Lecture Notes in Computer Science　　8124

Commenced Publication in 1973
Founding and Former Series Editors:
Gerhard Goos, Juris Hartmanis, and Jan van Leeuwen

W0235286

Christian Schulte (Ed.)

Principles and Practice of Constraint Programming

19th International Conference, CP 2013
Uppsala, Sweden, September 16-20, 2013
Proceedings

 Springer

Volume Editor

Christian Schulte
KTH Royal Institute of Technology
School of Information and Communication Technology
P.O. Box Forum 120
16440 Kista, Sweden
E-mail: cschulte@kth.se

ISSN 0302-9743   e-ISSN 1611-3349
ISBN 978-3-642-40626-3   e-ISBN 978-3-642-40627-0
DOI 10.1007/978-3-642-40627-0
Springer Heidelberg New York Dordrecht London

Library of Congress Control Number: 2013947084

CR Subject Classification (1998): F.4.1, G.1.6, F.2.2, F.3, G.2.2, D.3.2, F.1, E.1, I.2.8

LNCS Sublibrary: SL 2 – Programming and Software Engineering

Typesetting: Camera-ready by author, data conversion by Scientific Publishing Services, Chennai, India

Printed on acid-free paper

Springer is part of Springer Science+Business Media (www.springer.com)

Preface

This volume contains the proceedings of the 19th International Conference on Principles and Practice of Constraint Programming (CP 2013) that was held in Uppsala, Sweden, September 16–20, 2013. Detailed information on the conference can be found on its website cp2013.a4cp.org.

The CP conference is the annual international conference on constraint programming. It is concerned with all aspects of computing with constraints, including theory, algorithms, environments, languages, models, systems, and applications such as decision making, resource allocation, and agreement technologies. Besides the main technical track, CP 2013 featured an application track, which focused on uses of constraint technology and its comparison and integration with other optimization techniques (MIP, local search, SAT, . . .).

The interest of the research community in this conference was witnessed by the large number of submissions received this year. We received 170 (long and short) papers as follows: 138 papers submitted to the main track and 32 to the application track. Both long and short papers were reviewed to the same high standards of quality and no long papers were accepted as short papers. The reviewing process for the main track used a two-level Program Committee, consisting of senior Program Committee members and Program Committee members. Senior Program Committee members were responsible for managing a set of papers in their respective areas of expertise. They met in Stockholm during June 8–9, 2013. Each paper received at least three reviews, was extensively discussed, and additional reviews were added when needed. At the end of the reviewing process, we accepted 47 papers for the main technical track and 12 papers for the application track. All papers were presented at the conference.

Amongst the accepted papers, Jimmy Lee, Toby Walsh, and I selected a best technical track paper ("Parallel Discrepancy-Based Search" by Thierry Moisan, Jonathan Gaudreault, and Claude-Guy Quimper), a best application track paper ("Bin Packing with Linear Usage Costs – An Application to Energy Management in Data Centres" by Hadrien Cambazard, Deepak Mehta, Barry O'Sullivan, and Helmut Simonis), and a best student paper ("Filtering AtMostNValue with Difference Constraints: Application to the Shift Minimisation Personnel Task Scheduling Problem" by Jean-Guillaume Fages and Tanguy Lapègue). I am grateful to Jimmy Lee and Toby Walsh for their expert help in the selection process.

The conference program featured three invited talks, an invited public lecture, and an invited system presentation by distinguished scientists. This volume includes abstracts for the invited talks by Michela Milano, Torsten Schaub, and Peter Stuckey and the invited public lecture by Pascal Van Hentenryck. It also includes a paper for the invited system presentation by Pascal Van Hentenryck and Laurent Michel. The conference program included four tutorials: "MaxSAT

Latest Developments" by Carlos Ansótegui, "Replication and Recomputation in Scientific Experiments" by Ian Gent and Lars Kotthoff, "Constraint Programming for Vehicle Routing Problems" by Phil Kilby, and "Constraint Programming for the Control of Discrete Event Dynamic Systems" by Gérard Verfaille. As an additional part of the program, Peter Stuckey and Håkan Kjellerstrand organized the first international "Lightning Model and Solve Competition". The winners of the 2013 ACP Research Excellence Award and Doctoral Research Award presented their award talks. Many thanks to all of them for making essential contributions to an exciting conference program!

The conference included a Doctoral Program, which allowed doctoral students to come to the conference, present their work, and meet a mentor with similar research interests. I am very grateful to Christopher Mears and Nina Narodytska for doing a wonderful job in organizing the Doctoral Program.

The conference would not have been possible without the high-quality and interesting submissions from authors, which made the decision process so challenging. I would like to thank the whole Program Committee for the time spent in reviewing papers and in discussions. I am grateful to the additional reviewers, often recruited on very short notice. A special thank you goes to the senior Program Committee members for driving discussions, writing metareviews, and coming to the meeting in Stockholm. I would like to thank Gilles Pesant, who handled papers where I had a conflict of interest.

The conference would not have been possible without the great job done by Mats Carlsson, Pierre Flener, and Justin Pearson as Conference Chairs. They expertly took care of the local organization and I enjoyed our smooth and efficient collaboration. For conference publicity, I very much thank Guido Tack, who did a great job in advertising the conference and for a professional and always up-to-date website. I am very grateful to Laurent Michel, who acted as Workshop and Tutorial Chair and put together an exciting workshop and tutorial program. I am also grateful to Jimmy Lee and Peter Stuckey for sharing their past experiences as CP Program Chairs with me. I gratefully acknowledge local help in sponsoring matters by Karin Fohlstedt, Charlotta Jörsäter, and Victoria Knopf as well as in organizing the physical Senior Program Committee meeting by Sandra Gustavsson Nylén.

The Conference Chairs and I took on the task of soliciting sponsors for CP 2013. We would like to thank our many sponsors for their extraordinarily generous support; they are prominently—and deservedly so—listed in alphabetical order on a following page.

Last but not least, I want to thank the ACP Executive Committee for honoring me with the invitation to serve as Program Chair of CP 2013.

June 2013   Christian Schulte

Conference Organization

Conference Chairs

Mats Carlsson   SICS, Sweden
Pierre Flener   Uppsala University, Sweden
Justin Pearson   Uppsala University, Sweden

Program Chair and Application Track Chair

Christian Schulte   KTH Royal Institute of Technology, Sweden

Workshop and Tutorial Chair

Laurent Michel   University of Connecticut, USA

Doctoral Program Chairs

Christopher Mears   Monash University, Australia
Nina Narodytska   University of Toronto, Canada, and University of New South Wales, Australia

Publicity Chair

Guido Tack   Monash University, Australia

Senior Program Committee

Yves Deville   UCLouvain, Belgium
Pierre Flener   Uppsala University, Sweden
George Katsirelos   INRA, Toulouse, France
Christophe Lecoutre   CRIL, University of Artois, France
Jimmy Lee   The Chinese University of Hong Kong
Amnon Meisels   Ben-Gurion University of the Negev, Israel
Pedro Meseguer   IIIA-CSIC, Spain
Laurent Michel   University of Connecticut, USA
Barry O'Sullivan   4C, University College Cork, Ireland
Gilles Pesant   École Polytechnique de Montreal, Canada
Michel Rueher   University of Nice - Sophia Antipolis, France
Stefan Szeider   Vienna University of Technology, Austria

Michael Trick	Carnegie Mellon University, USA
Willem-Jan Van Hoeve	Carnegie Mellon University, USA
Toby Walsh	NICTA and UNSW, Australia
Roland Yap	National University of Singapore

Technical Program Committee

Fahiem Bacchus	University of Toronto, Canada
Chris Beck	University of Toronto, Canada
Nicolas Beldiceanu	École des Mines de Nantes, France
Christian Bessiere	Université Montpellier, France
Mats Carlsson	SICS, Sweden
Hubie Chen	Universidad del País Vasco and Ikerbasque, Spain
Geoffrey Chu	University of Melbourne, Australia
Ivan Dotu	Boston College, USA
Thibaut Feydy	University of Melbourne, Australia
Ian Gent	University of St. Andrews, UK
Alexandre Goldsztejn	CNRS, France
Emmanuel Hebrard	LAAS, CNRS, France
John Hooker	Carnegie Mellon University, USA
Said Jabbour	CRIL, University of Artois, France
Peter Jeavons	University of Oxford, UK
Christopher Jefferson	University of St. Andrews, UK
Narendra Jussien	École des Mines de Nantes, France
Serdar Kadıoğlu	Oracle Corporation, USA
Lars Kotthoff	University College Cork, Ireland
Michele Lombardi	University of Bologna, Italy
Ines Lynce	Technical University of Lisbon, Portugal
Christopher Mears	Monash University, Australia
Ian Miguel	University of St. Andrews, UK
Michela Milano	University of Bologna, Italy
Jean-Noël Monette	Uppsala University, Sweden
Peter Nightingale	University of St. Andrews, UK
Justin Pearson	Uppsala University, Sweden
Thierry Petit	École des Mines de Nantes, France
Claude-Guy Quimper	Université Laval, Canada
Louis-Martin Rousseau	École Polytechnique de Montreal, Canada
Jean-Charles Régin	Université Nice Sophia Antipolis, France
Ashish Sabharwal	IBM Research, USA
Thomas Schiex	INRA, France
Meinolf Sellmann	IBM Research, USA

Paul Shaw   IBM, France
Helmut Simonis   4C, Ireland
Christine Solnon   INSA Lyon, France
Peter Stuckey   NICTA and University of Melbourne, Australia
Guido Tack   Monash University, Australia
Peter Van Beek   University of Waterloo, Canada
Pascal Van Hentenryck   NICTA and University of Melbourne, Australia
Brent Venable   Tulane University, USA
Gérard Verfaillie   ONERA, France
Roie Zivan   Ben-Gurion University of the Negev, Israel
Stanislav Živný   University of Warwick, UK

Application Track Program Committee

Claire Bagley   Oracle Corporation, USA
Pedro Barahona   Universidade Nova de Lisboa, Portugal
Ken Brown   University College Cork, Ireland
Hadrien Cambazard   Grenoble University, France
Philippe Codognet   CNRS / UPMC / University of Tokyo, Japan
Sophie Demassey   École des Mines de Nantes, France
Pierre Flener   Uppsala University, Sweden
Maria Garcia de la Banda   Monash University, Australia
Arnaud Gotlieb   SIMULA Research Laboratory, Norway
Jimmy Lee   The Chinese University of Hong Kong
Michele Lombardi   University of Bologna, Italy
Laurent Michel   University of Connecticut, USA
Tomas Eric Nordlander   SINTEF ICT, Norway
Barry O'Sullivan   4C, University College Cork, Ireland
Federico Pecora   Örebro University, Sweden
Laurent Perron   Google, France
Michel Rueher   Université Nice Sophia Antipolis, France
Jean-Charles Régin   Université Nice Sophia Antipolis, France
Martin Sachenbacher   Technische Universität München, Germany
Pierre Schaus   UCLouvain, Belgium
Thomas Schiex   INRA, France
Paul Shaw   IBM, France
Helmut Simonis   4C, Ireland
Peter Stuckey   NICTA and University of Melbourne, Australia
Guido Tack   Monash University, Australia
Gérard Verfaillie   ONERA, France
Mark Wallace   Monash University, Australia

Additional Reviewers

Alejandro Arbelaez
James Bailey
Anton Belov
David Bergman
Christoph Berkholz
Christian Bessiere
Manuel Bodirsky
Alessio Bonfietti
Simone Bova
Simon Brockbank
Clément Carbonnel
Gilles Chabert
Jeff Choi
Andre Cire
David Cohen
Remi Coletta
Martin Cooper
Jorge Cruz
Veronica Dahl
Alessandro Dal Palù
Jessica Davies
Simon de Givry
Alban Derrien
Gregory Duck
Renaud Dumeur
Uwe Egly
Stefano Ermon
Pierre Flener
María Andreína Francisco Rodríguez
Maurizio Gabbrielli
Graeme Gange
Marco Gavanelli
Vibhav Gogate
Arnaud Gotlieb
Laurent Granvilliers
Diarmuid Grimes
Stefano Gualandi
Evgeny Gurevsky
Hossein Seyed Hashemi Doulabi
Patrik Haslum
Farshid Hassani Bijarbooneh
Benoît Hoessen
Marie-José Huguet

Barry Hurley
Siddhartha Jain
Mikoláš Janota
Nicolas Jozefowiez
Narendra Jussien
George Katsirelos
Zeynep Kiziltan
William Klieber
Arun Konagurthu
Marco Kuhlmann
Uwe Köckemann
Arnaud Lallouet
Javier Larrosa
Yat Chiu Law
Nadjib Lazaar
Kevin Leo
Olivier Lhomme
Chu-Min Li
Jerry Lonlac
Florian Lonsing
Xavier Lorca
Jean-Baptiste Mairy
Terrence W.K. Mak
Arnaud Malapert
Yuri Malitsky
Vasco Manquinho
Masoumeh Mansouri
Joao Marques-Silva
Barnaby Martin
Nicholas Mattei
Jacopo Mauro
Christopher Mears
Pedro Meseguer
Claude Michel
Andrea Micheli
Michela Milano
Thierry Moisan
Eric Monfroy
Jorge A. Navas
Samba Ndojh Ndiaye
Robert Nieuwenhuis
Todd Niven
Alexandre Papadopoulos

Federico Pecora
Laurent Perron
Justyna Petke
Karen Petrie
Nathalie Peyrard
Cédric Piette
Maria Silvia Pini
Charles Prud'Homme
Claude-Guy Quimper
Raghuram Ramanujan
Philippe Refalo
Jean-Charles Régin
Florian Richoux
Roberto Rossi
Olivier Roussel
Lakhdar Sais
András Salamon
Horst Samulowitz
Scott Sanner

Prateek Saxena
Tom Schrijvers
Andreas Schutt
Joseph Scott
Martina Seidl
Sagar Sen
Mohamed Siala
Laurent Simon
Friedrich Slivovsky
Kostas Stergiou
Sebastien Tabary
Johan Thapper
Evgenij Thorstensen
Gilles Trombettoni
Julien Vion
Mohamed Wahbi
Siert Wieringa
Lebbah Yahia
Alessandro Zanarini

Conference Sponsors

We would like to thank our sponsors (in alphabetical order) for their generous support: Association for Constraint Programming, Association Française pour la Programmation par Contraintes, AIMMS, AMPL, Artificial Intelligence – An International Journal, Kjell och Märta Beijers Stiftelse, Cadence, Certus, COSYTEC, FICO, IBM Research, ICS, Jeppesen, KTH Royal Institute of Technology, Microsoft Research – INRIA Joint Centre, NICTA, ONERA, Quintiq, SICS, SINTEF, Trade Extensions, Uppsala University, Världsklass Uppsala, Örebro University.

Table of Contents

Invited Talks

Optimization for Policy Making: The Cornerstone for an Integrated
Approach 1
    Michela Milano

Answer Set Programming: Boolean Constraint Solving for Knowledge
Representation and Reasoning 3
    Torsten Schaub

Those Who Cannot Remember the Past Are Condemned to Repeat
It 5
    Peter J. Stuckey

Invited Public Lecture

Decide Different! 7
    Pascal Van Hentenryck

Invited System Presentation

The Objective-CP Optimization System 8
    Pascal Van Hentenryck and Laurent Michel

Best Technical Track Paper

Parallel Discrepancy-Based Search 30
    Thierry Moisan, Jonathan Gaudreault, and Claude-Guy Quimper

Best Application Track Paper

Bin Packing with Linear Usage Costs – An Application to Energy
Management in Data Centres 47
    Hadrien Cambazard, Deepak Mehta, Barry O'Sullivan, and
    Helmut Simonis

Best Student Paper

Filtering AtMostNValue with Difference Constraints: Application to the
Shift Minimisation Personnel Task Scheduling Problem 63
    Jean-Guillaume Fages and Tanguy Lapègue

Technical Track Papers

A Parametric Approach for Smaller and Better Encodings of
Cardinality Constraints   80
   Ignasi Abío, Robert Nieuwenhuis, Albert Oliveras, and
   Enric Rodríguez-Carbonell

To Encode or to Propagate? The Best Choice for Each Constraint
in SAT   97
   Ignasi Abío, Robert Nieuwenhuis, Albert Oliveras,
   Enric Rodríguez-Carbonell, and Peter J. Stuckey

Automated Symmetry Breaking and Model Selection in CONJURE   107
   Ozgur Akgun, Alan M. Frisch, Ian P. Gent, Bilal Syed
   Hussain, Christopher Jefferson, Lars Kotthoff, Ian Miguel, and
   Peter Nightingale

Improving WPM2 for (Weighted) Partial MaxSAT   117
   Carlos Ansótegui, Maria Luisa Bonet, Joel Gabàs, and Jordi Levy

MinSAT versus MaxSAT for Optimization Problems   133
   Josep Argelich, Chu-Min Li, Felip Manyà, and Zhu Zhu

Adaptive Parameterized Consistency   143
   Amine Balafrej, Christian Bessiere, Remi Coletta, and
   El Houssine Bouyakhf

Global Inverse Consistency for Interactive Constraint Satisfaction   159
   Christian Bessiere, Hélène Fargier, and Christophe Lecoutre

Counting Spanning Trees to Guide Search in Constrained Spanning
Tree Problems   175
   Simon Brockbank, Gilles Pesant, and Louis-Martin Rousseau

On the Reduction of the CSP Dichotomy Conjecture to Digraphs   184
   Jakub Bulín, Dejan Delić, Marcel Jackson, and Todd Niven

A Scalable Approximate Model Counter   200
   Supratik Chakraborty, Kuldeep S. Meel, and Moshe Y. Vardi

Dominance Driven Search   217
   Geoffrey Chu and Peter J. Stuckey

Tractable Combinations of Global Constraints   230
   David A. Cohen, Peter G. Jeavons, Evgenij Thorstensen, and
   Stanislav Živný

Postponing Optimization to Speed Up MAXSAT Solving   247
   Jessica Davies and Fahiem Bacchus

Dead-End Elimination for Weighted CSP 263
  Simon de Givry, Steven D. Prestwich, and Barry O'Sullivan

Solving Weighted CSPs by Successive Relaxations.................... 273
  Erin Delisle and Fahiem Bacchus

Constraint-Based Program Reasoning with Heaps and Separation 282
  Gregory J. Duck, Joxan Jaffar, and Nicolas C.H. Koh

Model Combinators for Hybrid Optimization 299
  Daniel Fontaine, Laurent Michel, and Pascal Van Hentenryck

Modelling Destructive Assignments................................ 315
  Kathryn Francis, Jorge Navas, and Peter J. Stuckey

An Improved Search Algorithm for Min-Perturbation 331
  Alex Fukunaga

Explaining Propagators for Edge-Valued Decision Diagrams 340
  Graeme Gange, Peter J. Stuckey, and Pascal Van Hentenryck

A Simple and Effective Decomposition for the Multidimensional
Binpacking Constraint 356
  Stefano Gualandi and Michele Lombardi

Maintaining Soft Arc Consistencies in BnB-ADOPT$^+$ during Search.... 365
  Patricia Gutierrez, Jimmy H.M. Lee, Ka Man Lei,
  Terrence W.K. Mak, and Pedro Meseguer

Solving String Constraints: The Case for Constraint Programming 381
  Jun He, Pierre Flener, Justin Pearson, and Wei Ming Zhang

Blowing Holes in Various Aspects of Computational Problems, with
Applications to Constraint Satisfaction 398
  Peter Jonsson, Victor Lagerkvist, and Gustav Nordh

Solving QBF with Free Variables................................. 415
  William Klieber, Mikoláš Janota, Joao Marques-Silva, and
  Edmund Clarke

Globalizing Constraint Models................................... 432
  Kevin Leo, Christopher Mears, Guido Tack, and
  Maria Garcia de la Banda

A New Propagator for Two-Layer Neural Networks in Empirical Model
Learning 448
  Michele Lombardi and Stefano Gualandi

Bandit-Based Search for Constraint Programming.................... 464
  Manuel Loth, Michèle Sebag, Youssef Hamadi, and Marc Schoenauer

Focused Random Walk with Configuration Checking and Break
Minimum for Satisfiability 481
   Chuan Luo, Shaowei Cai, Wei Wu, and Kaile Su

Multi-Objective Constraint Optimization with Tradeoffs 497
   Radu Marinescu, Abdul Razak, and Nic Wilson

Multidimensional Bin Packing Revisited 513
   Michael D. Moffitt

A Parametric Propagator for Discretely Convex Pairs of SUM
Constraints 529
   *Jean-Noël Monette, Nicolas Beldiceanu, Pierre Flener, and
Justin Pearson*

Breaking Symmetry with Different Orderings 545
   Nina Narodytska and Toby Walsh

Time-Table Extended-Edge-Finding for the Cumulative Constraint 562
   Pierre Ouellet and Claude-Guy Quimper

Revisiting the Cardinality Reasoning for BinPacking Constraint........ 578
   François Pelsser, Pierre Schaus, and Jean-Charles Régin

Value Interchangeability in Scenario Generation...................... 587
   Steven D. Prestwich, Marco Laumanns, and Ban Kawas

Embarrassingly Parallel Search 596
   Jean-Charles Régin, Mohamed Rezgui, and Arnaud Malapert

Multi-Objective Large Neighborhood Search........................ 611
   Pierre Schaus and Renaud Hartert

Scheduling Optional Tasks with Explanation 628
   Andreas Schutt, Thibaut Feydy, and Peter J. Stuckey

Residential Demand Response under Uncertainty 645
   *Paul Scott, Sylvie Thiébaux, Menkes van den Briel, and
Pascal Van Hentenryck*

Lifting Structural Tractability to CSP with Global Constraints 661
   Evgenij Thorstensen

Empirical Study of the Behavior of Conflict Analysis in CDCL
Solvers 678
   Djamal Habet and Donia Toumi

Primal and Dual Encoding from Applications into Quantified Boolean
Formulas.. 694
   Allen Van Gelder

Asynchronous Forward Bounding Revisited........................   708
    Mohamed Wahbi, Redouane Ezzahir, and Christian Bessiere

Optimizing STR Algorithms with Tuple Compression.................   724
    Wei Xia and Roland H.C. Yap

Application Track Papers

Describing and Generating Solutions for the EDF Unit Commitment
Problem with the ModelSeeker   733
    *Nicolas Beldiceanu, Georgiana Ifrim, Arnaud Lenoir, and
    Helmut Simonis*

Solving the Agricultural Land Allocation Problem by Constraint-Based
Local Search   749
    Quoc Trung Bui, Quang Dung Pham, and Yves Deville

Constraint-Based Approaches for Balancing Bike Sharing Systems......   758
    Luca Di Gaspero, Andrea Rendl, and Tommaso Urli

Constraint Based Computation of Periodic Orbits of Chaotic Dynamical
Systems   774
    Alexandre Goldsztejn, Laurent Granvilliers, and Christophe Jermann

Laser Cutting Path Planning Using CP   790
    Mikael Z. Lagerkvist, Martin Nordkvist, and Magnus Rattfeldt

Atom Mapping with Constraint Programming   805
    *Martin Mann, Feras Nahar, Heinz Ekker, Rolf Backofen,
    Peter F. Stadler, and Christoph Flamm*

Beyond Feasibility: CP Usage in Constrained-Random Functional
Hardware Verification   823
    Reuven Naveh and Amit Metodi

Stochastic Local Search Based Channel Assignment in Wireless Mesh
Networks   832
    *M.A. Hakim Newton, Duc Nghia Pham, Wee Lum Tan,
    Marius Portmann, and Abdul Sattar*

Automatic Generation and Delivery of Multiple-Choice Math
Quizzes   848
    Ana Paula Tomás and José Paulo Leal

Constrained Wine Blending 864
    Philippe Vismara, Remi Coletta, and Gilles Trombettoni

The Berth Allocation and Quay Crane Assignment Problem Using
a CP Approach 880
    Stéphane Zampelli, Yannis Vergados, Rowan Van Schaeren,
    Wout Dullaert, and Birger Raa

Author Index 897

Optimization for Policy Making:
The Cornerstone for an Integrated Approach

Michela Milano

DISI, University of Bologna
V.le Risorgimento 2, 40136, Bologna, Italy

Abstract. Policy making is a very complex task taking into account several aspects related to sustainability, namely impact on the environments, health of productive sectors, economic implications and social acceptance. Optimization methods could be extremely useful for analysing alternative policy scenarios, but should be complemented with several other techniques such as machine learning, agent-based simulation, opinion mining and visualization to come up with an integrated system able to support decision making in the overall policy design life cycle. I will discuss how these techniques could be merged with optimization and I will identity some open research directions.

Policy making is the formulation of ideas or plans that are used by an organization or government as a basis for making decisions. Public policy issues cover a wide variety of fields such as economy, education, environment, health, social welfare, national and foreign affairs. They are extremely complex, occur in rapidly changing environments characterized by uncertainty, and involve conflicts among different interests and affect the three pillars of sustainable development, namely society, economy and the environment.

The government of a region or a nation should therefore take complex decisions on the basis of the available data (for example coming from the monitoring of previous policies), of the current economic situation, of the current level of environmental indicators, and on available resources. Basically the planning activity of a policy maker can be easily casted in a multi-criteria combinatorial optimization problem possibly under uncertainty, where Pareto optimal solutions are alternative political scenarios. Each scenario has its own cost and impact on environmental and economic indicators. Depending on the strategic political objectives, the policy maker might prefer one alternative among others. Thus, optimization can play a crucial role for improving the policy making process.

However, optimization is only one - yet important - cornerstone for the improvement of the overall policy making process. There are a number of techniques that could and should be merged with optimization to come up with integrated software tools aiding the policy maker in the overall policy design life cycle. One example is agent-based simulation [3] to mimic the social reaction to policy instruments. Another important technique is opinion mining [5] that basically extracts opinions and sentiments on specific policy topics from blogs and forums enabling e-participation in the policy design. Data mining and machine learning

C. Schulte (Ed.): CP 2013, LNCS 8124, pp. 1–2, 2013.

in general would also be extremely important for processing the always increasing amount of data coming from sensors, extracting relations between these data and the political interventions and possibly insert this extracted model into the optimization model.

Finally, policy makers are not ICT experts and should be aided in the use of the above-mentioned technology. Advanced visualization techniques should play an important role in the human-machine interaction.

Despite a number of research papers have been published in each above mentioned area, what is totally missing at present is a comprehensive tool that assists the policy maker in all phases of the decision making process. The tool should compute alternative scenarios each comprising both a well assessed plan and the corresponding implementation strategies to achieve its objective, its cost and its social acceptance. We need a tool that is able to integrate and consider at the same time global objectives and individual/social reactions. These two aspects could be (and often are) in conflict and possibly game theory could be used to find an equilibrium between the two parts.

During the talk I will present some recent work developed under the EU FP7 project called ePolicyn - *Engineering the policy making life cycle* - aimed at developing decision support systems aiding the policy maker across all phasees of the policy making process [1], [4], [2]. The case study will be on the regional energy plan of the Emilia Romagna region of Italy. We will show the different phases of the policy making process and explain where optimization could play a role and how other techniques should be integrated with it.

Acknowledgment. The author is partially supported by the European Union Seventh Framework Programme (FP7/2007-2013) under grant agreement n. 288147.

References

1. Gavanelli, M., Riguzzi, F., Milano, M., Cagnoli, P.: Logic-Based Decision Support for Strategic Environmental Assessment. In: Theory and Practice of Logic Programming, 26th Int'l. Conference on Logic Programming, ICLP 2010, vol. 10(4-6), pp. 643–658 (July 2010), Special Issue
2. Gavanelli, M., Riguzzi, F., Milano, M., Cagnoli, P.: Constraint and optimization techniques for supporting policy making. In: Yu, T., Chawla, N., Simoff, S. (eds.) Computational Intelligent Data Analysis for Sustainable Development, Routledge (2013)
3. Gilbert, N.: Agent based models. Sage Publications Inc. (2007)
4. Milano, M.: Sustainable energy policies: research challenges and opportunities. In: Design, Automation and Test in Europe, DATE, pp. 1143–1148 (2013)
5. Pang, B., Lee, L.: Opinion mining and sentiment analysis. Foundations and Trends in Information Retrieval 2(1-2), 1–135 (2008)

Answer Set Programming:
Boolean Constraint Solving
for Knowledge Representation and Reasoning

Torsten Schaub[*]

University of Potsdam, Germany
`torsten@cs.uni-potsdam.de`

Answer Set Programming (ASP; [1,2,3]) is a declarative problem solving approach, combining a rich yet simple modeling language with high-performance Boolean constraint solving capacities. ASP is particularly suited for modeling problems in the area of Knowledge Representation and Reasoning involving incomplete, inconsistent, and changing information. As such, it offers, in addition to satisfiability testing, various reasoning modes, including different forms of model enumeration, intersection or unioning, as well as multi-criteria and -objective optimization. From a formal perspective, ASP allows for solving all search problems in NP (and NP^{NP}) in a uniform way. Hence, ASP is well-suited for solving hard combinatorial search problems, like system design and timetabling. Prestigious applications of ASP include composition of Renaissance music [4], decision support systems for NASA shuttle controllers [5], reasoning tools in systems biology [6,7,8] and robotics [9,10], industrial team-building [11], and many more. The versatility of ASP is nicely reflected by the ASP solver *clasp* [12], winning first places at various solver competitions, such as ASP, MISC, PB, and SAT competitions. The solver *clasp* is at the heart of the open source platform *Potassco* hosted at `potassco.sourceforge.net`. Potassco stands for the "Potsdam Answer Set Solving Collection" [13] and has seen more than 30000 downloads world-wide since its inception at the end of 2008.

The talk will start with an introduction to ASP, its modeling language and solving methodology, and portray some distinguished ASP systems.

References

1. Gelfond, M., Lifschitz, V.: The stable model semantics for logic programming. In: Kowalski, R., Bowen, K. (eds.) Proceedings of the Fifth International Conference and Symposium of Logic Programming (ICLP 1988), pp. 1070–1080. MIT Press (1988)
2. Baral, C.: Knowledge Representation, Reasoning and Declarative Problem Solving. Cambridge University Press (2003)
3. Gebser, M., Kaminski, R., Kaufmann, B., Schaub, T.: Answer Set Solving in Practice. Synthesis Lectures on Artificial Intelligence and Machine Learning. Morgan and Claypool Publishers (2012)

---

[*] Affiliated with Simon Fraser University, Canada, and Griffith University, Australia.

C. Schulte (Ed.): CP 2013, LNCS 8124, pp. 3–4, 2013.

4. Boenn, G., Brain, M., De Vos, M., ffitch, J.: Automatic composition of melodic and harmonic music by answer set programming. In: Garcia de la Banda, M., Pontelli, E. (eds.) ICLP 2008. LNCS, vol. 5366, pp. 160–174. Springer, Heidelberg (2008)
5. Nogueira, M., Balduccini, M., Gelfond, M., Watson, R., Barry, M.: An A-prolog decision support system for the space shuttle. In: Ramakrishnan, I.V. (ed.) PADL 2001. LNCS, vol. 1990, pp. 169–183. Springer, Heidelberg (2001)
6. Erdem, E., Türe, F.: Efficient haplotype inference with answer set programming. In: Fox, D., Gomes, C. (eds.) Proceedings of the Twenty-Third National Conference on Artificial Intelligence (AAAI 2008), pp. 436–441. AAAI Press (2008)
7. Gebser, M., Schaub, T., Thiele, S., Veber, P.: Detecting inconsistencies in large biological networks with answer set programming. Theory and Practice of Logic Programming 11(2-3), 323–360 (2011)
8. Gebser, M., Guziolowski, C., Ivanchev, M., Schaub, T., Siegel, A., Thiele, S., Veber, P.: Repair and prediction (under inconsistency) in large biological networks with answer set programming. In: Lin, F., Sattler, U. (eds.) Proceedings of the Twelfth International Conference on Principles of Knowledge Representation and Reasoning (KR 2010), pp. 497–507. AAAI Press (2010)
9. Chen, X., Ji, J., Jiang, J., Jin, G., Wang, F., Xie, J.: Developing high-level cognitive functions for service robots. In: van der Hoek, W., Kaminka, G., Lespérance, Y., Luck, M., Sen, S. (eds.) Proceedings of the Ninth International Conference on Autonomous Agents and Multiagent Systems (AAMAS 2010), pp. 989–996. IFAAMAS (2010)
10. Erdem, E., Haspalamutgil, K., Palaz, C., Patoglu, V., Uras, T.: Combining high-level causal reasoning with low-level geometric reasoning and motion planning for robotic manipulation. In: Proceedings of the IEEE International Conference on Robotics and Automation (ICRA 2011), pp. 4575–4581. IEEE (2011)
11. Grasso, G., Iiritano, S., Leone, N., Lio, V., Ricca, F., Scalise, F.: An ASP-based system for team-building in the Gioia-Tauro seaport. In: Carro, M., Peña, R. (eds.) PADL 2010. LNCS, vol. 5937, pp. 40–42. Springer, Heidelberg (2010)
12. Gebser, M., Kaufmann, B., Schaub, T.: Conflict-driven answer set solving: From theory to practice. Artificial Intelligence 187-188, 52–89 (2012)
13. Gebser, M., Kaminski, R., Kaufmann, B., Ostrowski, M., Schaub, T., Schneider, M.: Potassco: The Potsdam answer set solving collection. AI Communications 24(2), 107–124 (2011)

Those Who Cannot Remember
the Past Are Condemned to Repeat It

Peter J. Stuckey

National ICT Australia, Victoria Laboratory
Department of Computing and Information Systems,
University of Melbourne, Australia
pstuckey@unimelb.edu.au

Abstract. Constraint programming is a highly successful technology for tackling complex combinatorial optimization problems. Any form of combinatorial optimization involves some form of search, and CP is very well adapted to make use of programmed search and strong inference to solve some problems that are out of reach of competing technologies. But much of the search that happens during a CP execution is effectively repeated. This arises from the combinatorial nature of the problems we are tackling. Learning about past unsuccessful searches and remembering this in an effective way can exponentially reduce the size of the search space. In this talk I will explain lazy clause generation, which is a hybrid constraint solving technique that steals all the best learning ideas from Boolean satisfiability solvers, but retains all the advantages of constraint programming. Lazy clause generation provides the state of the art solutions to a wide range of problems, and consistently outperforms other solving approaches in the MiniZinc challenge.

1 Introduction

In the early days of constraint programming there was considerable interest in learning from failure via look-back methods [1] and intelligent backtracking [2]. But this research faded out as propagation approaches proved to be more successful at tackling complex problems [3].

The SAT community revitalized learning, which is now the most critical component in a modern Davis-Putnam-Logemann-Loveland SAT solver, essentially because they devised data structures to efficiently store and propagate hundreds of thousands of learnt nogoods [4]. This technology has been incorporated in constraint programming solvers, first by Katsirelos and Bacchus [5] who used literals of the form $x = d$ and $x \neq d$ to represent integer variables. This was extended in the *Lazy Clause Generation* (LCG) approach [6] by using literals of the form $x \leq d$ and $x \geq d$. By storing nogoods that record the reason why a subtree search has failed, constraint programming solvers with learning can *exponentially* reduce the search required to find and prove optimal solutions.

In Lazy Clause Generation each constraint propagator is extended to be able to explain its propagation. Lazy Clause Generation has proved remarkably successful in tackling hard combinatorial optimization problems. It defines the state

C. Schulte (Ed.): CP 2013, LNCS 8124, pp. 5–6, 2013.

of the art complete method in many well studied scheduling problems, such as resource constraint project scheduling (RCPSP) [7], and variations like RCPSP with generalized precedences [8]. LCG has lead to substantial benefits in real life packing problems, such as carpet cutting [9]. LCG solvers have dominated the MiniZinc challenge competition www.minizinc.org since 2010 (although they are not eligible for prizes) illustrating the approach is applicable over a wide range of problem classes.

In this presentation, I will explain how lazy clause generation solvers work, some of the challenging algorithmic decisions that arise in creating explaining propagators, and some of the important emerging research directions such as *optimized Boolean encoding* [10], *lazy decomposition* [11], and *lifelong learning* [12].

References

1. Dechter, R.: Enhancement schemes for constraint processing: Backjumping, learning, and cutset decomposition. Artificial Intelligence 41, 273–312 (1990)
2. Prosser, P.: MAC-CBJ: Maintaining arc consistency with conflict-directed backjumping. Technical Report Research Report 177, University of Strathclyde (1995)
3. Bessiere, C., Regin, J.C.: MAC and combined heuristics: Two reasons to forsake FC (and CBJ?) on hard problems. In: Freuder, E.C. (ed.) CP 1996. LNCS, vol. 1118, pp. 61–75. Springer, Heidelberg (1996)
4. Moskewicz, M., Madigan, C., Zhao, Y., Zhang, L., Malik, S.: Chaff: Engineering an efficient SAT solver. In: Proceedings of the 39th Design Automation Conference, DAC 2001 (2001)
5. Katsirelos, G., Bacchus, F.: Generalized nogoods in CSPs. In: Proceedings of the 20th AAAI Conference on Artificial Intelligence, AAAI 2005, pp. 390–396 (2005)
6. Ohrimenko, O., Stuckey, P., Codish, M.: Propagation via lazy clause generation. Constraints 14(3), 357–391 (2009)
7. Schutt, A., Feydy, T., Stuckey, P., Wallace, M.: Explaining the cumulative propagator. Constraints 16(3), 250–282 (2011)
8. Schutt, A., Feydy, T., Stuckey, P., Wallace, M.: Solving RCPSP/max by lazy clause generation. Journal of Scheduling (2012), online first: http://dx.doi.org/10.1007/s10951-012-0285-x (August 2012)
9. Schutt, A., Stuckey, P., Verden, A.: Optimal carpet cutting. In: Lee, J. (ed.) CP 2011. LNCS, vol. 6876, pp. 69–84. Springer, Heidelberg (2011)
10. Metodi, A., Codish, M., Stuckey, P.J.: Boolean equi-propagation for concise and efficient SAT encodings of combinatorial problems. Journal of Artificial Intelligence Research 46, 303–341 (2013), http://www.jair.org/papers/paper3809.html
11. Abío, I., Stuckey, P.J.: Conflict directed lazy decomposition. In: Milano, M. (ed.) CP 2012. LNCS, vol. 7514, pp. 70–85. Springer, Heidelberg (2012)
12. Chu, G., Stuckey, P.J.: Inter-instance nogood learning in constraint programming. In: Milano, M. (ed.) CP 2012. LNCS, vol. 7514, pp. 238–247. Springer, Heidelberg (2012)

Decide Different!

Pascal Van Hentenryck

National ICT Australia, Victoria Laboratory
Department of Computing and Information Systems,
University of Melbourne, Australia
pvh@nicta.com.au

We live in a period where Information and Communication Technologies (ICT) has revolutionized the way we communicate, learn, work, and entertain ourselves. But we also live in challenging times, from climate change and natural disasters of increased intensity to rapid urbanization, pollution, economic stagnation, and a shrinking middle class in Western countries. In this lecture, we argue that ICT now has the opportunity to radically change the way we take decisions as a society, exploiting the wealth of data available to understand physical, biological, business, and human behaviors with unprecedented accuracy and speed. We illustrate this vision with challenging problems in disaster management, energy, medicine, and transportation.

Acknowledgments. NICTA is funded by the Australian Government as represented by the Department of Broadband, Communications and the Digital Economy and the Australian Research Council through the ICT Centre of Excellence program.

The Objective-CP Optimization System

Pascal Van Hentenryck[1] and Laurent Michel[2]

[1] NICTA, Australia
[2] University of Connecticut, Storrs, CT 06269-2155

Abstract. OBJECTIVE-CP is an optimization system that views an optimization program as the combination of a model, a search, and a solver. Models in OBJECTIVE-CP follow the modeling style of constraint programming and are concretized into specific solvers. Search procedures are specified in terms of high-level nondeterministic constructs, search combinators, and node selection strategies. OBJECTIVE-CP supports fully transparent parallelization of multi-start and branch & bound algorithms. The implementation of OBJECTIVE-CP is based on a sequence of model transformations, followed by a concretization step. Moreover, OBJECTIVE-CP features a constraint-programming solver following a micro-kernel architecture for ease of maintenance and extensibility. Experimental results show the practicability of the approach.

1 Introduction

This paper presents an overview of OBJECTIVE-CP, an optimization system written in OBJECTIVE-C (an object-oriented layer on top of C). OBJECTIVE-CP builds on more than two decades of research on the design and implementation of constraint-programming systems, from CHIP to systems such as ILOG SOLVER, OPL, ILOG CONCERT, COMET, GECODE, and MINIZINC which have probably had the strongest influence on its design and implementation. The design of OBJECTIVE-CP takes the the view that

Optimization Program = Model + Search + Solver

or, in other words, that an optimization program consists of a model, a search, and an underlying constraint solver.

Models are first-class objects in OBJECTIVE-CP; they also follow the style of constraint programming and are solver-independent. This allows for easy experimentation with different technologies and smooth hybridizations [5,4]. Models can be concretized into a specific solver to obtain an optimization program, (e.g., a constraint programs or a mixed-integer program). The resulting optimization program can be solved using a black-box search or a dedicated search procedure expressed in terms of the model variables. Search procedures in OBJECTIVE-CP are specified in terms of high-level nondeterministic constructs, search combinators, and node selection strategies, merging the benefits of search controllers and continuations [22] on the one hand and compositional combinators (e.g., [17]) on the other hand. The search language is generic and independent of

C. Schulte (Ed.): CP 2013, LNCS 8124, pp. 8–29, 2013.

the underlying solver, although obviously search procedures call the underlying solver for adding constraints, binding variables, and querying the search state. OBJECTIVE-CP transparently supports the parallelization of optimization programs, supporting parallel multi-start algorithms and parallel branch & bound.

The implementation of OBJECTIVE-CP performs a series of model transformations, including a flattening of the model, followed by a concretization of the final model into a specific solver. OBJECTIVE-CP also features a constraint-programming solver inspired by the micro-kernel approach to operating systems. It features small components, such as a propagation engine, a variable library, and a constraint library, that are separated and have minimal interfaces.

It is difficult to summarize the contributions of a large system. However, the following features of OBJECTIVE-CP are worth highlighting:

1. OBJECTIVE-CP enables the model and the search to be expressed in terms of the model variables, although the model can be concretized into different solvers;
2. OBJECTIVE-CP offers a rich, generic search language. The search language is independent of the underlying solver and merges the benefits of two high-level approaches to search: search controllers and search combinators. In particular, OBJECTIVE-CP provides a small set of abstractions that naturally combine to build complex search procedures.
3. OBJECTIVE-CP achieves a strong symbiosis with the underlying host language, i.e., OBJECTIVE-C. In particular, it allows for an iterative style in search procedures and makes heavy use of closures and first-order functions.
4. OBJECTIVE-CP provides first-class models which make it possible to offer model combinators and an implementation approach based on model transformations and concretizations.
5. OBJECTIVE-CP provides an automatic and transparent parallelization of optimization programs, even when the program feature a search procedure (i.e., not a black-box search).
6. OBJECTIVE-CP features a constraint-programming solver based on the concept of micro-kernel in operating systems, i.e., it strives to define small components with minimal interfaces.

This paper reviews the design and implementation of OBJECTIVE-CP. Section 2 gives a brief overview of OBJECTIVE-C. Section 3 then presents an overview of OBJECTIVE-CP, including models, search, and transparent parallelization. The implementation methodology and the experimental results are presented in Sections 4–5. Sections 6–7 discuss the related work and the conclusion.

2   The Host Language

OBJECTIVE-CP is written on top of OBJECTIVE-C, a high-level programming language that adds an object-oriented layer on top of C. OBJECTIVE-C marries the elegance of a fully dynamic object-oriented runtime based on dynamic message dispatching (a la SMALLTALK with the performance of C). OBJECTIVE-C features syntactic extensions over C to model classes (called @interface), interfaces (called @protocol) and categories that provide the ability to extend the API of a class with new methods without requiring access to the source code. OBJECTIVE-C inherits the static typing of C, yet it offers the ability to be loosely typed for the object-oriented extensions. The SMALLTALK heritage is significant. For instance, OBJECTIVE-C separates the notion of message and response behavior. It also offers introspection, message interception, and rerouting. The syntax of OBJECTIVE-C may seem peculiar at first. A method call cp.label(x) is written as [cp label: x] where label: is the method name. When using multiple arguments, OBJECTIVE-C "names" each of them. For instance, a method call cp.labelWith(x,v) could be become [cp label: x with: v], where the method name is label:with:. OBJECTIVE-C also features closures and first-order functions. For instance,

```
1 [cp onSolution: ^{printf(''found a solution \n'');}];
```

uses a closure ^{printf(''found a solution \n'')} which can be called subsequently, The snippet

```
1 [S enumerateWithBlock:^(int i) {printf(''\%d  '',i); }];
```

depicts the use of a first-order function. The code enumerates the elements of set S and calls the first-order function passing each element to parameter i. The body of the function prints the value of the set elements. Overall, OBJECTIVE-C is a nice compromise between the flexibility of SMALLTALK and the efficiency of C. It is particularly well-adpated for developing complex systems.

3   The Design of Objective-CP

This section reviews the design of OBJECTIVE-CP and its main concepts: models, programs, search procedures, and transparent parallelization. The focus is on introducing the concepts informally and conveying a sense of the global design. Model composition is covered in detail in [4].

3.1   Models

Figure 1 illustrates several features of OBJECTIVE-CP: It depicts a program which solves a capacitated warehouse location problem with a CP solver and a MIP solver. Lines 1–25 declare a model (line 1), its data (lines 2–7 where constants are omitted), its decision variables (lines 9–14), its constraints (lines 16–23), and its objective function (lines 24–25). The capacity constraints in line 17 feature reified constraints, while lines 21–22 feature element constraints to link the warehouse and store variables and to compute the transportation cost for each store. The objective function is stated in lines 24–25 and sums the fixed and transportation costs. This is a standard constraint-programming model for this problem, the only peculiarity being the syntax of OBJECTIVE-C.

```
 1 id<ORModel> model = [ORFactory createModel];
 2 ORInt fixed = ...;
 3 ORInt maxCost = ...;
 4 id<ORIntRange> Stores     = ...;
 5 id<ORIntRange> Warehouses = ...;
 6 ORInt* cap = ...;
 7 ORInt** tcost = ...;
 8
 9 id<ORIntVarArray> cost =
10    [ORFactory intVarArray: model range:Stores domain: RANGE(model,0,maxCost)];
11 id<ORIntVarArray> supp =
12    [ORFactory intVarArray: model range:Stores domain: Warehouses];
13 id<ORIntVarArray> open =
14    [ORFactory intVarArray: model range:Warehouses domain: RANGE(model,0,1)];
15
16 for(ORInt i=Warehouses.low;i <= Warehouses.up;i++)
17    [model add: [Sum(model,s,Stores,[supp[s] eq: @(i)]) leq: @(cap[i])]];
18 for(ORInt i=Stores.low;i <= Stores.up; i++) {
19    id<ORIntArray> row = [ORFactory intArray: model range: Warehouses with:
20      ^ORInt(ORInt j) { return tcost[i][j];}];
21    [model add: [[open elt: supp[i]] eq: @1]];
22    [model add: [cost[i] eq: [row elt: supp[i]]]];
23 }
24 [model minimize: Sum(model,s,Stores,cost[s]) plus:
25                  Sum(model,w,Warehouses,[@(fixed) mul: open[w]])];
26
27 id<CPProgram> cp = [ORFactory createCPProgram: model];
28 id<MIPProgram> mip = [ORFactory createMIPProgram: model];
29 [cp solve];
30 [mip solve];
```

Fig. 1. Capacitated Warehouse Location in OBJECTIVE-CP

This model is a specification and cannot be executed. No data structures are allocated for the variables, the constraints, and the objectives. For instance, a variable contains its domains but not in a form that can be used for computation and a constraint only collects its variables or expressions. This is similar to models in modeling languages and in the ILOG CONCERT library.

Models are first-class objects in OBJECTIVE-CP: They can be cloned and transformed and it is possible to retrieve their variables, constraints, and objective. They also support the definition of model combinators, an abstraction to build hybrid optimization algorithms compositionally [4].

3.2 Programs

To execute a model in OBJECTIVE-CP, it is necessary to create an optimization program. Lines 27–28 from Figure 1

```
1 id<CPProgram> cp  = [ORFactory createCPProgram: model];
2 id<MIPProgram> mip = [ORFactory createMIPProgram: model];
```

create a CP and a MIP program. In other words, these lines concretize the warehouse location model into two executable programs. These two programs are then solved in lines 29–30 with both technologies.

The implementation of lines 27–28 involves a series of model transformations followed by an actual concretization which creates the solver variables, constraints, and objectives. Like in modern modeling languages such as MINIZINC,

```
1 id<ORModel> m = [ORFactory createModel];
2
3 // data declarations and reading
4
5 id<ORIntVarArray> slab =
6   [ORFactory intVarArray: m range: SetOrders domain: Slabs];
7 id<ORIntVarArray> load =
8   [ORFactory intVarArray: m range: Slabs domain: Capacities];
9
10 [m add: [ORFactory packing: slab itemSize: weight load: load]];
11 for(ORInt s = Slabs.low; s <= Slabs.up; s++)
12   [m add:[Sum(m,c,Colors,Or(m,o,coloredOrder[c],[slab[o] eq:@(s)])) leq:@2]];
13 [m minimize: Sum(m,s,Slabs,[loss elt: [load at: s]])];
14
15 id<CPProgram> cp = [ORFactory createCPProgram: m];
16
17 [cp solve: ^{
18   for(ORInt i = SetOrders.low; i <= SetOrders.up; i++) {
19     ORInt ms = max(0,[cp maxBound: slab]);
20     [cp tryall: Slabs suchThat: ^bool(ORInt s) { return s <= ms+1; }
21                 in: ^void(ORInt s) { [cp label: slab[i] with: s]; }
22     ];
23   }
24 }]
```

Fig. 2. The Steel Mill Slab Problem in OBJECTIVE-CP

models are flattened to avoid potential redundant normalizations of expressions, constraints and objectives by several solvers. Solvers can then be kept small and compact. The flattened model must also be linearized for MIP solvers that do not support reification and element constraints. Observe that a model can be concretized multiple times as Figure 1 demonstrates. This feature is useful when designing hybrid optimization models such as those implemented in CML [5], to implement multi-start solvers, or portfolios of algorithms.

3.3 Search

The warehouse location programs use black-box searches. However, OBJECTIVE-CP lets optimizers state their own search algorithms. Figure 2 depicts a program for the steel mill slab problem (the data declaration and reading are omitted for brevity). The model (lines 5–13) is the same as in [6,8]: Line 10 states a global packing constraint and line 12 imposes reified constraints for the color requirements. The objective function in line 13 features element constraints. The model is concretized into a CP program on line 15. The search procedure is specified in lines 18–24: It combines a C for-loop with a nondeterministic tryall instruction (first introduced in OPL [20]). The search procedure breaks value symmetries on the slabs as in [8].

There are three features of OBJECTIVE-CP that deserve to be highlighted here. First, an optimization program in OBJECTIVE-CP is the combination of a propagation engine and a (solver-independent) search explorer as shown in Figure 3. It follows that the search in OBJECTIVE-CP is generic: It only features generic abstractions such as search controllers, nondeterministic constructs, and combinators, which do not depend on a particular solver technology. Once again,

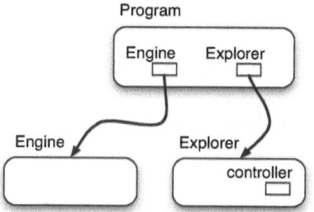

Fig. 3. The Structure of a CP Program in Objective-CP

this is an interesting property from a software engineering standpoint since it provides reusability and separation of concerns.

Second, the search procedure operates on the model variables, not on concrete variables of the underlying solver (in contrast to ILOG CONCERT). Programmers in OBJECTIVE-CP do not need to know the existence of concrete variables: The optimization program is responsible for performing the mapping from model to concrete variables.

Third, OBJECTIVE-CP achieves a strong symbiosis with its host language, since native iterative control structures in OBJECTIVE-C (e.g., for-loops) and non-deterministic constructs in OBJECTIVE-CP (e.g., `tryall`) are interleaved to implement search procedures. It is useful to dive into a simple example to show this symbiosis. The excerpt

```
1 -(void) labelArray: (id<ORIntVarArray>) x {
2   for (int i=x.range.low;i <= x.range.up; i++)
3     while (![cp bound: x[i]]) {
4       int v = [cp min: x[i]];
5       [cp try: ^{ [cp label:x[i] with:v];} or: ^{ [cp diff:x[i] with: v];}];
6     }
7 }
```

depicts how to label an array of variables in OBJECTIVE-CP. Line 2 iterates over the indices of the array, while lines 3–6 label variable x[i]. The variable labeling is a while-loop that terminates when x[i] is bound. Each iteration takes the smallest value in the domain of x[i] and either assigns it to x[i] or remove it from its domain. This nondeterministic step uses the `try` nondeterministic construct ([20,3]). What is interesting in this snippet is the ability to use a for-loop over local variables (e.g., i and v) and to combine it with nondeterminism. This is possible because, when the OBJECTIVE-CP implementation selects an alternative to the `try` instruction, it restores the values of i and v to their states before the execution of the `try`. This symbiosis enables an iterative style for writing search procedures that avoid the need to use goals and the *and* combinator in [10,17]. This symbiosis provides significant software engineering benefits, including a small and modular implementation, and the ability to avoid decoupling the syntactic structure of the search and the control flow of the execution. Even more important perhaps are the benefits for debugging. Users can place breakpoints anywhere in the code and use native debugging tools to inspect the computation state. With goals or combinators for sequential composition, users, when debugging, are in the interpreter code and have lost the connection to the original control flow.

The design of the search component in OBJECTIVE-CP is based on the motto

Search = Continuations + Controllers

or, in other words, the belief that a rich and extensible search language can be built bottom-up from continuations and search controllers. These abstractions make it possible to offer a rich collection of high-level nondeterministic constructs and search combinators. Moreover, controllers can also be used to specify node selection strategies. As a result, OBJECTIVE-CP unifies the search controllers of COMET [22] and the search combinators of [17] (and hence the goals of ILOG SOLVERS). We illustrate a few of these combinators and present a large neighborhood search based on them.

Limits. Limit combinators (e.g., [14]) impose conditions on a search procedure or, equivalently, on the exploration of a search subtree. These limits may concern a variety of measures including computation times and the number of solutions, as well as any condition provided by users. For instance,

```
1 [cp limitSolutions: k in: ^{ [cp labelArray: x]; }];
```

ensures that only k labelings of array x are returned. Once the k-th labeling has been obtained, the instruction fails. Limits can be composed naturally as in

```
1 [cp limitFailures: 200 in ^{
2    [cp limitSolutions: k in: ^{ [cp labelArray:x]; }];
3 }];
```

The resulting code also limits the number of failures in the labeling to 200.

The Repeat Combinator. The repeat combinator repeatedly executes a body until it succeeds, interleaving each iteration with the execution of a closure. In Prolog, it would be written as

```
1 repeat(Body,Restart) :- call(Body).
2 repeat(Body,Restart) :- call(Restart), repeat(Body,Restart).
```

It is particularly useful with various forms of randomization and local search algorithms. For instance, the fragment

```
1 [cp repeat: ^{ [cp limitFailures: nbFailures.value
2                          in: ^{ [cp labelHeuristic: h];}] }
3   onRepeat: ^{ [nbFailures setValue: nbFailures.value * f]; }
4 }];
```

depicts a (randomized) search procedure with restarts. In this example, each iteration is allowed `nbFailures.value` failures and the number of failures is increased geometrically after each search. The same excerpt can be transformed in a generic restart combinator that repeatedly executes a body until a condition c is met and executes **restart** when the body fails, i.e.,

```
1  [cp solve: ^{
2    [cp limitTime: timeLimit in: ^{
3      [cp repeat:^{
4        [cp limitFailures: failureLimit in:^{
5          for(ORInt i = SetOrders.low; i <= SetOrders.up; i++) {
6            ORInt ms = max(0,[cp maxBound: slab]);
7            [cp tryall: Slabs suchThat: ^bool(ORInt s) { return s <= ms+1;}
8                      in: ^void(ORInt s) {
9                [cp label: slab[o] with: s];
10               }];
11         }
12       }];
13     }]
14     onRepeat:^{
15       id<ORSolution> s = [[cp solutionPool] best];
16       if (s != nil)
17         [Orders enumerateWithBlock:^(ORInt i) {
18           if ([d next] <= Pr)
19             [cp label: x[i] with: [s intValue:x[i]]];
20         }];
21     }];
22   }];
23 }];
```

Fig. 4. A Large Neighborhood Search for the Steel Mill Slab Problem

```
1  -(void) restart: (void^()) body when: (BOOL^()) c onRestart: (void^())restart
2  {
3    [self repeat: [self limitCondition: c in: body()]; onRepeat:^{restart();}];
4  }
```

The earlier restart search, together with a time limit, can be written

```
1  [cp limitTime: 300 in: ^ {
2      [cp restart: ^{ [cp labelHeuristic: h]; }
3           when:   ^BOOL() { return cp.nbFailures > nbFailures.value; }
4           onRestart:  ^{ [nbFailures setValue: nbFailures.value * f]; }
5      ];
6  }];
```

These code fragments demonstrate compositionality. Limit, repeat, and nondeterministic combinators are interleaved in arbitrary ways.

Large Neighborhood Search. Figure 4 illustrates how search combinators can be used to implement a large neighborhood search (LNS) in OBJECTIVE-CP. The LNS search uses two limit combinators and a repeat combinator to transform the core search depicted previously into a large neighborhood search. The outermost limit combinator ensures that the overall search does not exceed the time limit. In the search, the body of the repeat combinator is the basic search (lines 5–10) presented in Figure 2, enclosed in a limit combinator that limits the number of failures (line 4). The restart code in lines 15–20 defines the neighborhood structure: It takes the best solution found so far, and fixes the value of a variable in this solution with probability Pr (d is a uniform distribution in [0,1]).

Implementation of Search Combinators. How does OBJECTIVE-CP implement combinators compositionally? *The key is to use controllers and to support the*

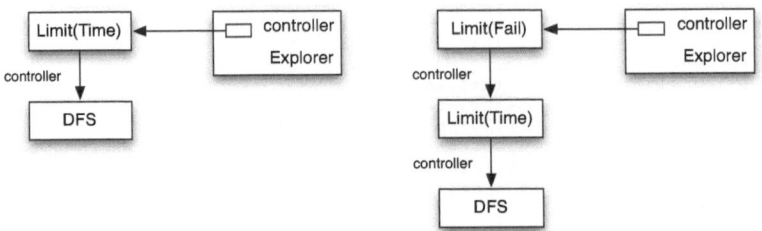

Fig. 5. Chaining Controllers

```
1 -(void) limitFailures: (int) max in: (void^()) body
2 {
3     id<ORSearchController> c = [[ORLimitFailures alloc] initWithLimit: max];
4     [self pushController: c];
5     body();
6     [c succeeds];
7     [self popController];
8 }
```

Fig. 6. The Implementation of a Limit Combinator

ability to chain controllers. Figure 6 depicts the implementation of the limit combinator for failures. Lines 3–4 create a limit controller (line 3) and push it as the top-level controller (line 4). Line 7 pops it once the body of the combinator is executed. Figure 5 depicts the chains of controllers for one and two limits, highlighting the compositionality of the approach. The effect of the **pushController** instruction is undone upon backtracking or, more generally, when the search jumps to other nodes in the search tree. The ability to chain controllers is the key for the modular and compositional design of search combinators. A controller for limiting the number of solutions is shown in Figure 7. It maintains the number of solutions produced so far (_nbSol), which is updated every time the body succeeds (line 11). Methods **startTryLeft** and **startTryRight** tests whether the limit is reached, in which case it fails. Otherwise, they delegate the call to the subcontroller, i.e., the next controller _controller in the chain. All other methods in the controller are inherited from the superclass and simply delegate the call to the subcontroller. Figure 8 depicts the implementation of the **repeat** combinator. Lines 2–3 create a continuation and a choice point. The first execution runs the body (line 6). Later executions creates a choice point, executes the **onRepeat** code, then the body.

3.4   Transparent Parallelization in Objective-CP

OBJECTIVE-CP supports fully transparent parallelizations of its programs. For instance, the concretization

```
1 id<CPProgram> cp = [ORFactory createCPProgram: model];
```

can be replaced by

```
1 id<CPProgram> cp = [ORFactory createCPMultiStartProgram: model nb: 4];
```

```
1  @implementation ORLimitSolution {
2      int _max;
3      int _nbSol;
4  }
5  -(ORLimitSolution*) initWithLimit: (int) m
6  {
7      self = [super init];
8      _max = m; _nbSol = 0;
9      return self;
10 }
11 -(void)  succeeds           { _nbSol++;}
12 -(BOOL) hasReachedLimit { return _nbSol >= _max; }
13 -(void) startTryLeft
14 {
15     if ([self hasReachedLimit]) [_controller fail];
16     else return [_controller startLeft];
17 }
18 -(void) startTryRight
19 {
20     if ([self hasReachedLimit]) [_controller fail];
21     else return [_controller startRight];
22 }
23 @end
```

Fig. 7. The Limit Solution Controller

```
1  -(void) repeat: (ORClosure) body onRepeat: (ORClosure) onRepeat {
2      NSCont* enter = [NSCont takeContinuation];
3      [_controller._val addChoice: enter];
4      if ([enter nbCalls]!=0)
5          if (onRepeat) onRepeat();
6      body();
7  }
```

Fig. 8. The Implementation of the **repeat** Combinator

to obtain a multistart search procedure capable of executing four searches on four threads with different random seeds: *The model and the search procedure are left unchanged.* At the implementation level, OBJECTIVE-CP concretizes the model four times to obtain four different CP programs. The search is executed on each CP program. Since the search is expressed in terms of the model objects, a particular thread executing the search will retrieve its concrete solver from the multistart program and then access the concrete objects in that solver.

OBJECTIVE-CP also supports a full transparent parallelization of a branch & bound search using a work-stealing model. For instance, the code

```
1 id<CPProgram> cp = [ORFactory createParCPProgram: model];
```

creates a parallel branch & bound that can execute search algorithms such as those described previously. These algorithms can be defined in terms of search combinators and nondeterministic constructs. The parallel implementation follows the computational model described in [11,12] which exploits search controllers (see Figure 9). A parallel constraint program consists of a set of workers, a problem pool, and a template to create the node selection strategy. Each worker is a solver with its own engine and search explorer. The explorer has its traditional chain of controllers and a parallel adapter that encapsulates two

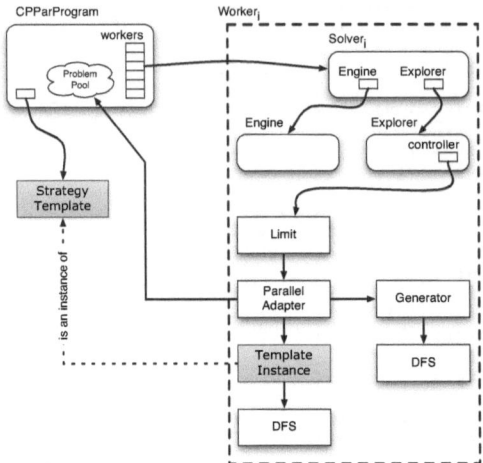

Fig. 9. The Parallel Solver Architecture

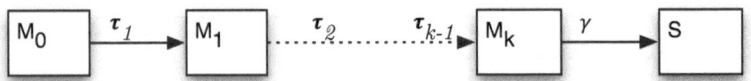

Fig. 10. Model Transformations and Concretizations

other controllers: an instance of the strategy template that is used to perform the search and a generator to produce subproblems into the problem pool.

4   The Implementation of Objective-CP

The OBJECTIVE-CP implementation receives a model M_0 as input, performs a number of model transformations $\tau_1, \ldots, \tau_{k-1}$ to obtain models $M_1, \ldots, M_k$ and then concretizes the final model M_k to obtain a concrete solver S (see Figure 10). Each model in this sequence is of the form $\langle X_i, C_i, O_i \rangle$, where X_i are the model variables, C_i is the set of constraints, O_i is the objective function, and $X_i \subseteq X_{i+1}$ ($1 \leq i < k$). This section reviews some of these steps.

4.1   Model Transformations

OBJECTIVE-CP supports a number of model transformations, including flattenings, linearizations, and relaxations. Flattenings are becoming a standard tool in optimization systems, as examplified by systems such as FLAT ZINC [13]. It removes the need to manipulate expressions in solvers that can then focus on implementing the core constraints. This methodology is also advocated in [18] to minimize the size of a kernel, possibly using views to minimize or eliminate most of the induced overhead. Figure 11 describes some flattening rules

$$\tau_f(\langle X, \{C_1, \ldots, C_k\}, O \rangle) = \langle X \cup X', C', O' \rangle \quad \textbf{where}$$

$$\begin{aligned}
\tau_c(C_i) &= \langle C_i', X_i' \rangle \quad (1 \le i \le k), \\
\tau_e(O) &= \langle O', C_O, X_o \rangle, \\
X' &= \left(\bigcup_{i=1}^k X_i' \right) \cup X_o, \\
C' &= \left(\bigcup_{i=1}^k C_i' \right) \cup C_o.
\end{aligned}$$

$$\tau_e(e_l * e_r) = \langle z, C, X \rangle \quad \textbf{where}$$

$$\begin{aligned}
\tau_e(e_l) &= \langle z_l, C_l, X_l \rangle, \\
\tau_e(e_r) &= \langle z_r, C_r, X_r \rangle, \\
C &= \{\texttt{mult}(z_l, z_r, z)\} \cup C_l \cup C_r, \\
X &= \{z, z_l, z_r\} \cup X_l \cup X_r.
\end{aligned}$$

$$\tau_c(\texttt{alldifferent}(e_1, \ldots e_n)) = \langle C, X \rangle \quad \textbf{where}$$

$$\begin{aligned}
\tau_e(e_i) &= \langle z_i, C_i, X_i \rangle \quad (1 \le i \le n) \\
C &= \{\texttt{alldifferent}(z_1, \ldots z_n)\} \cup \bigcup_{i=1}^n C_i \\
X &= \{z_1, \ldots, z_n\} \cup \bigcup_{i=1}^n X_i.
\end{aligned}$$

Fig. 11. Excerpts of The Flattening Transformation in OBJECTIVE-CP

```
1  τMIP(alldifferent(x1,...xn))  =  {
2      [[x1 = 1]] +...+ [[xn = 1]] ≤ 1,
3      ...
4      [[x1 = k]] +...+ [[xn = k]] ≤ 1,
5  }
```

Fig. 12. A Linearization of the Alldifferent Constraint

in OBJECTIVE-CP in terms of the function τ_f, τ_c, and τ_e to flatten a model, a constraint, and an expression. In these rules, the z_i's are brand new variables not used anywhere in the model. For instance, the figure illustrates the flattening of the **alldifferent** constraint, which flattens the expressions $e_1, \ldots, e_n$ to obtain the variables $z_1, \ldots, z_n$, the constraints $C_1, \ldots, C_n$, and the new variables $X_1, \ldots, X_n$. The resulting alldifferent constraint is solely expressed in terms of variables, not expressions. Finally, τ_e illustrates the flattening of a multiplication expression to obtain very simple constraints in the solver.

Models can be linearized for use in a MIP solver and there is considerable literature on how to perform such transformations (e.g., [16,9]). Figure 12 describes the linearization of the alldifferent constraint assuming that the variables take values in $1..k$. The linearization defines a number of inequalities over the literals $[\![x_i = j]\!]$, i.e., 0/1 variables that denote whether x_i is assigned the value j. The linearization of the variables also generates constraints of the form

$$x_i = \sum_{j=1}^k j * [\![x_i = j]\!].$$

As suggested in [16], these constraints may be enforced lazily when the model only uses the literals.

As indicated earlier, models are first-class objects in OBJECTIVE-CP and the implementation can include code of the form

```
id<ORModel> lfm = [[[m copy] flatten] linearize];
```

Another benefit of the OBJECTIVE-CP architecture is the factorization of transformations (flattenings, normalizations, and linearizations) across multiple solvers. Solvers do not have to be concerned with these transformations which are now performed at the model level. Obviously it does not mean that solvers cannot manipulate constraint globally as is necessary, say in hull and box-consistency [1,23]. Rather it simply means that the transformations provide a normalization of the model expressions in a form appropriate for the solvers.

4.2   Concretization

The concretization γ takes a model m in a flattened form appropriate for a solver s and concretizes m into s, i.e., $s = \gamma(m)$. The concretization associates a concrete variable with every model variable and a concrete object to every model object (e.g., a constraint). The concretization γ is used at various places. For instance, an instruction

```
[cp label: x with: v]
```

that labels variable x with value v is implemented by a call

```
[engine label: γ(x) with: v]
```

that concretizes variable x and calls the same method on the constraint engine. The literals of variable x can be accessed through

```
[cp literal: x for: v]
```

which returns $\gamma(\llbracket x = v \rrbracket)$. The instruction

```
[parcp label: x with: v]
```

in a parallel solver is implemented by the call

```
[cp_k label: x with: v]
```

which itself becomes

```
[engine_k label: γ_k(x) with: v]
```

where $\mathtt{cp}_k$ is the kth solver, $\mathtt{engine}_k$ is its engine, and γ_k is its concretization.

Adding a constraint c during the search requires some care, since these constraints are expressed in terms of the original model. The optimization program must preserve the chain of transformations $\tau_1, \ldots, \tau_k$ and apply them to constraint c to obtain a tuple $\langle X, C \rangle$, where X is a set of new variables and C is a set of new constraints. Both X and C can now be concretized through γ and posted in the solver. In other words, the addition of a constraint c executes

$$\gamma(\tau_n(\ldots(\tau_2(\tau_1(\langle\{\}, \{c\}\rangle)))\ldots)).$$

```
1 @protocol CPEngine
2 -(ORStatus)          add: (id<CPConstraint>) c;
3 -(void) setObjective: (id<ORObjective>) obj;
4 @end
```

Fig. 13. The `CPEngine` Interface

```
1 @protocol CPUKernel
2 -(void) scheduleCtrEvt: (id<CPCtrEvent>) list;
3 -(void) scheduleValEvt: (id<CPValEvent>) list;
4 -(void) triggerLossEvt: (id<CPTriggerMap>) map;
5 -(void) triggerBindEvt: (id<CPTriggerMap>) map;
6 -(ORStatus) propagate;
7 @end
```

Fig. 14. The Micro-Kernel Interface

4.3   A Micro-Kernel Architecture

The constraint-programming solver of OBJECTIVE-CP is based on a micro-kernel architecture inspired by ideas from operating systems. Micro-kernel architectures have become popular in operating systems as they favor extensibility, maintenance, and easier proofs of correctness. This section briefly reviews the main ideas behind the micro-kernel architecture.

The CP solver in OBJECTIVE-CP is only concerned with constraint propagation, as the search in OBJECTIVE-CP is solver-independent and lies in a separate library. The solver itself consists of two objects: an *engine* that defines the API to add constraints and objectives and a *kernel* that implements the propagation and provide minimal functionalities to define new propagators. Their interfaces are sketched in Figures 13 and 14.

The engine interface is used to register native constraints and objectives. No decomposition or rewriting is necessary at this stage, since these transformations took place earlier in the pipeline, The constraint interface

```
1 @protocol CPConstraint
2 -(ORUInt) getId;
3 -(ORStatus) post;
4 @end
```

is minimalist and only requires each constraint to be uniquely identified and to support a **post** method. Objective function are similar and abstract away the nature of the objective function.

The micro-kernel provides the interface to schedule and propagate events. It supports four types of events: constraint, variable, value, and trigger events. Intuitively, constraint events are used to propagate a constraint, variable events to execute a closure (e.g., upon a variable modification), value events to apply a first-order function (e.g., when a value is removed from a domain), and trigger events to update triggers. For space reasons, we do not discuss triggers in the rest of the paper: It suffices to say that they provide functionalities related to the watched literals in MINION [7]. The interfaces `CPCtrEvent` and `CPValEvent` represent simple lists of constraints, closures, and first-order function applications (`CPCtrEvent` contains both constraints and variable events for efficiency and simplicity reasons). Observe that the micro-kernel is agnostic with respect to the events themselves, which are in the realm of the variable definitions in other libraries. Hence, the micro-kernel architecture entirely separates the propagation from the variables and makes it possible to add new variable types compositionally without upgrading the kernel. In particular, the list of events are built and maintained outside the kernel.

The micro-kernel maintains an array of $P + 1$ queues to track the scheduled closures and first-order function applications. By default, value events are at priority P, variable events at priority $P - 1$, and constraint events at priorities $1..P - 2$. Priority 0 has a special role to be discussed shortly. Conceptually, method scheduleCtrEvt receives a list of k pairs $\langle c_i, p_i \rangle$ where c_i is a constraint and p_i is an integer priority in $1..P - 2$ and updates the queues with

$$Q_{p_i} = \texttt{enQueue}(Q_{p_i}, c_i) \ 0 \le i \le k - 1.$$

The other scheduling functions are similar.

Method propagate executes the propagation loop and its implementation is shown in Figure 15. The algorithm processes each non-empty queue in turn from the highest (P) to the lowest (1) priority. Line 5 finds the index of the highest priority queue with some events. Lines 6–9 pick the first highest priority event, execute it (line 7) and carry on until $p = 0$ which indicates that all queues in the $1..P$ range are empty. Finally, lines 12 and 15 unconditionally execute the events

```
1 -(ORStatus)propagate {
2   BOOL done = NO;
3   return tryfail(ORStatus^{
4     while (!done) {
5       p = max_{i=1}^{P} i · (Q_i ≠ ∅)
6       while (p ≠ 0) {
7         execute(deQueue(Q_p));
8         p = max_{i=1}^{P} i · (Q_i ≠ ∅);
9       }
10      done = Q_p = ∅;
11    }
12    while (Q_0 ≠ ∅) execute(deQueue(Q_0));
13    return ORSuspend;
14   }, ^ {
15    while (Q_0 ≠ ∅) execute(deQueue(Q_0));
16    return ORFail;
17   });
18 }
```

Fig. 15. The Micro-Kernel Propagation

held in Q_0 even after a failure has been discovered. As is customary, the dispatching of messages may schedule additional events that will be handled during this cycle. Lines 4–13 are the body of a closure which is passed to the micro-kernel function tryfail together with a failure handler (line 15–16). Function tryfail executes the first closure and diverts the control flow to the second closure when the fail function is called during the first closure execution. The events in Q_0 are always executed: They are typically used to collect and update monitoring information, which can then be used to implement heuristics or learning techniques. The micro-kernel does not have any reference to variable, domain, or even the nature of constraints: It only manipulates closures, first-order applications, and information about constraints (e.g., whether they have been propagated or not).

4.4   A Finite-Domain Integer Variable

We now sketch the implementation of a finite-domain variable, whose class CPIntVar is outlined as follows:

```
 1 @interface CPIntVar {
 2   id<CPEngine>                      _engine;
 3   id<CPDom>                         _dom;
 4   id<CPCtrEvent>  _min,_max,_bounds,_bind;
 5   id<CPValEvent>                    _loss;
 6 }
 7 -(id)initVar: (CPEngine*) engine low: (ORInt) low  up: (ORInt) up;
 8 -(void) whenChangeMinDo: (void^()) f;
 9 -(void) whenLoseValueDo: (void^(ORInt)) f;
10 -(void) whenChangePropagate: (id<CPConstraint>) c;
11 -(ORStatus) updateMin: (ORInt) newMin;
12 -(ORStatus) removeValue: (ORInt) value;
13 @end
```

Instance variable _dom points to a domain representation such as a range, a bit-vector, or a list of intervals. Methods whenChangeMinDo and whenLoseValueDo are used to register variable events and value events respectively. Their implementation is simple:

```
1 -(void) whenChangeMinDo: (void^()) f { [_min insert: f]; }
2 -(void) whenLoseValueDo: (void^(ORInt)) f { [_loss insert: f]; }
```

Method whenChangePropagate is slightly more involved: It creates a closure cl to propagate a constraint ctr and inserts a pair (cl,ctr) in the list.

The update methods of a variable are expected to schedule relevant events. Consider method updateMin:

```
1 -(ORStatus)updateMin:(ORInt)newMin {
2   BOOL changed = [_dom updateMin:newMin];
3   if (changed) {
4     [_engine scheduleCtrEvt:_min];
5     [_engine scheduleCtrEvt:_bounds];
6     if ([_dom size] == 1) [_engine scheduleCtrEvt:_bind];
7   }
8   return ORSuspend;
9 }
```

It updates the domain, and when the domain is modified, it schedules the execution of the events registered on the _min and _bounds lists. If the domain is a singleton, line 6 also schedules the _bind list. Note that method updateMin on the domain may raise a failure, which is captured in the tryfail construct.

4.5 Propagators

We now illustrate a few propagators. Figure 16 depicts a domain-consistent propagator for constraint $x = y + c$ using value events. Lines 3–4 cover the cases where one of the variables is bound and the other variable is updated accordingly. Lines 6–7 initiate the filtering of x and y by tightening their respective bounds. Lines 8–11 prune the domain of the variables. Lines 13–14 associates first-order functions with x and y to respond to the loss of value v from their domains. Figure 17 sketches the implementation of a global constraint using variable and constraint events. The post method scans all the variables (lines 4–6) and registers a closure to update internal data structures when a variable

```
 1 @implementation CPEqualDC   // x = y + c
 2 -(ORStatus) post {
 3    if (bound(x))        [y bind: x.min - c];
 4    else if (bound(y)) [x bind: y.min + c];
 5    else {
 6        [x updateMin:y.min + c andMax:y.max + c];
 7        [y updateMin:x.min - c andMax:x.max - c];
 8        for(ORInt i = x.min;i <= x.max; i++)
 9            if (![x member:i]) [y remove:i - c];
10        for(ORInt i = y.min; i <= y.max; i++)
11            if (![y member:i]) [x remove:i + c];
12
13        [x whenLoseValueDo:^(ORInt v) { [y remove: v - c];}];
14        [y whenLoseValueDo:^(ORInt v) { [x remove: v + c];}];
15    }
16    return ORSuspend;
17 }
```

Fig. 16. A Domain Consistency Propagator for $x = y + c$

```
 1 @implementation CPAllDifferent // on array _x
 2 -(ORStatus) post {
 3    [self initDataStructures];
 4    for(ORInt i = _x.low; i <= _x.up; i++)
 5        if (!_x[i].bound])
 6            [_var[i] whenBindDo: ^{ [self removeOnBind:i]; } ];
 7    [self propagate];
 8    for(ORInt k = _x.low; k <= _x.up; k++)
 9        if (!_x[k].bound) [_x[k] whenChangePropagate: self];
10    return ORSuspend;
11 }
12 -(void) propagate {
13    if (![self feasible])  fail();
14    [self prune];
15 }
16 @end
```

Fig. 17. The AllDifferent Skeleton

is bound. It propagates the constraint and registers itself with each variable to propagate whenever the domain of the source variable _x[k] changes (lines 8–9). Observe that this implementation combines variable-based and constraint-based propagation.

5   Experimental Results

Every programming system strikes a trade-off between expressiveness and efficiency. This section examines the efficiency of OBJECTIVE-CP to quantify this trade-off more precisely. It describes the compilation, runtime, and search efficiency of OBJECTIVE-CP. All experiments were carried out on MacOS X 10.8.3 running on a Core i7 at 2.6Ghz. All the results are based on 10 runs of the program (given that tie-breaks are randomized). All the times are reported in milliseconds. Columns $\mu(T)$ and $\sigma(T)$ report the average and standard deviation for the total runtime (measured from the moment the program starts to the moment it terminates). Column $\mu(|M|)$ reports the memory usage in kilobytes for the entire executable (it is measured at the `malloc` interface).

Table 1. Performance of OBJECTIVE-CP

Bench	$\mu(T)$	$\sigma(T)$	$\mu(|M|)$	$|X|$	$|C|$	$G(|X|)$	$G(|C|)$	$\mu(T_c)$	$\sigma(T_c)$
ais	1,444.3	111.8	1,336	59	33	1.49	1.88	1.9	0.6
costas	5,298.3	7,210.7	1,544	240	424	1.65	1	7.7	1.8
fdmul	427.3	64.2	674	29	15	1.83	1.6	1.4	0.5
magicserie	9,808.4	86.1	122,050	300	301	302.99	1	961.8	23.9
bibd	793.2	104.3	16,836	2,205	663	4	10.98	123	13.1
debruijn	7,619.3	207	554,915	57,346	57,346	2	1	1,535.3	121.9
golomb	28,068.2	238.8	904	132	134	1	1	2.2	0.4
coloringModel	14,872.7	248.2	1,818	81	1,309	1	1	10.9	2.4
eq20	23.6	2.9	986	7	20	21	1	2.3	0.7
latinSquare	498.7	55	1,660	147	182	1.33	1	4.5	1.6
slab	2,937.6	154.2	52,313	223	113	179.2	26.54	465.3	7.1
sport	5,987.8	380.3	2,018	287	113	1	1	4.2	1.2

Table 2. Comparison of OBJECTIVE-CP and COMET

System	Perfect $\mu(T)$	$\sigma(T)$	Sport $\mu(T)$	$\sigma(T)$	Slab $\mu(T)$	$\sigma(T)$	Slab-LNS $\mu(T)$	$\sigma(T)$
comet	6,145.64	180.8	4,655.7	90.9	5,205.5	102.6	17,241.7	10,644
ocp	5,884.16	204.8	4,712.1	76.3	2,213.2	59.1	6,308.2	3,632.5

Compilation Efficiency. Table 1 discusses model compilation efficiency. Columns $|X|$ and $|C|$ report the size of the model (number of variables and constraints), while $G(|X|)$ and $G(|C|)$ give the growth rate when the model has been flattened and concretized. $G(|X|) = 2.0$ states that the concretized model has twice as many variables as the original. $\mu(T_c)$ and $\sigma(T_c)$ report the average and standard deviation of the time (in milliseconds) needed to compile and concretize the high-level model. The compilation time is almost always insignificant rarely passing 10 milliseconds. For `Debruijn`, the compilation is the heaviest with about 1.5 seconds for a large instance with some 57,346 variables and constraints modeled as algebraic expressions. The next two most expensive are the `magicserie` and `slab` benchmarks that heavily rely on reifications.

Runtime Efficiency. Table 2 offers an insight on the performance of OBJECTIVE-CP relative to the COMET platform and shows that the OBJECTIVE-CP architecture competes with the COMET platform. The statistics for both systems were collected on the same (slightly faster) machine over 50 runs of each benchmark. `Slab-LNS` relies on a complex search heuristic with dynamic symmetry breaking and a large neighborhood search component (i.e., it uses several search combinators). Care was taken to implement the exact same search in COMET.

Search Efficiency. Table 3 provides an overview of several black-box search procedures embedded in combinators. Space limitations do not allow for a comprehensive suite of tests but, instead, we focus on demonstrating sound performance on a number of interesting search procedures. The magic square benchmark (size 12) was executed with both ABS and IBS within a restarting search with a failure limit doubling at each restart. The satisfaction knapsack (instance 4) results are

Table 3. Search with OBJECTIVE-CP

Bench	$\|Method\|$	$\mu(T_{cpu})$	$\sigma(T_{cpu})\|$	$\mu(T_{wc})$	$\sigma(T_{wc})\|$	$\mu(Choices)$	$\sigma(Choices)\|$	$\mu(\|M\|)$
magicsquareModel	ABS	7,871.8	3,251.6	8,043.9	3,335.7	57,952.5	37,181.1	4,850.1
magicsquareModel	IBS	15,021.2	7,581.3	15,228.9	7,816.8	101,308.1	144,010.7	2,289.2
knapsack	ABS	401.9	115.3	409.4	116.5	8,975.5	2,493.2	1,413
progressive	ABS	2,763.8	695.4	2,797.3	703.5	1,226.1	822.1	17,303.5
progressive	IBS	958.1	281.1	984.9	287.9	2,988.4	2,139.8	15,511.8
progressive	WDeg	1,587.3	1,068.2	1,630.4	1,087.9	8,769.2	6,724.9	15,459.4

for ABS alone (the others being too long). For the progressive party problem, the configuration is (2,8) and three black-box searches were compared. The search were all embedded in the same failure-limited restarting search. The results show that OBJECTIVE-CP achieves strong results on these benchmarks using state-of-the-art black-box algorithms. Note that, while ABS is the slowest of the three on progressive, it does far fewer choices than the other two and the bulk of the running time is spent in the probing. (No attempts were made to tweak the parameters which were identical on all benchmarks.).

6   Related Work

Obviously, OBJECTIVE-CP builds on top of over two decades of research in constraint programming and modeling systems. This section reviews some of the most relevant work from an overall system design standpoint. OPL, ILOG CONCERT, COMET, the G12 project, and GECODE probably had the most influences on its design and implementation. At some level, OBJECTIVE-CP can be viewed as a synthesis of the salient features of each of these systems and delivers, what we believe, is a sweet spot in the trade-off between expressiveness and efficiency. It goes without saying that different users may prefer different abstraction levels (e.g., a modeling language or a low-level library) based on their experience and the nature of their applications.

OPL [20] is a modeling language implemented on top of ILOG SOLVER which features high-level modeling and search abstractions. OPL introduced the `try` and `tryall` constructs but did not support compositional combinators and its implementation of node selection strategies was somewhat ad-hoc. Its implementation was in terms of goals. Models were not first-class objects and a dedicated language OPL SCRIPT [21] was needed to compose models. OBJECTIVE-CP borrowed the nondeterministic constructs of OPL but boosted almost all other aspects of the system (which is expected given the progress in the last 15 years).

ILOG CONCERT [2] introduced models into C++ libraries. ILOG CONCERT makes it possible to define models that can be then extracted by algorithms. The extraction process associates modeling objects (prefixed by `Ilo`) with implementation objects (prefixed by `Ilc` for the CP solver). Search goals can be expressed on the implementation objects in a way similar to ILOG SOLVER. OBJECTIVE-CP does not expose implementation objects and the search is implemented solely based on model variables. The search language of OBJECTIVE-CP is fundamentally different to those of ILOG CONCERT/ILOG SOLVER. At the functionality level, OBJECTIVE-CP enjoys a nice symbiosis with the underlying host languages, support an iterative style for search, and extensibility of the search languages. At the implementation level, it is based on continuations and controllers, not on recursive goals [15] which are close to the original implementation of constraint logic programming [19].

COMET [3] is a domain-specific language for hybrid optimization, featuring solvers for constraint programming, mathematical programming, and local search. Its search language for constraint programming is based on continuations and controllers [22] but it did not support compositional combinators as argued in [17]. COMET does not have the concept of models, which led to the development of CML [5]. OBJECTIVE-CP merge the controllers and the continuations of COMET with the concept of models and a more compositional search language into a low-level host language, achieving a strong symbiosis between the extensions and the host.

The G12 project at NICTA introduced the MINIZINC modeling language and its implementation through a series of transformations to FLATZINC [13]. MINIZINC has a lot of similarities with OPL but took a much more systematic implementation approach based on model transformations. OBJECTIVE-CP uses a similar strategy but exposes models as first-class objects and enables hybrid optimization through model combinators [4]. It also features a search language merging the benefits of search controllers [22] and search combinators [17].

7 Conclusion

This paper presented an overview of OBJECTIVE-CP based on the vision

Optimization Program = Model + Search + Solver

to strike a good balance between expressiveness, extensibility, and efficiency. From an expressiveness standpoint, OBJECTIVE-CP features high-level modeling and search languages which never refer to implementation objects. The search language of OBJECTIVE-CP is based on the motto

Search = Continuations + Controllers

or, in other words, the belief that a rich and extensible search language can be built bottom-up and compositionally from continuations and search controllers. OBJECTIVE-CP provides the ability to transparently parallelize multi-start and branch & bound algorithms. At the implementation level, OBJECTIVE-CP performs a number of model transformations of the user model before concretization into a specific solver. OBJECTIVE-CP features an efficient constraint-programming solver based on a micro-kernel architecture, separating propagation, variables, and constraints.

Acknowledgments. NICTA is funded by the Australian Government as represented by the Department of Broadband, Communications and the Digital Economy and the Australian Research Council through the ICT Centre of Excellence program.

References

1. Benhamou, F., Goualard, F., Granvilliers, L., Puget, J.-F.: Revising hull and box consistency. In: ICLP, pp. 230–244 (1999)
2. Concert Technology. Reference Manual, version 12.1 (2009)
3. Inc. Dynadec. Comet v2.1 user manual. Technical report, Providence, RI (2009)
4. Fontaine, D., Michel, L., Van Hentenryck, P.: Model Combinators for Hybrid Optimization. In: Schulte, C. (ed.) CP 2013, vol. 8124, pp. 300–315. Springer, Heidelberg (2013)
5. Fontaine, D., Michel, L.: A High Level Language for Solver Independent Model Manipulation and Generation of Hybrid Solvers. In: Beldiceanu, N., Jussien, N., Pinson, É. (eds.) CPAIOR 2012. LNCS, vol. 7298, pp. 180–194. Springer, Heidelberg (2012)
6. Gargani, A., Refalo, P.: An efficient model and strategy for the steel mill slab design problem. In: Bessière, C. (ed.) CP 2007. LNCS, vol. 4741, pp. 77–89. Springer, Heidelberg (2007)
7. Gent, I.P., Jefferson, C., Miguel, I.: Watched literals for constraint propagation in minion. In: Benhamou, F. (ed.) CP 2006. LNCS, vol. 4204, pp. 182–197. Springer, Heidelberg (2006)
8. Van Hentenryck, P., Michel, L.: The Steel Mill Slab Design Problem Revisited. In: Trick, M.A. (ed.) CPAIOR 2008. LNCS, vol. 5015, pp. 377–381. Springer, Heidelberg (2008)
9. Hooker, J.N.: Logic-Based Methods for Optimization: Combining Optimization and Constraint Satisfaction. John Wiley and Sons (2000)
10. Ilog Solver 4.4. Reference Manual. Ilog SA, Gentilly, France (1998)
11. Michel, L., See, A., Van Hentenryck, P.: Parallelizing constraint programs transparently. In: Bessière, C. (ed.) CP 2007. LNCS, vol. 4741, pp. 514–528. Springer, Heidelberg (2007)
12. Michel, L., See, A., Van Hentenryck, P.: Transparent Parallelization of Constraint Programming. INFORMS Journal on Computing 21(3), 363–382 (2009)
13. Nethercote, N., Stuckey, P.J., Becket, R., Brand, S., Duck, G.J., Tack, G.: MiniZinc: Towards a standard CP modelling language. In: Bessière, C. (ed.) CP 2007. LNCS, vol. 4741, pp. 529–543. Springer, Heidelberg (2007)
14. Perron, L.: Search Procedures and Parallelism in Constraint Programming. In: Jaffar, J. (ed.) CP 1999. LNCS, vol. 1713, pp. 346–361. Springer, Heidelberg (1999)
15. Puget, J.-F.: A C++ Implementation of CLP. In: Proceedings of SPICIS 1994, Singapore (November 1994)
16. Refalo, P.: Linear Formulation of Constraint Programming Models and Hybrid Solvers. In: Dechter, R. (ed.) CP 2000. LNCS, vol. 1894, pp. 369–383. Springer, Heidelberg (2000)
17. Schrijvers, T., Tack, G., Wuille, P., Samulowitz, H., Stuckey, P.: Search Combinators. Constraints 18(2), 269–305 (2013)
18. Schulte, C., Tack, G.: View-Based Propagator Derivation. Constraints 18(1), 75–107 (2013)

19. Van Hentenryck, P.: Constraint Satisfaction in Logic Programming. The MIT Press, Cambridge (1989)
20. Van Hentenryck, P.: The OPL Optimization Programming Language. The MIT Press, Cambridge (1999)
21. Van Hentenryck, P., Michel, L.: OPL Script: Composing and Controlling Models. In: Apt, K.R., Kakas, A.C., Monfroy, E., Rossi, F. (eds.) Compulog Net WS 1999. LNCS (LNAI), vol. 1865, pp. 75–90. Springer, Heidelberg (2000)
22. Van Hentenryck, P., Michel, L.: Nondeterministic Control for Hybrid Search. In: Barták, R., Milano, M. (eds.) CPAIOR 2005. LNCS, vol. 3524, pp. 380–395. Springer, Heidelberg (2005)
23. Van Hentenryck, P., Michel, L., Deville, Y.: Numerica: a Modeling Language for Global Optimization. The MIT Press, Cambridge (1997)

Parallel Discrepancy-Based Search

Thierry Moisan, Jonathan Gaudreault, and Claude-Guy Quimper

FORAC Research Consortium, Université Laval, Québec, Canada
Thierry.Moisan.1@ulaval.ca,
{Jonathan.Gaudreault,Claude-Guy.Quimper}@ift.ulaval.ca

Abstract. Backtracking strategies based on the computation of discrepancies have proved themselves successful at solving large problems. They show really good performance when provided with a high-quality domain-specific branching heuristic (variable and value ordering heuristic), which is the case for many industrial problems. We propose a novel approach (PDS) that allows parallelizing a strategy based on the computation of discrepancies (LDS). The pool of processors visits the leaves in exactly the same order as the centralized algorithm would do. The implementation allows for a natural/intrinsic load balancing to occur (filtering induced by constraint propagation would affect each processor pretty much in the same way), although there is no communication between processors. These properties make PDS a scalable algorithm that was used on a massively parallel supercomputer with thousands of cores. PDS improved the best known performance on an industrial problem.

1 Introduction

Constraint solvers have been used for decades and were successful at solving numerous operations research problems. For instance, these solvers are used for optimizing computer networks by better routing the traffic [1,2], and for planning and scheduling problems [3] in different industries, among them the forest products industry [4,5]. A solver accepts as input a combinatorial problem defined by a set of variables and a set of constraints posted on these variables. The solver usually explores the candidate solutions by doing a backtracking search in a tree.

With the rise of multi-core servers, there has been an increase in research for parallelizing constraint solvers. Parallelization is not trivial as there is need for a trade-off between the workload balance, the communication cost, and the duplication (redundancy) of work between the processors.

The choice of an efficient search strategy is instrumental in solving large industrial problems, even in a centralized environment (for performance reasons, it is essential to explore the most promising leaves first). Among others, backtracking strategies based on the analysis of discrepancies such as LDS [6], DDS [7], and DBDFS [8] have proved themselves successful at solving large problems. They show really good performance when provided with a high-quality branching heuristic (that is, variable and value ordering heuristic), which is the case for many industrial problems (e.g. [5]).

C. Schulte (Ed.): CP 2013, LNCS 8124, pp. 30–46, 2013.

In this article, we propose a novel approach (PDS) that allows parallelizing a strategy based on the computation of discrepancies (i.e. LDS). The proposed approach shows the following characteristics:

- The pool of processors globally visits the leaves in exactly the same order as the centralized version of LDS would do.
- There is no need for communication between the processors.
- The implementation allows for a natural/intrinsic load balancing to occur (filtering induced by constraint propagation would affect each processor pretty much in the same way).
- The method provides robustness (if a processor dies, it can be replaced by a new one that must however restart the work allocated to this processor).
- It offers good scaling: adding additional processors can never slow down the global process, unlike approaches using communication.

These properties make PDS a scalable algorithm that we used to solve industrial problems from the forest products industry (see [4,5]) using a massively parallel supercomputer called Colosse deployed at Université Laval.

The remainder of this paper is organized as follows. Section 2 reviews basic concepts related to parallel tree search. Sections 3 and 4 describe the original algorithm and the parallel version. Section 5 reports theoretical results and evaluates statistically the performance of the algorithm in order to illustrate different characteristics. Section 6 describes the experiments run on industrial problems and their results. Section 7 concludes the paper.

2   Basic Concepts

This section provides an overview of the main approaches regarding parallel tree search. We then give an overview of previous attempts that were made in order to parallelize discrepancy-based strategies.

2.1   Search Space in Shared Memory

The simplest method for parallel tree search is implemented by having many cores share a list of open nodes (nodes for which there is at least one of the children that is still unvisited). Starved processors just pick up the most promising node in the list and expand it. By defining different node evaluation functions, one can implement different strategies (DFS, BFS and others). A comprehensive framework based on this idea was proposed in [9]. Good performance is often reported, as in [10] where a parallel Best First Search was implemented, and evaluated up to 64 processors.

Although this kind of mechanism intrinsically provides excellent load balancing, it is known not to scale beyond a certain number of processors; beyond that point, performance starts to decrease. For this reason, the approach cannot easily be adapted for massively parallel supercomputers with thousands of cores.

2.2   Search Space Splitting / Work Stealing

This family of approaches is often reported as the most frequently seen in the literature [11]. The main idea is to have the search tree split into different regions allocated to processors (e.g. one processor branches to the left, the other processor branches to the right). As it is unlikely those subtrees will be of equal size, a *work stealing* mechanism (see [12,13]) is needed. Because it uses both communication and computation time, this cannot easily be scaled up to thousands of processors. In practice, we observe a decrease in performance when reaching a certain number of processors. However, interesting work was reported in [14]; the authors allocated specific processors to coordination tasks, allowing more processors to be used before performance starts to decline.

Another promising approach is reported in [11]. The authors used a search space splitting mechanism allowing good load balancing without needing a work stealing approach. They use a hashing function allocating implicitly the leaves to the processors. Each processor applies the same search strategy in its allocated search space, which solves the load balancing problem. However, like previous approaches, leaves are globally visited in a different order than they would be on a single-processor system. This could be a pity in situations where we know a really good domain-oriented search strategy, a strategy that the parallel algorithm failed to exploit to its full potential.

2.3   Las Vegas Algorithms / Portfolios

This approach consists in allocating the same search space to each processor. Each processor explores it using a different strategy, leading to a different visiting order of the leaves. No communication is required and an excellent level of load balancing is achieved (they all search the same search space). Even if this approach causes a high level of redundancy between processors, it shows really good performance in practice. Shylo et al. [15] greatly improves the method using randomized restart [16,17,18] on each processor.

As there is no communication between processors, this approach is fully scalable, although on small multi-core computers some authors increase the efficiency of the method by allowing processors to share information learned during the search (e.g. nogoods, see [19]).

In general, the main advantage of the algorithm portfolio approach is that one does not need to know a good search strategy beforehand: many strategies will be automatically tried at the same time by the parallel system, thanks to randomization. This is very useful because, as mentioned by [20] and [21], defining good domain-specific labelling strategies (that is, variable and value ordering heuristic) is a difficult task.

However, for complex applications where general strategies are inefficient and where very good domain-specific strategies are known (e.g. [4,5]) one would like to have the parallel algorithm exploit the domain-specific strategy.

To the best of our knowledge, it is the first time that LDS is parallelized this way. In [14] LDS was used locally by processors to search in the trees allocated to them (by a tree splitting / work stealing algorithm) but the global system did not replicate an LDS strategy. The original centralized LDS being an iterative algorithm, Boivin [22] tried running the first k iterations at the same time on k processors. The approach did not prove to be efficient for the following reason: when LDS is provided with a good labelling strategy, the kth iteration of LDS visit leaves that have considerably less expected probability of success than those in the first iterations. For domain-specific problems where centralized LDS is known to be good, only the first few processors were really helpful in the parallel implementation. Moreover, they were experiencing load balancing problems.

Finally, LDS was adapted for distributed optimization in [23,24]. However, distributed problems (DisCSP [25], DCOP [26] and HDCOP [27]) refers to a different context than parallel computing. These are problems that are distributed by nature; different agents are responsible for establishing the value of distinct variables and communication/coordination are inherent to those approaches. Therefore, the algorithm called MacDS we proposed in [23,24,27] could not serve as a basis for a scalable parallel LDS algorithm.

The next section provides a comprehensive description of the centralized version of LDS that will be parallelized in Section 4.

3   LDS

Harvey and Ginsberg [6] describe LDS with binary search trees, i.e. trees where each non-leaf node has two children. We present a generalization of LDS to n-ary trees which includes a modification by Walsh [7] that prevents visiting a leaf more than once. The search space of a problem can be represented as a tree where each node corresponds to a partial assignment. The root is the empty partial assignment and the leaves are complete assignments (also called solutions). Each child has one more variable assigned than its parent.

The value ordering heuristic is a function that orders the children of a node from the most likely one to lead to a solution to the least likely one. When represented graphically, the left child is the most likely one to lead to a solution and the right child is the least likely one. A *discrepancy* is a deviation from the first choice of the heuristic. We say that the first choice of the heuristic has zero discrepancy, the second choice has one discrepancy, the third choice has two discrepancies and so on. The discrepancy of a node is the sum of the discrepancies associated to each choice on the path from the root of the tree to the node. Figure 1 shows a search tree where the number of discrepancies is shown for each node.

Harvey and Ginsberg demonstrated that, with a good value ordering heuristic, the expected quality of a leaf decreases as the number of discrepancies increases. For that reason, they proposed to visit the leaves with the fewest discrepancies first and to keep the leaves with the most discrepancies for the end. Algorithm 2 visits all the leaves that have exactly k discrepancies. Algorithm 1 launches the search to visit all leaves in increasing number of discrepancies.

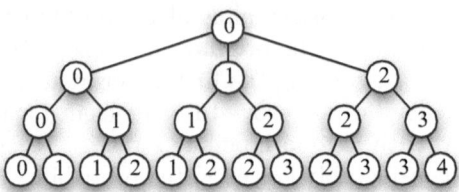

Fig. 1. Search tree. The discrepancy of each node is written inside the node.

4 PDS

We want to run an LDS search over multiple processors. Parallelization can be achieved in multiple ways but we set four goals that will influence our choices.

1. **Search Strategy Preservation.** We want the leaves of the search tree to be visited in the same order as they are on a single processor. Suppose that we mark each leaf of the tree with the time as it appears on a wall clock at the moment the leaf is visited. We assume that the clock is precise enough to break any ties. The ordering of the leaves by their visiting time should be the same regardless of the number of processors used.
2. **Workload Balancing.** We want the amount of work assigned to each processor to be evenly spread. This goal is particularly difficult to reach when the constraints filter the variable domains and make the search tree unbalanced.
3. **Robustness.** We aim at running the search on a large cluster of computers. It is frequent on those computers that a processor fails for different reasons and that the program must be restarted on another processor. It must be possible to identify which part of the search tree must be reassigned to another processor.
4. **Minimizing the Communication.** We aim at minimizing the communication between the processors. We actually want to avoid any communication. We make no assumptions about the geographical location of the processors and their ability to communicate. Communication should be limited to the broadcast of a solution.

Algorithm 1. LDS($[\mathrm{dom}(X_1), \ldots, \mathrm{dom}(X_n)]$)

---

for $k = 0..n$ **do**
  $s \leftarrow$ LDS-Probe($[\mathrm{dom}(X_1), \ldots, \mathrm{dom}(X_n)], k$)
  if $s \neq \emptyset$ **then return** s
return $\emptyset$

Algorithm 2. LDS-Probe($[\text{dom}(X_1), \ldots, \text{dom}(X_n)]$, k)

$Candidates \leftarrow \{X_i \mid |\text{dom}(X_i)| > 1\}$
if $Candidates = \emptyset$ **then**
   if $\text{dom}(X_1), \ldots, \text{dom}(X_n)$ *satisfies all the constraints* **then**
      return $\text{dom}(X_1), \ldots, \text{dom}(X_n)$
   return $\emptyset$
Choose a variable $X_i \in Candidates$
Let $v_0, \ldots, v_{|\text{dom}(X_i)|-1}$ be the values in $\text{dom}(X_i)$ sorted by the heuristic.
$\underline{d} \leftarrow \max(0, k - \sum_{X_a \in Candidates \setminus \{X_i\}}(|\text{dom}(X_a)| - 1))$
$\overline{d} \leftarrow \min(|\text{dom}(X_i)| - 1, k)$
for $d = \underline{d}..\overline{d}$ **do**
   $s \leftarrow$ LDS-Probe($[\text{dom}(X_1), \ldots, \text{dom}(X_{i-1}), \{v_d\},$
                $\text{dom}(X_{i+1}), \ldots, \text{dom}(X_n)], k - d$)
   if $s \neq \emptyset$ **then return** s
return $\emptyset$

We define a variation of LDS that we call PDS. We label ρ processors with an integer between 0 and $\rho - 1$ called the *processor id*. There is exactly one process running on each processor. The number of processors ρ and the processor id are given as input to each process. These two parameters are sufficient to identify which nodes of the search tree will be explored by each process.

We label each leaf s of the search tree by its visit time $t(s)$ in a centralized LDS. The first leaf to be visited has a visit time of $t(s_0) = 0$, the second leaf has a visit time of 1 and so on. We assign each leaf to a processor in a round-robin way by assigning a leaf s to processor $t(s) \bmod \rho$. A processor j is only allowed to visit a leaf s that satisfies $t(s) \bmod \rho = j$ or an ancestor of such a leaf. Consequently, before branching on a child node, a processor j has to check whether this child leads to a leaf it can visit. We show how to perform this test.

Let $C(X_1, \ldots, X_n, k)$ be the number of leaves with exactly k discrepancies in a search tree formed by the variables $X_1, \ldots, X_n$. The function $C(X_1, \ldots, X_n, k)$ is recursively defined as follows.

$$C(X_1, \ldots, X_n, k) = \begin{cases} 0 & \text{if } k < 0 \\ 1 & \text{if } k = 0 \\ \sum_{i=0}^{|\text{dom}(X_n)|-1} C(X_1, \ldots, X_{n-1}, k - i) & \text{otherwise} \end{cases} \quad (1)$$

When all domains have cardinality two, the recursion becomes $C(X_1, \ldots, X_n, k) = C(X_1, \ldots, X_{n-1}, k) + C(X_1, \ldots, X_{n-1}, k - 1)$. This recursion is the same that appears in Pascal's triangle to compute the binomial coefficients. We therefore have $C(X_1, \ldots, X_n, k) = \binom{n}{k}$ when $|\text{dom}(X_i)| = 2$. Intuitively, since each variable generates at most one discrepancy, the number of solutions with k discrepancies is the number of ways one can choose k variables among the n variables. When the domains have cardinalities greater than two, the recursion can be understood as follows: the variable X_n can generate a number of discrepancies i between 0 and $|\text{dom}(X_n)| - 1$. For each possible value of i, we count the number of solutions in the subtree of height $n - 1$ that have exactly $k - i$ discrepancies.

In equation (1), it seems that we consider a fixed ordering of the variables $X_1, \ldots, X_n$. However, the variable ordering imposed by the heuristic does not need to be static, but is required to be deterministic.

Consider a node a where a value is going to be assigned to X_n and none of the variables $X_1, \ldots, X_n$ are assigned. The node a has for children the nodes $c_0, \ldots, c_{|\text{dom}(X_n)|-1}$. Let $l(a, k)$ be the processor assigned to the left-most leaf with k discrepancies in the subtree rooted at a. From this construction, we have $l(a, k) = l(c_0, k)$ since branching from a to c_0 adds no discrepancies to the partial assignment and that both expressions refer to the same leaf. There are $C(X_1, \ldots, X_{n-1}, k)$ leaves with k discrepancies in the subtree rooted at c_0. Since each of these leaves are assigned to the processors in a round-robin way, the processor assigned to the first leaf in the subtree rooted at c_1 is therefore $(l(c_0, k) + C(X_1, \ldots, X_{n-1}, k)) \bmod \rho$. The same reasoning applies for the other children leading to the following recursion.

$$l(c_i, k - i) = \begin{cases} l(a, k) & \text{if } i = 0 \\ (l(c_{i-1}, k - i + 1) + C(X_1, \ldots, X_{n-1}, k - i + 1)) \bmod \rho & \text{otherwise} \end{cases}$$

We now have all the tools to present how the search strategy PDS proceeds. Each call to Algorithm 4 corresponds to the visit of a node in the search tree. The parameter k corresponds to the number of discrepancies that must lie on the path between this node and the leaves. Each processor visits only the nodes that lead to one of its assigned leaves. For each node a with children $c_0, c_1, \ldots$, Algorithm 4 computes which processor will treat the left-most leaf of the subtree rooted at c_i. This allows computing a range of processors that will visit each child. If the current processor is among that range, then it branches to the child.

5 Analysis

This Section provides an analysis of PDS in order to illustrate different properties of the algorithm. Section 4 showed how parallel cores can globally visit the leaves in the same order as the centralized algorithm would do. We now demonstrate the quality of the intrinsic workload balance that is achieved. First, when exploring

---

Algorithm 3. PDS($[\text{dom}(X_1), \ldots, \text{dom}(X_n)]$)

$l \leftarrow 0$
for $k = 0..n$ **do**
    $Candidates \leftarrow \{X_i \mid |\text{dom}(X_i)| > 1\}$
    $z \leftarrow C(Candidates \setminus \{X_i\}, k)$
    if $(currentProcessor - l) \bmod \rho < z$ **then**
        $s \leftarrow$ PDS-Probe($[\text{dom}(X_1), \ldots, \text{dom}(X_n)], k, l$)
        $l \leftarrow l + C(\{X_1, \ldots, X_n\}, k) \bmod \rho$
        if $s \neq \emptyset$ **then return** s
return $\emptyset$

Algorithm 4. PDS-Probe($[\text{dom}(X_1), \ldots, \text{dom}(X_n)]$, k, l)

$Candidates \leftarrow \{X_i \mid |\text{dom}(X_i)| > 1\}$
if $Candidates = \emptyset$ **then**
    if $dom(X_1), \ldots, dom(X_n)$ *satisfies all the constraints* **then**
        └ **return** $\text{dom}(X_1), \ldots, \text{dom}(X_n)$
    └ **return** $\emptyset$
Choose a variable $X_i \in Candidates$
Let $v_0, \ldots, v_{|\text{dom}(X_i)|-1}$ be the values in $\text{dom}(X_i)$ sorted by the heuristic.
$\underline{d} \leftarrow \max(0, k - \sum_{X_j \in Candidates \setminus \{X_i\}} (|\text{dom}(X_j)| - 1))$
$\overline{d} \leftarrow \min(|\text{dom}(X_i)| - 1, k)$
for $d = \underline{d}..\overline{d}$ **do**
    $z \leftarrow C(Candidates \setminus \{X_i\}, k - d)$
    if $(currentProcessor - l) \bmod \rho < z$ **then**
        $s \leftarrow$ PDS-Probe($[\text{dom}(X_1), \ldots, \text{dom}(X_{i-1}), \{v_d\},$
                    $\text{dom}(X_{i+1}), \ldots, \text{dom}(X_n)], k - d, l$)
        └ **if** $s \neq \emptyset$ **then return** s
    └ $l \leftarrow l + z \bmod \rho$
return $\emptyset$

---

the whole tree, the round-robin assignation of the processors ensures that the difference between the number of leaves visited by two processors is at most one. Workload balancing is easy to achieve when considering a complete search tree. However, it becomes harder to evenly divide the work among the processors when the tree is unbalanced. Search trees are often unbalanced when domain filtering and consistency technique are applied. We prove that when a value is filtered out of a variable domain and that a branch is cut from the tree, the workload is evenly reduced among all processors.

Theorem 1. *Let n be the number of variables in the problem. If a branch is cut from the search tree, the number of leaves removed from the workload of each processor differs by at most $n + 1$.*

Proof. The round-robin affection of the leaves with k discrepancies in a subtree guarantees that the number of leaves for each processor differs by at most one. Since we explore a subtree $n + 1$ times for solutions with 0, 1, ..., n discrepancies, the difference of workload between the processors is at most $n + 1$ leaves. □

5.1 Overhead

We do an overhead comparison of PDS, LDS, and DFS by counting the number of times a node is visited in a complete search tree associated to n binary variables.

A DFS in a tree with n variables visits the root and performs a DFS on two subtrees of $n - 1$ variables. Let $\text{DFS}(n)$ be the number of visited nodes.

$$\text{DFS}(n) = \begin{cases} 1 & \text{if } n = 0 \\ 2\text{DFS}(n) + 1 & \text{if } n > 0 \end{cases} \quad (2)$$

This non-homogeneous linear recurrence of first order solves to $\text{DFS}(n) = 2^{n+1} - 1$, i.e. the number of nodes in a complete binary tree of height n.

We consider a PDS with ρ processors. Let $\text{PDSprobe}_\rho(n, k, j)$ be the number of nodes visited by processors $j \in \{0, 1, \ldots, \rho - 1\}$ on a tree with n binary variables for which we seek leaves of k discrepancies. We assume that the left-most leaf must be visited by processor 0. If the left-most leaf has to be visited by processor a, one can retrieve the number of visited nodes by relabeling the processors and computing $\text{PDSprobe}_\rho(n, k, j - a \bmod \rho)$. When $k \in \{0, n\}$, the tree has a unique leaf with k discrepancies and only the processor $j = 0$ visits the $n + 1$ nodes between the root and the leaf. If the number of leaves with k discrepancies, $\binom{n}{k}$, is smaller than or equal to j, then the processor j does not have to visit the tree. In all other cases, the number of visited nodes depends on the number of visited nodes in the left and right subtrees. We have the following recurrence.

$$
\text{PDSprobe}_\rho(n, k, j) = \begin{cases} n + 1 & \text{if } j = 0 \wedge k \in \{0, n\} \\ 0 & \text{if } \binom{n}{k} \le j \\ \text{PDSprobe}_\rho(n - 1, k, j) & \text{otherwise} \\ \quad + \text{PDSprobe}_\rho(n - 1, k - 1, \\ \qquad j - \binom{n-1}{k} \bmod \rho) + 1 \end{cases}
$$

Let $\text{PDS}_\rho(n)$ be the total number of nodes visited by the ρ processors.

$$
\text{PDS}_\rho(n) = \sum_{k=0}^{n} \sum_{j=0}^{\rho-1} \text{PDSprobe}_\rho(n, k, j) = \sum_{k=1}^{n-1} \sum_{j=0}^{\rho-1} \text{PDSprobe}_\rho(n, k, j) + 2(n + 1)
$$

$$
= \sum_{k=1}^{n-1} \sum_{j=0}^{\rho-1} \text{PDSprobe}_\rho(n - 1, k, j)
$$

$$
+ \sum_{k=1}^{n-1} \sum_{j=0}^{\rho-1} \text{PDSprobe}_\rho\left(n - 1, k - 1, j - \binom{n-1}{k} \bmod \rho\right)
$$

$$
+ \sum_{k=1}^{n-1} \sum_{j=0}^{\min(\rho, \binom{n}{k})-1} 1 + 2(n + 1)
$$

One can replace $j - \binom{n-1}{k} \bmod \rho$ by j since we sum over $j = 0..\rho - 1$. We also perform a change of indices for k in the same summation.

$$
\text{PDS}_\rho(n) = \sum_{k=1}^{n-1} \sum_{j=0}^{\rho-1} \text{PDSprobe}_\rho(n - 1, k, j) + \sum_{k=0}^{n-2} \sum_{j=0}^{\rho-1} \text{PDSprobe}_\rho(n - 1, k, j)
$$

$$
+ \sum_{k=1}^{n-1} \min\left(\rho, \binom{n}{k}\right) + 2n + 2
$$

$$
= 2\text{PDS}_\rho(n - 1) + \sum_{k=1}^{n-1} \min\left(\rho, \binom{n}{k}\right) + 2
$$

Using backward substitutions solves the recurrence.

$$\text{PDS}_\rho(n) = 2^n + 2^n \sum_{i=1}^{n} \sum_{k=0}^{i} \frac{1}{2^i} \min(\rho, \binom{i}{k})$$

When solved for $\rho = 1$, we retrieve the number of visited nodes with LDS. We also simplify for $\rho \in \{2, 3\}$ assuming $n \geq 3$.

$$\text{LDS}(n) = 2^{n+2} - n - 3 \quad \text{PDS}_2(n) = 5 \cdot 2^n - 2n - 4 \quad \text{PDS}_3(n) = \frac{23}{4} 2^n - 3n - 5$$

We observe that, as the number of variables grows, a LDS visits twice the number of nodes than a DFS. Therefore, when DFS finishes to visit the entire tree, LDS visited half of the leaves. However, these leaves have fewer than $\frac{n}{2}$ discrepancies. So if the heuristic makes no mistakes at least half of the time, LDS finds a solution by the time DFS visits the entire tree. The overhead of LDS compared to DFS is therefore compensated by the search of more promising parts of the search tree.

As n grows, the ratios $\frac{\text{PDS}_2(n)}{\text{LDS}(n)}$ and $\frac{\text{PDS}_3(n)}{\text{LDS}(n)}$ tend to 1.25 and 1.43. These overheads of 25% and 43% grow slower than the number of processors and implies that 2 and 3 processors will visit the search tree in 62% and 48% of the time taken by one processor. Should the search visit the entire search tree, Figure 2 shows the speedup of PDS over LDS as the number of processors increases. We see that the speedup grows linearly except in the degenerate case where the number of leaves equal the number of processors.

To get a more accurate idea of the speedup, one needs to consider the quality of the solution (in an optimization problem) or the probability of finding a solution (in a satisfaction problem). This is done in the next section.

5.2   Statistical Analysis

We provide statistical results showing that the performance of the algorithm never declines, except in the degenerated case where there are more processors than leaves. It is therefore a worst-case analysis where the entire tree is explored.

Harvey and Ginsberg [6] showed, by analyzing binary SCP search trees from different problems, that the quality of a heuristic can be approximated/described by the probability p of finding a solution in the left subtree if no mistakes were made in the current partial assignment. Similarly, we say that the probability of finding a solution in the right subtree is q. If the solution is unique, we have $p + q = 1$. If there is more than one solution, we have $p + q \geq 1$ since there is a probability of having a solution both in the left subtree and the right subtree.

The better a heuristic is, the greater the ratio $\frac{p}{q}$ is. The extreme situation where $\frac{p}{q} = 1$ corresponds to a heuristic that does no better than random variable/value selection (all leaves share the same probability of being a solution, and using an LDS would not be a logical choice). The probability that a leaf

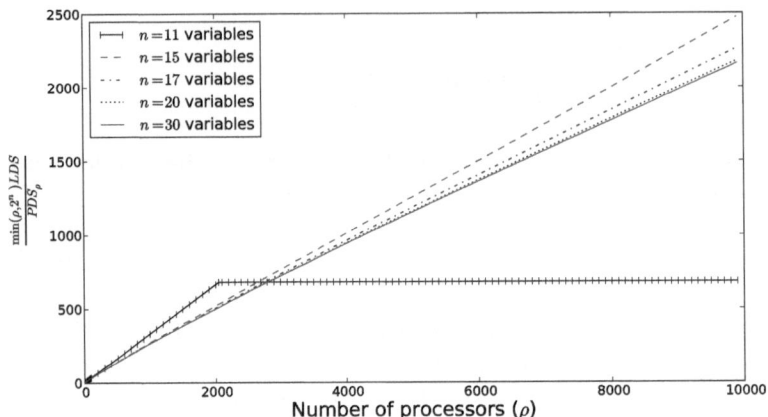

Fig. 2. Speedup for some number of processors

with k discrepancies is a solution is $p^{n-k}q^k$ since it involves branching k times on the right and $n - k$ times on the left.

Figure 3 shows the probability that a solution is found according to the number of visited nodes per processor. The probability that a leaf s_i with k deviations is a solution is $P(s_i) = p^{n-k}q^k$. The probability of finding a solution after visiting the leaves $s_1, \ldots, s_m$ is $1 - \prod_{i=1}^{m}(1 - P(s_i))$. For a given computation time, increasing the number of processors increases the probability that a solution has been found.

This clearly illustrates that increasing the number of processors increases the performance until ρ reaches the number of leaves in the search tree. From that point there is no more gain.

As the expected quality of a leaf decreases exponentially with its number of discrepancies (recall Section 3), adding more processors makes us visit additional leaves in the same computation time, but those leaves have smaller probability of success than the previous ones. This is a natural (and desired) consequence of using a good variable/value selection heuristics and a backtracking strategy visiting leaves in order of expected quality.

Figure 4 presents the speedup obtained for some probability that a solution is found. A lot of variation is present for low probability values. This is due to the use of the heuristic that points toward good solutions quickly. The speedup then converges toward a single value as the probability that a solution is found increases.

The next experiment studies the performance of the algorithm according to the quality of the variable/value selection heuristics used. We recall from Section 3 that the higher the $\frac{p}{q}$ ratio is, the more likely the solutions will be concentrated in leaves having few discrepancies. In contrast, the extreme situation where $p = q$ simulates the use of a heuristic that does no better than random variable/value selection (all leaves share the same probability of being a solution, and using an LDS would not be a logical choice).

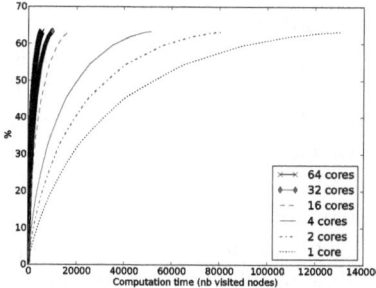
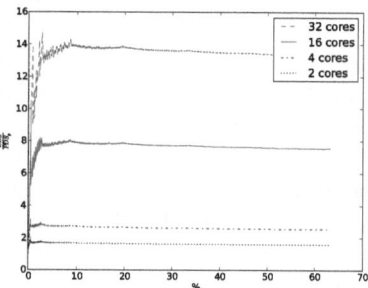

Fig. 3. Probability that a solution is found after some computation time $[n = 15$ vars; $p = 0.6$; $q = 0.4]$

Fig. 4. Speedup for some probability that a solution is found $[n = 15$ vars; $p = 0.6$; $q = 0.4]$

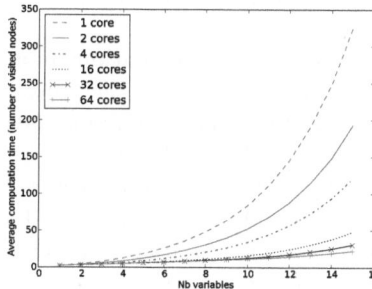
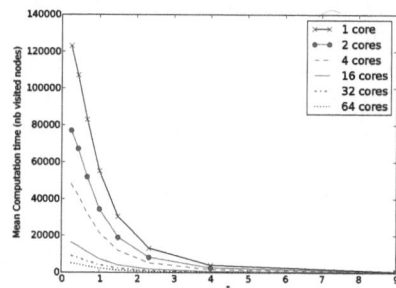

Fig. 5. Average computation time to find a solution according to the number of variables $[p = 0.6$; $q = 0.4]$

Fig. 6. Average computation time to find a solution according to the $\frac{p}{q}$ ratio where $p + q = 1$ with 15 variables

On Figure 6, the curve for 1 processor shows that computation time decreases exponentially when $\frac{p}{q}$ increases. Other curves show that when we provide additional processors, the computation time still decreases exponentially, but much more quickly.

6 Experimentation with Industrial Data

In a lumber finishing facility, lumbers are planed and sorted according to their grade (i.e. quality). It may be trimmed in order to produce a shorter lumber of a higher grade and value. The operation that improves a piece of lumber only depends on the piece of lumber itself with no consideration for the actual customer demand. This causes the production of multiple finished products at the same time (co-production) from a single raw product (divergence). This makes the production very difficult to plan according to the customer demand. There is a finite set of processes that can be used to transform one raw product into many finished products. The plant can only process lumber of a single category in a given production shift. Mills prefer long campaigns of a single category as it reduces costs: once the mill is configured for a given setup, they want to stay in this configuration for as many shifts as possible. The plant maintains an inventory of raw and finished products. For each customer order, a given quantity of a finished product has to be delivered at a specific time.

To sum up, the decisions that must be taken in order to plan the finishing operations are the following: (1) select a lumber category to process during a campaign, (2) decide when the campaign starts and for how long it lasts, and (3) for each campaign, decide the quantities of each compatible products to process. It is a single machine planning and scheduling problem. Each planning period corresponds to one "production shift" (approximatively 4 hours). The objective is to minimize orders lateness (modelled as a penalty cost) and production costs. The problem is fully described in [4] which provides a good heuristic for this problem. In [5], the heuristic is used to guide the search using constraint programming (applying LDS) and it outperforms the DFS and the mathematical programming approach.

Industrial instances are huge and there is a need for good solutions in shorter computation time. The instances have 65,142 variables and 50,238 constraints. Among them, there are 42 discrete decision variables whose domains have cardinality 6 and 4200 continuous decision variables. As we have a really good branching heuristic for which LDS works really well, this problem is an ideal candidate for PDS. This search heuristic first branches on variable/values for the integer variables (decisions 1 and 2 in the previous paragraph). Once the values for these variables are known, the remaining continuous variables (3) define a linear program that can be easily solved to optimality using the simplex method. Therefore, each time we have a valid assignment of the integer variables, we consider we have reached a leaf and we solve a linear program to evaluate the value of this solution. This implies that the leaves have a heavier computation time than the inner nodes. This situation differs from Section 5.1 where all nodes have the same computation time.

We implemented PDS and ran it on Colosse, a supercomputer with more than 8000 cores (dual, quad-core Intel Nehalem CPUs, 2.8 GHz with 24 GB RAM). Two Canadian lumber companies involved in the project provided the industrial instances. The four datasets have from 30 to 42 production shifts, from 20 to 133 processes, from 60 to 308 customer orders, from 20 to 68 raw products, and from 60 to 222 finished products.

6.1   Results and Discussion

Figures 7 to 10 show the objective value according to the computation time (maximum of one hour) for different numbers of processors. We also computed the best solution for LDS on these instances and the curve is indistinguishable from the PDS(1) curve. In Figure 10, we can see that a solution of quality of 1.1×10^7 is not found with 1 processor even after an hour but can be found in 10 minutes with 4096 processors. Furthermore, 1 processor obtained a solution of 1.5×10^7 in one hour while the same solution is found in a few seconds with

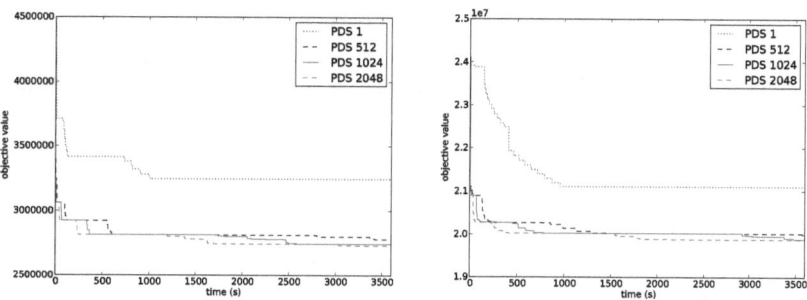

Fig. 7. Best solution found for K1 dataset

Fig. 8. Best solution found for M3 dataset

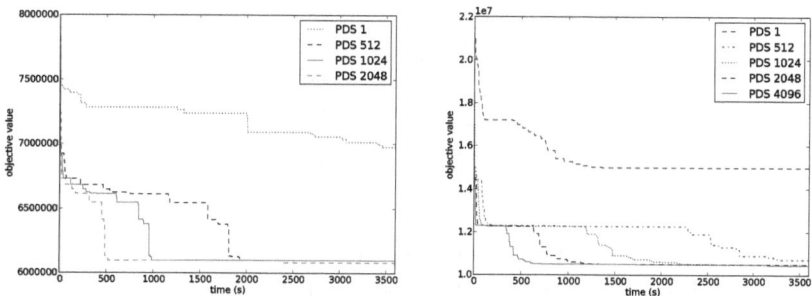

Fig. 9. Best solution found for M2 dataset

Fig. 10. Best solution found for M1 dataset

512 processors. This is a major improvement from an industrial point of view where computation time is the real constraint.

The harder instances are those where the heuristic has more difficulties and a good solution is obtained later in the search (Figure 7 is the easiest instance, Figure 10 is the hardest one). The absolute time saving is greater on harder instances when using PDS.

For each Figure from 7 to 10, the curves for 512, 1024, and 2048 processors have the same shape but get more compressed over time as the number of processors increases. This shows that the heuristic and the search strategies remains the same even in its parallelized version.

Table 1 lists statistics we computed during these experiments. The speedup is the ratio of the number of leaves visited by multiple processors over the number of leaves visited by one processor. PDS scales well: even with 4096 processors, the speedup is still increasing almost linearly.

One hour was insufficient to visit the entire search tree. With 4096 processors, we reached solutions with 6 discrepancies but did not visit all of them. Therefore, there is no idle time. However, we want to measure how the workload, in terms of visited leaves, differs between processors. Let χ_j be the number of leaves

Table 1. Industrial datasets experiments statistics. The column $\overline{\chi}$ is the average number of leaves visited by each processor. The column $max(\chi) - min(\chi)$ is the maximum difference of leaves treated between processors. The column $\frac{\overline{\chi}}{min(\chi)}$ is the average percentage of leaves differences between each processor and the minimal number of leaves treated by one processor.

dataSet	ρ	speedup	$\overline{\chi}$	$max(\chi) - min(\chi)$	$\frac{\overline{\chi}}{min(\chi)}(\%)$
K1	512	338.9	4668.6	550	8.6
K1	1024	585.9	4036.0	441	7.97
K1	2048	941.9	3244.4	363	7.79
M3	512	446.4	725.4	24	1.89
M3	1024	863.1	701.3	57	6.09
M3	2048	1601.1	650.4	25	2.92
M2	512	432.7	920.3	53	3.98
M2	1024	823.2	875.5	40	3.49
M2	2048	1604.7	853.3	23	1.94
M1	512	447.7	729.3	42	4.03
M1	1024	869.1	707.9	42	4.25
M1	2048	1656.1	674.4	79	11.1
M1	4096	3152.3	641.9	57	6.44

processed by processor j. Let $min(\chi)$ be the minimum value of χ_j for every $j \in 0, 1, \ldots, \rho-1$. Let $\overline{\chi}$ be the average number of leaves visited by each processor. The relative difference between $min(\chi)$ and $\overline{\chi}$ is $\frac{\overline{\chi}}{min(\chi)}$. This measure shows processors have visited roughly the same number of leaves.

We had hardware failures during the experiments and we have been able to restart a single processor while leaving the other ones running.

7 Conclusion

The contributions of this paper are twofold. First, we proposed a new parallelization scheme based on the LDS backtracking strategy. This parallelization does not alter the strategy since the visit order of the nodes remains unchanged. Moreover, PDS provides an intrinsic workload balancing, it scales on multiple processors, and it is robust to hardware failures. We provided a theoretical analysis that evaluated the performance of PDS based on the quality of the heuristic. This showed that adding more processors always provides a speedup.

Secondly, we experimented with a difficult industrial problem from the forest products industry for which an excellent problem-specific variable/value selection heuristics is known. This has been done by using as many as 4096 processors on a supercomputer. It shows the great potential of constraint programming in a massively parallel environment for which good search strategies are known.

References

1. Chabrier, A., Danna, E., Le Pape, C., Perron, L.: Solving a network design problem. Annals of Operations Research 130, 217–239 (2004)
2. Le Pape, C., Régin, J.-C., Shaw, P.: Robust and parallel solving of a network design problem. In: Van Hentenryck, P. (ed.) CP 2002. LNCS, vol. 2470, pp. 633–648. Springer, Heidelberg (2002)
3. Le Pape, C., Baptiste, P.: Heuristic control of a constraint-based algorithm for the preemptive job-shop scheduling problem. Journal of Heuristics 5, 305–325 (1999)
4. Gaudreault, J., Forget, P., Frayret, J.M., Rousseau, A., Lemieux, S., D'Amours, S.: Distributed operations planning in the lumber supply chain: Models and co-ordination. International Journal of Industrial Engineering: Theory, Applications and Practice 17 (2010)
5. Gaudreault, J., Frayret, J.M., Rousseau, A., D'Amours, S.: Combined planning and scheduling in a divergent production system with co-production: A case study in the lumber industry. Computers Operations Research 38, 1238–1250 (2011)
6. Harvey, W.D., Ginsberg, M.L.: Limited discrepancy search. In: Proceedings of the Fourteenth International Joint Conference on Artificial Intelligence (IJCAI 1995), pp. 607–613 (1995)
7. Walsh, T.: Depth-bounded discrepancy search. In: Proceedings of the Fifteenth International Joint Conference on Artificial Intelligence (IJCAI 1997), pp. 1388–1393 (1997)
8. Beck, J.C., Perron, L.: Discrepancy-bounded depth first search. In: Proceedings of the Second International Workshop on Integration of AI and OR Techniques in Constraint Programming for Combinatorial Optimization Problems (CP-AI-OR 2000), pp. 8–10 (2000)
9. Perron, L.: Search procedures and parallelism in constraint programming. In: Jaffar, J. (ed.) CP 1999. LNCS, vol. 1713, pp. 346–361. Springer, Heidelberg (1999)
10. Vidal, V., Bordeaux, L., Hamadi, Y.: Adaptive k-parallel best-first search: A simple but efficient algorithm for multi-core domain-independent planning. In: Proceedings of the Third International Symposium on Combinatorial Search (SOCS 2010) (2010)
11. Bordeaux, L., Hamadi, Y., Samulowitz, H.: Experiments with massively parallel constraint solving. In: Proceedings of the Twenty-First International Joint Conference on Artificial Intelligence (IJCAI 2009), pp. 443–448 (2009)
12. Michel, L., See, A., Van Hentenryck, P.: Transparent parallelization of constraint programming. INFORMS J. on Computing 21, 363–382 (2009)
13. Chu, G., Schulte, C., Stuckey, P.J.: Confidence-based work stealing in parallel constraint programming. In: Gent, I.P. (ed.) CP 2009. LNCS, vol. 5732, pp. 226–241. Springer, Heidelberg (2009)
14. Xie, F., Davenport, A.: Massively parallel constraint programming for supercomputers: Challenges and initial results. In: Lodi, A., Milano, M., Toth, P. (eds.) CPAIOR 2010. LNCS, vol. 6140, pp. 334–338. Springer, Heidelberg (2010)
15. Shylo, O.V., Middelkoop, T., Pardalos, P.M.: Restart strategies in optimization: Parallel and serial cases. Parallel Computing 37, 60–68 (2010)
16. Luby, M., Sinclair, A., Zuckerman, D.: Optimal speedup of las vegas algorithms. Information Processing Letters 47, 173–180 (1993)
17. Gomes, C.P.: Boosting combinatorial search through randomization. In: Proceedings of the Fifteenth National/Tenth Conference on Artificial Intelligence/Innovative Applications of Artificial Intelligence (AAAI 1998/IAAI 1998), pp. 431–437 (1998)

18. Gomes, C.P.: Complete randomized backtrack search. In: Constraint and Integer Programming: Toward a Unified Methodology, pp. 233–283 (2003)
19. Hamadi, Y., Sais, L.: Manysat: a parallel sat solver. Journal on Satisfiability, Boolean Modeling and Computation 6, 245–262 (2009)
20. Hamadi, Y., Ringwelski, G.: Boosting distributed constraint satisfaction. Journal of Heuristics, 251–279 (2010)
21. Puget, J.F.: Constraint programming next challenge: Simplicity of use. In: Wallace, M. (ed.) CP 2004. LNCS, vol. 3258, pp. 5–8. Springer, Heidelberg (2004)
22. Boivin, S., Gendron, B., Pesant, G.: Parallel constraint programming discrepancy-based search decomposition. Optimization days, Montréal, Canada (2007)
23. Gaudreault, J., Frayret, J.M., Pesant, G.: Discrepancy-based method for hierar-chical distributed optimization. In: Nineteenth International Conference on Tools with Artificial Intelligence (ICTAI 2007), pp. 75–81 (2007)
24. Gaudreault, J., Frayret, J.M., Pesant, G.: Distributed search for supply chain co-ordination. Computers in Industry 60, 441–451 (2009)
25. Yokoo, M.: Distributed constraint satisfaction: foundations of cooperation in multi-agent systems. Springer, London (2001)
26. Modi, P.J., Shen, W.M., Tambe, M., Yokoo, M.: Adopt: Asynchronous distributed constraint optimization with quality guarantees. Artificial Intelligence 161, 149–180 (2006)
27. Gaudreault, J., Frayret, J.M., Pesant, G.: Discrepancy-based optimization for dis-tributed supply chain operations planning. In: Proceeding of the Ninth Interna-tional Workshop on Distributed Constraint Reasoning (DCR 2007) (2007)

Bin Packing with Linear Usage Costs – An Application to Energy Management in Data Centres★

Hadrien Cambazard[1], Deepak Mehta[2], Barry O'Sullivan[2], and Helmut Simonis[2]

[1] G-SCOP, Université de Grenoble, Grenoble INP; UJF Grenoble 1, CNRS, France
hadrien.cambazard@grenoble-inp.fr
[2] Cork Constraint Computation Centre, University College Cork, Ireland
{d.mehta,b.osullivan,h.simonis}@4c.ucc.ie

Abstract. EnergeTIC is a recent industrial research project carried out in Grenoble on optimizing energy consumption in data-centres. The efficient management of a data-centre involves minimizing energy costs while ensuring service quality. We study the problem formulation proposed by EnergeTIC. First, we focus on a key sub-problem: a bin packing problem with linear costs associated with the use of bins. We study lower bounds based on Linear Programming and extend the bin packing global constraint with cost information. Second, we present a column generation model for computing the lower bound on the original energy management problem where the pricing problem is essentially a cost-aware bin packing with side constraints. Third, we show that the industrial benchmark provided so far can be solved to near optimality using a large neighborhood search.

1 Introduction

Energy consumption is one of the most important sources of expense in data centers. The ongoing increase in energy prices (a 50% increase is forecasted by the French senate by 2020) and the growing market for cloud computing are the main incentives for the design of energy efficient centers. We study a problem associated with the EnergeTIC[1] project which was accredited by the French government (FUI) [2]. The objective is to control the energy consumption of a data center and ensure that it is consistent with application needs, economic constraints and service level agreements. We focus on how to reduce energy cost by taking variable cpu requirements of client applications, IT equipment and virtualization techniques into account.

There are a variety of approaches to energy management in data centres, the most well-studied of which is energy-aware workload consolidation. A Mixed Integer Programming (MIP) approach to dynamically configuring the consolidation of multiple services or applications in a virtualised server cluster has been proposed in [16]. That work focused on power efficiency and considered the costs of turning on/off the servers.

★ The authors acknowledge their industrial partners (Bull, Schneider Electric, Business & Decision and UXP) as well as several public research institutions (G2Elab, G-SCOP and LIG). The authors from UCC are supported by Science Foundation Ireland Grant No. 10/IN.1/I3032.
[1] Minalogic EnergeTIC is a Global competitive cluster located in Grenoble France and fostering research-led innovation in intelligent miniaturized products and solutions for industry.

C. Schulte (Ed.): CP 2013, LNCS 8124, pp. 47–62, 2013.
© Springer-Verlag Berlin Heidelberg 2013

However, workloads were entirely homogeneous and there was little uncertainty around the duration of tasks. Constraint Programming is used in [8] with a different cost model.

A combinatorial optimisation model for the problem of loading servers to a desired utilisation level has, at its core, a bin packing (BP) problem [20]. In such a model each server is represented by a bin with a capacity equal to the amount of resource available. Bin packing is a very well studied NP-Hard problem. A significant amount of work has been conducted on lower bounds [13], approximation and exact algorithms. Although this research is still very active as demonstrated by the recent progress [17], researchers have started to look at variants involved in industrial applications.

In Section 2 we present an extension of bin packing which is a key sub-problem of the application domain and we show how to handle it efficiently with Constraint Programming (CP). In Section 3 we study the formulation of the EnergeTIC problem. In particular a lower bound computation technique is designed to assert the quality of the upper bounds found by a large neighborhood search. Section 4 reports the experiments on a real data-set followed by conclusions in Section 5.

2　Bin Packing with Linear Usage Costs

We consider a variant of the Bin Packing problem (BP) [20], which is the key sub-problem of the application investigated here. We denote by $S = \{w_1, \ldots, w_n\}$ the integer sizes of the n items such that $w_1 \leq w_2 \leq \ldots w_n$. A bin j is characterized by an integer capacity C_j, a non-negative fixed cost f_j and a non-negative cost c_j for each unit of used capacity. We denote by $B = \{\{C_1, f_1, c_1\}, \ldots, \{C_m, f_m, c_m\}\}$ the characteristics of the m bins. A bin is used when it contains at least one item. Its cost is a linear function $f_j + c_j l_j$, where l_j is the total size of the items in bin j. The total load is denoted by $W = \sum_{i=1}^{n} w_i$ and the maximum capacity by $C_{max} = max_{1 \leq j \leq m} C_j$. The problem is to assign each item to a bin subject to the capacity constraints so that we minimize the sum of the costs of all bins. We refer to this problem as the *Bin Packing with Usage Cost* problem (BPUC). BP is a special case of BPUC where all f_j are set to 1 and all c_j to 0. The following example shows that a good solution for BP might not yield a good solution for BPUC.

Example 1. In Figure 1, Scenario 1, B =$\{(9,0,1),(3,0,2),(3,0,2),(3,0,2),(3,0,2)\}$ and S = $\{2,2,2,2,3,3,3\}$. Notice that $\forall j, f_j = 0$. The packing (P_1) : $\{\{2,2,2,2\}, \{3\}, \{3\}, \{3\}, \{\}\}$ is using the minimum number of bins and has a cost of 26 ($8*1 + 3*2 + 3*2 + 3*2$). The packing (P_2): $\{\{3,3,3\}, \{2\}, \{2\}, \{2\}, \{2\}\}$ is using one more bin but has a cost of 25 ($9 + 2*2 + 2*2 + 2*2 + 2*2$). Here, (P_2) is better than (P_1) and using the minimum number of bins is not a good strategy. Now change the last unit cost to $c_5 = 3$ (see Figure 1, Scenario 2). The cost of (P_1) remains unchanged since it does not use bin number 5 but the cost of (P_2) increases to 27, and thus (P_1) is now better than (P_2).

Literature Review. A first relevant extension of BP for the current paper is called Variable Size Bin-Packing, where bins have different capacities and the problem is to minimize the sum of the wasted space over all used bins [15]. It can be seen as a special case of BPUC where all $f_j = C_j$ and $c_j = 0$. Recent lower bounds and an exact approach are examined in [11]. A generalization to any kind of fixed cost is presented in [5], which can be seen as a special case of BPUC where all $c_j = 0$. Concave costs of bin utilization

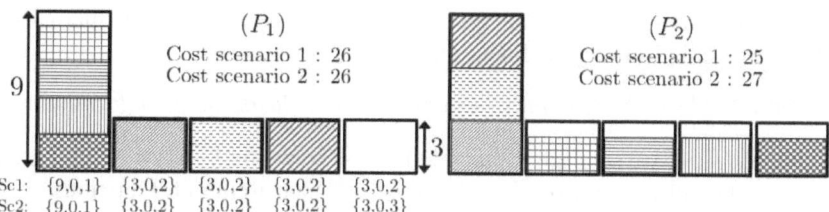

Fig. 1. Example of optimal solutions in two scenarios of costs. In Scenario 1, the best solution has no waste on the cheapest bin. In Scenario 2, it does not fill completely the cheapest bin.

studied in [12] are more general than the linear cost functions of BPUC. However [12] does not consider bins of different capacities and deals with the performance of classical BP heuristics whereas we are focusing on lower bounds and exact algorithms. Secondly, BP with general cost structures have been introduced [3] and studied [9]. The authors investigated BP with non-decreasing and concave cost functions of the number of items put in a bin. They extend it with profitable optional items in [4]. Their framework can model a fixed cost but does not relate to bin usage.

2.1 Basic Formulation and Lower Bounds

Numerous linear programming models have been proposed for BP [7]. We first present a formulation for BPUC. For each bin a binary variable y_j is set to 1 if bin j is used in the packing, and 0 otherwise. For each item $i \in \{1, \ldots, n\}$ and each bin $j \in \{1, \ldots, m\}$ a binary variable x_{ij} is set to 1 if item i is packed into bin j, and 0 otherwise. We add non-negative variables l_j representing the load of each bin j. The model is as follows:

$$
\begin{aligned}
\textbf{Minimize} \quad & z_1 = \sum_{j=1}^{m}(f_j y_j + c_j l_j) & & \\
(1.1) \quad & \sum_{j=1}^{m} x_{ij} = 1, & & \forall i \in \{1, \ldots, n\} \\
(1.2) \quad & \sum_{i=1}^{n} w_i x_{ij} = l_j, & & \forall j \in \{1, \ldots, m\} \\
(1.3) \quad & l_j \leq C_j y_j, & & \forall j \in \{1, \ldots, m\} \\
(1.4) \quad & x_{ij} \in \{0,1\}, y_j \in \{0,1\}, l_j \geq 0 & & \forall j \in \{1, \ldots, m\}, \forall i \in \{1, \ldots, n\}
\end{aligned}
\tag{1}
$$

Constraint (1.1) states that each item is assigned to one bin whereas (1.2) and (1.3) enforce the capacity of the bins. We now characterize the linear relaxation of the model. Let $r_j = f_j/C_j + c_j$ be a real number associated with bin j. If bin j is filled completely, r_j is the cost of one unit of space in bin j. We sort the bins by non-decreasing r_j: $r_{a_1} \leq r_{a_2} \leq \ldots \leq r_{a_m}$; $a_1, \ldots, a_m$ is a permutation of the bin indices $1, \ldots, m$. Let k be the minimum number of bins such that $\sum_{j=1}^{k} C_{a_j} \geq W$.

Proposition 1. *Let z_1^* be the optimal value of the linear relaxation of the formulation (1). We have $z_1^* \geq Lb_1$ with $Lb_1 = \sum_{j=1}^{k-1} C_{a_j} r_{a_j} + (W - \sum_{j=1}^{k-1} C_{a_j}) r_{a_k}$.*

Proof. $z_1^* = \sum_{j=1}^{m}(f_j y_j + c_j l_j) \geq \sum_{j=1}^{m}(f_j \frac{l_j}{C_j} + c_j l_j)$ because of the constraint $l_j \leq C_j y_j$, so $z_1^* \geq \sum_{j=1}^{m}(\frac{f_j}{C_j} + c_j)l_j \geq \sum_{j=1}^{m} r_j l_j$. Lb_1 is the quantity minimizing $\sum_{j=1}^{m} r_j l_j$ under the constraints $\sum_j l_j = W$ where each $l_j \leq C_j$. To minimize the quantity we must split W over the l_j related to the smallest coefficients r_j. Hence, $z_1^* \geq \sum_{j=1}^{m} r_j l_j \geq Lb_1$. □

Lb_1 is a lower bound of BPUC that can be easily computed. Also notice that Lb_1 is the bound that we get by solving the linear relaxation of formulation (1).

Proposition 2. *Lb_1 is the optimal value of the linear relaxation of the formulation (1).*

Proof. For all $j < k$, we set each y_{a_j} to 1 and l_{a_j} to C_j. We fix l_{a_k} to $(W - \sum_{j=1}^{k-1} C_{a_j})$ and y_{a_k} to l_{a_k}/C_{a_k}. For all $j > k$ we set $y_{a_j} = 0$ and $l_{a_j} = 0$. Constraints (1.3) are thus satisfied. Finally we fix $x_{i,a_j} = \frac{l_{a_j}}{W}$ for all i, j so that constraints (1.2) and (1.1) are satisfied. This is a feasible solution of the linear relaxation of (1) achieving an objective value of Lb_1. We have, therefore, $Lb_1 \geq z_1^*$ and consequently $z_1^* = Lb_1$ from Proposition 1. $\qquad\square$

Adding the constraint $x_{ij} \leq y_j$ for each item i and bin j, strengthens the linear relaxation only if $W < C_{a_k}$. Indeed, the solution given in the proof is otherwise feasible for the constraint, $(\forall j < k, x_{i,a_j} = \frac{l_{a_j}}{W} \leq y_{a_j} = 1$ and for $j = k$ we have $\frac{l_{a_k}}{W} \leq \frac{l_{a_k}}{C_{a_k}}$ if $W \geq C_{a_k}$).

2.2 Two Extended Formulations of BPUC

The Cutting Stock Model. The formulation of Gilmore and Gomory for the cutting stock problem [10] can be adapted for BPUC. The items of equal size are now grouped and for $n' \leq n$ different sizes we denote the number of items of sizes $w'_1, \ldots, w'_{n'}$ by $q_1, \ldots, q_{n'}$ respectively. A cutting pattern for bin j is a combination of item sizes that fits into bin j using no more than q_d items of size w'_d. In the i-th pattern of bin j, the number of items of size w'_d that are in the pattern is denoted g_{dij}. Let I_j be the set of all patterns for bin j. The cost of the i-th pattern of bin j is therefore equal to $co_{ij} = f_j + (\sum_{d=1}^{n'} g_{dij} w'_d) c_j$. The cutting stock formulation is using a variable p_{ij} for the i-th pattern of bin j:

$$
\begin{aligned}
\textbf{Minimize} \quad & z_2 = \sum_{j=1}^{m} \sum_{i \in I_j} co_{ij} p_{ij} \\
(2.1) \quad & \sum_{j=1}^{m} \sum_{i \in I_j} g_{dij} p_{ij} = q_d & \forall d \in \{1, \ldots, n'\} \\
(2.2) \quad & \sum_{i \in I_j} p_{ij} = 1 & \forall j \in \{1, \ldots, m\} \\
(2.3) \quad & p_{ij} \in \{0, 1\} & \forall j \in \{1, \ldots, m\}, i \in I_j
\end{aligned}
\tag{2}
$$

Constraint (2.1) states that each item has to appear in a pattern (thus in a bin) and (2.2) enforces one pattern to be designed for each bin (convexity constraints). A pattern p_{ij} for bin j is valid if $\sum_{d=1}^{n'} g_{dij} w'_d \leq C_j$ and all g_{dij} are integers such that $q_d \geq g_{dij} \geq 0$. The sets I_j have an exponential size so the linear relaxation of this model can be solved using column generation. The pricing step is a knapsack problem that can be solved efficiently by dynamic programming if the capacities are small enough.

The Arc-Flow Model. Carvalho introduced an elegant Arc-Flow model for BP [6,7]. His model explicitly uses each unit of capacity of the bins. In the following we show how to adapt it for BPUC. Consider a multi-graph $G(V, A)$, where $V = \{0, 1, ..., C_{max}\} \cup \{F\}$ is the set of $C_{max} + 2$ nodes labeled from 0 to C_{max} and a final node labeled F, and $A = I \cup J$ is the set of two kinds of edges. An edge $(a, b) \in I$ between two nodes labelled $a \leq C_{max}$ and $b \leq C_{max}$ represents the use of an item of size $b - a$. An edge of $(a, F) \in J$ for each bin j represents the usage a of the bin j, and therefore $a \leq C_j$. An example of

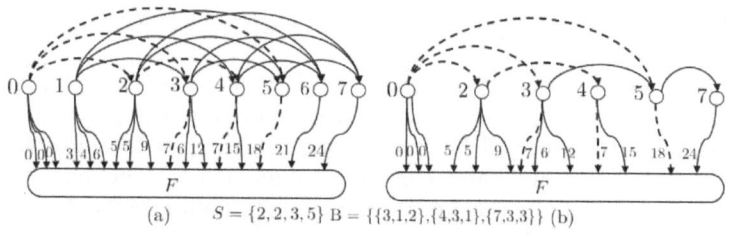

(a) $S = \{2,2,3,5\}$ B $= \{\{3,1,2\},\{4,3,1\},\{7,3,3\}\}$ (b)

Fig. 2. (a) An example of the graph underlying the ARC-FLOW model for $S = \{2,2,3,5\}$, $B = \{\{3,1,2\},\{4,3,1\},\{7,3,3\}\}$ so that $C_{max} = 7$. A packing is shown using a dotted line: $\{3\}$ is put in the first bin for a cost of 7, $\{2,2\}$ is in the second bin for a cost of 7 and $\{5\}$ in the last bin for a cost of 18. (b) The graph underlying the ARC-FLOW model after the elimination of symmetries.

such a graph is shown in Figure 2(a). Notice that this formulation has symmetries since a packing can be encoded by many different paths. Some reduction rules were given by Carvalho [6], which help in reducing such symmetries (see Figure 2(b)).

BPUC can be seen as a minimum cost flow between 0 and F with constraints enforcing the number of edges of a given length used by the flow to be equal to the number of items of the corresponding size. We have variables x_{ab} for each edge $(a, b) \in I$ as well as variables y_{aj} for each pair of bin $j \in \{1, \ldots, m\}$ and $a \in V$. The cost of using an edge $(a, F) \in J$ for bin j with $a > 0$ is $co_{aj} = f_j + a \cdot c_j$ and $co_{0j} = 0$. The model is as follows:

$$\text{Minimize} \qquad z_3 = \sum_{j=1}^{m} \sum_{k=0}^{k=C_{max}} co_{kj} y_{kj}$$

$$
\begin{align}
&(3.1) \quad \textstyle\sum_{(a,b)\in A} x_{ab} - \sum_{(b,c)\in A} x_{bc} - \sum_{j=1}^{m} y_{bj} = \begin{cases} 0 & \forall b \in \{1,2,\ldots,C_{max}\} \\ -m & \text{for } b = 0 \end{cases} \\
&(3.2) \quad \textstyle\sum_{a=0}^{C_j} y_{aj} = 1 && \forall j \in \{1,\ldots,m\} \\
&(3.3) \quad \textstyle\sum_{(k,k+w'_d)\in A} x_{k,k+w'_d} = q_d && \forall d \in \{1,2,\ldots,n'\} && (3) \\
&(3.4) \quad y_{aj} = 0 && \forall (j,a) \in \{1,\ldots,m\} \times \{C_j+1,\ldots,C_{max}\} \\
&(3.5) \quad x_{ab} \in \mathbb{N} && \forall (a,b) \in A \\
&(3.6) \quad y_{aj} \in \{0,1\} && \forall (j,a) \in \{1,\ldots,m\} \times \{0,\ldots,C_{max}\}
\end{align}
$$

Constraint (3.1) enforces the flow conservation at each node, and Constraint (3.2) states that each bin should be used exactly once. Constraint (3.3) ensures that all the items are packed, while Constraint (3.4) enforces that bin j is not used beyond its capacity C_j. A solution can be obtained again by decomposing the flow into paths. The number of variables in this model is in $O((n' + m) \cdot C_{max})$ and the number of constraints is $O(C_{max} + m + n')$. Although its LP relaxation is stronger than that of Model (1), it remains dominated by that of Model (2).

Proposition 3. $z_3^* \leq z_2^*$. *The optimal value of the linear relaxation of (3) is less than the optimal value of the linear relaxation of (2).*

Proof. Let (p^*) be a solution of the linear relaxation of (2). Each pattern p_{ij}^* is mapped to a path of the ARC-FLOW model. A fractional value p_{ij}^* is added on the arcs corresponding to the item sizes of the pattern (the value of the empty patterns for which all $g_{dij} = 0$ is put on the arcs y_{0j}). The flow conservation (3.1) is satisfied by construction, so is (3.2) because of (2.2) and so are the demand constraints (3.3) because of (2.1). Any solution of (2) is thus encoded as a solution of (3) for the same cost so $z_3^* \leq z_2^*$. □

Proposition 4. z_2^* *can be stronger than* z_3^* *i.e there exist instances such that* $z_2^* > z_3^*$.

Proof. Consider the following instance: $S = \{1, 1, 2\}$ and $B = \{\{3, 1, 1\}, \{3, 4, 4\}\}$. Two items of size 1 occurs so that $n' = 2$, $q_1 = 2$, $q_2 = 1$ corresponding to $w_1' = 1$, $w_2' = 2$. The two bins have to be used and the first *dominates* the second (the maximum possible space is used in bin 1 in any optimal solution) so the optimal solution is the packing $\{\{2, 1\}, \{1\}\}$ (cost of 12). Let's compute the value of z_2^*. It must fill the first bin with the pattern $[g_{111}, g_{211}] = [1, 1]$ for a cost of 4. Only three possible patterns can be used to fill the second bin: $[0, 0]$, $[1, 0]$ and $[2, 0]$ (a valid pattern p_{i2} is such that $g_{1i2} \leq 2$). The best solution is using $[g_{112}, g_{212}] = [2, 0]$ and $[g_{122}, g_{222}] = [0, 0]$ taking both a 0.5 value to get a total cost $z_2^* = 4 + 6 = 10$. The ARC-FLOW model uses a path to encode the same first pattern $[1,1]$ for bin 1. But it can build a path for bin 2 with a $\frac{1}{3}$ unit of flow taking three consecutive arcs of size 1 to reach a better cost of $\frac{1}{3} * 16 \approx 5, 33$. This path would be a pattern $[3,0]$ which is not valid for (2). So $z_3^* \approx 9.33$ and $z_2^* > z_3^*$.   □

The ARC-FLOW model may use a path containing more than q_d arcs of size w_d' with a positive flow whereas no such patterns exist in (3) **because** the sub-problem is subject to the constraint $0 \leq g_{dij} \leq q_d$. The cutting stock formulation used in [6] ignores this constraint and therefore the bounds are claimed to be equivalent.

2.3   Extending the Bin Packing Global Constraint

A bin packing global constraint was introduced in CP by [19]. We present an extension of this global constraint to handle BPUC. The scope and parameters are as follows:

$$\textsc{BinPackingUsageCost}([x_1, \ldots, x_n], [l_1, \ldots, l_m], [y_1, \ldots, y_m], b, z, S, B)$$

Variables $x_i \in \{1, \ldots, m\}$, $l_j \in [0, \ldots, C_j]$ and $b \in \{1, \ldots, m\}$ denote the bin assigned to item i, the load of bin j, and the number of bins used, respectively. These variables are also used by the BINPACKING constraint. Variables $y_i \in \{0, 1\}$ and $z \in \mathbb{R}$ are due to the cost. They denote whether bin j is open, and the cost of the packing. The last two arguments refers to BPUC and give the size of the items as well as the costs (fixed and unit). In the following, $\underline{x}$ (resp. $\overline{x}$) denotes the lower (resp. upper) bound of variable x.

Cost-Based Propagation Using Lb_1. The characteristics of the bins of the restricted BPUC problem based on the current state of the domains of the variables is denoted by B', and defined by $B' = \{\{C_1', f_1', c_1\}, \ldots, \{C_m', f_m', c_m\}\}$ where $C_j' = \overline{l_j} - \underline{l_j}$, and $f_j' = (1 - y_j) f_j$. The total load that remains to be allocated to the bins is denoted $W' = W - \sum_{j=1}^{m} \underline{l_j}$. Notice that we use the lower bounds of the loads rather than the already packed items. We assume it is strictly better due to the reasonings of the bin packing constraint.

Lower Bound of z. The first propagation rule is the update of the lower bound $\underline{z}$ of z. The bound is summing the cost due to open bins and minimum loads with the value of Lb_1 on the remaining problem. It gives a maximum possible cost increase *gap*:

$$Lb_1' = \sum_{j=1}^{m} (\underline{l_j} c_j + y_j f_j) + Lb_1(W', B'); \qquad \underline{z} \leftarrow max(\underline{z}, Lb_1'); \qquad gap = \overline{z} - Lb_1' \quad (4)$$

Bounds of the load variables. We define the bin packing problem B'' obtained by excluding the space supporting the lower bound $Lb_1(W', B')$. Lb_1 is using L_j units of space on bin a_j. The bins $a_1, \ldots, a_{k-1}$ are fully used so $\forall j < k, L_j = C'_{a_j}$, for bin a_k we have $L_k = W' - \sum_{j=1}^{k-1} C'_{a_j}$ and $\forall j > k, L_j = 0$. The resulting bins are defined as $B'' = \{\{C''_1, f'_1, c_1\}, \ldots, \{C''_m, f'_m, c_m\}\}$ where $C''_{a_j} = 0$ for all $j < k$, $C''_{a_k} = C'_{a_k} - L_k$ and $C''_{a_j} = C'_{a_j}$ for all $j > k$. Lower and upper bounds of loads are adjusted with rules (5).

Let $q^1_{a_j}$ be the largest quantity that can be removed from a bin a_j, with $j \leq k$, and put at the other cheapest possible place without overloading $\bar{z}$. Consequently, when $j < k$, $q^1_{a_j}$ is the largest value in $[1, L_j]$ such that $(Lb_1(q^1_{a_j}, B'') - q^1_{a_j} r_{a_j}) \leq gap$. When $j = k$, the same reasonning can be done by setting $C''_{a_k} = 0$ in B''.

Similarly, let $q^2_{a_j}$ be the largest value in $[1, C'_{a_j}]$ that can be put on a bin a_j, with $j \geq k$, without triggering a contradiction with the remaining gap of cost. $q^2_{a_j}$ is thus the largest value in $[1, C'_{a_j}]$ such that $(q^2_{a_j} r_{a_j} - (Lb_1(W', B') - Lb_1(W' - q^2_{a_j}, B'))) \leq gap$.

$$\forall j \leq k, \quad \underline{l_{a_j}} \leftarrow \underline{l_{a_j}} + L_j - q^1_{a_j}; \qquad \qquad \forall j \geq k, \quad \overline{l_{a_j}} \leftarrow \underline{l_{a_j}} + q^2_{a_j}. \quad (5)$$

Channeling. The constraint ensures two simple rules relating the load and open-close variables (a bin of zero load can be open): $y_j = 0 \implies l_j = 0$ and $l_j > 0 \implies y_j = 1$.

Bounds of the open-close variables. The propagation rule for l_j can derive $l_j > 0$ from (5), which in turn (because of the channeling between y and l) will force a bin to open *i.e* $y_{a_j} \in \{0, 1\}$ will change to $y_{a_j} = 1$. To derive that a y_j has to be fixed to 0, we can use Lb_1 similarly to the reasonings presented for the load variables (checking that the increase of cost for opening a bin remains within the gap).

Tightening the bounds of the load variables can trigger the existing filtering rules of the bin packing global constraint thus forbidding or committing items to bins. Notice that items are only increasing the cost indirectly by increasing the loads of the bins because the cost model is defined by the state of the bins (rather than the items). The cost-based propagation on x is thus performed by the bin packing global constraint solely as a consequence of the updates on the bin related variables, i.e. l and y.

Algorithms and Complexity. Assuming that B' and W' are available, $Lb_1(W', B')$ can be computed in $O(m \log(m))$ time. Firstly we compute the r_j values corresponding to B' for all bins. Secondly, we sort the bins in non-decreasing r_j. Finally, the bound is computed by iterating over the sorted bins and the complexity is dominated by the sorting step. After computing $Lb_1(W', B')$, the values a_j (the permutation of the bins) such that $r_{a_1} \leq r_{a_2} \leq \ldots \leq r_{a_m}$ are available as well as the critical k and $L_k = W' - \sum_{j=1}^{k-1} C'$. The propagation of $\underline{l_{a_j}}$ and $\overline{l_{a_j}}$ can then be done in $O(m)$ as shown in Figure 3.

3 Application – Energy Optimization in a Data Centre

The system developed by EnergeTIC is based on a model of the energy consumption of the various components in a data centre, a prediction system to forecast the demand and an optimization component computing the placement of virtual machines onto servers.

Algorithm 1: UpdateMinimumLoad

Input: a_j with $j \le k$, B', gap

Output: a lower bound of l_{a_j}

1. $costInc = 0$; $q = 0$; $b = k$;
2. **If** $(j == k)$ $\{b = k + 1;\}$
3. **While** $(q < L_j$ && $b \le m)$
4. $loadAdd = \min(L_j - q, C'_{a_b} - L_b)$;
5. $costIncb = loadAdd \times (r_{a_b} - r_{a_j})$;
6.   **If** $((costIncb + costInc) > gap)$
7. $q = q + \lfloor \frac{gap - costInc}{r_{a_b} - r_{a_j}} \rfloor$;
8.   **return** $L_j + \underline{l_{a_j}} - q$;
9. $costInc = costInc + costIncb$;
10. $q = q + loadAdd$; $b = b + 1$;
11. **return** $\underline{l_{a_j}}$

Algorithm 2: UpdateMaximumLoad

Input: a_j with $j \ge k$, B', gap

Output: an upper bound of l_{a_j}

1. $costInc = 0$; $q = 0$; $b = k$;
2. **If** $(j == k)$ $\{q = L_k;$ $b = k - 1;\}$
3. **While** $(q < C'_{a_j}$ && $b \ge 0)$
4. $loadAdd = \min(L_b, C'_{a_j} - q)$;
5. $costIncb = loadAdd \times (r_{a_j} - r_{a_b})$;
6.   **If** $((costIncb + costInc) > gap)$
7. $q = q + \lfloor \frac{gap - costInc}{r_{a_j} - r_{a_b}} \rfloor$;
8.   **return** $\overline{l_{a_j}} + q$;
9. $costInc = \overline{costInc} + costIncb$;
10. $q = q + loadAdd$; $b = b - 1$;
11. **return** $\overline{l_{a_j}}$

Fig. 3. Propagation algorithms for updating the lower and upper bounds of the load variables

Energy Model. In the last decade green data centres have focused on limiting the amount of energy that is not used for running the client's applications. The Power Usage Effectiveness (PUE) is a key indicator introduced by the Green Grid consortium [1] which measures the ratio between the total energy consumption of the data centre and the energy used by its IT systems (e.g., servers, networks, etc.). A value of 1 is the perfect score. The current average in industry is around 1.7 and the most efficient data centres are reaching 1.4 or even 1.2. As not all electrical power consumed by the IT equipment is transformed into a useful work product, the need to refine such a metric arose quickly. Therefore, the Green Grid proposed a very fine grained indicator for that purpose [1]. This metric, although very accurate, is not really used in practice because of its complexity and no consensus has been reached for a practical and relevant indicator. The EnergeTIC project introduced a new energy indicator which is defined as the ratio between the total energy consumed and the energy really used to run clients' applications. This indicator however relies on a model of the energy consumption of each equipment. A system, based on three different servers (quad-, bi- and mono- processor) with different energy behaviors, was provided by Bull to perform the measurements. As an example, the energy cost of the power consumption of three different servers at different CPU loads taken from one of the problem instances is shown in Figure 4.

Demand Model. The demands of the real benchmarks used in the experimental section are coming from the Green Data Centre of Business & Decision Eolas located in Grenoble. It was used to study and validate the system operationally. It is instrumented with thousands of sensors spread over the site to monitor the energy consumption of the centre and claims a PUE between 1.28 and 1.34. It deals with an heterogeneous demand: web applications, e-commerce, e-business, e-administrations. An example showing variable requirements of CPU usage over 24 time-periods for multiple virtual machines taken from one of the problem instances is shown in Figure 5.

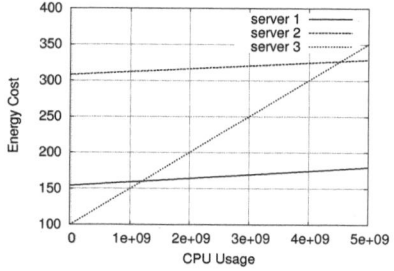
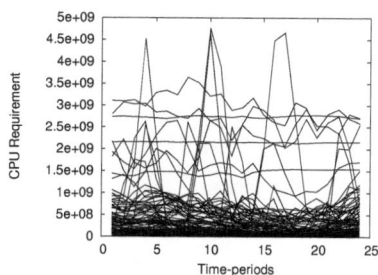

Fig. 4. Energy cost vs CPU Usage for 3 servers   **Fig. 5.** Variable demands of virtual machines

3.1 Problem Description and Notation

The problem is to place a set of virtual machines on a set of servers over multiple time-periods to minimize the energy cost of the data center. The CPU usage of a VM is changing over time. At each period, we must ensure that the virtual machines have enough resources (CPU and memory). Let $VM = \{v_1, \ldots, v_n\}$ be the set of virtual machines, $SE = \{s_1, \ldots, s_m\}$ be the set of servers and $T = \{p_1, \ldots, p_h\}$ be the set of periods.

Virtual Machines. A virtual machine v_i is characterized by a memory consumption M_i independent of the time-period, a set $SA_i \subseteq S$ of allowed servers where it can be hosted, and a potential initial server (for time-period p_0) denoted by $Iserv_i$ (which might be unknown). A virtual machine v_i has a CPU consumption U_{it} at time-period t.

Servers. A server s_j can be in two different states: ON=1 or STBY=0 (stand-by). It is characterized by: a CPU capacity $Umax_j$; a memory capacity $Mmax_j$; a fixed cost of usage $Emin_j$ (in Watts); a unit cost τ_j per unit of used capacity; a basic CPU consumption Ca_j when it is ON (to run the operating system and other permanent tasks); an energy consumption $Esby_j$ when it is in state STBY; an energy consumption $Esta_j$ to change the state of the server from STBY to ON; an energy consumption $Esto_j$ to change the state of the server from ON to STBY; a maximum number $Nmax_j$ of virtual machines that can be allocated to it at any time period; a set of periods $P_j \subseteq T$ during which s_j is forced to be ON; and a potential initial state $Istate_j \in \{0, 1\}$.

If a server is ON, its minimum cost is $Emin_j + \tau_j Ca_j$, and if it is STBY, its cost is $Esby_j$. For the sake of simplicity, to compute the fixed energy cost of an active server we include the basic consumption Ca_j and the standby energy $Estby_j$ in $Emin_j$ so that $Emin'_j = Emin_j - Estby_j + \tau_j Ca_j$. This way we can state the BINPACKINGUSAGECOST directly with the semantic given earlier by adding the constant $\sum_{s_j \in SE} Estby_j$ in the final objective value. We also shift the CPU capacity of the servers: $Umax'_j = Umax_j - Ca_j$.

Migrations. The maximum number of changes of servers among all virtual machines from one period to the next is denoted by N and the cost of a migration by $Cmig$.

The problem can be seen as a series of cost-aware bin packing problems (one per period) in two dimensions (CPU and memory) that are coupled by the migration constraints and the cost for changing the state of a server. Figure 6 gives an overview of the problem. This example has four servers, each shown by a rectangle whose dimensions

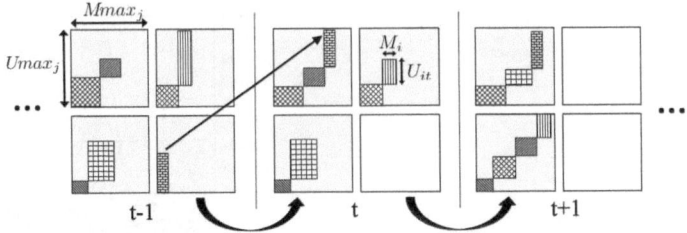

Fig. 6. A solution over three time periods. Virtual machines migrate to turn off two servers at $t+1$.

Minimize $\sum_{s_j \in SE} \sum_{t \in T} (Esta_j bto_{jt} + Esto_j otb_{jt} + \tau_j cpu_{jt} + Emin'_j o_{jt}) + Cmig(\sum_{v_i \in VM} \sum_{t \in T} a_{it})$

(6.1)	$\sum_{s_j \in SE} x_{ijt}$	$= 1$	$(\forall\, v_i \in VM, p_t \in T)$	
(6.2)	x_{ijt}	$= 0$	$(\forall\, v_i \in VM, p_t \in T, s_j \notin SA_i)$	
(6.3)	x_{ijt}	$\leq o_{jt}$	$(\forall\, v_i \in VM, p_t \in T, s_j \in SE)$	
(6.4)	cpu_{jt}	$= \sum_{v_i \in VM} U_{it} x_{ijt}$	$(\forall\, s_j \in SE, p_t \in T)$	
(6.5)	cpu_{jt}	$\leq Umax'_j o_{jt}$	$(\forall\, s_j \in SE, p_t \in T)$	
(6.6)	$\sum_{v_i \in VM} M_i x_{ijt}$	$\leq Mmax_j o_{jt}$	$(\forall\, s_j \in SE, p_t \in T)$	
(6.7)	$\sum_{v_i \in VM} x_{ijt}$	$\leq Nmax_j o_{jt}$	$(\forall\, s_j \in SE, p_t \in T)$	
(6.8)	mig_{ijt}	$\geq x_{ijt} - x_{ijt-1}$	$(\forall\, v_i \in VM, s_j \in SE, p_t \in T)$	(6)
(6.8')	a_{it}	$\geq \sum_{s_j \in SE} mig_{ijt}$	$(\forall\, v_i \in VM, p_t \in T)$	
(6.9)	$\sum_{v_i \in VM} a_{it}$	$\leq N$	$(\forall\, p_t \in T)$	
(6.10)	bto_{jt}	$\geq o_{jt} - o_{jt-1}$	$(\forall\, s_j \in SE, p_t \in T)$	
(6.11)	otb_{jt}	$\geq o_{jt-1} - o_{jt}$	$(\forall\, s_j \in SE, p_t \in T)$	
(6.12)	o_{jt}	$= 1$	$(\forall\, s_j \in SE, p_t \in P_j)$	
(6.13)	x_{ij0}	$= 0$	$(\forall\, v_i \in VM, s_j \in SE - \{Iserv_i\})$	
(6.14)	$x_{i,Iserv_i,0}$	$= 1$	$(\forall\, v_i \in VM)$	
(6.15)	o_{j0}	$= Istate_j$	$(\forall\, s_j \in SE)$	

are representing the cpu and memory capacities. A virtual machine is a small rectangle whose height (its cpu) varies from one period to the next. In this scenario, the cpu needs of some virtual machines decrease allowing to find better packings and turn off two servers at $t + 1$.

3.2 An Integer Linear Model

We present the integer linear model of the problem in which the following variables are used: $x_{ijt} \in \{0, 1\}$ indicates whether virtual machine v_i is placed on server s_j at time t. $cpu_{jt} \in [0, Umax'_j]$ gives the cpu consumption of s_j at period t. $o_{jt} \in \{0, 1\}$ is set to 1 if s_j is ON at time t, 0 otherwise. $bto_{jt} \in \{0, 1\}$ is set to 1 if s_j was in STBY at $t-1$ and is turned ON at t. $otb_{jt} \in \{0, 1\}$ is set to 1 if s_j was in ON at $t-1$ and is put STBY at t. $mig_{ijt} \in \{0, 1\}$ is set to 1 if v_i is on s_j at time t and was on a different server at $t-1$. $a_{it} \in \{0, 1\}$ is set to 1 if v_i is on a different server at t than the one it was using at $t-1$.

The initial state is denoted by $t = 0$. The model is summarized in Model (6). We omit the constant term $\sum_{s_j \in SE} Estby_j$ from the objective function. Constraint (6.1) states that a virtual machine has to be on a server at any time; (6.2) enforces the forbidden servers for each machine; (6.3) enforces a server to be ON if it is hosting at least one virtual machine; (6.4) links the CPU load of a server to the machines assigned to it. (6.5–6.7) are the resource constraints (CPU, memory and cardinality) of each server; (6.8,6.8′,6.9) allow us to count the number of migrations and state the limit on N (6.8 and 6.8′ together give a stronger linear relaxation than the single $a_{it} \geq x_{ijt} - x_{ijt-1}$); (6.10-6.11) keeps track of the change of states of the servers; (6.12) states the periods where a server has to be ON; (6.13–6.15) enforce the initial state ($t = 0$). The number of constraints of this model is dominated by the $n \times m \times h$ number of (6.8) and (6.3).

3.3   Lower Bound – An Extended Formulation

Solving large-sized instances of this application domain within short time limits is beyond the capability of exact algorithms. Therefore, one is generally forced to use an incomplete approach. Although an incomplete approach like large neighborhood search can usually find feasible solutions quickly, their qualities are often not evaluated as no bounds or provable approximation ratio can be found in the literature. Hence, it is important to be able to compute tighter lower bounds. In this section we present a column generation-based approach for computing a lower bound. Although we focus on a lower bound for the particular formulation (6), we believe it is generic enough to be relevant to other closely related problems of the literature that have at their core a series of cost-aware bin packing problems coupled with cost/migration constraints.

Let $b_{kt} \in \{0, 1\}$ be a variable for each bin packing of each time period to know whether the packing k is used for time period t. The set of all packings for period t is denoted by Ω_t. The packing k of period t is characterized by its cost c_{kt}, the server where each virtual machine is run and the state of each server. We use $x_{kijt} = 1$ if v_i is placed on s_j in the packing k at time period t and $o_{kjt} = 1$ if server s_j is ON in the packing k. In addition to b_{kt}, the variables bto_{jt}, otb_{jt}, a_{it} and mig_{ijt} that we have already introduced for (6) are used in the column generation model (7). The restricted master problem is defined for a restricted number of packing variables ($\forall t \leq m$, $b_{kt} \in \Omega'_t \subset \Omega_t$):

Minimize $z_4 = \sum_{t \in T}(\sum_{s_j \in SE}(Esta_j bto_{jt} + Esto_j otb_{jt}) + \sum_{k \in \Omega_t} c_{kt} b_{kt} + \sum_{v_i \in VM} Omiga_{it})$

$$
\begin{array}{llll}
(7.1) & \sum_{k \in \Omega_t} b_{kt} = 1 & (\forall\, p_t \in T) & (\lambda_t) \\
(7.2) & mig_{ijt} \geq \sum_{k \in \Omega_t} x_{kijt} b_{kt} - \sum_{k \in \Omega_t} x_{k,i,j,t-1} b_{k,t-1} & (\forall\, v_i \in VM, s_j \in SE, p_t \in T) & (\pi_{ijt}) \\
(7.3) & a_{it} \geq \sum_{s_j \in SE} mig_{ijt} & (\forall\, v_i \in VM, p_t \in T) & \\
(7.4) & \sum_{v_i \in VM} a_{it} \leq N & (\forall\, p_t \in T) & \\
(7.5) & bto_{jt} \geq \sum_{k \in \Omega_t} o_{kjt} b_{kt} - \sum_{k \in \Omega_t} o_{k,j,t-1} b_{k,t-1} & (\forall\, s_j \in SE, p_t \in T) & (\alpha_{jt}) \\
(7.6) & otb_{jt} \geq \sum_{k \in \Omega_t} o_{k,j,t-1} b_{k,t-1} - \sum_{k \in \Omega_t} o_{kjt} b_{kt} & (\forall\, s_j \in SE, p_t \in T) & (\beta_{jt})
\end{array}
\qquad (7)
$$

Let λ_t, π_{ijt}, α_{jt} and β_{jt} be the dual variables of constraints (7.1), (7.2), (7.5) and (7.6) respectively. We have h independent pricing problems and for each time period t we are looking for a negative reduced cost packing. The number of constraints (7.2) can prevent us from solving the relaxation of the master problem alone. We therefore turned to a relaxation of the migration constraints. The rationale is that the migration cost is really dominated by the server costs. Let $nmig_{jt} \in \mathbb{N}$ be the number of migrations occurring on server j and $u_{kjt} = \sum_{i \in VM} x_{kijt}$ the number of virtual machines allocated to server j in the k-th packing of time t. We suggest removing the a and mig variables from formulation (7), adding the $nmig$ variables instead and replacing (7.2)–(7.4) by :

$$(7.2')\ nmig_{jt} \quad \geq \sum_{k \in \Omega_t} u_{kjt} b_{kt} - \sum_k u_{k,j,t-1} b_{k,t-1}\ (s_j \in SE, p_t \in T) \quad (\pi_{jt})$$
$$(7.3')\ \sum_{j \in SE} nmig_{jt} \leq N \qquad\qquad\qquad\qquad\qquad (\forall\, p_t \in T) \qquad\qquad (\gamma_t)$$

The last term in the objective is replaced by $Cmig(\sum_{t \in T} \sum_{j \in SE} nmig_{jt})$. The pricing problem for period t can now be seen as a cost-aware bin packing problem with an extra cost related to the number of items assigned to a bin and two side constraints: a cardinality and memory capacity constraint. The reduced cost r_{kt} of packing b_{kt} is equal to

$$r_{kt} = c_{kt} - \sum_{j \in SE} (o_{kjt}(-\alpha_{jt} + \alpha_{j,t+1} + \beta_{jt} - \beta_{j,t+1}) + u_{kjt}(-\pi_{jt} + \pi_{j,t+1})) - \lambda_t \qquad (8)$$

For each bin j, the fixed and unit costs can be set to $f_j = Emin'_j - (-\alpha_{jt} + \alpha_{j,t+1} + \beta_{jt} - \beta_{j,t+1})$ and $c_j = \tau_j$ respectively. The cost depending on the number of items placed in bin j is denoted $\tau c_j = -(-\pi_{jt} + \pi_{j,t+1})$. Ignoring the constant term $-\lambda_t$ of the objective function, we summarize the pricing problem of period t by a CP model:

$$
\begin{aligned}
&\textbf{Minimize } r_t = ccpu + ccard \\
&(9.1)\ \textsc{BinPackingUsageCost}([x_1,\ldots,x_n],[cpu_1,\ldots,cpu_m],[y_1,\ldots,y_m],nbb,ccpu, \\
&\qquad\qquad [U_{1t},\ldots,U_{1n}],[(f_1,\tau_1),\ldots,(f_m,\tau_m)]) \\
&(9.2)\ \textsc{BinPackingUsageCost}([x_1,\ldots,x_n],[oc_1,\ldots,oc_m],\quad [y_1,\ldots,y_m],nbb,ccard, \\
&\qquad\qquad [1,\ldots,1],[(0,\tau c_1),\ldots,(0,\tau c_m)]) \\
&(9.3)\ \textsc{BinPacking}([x_1,\ldots,x_n],[mem_{1t},\ldots,mem_{mt}],nbb',[M_{1t},\ldots,M_{nt}]) \\
&(9.4)\ nbb \geq nbb' \\
&(9.5)\ \textsc{GlobalCardinality}([x_1,\ldots,x_n],[oc_1,\ldots,oc_m]) \\
&(9.6)\ y_j \begin{cases} = 1 & \text{if } p_t \in P_j \text{ or } f_j \leq 0 \\ \in \{0,1\} & \text{otherwise} \end{cases} \qquad (\forall\, s_j \in SE)
\end{aligned}
\qquad (9)
$$

Each variable $x_i \in SA_i$ gives the bin where item v_i is placed. $cpu_j \in [0, Umax'_j]$ and $mem_j \in [0, Mmax_j]$ encode the cpu and memory load of bin j, respectively. The number of items placed in bin j is given by $oc_j \in \{0, \ldots, n\}$ and $y_j \in \{0, 1\}$ indicates if bin j is ON or not. The number of bins used is $nbb \in \{1, ..., m\}$ (nbb' is an intermediate variable). Finally $ccpu \geq 0$ and $ccard \geq 0$ are real variables representing the costs related to CPU and cardinality. The costs are expressed by the state of the bins, thus matching the model of Section 2. A negative f_j is handled by pre-fixing y_j to 1 (constraint (9.6)).

Dual Bound. The bottleneck of this method is the hardness of the pricing step, as proving that no negative reduced cost packing exist is unlikely to be tractable. At any iteration, if the optimal reduced costs $r^* = (r_1^*, \ldots, r_h^*)$ are known, a well-known lower bound of the linear relaxation of the master is $w_4 = z_4^* + \sum_{t \in P_t} r_t^*$ where z_4^* is the current optimal value of the **restricted** master at this iteration. Indeed, since r_t^* is the best reduced cost for period t, $\forall k \in \Omega_t$, $r_{kt} \geq r_t^*$, and using (8) we have the following:

$$\forall k \in \Omega_t,\ c_{tk} \geq r_t^* + \sum_{j \in SE} (o_{kjt}(-\alpha_{jt} + \alpha_{j,t+1} + \beta_{jt} - \beta_{j,t+1}) + u_{kjt}(-\pi_{jt} + \pi_{j,t+1})) + \lambda_t.$$

This shows that the solution $(\gamma, \pi, \alpha, \beta, \lambda + r^*)$ is dual feasible for the master which explains w_4. Now this reasoning also holds for any value smaller than r_t^*. Therefore we still get a valid lower bound $\underline{w_4}$ if we use a lower bound $\underline{r_t^*}$ of each r_t^* and $w_4 = z_4^* + \sum_{t \in P_t} r_t^* \leq w_4$. We note that this algorithm can therefore return a valid bound without succeeding in solving a single pricing problem to optimality. At the moment, the pricing problem is solved using a linear solver with a time-limit of three seconds so the best bound is used for $\underline{r_t^*}$ if the time limit is reached. This is critical for scaling with sub-problem size. We can always return the best $\overline{w_4}$ found over all iterations. In practice we terminate when the gap between $\underline{w_4}$ and z_4^* is less than 0.1%.

3.4 Upper Bounds

The EnergeTIC team initially designed a MIP model that was embedded in their platform but it failed to scale. The details of this model are not reported here. We investigated three different approaches for computing upper bounds. The first approach is the MIP model (6) of Section 3.2 which is an improvement of the model designed by EnergeTIC. The second approach which we call Temporal Greedy (TG) is currently employed in their platform. It proceeds by decomposing time and is more scalable. It greedily solves the problem period by period using model (6) restricted to one period (enforcing the known assignment of the previous period). Each time-period is used as a starting period as long as there is time left, and therefore, if required, the assignment is extended in both directions (toward the beginning and toward the end). The last one is a large neighborhood search (LNS)[14], which was originally developed for the machine reassignment problem of 2012 ROADEF Challenge which had only 1 time-period. Therefore we extended it in order to handle multiple time-periods.

4 Experimental Results

Cost-Aware Bin Packing Benchmarks. We first compare on randomly generated instances the lower bounds z_1^*, z_2^*, z_3^* as well as exact algorithms: Model (1), ARC-FLOW Model (3) and a CP model using the BINPACKINGUSAGECOST constraint. Standard symmetry/dominance breaking techniques for BP are applied to the MIP [18] of Model (1) and CP [19]. A random instance is defined by (n, m, X), where n is the number of items ($n \in \{15, 25, 200, 250, 500\}$), m is the number of bins ($m \in \{10, 15, 25, 30\}$), and parameter $X \in \{1, 2, 3\}$ denotes that the item sizes are uniformly randomly generated in the intervals [1, 100], [20, 100], and [50, 100] respectively. The capacities of the bins are picked randomly from the sets {80, 100, 120, 150, 200, 250} and {800, 1000, 1200, 1500, 2000, 2500} when $n \in \{15, 25\}$ and $n \in \{200, 250, 500\}$ respectively. The fixed cost of each bin is set to its capacity and the unit cost is randomly picked from the interval [0, 1[. For each combination of $(n, m) \in \{(15, 10), (25, 15), (25, 25), (200, 10), (250, 15), (500, 30)\}$ and $X \in \{1, 2, 3\}$ we generated 10 instances giving 180 instances in total.

The time-limit was 600 seconds. If an approach failed to solve an instance within the time-limit then 600 was recorded as its solution time. All the experiments were carried out on a Dual Quad Core Xeon CPU, running Linux 2.6.25 x64, with 11.76 GB of RAM, and 2.66 GHz processor speed. The LP solver used was CPLEX 12.5 (default parameters) and the CP solver was Choco 2.1.5. Table 1 reports results for some classes due to lack of space. We report the average cpu time (denoted cpu) and the average

Table 1. Comparison of bounds obtained using MIP, Arc-Flow, CP, and Cutting-Stock approaches on random bin packing with usage cost problem instances with 600 seconds time-limit

n	m	X	best ub	MIP				CP			Arc-Flow				Cutting-Stock
				z_1^*	ub	#nu	cpu	ub	#nu	cpu	z_3^*	ub	#nu	cpu	z_2^*
15	10	1	1005.2	956.8	1005.2	0 (0)	1.2	1005.2	0 (0)	**0.5**	959.6	1005.2	0 (0)	2.1	**960.3**
15	10	2	1267.4	1230.5	1267.4	0 (0)	1.1	1267.4	0 (0)	**0.2**	1244.5	1267.4	0 (0)	0.7	**1245.0**
15	10	3	1574.5	1522.3	1574.5	0 (0)	0.8	1574.5	0 (0)	0.7	1553.0	1574.5	0 (0)	**0.6**	**1553.5**
25	15	1	1665.6	1636.3	1665.6	0 (0)	35.1	1665.6	0 (0)	24.0	1638.9	1665.6	1 (0)	42.7	**1639.0**
25	15	2	2127.1	2086.4	2127.1	0 (0)	74.2	2127.1	0 (0)	**12.9**	2094.6	2127.1	0 (0)	61.2	**2094.9**
25	15	3	2682.8	2613.1	2682.8	0 (0)	22.6	2685.6	2 (0)	144.0	2657.9	2682.8	0 (0)	**11.3**	2657.9
500	30	1	32387.2	32187.0	32387.2	0 (0)	**18.1**	32387.2	0 (0)	57.6	32187.0	-	10 (10)	600	32187.0
500	30	2	40422.7	40235.8	40513.5	3 (0)	301.2	40422.7	**0 (0)**	34.2	40235.8	-	10 (10)	600	40235.8
500	30	3	53395.6	53236.3	-	9 (2)	558.5	53395.6	**3 (0)**	201.3	53236.3	-	10 (10)	600	53236.3

value of upper/lower-bounds found (denoted ub / z_x^*) (when a value is found for each instance of the class). Column **#nu** is a pair $x(y)$ giving the number of instances x (resp. y) for which an approach failed to prove optimality (resp. to find a feasible solution).

For the cutting-stock approach upper-bounds are not shown as the branch-and-price algorithm was not implemented. The CP approach shows better performance when scaling to larger size instances (and capacities) than the MIP and Arc-Flow models.

EnergeTIC Benchmarks. The industry partners provided 74 instances, where the maximum number of virtual machines (items), servers (bins), and time-periods are 242, 20 and 287 respectively. [2] The time-limit is 600 seconds. As mentioned previously, we compared three approaches for computing upper bounds: the MIP model, the Temporal Greedy approach (TG) currently used by EnergeTIC, and large neighborhood search (LNS) [14]. We also analyzed the lower bounds provided by the linear relaxation of the MIP model (LP), the best lower bound established by MIP when reaching the time-limit (MIP LB), and the bound provided by the linear relaxation of formulation (7) (CG). Table 2 gives an overview of the results by reporting (over the 74 instances) the average/median/max values of the **gap** to the best known bound[3], the **cpu** time, and the number of instances **#nu** when an approach fails to return any results within the time-limit. Table 2 also gives the results for a few hard instances.

Upper Bounds. Out of 74 instances, MIP was able to find solutions for 71 instances within the time-limit out of which 54 are proved optimal. It thus failed for 3 instances where the space requirement for CPLEX exceeded 11GB. Notice that the largest size instance has $1,389,080$ decision variables. Clearly, MIP-based systematic search cannot scale in terms of time and memory. TG is able to find solutions for 73 instances (so it failed on one instance), out of which 26 are optimal. Its quality deteriorates severely when one should anticipate expensive peaks in demand by placing adequately virtual machines several time periods before the peak. This can be seen in Table 2 where the maximum gap is 119.35%. LNS succeeds to find feasible solutions for all instances within 2 seconds, on average, but it was terminated after 600 seconds and for 41

---

[2] The benchmarks are available from http://www.4c.ucc.ie/$\sim$dm6/energetic.tar.gz
[3] The gaps for lower and upper bounds are computed as $\frac{100 \times (best\_ub - lb)}{best\_ub}$ and $\frac{100 \times (ub - best\_lb)}{best\_lb}$) respectively. To compute mean/average/max values of gaps or time of a given approach, we exclude the instances where it fails to return any value (no feasible solution or a zero lower bound).

Table 2. Comparison between lower and upper bounds of the various approaches with 600 seconds time-limit (over 74 instances in the first part of the table and on a few specific instances in the second part)

			Lower bounds						Upper bounds				
			LP		CG		MIP LB		LNS	MIP		TG	
			gap	cpu	gap	cpu	gap	cpu	gap	gap	cpu	gap	cpu
	Mean		9,64	3,13	**0,32**	23,31	0,90	191,92	0,51	0,03	191,92	7,00	42,50
	Median		8,33	0,23	**0,10**	1,3	0	2,67	0	0	2,67	0,06	1,45
	Max		58,36	95,66	**7,14**	600	26,42	600	4,58	0,74	600	119,35	600
	#nu		3		**0**		3		**0**	3		1	
n	m	h	value	cpu	value	cpu	value	cpu	value	value	cpu	value	cpu
32	3	96	23492,8	9,5	**25404,7**	15,2	25043,6	600	25586,7	25575,7	600	36049,7	112,3
36	3	287	122831,3	4,5	**126716,8**	132,2	126597,9	600	127018,6	127654,4	600	127036,6	600
242	20	24	0	600	**37482,5**	600	0	600	40362,5	-	600	43027,6	14,2
242	20	24	0	600	**36890,8**	24,2	0	600	37701,6	-	600	36897,4	600
242	20	287	0	600	**431704,0**	600	0	600	439926,2	-	600	-	600
90	7	8	10420,7	14,4	**11431,9**	0,2	11236,3	600	11728,2	11435,3	600	11435,5	1,5

instances it found optimal solutions. Its average gap to the best known lower bound is less than 0.5% showing that LNS scales very well both in quality and problem size.

Lower bounds. The LP bound can be very far from the optimal value (its maximum gap is 58.36%) and does not scale since it fails on 3 instances even with 2 hour time-limit. The MIP obviously fails if the LP has failed. However, when solving the MIP, CPLEX automatically strengthens the formulation which allows us to solve many instances optimally where the LP bound was initially quite bad. Nevertheless, even after search there are cases where the gap can remain quite large (maximum of 26.42%). CG exhibits very good behaviour. Firstly, its gap clearly outperforms other bounds. Secondly, it can be stopped at any time and returns its current best master dual bound which is why #nu is 0 even though the time limit is reached on two cases (shown in Table 2). The first would improve to 38614.3 in 2000s whereas the second converges in 700 seconds without any improvement. Tables 2 shows that CG scales well both in terms of quality and size.

5   Conclusion and Future Work

Many optimisation problems in data centres can be described as a series of consecutive multi-dimensional Bin Packing with Usage Costs (BPUC) problems coupled by migration constraints and costs. First, we studied the lower bounds of a critical variant of bin packing for this domain that includes linear usage costs. We designed a CP approach that gives, so far, the best algorithm to solve BPUC exactly. Secondly, the usefulness of the exact algorithm and the efficient bounds for BPUC is shown within a column generation approach for the energy cost optimisation problem arising in data centres. These bounds are evaluated experimentally on real benchmarks and they assert the efficiency of the LNS approach [14] which was extended to handle consecutive BPUC problems.

The next step is to generalize the Martello and Toth bound L_2 [13] to the linear cost function which should improve the BINPACKINGUSAGECOST global constraint. We also plan to evaluate both column generation and LNS approaches on even larger size instances. We intend to solve the pricing problem with CP as we believe it can scale better for larger size problems.

References

1. A framework for data center energy productivity. Technical report, The green grid (2008)
2. Efficience des Data Centers, les retombées du projet EnergeTIC. Technical report (2013), http://www.vesta-system.cades-solutions.com/images/vestalis/4/energetic_white2@paper.pdf
3. Anily, S., Bramel, J., Simchi-Levi, D.: Worst-case analysis of heuristics for the bin packing problem with general cost structures. Operational Research 42 (1994)
4. Baldi, M.M., Crainic, T.G., Perboli, G., Tadei, R.: The generalized bin packing problem. Transportation Research Part E: Logistics and Transportation Review 48(6), 1205–1220 (2012)
5. Crainic, T.G., Perboli, G., Rei, W., Tadei, R.: Efficient lower bounds and heuristics for the variable cost and size bin packing problem. Comput. Oper. Res. 38(11), 1474–1482 (2011)
6. de Carvalho, J.M.V.: Exact solution of bin packing problems using column generation and branch-and-bound. Annals of Operations Research 86(0), 629–659 (1999)
7. de Carvalho, J.M.V.: LP models for bin packing and cutting stock problems. European Journal of Operational Research 141(2), 253–273 (2002)
8. Dupont, C., Schulze, T., Giuliani, G., Somov, A., Hermenier, F.: An energy aware framework for virtual machine placement in cloud federated data centres. In: Proceedings of the 3rd International Conference on Future Energy Systems: Where Energy, Computing and Communication Meet, e-Energy 2012, pp. 4:1–4:10. ACM, New York (2012)
9. Epstein, L., Levin, A.: Bin packing with general cost structures. Math. Program. 132(1-2), 355–391 (2012)
10. Gilmore, P.C., Gomory, R.E.: A linear programming approach to the cutting-stock problem. Operations research 11, 863–888 (1963)
11. Haouari, M., Serairi, M.: Relaxations and exact solution of the variable sized bin packing problem. Comput. Optim. Appl. 48(2), 345–368 (2011)
12. Li, C.L., Chen, Z.L.: Bin-packing problem with concave costs of bin utilization. Naval Research Logistics 53, 298–308 (2006)
13. Martello, S., Toth, P.: Lower bounds and reduction procedures for the bin packing problem. Discrete Applied Mathematics 28(1), 59–70 (1990)
14. Mehta, D., O'Sullivan, B., Simonis, H.: Comparing solution methods for the machine reassignment problem. In: Milano, M. (ed.) CP 2012. LNCS, vol. 7514, pp. 782–797. Springer, Heidelberg (2012)
15. Monaci, M.: Algorithms for Packing and Scheduling Problems. PhD thesis, Universit di Bologna (2012)
16. Petrucci, V., Loques, O., Mosse, D.: A dynamic configuration model for power-efficient virtualized server clusters. In: Proceedings of the 11th Brazilian Workshop on Real-Time and Embedded Systems (2009)
17. Rothvoss, T.: Approximating bin packing within $O(\log OPT \cdot \log\log OPT)$ bins. Technical report, MIT (2013)
18. Salvagnin, D.: Orbital shrinking: A new tool for hybrid MIP/CP methods. In: Gomes, C., Sellmann, M. (eds.) CPAIOR 2013. LNCS, vol. 7874, pp. 204–215. Springer, Heidelberg (2013)
19. Shaw, P.: A constraint for bin packing. In: Wallace, M. (ed.) CP 2004. LNCS, vol. 3258, pp. 648–662. Springer, Heidelberg (2004)
20. Srikantaiah, S., Kansal, A., Zhao, F.: Energy aware consolidation for cloud computing. In: Proceedings of HotPower (2008)

Filtering `AtMostNValue` with Difference Constraints: Application to the Shift Minimisation Personnel Task Scheduling Problem

Jean-Guillaume Fages[1] and Tanguy Lapègue[2]

[1] École des Mines de Nantes, LINA (UMR CNRS 6241), LUNAM Université,
[2] École des Mines de Nantes, IRCCyN (UMR CNRS 6597), LUNAM Université,
4 rue Alfred Kastler, La Chantrerie, BP20722, 44307 Nantes Cedex 3, France
{jean-guillaume.fages,tanguy.lapegue}@mines-nantes.fr

Abstract. This paper introduces a propagator which filters a conjunction of difference constraints and an `AtMostNValue` constraint. This propagator is relevant in many applications such as the Shift Minimisation Personnel Task Scheduling Problem, which is used as a case study all along this paper. Extensive experiments show that it significantly improves a straightforward CP model, so that it competes with best known approaches from Operational Research.

Keywords: `AtMostNValue`, Constraints Conjunction, Global Constraints, Shift Minimisation Personnel Task Scheduling Problem.

1 Introduction

The problem of minimising the number of distinct values among a set of variables subject to difference constraints occurs in many real-life contexts, where an assignment of resources to tasks has to be optimised. For instance, in transports, crews have to be assigned to trips [25]. In schools, classes have to be assigned to rooms [6]. In airports, maintenance tasks have to be assigned to ground crew employees [9,10]. In some factories, fixed jobs have to be assigned to machines [12,13,22]. In a more theoretical context, one may need to color a graph, such that adjacent vertices have distinct colors and not every color can be taken by every node [15,16].

In order to illustrate our contribution, we consider the Shift Minimisation Personnel Task Scheduling Problem (SMPTSP). This problem belongs to the set of personnel scheduling problems (see [11,32] for an overview). It arises when a set of tasks, fixed in time, have to be assigned to a set of shifts so that overlapping tasks should not be assigned to the same shift. Each shift is associated with a given subset of assignable tasks. The objective is to minimise the number of used shifts. This problem typically occurs as the second step of decomposition methods which handle the creation of rosters in a first step and the assignment of tasks in a second one. With this kind of methods, side constraints, related to

C. Schulte (Ed.): CP 2013, LNCS 8124, pp. 63–79, 2013.

personnel roster design, are considered in the first step only, hence the simplicity of the SMPTSP formulation. Nonetheless, current exact approaches from Operational Research fail to solve large scale instances. This is the main motivation for investigating a Constraint-Programming (CP) approach.

The core idea of CP is to design independent constraints that can be combined through common variables, in order to model constrained problems. However, in practice, it is often more interesting to design global constraints [3]. These constraints are able to consider a larger part of the problem, hence their filtering impact is increased. For instance, the `AllDifferent` global constraint is the conjunction over a clique of difference binary constraints, and it has been proved highly relevant within CP solvers [29]. However, developing effective global constraints is often difficult, and it also tends to make CP solver maintenance more expensive, which is one of the greatest concerns of the CP community [27]. Consequently, from a practical point of view, one would rather adapt existing constraints than to implement brand new ones, in order to capitalise over previous work. In this paper we investigate the interest of considering difference constraints when filtering the well known `AtMostNValue` constraint [2,4,28]. We introduce a new propagator whose implementation is based on the state-of-the-art `AtMostNValue` propagator [4]. A wide range of experiments show that our propagator significantly improves the CP model, so that it competes with the most recent SMPTSP dedicated approaches.

The remainder of the paper is organised as follows: Section 2 is devoted to the description of the SMPTSP, in Section 3 we show how the straightforward CP model of the SMPTSP can be improved with a new propagator. Our approach is validated by an extensive experimental study in Section 4, followed by our conclusions.

2   Description of the SMPTSP

In the following, $\mathcal{T}$ and $\mathcal{W}$ refer respectively to the set of tasks and workers (shifts may be seen as workers with specific skills). Given a task $t \in \mathcal{T}$, we refer to the set of workers that can be assigned to t as $\mathcal{W}_t \subseteq \mathcal{W}$. Since tasks are fixed, it is easy to find the set of maximal sets of overlapping tasks, which is referred to as $\mathcal{C}$. Actually, it amounts to finding the set of maximal cliques in an interval graph. The size of the largest clique, which provides a trivial lower bound on the required number of workers, is referred to as $LB_{\neq}$. For instance, if we consider the example given of Figure 1, we have $\mathcal{C} = \{\mathcal{K}_1, \mathcal{K}_2, \mathcal{K}_3\}$ with $\mathcal{K}_1 = \{t_1, t_2, t_3\}$, $\mathcal{K}_2 = \{t_1, t_3, t_4\}$, $\mathcal{K}_3 = \{t_4, t_5\}$ and $LB_{\neq} = 3$. This example will be used all along the article to illustrate our points.

The SMPTSP may be stated in Mathematical Programming, by using binary variables $x_{t,w}$ and y_w which specify respectively if the task t is assigned to the worker w and if the worker w is used. Based on these variables, the number of used workers is given by (1) and the assignment of tasks to qualified workers

$$\mathcal{W}_{t_3} = \{w_1, w_3\} \qquad \mathcal{W}_{t_5} = \{w_1, w_2, w_5\}$$

$$\mathcal{W}_{t_2} = \{w_1, w_2, w_3\} \quad \mathcal{W}_{t_4} = \{w_3, w_4, w_5\}$$

$$\mathcal{W}_{t_1} = \{w_2, w_3, w_4\}$$

$\longrightarrow$ Time

(a) Input data.

Task	Worker
t_1	w_2
t_2	w_3
t_3	w_1
t_4	w_3
t_5	w_1

(b) Optimal solution with three workers.

Fig. 1. A basic example with 5 workers ($w_{1..5}$) and 5 tasks ($t_{1..5}$)

is ensured by (2). The purpose of the constraint (3) is twofold: first it prevents workers to work on overlapping tasks, then it ensures that workers assigned to at least one task are counted as used. We refer to this model as *MIP model*:

$$\begin{aligned}
& \textit{minimise} \ \sum_{w \in \mathcal{W}} y_w & & (1) \\
& \textit{subject to:} \ \sum_{w \in \mathcal{W}_t} x_{t,w} = 1 & , \forall t \in \mathcal{T} & (2) \\
& \qquad\quad \sum_{t \in \mathcal{K}} x_{t,w} \leq y_w & , \forall w \in \mathcal{W}, \forall \mathcal{K} \in \mathcal{C} & (3) \\
& \qquad\quad x_{t,w} \in \{0,1\} & , \forall t \in \mathcal{T}, \forall w \in \mathcal{W}_t & (4) \\
& \qquad\quad y_w \in \{0,1\} & , \forall w \in \mathcal{W} & (5)
\end{aligned}$$

3 A CP Model Based on the AtMostNValue Constraint

This section first introduces a straightforward CP formulation of the SMPTSP. Next, it recalls the former AtMostNValue propagator our approach is based on, and provides a new formalism to define propagators of the same family. Then, it introduces a new propagator which filters AtMostNValue while considering a set of difference constraints. We show how to improve and diversify its impact on variables. Finally, we discuss the case of dynamic difference constraints and provide some implementation guidelines.

3.1 A Straightforward CP Formulation

The SMPTSP can be formulated within CP with a set of $|\mathcal{T}|$ integer variables $\mathcal{X}$ and one objective variable z. For each task $t_i \in \mathcal{T}$, the variable x_i gives the worker assigned to the task t_i. The objective variable z gives the number of workers assigned to at least one task. We refer to this model as *CP model*:

$$\begin{aligned}
& \textit{minimise} \ z & & (6) \\
& \textit{subject to:} \ \texttt{AllDifferent}(\{x_i \mid i \in \mathcal{K}\}) , \forall \mathcal{K} \in \mathcal{C} & & (7) \\
& \qquad\quad \texttt{AtMostNValue}(\mathcal{X}, z) & & (8) \\
& \qquad\quad Dom(z) = [LB_{\neq}, |\mathcal{W}|] & & (9) \\
& \qquad\quad Dom(x_i) = \mathcal{W}_{t_i} & , \forall t_i \in \mathcal{T} & (10)
\end{aligned}$$

The expressive language offered by CP enables to model the problem through two global constraints. In (7), AllDifferent constraints [29] are used to forbid workers ubiquity, *i.e.*, tasks which overlap in time need to be assigned to different workers. In (8), the `AtMostNValue` constraint [4] is used to restrict the number of workers that are involved in the schedule. Then, variable initial domain definitions are given by (9) and (10). More precisely, trivial lower and upper bounds for z are respectively the maximum number of overlapping tasks and the number of available workers.

3.2   State-of-the-Art Filtering of the `AtMostNValue` Constraint

The `AtMostNValue` constraint belongs to the *Number of Distinct Values* constraint family [2]. It has been introduced in [28] to specify music programs but the first filtering algorithm was provided in [2]. Then, the `AtMostNValue` constraint has been widely investigated in [4], where the authors proved that ensuring the generalised arc consistency (GAC) of `AtMostNValue` is NP-hard, and provide various filtering algorithms. According to this study, the *greedy* propagator they introduced provides a good tradeoff between filtering and runtime. Thus, we use it as the reference propagator for filtering the `AtMostNValue` constraint. Before describing this propagator we need to recall a few definitions:

Definition 1. *The intersection graph of a set of variables $\mathcal{X}$, $G_\mathcal{I}(\mathcal{X}) = (V, E_\mathcal{I})$, is defined by a vertex set V where each variable $x_i \in \mathcal{X}$ is associated with a vertex $i \in V$, and an edge set $E_\mathcal{I}$ representing domain intersections: for any $(i, j) \in V^2$, there is an edge $(i, j) \in E_\mathcal{I}$ if and only if $Dom(x_i) \cap Dom(x_j) \neq \emptyset$.*

Definition 2. *An independent set of a graph $G = (V, E)$ is a subset, $A \subseteq V$, of disjoint vertices, i.e., for any $(i, j) \in A^2$ such that $i \neq j$, $(i, j) \notin E$.*

Definition 3. *A maximum independent set of a graph G is an independent set whose cardinality is maximal. The cardinality of a maximum independent set of a graph G is noted $\alpha(G)$.*

In the following, the set of all independent sets of a graph G, is referred to as $\mathcal{IS}(G)$. The filtering algorithm proposed in [4] stems from the search of a maximum independent set in $G_\mathcal{I}(\mathcal{X})$. Since this problem is NP-Hard [14], it actually computes an independent set A of $G_\mathcal{I}(\mathcal{X})$, in a greedy way, by selecting nodes of minimum degree first [17]. This heuristic is referred to as `MD`. Then, the propagator filters according to the following rules:

- $\mathcal{R}_1$: $\underline{z} \leftarrow max(\underline{z}, |A|)$
- $\mathcal{R}_2$: $|A| = \overline{z} \Rightarrow \forall i \in V, Dom(x_i) \leftarrow Dom(x_i) \cap \bigcup_{a \in A} Dom(x_a)$

Where $\underline{z}$ and $\overline{z}$ respectively refer to the lower bound and the upper bound of the variable z. $\mathcal{R}_1$ states that the cardinality of A is a valid lower bound for z. $\mathcal{R}_2$ states that whenever the cardinality of the independent set A is equal to the upper bound of z, then variables in $\mathcal{X}$ have to take their values among the subset of values induced by A. Indeed, variables associated with an independent set of an intersection graph take different values, by definition. Thus, using a value outside of this subset of values would lead to use at least $|A| + 1$ values, which is a contradiction.

Thus, the greedy propagator of `AtMostNValue` takes a graph G as input, calls a function F to compute independent sets in G and then filters variable domains with a set of rules $\mathcal{R}$. Therefore, we introduce the notation $\text{AMNV}\langle G|F|\mathcal{R}\rangle$ to define such a family of propagators. Consequently, the greedy propagator introduced in [4] is referred to as

$\text{AMNV}\langle G_{\mathcal{I}}|\text{MD}|\mathcal{R}_{1,2}\rangle$. In the following, we suggest improvements for G, F and $\mathcal{R}$, leading to a new propagator which filters `AtMostNValue` and a set of difference constraints.

To illustrate the state-of-the-art propagator, we now apply it to our example (cf. Figure 2). Because of variables domain definition, the intersection graph corresponding to our example, is a complete graph. Thus, MD select only one node, x_3 for instance. Then $\mathcal{R}_1$ states that the number of workers required is at least one. If we now assume that the number of workers required is at most one, then $\mathcal{R}_2$ states that values w_2, w_4 and w_5 must be removed from the domain of the variables (cf. Figure 2a). Consequently, the edges (x_1, x_2), (x_1, x_5) and (x_5, x_4) have to be removed in order to obtain the new intersection graph. Based on this new graph, if we assume that x_1 and x_5 are then used as a new independent set (cf. Figure 2b) then $\mathcal{R}_1$ states that the number of required workers is at least two, leading to fail, since $z = 1$.

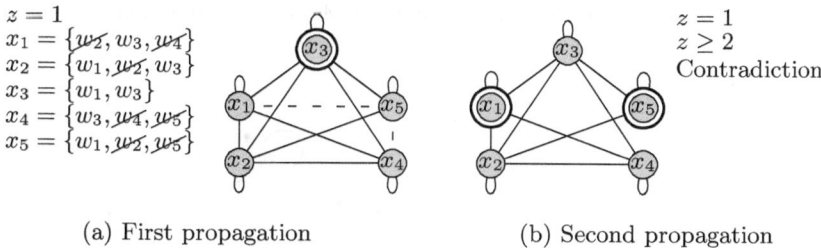

$z = 1$
$x_1 = \{\cancel{w_2}, w_3, \cancel{w_4}\}$
$x_2 = \{w_1, \cancel{w_2}, w_3\}$
$x_3 = \{w_1, w_3\}$
$x_4 = \{w_3, \cancel{w_4}, \cancel{w_5}\}$
$x_5 = \{w_1, \cancel{w_2}, \cancel{w_5}\}$

$z = 1$
$z \geq 2$
Contradiction

(a) First propagation   (b) Second propagation

Fig. 2. Applying $\text{AMNV}\langle G_{\mathcal{I}}|\text{MD}|\mathcal{R}_{1,2}\rangle$ to our example when $z = 1$

3.3 Embedding Difference Constraints into `AtMostNValue`

As the SMPTSP only considers two kind of constraints (`AtMostNValue` and `AllDifferent`), filtering their conjunction may be very profitable. For that purpose, this section introduces an implied propagator, in the form $\text{AMNV}\langle G|F|\mathcal{R}\rangle$, which considers difference constraints. As a first step, we suggest to consider a new graph, referred to as a *constrained intersection graph* (Definition 4), instead of the intersection graph of variables.

Definition 4. *Given a set of variables $\mathcal{X}$ and a set of difference constraints $\mathcal{D}$, the constrained intersection graph, $G_{\mathcal{CI}}(\mathcal{X}, \mathcal{D}) = (V, E_{\mathcal{CI}})$, of $\mathcal{X}$ and $\mathcal{D}$ is defined by a vertex set V where each variable $x_i \in \mathcal{X}$ is associated with a vertex $i \in V$, and an edge set $E_{\mathcal{CI}}$ representing possible classes of equivalence: for any $(i, j) \in V^2$, there is an edge $(i, j) \in E_{\mathcal{CI}}$ if and only if $Dom(x_i) \cap Dom(x_j) \neq \emptyset$ and $\textbf{neq}(x_i, x_j) \notin \mathcal{D}$.*

In this paper, we consider a single set of variables $\mathcal{X}$ and a single set of difference constraints $\mathcal{D}$, thus, for the sake of clarity $G_{\mathcal{CI}}(\mathcal{X}, \mathcal{D})$ and $G_{\mathcal{I}}(\mathcal{X})$ will be respectively noted $G_{\mathcal{CI}}$ and $G_{\mathcal{I}}$. It is worth noticing that $G_{\mathcal{CI}} \subseteq G_{\mathcal{I}}$.

Proposition 1. $\mathcal{IS}(G_{\mathcal{I}}) \subseteq \mathcal{IS}(G_{\mathcal{CI}})$, hence $\alpha(G_{\mathcal{I}}) \leq \alpha(G_{\mathcal{CI}})$.

Proof. Let $A_{\mathcal{I}}$ be an independent set in $G_{\mathcal{I}}$. Since $G_{\mathcal{CI}}$ and $G_{\mathcal{I}}$ are based on the same variable set, they share the same vertex set, so $A_{\mathcal{I}}$ is also a subset of vertices of $G_{\mathcal{CI}}$. Since vertices of $A_{\mathcal{I}}$ are pairwise disjoint in $G_{\mathcal{I}}$ (by assumption) and since all edges of $G_{\mathcal{CI}}$ also belong to $G_{\mathcal{I}}$, then vertices of $A_{\mathcal{I}}$ are also pairwise disjoint in $G_{\mathcal{CI}}$. Consequently, $A_{\mathcal{I}}$ is an independent set of $G_{\mathcal{CI}}$. Thus, all independent sets of $G_{\mathcal{I}}$ are independent sets of $G_{\mathcal{CI}}$, so $\max\limits_{I \in \mathcal{IS}(G_{\mathcal{I}})} |I| \leq \max\limits_{I_c \in \mathcal{IS}(G_{\mathcal{CI}})} |I_c|$, hence $\alpha(G_{\mathcal{I}}) \leq \alpha(G_{\mathcal{CI}})$. $\square$

Note that a maximum independent set in $G_{\mathcal{I}}$, is not necessarily maximal in $G_{\mathcal{CI}}$. For instance, one may consider a non-empty set of variables with identical domains and with a difference constraint over each pair of distinct variables. Then $G_{\mathcal{I}}$ is a complete graph, whereas there are no edges, but loops, in $G_{\mathcal{CI}}$. Consequently, $\alpha(G_{\mathcal{I}}) = 1$ whereas $\alpha(G_{\mathcal{CI}}) = |V|$. It is worth noticing that in our context, the bigger the independent set, the higher the chance to filter variable domains. Thus, using $G_{\mathcal{CI}}$ is apriori better than using $G_{\mathcal{I}}$ to filter AtMostNValue when difference constraints figure in the model (Proposition 2).

To illustrate the interest of $G_{\mathcal{CI}}$, we now use it on our example (cf. Figure 3). Because of difference constraints, $G_{\mathcal{CI}}$ is sparser than $G_{\mathcal{I}}$ (cf. 3a). Thus, MD is now able to compute a larger independent set, leading to find a better lower bound of z. For instance, if we consider the independent set $\{x_1, x_2, x_3\}$ (cf. Figure 3b) then AMNV$\langle G_{\mathcal{CI}}|MD|\mathcal{R}_{1,2}\rangle$ states that the number of required workers is at least three. Moreover, if we assume that the number of required workers is at most three, then the value w_5 has to be removed from the variable domains.

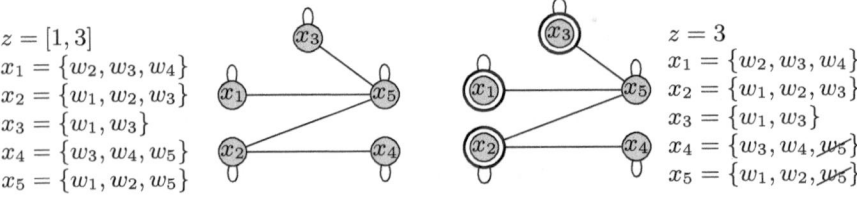

$$z = [1,3]$$
$$x_1 = \{w_2, w_3, w_4\}$$
$$x_2 = \{w_1, w_2, w_3\}$$
$$x_3 = \{w_1, w_3\}$$
$$x_4 = \{w_3, w_4, w_5\}$$
$$x_5 = \{w_1, w_2, w_5\}$$

(a) Constrained intersection graph

$$z = 3$$
$$x_1 = \{w_2, w_3, w_4\}$$
$$x_2 = \{w_1, w_2, w_3\}$$
$$x_3 = \{w_1, w_3\}$$
$$x_4 = \{w_3, w_4, \cancel{w_5}\}$$
$$x_5 = \{w_1, w_2, \cancel{w_5}\}$$

(b) Filtering from an independent set

Fig. 3. Use of AMNV$\langle G_{\mathcal{CI}}|MD|\mathcal{R}_{1,2}\rangle$ on our example

Proposition 2. *Given an oracle $\mathcal{O}$ which computes all maximum independent sets of any graph, then* AMNV$\langle G_{\mathcal{CI}}|\mathcal{O}|\mathcal{R}_{1,2}\rangle$ *dominates* AMNV$\langle G_{\mathcal{I}}|\mathcal{O}|\mathcal{R}_{1,2}\rangle$

Proof. First of all, since $\mathcal{O}$ is able to compute all maximum independent sets of any graph, then the lower bound given by $\mathcal{R}_1$ in $G_{\mathcal{CI}}$ is equal to $\alpha(G_{\mathcal{CI}})$ whereas the lower bound given by $\mathcal{R}_1$ in $G_{\mathcal{I}}$ is equal to $\alpha(G_{\mathcal{I}})$. Since $\alpha(G_{\mathcal{I}}) \leq \alpha(G_{\mathcal{CI}})$ (Proposition 1), then AMNV$\langle G_{\mathcal{CI}}|\mathcal{O}|\mathcal{R}_1\rangle$ dominates AMNV$\langle G_{\mathcal{I}}|\mathcal{O}|\mathcal{R}_1\rangle$. Second, since $\mathcal{O}$ is able to compute all maximum independent sets of any graph, and since $\mathcal{IS}(G_{\mathcal{I}}) \subseteq \mathcal{IS}(G_{\mathcal{CI}})$ (Proposition 1), then values filtered by AMNV$\langle G_{\mathcal{I}}|\mathcal{O}|\mathcal{R}_2\rangle$ are also filtered by AMNV$\langle G_{\mathcal{CI}}|\mathcal{O}|\mathcal{R}_2\rangle$. Consequently, AMNV$\langle G_{\mathcal{CI}}|\mathcal{O}|\mathcal{R}_2\rangle$ dominates AMNV$\langle G_{\mathcal{I}}|\mathcal{O}|\mathcal{R}_2\rangle$ and thus AMNV$\langle G_{\mathcal{CI}}|\mathcal{O}|\mathcal{R}_{1,2}\rangle$ dominates AMNV$\langle G_{\mathcal{I}}|\mathcal{O}|\mathcal{R}_{1,2}\rangle$. $\square$

Proposition 3. *Given an independent set A in $G_{\mathcal{CI}}$ such that $|A| = \overline{z}$, any solution of the conjunction of* AtMostNValue *and $\mathcal{D}$ satisfies the following formula: $\forall i \in V \backslash A$, $\exists a \in A_i$ s.t. $x_i = x_a$, where A_i denotes $\{a \in A | (i,a) \in E_{\mathcal{CI}}\}$.*

Proof. Given an independent set A in $G_{\mathcal{CI}}$ such that $|A| = \overline{z}$. Let's assume that there exists a solution S to the conjunction of AtMostNValue and $\mathcal{D}$ such that there exists a vertex $i \in V \backslash A$ for which $\forall a \in A_i, x_i \neq x_a$. Thus, S is solution of the conjunction of AtMostNValue and $\mathcal{D} \cup \{$neq$(x_i, x_a)|a \in A_i\}$. Consequently, $A \cup \{i\}$ is a valid independent set in $G_{\mathcal{CI}}$. Then $\mathcal{R}_1$ states that $\overline{z} \leftarrow |A \cup \{i\}|$, i.e., $\overline{z} \leftarrow \overline{z} + 1$ which is not possible. Consequently, such a solution S does not exist, hence Proposition 3 holds. $\square$

From a filtering perspective, Proposition 3 leads to consider the following rule:

- $\mathcal{R}_3$: $|A| = \bar{z} \Rightarrow \forall i \in V \backslash A$ $\begin{cases} Dom(x_i) \leftarrow Dom(x_i) \cap \bigcup_{a \in A_i} Dom(x_a) \\ A_i = \{a\} \Rightarrow Dom(x_a) \leftarrow Dom(x_a) \cap Dom(x_i) \end{cases}$

This rule is actually a refined variant of $\mathcal{R}_2$. While this change is quite simple, it may have a significant impact in practice, especially on large scale problems were, presumably, $\forall i \in V \backslash A$, $|A_i| << |A|$. Note the particular case that occurs when, for some node $i \in V \backslash A$, $A_i = \{a\}$ enables to learn the valid equality $x_i = x_a$. This way, it is possible to filter the domain of the variable x_a of the independent set as well. From a theoretical point of view $\mathcal{R}_3$ is also stronger than $\mathcal{R}_2$ (cf Proposition 4).

Proposition 4. *Given a deterministic heuristic H which computes a set of independent sets in G_{CI}, then AMNV$\langle G_{CI}|H|\mathcal{R}_3 \rangle$ dominates AMNV$\langle G_{CI}|H|\mathcal{R}_2 \rangle$*

Proof. Since AMNV$\langle G_{CI}|\text{H}|\mathcal{R}_3 \rangle$ and AMNV$\langle G_{CI}|\text{H}|\mathcal{R}_2 \rangle$ use the same deterministic heuristic H in the same graph G_{CI}, then they filter with the same independent set A. Since, for any node $i \in V$, $A_i = \{a \in A | (i, a) \in E_{CI}\}$, then $A_i \subseteq A$. Consequently, any value filtered by $\mathcal{R}_2$ is also filtered by $\mathcal{R}_3$. ☐

To illustrate this rule, we now apply it on our example (cf. Figure 4). If we consider the independent set $\{x_1, x_3, x_4\}$ (cf. Figure 4a) and an objective $z = [1, 3]$, then AMNV$\langle G_{CI}|\text{MD}|\mathcal{R}_{1,2} \rangle$ deduces that $z = 3$ but cannot filter $\mathcal{X}$ domains. However, AMNV$\langle G_{CI}|\text{MD}|\mathcal{R}_{1,3} \rangle$ enables to filter the value w_5, which is not included in $Dom(x_1) \cup Dom(x_3) = \{w_1, w_2, w_3, w_4\}$, from the domain of x_5 and values w_1 and w_2 for variable x_2 because they do not figure in $Dom(x_4) = \{w_3, w_4, w_5\}$ (cf. Figure 4b). Moreover, AMNV$\langle G_{CI}|\text{MD}|\mathcal{R}_{1,3} \rangle$ also enables to learn the valid equality $x_2 = x_4$, which enables to remove values w_4 and w_5 from the domain of x_4.

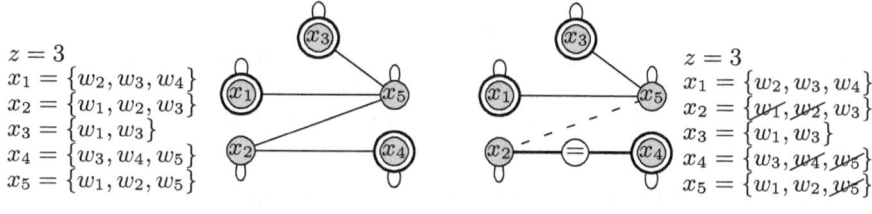

$z = 3$
$x_1 = \{w_2, w_3, w_4\}$
$x_2 = \{w_1, w_2, w_3\}$
$x_3 = \{w_1, w_3\}$
$x_4 = \{w_3, w_4, w_5\}$
$x_5 = \{w_1, w_2, w_5\}$

(a) Filtering with AMNV$\langle G_{CI}|\text{MD}|\mathcal{R}_{1,2} \rangle$

$z = 3$
$x_1 = \{w_2, w_3, w_4\}$
$x_2 = \{\not{w_1}, \not{w_2}, w_3\}$
$x_3 = \{w_1, w_3\}$
$x_4 = \{w_3, \not{w_4}, \not{w_5}\}$
$x_5 = \{w_1, w_2, \not{w_5}\}$

(b) Filtering with AMNV$\langle G_{CI}|\text{MD}|\mathcal{R}_{1,3} \rangle$

Fig. 4. Using AMNV$\langle G_{CI}|\text{MD}|\mathcal{R}_{1,3} \rangle$ on our example

3.4 Diversifying Filtering

CP frameworks traditionally perform a fix point over constraints at each branching node [30]. This implies that our model may compute thousands of independent sets during the search process. Thus, it is advised to use a greedy algorithm, such as MD, to filter the `AtMostNValue` constraint [4]. This heuristic is quite efficient but it lacks diversification for both its bound and the resulting filtering, which depend on the computed independent set. This may lead to unfortunate results when the considered graph does not suit this deterministic heuristic.

Proposition 5. *Given two functions F_1 and F_2 which compute a set of independent sets in G_{CI}, respectively noted $\mathcal{A}_1$ and $\mathcal{A}_2$. Let a set $\mathcal{A}_{22}$ denote $\mathcal{A}_2 \setminus \mathcal{A}_1$. If $\max_{I_2 \in \mathcal{A}_{22}} |I_2| < \max_{I_1 \in \mathcal{A}_1} |I_1|$, then $\mathtt{AMNV}\langle G_{CI}|F_1|\mathcal{R}_{1,3}\rangle$ dominates $\mathtt{AMNV}\langle G_{CI}|F_2|\mathcal{R}_{1,3}\rangle$.*

Proof. First, $\max_{I_2 \in \mathcal{A}_{22}} |I_2| < \max_{I_1 \in \mathcal{A}_1} |I_1|$ implies that $\max_{I_2 \in \mathcal{A}_2} |I_2| \leq \max_{I_1 \in \mathcal{A}_1} |I_1|$, hence $\mathtt{AMNV}\langle G_{CI}|F_1|\mathcal{R}_1\rangle$ dominates $\mathtt{AMNV}\langle G_{CI}|F_2|\mathcal{R}_1\rangle$. Second, let's now assume that a value is removed from a domain of a variable in $\mathcal{X}$, by $\mathtt{AMNV}\langle G_{CI}|F_2|\mathcal{R}_3\rangle$. If this filtering occurs while considering an independent set of $\mathcal{A}_1 \cap \mathcal{A}_2$, then the same filtering is performed by $\mathtt{AMNV}\langle G_{CI}|F_1|\mathcal{R}_{1,3}\rangle$. Else, this filtering occurs when considering an independent set I_2 of $\mathcal{A}_{22}$. A necessary condition so that $\mathcal{R}_3$ triggers filtering is that, $|I_2| = \overline{z}$. As $|I_2| < \max_{I_1 \in \mathcal{A}_1} |I_1|$, then the problem is actually infeasible, which is captured by $\mathtt{AMNV}\langle G_{CI}|F_1|\mathcal{R}_1\rangle$. Thus, $\mathtt{AMNV}\langle G_{CI}|F_1|\mathcal{R}_{1,3}\rangle$ dominates $\mathtt{AMNV}\langle G_{CI}|F_2|\mathcal{R}_{1,3}\rangle$. □

As highlighted by the Proposition 5, it may be interesting to obtain several independent sets to apply filtering rules over a greater set of variables. The simplest way to get diversification would be to randomly break ties of MD. However, it turns out that this does not bring enough diversification to improve results. Thus, we introduce the $\mathtt{R}^k$ algorithm (cf. Algorithm 1), which keeps the philosophy of a fast and simple greedy algorithm, while providing diversification. This heuristic performs k iterations, each one computes an independent set by selecting nodes randomly, with a uniform probability distribution. It thus provides a set of k independent sets. Note that we also tried to use a weighted random that favors nodes with a small degree, but it does not outperform $\mathtt{R}^k$, while being more complicated, and thus it is not presented in this paper. From a theoretical point of view, this approach presents several interesting properties. First, it offers control over its runtime complexity and its expected quality. Second, computing independent sets randomly tends to impact variable domains homogeneously. Note that, when $k \to \infty$, then the method tends to enumerate and filter with all maximum independent sets, which is optimal with the given filtering rules. However, from a practical point of view, one has to bound the number of iterations in order to get a reasonable runtime. We suggest the default setting $k = 30$, which seems to perform better than MD, on average, without introducing a significant overhead. However, MD offers a good approximation ratio [17], hence we recommend to use both.

3.5 The Case of Dynamic Difference Constraints

Our approach focuses on the SMPTSP which considers a set of tasks that are already fixed in time. Thus, difference constraints are given as input through a set of AllDifferent constraints. However, one may be interested in filtering AtMostNValue when difference constraints appear dynamically during the search. For instance, in disjunctive scheduling problems where tasks have to be fixed in time, difference constraints implicitly appear as variable domains are reduced. It is worth noticing that in this situation, difference constraints would no longer be well propagated because of the absence of AllDifferent constraints in the model. Fortunately, the AtMostNValue propagator can help to get back a global view of the problem and thus a powerful filtering thanks to Proposition 6.

Proposition 6. *If A is an independent set of G_{CI}, let $\mathcal{X}_A$ denote $\{x_i \in \mathcal{X} | i \in A\}$, then the constraint $AllDifferent(\mathcal{X}_A)$ holds.*

Algorithm 1. R^k algorithm to compute independent sets

Global variables:

 $G_{\mathcal{CI}}$, the constrained intersection graph of which independent sets have to be computed

 k, the number of iterations (and thus the number of independent sets to compute)

1: $\mathcal{A} \leftarrow \emptyset$ // Creates a set of independent sets, initially empty
2: $count \leftarrow k$
3: **while** $(count \geq 0)$ **do**
4:   $count \leftarrow count - 1$
5:   $G \leftarrow G_{\mathcal{CI}}.copy()$ // Copies the constrained intersection graph
6:   $A \leftarrow \emptyset$ // Creates an independent set A, initially empty
7:   **while** $(G \neq \emptyset)$ **do**
8:    $x \leftarrow randomNode(G)$// Randomly selects a node x in G
9:    $A \leftarrow A \cup \{x\}$// Adds x to the independent set A
10:    $G \leftarrow G \setminus \{y | (x, y) \in G\}$ // Removes x's neighbors from G (including x itself)
11:   **end while**
12:   $\mathcal{A} \leftarrow \mathcal{A} \cup \{A\}$ // adds the independent set A to $\mathcal{A}$
13: **end while**
14: **return** $\mathcal{A}$ // returns a set of independent sets of $G_{\mathcal{CI}}$

Proof. By definition of a constrained intersection graph, an independent set represents variables that are either already different or constrained to be. □

From Proposition 6, we derive a new filtering rule for AtMostNValue which calls a filtering algorithm of AllDifferent over variables corresponding to the independent set A it has computed:

 – $\mathcal{R}_4$: AllDifferent$(\{x_i \in \mathcal{X} | i \in A\})$

Note that this rule can either directly call a filtering algorithm of AllDifferent over the appropriate subset of variables or post dynamically an AllDifferent constraint into the solver. The first option is the simplest one, but cannot filter incrementally, so a *bound-consistent* filtering algorithm of AllDifferent, such as the one presented in [26], would presumably be more relevant than the GAC one [29]. Moreover, the subset of variables to be different is lost right after filtering and the next propagation may involve a worse independent set. Instead, the second option can involve incremental GAC AllDifferent constraints which would remain in the current search branch. However, since each independent set can post an AllDifferent constraint, it has the drawback of leading to a potential explosion over the number of such constraints. Thus, one needs to set up a constraint pool to manage those constraints, by retaining only the most interesting constraints and removing dominated ones. One can see a parallel with *Cut pools* of MIP solvers [1]. For solver maintenance simplicity purposes, we recommend the first option.

3.6   Implementation

So far, we have seen that it was possible to tune the original greedy AtMostNValue propagator. We now suggest a simple and non-intrusive implementation (see Algorithm 2). In this implementation, no assumption is made on the set of difference constraints $\mathcal{D}$ which can thus grow during the resolution process. For instance, $\mathcal{D}$ can be modified by the user or other constraints. Note also that no assumption is made about the kind of variables in $\mathcal{X}$, which can then be multidimensional continuous variables, as in [7]. The constrained intersection graph is stored as a backtrackable structure which is updated at the beginning of the filtering algorithm, *i.e.*, after potentially many domain modifications. This turns out to be much faster than updating G_{CI} incrementaly on each domain modification or rebuilding it from scratch. Once the graph has been updated, the propagator computes a set of independent sets with a function F. For each independent set, the propagator filters the lower bound of z and filters $\mathcal{X}$ with a set of rules $\mathcal{R}$. Thus, one can see that this propagator is flexible and not intrusive.

Algorithm 2. AMNV$\langle G_{CI}|F|\mathcal{R}\rangle$ - filtering algorithm

---

Global variables:
  z, an integer variable
  $\mathcal{X}$, a set of variables
  $\mathcal{D}$, a set of difference constraints
  $G_{CI} = (V, E_{CI})$, a backtrackable graph
  F, a function that computes a set of independent sets in a given graph
  $\mathcal{R}$, a set of filtering rules

1: **if** (first call of the propagator) **then**
2:      // graph generation
3:      $V \leftarrow [1, |\mathcal{X}|]$
4:      $E_{CI} \leftarrow \emptyset$
5:      **for** $((i, j) \in V^2)$ **do**
6:          **if** $(Dom(x_i) \cap Dom(x_j) \neq \emptyset \wedge \text{neq}(x_i, x_j) \notin \mathcal{D})$ **then**
7:              $E_{CI} \leftarrow E_{CI} \cup \{(i, j)\}$
8:          **end if**
9:      **end for**
10:     $G_{CI} \leftarrow (V, E_{CI})$
11: **else**
12:        //graph lazy update
13:        **for** $((i, j) \in E_{CI})$ **do**
14:            **if** $(Dom(x_i) \cap Dom(x_j) = \emptyset \vee \text{neq}(x_i, x_j) \in \mathcal{D})$ **then**
15:                $E_{CI} \leftarrow E_{CI} \backslash \{(i, j)\}$
16:            **end if**
17:        **end for**
18: **end if**
19: $\mathcal{A} \leftarrow F(G_{CI})$// computes a set $\mathcal{A}$ of independent sets
20: **for** $(A \in \mathcal{A})$ **do**
21:     $filter(\mathcal{X}, z, G_{CI}, A, \mathcal{R})$ // rule-based filtering
22: **end for**

---

4   Experimental Study

In order to evaluate the interest of our contribution, we have performed extensive tests. First, Section 4.1 introduces and motivates a new benchmark data set for the SMPTSP. Second, Section 4.2 describes the search strategies that were used in this study. Next, in Section 4.3, we focus on the objective lower bound quality at root node. We show that it changes completely depending on the model that is used. Section 4.4 highlights the potential benefit of strengthening the filtering with $\mathbf{R}^k$. Section 4.5 focuses on the scalability issue, *i.e.*, it investigates the ability of our approach to compute tight bounds, even on large instances. Finally, in Section 4.6 we compare our approach to the best known results on the state-of-the-art SMPTSP instances.

Our algorithms have been implemented in JAVA, we have used Cplex 12.4 with default settings to test *MIP model* and Choco 13-03 [8] to perform every test related to *CP model*. All our implementations are available online at [23]. In the following line-graphs, instances are sorted in order to ease graphs reading. Moreover, in order to simplify result analysis, we note respectively z^* and $\underline{z}_r$ the optimal objective and the objective lower bound at root node. Finally, a CP model using a propagator $\text{AMNV}\langle G|\text{F}|R\rangle$, is noted $\downarrow\text{AMNV}\langle G|\text{F}|R\rangle$ when used within a top-down minimisation strategy and $\uparrow\text{AMNV}\langle G|\text{F}|R\rangle$ when used within a bottom-up minimisation strategy.

4.1   A New Set of Challenging SMPTSP Instances

Very recently, Smet *et. al.* have shown in [31] that the 137 instances provided by Krishnamoorthy *et. al.* in [20] admit a feasible solution with an objective value equal to $LB_{\neq}$. Since worker skills are taken into account in [20], this result is somehow surprising: one may expect that considering worker skills would have an impact on the optimum, but actually, it only makes it harder to find a feasible solution. Consequently, these instances do not provide much challenge regarding to the search of interesting lower bounds. Therefore, we propose a new set of challenging instances whose maximal number of overlapping tasks does not provide a good lower bound.

In order to generate this new benchmark we used a dedicated procedure based on some of the empirical results presented in [20] and [31]. First of all, it is specified in [20] that the average tightness, defined as the sum of processing times over the sum of shift lenghts should be close to 90% in order to obtain challenging instances. Another hardness analysis [31] shows that the smaller the average task processing time, the more difficult instances are to solve. Based on these two conclusions, we designed a dedicated procedure able to generate instances for which the maximum number of overlapping tasks does not provide a good lower bound.

The procedure which is given in details in [23] generates randomly a set of tasks ranging from 15 minutes to 2 hours. We consider six different kinds of shift. The first three aim at splitting a working day into shifts of 8 hours which is very common in personnel scheduling [32]. The other three are obtained from the previous ones by introducing an offset of 2 hours to their starting time, so that each task is entirely contained in at least one shift. Based on this simple procedure, we provide 100 new instances with a number of tasks ranging from 70 to 1600 and a number of available workers ranging from 60 to 950. These instances along with our generator are available online [23]. In the following, we refer to the instances of Krishnamoorthy *et. al.* as Data_137, and our generated instances as Data_100.

4.2   Search Strategies

Our contribution mainly concerns the lower bounding of the problem, and the lower the objective upper bound, the stronger our filtering. Hence, we naturally employed a bottom-up minimisation approach. It tries to compute a solution involving k workers, where k is initialised to the root lower bound of z and incremented by one each time the solver proved unsatisfiability. Thus, the first solution found is optimal.

It is worth noticing that when using a bottom-up approach, most part of the search concerns infeasible regions of the search space. In order to get out of them as quickly as possible, it seems wiser to use intensification processes within the branching heuristic, following the *Fail First* principle [18]. Thus, our bottom-up approach uses a sequential heuristic: a first heuristic randomly determines which subset of workers will be used in the solution, then a `first-fail` heuristic is used to assign tasks to workers: the variable associated with the smallest domain is selected first, and then fixed to its lower bound. The overall branching scheme is reinforced by the `Last Conflict` heuristic of [24] which aims at identifying critical variables.

4.3   Impact of Suggested Improvements on the Root Lower Bound

In order to evaluate the pratical interest of our contribution, we have compared the value of $\underline{z}_r$ for several settings, on Data_100. Results are reported on Figure 5. Using $\text{AMNV}\langle G_{\mathcal{CI}}|\text{MD},\text{R}^{30}|\mathcal{R}_{1,3}\rangle$ instead of $\text{AMNV}\langle G_{\mathcal{I}}|\text{MD}|\mathcal{R}_{1,2}\rangle$ dramatically improves the value of $\underline{z}_r$: on average, our approach is able to increase $LB_{\neq}$ by a factor two, whereas the former approach is not able to reach this trivial lower bound. Moreover, our approach scales much better than the classical one which is not able to increase $\underline{z}_r$ whatever the size of the instance. Actually, this is not really surprising since $\text{AMNV}\langle G_{\mathcal{I}}|\text{MD}|\mathcal{R}_{1,2}\rangle$ is blind to AllDifferent constraints, which represent a big part of the problem. Using $G_{\mathcal{CI}}$ instead of $G_{\mathcal{I}}$ makes these AllDifferent constraints visible to AtMostNValue, hence the increased of $\underline{z}_r$. Finally, to a lesser extent, combining R^{30} and MD also contributes to improves $\underline{z}_r$ by 7%. Given the high quality of the lower bound reached by $\text{AMNV}\langle G_{\mathcal{CI}}|\text{MD}|\mathcal{R}_{1,3}\rangle$, an increase of 7% is already significant. Moreover, considering the simplicity of this modification, it is very interesting.

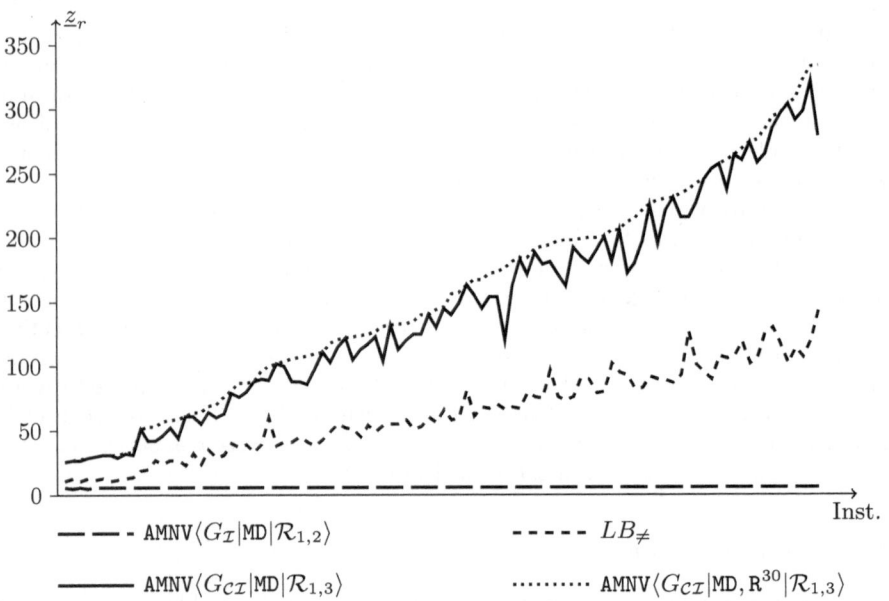

Fig. 5. Value of $\underline{z}_r$ depending on the propagator of AtMostNValue.

One may wonder if the straightforward CP model is able to catch up with our approach after a few minutes. Hence, we have evaluated the evolution of $\underline{z}$ after five minutes of bottom-up resolution. It turned out that its best improvement was by 7 units only, which is far too small to catch up with $LB_{\neq}$.

4.4   Managing the Tradeoff between Filtering and Runtime

As explained in Section 3.4, using the heuristic R^k is a simple and effective way to obtain diversification in filtering. The parameter k enables to manage the tradeoff between expected filtering power and runtime. Since R^k is suggested as a complement to

MD, it must be seen as an improvement opportunity. Back to our case, many values of k have been tested, within a time limit of 5 minutes (Table 1). Interestingly, the optimal setting changes quite a lot from one data set to another, hence using an automatic algorithm configuration program [19] seems relevant to get the best results. More precisely, it can be seen that for Data_100 instances, the more filtering the better. The time spent in $\mathbf{R}^k$ is compensated by the search space reduction it provides. Thus, it may be worth strengthening even more the filtering, by filtering conjunctions of `AllDifferent` constraints for instance [5]. Instead, Data_137 instances are well solved by our model, hence the faster $\mathbf{R}^k$ the better. We report that using $\mathbf{R}^{800}$ with $\mathcal{R}_{1,2}$ instead of $\mathcal{R}_{1,3}$, leads to finding 7 optima only. Thus, the interest of $\mathbf{R}^k$ lies not only in finding large independent sets ($\mathcal{R}_1$), but mostly in diversifying the filtering ($\mathcal{R}_3$).

Table 1. Number of optima found with $\uparrow$AMNV$\langle G_{\mathcal{CI}}|MD,\mathbf{R}^k|\mathcal{R}_{1,3}\rangle$. For reference, *MIP model* finds respectively 65 and 46 optima on Data_137 and Data_100.

Benchmark	Data_137					Data_100				
k	10	30	50	70	90	100	200	400	800	1600
Nb Opt.	101	**109**	106	99	95	19	22	28	**35**	31

4.5   Scalability Study

We now evaluate the ability of our approach to provide good bounds on large scale instances. For that, we compare the values of $\underline{z}$ and $\overline{z}$ obtained with *MIP model*, with those obtained with $\uparrow$AMNV$\langle G_{\mathcal{CI}}|MD,\mathbf{R}^{800}|\mathcal{R}_{1,3}\rangle$ and $\downarrow$AMNV$\langle G_{\mathcal{CI}}|MD,\mathbf{R}^{30}|\mathcal{R}_{1,3}\rangle$, after 5 minutes of resolution. For this last one, we did not use the branching heuristic described in section 4.2 which is no longer relevant in a top-down mode. Instead we used `first-fail` to select a task and then we assign it to any worker, preferably already used in the schedule. Moreover, our propagator is unlikely to filter significantly in a top-down approach, since the upper bound is presumably too high to enable filtering. Consequently, we set k to 30 so that it improves MD at least expense.

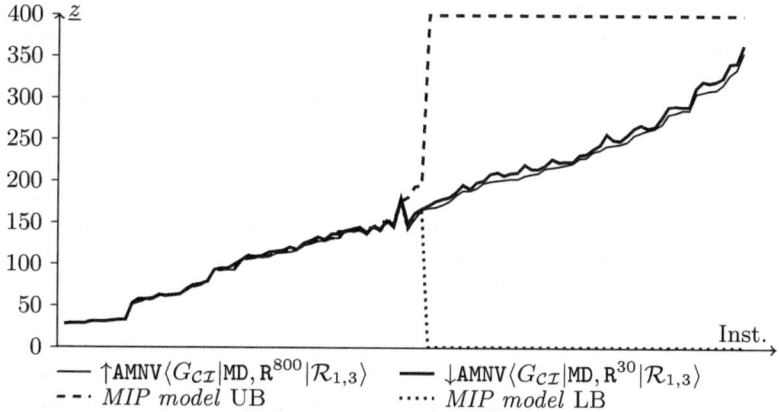

Fig. 6. Objective value after a 5 minutes resolution, on Data_100

Figure 6 gives general trends regarding the ability to scale of *MIP model* and *CP model*. It shows that *MIP model* fails to solve about half of instances: it is not able to provide either lower bounds or upper bounds. Regarding this scalability issue, the top-down and bottom-up CP approaches perform very well. Their relative gap varies around 2%. Moreover, it does not increase much with the instance size. Consequently, using $\uparrow\texttt{AMNV}\langle G_{\mathcal{CI}}|\texttt{MD},\texttt{R}^{800}|\mathcal{R}_{1,3}\rangle$ and $\downarrow\texttt{AMNV}\langle G_{\mathcal{CI}}|\texttt{MD},\texttt{R}^{30}|\mathcal{R}_{1,3}\rangle$ leads to finding tight bounds for z, even on large instances.

4.6 A Competitive Approach

In order to estimate the quality of our approach, we consider state-of-the-art SMPTSP instances [21]. We compare the performances of $\uparrow\texttt{AMNV}\langle G_{\mathcal{CI}}|\texttt{MD},\texttt{R}^{30}|\mathcal{R}_{1,3}\rangle$ and *MIP model*, but we also compare our results to those of Krishnamoorthy *et. al.* and Smet *et. al.* who use two different metaheuristics. As illustrated on Figure 7, the CP model clearly dominates *MIP model* on this data set, since it is able to reach the optimum of all but one instance within the time limit. It is important to notice that the only other method that is able to solve these instances to optimality is the metaheuristic given in [31]. This metaheuristic is actually faster than our approach, since it is able to reach all optima in 5 minutes, whereas our approach, after 5 minutes, has proved 80% of optima. However, it provides no lower bound, hence it does not prove optimality, except when $LB_{\neq}$ equals z^*.

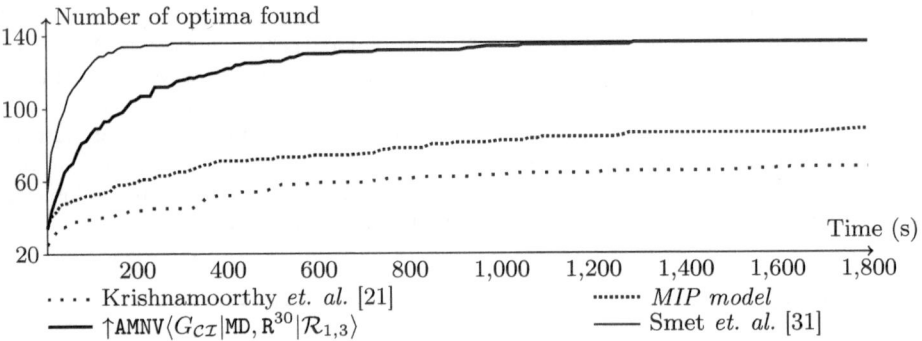

Fig. 7. Number of optima found, depending on runtime, on Data_137

As the SMPTSP often occurs as a subproblem of a more general method, it seems critical to provide a good lower bound in a short time. To evaluate the quality of $\underline{z}_r$, we compare its value and its required runtime with those of the lower bounding procedure given in [21]. This procedure is based on a Lagrangean relaxation, therefore, we refer to its lower bound as $\underline{z}_{\mathcal{L}}$. To have a fair comparison, we do not use $LB_{\neq}$ in our model, since it gives the optimal value on the state-of-the-art instances. As illustrated by Figure 8a, the absolute gap between $\underline{z}_r$ and z^* is on average smaller than the gap between $\underline{z}_{\mathcal{L}}$ and z^*. More precisely $\underline{z}_{\mathcal{L}}$ is on average seven times larger than $\underline{z}_r$. Moreover the empirical worst-case gap of $\underline{z}_r$ is also much better than the one of $\underline{z}_{\mathcal{L}}$, since there is a factor 6 between these two values. Finally, as illustrated by Figure 8b the runtime of $\underline{z}_r$ is much smaller than the one of $\underline{z}_{\mathcal{L}}$: the average runtime of our lower bounding procedure is thirty times quicker than the one presented in [21].

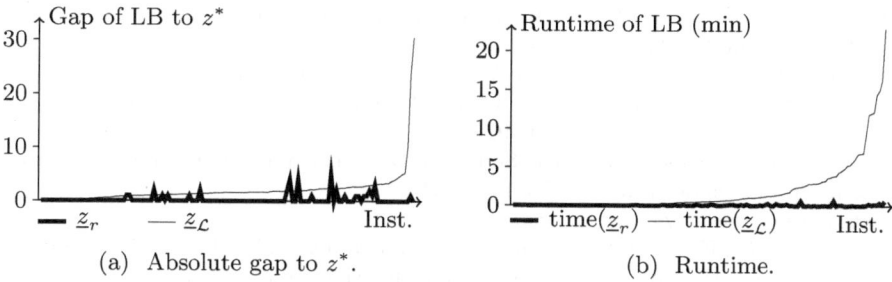

(a) Absolute gap to z^*.   (b) Runtime.

Fig. 8. Overall quality of $\underline{z}_r$ and $\underline{z}_{\mathcal{L}}$ on Data_137

5   Conclusions

In this paper, we have presented a new propagator to filter the conjunction of an `AtMostNValue` constraint and difference constraints. This propagator relies on a more appropriate graph structure, as well as refined filtering rules. We provide a simple way to diversify filtering, in order to improve the overall approach. Moreover, this propagator gives control over the tradeoff between filtering and runtime. Since it is simple to implement, effective and relevant for many applications, we believe that this propagator would benefit the CP community.

Furthermore, we have introduced a CP approach to solve the SMPTSP. This model outperforms previous exact approaches on state-of-the-art benchmarks. Furthermore, it provides very good lower bounds, even on large instances, in a short time. Thus, the CP approach competes with both exact and heuristic state-of-the-art approaches.

Future work may focus on the application of this propagator in the context of dynamic difference constraints, which occurs in many disjunctive scheduling problems. Another research perspective would be to investigate how to adapt the parameter k of R^k during the search, to improve even more the tradeoff between filtering and runtime.

Acknowledgements. The authors thank the anonymous referees as well as Xavier Lorca, Damien Prot and Thierry Petit for their useful comments.

References

1. Andreello, G., Caprara, A., Fischetti, M.: Embedding $\{0, 1/2\}$-cuts in a branch-and-cut framework: A computational study. INFORMS J. on Computing 19(2), 229–238 (2007)
2. Beldiceanu, N.: Pruning for the *minimum* constraint family and for the *numberofdistinctvalues* constraint family. In: Walsh, T. (ed.) CP 2001. LNCS, vol. 2239, pp. 211–224. Springer, Heidelberg (2001)
3. Beldiceanu, N., Carlsson, M., Demassey, S., Petit, T.: Global constraint catalogue: Past, present and future. Constraints 12(1), 21–62 (2007)

4. Bessière, C., Hebrard, E., Hnich, B., Kiziltan, Z., Walsh, T.: Filtering Algorithms for the NValue Constraint. Constraints 11(4), 271–293 (2006)
5. Bessiere, C., Katsirelos, G., Narodytska, N., Quimper, C.G., Walsh, T.: Propagating conjunctions of alldifferent constraints. CoRR abs/1004.2626 (2010)
6. Carter, M.W., Tovey, C.A.: When is the classroom assignment problem hard? Operations Research 40(1), 28–39 (1992)
7. Chabert, G., Jaulin, L., Lorca, X.: A constraint on the number of distinct vectors with application to localization. In: Gent, I.P. (ed.) CP 2009. LNCS, vol. 5732, pp. 196–210. Springer, Heidelberg (2009)
8. CHOCO Team: choco: an Open Source Java Constraint Programming Library. Tech. rep., Ecole des Mines de Nantes (2010), http://www.emn.fr/z-info/choco-solver/
9. Dijkstra, M.C., Kroon, L.G., Salomon, M., Van Nunen, J.A.E.E., van Wassenhove, L.N.: Planning the Size and Organization of KLM's Aircraft Maintenance Personnel. Interfaces 24(6), 47–58 (1994)
10. Dowling, D., Krishnamoorthy, M., Mackenzie, H., Sier, D.: Staff rostering at a large international airport. Annals of Operations Research 72, 125–147 (1997)
11. Ernst, A.T., Jiang, H., Krishnamoorthy, M., Sier, D.: Staff scheduling and rostering: A review of applications, methods and models. European Journal of Operational Research 153(1), 3–27 (2004)
12. Fischetti, M., Martello, S., Toth, P.: The fixed job schedule problem with spread-time constraints. Operations Research 35(6), 849–858 (1987)
13. Fischetti, M., Martello, S., Toth, P.: The fixed job schedule problem with working-time constraints. Operations Research 37(3), 395–403 (1989)
14. Garey, M.R., Johnson, D.S.: Computers and Intractability: A Guide to the Theory of NP-Completeness, 1st edn. W. H. Freeman (January 1979)
15. Golumbic, M.C.: Algorithmic Graph Theory and Perfect Graphs, 2nd edn. Annals of Discrete Mathematics, vol. 57. Elsevier (2004)
16. Gualandi, S., Malucelli, F.: Exact solution of graph coloring problems via constraint programming and column generation. INFORMS J. Comput. Sc. 24, 81–100 (2012)
17. Halldórsson, M.M., Radhakrishnan, J.: Greed is good: Approximating independent sets in sparse and bounded-degree graphs. Algorithmica 18(1), 145–163 (1997)
18. Haralick, R.M., Elliott, G.L.: Increasing tree search efficiency for constraint satisfaction problems. In: Proceedings of the 6th International Joint Conference on Artificial Intelligence, IJCAI 1979, vol. 1, pp. 356–364. Morgan Kaufmann Publishers Inc. (1979)
19. Kadioglu, S., Malitsky, Y., Sellmann, M., Tierney, K.: Isac - instance-specific algorithm configuration. In: ECAI. Frontiers in Artificial Intelligence and Applications, vol. 215, pp. 751–756. IOS Press (2010)
20. Krishnamoorthy, M., Ernst, A.T.: The personnel task scheduling problem. In: Optimization Methods and Applications, pp. 343–368. Springer, US (2001)
21. Krishnamoorthy, M., Ernst, A., Baatar, D.: Algorithms for large scale shift minimisation personnel task scheduling problems. European Journal of Operational Research 219, 34–48 (2012)
22. Kroon, L.G., Salomon, M., Wassenhove, L.N.V.: Exact and approximation algorithms for the tactical fixed interval scheduling problem. Operations Research 45(4) (1997)
23. Lapègue, T., Fages, J.G., Prot, D., Bellenguez-Morineau, O.: Personnel Task Scheduling Problem Library (2013), https://sites.google.com/site/ptsplib/smptsp/home

24. Lecoutre, C., Sais, L., Tabary, S., Vidal, V.: Reasoning from last conflict(s) in constraint programming. Artif. Intell. 173(18), 1592–1614 (2009)
25. Leone, R., Festa, P., Marchitto, E.: A Bus Driver Scheduling Problem: a new mathematical model and a GRASP approximate solution. Journal of Heuristics 17(4), 441–466 (2010)
26. López-Ortiz, A., Quimper, C.G., Tromp, J., van Beek, P.: A fast and simple algorithm for bounds consistency of the alldifferent constraint. In: IJCAI, pp. 245–250. Morgan Kaufmann (2003)
27. Monette, J.N., Flener, P., Pearson, J.: Towards solver-independent propagators. In: Milano, M. (ed.) CP 2012. LNCS, vol. 7514, pp. 544–560. Springer, Heidelberg (2012)
28. Pachet, F., Roy, P.: Automatic generation of music programs. In: Jaffar, J. (ed.) CP 1999. LNCS, vol. 1713, pp. 331–345. Springer, Heidelberg (1999)
29. Régin, J.C.: A Filtering Algorithm for Constraints of Difference in CSPs. In: National Conference on Artificial Intelligence, pp. 362–367. AAAI (1994)
30. Schulte, C., Stuckey, P.J.: Efficient constraint propagation engines. CoRR abs/cs/0611009 (2006)
31. Smet, P., Wauters, T., Mihaylow, M., Vanden Berghe, G.: The shift minimisation personnel task scheduling problem: a new hybrid approach and computational insights. Technical report (2013)
32. Van den Bergh, J., Beliën, J., De Bruecker, P., Demeulemeester, E., De Boeck, L.: Personnel scheduling: A literature review. European Journal of Operational Research 226(3), 367–385 (2013)

A Parametric Approach for Smaller and Better Encodings of Cardinality Constraints

Ignasi Abío[1], Robert Nieuwenhuis[2], Albert Oliveras[2], and Enric Rodríguez-Carbonell[2]

[1] Theoretical Computer Science, TU Dresden, Germany
[2] Technical University of Catalonia, Barcelona

Abstract. Adequate encodings for high-level constraints are a key ingredient for the application of SAT technology. In particular, *cardinality constraints* state that at most (at least, or exactly) k out of n propositional variables can be true. They are crucial in many applications. Although sophisticated encodings for cardinality constraints exist, it is well known that for small n and k straightforward encodings without auxiliary variables sometimes behave better, and that the choice of the right trade-off between minimizing either the number of variables or the number of clauses is highly application-dependent. Here we build upon previous work on Cardinality Networks to get the best of several worlds: we develop an arc-consistent encoding that, by recursively decomposing the constraint into smaller ones, allows one to decide which encoding to apply to each sub-constraint. This process minimizes a function $\lambda \cdot num\_vars + num\_clauses$, where λ is a parameter that can be tuned by the user. Our careful experimental evaluation shows that (e.g., for $\lambda = 5$) this new technique produces much smaller encodings in variables *and* clauses, and indeed strongly improves SAT solvers' performance.

1 Introduction

This paper presents a new encoding into SAT of *cardinality constraints*, that is, constraints of the form $x_1 + \cdots + x_n \mathbin{\#} k$, where k is a natural number, the x_i are propositional variables, and the relation operator # belongs to $\{<, >, \leqslant, \geqslant, =\}$. Cardinality constraints are present in many practical SAT applications, such as cumulative scheduling [17] or timetabling [4]. They also arise as components of some SAT-based techniques, e.g., for MaxSAT [11].

Here we are interested in encoding a cardinality constraint C with a clause set S (possibly with auxiliary variables) that is not only equisatisfiable, but also *arc-consistent*: given a partial assignment A, if x_i is true (false) in every extension of A satisfying C, then unit propagating A on S sets x_i to true (false)[1]. Enforcing arc-consistency by unit propagation in this way has of course an important positive impact on the practical efficiency of SAT solvers.

A straightforward encoding of a cardinality constraint $x_1 + \cdots + x_n \leqslant k$ is to state, for each subset Y of $\{x_1, \ldots, x_n\}$ with $|Y| = k + 1$, that at least one variable of Y must be false. This can be done by asserting $\binom{n}{k+1}$ clauses of the form $\overline{x_{i_1}} \vee \ldots \vee \overline{x_{i_{k+1}}}$. This kind of construction frequently works well, although it is of course not reasonable for

---

[1] Sometimes this notion is called *generalized arc-consistency*.

C. Schulte (Ed.): CP 2013, LNCS 8124, pp. 80–96, 2013.
© Springer-Verlag Berlin Heidelberg 2013

large n and k, which is our aim in this work. Successively more sophisticated encodings using auxiliary variables have been defined that require fewer clauses (see Section 2). But still, for small n and k the straightforward encodings may behave better in practice. An additional issue is that, for the efficiency of the SAT solver, the choice of the right trade-off between minimizing either the number of auxiliary variables or the number of clauses is highly application-dependent.

Here we build upon and improve previous work on encoding cardinality constraints with Cardinality Networks [2,3], which use $O(n \log^2 k)$ variables and clauses (see Section 3). The idea is to get the best of several worlds: we develop a hybrid arc-consistent encoding that, by recursively decomposing the constraint into smaller ones, allows one to decide whether to apply a recursive (see Section 4) or a direct (see Section 5) encoding to each sub-constraint. This process minimizes a function $\lambda \cdot num\_vars + num\_clauses$, where λ is a parameter that can be tuned by the user (see Section 6). Our experimental evaluation shows that (e.g., for $\lambda = 5$) this new technique produces much smaller encodings in variables *and* clauses, and indeed strongly improves the performance of SAT solvers (see Section 7).

2 Related Work

Because of their practical importance, encodings of cardinality constraints into SAT have been thoroughly studied over the last few years. In this section we review some of the most important works in the literature.

In [20], Warners considered the more general pseudo-Boolean case, where constraints are of the form $a_1 x_1 + \ldots + a_n x_n \leq k$, being the a_i's and the k integer coefficients and the x_i's Boolean variables. The encoding is based on using adders for numbers represented in binary. For cardinality constraints the encoding uses $O(n)$ clauses and variables, but does not preserve arc consistency.

Bailleux and Boufkhad presented in [5] an arc-consistent encoding of cardinality constraints that uses $O(n \log n)$ variables and $O(n^2)$ clauses. The encoding consists of a *totalizer* and a *comparator*. The totalizer can be seen as a binary tree, where the leaves are the x_i's variables. Each inner node is labeled with a number s and uses s auxiliary variables to represent, in unary, the sum of the leaves of the corresponding subtree. As for the comparator, it is easily encoded thanks to the unary representation, which also allows handling constraints of the form $k_1 \leq x_1 + \ldots + x_n \leq k_2$ without splitting.

A more applied work is the one of Büttner and Rintanen [19]. Although their main interest was in planning, they suggested two encodings of cardinality constraints. The first one is based on encoding an injective mapping between the true x_i's variables and k elements. It uses $O(nk)$ clauses and variables and is not arc-consistent. The other encoding is a small modification of [5]. Based on the observation that counting up to $k + 1$ suffices, they can reduce the number of variables and clauses used in each node. The resulting encoding requires $O(nk)$ variables and $O(nk^2)$ clauses, which improves on [5] if k is small enough.

In [18], Sinz proposed two different encodings, both based on counters. The first encoding uses a sequential counter where numbers are represented in unary. It needs $O(nk)$ clauses and variables and is arc-consistent. The second one is based on a parallel counter, where numbers are represented in binary. It requires $O(n)$ clauses and variables, but is not arc-consistent.

Another kind of encoding was used in [6], where a BDD-like technique was proposed for pseudo-Boolean constraints. The encoding is arc-consistent, and uses $O(n^2)$ clauses and variables when applied to cardinality constraints. The idea is as follows: given a pseudo-Boolean constraint $a_1 x_1 + \ldots + a_n x_n \leq k$, the root of the BDD is labeled with variable $D_{n,k}$, expressing that the sum of the first n terms is at most k. The two corresponding children are $D_{n-1,k}$ and $D_{n-1,k-a_n}$, indicating the two cases that correspond to setting x_n to false and true, respectively. Then the necessary clauses are added to express the relationship between the variables, and trivial cases are treated accordingly.

The same authors presented in [7] a polynomial and arc-consistent encoding of pseudo-Boolean constraints. When restricted to cardinality constraints it is similar to [5], but the latter is better in terms of size.

Yet another approach for encoding cardinality constraints was suggested in [1]. The authors revisit the idea of using totalizers, and realize that totalizers require two parameters: the encoding used (unary or binary) and the way the totalizers are grouped (e.g. $(a + b) + (c + d)$ or $(((a + b) + c) + d)$). A thorough experimentation is performed, to which they add two extra aspects: (i) how to order the variables; and (ii), the use of encodings in parallel, hoping the SAT solver will focus on the most appropriate one for each problem.

Finally, Eén and Sörensson [10] presented three encodings for pseudo-Boolean constraints. The first encoding is BDD-based, similar to [6]. The second one, based on adder networks, improves that of [20] in that it uses less adders, but is still linear and does not preserve arc consistency. Finally, their third encoding uses *Sorting Networks* [8]. A Sorting Network takes input variables $(x_1 \ldots x_n)$ and returns as outputs $(y_1 \ldots y_n)$ the sorted input values in decreasing order. Hence, an output variable y_k will become true iff there are at least k true input variables, and false iff there are at least $n - k + 1$ false ones. Now, to express $x_1 + \cdots + x_n \geq k$, it suffices to add a unit clause y_k; similarly, for $x_1 + \cdots + x_n \leq k$ one adds $\overline{y_{k+1}}$, and both are added if the relation is =. This encoding, when restricted to cardinality constraints, preserves arc consistency and requires $O(n \log^2 n)$ clauses and variables. The Cardinality Networks of [2,3] reduce this to $O(n \log^2 k)$, which is important as often $n \gg k$.

A similar approach uses so-called Pairwise Cardinality Networks [9], which are based on Pairwise Sorting Networks [15] instead of Sorting Networks. By means of partial evaluation, this method also achieves $O(n \log^2 k)$ variables and clauses. Finally, we were recently informed that a hybrid approach based on Pairwise Cardinality Networks similar to that presented here was implemented in the BEE system [13]. However, no detailed description or experimental evaluation is available. Moreover, our proposal in this paper is more general, in the sense that it allows the user to tune the parameter λ when minimizing the objective function $\lambda \cdot num\_vars + num\_clauses$.

3 Preliminaries

In this work we describe a method for producing cardinality networks that generalizes the construction of [3]. The core idea of these approaches, which dates back to [8], consists in encoding a circuit that implements mergesort by means of a set of clauses. The most basic components of these circuits are 2-comparators.

A *2-comparator* is a sorting network of size 2, i.e., it has 2 input variables (x_1 and x_2) and 2 output variables (y_1 and y_2) such that y_1 is true if and only if at least one of the input variables is true, and y_2 is true if and only if both two input variables are true. In the following, 2-comparators are denoted by $(y_1, y_2) = 2\text{-Comp}(x_1, x_2)$. As pointed out in [3], for encoding $\leqslant$-constraints, only the three clauses on the first row of Fig. 1 are needed to guarantee arc-consistency. The three clauses on the second row suffice for $\geqslant$-constraints and all six must be present when encoding =-constraints. Note that the usual polarity argument [16] cannot be directly applied here, as we are interested not only in preserving satisfiability, but also arc-consistency under unit propagation.

$$x_1 \rightarrow y_1, \quad x_2 \rightarrow y_1, \quad x_1 \wedge x_2 \rightarrow y_2,$$
$$\overline{x_1} \rightarrow \overline{y_2}, \quad \overline{x_2} \rightarrow \overline{y_2}, \quad \overline{x_1} \wedge \overline{x_2} \rightarrow \overline{y_1}$$

Fig. 1. A 2-comparator: clauses (left) and graphical representation (right)

4 Arbitrary-Sized Recursive Cardinality Networks

In this section we generalize the recursive construction of cardinality networks given in [3] by allowing inputs and outputs of any size, not necessarily a power of two. Not only does this avoid adding dummy variables that are not actually needed (which, as will be seen in Section 7, has an impact on performance on its own), but also becomes useful when combining with the direct (non-recursive) constructions of Section 5.

In what follows, we denote by $\lfloor r \rfloor$ and $\lceil r \rceil$ the floor and ceiling functions respectively. Moreover, for simplicity, we will assume that the constraint to be encoded is a $\leqslant$-constraint. However, similar constructions for the other constraints can be devised.

4.1 Merge Networks

A *merge network* takes as input two (decreasingly) ordered sets of sizes a and b and produces a (decreasingly) ordered set of size $a + b$. We can build a merge network with inputs $(x_1, \ldots, x_a)$ and $(x'_1, \ldots, x'_b)$ in a recursive way as follows[2]:

– If $a = b = 1$, a merge network is a 2-comparator:

$$\text{Merge}(x_1; x'_1) := 2\text{-Comp}(x_1, x'_1).$$

– If $a = 0$, a merge network returns the second input:

$$\text{Merge}(; x'_1, x'_2, \ldots, x'_b) := (x'_1, x'_2, \ldots, x'_b).$$

---

[2] Notice we use the notation $\text{Merge}(X; X')$ instead of $\text{Merge}((X), (X'))$ for simplicity.

- If a and b are even, $a > 0$, $b > 0$ and either $a > 1$ or $b > 1$, let us define

$$\begin{aligned}
(z_1, z_3, \ldots, z_{a-3}, z_{a-1}, \\ z_{a+1}, z_{a+3}, \ldots, z_{a+b-1}) &= \begin{aligned}\text{Merge}(x_1, x_3, \ldots, x_{a-1}; \\ x_1', x_3', \ldots, x_{b-1}'),\end{aligned} \\[4pt]
(z_2, z_4, \ldots, z_{a-2}, z_a, \\ z_{a+2}, z_{a+4}, \ldots, z_{a+b}) &= \begin{aligned}\text{Merge}(x_2, x_4, \ldots, x_a; \\ x_2', x_4', \ldots, x_b'),\end{aligned} \\[4pt]
(y_2, y_3) &= 2\text{-Comp}(z_2, z_3), \\
&\cdots \\
(y_{a+b-2}, y_{a+b-1}) &= 2\text{-Comp}(z_{a+b-2}, z_{a+b-1}).
\end{aligned}$$

Then,

$$\text{Merge}(x_1, x_2, \ldots, x_a; x_1', x_2', \ldots, x_b') := (z_1, y_2, y_3, \ldots, y_{a+b-1}, z_{a+b}).$$

- If a is even, b is odd, $a > 0$, $b > 0$ and either $a > 1$ or $b > 1$, let us define

$$\begin{aligned}
(z_1, z_3, \ldots, z_{a-1}, \\ z_{a+1}, z_{a+3}, \ldots, z_{a+b}) &= \begin{aligned}\text{Merge}(x_1, x_3, \ldots, x_{a-1}; \\ x_1', x_3', \ldots, x_b'),\end{aligned} \\[4pt]
(z_2, z_4, \ldots, z_a, z_{a+2}, \\ z_{a+4}, \ldots, z_{a+b-1}) &= \begin{aligned}\text{Merge}(x_2, x_4, \ldots, x_a; \\ x_2', x_4', \ldots, x_{b-1}'),\end{aligned} \\[4pt]
(y_2, y_3) &= 2\text{-Comp}(z_2, z_3), \\
&\cdots \\
(y_{a+b-1}, y_{a+b}) &= 2\text{-Comp}(z_{a+b-1}, z_{a+b}).
\end{aligned}$$

Then,

$$\text{Merge}(x_1, x_2, \ldots, x_a; x_1', x_2', \ldots, x_b') := (z_1, y_2, y_3, \ldots, y_{a+b-1}, y_{a+b}).$$

- If a and b are odd, $a > 0$, $b > 0$ and either $a > 1$ or $b > 1$, let us define

$$\begin{aligned}
(z_1, z_3, \ldots, z_{a-2}, z_a, \\ z_{a+1}, z_{a+3}, \ldots, z_{a+b}) &= \begin{aligned}\text{Merge}(x_1, x_3, \ldots, x_a; \\ x_1', x_3', \ldots, x_b'),\end{aligned} \\[4pt]
(z_2, z_4, \ldots, z_{a-3}, z_{a-1}, \\ z_{a+2}, z_{a+4}, \ldots, z_{a+b-1}) &= \begin{aligned}\text{Merge}(x_2, x_4, \ldots, x_{a-3}, x_{a-1}; \\ x_2', x_4', \ldots, x_{b-1}'),\end{aligned} \\[4pt]
(y_2, y_3) &= 2\text{-Comp}(z_2, z_3), \\
&\cdots \\
(y_{a+b-2}, y_{a+b-1}) &= 2\text{-Comp}(z_{a+b-2}, z_{a+b-1}).
\end{aligned}$$

Then,

$$\text{Merge}(x_1, x_2, \ldots, x_a; x_1', x_2', \ldots, x_b') := (z_1, y_2, y_3, \ldots, y_{a+b-1}, z_{a+b}).$$

- The remaining cases are defined thanks to the symmetry of the merge function, i.e., due to $\text{Merge}(X, X') = \text{Merge}(X', X)$.

The base cases do not require any explanation. As regards the recursive ones, first notice that the set of values $x_1, x_2, \ldots, x_a, x_1', x_2', \ldots, x_b'$ is always preserved. Further, the

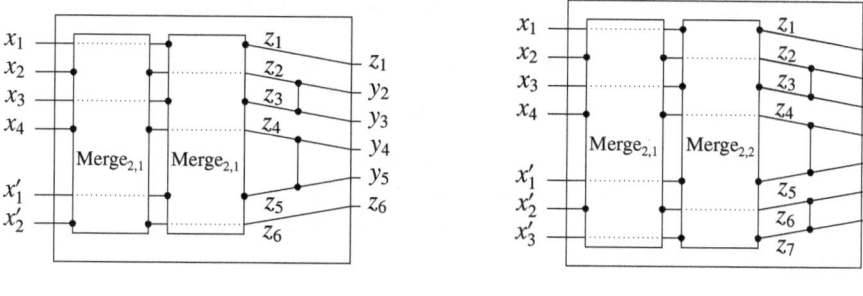

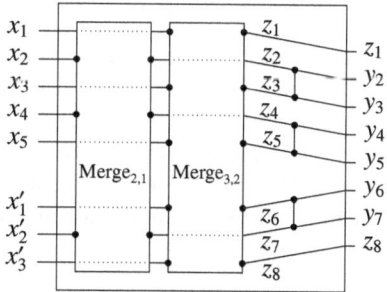

Fig. 2. Different examples of merge networks

output bits are sorted, as $z_{2i} \geq z_{2(i+1)}$, $z_{2i} \geq z_{2(i+1)+1}$, $z_{2i+1} \geq z_{2(i+1)}$ and $z_{2i+1} \geq z_{2(i+1)+1}$ imply that $\min(z_{2i}, z_{2i+1}) \geq \max(z_{2(i+1)}, z_{2(i+1)+1})$. Figure 2 shows examples of some of these recursive cases.

The number of auxiliary variables and clauses of a merge network defined in this way can be recursively computed. A merge network with inputs of size $(1, 1)$ needs 2 variables and 3 clauses. A merge network with inputs of size $(0, b)$ needs no variables and clauses. A merge network with inputs of size (a, b) with $a > 1$ or $b > 1$ needs $V_1 + V_2 + 2 \left\lfloor \frac{a+b-1}{2} \right\rfloor$ variables and $C_1 + C_2 + 3 \left\lfloor \frac{a+b-1}{2} \right\rfloor$ clauses, where V_1 and C_1 are the number of variables and clauses in a merge network with inputs of size $\left(\left\lceil \frac{a}{2} \right\rceil, \left\lceil \frac{b}{2} \right\rceil \right)$, and V_2, C_2 are idem in a merge network with inputs of size $\left(\left\lfloor \frac{a}{2} \right\rfloor, \left\lfloor \frac{b}{2} \right\rfloor \right)$.

In comparison to [3], in that work it was assumed that $a = b = 2^m$ for some $m \geq 0$. Thanks to this, only one base case ($a = b = 1$) and one recursive case (a, b even) were considered there. All the other cases introduced here are needed for arbitrary a and b.

4.2 Sorting Networks

A *sorting network* takes an input of size n and sorts it. It can be built in a recursive way as follows, using the same strategy as in mergesort:

– If $n = 1$, the output of the sorting network is its input:

$$\text{Sorting}(x_1) := x_1$$

- If $n = 2$, a sorting network is a single merge (i.e., a 2-comparator):

$$\text{Sorting}(x_1, x_2) := \text{Merge}(x_1; x_2).$$

- For $n > 2$, take l with $1 \leqslant l < n$: Let us define

$$(z_1, z_2, \ldots, z_l) = \text{Sorting}(x_1, x_2, \ldots, x_l),$$
$$(z_{l+1}, z_{l+2}, \ldots, z_n) = \text{Sorting}(x_{l+1}, x_{l+2}, \ldots, x_n),$$
$$(y_1, y_2, \ldots, y_n) = \text{Merge}(z_1, z_2, z_l; z_{l+1}, \ldots, z_n).$$

Then,

$$\text{Sorting}(x_1, x_2, \ldots, x_n) := (y_1, y_2, \ldots, y_n).$$

Again, the number of auxiliary variables and clauses needed in these networks can be recursively computed. A sorting network of input size 1 needs no variables and clauses. A sorting network of input size 2 needs 2 variables and 3 clauses. A sorting network of input size n composed by a sorting network of size l and a sorting network of size $n - l$ needs $V_1 + V_2 + V_3$ variables and $C_1 + C_2 + C_3$ clauses, where $(V_1, C_1), (V_2, C_2)$ are the number of variables and clauses used in the sorting networks of sizes l and $n - l$, and (V_3, C_3) are the number of variables and clauses needed in the merge network with inputs of sizes $(l, n - l)$.

In comparison to [3], in that work n is assumed to be a power of two. Moreover, in the recursive case l is always chosen to be $n/2$, while here we can build sorting networks of any size, and have the additional freedom of choosing the sizes of the two sorting network components.

4.3 Simplified Merge Networks

A *simplified merge* is a reduced version of a merge, used when we are only interested in some of the outputs, but not all. Recall that we want to encode a constraint of the form $x_1 + \ldots + x_n \leqslant k$, and hence we are only interested in the first $k + 1$ bits of the sorted output. Thus, in a c-simplified merge network, the inputs are two sorted sequences of variables $(x_1, x_2, \ldots, x_a; x'_1, x'_2, \ldots, x'_b)$, and the network produces a sorted output of the desired size, c, $(y_1, y_2, \ldots, y_c)$. The network satisfies that y_r is true if there are at least r true inputs. We can build a recursive simplified merge as follows:

- If $a = b = c = 1$, let us add the clauses $x_1 \rightarrow y$, $x'_1 \rightarrow y$[3]. Then:

$$\text{SMerge}_1(x_1; x'_1) := y.$$

- If $a > c$, we can ignore the last $a - c$ bits of the first input (similarly if $b > c$):

$$\text{SMerge}_c(x_1, x_2, \ldots, x_a; x'_1, \ldots, x'_b) = \text{SMerge}_c(x_1, x_2, \ldots, x_c; x'_1, \ldots, x'_b).$$

- If $a + b \leqslant c$, the simplified merge is a merge:

$$\text{SMerge}_c(x_1, \ldots, x_a; x'_1, \ldots, x'_b) = \text{Merge}(x_1, \ldots, x_a; x'_1, \ldots, x'_b).$$

---

[3] Notice that these clauses correspond to the bit of the 2-comparator with lower index. Clause $\overline{x_1} \wedge \overline{x_2} \rightarrow \overline{y}$ does not need to be included here following the reasoning given in Section 3.

– If $a, b \leqslant c$, $a + b > c$ and c is even: Let us define

$$(z_1, z_3, \ldots, z_{c+1}) = \mathrm{SMerge}_{c/2+1}(x_1, x_3, \ldots; x'_1, x'_3, \ldots),$$
$$(z_2, z_4, \ldots, z_c) = \mathrm{SMerge}_{c/2}(x_2, x_4, \ldots; x'_2, x'_4, \ldots),$$
$$(y_2, y_3) = 2\text{-}\mathrm{Comp}(z_2, z_3),$$

$$\cdots$$

$$(y_{c-2}, y_{c-1}) = 2\text{-}\mathrm{Comp}(z_{c-2}, z_{c-1}).$$

and add the clauses $z_c \to y_c$, $z_{c+1} \to y_c$. Then,

$$\mathrm{SMerge}_c(x_1, x_2, \ldots, x_a; x'_1, x'_2, \ldots, x'_b) := (z_1, y_2, y_3, \ldots, y_c),$$

– If $a, b \leqslant c$, $a + b > c$ and $c > 1$ is odd: Let us define

$$(z_1, z_3, \ldots, z_c) = \mathrm{SMerge}_{\frac{c+1}{2}}(x_1, x_3, \ldots; x'_1, x'_3, \ldots),$$
$$(z_2, z_4, \ldots, z_{c-1}) = \mathrm{SMerge}_{\frac{c-1}{2}}(x_2, x_4, \ldots; x'_2, x'_4, \ldots),$$
$$(y_2, y_3) = 2\text{-}\mathrm{Comp}(z_2, z_3),$$

$$\cdots$$

$$(y_{c-1}, y_c) = 2\text{-}\mathrm{Comp}(z_{c-1}, z_c).$$

Then,

$$\mathrm{SMerge}_c(x_1, x_2, \ldots, x_a; x'_1, x'_2, \ldots, x'_b) := (z_1, y_2, y_3, \ldots, y_c).$$

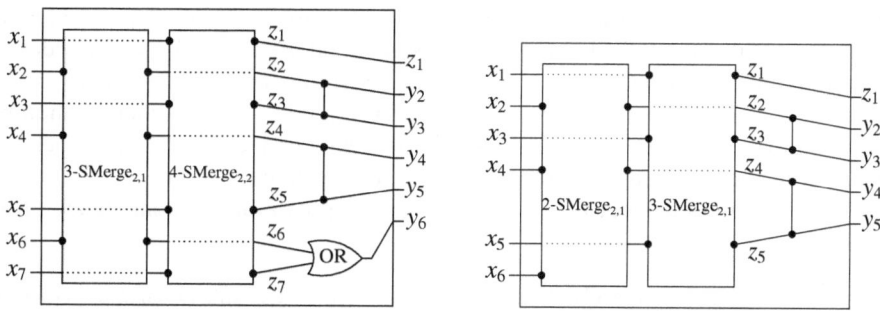

Fig. 3. Two examples of simplified merge networks

Figure 3 shows two examples of simplified merges: The first one shows a 6-simplified merge with inputs of sizes 3 and 4. The second one corresponds to a 5-simplified merge with inputs of sizes 2 and 4.

We can recursively compute the auxiliary variables and clauses needed in simplified merge networks. In the recursive case, we need $V_1 + V_2 + c - 1$ variables and $C_1 + C_2 + C_3$ clauses, where $(V_1, C_1), (V_2, C_2)$ are the number of clauses and variables needed in simplified merge networks of sizes $\left(\left\lceil \frac{a}{2} \right\rceil, \left\lceil \frac{b}{2} \right\rceil, \left\lfloor \frac{c}{2} \right\rfloor + 1 \right), \left(\left\lfloor \frac{a}{2} \right\rfloor, \left\lfloor \frac{b}{2} \right\rfloor, \left\lfloor \frac{c}{2} \right\rfloor \right)$, and

$$C_3 = \begin{cases} \frac{3c-3}{2} & \text{if } c \text{ is odd,} \\ \frac{3c-2}{2} + 2 & \text{if } c \text{ is even.} \end{cases}$$

Compared to [3], there it was assumed that $a = b = 2^m$ for some $m \geq 0$, and $c = 2^m + 1$. Similarly to merge networks, only one base case and one recursive case were considered. All the other cases introduced here are needed for arbitrary a, b and c.

4.4   *m*-Cardinality Networks

An *m-cardinality network* takes an input of size n and outputs the first m sorted bits. Recursively, an *m*-cardinality network with input $x_1, x_2, \ldots, x_n$ can be defined as follows:

- If $n \leqslant m$, a cardinality network is a sorting network:

$$\mathrm{Card}_m(x_1, x_2, \ldots, x_n) := \mathrm{Sorting}(x_1, x_2, \ldots, x_n).$$

- If $n > m$, take l with $1 \leqslant l < n$. Let us define

$$
\begin{aligned}
(z_1, z_2, \ldots, z_A) &= \mathrm{Card}_m(x_1, x_2, \ldots, x_l), \\
(z_1', z_2', \ldots, z_B') &= \mathrm{Card}_m(x_{l+1}, x_{l+2}, \ldots, x_n), \\
(y_1, y_2, \ldots, y_m) &= \mathrm{SMerge}_m(z_1, z_2, \ldots, z_A; z_1', z_2', \ldots, z_B'),
\end{aligned}
$$

where $A = \min\{l, m\}$ and $B = \min\{n - l, m\}$. Then,

$$\mathrm{Card}_m(x_1, x_2, \ldots, x_n) := (y_1, y_2, \ldots, y_m).$$

Again, the number of auxiliary variables and clauses needed in these networks can be recursively computed. An *m*-cardinality network of size n composed by an *m*-cardinality network of size l and an *m*-cardinality network of size $n - l$ needs $V_1 + V_2 + V_3$ variables and $C_1 + C_2 + C_3$ clauses, where $(V_1, C_1), (V_2, C_2)$ are the number of variables and clauses used in the *m*-cardinality networks of sizes l and $n - l$, and (V_3, C_3) are idem in the *m*-simplified merge network with inputs of sizes $(\min\{l, m\}, \min\{n - l, m\})$.

Compared to [3], in that work m is assumed to be a power of two, and n a multiple of m. Moreover, similarly to sorting networks, in the recursive case l is always chosen to be m, while here we have an additional degree of freedom.

Using the same techniques in [3] one can easily prove the arc-consistency of the encoding.

Theorem 1. *The Recursive Cardinality Network encoding is arc-consistent: consider a cardinality constraint $x_1 + \ldots + x_n \leqslant k$, its corresponding cardinality network $(y_1, y_2, \ldots, y_{k+1}) = \mathrm{Card}_{k+1}(x_1, x_2, \ldots, x_n)$, and the unit clause $\neg y_{k+1}$. If we now set to true k input variables, then unit propagation sets to false the remaining $n - k$ input variables.*

Proof (sketch). The proof relies on the following lemmas, which formalize the propagation properties of the building blocks of cardinality networks:

Lemma 1 (Merge Networks). *Let S be the set of clauses of*

$$(y_1, y_2, \ldots, y_{a+b}) = \mathrm{Merge}(x_1, x_2, \ldots, x_a; x_1', x_2', \ldots, x_b').$$

Let $p, q \in \mathbb{N}$ with $0 \leq p \leq a$ and $0 \leq q \leq b$. Then:

1. $S \cup \{x_1, \ldots, x_p, x_1', \ldots, x_q'\} \models_{\mathrm{UP}} y_1, \ldots, y_{p+q}.$
2. *If $p < a$ and $q < b$ then* $S \cup \{x_1, \ldots, x_p, x_1', \ldots, x_q', \overline{y_{p+q+1}}\} \models_{\mathrm{UP}} \overline{x_{p+1}}, \overline{x_{q+1}'}.$
3. *If $p = a$ and $q < b$ then* $S \cup \{x_1, \ldots, x_p, x_1', \ldots, x_q', \overline{y_{p+q+1}}\} \models_{\mathrm{UP}} \overline{x_{q+1}'}.$
4. *If $p < a$ and $q = b$ then* $S \cup \{x_1, \ldots, x_p, x_1', \ldots, x_q', \overline{y_{p+q+1}}\} \models_{\mathrm{UP}} \overline{x_{p+1}}.$

Lemma 2 (Sorting Networks). *Let* $X = (x_1, x_2, \ldots, x_n)$, $X' \subseteq X$ *and* S *be the set of clauses of* $(y_1, y_2, \ldots, y_n) = \text{Sorting}(X)$. *Let* $p = |X'|$. *Then:*

1. $S \cup X' \models_{\text{UP}} y_1, \ldots, y_p$.
2. *If* $p = |X'| < n$, *then* $S \cup X' \cup \{\overline{y_{p+1}}\} \models_{\text{UP}} \overline{x_i}$ *for all* $x_i \notin X'$.

Lemma 3 (Simplified Merge Networks). *Let* S *be the set of clauses of*

$$(y_1, y_2, \ldots, y_c) = \text{SMerge}_c(x_1, x_2, \ldots, x_a; x'_1, x'_2, \ldots, x'_b).$$

Let $p, q \in \mathbb{N}$ *be such that* $0 \le p \le a$, $0 \le q \le b$. *Then:*

1. *If* $p + q \le c$, *then* $S \cup \{x_1, \ldots, x_p, x'_1, \ldots, x'_q\} \models_{\text{UP}} y_1, \ldots, y_{p+q}$.
2. *If* $p < a$, $q < b$ *and* $p+q < c$, *then* $S \cup \{x_1, \ldots, x_p, x'_1, \ldots, x'_q, \overline{y_{p+q+1}}\} \models_{\text{UP}} \overline{x_{p+1}}, \overline{x'_{q+1}}$.
3. *If* $p = a$, $q < b$ *and* $p + q < c$, *then* $S \cup \{x_1, \ldots, x_p, x'_1, \ldots, x'_q, \overline{y_{p+q+1}}\} \models_{\text{UP}} \overline{x'_{q+1}}$.
4. *If* $p < a$, $q = b$ *and* $p + q < c$, *then* $S \cup \{x_1, \ldots, x_p, x'_1, \ldots, x'_q, \overline{y_{p+q+1}}\} \models_{\text{UP}} \overline{x_{p+1}}$.

Lemma 4 (Cardinality Networks). *Let* $X = (x_1, x_2, \ldots, x_n)$, $X' \subseteq X$ *and* S *be the set of clauses of* $(y_1, y_2, \ldots, y_m) = \text{Card}_m(X)$. *Let* $p = |X'|$. *Then:*

1. *If* $p \le m$, *then* $S \cup X' \models_{\text{UP}} y_1, \ldots, y_p$.
2. *If* $p < m$, *then* $S \cup X' \cup \{\overline{y_{p+1}}\} \models_{\text{UP}} \overline{x_i}$ *for all* $x_i \notin X'$.

Each lemma is proved by induction and using the corresponding lemmas of the inner building blocks. The proofs of Lemmas 1 and 3 require considering four cases according to the parities of p and q. Finally, the theorem follows as a corollary of Lemma 4.

For the sake of illustration, let us prove the case $a, b \le c$, $a + b > c$, with c even, of the inductive case of property 1 in Lemma 3. So, let us consider the set of clauses of

$$(z_1, y_2, y_3, \ldots, y_c) = \text{SMerge}_c(x_1, x_2, \ldots, x_a; x'_1, x'_2, \ldots, x'_b)$$

consisting of the clauses $z_c \to y_c$, $z_{c+1} \to y_c$ and those in

$$
\begin{aligned}
(z_1, z_3, \ldots, z_{c+1}) &= \text{SMerge}_{c/2+1}(x_1, x_3, \ldots; x'_1, x'_3, \ldots), \\
(z_2, z_4, \ldots, z_c) &= \text{SMerge}_{c/2}(x_2, x_4, \ldots; x'_2, x'_4, \ldots), \\
(y_2, y_3) &= 2\text{-Comp}(z_2, z_3), \\
&\cdots \\
(y_{c-2}, y_{c-1}) &= 2\text{-Comp}(z_{c-2}, z_{c-1}).
\end{aligned}
$$

Let $p, q \in \mathbb{N}$ such that $0 \le p \le a$, $0 \le q \le b$ and $p + q \le c$. If $p = q = 0$ there is nothing to prove. Otherwise let us show $S \cup \{x_1, \ldots, x_p, x'_1, \ldots, x'_q\} \models_{\text{UP}} z_1, y_i$ for all $2 \le i \le p + q$.

Here we focus on the subcase p and q even, being the other three cases analogous. Hence, let $p = 2p'$ and $q = 2q'$. In $x_1, x_2, \ldots, x_p$ there are p' odd indices and p' even indices. Similarly, in $x'_1, x'_2, \ldots, x'_q$ there are q' odd indices and q' even indices. Thus, using the IH (note $p' + q' \le c/2 < c/2 + 1$), we have that the clauses of the subnetwork $(z_1, z_3, \ldots, z_{c+1}) = \text{SMerge}_{c/2+1}(x_1, x_3, \ldots; x'_1, x'_3, \ldots)$ propagate by unit propagation the literals $z_1, \ldots, z_{2(p'+q')-1}$; and that the clauses of $(z_2, z_4, \ldots, z_c) = \text{SMerge}_{c/2}(x_2, x_4, \ldots; x'_2, x'_4, \ldots)$ propagate by unit propagation the literals $z_2, \ldots, z_{2(p'+q')}$. Altogether, all literals z_j with $1 \le j \le p + q$ can be propagated by unit propagation.

Let us take $2 \le i \le p + q$. If i is odd then, thanks to literals z_{i-1} and z_i and clause $z_{i-1} \wedge z_i \to y_i$ of the 2-comparator $(y_{i-1}, y_i) = 2\text{-Comp}(z_{i-1}, z_i)$, literal y_i is propagated. If i is even, then thanks to literal z_i and clause $z_i \to y_i$, literal y_i is propagated too.

5   Direct Cardinality Networks

In this section we introduce an alternative technique for building cardinality networks which we call *direct*, as it is non-recursive. This method uses many fewer auxiliary variables than the recursive approach explained in Section 4. On the other hand, the number of clauses of this construction makes it competitive only for small sizes. However, this is not a problem as we will see in Section 6, as the two techniques can be combined.

As in the recursive construction described in Section 4, the building blocks of direct cardinality networks are merge, sorting and simplified merge networks:

- **Merge Networks.** They are defined as follows[4]:

$$\text{Merge}(x_1, x_2, \ldots, x_a; x'_1, x'_2, \ldots, x'_b) := (y_1, y_2, y_3, \ldots, y_{a+b-1}, y_{a+b}),$$

with clauses $\{x_i \to y_i, \ x'_j \to y_j, \ x_i \wedge x'_j \to y_{i+j} \ : \ 1 \leqslant i \leqslant a, 1 \leqslant j \leqslant b\}$. Notice we need $a + b$ variables and $ab + a + b$ clauses.

- **Sorting Networks.** A sorting network can be built as follows:

$$\text{Sorting}(x_1, x_2, \ldots, x_n) := (y_1, y_2, \ldots, y_n),$$

with clauses $\{x_{i_1} \wedge x_{i_2} \wedge \cdots \wedge x_{i_k} \to y_k \ : \ 1 \leqslant k \leqslant n, 1 \leqslant i_1 < i_2 < \cdots < i_k \leqslant n\}$. Therefore, we need n auxiliary variables and $2^n - 1$ clauses.

- **Simplified Merge Networks.** The definition of c-simplified merge is the same as in Section 4, except for the cases in which $a, b \leqslant c$ and $a + b > c$, where:

$$\text{SMerge}_c(x_1, x_2, \ldots, x_a; x'_1, x'_2, \ldots, x'_b) := (y_1, y_2, \ldots, y_c),$$

with clauses $\{x_i \to y_i, \ x'_j \to y_j, \ x_i \wedge x'_j \to y_{i+j} \ : \ 1 \leqslant i \leqslant a, 1 \leqslant j \leqslant b, i + j \leqslant c\}$. This approach needs c variables and $(a + b)c - \frac{c(c-1)}{2} - \frac{a(a-1)}{2} - \frac{b(b-1)}{2}$ clauses.

- **m-Cardinality Networks.** As in Section 4, except for the case $n > m$, where:

$$\text{Card}_m(x_1, x_2, \ldots, x_n) := (y_1, y_2, \ldots, y_m)$$

with clauses $\{x_{i_1} \wedge x_{i_2} \wedge \cdots \wedge x_{i_k} \to y_k \ : \ 1 \leqslant k \leqslant m, 1 \leqslant i_1 < i_2 < \cdots < i_k \leqslant n\}$. This approach needs m variables and $\binom{n}{1} + \binom{n}{2} + \cdots + \binom{n}{m}$ clauses.

As regards the arc-consistency of the encoding, the following can be easily proved:

Theorem 2. *The Direct Cardinality Network encoding is arc-consistent.*

Proof (sketch). The proof uses lemmas analogous to Lemmas 1, 2, 3 and 4. For illustration purposes, let us show property 1 in Lemma 3. Let us consider the clause set of $(y_1, y_2, \ldots, y_c) = \text{SMerge}_c(x_1, x_2, \ldots, x_a; x'_1, x'_2, \ldots, x'_b)$, i.e.,

$$\{x_i \to y_i, \ x'_j \to y_j, \ x_i \wedge x'_j \to y_{i+j} \ : \ 1 \leqslant i \leqslant a, 1 \leqslant j \leqslant b, i + j \leqslant c\}.$$

Let $p, q \in \mathbb{N}$ be such that $0 \leq p \leq a, 0 \leq q \leq b$ and $p + q \leq c$. If $p = q = 0$ there is nothing to prove. Otherwise let us consider $1 \leq k \leq p + q$. Let $0 \leq i \leq p$ and $0 \leq j \leq q$ be such that $i + j = k$. If $i = 0$ then $j = k$ and the clause $x'_j \to y_j$ propagates y_k. Similarly, if $j = 0$ then $i = k$ and the clause $x_i \to y_i$ propagates y_k. Finally, if $i \geq 1$ and $j \geq 1$ the clause $x_i \wedge x'_j \to y_{i+j}$ propagates y_k.

---

[4] Direct merge networks are similar to the totalizers of [7].

6 Combining Recursive and Direct Cardinality Networks

The recursive approach produces shorter networks than the direct approach when the input is middle-sized. Still, the recursive method for building a network needs to inductively produce networks for smaller and smaller input sizes. At some point, the networks we need have a sufficiently small number of inputs such that the direct method can build them using fewer clauses and variables than the recursive approach. Here a *mixed encoding* is presented: large cardinality networks are build with the recursive approach but their components are produced with the direct approach if their size is small enough.

In more detail, assume a merge of input sizes a and b is needed. We can use the direct approach, which needs $V_D = a + b$ auxiliary variables and $C_D = ab + a + b$ clauses; or we could use the recursive approach. With the recursive approach, we have to built two merge networks of sizes $\left(\left\lceil \frac{a}{2} \right\rceil, \left\lceil \frac{b}{2} \right\rceil\right)$ and $\left(\left\lfloor \frac{a}{2} \right\rfloor, \left\lfloor \frac{b}{2} \right\rfloor\right)$. These networks are also built with this mixed approach. Then, we compute the clauses and variables needed in the recursive approach, V_R and C_R, with the formula of Section 4.1: $V_R = V_1 + V_2 + 2\left\lfloor \frac{a+b-1}{2} \right\rfloor$, $C_R = C_1 + C_2 + 3\left\lfloor \frac{a+b-1}{2} \right\rfloor$, where (V_1, C_1) and (V_2, C_2) are, respectively, the number of variables and clauses needed in the recursive merge networks.

Finally, we compare the values of V_R, V_D, C_R and C_D, and decide which method is better for building the merge network. Notice that we cannot minimize both the number of variables and clauses; therefore, here we try to minimize the function $\lambda \cdot V + C$, for some fixed value $\lambda > 0$.[5] The parameter λ allows us to adjust the relative importance of the number of variables with respect to the number of clauses of the encoding. Notice that this algorithm for building merge networks (and similarly, sorting, simplified merge and cardinality networks) can easily be implemented with dynamic programming. See Section 7 for an experimental evaluation of the numbers of variables and clauses in cardinality networks built with this mixed approach.

The arc-consistency of the mixed encoding easily follows from the arc-consistency of the two encodings it is based on.

Theorem 3. *The Mixed Cardinality Network encoding is arc-consistent.*

Proof (sketch). The proof uses lemmas analogous to Lemmas 1, 2, 3 and 4. In turn, these lemmas are proved by combining the proofs outlined in Theorems 1 and 2.

7 Experimental Evaluation

In previous work [3], it was shown that power-of-two (Recursive) Cardinality Networks have overall better performance than other well-known methods such as Sorting Networks [10], Adders [10] and the BDD-based encoding of [6]. In what follows we will show that the generalization of Cardinality Networks to arbitrary size and their combination with Direct Encodings, yielding what we have called the **Mixed** approach, makes them significantly better, both in the size of the encoding and the SAT solver runtime.

We start the evaluation focusing on the size of the resulting encoding. In Figure 4 we show a representative graph, which indicates the size, in terms of variables and clauses, of the encoding of a cardinality network with input size 100 and varying output size m.

---

[5] This function can be replaced by any other monotone function that can be efficiently evaluated.

It can be seen that, since we minimize the function $\lambda \cdot V + C$, where V is the number of variables and C the number of clauses, the bigger λ is, the fewer variables we obtain, at the expense of a slight increase in the number of clauses. Also, it can be seen that using power-of-two Cardinality Networks as in [3] is particularly harmful when m is slightly larger than a power of two.

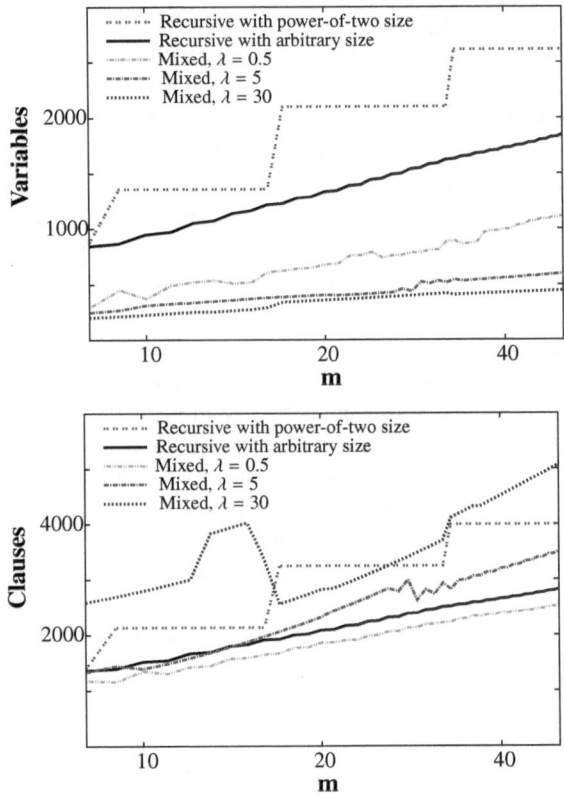

Fig. 4. Number of variables and clauses generated by **Mixed** and the Recursive Cardinality Networks approaches with input size 100 and different output sizes m

Although having a smaller encoding is beneficial, this should be accompanied with a reduction in SAT solver runtime. Hence, let us now move to assess how our new encoding affects the performance of SAT solvers. In this evaluation, in addition to considering the power-of-two Recursive Cardinality Networks in [3] (**Power-of-two CN**), the (arbitrary-size) Recursive Cardinality Networks presented in Section 4 (**Arbitrary-sized CN**) and the **Mixed** approach of Section 6, we have also included other well-known encodings in the literature: the adder-based encoding (**Adder**) of [10] and the BDD-based encoding (**BDD**) of [6]. We believe these encodings are representative of all different approaches that have been used to deal with cardinality constraints. Other works, like the adder-based encoding of [20], the BDD-based one of [10] or the work by

Anbulagan and Grastien [1], are small variations or combinations of the encodings we have chosen. Moreover, we have implemented an SMT-based approach (**SMT**) to Cardinality Constraints. In a nutshell, we have coupled a SAT solver with a theory solver that handles all cardinality constraints. As soon as a cardinality constraint is violated by the current partial assignment, the SAT solver is forced to backtrack and, when the value of a variable can be propagated thanks to a cardinality constraint, this information is passed to the SAT solver. In other words, cardinality constraints are not translated into SAT, but rather tackled by a dedicated algorithm, similar in nature to what some pseudo-Boolean solvers do. See [14] for more information about SMT.

The SAT solver we have used in this evaluation is Lingeling version *ala*, a state-of-the-art CDCL (Conflict-Driven Clause Learning) SAT solver that implements several in/preprocessing techniques. All experiments were conducted on a 2Ghz Linux Quad-Core AMD with the three following sets of benchmarks:

1.-MSU4 Suite. These benchmarks are intermediate problems generated by an implementation of the *msu4* algorithm [12], which reduces a Max-SAT problem to a series of SAT problems with cardinality constraints. The *msu4* implementation was run of a variety of problems (filter design, logic synthesis, minimum-size test pattern generation, haplotype inference and maximum-quartet consistency) from the Partial Max-SAT division of the Third Max-SAT evaluation[6]. The suite consists of about 14000 benchmarks, each of which contains multiple $\leqslant$-cardinality constraints.

2.-Discrete-Event System Diagnosis Suite. The second set of benchmarks we have used is the one introduced in [1]. These problems come from discrete-event system (DES) diagnosis. As it happened with the Max-SAT problems, a single DES problem produced a family of "SAT + cardinality constraints" problems. This way, out of the roughly 600 DES problems, we obtained a set of around 6000 benchmarks, each of which contained a single very large $\leqslant$-cardinality constraint.

3.-Tomography Suite. The last set of benchmarks we have used is the one introduced in [5]. The idea is to first generate an $N \times N$ grid in which some cells are filled and some others are not. The problem consists in finding out which are the filled cells using only the information of how many filled cells there are in each row, column and diagonal. For that purpose, variables x_{ij} are used to indicate whether cell (i, j) is filled and several =-cardinality constraints impose how many filled cells there are in each row, column and diagonal. We generated 2600 benchmarks (100 instances for each size $N = 15 \dots 40$).

Results are summarized in Table 1, which compares the **Mixed** (with $\lambda = 5$) encoding with the aforementioned encodings. The time limit was set to 600 seconds per benchmark and we only considered benchmarks for which at least one of the methods took more than 5 seconds. There are three tables, one for each benchmark suite. In each table, columns indicate in how many benchmarks the **Mixed** encoding exhibits the corresponding speed-up or slow-down factor with respect to the method indicated in each row. For example, in the table for the **MSU4 suite**, the first row indicates that in 43 benchmarks, Power-of-two Cardinality Networks timed out (**TO**) whereas our new encoding did not. The columns next to it indicate that in 732 benchmarks the novel encoding was at least 4 times faster, in 2957 between 2 and 4 times faster, etc.

---

[6] See http://www.maxsat.udl.cat/08/index.php?disp=submitted-benchmarks

Table 1. Comparison of SAT solver runtime. Figures show number of benchmarks in which **Mixed** shows the corresponding speed-up/slow-down factor w.r.t. other methods.

	Speed-up factor of Mixed					Slow-down factor of Mixed				
	TO	4	2	1.5	TOT.	1.5	2	4	TO	TOT.
MSU4 suite										
Power-of-two CN	43	732	2957	1278	*5010*	1	23	13	11	*48*
Arbitrary-sized CN	10	149	544	726	*1429*	3	106	43	80	*232*
Adder	985	1207	1038	1250	*4480*	0	13	36	40	*89*
BDD	187	1139	1795	1292	*4413*	4	10	31	36	*81*
SMT	1143	323	102	53	*1621*	0	1417	211	63	*1691*
DES suite										
Power-of-two CN	13	21	265	638	*937*	6	12	7	46	*71*
Arbitrary-sized CN	19	21	75	404	*519*	5	12	11	45	*73*
Adder	218	235	611	1283	*2347*	0	5	3	42	*50*
BDD	705	3944	759	51	*5459*	0	0	0	0	*0*
SMT	3003	1134	262	73	*4472*	0	15	19	15	*49*
Tomography suite										
Power-of-two CN	118	388	408	175	*1089*	64	82	159	121	*426*
Arbitrary-sized CN	104	430	432	169	*1135*	67	81	158	11	*417*
Adder	492	591	371	143	*1597*	14	20	39	35	*108*
BDD	0	0	0	0	*0*	112	1367	184	51	*1714*
SMT	0	10	25	11	*46*	112	1250	155	68	*1585*

We can see from the table that in the **MSU4** and **DES** suites, which contain benchmarks coming from real-world applications, our new encoding in general outperforms the other methods (except for some instances in which **Mixed** times out and the other cardinality network-based encodings do not; also, in **MSU4**, SMT and **Mixed** obtain comparable results). We want to remark that the gain comes both from using arbitrary-sized networks as well as from combining them with direct encodings, as can be seen from the second row of each table. In particular, this shows the negative impact of the dummy variables of [3], which hinder the performance in spite of the unit propagation of the SAT solver. Finally, in the **Tomography** suite, the BDD-based encoding and the SMT system outperform all other methods, but among the rest of the approaches the **Mixed** encoding exhibits the best performance. Altogether, the **Mixed** encoding is the most robust technique according to the results of this evaluation.

8   Conclusion and Future Work

The contributions of this paper are: (*i*) an extension of the recursive cardinality networks of [3] to arbitrary input and output sizes; (*ii*) a non-recursive construction of cardinality networks that is competitive for small sizes; (*iii*) a parametric combination of these two approaches for producing cardinality networks that not only improves on the size of the encoding, but also yields significant speedups in SAT solver performance.

As regards future work, we plan to develop encoding techniques for cardinality constraints that do not process constraints one-at-a-time but simultaneously, in order to exploit their similarities. We foresee that the flexibility of the approach presented here with respect to the original construction in [3], will open the door to sharing the internal networks among the cardinality constraints present in a SAT problem.

Acknowledgments. Abío is supported by DFG Graduiertenkolleg 1763 (QuantLA). All other authors are partially supported by Spanish MEC/MICINN under SweetLogics project (TIN 2010-21062-C02-01). We also thank the reviewers for their comments.

References

1. Anbulagan, A.G.: Importance of Variables Semantic in CNF Encoding of Cardinality Constraints. In: Bulitko, V., Beck, J.C. (eds.) Eighth Symposium on Abstraction, Reformulation, and Approximation, SARA 2009. AAAI (2009)
2. Asín, R., Nieuwenhuis, R., Oliveras, A., Rodríguez-Carbonell, E.: Cardinality networks and their applications. In: Kullmann, O. (ed.) SAT 2009. LNCS, vol. 5584, pp. 167–180. Springer, Heidelberg (2009)
3. Asín, R., Nieuwenhuis, R., Oliveras, A., Rodríguez-Carbonell, E.: Cardinality Networks: a theoretical and empirical study. Constraints 16(2), 195–221 (2011)
4. Achá, R.A., Nieuwenhuis, R.: Curriculum-based course timetabling with SAT and MaxSAT. Annals of Operations Research, 1–21 (February 2012)
5. Bailleux, O., Boufkhad, Y.: Efficient CNF Encoding of Boolean Cardinality Constraints. In: Rossi, F. (ed.) CP 2003. LNCS, vol. 2833, pp. 108–122. Springer, Heidelberg (2003)
6. Bailleux, O., Boufkhad, Y., Roussel, O.: A translation of pseudo boolean constraints to sat. JSAT 2(1-4), 191–200 (2006)
7. Bailleux, O., Boufkhad, Y., Roussel, O.: New Encodings of Pseudo-Boolean Constraints into CNF. In: Kullmann, O. (ed.) SAT 2009. LNCS, vol. 5584, pp. 181–194. Springer, Heidelberg (2009)
8. Batcher, K.E.: Sorting Networks and their Applications. In: AFIPS Spring Joint Computing Conference, pp. 307–314 (1968)
9. Codish, M., Zazon-Ivry, M.: Pairwise cardinality networks. In: Clarke, E.M., Voronkov, A. (eds.) LPAR-16 2010. LNCS, vol. 6355, pp. 154–172. Springer, Heidelberg (2010)
10. Eén, N., Sörensson, N.: Translating Pseudo-Boolean Constraints into SAT. Journal on Satisfiability, Boolean Modeling and Computation 2, 1–26 (2006)
11. Fu, Z., Malik, S.: Solving the minimum-cost satisfiability problem using SAT based branch-and-bound search. In: Proceedings of the 2006 IEEE/ACM International Conference on Computer-Aided Design, ICCAD 2006, pp. 852–859. ACM, New York (2006)
12. Marques-Silva, J., Planes, J.: Algorithms for Maximum Satisfiability using Unsatisfiable Cores. In: 2008 Conference on Design, Automation and Test in Europe Conference, DATE 2008, pp. 408–413. IEEE Computer Society (2008)
13. Metodi, A., Codish, M., Stuckey, P.J.: Boolean equi-propagation for concise and efficient sat encodings of combinatorial problems. J. Artif. Intell. Res., JAIR 46, 303–341 (2013)
14. Nieuwenhuis, R., Oliveras, A., Tinelli, C.: Solving SAT and SAT Modulo Theories: From an abstract Davis–Putnam–Logemann–Loveland procedure to DPLL(T). Journal of the ACM, JACM 53(6), 937–977 (2006)
15. Parberry, I.: The pairwise sorting network. Parallel Processing Letters 2, 205–211 (1992)
16. David, A.: Plaisted and Steven Greenbaum. A structure-preserving clause form translation. J. Symb. Comput. 2(3), 293–304 (1986)
17. Schutt, A., Feydy, T., Stuckey, P.J., Wallace, M.G.: Why cumulative decomposition is not as bad as it sounds. In: Gent, I.P. (ed.) CP 2009. LNCS, vol. 5732, pp. 746–761. Springer, Heidelberg (2009)

18. Sinz, C.: Towards an optimal CNF encoding of boolean cardinality constraints. In: van Beek, P. (ed.) CP 2005. LNCS, vol. 3709, pp. 827–831. Springer, Heidelberg (2005)
19. Büttner, M., Rintanen, J.: Satisfiability planning with constraints on the number of actions. In: Biundo, S., Myers, K.L., Rajan, K. (eds.) 15th International Conference on Automated Planning and Scheduling, ICAPS 2005, pp. 292–299. AAAI (2005)
20. Warners, J.P.: A Linear-Time Transformation of Linear Inequalities into Conjunctive Normal Form. Information Processing Letters 68(2), 63–69 (1998)

To Encode or to Propagate?
The Best Choice for Each Constraint in SAT

Ignasi Abío[1], Robert Nieuwenhuis[2],
Albert Oliveras[2], Enric Rodríguez-Carbonell[2], and Peter J. Stuckey[3]

[1] Theoretical Computer Science, TU Dresden, Germany
[2] Technical University of Catalonia, Barcelona
[3] National ICT Australia and the University of Melbourne

Abstract. Sophisticated compact SAT encodings exist for many types of constraints. Alternatively, for instances with many (or large) constraints, the SAT solver can also be extended with built-in propagators (the SAT Modulo Theories approach, SMT). For example, given a cardinality constraint $x_1 + \ldots + x_n \leq k$, as soon as k variables become true, such a propagator can set the remaining variables to false, generating a so-called explanation clause of the form $x_1 \wedge \ldots \wedge x_k \rightarrow \overline{x_i}$. But certain "bottle-neck" constraints end up generating an exponential number of explanations, equivalent to a naive SAT encoding, much worse than using a compact encoding with auxiliary variables from the beginning. Therefore, Abío and Stuckey proposed starting off with a full SMT approach and *partially* encoding, on the fly, only certain "active" parts of constraints. Here we build upon their work. Equipping our solvers with some additional bookkeeping to monitor constraint activity has allowed us to shed light on the effectiveness of SMT: many constraints generate very few, or few different, explanations. We also give strong experimental evidence showing that it is typically unnecessary to consider partial encodings: it is competitive to encode the few really active constraints entirely. This makes the approach amenable to any kind of constraint, not just the ones for which *partial* encodings are known.

1   Introduction

The "SAT revolution" [Var09] has made SAT solvers a very appealing tool for solving constraint satisfaction and optimization problems. Apart from their efficiency, SAT tools are push-button technology, with a single fully automatic variable selection heuristic. For many types of constraints, sophisticated compact SAT encodings exist. Such encodings usually introduce auxiliary variables, which allows one to obtain succinct formulations. Auxiliary variables frequently also have a positive impact on the size and reusability of the *learned* clauses (lemmas), and, in combination with the possibility of deciding (splitting) on them, on the quality of the search.

Building in Constraints: SAT Modulo Theories (SMT). On problem instances with many (or very large) constraints, where encodings lead to huge numbers of clauses and variables, it may be preferable to follow an alternative approach: in SMT [NOT06,BHvMW09], the SAT solver is extended with a built-in

C. Schulte (Ed.): CP 2013, LNCS 8124, pp. 97–106, 2013.

propagator for each constraint, making it amenable to sophisticated constraint-specific reasoning (as in Constraint Programming). For example, given a cardinality constraint $x_1 + \ldots + x_n \leq k$, as soon as k of its variables become true, such a propagator can set any other variable x_j to false. If at some later point this propagated literal $\overline{x_j}$ takes part in a conflict, a so-called *explanation* clause of the form $x_{i_1} \wedge \ldots \wedge x_{i_k} \rightarrow \overline{x_j}$ is used, thus fully integrating such propagators in the SAT solver's conflict analysis and backjumping mechanisms. As usual in SMT, here we consider that such explanations are (i) only produced when needed during conflict analysis and (ii) are not learned (only the resulting lemma is).

The Remarkable Effectiveness of SMT. SMT is remarkably effective. The intuitive reason is that, while searching for a solution for a given problem instance, some constraints only block the current solution candidate very few times, and moreover they do this almost always in the same way. In this paper we shed some more light on this intuitive idea. We perform experiments with a number of notions of *constraint activity* in this sense, that is, the (recent or total) number of (different or all) explanations that each constraint generates. Indeed, as we will see: A) many constraints generate very few, or few different, explanations, and B) generating only these explanations can be much more effective than dealing with a full encoding of the constraint.

The Dark Side of SMT. Frequently, there are also certain "bottle-neck" constraints that end up generating an exponential number of explanations, equivalent to a naive SAT encoding. A theoretical but illustrative example is:

Lemma 1. *An SMT solver will generate an exponential number of explanations when proving the unsatisfiability of the input problem consisting of only the two cardinality constraints $x_1 + \ldots + x_n \leq n/2$ and $x_1 + \ldots + x_n > n/2$.*

This lemma holds because any SMT solver, when proving unsatisfiability, generates a propositionally unsatisfiable set of clauses (the input ones plus the lemmas), and if a single one of the all $\binom{n}{k+1} + \binom{n}{n-k}$ explanations (where $k = n/2$) has not been generated, say, the explanation $\overline{x_1} \vee \ldots \vee \overline{x_{k+1}}$, then the assignment that sets $x_1, \ldots, x_{k+1}$ to true and the remaining $n - k - 1$ variables to false is a model. Such situations indeed happen in practice: for some constraints SMT ends up generating a full or close to full encoding, which is moreover a very naive exponential one, with no auxiliary variables. If a polynomial-size encoding for such a constraint exists (possibly with auxiliary variables), using it right from the beginning is a much better alternative. This is shown in the following figure:

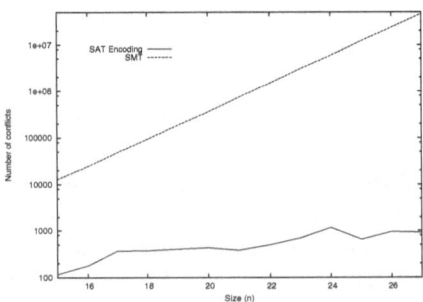

It gives the number of conflicts needed to prove unsatisfiability of the previous example, varying n, with our Barcelogic solver in SMT mode and with a SAT encoding based on Cardinality Networks. SMT exhibits exponential behavior (note the logarithmic scale). The encoding-based version scales up much better; in fact, a polynomial-size refutation for it exists, although it is not clear from the figure whether the solver always finds it or not.

Getting the Best of Both. In their *conflict-directed lazy decomposition (LD)* approach [AS12], Abío and Stuckey proposed starting off the solver using an SMT approach for all constraints of the problem instance, and *partially* encoding (or *decomposing*), on the fly, only the "active" *parts of* some constraints. The decision of when and which auxiliary variables to introduce during the solving process is taken with a particular concrete encoding in mind: if, according to the explanations that are being generated, it is observed that an auxiliary variable of the encoding and its corresponding part of the encoding would have been "active", then it is added to the formula, together with all of the involved clauses of the encoding. In this way, fully active constraints end up being completely encoded using the compact encoding with auxiliary variables, and less active ones are handled by SMT. In [AS12] it is shown that this can be done for the Cardinality/Sorting Network encoding of cardinality constraints, and, although in a complicated way, for BDD-encodings of pseudo-Boolean constraints, performing essentially always at least as well as the best of SMT and encoding.

Going Beyond. A shortcoming of [AS12] is that it is highly dependent on the constraint to be dealt with and the chosen encoding, making it unlikely to be applicable to other more complex constraints, and in any case equipping the theory solver with the required features is a highly non-trivial task. Here we propose another technique that is much simpler. It does not depend on the concrete constraint under consideration and can in fact be applied to any class of constraints that can be either encoded or built in. As mentioned previously, we have devised and analyzed bookkeeping methods for different notions of constraint activity that are cheap enough not to slow down solving appreciably. As a result, here we show, giving strong experimental evidence, that it is typically unnecessary to consider partial encodings: the few really active constraints can usually be encoded –on the fly– entirely. This makes the approach amenable to any kind of constraint, not just the ones for which *partial* encodings are known. Results on problems containing cardinality and pseudo-Boolean constraints are comparable, and frequently outperform all three of its competitors: SMT, encoding, and the partial lazy decomposition method of [AS12].

2   SAT and SAT Encoding

Let $\mathcal{X} = \{x_1, x_2, \ldots x_n\}$ be a finite set of propositional *variables*. If $x \in \mathcal{X}$ then x and $\bar{x}$ are *literals*. The *negation* of a literal l, written $\bar{l}$, denotes $\bar{x}$ if l is x, and x if l is $\bar{x}$. A *clause* is a disjunction of literals $l_1 \vee \ldots \vee l_n$. A *(CNF) formula* is a conjunction of clauses.

An *assignment* A is a set of literals such that $\{x, \overline{x}\} \subseteq A$ for no x. A literal l is *true* in A if $l \in A$, is *false* if $\overline{l} \in A$, and is *undefined* otherwise. A clause C is true in A if at least one of its literals is true in A. A formula F is true in A if all its clauses are true in A, and then A is a *model* of F. The *satisfiability (SAT) problem* consists in, given a formula F, to decide if it has a model. Systems that decide the SAT problem are called *SAT solvers*.

A function $C : \{0,1\}^n \to \{0,1\}$ is called a *constraint*. Given a constraint C, a *(SAT) encoding* for it is a formula F (possibly with auxiliary variables) that is equisatisfiable. An important class of constraints are *cardinality constraints*, which state that at most (or at least, or exactly) k out of n variables can be true. Common encodings for it are based on networks of adders [BB03,uR05,AG09,Sin05], or Sorting Networks [ES06,CZI10,ANORC09,ANORC11]. Cardinality constraints are generalized by *pseudo-Boolean constraints*, of the form $a_1 x_1 + \cdots + a_n x_n \# k$, where the a_i and k are integer coefficients, and $\#$ belongs to $\{\leq, \geq, =\}$. Again, several encodings exist, based on performing arithmetic [War98,BBR09,ES06] or computing BDD's [BBR06,ES06,ANO+12]. Most convenient encodings are the ones for which the SAT solver's unit propagation mechanism preserves domain-consistency.

3   To Encode or Not to Encode?

In this section we will discuss situations where encoding a constraint is better than using a propagator for it or vice versa, and how to detect them. The reasoning will consist of both theoretical insights and experimental evaluation. For the latter, 5 benchmarks suites will be used, in which all benchmarks solvable in less than 5 seconds by both methods have been removed.

1.-MSU4: 5729 problems generated in the execution of the *msu4* algorithm [MSP08] for Max-SAT. Each benchmark contains very few $\leq$-cardinality constraints.

2.-Discrete-Event System Diagnosis: 4526 discrete-event system (DES) diagnosis [AG09] problems. Each benchmark contains a single very large $\leq$-cardinality constraint.

3.-Tomography: 2021 tomography problems introduced in [BB03]. Each problem contains many =-cardinality constraints.

4.-PB Evaluation: 669 benchmarks from the pseudo-Boolean Competition[1] 2011 (category DEC-SMALLINT-LIN), with multiple cardinality and pseudo-Boolean constraints.

5.-RCPSP: 577 benchmarks coming from the PSP-Lib[2]. These are scheduling problems with a fixed makespan. Several pseudo-Boolean constraints are present.

To start with, let us experimentally confirm that SMT and encoding-based methods are complementary, and so a hybrid method getting the best of both is worth pursuing. For this purpose, we implemented an SMT-based system

---

[1] http://www.cril.univ-artois.fr/PB11/
[2] http://webserver.wi.tum.de/psplib

and encodings into SAT. For cardinality constraints, we used the Cardinality Networks encoding of [ANORC11], whereas for pseudo-Boolean constraints, the BDD-based one of [ANO+12]. The reason for this choice is that, according to the experimental results of [ANORC11,ANO+12], these two encodings are the globally most robust ones in practice. However, any other choice would have been possible, since the approach we will present is encoding-independent. A time limit of 600 seconds was set per benchmark and, in order to have a fair comparison, in both systems the same underlying SAT solver was used (Barcelogic). Results can be seen in Table 1, where one can observe that the encoding performs very well in the MSU4 and DES suite, and is significatively worse in the other three.[3]

Table 1. Comparison between encoding and SMT. Table on the left indicate the percentage of benchmarks where each method outperforms (is at least 1.5 times faster than) the other. On the right, the geometric mean (in seconds) of the instances solves by both methods.

Benchmark suite	Encoding $\geq$ 1.5x faster	SMT $\geq$ 1.5x faster	Geometric mean	
			Encoding	SMT
MSU4	39.37%	15.39%	1.71	23.53
DES	92.06%	0.28%	2.3	56.02
Tomography	5.93%	86.49%	46.95	4.37
PB evaluation	7.02%	43.49%	25.53	3.79
RCPSP	0.69%	46.62%	106.65	5.8

Lemma 1 explains why SMT is worse in some suites, but not why it is better in some others. The latter happens on benchmarks with many constraints. A possible explanation could be that many of these constraints are not very active, i.e. they produce very few, if any, explanations. If this is the case, SMT has an advantage over an encoding: only active constraints will generate explanations, whereas an encoding approach would also have to encode all low-activity constraints right from the beginning. This notion of constraint activity, counting the number of times the propagator generates an explanation, is very similar to earlier activity-based lemma deletion policies in SAT solvers [GN02]. In order to evaluate how often this situation happens, we ran our SMT system computing the number of explanations each constraint generates. Results can be seen in Table 2, where we considered a constraint to have low activity if it generates less than 100 (possibly repeated) explanations.

Each row contains the data for each suite: e.g., in 74.6% of the MSU4 benchmarks between 0 and 5% of the constraints had low activity. In the PB evaluation and in the RCPSP benchmarks, the number of low-activity constraints is high and hence, this might explain why SMT behaves better than the encoding on these suites. However, in the Tomography suite, constraints tend to be very active, which refutes our conjecture of why SMT performs so well on these benchmarks.

---

[3] Note that rows do not add up to 100 % as benchmarks in which the two methods are comparable are not shown.

Table 2. Number of low-activity constraints in distinct benchmarks suites

Suite	Perc. of benchs with this perc. of low-act. constr.						
	0-5%	5-10%	10-20%	20-40%	40-60%	60-80%	80-95%	95-100%
MSU4	74.6	0	0	0	24.9	0.5	0	0
DES	99.9	0	0	0	0	0	0	0.1
Tomography	100	0	0	0	0	0	0	0
PB evaluation	54	21.6	20.5	0.6	1.1	0.6	1.7	20.5
RCPSP	0	0	2.2	13.2	51.1	31.3	2.2	0

What happens in the Tomography suite is that although constraints are very active, the SMT solver does not end up generating the whole naive encoding because many explanations are repeated. Hence, a sophisticated encoding would probably generate many more clauses, as the whole constraint would be decomposed, even irrelevant parts.

To confirm this hypothesis we ran our SMT solver counting repeated explanations. Results can be seen in Table 3. Each row[4] corresponds to a different suite: e.g., the 100 in the third row indicates that all benchmarks in the Tomography suite had at least half of its constraints producing between 80 and 95% of repeated explanations. In general, if a constraint produces many repeated explanations, it is unlikely that it might end up generating its whole naive encoding. This explains why SMT has good results in this suite, as well as in PB evaluation and RCPSP. Hence, the number of repeated explanations seems to be a robust indicator of whether we should encode a constraint or use a propagator.

Table 3. The percentage of benchmark instances where more than half the constraints have a given percentage of repeated explanations

Suite	Benchs with >50% of the ctrs. w./ this perc. of rep. expl.						
	0-5%	5-10%	10-20%	20-40%	40-60%	60-80%	80-95%	95-100%
MSU4	53.8	9.1	11.6	8.5	2	0.8	0.2	0
DES	21.4	29.8	35.2	13.6	0	0	0	0
Tomography	0	0	0	0	0	0	100	0
PB evaluation	6.2	0	0	0	0	0.6	14.2	51.7
RCPSP	0	0	0	0	0	5.5	52.7	1.1

4   Implementation and Experimental Evaluation

Taking into account Section 3, we implemented a system that processes SAT problems augmented with cardinality and pseudo-Boolean constraints. Although our approach is easily applicable much more generally, here we focus on these two types of constraints in order to be able to compare with [AS12]. Our aim is to show that a very simple approach gets the best of SMT and encoding methods. The starting point for our implementation is an SMT solver equipped with the ability of encoding cardinality constraints via Cardinality Networks and pseudo-Boolean constraints via BDDs.

In order to know which constraints to encode we need to keep track of the percentage of different explanations that the constraints generate. To do this we

---

[4] Note that the percentages in each row do not need to add 100.

attach to each constraint all the different explanations it produces. When an explanation is generated, we traverse the list of previous explanations, checking if it already exists. To speed up comparison, we first compare the size and only if they are equal, we compare the explanations, which are sorted to make comparison faster. This would be very expensive if constraints with many different explanations existed, but those constraints end up being encoded and after that do not cause any further bookkeeping overhead. Hence, more complex data structures would not help here. In our implementation, we only collect information during the first 2 minutes, since, according to our experiments, after that the information stabilizes.

Another important source of information to consider is how large the ad-hoc encoding of each constraint would be. If the number of generated explanations becomes close to the number of clauses the encoding requires, according to our experiments then it is advantageous to encode the constraint. Besides, if a constraint is producing many different explanations, we found that it is likely to end up generating the full (or a large part of the) naive encoding. Discovering and avoiding this situation is highly beneficial.

We also experimented with different ways of counting the number of *recent* occurrences of a given explanation in conflicts, without any significant findings.

Finally, following all previous observations, we encode a constraint if at least one of two conditions holds: (i) the number of different explanations is more than half the number of clauses of the compact, sophisticated encoding, (ii) more than 70% of the explanations are new and more than 5000 explanations have already been generated.

We compared the resulting system (**New** in the tables) with an SMT system, another one which encodes all constraints from the start (**Enc.**) and Lazy Decomposition [AS12] (**LD**). Results can be seen in Table 4. Each cell contains the

Table 4. Comparison among different methods on all benchmarks suites

MSU4

	<10s	<30s	<60s	<120s	<300s	<600s
Enc.	**5374**	**5525**	5578	5621	5659	5677
SMT	4322	4530	4603	4667	4737	4767
LD	5196	5414	5528	5598	5655	5674
New	5222	5479	**5585**	**5636**	**5666**	**5679**

DES

	<10s	<30s	<60s	<120s	<300s	<600s
Enc.	**2521**	**3333**	**3692**	**3903**	**4102**	**4228**
SMT	362	654	850	1023	1256	1452
LD	570	1230	1761	2525	3558	4019
New	836	2156	3293	3800	4053	4166

Tomography

	<10s	<30s	<60s	<120s	<300s	<600s
Enc.	773	1112	1314	1501	1759	1932
SMT	1457	1748	1858	1962	**2014**	**2021**
LD	1027	1239	1399	1561	1763	1918
New	**1556**	**1818**	**1935**	**1971**	2012	**2021**

PB evaluation

	<10s	<30s	<60s	<120s	<300s	<600s
Enc.	268	337	358	376	399	414
SMT	**364**	**377**	**386**	**392**	**409**	414
LD	352	371	379	388	403	**416**
New	269	341	360	381	404	415

RCPSP

	<10s	<30s	<60s	<120s	<300s	<600s
Enc.	7	22	52	91	139	175
SMT	**132**	**179**	**206**	224	**249**	**272**
LD	114	160	178	189	216	228
New	111	169	202	**225**	**249**	271

number of problems that could be solved in less than the number of seconds of the corresponding column.

The first important conclusion is that we can obtain comparable, in some cases better, results than the LD method. This is worth mentioning since our approach is much simpler to implement and does not pose any requirement on the encodings to be used. Secondly, our approach always solves a very similar number of problems to the best option for each suite. Only in the DES suite, there is some difference that can be explained by the fact that SMT has extraordinarily poor performance on those benchmarks. Thus, just running the system in SMT mode for few seconds before encoding the constraints, as our new system does on these instances, has a strong negative impact because the many explanations generated in this early stage hinder the search later on. This could be mitigated by using more aggressive lemma deletion policies.

5   Conclusions and Other Related Work

This work is part of a project with the aim of deepening our understanding of what choices between SMT and encodings are optimal in practical problems. Here we have seen that the use of adaptive strategies is clearly advantageous. Moreover, we have given a simpler approach for which it becomes possible to handle many other types of constraints, as we plan to investigate next.

Another possibility for future work concerns the version of SMT in which explanation clauses are generated and learned immediately when a constraint propagates, as in the initial version of Lazy Clause Generation [OSC07], which worked remarkably well on resource-constrained project scheduling problems (RCPSPs). Indeed, we have now discovered that in these problems the number of different explanations is specially low. It may turn out to be advantageous to handle these constraints with clauses, which are prioritized in the solver with respect to constraint propagators.

Related Work. Apart from [AS12], another related proposal is [MP11]: to solve a propositional formula F plus additional pseudo-Boolean constraints, the SAT solver first finds a model M for F ("unsat" if there is none); then a few of the constraints that are false in M are picked ("sat" if there is none), simplified using the unit clauses found so far, encoded, and added to F; and the process is iterated. A drawback of this method is that it may fully encode low-activity constraints just because they happen to be false in M, whereas we really monitor activity. Also, we only need one run of the solver. Finally, it is clear that any method that encodes on the fly (including ours) can simplify constraints with the unit clauses available at that point.

Acknowledgments. First four authors partially supported by MEC/MICINN under SweetLogics project (TIN 2010-21062-C02-01). Abío is also supported by DFG Graduiertenkolleg 1763 (QuantLA). NICTA is funded by the Australian Government as represented by the Department of Broadband, Communications and the Digital Economy and the Australian Research Council through the ICT Centre of Excellence program.

References

AG09.   Anbulagan, Grastien, A.: Importance of Variables Semantic in CNF Encoding of Cardinality Constraints. In: Bulitko, V., Beck, J.C. (eds.) Eighth Symposium on Abstraction, Reformulation, and Approximation, SARA 2009. AAAI (2009)

ANO+12.   Abío, I., Nieuwenhuis, R., Oliveras, A., Rodríguez-Carbonell, E., Mayer-Eichberger, V.: A new look at bdds for pseudo-boolean constraints. J. Artif. Intell. Res (JAIR) 45, 443–480 (2012)

ANORC09.   Asín, R., Nieuwenhuis, R., Oliveras, A., Rodríguez-Carbonell, E.: Cardinality networks and their applications. In: Kullmann, O. (ed.) SAT 2009. LNCS, vol. 5584, pp. 167–180. Springer, Heidelberg (2009)

ANORC11.   Asín, R., Nieuwenhuis, R., Oliveras, A., Rodríguez-Carbonell, E.: Cardinality Networks: a theoretical and empirical study. Constraints 16(2), 195–221 (2011)

AS12.   Abío, I., Stuckey, P.J.: Conflict directed lazy decomposition. In: Milano, M. (ed.) CP 2012. LNCS, vol. 7514, pp. 70–85. Springer, Heidelberg (2012)

BB03.   Bailleux, O., Boufkhad, Y.: Efficient CNF Encoding of Boolean Cardinality Constraints. In: Rossi, F. (ed.) CP 2003. LNCS, vol. 2833, pp. 108–122. Springer, Heidelberg (2003)

BBR06.   Bailleux, O., Boufkhad, Y., Roussel, O.: A translation of pseudo boolean constraints to sat. JSAT 2(1-4), 191–200 (2006)

BBR09.   Bailleux, O., Boufkhad, Y., Roussel, O.: New Encodings of Pseudo-Boolean Constraints into CNF. In: Kullmann, O. (ed.) SAT 2009. LNCS, vol. 5584, pp. 181–194. Springer, Heidelberg (2009)

BHvMW09.   Biere, A., Heule, M.J.H., van Maaren, H., Walsh, T. (eds.): *Handbook of Satisfiability*. Frontiers in Artificial Intelligence and Applications, vol. 185. IOS Press (February 2009)

CZI10.   Codish, M., Zazon-Ivry, M.: Pairwise cardinality networks. In: Clarke, E.M., Voronkov, A. (eds.) LPAR-16 2010. LNCS, vol. 6355, pp. 154–172. Springer, Heidelberg (2010)

ES06.   Eén, N., Sörensson, N.: Translating Pseudo-Boolean Constraints into SAT. Journal on Satisfiability, Boolean Modeling and Computation 2, 1–26 (2006)

GN02.   Goldberg, E., Novikov, Y.: BerkMin: A Fast and Robust SAT-Solver. In: 2002 Conference on Design, Automation, and Test in Europe, DATE 2002, pp. 142–149. IEEE Computer Society (2002)

MP11.   Manolios, P., Papavasileiou, V.: Pseudo-Boolean solving by incremental translation to SAT. In: Formal Methods in Computer-Aided Design, FMCAD 2011 (2011)

MSP08.   Marques-Silva, J., Planes, J.: Algorithms for Maximum Satisfiability using Unsatisfiable Cores. In: 2008 Conference on Design, Automation and Test in Europe Conference, DATE 2008, pp. 408–413. IEEE Computer Society (2008)

NOT06.   Nieuwenhuis, R., Oliveras, A., Tinelli, C.: Solving SAT and SAT Modulo Theories: From an abstract Davis–Putnam–Logemann–Loveland procedure to DPLL(T). Journal of the ACM, JACM 53(6), 937–977 (2006)

OSC07.   Ohrimenko, O., Stuckey, P.J., Codish, M.: Propagation = lazy clause generation. In: Bessière, C. (ed.) CP 2007. LNCS, vol. 4741, pp. 544–558. Springer, Heidelberg (2007)

Sin05.   Sinz, C.: Towards an optimal CNF encoding of boolean cardinality con-
            straints. In: van Beek, P. (ed.) CP 2005. LNCS, vol. 3709, pp. 827–831.
            Springer, Heidelberg (2005)
uR05.   Büttner, M., Rintanen, J.: uttner and J. Rintanen. Satisfiability planning
            with constraints on the number of actions. In: Biundo, S., Myers, K.L.,
            Rajan, K. (eds.) 15th International Conference on Automated Planning
            and Scheduling, ICAPS 2005, pp. 292–299. AAAI (2005)
Var09.   Vardi, M.Y.: Symbolic techniques in propositional satisfiability solving.
            In: Kullmann, O. (ed.) SAT 2009. LNCS, vol. 5584, pp. 2–3. Springer,
            Heidelberg (2009)
War98.   Warners, J.P.: A Linear-Time Transformation of Linear Inequalities into
            Conjunctive Normal Form. Information Processing Letters 68(2), 63–69
            (1998)

Automated Symmetry Breaking
and Model Selection in CONJURE

Ozgur Akgun[1], Alan M. Frisch[2], Ian P. Gent[1], Bilal Syed Hussain[1],
Christopher Jefferson[1], Lars Kotthoff[3], Ian Miguel[1], and Peter Nightingale[1]

[1] University of St. Andrews
[2] University of York
[3] Cork Constraint Computation Centre

Abstract. Constraint modelling is widely recognised as a key bottleneck in
applying constraint solving to a problem of interest. The CONJURE automated
constraint modelling system addresses this problem by automatically refining
constraint models from problem specifications written in the ESSENCE language.
ESSENCE provides familiar mathematical concepts like sets, functions and relations nested to any depth. To date, CONJURE has been able to produce a set
of alternative model kernels (i.e. without advanced features such as symmetry
breaking or implied constraints) for a given specification. The first contribution
of this paper is a method by which CONJURE can break symmetry in a model as
it is introduced by the modelling process. This works at the problem class level,
rather than just individual instances, and does not require an expensive detection
step after the model has been formulated. This allows CONJURE to produce a
higher quality set of models. A further limitation of CONJURE has been the lack
of a mechanism to select among the models it produces. The second contribution
of this paper is to present two such mechanisms, allowing effective models to be
chosen automatically.

1 Introduction and Background

For constraint programming to achieve its potential widespread industrial and academic
use, reducing the *modelling bottleneck* [29] is of central importance. This is the problem
of formulating a problem of interest as a constraint model suitable for input to a constraint solver. There are typically many possible models for a given problem, and the
model chosen can dramatically affect the efficiency of constraint solving. This presents
a serious obstacle for non-expert users, who have difficulty in formulating a good (or
even correct) model from among the many possible alternatives. Therefore, automating constraint modelling is a desirable goal. Numerous approaches have been taken to
automate aspects of constraint modelling, including: case-based reasoning [23]; theorem proving [6]; automated transformation of medium-level solver-independent constraint models [27, 28, 30, 33]; and refinement of abstract constraint specifications [9]
in languages such as ESRA [8], ESSENCE [10], $\mathcal{F}$ [18] or Zinc [21, 25]. Some systems [2–4, 7, 22] aim to learn constraint models from positive or negative examples.

This paper focuses on the refinement-based approach, in which a user writes abstract constraint specifications to describe a problem at a higher level than that where

C. Schulte (Ed.): CP 2013, LNCS 8124, pp. 107–116, 2013.

modelling decisions are normally made. Abstract constraint specification languages, e.g. ESSENCE and Zinc, support abstract variables with types for common mathematical structures such as sets, multisets, functions, and relations, as well as nested types, such as set of sets and multiset of functions. Problems can often be specified very concisely in this way. For example, the Social Golfers Problem [17], which is to find a set of partitions of golfers subject to some constraints, can be specified directly (see Fig. 1) without the need to model the sets or partitions as matrices of Integer variables.

We use ESSENCE in this paper [10]. An ESSENCE specification, such as that in Fig. 1, identifies: the input parameters of the problem class (given), whose values define an instance; the combinatorial objects to be found (find); and the constraints the objects must satisfy (such that). An objective function may also be specified (min/maximising) and identifiers may be declared (letting). Abstract constraint specifications must be *refined* into concrete constraint models for existing constraint solvers. Our CONJURE system[1] [1] uses refinement rules to convert an ESSENCE specification into the solver-independent constraint modelling language ESSENCE' [30]. From ESSENCE' we use SAVILE ROW[2] to translate the model into input for a particular constraint solver while performing solver-specific model optimisations.

CONJURE has been able to produce the *kernels* of constraint models, without advanced features like symmetry breaking often used by experts to improve model performance. The first contribution of this paper is to automate the generation of symmetry-breaking constraints. Much symmetry enters constraint models through the process of constraint modelling [11]. CONJURE exploits this by breaking symmetry as it enters the model. This obviates the need for an expensive symmetry detection step following model formulation, as used by other approaches [24,26]. The added symmetry breaking constraints hold for the entire parameterised problem class — not just a single problem instance — without the need to identify graph automorphisms.

The second contribution of this paper is to automate model selection. Previously, CONJURE has been able to produce a (typically large) set of alternative models through the application of alternative refinement rules, but not to select among these models. CONJURE can now automatically select the best models for a problem class.

2   Automated Symmetry Breaking

Symmetry enters constraint models in two ways. Some problems have inherent symmetries, which if not broken get reflected in the model. Other symmetries are introduced by the modelling process; in this case a single solution to the problem corresponds to multiple assignments to the variables of the model. We call these *model* symmetries. As an example, consider the Social Golfers Problem (Fig. 1), which requires finding a set of w partitions. If this set is modelled as an array indexed by $1..w$ then all $w!$ permutations of the array correspond to the same set. This symmetry is introduced when an arbitrary decision is made about which set element goes in which cell of the array. Similarly, if the $g * s$ *Golfers* are modelled by the Integers $1..g * s$ then $g * s$ symmetries are introduced because of the arbitrary decision of which golfer corresponds to which

---

[1] https://bitbucket.org/stacs_cp/conjure-public/
[2] http://savilerow.cs.st-andrews.ac.uk

```
given    w, g, s : int(1..)
letting Golfers be new type of size g * s
find     sched  : set (size w) of partition (regular, size g)
                    from Golfers
such that
 forAll week1, week2 in sched, week1 != week2 .
   forAll group1 in parts(week1) .
     forAll group2 in parts(week2) .
       |group1 intersect group2| < 2
```

Fig. 1. ESSENCE specification of the Social Golfers Problem

Integer. The problem-specification language ESSENCE has been designed such that, unlike other modelling languages, problems can be specified without having to make the arbitrary decisions that introduce model symmetries.

Frisch et al. [11] show how each modelling rule of CONJURE can be extended to generate a description of the symmetries it introduces and how the generated descriptions can be composed to form a description of the symmetries introduced into the model. The intention was that this could then be used to generate symmetry-breaking constraints, though these descriptions were never fully developed into a method for automatically generating symmetry-breaking constraints.

The current version of CONJURE takes a different approach to generating symmetry breaking constraints: every rule that introduces symmetries also generates a constraint to break those symmetries. CONJURE has 28 such rules. There is only one rule which does not break all symmetries it introduces – the rule that refines an unnamed type, such as *Golfers*, to a range of Integers. This is because each unnamed type can be used in multiple places, and the symmetry of an unnamed type must be broken in a globally consistent way. All the other symmetries we introduce are independent, so we can add constraints which immediately break each introduced group of symmetries in a valid and complete manner. This leads to globally valid and complete symmetry breaking. We plan to handle unnamed types in the future.

To illustrate how CONJURE rules can be extended to generate symmetry-breaking constraints, consider the rule given below to build the explicit representation of a set.

```
Representation: Set~Explicit~Sym
Matches:        set (size &n, ..) of &tau
Produces:       refn : matrix indexed by [int(1..&n)] of &tau
Constraint:     allDiff(refn)
```

This rule transforms a set of a size n into a matrix with n index values, where each value in the matrix is a member of the set. A constraint is imposed to ensure that the cells of the matrix are all different. For any tau other than Integers or Booleans, CONJURE has to further decompose the allDiff constraint into $O(n^2)$ not-equal constraints.

Now consider extending this rule to generate a constraint to break the symmetry it introduces, that the index values of the matrix can be permuted in any way. The simplest way to break this symmetry is to impose a total order on the elements of the matrix. As the elements of the matrix can be any type tau we introduce two new operators, $\dot{\leq}$ and $\dot{<}$. These operators provide a total ordering (and a strict version of the same total ordering) for all types in CONJURE. These orderings are not intended to be "natural" and are not available to ESSENCE users. They are used only in refinement rules to generate effective symmetry-breaking constraints. Using these orderings, the Set~Explicit~Sym rule is modified to a rule that breaks all the symmetries it introduces:

```
Representation: Set~Explicit
Matches:        set (size &n, ..) of &tau
Produces:       refn : matrix indexed by [int(1..&n)] of &tau
Constraint:     forAll i : int(1..&n-1) . refn[i] .< refn[i+1]
```

Rather than introducing a chain of $\leq$ constraints, this rule exploits the fact that the elements of the set are required to be all different and strengthens the ordering to $\dot{<}$ constraint. This replaces $O(n^2)$ not-equal constraints with only $O(n)$ $\dot{<}$ constraints.

Other refinement rules can exploit the fact that symmetry breaking is performed immediately to produce more efficient refinements. Consider refining the constraint $S = T$ by representing the sets S and T as matrices S' and T' with the Set~Explicit~Sym representation. To find if S' and T' represent the same set we must check if each element of S' is equal to *any* element of T' and whether the two sets have the same cardinality, since the order of elements in the matrices can be different. However, when the Set~Explicit representation is used we can refine $S = T$ to the constraint $S' = T'$, because each assignment of S is represented by exactly one assignment to S', which satisfies the symmetry breaking constraint. This gives a much smaller constraint, which propagates much more efficiently.

We illustrate the new approach to symmetry-breaking by showing how the SGP specification (Fig. 1) is refined into a model with symmetry-breaking constraints. We consider generating only one model. We will consider only how the decision variables are refined, ignoring all constraints other than symmetry-breaking constraints. First, CONJURE replaces type of size g*s with int(1..g*s):

```
given  w, g, s : int(1..)
find   sched'  : set (size w) of partition (regular, size g) from int(1..g*s)
```

After this, CONJURE refines the type of the decision variable by rewriting the outer set constructor using the Set~Explicit rule given in the previous section. This generates the following refinement.

```
given      w, g, s : int(1..)
find       sched'  : matrix indexed by [int(1..w)] of
                         partition (regular, size g) from int(1..g*s)
such that  forAll i : int(1..w-1). sched'[i] .< sched'[i+1]
```

This refinement step shows all of the important features of our method. CONJURE has introduced a new, compact constraint which both breaks symmetry, and ensures all members of the matrix are distinct. Next, it transforms the partition into a set of sets:

```
given      w, g, s : int(1..)
find       sched'' : matrix indexed by [int(1..w)] of
                         set (size g) of set (size (g*s)/g) of int(1..g*s)
such that  forAll i : int(1..w-1). sched''[i] .< sched''[i+1],
           forAll j : int(1..w).
               forAll k1,k2 in sched''[j], k1 != k2. | k1 intersect k2 | = 0
```

This refinement does not appear to have changed the symmetry-breaking constraint but it has in fact been refined from a partition to a set of sets. CONJURE has also added a constraint to impose that the cells of the partition are distinct. This structural constraint constrains the sets to be disjoint. CONJURE now applies Set~Explicit again.

```
given       w, g, s : int(1..)
find        sched''' : matrix indexed by [int(1..w), int(1..g)]
                of set (size (g*s)/g) of int(1..g*s)
such that   forAll i : int(1..w-1). sched'''[i,..] .< sched'''[i+1,..],
            forAll j : int(1..w). forAll k : int(1..g-1).
                sched'''[j,k] .< sched'''[j,k+1]
```

The first constraint here is the refined version of the already existing symmetry-breaking constraint. Once again by design the $\dot{<}$ constraint maps naturally to the matrices used in refinement. The second constraint is the symmetry breaking on matrix of sets, now transformed into a matrix of matrices. CONJURE uses the same refinement rule, even though we are now refining a set inside a matrix. CONJURE automatically deals with the array indices and inserts the outer `forAll j : int(1..w)` in a process called *lifting*. We finally apply `Set~Explicit` once more and change the $\dot{<}$ and $\le$ constraints to their final form – lexicographic ordering constraints on matrices and ordering on Integers.

```
given       w, g, s : int(1..)
find        sched''' : matrix indexed by [int(1..w), int(1..g),int(1..(g*s/g))]
                of int(1..g*s)
such that   forAll i : int(1..w-1). sched'''[i,..,..] <lex sched'''[i+1,..,..],
            forAll j : int(1..w). forAll k : int(1..(g*s)/g-1).
                sched'''[j,k,..] <lex sched'''[j,k+1,..]
            forAll j : int(1..w). forAll k : int(1..g)
                forAll l : int(1..(g*s)/g-1). sched'''[j,k,l] < sched'''[j,k,l+1]
```

If CONJURE had not the broken symmetries immediately, but instead used the `Set~Explicit~Sym` representation, the constraints requiring each partition in the outermost set to be different would now be very complex. This shows the benefit of breaking symmetries as soon as they are introduced, rather than delaying and using a general technique for symmetry breaking after model generation is finished.

3   Automated Model Selection

Our previous work [1] shows that CONJURE can successfully refine a set of model kernels (i.e. excluding symmetry breaking and implied constraints) from a given specification, and that this set contains the kernels of *effective* models. However, without symmetry breaking the performance of these model kernels is poor since refinement of abstract types naturally introduces a great deal of symmetry. Therefore, the symmetry breaking approach described above is a necessary step in producing practically useful models. Having thus enhanced CONJURE the natural next step is to provide a means to select an effective model automatically. We propose and evaluate two such approaches: a lightweight heuristic based purely on an analysis of model structure and an approach that uses a set of training instances to perform model selection by means of a race.

3.1   The Compact Heuristic

If time is limited it is sensible to provide a rapid model selection method, avoiding both generating all models and training using instance data. Our solution is a heuristic employed during refinement to commit greedily to promising modelling choices at each point where an abstract type or a constraint expression may be refined in multiple ways.

It is named Compact since it favours transformations that produce smaller expressions. For an abstract type, we define an ordering as follows: concrete domains (such as `bool`, `matrix`) are smaller than abstract domains; within concrete domains, `bool` is smaller than `int` and `int` is smaller than `matrix`. These rules are applied recursively, so that a one-dimensional matrix of `int` is smaller than any two-dimensional matrix. Abstract type constructors have the ordering `set` < `mset` < `function` < `relation` < `partition`, which is also applied recursively. Compact will select the smallest domain according to this order. For a constraint expression (and the objective), Compact chooses the refinement with the most shallow abstract syntax tree.

3.2   Racing

Our second selection method takes as input a set of instances representative of the distribution of instances a user wishes to solve. Our measure of quality of a model with respect to an instance is the time taken for SAVILE ROW to instantiate the model and translate for input to the MINION constraint solver [13] plus the time taken for MINION to solve the instance. We include the time taken by SAVILE ROW since it adds desirable instance-specific optimisations to the model, such as common subexpression elimination [14]. Given a parameter $\rho \geq 1$, a model is ρ-*dominated* on an instance by another model if the measure for the second model is at least ρ times faster than the first.

We iterate over the set of instances and conduct a *race* [5] for each. The set of models entered into the race for instance i are the winners of the race for instance $i - 1$, with all models entered in the first race. The 'winners' of an instance race are the models not ρ-dominated by any other model. After we have iterated over all of the supplied instances, the subset of models remaining is selected for the specified class. This naturally suggests the notion of a *model portfolio*, analogous to algorithm portfolios [16, 19].

A set of instances is ρ-*fractured* if every model is ρ-dominated on at least one instance. If the supplied set of instances is fractured, races run with different instance orderings can produce disjoint sets of models. We observe this experimentally in Section 3.4 and discuss its consequences.

3.3   Case Study: Equidistant Frequency Permutation Arrays

We illustrate the model selection process using the Equidistant Frequency Permutation Array (EFPA) problem [20]: 'The problem has parameters v, q, λ, d and it is to find a set E of size v, of sequences of length $q\lambda$, such that each sequence contains λ of each symbol in the set $\{1, \ldots, q\}$. For each pair of sequences in E, the pair are Hamming distance d apart (i.e. there are d places where the sequences disagree)'.

This problem is specified in ESSENCE (see Fig. 2) with a single abstract decision variable E and two constraints. The first ensures that each codeword must contain each symbol λ times, the second that each pair of codewords must differ in exactly d places. CONJURE refines this specification into 45 models. The type of E is a fixed size set of total functions. The outer set is always modelled using the explicit representation (as a vector of the inner type) and the symmetry is broken by constraining the elements of the vector to be in increasing order according to $\dot{<}$. The total function is refined in two ways: to a vector, or to a relation. In the latter case the relation is refined in four different

```
given      d, lambda, q, v : int(1..)
letting    Character    be domain int(1..q)
letting    Index        be domain int(1..lambda * q)
letting    String       be domain function (total)
                        Index --> Character
find       E            : set (size v) of String
such that  forAll s in E . forAll a : Character .
               (sum i : Index . toInt(s(i) = a)) = lambda,
           forAll s1, s2 in E, s1 != s2 .
               (sum i : Index . toInt(s1(i) != s2(i))) = d
```

Fig. 2. ESSENCE specification of the EFPA Problem

ways, giving five representations of E in total. Subsets of these five are channelled and constraints are stated on different representations to create 45 models.

For EFPA we use 24 instances from Huczynska et al. [20], and 12 casier instances that were created by taking the satisfiable instances from Huczynska et al. and reducing v by one. Identifying instances by the tuple $\langle d, \lambda, q, v \rangle$, the first instance we race is $\langle 3, 7, 7, 5 \rangle$. This instance is exceptionally discriminating. The number of winners is 4, so we have eliminated 41 models at this stage. We will see in Section 3.4 that not all problems converge so quickly. Second, the remaining models are raced on the instance $\langle 3, 8, 8, 6 \rangle$. This does not eliminate any models, although they are ranked in a different order. This process is continued for another 30 instances that eliminate no models. Instance $\langle 6, 4, 3, 12 \rangle$ eliminates one model, leaving three. Finally, the last three instances eliminate no more models so the final winning set has three models.

All of the final set of models contain the vector representation of the total function. Two of the models refine the function to a relation, then to a two-dimensional matrix of Boolean variables (which is channelled with the vector). These two models differ in one constraint. The relative similarity of these three models shows that on this problem there is a clear cluster of similar winners among a more diverse set of models.

For this problem, Compact generates the model which uses the vector representation for the function variable without any channelling. Although it uses far less information and is very quick in comparison to racing, it manages to find one of the 'winner' models.

3.4   Experimental Evaluation

In this section, we present the results of model selection for the five problem classes presented in Table 1: EFPA [20], Social Golfers Problem (SGP) [15], Progressive Party Problem (PPP) [32], the SONET network design problem [31], and Error Correcting Codes (ECC) [12]. Although not generally feasible in practice, for the purpose of this experiment we ran a race for every model on every instance with no pruning of models between races. We set $\rho = 2$ and a timeout of one hour. Furthermore, a model that solves an instance within ten seconds is considered to be non-dominated on that instance. The results presented in the *Winner set size* column of Table 1 show the number of non-dominated models in each case. For the second problem class, SGP, the set of instances is fractured; every winner set contains either model 2 or model 3 but not both.

Now consider the performance of the racing scheme. Notice that the winner set of a race must contain all the non-dominated models. It may contain further models that

Table 1. Experimental results

Inputs			Steps to convergence		Results	
Problem	Models	Instances	Mean	Std. Dev.	Winner set size	Compact
EFPA	45	36	9.64	4.65	1	Yes
SGP	4	37	5.14	1.78	Fractured	Yes
PPP	81	11	5.67	1.47	4	No
ECC	108	26	2.75	0.43	4	No
SONET	27	47	4.30	1.93	1	Yes

were not eliminated because of the eager pruning policy. Such models are dominated on some instance by models that were eliminated earlier in the racing process. The number and identity of these extra models is dependent on the order that the instances are considered. Also dependent on this order is the rate of convergence.

In order to test the importance of instance order, we ran 50 races with randomly-selected instance orders. The racing scheme does not know if the problem instances are fractured, though in some cases it may detect that it is. We make the distinction solely for the sake of this study. For the four non-fractured problem classes, all 50 sample races yielded a winner set comprising exactly the non-dominated models. In contrast, the SGP does exhibit fracturing. On the 50 runs every winner set is a singleton comprising either model 1, 2 or 3. The mean and standard deviation of the number of instances raced before reaching the final model set are given in Table 1 under the *Steps to convergence* heading.

Table 1 also presents whether Compact manages to generate a model that is in one of the winner sets found by racing. It finds a winner model for two out of four non-fractured problem classes. Moreover, it finds a winner model for one of the subdivisions in the fractured class, SGP. This is a promising result, considering that Compact works with far less information than racing and is very cheap.

4   Conclusions

This paper has demonstrated significant progress towards the goal of automated constraint modelling. We have shown how symmetry can be broken cheaply and automatically as it enters the model through the modelling process, increasing the quality of the models that CONJURE can produce beyond model kernels. Furthermore, we have shown how CONJURE can select *effective* models using a racing process and the Compact heuristic.

Acknowledgements. We thank the anonymous reviewers for their comments. This research is supported by UK EPSRC grants no EP/H004092/1 and EP/K015745/1 and EU FP7 grant 284715.

References

1. Akgun, O., Miguel, I., Jefferson, C., Frisch, A.M., Hnich, B.: Extensible automated constraint modelling. In: AAAI 2011: Twenty-Fifth Conference on Artificial Intelligence (2011)
2. Beldiceanu, N., Simonis, H.: A model seeker: Extracting global constraint models from positive examples. In: Milano, M. (ed.) CP 2012. LNCS, vol. 7514, pp. 141–157. Springer, Heidelberg (2012)
3. Bessière, C., Coletta, R., Freuder, E.C., O'Sullivan, B.: Leveraging the learning power of examples in automated constraint acquisition. In: Wallace, M. (ed.) CP 2004. LNCS, vol. 3258, pp. 123–137. Springer, Heidelberg (2004)
4. Bessiere, C., Coletta, R., Koriche, F., O'Sullivan, B.: Acquiring constraint networks using a SAT-based version space algorithm. In: AAAI 2006, pp. 1565–1568 (2006)
5. Birattari, M., Stützle, T., Paquete, L., Varrentrapp, K.: A racing algorithm for configuring metaheuristics. In: Proceedings of the Genetic and Evolutionary Computation Conference, pp. 11–18. Morgan Kaufmann (2002)
6. Charnley, J., Colton, S., Miguel, I.: Automatic generation of implied constraints. In: Proc. of ECAI 2006, pp. 73–77. IOS Press (2006)
7. Coletta, R., Bessière, C., O'Sullivan, B., Freuder, E.C., O'Connell, S., Quinqueton, J.: Semi-automatic modeling by constraint acquisition. In: Rossi, F. (ed.) CP 2003. LNCS, vol. 2833, pp. 812–816. Springer, Heidelberg (2003)
8. Flener, P., Pearson, J., Ågren, M.: Introducing ESRA, a relational language for modelling combinatorial problems (Abstract). In: Rossi, F. (ed.) CP 2003. LNCS, vol. 2833, pp. 971–971. Springer, Heidelberg (2003)
9. Frisch, A.M., Jefferson, C., Hernandez, B.M., Miguel, I.: The rules of constraint modelling. In: Proc. of the IJCAI 2005, pp. 109–116 (2005)
10. Frisch, A.M., Harvey, W., Jefferson, C., Martínez-Hernández, B., Miguel, I.: Essence: A constraint language for specifying combinatorial problems. Constraints 13(3), 268–306 (2008), http://dx.doi.org/10.1007/s10601-008-9047-y
11. Frisch, A.M., Jefferson, C., Martinez-Hernandez, B., Miguel, I.: Symmetry in the generation of constraint models. In: Proceedings of the International Symmetry Conference (2007)
12. Frisch, A., Jefferson, C., Miguel, I.: Constraints for breaking more row and column symmetries. In: Rossi, F. (ed.) CP 2003. LNCS, vol. 2833, pp. 318–332. Springer, Heidelberg (2003)
13. Gent, I.P., Jefferson, C., Miguel, I.: Minion: A fast scalable constraint solver. In: Proceedings ECAI 2006, pp. 98–102 (2006)
14. Gent, I.P., Miguel, I., Rendl, A.: Common subexpression elimination in automated constraint modelling. In: Workshop on Modeling and Solving Problems with Constraints, pp. 24–30 (2008)
15. Gent, I.P., Walsh, T.: CSPlib: a benchmark library for constraints. Tech. rep. (April 16, 1999), http://citeseer.ist.psu.edu/5054.html, http://www.cs.strath.ac.uk/~apes/reports/apes-09-1999.ps.gz
16. Gomes, C.P., Selman, B.: Algorithm portfolios. Artificial Intelligence 126(1-2), 43–62 (2001)
17. Harvey, W.: Symmetry breaking and the social golfer problem. In: Proc. SymCon 2001: Symmetry in Constraints, co-located with CP 2001, pp. 9–16 (2001)
18. Hnich, B.: Thesis: Function variables for constraint programming. AI Commun. 16(2), 131–132 (2003)
19. Huberman, B.A., Lukose, R.M., Hogg, T.: An economics approach to hard computational problems. Science 275(5296), 51–54 (1997)

20. Huczynska, S., McKay, P., Miguel, I., Nightingale, P.: Modelling equidistant frequency permutation arrays: An application of constraints to mathematics. In: Gent, I.P. (ed.) CP 2009. LNCS, vol. 5732, pp. 50–64. Springer, Heidelberg (2009)
21. Koninck, L.D., Brand, S., Stuckey, P.J.: Data independent type reduction for zinc. In: ModRef 2010 (2010)
22. Lallouet, A., Lopez, M., Martin, L., Vrain, C.: On learning constraint problems. In: 22nd IEEE International Conference on Tools with Artificial Intelligence (ICTAI), vol. 1, pp. 45–52 (2010)
23. Little, J., Gebruers, C., Bridge, D.G., Freuder, E.C.: Using case-based reasoning to write constraint programs. In: Rossi, F. (ed.) CP 2003. LNCS, vol. 2833, pp. 983–983. Springer, Heidelberg (2003)
24. Mancini, T., Cadoli, M.: Detecting and breaking symmetries by reasoning on problem specifications. In: Zucker, J.-D., Saitta, L. (eds.) SARA 2005. LNCS (LNAI), vol. 3607, pp. 165–181. Springer, Heidelberg (2005)
25. Marriott, K., Nethercote, N., Rafeh, R., Stuckey, P.J., de la Banda, M.G., Wallace, M.: The design of the zinc modelling language. Constraints 13(3) (2008), http://dx.doi.org/10.1007/s10601-008-9041-4
26. Mears, C., Niven, T., Jackson, M., Wallace, M.: Proving symmetries by model transformation. In: Lee, J. (ed.) CP 2011. LNCS, vol. 6876, pp. 591–605. Springer, Heidelberg (2011)
27. Mills, P., Tsang, E., Williams, R., Ford, J., Borrett, J.: EaCL 1.5: An easy abstract constraint optimisation programming language. Tech. rep., University of Essex, Colchester, UK (December 1999)
28. Nethercote, N., Stuckey, P.J., Becket, R., Brand, S., Duck, G.J., Tack, G.R.: MiniZinc: Towards a standard CP modelling language. In: Bessière, C. (ed.) CP 2007. LNCS, vol. 4741, pp. 529–543. Springer, Heidelberg (2007)
29. Puget, J.F.: Constraint programming next challenge: Simplicity of use. In: Wallace, M. (ed.) CP 2004. LNCS, vol. 3258, pp. 5–8. Springer, Heidelberg (2004)
30. Rendl, A.: Thesis: Effective Compilation of Constraint Models. Ph.D. thesis, University of St. Andrews (2010)
31. Smith, B.M.: Search strategies for optimization: Modelling the sonet problem. Tech. Rep. Research Report APES-70-2003. presented at 2nd International Workshop on Reformulating Constraint Satisfaction Problems (2003)
32. Smith, B.M., Brailsford, S.C., Hubbard, P.M., Williams, H.P.: The progressive party problem: Integer linear programming and constraint programming compared. In: Montanari, U., Rossi, F. (eds.) CP 1995. LNCS, vol. 976, pp. 36–52. Springer, Heidelberg (1995)
33. Van Hentenryck, P.: The OPL Optimization Programming Language. MIT Press, Cambridge (1999)

Improving WPM2
for (Weighted) Partial MaxSAT[*]

Carlos Ansótegui[1], Maria Luisa Bonet[2], Joel Gabàs[1], and Jordi Levy[3]

[1] DIEI, Univ. de Lleida
{carlos,joel.gabas}@diei.udl.cat
[2] LSI, UPC
bonet@lsi.upc.edu
[3] IIIA-CSIC
levy@iiia.csic.es

Abstract. Weighted Partial MaxSAT (WPMS) is an optimization variant of the Satisfiability (SAT) problem. Several combinatorial optimization problems can be translated into WPMS. In this paper we extend the state-of-the-art WPM2 algorithm by adding several improvements, and implement it on top of an SMT solver. In particular, we show that by focusing search on solving to optimality subformulas of the original WPMS instance we increase the efficiency of WPM2. From the experimental evaluation we conducted on the PMS and WPMS instances at the 2012 MaxSAT Evaluation, we can conclude that the new approach is both the best performing for industrial instances, and for the union of industrial and crafted instances.

1 Introduction

In the last decade Satisfiability (SAT) solvers have progressed dramatically in performance due to new algorithms, such as, conflict directed clause learning [36], and better implementation techniques. Thanks to these advances, nowadays the best SAT solvers can tackle hard decision problems. Our aim is to push this technology forward to deal with optimization problems.

The Maximum Satisfiability (MaxSAT) problem is the optimization version of SAT. The idea behind this formalism is that sometimes not all constraints of a problem can be satisfied, and we try to satisfy the maximum number of them. The MaxSAT problem can be further generalized to the Weighted Partial MaxSAT (WPMS) problem.

In the MaxSAT community, we find two main classes of algorithms: branch and bound [17, 22, 24, 26, 27] and SAT-based [2, 14, 19–21, 31–33]. The latter clearly dominate on industrial and some crafted instances, as we can see in the results of the last 2012 MaxSAT Evaluation. SAT-based MaxSAT algorithms basically reformulate a MaxSAT instance into a sequence of SAT instances. By solving these SAT instances the MaxSAT problem can be solved [6].

---

[*] This research has been partially founded by the CICYT research projects TASSAT (TIN2010-20967-C04-01/03/04) and ARINF (TIN2009-14704-C03-01).

C. Schulte (Ed.): CP 2013, LNCS 8124, pp. 117–132, 2013.

In this paper we revisit the SAT-based MaxSAT algorithm WPM2 [5] which belongs to a family of algorithms that exploit the information from the unsatisfiable cores the underlying SAT solver provides. This algorithm is the natural extension to the weighted case of the Partial MaxSAT algorithm PM2 [3, 4]. In our experimental investigation the original WPM2 algorithm solves 796 out of 1474 from the whole benchmark of PMS and WPMS industrial and crafted instances at the 2012 MaxSAT Evaluation. We have extended WPM2 with several complementary improvements. First of all, we apply the stratification approach described in [2], what results in solving 74 additional instances. Secondly, we introduce a new criteria to decide when soft clauses can be hardened, that provides 66 additional solved instances. The hardening of soft clauses in MaxSAT SAT-based solvers has been previously studied in [2, 33]. Finally, our most effective contribution is to introduce a new strategy that focuses search on solving to optimality subformulas of the original MaxSAT instance. Actually, the new WPM2 algorithm is parametric on the approach we use to optimize these subformulas. This allows to combine the strength of exploiting the information extracted from unsatisfiable cores and other optimization approaches. By solving these smaller optimization problems we get the most significant boost in our new WPM2 algorithm. In particular, we experiment with three approaches: (i) refine the lower bound on these subformulas with the subsetsum function [5, 13], (ii) refine the upper bound with the strategy applied in minisat+ [15], SAT4J [10], qmaxsat [21] or ShinMaxSat [20], and (iii) a binary search scheme where the lower bound and upper bound are refined as in the previous approaches. The best performing approach in our experimental analysis is the second one and it allows to solve up to 238 additional instances. As a summary, the overall speed-up we achieved on the original WPM2 solver is about 378 additional solved instances, a 47% more.

As we mentioned, SAT-based MaxSAT algorithms reformulate a MaxSAT instances into a sequence of SAT instances. Obviously, it is important to use an efficient SAT solver. Also, most SAT-based MaxSAT algorithms require the addition of Pseudo-Boolean (PB) linear constraints as a result of the reformulation process. These PB constraints are used to bound the cost of the optimal assignment. Currently, in most state-of-the-art SAT-based MaxSAT solvers, PB constraints are translated into SAT. However, there is no known SAT encoding which can guarantee the original propagation power of the constraint, i.e, what we call arc-consistency, while keeping the translation low in size. The best approach so far, has a cubic complexity [8]. This can be a bottleneck for WPM2 [5] and also for other algorithms such as, BINCD [19] or SAT4J [10].

In order to treat PB constraints with specialized inference mechanisms and a moderate cost in size, while preserving the strength of SAT techniques for the rest of the formula, we use the Satisfiability Modulo Theories (SMT) technology [35]. Related work in this sense can be found in [34]. Also, in [1] a Weighted Constraint Satisfaction Problems (WCSP) solver implementing the original WPM1 [4] algorithm is presented.

An SMT instance is a generalization of a Boolean formula in which some propositional variables have been replaced by predicates with predefined interpretations from background theories such as, e.g., linear integer arithmetic. Most modern SMT solvers integrate a SAT solver with decision procedures (theory solvers) for sets of literals belonging to each theory. This way, we can hopefully get the best of both worlds: in particular, the efficiency of the SAT solver for the Boolean reasoning and the efficiency of special-purpose algorithms for the theory reasoning.

Another reasonable choice would be to use a PB solver, which can be seen as a particular case of an SMT solver specialized on the theory of PB constraints [28, 29]. However, if we also want to solve problems modeled with richer formalisms like WCSP, the SMT approach seems a better choice since we can take advantage of a wide range of theories [1].

In this work, we implemented both the last version of the WPM1 algorithm [2] and the revisited version of the WPM2 algorithm on top the of the SMT solver Yices. Then, we performed an extensive experimental evaluation comparing them with the best two solvers for PMS and WPMS categories at the 2012 MaxSAT Evaluation and with three additional solvers that did not take part but have been reported to exhibit good performance: *bincd2*, which is the new version of the BINCD algorithm [19] described in [33], with the best configuration reported by authors, *maxhs* from [14], which consists in an hybrid SAT and Integer Linear Programming (ILP) approach, and *ilp* which performs a translation of WPMS into ILP solved with IBM-CPLEX studio124 [7].

We observe that the implementation on SMT of our new WPM2 algorithm with the second approach for optimizing the subformulas is the best performing solver for both PMS and WPMS industrial instances. We also observe that it is the best performing for the union of PMS and WPMS industrial and crafted instances, what shows this is a robust approach. These results make us conjecture that by improving the interaction of our new WPM2 algorithm with diverse optimization techniques applied on the subformulas we can get additional speed-ups.

This paper proceeds as follows. Section 2 presents some preliminary concepts. Section 3 describes WPM2 [5] and the new improvements. Section 4 describes the SMT problem and discuss some implementation details of the SMT-based MaxSAT algorithms. Section 5 presents the experimental evaluation. Finally, Section 6 shows the conclusions and the future work.

2   Preliminaries

A *literal* is either a Boolean variable x or its negation $\overline{x}$. A *clause* C is a disjunction of literals. A *weighted clause* is a pair (C, w), where C is a clause and w is a natural number or infinity, indicating the penalty for falsifying the clause C. A *Weighted Partial MaxSAT formula* is a multiset of weighted clauses

$$\varphi = \{(C_1, w_1), \ldots, (C_m, w_m), (C_{m+1}, \infty), \ldots, (C_{m+m'}, \infty)\}$$

where the first m clauses are soft and the last m' clauses are hard. The set of variables occurring in a formula φ is noted as $\mathrm{var}(\varphi)$.

A *(total) truth assignment* for a formula φ is a function $I : \mathrm{var}(\varphi) \to \{0,1\}$, that can be extended to literals, clauses and SAT formulas. For MaxSAT formulas is defined as $I(\{(C_1, w_1), \ldots, (C_m, w_m)\}) = \sum_{i=1}^{m} w_i\,(1 - I(C_i))$. The *optimal cost* of a formula is $\mathrm{cost}(\varphi) = \min\{I(\varphi) \mid I : \mathrm{var}(\varphi) \to \{0,1\}\}$ and an *optimal assignment* is an assignment I such that $I(\varphi) = \mathrm{cost}(\varphi)$.

The *Weighted Partial MaxSAT* problem for a Weighted Partial MaxSAT formula φ is the problem of finding an *optimal assignment*.

3   WPM2 Algorithm

The WPM2 algorithm [5] is described in Algorithm 1. The fragments in gray (lines 4, 10, 11, 13-18 and 20) correspond to the new improvements we have incorporated.

In the WPM2 algorithm, we extend soft clauses C_i with a unique fresh auxiliary blocking variable b_i obtaining $\varphi_w = \{C_i \vee b_i\}_{i=1\ldots m} \cup \{C_{m+i}\}_{i=1\ldots m'}$. Notice that b_i will be set to true by a SAT solver on φ_w if C_i is false. We also work with a set AL of at-least PB constraints of the form $\sum_{i\in A} w_i\, b_i \geq k$ on the variables b_i, and a similar set AM of at-most constraints of the form $\sum_{i\in A} w_i\, b_i \leq k$, that are modified at every iteration of the algorithm.

Intuitively, the WPM2 algorithm refines at every iteration the lower bound on φ till it reaches the optimum $\mathrm{cost}(\varphi)$. The AM constraints are used to bound the cost of the falsified clauses. The AL constraints are used to impose that subsets of soft clauses have a minimum cost and to compute the AM constraints, as we will see later. The algorithm ends when $\varphi_w \cup CNF(AL \cup AM)$ becomes satisfiable[1], where CNF is the translation to SAT of the PB constraints.

Technically speaking, the AL constraints give lower bounds on $\mathrm{cost}(\varphi)$. The AM constraints enforce that all solutions of the set of constraints $AL \cup AM$ are the solutions of AL of minimal cost. This ensures that any solution of the formula sent to the solver, $\varphi_w \cup CNF(AL \cup AM)$, if there is any, is an optimal assignment of φ. Therefore, given a set of at-least constraints AL we compute a *corresponding* set of at-most constraints AM as follows. First, we need to introduce the notion of *core* and *cover*. A core is a set of indexes A such that $\sum_{i\in A} w_i\, b_i \geq k \in AL$. Function $core(\sum_{i\in A} w_i\, b_i \geq k)$ returns the core A and function $cores(AL)$ returns $\{core(al) \mid al \in AL\}$. Covers are defined from cores as follows.

Definition 1. *Given a set of cores L, we say that the set of indexes A is a* cover *of L, if it is a minimal non-empty set such that, for every $A' \in L$, if $A' \cap A \neq \emptyset$, then $A' \subseteq A$. Given a set of cores L, we denote the set of covers of L as $SC(L)$.*

---

[1] The AL constraints are redundant, i.e., not required to be sent to the SAT solver for the soundness of the algorithm but help to speed up the search.

Algorithm 1. Revisited WPM2 algorithm.

Input: $\varphi = \{(C_1, w_1), \ldots, (C_m, w_m), (C_{m+1}, \infty), \ldots, (C_{m+m'}, \infty)\}$

1: **if** $sat(\{C_i \in \varphi \mid w_i = \infty\}) = (UNSAT, \_, \_)$ **then return** $(\infty, \emptyset)$
2: $\varphi_w := \{C_1 \vee b_1, \ldots, C_m \vee b_m, C_{m+1}, \ldots, C_{m+m'}\}$ ▷Extend all soft clauses
3: $AL := \{w_1 b_1 \geq 0, \ldots, w_m b_m \geq 0\}$ ▷Set of at-least constraints
4: $w_{max} := \infty$
5: **while** $true$ **do**
6:     $AM := \emptyset$ ▷Set of at-most constraints
7:     **foreach** $(\sum_{i \in A} w_i b_i \geq k) \in AL$ **do**
8:         **if** $A \in SC(cores(AL))$ **then**
9:             $AM := AM \cup \{\sum_{i \in A} w_i b_i \leq k\}$
10:     $(st, \varphi_c, \mathcal{I}) := sat(\varphi_w \backslash \{C_i \vee b_i \mid (C_i, w_i) \in \varphi \wedge w_i < w_{max}\} \cup CNF(AL \cup AM))$
11:     **if** $st =$ SAT and $w_{max} = 0$ **then return** $(\mathcal{I}(\varphi), \mathcal{I})$
12:     **else**
13:         **if** $st =$ SAT **then**
14:             $W := \sum \{w_i \mid (C_i, w_i) \in \varphi \wedge w_i < w_{max}\}$
15:             $\varphi_h := harden(\varphi, AM, W)$
16:             $w_{max} := decrease(w_{max}, \varphi)$
17:         **else**
18:             $A := \{i \mid (C_i \vee b_i) \in (\varphi_c \backslash \varphi_h)\}$ ▷New core
19:             $A := \bigcup_{\substack{A' \in cores(AL) \\ A' \cap A \neq \emptyset}} A'$ ▷New cover
20:             $k := newbound(AL \cup \varphi_w, A)$
21:             $AL := \{al \in AL \mid core(al) \neq A\} \cup \{\sum_{i \in A} w_i b_i \geq k\}$

Given a set AL, the set AM is the set of at-most constrains $\sum_{i \in A} w_i b_i \leq k$ such that $A \in SC(cores(AL))$ and k is the solution of minimizing $\sum_{i \in A} w_i b_i$ subject to AL and $b_i \in \{0, 1\}$.

The algorithm starts with $AL = \{w_1 b_1 \geq 0, \ldots, w_m b_m \geq 0\}$ and the corresponding $AM := \{w_1 b_1 \leq 0, \ldots, w_m b_m \leq 0\}$ that ensures that the unique solution of $AL \cup AM$ is $b_1 = \cdots = b_m = 0$ with cost 0^2. At every iteration, the algorithm calls a SAT solver with $\varphi_w \cup CNF(AL \cup AM)$. If it returns SAT, then the interpretation $\mathcal{I}$ is a MaxSAT solution of φ and we return the optimal cost $\mathcal{I}(\varphi)$. If it returns UNSAT, then we use the information of the unsatisfiable core φ_c obtained by the SAT solver to enlarge the set AL, excluding more interpretations on the b_i's that are not partial solutions of φ_w. Before calling again the SAT solver, we update AM conveniently, to ensure that solutions to the new constraints $AL \cup AM$ are still minimal solutions of the new AL constraint set. Notice that in every iteration the set of solutions of $\{b_1, \ldots, b_m\}$ defined by AL is decreased, whereas the set of solutions of AM is increased.

---

[2] In the implementation, we do not add a blocking variable to a soft clause till it appears into a core.

One key point in WPM2 is to compute the *newbound*(AL, A) (line 20) which corresponds to the following optimization problem:

$$\text{minimize} \sum_{i \in A} w_i \cdot b_i \quad \text{subject to } \{\sum_{i \in A} w_i \cdot b_i \geq k\} \cup AL \tag{1}$$

where $k = 1 + \sum \{k' \mid \sum_{i \in A'} w_i b_i \leq k' \in AM \wedge A' \subseteq A\}$.

Notice that by removing the AL constraints in (1), we get the subsetsum problem [13]. In the original WPM2 algorithm [5], the subsetsum problem is progressively solved until we get a solution that also satisfies the AL constraints. This satisfiability check in the original WPM2 is performed with a SAT solver.

In what follows, we present how we have modified the original WPM2 algorithm (fragments in gray in Algorithm 1) by incorporating several improvements: the application of a stratified approach, the hardening of soft clauses and the optimization of the subformulas defined by the covers.

3.1   Stratified Approach

As in [4] for WPM1, we apply a stratified approach. The stratified approach (lines 4, 10, 11 and 16) consists in sending to the SAT solver only those soft clauses with weight $w_i \geq w_{max}$. Then, when the SAT solver returns SAT, if there are still unsent clauses, we decrease w_{max} to include additional clauses to the formula. From [4], we also apply the diversity heuristic (line 16) which supplies us with an efficient method to calculate how we have to reduce the value of w_{max} in the stratified approach, so that, when there is a big variety of distinct weights, w_{max} decreases faster, and, when there is a low diversity, w_{max} is decreased to the following value of w_i. Similar approach with an alternative heuristic for grouping clauses can be found in [32].

3.2   Clause Hardening

The hardening of soft clauses in MaxSAT SAT-based solvers has been previously studied in [2, 11, 18, 23, 25, 30, 33]. Inspired by these works we study a hardening scheme for WPM2. While clause hardening was reported to have no positive effect in WPM1 [2], we will see that it boosts efficiency in WPM2.

The clause hardening (lines 14, 15 and 18) consists in considering hard those soft clauses whose satisfiability we know does not need to be reconsidered. We need some lemma ensuring that falsifying those soft clauses would lead us to suboptimal solutions. In the case of WPM1, all soft clauses satisfying $w_i > W$, where $W = \sum \{w_i \mid (C_i, w_i) \in \varphi \wedge w_i < w_{max}\}$ is the sum of weights of clauses not sent to the SAT solver, can be hardened. The correctness of this transformation is ensured by the following lemma:

Lemma 1 (Lemma 24 in [6])
Let $\varphi_1 = \{(C_1, w_1), \ldots, (C_m, w_m), (C_{m+1}, \infty), \ldots, (C_{m+m'}, \infty)\}$ be a MaxSAT formula with cost zero, let $\varphi_2 = \{(C'_1, w'_1), \ldots, (C'_r, w'_r)\}$ be a MaxSAT formula without hard clauses and $W = \sum_{j=1}^{r} w'_j$. Let

$$harden(w) = \begin{cases} w & if\ w \leq W \\ \infty & if\ w > W \end{cases}$$

and $\varphi_1' = \{(C_i, harden(w_i)) \mid (C_i, w_i) \in \varphi_1\}$. Then, $cost(\varphi_1 \cup \varphi_2) = cost(\varphi_1' \cup \varphi_2)$, and any optimal assignment for $\varphi_1' \cup \varphi_2$ is an optimal assignment of $\varphi_1 \cup \varphi_2$.

However, this lemma is not useful in the case of WPM2 because we do not proceed by transforming the formula, like in WPM1. Therefore, we generalize this lemma. For this, we need to introduce the notion of *optimal* of a formula.

Definition 2. *Given a MaxSAT formula* $\varphi = \{(C_1, w_1), \ldots, (C_m, w_m), (C_{m+1}, \infty), \ldots, (C_{m+m'}, \infty)\}$, *we say that* k *is a (possible) optimal of* φ *if there exists a subset* $A \subseteq \{1, \ldots, m\}$ *such that* $\sum_{i \in A} w_i = k$.

Notice that, for any interpretation I of the variables of φ, we have that $I(\varphi)$ is an optimal of φ. However, if k is an optimal, there does not exist necessarily an interpretation I satisfying $I(\varphi) = k$. Notice also that, given φ and k, finding the *next optimal*, i.e. finding the smallest $k' > k$ such that k' is an optimal of φ is equivalent to the subset sum problem.

Lemma 2. *Let* $\varphi_1 \cup \varphi_2$ *be a MaxSAT formula and* k_1 *and* k_2 *values such that:* $cost(\varphi_1 \cup \varphi_2) = k_1 + k_2$ *and any assignment* I *satisfies* $I(\varphi_1) \geq k_1$ *and* $I(\varphi_2) \geq k_2$. *Let* k' *be the smallest possible optimal of* φ_2 *such that* $k' > k_2$. *Let* φ_3 *be a set of soft clauses with* $W = \sum\{w_i \mid (C_i, w_i) \in \varphi_3\}$.

Then, if $W < k' - k_2$, *then any optimal assignment* I' *of* $\varphi_1 \cup \varphi_2 \cup \varphi_3$ *assigns* $I'(\varphi_2) = k_2$

Proof. Let I' be any optimal assignment of $\varphi_1 \cup \varphi_2 \cup \varphi_3$. On the one hand, as for any other assignment, we have $I'(\varphi_2) \geq k_2$.

On the other hand, any of the optimal assignments I of $\varphi_1 \cup \varphi_2$ can be extended (does not matter how) to the variables of $var(\varphi_3) \setminus var(\varphi_1 \cup \varphi_2)$, such that

$$I(\varphi_1 \cup \varphi_2 \cup \varphi_3) = I(\varphi_1) + I(\varphi_2) + I(\varphi_3) \leq k_1 + k_2 + W < k_1 + k' \quad (2)$$

Now, assume that $I'(\varphi_2) \neq k_2$, then $I'(\varphi_2) \geq k'$. As any other assignment, $I'(\varphi_1) \geq k_1$. Hence, $I'(\varphi_1 \cup \varphi_2 \cup \varphi_3) \geq k_1 + k' > I(\varphi_1 \cup \varphi_2 \cup \varphi_3)$, but this contradicts the optimality of I'. Therefore, $I'(\varphi_2) = k_2$.

$\square$

In order to apply this lemma we have to consider partitions of the formula $\varphi_1 \cup \varphi_2$ ensuring $cost(\varphi_1 \cup \varphi_2) = k_1 + k_2$ and $I(\varphi_1) \geq k_1$ and $I(\varphi_2) \geq k_2$, for any assignment I. This can be easily ensured, in the case of WPM2, if both φ_1 and φ_2 are unions of covers. Then, we only have to check if the next possible optimal k' of φ_2 exceeds the previous one k_2 more than the sum W of the weights of the clauses not sent to the SAT solver. In such a case, we can consider all soft clauses of φ_2 and their corresponding AM constraint with k_2 as hard clauses. In other words, we do not need to recompute the partial optimal k_2 of φ_2.

Finally, in line 15 of Algorithm 1, function $harden(\varphi, AM, W)$ returns the set of soft clauses φ_h that needs to be considered hard based on the previous analysis according to: the current set of covers AM, the next optimals of these covers and the sum of the weights W of soft clauses beyond the current w_{max}, i.e., not yet sent to the SAT solver.

3.3   Cover Optimization

As we have mentioned earlier, one key point in WPM2 is how to compute the $newbound(AL, A)$ (line 20). Actually, we can solve to optimality the subformulas defined by the union of the soft clauses related to the cover A and the hard clauses.

Definition 3. *Given a MaxSAT formula* $\varphi = \{(C_1, w_1), \ldots, (C_m, w_m), (C_{m+1}, \infty), \ldots, (C_{m+m'}, \infty)\}$ *and a set of indexes* A, *we define the subformula,* $\varphi[A]$, *as follows:* $\varphi[A] = \{(C_i, w_i) \in \varphi \mid i \in A \vee w_i = \infty)\}$

Solving to optimality $\varphi[A]$ give us the optimal value $k = cost(\varphi[A])$ for the AM constraint related to cover A. In order to do this, while taking advantage of the AL constraints generated so far, we only have to extend the minimization problem corresponding to the *newbound* (1) function, by adding φ_w to the constraints, i.e, $newbound(AL \cup \varphi_w, A)$[3]. Notice that $newbound(AL \cup \varphi_w, A) \geq newbound(AL, A)$.

In order to optimize $\varphi[A]$, we can use any exact approach related to MaxSAT, such as, MaxSAT branch and bound algorithms, MaxSAT SAT-based algorithms, saturation under the MaxSAT resolution rule, or we can use other solving techniques such as PB solvers or ILP techniques, etc. Our new WPM2 algorithm is parametric on any suitable optimization solving approach. In this work, we present three approaches.

The first and natural approach consists in iteratively refining (increasing) the lower bound on the optimal k for $\varphi[A]$ by applying the subsetsum function as in the original WPM2. The procedure stops when we satisfy the constraints $AL \cup \varphi_w$. Notice that since we have included φ_w into the set of constraints, the solution we will eventually get has to be optimal for $\varphi[A]$.

The second approach consists in iteratively refining (decreasing) the upper bound following the strategy applied in minisat+ [15], SAT4J [10], qmaxsat [21] or ShinMaxSat [20]. The upper bound ub is initially set to the sum of the weights w_i of the soft clauses in $\varphi[A]$. Then, we iteratively test whether $k = ub - 1$ is feasible or not. Whenever we get a satisfying assignment, we update ub to the sum of the weights w_i of those soft clauses where b_i evaluates to true under the satisfying assignment. If we get an unsatisfiable answer, the previous ub is the optimal value for $\varphi[A]$.

The third approach applies a binary search scheme [12, 16, 19]. We additionally refine the lower bound as in our first approach and the upper bound as in the second approach.

---

[3] We can actually exclude from φ_w all the soft clauses not in $\varphi[A]$.

The worst case complexity, in terms of the number of calls to the SAT solver, of the new WPM2 algorithm is the number of times that the newbound function is called (bounded by the number of clauses) multiplied by the number of SAT calls needed in each call to the newbound function. This latter number is logarithmic on the sum of the weights of the clauses of the core if we use a binary search, hence essentially the number of clauses. Therefore, the worst case complexity, when using a binary search to solve to optimality the subformulas, is quadratic on the number of soft clauses.

In order to see that the number of calls to the newbound function is bounded by the number of clauses we just need to recall that WPM2 merges the covers. Consider a binary tree where the soft clauses are the leaves, and the internal nodes represent the merges (calls to the newbound function). A binary tree of n leaves has n-1 internal nodes.

Solving to optimality all the covers can be very costly since these are NP-hard problems. Depending on the unsatisfiable cores we get in the general loop of the WPM2 algorithm some covers have to be merged. Therefore, we may argue that part of the work we did in order to optimize these covers can be useless[4]. For example, a reasonable strategy is to optimize the current cover only if it was not the result of merging other covers, i.e., when the last unsatisfiable core is contained into a cover. In the experimental evaluation, we will see that although the number of solved instances does not vary too much, the mean time for solving some families can be decreased.

4   Engineering Efficient SMT-Based MaxSAT Solvers

We have implemented both the last version of the WPM1 algorithm [2] and the revisited version of the WPM2 algorithm on top the of the SMT solver Yices.

As we have said, an SMT instance is a generalization of SAT where some propositional variables are replaced by predicates with predefined interpretations from background theories. Among the theories considered in the SMT library [9] we are interested in QF_LIA (Quantifier-Free *Linear Integer Arithmetic*). With the QF_LIA theory we can model the PB constraints that SAT-based MaxSAT algorithms generate during their execution. Therefore, for the SMT-based MaxSAT algorithm, we just need to replace the conversion to CNF (line 10 in Algorithm 1) by the proper linear integer arithmetic predicates.

As suggested in [16, 31], we can preserve some learned lemmas from previous iterations that may help to reduce the search space. In order to do that, we execute the SMT solver in incremental mode. Within this mode, we can call the solve routine and add new clauses (assertions) on demand, while preserving learned lemmas. However, notice that our algorithms delete parts of the formula between iterations. For example, in lines 7 to 9 of Algorithm 1 we recompute the set AM, possibly erasing some of the at-most constraints. Therefore, we have to take care also of any learned lemma depending on them.

---

[4] The related *AL* constraints can still be kept.

The SMT solver Yices gives the option of marking assertions as *retractable*. If the SMT solver does not support the deletion of assertions but supports the usage of assumptions, we can replace every retractable assertion C, with $a \rightarrow C$, where a is an assumption. Before each call, we activate the assumptions of assertions that have not been retracted by the algorithm. Notice that assertions that do have been retracted will have a pure literal ($\overline{a}$) such that a has not been activated. Therefore, the solver can safely set to false a deactivating the clause. Moreover, any learned lemma on those assertions will also include $\overline{a}$. For example, Z3 and Mathsat SMT solvers do not allow to delete clauses, but they allow the use of assumptions.

5   Experimental Results

In this section we present an intensive experimental investigation on the PMS and WPMS industrial and crafted instances from the 2012 MaxSAT Evaluation. We provide results for our new WPM2 SMT-based MaxSAT solver, for a WPM1 [2] SMT-based MaxSAT solver, the best two solvers for each category of the 2012 MaxSAT Evaluation, and three solvers which did not participate but the authors have reported to exhibit good performance. We run our experiments on a cluster featured with 2.27 GHz processors, memory limit of 3.9 GB and a timeout of 7200 seconds per instance.

The experimental results are presented in Tables 1 and 2 following the same classification criteria as in the 2012 MaxSAT Evaluation. For each solver and family of instances, we present the number of solved instances in parenthesis and the mean solving time. Solvers are ordered from left to right according to the total number of solved instances. The results for the best performing solver in each family are presented in bold. The number of instances of every family is specified in the column under the sign '#'. Since different families may have different number of instances, we also include for each solver the mean ratio of solved instances.

Our new WPM2 algorithm is implemented on top of the Yices SMT solver (version 1.0.29). The different versions of WPM2 and corresponding implementations are named $wpm2$ where subindexes can be $_s$ that stands for stratified approach with diversity heuristic and $_h$ for hardening. Regarding to how we perform the cover optimization, $_l$ stands for lower bound refinement based on subsetsum, $_u$ for upper bound refinement based on satisfying truth assignment, and $_b$ for binary search. Finally, $_a$ stands for optimizing all the covers and $_c$ for optimizing only covers that contain the last unsatisfiable core.

Table 1 shows our first experiment, where we evaluate the impact of each variation on the original $wpm2$. By using a stratified approach with the diversity heuristic ($wpm2_s$) we solve some additional instances in all categories having the best improvement in WPMS crafted. Overall, we solve 74 more instances. By adding hardening ($wpm2_{sh}$) we solve 66 more instances, mainly in WPMS industrial family *haplotyping-pedigrees*.

Table 1. Experimental results of different versions of *wpm2*

Instance set	#	wpm2	wpm2_s	wpm2_sh	wpm2_shlc	wpm2_shlq	wpm2_shbc	wpm2_shbq	wpm2_shuc	wpm2_shuq
PMS-Industrial										
aes	7	0.00(0)	0.00(0)	0.00(0)	0.00(0)	0.00(0)	0.00(0)	**10.34(1)**	1836.54(1)	514.14(1)
bcp-fir	59	83.29(57)	23.44(57)	23.44(57)	40.87(57)	104.75(57)	113.13(57)	146.03(57)	60.91(58)	**57.36(58)**
bcp-hipp-yRa1-simp	17	504.81(13)	155.95(12)	155.95(12)	95.16(12)	239.17(13)	151.30(13)	813.36(16)	107.24(16)	160.55(16)
bcp-hipp-yRa1-su	38	380.47(19)	164.70(16)	164.70(16)	181.49(18)	667.59(19)	226.66(25)	585.82(28)	555.92(33)	**315.87(34)**
bcp-msp	64	821.88(25)	756.59(26)	756.59(26)	283.50(28)	606.35(28)	283.78(31)	464.47(30)	711.08(36)	912.70(34)
bcp-mtg	40	1363.97(18)	852.58(23)	852.58(23)	1320.80(28)	786.09(34)	1296.44(32)	940.62(37)	997.38(35)	**578.75(39)**
bcp-syn	74	60.29(41)	302.54(41)	302.54(41)	296.12(41)	83.15(39)	83.15(39)	69.23(42)	103.21(43)	78.22(42)
circuit-trace-compaction	4	285.74(3)	835.72(4)	835.72(4)	230.85(3)	134.79(4)	145.72(4)	151.41(4)	**118.24(4)**	129.10(4)
haplotype-assembly	6	2.87(5)	7.17(5)	7.17(5)	9.80(5)	42.86(5)	15.53(5)	15.53(5)	18.28(4)	65.94(5)
pbo-mqc-nencdr	84	866.38(84)	821.16(84)	821.16(84)	142.43(84)	107.70(84)	130.90(84)	127.99(84)	245.77(84)	257.06(84)
pbo-mqc-nlogencdr	84	362.20(84)	353.14(84)	353.14(84)	24.48(84)	17.19(84)	58.42(84)	65.79(84)	124.43(84)	140.44(84)
pbo-routing	15	0.46(15)	2.14(15)	2.14(15)	3.89(15)	6.59(15)	4.88(15)	6.72(15)	5.42(15)	6.73(15)
protein-ins	12	2626.34(10)	2162.13(9)	2162.13(9)	476.03(12)	552.46(12)	360.76(12)	284.63(12)	**234.31(12)**	333.85(12)
Total	504	374 / 69.6%	376 / 70.9%	376 / 70.9%	387 / 72.5%	394 / 76.0%	404 / 77.5%	415 / 81.4%	425 / 81.7%	**428 / 83.6%**
WPMS-Industrial										
haplotyping-pedigrees	100	16.01(22)	292.29(25)	242.08(90)	154.12(92)	142.59(92)	199.61(96)	86.67(95)	202.70(98)	**176.02(98)**
timetabling	26	1017.27(8)	720.89(8)	651.84(8)	1008.59(9)	**430.67(9)**	1544.43(9)	931.17(8)	932.77(8)	1438.59(9)
upgradeability-problem	100	19.19(100)	19.67(100)	15.27(100)	96.88(100)	375.70(100)	99.51(100)	365.70(100)	97.83(100)	371.96(100)
Total	226	130 / 50.9%	133 / 51.9%	198 / 73.6%	201 / 75.5%	201 / 75.5%	205 / 76.9%	203 / 75.3%	206 / 76.3%	**207 / 77.5%**
Total Industrial	730	504 / 66.1%	509 / 67.3%	574 / 71.4%	588 / 73.0%	595 / 75.9%	609 / 77.4%	618 / 80.2%	631 / 80.7%	**635 / 82.5%**
PMS-Crafted										
frb	25	0.00(0)	0.00(0)	0.00(0)	0.00(0)	0.00(0)	0.00(0)	0.00(0)	0.00(0)	0.00(0)
job-shop	3	68.31(3)	63.37(3)	63.37(3)	50.84(3)	53.11(3)	51.29(3)	49.87(3)	88.41(3)	58.75(3)
maxclique-random	96	478.65(76)	471.97(76)	471.97(76)	435.07(77)	533.97(78)	423.89(83)	392.04(82)	565.44(87)	512.17(89)
maxclique-structured	62	779.71(21)	592.07(21)	592.07(21)	511.52(22)	783.37(24)	650.87(25)	838.38(26)	802.09(26)	**876.90(27)**
maxone-3sat	80	16.42(80)	18.11(80)	18.11(80)	10.90(80)	9.39(80)	5.34(80)	**5.29(80)**	6.65(80)	6.47(80)
maxone-structured	60	156.20(58)	89.15(59)	89.15(59)	14.42(60)	32.46(60)	13.11(60)	46.68(60)	**9.24(60)**	50.43(60)
min-enc-kbtree	42	1607.47(4)	707.46(4)	707.46(4)	2057.16(5)	2086.38(5)	2254.64(6)	1777.41(5)	**921.09(6)**	1906.23(5)
pseudo-miplib	4	122.23(4)	93.95(4)	93.95(4)	109.21(4)	64.35(4)	27.35(4)	37.81(4)	54.30(4)	34.23(4)
Total	372	246 / 64.9%	247 / 65.1%	247 / 65.1%	251 / 65.9%	254 / 66.5%	261 / 67.6%	260 / 67.4%	266 / 68.4%	**268 / 68.5%**
WPMS-Crafted										
auc-paths	86	0.00(0)	376.67(1)	370.25(1)	706.38(33)	546.57(33)	1019.93(74)	948.64(75)	435.71(82)	**271.16(82)**
auc-scheduling	84	0.00(0)	133.39(51)	136.89(51)	71.37(84)	69.59(84)	2.73(84)	1.66(84)	1.25(84)	**1.07(84)**
min-enc-planning	56	491.39(25)	141.80(38)	138.29(38)	3.24(56)	2.47(56)	0.75(56)	0.77(56)	0.74(56)	0.80(56)
min-enc-warehouses	18	11.36(1)	3.36(1)	3.43(1)	0.07(1)	0.08(1)	0.04(1)	0.05(1)	**1227.86(2)**	0.05(1)
pseudo-miplib	12	2936.02(3)	1533.63(3)	1487.40(3)	13.72(3)	11.81(3)	471.93(4)	990.65(5)	299.53(4)	**695.31(5)**
random-net	74	187.40(9)	0.00(0)	0.00(0)	1041.56(15)	1943.41(17)	1633.69(33)	**1529.94(33)**	3369.16(17)	3151.03(15)
wcsp-spot5-dir	21	1067.69(8)	310.45(10)	248.53(12)	420.10(14)	428.10(14)	495.17(14)	470.92(14)	113.90...	113.90(15)
wcsp-spot5-log	21	—	469.38(10)	120.60(9)	100.07(13)	59.16(13)	133.75(14)	48.91(13)	23.84(14)	**16.50(14)**
Total	372	114 / 31.9%	114 / 31.9%	115 / 32.5%	219 / 52.2%	221 / 52.6%	280 / 62.9%	**281 / 63.4%**	273 / 62.0%	271 / 62.0%
Total Crafted	744	292 / 42.2%	361 / 48.5%	362 / 48.8%	470 / 59.1%	475 / 59.5%	541 / 65.2%	**541 / 65.4%**	539 / 65.2%	539 / 65.3%
Total (W)PMS	1474	796 / 54.1%	870 / 57.9%	936 / 60.1%	1058 / 66.1%	1070 / 67.7%	1150 / 71.3%	1159 / 72.8%	1170 / 72.9%	**1174 / 73.9%**

Table 2. Experimental results of best *wpm2* version compared with other solvers

(a) Partial Industrial

Instance set	#	wpm2_shau	bincd2	qms0.21g2	pwbo2.1	shinms	wpm1	ilp
aes	7	514.44(1)	453.22(1)	3154.99(1)	68.10(56)	0.00(0)	3073.19(1)	**1310.95(3)**
bcp-fir	59	57.36(58)	44.09(58)	108.17(56)	174.98(15)	13.54(22)	10.87(57)	**62.86(59)**
bcp-hipp-yRa1_simp	17	160.55(16)	170.49(16)	**358.14(17)**	97.91(25)	40.66(16)	70.27(16)	666.94(6)
bcp-hipp-yRa1_su	38	315.87(34)	244.97(32)	**105.60(35)**	96.14(26)	282.26(34)	244.23(28)	0.00(0)
bcp-msp	64	912.79(34)	1.15(40)	451.50(30)	0.57(40)	281.37(22)	1053.47(7)	855.96(37)
bcp-mtg	40	578.75(39)	**213.47(38)**	283.64(35)	21.82(39)	0.60(40)	8.54(40)	769.27(29)
bcp-syn	74	78.22(42)	28.56(43)	**45.01(44)**	200.11(2)	86.98(33)	59.30(45)	**18.95(71)**
circuit-trace-compaction	4	129.10(4)	109.31(4)	153.22(5)	9.09(5)	52.22(4)	118.43(4)	6921.80(1)
haplotype-assembly	6	65.94(5)	728.30(5)	153.01(4)	0.00(0)	0.00(0)	**2.63(5)**	2124.72(5)
pbo-mqc-nencdr	84	257.06(84)	278.45(84)	**58.78(84)**	222.19(68)	145.71(84)	804.25(54)	1109.89(6)
pbo-mqc-nlogencdr	84	140.44(84)	78.58(84)	**23.69(84)**	71.86(82)	180.37(79)	403.38(55)	508.21(6)
pbo-routing	15	6.73(15)	1.14(15)	3.61(15)	27.67(15)	4.80(15)	1.75(15)	19.68(15)
protein-ins	12	333.85(12)	314.09(3)	**128.58(12)**	0.11(1)	206.51(4)	1812.03(3)	2.72(1)
Total	504	**428**	423	418	374	353	330	239
Mean ratio		**83.6%**	78.2%	83.0%	66.3%	63.6%	68.3%	48.9%

(b) Weighted Partial Industrial

Instance set	#	wpm2_shau	wpm1	pwbo2.1	bincd2	mazhs	ilp	shinms
haplotyping-pedigrees	100	**176.02(98)**	212.76(93)	123.00(87)	544.80(73)	1089.24(39)	1892.16(18)	1203.99(47)
timetabling	26	0.00(0)	1347.39(11)	—	168.55(8)	1249.85(6)	0.00(0)	2261.00(5)
upgradeability-problem	100	**371.96(100)**	4.57(100)	32.67(100)	76.40(100)	13.41(100)	19.26(100)	0.00(0)
Total	226	**207**	204	181	145	—	118	52
Mean ratio		77.5%	78.4%	71.3%	67.9%	54.0%	39.3%	22.1%

(c) Partial Crafted

Instance set	#	wpm2_shau	ilp	akms_ls	wpm1	qms.s0.21	pwbo2.1	shinms	mazhs	bincd2
frb	25	**346.73(25)**	1152.97(13)	159.47(5)	—	—	—	43.52(23)	—	—
job-shop	3	**41.51(3)**	0.00(0)	0.00(0)	1.56(3)	—	—	36.44(3)	58.75(3)	—
maxclique-random	96	269.93(83)	45.13(96)	**1.09(96)**	2.61(53)	—	—	339.30(76)	512.17(89)	76.40(100)
maxclique-structured	62	800.18(30)	326.51(38)	**281.51(41)**	78.89(17)	—	—	401.45(23)	876.90(27)	—
maxone-3sat	80	198.39(80)	13.12(80)	0.47(80)	157.33(4)	—	**6.35(80)**	694.52(78)	6.47(80)	—
maxone-structured	60	**6.35(60)**	337.60(59)	482.29(38)	**4.03(70)**	—	—	3.53(59)	50.43(60)	—
min-enc-kbtree	42	248.02(6)	**162.89(42)**	3199.18(34)	37.70(13)	—	—	513.97(5)	1906.23(5)	—
pseudo_miplib	4	**1.84(4)**	34.12(4)	258.91(3)	399.59(4)	—	—	3.44(4)	34.23(4)	—
Total	372	291	332	297	318	204	271	268	245	228
Mean ratio		**81.1%**	76.5%	63.2%	77.4%	71.3%	77.0%	68.5%	65.3%	59.3%

(d) Weighted Partial Crafted

Instance set	#	wpm1	ilp	akms_ls	shinms	wpm2_shau	pwbo2.1	mazhs	pwbo2.1	bincd2
auc-paths	86	**0.49(86)**	0.38(84)	2.57(86)	317.62(84)	271.16(84)	110.67(19)	35.41(86)	0.00(0)	1414.73(12)
auc-scheduling	84	**0.38(84)**	1.56(84)	68.27(84)	5.81(84)	1.07(84)	7.65(81)	965.10(78)	141.77(81)	—
min-enc-planning	56	296.52(56)	2.61(53)	20.11(2)	8.15(52)	0.80(56)	**0.46(56)**	459.18(31)	32.74(54)	—
min-enc-warehouses	18	**0.49(18)**	78.89(17)	141.21(40)	0.43(1)	0.05(1)	3.78(14)	0.18(1)	2.10(1)	—
pseudo-miplib	12	82.82(3)	157.33(4)	0.26(2)	**127.99(5)**	695.31(5)	3.97(3)	0.03(1)	1072.69(4)	3151.03(15)
random-net	74	532.74(59)	**4.03(70)**	4060.60(8)	0.00(0)	113.90(14)	42.15(35)	2770.76(10)	0.00(0)	—
wcsp-spot5-dir	21	42.88(4)	37.70(13)	1555.46(6)	**743.63(21)**	16.50(14)	61.89(8)	101.11(6)	127.73(12)	—
wcsp-spot5-log	21	322.93(8)	399.59(14)	108.70(5)	**200.73(17)**	—	1.71(6)	357.32(6)	299.32(13)	—
Total	372	**332**	318	233	264	271	222	222	219	177
Mean ratio		**78.6%**	77.4%	45.3%	64.8%	62.0%	54.4%	41.6%	59.3%	45.6%

Table 3. Summary of solved instances and mean ratio % for best solvers

solvers	pms	wpms	Ind.	pms	wpms	Cra.	Total
	428	**207**	**635**	268	271	539	**1174**
$wpm2_{shua}$	**83.6 %**	77.5 %	**82.5 %**	68.5 %	62.0 %	65.3 %	**73.9 %**
	330	204	534	207	318	525	1059
$wpm1$	68.3 %	**78.4 %**	70.2 %	58.3 %	77.4 %	67.9 %	69.0 %
	423	181	604	245	177	422	1026
$bincd2$	78.2 %	67.9 %	76.3 %	65.3 %	45.6 %	55.5 %	65.9 %
	239	118	357	**332**	**332**	**664**	1021
ilp	48.9 %	39.3 %	47.1 %	76.5 %	**78.6 %**	**77.6 %**	62.3 %
	374	194	568	228	222	450	1018
$pwbo2.1$	66.3 %	71.3 %	67.2 %	59.3 %	54.5 %	56.9 %	62.1 %
	353	52	405	271	264	535	940
$shinms$	63.6 %	22.1 %	55.8 %	77.0 %	64.8 %	70.9 %	63.4 %
	418			291			
qms	83.0 %			**81.1 %**			

Regarding our three approaches for optimizing the covers, we can see that by optimizing with subsetsum ($wpm2_{shla}$) we solve some additional instances in all categories having the best improvement in WPMS industrial with 18 more and in WPMS crafted with 106 more. It is important to highlight that optimizing covers with subsetsum, instead of applying the subsetsum as in the original WPM2 algorithm, leads to a total improvement of 134 additional solved instances, with respect to $wpm2_{sh}$.

Optimizing all covers by refining the upper bound ($wpm2_{shua}$), we get an additional boost with respect to $wpm2_{shla}$. We can see that we solve some additional instances in all categories. We get the best improvement for PMS industrial, solving 34 additional instances, and for WPMS crafted, 50 more. Notice that the overall increase with respect to $wpm2_{sh}$ is of 238 additional solved instances.

Binary search ($wpm2_{shba}$) improves 10 instances in WPMS crafted with respect to $wpm2_{shua}$. But the global performance with respect to $wpm2_{sh}$, 223, is not as good as only refining the upper bound ($wpm2_{shua}$).

Optimizing only covers that contain the last unsatisfiable core solves almost the same instances as optimizing all covers but improves the average running time in the WPMS industrial family *upgradeability-problem* by a factor of 4.

Table 2 shows the results of our second experiment where we compare the best variation and implementation of our new WPM2 algorithm ($wpm2_{shua}$) with several solvers. In particular, we compare with the best two solvers for the PMS and WPMS industrial and crafted instances of the 2012 MaxSAT Evaluation: PMS industrial ($qms0.21g2$, $pwbo2.1$), WPMS industrial ($pwbo2.1$ [31, 32], $wpm1$ [2][5]), PMS crafted ($qms0.21$ [21], $akms\_ls$ [22] and WPMS crafted ($wpm1$, $shinms$ [20]). We also compare with three additional MaxSAT solvers: $bincd2$, which is the new version of the BINCD algorithm [19] described in [33], with the best configuration reported by authors, $maxhs$ from [14], which consists in an hybrid SAT-ILP approach, and ilp, which translates WPMS into ILP and applies the MIP solver IBM-CPLEX studio124 [7].

---

[5] We present in this paper a version implemented on top of the Yices SMT solver.

Table 2(a) presents the results for the PMS industrial instances. Our $wpm2_{shua}$ is the first one in solved instances with 428 and mean ratio with 93.6%, closely followed by $bincd2$ and $qms0.21g2$.

Table 2(b) presents the results for the WPMS industrial instances. As we can see, our $wpm2_{shua}$ and $wpm1$ dominate this category with 207 and 204 solved instances and 77.5% and 78.4% mean ratio, resp.

As a summary of industrial instances, we can conclude that our $wpm2_{shua}$ is the best performing solver with a total of 635 solved instances, followed by $bincd2$ with a total of 604. We do not have results for any version of qms since it only works for PMS instances. The closest solver to the search scheme of qms would be $shinms$ but it does not perform well for WPMS industrial.

Table 2(c) presents the results for the PMS crafted instances. The ilp approach solves 332 of 372 instances, 35 more than $akms\_ls$. This is remarkable since branch and bound solvers, like $akms\_ls$, have always dominated this category since 2006. PMS solver $qms0.21$ is the third in solved instances but the first in mean ratio with 81.1%. Our $wpm2_{shua}$ is the fifth in solved instances with 268 and the fourth in mean ratio with 68.5%.

Table 2(d) presents the results for the WPMS crafted instances. Again, the ilp approach is the best one, solving 332 of 372 instances, 14 more than the second one, $wpm1$. Our $wpm2_{shua}$ is the third in solved instances with 271 and the fourth in mean ratio with 62.0%.

As a summary of crafted instances, we can conclude that ilp is the best performing approach, and our $wpm2_{shua}$ is the second in total solved instances.

In Table 3 we can see a summary of the solved instances and mean ratio per category for best solvers. We recall that all solvers accept weights except qms that is only for PMS. Our $wpm2_{shua}$ is the first in solved instances for both PMS industrial and WPMS industrial. In crafted categories it is the second in total solved instances. However, for both PMS crafted and WPMS crafted categories ilp is the first in solved instances. We can conclude that our $wpm2_{shau}$ is the most robust solver across all four PMS and WPMS industrial and crafted categories, followed by $wpm1$ and $bincd2$.

6   Conclusions and Future Work

From the experimental evaluation, we conclude that the new WPM2 solver is the best performing solver for PMS and WPMS industrial instances and the best on the union of PMS and WPMS industrial and crafted instances. In particular, we have shown that solving to optimality the subformulas defined by covers really works in practice. As future work, we will study how to improve the interaction with the optimization of the subformulas. A portfolio that selects the most suitable optimization approach depending on the structure of the subformula seems another way of achieving additional speed-ups. Finally, we have also shown that SMT technology is an underlying efficient technology for solving the MaxSAT problem.

References

1. Ansótegui, C., Bofill, M., Palahí, M., Suy, J., Villaret, M.: A Proposal for Solving Weighted CSPs with SMT. In: Proceedings of the 10th International Workshop on Constraint Modelling and Reformulation (ModRef 2011), pp. 5–19 (2011)
2. Ansótegui, C., Bonet, M.L., Gabàs, J., Levy, J.: Improving sat-based weighted maxsat solvers. In: Milano, M. (ed.) CP 2012. LNCS, vol. 7514, pp. 86–101. Springer, Heidelberg (2012)
3. Ansótegui, C., Bonet, M.L., Levy, J.: On solving MaxSAT through SAT. In: Proc. of the 12th Int. Conf. of the Catalan Association for Artificial Intelligence (CCIA 2009), pp. 284–292 (2009)
4. Ansótegui, C., Bonet, M.L., Levy, J.: Solving (weighted) partial MaxSAT through satisfiability testing. In: Kullmann, O. (ed.) SAT 2009. LNCS, vol. 5584, pp. 427–440. Springer, Heidelberg (2009)
5. Ansotegui, C., Bonet, M.L., Levy, J.: A new algorithm for weighted partial maxsat. In: Proc. the 24th National Conference on Artificial Intelligence (AAAI 2010) (2010)
6. Ansótegui, C., Bonet, M.L., Levy, J.: Sat-based maxsat algorithms. Artif. Intell. 196, 77–105 (2013)
7. Ansotegui, C., Gabas, J.: Solving maxsat with mip. In: CPAIOR (2013)
8. Bailleux, O., Boufkhad, Y., Roussel, O.: New encodings of pseudo-boolean constraints into CNF. In: Kullmann, O. (ed.) SAT 2009. LNCS, vol. 5584, pp. 181–194. Springer, Heidelberg (2009)
9. Barrett, C., Stump, A., Tinelli, C.: The Satisfiability Modulo Theories Library (SMT-LIB) (2010), http://www.SMT-LIB.org
10. Berre, D.L.: Sat4j, a satisfiability library for java (2006), http://www.sat4j.org
11. Borchers, B., Furman, J.: A two-phase exact algorithm for max-sat and weighted max-sat problems. J. Comb. Optim. 2(4), 299–306 (1998)
12. Cimatti, A., Franzén, A., Griggio, A., Sebastiani, R., Stenico, C.: Satisfiability modulo the theory of costs: Foundations and applications. In: Esparza, J., Majumdar, R. (eds.) TACAS 2010. LNCS, vol. 6015, pp. 99–113. Springer, Heidelberg (2010)
13. Cormen, T.H., Leiserson, C.E., Rivest, R.L., Stein, C.: Introduction to Algorithms, 3rd edn. MIT Press (2009)
14. Davies, J., Bacchus, F.: Solving MAXSAT by solving a sequence of simpler SAT instances. In: Lee, J. (ed.) CP 2011. LNCS, vol. 6876, pp. 225–239. Springer, Heidelberg (2011)
15. Eén, N., Sörensson, N.: Translating pseudo-boolean constraints into SAT. JSAT 2(1-4), 1–26 (2006)
16. Fu, Z., Malik, S.: On solving the partial MAX-SAT problem. In: Biere, A., Gomes, C.P. (eds.) SAT 2006. LNCS, vol. 4121, pp. 252–265. Springer, Heidelberg (2006)
17. Heras, F., Larrosa, J., Oliveras, A.: MiniMaxSat: A new weighted Max-SAT solver. In: Marques-Silva, J., Sakallah, K.A. (eds.) SAT 2007. LNCS, vol. 4501, pp. 41–55. Springer, Heidelberg (2007)
18. Heras, F., Larrosa, J., Oliveras, A.: Minimaxsat: An efficient weighted max-sat solver. J. Artif. Intell. Res (JAIR) 31, 1–32 (2008)
19. Heras, F., Morgado, A., Marques-Silva, J.: Core-guided binary search algorithms for maximum satisfiability. In: Proc. the 25th National Conference on Artificial Intelligence (AAAI 2011) (2011)

20. Honjyo, K., Tanjo, T.: Shinmaxsat, a Weighted Partial Max-SAT solver inspired by MiniSat+, Information Science and Technology Center, Kobe University
21. Koshimura, M., Zhang, T., Fujita, H., Hasegawa, R.: Qmaxsat: A partial max-sat solver. JSAT 8(1/2), 95–100 (2012)
22. Kügel, A.: Improved exact solver for the weighted max-sat problem (to appear)
23. Larrosa, J., Heras, F., de Givry, S.: A logical approach to efficient max-sat solving. Artif. Intell. 172(2-3), 204–233 (2008)
24. Li, C.M., Manyà, F., Mohamedou, N., Planes, J.: Exploiting cycle structures in Max-SAT. In: Kullmann, O. (ed.) SAT 2009. LNCS, vol. 5584, pp. 467–480. Springer, Heidelberg (2009)
25. Li, C.M., Manyà, F., Planes, J.: New inference rules for Max-SAT. J. Artif. Intell. Res (JAIR) 30, 321–359 (2007)
26. Lin, H., Su, K.: Exploiting inference rules to compute lower bounds for Max-SAT solving. In: IJCAI 2007, pp. 2334–2339 (2007)
27. Lin, H., Su, K., Li, C.M.: Within-problem learning for efficient lower bound computation in Max-SAT solving. In: Proc. the 23rd National Conference on Artificial Intelligence (AAAI 2008), pp. 351–356 (2008)
28. Manquinho, V., Marques-Silva, J., Planes, J.: Algorithms for weighted boolean optimization. In: Kullmann, O. (ed.) SAT 2009. LNCS, vol. 5584, pp. 495–508. Springer, Heidelberg (2009)
29. Manquinho, V.M., Martins, R., Lynce, I.: Improving unsatisfiability-based algorithms for boolean optimization. In: Strichman, O., Szeider, S. (eds.) SAT 2010. LNCS, vol. 6175, pp. 181–193. Springer, Heidelberg (2010)
30. Marques-Silva, J., Argelich, J., Graça, A., Lynce, I.: Boolean lexicographic optimization: algorithms & applications. Ann. Math. Artif. Intell. 62(3-4), 317–343 (2011)
31. Martins, R., Manquinho, V.M., Lynce, I.: Exploiting cardinality encodings in parallel maximum satisfiability. In: ICTAI, pp. 313–320 (2011)
32. Martins, R., Manquinho, V., Lynce, I.: Clause sharing in parallel MaxSAT. In: Hamadi, Y., Schoenauer, M. (eds.) LION 2012. LNCS, vol. 7219, pp. 455–460. Springer, Heidelberg (2012)
33. Morgado, A., Heras, F., Marques-Silva, J.: Improvements to core-guided binary search for MaxSAT. In: Cimatti, A., Sebastiani, R. (eds.) SAT 2012. LNCS, vol. 7317, pp. 284–297. Springer, Heidelberg (2012)
34. Nieuwenhuis, R., Oliveras, A.: On SAT modulo theories and optimization problems. In: Biere, A., Gomes, C.P. (eds.) SAT 2006. LNCS, vol. 4121, pp. 156–169. Springer, Heidelberg (2006)
35. Sebastiani, R.: Lazy Satisfiability Modulo Theories. Journal on Satisfiability, Boolean Modeling and Computation 3(3-4), 141–224 (2007)
36. Silva, J.P.M., Sakallah, K.A.: Grasp: A search algorithm for propositional satisfiability. IEEE Trans. Computers 48(5), 506–521 (1999)

MinSAT versus MaxSAT for Optimization Problems

Josep Argelich[1], Chu-Min Li[2], Felip Manyà[3], and Zhu Zhu[2]

[1] Dept. of Computer Science, Universitat de Lleida, Lleida, Spain
[2] MIS, Université de Picardie Jules Verne, Amiens, France
[3] Artificial Intelligence Research Institute (IIIA, CSIC), Bellaterra, Spain

Abstract. Despite their similarities, MaxSAT and MinSAT use different encodings and solving techniques to cope with optimization problems. In this paper we describe a new weighted partial MinSAT solver, define original MinSAT encodings for relevant combinatorial problems, propose a new testbed for evaluating MinSAT, report on an empirical investigation comparing MinSAT with MaxSAT, and provide new insights into the duality between MinSAT and MaxSAT.

1 Introduction

MinSAT is the problem of finding a truth assignment that minimizes the number of satisfied clauses in a CNF formula, and MaxSAT is the problem of finding a truth assignment that maximizes the number of satisfied clauses. When hard and soft clauses, and weights are considered, the problem is known as weighted partial MinSAT/MaxSAT.

The promising results on MaxSAT as a generic approach to solving combinatorial optimization problems [6,16] led us to investigate the opportunities that MinSAT offers in optimization [19,20]. At first sight, it may seem that MaxSAT and MinSAT are so close that it does not pay off to devote efforts to MinSAT, but this is not completely certain. In [19,20], branch-and-bound MinSAT solvers apply upper bounding techniques not applicable in MaxSAT solvers, and this provides a competitive advantage to MinSAT, which solves MaxClique and combinatorial auction instances (that are beyond the reach of current MaxSAT solvers) even faster than using dedicated algorithms. Interestingly, the MinSAT and MaxSAT encodings of these problems are almost identical (the only difference is that the literals in the soft clauses, which are unitary, have opposite polarity). The superiority of MinSAT in [19,20] is due to the MinSAT solving techniques, not to the encoding. The compared MinSAT and MaxSAT solvers use the same data structures but implement different techniques. Since performance depends on both solvers and encodings, our first goal is to define suitable MinSAT encodings of NP-hard problems, and improve the performance by using more efficient encodings.

All the interesting genuine MinSAT encodings defined so far, except for random Min-kSAT instances, only contain unit clauses in the soft part. In the literature we can find experimental investigations solving instances with non-unit soft clauses, but the optimization problem that is actually being solved is a genuine or transformed MaxSAT instance. The first contribution of the paper is the definition of genuine MinSAT encodings of practical optimization problems, containing non-unit soft clauses, that are completely different from their MaxSAT counterparts, and a comparison that provides empirical evidence of the gains that could be achieved. More specifically, we define

C. Schulte (Ed.): CP 2013, LNCS 8124, pp. 133–142, 2013.
© Springer-Verlag Berlin Heidelberg 2013

novel encodings from weighted MaxCSP to MinSAT that fulfill our requirements. We use weighted MaxCSP as a good source of benchmarks for MinSAT solvers, but not with the aim of competing with weighted MaxCSP solvers in their own territory.

The second contribution is the incorporation of the inference rules defined in [17] into MinSatz [19,20]. It turns out that MaxSAT inference rules can be applied to Min-SAT too, because they preserve the cost distribution over all the models. As a result, we have implemented a new version of MinSatz that incorporates inference rules, and provided empirical evidence that inference rules may produce significant speedups.

Last but not least, we would like to highlight that MinSAT solving is a novel and emerging technology with a remarkable potential in optimization. It allows to look at problems from a different perspective because, in contrast to existing Boolean optimization approaches, it works by maximizing the number of violated constraints instead of working by minimizing that number. This fact leads to the definition of different encodings and solving techniques that exploit the duality between MinSAT and MaxSAT.

Related Work: The work on MinSAT for solving optimization problems may be divided into three categories: (I) Transformation between MinSAT and MaxSAT: Reductions from MinSAT to PMaxSAT were defined in [18], but these reductions do not generalize to WPMinSAT. This drawback was overcome with the definition of the natural encoding [13], which was improved in [23]. Reductions of WPMinSAT to Group MaxSAT were evaluated in [11]. (II) Branch-and-bound solvers: The only existing WPMinSAT solver, MinSatz [19,20], is based on MaxSatz [17], and implements upper bounds that exploit clique partition algorithms and MaxSAT technology. (III) SAT-based solvers: There exist two WPMinSAT solvers of this class [2,11]. The main difference with SAT-based MaxSAT solvers lies in the way of relaxing soft clauses.

2   Preliminaries

A weighted clause is a pair (c, w), where c is a is a disjunction of literals and w, its weight, is a natural number or infinity. A clause is hard if its weight is infinity (for simplicity, we omit infinity weights); otherwise it is soft. A Weighted Partial Min-SAT (MaxSAT) instance is a multiset of hard clauses and weighted soft clauses. A truth assignment assigns to each propositional variable either 0 or 1. The Weighted Partial MinSAT (MaxSAT) problem, or WPMinSAT (WPMaxSAT), for an instance ϕ consists in finding an assignment in which the sum of weights of the satisfied (falsified) soft clauses is minimal, and all the hard clauses are satisfied. The Weighted Min-SAT (MaxSAT) problem, or WMinSAT (WMaxSAT), is the WPMinSAT (WPMaxSAT) problem when there are no hard clauses. The Partial MinSAT (MaxSAT) problem, or PMinSAT (PMaxSAT), is the WPMinSAT (WPMaxSAT) problem when all the soft clauses have the same weight. The MinSAT (MaxSAT) problem is the Partial MinSAT (MaxSAT) problem when there are no hard clauses.

A CSP instance is a triple $\langle \mathcal{X}, \mathcal{D}, \mathcal{C} \rangle$, where $\mathcal{X} = \{X_1, \ldots, X_n\}$ is a set of variables, $\mathcal{D} = \{d(X_1), \ldots, d(X_n)\}$ is a set of finite domains, and $\mathcal{C} = \{C_1, \ldots, C_m\}$ is a set of constraints. Each $C_i = \langle S_i, R_i \rangle$ in $\mathcal{C}$ is a relation R_i over a subset of $S_i = \{X_{i_1}, \ldots, X_{i_k}\} \subseteq \mathcal{X}$, called the scope. R_i may be represented extensionally as a subset of the Cartesian product $d(X_{i_1}) \times \cdots \times d(X_{i_k})$. The tuples belonging to R_i represent the allowed values and are called goods, and the rest of tuples represent the forbidden values and are called nogoods. An assignment v for a CSP instance $\langle \mathcal{X}, \mathcal{D}, \mathcal{C} \rangle$ is a mapping that assigns to every variable $X_i \in \mathcal{X}$ an element $v(X_i) \in d(X_i)$. It satisfies a constraint $\langle \{X_{i_1}, \dots, X_{i_k}\}, R_i \rangle \in C$ iff $\langle v(X_{i_1}), \dots, v(X_{i_k}) \rangle \in R_i$. The Constraint Satisfaction Problem (CSP) for an instance P consists in finding a satisfying assignment for P.

A Weighted MaxCSP (WMaxCSP) instance is defined as a triple $\langle \mathcal{X}, \mathcal{D}, \mathcal{C} \rangle$, where $\mathcal{X}$ and $\mathcal{D}$ are variables and domains as in CSP, and $\mathcal{C}$ is a set of weighted constraints. A weighted constraint $\langle c, w \rangle$ is just a classical constraint c plus a weight w over natural numbers. The cost of an assignment v is the sum of the weights of all constraints violated by v. An optimal solution is an assignment with minimal cost. In the particular case where all the constraints have the same weight, it is called the MaxCSP problem.

3 Encodings from WMaxCSP to WPMinSAT

We define the MinSAT counterparts of the direct, minimal support and interval-based minimal support encodings from WMaxCSP to WPMaxSAT defined in [5]. All these encodings are correct: solving a WMaxCSP instance is equivalent to solving the WPMinSAT instance derived by any of our encodings. Other variants and encodings could be tried. We selected them because they were superior in most of our tests. We assume binary constraints, but the direct encoding is valid for non-binary constraints too.

Direct Encoding. A Boolean variable x_i is associated with each value i that the CSP variable X can take. If the domain $d(X)$ has size m, the *ALO* clause of X is $x_1 \lor \cdots \lor x_m$, and ensures that X is given a value. The *AMO* clauses of X are the set of clauses $\{\neg x_i \lor \neg x_j | i, j \in d(X), i < j\}$, and ensure that X takes no more than one value.

Definition 1. *The direct encoding of a WMaxCSP instance $\langle \mathcal{X}, \mathcal{D}, \mathcal{C} \rangle$ is the WPMinSAT instance that contains as hard clauses the ALO and AMO clauses for every CSP variable in $\mathcal{X}$, and a soft clause $(\neg x_i \lor \neg y_j, w)$ for every good $(X = i, Y = j)$ of every weighted constraint $\langle C, w \rangle$ of $\mathcal{C}$ with scope $\{X, Y\}$.*

Since PMinSAT maximizes the number of falsified soft clauses, the idea behind the PMinSAT encoding of a MaxCSP instance is that we want to force the violation of one soft clause for every satisfied constraint, because of that we negate the goods instead of the nogoods as is done for PMaxSAT encodings of MaxCSP instances in [5]. If an interpretation is compatible with a good of a constraint, then one soft clause of the clauses encoding the constraint is falsified, and if it is not compatible with any good, then all the clauses are satisfied. The idea is the same when weights are added.

Example 1. Let $\langle \mathcal{X}, \mathcal{D}, \mathcal{C} \rangle$ be the WMaxCSP instance where $\mathcal{X} = \{X, Y\}$, $d(X) = d(Y) = \{1, 2, 3\}$, $\mathcal{C} = \{X = Y\}$, and the weight of $X = Y$ is 3. The direct encoding from WMaxCSP to WPMinSAT is formed by the hard clauses encoding the ALO and AMO conditions of X and Y, and the soft clauses $(\neg x_1 \lor \neg y_1, 3)$, $(\neg x_2 \lor \neg y_2, 3)$, and $(\neg x_3 \lor \neg y_3, 3)$. In contrast to the WPMaxSAT direct encoding, which needs a quadratic number of binary soft clauses in the domain size for encoding the equality constraint, the WPMinSAT direct encoding just needs a linear number of binary clauses. So, this example shows that certain constraints can be more compactly defined with MinSAT.

Minimal Support Encoding. In the support encoding from CSP to SAT [12,9,22], besides the ALO and AMO clauses, there are clauses that encode the support for a value instead of encoding conflicts. The support for a value i of a variable X across a binary constraint with scope $\{X, Y\}$ is the set of values of Y which allow $X = i$. If $v_1, \ldots, v_k$ are the supporting values of variable Y for $X = i$, the clause $\neg x_i \vee y_{v_1} \vee \cdots \vee y_{v_k}$ (called support clause) is added. There is one support clause for each value in the domain and for each pair of variables X, Y involved in a constraint. In the support encoding, a clause in each direction is used: one for the pair X, Y and one for Y, X, while in the minimal support encoding [5], for every constraint with scope $\{X, Y\}$, the added clauses are the support clauses either for all the domain values of X or for all the domain values of Y. The minimal support encoding does not maintain arc consistency through unit propagation but usually is better for SAT solvers with learning and for MaxSAT solvers [5]. In our MinSAT setting, we focus on the minimal support encoding because it greatly outperformed the support encoding in our tests. Before defining the encoding, we need to define the *negative support* for a value j of a CSP variable X across a binary constraint with scope $\{X, Y\}$ as the set of values of Y which forbid $X = j$.

Definition 2. *The minimal support encoding of a WMaxCSP instance $\langle \mathcal{X}, \mathcal{D}, \mathcal{C} \rangle$ is the WPMinSAT instance that contains as hard clauses the ALO and AMO clauses for every CSP variable in $\mathcal{X}$ and, for every constraint with scope $\{X, Y\}$ and weight w, either the soft clause $(\neg x_i \vee y_{v_1} \vee \cdots \vee y_{v_n}, w)$ for every value $i \in d(X)$, where $v_1, \ldots, v_n$ is the negative support for value i, or the soft clause $(\neg y_j \vee x_{u_1} \vee \cdots \vee x_{u_m}, w)$ for every value $j \in d(Y)$, where $u_1, \ldots, u_m$ is the negative support for value j.*

The main difference with the minimal support encoding from WMaxCSP to WP-MaxSAT [5] is that the support clauses now include the negative support instead of the positive support. This allows the violation of a soft clause for every satisfied constraint. In the experiments, we select the variable that produces support clauses of smaller size. To this end, we give a score of 16 to unit clauses, a score of 4 to binary clauses and a score of 1 to ternary clauses, and select the variable with higher sum of scores.

Example 2. Let $\langle \mathcal{X}, \mathcal{D}, \mathcal{C} \rangle$ be the MaxCSP instance where $\mathcal{X} = \{X, Y\}$, $d(X) = d(Y) = \{1, 2, 3\}$, $\mathcal{C} = \{X \neq Y\}$, and the constraint weight is 2. The minimal support encoding from WMaxCSP to WPMinSAT is formed by the ALO and AMO hard clauses of X and Y, and the soft clauses $(\neg x_1 \vee y_1, 2)$, $(\neg x_2 \vee y_2, 2)$, and $(\neg x_3 \vee y_3, 2)$. Notice that we could also define the minimal support encoding by replacing the previous soft clauses with $(x_1 \vee \neg y_1, 2)$, $(x_2 \vee \neg y_2, 2)$, and $(x_3 \vee \neg y_3, 2)$ if we add the support clauses for the domain values of Y instead of the domain values of X. The length of the soft clauses of the minimal support WPMaxSAT encoding is linear in the domain size for the inequality constraint whereas in MinSAT all the soft clauses are binary.

If we replace $\mathcal{C} = \{X \neq Y\}$ with $\mathcal{C} = \{X \leq Y\}$, we can derive the minimal support encoding containing the soft clauses $(\neg x_1, 2)$, $(\neg x_2 \vee y_1, 2)$, and $(\neg x_3 \vee y_1 \vee y_2, 2)$. Interestingly, we derive a unit clause that can be very useful for applying inference rules. In the minimal support encoding from WMaxCSP to WPMaxSAT no unit clause is derived. It contains the soft clauses $(\neg x_2 \vee y_2 \vee y_3, 2)$, and $(\neg x_3 \vee y_3, 2)$.

Interval-Based Minimal Support Encoding. An alternative for modeling a CSP variable X is the use of Boolean variables of the form $x_i^{\geq}$ (instead of $x = i$), assuming a total ordering on the domain. Variables of the form $x_i^{\geq}$ are called regular variables, and their intended meaning is that $x_i^{\geq}$ is true iff $X \geq i$ [7,3,10]. For simplicity, we assume here that domains are subsets of natural numbers.

One advantage of introducing regular literals (i.e.; regular variables and their negations) is that the number of clauses needed for each CSP variable is linear rather than quadratic in the domain size, they may produce more compact encodings, and are particularly useful for dealing with large domains. We focus on the interval-based minimal support encoding [5] because it outperformed other encodings with regular literals.

Given a MinSAT support clause $\neg x_j \vee y_{v_1} \vee y_{v_2} \vee \cdots \vee y_{v_k}$, the negative support of variable Y for $X = j$ is now encoded in intervals using regular literals. Let us see an example: If the domain of Y is $\{1, 2, \ldots, 10\}$ and we are given the support clause $\neg x_2 \vee y_2 \vee y_3 \vee y_6 \vee y_8 \vee y_9$, then the negative support is represented by the following intervals: $[2, 3], [6, 6]$, and $[8, 9]$. The interval-based encoding for this clause is as follows: $x_2^{\geq} \wedge \neg x_3^{\geq} \rightarrow y_2^{\geq}, x_2^{\geq} \wedge \neg x_3^{\geq} \wedge y_4^{\geq} \rightarrow y_6^{\geq}, x_2^{\geq} \wedge \neg x_3^{\geq} \wedge y_7^{\geq} \rightarrow y_8^{\geq}, x_2^{\geq} \wedge \neg x_3^{\geq} \rightarrow \neg y_{10}^{\geq}$. Notice that the number of clauses needed for each support clause is linear in the largest domain size, and each clause has at most four regular literals.

Definition 3. *The interval-based minimal support encoding of a WMaxCSP instance $\langle \mathcal{X}, \mathcal{D}, \mathcal{C} \rangle$ is the WPMinSAT instance obtained from the minimal support encoding by (i) replacing the ALO and AMO clauses for every variable $X \in \mathcal{X}$ with domain size m with the clauses $x_m^{\geq} \rightarrow x_{m-1}^{\geq}, x_{m-1}^{\geq} \rightarrow x_{m-2}^{\geq}, \ldots, x_3^{\geq} \rightarrow x_2^{\geq}$; and (ii) replacing every support clause with the corresponding interval-based regular clauses using the negative support; i.e, if the negative support of variable Y for $X = i$ can be represented by the intervals $[l_1, u_1], [l_2, u_2], \ldots [l_k, u_k]$, we add the clauses $x_i^{\geq} \wedge \neg x_{i+1}^{\geq} \rightarrow y_{l_1}^{\geq}, x_i^{\geq} \wedge \neg x_{i+1}^{\geq} \wedge y_{u_1+1}^{\geq} \rightarrow y_{l_2}^{\geq}, \ldots, x_i^{\geq} \wedge \neg x_{i+1}^{\geq} \wedge y_{u_{k-1}+1}^{\geq} \rightarrow y_{l_k}^{\geq}, x_i^{\geq} \wedge \neg x_{i+1}^{\geq} \rightarrow \neg y_{u_k+1}^{\geq}$.*

Example 3. The WPMinSAT interval-based minimal support encoding for the WMaxCSP instance of Example 2 with the constraint $X \neq Y$ is formed by the hard clauses $\neg x_3^{\geq} \vee x_2^{\geq}$ and $\neg y_3^{\geq} \vee y_2^{\geq}$, and the soft clauses $(x_2^{\geq} \vee \neg y_2^{\geq}, 2), (\neg x_2^{\geq} \vee x_3^{\geq} \vee y_2^{\geq}, 2), (\neg x_2^{\geq} \vee x_3^{\geq} \vee \neg y_3^{\geq}, 2)$, and $(\neg x_3^{\geq} \vee y_3^{\geq}, 2)$.

4   A New WPMinSAT Solver

We implemented a new version of MinSatz [20], which is the only existing branch-and-bound WPMinSAT solver, by incorporating all the inference rules of MaxSatz [17]. Such rules preserve the number of falsified clauses over all the models and can be applied in MinSatz, to help detect contradictions earlier, because it works by maximizing the number of falsified clauses (instead of minimizing that number as in MaxSAT).

For lack of space we just define one rule of [17]: The clauses $x, y, \neg x \vee \neg y$ are replaced with $\square, x \vee y$. The clause $x \vee y$ ensures that the number of falsified clauses is preserved also when $x = y = 0$. In this case, two clauses are falsified.

Table 1. Experimental results for random binary MaxCSP, graph coloring, and kbtree instances. The first column displays the parameters provided to the generators, the second column displays the number of tested instances, and the rest of columns display, for the encoding and solver indicated in the first row, the mean time (seconds) needed to solve an instance among the instances that were solved within a cutoff time of 1800 seconds, followed by the total number of solved instances in parentheses. The best results are in bold.

Random binary CSP

(n, d, p_1, p_2)	#	Toulbar2	MinSatz (supc)	MaxSatz (isupc)	MinSatz (isupc)	Abscon	MaxSatz (supc)	MinSatz (dir)	MaxSatz (dir)	WPM1 (supc)
(22, 5, 180, 2)	100	**0.01(100)**	0.37(100)	**0.01(100)**	438.21(98)	11.66(100)	**0.01(100)**	937.91(27)	**0.01(100)**	**0.01(100)**
(22, 5, 180, 4)	100	**0.01(100)**	7.29(100)	0.93(100)	187.77(100)	11.59(100)	1.65(100)	1084.66(27)	0.02(100)	0.02(100)
(22, 5, 180, 6)	100	**0.58(100)**	44.29(100)	62.21(100)	377.95(100)	49.73(100)	23.20(100)	1179.06(2)	7.40(100)	730.03(32)
(22, 5, 180, 8)	100	**2.20(100)**	99.62(100)	217.00(100)	522.45(100)	68.60(100)	59.37(100)	1127.71(2)	111.02(100)	0.00(0)
(22, 5, 180, 10)	100	**4.96(100)**	173.71(100)	362.44(100)	586.06(100)	114.64(100)	129.01(100)	1715.14(3)	854.61(93)	0.00(0)
(22, 5, 180, 12)	100	**7.48(100)**	247.97(100)	441.82(100)	634.37(100)	202.55(100)	237.51(100)	1314.50(5)	1487.47(1)	0.00(0)
(22, 5, 180, 14)	100	**9.52(100)**	285.39(100)	359.80(100)	513.48(100)	350.16(100)	422.87(100)	1390.58(23)	0.00(0)	0.00(0)
(22, 5, 180, 16)	100	**9.74(100)**	276.80(100)	224.28(100)	366.35(100)	516.14(100)	696.88(99)	1033.17(92)	0.00(0)	0.00(0)
(22, 5, 180, 18)	100	**8.49(100)**	225.65(100)	124.22(100)	173.55(100)	639.40(100)	1021.36(89)	334.16(100)	0.00(0)	0.00(0)
(22, 5, 180, 20)	100	**6.47(100)**	176.76(100)	54.13(100)	72.30(100)	764.34(96)	1122.14(62)	73.99(100)	0.00(0)	0.00(0)
(22, 5, 180, 22)	100	**2.69(100)**	61.97(100)	14.75(100)	9.34(100)	669.01(99)	989.45(92)	5.41(100)	0.00(0)	0.00(0)
(22, 5, 180, 24)	100	0.08(100)	0.25(100)	0.93(100)	0.04(100)	195.24(100)	44.76(100)	**0.01(100)**	0.00(0)	319.82(1)
Total	1200	**1200**	1200	1200	1198	1195	1142	581	494	233

Graph Coloring

(n, k, c)	#	Toulbar2	Abscon	MinSatz (supc)	MaxSatz (dir)	MaxSatz (isupc)	MinSatz (isupc)	MaxSatz (supc)	MinSatz (dir)	WPM1 (dir)
(16, 16, 6)	100	**10.48(100)**	85.65(100)	106.75(100)	84.99(100)	167.05(100)	615.50(75)	688.32(63)	557.89(30)	86.48(91)
(16, 16, 4)	100	1.98(100)	49.32(100)	7.22(100)	4.92(100)	**1.60(100)**	26.07(100)	22.66(100)	88.79(100)	668.69(24)
(18, 18, 6)	100	**132.01(100)**	502.96(81)	649.46(74)	576.50(67)	1036.78(37)	1275.15(2)	0.00(0)	0.00(0)	247.14(40)
(18, 18, 4)	100	16.35(100)	130.90(100)	65.48(100)	50.95(100)	**12.44(100)**	290.03(100)	262.91(100)	692.70(84)	0.00(0)
(20, 20, 7)	100	**514.20(63)**	588.24(40)	886.25(5)	757.93(11)	1169.70(12)	0.00(0)	0.00(0)	0.00(0)	234.16(34)
(20, 20, 5)	100	**478.25(95)**	1024.63(30)	872.03(46)	899.19(22)	0.00(0)	0.00(0)	0.00(0)	0.00(0)	437.06(1)
Total	600	**558**	451	425	400	349	277	263	214	190

Kbtrees

t	#	Toulbar2	MinSatz (supc)	MinSatz (isupc)	Abscon	MaxSatz (isupc)	MaxSatz (supc)	MinSatz (dir)	MaxSatz (dir)	WPM1 (dir)
10	50	**0.01(50)**	1.46(50)	39.81(50)	18.72(50)	489.89(50)	47.07(50)	65.93(50)	0.01(50)	0.01(50)
20	50	**0.01(50)**	2.93(50)	27.02(50)	25.42(50)	485.40(48)	244.92(50)	34.26(50)	1.42(50)	0.30(50)
30	50	**0.19(50)**	4.58(50)	34.21(50)	50.73(50)	444.69(49)	800.77(35)	260.17(50)	209.61(50)	541.71(11)
40	50	**1.18(50)**	11.21(50)	47.04(50)	77.81(50)	213.32(50)	930.23(21)	925.71(41)	0.00(0)	0.00(0)
50	50	**2.05(50)**	11.33(50)	45.08(50)	104.27(50)	90.48(50)	907.35(3)	1040.50(18)	0.00(0)	0.00(0)
60	50	**2.77(50)**	15.04(50)	31.25(50)	177.85(50)	65.45(50)	998.99(37)	1200.68(7)	0.00(0)	0.00(0)
70	50	**3.49(50)**	19.10(50)	28.29(50)	306.46(50)	8.39(50)	1509.73(2)	0.00(0)	0.00(0)	0.00(0)
80	50	**1.60(50)**	18.88(50)	5.79(50)	321.73(50)	1.83(50)	0.00(0)	0.00(0)	0.00(0)	0.00(0)
90	50	**0.82(50)**	33.60(50)	1.46(50)	370.69(50)	0.00(0)	0.00(0)	0.00(0)	0.00(0)	0.00(0)
Total	450	**450**	450	450	450	397	298	216	150	111

MinSatz is also characterized by incorporating the upper bounds, described in [20], that exploit clique partition algorithms and MaxSAT technology. In a sense, the new MinSatz combines efficient MaxSAT techniques with genuine MinSAT techniques.

To show the impact of inference rules in performance, Figure 1 compares the old and new MinSatz solvers on weighted Min2SAT instances, generated uniformly at random, having 2000 variables, and with a number of clauses ranging from 1000 to 3000, and the weights ranging from 1 to 10. We used the computer described below, and solved 100 instances per point. MinSatz-ir refers to MinSatz without inference rules.

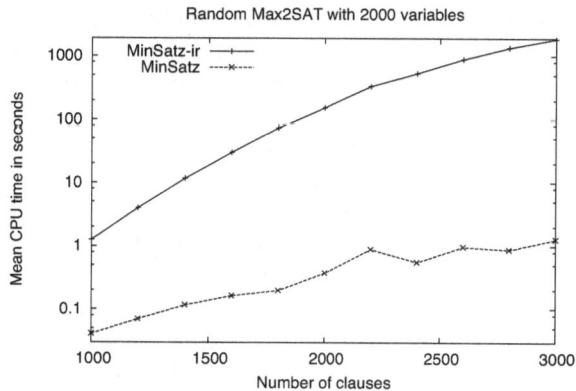

Fig. 1. Comparison of Minsatz with and without inference rules on weighted Min2SAT instances

5   Experimental Results

We compared the performance of the direct encoding (*dir*), the minimal support encoding (*supc*), and the interval-based support encoding (*isupc*) of different optimization problems with WCSP, WPMinSAT and WPMaxSAT solvers. Experiments were performed on a cluster with Intel Xeon 2.67GHz processors with 4GB of memory.

The solved problems (that were first derived as MaxCSP instances and then translated to the mentioned WPMinSAT and WPMaxSAT encodings) are: *(I) Random binary MaxCSP (model B [21]):* In the class $\langle n, d, p_1, p_2 \rangle$ with n variables of domain size d, we choose a random subset of exactly $p_1 n(n-1)/2$ constraints (rounded to the nearest integer), each with exactly $p_2 d^2$ conflicts; p_1 may be thought of as the density and p_2 as the tightness. *(II) Graph coloring:* unsatisfiable graph coloring instances of Culberson [8] with option IID (independent random edge assignment). The parameters of the generator are: number of vertices (n), optimum number of colors to get a valid coloring (k), and number of colors we use to color the graph (c). We solved the problem of finding a coloring that minimizes the number of adjacent vertices with the same color. *(III) Kbtrees:* Clique tree instances with different constraint tightness (t), tested in [4].

We used the MaxSAT solvers MaxSatz and WPM1 [1] from the last MaxSAT Evaluation, the new version of MinSatz, and the WCSP solvers Abscon [15] and Toulbar2 [14]. For WPM1 we just show results with the best performing encoding. We also tested WPMin1 [2] and the WPMinSAT instances produced by transforming the tested

WPMaxSAT instances using the natural flow network encoding [23], but their performance was not competitive.

Table 1 shows the results obtained. As a general comment for the solved testbed, we should say that MinSAT outperforms MaxSAT, the minimal support and interval-based minimal support encodings are better than the direct encoding in MinSAT, and encodings containing fewer and smaller clauses are usually more efficient. Even when our aim is to study the duality and complementarity between MaxSAT and MinSAT, and not to compete with WCSP solvers in their own territory, it is worth mentioning the good behavior of MinSatz wrt Abscon, as well as the still existing gap between MinSatz and Toulbar2 that should stimulate further research on MinSAT. The efforts devoted to WCSP solvers cannot be compared with the recent efforts on WPMinSAT solvers. Recall that WCSP is used here as a source for getting benchmarks for WPMinSAT solvers, and for advancing in the construction of a challenging testbed of MinSAT instances.

In random MaxCSP, the performance of MinSAT versus MaxSAT depends on the number of conflicts. The number of conflicts is related to the clause size in the support encodings, and is related to the number of clauses in the direct encoding. Also observe that *isupc* is superior to *supc* on some subsets of instances on both MaxSAT and MinSAT solvers, showing the relevance of using regular literals in some cases.

In graph coloring, MinSAT solves more instances than MaxSAT but MaxSAT is faster in some subsets, providing again evidence of the complementarity between Min-SAT and MaxSAT. We also observe that regular literals are relevant in some cases. There are a substantial number of instances beyond the reach of the tested solvers, and are a challenge for future experimental investigations. Observe that MinSAT outperforms Abscon on some sets.

In the clique tree results, all the MinSAT options are superior to the best MaxSAT option, solving 152 additional instances. Our encodings allowed to solve for the first time a number of instances not solved in [4].

6   Conclusions

We have defined original encodings from WMaxCSP to WPMinSAT, created a new testbed for evaluating MinSAT solvers, developed a new WPMinSAT solver, and provided new insights into MinSAT such as the duality between MinSAT and MaxSAT encodings. Our experiments indicate that MaxSAT solving and MinSAT solving are complementary. Depending on the structure of the instances, more compact and efficient encodings are produced either with MinSAT or with MaxSAT. In a similar way, the solving techniques of MinSAT outperform the techniques of MaxSAT for some instances, and vice versa.

MaxSAT and MinSAT are important optimization problems that deserve to be compared, and their duality and complementarity deserve to be analyzed. Moreover, as MinSAT-based problem solving is a new research topic, there are not as many available benchmarks as in MaxSAT for evaluating solvers and stimulating the development of new ones. We hope the results of the paper could contribute to gain new insights on these problems, advance the state of the art of MinSAT-based problem solving, and provide tools for developing and evaluating MinSAT solvers.

References

1. Ansótegui, C., Bonet, M.L., Levy, J.: A new algorithm for weighted partial MaxSAT. In: Proceedings of the 24th AAAI Conference on Artificial Intelligence, AAAI 2008, Atlanta, USA, pp. 3–8 (2010)
2. Ansótegui, C., Li, C.M., Manyà, F., Zhu, Z.: A SAT-based approach to MinSAT. In: Proceedings of the 15th International Conference of the Catalan Association for Artificial Intelligence, CCIA 2012, Alacant, Spain, pp. 185–189. IOS Press (2012)
3. Ansótegui, C., Manyà, F.: Mapping Problems with Finite-Domain Variables to Problems with Boolean Variables. In: Hoos, H.H., Mitchell, D.G. (eds.) SAT 2004. LNCS, vol. 3542, pp. 1–15. Springer, Heidelberg (2005)
4. Argelich, J., Cabiscol, A., Lynce, I., Manyà, F.: Sequential Encodings from Max-CSP into Partial Max-SAT. In: Kullmann, O. (ed.) SAT 2009. LNCS, vol. 5584, pp. 161–166. Springer, Heidelberg (2009)
5. Argelich, J., Cabiscol, A., Lynce, I., Manyà, F.: Efficient encodings from CSP into SAT, and from MaxCSP into MaxSAT. Multiple-Valued Logic and Soft Computing 19(1-3), 3–23 (2012)
6. Argelich, J., Li, C.M., Manyà, F., Planes, J.: The first and second Max-SAT evaluations. Journal on Satisfiability, Boolean Modeling and Computation 4, 251–278 (2008)
7. Béjar, R., Hähnle, R., Manyà, F.: A modular reduction of regular logic to classical logic. In: Proceedings, 31st International Symposium on Multiple-Valued Logics (ISMVL), Warsaw, Poland, pp. 221–226. IEEE CS Press, Los Alamitos (2001)
8. Culberson, J.: Graph coloring page: The flat graph generator (1995), http://web.cs.ualberta.ca/~joe/Coloring/Generators/flat.html
9. Gent, I.P.: Arc consistency in SAT. In: Proceedings of the 15th European Conference on Artificial Intelligence, ECAI, Lyon, France, pp. 121–125 (2002)
10. Gent, I.P., Nightingale, P.: A new encoding of AllDifferent into SAT. In: Proceedings of the 3rd International Workshop on Modelling and Reformulating Constraint Satisfaction Problems, CP 2004 Workshop, Toronto, Canada, pp. 95–110 (2004)
11. Heras, F., Morgado, A., Planes, J., Marques-Silva, J.: Iterative SAT solving for minimum satisfiability. In: Proceedings of the IEEE 24th International Conference on Tools with Artificial Intelligence, ICTAI 2012, Athens, Greece, pp. 922–927 (2012)
12. Kasif, S.: On the parallel complexity of discrete relaxation in constraint satisfaction networks. Artificial Intelligence 45, 275–286 (1990)
13. Kügel, A.: Natural Max-SAT encoding of Min-SAT. In: Hamadi, Y., Schoenauer, M. (eds.) LION 6. LNCS, vol. 7219, pp. 431–436. Springer, Heidelberg (2012)
14. Larrosa, J., Schiex, T.: In the quest of the best form of local consistency for weighted CSP. In: Proceedings of the International Joint Conference on Artificial Intelligence, IJCAI 2003, Acapulco, México, pp. 239–244 (2003)
15. Lecoutre, C., Tabary, S.: Abscon 112: Towards more robustness. In: Proceedings of the 3rd International Constraint Solver Competition in CP 2008 (2008)
16. Li, C.M., Manyà, F.: MaxSAT, hard and soft constraints. In: Biere, A., van Maaren, H., Walsh, T. (eds.) Handbook of Satisfiability, pp. 613–631. IOS Press (2009)
17. Li, C.M., Manyà, F., Planes, J.: New inference rules for Max-SAT. Journal of Artificial Intelligence Research 30, 321–359 (2007)
18. Li, C.M., Manyà, F., Quan, Z., Zhu, Z.: Exact MinSAT solving. In: Strichman, O., Szeider, S. (eds.) SAT 2010. LNCS, vol. 6175, pp. 363–368. Springer, Heidelberg (2010)
19. Li, C.M., Zhu, Z., Manyà, F., Simon, L.: Minimum satisfiability and its applications. In: Proceedings of the 22nd International Joint Conference on Artificial Intelligence, IJCAI 2011, Barcelona, Spain, pp. 605–610 (2011)

20. Li, C.M., Zhu, Z., Manyà, F., Simon, L.: Optimizing with minimum satisfiability. Artificial Intelligence 190, 32–44 (2012)
21. Smith, B.M., Dyer, M.E.: Locating the phase transition in binary constraint satisfaction problems. Artificial Intelligence 81, 155–181 (1996)
22. Walsh, T.: SAT v CSP. In: Dechter, R. (ed.) CP 2000. LNCS, vol. 1894, pp. 441–456. Springer, Heidelberg (2000)
23. Zhu, Z., Li, C.-M., Manyà, F., Argelich, J.: A New Encoding from MinSAT into MaxSAT. In: Milano, M. (ed.) CP 2012. LNCS, vol. 7514, pp. 455–463. Springer, Heidelberg (2012)

Adaptive Parameterized Consistency*

Amine Balafrej[1,2], Christian Bessiere[1],
Remi Coletta[1], and El Houssine Bouyakhf[2]

[1] CNRS, University of Montpellier, France
[2] LIMIARF/FSR, University Mohammed V Agdal, Rabat, Morocco
{balafrej,bessiere,coletta}@lirmm.fr, bouyakhf@fsr.ac.ma

Abstract. State-of-the-art constraint solvers uniformly maintain the same level of local consistency (usually arc consistency) on all the instances. We propose *parameterized local consistency,* an original approach to adjust the level of consistency depending on the instance and on which part of the instance we propagate. We do not use as parameter one of the features of the instance, as done for instance in portfolios of solvers. We use as parameter the *stability* of values, which is a feature based on the state of the arc consistency algorithm during its execution. Parameterized local consistencies choose to enforce arc consistency or a higher level of local consistency on a value depending on whether the stability of the value is above or below a given threshold. We also propose a way to dynamically adapt the parameter, and thus the level of local consistency, during search. This approach allows us to get a good trade-off between the number of values pruned and the computational cost. We validate our approach on various problems from the CSP competition.

1 Introduction

Enforcing constraint propagation by applying local consistency during search is one of the strengths of constraint programming (CP). It allows the constraint solver to remove locally inconsistent values. This leads to a reduction of the search space. Arc consistency is the oldest and most well-known way of propagating constraints [2]. It has the nice feature that it does not modify the structure of the constraint network. It just prunes infeasible values. Arc consistency is the standard level of consistency maintained in constraint solvers. Several other local consistencies pruning only values and stronger than arc consistency have been proposed, such as max restricted path consistency or singleton arc consistency [7]. These local consistencies are seldom used in practice because of the high computational cost of maintaining them during search. However, on some problems, maintaining arc consistency is not a good choice because of the high number of ineffective revisions of constraints that penalize the CPU time. For instance, Stergiou observed that when solving the scen11 radio link frequency assignment problem (RLFAP) with an algorithm maintaining arc consistency, only 27 out of the 4103 constraints of the problem were identified as causing a domain wipe out and 1921 constraints did not prune any value [10].

---

* This work has been funded by the EU project ICON (FP7-284715).

C. Schulte (Ed.): CP 2013, LNCS 8124, pp. 143–158, 2013.
© Springer-Verlag Berlin Heidelberg 2013

Choosing the right level of local consistency for solving a problem requires finding the good trade-off between the ability of this local consistency to remove inconsistent values, and the cost of the algorithm that enforces it. Stergiou suggests to take advantage of the power of strong consistencies to reduce the search space while avoiding the high cost of maintaining them in the whole network. His method results in a heuristic approach based on the monitoring of propagation events to dynamically adapt the level of local consistency (arc consistency or max restricted path consistency) to individual constraints. This prunes more values than arc consistency and less than max restricted path consistency. The level of propagation obtained is not characterized by a local consistency property. Depending on the order of propagation we can converge on different closures. When dealing with global constraints, some work propose to weaken arc consistency instead of strengthening it. In [8], Katriel et al. proposed a randomized filtering scheme for AllDifferent and Global Cardinality Constraint. In [9], Sellmann introduced the concept of approximated consistency for optimization constraints and provided filtering algorithms for Knapsack Constraints based on bounds with guaranteed accuracy.

In this paper we define the notion of stability of values. This is an original notion not based on characteristics of the instance to solve but based on the state of the arc consistency algorithm during its propagation. Based on this notion, we propose *parameterized consistencies*, an original approach to adjust the level of consistency inside a given instance. The intuition is that if a value is hard to prove arc consistent (i.e., the value is not stable for arc consistency), this value will perhaps be pruned by a stronger local consistency. The parameter p specifies the threshold of stability of a value v below which we will enforce a higher consistency to v. A parameterized consistency p-LC is thus an intermediate level of consistency between arc consistency and another consistency LC, stronger than arc consistency. The strength of p-LC depends on the parameter p. This approach allows us to find a trade-off between the pruning power of the local consistency and the computational cost of the algorithm that achieves it. We apply p-LC to the case where LC is max restricted path consistency. We describe the algorithm p-maxRPC3 (based on maxRPC3 [1]) that achieves p-max restricted path consistency. Then, we propose ap-LC, an adaptive variant of p-LC which adapts dynamically and locally the level of local consistency during search. Finally, we experimentally assess the practical relevance of parameterized local consistency. We show that by making good choices for the parameter p we take advantage of both arc consistency light computational cost and LC effectiveness of pruning. In the best cases, a solver using p-LC explores the same number of nodes as LC with a number of constraint checks lower than LC, resulting in a CPU-time lower than both arc consistency-based or LC-based solvers.

2   Background

A *constraint network* is defined as a set of n variables $X = \{x_1, ..., x_n\}$, a set of ordered domains $D = \{D(x_1), ..., D(x_n)\}$, and a set of e constraints

$C = \{c_1, ..., c_e\}$. Each constraint c_k is defined by a pair $(var(c_k), sol(c_k))$, where $var(c_k)$ is an ordered subset of X, and $sol(c_k)$ is a set of combinations of values (tuples) satisfying c_k. In the following, we restrict ourselves to binary constraints because the local consistency (maxRPC) we use here to instantiate our approach is defined on the binary case only. However, the notions we introduce can be extended to non-binary constraints, by using maxRPWC for instance [4]. A binary constraint c between x_i and x_j will be denoted by c_{ij}, and $\Gamma(x_i)$ will denote the set of variables x_j involved in a constraint with x_i.

A value $v_j \in D(x_j)$ is called an *arc consistent support (AC support)* for $v_i \in D(x_i)$ on c_{ij} if $(v_i, v_j) \in sol(c_{ij})$. A value $v_i \in D(x_i)$ is *arc consistent (AC)* if and only if for all $x_j \in \Gamma(x_i)$ v_i has an AC support $v_j \in D(x_j)$ on c_{ij}. A domain $D(x_i)$ is arc consistent if it is non empty and all values in $D(x_i)$ are arc consistent. A network is arc consistent if all domains in D are arc consistent. If enforcing arc consistency on a network N leads to a domain wipe out, we say that N is arc inconsistent.

A tuple $(v_i, v_j) \in D(x_i) \times D(x_j)$ is *path consistent (PC)* if and only if for any third variable x_k there exists a value $v_k \in D(x_k)$ such that v_k is an AC support for both v_i and v_j. In such a case, v_k is called *witness* for the path consistency of (v_i, v_j).

A value $v_j \in D(x_j)$ is a *max restricted path consistent (maxRPC)* support for $v_i \in D(x_i)$ on c_{ij} if and only if it is an AC support and the tuple (v_i, v_j) is path consistent. A value $v_i \in D(x_i)$ is max restricted path consistent on a constraint c_{ij} if and only if $\exists v_j \in D(x_j)$ maxRPC support for v_i on c_{ij}. A value $v_i \in D(x_i)$ is max restricted path consistent iff for all $x_j \in \Gamma(x_i)$ v_i has a maxRPC support $v_j \in D(x_j)$ on c_{ij}. A domain $D(x_i)$ is maxRPC if it is non empty and all values in $D(x_i)$ are maxRPC. A network is maxRPC if all domains in D are maxRPC.

We say that a local consistency LC_1 is stronger than a local consistency LC_2 $(LC_2 \preceq LC_1)$ if LC_2 holds on any constraint network on which LC_1 holds.

The problem of deciding whether a constraint network has solutions is called the *constraint satisfaction problem (CSP)*, and it is NP-complete. Solving a CSP is done by backtrack search that maintains some level of consistency between each branching step.

3   Parameterized Consistency

In this section we present an original approach to parameterize a level of consistency LC stronger than arc consistency so that it degenerates to arc consistency when the parameter equals 0, to LC when the parameters equals 1, and to levels in between when the parameter is between 0 and 1. The idea behind this is to be able to adjust the level of consistency to the instance to be solved, hoping that such an adapted level of consistency will prune significantly more values than arc consistency while being less time consuming than LC.

Parameterized consistency is based on the concept of stability of values. We first need to define the 'distance to end' of a value in a domain. This captures how far a value is from the last in its domain. In the following, $rank(v, S)$ is the position of value v in the ordered set of values S.

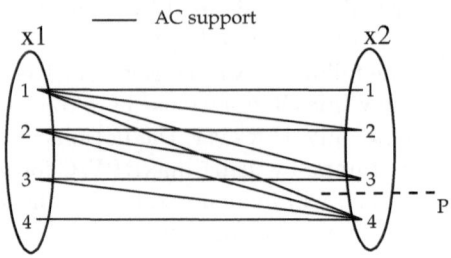

Fig. 1. Stability of supports on the example of the constraint $x_1 \leq x_2$ with the domains $D(x_1) = D(x_2) = \{1, 2, 3, 4\}$. $(x_1, 4)$ is not p-stable for AC.

Definition 1 (Distance to end of a value). *The* distance to end *of a value* $v_i \in D(x_i)$ *is the ratio*

$$\Delta(x_i, v_i) = (|D_o(x_i)| - rank(v_i, D_o(x_i)))/|D_o(x_i)|,$$

where $D_o(x_i)$ is the initial domain of x_i.

We see that the first value in $D_o(x_i)$ has distance $(|D_o(x_i)| - 1)/|D_o(x_i)|$ and the last one has distance 0. Thus, $\forall v_i \in D(x_i), 0 \leq \Delta(x_i, v_i) < 1$.

We can now give the definition of what we call the parameterized stability of a value for arc consistency. The idea is to define stability for values based on the distance to the end of their AC supports. For instance, consider the constraint $x_1 \leq x_2$ with the domains $D(x_1) = D(x_2) = \{1, 2, 3, 4\}$ (see Figure 1). $\Delta(x_2, 1) = (4 - 1)/4 = 0.75$, $\Delta(x_2, 2) = 0.5$, $\Delta(x_2, 3) = 0.25$ and $\Delta(x_2, 4) = 0$. If $p = 0.2$, the value $(x_1, 4)$ is not p-stable for AC, because the first and only AC support of $(x_1, 4)$ in the ordering used to look for supports, that is $(x_2, 4)$, has a distance to end smaller than the threshold p. Proving that the pair $(4, 4)$ is inconsistent (by a stronger consistency) could lead to the pruning of $(x_1, 4)$. In other words, applying a stronger consistency on $(x_1, 4)$ has more chances to lead to its removal than applying it on for instance $(x_1, 1)$, which had no difficulty to find its first AC support (distance to en of $(x_2, 1)$ is 0.75).

Definition 2 (p-stability for AC). *A value $v_i \in D(x_i)$ is p-stable for AC on c_{ij} iff v_i has an AC support $v_j \in D(x_j)$ on c_{ij} such that $\Delta(x_j, v_j) \geq p$. A value $v_i \in D(x_i)$ is p-stable for AC iff $\forall x_j \in \Gamma(x_i)$, v_i is p-stable for AC on c_{ij}.*

We are now ready to give the first definition of parameterized local consistency. This first definition can be applied to any local consistency LC for which the consistency of a value on a constraint is well defined. This is the case for instance for all triangle-based consistencies [6,2].

Definition 3 (Constraint-based p-LC). *Let LC be a local consistency stronger than AC for which the LC consistency of a value on a constraint is defined. A value $v_i \in D(x_i)$ is constraint-based p-LC on c_{ij} iff it is p-stable for*

AC on c_{ij}, or it is LC on c_{ij}. A value $v_i \in D(x_i)$ is constraint-based p-LC iff $\forall c_{ij}$, v_i is constraint-based p-LC on c_{ij}. A constraint network is constraint-based p-LC iff all values in all domains in D are constraint-based p-LC.

Theorem 1. *Let LC be a local consistency stronger than AC for which the LC consistency of a value on a constraint is defined. Let p_1 and p_2 be two parameters in $[0..1]$. If $p_1 < p_2$ then $AC \preceq$ constraint-based p_1-$LC \preceq$ constraint-based p_2-$LC \preceq LC$.*

Proof. Suppose that there exist two parameters p_1, p_2 such that $0 \le p_1 < p_2 \le 1$, and suppose that there exists a p_2-LC constraint network N that contains a p_2-LC value (x_i, v_i) that is p_1-LC inconsistent. Let c_{ij} be the constraint on which (x_i, v_i) is p_1-LC inconsistent. Then, $\nexists v_j \in D(x_j)$ that is an AC support for (x_i, v_i) on c_{ij} such that $\Delta(x_j, v_j) \ge p_1$. Thus, v_i is not p_2-stable for AC on c_{ij}. In addition, v_i is not LC on c_{ij}. Therefore, v_i is not p_2-LC, and N is not p_2-LC. ∎

Definition 3 can be modified to a more coarse-grained version that is not dependent on the consistency of values on a constraint. It will have the advantage to apply to any type of strong local consistency, even those, like singleton arc consistency, for which the consistency of a value on a constraint is not defined.

Definition 4 (Value-based p-LC). *Let LC be a local consistency stronger than AC. A value $v_i \in D(x_i)$ is value-based p-LC if and only if it is p-stable for AC or it is LC. A constraint network is value-based p-LC if and only if all values in all domains in D are value-based p-LC.*

Theorem 2. *Let LC be a local consistency stronger than AC. Let p_1 and p_2 be two parameters in $[0..1]$. If $p_1 < p_2$ then $AC \preceq$ value-based p_1-$LC \preceq$ value-based p_2-$LC \preceq LC$.*

Proof. Suppose that there exist two parameters p_1, p_2 such that $0 \le p_1 < p_2 \le 1$, and suppose that there exists a p_2-LC constraint network N that contains a p_2-LC value (x_i, v_i) that is p_1-LC-inconsistent. v_i is p_1-LC-inconsistent means that:

1. v_i is not p_1-stable for AC: $\exists c_{ij}$ on which v_i is not p_1-stable for AC. Then $\nexists v_j \in D(x_j)$ that is an AC support for (x_i, v_i) on c_{ij} such that $\Delta(x_j, v_j) \ge p_1$. Therefore, v_i is not p_2-stable for AC on c_{ij}, then v_i is not p_2-stable for AC.
2. v_i is LC inconsistent

(1) and (2) imply that v_i is not p_2-LC, and N is not p_2-LC. ∎

For both types of definitions of p-LC, we have the following property on the extreme cases ($p = 0, p = 1$).

Corollary 1. *Let LC_1 and LC_2 be two local consistencies stronger than AC. We have: value-based 0-$LC_2 = AC$ and value-based 1-$LC_2 = LC$. If the LC_1 consistency of a value on a constraint is defined, we also have: constraint-based 0-$LC_1 = AC$ and constraint-based 1-$LC_1 = LC$.*

4   Parameterized maxRPC: p-maxRPC

To illustrate the benefit of our approach, we apply *parameterized consistency* to maxRPC to obtain the p-maxRPC level of consistency that achieves a consistency level between AC and maxRPC.

Definition 5 (p-maxRPC). *A value, a network, are p-maxRPC if and only if they are constraint-based p-maxRPC.*

From Theorem 1 and Corollary 1 we derive the following corollary.

Corollary 2. *For any two parameters p_1, p_2, $0 \leq p_1 < p_2 \leq 1$, $AC \preceq p_1$-maxRPC $\preceq p_2$-maxRPC $\preceq$ maxRPC. 0-maxRPC = AC and 1-maxRPC = maxRPC.*

Algorithm 1. Initialization(X, D, C, Q)

```
 1  begin
 2      foreach x_i ∈ X do
 3          foreach v_i ∈ D(x_i) do
 4              foreach x_j ∈ Γ(x_i) do
 5                  p-support ← false
 6                  foreach v_j ∈ D(x_j) do
 7                      if (v_i, v_j) ∈ c_ij then
 8                          LastAC_{x_i,v_i,x_j} ← v_j
 9                          if Δ(x_j, v_j) ≥ p then
10                              p-support ← true
11                              LastPC_{x_i,v_i,x_j} ← v_j
12                              break;
13                      if searchPCwit(v_i, v_j) then
14                          p-support ← true
15                          LastPC_{x_i,v_i,x_j} ← v_j
16                          break;
17                  if ¬p-support then
18                      remove v_i from D(x_i)
19                      Q ← Q ∪ {x_i}
20                      break;
21              if D(x_i) = ∅ then return false
22      return true
```

We propose an algorithm for p-maxRPC, based on maxRPC3, the best existing maxRPC algoritm. We do not describe maxRPC3 in full detail as it can be found in [1]. We only describe procedures where changes to maxRPC3 are necessary to design p-maxRPC3, a coarse grained algorithm that performs p-maxRPC. We use light grey to emphasize the modified parts of the original maxRPC3 algorithm.

Algorithm 2. checkPCsupLoss(v_j, x_i)

---

1 **begin**
2   **if** $LastAC_{x_j,v_j,x_i} \in D(x_i)$ **then** $b_i \leftarrow max(LastPC_{x_j,v_j,x_i}+1, LastAC_{x_j,v_j,x_i})$
3   **else** $b_i \leftarrow max(LastPC_{x_j,v_j,x_i}+1, LastAC_{x_j,v_j,x_i}+1)$
4   **foreach** $v_i \in D(x_i), v_i \geq b_i$ **do**
5   **if** $(v_j, v_i) \in c_{ji}$ **then**
6   **if** $LastAC_{x_j,v_j,x_i} \notin D(x_i)$ & $LastAC_{x_j,v_j,x_i} > LastPC_{x_j,v_j,x_i}$ **then**
7          $LastAC_{x_j,v_j,x_i} \leftarrow v_i$
8   **if** $\Delta(x_i, v_i) \geq p$ **then** $LastPC_{x_j,v_j,x_i} \leftarrow v_i$ **return** *true*
9   **if** searchPCwit(v_j, v_i) **then** $LastPC_{x_j,v_j,x_i} \leftarrow v_i$ **return** *true*
10   **return** *false*

Algorithm 3. checkPCwitLoss(x_j, v_j, x_i)

---

1 **begin**
2   **foreach** $x_k \in \Gamma(x_j) \cap \Gamma(x_i)$ **do**
3   witness $\leftarrow$ *false*
4   **if** $v_k \leftarrow LastPC_{x_j,v_j,x_k} \in D(x_k)$ **then**
5   **if** $\Delta(x_k, v_k) \geq p$ **then** witness $\leftarrow$ *true*
6   **else**
7   **if** $LastAC_{x_j,v_j,x_i} \in D(x_i)$ & $LastAC_{x_j,v_j,x_i} = LastAC_{x_k,v_k,x_i}$
8   **OR** $LastAC_{x_j,v_j,x_i} \in D(x_i)$ & $(LastAC_{x_j,v_j,x_i}, v_k) \in c_{ik}$
9   **OR** $LastAC_{x_k,v_k,x_i} \in D(x_i)$ & $(LastAC_{x_k,v_k,x_i}, v_j) \in c_{ij}$
10   **then** witness $\leftarrow$ *true*
11   **else**
12   **if** searchACsup(x_j, v_j, x_i) & searchACsup(x_k, v_k, x_i) **then**
13   **foreach** $v_i \in D(x_i), v_i \geq max(LastAC_{x_j,v_j,x_i}, LastAC_{x_k,v_k,x_i})$ **do**
14   **if** $(v_j, v_i) \in c_{ji}$ & $(v_k, v_i) \in c_{ki}$ **then**
15   witness $\leftarrow$ *true*
16   break;
17   **if** ¬witness & ¬checkPCsupLoss(v_j, x_k) **then** **return** *false*
18   **return** *true*

---

maxRPC3 uses a propagation list Q where it inserts the variables whose domains have changed. It also uses two other data structures: LastAC and LastPC. For each value (x_i, v_i) $LastAC_{x_i,v_i,x_j}$ stores the smallest AC support for (x_i, v_i) on c_{ij} and $LastPC_{x_i,v_i,x_j}$ stores the smallest PC support for (x_i, v_i) on c_{ij} (i.e., the smallest AC support (x_j, v_j) for (x_i, v_i) on c_{ij} such that (v_i, v_j) is PC). This algorithm consists in two phases: initialization and propagation.

In the initialization phase (algorithm 1) maxRPC3 checks if each value (x_i, v_i) has a maxRPC-support (x_j, v_j) on each constraint c_{ij}. If not, it removes v_i from $D(x_i)$ and inserts x_i in Q. To check if a value (x_i, v_i) has a maxRPC-support on a constraint c_{ij}, maxRPC3 looks first for an AC-support (x_j, v_j) for (x_i, v_i) on c_{ij}, then it checks if (v_i, v_j) is PC. In this last step, changes were necessary to obtain p-maxRPC3 (lines 9-12). We check if (v_i, v_j) is PC (line 13) only if $\Delta(x_j, v_j)$ is smaller than the parameter p (line 9).

The propagation phase of maxRPC3 consists in propagating the effect of deletions. While Q is non empty, maxRPC3 extracts a variable x_i from Q and checks for each value (x_j, v_j) of each neighboring variable $x_j \in \Gamma(x_i)$ if it is not maxRPC because of deletions of values in $D(x_i)$. A value (x_j, v_j) becomes maxRPC inconsistent in two cases: if its unique PC-support (x_i, v_i) on c_{ij} has been deleted, or if we deleted the unique witness (x_i, v_i) for a pair (v_j, v_k) such that (x_k, v_k) is the unique PC-support for (x_j, v_j) on c_{jk}. So, to propagate deletions, maxRPC3 checks if the last maxRPC support (last known support) of (x_j, v_j) on c_{ij} still belongs to the domain of x_i, otherwise it looks for the next support (algorithm 2). If such a support does not exist, it removes the value v_j and adds the variable x_j to Q. Then if (x_j, v_j) has not been removed in the previous step, maxRPC3 checks (algorithm 3) whether there is still a witness for each pair (v_j, v_k) such that (x_k, v_k) is the PC support for (x_j, v_j) on c_{jk}. If not, it looks for the next maxRPC support for (x_j, v_j) on c_{jk}. If such a support does not exist, it removes v_j from $D(x_j)$ and adds the variable x_j to Q.

In the propagation phase also, we modified maxRPC3 to check if the values are still p-maxRPC instead of checking if they are maxRPC. In p-maxRPC3, the last p-maxRPC support for (x_j, v_j) on c_{ij} is the last AC support if (x_j, v_j) is p-stable for AC on c_{ij}. If not, it is the last PC support. Thus, p-maxRPC3 checks if the last p-maxRPC support (last known support) of (x_j, v_j) on c_{ij} still belongs to the domain of x_i. If not, it looks (algorithm 2) for the next AC support (x_i, v_i) on c_{ij}, and checks if (v_i, v_j) is PC (line 9) only when $\Delta(x_i, v_i) < p$ (line 8). If no p-maxRPC support exists, p-maxRPC3 removes the value and adds the variable x_j to Q. If the value (x_j, v_j) has not been removed in the previous phase, p-maxRPC3 checks (algorithm 3) whether there is still a witness for each pair (v_j, v_k) such that (x_k, v_k) is the p-maxRPC support for v_j on c_{jk} and $\Delta(x_k, v_k) < p$. If not, it looks for the next p-maxRPC support for v_j on c_{jk}. If such a support does not exist, it removes v_j from $D(x_j)$ and adds the variable x_j to Q.

5   Experimental Validation of p-maxRPC

To validate the approach of parameterized local consistency, we made a first basic experiment. The purpose of this experiment is to see if there exist problems on which a given level of p-maxRPC, with a p uniform on all the constraint network and static during the whole search is more efficient than AC or maxRPC, or both.

We have implemented the algorithms that achieve p-maxRPC as described in the previous section in our own binary constraint solver, in addition to maxRPC (maxRPC3 version [1]) and AC (AC2001 version [3]). All these algorithms are maintained during search. We tested these algorithms on several classes of problems of the International Constraint Solver Competition 09[1]. We have only selected problems involving binary constraints. To isolate the effect of propagation, we used the lexicographic ordering for variables and values. We set the CPU timeout to one hour. Our experiments were conducted on a 12-core Genuine Intel machine with 16Gb of RAM running at 2.92GHz.

On each instance of our experiment, we ran AC, max-RPC, and p-maxRPC for all values of p in $\{0.1, 0.2, \ldots, 0.9\}$. Performance has been measured in terms of CPU time in seconds, the number of visited nodes (NODE) and the number of constraint checks (CCK). Results are presented as "CPU time (p)", where p is the parameter for which p-maxRPC gives the best result.

Table 1 reports the performance of AC, maxRPC, and p-maxRPC for the value of p producing the best CPU time, on Radio Link Frequency Assignment Problems (RLFAPs), Geom problems, and queens knights problems. The CPU time of the best algorithm is highlighted with bold. On RLFAP and Geom, we observe the existence of a parameter p where p-maxRPC is faster than *both* AC and maxRPC for most instances of these two classes of problems. On the queens-knight problem, however, AC is always the best algorithm. In Figures 2 and 3, we try to understand more closely what makes p-maxRPC better or worse than AC and maxRPC. Figures 2 and 3 plot the performance (CPU, NODE and CCK) of p-maxRPC for all values of p from 0 to 1 by steps of 0.1 against performance of AC and maxRPC. Figure 2 shows an instance where p-maxRPC solves the problem faster than AC and maxRPC for values of p in the range $[0.3..0.8]$. We observe that p-maxRPC is faster than AC and maxRPC when it reduces the size of the search space as much as maxRPC (same number of nodes visited) with a number of CCK closer to the number of CCK produced by AC. Figure 3 shows an instance where the CPU time for p-maxRPC is never better than *both* AC and maxRPC. We see that if the CPU time for p-maxRPC is two to three times better than maxRPC, it fails to improve AC because the number of constraint checks performed by p-maxRPC is much higher than the number of constraint checks performed by AC, whereas the number of nodes visited by p-maxRPC is not significantly reduced compared to the number of nodes visited by AC. From these observations, it thus seems that p-maxRPC outperforms AC and maxRPC when it finds a compromise between the number of nodes visited (the power of maxRPC) and the number of CCK needed to maintain (the light cost of AC).

In Figures 2 and 3 we can see that the CPU time for 1-maxRPC (respectively 0-maxRPC) is greater than the CPU time for maxRPC (respectively AC) although the two consistencies are equivalent. The reason is that p-maxRPC performs tests on the distances. For $p = 0$, we also explain this difference by the fact that p-maxRPC maintains data structures that AC does not use.

---

[1] http://cpai.ucc.ie/09/

Table 1. Performance (CPU time, nodes and constraint checks) of AC, p-maxRPC, and maxRPC on various instances

		AC	p-maxRPC	maxRPC
scen1-f8	CPU(s)	Time-out	**1.39** (0.2)	6.10
	#nodes	–	927	917
	#ccks	–	1,397,440	26,932,990
scen2-f24	CPU(s)	Time-out	**0.13** (0.3)	0.65
	#nodes	–	201	201
	#ccks	–	296,974	3,462,070
scen3-f10	CPU(s)	Time-out	**0.89** (0.5)	2.80
	#nodes	–	469	408
	#ccks	–	874,930	13,311,797
geo50-20d4-75-26	CPU(s)	111.48	17.80 (1.0)	**15.07**
	#nodes	477,696	3,768	3,768
	#ccks	96,192,822	40,784,017	40,784,017
geo50-20d4-75-43	CPU(s)	1,671.35	**1,264.36** (0.5)	1,530.02
	#nodes	4,118,134	555,259	279,130
	#ccks	1,160,664,461	1,801,402,535	3,898,964,831
geo50-20d4-75-46	CPU(s)	1,732.22	**371.30** (0.6)	517.35
	#nodes	3,682,394	125,151	64,138
	#ccks	1,516,856,615	584,743,023	1,287,674,430
geo50-20d4-75-84	CPU(s)	404.63	**0.44** (0.6)	0.56
	#nodes	2,581,794	513	333
	#ccks	293,092,144	800,657	1,606,047
queensK10-5-add	CPU(s)	**27.14**	30.79 (0.2)	98.44
	#nodes	82,208	81,033	78,498
	#ccks	131,098,933	148,919,686	954,982,880
queensK10-5-mul	CPU(s)	**43.89**	83.27 (0.1)	300.74
	#nodes	74,968	74,414	70,474
	#ccks	104,376,698	140,309,576	1,128,564,278

6   Adaptative Parameterized Consistency: ap-maxRPC

In the previous section, we have defined p-maxRPC, a version of parameterized consistency where the strong local consistency is maxRPC. We have performed some initial experiments where p has the same value during the whole search and everywhere in the constraint network. However, the algorithm we proposed to enforce p-maxRPC does not specify how p is chosen. In this section, we propose two possible ways to dynamically and locally adapt the parameter p in order to solve the problem faster than both AC and maxRPC. Instead of using a single parameter p during the whole search and for the whole constraint network, we propose to use several local parameters and to adapt the level of local consistency by dynamically adjusting the value of the different local parameters during search. The idea is to concentrate the effort of propagation by increasing the level of consistency in the most difficult parts of the instance. We can determine these

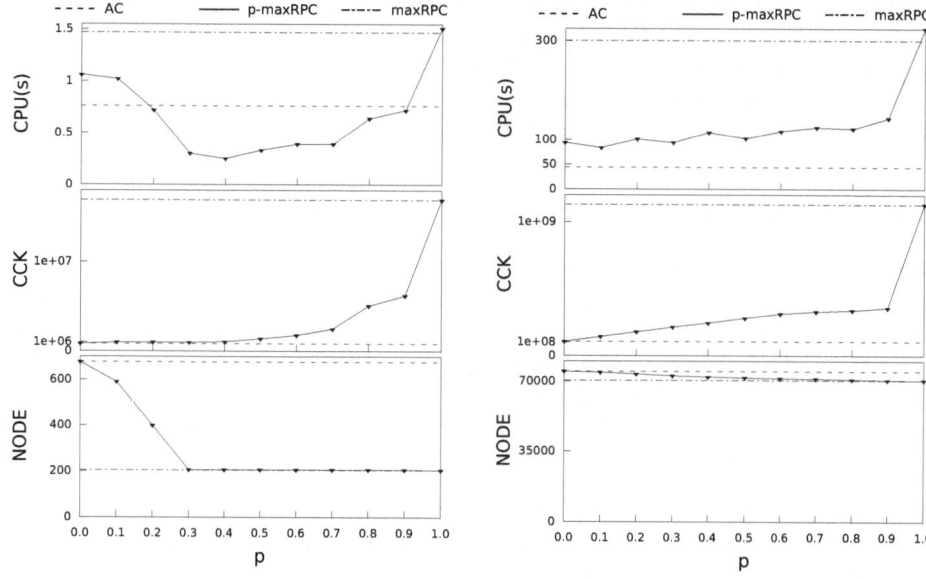

Fig. 2. Instance where p-maxRPC outperforms both AC and maxRPC

Fig. 3. Instance where AC outperforms p-maxRPC

difficult parts using heuristics based on conflicts in the same vein as the weight of a constraint or the weighted degree of a variable in [5].

6.1 Constraint-Based ap-maxRPC : apc-maxRPC

The first technique we propose, called constraint-based ap-maxRPC, assigns a parameter $p(c_k)$ to each constraint c_k in C. We define this parameter to be correlated to the *weight* of the constraint. The idea is to apply a higher level of consistency in parts of the problem where the constraints are the most active.

Definition 6 (The weight of a constraint [5]). *The weight $w(c_k)$ of a constraint $c_k \in C$ is an integer that is incremented every time a domain wipe out occurs while performing propagation on this constraint.*

We define the adaptive parameter $p(c_k)$ local to constraint c_k in such a way that it is greater when the weight $w(c_k)$ is higher wrt to other constraints.

$$\forall c_k \in C, p(c_k) = \frac{w(c_k) - min_{c \in C}(w(c))}{max_{c \in C}(w(c)) - min_{c \in C}(w(c))} \tag{1}$$

Equation 1 is normalized so that we are guaranteed that $0 \leq p(c_k) \leq 1$ for all $c_k \in C$ and that there exists c_{k_1} with $p(c_{k_1}) = 0$ (the constraint with lowest weight) and c_{k_2} with $p(c_{k_2}) = 1$ (the constraint with highest weight).

We are ready to define adaptive parameterized consistency based on constraints.

Definition 7 (constraint-based ap-maxRPC). *A value $v_i \in D(x_i)$ is constraint-based ap-maxRPC (or apc-maxRPC) on a constraint c_{ij} if and only if it is constraint-based $p(c_{ij})$-maxRPC. A value $v_i \in D(x_i)$ is apc-maxRPC iff $\forall c_{ij}$, v_i is apc-maxRPC on c_{ij}. A constraint network is apc-maxRPC iff all values in all domains in D are apc-maxRPC.*

6.2   Variable-Based ap-maxRPC: apx-maxRPC

The technique proposed in Section 6.1 can only be used on consistencies where the consistency of a value on a constraint is defined. We give a second technique which can be used on constraint-based or variable-based local consistencies indifferently. We instantiate our definitions to maxRPC but the extension to other consistencies is direct. We call this new technique variable-based ap-maxRPC. We need to define the weighted degree of a variable as the aggregation of the weights of all constraints involving it.

Definition 8 (The weighted degree of a variable [5]). *The weighted degree $wdeg(x_i)$ of a variable x_i is the sum of the weights of the constraints involving x_i and one other uninstantiated variable.*

We associate each variable with an adaptive local parameter based on its weighted degree.

$$\forall x_i \in X, p(x_i) = \frac{wdeg(x_i) - min_{x \in X}(wdeg(x))}{max_{x \in X}(wdeg(x)) - min_{x \in X}(wdeg(x))} \tag{2}$$

As in Equation 1, we see that the local parameter is normalized so that we are guaranteed that $0 \leq p(x_i) \leq 1$ for all $x_i \in X$ and that there exists x_{k_1} with $p(x_{k_1}) = 0$ (the variable with lowest weighted degree) and x_{k_2} with $p(x_{k_2}) = 1$ (the variable with highest weighted degree).

Definition 9 (variable-based ap-maxRPC). *A value $v_i \in D(x_i)$ is variable-based ap-maxRPC (or apx-maxRPC) if and only if it is value-based $p(x_i)$-maxRPC. A constraint network is apx-maxRPC iff all values in all domains in D are apx-maxRPC.*

7   Experimental Evaluation of ap-maxRPC

In Section 5 we have shown that maintaining a static form of p-maxRPC during the whole search can lead to a promising trade-off between computational effort and pruning when all algorithms follow the same static variable ordering. In this section, we want to put our contributions in the real context of a solver using the best known variable ordering heuristic, $dom/wdeg$, though it is known that this heuristic is so good that it reduces a lot the differences in performance that other features of the solver could provide. We have compared the two variants of adaptive parameterized consistency introduced in Section 6 to AC and maxRPC.

We ran the four algorithms on radio link frequency assignment problems, geom problems, and queens knights problems.

Table 2 reports some representative results. A first observation is that, thanks to the *dom/wdeg* heuristic, we were able to solve more instances before the cutoff of one hour, especially the scen11 variants of RLFAP. A second observation is that *apc*-maxRPC and *apx*-maxRPC are both faster than at least one of the two extreme consistencies (AC and maxRPC) on all instances except scen7-w1-f4 and geo50-20-d4-75-30. Third, when *apx*-maxRPC and/or *apc*-maxRPC are faster than both AC and maxRPC (scen1-f9, scen2-f25, scen11-f9, scen11-f10 and scen11-f11), we observe that the gap in performance in terms of nodes and CCKs between AC and maxRPC is significant. Except for scen7-w1-f4, the number of nodes visited by AC is three to five times greater than the number of nodes visited by maxRPC and the number of constraint checks performed by maxRPC is twelve to sixteen times greater than the number of constraint checks performed by AC. For the Geom instances the CPU time of the *ap*-maxRPC algorithms is between AC and maxRPC, and it is never lower than the CPU time of AC. This probably means that when solving these instances with the *dom/wdeg* heuristic, there is no need for sophisticated local consistencies. In general we see that the *ap*-maxRPC algorithms fail to improve both the two extreme consistencies simultaneously for the instances where the performance gap between AC and maxRPC is low.

If we compare *apx*-maxRPC to *apc*-maxRPC, we observe that although *apx*-maxRPC is coarser in its design than *apc*-maxRPC, *apx*-maxRPC is often faster than *apc*-maxRPC. We can explain this by the fact that the constraints initially all have the same weight equal to 1. Hence, all local parameters $ap(c_k)$ initially have the same value 0, so that *apc*-maxRPC starts resolution by applying AC everywhere. It will start enforcing some amount of maxRPC only after the first wipe-out occurred. On the contrary, in *apx*-maxRPC, when constraints all have the same weight, the local parameter $p(x_i)$ is correlated to the degree of the variable x_i. As a result, *apx*-maxRPC benefits from the filtering power of maxRPC even before the first wipe-out.

In Table 2, we reported only the results on a few representative instances. Table 3 summarizes the whole set of experiments. It shows the average CPU time for each algorithm on all instances of the different classes of problems tested. We considered only the instances solved before the cutoff of one hour by at least one of the four algorithms. To compute the average CPU time of an algorithm on a class of problems, we add the CPU time needed to solve each instance solved before the cutoff of one hour, and for the instances not solved before the cutoff, we add one hour. We observe that the adaptive approach is, on average, faster than the two extreme consistencies AC and maxRPC, except on the Geom class.

In *apx*-maxRPC and *apc*-maxRPC, we update the local parameters $p(x_i)$ or $p(c_k)$ at each node in the search tree. We could wonder if such a frequent update does not produce too much overhead. To answer this question we performed a simple experiment in which we update the local parameters every 10 nodes only.

Table 2. Performance (CPU time, nodes and constraint checks) of AC, variable-based ap-maxRPC (apx-maxRPC), constraint-based ap-maxRPC (apc-maxRPC), and maxRPC on various instances

		AC	apx-maxRPC	apc-maxRPC	maxRPC
scen1-f9	CPU(s)	90.34	**31.17**	33.40	41.56
	#nodes	2,291	1,080	1,241	726
	#ccks	3,740,502	3,567,369	2,340,417	50,045,838
scen2-f25	CPU(s)	70.57	46.40	**27.22**	81.40
	#nodes	12,591	4,688	3,928	3,002
	#ccks	15,116,992	38,239,829	8,796,638	194,909,585
scen6-w2	CPU(s)	7.30	1.25	2.63	**0.01**
	#nodes	2,045	249	610	0
	#ccks	2,401,057	1,708,812	1,914,113	85,769
scen7-w1-f4	CPU(s)	0.28	**0.17**	0.54	0.30
	#nodes	567	430	523	424
	#ccks	608,040	623,258	584,308	1,345,473
scen11-f9	CPU(s)	2,718.65	**1,110.80**	1,552.20	2,005.61
	#ccks	103,506	40,413	61,292	32,882
	#nodes	227,751,301	399,396,873	123,984,968	3,637,652,122
scen11-f10	CPU(s)	225.29	**83.89**	134.46	112.18
	#ccks	9,511	3,510	4,642	2,298
	#nodes	12,972,427	17,778,458	6,717,485	156,005,235
scen11-f11	CPU(s)	156.76	**39.39**	93.69	76.95
	#ccks	7,050	2,154	3,431	1,337
	#nodes	7,840,552	10,006,821	5,143,592	91,518,348
scen11-f12	CPU(s)	139.91	69.50	88.76	**61.92**
	#ccks	7,050	2,597	3,424	1,337
	#nodes	7,827,974	11,327,536	5,144,835	91,288,023
geo50-20d4-75-19	CPU(s)	**242.13**	553.53	657.72	982.34
	#nodes	195,058	114,065	160,826	71,896
	#ccks	224,671,319	594,514,132	507,131,322	2,669,750,690
geo50-20d4-75-30	CPU(s)	**0.84**	1.01	1.07	1.02
	#nodes	359	115	278	98
	#ccks	261,029	432,705	313,168	1,880,927
geo50-20d4-75-84	CPU(s)	**0.02**	0.09	0.05	0.29
	#nodes	59	54	59	52
	#ccks	33,876	80,626	32,878	697,706
queensK20-5-mul	CPU(s)	787.35	2,345.43	**709.45**	Time-out
	#codes	55,596	40,606	41,743	–
	#ccks	347,596,389	6,875,941,876	379,826,516	–
queensK15-5-add	CPU(s)	24.69	17.01	**14.98**	35.05
	#codes	24,639	12,905	12,677	11,595
	#ccks	90,439,795	91,562,150	58,225,434	394,073,525

Table 3. Average CPU time of AC, variable-based ap-maxRPC (apx-maxRPC), constraint-based ap-maxRPC (apc-maxRPC), and maxRPC on all instances of each class of problems tested, when the local parameters are updated at each node

		AC	apx-**maxRPC**	apc-**maxRPC**	**maxRPC**
Average(CPU)	**geom**	**69.28**	180.57	191.03	279.30
	scen	18.95	9.63	**8.30**	13.94
	scen11	810.15	**325.90**	467.28	564.17
	queensK	135.95	395.41	**121.75**	610.51

Table 4. Average CPU time of AC, variable-based ap-maxRPC (apx-maxRPC), constraint-based ap-maxRPC (apc-maxRPC), and maxRPC on all instances of each class of problems tested, when the local parameters are updated every 10 nodes

		AC	apx-**maxRPC**	apc-**maxRPC**	**maxRPC**
Average(CPU)	**geom**	**69.28**	147.20	189.42	279.30
	scen	18.95	**7.40**	8.86	13.94
	scen11	810.15	**311.74**	417.97	564.17
	queensK	135.95	269.51	**117.18**	610.52

We re-ran the whole set of experiments with this new setting. Table 4 reports the CPU time average results. We observe that when the local parameters are updated every 10 nodes, the gain for the adaptive approach is, on average, greater than when the local parameters are updated at each node. This gives room for improvement, by trying to adapt the frequency of update of these parameters.

8   Conclusion

We have introduced the notion of stability of values for arc consistency, a notion based on the depth of their supports in their domain. We have used this notion to propose parameterized consistency, a technique that allows to define levels of local consistency of increasing strength between arc consistency and a given strong local consistency. We have instantiated the generic parameterized consistency approach to max restricted path consistency. We have experimentally shown that the concept of parameterized consistency is viable. Then we have introduced two techniques which allow us to make the parameter adaptable dynamically and locally during search. We have evaluated these two techniques experimentally and we have observed that adapting the level of local consistency during search using the parameterized consistency concept is a promising approach that can outperform both MAC and a strong local consistency on many problems.

References

1. Balafoutis, T., Paparrizou, A., Stergiou, K., Walsh, T.: New algorithms for max restricted path consistency. Constraints 16(4), 372–406 (2011)
2. Bessiere, C.: Constraint propagation. In: Rossi, F., van Beek, P., Walsh, T. (eds.) Handbook of Constraint Programming, ch. 3, Elsevier (2006)
3. Bessiere, C., Régin, J.C., Yap, R.H.C., Zhang, Y.: An optimal coarse-grained arc consistency algorithm. Artif. Intell. 165(2), 165–185 (2005)
4. Bessiere, C., Stergiou, K., Walsh, T.: Domain filtering consistencies for non-binary constraints. Artif. Intell. 172(6-7), 800–822 (2008)
5. Boussemart, F., Hemery, F., Lecoutre, C., Sais, L.: Boosting systematic search by weighting constraints. In: ECAI, pp. 146–150 (2004)
6. Debruyne, R., Bessiere, C.: Domain filtering consistencies. Journal of Artificial Intelligence Research 14, 205–230 (2001)
7. Debruyne, R., Bessière, C.: Some practicable filtering techniques for the constraint satisfaction problem. In: IJCAI (1), pp. 412–417 (1997)
8. Katriel, I., Van Hentenryck, P.: Randomized filtering algorithms. Technical Report CS-06-09, Brown University (June 2006)
9. Sellmann, M.: Approximated consistency for knapsack constraints. In: Rossi, F. (ed.) CP 2003. LNCS, vol. 2833, pp. 679–693. Springer, Heidelberg (2003)
10. Stergiou, K.: Heuristics for dynamically adapting propagation in constraint satisfaction problems. AI Commun. 22, 125–141 (2009)

Global Inverse Consistency
for Interactive Constraint Satisfaction*

Christian Bessiere[1], Hélène Fargier[2], and Christophe Lecoutre[3]

[1] LIRMM-CNRS, University of Montpellier, France
[2] IRIT-CNRS, University of Toulouse, France
[3] CRIL-CNRS, University of Artois, Lens, France
bessiere@lirmm.fr, fargier@irit.fr, lecoutre@cril.fr

Abstract. Some applications require the interactive resolution of a constraint problem by a human user. In such cases, it is highly desirable that the person who interactively solves the problem is not given the choice to select values that do not lead to solutions. We call this property *global inverse consistency*. Existing systems simulate this either by maintaining arc consistency after each assignment performed by the user or by compiling offline the problem as a multi-valued decision diagram. In this paper, we define several questions related to global inverse consistency and analyse their complexity. Despite their theoretical intractability, we propose several algorithms for enforcing global inverse consistency and we show that the best version is efficient enough to be used in an interactive setting on several configuration and design problems. We finally extend our contribution to the inverse consistency of tuples.

1 Introduction

Constraint Programming (CP) is widely used to express and solve combinatorial problems. Once the problem is modelled as a constraint network, efficient solving techniques generate a solution satisfying the constraints, if such a solution exists. However, there are situations where the user has strong opinions about the way to build good solutions to the problem but some of the desirable/undesirable combinations will become clear only once some of the variables are assigned. In this case, the constraint solver should be there to assist the user in the solution design and to ensure her choices remain in the feasible space, removing the combinatorial complexity from her shoulders. See the Synthia system for protein design as an early example of using CP to interactively solve a problem [12]. Another well known example of such an interactive solving of constraint-based models is product configuration [7,1]. The person modelling the product as a constraint network for the company knows its technical and marketing requirements. She models the feasibility, availability and/or marketing constraints about the product. This constraint network captures the catalog of possible products, which may contain billions of solutions, but in an intentional and compact way. Nevertheless, the modeller does not know the constraints or preferences of the customer(s). Now, this

---

* This work has been funded by the ANR ("Agence Nationale de la Recherche") project BR4CP (ANR-11-BS02-008).

C. Schulte (Ed.): CP 2013, LNCS 8124, pp. 159–174, 2013.

is the customer who will look for solutions, with her own constraints and preferences on the price, the colour, or any other configurable feature.

These applications refer to an interactive solving process where the user selects values for variables according to her own preferences and the system checks the constraints of the network, until all variables are assigned and satisfy all constraints of the network. This solving policy raises an important issue: the person who interactively solves the problem should not be led to a dead-end where satisfying all constraints of the network is impossible. Existing interactive solving systems address this issue either by compiling the constraint network into a multivalued decision diagram (MDD) at the modelling phase [1,9,10] or by enforcing arc consistency on the network after each assignment performed by the user [12]. Compiling the constraint network as a MDD can require a significant amount of time and space. That is why compilation is performed offline (before the solving session). As a consequence, configurators based on a MDD compilation are restricted to static constraint networks: non-unary constraints can neither be added nor removed once the network compiled. It is thus not possible for the user to perform complex requirements, e.g., she is interested in travelling to Venezia only during the carnival period. Arc and dynamic arc consistencies require a lighter computational effort but the user can be trapped in dead-ends, which is very risky from a commercial point of view. It has been shown in [5] that arc consistency (and even higher levels of local consistency) can be very bad approximations of the ideal state where all values remaining in the network can be extended to solutions.

The message of our paper is that for many of the problems that require interactive solving of the problem, and especially for real problems, it is computationally feasible to maintain the domains of the variables in a state where they only contain those values which belong to a complete solution extending the current choices of the user. Inspired by the nomenclature used in [6] and [15], we call this level of consistency *global inverse consistency* (GIC).

Our contribution addresses several aspects. First, we formally characterise the questions that underlie the interactive constraint solving loop and we show that they are all NP-hard. Second, we provide several algorithms with increasing sophistication to address those tasks and we experimentally show that the most efficient one is efficient enough to be used in an interactive constraint solving loop of several non trivial configuration and design problems. Third, we finally extend all these contributions to the *positive consistency* of constraints, which is a problem closely related to GIC that appears in configuration.

2   Background

A (discrete) constraint network (CN) N is composed of a finite set of n variables, denoted by $vars(N)$, and a finite set of e constraints, denoted by $cons(N)$. Each variable x has a domain which is the finite set of values that can be assigned to x. The initial domain of a variable x is denoted by $dom^{init}(x)$ whereas the current domain of x is denoted by $dom(x)$; we always have $dom(x) \subseteq dom^{init}(x)$. The maximum domain size for a given CN will be denoted by d. To simplify, a variable-value pair (x, a) such that $x \in vars(N)$ and $a \in dom(x)$ is called a value of N. Each constraint c involves an ordered set of variables, called the *scope* of c and denoted by $scp(c)$, and is semantically defined by a relation, denoted by $rel(c)$, which contains the set of tuples allowed for the variables involved in c. The *arity* of a constraint c is the size of $scp(c)$, and will usually be denoted by r.

An *instantiation* I of a set $X = \{x_1, \ldots, x_k\}$ of variables is a set $\{(x_1, a_1), \ldots, (x_k, a_k)\}$ such that $\forall i \in 1..k, a_i \in dom^{init}(x_i)$; X is denoted by $vars(I)$ and each a_i is denoted by $I[x_i]$. An instantiation I *on* a CN N is an instantiation of a set $X \subseteq vars(N)$; it is *complete* if $vars(I) = vars(N)$. I is *valid* on N iff $\forall (x, a) \in I, a \in dom(x)$. I *covers* a constraint c iff $scp(c) \subseteq vars(I)$, and I *satisfies* a constraint c with $scp(c) = \{x_1, \ldots, x_r\}$ iff (i) I covers c and (ii) the tuple $(I[x_1], \ldots, I[x_r]) \in rel(c)$. An instantiation I on a CN N is *locally consistent* iff (i) I is valid on N and (ii) every constraint of N covered by I is satisfied by I. A *solution* of N is a complete locally consistent instantiation on N; $sols(N)$ denotes the set of solutions of N. A CN N is *satisfiable* iff $sols(N) \neq \emptyset$.

The ubiquitous example of constraint propagation is enforcement of *generalised arc consistency* (GAC) which removes values from domains without reducing the set of solutions of the constraint network. A value (x, a) of a CN N is GAC on N iff for every constraint c of N involving x, there exists a valid instantiation I of $scp(c)$ such that I satisfies c and $I[x] = a$. N is GAC iff every value of N is GAC. Enforcing GAC means removing GAC-inconsistent values from domains until the constraint network is GAC. In this paper, we shall refer to MAC which is an algorithm considered to be among the most efficient generic approaches for the solution of CNs. MAC [17] explores the search space depth-first and enforces (generalised) arc consistency after each decision taken (variable assignment or value refutation) during search. A *past* variable is a variable explicitly assigned by the search algorithm whereas a *future* variable is a variable not (explicitly) assigned. The set of future variables of a CN N is denoted by $vars^{fut}(N)$.

3   Problems Raised by Interactive Constraint Solving

In this section we formally characterise the questions that underlie the interactive constraint solving loop and we study their theoretical complexity.

3.1   Formalization

We first need to define global inverse consistency.

Definition 1 (Global Inverse Consistency). *A value (x, a) of a CN N is* globally inverse consistent *(GIC) iff $\exists I \in sols(N) \mid I[x] = a$. A CN N is GIC iff every value of N is GIC.*

The *GIC closure* of N is the CN obtained from N by removing all the values that do not belong to a solution of N. The obvious problems that follow are to check whether a constraint network is GIC or not, and to enforce GIC.

Problem 1 (Deciding GIC). *Given a CN N, is N GIC?*

Problem 2 (Computing GIC). *Given a CN N, compute the GIC closure of N.*

As we are interested in interactive solving, we define the problem of restoring (maintaining) GIC after the user has performed a variable assignment.

Problem 3 (Restoring GIC). *Given a CN N that is GIC, and a value (x, a) of N, restore GIC after the assignment $x = a$ has been performed.*

In a configuration setting, as soon as some mandatory variables have been set, the user can ask for an automatic completion of the remaining variables. Hence the definition of following problem:

Problem 4 (Solving a GIC network). *Given a CN N that is GIC, find a solution to N.*

3.2   Complexity Results

Not surprisingly, the basic questions related to GIC (Problems 1 and 2) are intractable.

Theorem 1 (Problem 1). *Deciding whether a constraint network N is GIC is NP-complete, even if N is satisfiable.*

Proof. We first prove membership to NP. For each value (x, a) of N, it is sufficient to provide a solution I of N such that the projection $I[x]$ of I on variable x is equal to a. This certificate has size $n \cdot n \cdot d$ and can be checked in polynomial time.

Completeness for NP is proved by reducing 3COL to the problem of deciding whether a satisfiable CN is GIC. Take any instance of the 3COL problem, that is, a graph $G = (V, E)$. Consider the CN N where $vars(N) = \{x_i \mid i \in V\}$, $dom(x_i) = \{0, 1, 2, 3\}, \forall i \in V$, and $cons(N) = \{(x_i \neq x_j) \vee (x_i = 0 \wedge x_j = 0) \mid (i, j) \in E\}$. Clearly $[0, \ldots, 0]$ is a solution of N, and by construction, N has other solutions iff G is 3-colourable. Now, if G is 3-colourable, N is GIC because colours are completely interchangeable. Therefore, N is GIC iff G is 3-colourable.   □

Our proof shows that hardness for deciding GIC holds for binary CNs (i.e., CNs only involving binary constraints). We have another proof, inspired from that used in Theorem 3 in [2], that shows that deciding GIC is still hard for Boolean domains and quaternary constraints.

Theorem 2 (Problem 2). *Computing the GIC closure of a constraint network N is NP-hard and NP-easy, even if N is satisfiable.*

Proof. We prove NP-easiness by showing that a polynomial number of calls to a NP oracle are sufficient to build the GIC closure of N. For each value (x, a) of N, we ask the NP oracle whether N with the extra constraint $x = a$ is satisfiable (we call this an *inverse check*). Once all values have been tested, we build the GIC closure of N by removing from each $dom(x)$ all values a for which the oracle test returned 'no'. Hardness is a direct corollary of Theorem 1.   □

Notice that the two previous intractability results are still valid when the CN is satisfiable, as is the case at the beginning of an interactive resolution session.

We finally show that Problems 3 and 4 are unfortunately not easier than checking GIC or enforcing GIC from scratch. But they are not harder.

Theorem 3 (Problem 3). *Given a CN N that is GIC, and a value (y, b) of N, computing the GIC closure of the CN N', where $vars(N') = vars(N)$ and $cons(N') = cons(N) \cup \{y = b\}$ is NP-hard and NP-easy.*

Proof. NP-easiness is proved as in the proof of Theorem 2 by showing that a polynomial number of calls to a NP oracle are sufficient to build the GIC closure of N'. For each value (x, a) of N (except values (y, a) with $a \neq b$), we ask the NP oracle whether N' with the extra constraint $x = a$ is satisfiable. Once all values have been tested, we build the closure of N' by removing from $dom(y)$ all values $a \neq b$ and removing from each $dom(x)$ all values a for which the oracle test returned 'no'. Hardness is a direct corollary of Theorem 7 in [2]. □

Theorem 4 (Problem 4). *Generating a solution to a GIC constraint network cannot be done in polynomial time, unless $P = NP$.*

Proof. The following proof is derived from [16]. But it is also a corollary of the recent and more complex Theorem 3.1 in [8].

Suppose we have an algorithm A that generates a solution to a GIC constraint network N in time bounded by a polynom $p(|N|)$. Take any instance of the 3COL problem, that is, a graph $G = (V, E)$. Consider the CN N where $vars(N) = \{x_i \mid i \in V\}$, $dom(x_i) = \{0, 1, 2\}, \forall i \in V$, and $cons(N) = \{x_i \neq x_j \mid (i, j) \in E\}$. N has a solution iff G is 3-colourable. Now, if G is 3-colourable, N is GIC because colours are completely interchangeable. Thus, it is sufficient to run A during $p(|N|)$ steps. If it returns a solution to N, then the 3COL instance is satisfiable. Otherwise, the 3COL instance is unsatisfiable. Therefore, as 3COL is NP-complete, there cannot exist a polynomial algorithm for generating a solution to a GIC constraint network, unless $P = NP$. □

4   GIC Algorithms

In this section, we introduce four algorithms to enforce global inverse consistency. These GIC algorithms use increasingly sophisticated data structures and techniques that have recently proved their worth in filtering algorithms proposed in the literature; e.g., see [14,18]. To simplify our presentation, we assume that the CNs are satisfiable, which is the case in interactive resolution, allowing us to avoid handling domain wipe-outs in the GIC procedures. Note that these algorithms can be used to enforce GIC, but also to maintain it during a user-driven search. This is why we refer to the set $vars^{fut}(N)$ of future variables in some instructions.

The first algorithm, GIC1, described in Algorithm 1, is really basic: it will be used as our baseline during our experiments. For each value a in the domain of a future variable x, a solution for the CN N where x is assigned the value a, denoted by $N|_{x=a}$, is sought using a complete search algorithm. This search algorithm, called here searchSolutionFor, either returns the first solution that can be found, or the special value nil. Our implementation choice will be the algorithm MAC that maintains (G)AC during a backtrack search [17]. Hence, in Algorithm 1, when it is proved with searchSolutionFor that no solution exists, i.e., $I = nil$, the value a can be deleted. Note that, in contrary to weaker forms of consistency, when a value is pruned there is no need for GIC to repeat the process of iterating over the values remaining in the CN.

Algorithm 1. GIC1(N: CN)

---

1 **foreach** *variable* $x \in vars^{fut}(N)$ **do**
2   **foreach** *value* $a \in dom(x)$ **do**
3        $I \leftarrow$ searchSolutionFor($N|_{x=a}$)
4   **if** $I = nil$ **then**
5   remove a from $dom(x)$

Algorithm 2. handleSolution2/3(x: variable, I: instantiation)

---

1 **foreach** *variable* $y \in vars^{fut}(N) \mid y$ *is revised after* x **do**
2   **if** $\text{stamp}[y][I[y]] \neq \text{time}$ **then**
3        $\text{stamp}[y][I[y]] \leftarrow \text{time}$
4        $\text{nbGic}[y] + +$

Algorithm 3. isValid(X : set of variables, I : instantiation): Boolean

---

1 **foreach** *variable* $x \in X$ **do**
2   **if** $I[x] \notin dom(x)$ **then**
3   **return** *false*

4 **return** *true*

Algorithm 4. GIC2/3(N: CN)

Data: GIC3 is obtained by considering light grey coloured instructions between lines 5
      and 6, and after line 10

1 $\text{time} + +$
2 **foreach** *variable* $x \in vars^{fut}(N)$ **do**
3     $\text{nbGic}[x] \leftarrow 0$

4 **foreach** *variable* $x \in vars^{fut}(N) \mid \text{nbGic}[x] < |dom(x)|$ **do**
5   **foreach** *value* $a \in dom(x) \mid \text{stamp}[x][a] < \text{time}$ **do**
       if isValid($vars(N)$,$residue[x][a]$) **then**
          handleSolution2/3(x,$residue[x][a]$)
          continue
6        $I \leftarrow$ searchSolutionFor($N|_{x=a}$)
7   **if** $I = nil$ **then**
8   remove a from $dom(x)$
9   **else**
10   handleSolution2/3(x,I)
          $residue[x][a] \leftarrow I$

The second algorithm, GIC2 described in Algorithm 4 (ignoring light grey lines), uses timestamping. This is useful when GIC is maintained during a user-driven search. We use an integer variable time for counting time, and we introduce a two-dimensional array stamp that associates with each value (x, a) of the CN the last time (value of stamp$[x][a]$) a solution was found for that value. We also assume that variables are implicitly totally ordered (for example, in lexicographic order). Then, the idea is to increment the value of the variable time whenever a new call to GIC2 is performed (see line 1) and to test time against each value (x, a) of the CN (see line 5) to determine whether it is necessary or not to search for a solution for (x, a). When a solution I is found, function handleSolution2/3 is called at line 10 in order to update stamps. Actually, we only update the stamps of values in I corresponding to variables that are processed after x in the loop of revisions (line 4) in Algorithm 4. These are the variables that have not been processed yet by the loop at line 4 of Algorithm 4. Finally, by further introducing a one-dimensional array nbGic that associates with each variable x of the CN the number of values in $dom(x)$ that have been proved to be GIC, it is possible to avoid some iterations of loop 5; see initialization at lines 2-3, testing at line 4 and update at line 4 of Algorithm 2.

The third algorithm, GIC3, described in Algorithm 4 when considering light grey lines, can be seen as a refinement of GIC2 obtained by exploiting residues, which correspond to solutions that have been previously found. Here, we introduce a two-dimensional array residue that associates with each value (x, a) of the CN the last solution found for this value (potentially, during another call to GIC3). Because residual solutions may not be valid anymore, for each value (x, a) we need to test the validity of residue$[x][a]$ by calling the function isValid; see instructions between lines 5 and 6. If the residue is valid, we call handleSolution2/3 to update the other data structures, and we continue with the next value in the domain of x. A validity test, Algorithm 3, only checks that all values in a given complete instantiation are still present in the current domains. Of course, when a new solution is found, we record it as a residue; see instruction after line 10.

Our last algorithm, GIC4 described in Algorithm 6, is based on an original use of simple tabular reduction [18]. The principle is to record all solutions found during the enforcement of GIC in a table, so that an (adaptation of an) algorithm such as STR2 [13] can be applied. The current table is given by all elements of an array solutions at indices ranging from 1 to nbSolutions. As for STR2, we introduce two sets of variables called S^{val} and S^{sup}. The former allows us to limit validity control of solutions to the variables whose domains have changed recently (i.e., since the last execution of GIC4). This is made possible by reasoning from domain cardinalities, as performed at lines 3 and 26–27 with the array lastSize. The latter (S^{sup}) contains any future variable x for which at least one value is not in the array gicValues[x], meaning that it has still to be proved GIC. Related details can be found in [13]. After the initialization of S^{val} and S^{sup} (lines 1–8), each instantiation solutions[i] of the current table is processed (lines 11–16). If it remains valid (hence, a solution), we update structures gicValues and S^{sup} by calling the function handleSolution4. Otherwise, this instantiation is deleted by swapping it with the last one. The rest of the algorithm (lines 17–25) just tries to find a solution support for each value not present in gicValues. When a

---

Algorithm 5. handleSolution4(I : instantiation)

---

1 **foreach** *variable* $x \in S^{sup}$ **do**
2   **if** $I[x] \notin$ gicValues$[x]$ **then**
3   gicValues$[x] \leftarrow$ gicValues$[x] \cup \{I[x]\}$
4   **if** $|$gicValues$[x]| = |dom(x)|$ **then**
5          $S^{sup} \leftarrow S^{sup} \setminus \{x\}$

Algorithm 6. GIC4(N: CN)

```
// Initialization of structures
```
1 $S^{val} \leftarrow \emptyset$
2 **foreach** *variable* $x \in vars(N)$ **do**
3   **if** $|dom(x)| \neq$ lastSize$[x]$ **then**
4       $S^{val} \leftarrow S^{val} \cup \{x\}$

5 $S^{sup} \leftarrow \emptyset$
6 **foreach** *variable* $x \in vars^{fut}(N)$ **do**
7   gicValues$[x] \leftarrow \emptyset$
8    $S^{sup} \leftarrow S^{sup} \cup \{x\}$

```
// The table of current solutions is traversed
```
9 $i \leftarrow 1$
10 **while** $i \leq$ nbSolutions **do**
11   **if** isValid(S^{val},solutions$[i]$) **then**
12   handleSolution4(solutions$[i]$)
13       $i++$
14   **else**
15   solutions$[i] \leftarrow$ solutions$[$nbSolutions$]$
16   nbSolutions $--$

```
// Search for values not currently supported is performed
```
17 **foreach** *variable* $x \in S^{sup}$ **do**
18   **foreach** *value* $a \in dom(x) \setminus$ gicValues$[x]$ **do**
19       $I \leftarrow$ searchSolutionFor($N|_{x=a}$)
20   **if** $I = nil$ **then**
21   remove a from $dom(x)$
22   **else**
23   nbSolutions $++$
24   solutions$[$nbSolutions$] \leftarrow I$
25   handleSolution4(I)

26 **foreach** *variable* $x \in vars^{fut}(N)$ **do**
27   lastSize$[x] \leftarrow |dom(x)|$

new solution is found, it is recorded in the current table (lines 23–24) and handleSolution4 is called (line 25).

Theorem 5. *Algorithms GIC1, GIC2, GIC3 and GIC4 enforce GIC.*

Proof. (sketch) This is immediate for GIC1. For GIC2 and GIC3, the use of timestamps and residues permits us to avoid useless inverse checks. For GIC4, the same arguments as those used for proving that STR2 enforces GAC hold. Simply, additional inverse checks are performed for values not collected (in `gicValues`) during the traversal of the current table.   □

The worst-case space complexity (for the specific data structures) of GIC1 is $O(1)$. For GIC2 and GIC3, this is $O(nd)$ because `nbGic` is $O(n)$, `stamp` and `residue` are $O(nd)$. For GIC4, S^{val}, S^{sup} and `lastSize` are $O(n)$, `gicValues` is $O(nd)$, and the structure `solutions` is $O(n^2 d)$. The time complexity of the GIC algorithms can be expressed in term of the number of calls to the (oracle) searchSolutionFor. For GIC1, this is $O(nd)$. For GIC2, in the *best-case*, only d calls are necessary, one call permitting to prove (through timestamping) that n values are GIC. For GIC3 and GIC4, still in the *best-case* and assuming the case of maintaining GIC (i.e., after the assignment of a variable by the user), no call to the oracle is necessary (residues and the current table permit alone to prove that all values are GIC). This rough analysis of time complexity suggests that GIC3 and GIC4 might be the best options.

5   Tuple Inverse Consistency

Up to this point, we have based our analysis on the last part of the interactive resolution process, i.e., the specification of a solution of the constraint network by the user. This allowed us to make the simplifying assumptions that the user is only looking at the domains of the variables. After each variable assignment, she just wants to know which values remain feasible for non assigned variables.

The situation is different at the modelling phase, e.g., the engineers of the company dynamically build the set of constraints that define the configurable product. At this point, GIC is also a crucial functionality, not for deriving a solution (a end product), but to ensure that each of the options proposed in the catalog (each of the values in the domains of the constraint network) is present in at least one end product. It is meaningless to propose (and advertise on) a sophisticated air bag system when it cannot equip any car in practice.

In that modelling phase, the need for information on the extensibility to solutions is not restricted to domains, but extends to (some of) the constraints of the model. Many constraints have actually a double meaning. Following the standard semantics of constraints, the first one is negative: technical constraints forbid combinations of variables. The second one is positive: the possibilities that are left by some (generally, table) constraints *have to* be effective. Let us assume, for instance, that a constraint means 'The level of equipment of vehicles with type M3 engine can be middle level or luxurious'. If some other constraint excludes the vehicles M3 engine for luxurious level of equipment, the specification of the product is considered as inconsistent. This property has been called *positive consistency* in [2] and actually refers to the extensibility to a solution of each of the tuples allowed by the constraint of interest:

Definition 2 (Tuple Inverse Consistency). *Given a CN N, a tuple τ on a set of variable X is said to be inverse consistent (TIC) in N iff there exists a solution I of N such that $\forall x \in X, I[x] = \tau[x]$.*

Definition 3 (Positive Consistency). *A constraint c is positively consistent in N iff for any valid tuple $\tau \in rel(c)$, τ is TIC.*

The positive closure of a constraint c is the constraint obtained from c by removing from $rel(c)$ all the valid tuples that are not TIC in N. The obvious problem that follows is to check whether a constraint is positively consistent or not.

Problem 5 (Deciding Positive Consistency). *Given a CN N and a constraint c of N, is c positively consistent in N?*

Deciding positive consistency has been shown to be NP-hard, even when the constraint network is known to be satisfiable ([2]). The other problem of interest is to restore positive consistency on a constraint after the user has refined her model by adding a constraint to the network.

Problem 6 (Restoring Positive Consistency). *Given a CN N, given a positive consistent constraint c in N, given any extra constraint c' not in cons(N), compute the new positive closure of c in the network obtained from N by adding c' to cons(N).*

6 Experiments

In order to show the practical interest of our approach, we have performed several experiments mainly using a computer with processors Intel(R) Core(TM) i7-2820QM CPU 2.30GHz; for random instances, we used a cluster of Xeon 3.0GHz with 13GB of RAM. Our main purpose was to determine whether maintaining GIC is a viable option for configuration-like problem instances and for interactive puzzle creation, as well as to compare the relative efficiency of the four GIC algorithms described in Section 4.

Table 1. Features of six Renault configuration instances

	n	d	e	r	t	D	T
souffleuse	32	12	35	3	55	145	350
megane	99	42	113	10	48,721	396	194,838
master	158	324	195	12	26,911	732	183,701
small	139	16	147	8	222	340	3,044
medium	148	20	174	10	2,718	424	9,532
big	268	324	332	12	26,881	1,273	225,989

In Table 1, we show relevant features of car configuration instances, generated with the help of our industrial partner Renault. For each of the six instances currently available,[1] we indicate

---

[1] See http://www.irit.fr/~Helene.Fargier/BR4CP/benches.html

- the number of variables (n),
- the size of the greatest domain (d),
- the number of constraints (e),
- the greatest constraint arity (r),
- the size of the greatest table (t),
- the total number of values ($D = \sum_{x \in vars(N)} |dom(x)|$),
- and the total number of tuples ($T = \sum_{c \in cons(N)} |rel(c)|$).

The left part of Table 2 presents the CPU time required to establish GIC on the six Renault configuration instances. Clearly GIC1 is outperformed by the three other algorithms, which have here rather similar efficiency. The right part of Table 2 aims at simulating the behaviour of a configuration software user who makes the variable choices and value selections. It presents the CPU time required to maintain GIC along a single branch built by performing random variable assignments (random variable assignment simulates the user, who chooses the variables and the values according to her preference). Specifically, variables and values are randomly selected in turn, and after each assignment, GIC is systematically enforced to maintain this property. Of course, no conflict (dead-end) can occur along the branch due to the strength of GIC, which is why we use the term of greedy executions. CPU times are given on average for 100 executions (different random orderings). For all instances, GIC3 and GIC4 are maintained very fast, whereas on the biggest instances, GIC2 requires a few seconds and GIC1 around ten seconds.

Table 2. CPU time (in seconds) to establish GIC on Renault configuration instances, and to maintain it (average over 100 random greedy executions)

	Establishing GIC with				Maintaining GIC with			
	GIC1	GIC2	GIC3	GIC4	GIC1	GIC2	GIC3	GIC4
souffleuse	0.02	0.01	0.01	0.01	0.13	0.07	**0.02**	**0.02**
megane	2.94	0.71	0.72	0.71	4.26	1.18	0.05	**0.04**
master	2.45	1.35	1.33	1.33	9.81	3.57	0.07	**0.06**
small	0.14	0.02	0.03	0.03	0.32	0.05	**0.01**	**0.01**
medium	0.26	0.04	0.05	0.04	0.35	0.04	**0.01**	**0.01**
big	4.19	1.16	1.10	1.10	12.6	2.60	**0.05**	**0.05**

One great advantage of GIC is that it guarantees that a conflict can never occur during a configuration session. However, one may wonder whether the risk of failure(s) is really important in user-driven searches that use a weaker consistency such as GAC or a partial form of it (Forward Checking). Table 3 shows the number of conflicts (sum over 100 executions using random orderings) encountered when following a MAC or a nFC2 [3] strategy. The number of conflict situations can be very large with nFC2 (for two instances, we even report the impossibility of finding a solution within 10 minutes with some random orderings). For MAC, the number of failures is rather small but the risk is not null (for example, the risk is equal to 5% for megane).

Table 3. Number of conflicts encountered when running nFC2 and MAC (sum over 100 random executions)

	souffleuse	megane	master	small	medium	big
nFC2	252,605	313,910	time-out	3,728	7,824	time-out
MAC	0	7	5	0	3	3

The encouraging results obtained on Renault configuration instances led us to test other problems, in particular to get a better picture of the relative efficiency of the various GIC algorithms. For example, on classical Crossword instances (see Table 4), GIC1 is once again clearly outperformed while the three other algorithms are quite close, where there is still a a small benefit of using GIC4.

Table 4. CPU time (in seconds) to establish GIC on some Crosswords instances, and to maintain it on average over 100 random greedy executions

	Establishing GIC with				Maintaining GIC with			
	GIC1	GIC2	GIC3	GIC4	GIC1	GIC2	GIC3	GIC4
ogd-vg5-5	2.25	0.67	0.67	0.67	2.34	0.79	0.73	**0.70**
ogd-vg5-6	6.40	2.18	2.19	2.19	7.42	2.82	2.58	**2.48**
ogd-vg5-7	25.8	9.91	9.87	9.84	33.4	15.2	14.3	**13.8**

Table 5. CPU time (in seconds) to establish GIC on Puzzle instances, and to maintain it on average over 100 random greedy executions until a unique solution is found

	Establishing GIC with				Maintaining GIC with			
	GIC1	GIC2	GIC3	GIC4	GIC1	GIC2	GIC3	GIC4
sudoku-9x9	1.58	0.32	0.32	0.31	15.3	2.71	2.10	**1.74**
sudoku-16x16	6.04	0.51	0.50	0.50	246	25.5	26.5	**18.9**
magicSquare-4x4	0.96	0.26	0.28	0.28	1.63	**0.69**	0.71	0.71
magicSquare-5x5	14.7	3.01	3.10	2.99	55.1	15.9	15.6	**13.7**

It is worthwhile to note that GIC is a nice property that can be useful when puzzles, where hints are specified, have to be conceived. Typically, one looks for puzzles where only one solution exists. One way of building such puzzles is to add hints in sequence, while maintaining GIC, until all domains become singleton. For example, this is a possible approach for constructing Sudoku and Magic Square grids, with the advantage that the user can choose freely the position of the hints.[2] On the left part of Table 5, we report the time to enforce GIC on empty Sudoku grids of size 9x9 and 16x16, and on empty Magic square of size 4x4 and 5x5, and on the right part, the average time required to maintain GIC until a fixed point is reached, meaning that after several hints have been randomly selected and propagated, we have the guarantee of having a one-solution puzzle. GIC4 is a clear winner, with for example, a 30% speedup over GIC2

---

[2] However, we are not claiming that maintaining GIC is the unique answer to this problem.

and GIC3 on sudoku-16x16, and more than one order of magnitude over GIC1. Overall, the results we obtain show that MIC, i.e., maintaining GIC, is a practicable solution (at least for some problems) as the average time between each decision of the user is small with GIC4.

The efficiency of MIC on structured under-constrained instances piqued our curiosity. So we decided to compare MIC (embedding GIC4) and MAC on series of binary random instances generated from Model RB [20]. For the class $RB(2, 30, 0.8, 3, t)$, see [19], we obtain instances with 30 variables, 15 values per domain and 306 binary constraints of tightness t, and for the class $RB(2, 40, 0.8, 3, t)$, instances with 40 variables, 19 values per domain and 443 binary constraints of tightness t. For each value of t ranging from 0.01 to 0.50 (step of 0.01), a series of 100 instances was generated so as to observe the behaviour of MIC on both under-constrained instances and over-constrained instances; the theoretical threshold is around 0.23. Figure 1 shows the average CPU time of MIC and MAC on series of class $RB(2, 30, 0.8, 3, t)$. On the left, Figure 1(a), the ordering of variables and values is random (simulating a free user-driven search). MIC outperforms MAC when the ordering is random and the tightness is greater than or equal to 0.23. That means that the strong inference capability of MIC do pay off for the unsatisfiable instances. On the right, Figure 1(b), the variable ordering heuristic is *dom/wdeg* [4] and the value ordering heuristic is lexico. Obviously, MAC with *dom/wdeg* is clearly faster than MIC. However, if used in a context of interactive resolution, the *dom/wdeg* ranking of the variables drives the user, who is not free anymore in the choices of its variables. It may ask her to assign first variables that are meaningless to her, restricting her future choices on important variables. The outcome will be a solution which is very bad with respect to the preferences of the user. All of this suggests that MIC can be efficient enough to be used in practice, except for a (small) region of satisfiable instances lying at the left of the threshold point. Figure 2 shows similar results with respect to series of class $RB(2, 40, 0.8, 3, t)$.

One other practical issue we are interested in is the effectiveness of positive consistency. Hence, we tested to establish positive consistency on existing constraints of the Renault configuration instances, see Table 6. The algorithm we used here is a simple adaptation of GIC1 to tuples (so, certainly, several optimizations are possible). A few hundreds of seconds are necessary to ensure the positive consistency of all existing constraints of the biggest instances.

Table 6. CPU time (in seconds) and filtering in term of the number of tuples deleted when establishing positive consistency on Renault configuration instances

	souffleuse	megane	master	small	medium	big
CPU	0.68	352	368	2.6	4.2	613
# tuples removed	0	138,493	90,874	240	5,425	105,020

Finally, in our last experiment, for each constraint network, we randomly select a constraint of interest c_i for which positive consistency must be ensured (as if the modeller were asking for the positive consistency of this constraint), and we randomly select a set C containing 10% of the set of constraints. We initially consider the CN without

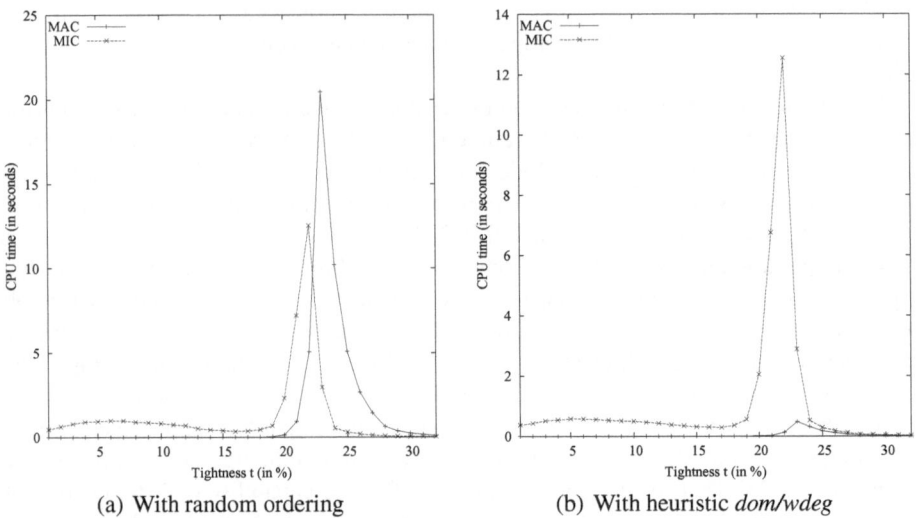

(a) With random ordering   (b) With heuristic *dom/wdeg*

Fig. 1. Mean search cost (100 instances) of solving instances in class $RB(2, 30, 0.8, 3, t)$ with MAC and MIC

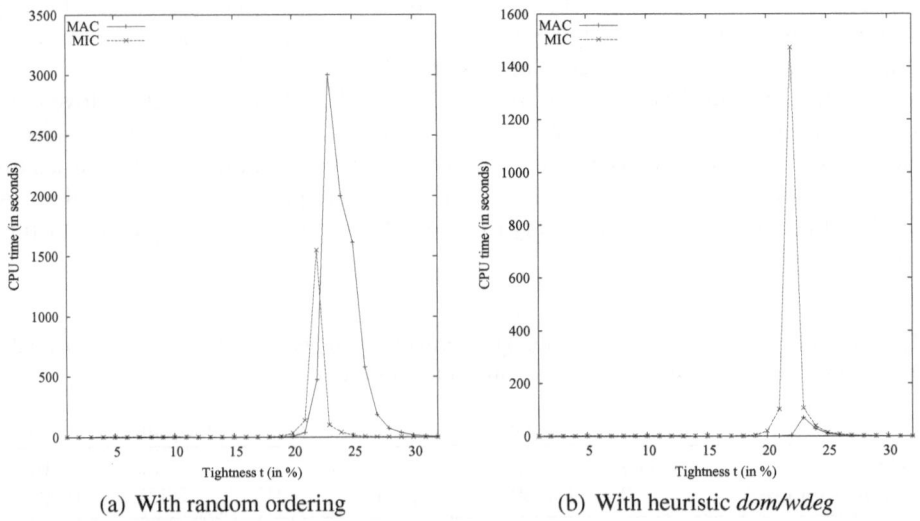

(a) With random ordering   (b) With heuristic *dom/wdeg*

Fig. 2. Mean search cost (100 instances) of solving instances in class $RB(2, 40, 0.8, 3, t)$ with MAC and MIC

the constraints in C, and we enforce positive consistency on c_i. Then we simulate a session of product modelling: we post each constraint in C in turn and maintain positive consistency on c_i. In our implementation (not detailed here due to lack of space), we use residues, i.e., a solution stored for each tuple of c_i. The first line of Table 7 shows the average CPU time to maintain positive consistency on the constraint of interest. For the second line, the constraint of interest is not randomly chosen but set to the constraint with the largest table. The obtained results are rather promising (except for the instance megane).

Table 7. Dynamic positive consistency filtering on Renault configuration instances (average CPU time over 100 executions)

	megane	master	big
random	9.97	10.1	36.4
largest	106.6	11.4	20.6

7 Conclusion

We have analysed the problems that arise in applications that require the interactive resolution of a constraint problem by a human user. The central notion is global inverse consistency of the network because it ensures that the person who interactively solves the problem is not given the choice to select values that do not lead to solutions. We have shown that deciding, computing, or restoring global inverse consistency, and other related problems are all NP-hard. We have proposed several algorithms for enforcing global inverse consistency and we have shown that the best version is efficient enough to be used in an interactive setting on several configuration and design problems. This is a great advantage compared to existing techniques usually used in configurators. As opposed to techniques maintaining arc consistency, our algorithms give an exact picture of the values remaining feasible. As opposed to compiling offline the problem as a multi-valued decision diagram, our algorithms can deal with constraint networks that change over time (e.g., an extra non-unary constraint posted by a customer who does not want to buy a car with more than 100,000 miles except if it is a Volvo). We have finally extended our contribution to the inverse consistency of tuples, which is useful at the modelling phase of configuration problems.

One direct perspective of this work is to try computing diverse solutions when enforcing GIC. This should permit, on average, to reduce the number of search runs. Some techniques developed in [11] might be useful.

References

1. Amilhastre, J., Fargier, H., Marquis, P.: Consistency restoration and explanations in dynamic CSPs - application to configuration. Artificial Intelligence 135(1-2), 199–234 (2002)
2. Astesana, J.M., Cosserat, L., Fargier, H.: Constraint-based vehicle configuration: A case study. In: Proceedings of ICTAI 2010, pp. 68–75 (2010)

3. Bessiere, C., Meseguer, P., Freuder, E.C., Larrosa, J.: On Forward Checking for non-binary constraint satisfaction. Artificial Intelligence 141, 205–224 (2002)
4. Boussemart, F., Hemery, F., Lecoutre, C., Sais, L.: Boosting systematic search by weighting constraints. In: Proceedings of ECAI 2004, pp. 146–150 (2004)
5. Debruyne, R., Bessiere, C.: Domain filtering consistencies. Journal of Artificial Intelligence Research 14, 205–230 (2001)
6. Freuder, E.C., Elfe, C.D.: Neighborhood inverse consistency preprocessing. In: Proceedings of AAAI 1996, Portland, Oregon, pp. 202–208 (1996)
7. Gelle, E., Weigel, R.: Interactive configuration using constraint satisfaction techniques. In: Proceedings of PACT 1996, pp. 37–44 (1996)
8. Gottlob, G.: On minimal constraint networks. Artificial Intelligence 191-192, 42–60 (2012)
9. Hadzic, T., Andersen, H.R.: Interactive reconfiguration in power supply restoration. In: van Beek, P. (ed.) CP 2005. LNCS, vol. 3709, pp. 767–771. Springer, Heidelberg (2005)
10. Hadzic, T., Hansen, E.R., O'Sullivan, B.: Layer compression in decision diagrams. In: Proceedings of ICTAI 2008, pp. 19–26 (2008)
11. Hebrard, E., Hnich, B., O'Sullivan, B., Walsh, T.: Finding diverse and similar solutions in constraint programming. In: Proceedings of AAAI 2005, pp. 372–377 (2005)
12. Janssen, P., Jégou, P., Nouguier, B., Vilarem, M.C., Castro, B.: SYNTHIA: Assisted design of peptide synthesis plans. New Journal of Chemistry 14(12), 969–976 (1990)
13. Lecoutre, C.: STR2: Optimized simple tabular reduction for table constraints. Constraints 16(4), 341–371 (2011)
14. Lecoutre, C., Hemery, F.: A study of residual supports in arc consistency. In: Proceedings of IJCAI 2007, pp. 125–130 (2007)
15. Martinez, D.: Résolution interactive de problemes de satisfaction de contraintes. PhD thesis, Supaero, Toulouse, France (1998)
16. Papadimitriou, C.: Private communication (1999)
17. Sabin, D., Freuder, E.C.: Contradicting conventional wisdom in constraint satisfaction. In: Borning, A. (ed.) PPCP 1994. LNCS, vol. 874, pp. 10–20. Springer, Heidelberg (1994)
18. Ullmann, J.R.: Partition search for non-binary constraint satisfaction. Information Science 177, 3639–3678 (2007)
19. Xu, K., Boussemart, F., Hemery, F., Lecoutre, C.: Random constraint satisfaction: easy generation of hard (satisfiable) instances. Artificial Intelligence 171(8-9), 514–534 (2007)
20. Xu, K., Li, W.: Exact phase transitions in random constraint satisfaction problems. Journal of Artificial Intelligence Research 12, 93–103 (2000)

Counting Spanning Trees to Guide Search in Constrained Spanning Tree Problems

Simon Brockbank, Gilles Pesant, and Louis-Martin Rousseau

[1] École polytechnique de Montréal, Montreal, Canada
[2] CIRRELT, Université de Montréal, Montreal, Canada
{simon.brockbank,gilles.pesant,louis-martin.rousseau}@polymtl.ca

Abstract. Counting-based branching heuristics such as maxSD were shown to be effective on a variety of constraint satisfaction problems. These heuristics require that we equip each family of constraints with a dedicated algorithm to compute the local solution density of variable assignments, much as what has been done with filtering algorithms to apply local inference. This paper derives an exact polytime algorithm to compute solution densities for a spanning tree constraint, starting from a known result about the number of spanning trees in a graph. We then empirically compare branching heuristics based on that result with other generic heuristics.

1 Introduction

Constraint programming is a powerful approach that can be used to solve combinatorial problems. However its success depends heavily on heuristics that can guide the search toward promising areas of the search tree. One can design a heuristic dedicated to the particular problem at hand or rely on out-of-the-box generic heuristics that have shown good performance on a variety of problems. The last decade has witnessed renewed interest in the design of robust generic branching heuristics (e.g. [9,6]). In particular Zanarini and Pesant[15] introduced branching heuristics based on the concept of *solution density*, i.e. the proportion of solutions local to a constraint featuring a given variable-value assignment.

Definition 1 (solution density). *Given a constraint* $c(x_1, \ldots, x_n)$, *its number of solutions* $\#c(x_1, \ldots, x_n)$, *respective finite domains* D_i $_{1 \leq i \leq n}$, *a variable* x_i *in the scope of* c, *and a value* $d \in D_i$, *we will call*

$$\sigma(x_i, d, c) = \frac{\#c(x_1, \ldots, x_{i-1}, d, x_{i+1}, \ldots, x_n)}{\#c(x_1, \ldots, x_n)}$$

the solution density *of pair* (x_i, d) *in c. It measures how often a certain assignment is part of a solution to* c.

Specialized algorithms have been designed to compute solution densities for several families of constraints[8]. In this paper we propose an exact polytime algorithm that computes solution densities for a spanning tree constraint.

C. Schulte (Ed.): CP 2013, LNCS 8124, pp. 175–183, 2013.

Definition 2 (Spanning Tree Constraint (adapted from [4])). *Given an undirected graph $G(V, E)$ and set variable $T \subseteq E$, constraint spanningTree(G, T) restricts T to be a spanning tree of G.*

For the sake of conforming to the previous definition of solution density, especially important if we are to allow the combination of solution density information from different constraints, we instead represent T as an array of boolean variables.

The rest of the paper is organized as follows: Section 2 exposes the related work, Section 3 describes our algorithm to compute solution densities for the spanning tree constraint, Section 4 discusses how our data structures are updated in the course of backtrack search, and Section 5 provides supporting empirical evidence of branching based on solution density.

2   Related Work

Research in the CP community about imposed tree structures has focused so far on filtering algorithms and not on branching heuristics. Beldiceanu et al.[2] introduced the tree constraint, which addresses the digraph partitioning problem from a constraint programming perspective. In their work a constraint that enforces a set of vertex-disjoint anti-arborescences is proposed. They achieve domain consistency in $\mathcal{O}(nm)$ time, where n is the number of vertices and m is the number of edges in the graph. Their pruning relies on the identification of strong articulation points in the graph and of roots and sinks (to evaluate the minimum and maximum number of trees required to partition the graph).

Dooms and Katriel[3] introduced the MST constraint, requiring the tree variable to represent a minimum spanning tree of the graph on which the constraint is defined. Many variants of the minimum spanning tree problem, such as minimum k-spanning tree and Steiner tree are known to be NP-hard, even though its basic version can be solved in polynomial time. Those problems can be modeled by combining the MST constraint and other constraints. The authors proposed polytime bound consistent filtering algorithms for several restrictions of this constraint. They proceed by classifying edges in three sets: mandatory, possible, and forbidden. Afterwards Dooms and Katriel[4] proposed a weighted spanning tree constraint, in which both the tree and the weight of the edges are variables, and considered several filtering algorithms. In their work a set variable is used, indicating which edges are tree edges.

The filtering proposed by Dooms and Katriel[4] was then simplified and improved by Régin[10], who proposed an incremental filtering algorithm by maintaining a connected component tree which represents disjoint trees merging operations in Kruskal's algorithm, and by computing lowest common ancestors on that tree. Domain consistency was thus achieved in $\mathcal{O}(m + n \log n)$ time. Subsequently, Régin et al.[11] improved the time complexity of that filtering and also considered mandatory edges.

3   Computing Solution Densities

The *Laplacian matrix* $L(G)$ of a graph G is formed by subtracting the adjacency matrix of G from the diagonal matrix whose i^{th} entry is equal to the degree of vertex i in G. Henceforth for notational convenience we will refer to it simply as L. For example Figure 1 shows a graph and its Laplacian matrix.

$$L = \begin{pmatrix} 3 & -1 & -1 & -1 \\ -1 & 2 & -1 & 0 \\ -1 & -1 & 3 & -1 \\ -1 & 0 & -1 & 2 \end{pmatrix}$$

Fig. 1. The kite graph and its Laplacian matrix

The (i, j)-*minor* of a square matrix M, denoted M_{ij}, is the determinant of the sub-matrix obtained by removing from M its ith row and j^{th} column. The Laplacian matrix has the interesting property that its (i, j)-minor, for any row i and column j, is equal to the number of spanning trees of the corresponding graph.

Theorem 1 (Kirchhoff's Matrix-Tree Theorem [13]). *Denote by $\tau(G)$ the number of spanning trees of graph G on n vertices. For any $1 \leq i, j \leq n$,*

$$\tau(G) = L_{ij}.$$

So the number of solutions to a `spanningTree` constraint can be computed as the determinant of a $(n - 1) \times (n - 1)$ matrix, in $\mathcal{O}(n^3)$ time.

If we remove the first row and column of the Laplacian matrix at Figure 1, the resulting minor is $2 \times (3 \times 2 - (-1) \times (-1)) - (-1) \times (-1 \times 2 - (-1) \times 0) = 8$ and one can easily verify that there are eight possible spanning trees for that graph.

But we are interested in computing the solution density of an edge $(i, j) \in E$. One way to approach this is by counting the number of spanning trees not using that edge, $\tau(G \backslash \{(i, j)\})$, and then dividing that by the total number of spanning trees, yielding the solution density of the corresponding variable being assigned value 0 (i.e. $(i, j) \notin T$):

$$\sigma((i, j), 0, \texttt{spanningTree}(G, T)) = \frac{\tau(G \setminus \{(i, j)\})}{\tau(G)}.$$

Let $L' = L(G \backslash \{(i, j)\})$. How different is L' from L? It will be identical except for entries ℓ_{ii}, ℓ_{jj}, ℓ_{ij}, and ℓ_{ji}. Since we can choose any row and column of L to compute our minor, consider removing row and column i. Then the only difference is $\ell'_{jj} = \ell_{jj} - 1$. The *Sherman-Morrison formula* [12] tells us that if

M' is obtained from matrix M by replacing its jth column, $(M)_j$, by column vector u then

$$\det(M') = (1 + e_j^\top M^{-1}(u - (M)_j))\det(M).$$

In our case $(u - (M)_j) = -e_j$ so the right-hand side of the previous equation simplifies to $(1 - e_j^\top M^{-1} e_j)\det(M) = (1 - m_{jj}^{-1})\det(M)$. So finally we have

$$\sigma((i,j), 0, \mathtt{spanningTree}(G,T)) = \frac{L'_{ii}}{L_{ii}} = \frac{(1 - m_{jj}^{-1})L_{ii}}{L_{ii}} = 1 - m_{jj}^{-1},$$

and of course

$$\sigma((i,j), 1, \mathtt{spanningTree}(G,T)) = m_{jj}^{-1}.$$

Computing solution densities turns out to be quite simple: for each edge (i,j) incident with vertex i such that $j < i$ (respectively $j > i$), the corresponding value is the jth (respectively $(j-1)^{th}$) entry on the diagonal of the inverse of M, the sub-matrix of Laplacian matrix L obtained by removing its ith row and column. Repeating this from every vertex of a vertex cover of G provides solution densities for every edge. If γ is the size of the vertex cover used then the whole procedure takes $\mathcal{O}(\gamma n^3)$ time.

Example 1. Let M be the sub-matrix of L obtained by removing its first row and column as before. Then

$$M^{-1} = \begin{pmatrix} 5/8 & 2/8 & 1/8 \\ 2/8 & 4/8 & 2/8 \\ 1/8 & 2/8 & 5/8 \end{pmatrix}$$

and the solution density of edges $(1,2)$, $(1,3)$, and $(1,4)$ being used in T is respectively $\frac{5}{8}$, $\frac{4}{8}$, and $\frac{5}{8}$.

4   Integration into Backtrack Search

In this section we describe some of the implementation details and issues. As branching decisions are made and domain filtering is applied, some edges of G will be required in T and others, forbidden. These changes must be reflected in our data structures. Our data structures are reversible so that they are restored upon backtracking.

We use a heuristic greedy algorithm to compute our initial vertex cover — it may be worthwhile spending the time to compute a minimum vertex cover but that cover will need to be revised as vertices are merged following edge contractions. A simple way to update a vertex cover containing vertex j when edge (i,j) is contracted is to replace it with vertex i.

4.1   Updating the Laplacian Matrix

If edge (i, j) is forbidden it is simply removed from the graph. To reflect that in the Laplacian matrix we simply add one to entries ℓ_{ij} and ℓ_{ji}. The degree of each endpoint is also updated by subtracting one to ℓ_{ii} and ℓ_{jj}.

If edge (i, j) is required we contract it in the graph, so that (i, j) is implicitly part of the spanning tree. To update the Laplacian matrix we start by adding to vertex i all the edges (j, k): $\ell_{ik} \leftarrow \ell_{ik} + \ell_{jk}$. This may create multiple edges. The degree of vertex i, ℓ_{ii}, is also updated accordingly. Then, since vertex j is now merged with i, we remove all the edges connected to it, by setting to zero row and column j of the Laplacian matrix. Finally we set ℓ_{jj} to 1 so that minors will be computed correctly when they include row and column j.

Example 2. Recall Figure 1 and suppose edge $(1, 2)$ is now required for the spanning tree: we contract it and merge vertex 2 with 1. The new Laplacian matrix will be (note the double edge $(1, 3)$):

$$L = \begin{pmatrix} 3 & 0 & -2 & -1 \\ 0 & 1 & 0 & 0 \\ -2 & 0 & 3 & -1 \\ -1 & 0 & -1 & 2 \end{pmatrix}$$

4.2   Updating Solution Densities

The solution densities will change and we would like to avoid recomputing them from scratch. Given the inverse of matrix M can we incrementally compute the inverse of a slightly different matrix M'? The *Sherman-Morrison formula* further reveals that if M' is obtained from M by replacing its ith column, $(M)_i$, by column vector u as before then

$$M'^{-1} = M^{-1} - \frac{(M^{-1}(u - (M)_i))(e_i^\top M^{-1})}{1 + e_i^\top M^{-1}(u - (M)_i)}.$$

This can be computed in $\mathcal{O}(n^2)$ time.

In some cases we can lower that time complexity considerably. Consider forbidden edge (i, j). For any edge (i, k) whose solution density was obtained through the inverse of a sub-matrix removing row and column i from L, removing edge (i, j) only changes one entry in that sub-matrix, as we saw before, and the previous formula simplifies to

$$M'^{-1} = M^{-1} - \frac{(M^{-1} \cdot (-e_j)) \cdot (e_j^\top \cdot M^{-1})}{1 - m_{jj}^{-1}} = M^{-1} + \frac{1}{1 - m_{jj}^{-1}} \cdot Q$$

where $Q = (q_{hk})$ is an $(n-1) \times (n-1)$ matrix with $q_{hk} = m_{hj}^{-1} \cdot m_{jk}^{-1}$. Because we only need the kth entry on the diagonal, $m_{kk}^{-1} + (m_{kj}^{-1})^2/(1 - m_{jj}^{-1})$, the update for that edge takes constant time. What preceded equally applies for any edge (j, k) with a sub-matrix removing row and column j from L.

Example 3. Recall that for the graph at Figure 1 the solution density of edge $(1, 4)$ is $\frac{5}{8}$. Suppose edge $(1, 2)$ is now forbidden in the spanning tree. The updated solution density will be $\frac{5}{8} + (m_{42}^{-1})^2/(1 - m_{22}^{-1}) = \frac{5}{8} + (\frac{1}{8})^2/(1 - \frac{5}{8}) = \frac{2}{3}$.

5   Experiments

To demonstrate the effectiveness of using solution density information from a spanningTree constraint to guide a branching heuristic on some constrained spanning tree problems, we consider finding degree-constrained spanning trees of a graph. Note that the special case of a maximum degree of 2 corresponds to the Hamiltonian path problem. We created some graphs using a generator designed to produce hard Hamiltonian path instances for backtracking algorithms [14]. We used the IBM ILOG CP v1.6 solver for our implementation and performed our experiments on a AMD Opteron 2.2GHz with 1GB of memory. Our current implementation does not include the incremental algorithm described in Section 4.2 so the times reported are with matrix inversions computed from scratch at every search tree node. We report comparative results between maxSD, impact-based search (IBS), and random variable and value selection (random). Heuristic maxSD considers solution density information from each constraint and branches on the variable-value pair corresponding to the highest solution density observed. For IBS impacts are initialized by probing at the root node. At a search tree node the five best variables according to the approximated impact are identified. For that subset, we compute node impacts and branch on the best variable (highest impact) and value (lowest impact). This is consistent with what is suggested in the IBM ILOG solver documentation. For random we report the average of ten runs.

We used simple filtering rules for our constraint — our objective is not to solve that problem in the best way possible but rather to evaluate a counting-based branching heuristic. The first one forces each vertex to have degree at least one in the tree by lower bounding the sum of the variables corresponding to the edges incident to it. The second one fixes the number of edges that can be part of the spanning tree: as a spanning tree is formed by $n - 1$ edges, the sum of all variables must equal that value. Finally, since a tree is acyclic, we maintain the connected component in which each vertex lies, removing any extraneous intra connected component edges. In addition to the spanningTree constraint, we add to our model for each vertex i an upper bound on the sum of the variables corresponding to the edges incident to i.

We first generated random graphs of 15, 20, 25, 30, and 35 vertices (10 instances each). The generator ensures the existence of a Hamiltonian path. Turning first to a degree-2 bound, Table 1 left indicates that using maxSD effectively guides the search to a solution in several orders of magnitude fewer backtracks than the other two branching heuristics. Even though maxSD appears slower on small graphs, as displayed in Table 1 right, as the graphs become larger, this approach becomes faster than IBS and random.

Table 1. Number of backtracks (left) and time in seconds (right) before finding a spanning tree of maximum degree 2. Each line represents an average over 10 instances.

n	maxSD	IBS	random
15	0.2	229.8	49.0
20	1.5	533.0	976.6
25	2.1	1772.3	5919.6
30	71.7	12517.1	91454.4
35	112.2	18405.4	139861.3

n	maxSD	IBS	random
15	0.029	0.001	0.001
20	0.080	0.012	0.020
25	0.187	0.085	0.173
30	0.815	0.897	1.873
35	1.769	4.742	14.646

Table 2. Number of backtracks (left) and time in seconds (right) before finding a spanning tree of maximum degree 3. Each line represents an average over 10 instances.

n	maxSD	IBS	random
15	0.0	225.6	1.3
20	0.0	315.2	53.2
25	0.0	446.7	882.0
30	0.0	495.1	18589.8
35	0.0	566.8	20001.4

n	maxSD	IBS	random
15	0.039	0.002	0.001
20	0.100	0.013	0.001
25	0.222	0.021	0.311
30	0.441	0.039	0.093
35	0.852	0.063	2.333

We then turn to a degree-3 bound (see Table 2). It clearly demonstrates that using solution densities to find spanning trees in random graphs is a very effective approach. A maximum degree of 3 is much less restrictive than a maximum degree of 2 and more spanning trees in that graph will have that property. Therefore the first few spanning trees found satisfy all constraints. For all graphs, the solution density branching heuristic finds a spanning tree without any backtrack, unlike the other approaches. Despite not having to backtrack, maxSD remains slower than IBS on these instances since the latter only requires a few hundred backtracks.

Table 3. Number of backtracks (left) and time in seconds (right) before finding a Hamiltonian path in crossroad graphs. Each line represents an average over 10 instances.

n	maxSD	IBS	random
3	0.2	7721.9	8530.5
4	0.1	262011.7	191195.8
5	0.4	162353.0	-

n	maxSD	IBS	random
3	0.085	0.255	0.062
4	0.280	26.379	3.674
5	0.676	586.679	-

We also generated *crossroad graphs* using the same graph generator. These graphs are made up of small subgraphs only connected to each other via "bridge" edges. We generated 10 instances each of crossroad graphs containing 3, 4, and 5 subgraphs (with up to 35 vertices in total) and then tried to find a Hamiltonian path (spanning tree of degree 2). Results are shown in Table 3.

Using maxSD on these hard graphs is very effective, always finding a solution in much fewer backtracks than the other approaches. For the instances made up of 5 subgraphs, random could not solve a single instance within 2 hours of computing time. Here maxSD is also orders of magnitude faster than the other two branching heuristics.

6   Conclusion

We presented a new algorithm that computes exact solution densities for the spanning tree constraint in $\mathcal{O}(\gamma n^3)$ time, where γ is the size of a vertex cover for the graph, and updates solution densities in $\mathcal{O}(\gamma n^2)$ time, even in some cases achieving constant time updates per edge. Building the Laplacian matrix of a graph and inverting selected sub-matrices, the proportion of spanning trees including a certain edge of the graph can be calculated. By relying on that information, search can be oriented towards areas of the search space with high solution density with respect to the spanning tree structure and we gave some empirical evidence that this helps solve constrained spanning tree problems.

As future work we would like to try other types of constrained spanning tree problems. There are several application areas that involve finding spanning trees, such as network design, telecommunication, or transportation. Examples of these problems are the degree-constrained problem [7], the hop-constrained problem [5] or the diameter-constrained minimum spanning tree [1]. We also plan to investigate the compatibility of our solution density algorithm with more powerful filtering algorithms and variants of the constraint as proposed in the literature. For example the Matrix-Tree Theorem to count the number of spanning trees has already been generalized to directed graphs.

References

1. Abdalla, A., Deo, N.: Random-tree diameter and the diameter-constrained mst. Int. J. Comput. Math. 79(6), 651–663 (2002)
2. Beldiceanu, N., Flener, P., Lorca, X.: The tree constraint. In: Barták, R., Milano, M. (eds.) CPAIOR 2005. LNCS, vol. 3524, pp. 64–78. Springer, Heidelberg (2005)
3. Dooms, G., Katriel, I.: The minimum spanning tree constraint. In: Benhamou, F. (ed.) CP 2006. LNCS, vol. 4204, pp. 152–166. Springer, Heidelberg (2006)
4. Dooms, G., Katriel, I.: The "not-too-heavy spanning tree" constraint. In: Van Hentenryck, P., Wolsey, L.A. (eds.) CPAIOR 2007. LNCS, vol. 4510, pp. 59–70. Springer, Heidelberg (2007)
5. Gouveia, L., Simonetti, L., Uchoa, E.: Modeling hop-constrained and diameter-constrained minimum spanning tree problems as steiner tree problems over layered graphs. Math. Program. 128(1-2), 123–148 (2011)
6. Michel, L., Van Hentenryck, P.: Activity-based search for black-box constraint programming solvers. In: Beldiceanu, N., Jussien, N., Pinson, É. (eds.) CPAIOR 2012. LNCS, vol. 7298, pp. 228–243. Springer, Heidelberg (2012)
7. Narula, S.C., Ho, C.A.: Degree-constrained minimum spanning tree. Computers & OR 7(4), 239–249 (1980)

8. Pesant, G., Quimper, C.-G., Zanarini, A.: Counting-based search: Branching heuristics for constraint satisfaction problems. J. Artif. Intell. Res. (JAIR) 43, 173–210 (2012)
9. Refalo, P.: Impact-Based Search Strategies for Constraint Programming. In: Wallace, M. (ed.) CP 2004. LNCS, vol. 3258, pp. 557–571. Springer, Heidelberg (2004)
10. Régin, J.-C.: Simpler and incremental consistency checking and arc consistency filtering algorithms for the weighted spanning tree constraint. In: Trick, M.A. (ed.) CPAIOR 2008. LNCS, vol. 5015, pp. 233–247. Springer, Heidelberg (2008)
11. Régin, J.-C., Rousseau, L.-M., Rueher, M., van Hoeve, W.-J.: The weighted spanning tree constraint revisited. In: Lodi, A., Milano, M., Toth, P. (eds.) CPAIOR 2010. LNCS, vol. 6140, pp. 287–291. Springer, Heidelberg (2010)
12. Sherman, J., Morrison, W.J.: Adjustment of an inverse matrix corresponding to a change in one element of a given matrix. Annals of Mathematical Statistics 21, 124–127 (1950)
13. Tutte, W.T.: Graph Theory. Encyclopedia of Mathematics and Its Applications, vol. 21, 333 pages. Cambridge University Press (2001)
14. Vandegriend, B.: Finding hamiltonian cycles: Algorithms, graphs and performance. Master's thesis, University of Alberta (1998), http://webdocs.cs.ualberta.ca/~joe/Theses/HCarchive/main.html
15. Zanarini, A., Pesant, G.: Solution counting algorithms for constraint-centered search heuristics. In: Bessière, C. (ed.) CP 2007. LNCS, vol. 4741, pp. 743–757. Springer, Heidelberg (2007)

On the Reduction of the CSP Dichotomy Conjecture to Digraphs*

Jakub Bulín[1], Dejan Delić[2], Marcel Jackson[3], and Todd Niven[3]

[1] Faculty of Mathematics and Physics,
Charles University in Prague, Czech Republic
[2] Department of Mathematics, Ryerson University, Canada
[3] Department of Mathematics, La Trobe University, Australia
m.g.jackson@latrobe.edu.au, ddelic@ryerson.ca,
{jakub.bulin,toddniven}@gmail.com

Abstract. It is well known that the constraint satisfaction problem over general relational structures can be reduced in polynomial time to digraphs. We present a simple variant of such a reduction and use it to show that the algebraic dichotomy conjecture is equivalent to its restriction to digraphs and that the polynomial reduction can be made in logspace. We also show that our reduction preserves the bounded width property, i.e., solvability by local consistency methods. We discuss further algorithmic properties that are preserved and related open problems.

1 Introduction

A fundamental problem in constraint programming is to understand the computational complexity of constraint satisfaction problems (CSPs). While it is well known that the class of all constraint problems is NP-complete, there are many subclasses of problems for which there are efficient solving methods. One way to restrict the instances is to only allow a fixed set of constraint relations, often referred to as a *constraint language* [5] or *fixed template*. Classifying the computational complexity of fixed template CSPs has been a major focus in the theoretical study of constraint satisfaction. In particular it is of interest to know which templates produce polynomial time solvable problems to help provide more efficient solution techniques.

The study of fixed template CSPs dates back to the 1970's with the work of Montanari [18] and Schaefer [19]. A standout result from this era is that of Schaefer who showed that the CSPs arising from constraint languages over 2-element domains satisfy a *dichotomy*. The decision problem for fixed template

---

* The first author was supported by the grant projects GAČR 201/09/H012, GA UK 67410, SVV-2013-267317; the second author gratefully acknowledges support by the Natural Sciences and Engineering Research Council of Canada in the form of a Discovery Grant; the third and fourth were supported by ARC Discovery Project DP1094578; the first and fourth authors were also supported by the Fields Institute.

C. Schulte (Ed.): CP 2013, LNCS 8124, pp. 184–199, 2013.

CSPs over finite domains belong to the class NP, and Schaefer showed that in the 2-element domain case, a constraint language is either solvable in polynomial time or NP-complete. Dichotomies cannot be expected for decision problems in general, since (under the assumption that P≠NP) there are many problems in NP that are neither solvable in polynomial time, nor NP-complete [15]. Another important dichotomy was proved by Hell and Nešetřil [9]. They showed that if a fixed template is a finite simple graph (the vertices make up the domain and the edges make up the only allowable constraints), then the corresponding CSP is either polynomial time solvable or NP-complete. The decision problem for a graph constraint language can be rephrased as graph homomorphism problem (a graph homomorphism is a function from the vertices of one graph to another such that the edges are preserved). Specifically, given a fixed graph $\mathcal{H}$ (the constraint language), an instance is a graph $\mathcal{G}$ together with the question "Is there a graph homomorphism from $\mathcal{G}$ to $\mathcal{H}$?". In this sense, 3-colorability corresponds to $\mathcal{H}$ being the complete graph on 3 vertices. The notion of graph homomorphism problems naturally extends to directed graph (digraph) homomorphism problems and to relational structure homomorphism problems.

These early examples of dichotomies, by Schaefer, Hell and Nešetřil, form the basis of a larger project of classifying the complexity of fixed template CSPs. Of particular importance in this project is to prove the so-called *CSP Dichotomy Conjecture* of Feder and Vardi [8] dating back to 1993. It states that the CSPs relating to a fixed constraint language over a finite domain are either polynomial time solvable or NP-complete. To date this conjecture remains unanswered, but it has driven major advances in the study of CSPs.

One such advance is the algebraic connection revealed by Jeavons, Cohen and Gyssens [13] and later refined by Bulatov, Jeavons and Krokhin [5]. This connection associates with each finite domain constraint language $\mathbb{A}$ a finite algebraic structure $\mathbf{A}$. The properties of this algebraic structure are deeply linked with the computational complexity of the constraint language. In particular, for a fixed constraint language $\mathbb{A}$, if there does not exist a particular kind of operation, known as a Taylor polymorphism, then the class of problems determined by $\mathbb{A}$ is NP-complete. Bulatov, Jeavons and Krokhin [5] go on to conjecture that all other constraint languages over finite domains determine polynomial time CSPs (a stronger form of the CSP Dichotomy Conjecture, since it describes where the split between polynomial time and NP-completeness lies). This conjecture is often referred to as the *Algebraic CSP Dichotomy Conjecture*. Many important results have been built upon this algebraic connection. Bulatov [6] extended Schaefer's [19] result on 2-element domains to prove the CSP Dichotomy Conjecture for 3-element domains. Barto, Kozik and Niven [3] extended Hell and Nešetřil's result [9] on simple graphs to constraint languages consisting of a finite digraph with no sources and no sinks. Barto and Kozik [2] gave a complete algebraic description of the constraint languages over finite domains that are solvable by local consistency methods (these problems are said to be of *bounded width*) and as a consequence it is decidable to determine whether a constraint language can be solved by such methods.

In their seminal paper, Feder and Vardi [8] not only conjectured a dichotomy, they also reduced the problem of proving the dichotomy conjecture to the particular case of digraph homomorphism problems, and even to digraph homomorphism problems where the digraph is balanced (here balanced means that its vertices can be partitioned into levels). Specifically, for every template $\mathbb{A}$ (a finite relational structure of finite type) there is a balanced digraph (digraphs are particular kinds of relational structures) $\mathcal{D}(\mathbb{A})$ such that the CSP over $\mathbb{A}$ is polynomial time equivalent to that over $\mathcal{D}(\mathbb{A})$.

2   The Main Results

In general, fixed template CSPs can be modelled as relational structure homomorphism problems [8]. For detailed formal definitions of relational structures, homomorphisms and polymorphisms, see Section 3.

Let $\mathbb{A}$ be a finite structure with signature $\mathcal{R}$ (the fixed template), then the *constraint satisfaction problem for* $\mathbb{A}$ is the following decision problem.

Constraint Satisfaction Problem for $\mathbb{A}$.

CSP$(\mathbb{A})$
INSTANCE: A finite $\mathcal{R}$-structure $\mathbb{X}$.
QUESTION: Is there a homomorphism from $\mathbb{X}$ to $\mathbb{A}$?

The dichotomy conjecture [8] can be stated as follows:

CSP Dichotomy Conjecture. *Let $\mathbb{A}$ be a finite relational structure. Then* CSP$(\mathbb{A})$ *is solvable in polynomial time or NP-complete.*

The dichotomy conjecture is equivalent to its restriction to digraphs [8], and thus can be restated as follows:

CSP Dichotomy Conjecture. *Let $\mathbb{H}$ be a finite digraph. Then* CSP$(\mathbb{H})$ *is solvable in polynomial time or NP-complete.*

Every finite relational structure $\mathbb{A}$ has a unique *core* substructure $\mathbb{A}'$ (see Section 3.3 for the precise definition) such that CSP$(\mathbb{A})$ and CSP$(\mathbb{A}')$ are identical problems, i.e., the "yes" and "no" instances are precisely the same. The algebraic dichotomy conjecture [5] is the following:

Algebraic CSP Dichotomy Conjecture. *Let $\mathbb{A}$ be a finite relational structure that is a core. If $\mathbb{A}$ has a Taylor polymorphism then* CSP$(\mathbb{A})$ *is solvable in polynomial time, otherwise* CSP$(\mathbb{A})$ *is NP-complete.*

Indeed, perhaps the above conjecture should be called the *algebraic tractability conjecture* since it is known that if a core $\mathbb{A}$ does not possess a Taylor polymorphism, then CSP$(\mathbb{A})$ is NP-complete [5].

Feder and Vardi [8] proved that every fixed template CSP is polynomial time equivalent to a digraph CSP. This article will provide the following theorem, which replaces "polynomial time" with "logspace" and reduces the algebraic dichotomy conjecture to digraphs.

Theorem 1. *Let $\mathbb{A}$ be a finite relational structure. There is a finite digraph $\mathcal{D}(\mathbb{A})$ such that*

(i) $\mathrm{CSP}(\mathbb{A})$ and $\mathrm{CSP}(\mathcal{D}(\mathbb{A}))$ are logspace equivalent,
(ii) $\mathbb{A}$ has a Taylor polymorphism if and only if $\mathcal{D}(\mathbb{A})$ has a Taylor polymorphism, and
(iii) $\mathbb{A}$ is a core if and only if $\mathcal{D}(\mathbb{A})$ is a core.

Furthermore, if $\mathbb{A}$ is a core, then $\mathrm{CSP}(\mathbb{A})$ has bounded width if and only if $\mathrm{CSP}(\mathcal{D}(\mathbb{A}))$ has bounded width.

Proof. To prove (i), one reduction follows from Lemma 3 and Lemma 1. The other reduction is Lemma 4.

To prove (ii) we employ Theorem 2; it suffices to show that $\mathbb{A}$ has a WNU polymorphism if and only if $\mathcal{D}(\mathbb{A})$ has a WNU polymorphism. The forward implication (which is the crucial part of our proof) is proved in Lemma 8 and the converse follows from Lemma 3 and Lemma 2. Item (iii) is Corollary 1.

The preservation of bounded width follows from Corollary 1, Lemma 8 and Theorem 3.   □

See Remark 1 in Section 4 for the size of $\mathcal{D}(\mathbb{A})$. The "Taylor polymorphism" in Theorem 1 (ii) can be replaced by many other polymorphism properties, but space constraints do not allow us to elaborate here.

As a direct consequence of Theorem 1 (ii) and (iii) above, we can restate the algebraic dichotomy conjecture:

Algebraic CSP Dichotomy Conjecture. *Let $\mathbb{H}$ be a finite digraph that is a core. If $\mathbb{H}$ has a Taylor polymorphism then $\mathrm{CSP}(\mathbb{H})$ is solvable in polynomial time, otherwise $\mathrm{CSP}(\mathbb{H})$ is NP-complete.*

3   Background and Definitions

We approach fixed template constraint satisfaction problems from the "homomorphism problem" point of view. For background on the homomorphism approach to CSPs, see [8], and for background on the algebraic approach to CSP, see [5].

A *relational signature* $\mathcal{R}$ is a (in our case finite) set of *relation symbols* R_i, each with an associated arity k_i. A (finite) *relational structure* $\mathbb{A}$ over *relational signature* $\mathcal{R}$ (called an $\mathcal{R}$-*structure*) is a finite set A (the *domain*) together with a relation $R_i \subseteq A^{k_i}$, for each relation symbol R_i of arity k_i in $\mathcal{R}$. A *CSP template* is a fixed finite $\mathcal{R}$-structure, for some signature $\mathcal{R}$.

For simplicity we do not distinguish the relation with its associated relation symbol, however to avoid ambiguity, sometimes we write $R^{\mathbb{A}}$ to indicate that R belongs to $\mathbb{A}$. We will often refer to the domain of relational structure $\mathbb{A}$ simply by A. When referring to a fixed relational structure, we may simply specify it as $\mathbb{A} = (A; R_1, R_2, \ldots, R_k)$. For technical reasons we require that all the relations of a relational structure are nonempty.

3.1   Notation

For a positive integer n we denote the set $\{1, 2, \ldots, n\}$ by $[n]$. We write tuples using boldface notation, e.g. $\mathbf{a} = (a_1, a_2, \ldots, a_k) \in A^k$ and when ranging over tuples we use superscript notation, e.g. $(\mathbf{r}^1, \mathbf{r}^2, \ldots, \mathbf{r}^l) \in R^l \subseteq (A^k)^l$, where $\mathbf{r}^i = (r_1^i, r_2^i, \ldots, r_k^i)$, for $i = 1, \ldots, l$.

Let $R_i \subseteq A^{k_i}$ be relations of arity k_i, for $i = 1, \ldots, n$. Let $k = \sum_{i=1}^n k_i$ and $l_i = \sum_{j<i} k_j$. We write $R_1 \times \cdots \times R_n$ to mean the k-ary relation

$$\{(a_1, \ldots, a_k) \in A^k \mid (a_{l_i+1}, \ldots, a_{l_i+k_i}) \in R_i \text{ for } i = 1, \ldots, n\}.$$

An n-ary operation on a set A is simply a mapping $f : A^n \to A$; the number n is the arity of f. Let f be an n-ary operation on A and let $k > 0$. We define $f^{(k)}$ to be the n-ary operation obtained by applying f coordinatewise on A^k. That is, we define the n-ary operation $f^{(k)}$ on A^k by

$$f^{(k)}(\mathbf{a}^1, \ldots, \mathbf{a}^n) = (f(a_1^1, \ldots, a_1^n), \ldots, f(a_k^1, \ldots, a_k^n)),$$

for $\mathbf{a}^1, \ldots, \mathbf{a}^n \in A^k$.

We will be particularly interested in so-called idempotent operations. An n-ary operation f is said to be *idempotent* if it satisfies the equation

$$f(x, x, \ldots, x) = x.$$

3.2   Homomorphisms, Cores and Polymorphisms

We begin with the notion of a relational structure homomorphism.

Definition 1. *Let $\mathbb{A}$ and $\mathbb{B}$ be relational structures in the same signature $\mathcal{R}$. A homomorphism from $\mathbb{A}$ to $\mathbb{B}$ is a mapping φ from A to B such that for each n-ary relation symbol R in $\mathcal{R}$ and each n-tuple $\mathbf{a} \in A^n$, if $\mathbf{a} \in R^{\mathbb{A}}$, then $\varphi(\mathbf{a}) \in R^{\mathbb{B}}$, where φ is applied to $\mathbf{a}$ coordinatewise.*

We write $\varphi : \mathbb{A} \to \mathbb{B}$ to mean that φ is a homomorphism from $\mathbb{A}$ to $\mathbb{B}$, and $\mathbb{A} \to \mathbb{B}$ to mean that there exists a homomorphism from $\mathbb{A}$ to $\mathbb{B}$.

An *isomorphism* is a bijective homomorphism φ such that φ^{-1} is a homomorphism. A homomorphism $\mathbb{A} \to \mathbb{A}$ is called an *endomorphism*. An isomorphism from $\mathbb{A}$ to $\mathbb{A}$ is an *automorphism*. It is an easy fact that if $\mathbb{A}$ is finite, then every surjective endomorphism is an automorphism.

A finite relational structure $\mathbb{A}'$ is a *core* if every endomorphism $\mathbb{A}' \to \mathbb{A}'$ is surjective (and therefore an automorphism). For every $\mathbb{A}$ there exists a relational structure $\mathbb{A}'$ such that $\mathbb{A} \to \mathbb{A}'$ and $\mathbb{A}' \to \mathbb{A}$ and $\mathbb{A}'$ is minimal with respect to these properties. The structure $\mathbb{A}'$ is called the *core of* $\mathbb{A}$. The core of $\mathbb{A}$ is unique (up to isomorphism) and $\mathrm{CSP}(\mathbb{A})$ and $\mathrm{CSP}(\mathbb{A}')$ are the same decision problems. Equivalently, the core of $\mathbb{A}$ can be defined as a minimal induced substructure that $\mathbb{A}$ retracts onto. (See [10] for details on cores for graphs, cores for relational structures are a natural generalisation.)

The notion of *polymorphism* is central in the so called algebraic approach to CSP. Polymorphisms are a natural generalization of endomorphisms to higher arity operations.

Definition 2. *Given an $\mathcal{R}$-structure $\mathbb{A}$, an n-ary polymorphism of $\mathbb{A}$ is an n-ary operation f on A such that f preserves the relations of $\mathbb{A}$. That is, if $\mathbf{a}^1, \dots, \mathbf{a}^n \in R$, for some k-ary relation R in $\mathcal{R}$, then $f^{(k)}(\mathbf{a}^1, \dots, \mathbf{a}^n) \in R$.*

Thus, an endomorphism is a 1-ary polymorphism.

In this paper we will be interested in the following kind of polymorphisms.

Definition 3. *A* weak near-unanimity *(WNU) polymorphism is an n-ary idempotent polymorphism ω, for some $n \geq 3$, that satisfies the following equations (for all x, y):*

$$\omega(x, \dots, x, y) = \omega(x, \dots, x, y, x) = \dots = \omega(y, x, \dots, x).$$

We call the above WNU *equations.*

Note that since we assume that a WNU polymorphism ω is idempotent it also satisfies the equation

$$\omega(x, x, \dots, x) = x.$$

Of particular interest, with respect to the algebraic dichotomy conjecture, are Taylor polymorphisms. We will not need to explicitly define Taylor polymorphisms (and only need consider WNU polymorphisms) by the following theorem.

Theorem 2. [17] *A finite relational structure $\mathbb{A}$ has a Taylor polymorphism if and only if $\mathbb{A}$ has a WNU polymorphism.*

Weak near-unanimity polymorphisms can be also used to characterise CSPs of bounded width (see [2] for a detailed explanation of the bounded width algorithm).

Theorem 3. [2,17] *Let $\mathbb{A}$ be a finite relational structure that is a core. Then $\mathrm{CSP}(\mathbb{A})$ is of bounded width if and only if $\mathbb{A}$ has WNU polymorphisms of all but finitely many arities.*

3.3 Primitive Positive Definability

A first order formula is called *primitive positive* if it is an existential conjunction of atomic formulæ. Since we only refer to relational signatures, a primitive positive formula is simply an existential conjunct of formulæ of the form $x = y$ or $(x_1, x_2, \dots, x_n) \in R$, where R is a relation symbol of arity n.

For example, if we have a binary relation symbol E in our signature, then the formula

$$\psi(x, y) = (\exists z)((x, z) \in E \ \wedge \ (z, y) \in E),$$

pp-defines a binary relation in which elements a, b are related if there is a directed path of length 2 from a to b in E.

Definition 4. *A relational structure $\mathbb{B}$ is* primitive positive definable *in $\mathbb{A}$ (or $\mathbb{A}$ pp-defines $\mathbb{B}$) if*

(i) the set B is a subset of A and is definable by a primitive positive formula interpreted in $\mathbb{A}$, and

(ii) each relation R in the signature of $\mathbb{B}$ is definable on the set B by a primitive positive formula interpreted in $\mathbb{A}$.

The following result relates the above definition to the complexity of CSPs.

Lemma 1. [13] *Let $\mathbb{A}$ be a finite relational structure that pp-defines $\mathbb{B}$. Then, $CSP(\mathbb{B})$ is polynomial time (indeed, logspace) reducible to $\mathrm{CSP}(\mathbb{A})$.*

It so happens that, if $\mathbb{A}$ pp-defines $\mathbb{B}$, then $\mathbb{B}$ inherits the polymorphisms of $\mathbb{A}$. See [5] for a detailed explanation.

Lemma 2. [5] *Let $\mathbb{A}$ be a finite relational structure that pp-defines $\mathbb{B}$. If φ is a polymorphism of $\mathbb{A}$, then its restriction to B is a polymorphism of $\mathbb{B}$.*

In particular, if $\mathbb{A}$ pp-defines $\mathbb{B}$ and $\mathbb{A}$ has a WNU polymorphism ω, then ω restricted to B is a WNU polymorphism of $\mathbb{B}$.

In the case that $\mathbb{A}$ pp-defines $\mathbb{B}$ and $\mathbb{B}$ pp-defines $\mathbb{A}$, we say that $\mathbb{A}$ and $\mathbb{B}$ are *pp-equivalent*. In this case, CSP($\mathbb{A}$) and CSP($\mathbb{B}$) are essentially the same problems (they are logspace equivalent) and $\mathbb{A}$ and $\mathbb{B}$ have the same polymorphisms.

Example 1. Let $\mathbb{A} = (A; R_1, \ldots, R_n)$, where each R_i is k_i-ary, and define $R = R_1 \times \cdots \times R_n$. Then the structure $\mathbb{A}' = (A; R)$ is pp-equivalent to $\mathbb{A}$.

Indeed, let $k = \sum_{i=1}^{n} k_i$ be the arity of R and $l_i = \sum_{j<i} k_j$ for $i = 1, \ldots, n$. The relation R is pp-definable from $R_1, \ldots, R_n$ using the formula

$$\Psi(x_1, \ldots, x_k) = \bigwedge_{i=1}^{n} (x_{l_i+1}, \ldots, x_{l_i+k_i}) \in R_i.$$

The relation R_1 can be defined from R by the primitive positive formula

$$\Psi(x_1, \ldots, x_{k_1}) = (\exists y_{k_1+1}, \ldots, \exists y_k)((x_1, \ldots, x_{k_1}, y_{k_1+1}, \ldots, y_k) \in R)$$

and the remaining R_i's can be defined similarly.

Example 1 shows that when proving Theorem 1 we can restrict ourselves to relational structures with a single relation.

3.4 Digraphs

A *directed graph*, or *digraph*, is a relational structure $\mathbb{G}$ with a single binary relation symbol E as its signature. We typically call the members of G and $E^{\mathbb{G}}$ *vertices* and *edges*, respectively. We usually write $a \to b$ to mean $(a, b) \in E^{\mathbb{G}}$, if there is no ambiguity.

A special case of relational structure homomorphism (see Definition 1), is that of digraph homomorphism. That is, given digraphs $\mathbb{G}$ and $\mathbb{H}$, a function $\varphi : G \to H$ is a homomorphism if $(\varphi(a), \varphi(b)) \in E^{\mathbb{H}}$ whenever $(a, b) \in E^{\mathbb{G}}$.

Definition 5. *For* $i = 1, \ldots, n$, *let* $\mathbb{G}_i = (G_i, E_i)$ *be digraphs. The* direct product *of* $\mathbb{G}_1, \ldots, \mathbb{G}_n$, *denoted by* $\prod_{i=1}^{n} \mathbb{G}_i$, *is the digraph with vertices* $\prod_{i=1}^{n} G_i$ *(the cartesian product of the sets* G_i*) and edge relation*

$$\{(\mathbf{a}, \mathbf{b}) \in (\prod_{i=1}^{n} G_i)^2 \mid (a_i, b_i) \in E_i \text{ for } i = 1 \ldots, n\}.$$

If $\mathbb{G}_1 = \cdots = \mathbb{G}_n = \mathbb{G}$ *then we write* $\mathbb{G}^n$ *to mean* $\prod_{i=1}^{n} \mathbb{G}_i$.

With the above definition in mind, an n-ary polymorphism on a digraph $\mathbb{G}$ is simply a digraph homomorphism from $\mathbb{G}^n$ to $\mathbb{G}$.

Definition 6. *A digraph* $\mathbb{P}$ *is an* oriented path *if it consists of a sequence of vertices* $v_0, v_1, \ldots, v_k$ *such that precisely one of* $(v_{i-1}, v_i), (v_i, v_{i-1})$ *is an edge, for each* $i = 1, \ldots, k$. *We require oriented paths to have a direction; we denote the* initial vertex v_0 *and the* terminal vertex v_k *by* $\iota\mathbb{P}$ *and* $\tau\mathbb{P}$, *respectively.*

Given a digraph $\mathbb{G}$ and an oriented path $\mathbb{P}$, we write $a \xrightarrow{\mathbb{P}} b$ to mean that we can walk in $\mathbb{G}$ from a following $\mathbb{P}$ to b, i.e., there exists a homomorphism $\varphi : \mathbb{P} \to \mathbb{G}$ such that $\varphi(\iota\mathbb{P}) = a$ and $\varphi(\tau\mathbb{P}) = b$. Note that for every $\mathbb{P}$ there exists a primitive positive formula $\psi(x, y)$ such that $a \xrightarrow{\mathbb{P}} b$ if and only if $\psi(a, b)$ is true in $\mathbb{G}$. If there exists an oriented path $\mathbb{P}$ such that $a \xrightarrow{\mathbb{P}} b$, we say that a and b are *connected.* If vertices a and b are connected, then the *distance* from a to b is the number of edges in the shortest oriented path connecting them. Connectedness forms an equivalence relation on G; its classes are called the *connected components* of $\mathbb{G}$. We say that a digraph is connected if it consists of a single connected component.

A connected digraph is *balanced* if it admits a *level function* $\mathrm{lvl} : G \to \mathbb{N}$, where $\mathrm{lvl}(b) = \mathrm{lvl}(a) + 1$ whenever (a, b) is an edge, and the minimum level is 0. The maximum level is called the *height* of the digraph. Oriented paths are natural examples of balanced digraphs.

By a *zigzag* we mean the oriented path $\bullet \to \bullet \leftarrow \bullet \to \bullet$ and a *single edge* is the path $\bullet \to \bullet$. For oriented paths $\mathbb{P}$ and $\mathbb{P}'$, the *concatenation of* $\mathbb{P}$ *and* $\mathbb{P}'$, denoted by $\mathbb{P} + \mathbb{P}'$, is the oriented path obtained by identifying $\tau\mathbb{P}$ with $\iota\mathbb{P}'$.

Our digraph reduction as described in Section 4 relies on oriented paths obtained by concatenation of zigzags and single edges. For example, the path in Figure 1 is a concatenation of a single edge followed by two zigzags and two more single edges (for clarity, we organise its vertices into levels).

4   The Reduction to Digraphs

In this section we take an arbitrary finite relational structure $\mathbb{A}$ and construct a balanced digraph $\mathcal{D}(\mathbb{A})$ such that $\mathrm{CSP}(\mathbb{A})$ and $\mathrm{CSP}(\mathcal{D}(\mathbb{A}))$ are logspace equivalent.

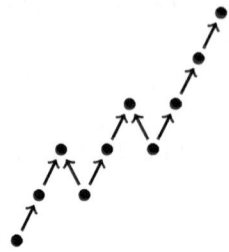

Fig. 1. A minimal oriented path

Let $\mathbb{A} = (A; R_1, \ldots, R_n)$ be a finite relational structure, where R_i is of arity k_i, for $i = 1, \ldots, n$. Let $k = \sum_{i=1}^{n} k_i$ and let R be the k-ary relation $R_1 \times \cdots \times R_n$. For $\mathcal{I} \subseteq [k]$ define $\mathbb{Q}_{\mathcal{I},l}$ to be a single edge if $l \in \mathcal{I}$, and a zigzag if $l \in [k] \setminus \mathcal{I}$.

We define the oriented path $\mathbb{Q}_{\mathcal{I}}$ (of height $k + 2$) by

$$\mathbb{Q}_{\mathcal{I}} = \bullet \to \bullet + \mathbb{Q}_{\mathcal{I},1} + \mathbb{Q}_{\mathcal{I},2} + \ldots + \mathbb{Q}_{\mathcal{I},k} + \bullet \to \bullet$$

Instead of $\mathbb{Q}_{\emptyset}, \mathbb{Q}_{\emptyset,l}$ we write just $\mathbb{Q}, \mathbb{Q}_l$, respectively. For example, the oriented path in Figure 1 is $\mathbb{Q}_{\mathcal{I}}$ where $k = 3$ and $\mathcal{I} = \{3\}$. We will need the following observation.

Observation. *Let $\mathcal{I}, \mathcal{J} \subseteq [k]$. A homomorphism $\varphi : \mathbb{Q}_{\mathcal{I}} \to \mathbb{Q}_{\mathcal{J}}$ exists, if and only if $\mathcal{I} \subseteq \mathcal{J}$. In particular $\mathbb{Q} \to \mathbb{Q}_{\mathcal{I}}$ for all $\mathcal{I} \subseteq [k]$. Moreover, if φ exists, it is unique and surjective.*

We are now ready to define the digraph $\mathcal{D}(\mathbb{A})$.

Definition 7. *For every $e = (a, \mathbf{r}) \in A \times R$ we define $\mathbb{P}_e$ to be the path $\mathbb{Q}_{\{i : a = r_i\}}$. The digraph $\mathcal{D}(\mathbb{A})$ is obtained from the digraph $(A \cup R; A \times R)$ by replacing every $e = (a, \mathbf{r}) \in A \times R$ by the oriented path $\mathbb{P}_e$ (identifying $\iota\mathbb{P}_e$ with a and $\tau\mathbb{P}_e$ with $\mathbf{r}$).*

(We often write $\mathbb{P}_{e,l}$ to mean $\mathbb{Q}_{\mathcal{I},l}$ where $\mathbb{P}_e = \mathbb{Q}_{\mathcal{I}}$.)

Example 2. Consider the relational structure $\mathbb{A} = (\{0, 1\}; R)$ where $R = \{(0, 1), (1, 0)\}$, i.e., $\mathbb{A}$ is the 2-cycle. Figure 2 is a visual representation of $\mathcal{D}(\mathbb{A})$.

Remark 1. The number of vertices in $\mathcal{D}(\mathbb{A})$ is $(3k + 1)|R||A| + (1 - 2k)|R| + |A|$ and the number of edges is $(3k + 2)|R||A| - 2k|R|$. The construction of $\mathcal{D}(\mathbb{A})$ can be performed in logspace (under any reasonable encoding).

Proof. The vertices of $\mathcal{D}(\mathbb{A})$ consist of the elements of $A \cup R$, along with vertices from the connecting paths. The number of vertices lying strictly within the connecting paths would be $(3k + 1)|R||A|$ if every $\mathbb{P}_e$ was $\mathbb{Q}$. We need to deduct 2 vertices whenever there is a single edge instead of a zigzag and there are $\sum_{(a,\mathbf{r}) \in A \times R} |\{i : a = r_i\}| = k|R|$ such instances. The number of edges is counted very similarly. $\qquad\square$

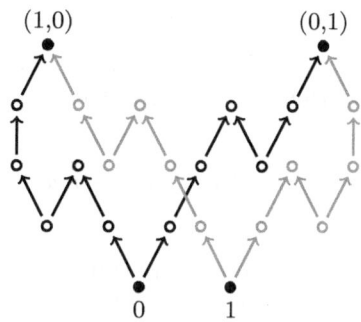

Fig. 2. $\mathcal{D}(\mathbb{A})$ where $\mathbb{A}$ is the 2-cycle

Remark 2. Note that if we apply this construction to itself (that is, $\mathcal{D}(\mathcal{D}(\mathbb{A}))$) then we obtain balanced digraphs of height 4. When applied to digraphs, the $\mathcal{D}$ construction is identical to that given by Feder and Vardi [8, Theorem 13].

The following lemma, together with Lemma 1, shows that $\mathrm{CSP}(\mathbb{A})$ reduces to $\mathrm{CSP}(\mathcal{D}(\mathbb{A}))$ in logspace.

Lemma 3. $\mathbb{A}$ *is pp-definable from* $\mathcal{D}(\mathbb{A})$.

Proof. Example 1 demonstrates that $\mathbb{A}$ is pp-equivalent to $(A; R)$. We now show that $\mathcal{D}(\mathbb{A})$ pp-defines $(A; R)$, from which it follows that $\mathcal{D}(\mathbb{A})$ pp-defines $\mathbb{A}$.

Note that $\mathbb{Q} \to \mathbb{P}_e$ for all $e \in A \times R$, and $\mathbb{Q}_{\{i\}} \to \mathbb{P}_{(a,\mathbf{r})}$ if and only if $a = r_i$. The set A is pp-definable in $\mathcal{D}(\mathbb{A})$ by $A = \{x \mid (\exists y)(x \xrightarrow{\mathbb{Q}} y)\}$ and the relation R can be defined as the set $\{(x_1, \ldots, x_k) \mid (\exists y)(x_i \xrightarrow{\mathbb{Q}_{\{i\}}} y \text{ for all } i \in [k])\}$, which is also a primitive positive definition. $\qquad\square$

It is not, in general, possible to pp-define $\mathcal{D}(\mathbb{A})$ from $\mathbb{A}$.[1] Nonetheless the following lemma is true.

Lemma 4. $\mathrm{CSP}(\mathcal{D}(\mathbb{A}))$ *reduces in logspace to* $\mathrm{CSP}(\mathbb{A})$.

The proof of Lemma 4 is rather technical, though broadly follows the polynomial process described in the proof of [8, Theorem 13] (as mentioned, our construction coincides with theirs in the case of digraphs). Details of the argument will be presented in a subsequent expanded version of this article.

Lemma 3 and Lemma 4 complete the proof of part (i) of Theorem 1. As this improves the oft-mentioned polynomial time equivalence of general CSPs with digraph CSPs, we now present it as stand-alone statement.

Theorem 4. *Every fixed finite template CSP is logspace equivalent to the CSP over some finite digraph.*

---

[1] Using the definition of pp-definability as described in this paper, this is true for cardinality reasons. However, a result of Kazda [14] can be used to show that the statement remains true even for more general definitions of pp-definability.

5 Preserving Cores

In what follows, let $\mathbb{A}$ be a fixed finite relational structure. Without loss of generality we may assume that $\mathbb{A} = (A; R)$, where R is a k-ary relation (see Example 1).

Lemma 5. *The endomorphisms of $\mathbb{A}$ and $\mathcal{D}(\mathbb{A})$ are in one-to-one correspondence.*

Proof. We first show that every endomorphism φ of $\mathbb{A}$ can be extended to an endomorphism $\overline{\varphi}$ of $\mathcal{D}(\mathbb{A})$. Let $\overline{\varphi}(a) = \varphi(a)$ for $a \in A$, and let $\overline{\varphi}(\mathbf{r}) = \varphi^{(k)}(\mathbf{r})$ for $\mathbf{r} \in R$. Note that $\varphi^{(k)}(\mathbf{r}) \in R$ since φ is an endomorphism of $\mathbb{A}$.

Let $c \in \mathcal{D}(\mathbb{A}) \setminus (A \cup R)$ and let $e = (a, \mathbf{r})$ be such that $c \in \mathbb{P}_e$. Define $e' = (\varphi(a), \varphi^{(k)}(\mathbf{r}))$. If $\mathbb{P}_{e,l}$ is a single edge for some $l \in [k]$, then $r_l = a$ and $\varphi(r_l) = \varphi(a)$, and therefore $\mathbb{P}_{e',l}$ is a single edge. Thus there exists a (unique) homomorphism $\mathbb{P}_e \to \mathbb{P}_{e'}$. Define $\overline{\varphi}(c)$ to be the image of c under this homomorphism, completing the definition of $\overline{\varphi}$.

We now show that every endomorphism Φ of $\mathcal{D}(\mathbb{A})$ is of the form $\overline{\varphi}$, for some endomorphism φ of $\mathbb{A}$. Let Φ be an endomorphism of $\mathcal{D}(\mathbb{A})$. Let φ be the restriction of Φ to A. By Lemma 2 and Lemma 3, φ is an endomorphism of $\mathbb{A}$. For every $e = (a, \mathbf{r})$, the endomorphism Φ maps $\mathbb{P}_e$ onto $\mathbb{P}_{(\varphi(a), \Phi(\mathbf{r}))}$. If we set $a = r_l$, then $\mathbb{P}_{e,l}$ is a single edge. In this case it follows that $\mathbb{P}_{(\varphi(a), \Phi(\mathbf{r})),l}$ is also a single edge. Thus, by the construction of $\mathcal{D}(\mathbb{A})$ the lth coordinate of $\Phi(\mathbf{r})$ is $\varphi(a) = \varphi(r_l)$. This proves that the restriction of Φ to R is $\varphi^{(k)}$ and therefore $\Phi = \overline{\varphi}$. $\qquad\square$

The following corollary is Theorem 1 (iii).

Corollary 1. $\mathbb{A}$ *is a core if and only if* $\mathcal{D}(\mathbb{A})$ *is a core.*

Proof. To prove the corollary we need to show that an endomorphism φ of $\mathbb{A}$ is surjective if and only if $\overline{\varphi}$ (from Lemma 5) is surjective. Clearly, if $\overline{\varphi}$ is surjective then so is φ.

Assume φ is surjective (and therefore an automorphism of $\mathbb{A}$). It follows that $\varphi^{(k)}$ is surjective on R and therefore $\overline{\varphi}$ is a bijection when restricted to the set $A \cup R$. Let $a \in A$ and $\mathbf{r} \in R$. By definition we know that $\overline{\varphi}$ maps $\mathbb{P}_{(a,\mathbf{r})}$ homomorphically onto $\mathbb{P}_{(\varphi(a), \varphi^{(k)}(\mathbf{r}))}$. Since φ has an inverse φ^{-1}, it follows that $\overline{\varphi^{-1}}$ maps $\mathbb{P}_{(\varphi(a), \varphi^{(k)}(\mathbf{r}))}$ homomorphically onto $\mathbb{P}_{(a,\mathbf{r})}$. Thus $\mathbb{P}_{(a,\mathbf{r})}$ and $\mathbb{P}_{(\varphi(a), \varphi^{(k)}(\mathbf{r}))}$ are isomorphic, completing the proof. $\qquad\square$

To complete the proof of Theorem 1, it remains to show that our reduction preserves WNUs.

6 The Reduction Preserves WNUs

For $m > 0$ let Δ_m denote the connected component of the digraph $\mathcal{D}(\mathbb{A})^m$ containing the *diagonal* (i.e., the set $\{(c, c, \ldots, c) \mid c \in \mathcal{D}(\mathbb{A})\}$).

Lemma 6. *The elements of the diagonal are all connected in $\mathcal{D}(\mathbb{A})^m$. Furthermore, $A^m \subseteq \Delta_m$ and $R^m \subseteq \Delta_m$.*

Proof. The first statement follows from the fact that $\mathcal{D}(\mathbb{A})$ is connected. To see that $A^m \subseteq \Delta_m$ and $R^m \subseteq \Delta_m$, fix $a \in A$, and so by definition, $(a, a, \ldots, a) \in \Delta_m$. Let $(\mathbf{r}^1, \ldots, \mathbf{r}^m) \in R^m$ and for every $i \in [m]$ let $\varphi_i : \mathbb{Q} \to \mathbb{P}_{(a, \mathbf{r}^i)}$. The homomorphism defined by $x \mapsto (\varphi_1(x), \ldots, \varphi_m(x))$ witnesses $(a, \ldots, a) \xrightarrow{\mathbb{Q}} (\mathbf{r}^1, \ldots, \mathbf{r}^m)$ in $\mathcal{D}(\mathbb{A})^m$. Hence $R^m \subseteq \Delta_m$. A similar argument gives $A^m \subseteq \Delta_m$. $\square$

The following lemma shows that there is only one non-trivial connected component of $\mathcal{D}(\mathbb{A})^m$ that contains tuples (whose entries are) on the same level in $\mathcal{D}(\mathbb{A})$; namely Δ_m.

Lemma 7. *Let $m > 0$ and let Γ be a connected component of $\mathcal{D}(\mathbb{A})^m$ containing an element $\mathbf{c}$ such that $\mathrm{lvl}(c_1) = \cdots = \mathrm{lvl}(c_m)$. Then every element $\mathbf{d} \in \Gamma$ is of the form $\mathrm{lvl}(d_1) = \cdots = \mathrm{lvl}(d_m)$ and the following hold.*

(i) If $\mathbf{c} \to \mathbf{d}$ is an edge in Γ such that $\mathbf{c} \notin A^m$ and $\mathbf{d} \notin R^m$, then there exist $e_1, \ldots, e_m \in A \times R$ and $l \in [k]$ such that $\mathbf{c}, \mathbf{d} \in \prod_{i=1}^m \mathbb{P}_{e_i, l}$.
(ii) Either $\Gamma = \Delta_m$ or Γ is one-element.

Proof. First observe that if an element $\mathbf{d}$ is connected in $\mathcal{D}(\mathbb{A})^m$ to an element $\mathbf{c}$ with $\mathrm{lvl}(c_1) = \cdots = \mathrm{lvl}(c_m)$, then there is an oriented path $\mathbb{Q}'$ such that $\mathbf{c} \xrightarrow{\mathbb{Q}'} \mathbf{d}$ from which it follows that $\mathrm{lvl}(d_1) = \cdots = \mathrm{lvl}(d_m)$.

To prove (i), let $\mathbf{c} \to \mathbf{d}$ be an edge in Γ such that $\mathbf{c} \notin A^m$ and $\mathbf{d} \notin R^m$. For $i = 1, \ldots, m$ let e_i be such that $c_i \in \mathbb{P}_{e_i}$ and let $l = \mathrm{lvl}(c_1)$. The claim now follows immediately from the construction of $\mathcal{D}(\mathbb{A})$.

It remains to prove (ii). If $|\Gamma| > 1$, then there is an edge $\mathbf{c} \to \mathbf{d}$ in Γ. If $\mathbf{c} \in A^m$ or $\mathbf{d} \in R^m$, then the result follows from Lemma 6. Otherwise, from (i), there exists $l \in [k]$ and $e_i = (a_i, \mathbf{r}^i)$ such that $\mathbf{c}, \mathbf{d} \in \prod_{i=1}^m \mathbb{P}_{e_i, l}$.

For every $i \in [m]$ we can walk from c_i to $\iota \mathbb{P}_{e_i, l}$ following the path $\bullet \to \bullet \leftarrow \bullet$; and so $\mathbf{c}$ and $(\iota \mathbb{P}_{e_1, l}, \ldots, \iota \mathbb{P}_{e_m, l})$ are connected. For every $i \in [m]$ there exists a homomorphism $\varphi_i : \mathbb{Q} \to \mathbb{P}_{e_i}$ such that $\varphi_i(\iota \mathbb{Q}) = a_i$ and $\varphi_i(\iota \mathbb{Q}_l) = \iota \mathbb{P}_{e_i, l}$. The homomorphism $\mathbb{Q} \to \mathcal{D}(\mathbb{A})^m$ defined by $x \mapsto (\varphi_1(x), \ldots, \varphi_m(x))$ shows that $(a_1, \ldots, a_m)$ and $(\iota \mathbb{P}_{e_1, l}, \ldots, \iota \mathbb{P}_{e_m, l})$ are connected. By transitivity, $(a_1, \ldots, a_m)$ is connected to $\mathbf{c}$ and therefore $(a_1, \ldots, a_m) \in \Gamma$. Using (i) we obtain $\Gamma = \Delta_m$. $\square$

We are now ready to prove the main ingredient of Theorem 1 (ii). The proof of Lemma 8 is similar in essence to the proof of Lemma 5, although more complicated.

Lemma 8. *If $\mathbb{A}$ has an m-ary WNU polymorphism, then $\mathcal{D}(\mathbb{A})$ has an m-ary WNU polymorphism.*

Proof. Let ω be an m-ary WNU polymorphism of $\mathbb{A}$. We construct an m-ary operation $\overline{\omega}$ on $\mathcal{D}(\mathbb{A})$. We split the definition into several cases and subcases. Let $\mathbf{c} \in \mathcal{D}(\mathbb{A})^m$.

Case 1. $|\{\mathrm{lvl}(c_1), \ldots, \mathrm{lvl}(c_m)\}| > 1$.

<u>1a</u>   If there exists $i \in [m]$ such that $|\{\mathrm{lvl}(c_j) \mid j \neq i\}| = 1$, we define $\overline{\omega}(\mathbf{c}) = c_i$.

<u>1b</u>   Else let $\overline{\omega}(\mathbf{c}) = c_1$.

Case 2. $\mathrm{lvl}(c_1) = \cdots = \mathrm{lvl}(c_m)$, but $\mathbf{c} \notin \Delta_m$.

<u>2a</u>   If there exists $i \in [m]$ such that $|\{c_j \mid j \neq i\}| = 1$, we define $\overline{\omega}(\mathbf{c}) = c_i$.

<u>2b</u>   Else let $\overline{\omega}(\mathbf{c}) = c_1$.

Case 3. $\mathbf{c} \in \Delta_m$.

<u>3a</u>   If $\{c_1, \ldots, c_m\} \subseteq A$, we define $\overline{\omega}(\mathbf{c}) = \omega(\mathbf{c})$.

<u>3b</u>   If $\{c_1, \ldots, c_m\} \subseteq R$, we define $\overline{\omega}(\mathbf{c}) = \omega^{(k)}(\mathbf{c})$.

<u>3c</u>   Else, there exists $\mathbf{d} \in \Delta_m \setminus (A^m \cup R^m)$ such that $\mathbf{c} \to \mathbf{d}$ or $\mathbf{d} \to \mathbf{c}$ in $\mathcal{D}(\mathbb{A})^m$. By Lemma 7 (ii), there exist $l \in [k]$ and $e_i = (a_i, \mathbf{r}^i)$ such that $\mathbf{c} \in \prod_{i=1}^{m} \mathbb{P}_{e_i, l}$. Let $e = (a, \mathbf{r})$, where $a = \omega(a_1, \ldots, a_m)$ and $\mathbf{r} = \omega^{(k)}(\mathbf{r}^1, \ldots, \mathbf{r}^m)$. We set $\overline{\omega}(\mathbf{c}) = \Phi(\mathbf{c})$, where $\Phi : \prod_{i=1}^{m} \mathbb{P}_{e_i, l} \to \mathbb{P}_{e, l}$ is defined as follows.

1. If $\mathbb{P}_{e,l}$ is a single edge, then we set

$$\Phi(\mathbf{u}) = \begin{cases} \iota \mathbb{P}_{e,l} & \text{if } \mathrm{lvl}(u_1) = \cdots = \mathrm{lvl}(u_m) = \mathrm{lvl}(\iota \mathbb{P}_{e,l}) \\ \tau \mathbb{P}_{e,l} & \text{otherwise.} \end{cases}$$

2. If $\mathbb{P}_{e,l}$ is a zigzag, then let $I = \{i \in [m] \mid \mathbb{P}_{e_i, l} \text{ is a zigzag}\}$. For every $i \in I$ let $\phi_i : \mathbb{P}_{e_i, l} \to \mathbb{P}_{e,l}$ be the unique isomorphism. We define $\Phi(\mathbf{u})$ to be the vertex from $\{\phi_i(u_i) : i \in I\}$ with minimal distance from $\iota \mathbb{P}_{e,l}$.

Let us first comment on correctness of the definition. In subcase <u>3b</u>, $\omega^{(k)}(\mathbf{c}) \in R$ follows from the fact that ω preserves R. In subcase <u>3c</u>, if $\mathbb{P}_{e,l}$ is a zigzag, then $I \neq \emptyset$. Indeed, if all the $\mathbb{P}_{e_i, l}$'s were single edges, then $r_l = \omega(r_l^1, \ldots, r_l^m) = \omega(a_1, \ldots, a_m) = a$ and so $\mathbb{P}_{e,l}$ would also be a single edge. The e_i's are uniquely determined by $\mathbf{c}$, and the choice of l is unique as well, with one exception: if $l < k$ and $c_i = \tau \mathbb{P}_{e_i, l} = \iota \mathbb{P}_{e_i, l+1}$ for every $i \in [m]$, then we have $\mathbf{d} \to \mathbf{c} \to \mathbf{d}'$ for some $\mathbf{d}, \mathbf{d}' \in \mathcal{D}(\mathbb{A})^m$ and we can choose $l+1$ instead of l. However, it is not hard to see that the value assigned to $\overline{\omega}(\mathbf{c})$ is the same in both cases, namely it is the vertex $\tau \mathbb{P}_{e,l} = \iota \mathbb{P}_{e,l+1}$ (see property (b) below).

We need the following properties of the mapping Φ defined in <u>3c</u>.

(a) Φ is a homomorphism.
(b) $\Phi(\iota \mathbb{P}_{e_1, l}, \ldots, \iota \mathbb{P}_{e_m, l}) = \iota \mathbb{P}_{e,l}$ and $\Phi(\tau \mathbb{P}_{e_1, l}, \ldots, \tau \mathbb{P}_{e_m, l}) = \tau \mathbb{P}_{e,l}$.
(c) Φ does not depend on the ordering of the tuple $(e_1, \ldots, e_m)$.
(d) If $e_1 = \cdots = e_m = e$, then $\Phi : \mathbb{P}_{e,l}^m \to \mathbb{P}_{e,l}$ is idempotent, i.e., $\Phi(u, \ldots, u) = u$ for all $u \in \mathbb{P}_{e,l}$.

All of the above properties follow easily from the definition of Φ. We leave the verification to the reader. It remains to prove that $\overline{\omega}$ is a WNU polymorphism of $\mathcal{D}(\mathbb{A})$.

Claim. $\overline{\omega}$ is idempotent and satisfies the WNU equations.

Let $c, d \in \mathcal{D}(\mathbb{A})$. Note that all of the tuples $(c, \ldots, c, d), (c, \ldots, d, c), \ldots,$ $(d, c, \ldots, c)$ fall into the same subcase; the possibilities are <u>3a</u>, <u>3b</u>, <u>3c</u> and if $c \neq d$, then also <u>1a</u> or <u>2a</u>. In subcases <u>1a</u> and <u>2a</u> the definition does not depend on the ordering of the input tuple at all; therefore the WNU equations hold (and since $c \neq d$, idempotency does not apply).

In case 3 we use the fact that ω and $\omega^{(k)}$ are idempotent and satisfy the WNU equations. In <u>3a</u> and <u>3b</u> the claim follows immediately. In <u>3c</u> note that e is the same for all of the tuples $(c, \ldots, c, d), \ldots, (d, c, \ldots, c)$, and if $c = d$, then $e_1 = \cdots = e_m = e$. The WNU equations follow from property (c) of the mapping Φ, and idempotency follows from (d).

Claim. $\overline{\omega}$ is a polymorphism of $\mathcal{D}(\mathbb{A})$.

Let $\mathbf{c}, \mathbf{d} \in \mathcal{D}(\mathbb{A})^m$ be such that $\mathbf{c} \to \mathbf{d}$ is an edge in $\mathcal{D}(\mathbb{A})^m$, that is, $c_i \to d_i$ for all $i \in [m]$. Both the tuples $\mathbf{c}$ and $\mathbf{d}$ fall into the same case and, by Lemma 7 (iii), it cannot be case 2. If it is case 1, then they also fall into the same subcase and it is easily seen that $\overline{\omega}(\mathbf{c}) \to \overline{\omega}(\mathbf{d})$.

If $\mathbf{c}$ falls into subcase <u>3a</u>, then $\mathbf{d}$ falls into <u>3c</u>. Let e_i be such that $d_i \in \mathbb{P}_{e_i}$. As $c_i \to d_i$, it follows that $e_i = (c_i, \mathbf{r}^i)$ for some $\mathbf{r}^i \in R$ and $d_i = \iota \mathbb{P}_{e_i, 1}$. Let us define $c = \omega(\mathbf{c})$, $\mathbf{r} = \omega^{(k)}(\mathbf{r}^1, \ldots, \mathbf{r}^m)$ and $e = (c, \mathbf{r})$. Now $\overline{\omega}(\mathbf{d})$ is the result of the mapping Φ applied to the tuple of initial vertices of the $\mathbb{P}_{e_i, 1}$'s, which (by property (b)) is the initial vertex of $\mathbb{P}_{e, 1}$. So $\overline{\omega}(\mathbf{c}) = c \to \iota \mathbb{P}_{e, 1} = \overline{\omega}(\mathbf{d})$. The argument is similar if $\mathbf{d}$ falls into <u>3b</u>.

It remains to verify that $\overline{\omega}(\mathbf{c}) \to \overline{\omega}(\mathbf{d})$ if both $\mathbf{c}$ and $\mathbf{d}$ fall into subcase <u>3c</u>. Both $\overline{\omega}(\mathbf{c})$ and $\overline{\omega}(\mathbf{d})$ are defined using the mapping $\Phi : \prod_{i=1}^{m} \mathbb{P}_{e_i, l} \to \mathbb{P}_{e, l}$. Since Φ is a homomorphism, we have $\overline{\omega}(\mathbf{c}) = \Phi(\mathbf{c}) \to \Phi(\mathbf{d}) = \overline{\omega}(\mathbf{d})$, concluding the proof. $\qquad \square$

7 Discussion

The algebraic dichotomy conjecture proposes a polymorphism characterisation of tractability for core CSPs. A number of other algorithmic properties are also either proved or conjectured to correspond to the existence of polymorphisms with special equational properties. For instance, solvability by the few subpowers algorithm (a generalization of Gaussian elimination) as described in Idziak et al. [11] has a polymorphism characterisation [4], as well as problems of bounded width (see Theorem 3). The final statement in Theorem 1 already shows that $\mathbb{A}$ has bounded width if and only if $\mathcal{D}(\mathbb{A})$ has bounded width. Kazda [14] showed that every digraph with a Maltsev polymorphism must have a majority polymorphism, which is not the case for finite relational structures in general. In a later version of the present article we will show that Theorem 1(ii) extends to include almost all commonly encountered polymorphism properties aside from Maltsev. For instance the CSP over $\mathbb{A}$ is solvable by the few subpowers algorithm if and only if the same is true for $\mathcal{D}(\mathbb{A})$. Among other conditions preserved under our reduction to digraphs is that of arc consistency (or width 1 problems [7]) and problems of bounded strict width [8]. The following example is a powerful consequence of the result.

Example 3. Let $\mathbb{A}$ be the structure on $\{0, 1\}$ with a single 4-ary relation

$$\{(0, 0, 0, 1), (0, 1, 1, 1), (1, 0, 1, 1), (1, 1, 0, 1)\}.$$

Clearly $\mathbb{A}$ is a core. The polymorphisms of $\mathbb{A}$ can be shown to be the idempotent term functions of the two element group, and from this it follows that $\mathrm{CSP}(\mathbb{A})$ is solvable by the few subpowers algorithm of [11], but is not bounded width. Then the CSP over the digraph $\mathcal{D}(\mathbb{A})$ is also solvable by few subpowers but is not bounded width (that is, is not solvable by local consistency check).

Prior to the announcement of this example it had been temporarily conjectured by some researchers that solvability by the few subpowers algorithm implied solvability by local consistency check in the case of digraphs (this was the opening conjecture in Maróti's keynote presentation at the Second International Conference on Order, Algebra and Logics in Krakow 2011 for example). With 78 vertices and 80 edges, Example 3 also serves as a simpler alternative to the 368-vertex, 432-edge digraph whose CSP was shown by Atserias in [1, §4.2] to be tractable but not solvable by local consistency check.

There are also conjectured polymorphism classifications of the property of solvability within nondeterministic logspace and within logspace; see Larose and Tesson [16]. The required polymorphism conditions are among those we can show are preserved under the transition from $\mathbb{A}$ to $\mathcal{D}(\mathbb{A})$. It then follows that these conjectures are true provided they can be established in the restricted case of CSPs over digraphs.

Open Problems. We conclude our paper with some further research directions. It is possible to show that the logspace reduction in Lemma 4 *cannot* be replaced by first order reductions. Is there a different construction that translates general CSPs to digraph CSPs with first order reductions in both directions?

Feder and Vardi [8] and Atserias [1] provide polynomial time reductions of CSPs to digraph CSPs. We vigorously conjecture that their reductions preserve the properties of possessing a WNU polymorphism (and of being cores; but this is routinely verified). Do these or other constructions preserve the precise arity of WNU polymorphisms? What other polymorphism properties are preserved? Do they preserve the bounded width property? Can they preserve *conservative* polymorphisms (the polymorphisms related to list homomorphism problems)? The third and fourth authors with Kowalski [12] have recently shown that a minor variation of the $\mathcal{D}$ construction in the present article preserves k-ary WNU polymorphisms (and can serve as an alternative to $\mathcal{D}$ in Theorem 1), but always fails to preserve many other polymorphism properties (such as those witnessing strict width).

Acknowledgements. The authors would like to thank Libor Barto, Marcin Kozik, Miklós Maróti and Barnaby Martin for their thoughtful comments and discussions.

References

1. Atserias, A.: On digraph coloring problems and treewidth duality. European J. Combin. 29(4), 796–820 (2008), http://dx.doi.org/10.1016/j.ejc.2007.11.004

2. Barto, L., Kozik, M.: Constraint satisfaction problems of bounded width. In: 2009 50th Annual IEEE Symposium on Foundations of Computer Science (FOCS 2009), pp. 595–603. IEEE Computer Soc., Los Alamitos (2009), http://dx.doi.org/10.1109/FOCS.2009.32
3. Barto, L., Kozik, M., Niven, T.: The CSP dichotomy holds for digraphs with no sources and no sinks (a positive answer to a conjecture of Bang-Jensen and Hell). SIAM J. Comput. 38(5), 1782–1802 (2008/2009), http://dx.doi.org/10.1137/070708093
4. Berman, J., Idziak, P., Marković, P., McKenzie, R., Valeriote, M., Willard, R.: Varieties with few subalgebras of powers. Trans. Amer. Math. Soc. 362(3), 1445–1473 (2010), http://dx.doi.org/10.1090/S0002-9947-09-04874-0
5. Bulatov, A., Jeavons, P., Krokhin, A.: Classifying the complexity of constraints using finite algebras. SIAM J. Comput. 34(3), 720–742 (2005), http://dx.doi.org/10.1137/S0097539700376676
6. Bulatov, A.A.: A dichotomy theorem for constraint satisfaction problems on a 3-element set. J. ACM 53(1), 66–120 (2006),
   http://dx.doi.org/10.1145/1120582.1120584
7. Dalmau, V., Pearson, J.: Closure functions and width 1 problems. In: Jaffar, J. (ed.) CP 1999. LNCS, vol. 1713, pp. 159–173. Springer, Heidelberg (1999)
8. Feder, T., Vardi, M.Y.: The computational structure of monotone monadic SNP and constraint satisfaction: a study through Datalog and group theory. SIAM J. Comput. 28(1), 57–104 (electronic) (1999), http://dx.doi.org/10.1137/S0097539794266766
9. Hell, P., Nešetřil, J.: On the complexity of H-coloring. J. Combin. Theory Ser. B 48(1), 92–110 (1990), http://dx.doi.org/10.1016/0095-8956(90)90132-J
10. Hell, P., Nešetřil, J.: Graphs and homomorphisms. Oxford Lecture Series in Mathematics and its Applications, vol. 28. Oxford University Press, Oxford (2004), http://dx.doi.org/10.1093/acprof:oso/9780198528173.001.0001
11. Idziak, P., Marković, P., McKenzie, R., Valeriote, M., Willard, R.: Tractability and learnability arising from algebras with few subpowers. SIAM J. Comput. 39(7), 3023–3037 (2010), http://dx.doi.org/10.1137/090775646
12. Jackson, M., Kowalski, T., Niven, T.: Digraph related constructions and the complexity of digraph homomorphism problems. arXiv:1304.4986 [math.CO] (2013)
13. Jeavons, P., Cohen, D., Gyssens, M.: Closure properties of constraints. J. ACM 44(4), 527–548 (1997), http://dx.doi.org/10.1145/263867.263489
14. Kazda, A.: Maltsev digraphs have a majority polymorphism. European J. Combin. 32(3), 390–397 (2011), http://dx.doi.org/10.1016/j.ejc.2010.11.002
15. Ladner, R.E.: On the structure of polynomial time reducibility. J. Assoc. Comput. Mach. 22, 155–171 (1975)
16. Larose, B., Tesson, P.: Universal algebra and hardness results for constraint satisfaction problems. Theor. Comput. Sci. 410(18), 1629–1647 (2009)
17. Maróti, M., McKenzie, R.: Existence theorems for weakly symmetric operations. Algebra Universalis 59(3-4), 463–489 (2008),
    http://dx.doi.org/10.1007/s00012-008-2122-9
18. Montanari, U.: Networks of constraints: fundamental properties and applications to picture processing. Information Sci. 7, 95–132 (1974)
19. Schaefer, T.J.: The complexity of satisfiability problems. In: Conference Record of the Tenth Annual ACM Symposium on Theory of Computing (San Diego, Calif., 1978), pp. 216–226. ACM, New York (1978)

A Scalable Approximate Model Counter[*][**]

Supratik Chakraborty[1], Kuldeep S. Meel[2], and Moshe Y. Vardi[2]

[1] Indian Institute of Technology Bombay, India
[2] Department of Computer Science, Rice University

Abstract. *Propositional model counting* (#SAT), i.e., counting the number of satisfying assignments of a propositional formula, is a problem of significant theoretical and practical interest. Due to the inherent complexity of the problem, *approximate model counting*, which counts the number of satisfying assignments to within given tolerance and confidence level, was proposed as a practical alternative to exact model counting. Yet, approximate model counting has been studied essentially only theoretically. The only reported implementation of approximate model counting, due to Karp and Luby, worked only for DNF formulas. A few existing tools for CNF formulas are *bounding model counters*; they can handle realistic problem sizes, but fall short of providing counts within given tolerance and confidence, and, thus, are not approximate model counters.

We present here a novel algorithm, as well as a reference implementation, that is the first scalable approximate model counter for CNF formulas. The algorithm works by issuing a polynomial number of calls to a SAT solver. Our tool, ApproxMC, scales to formulas with tens of thousands of variables. Careful experimental comparisons show that ApproxMC reports, with high confidence, bounds that are close to the exact count, and also succeeds in reporting bounds with small tolerance and high confidence in cases that are too large for computing exact model counts.

1 Introduction

Propositional model counting, also known as #SAT, concerns counting the number of models (satisfying truth assignments) of a given propositional formula. This problem has been the subject of extensive theoretical investigation since its

[*] Authors would like to thank Henry Kautz and Ashish Sabhrawal for their valuable help in experiments, and Tracy Volz for valuable comments on the earlier drafts. Work supported in part by NSF grants CNS 1049862 and CCF-1139011, by NSF Expeditions in Computing project "ExCAPE: Expeditions in Computer Augmented Program Engineering," by BSF grant 9800096, by a gift from Intel, by a grant from Board of Research in Nuclear Sciences, India, and by the Shared University Grid at Rice funded by NSF under Grant EIA-0216467, and a partnership between Rice University, Sun Microsystems, and Sigma Solutions, Inc.

[**] A longer version of this paper is available at http://www.cs.rice.edu/CS/Verification/Projects/ApproxMC/

C. Schulte (Ed.): CP 2013, LNCS 8124, pp. 200–216, 2013.
© Springer-Verlag Berlin Heidelberg 2013

introduction by Valiant [35] in 1979. Several interesting applications of #SAT have been studied in the context of probabilistic reasoning, planning, combinatorial design and other related fields [24,4,9]. In particular, probabilistic reasoning and inferencing have attracted considerable interest in recent years [13], and stand to benefit significantly from efficient propositional model counters.

Theoretical investigations of #SAT have led to the discovery of deep connections in complexity theory [3,29,33]: #SAT is #P-complete, where #P is the set of counting problems associated with decision problems in the complexity class NP. Furthermore, $P^{\#SAT}$, that is, a polynomial-time machine with a #SAT oracle, can solve all problems in the entire polynomial hierarchy. In fact, the polynomial-time machine only needs to make one #SAT query to solve any problem in the polynomial hierarchy. This is strong evidence for the hardness of #SAT.

In many applications of model counting, such as in probabilistic reasoning, the exact model count may not be critically important, and approximate counts are sufficient. Even when exact model counts are important, the inherent complexity of the problem may force one to work with approximate counters in practice. In [31], Stockmeyer showed that counting models within a specified tolerance factor can be achieved in deterministic polynomial time using a Σ_2^p-oracle. Karp and Luby presented a fully polynomial randomized approximation scheme for counting models of a DNF formula [18]. Building on Stockmeyer's result, Jerrum, Valiant and Vazirani [16] showed that counting models of CNF formulas within a specified tolerance factor can be solved in random polynomial time using an oracle for SAT.

On the implementation front, the earliest approaches to #SAT were based on DPLL-style SAT solvers and computed exact counts. These approaches consisted of incrementally counting the number of solutions by adding appropriate multiplication factors after a partial solution was found. This idea was formalized by Birnbaum and Lozinkii [6] in their model counter CDP. Subsequent model counters such as Relsat [17], Cachet [26] and sharpSAT [32] improved upon this idea by using several optimizations such as component caching, clause learning, look-ahead and the like. Techniques based on Boolean Decision Diagrams and their variants [23,21], or d-DNNF formulae [8], have also been used to compute exact model counts. Although exact model counters have been successfully used in small- to medium-sized problems, scaling to larger problem instances has posed significant challenges in practice. Consequently, a large class of practical applications has remained beyond the reach of exact model counters.

To counter the scalability challenge, more efficient techniques for counting models approximately have been proposed. These counters can be broadly divided into three categories. Counters in the first category are called (ε, δ) counters, following Karp and Luby's terminology [18]. Let ε and δ be real numbers such that $0 < \varepsilon \leq 1$ and $0 < \delta \leq 1$. For every propositional formula F with $\#F$ models, an (ε, δ) counter computes a number that lies in the interval $[(1 + \varepsilon)^{-1}\#F, (1 + \varepsilon)\#F]$ with probability at least $1 - \delta$. We say that ε is the *tolerance* of the count, and $1 - \delta$ is its *confidence*. The counter described in this paper and also that due to Karp and Luby [18] belong to this category. The approximate-counting algorithm of Jerrum et al. [16] also belongs to this category; however, their algorithm does not lend itself to an implementation that scales in practice. Counters in the second category are called *lower (or upper) bounding counters*, and are parameterized by a confidence probability $1 - \delta$. For every propositional formula F with $\#F$ models, an upper (resp., lower) bounding counter computes a number that is at least as large (resp., as small) as $\#F$ with probability at least $1 - \delta$. Note that bounding counters *do not* provide any tolerance guarantees. The large majority of approximate counters used in practice are bounding counters. Notable examples include SampleCount [14], BPCount [20], MBound (and Hybrid-MBound) [12], and MiniCount [20]. The final category of counters is called *guarantee-less counters*. These counters provide no guarantees at all but they can be very efficient and provide good approximations in practice. Examples of guarantee-less counters include ApproxCount [36], SearchTreeSampler [10], SE [25] and SampleSearch [11].

Bounding both the tolerance and confidence of approximate model counts is extremely valuable in applications like probabilistic inference. Thus, designing (ε, δ) counters that scale to practical problem sizes is an important problem. Earlier work on (ε, δ) counters has been restricted largely to theoretical treatments of the problem. The only counter in this category that we are aware of as having been implemented is due to Karp and Luby [22]. Karp and Luby's original implementation was designed to estimate reliabilities of networks with failure-prone links. However, the underlying Monte Carlo engine can be used to approximately count models of DNF, *but not CNF*, formulas.

The counting problems for both CNF and DNF formulae are #P-complete. While the DNF representation suits some applications, most modern applications of model counting (e.g. probabilistic inference) use the CNF representation. Although *exact* counting for DNF and CNF formulae are polynomially inter-reducible, there is no known polynomial reduction for the corresponding *approximate* counting problems. In fact, Karp and Luby remark in [18] that it is highly unlikely that their randomized approximate algorithm for DNF formulae can be adapted to work for CNF formulae. Thus, there has been no prior implementation of (ε, δ) counters for CNF formulae *that scales in practice*. In this paper, we present the first such counter. As in [16], our algorithm runs in random polynomial time using an oracle for SAT. Our extensive experiments show that our algorithm scales, with low error, to formulae arising from several application domains involving tens of thousands of variables.

The organization of the paper is as follows. We present preliminary material in Section 2, and related work in Section 3. In Section 4, we present our algorithm, followed by its analysis in Section 5. Section 6 discusses our experimental methodology, followed by experimental results in Section 7. Finally, we conclude in Section 8.

2   Notation and Preliminaries

Let Σ be an alphabet and $R \subseteq \Sigma^* \times \Sigma^*$ be a binary relation. We say that R is an $\mathcal{NP}$-relation if R is polynomial-time decidable, and if there exists a polynomial $p(\cdot)$ such that for every $(x, y) \in R$, we have $|y| \leq p(|x|)$. Let L_R be the language $\{x \in \Sigma^* \mid \exists y \in \Sigma^*, (x, y) \in R\}$. The language L_R is said to be in $\mathcal{NP}$ if R is an $\mathcal{NP}$-relation. The set of all satisfiable propositional logic formulae in CNF is a language in $\mathcal{NP}$. Given $x \in L_R$, a *witness* or *model* of x is a string $y \in \Sigma^*$ such that $(x, y) \in R$. The set of all models of x is denoted R_x. For notational convenience, fix Σ to be $\{0, 1\}$ without loss of generality. If R is an $\mathcal{NP}$-relation, we may further assume that for every $x \in L_R$, every witness $y \in R_x$ is in $\{0, 1\}^n$, where $n = p(|x|)$ for some polynomial $p(\cdot)$.

Let $R \subseteq \{0, 1\}^* \times \{0, 1\}^*$ be an $\mathcal{NP}$ relation. The *counting problem* corresponding to R asks "Given $x \in \{0, 1\}^*$, what is $|R_x|$?". If R relates CNF propositional formulae to their satisfying assignments, the corresponding counting problem is called #SAT. The primary focus of this paper is on (ε, δ) counters for #SAT. The randomized (ε, δ) counters of Karp and Luby [18] for DNF formulas are *fully polynomial*, which means that they run in time polynomial in the size of the input formula F, $1/\varepsilon$ and $\log(1/\delta)$. The randomized (ε, δ) counters for CNF formulas in [16] and in this paper are however fully polynomial *with respect to a SAT oracle*.

A special class of hash functions, called *r-wise independent* hash functions, play a crucial role in our work. Let n, m and r be positive integers, and let $H(n, m, r)$ denote a family of r-wise independent hash functions mapping $\{0, 1\}^n$ to $\{0, 1\}^m$. We use $\Pr[X : \mathcal{P}]$ to denote the probability of outcome X when sampling from a probability space $\mathcal{P}$, and $h \xleftarrow{R} H(n, m, r)$ to denote the probability space obtained by choosing a hash function h uniformly at random from $H(n, m, r)$. The property of r-wise independence guarantees that for all $\alpha_1, \ldots \alpha_r \in \{0, 1\}^m$ and for all distinct $y_1, \ldots y_r \in \{0, 1\}^n$, $\Pr\left[\bigwedge_{i=1}^r h(y_i) = \alpha_i : h \xleftarrow{R} H(n, m, r)\right] = 2^{-mr}$. For every $\alpha \in \{0, 1\}^m$ and $h \in H(n, m, r)$, let $h^{-1}(\alpha)$ denote the set $\{y \in \{0, 1\}^n \mid h(y) = \alpha\}$. Given $R_x \subseteq \{0, 1\}^n$ and $h \in H(n, m, r)$, we use $R_{x,h,\alpha}$ to denote the set $R_x \cap h^{-1}(\alpha)$. If we keep h fixed and let α range over $\{0, 1\}^m$, the sets $R_{x,h,\alpha}$ form a partition of R_x. Following the notation in [5], we call each element of such a partition a *cell* of R_x induced by h. It was shown in [5] that if h is chosen uniformly at random from $H(n, m, r)$ for $r \geq 1$, then the expected size of $R_{x,h,\alpha}$, denoted $\mathsf{E}[|R_{x,h,\alpha}|]$, is $|R_x|/2^m$, for each $\alpha \in \{0, 1\}^m$.

The specific family of hash functions used in our work, denoted $H_{xor}(n, m, 3)$, is based on randomly choosing bits from $y \in \{0, 1\}^n$ and xor-ing them. This family of hash functions has been used in earlier work [12], and has been shown to be 3-independent in [15]. Let $h(y)[i]$ denote the ith component of the bit-vector obtained by applying hash function h to y. The family $H_{xor}(n, m, 3)$ is defined as $\{h(y) \mid (h(y))[i] = a_{i,0} \oplus (\bigoplus_{k=1}^n a_{i,k} \cdot y[k]), a_{i,j} \in \{0, 1\}, 1 \leq i \leq m, 0 \leq j \leq n\}$, where $\oplus$ denotes the xor operation. By randomly choosing the $a_{i,j}$'s, we can randomly choose a hash function from this family.

3   Related Work

Sipser pioneered a hashing based approach in [30], which has subsequently been used in theoretical [34,5] and practical [15,12,7] treatments of approximate counting and (near-)uniform sampling. Earlier implementations of counters that use the hashing-based approach are MBound and Hybrid-MBound [12]. Both these counters use the same family of hashing functions, i.e., $H_{xor}(n, m, 3)$, that we use. Nevertheless, there are significant differences between our algorithm and those of MBound and Hybrid-MBound. Specifically, we are able to exploit properties of the $H_{xor}(n, m, 3)$ family of hash functions to obtain a fully polynomial (ε, δ) counter with respect to a SAT oracle. In contrast, both MBound and Hybrid-MBound are bounding counters, and cannot provide bounds on tolerance. In addition, our algorithm requires no additional parameters beyond the tolerance ε and confidence $1 - \delta$. In contrast, the performance and quality of results of both MBound and Hybrid-MBound, depend crucially on some hard-to-estimate parameters. It has been our experience that the right choice of these parameters is often domain dependent and difficult.

Jerrum, Valiant and Vazirani [16] showed that if R is a self-reducible $\mathcal{NP}$ relation (such as SAT), the problem of generating models *almost uniformly* is polynomially inter-reducible with approximately counting models. The notion of almost uniform generation requires that if x is a problem instance, then for every $y \in R_x$, we have $(1 + \varepsilon)^{-1}\varphi(x) \leq \Pr[y \text{ is generated}] \leq (1 + \varepsilon)\varphi(x)$, where $\varepsilon > 0$ is the specified tolerance and $\varphi(x)$ is an appropriate function. Given an almost uniform generator $\mathcal{G}$ for R, an input x, a tolerance bound ε and an error probability bound δ, it is shown in [16] that one can obtain an (ε, δ) counter for R by invoking $\mathcal{G}$ polynomially (in $|x|$, $1/\varepsilon$ and $\log_2(1/\delta)$) many times, and by using the generated samples to estimate $|R_x|$. For convenience of exposition, we refer to this approximate-counting algorithm as the JVV algorithm (after the last names of the authors).

An important feature of the JVV algorithm is that it uses the almost uniform generator $\mathcal{G}$ as a black box. Specifically, the details of how $\mathcal{G}$ works is of no consequence. Prima facie, this gives us freedom in the choice of $\mathcal{G}$ when implementing the JVV algorithm. Unfortunately, while there are theoretical constructions of uniform generators in [5], we are not aware of any implementation of an almost uniform generator that scales to CNF formulas involving thousands of variables. The lack of a scalable and almost uniform generator presents a significant hurdle in implementing the JVV algorithm for practical applications. It is worth asking if we can make the JVV algorithm work without requiring $\mathcal{G}$ to be an almost uniform generator. A closer look at the proof of correctness of the JVV algorithm [16] shows it relies crucially on the ability of $\mathcal{G}$ to ensure that the probabilities of generation of any two distinct models of x differ by a factor in $O(\varepsilon^2)$. As discussed in [7], existing algorithms for randomly generating models either provide this guarantee but scale very poorly in practice (e.g., the algorithms in [5,37]), or scale well in practice without providing the above guarantee (e.g., the algorithms in [7,15,19]). Therefore, using an existing generator as a black box in the JVV algorithm would not give us an (ε, δ) model counter that scales in practice. The primary contribution of this paper is to show that a scalable (ε, δ) counter can indeed be designed by using the same insights that went into the design of a *near uniform* generator, UniWit [7], but without using the generator as a black box in the approximate counting algorithm. Note that near uniformity, as defined in [7], is an even more relaxed notion of uniformity than almost uniformity. We leave the question of whether a near uniform generator can be used as a black box to design an (ε, δ) counter as part of future work.

The central idea of UniWit, which is also shared by our approximate model counter, is the use of r-wise independent hashing functions to randomly partition the space of all models of a given problem instance into "small" cells. This idea was first proposed in [5], but there are two novel insights that allow UniWit [7] to scale better than other hashing-based sampling algorithms [5,15], while still providing guarantess on the quality of sampling. These insights are: (i) the use of computationally efficient linear hashing functions with low degrees of independence, and (ii) a drastic reduction in the size of "small" cells, from n^2 in [5] to $n^{1/k}$ (for $2 \leq k \leq 3$) in [7], and even further to a constant in the current paper. We continue to use these key insights in the design of our approximate model counter, although UniWit is not used explicitly in the model counter.

4 Algorithm

We now describe our approximate model counting algorithm, called ApproxMC. As mentioned above, we use 3-wise independent linear hashing functions from the $H_{xor}(n, m, 3)$ family, for an appropriate m, to randomly partition the set of models of an input formula into "small" cells. In order to test whether the generated cells are indeed small, we choose a random cell and check if it is non-empty and has no more than *pivot* elements, where *pivot* is a threshold that depends only on the tolerance bound ε. If the chosen cell is not small, we randomly partition the set of models into twice as many cells as before by choosing a random hashing function from the family $H_{xor}(n, m + 1, 3)$. The above procedure is repeated until either a randomly chosen cell is found to be non-empty and small, or the number of cells exceeds $\frac{2^{n+1}}{pivot}$. If all cells that were randomly chosen during the above process were either empty or not small, we report a counting failure and return $\perp$. Otherwise, the size of the cell last chosen is scaled by the number of cells to obtain an ε-approximate estimate of the model count.

The procedure outlined above forms the core engine of ApproxMC. For convenience of exposition, we implement this core engine as a function ApproxMCCore. The overall ApproxMC algorithm simply invokes ApproxMCCore sufficiently many times, and returns the median of the non-$\perp$ values returned by ApproxMCCore. The pseudocode for algorithm ApproxMC is shown below.

Algorithm ApproxMC(F, ε, δ)
1: $counter \leftarrow 0; C \leftarrow$ emptyList;
2: $pivot \leftarrow 2 \times$ ComputeThreshold(ε);
3: $t \leftarrow$ ComputeIterCount(δ);
4: **repeat**:
5:     $c \leftarrow$ ApproxMCCore($F, pivot$);
6:     $counter \leftarrow counter + 1$;
7:     **if** ($c \neq \perp$)
8:         AddToList(C, c);
9: **until** ($counter < t$);
10: $finalCount \leftarrow$ FindMedian(C);
11: **return** $finalCount$;

Algorithm ComputeThreshold(ε)
1: **return** $\left\lceil 3e^{1/2} \left(1 + \frac{1}{\varepsilon}\right)^2 \right\rceil$;

Algorithm ComputeIterCount(δ)
1: **return** $\lceil 35 \log_2(3/\delta) \rceil$;

Algorithm ApproxMC takes as inputs a CNF formula F, a tolerance ε ($0 < \varepsilon \leq 1$) and δ ($0 < \delta \leq 1$) such that the desired confidence is $1 - \delta$. It computes two key parameters: (i) a threshold $pivot$ that depends only on ε and is used in ApproxMCCore to determine the size of a "small" cell, and (ii) a parameter t (≥ 1) that depends only on δ and is used to determine the number of times ApproxMCCore is invoked. The particular choice of functions to compute the parameters $pivot$ and t aids us in proving theoretical guarantees for ApproxMC in Section 5. Note that $pivot$ is in $\mathcal{O}(1/\varepsilon^2)$ and t is in $\mathcal{O}(\log_2(1/\delta))$. All non-$\perp$ estimates of the model count returned by ApproxMCCore are stored in the list C. The function AddToList(C, c) updates the list C by adding the element c. The final estimate of the model count returned by ApproxMC is the median of the estimates stored in C, computed using FindMedian(C). We assume that if the list C is empty, FindMedian(C) returns $\perp$.

The pseudocode for algorithm ApproxMCCore is shown below.

Algorithm ApproxMCCore($F, pivot$)
/* Assume $z_1, \ldots z_n$ are the variables of F */
1: $S \leftarrow$ BoundedSAT($F, pivot + 1$);
2: **if** ($|S| \leq pivot$)
3:     **return** $|S|$;
4: **else**
5:     $l \leftarrow \lfloor \log_2(pivot) \rfloor - 1; i \leftarrow l - 1$;
6:     **repeat**
7:         $i \leftarrow i + 1$;
8:         Choose h at random from $H_{xor}(n, i - l, 3)$;
9:         Choose α at random from $\{0, 1\}^{i-l}$;
10:        $S \leftarrow$ BoundedSAT($F \wedge (h(z_1, \ldots z_n) = \alpha), pivot + 1$);
11:    **until** ($1 \leq |S| \leq pivot$) or ($i = n$);
12:    **if** ($|S| > pivot$ **or** $|S| = 0$) **return** $\perp$;
13:    **else return** $|S| \cdot 2^{i-l}$;

Algorithm ApproxMCCore takes as inputs a CNF formula F and a threshold *pivot*, and returns an ε-approximate estimate of the model count of F. We assume that ApproxMCCore has access to a function BoundedSAT that takes as inputs a proposition formula F' that is the conjunction of a CNF formula and xor constraints, as well as a threshold $v \geq 0$. BoundedSAT(F', v) returns a set S of models of F' such that $|S| = \min(v, \#F')$. If the model count of F is no larger than *pivot*, then ApproxMCCore returns the exact model count of F in line 3 of the pseudocode. Otherwise, it partitions the space of all models of F using random hashing functions from $H_{xor}(n, i - l, 3)$ and checks if a randomly chosen cell is non-empty and has at most *pivot* elements. Lines 8–10 of the repeat-until loop in the pseudocode implement this functionality. The loop terminates if either a randomly chosen cell is found to be small and non-empty, or if the number of cells generated exceeds $\frac{2^{n+1}}{pivot}$ (if $i = n$ in line 11, the number of cells generated is $2^{n-l} \geq \frac{2^{n+1}}{pivot}$). In all cases, unless the cell that was chosen last is empty or not small, we scale its size by the number of cells generated by the corresponding hashing function to compute an estimate of the model count. If, however, all randomly chosen cells turn out to be empty or not small, we report a counting error by returning $\perp$.

Implementation Issues: There are two steps in algorithm ApproxMCCore (lines 8 and 9 of the pseudocode) where random choices are made. Recall from Section 2 that choosing a random hash function from $H_{xor}(n, m, 3)$ requires choosing random bit-vectors. It is straightforward to implement these choices and also the choice of a random $\alpha \in \{0, 1\}^{i-l}$ in line 9 of the pseudocode, if we have access to a source of independent and uniformly distributed random bits. Our implementation uses pseudo-random sequences of bits generated from nuclear decay processes and made available at HotBits [2]. We download and store a sufficiently long sequence of random bits in a file, and access an appropriate number of bits sequentially whenever needed. We defer experimenting with sequences of bits obtained from other pseudo-random generators to a future study.

In lines 1 and 10 of the pseudocode for algorithm ApproxMCCore, we invoke the function BoundedSAT. Note that if h is chosen randomly from $H_{xor}(n, m, 3)$, the formula for which we seek models is the conjunction of the original (CNF) formula and xor constraints encoding the inclusion of each witness in $h^{-1}(\alpha)$. We therefore use a SAT solver optimized for conjunctions of xor constraints and CNF clauses as the back-end engine. Specifically, we use CryptoMiniSAT (version 2.9.2) [1], which also allows passing a parameter indicating the maximum number of witnesses to be generated.

Recall that ApproxMCCore is invoked t times with the same arguments in algorithm ApproxMC. Repeating the loop of lines 6–11 in the pseudocode of ApproxMCCore in each invocation can be time consuming if the values of $i - l$ for which the loop terminates are large. In [7], a heuristic called *leap-frogging* was proposed to overcome this bottleneck in practice. With leap-frogging, we register the smallest value of $i - l$ for which the loop terminates during the first few invocations of ApproxMCCore. In all subsequent invocations of ApproxMCCore with the same arguments, we start iterating the loop of lines 6–11 by initializing $i - l$

to the smallest value registered from earlier invocations. Our experiments indicate that leap-frogging is extremely efficient in practice and leads to significant savings in time after the first few invocations of ApproxMCCore. A theoretical analysis of leapfrogging is deferred to future work.

5   Analysis of ApproxMC

The following result, a minor variation of Theorem 5 in [28], about Chernoff-Hoeffding bounds plays an important role in our analysis.

Theorem 1. *Let Γ be the sum of r-wise independent random variables, each of which is confined to the interval $[0, 1]$, and suppose $\mathsf{E}[\Gamma] = \mu$. For $0 < \beta \leq 1$, if $r \leq \lfloor \beta^2 \mu e^{-1/2} \rfloor \leq 4$ then $\Pr\left[|\Gamma - \mu| \geq \beta \mu \right] \leq e^{-r/2}$.*

Let F be a CNF propositional formula with n variables. The next two lemmas show that algorithm ApproxMCCore, when invoked from ApproxMC with arguments F, ε and δ, behaves like an (ε, d) model counter for F, for a fixed confidence $1 - d$ (possibly different from $1 - \delta$). Throughout this section, we use the notations R_F and $R_{F,h,\alpha}$ introduced in Section 2.

Lemma 1. *Let algorithm ApproxMCCore, when invoked from ApproxMC, return c with i being the final value of the loop counter in ApproxMCCore. Then,*
$$\Pr\left[(1+\varepsilon)^{-1} \cdot |R_F| \leq c \leq (1+\varepsilon) \cdot |R_F| \;\middle|\; c \neq \perp \text{ and } i \leq \log_2 |R_F| \right] \geq 1 - e^{-3/2}.$$

Proof. Referring to the pseudocode of ApproxMCCore, the lemma is trivially satisfied if $|R_F| \leq pivot$. Therefore, the only non-trivial case to consider is when $|R_F| > pivot$ and ApproxMCCore returns from line 13 of the pseudocode. In this case, the count returned is $2^{i-l} \cdot |R_{F,h,\alpha}|$, where $l = \lfloor \log_2(pivot) \rfloor - 1$ and α, i and h denote (with abuse of notation) the values of the corresponding variables and hash functions in the final iteration of the repeat-until loop in lines 6–11 of the pseudocode.

For simplicity of exposition, we assume henceforth that $\log_2(pivot)$ is an integer. A more careful analysis removes this restriction with only a constant factor scaling of the probabilities. From the pseudocode of ApproxMCCore, we know that $pivot = 2\left\lceil 3e^{1/2}\left(1 + \frac{1}{\varepsilon}\right)^2 \right\rceil$. Furthermore, the value of i is always in $\{l, \ldots n\}$. Since $pivot < |R_F| \leq 2^n$ and $l = \lfloor \log_2 pivot \rfloor - 1$, we have $l < \log_2 |R_F| \leq n$. The lemma is now proved by showing that for every i in $\{l, \ldots \lfloor \log_2 |R_F| \rfloor\}$, $h \in H(n, i - l, 3)$ and $\alpha \in \{0, 1\}^{i-l}$, we have $\Pr\left[(1+\varepsilon)^{-1} \cdot |R_F| \leq 2^{i-l} |R_{F,h,\alpha}| \leq (1+\varepsilon) \cdot |R_F| \right] \geq (1 - e^{-3/2})$.

For every $y \in \{0, 1\}^n$ and for every $\alpha \in \{0, 1\}^{i-l}$, define an indicator variable $\gamma_{y,\alpha}$ as follows: $\gamma_{y,\alpha} = 1$ if $h(y) = \alpha$, and $\gamma_{y,\alpha} = 0$ otherwise. Let us fix α and y and choose h uniformly at random from $H(n, i - l, 3)$. The random choice of h induces a probability distribution on $\gamma_{y,\alpha}$, such that $\Pr\left[\gamma_{y,\alpha} = 1\right] = \Pr\left[h(y) = \alpha\right] = 2^{-(i-l)}$, and $\mathsf{E}\left[\gamma_{y,\alpha}\right] = \Pr\left[\gamma_{y,\alpha} = 1\right] = 2^{-(i-l)}$. In addition, the 3-wise independence of hash functions chosen from $H(n, i - l, 3)$ implies that for every distinct $y_a, y_b, y_c \in R_F$, the random variables $\gamma_{y_a,\alpha}$, $\gamma_{y_b,\alpha}$ and $\gamma_{y_c,\alpha}$ are 3-wise independent.

Let $\Gamma_\alpha = \sum_{y \in R_F} \gamma_{y,\alpha}$ and $\mu_\alpha = \mathsf{E}\,[\Gamma_\alpha]$. Clearly, $\Gamma_\alpha = |R_{F,h,\alpha}|$ and $\mu_\alpha = \sum_{y \in R_F} \mathsf{E}\,[\gamma_{y,\alpha}] = 2^{-(i-l)}|R_F|$. Since $|R_F| > pivot$ and $i \leq \log_2 |R_F|$, using the expression for $pivot$, we get $3 \leq \left\lfloor e^{-1/2}(1 + \frac{1}{\varepsilon})^{-2} \cdot \frac{|R_F|}{2^{i-l}} \right\rfloor$. Therefore, using Theorem 1, $\Pr\left[|R_F| \cdot \left(1 - \frac{\varepsilon}{1+\varepsilon}\right) \leq 2^{i-l}|R_{F,h,\alpha}| \leq (1 + \frac{\varepsilon}{1+\varepsilon})|R_F|\right] \geq 1 - e^{-3/2}$. Simplifying and noting that $\frac{\varepsilon}{1+\varepsilon} < \varepsilon$ for all $\varepsilon > 0$, we obtain $\Pr\left[(1 + \varepsilon)^{-1} \cdot |R_F| \leq 2^{i-l}|R_{F,h,\alpha}| \leq (1 + \varepsilon) \cdot |R_F|\right] \geq 1 - e^{-3/2}$.

Lemma 2. *Given $|R_F| > pivot$, the probability that an invocation of* ApproxMCCore *from* ApproxMC *returns non-$\bot$ with $i \leq \log_2 |R_F|$, is at least $1 - e^{-3/2}$.*

Proof. Let us denote $\log_2 |R_F| - l = \log_2 |R_F| - (\lfloor \log_2(pivot) \rfloor - 1)$ by m. Since $|R_F| > pivot$ and $|R_F| \leq 2^n$, we have $l < m + l \leq n$. Let p_i ($l \leq i \leq n$) denote the conditional probability that ApproxMCCore($F, pivot$) terminates in iteration i of the repeat-until loop (lines 6–11 of the pseudocode) with $1 \leq |R_{F,h,\alpha}| \leq pivot$, given $|R_F| > pivot$. Since the choice of h and α in each iteration of the loop are independent of those in previous iterations, the conditional probability that ApproxMCCore($F, pivot$) returns non-$\bot$ with $i \leq \log_2 |R_F| = m + l$, given $|R_F| > pivot$, is $p_l + (1 - p_l)p_{l+1} + \cdots + (1 - p_l)(1 - p_{l+1}) \cdots (1 - p_{m+l-1})p_{m+l}$. Let us denote this sum by P. Thus, $P = p_l + \sum_{i=l+1}^{m+l} \prod_{k=l}^{i-1}(1 - p_k)p_i \geq \left(p_l + \sum_{i=l+1}^{m+l-1} \prod_{k=l}^{i-1}(1 - p_k)p_i\right)p_{m+l} + \prod_{s=l}^{m+l-1}(1 - p_s)p_{m+l} = p_{m+l}$. The lemma is now proved by using Theorem 1 to show that $p_{m+l} \geq 1 - e^{-3/2}$.

It was shown in Lemma 1 that $\Pr\left[(1 + \varepsilon)^{-1} \cdot |R_F| \leq 2^{i-l}|R_{F,h,\alpha}| \leq (1 + \varepsilon) \cdot |R_F|\right] \geq 1 - e^{-3/2}$ for every $i \in \{l, \ldots \lfloor \log_2 |R_F| \rfloor\}$, $h \in H(n, i - l, 3)$ and $\alpha \in \{0,1\}^{i-l}$. Substituting $\log_2 |R_F| = m + l$ for i, re-arranging terms and noting that the definition of m implies $2^{-m}|R_F| = pivot/2$, we get $\Pr\left[(1 + \varepsilon)^{-1}(pivot/2) \leq |R_{F,h,\alpha}| \leq (1 + \varepsilon)(pivot/2)\right] \geq 1 - e^{-3/2}$. Since $0 < \varepsilon \leq 1$ and $pivot > 4$, it follows that $\Pr\left[1 \leq |R_{F,h,\alpha}| \leq pivot\right] \geq 1 - e^{-3/2}$. Hence, $p_{m+l} \geq 1 - e^{-3/2}$.

Theorem 2. *Let an invocation of* ApproxMCCore *from* ApproxMC *return c. Then* $\Pr\left[c \neq \bot \text{ and } (1 + \varepsilon)^{-1} \cdot |R_F| \leq c \leq (1 + \varepsilon) \cdot |R_F|\right] \geq (1 - e^{-3/2})^2 > 0.6$.

Proof sketch: It is easy to see that the required probability is at least as large as $\Pr\left[c \neq \bot \text{ and } i \leq \log_2 |R_F| \text{ and } (1 + \varepsilon)^{-1} \cdot |R_F| \leq c \leq (1 + \varepsilon) \cdot |R_F|\right]$. From Lemmas 1 and 2, the latter probability is $\geq (1 - e^{-3/2})^2$.

We now turn to proving that the confidence can be raised to at least $1 - \delta$ for $\delta \in (0,1]$ by invoking ApproxMCCore $\mathcal{O}(\log_2(1/\delta))$ times, and by using the median of the non-$\bot$ counts thus returned. For convenience of exposition, we use $\eta(t, m, p)$ in the following discussion to denote the probability of at least m heads in t independent tosses of a biased coin with $\Pr\,[heads] = p$. Clearly, $\eta(t, m, p) = \sum_{k=m}^{t} \binom{t}{k} p^k (1 - p)^{t-k}$.

Theorem 3. *Given a propositional formula F and parameters ε $(0 < \varepsilon \leq 1)$ and δ $(0 < \delta \leq 1)$, suppose $\mathsf{ApproxMC}(F, \varepsilon, \delta)$ returns c. Then $\Pr\left[(1+\varepsilon)^{-1} \cdot |R_F| \leq c \leq (1+\varepsilon) \cdot |R_F|\right] \geq 1 - \delta$.*

Proof. Throughout this proof, we assume that $\mathsf{ApproxMCCore}$ is invoked t times from $\mathsf{ApproxMC}$, where $t = \lceil 35 \log_2(3/\delta) \rceil$ (see pseudocode for $\mathsf{ComputeIterCount}$ in Section 4). Referring to the pseudocode of $\mathsf{ApproxMC}$, the final count returned by $\mathsf{ApproxMC}$ is the median of non-$\perp$ counts obtained from the t invocations of $\mathsf{ApproxMCCore}$. Let Err denote the event that the median is not in $\left[(1+\varepsilon)^{-1} \cdot |R_F|, (1+\varepsilon) \cdot |R_F|\right]$. Let "$\#non\perp = q$" denote the event that q (out of t) values returned by $\mathsf{ApproxMCCore}$ are non-$\perp$. Then, $\Pr[Err] = \sum_{q=0}^{t} \Pr[Err \mid \#non\perp = q] \cdot \Pr[\#non\perp = q]$.

In order to obtain $\Pr[Err \mid \#non\perp = q]$, we define a 0-1 random variable Z_i, for $1 \leq i \leq t$, as follows. If the ith invocation of $\mathsf{ApproxMCCore}$ returns c, and if c is either $\perp$ or a non-$\perp$ value that does not lie in the interval $[(1+\varepsilon)^{-1} \cdot |R_F|, (1+\varepsilon) \cdot |R_F|]$, we set Z_i to 1; otherwise, we set it to 0. From Theorem 2, $\Pr[Z_i = 1] = p < 0.4$. If Z denotes $\sum_{i=1}^{t} Z_i$, a necessary (but not sufficient) condition for event Err to occur, given that q non-$\perp$s were returned by $\mathsf{ApproxMCCore}$, is $Z \geq (t - q + \lceil q/2 \rceil)$. To see why this is so, note that $t - q$ invocations of $\mathsf{ApproxMCCore}$ must return $\perp$. In addition, at least $\lceil q/2 \rceil$ of the remaining q invocations must return values outside the desired interval. To simplify the exposition, let q be an even integer. A more careful analysis removes this restriction and results in an additional constant scaling factor for $\Pr[Err]$. With our simplifying assumption, $\Pr[Err \mid \#non\perp = q] \leq \Pr[Z \geq (t - q + q/2)] = \eta(t, t - q/2, p)$. Since $\eta(t, m, p)$ is a decreasing function of m and since $q/2 \leq t - q/2 \leq t$, we have $\Pr[Err \mid \#non\perp = q] \leq \eta(t, t/2, p)$. If $p < 1/2$, it is easy to verify that $\eta(t, t/2, p)$ is an increasing function of p. In our case, $p < 0.4$; hence, $\Pr[Err \mid \#non\perp = q] \leq \eta(t, t/2, 0.4)$.

It follows from above that $\Pr[Err] = \sum_{q=0}^{t} \Pr[Err \mid \#non\perp = q] \cdot \Pr[\#non\perp = q] \leq \eta(t, t/2, 0.4) \cdot \sum_{q=0}^{t} \Pr[\#non\perp = q] = \eta(t, t/2, 0.4)$. Since $\binom{t}{t/2} \geq \binom{t}{k}$ for all $t/2 \leq k \leq t$, and since $\binom{t}{t/2} \leq 2^t$, we have $\eta(t, t/2, 0.4) = \sum_{k=t/2}^{t} \binom{t}{k}(0.4)^k(0.6)^{t-k} \leq \binom{t}{t/2} \sum_{k=t/2}^{t}(0.4)^k(0.6)^{t-k} \leq 2^t \sum_{k=t/2}^{t}(0.6)^t(0.4/0.6)^k \leq 2^t \cdot 3 \cdot (0.6 \times 0.4)^{t/2} \leq 3 \cdot (0.98)^t$. Since $t = \lceil 35 \log_2(3/\delta) \rceil$, it follows that $\Pr[Err] \leq \delta$.

Theorem 4. *Given an oracle for SAT, $\mathsf{ApproxMC}(F, \varepsilon, \delta)$ runs in time polynomial in $\log_2(1/\delta)$, $|F|$ and $1/\varepsilon$ relative to the oracle.*

Proof. Referring to the pseudocode for $\mathsf{ApproxMC}$, lines 1–3 take time no more than a polynomial in $\log_2(1/\delta)$ and $1/\varepsilon$. The repeat-until loop in lines 4–9 is repeated $t = \lceil 35 \log_2(3/\delta) \rceil$ times. The time taken for each iteration is dominated by the time taken by $\mathsf{ApproxMCCore}$. Finally, computing the median in line 10 takes time linear in t. The proof is therefore completed by showing that $\mathsf{ApproxMCCore}$ takes time polynomial in $|F|$ and $1/\varepsilon$ relative to the SAT oracle.

Referring to the pseudocode for $\mathsf{ApproxMCCore}$, we find that $\mathsf{BoundedSAT}$ is called $\mathcal{O}(|F|)$ times. Each such call can be implemented by at most *pivot* + 1 calls to a SAT oracle, and takes time polynomial in $|F|$ and $pivot + 1$ relative to the oracle. Since $pivot + 1$ is in $\mathcal{O}(1/\varepsilon^2)$, the number of calls to the SAT oracle, and the total time taken by all calls to BoundedSAT in each invocation of ApproxMCCore is a polynomial in $|F|$ and $1/\varepsilon$ relative to the oracle. The random choices in lines 8 and 9 of ApproxMCCore can be implemented in time polynomial in n (hence, in $|F|$) if we have access to a source of random bits. Constructing $F \wedge h(z_1, \ldots z_n) = \alpha$ in line 10 can also be done in time polynomial in $|F|$.

6   Experimental Methodology

To evaluate the performance and quality of results of ApproxMC, we built a prototype implementation and conducted an extensive set of experiments. The suite of benchmarks represents problems from practical domains as well as problems of theoretical interest. In particular, we considered a wide range of model counting benchmarks from different domains including grid networks, plan recognition, DQMR networks, Langford sequences, circuit synthesis, random k-CNF and logistics problems [27,20]. The suite consisted of benchmarks ranging from 32 variables to 229100 variables in CNF representation. The complete set of benchmarks (numbering above 200) is available at http://www.cs.rice.edu/CS/Verification/Projects/ApproxMC/.

All our experiments were conducted on a high-performance computing cluster. Each individual experiment was run on a single node of the cluster; the cluster allowed multiple experiments to run in parallel. Every node in the cluster had two quad-core Intel Xeon processors with 4GB of main memory. We used 2500 seconds as the timeout for each invocation of BoundedSAT in ApproxMCCore, and 20 hours as the timeout for ApproxMC. If an invocation of BoundedSAT in line 10 of the pseudo-code of ApproxMCCore timed out, we repeated the iteration (lines 6–11 of the pseudocode of ApproxMCCore) without incrementing i. The parameters ε (tolerance) and δ (confidence being $1 - \delta$) were set to 0.75 and 0.1 respectively. With these parameters, ApproxMC successfully computed counts for benchmarks with upto $33,000$ variables.

We implemented leap-frogging, as described in [7], to estimate initial values of i from which to start iterating the repeat-until loop of lines 6–11 of the pseudocode of ApproxMCCore. To further optimize the running time, we obtained tighter estimates of the iteration count t used in algorithm ApproxMC, compared to those given by algorithm ComputeIterCount. A closer examination of the proof of Theorem 3 shows that it suffices to have $\eta(t, t/2, 0.4) \leq \delta$. We therefore pre-computed a table that gave the smallest t as a function of δ such that $\eta(t, t/2, 0.4) \leq \delta$. This sufficed for all our experiments and gave smaller values of t (we used $t{=}41$ for $\delta{=}0.1$) compared to those given by ComputeIterCount.

For purposes of comparison, we also implemented and conducted experiments with the exact counter Cachet [26] by setting a timeout of 20 hours on the same computing platform. We compared the running time of ApproxMC with that of Cachet for several benchmarks, ranging from benchmarks on which Cachet ran very efficiently to those on which Cachet timed out. We also measured the quality of approximation produced by ApproxMC as follows. For each benchmark on which Cachet did not time out, we obtained the approximate count from ApproxMC with parameters $\varepsilon = 0.75$ and $\delta = 0.1$, and checked if the approximate count was indeed within a factor of 1.75 from the exact count. Since the theoretical guarantees provided by our analysis are conservative, we also measured the relative error of the counts reported by ApproxCount using the L_1 norm, for all benchmarks on which Cachet did not time out. For an input formula F_i, let A_{F_i} (resp., C_{F_i}) be the count returned by ApproxCount (resp., Cachet). We computed the L_1 norm of the relative error as $\frac{\sum_i |A_{F_i} - C_{F_i}|}{\sum_i C_{F_i}}$.

Since Cachet timed out on most large benchmarks, we compared ApproxMC with state-of-the-art bounding counters as well. As discussed in Section 1, bounding counters do not provide any tolerance guarantees. Hence their guarantees are significantly weaker than those provided by ApproxMC, and a direct comparison of performance is not meaningful. Therefore, we compared the sizes of the intervals (i.e., difference between upper and lower bounds) obtained from existing state-of-the-art bounding counters with those obtained from ApproxMC. To obtain intervals from ApproxMC, note that Theorem 3 guarantees that if ApproxMC(F, ε, δ) returns c, then $\Pr[\frac{c}{1+\varepsilon} \leq |R_F| \leq (1+\varepsilon)\cdot c] \geq 1-\delta$. Therefore, ApproxMC can be viewed as computing the interval $[\frac{c}{1+\varepsilon}, (1+\varepsilon)\cdot c]$ for the model count, with confidence δ. We considered state-of-the-art lower bounding counters, viz. MBound [12], Hybrid-MBound [12], SampleCount [14] and BPCount [20], to compute a lower bound of the model count, and used MiniCount [20] to obtain an upper bound. We observed that SampleCount consistently produced better (i.e. larger) lower bounds than BPCount for our benchmarks. Furthermore, the authors of [12] advocate using Hybrid-MBound instead of MBound. Therefore, the lower bound for each benchmark was obtained by taking the maximum of the bounds reported by Hybrid-MBound and SampleCount.

We set the confidence value for MiniCount to 0.99 and SampleCount and Hybrid-MBound to 0.91. For a detailed justification of these choices, we refer the reader to the full version of our paper. Our implementation of Hybrid-MBound used the "conservative" approach described in [12], since this provides the best lower bounds with the required confidence among all the approaches discussed in [12]. Finally, to ensure fair comparison, we allowed all bounding counters to run for 20 hours on the same computing platform on which ApproxMC was run.

7   Results

The results on only a subset of our benchmarks are presented here for lack of space. Figure 1 shows how the running times of ApproxMC and Cachet compared on this subset of our benchmarks. The y-axis in the figure represents time in seconds, while the x-axis represents benchmarks arranged in ascending order of running time of ApproxMC. The comparison shows that although Cachet performed better than ApproxMC initially, it timed out as the "difficulty" of problems increased. ApproxMC, however, continued to return bounds with the specified tolerance and confidence, for many more difficult and larger problems. Eventually, however, even ApproxMC timed out for very large problem instances.

Our experiments clearly demonstrate that there is a large class of practical problems that lie beyond the reach of exact counters, but for which we can still obtain counts with (ε, δ)-style guarantees in reasonable time. This suggests that given a model counting problem, it is advisable to run Cachet initially with a small timeout. If Cachet times out, ApproxMC should be run with a larger timeout. Finally, if ApproxMC also times out, counters with much weaker guarantees but shorter running times, such as bounding counters, should be used.

Figure 2 compares the model count computed by ApproxMC with the bounds obtained by scaling the exact count obtained from Cachet by the tolerance factor (1.75) on a subset of our benchmarks. The y-axis in this figure represents the model count on a log-scale, while the x-axis represents the benchmarks arranged in ascending order of the model count. The figure shows that in all cases, the count reported by ApproxMC lies within the specified tolerance of the exact count. Although we have presented results for only a subset of our benchmarks

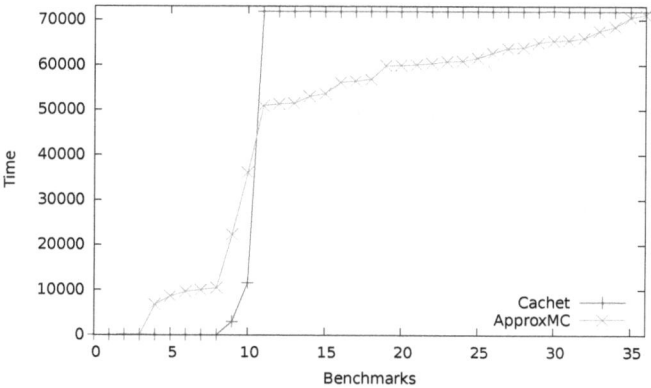

Fig. 1. Performance comparison between ApproxMC and Cachet. The benchmarks are arranged in increasing order of running time of ApproxMC.

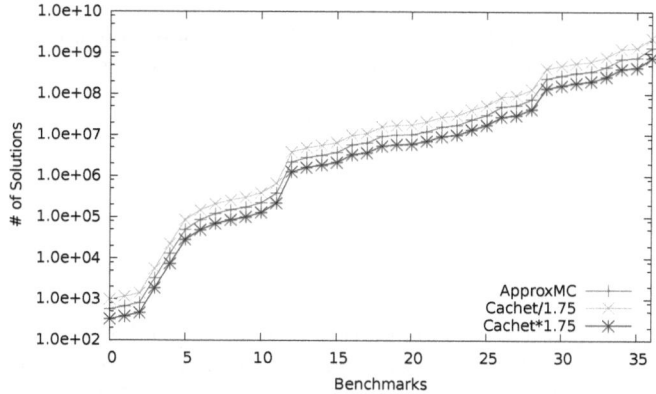

Fig. 2. Quality of counts computed by ApproxMC. The benchmarks are arranged in increasing order of model counts.

(37 in total) in Figure 2 for reasons of clarity, the counts reported by ApproxMC were found to be within the specified tolerance of the exact counts for *all* 95 benchmarks for which Cachet reported exact counts. We also found that the L_1 norm of the relative error, considering all 95 benchmarks for which Cachet returned exact counts, was 0.033. Thus, ApproxMC has approximately 4% error in practice – much smaller than the theoretical guarantee of 75% with $\varepsilon = 0.75$.

Figure 3 compares the sizes of intervals computed using ApproxMC and using state-of-the-art bounding counters (as described in Section 6) on a subset of our benchmarks. The comparison clearly shows that the sizes of intervals computed using ApproxMC are consistently smaller than the sizes of the corresponding intervals obtained from existing bounding counters. Since smaller intervals with comparable confidence represent better approximations, we conclude that ApproxMC computes better approximations than a combination of existing bounding counters. In all cases, ApproxMC improved the upper bounds from MiniCount significantly; it also improved lower bounds from SampleCount and MBound to a lesser extent. For details, please refer to the full version.

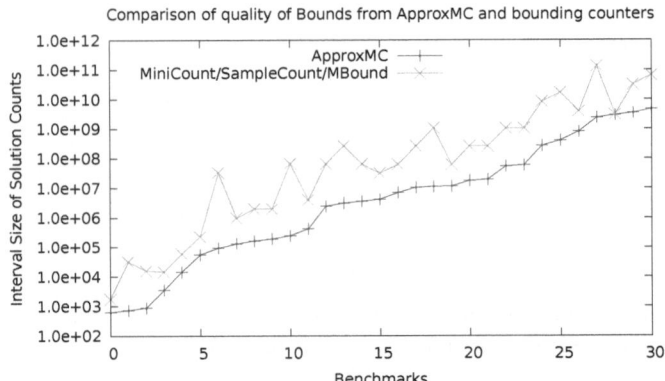

Fig. 3. Comparison of interval sizes from ApproxMC and those from bounding counters. The benchmarks are arranged in increasing order of model counts.

8   Conclusion and Future Work

We presented ApproxMC, the first (ε, δ) approximate counter for CNF formulae that scales in practice to tens of thousands of variables. We showed that ApproxMC reports bounds with small tolerance in theory, and with much smaller error in practice, with high confidence. Extending the ideas in this paper to probabilistic inference and to count models of SMT constraints is an interesting direction of future research.

References

1. CryptoMiniSAT, http://www.msoos.org/cryptominisat2/
2. HotBits, http://www.fourmilab.ch/hotbits
3. Angluin, D.: On counting problems and the polynomial-time hierarchy. Theoretical Computer Science 12(2), 161–173 (1980)
4. Bacchus, F., Dalmao, S., Pitassi, T.: Algorithms and complexity results for #SAT and bayesian inference. In: Proc. of FOCS, pp. 340–351 (2004)
5. Bellare, M., Goldreich, O., Petrank, E.: Uniform generation of NP-witnesses using an NP-oracle. Information and Computation 163(2), 510–526 (1998)
6. Birnbaum, E., Lozinskii, E.L.: The good old Davis-Putnam procedure helps counting models. Journal of Artificial Intelligence Research 10(1), 457–477 (1999)
7. Chakraborty, S., Meel, K.S., Vardi, M.Y.: A scalable and nearly uniform generator of SAT witnesses. In: Sharygina, N., Veith, H. (eds.) CAV 2013. LNCS, vol. 8044, pp. 608–623. Springer, Heidelberg (2013)
8. Darwiche, A.: New advances in compiling CNF to decomposable negation normal form. In: Proc. of ECAI, pp. 328–332. Citeseer (2004)
9. Domshlak, C., Hoffmann, J.: Probabilistic planning via heuristic forward search and weighted model counting. Journal of Artificial Intelligence Research 30(1), 565–620 (2007)
10. Ermon, S., Gomes, C.P., Selman, B.: Uniform solution sampling using a constraint solver as an oracle. In: Proc. of UAI (2012)
11. Gogate, V., Dechter, R.: Samplesearch: Importance sampling in presence of determinism. Artificial Intelligence 175(2), 694–729 (2011)
12. Gomes, C.P., Sabharwal, A., Selman, B.: Model counting: A new strategy for obtaining good bounds. In: Proc. of AAAI, pp. 54–61 (2006)
13. Gomes, C.P., Sabharwal, A., Selman, B.: Model counting. In: Biere, A., Heule, M., Maaren, H.V., Walsh, T. (eds.) Handbook of Satisfiability. Frontiers in Artificial Intelligence and Applications, vol. 185, pp. 633–654. IOS Press (2009)
14. Gomes, C.P., Hoffmann, J., Sabharwal, A., Selman, B.: From sampling to model counting. In: Proc. of IJCAI, pp. 2293–2299 (2007)
15. Gomes, C.P., Sabharwal, A., Selman, B.: Near-uniform sampling of combinatorial spaces using XOR constraints. In: Proc. of NIPS, pp. 670–676 (2007)
16. Jerrum, M.R., Valiant, L.G., Vazirani, V.V.: Random generation of combinatorial structures from a uniform distribution. Theoretical Computer Science 43(2-3), 169–188 (1986)
17. Bayardo Jr., R.J., Schrag, R.: Using CSP look-back techniques to solve real-world SAT instances. In: Proc. of AAAI, pp. 203–208 (1997)
18. Karp, R.M., Luby, M., Madras, N.: Monte-Carlo approximation algorithms for enumeration problems. Journal of Algorithms 10(3), 429–448 (1989)
19. Kitchen, N., Kuehlmann, A.: Stimulus generation for constrained random simulation. In: Proc. of ICCAD, pp. 258–265 (2007)
20. Kroc, L., Sabharwal, A., Selman, B.: Leveraging belief propagation, backtrack search, and statistics for model counting. In: Trick, M.A. (ed.) CPAIOR 2008. LNCS, vol. 5015, pp. 127–141. Springer, Heidelberg (2008)
21. Löbbing, M., Wegener, I.: The number of knight's tours equals 33,439,123,484,294 – counting with binary decision diagrams. The Electronic Journal of Combinatorics 3(1), R5 (1996)
22. Luby, M.G.: Monte-Carlo Methods for Estimating System Reliability. PhD thesis, EECS Department, University of California, Berkeley (June 1983)

23. Minato, S.: Zero-suppressed bdds for set manipulation in combinatorial problems. In: Proc. of Design Automation Conference, pp. 272–277 (1993)
24. Roth, D.: On the hardness of approximate reasoning. Artificial Intelligence 82(1), 273–302 (1996)
25. Rubinstein, R.: Stochastic enumeration method for counting np-hard problems. In: Methodology and Computing in Applied Probability, pp. 1–43 (2012)
26. Sang, T., Bacchus, F., Beame, P., Kautz, H., Pitassi, T.: Combining component caching and clause learning for effective model counting. In: Proc. of SAT (2004)
27. Sang, T., Bearne, P., Kautz, H.: Performing bayesian inference by weighted model counting. In: Prof. of AAAI, pp. 475–481 (2005)
28. Schmidt, J.P., Siegel, A., Srinivasan, A.: Chernoff-Hoeffding bounds for applications with limited independence. SIAM Journal on Discrete Mathematics 8, 223–250 (1995)
29. Simon, J.: On the difference between one and many. In: Salomaa, A., Steinby, M. (eds.) ICALP 1977. LNCS, vol. 52, pp. 480–491. Springer, Heidelberg (1977)
30. Sipser, M.: A complexity theoretic approach to randomness. In: Proc. of STOC, pp. 330–335 (1983)
31. Stockmeyer, L.: The complexity of approximate counting. In: Proc. of STOC, pp. 118–126 (1983)
32. Thurley, M.: sharpSAT – counting models with advanced component caching and implicit BCP. In: Biere, A., Gomes, C.P. (eds.) SAT 2006. LNCS, vol. 4121, pp. 424–429. Springer, Heidelberg (2006)
33. Toda, S.: On the computational power of PP and (+)P. In: Proc. of FOCS, pp. 514–519. IEEE (1989)
34. Trevisan, L.: Lecture notes on computational complexity. Notes written in Fall (2002), http://citeseerx.ist.psu.edu/viewdoc/download?doi=10.1.1.71.9877&rep=rep1&type=pdf
35. Valiant, L.G.: The complexity of enumeration and reliability problems. SIAM Journal on Computing 8(3), 410–421 (1979)
36. Wei, W., Selman, B.: A new approach to model counting. In: Bacchus, F., Walsh, T. (eds.) SAT 2005. LNCS, vol. 3569, pp. 324–339. Springer, Heidelberg (2005)
37. Yuan, J., Aziz, A., Pixley, C., Albin, K.: Simplifying boolean constraint solving for random simulation-vector generation. IEEE Trans. on CAD of Integrated Circuits and Systems 23(3), 412–420 (2004)

Dominance Driven Search

Geoffrey Chu[2] and Peter J. Stuckey[1,2]

[1] National ICT Australia, Victoria Laboratory
[2] Department of Computing and Information Systems,
University of Melbourne, Australia
{gchu,pjs}@csse.unimelb.edu.au

Abstract. Recently, a generic method for identifying and exploiting dominance relations using *dominance breaking constraints* was proposed. In this method, sufficient conditions for a solution to be dominated are identified and these conditions are used to generate dominance breaking constraints which prune off the dominated solutions. We propose to use these dominance relations in a different way in order to boost the search for good/optimal solutions. In the new method, which we call *dominance jumping*, when search reaches a point where all solutions in the current domain are dominated, rather than simply backtrack as in the original dominance breaking method, we jump to the subtree which dominates the current subtree. This new strategy allows the solver to move from a bad subtree to a good one, significantly increasing the speed with which good solutions can be found. Experiments across a range of problems show that the method can be very effective when the original search strategy was not very good at finding good solutions.

1 Introduction

Recently, a generic method for identifying and exploiting dominance relations using *dominance breaking constraints* was proposed in [3]. This method analyzes the effects of different assignment mappings on the satisfiability and objective value of solutions, and finds sufficient conditions under which a solution is dominated by (i.e., no better than) another one. Dominance breaking constraints are then generated which prune off these dominated solutions.

While symmetry and dominance breaking constraints are very powerful and can produce orders of magnitude speedup on a wide range of problems, it is well known that static symmetry breaking constraints (e.g., [20,4,14,21,8]) can conflict with the search strategy, leading to less speedup or even a slowdown [10]. Static dominance breaking constraints, which are a generalization of static symmetry breaking constraints, suffer from a similar problem. In optimization problems, dominance breaking constraints prevent the solver from finding any solution which is dominated. While such dominated solutions may not be optimal, they may nevertheless improve the current best solution and allow additional pruning through branch and bound. If the search strategy quickly guides the search to good/optimal non-dominated solutions, then there is no conflict between the search and the dominance breaking constraints. However, if we have a search strategy which keeps pushing the search into subtrees with only dominated solutions, then the search will potentially conflict with the dominance breaking constraints.

C. Schulte (Ed.): CP 2013, LNCS 8124, pp. 217–229, 2013.

Consider a situation where the next 100 solutions in the search tree are all dominated, but one or more of them does improve the current best solution. A search without dominance breaking constraints will go through that portion of the search space without the pruning provided by the dominance breaking constraints, but once one of these better solutions are found, it will gain some additional pruning from branch and bound. A search with dominance breaking constraints will have the pruning provided by the dominance breaking constraints, but none of the pruning from the better solution that it could have found among these 100 solutions. Thus it is possible that the solver is actually slower at finding the optimal solution with dominance breaking constraints than without.

In symmetry breaking, the potential conflict between search strategy and static symmetry breaking constraints can be overcome by using a dynamic symmetry breaking method such as symmetry breaking during search (SBDS) [1,11] or symmetry breaking by dominance detection (SBDD) [5]. In this paper, we propose a different way to overcome this problem in the dominance case. We propose an altered method called *dominance jumping*, which exploits the information contained in the dominance relations in a different way. When a dominance breaking constraint prunes a partial assignment (because it is dominated by another), instead of simply backtracking as a normal CP solver would do, we jump to the subtree represented by the partial assignment which dominates the current one. Then, rather than getting stuck in a subtree where all the solutions are dominated and therefore not discoverable by a search with dominance breaking constraints, the dominance jumping can take us to a subtree with non-dominated solutions, allowing the search to find good non-dominated solutions and benefit from the additional pruning provided by branch and bound.

The rest of the paper is organized as follows: in the next section we introduce notation, and recall the approach to creating dominance breaking constraint described in [3]. In Section 3 we describe dominance jumping. In Section 4 we give experimental results comparing dominance breaking and dominance jumping. In Section 5 we examine related work, and finally in Section 6 we conclude.

2   Definitions

2.1   Constraints, Literals, and COPs

Let $\equiv$ denote syntactical identity, $\Rightarrow$ denote logical implication and $\Leftrightarrow$ denote logical equivalence. We define variables and constraints in a problem independent way. A variable v is a mathematical quantity capable of assuming any value from a set of values called the *default domain* of v. Each variable is typed, e.g., Boolean or Integer, and its type determines its default domain, e.g., $\{0,1\}$ for Boolean variables and $\mathbb{Z}$ for Integer variables. Given a set of variables V, let Θ_V denote the set of valuations over V where each variable in V is assigned to a value in its default domain. A constraint c over a set of variables V is defined by a set of valuations $solns(c) \subseteq \Theta_V$. Given a valuation θ over $V' \supseteq V$, we say θ satisfies c if the restriction of θ onto V is in $solns(c)$. Otherwise, we say that θ violates c. A domain D over variables V is a set of *unary constraints*, one for each variable in V. In an abuse of notation, if a symbol A refers to a set of constraints $\{c_1,\ldots,c_n\}$, we will often also use the symbol A to refer to the constraint $c_1 \wedge \ldots \wedge c_n$. This allows us to avoid repetitive use of conjunction symbols.

A *Constraint Satisfaction Problem* (CSP) is a tuple $P \equiv (V,D,C)$, where V is a set of variables, D is a domain over V, and C is a set of n-ary constraints. A valuation θ over V is a *solution* of P if it satisfies every constraint in D and C. The aim of a CSP is to find a solution or to prove that none exist. In a *Constraint Optimization Problem* (COP) $P \equiv (V,D,C,f)$, we also have an objective function f mapping Θ_V to an ordered set, e.g., $\mathbb{Z}$ or $\mathbb{R}$, and we wish to minimize or maximize f over the solutions of P. In this paper, we deal with finite domain problems only, i.e., where the initial domain D constrains each variable to take values from a finite set of values.

CP solvers solve CSP's by interleaving search with inference. We begin with the original problem at the root of the search tree. At each node in the search tree, we propagate the constraints to try to infer variable/value pairs which can no longer be taken in any solution in this subtree. Such pairs are removed from the current domain. If some variable's domain becomes empty, then the subtree has no solution and the solver backtracks. If all the variables are assigned and no constraint is violated, then a solution has been found and the solver can terminate. If inference is unable to detect either of the above two cases, the solver further divides the problem into a number of more constrained subproblems and searches each of those in turn. COP's are typically solved via branch and bound where we solve a series of CSP's with increasingly tight bounds on the objective value.

2.2 Dominance Breaking

Without loss of generality, assume that we are dealing with a minimization problem. We say that assignment θ_1 dominates θ_2 if either: 1) θ_1 is a solution and θ_2 is a non-solution, or 2) they are both solutions or both non-solutions and $f(\theta_1) \leq f(\theta_2)$. In [3], a generic method for identifying and exploiting dominance relations via dominance breaking constraints was proposed. The method can be briefly outlined as follows:

Step 1 Choose a refinement of the objective function f' with the property that $\forall \theta_1, \theta_2, f(\theta_1) < f(\theta_2)$ implies $f'(\theta_1) < f'(\theta_2)$.

Step 2 Find mappings $\sigma : \Theta_V \to \Theta_V$ which are likely to map solutions to better solutions.

Step 3 For each σ, find a constraint $scond(\sigma)$ s.t. if $\theta \in solns(C \wedge D \wedge scond(\sigma))$, then $\sigma(\theta) \in solns(C \wedge D)$.

Step 4 For each σ, find a constraint $ocond(\sigma)$ s.t. if $\theta \in solns(C \wedge D \wedge ocond(\sigma))$, then $f'(\sigma(\theta)) < f'(\theta)$.

Step 5 For each σ, post the dominance breaking constraint $db(\sigma) \equiv \neg(scond(\sigma) \wedge ocond(\sigma))$.

The method analyzes the effects of different assignment mappings σ on the satisfiability and objective value of solutions, and finds sufficient conditions $scond(\sigma) \wedge ocond(\sigma)$ under which a solution is dominated by another one. Dominance breaking constraints $db(\sigma) \equiv \neg(scond(\sigma) \wedge ocond(\sigma))$ are then generated which prune off these dominated solutions. We now restate the main theorem from [3] showing the correctness of the dominance breaking constraints generated by this method.

Theorem 1. *Given a finite domain COP* $P \equiv (V,D,C,f)$, *a refinement of the objective function* f' *satisfying* $\forall \theta_1, \theta_2, f(\theta_1) < f(\theta_2)$ *implies* $f'(\theta_1) < f'(\theta_2)$, *a set of mappings S, and for each mapping* $\sigma \in S$ *constraints* $scond(\sigma)$ *and* $ocond(\sigma)$ *satisfying:* $\forall \sigma \in S,$

if $\theta \in solns(C \wedge D \wedge scond(\sigma))$, *then* $\sigma(\theta) \in solns(C \wedge D)$, *and:* $\forall \sigma \in S$, *if* $\theta \in solns(C \wedge D \wedge ocond(\sigma))$, *then* $f'(\sigma(\theta)) < f'(\theta)$, *we can add all of the dominance breaking constraints* $db(\sigma) \equiv \neg(scond(\sigma) \wedge ocond(\sigma))$ *to P without changing its satisfiability or optimal value.*

A proof of this theorem and more details on the method can be found in [3].

Example 1. Consider the 0-1 knapsack problem where x_i are 0-1 variables, we have constraint $\sum w_i x_i \leq W$ and we have objective $f = -\sum v_i x_i$, where w_i and v_i are constants.

Step 1 Initially, let us not refine the objective function leaving $f' = f$.

Step 2 Consider mappings which swap the values of two variables, i.e., $\forall i < j, \sigma_{i,j}$ swaps x_i and x_j.

Step 3 A sufficient condition for $\sigma_{i,j}$ to map the current solution to another solution is: $scond(\sigma_{i,j}) \equiv w_i x_j + w_j x_i \leq w_i x_i + w_j x_j$. Rearranging, we get: $(w_i - w_j)(x_i - x_j) \geq 0$.

Step 4 A sufficient condition for $\sigma_{i,j}$ to map the current solution to an assignment with a better objective value is: $ocond(\sigma_{i,j}) \equiv v_i x_j + v_j x_i > v_i x_i + v_j x_j$. Rearranging, we get: $(v_i - v_j)(x_i - x_j) < 0$.

Step 5 For each $\sigma_{i,j}$, we can post the dominance breaking constraint: $db(\sigma_{i,j}) \equiv \neg(scond(\sigma_{i,j}) \wedge ocond(\sigma_{i,j}))$. After simplifying, we have $db(\sigma_{i,j}) \equiv x_i \leq x_j$ if $w_i \geq w_j$ and $v_i < v_j$, $db(\sigma_{i,j}) \equiv x_i \geq x_j$ if $w_i \leq w_j$ and $v_i > v_j$, and $db(\sigma_{i,j}) \equiv true$ for all other cases.

These dominance breaking constraints ensure that if one item has worse value and greater or equal weight to another, then it cannot be chosen without choosing the other also.

We can refine the objective to get stronger dominance breaking constraints. In Step 1, we can tie break solutions with equal objective value by the weight used, and then lexicographically, i.e., $f' = lex(f, \sum w_i x_i, x_1, \ldots, x_n)$. In Step 4, we have: $\forall i < j, ocond(\sigma_{i,j}) \equiv \sigma(f') < f' \equiv ((v_i - v_j)(x_i - x_j) < 0) \vee ((v_i - v_j)(x_i - x_j) = 0 \wedge (w_i - w_j)(x_i - x_j) > 0) \vee ((v_i - v_j)(x_i - x_j) = 0 \wedge (w_i - w_j)(x_i - x_j) = 0 \wedge x_j < x_i)$. In Step 5, after simplifying, in addition to the dominance breaking constraints we had before, we would also have: $db(\sigma_{i,j}) \equiv x_i \leq x_j$ if $w_i > w_j$ and $v_i = v_j$, $db(\sigma_{i,j}) \equiv x_i \geq x_j$ if $w_i < w_j$ and $v_i = v_j$, and $db(\sigma_{i,j}) \equiv x_i \leq x_j$ if $w_i = w_j$ and $v_i = v_j$ which is a symmetry breaking constraint. □

3   Dominance Jumping

A propagator for a dominance breaking constraint can do two things: 1) it can check the consistency of the current domain w.r.t. the dominance breaking constraint, producing failure if it is inconsistent, and 2) it can prune off variable/value pairs which, if taken, will cause inconsistency. In the original method, the failure and propagation produced by these propagators are treated the same as any other propagator in the system. In our altered method, whenever dominance jumping is active, we modify this behavior as follows: 1) we check consistency only and never prune any values using the dominance breaking constraints, 2) when a failure is detected, we perform a *dominance jump* rather than a normal backtrack.

As can be seen from the definitions in Section 2, each dominance breaking constraint $db(\sigma)$ is generated from an assignment mapping σ. When a domain D is failed by $db(\sigma)$, it means that every solution θ in D is dominated by a corresponding solution $\sigma(\theta)$. Rather than simply failing and backtracking, we can instead perform a *dominance jump* to get to the part of the search tree that contains these better solutions. Let us extend σ to also map domains to domains via $solns(\sigma(D)) \equiv \{\sigma(\theta) \mid \theta \in solns(D)\}$. Then, if D is failed by $db(\sigma)$, the domain $\sigma(D)$ must contain solutions which dominate those in the current domain D. We want to calculate this new domain $\sigma(D)$ and jump to there. In this paper, we consider only σ which are literal mappings, i.e., assignment mappings which map each equality literal $x = v$ to the same or another equality literal $x' = v'$ in all assignments. All of the σ used in [3] are literal mappings, and indeed we expect that most practically useful mappings for the method proposed in [3] will be literal mappings. Let us extend σ to map equality literals to equality literals and disequality literals to disequality literals such that if $\sigma(x = v) = (x' = v')$, then $\sigma(x \neq v) = (x' \neq v')$.

Any domain D can be expressed as a set of disequality literals $lits_D$ representing which variable/value pairs from the original domain has been pruned. For example, suppose the initial domain D_{init} of x_1, x_2 were $\{1,2,3,4,5\}$ and the current domain D is $x_1 \in \{1,3,5\}, x_2 \in \{2,3,4\}$. Then $lits_D \equiv \{x_1 \neq 2, x_1 \neq 4, x_2 \neq 1, x_2 \neq 5\}$. Using the set of literals $\sigma(lits_D) \equiv \{\sigma(l) \mid l \in lits_D\}$ as decisions from the root node will take us to the search space $\sigma(D)$. We do this by backtracking to the deepest level such that all previous decisions in the current search path are in $\sigma(lits_D)$. We then suspend the normal search strategy and draw decisions from $\sigma(lits_D)$ until it is either exhausted, or some conflict occurs. After that, we resume the normal search strategy. If D had variables which were fixed, we can use those equality literals in $lits_D$ instead so we need to make fewer decisions to get to $\sigma(D)$. Similarly, if σ happens to also map inequality literals to inequality literals (e.g., in a mapping which swaps variables), we can use those to reduce the number of decisions.

Example 2. Consider the Photo problem. A group of people wants to take a group photo where they stand in one line. Each person has preferences regarding who they want to stand next to. We want to find the arrangement which satisfies the most preferences.

We can model this as follows. Let $x_i \in \{1, \ldots, n\}$ for $i = 1, \ldots, n$ be variables where x_i represent the person in the ith place. Let p be a 2d integer array where $p[i][j] = p[j][i] = 2$ if person i and j both want to stand next to each other, $p[i][j] = p[j][i] = 1$ if only one of them wants to stand next to the other, and $p[i][j] = p[j][i] = 0$ if neither want to stand next to each other. The only constraint is: $alldiff([x_1, \ldots, x_n])$. The objective function to be maximised is given by: $f = \sum_{i=1}^{n-1} p[x_i][x_{i+1}]$. In MiniZinc [18] it would be modelled as:

```
int: n;                               % number of people
set of int: Person = 1..n;
array[Person,Person] of 0..2: p;      % preferences

array[Person] of Person: x;           % person in position i;

constraint alldifferent(x);

solve maximize sum(i in 1..n-1)(p[x[i],x[i+1]]);
```

As described in [3], if we consider the mappings $\sigma_{i,j}$ which flip a subsequence of the x's from the ith position to the jth (i.e., map x_i to x_j, x_{i+1} to x_{j-1}, ..., x_j to x_i), we can generate a number of dominance breaking constraints. For $2 \geq i < j \leq n - 1$, $db(\sigma_{i,j}) \equiv p[x_{i-1}][x_i] + p[x_j][x_{j+1}] + (x_i < x_j) > p[x_{i-1}][x_j] + p[x_i][x_{j+1}]$. For $i = 1, 2 \leq j$, $db(\sigma_{i,j}) \equiv p[x_j][x_{j+1}] + (x_i < x_j) > p[x_i][x_{j+1}]$. For $i \leq n - 1, j = n$, $db(\sigma_{i,j}) \equiv p[x_{i-1}][x_i] + (x_i < x_j) > p[x_{i-1}][x_j]$. For $i = 1, j = n$, $db(\sigma_{i,j}) \equiv (x_i < x_j) > 0$.

We now illustrate the difference between dominance breaking and dominance jumping on a simple example. Suppose $n = 6$ and person 1 wants to stand next to person 6, person 6 wants to stand next to person 2, person 2 wants to stand next to person 5, person 5 wants to stand next to person 3, and person 3 wants to stand next to person 4. This is expressed by the MiniZinc data file:

```
n = 6;
p = [| 0, 0, 0, 0, 0, 1
     | 0, 0, 0, 0, 1, 1
     | 0, 0, 0, 1, 1, 0
     | 0, 0, 1, 0, 0, 0
     | 0, 1, 1, 0, 0, 0
     | 1, 1, 0, 0, 0, 0 |];
```

Suppose we use a naive search strategy such as labelling the x_i in order, trying the lowest value available in the domain first. With neither dominance breaking nor dominance jumping, it takes the search 51 conflicts to reach an optimal solution $x_1 = 1, x_2 = 6, x_3 = 2, x_4 = 5, x_5 = 3, x_6 = 4$. With dominance breaking, the search proceeds as follows. At the first decision level, we try $x_1 = 1$. At the second decision level, we try $x_2 = 2$. At this point, the constraint $p[x_1][x_2] + (x_2 < x_6) > p[x_1][x_6]$ propagates to force $p[x_1][x_6] = 0$, which forces $x_6 \neq 6$. At the third decision level, we try $x_3 = 3$. At this point, the constraint $p[x_2][x_3] + (x_3 < x_6) > p[x_2][x_6]$ propagates to force $p[x_2][x_6] = 0$, which forces $x_6 \neq 5$, which forces $x_6 = 4$. The constraint $p[x_3][x_4] + (x_4 < x_6) > p[x_3][x_6]$ now propagates and forces $p[x_3][x_4] \geq 2$ because $p[x_3][x_6] = p[3][4] = 1$, and x_4 is either 5 or 6 so $(x_4 < x_6) = 0$. But then $x_4 \neq 5$ and $x_4 \neq 6$ and we have a failure. We then backtrack and continue the search. After another 25 conflicts, we reach the optimal solution. The search tree is shown in Figure 1.

With dominance jumping, the search proceeds as follows. We make the 5 decisions $x_1 = 1, x_2 = 2, x_3 = 3, x_4 = 4, x_5 = 5$, which forces $x_6 = 6$. At this point, a number of dominance breaking constraints are detected to be violated, telling us that certain mappings can improve the solution. For example, $db(\sigma_{2,6})$ is violated. Using the mapping $\sigma_{2,6}$ to perform a jump means that we backtrack to decision level 1, and then try the decisions $x_2 = 6, x_3 = 5, x_4 = 4, x_5 = 3, x_6 = 2$. After these, we again detect that a dominance breaking constraint is violated, e.g., $db(\sigma_{3,6})$. Performing this jump means that we backtrack to decision level 2 and try $x_3 = 2, x_4 = 3, x_4 = 4, x_6 = 5$. This violates $db(\sigma_{4,6})$. Applying this jump causes us to backtrack to decision level 3 and try $x_4 = 5, x_5 = 4, x_6 = 3$. This violates $db(\sigma_{5,6})$. Applying this jump causes us to backtrack to decision level 4 and try $x_5 = 3, x_6 = 4$, finally giving a non-dominated solution of $x_1 = 1, x_2 = 6, x_3 = 2, x_4 = 5, x_5 = 3, x_6 = 4$. In this case, it only took 4 conflicts and 4 jumps to bring us to the optimal solution. The search tree is shown in Figure 2.   □

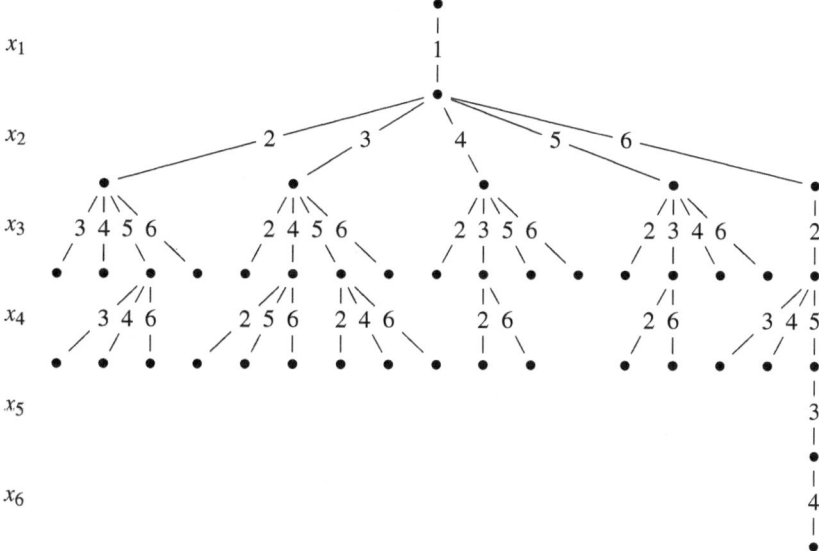

Fig. 1. Search tree with dominance breaking

The effect of dominance jumping is significantly different from dominance breaking. When a "bad" decision is made (e.g., $x_2 = 2$ after $x_1 = 1$ in Example 2), dominance breaking is incapable of "fixing" the problem. Instead, it just helps the solver to fail that bad subtree quicker so that it can backtrack out of the bad decision. However, in general, it still takes exponential time to undo the bad decision. Dominance jumping on the other hand, can potentially fix a bad decision and replace it with a good one very quickly. In Example 2, after $x_1 = 1, x_2 = 2$, if some x_i is set to 6, it is likely that flipping the subsequence from 2 to i improves the objective. Thus dominance jumping will jump to a subtree where $x_1 = 1, x_2 = 6$, immediately undoing the bad decision $x_2 = 2$.

Note that dominance jumping does not require all variables involved in the dominance breaking constraint to be fixed in order to jump.

Example 3. Consider a simple problem: $x_1 + x_2 + x_3 + x_4 \leq 9 \wedge alldiff(x_1, x_2, x_3, x_4)$ with $D(x_i) = [1..6]$. All variables are symmetric. So $\sigma_{i,j}$ which swaps the values of x_i and x_j is a mapping that preserves solutions. Using a lexicographic objective f' we can compute $db(\sigma_{i,j}) = x_i \leq x_j$ for $i < j$. Imagine we label $x_2 = 1$, then propagation causes $D(x_1) = [2..6]$ and $db(\sigma_{1,2})$ fails. We compute $\sigma_{1,2}(D)$ as $D(x_2) = [2..6]$ and $D(x_1) = \{1\}$. The dominance jump goes to the root and then sets $x_1 = 1$ (which will set $D(x_2) = [2..6]$) and then continues its search. Of course for this trivial example dominance breaking is clearly superior.   □

Both dominance breaking and dominance jumping are optimality preserving. As Theorem 1 states, the addition of dominance breaking constraints to the problem does not change its satisfiability or optimal value. Performing dominance jumping will not

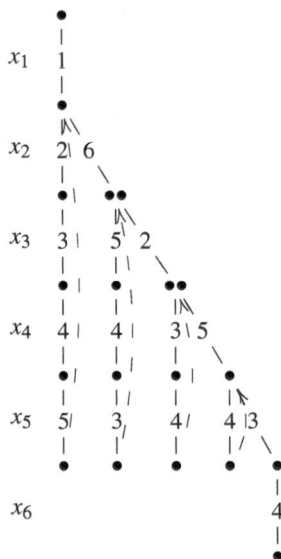

Fig. 2. Search tree with dominance jumping

change its satisfiability or optimal value either. When a dominance jump is performed, nothing is actually pruned. We are only doing a restart and temporarily using a different search strategy to guide the solver to another part of the search space. The subtree we jumped away from is not counted as fully searched, so dominance jumping can never prune off optimal solutions of the problem. However, there can be issues of termination when dominance jumping is used, as if we keep jumping, we may never complete a full search of the search tree. Nogood learning techniques such as [13,19,6]) can be used to overcome this problem. Such techniques record nogoods which tell us which parts of the search tree the solver has proved contains no solution better than the current best solution. Such nogoods allow the solver to keep track of which subtrees have been exhaustively searched. If we keep all such nogoods permanently, the search will always terminate and will be complete, guaranteeing that the correct optimal value is found. We implemented dominance jumping in the nogood learning solver CHUFFED which already has nogood learning built in, and thus it is capable of performing a complete search with dominance jumping.

The effectiveness of dominance jumping depends significantly on how good the original search strategy was. If the original search strategy was already good enough that it very quickly leads the search to an optimal solution, then dominance jumping is largely pointless. On the other hand, if the original search strategy was quite naive, or is not designed for finding good solutions, then dominance jumping can be very useful. It is quite common that the search strategies used in CP solvers are not good at finding good solutions. This is because many of them are designed to reduce the size of the search tree, rather than to order the subtrees so that good solutions are found first. Many of the common dynamic search strategies such as first fail [12], dom/wdeg and its variants [2], and activity based search [17] are purely designed to make the search tree smaller and makes no effort whatsoever to try to guide the solver to good solutions quickly. But when dominance jumping is used on top of them, even very naive search strategies which are bad at finding solutions can turn into good ones, as a significant amount of information about the objective and the structure of the problem is contained in the dominance relations. By exploiting this through dominance jumping, we can bring the solver to a good subtree even if the search strategy initially sent it to a bad one.

While dominance jumping can be effective in the solution finding phase of an optimization problem, it is completely useless in the proof of optimization phase. In that phase, we are no longer trying to find any solutions. Jumping around in the search space will simply make it more difficult to complete the proof of optimality. Nogood learning can prevent thrashing behavior caused by the jumping and allow a complete proof of optimality to be derived. However, for large/hard problems, this may require an unreasonably large number of nogoods to be stored, causing the solver to run out of memory. Ideally, if a proof of optimality is desired, we should use some dynamic method to switch off dominance jumping and go back to pure dominance breaking. Such strategies will be explored in future work.

4   Experiments

The experiments were performed on Xeon Pro 2.4GHz processors using the state of the art CP solver CHUFFED. We use a time out of 600 seconds. We compare the base solver vs dominance breaking and dominance jumping. We use the 7 optimization problems used in [3]: Photo-n the photo problem of Example 2 with n people, Knapsack-n 0-1 knapsack problem of Example 1 with n objects, Nurse-n nurse scheduling problem [15] with 15 nurses and n days, RCPSP resource constrained scheduling problem J120 instances, Talent-n talent scheduling problem [9] with n scenes, SteelMill-n steel mill scheduling [10] with n orders, and PCBoard-n-m PB board manufacturing problem with n components and m devices. We use basic inorder fixed search strategies that are not specifically designed to find good solutions. MiniZinc models and data for the problems can be found at: www.cs.mu.oz.au/~pjs/dom-jump/

Results are shown in Table 1: *opt* is average of the best solution found; *otime* is the geometric mean of time to find the best solution; *etime* is the geometric mean of time to find a solution at least as good as the worst of the best solution found by dominance breaking and the best solution found by dominance jumping, so we can directly compare *etime* to see how much time each took to get the same quality solution; *stime* is the geometric mean of the time to solve the instance completely (timeouts count as 600); and finally *svd* the number of instances solved to optimality is given in brackets. The best values out of the three methods are given in bold. When there is a tie on the best value, we tie-break on the time required to achieve the value.

The results show that both dominance breaking and dominance jumping substantially improve upon solving without dominance information. Dominance jumping is clearly better at finding good solutions faster. The average best solution found is almost always better. Dominance jumping almost always wins in *etime*, the only exception is in smaller knapsacks and in nurse scheduling where dominance breaking is obviously far superior. Notice how, as the difficulty of the instances grows with size, dominance jumping becomes more advantageous.

Table 1. Comparison of the solver with nothing (none), with dominance breaking constraints (dominance breaking), and with dominance jumping. dominance jumping.

Problem	none				dominance breaking					dominance jumping				
	opt	*otime*	*stime*	*svd*	*opt*	*otime*	*etime*	*stime*	*svd*	*opt*	*otime*	*etime*	*stime*	*svd*
Photo-16	19.5	16.43	50.13	18	19.7	1.03	1.03	**11.40**	**20**	**19.7**	0.10	**0.10**	16.64	**20**
Photo-18	21.4	39.45	182.7	15	21.5	2.27	2.27	**76.26**	**20**	**21.5**	0.29	**0.29**	80.62	19
Photo-20	21.4	179.6	393.3	5	23.15	31.06	31.06	262.4	7	**23.15**	1.16	**1.16**	**232.9**	**9**
Photo-22	22.8	185.1	368.7	4	25.15	51.56	38.9	294.7	6	**25.4**	2.18	**0.76**	**244.4**	7
Photo-24	21.55	213.5	596.5	1	26.75	147.9	147.9	586.7	3	**27.15**	2.00	**0.52**	**495.6**	**4**
Knapsack-100	1827.95	222.3	600	0	**2583.2**	0.30	**0.30**	**0.93**	**20**	2583.2	0.60	0.60	1.52	20
Knapsack-150	3605.75	240.6	600	0	**5810.35**	13.8	**13.82**	**47.62**	19	5810.35	27.71	27.71	82.26	19
Knapsack-200	5910.55	299.2	600	0	**10422.4**	180.1	**126.2**	**261.0**	3	10415.8	220.7	206.2	491.9	2
Knapsack-250	8732.25	342.4	600	0	16235.25	253.9	186.2	600	0	**16235.5**	327.1	**151.7**	600	0
Knapsack-300	12000	288.9	600	0	23212	272.6	157.2	600	0	**23294.9**	302.2	**123.0**	600	0
Nurse-14	149.2	61.35	61.82	18	**151.35**	50.99	**49.14**	**52.33**	19	150.2	72.45	72.45	74.06	18
Nurse-21	137.1	416.5	600	0	**172.7**	248.0	**217.6**	600	0	172.4	280.7	254.3	600	0
Nurse-28	161.5	400.5	600	0	222.4	245.1	**213.5**	600	0	**222.45**	310.2	277.6	600	0
Nurse-35	187.5	355.0	600	0	**275.75**	339.8	**164.7**	600	0	275.1	223.7	196.4	600	0
Nurse-42	213.1	382.3	600	0	**321.9**	377.4	**193.3**	600	0	320.55	332.9	332.9	600	0
RCPSP	110.9	31.37	41.62	72	114.17	7.51	7.51	13.97	57	**110.86**	12.44	**3.92**	**15.83**	**72**
Talent-16	106.1	7.58	19.03	20	106.05	0.92	0.92	3.25	20	**106.05**	0.41	**0.41**	**2.79**	**20**
Talent-18	149.45	127.1	239.3	16	147	5.22	5.22	17.26	20	**147**	2.73	**2.73**	**15.81**	**20**
Talent-20	270.2	321.4	497.3	5	184.45	27.22	27.22	81.48	20	**184.45**	14.85	**14.85**	**73.40**	**20**
Talent-22	387.1	369.2	600	0	270.9	34.32	34.32	353.2	13	**204.05**	51.4	**21.10**	**310.5**	14
Talent-24	566.65	323.8	600	0	322.25	161.6	154.9	547.1	2	**260.1**	123.4	**18.27**	**510.7**	**3**
SteelMill-40	5.45	71.02	129.3	10	1.6	42.30	40.80	54.60	15	**0.65**	20.24	**9.33**	**21.75**	17
SteelMill-45	8.2	105.7	269.6	6	1.55	121.7	121.7	134.1	14	**0.35**	46.44	**31.92**	**52.83**	18
SteelMill-50	16.35	319.6	560.1	1	9.85	91.00	90.73	332.1	10	**1.1**	142.1	**30.75**	**198.9**	16
SteelMill-55	25.95	305.0	600	0	16.05	67.26	60.19	419.3	6	**2.9**	212.2	**23.59**	**275.6**	13
SteelMill-60	32.55	274.4	584.8	1	19.9	59.56	54.65	497.0	2	**5**	224.7	**23.82**	**399.0**	8
PCBoard-6-8	206.25	298.8	341.1	8	217.7	17.50	17.50	**24.36**	**20**	217.7	1.29	**1.29**	68.19	14
PCBoard-6-9	225.9	440.2	532.8	2	**246.7**	84.81	80.04	**142.5**	**20**	246.6	8.39	**8.39**	438.5	3
PCBoard-7-9	242.15	418.6	600	0	277.7	123.5	111.4	**498.5**	7	**282.5**	21.95	**5.56**	600	0
PCBoard-7-10	266.9	383.7	600	0	282.55	6.32	6.32	600	0	**319.15**	28.48	**0.86**	600	0
PCBoard-8-10	293.9	379.2	600	0	308.4	4.35	4.35	**597**	1	**357.9**	30.32	**0.38**	600	0

Unsuprisingly dominance jumping is usually better at proving optimality having better *stime* in most of the smaller instances. Suprisingly dominance jumping actually turns out to be preferable to dominance breaking even for proving optimality on RCPSP, Talent and SteelMill. For these problems the proof of optimality is not the larger part of the search space, that is once we find the optimal solution for these problems it is often not too difficult to prove it optimal.

Next we compare the three methods using 3 different search heuristics for the Photo problem, to see how the search strategy affects the effectiveness of dominance breaking and dominance jumping. The first is the basic inorder fixed search used above (inorder). The second greedily finds a person who most wants to stand next to an already assigned person and assigns them next to them (greedy). The third uses the first fail heuristic to pick which variable to assign next (first-fail).

The results in Table 2 show that dominance jumping is still much better at reaching a good solution than dominance breaking, regardless of the search strategy. The results clearly illustrate that the biggest advantage of dominance jumping arises in the inorder search, which does not try to look for good solutions. But dominance jumping is still advantageous over dominance breaking using the greedy search, although to a lesser degree. Using first-fail dominance breaking is better at proving optimality, since first-fail search also concentrates on reducing the search space, but as the size of the problem grows, its advantage reduces, until for Photo-24 dominance jumping is superior in proving optimality as well.

Table 2. Comparison of dominance breaking and dominance jumping given different search heuristics

Problem	Search	none				dominance breaking					dominance jumping				
		opt	otime	stime	svd	opt	otime	etime	stime	svd	opt	otime	etime	stime	svd
	inorder	19.65	7.37	11.47	18	19.70	0.79	0.79	2.35	20	**19.70**	0.08	**0.08**	**0.98**	**20**
Photo-16	greedy	19.70	1.29	3.44	19	19.70	0.45	0.45	1.71	20	**19.70**	0.04	**0.04**	**0.56**	**20**
	first-fail	19.70	0.72	1.22	20	19.70	0.24	0.24	**0.65**	**20**	**19.70**	0.13	**0.13**	0.76	20
	inorder	21.45	18.47	45.25	19	21.50	1.42	1.42	9.71	20	**21.50**	0.20	**0.20**	**5.60**	**20**
Photo-18	greedy	21.50	0.69	7.80	19	21.50	0.39	0.39	4.09	20	**21.50**	0.08	**0.08**	**2.72**	**20**
	first-fail	21.50	1.05	2.79	20	21.50	0.28	0.28	**0.97**	**20**	**21.50**	0.24	**0.24**	1.62	20
	inorder	22.50	161.35	228.98	10	23.20	16.74	16.74	54.97	16	**23.20**	1.23	**1.23**	**15.92**	**16**
Photo-20	greedy	23.00	5.36	38.38	12	23.20	3.26	3.26	19.01	18	**23.20**	0.10	**0.10**	**3.29**	**18**
	first-fail	23.20	4.90	12.95	17	23.20	1.17	1.17	**3.45**	**20**	**23.20**	0.54	**0.54**	3.46	20
	inorder	23.70	137.69	254.99	55	25.35	44.46	44.46	102.18	13	**25.40**	2.09	**1.57**	**12.27**	**14**
Photo-22	greedy	25.15	8.28	60.94	11	25.40	9.54	9.54	36.68	15	**25.45**	0.60	**0.60**	**5.38**	**15**
	first-fail	25.45	13.94	19.71	16	25.50	3.36	3.36	**6.42**	**20**	**25.50**	2.33	**2.33**	6.55	20
	inorder	22.45	216.79	519.93	3	26.85	87.71	87.71	350.3	7	**27.10**	1.36	**0.71**	**61.88**	**8**
Photo-24	greedy	26.60	11.75	184.26	7	26.95	16.52	16.52	171.90	8	**27.35**	0.66	**0.23**	**27.71**	**12**
	first-fail	26.55	43.17	88.59	12	27.45	13.10	13.10	33.29	19	**27.45**	5.12	**5.12**	**20.97**	**19**

5   Related Work

Dominance breaking constraints were introduced only recently in [3] and there has been little work analyzing how they may conflict with the search or how that problem can be overcome. The closest related work is that for the special case of symmetry breaking. In this case, potential conflicts between the search and static symmetry breaking constraints can be overcome by using dynamic symmetry breaking techniques such as SBDS [1,11] or SBDD [5]. However, neither of these methods obviously generalize to the dominance case. In the case of symmetry, we have sets of equally good symmetric subtrees. The policy in SBDS/SBDD is to search the first member of each such set encountered during search, and to prune all other members as soon as they are encountered. In dominance breaking however, we have pairs of subtrees where one may be strictly better than the other (i.e., contains a strictly better solution). The ordering between them is not up to us to decide as it is determined by the search strategy, and we cannot simply decide to always search the first of the pair and prune the second. We could try a different policy such as: if we encounter the good one first, we prune the bad one later, and if we encounter the bad one first, we search both. Indeed, such a policy was proposed in [7]. However, checking whether a subtree is dominated by any of the previously searched subtrees is extremely complex in general, and is much harder than simply determining whether it is dominated by some (possibly not yet explored) subtree. In [7], it is shown how an incomplete version of such a dominance check can be performed using greedy local search for the Travelling Salesman Problem. However, it is not clear how complete it is or whether it can generalise to other problems. Also, even if the dominance check can be performed efficiently, such a method will still perform more search in general than the dominance jumping method presented here. In dominance jumping, regardless of the order in which we encounter the pair of subtrees, we will only search the good one and will always skip the bad one, because if we encounter the bad one first, we will simply immediately jump to the good one.

Dominance jumping shares some features with local search/repair methods such as min-conflict search [16]. However, such methods typically travel through the space of infeasible solutions and jump at every node. Dominance jumping on the other hand is a systematic search which remains within feasible space, and only occasionally jumps when we are guaranteed to get to a better subtree.

Dominance jumping is also related to best first search. When best first search reaches a node which is recognized as possibly dominated (since the lower bound on the objective is substantially worse than another part of the search tree), it jumps to what it thinks is the best node and explores from there. In this case the jump is a heuristic, unlike in dominance jumping where we have a proof that the current node is suboptimal and we jump to a strictly better node.

6   Conclusion

We have developed a new method called dominance jumping to exploit the dominance relations identified by the method proposed in [3]. Rather than failing and backtracking as in the original method, we use the dominance relation to jump to a different part of the search tree that dominates the current subtree. Unlike static dominance breaking constraints, the new method will not conflict with the search strategy. Experimental evidence shows that the method allows good/optimal solutions to be found much more quickly on a wide range of problems. Important future work is to examine how to automatically determine when during search to switch from dominance jumping to dominance breaking, so that we can take advantage of the strengths of both approaches.

Acknowledgments. NICTA is funded by the Australian Government as represented by the Department of Broadband, Communications and the Digital Economy and the Australian Research Council.

References

1. Backofen, R., Will, S.: Excluding symmetries in constraint-based search. In: Jaffar, J. (ed.) CP 1999. LNCS, vol. 1713, pp. 73–87. Springer, Heidelberg (1999)
2. Boussemart, F., Hemery, F., Lecoutre, C., Sais, L.: Boosting systematic search by weighting constraints. In: Procs. of ECAI 2004, pp. 146–150 (2004)
3. Chu, G., Stuckey, P.J.: A generic method for identifying and exploiting dominance relations. In: Milano, M. (ed.) CP 2012. LNCS, vol. 7514, pp. 6–22. Springer, Heidelberg (2012)
4. Crawford, J.M., Ginsberg, M.L., Luks, E.M., Roy, A.: Symmetry-breaking predicates for search problems. In: Proceedings of the 5th International Conference on Principles of Knowledge Representation and Reasoning, pp. 148–159. Morgan Kaufmann (1996)
5. Fahle, T., Schamberger, S., Sellmann, M.: Symmetry breaking. In: Walsh, T. (ed.) CP 2001. LNCS, vol. 2239, pp. 93–107. Springer, Heidelberg (2001)
6. Feydy, T., Stuckey, P.J.: Lazy clause generation reengineered. In: Gent, I.P. (ed.) CP 2009. LNCS, vol. 5732, pp. 352–366. Springer, Heidelberg (2009)
7. Focacci, F., Shaw, P.: Pruning sub-optimal search branches using local search. In: CPAIOR, vol. 2, pp. 181–189 (2002)
8. Frisch, A.M., Jefferson, C., Miguel, I.: Constraints for Breaking More Row and Column Symmetries. In: Rossi, F. (ed.) CP 2003. LNCS, vol. 2833, pp. 318–332. Springer, Heidelberg (2003)
9. de la Banda, M.G., Stuckey, P.J., Chu, G.: Solving talent scheduling with dynamic programming. INFORMS Journal on Computing 23(1), 120–137 (2011)
10. Gargani, A., Refalo, P.: An efficient model and strategy for the steel mill slab design problem. In: Bessière, C. (ed.) CP 2007. LNCS, vol. 4741, pp. 77–89. Springer, Heidelberg (2007)

11. Gent, I., Smith, B.M.: Symmetry breaking in constraint programming. In: 14th European Conference on Artificial Intelligence, pp. 599–603 (2000)
12. Haralick, R.M., Elliott, G.L.: Increasing tree search efficiency for constraint satisfaction problems. Artificial Intelligence (1980)
13. Lecoutre, C., Sais, L., Tabary, S., Vidal, V.: Nogood Recording from Restarts. In: Veloso, M.M. (ed.) Proc. of IJCAI 2007, pp. 131–136 (2007)
14. Luks, E.M., Roy, A.: The Complexity of Symmetry-Breaking Formulas. Ann. Math. Artif. Intell. 41(1), 19–45 (2004)
15. Miller, H.E., Pierskalla, W.P., Rath, G.J.: Nurse scheduling using mathematical programming. Operations Research, 857–870 (1976)
16. Minton, S., Johnston, M.D., Philips, A.B., Laird, P.: Minimizing conflicts: a heuristic repair method for constraint satisfaction and scheduling problems. Artificial Intelligence 58(1), 161–205 (1992)
17. Moskewicz, M., Madigan, C., Zhao, Y., Zhang, L., Malik, S.: Chaff: engineering an efficient SAT solver. In: Procs. of DAC 2001, pp. 530–535 (2001)
18. Nethercote, N., Stuckey, P.J., Becket, R., Brand, S., Duck, G.J., Tack, G.: MiniZinc: Towards a Standard CP Modelling Language. In: Bessière, C. (ed.) CP 2007. LNCS, vol. 4741, pp. 529–543. Springer, Heidelberg (2007)
19. Ohrimenko, O., Stuckey, P.J., Codish, M.: Propagation = Lazy Clause Generation. In: Bessière, C. (ed.) CP 2007. LNCS, vol. 4741, pp. 544–558. Springer, Heidelberg (2007)
20. Puget, J.F.: On the Satisfiability of Symmetrical Constrained Satisfaction Problems. In: Komorowski, J., Raś, Z.W. (eds.) ISMIS 1993. LNCS, vol. 689, pp. 350–361. Springer, Heidelberg (1993)
21. Puget, J.F.: Breaking symmetries in all different problems. In: Kaelbling, L.P., Saffiotti, A. (eds.) Proc. of IJCAI 2005, pp. 272–277. Professional Book Center (2005)

Tractable Combinations of Global Constraints

David A. Cohen[1], Peter G. Jeavons[2],
Evgenij Thorstensen[2,*], and Stanislav Živný[3,**]

[1] Department of Computer Science, Royal Holloway, University of London, UK
d.cohen@rhul.ac.uk
[2] Department of Computer Science, University of Oxford, UK
firstname.lastname@cs.ox.ac.uk
[3] Department of Computer Science, University of Warwick, UK
s.zivny@warwick.ac.uk

Abstract. We study the complexity of constraint satisfaction problems involving global constraints, i.e., special-purpose constraints provided by a solver and represented implicitly by a parametrised algorithm. Such constraints are widely used; indeed, they are one of the key reasons for the success of constraint programming in solving real-world problems.

Previous work has focused on the development of efficient propagators for individual constraints. In this paper, we identify a new tractable class of constraint problems involving global constraints of unbounded arity. To do so, we combine structural restrictions with the observation that some important types of global constraint do not distinguish between large classes of equivalent solutions.

1 Introduction

Constraint programming (CP) is widely used to solve a variety of practical problems such as planning and scheduling [23,30], and industrial configuration [1,22]. The theoretical properties of constraint problems, in particular the computational complexity of different types of problem, have been extensively studied and quite a lot is known about what restrictions on the general *constraint satisfaction problem* are sufficient to make it tractable [2,7,11,17,20,25].

However, much of this theoretical work has focused on problems where each constraint is represented *explicitly*, by a table of allowed assignments.

In practice, however, a lot of the success of CP is due to the use of special-purpose constraint types for which the software tools provide dedicated algorithms [28,16,31]. Such constraints are known as *global constraints* and are usually represented *implicitly* by an algorithm in the solver. This algorithm may take as a parameter a *description* that specifies exactly which kinds of assignments a particular instance of this constraint should allow.

Theoretical work on global constraints has to a large extent focused on developing efficient algorithms to achieve various kinds of local *consistency* for

---

* Supported by EPSRC grant EP/G055114/1.
** Supported by a Senior Research Fellowship from Warwick's DIMAP.

C. Schulte (Ed.): CP 2013, LNCS 8124, pp. 230–246, 2013.

individual constraints. This is generally done by pruning from the domains of variables those values that cannot lead to a satisfying assignment [5,29]. Another strand of research has explored when it is possible to replace global constraints by collections of explicitly represented constraints [6]. These techniques allow faster implementations of algorithms for *individual constraints*, but do not shed much light on the complexity of problems with multiple *overlapping* global constraints, which is something that practical problems frequently require.

As an example, consider the following family of constraint problems involving clauses and cardinality constraints of unbounded arity.

Example 1. Consider a family of constraint problems on a set of Boolean variables $\{x_1, x_2, \ldots, x_{3n}\}$ (where $n = 2, 3, 4, \ldots$), with the following five constraints:

- C_1 is the binary clause $x_1 \vee x_{2n+1}$;
- C_2 is a cardinality constraint on $\{x_1, x_2, \ldots, x_n\}$ specifying that exactly one of these variables takes the value 1;
- C_3 is a cardinality constraint on $\{x_{2n+1}, x_{2n+2}, \ldots, x_{3n}\}$ specifying that exactly one of these variables takes the value 1;
- C_4 is a cardinality constraint on $\{x_2, x_3, \ldots, x_{3n}\} - \{x_{2n+1}\}$ specifying that exactly $n + 1$ of these variables takes the value 1;
- C_5 is the clause $\neg x_{n+1} \vee \neg x_{n+2} \vee \cdots \vee \neg x_{2n}$.

This problem is illustrated in Figure 1.

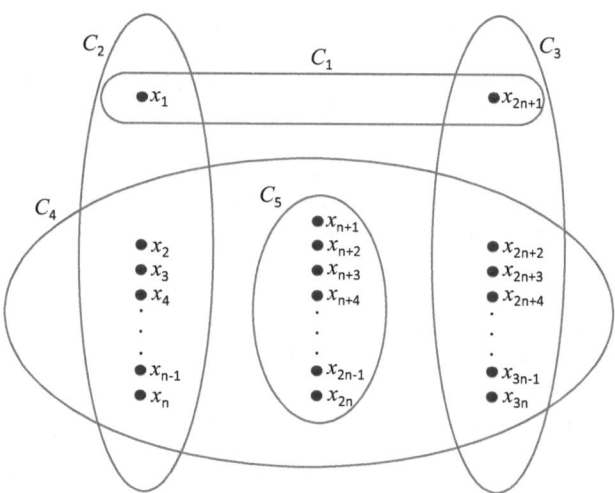

Fig. 1. The structure of the constraint problems in Example 1

This family of problems is not included in any previously known tractable class, but will be shown to be tractable using the results of this paper.

As discussed in [9], when the constraints in a family of problems have unbounded arity, the way that the constraints are *represented* can significantly
affect the complexity. Previous work in this area has assumed that the global
constraints have specific representations, such as propagators [19], negative constraints [10], or GDNF/decision diagrams [9], and exploited properties particular
to that representation. In contrast, here we investigate the conditions that yield
efficiently solvable classes of constraint problems with global constraints, without requiring any specific representation. Many global constraints have succinct
representations, so even problems with very simple structures are known to be
hard in some cases [24,29]. We will therefore need to impose some restrictions
on the properties of the individual global constraints, as well as on the problem
structure.

To obtain our results, we define a notion of equivalence on assignments and a
new width measure that identifies variables that are constrained in exactly the
same way. We then show that we can replace variables that are equated under our
width measure with a single new variable whose domain represents the possible
equivalence classes of assignments. Both of these simplification steps, merging
variables and equating assignments, can be seen as techniques for eliminating
symmetries in the original problem formulation. We describe some sufficient
conditions under which these techniques provide a polynomial-time reduction to
a known tractable case, and hence identify new tractable classes of constraint
problems involving global constraints.

2   Global Constraints and Constraint Problems

In order to be more precise about the way in which global constraints are represented, we will extend the standard definition of a constraint problem.

Definition 1 (Variables and assignments). *Let V be a set of variables, each
with an associated set of domain elements. We denote the set of domain elements
(the domain) of a variable v by $D(v)$. We extend this notation to arbitrary subsets
of variables, W, by setting $D(W) = \bigcup_{v \in W} D(v)$.*

An assignment *of a set of variables V is a function $\theta : V \to D(V)$ that maps
every $v \in V$ to an element $\theta(v) \in D(v)$. We denote the restriction of θ to a set
of variables $W \subseteq V$ by $\theta|_W$. We also allow the special assignment $\bot$ of the empty
set of variables. In particular, for every assignment θ, we have $\theta|_\emptyset = \bot$.*

Global constraints have traditionally been defined, somewhat vaguely, as constraints without a fixed arity, possibly also with a compact representation of
the constraint relation. For example, in [23] a global constraint is defined as "a
constraint that captures a relation between a non-fixed number of variables".

Below, we offer a precise definition similar to the one in [5], where the authors
define global constraints for a domain D over a list of variables σ as being
given intensionally by a function $D^{|\sigma|} \to \{0, 1\}$ computable in polynomial time.
Our definition differs from this one in that we separate the general *algorithm* of

a global constraint (which we call its *type*) from the specific description. This separation allows us a better way of measuring the size of a global constraint, which in turn helps us to establish new complexity results.

Definition 2 (Global constraints). *A global constraint type is a parametrised polynomial-time algorithm that determines the acceptability of an assignment of a given set of variables.*

Each global constraint type, e, has an associated set of descriptions, $\Delta(e)$. *Each description $\delta \in \Delta(e)$ specifies appropriate parameter values for the algorithm e. In particular, each $\delta \in \Delta(e)$ specifies a set of variables, denoted by* vars(δ).

A global constraint $e[\delta]$, where $\delta \in \Delta(e)$, is a function that maps assignments of vars(δ) *to the set $\{0, 1\}$. Each assignment that is allowed by $e[\delta]$ is mapped to 1, and each disallowed assignment is mapped to 0. The* extension *or* constraint relation *of $e[\delta]$ is the set of assignments, θ, of* vars(δ) *such that $e[\delta](\theta) = 1$. We also say that such assignments* satisfy *the constraint, while all other assignments* falsify *it.*

When we are only interested in describing the set of assignments that satisfy a constraint, and not in the complexity of determining membership in this set, we will sometimes abuse notation by writing $\theta \in e[\delta]$ to mean $e[\delta](\theta) = 1$.

As can be seen from the definition above, a global constraint is not usually explicitly represented by listing all the assignments that satisfy it. Instead, it is represented by some description δ and some algorithm e that allows us to check whether the constraint relation of $e[\delta]$ includes a given assignment. To stay within the complexity class NP, this algorithm is required to run in polynomial time. As the algorithms for many common global constraints are built into modern constraint solvers, we measure the *size* of a global constraint's representation by the size of its description.

Example 2 (EGC). A very general global constraint type is the *extended global cardinality* constraint type [26,29]. This form of global constraint is defined by specifying for every domain element a a finite set of natural numbers $K(a)$, called the cardinality set of a. The constraint requires that the number of variables which are assigned the value a is in the set $K(a)$, for each possible domain element a.

Using our notation, the description δ of an EGC global constraint specifies a function $K_\delta : D(\text{vars}(\delta)) \to \mathcal{P}(\mathbb{N})$ that maps each domain element to a set of natural numbers. The algorithm for the EGC constraint then maps an assignment θ to 1 if and only if, for every domain element $a \in D(\text{vars}(\delta))$, we have that $|\{v \in \text{vars}(\delta) \mid \theta(v) = a\}| \in K_\delta(a)$.

The cardinality constraint C_2 from Example 1 can be expressed as an EGC global constraint with description δ such that $K_\delta(1) = \{1\}$, and $K_\delta(0) = \{n-1\}$.

Example 3 (Clauses). We can view the disjunctive clauses used to define propositional satisfiability problems as a global constraint type in the following way.

The description δ of a clause is simply a list of the literals that it contains, and vars(δ) is the corresponding set of variables. The algorithm for the clause then maps any Boolean assignment θ of vars(δ) that satisfies the disjunction of the literals specified by δ to 1, and all other assignments to 0.

Note that a clause forbids precisely one assignment to vars(δ) (the one that falsifies all of the literals in the clause). Hence the extension of a clause contains $2^{|\text{vars}(\delta)|} - 1$ assignments, so the size of the constraint *relation* grows exponentially with the number of variables, but the size of the constraint *description* grows only linearly.

Example 4 (Table and negative constraints). A rather degenerate example of a a global constraint type is the *table* constraint.

In this case the description δ is simply a list of assignments of some fixed set of variables, vars(δ). The algorithm for a table constraint then decides, for any assignment of vars(δ), whether it is included in δ. This can be done in a time which is linear in the size of δ and so meets the polynomial time requirement.

Negative constraints are complementary to table constraints, in that they are described by listing *forbidden* assignments. The algorithm for a negative constraint $e[\delta]$ decides, for any assignment of vars(δ), whether it is *not* included in δ. Observe that the clauses described in Example 3 are a special case of the negative constraint type, as they have exactly one forbidden assignment.

We observe that any global constraint can be rewritten as a table or negative constraint. However, this rewriting will, in general, incur an exponential increase in the size of the description.

Definition 3 (CSP instance). *An instance of the constraint satisfaction problem (CSP) is a pair $\langle V, C \rangle$ where V is a finite set of variables, and C is a set of global constraints such that for every $e[\delta] \in C$, vars(δ) $\subseteq V$. In a CSP instance, we call vars(δ) the* scope *of the constraint $e[\delta]$.*

A solution *to a CSP instance $\langle V, C \rangle$ is an assignment θ of V which satisfies every global constraint, i.e., for every $e[\delta] \in C$ we have $\theta|_{\text{vars}(\delta)} \in e[\delta]$.*

The general constraint satisfaction problem is clearly NP-complete, so in the remainder of the paper we shall look for more restricted versions of the problem that are *tractable*, that is, solvable in polynomial time.

3 Restricted Classes of Constraint Problems

First, we are going to consider restrictions on the way that the constraints in a given instance interact with each other, or, in other words, the way that the constraint scopes overlap; such restrictions are known as *structural* restrictions [11,17,20].

Definition 4 (Hypergraph). *A hypergraph $\langle V, H \rangle$ is a set of vertices V together with a set of hyperedges $H \subseteq \mathcal{P}(V)$.*

Given a CSP instance $P = \langle V, C \rangle$, the hypergraph of P, denoted hyp(P), has vertex set V together with a hyperedge vars(δ) *for every $e[\delta] \in C$.*

One special class of hypergraphs that has received a great deal of attention is the class of *acyclic* hypergraphs [3]. This notion is a generalisation of the idea of tree-structure in a graph, and has been very important in the analysis of relational databases. A hypergraph is said to be acyclic if repeatedly removing all hyperedges contained in other hyperedges, and all vertices contained in only a single hyperedge, eventually deletes all vertices [3].

Solving a CSP instance P whose constraints are represented extensionally (i.e., as table constraints) is known to be tractable if the hypergraph of P, hyp(P), is acyclic [21]. Indeed, this has formed the basis for more general notions of "bounded cyclicity" [21] or "bounded hypertree width" [18], which have also been shown to imply tractability for problems with explicitly represented constraint relations. However, this is no longer true if the constraints are global, not even when we have a fixed, finite domain, as the following examples show.

Example 5. Any hypergraph containing only a single edge is clearly acyclic (and therefore has hypertree width one [18]), but the class of CSP instances consisting of a single EGC constraint over an unbounded domain is NP-complete [26].

Example 6. The NP-complete problem of 3-colourability [15] is to decide, given a graph $\langle V, E \rangle$, whether the vertices V can be coloured with three colours such that no two adjacent vertices have the same colour.

We may reduce this problem to a CSP with EGC constraints (cf. Example 2) as follows: Let V be the set of variables for our CSP instance, each with domain $\{r, g, b\}$. For every edge $\langle v, w \rangle \in E$, we post an EGC constraint with scope $\{v, w\}$, parametrised by the function K such that $K(r) = K(g) = K(b) = \{0, 1\}$. Finally, we make the hypergraph of this CSP instance acyclic by adding an EGC constraint with scope V parametrised by the function K' such that $K'(r) = K'(g) = K'(b) = \{0, \ldots, |V|\}$. This reduction clearly takes polynomial time, and the hypergraph of the resulting instance is acyclic.

These examples indicate that when dealing with implicitly represented constraints we cannot hope for tractability using structural restrictions alone. We are therefore led to consider *hybrid* restrictions, which restrict both the nature of the constraints and the structure at the same time.

Definition 5 (Constraint catalogue). *A constraint catalogue is a set of global constraints. A CSP instance $\langle V, C \rangle$ is said to be over a constraint catalogue $\mathcal{C}$ if for every $e[\delta] \in C$ we have $e[\delta] \in \mathcal{C}$.*

Previous work on the complexity of constraint problems has restricted the *extensions* of the constraints to a specified set of *relations*, known as a constraint *language* [7]. This is an appropriate form of restriction when all constraints are given explicitly, as table constraints. However, here we work with global constraints where the relations are often implicit, and this can significantly alter the complexity of the corresponding problem classes, as we will illustrate below. Hence we allow a more general form of restriction on the constraints by specifying a constraint catalogue containing all allowed constraints.

Definition 6 (Restricted CSP class). *Let $\mathcal{C}$ be a constraint catalogue, and let $\mathcal{H}$ be a class of hypergraphs. We define $\mathsf{CSP}(\mathcal{H}, \mathcal{C})$ to be the class of CSP instances over $\mathcal{C}$ whose hypergraphs are in $\mathcal{H}$.*

Using Definition 6, we will restate an earlier structural tractability result, which will form the basis for our results in Section 5.

Definition 7 (Treewidth). *A tree decomposition of a hypergraph $\langle V, H \rangle$ is a pair $\langle T, \lambda \rangle$ where T is a tree and λ is a labelling function from nodes of T to subsets of V, such that*

1. *for every $v \in V$, there exists a node t of T such that $v \in \lambda(t)$,*
2. *for every hyperedge $h \in E$, there exists a node t of T such that $h \subseteq \lambda(t)$, and*
3. *for every $v \in V$, the set of nodes $\{t \mid v \in \lambda(t)\}$ induces a connected subtree of T.*

The width *of a tree decomposition is $\max(\{|\lambda(t)| - 1 \mid t \text{ node of } T\})$. The* treewidth $\mathsf{tw}(G)$ *of a hypergraph G is the minimum width over all its tree decompositions.*

Let $\mathcal{H}$ be a class of hypergraphs, and define $\mathsf{tw}(\mathcal{H})$ to be the maximum treewidth over the hypergraphs in $\mathcal{H}$. If $\mathsf{tw}(\mathcal{H})$ is unbounded we write $\mathsf{tw}(\mathcal{H}) = \infty$; otherwise $\mathsf{tw}(\mathcal{H}) < \infty$.

We can now restate using the language of global constraints the following result, from Dalmau et al. [12], which builds on several earlier results [13,14].

Theorem 1 ([12]). *Let $\mathcal{C}$ be a constraint catalogue and $\mathcal{H}$ a class of hypergraphs. $\mathsf{CSP}(\mathcal{H}, \mathcal{C})$ is tractable if $\mathsf{tw}(\mathcal{H}) < \infty$.*

Observe that the family of constraint problems described in Example 1 is not covered by the above result, because the treewidth of the associated hypergraphs is unbounded.

4   Cooperating Constraint Catalogues

Whenever constraint scopes overlap, we may ask whether the possible assignments to the variables in the overlap are essentially different. It may be that some assignments extend to precisely the same satisfying assignments in each of the overlapping constraints. If so, we may as well identify such assignments.

Definition 8 (Disjoint union of assignments). *Let θ_1 and θ_2 be two assignments of disjoint sets of variables V_1 and V_2, respectively. The disjoint union of θ_1 and θ_2, denoted $\theta_1 \oplus \theta_2$, is the assignment of $V_1 \cup V_2$ such that $(\theta_1 \oplus \theta_2)(v) = \theta_1(v)$ for all $v \in V_1$, and $(\theta_1 \oplus \theta_2)(v) = \theta_2(v)$ for all $v \in V_2$.*

Definition 9 (Projection). *Let Θ be a set of assignments of a set of variables V. The projection of Θ onto a set of variables $X \subseteq V$ is the set of assignments $\pi_X(\Theta) = \{\theta|_X \mid \theta \in \Theta\}$.*

Note that when $\Theta = \emptyset$ we have $\pi_X(\Theta) = \emptyset$ for any set X, but when $X = \emptyset$ and $\Theta \neq \emptyset$, we have $\pi_X(\Theta) = \{\bot\}$.

Definition 10 (Assignment extension). *Let $e[\delta]$ be a global constraint, and $X \subseteq \text{vars}(\delta)$. For every assignment μ of X, let $\text{ext}(\mu, e[\delta]) = \pi_{\text{vars}(\delta)-X}(\{\theta \in e[\delta] \mid \theta|_X = \mu\})$.*

In other words, for any assignment μ of X, the set $\text{ext}(\mu, e[\delta])$ is the set of assignments of $\text{vars}(\delta) - X$ that extend μ to a satisfying assignment for $e[\delta]$; i.e., those assignments θ for which $\mu \oplus \theta \in e[\delta]$.

Definition 11 (Extension equivalence). *Let $e[\delta]$ be a global constraint, and $X \subseteq \text{vars}(\delta)$. We say that two assignments θ_1, θ_2 to X are extension equivalent on X with respect to $e[\delta]$ if $\text{ext}(\theta_1, e[\delta]) = \text{ext}(\theta_2, e[\delta])$. We denote this equivalence relation by $\text{equiv}[e[\delta], X]$; that is, $\text{equiv}[e[\delta], X](\theta_1, \theta_2)$ holds if and only if θ_1 and θ_2 are extension equivalent on X with respect to $e[\delta]$.*

In other words, two assignments to some subset of the variables of a constraint $e[\delta]$ are extension equivalent if every assignment to the rest of the variables combines with both of them to give either two assignments that satisfy $e[\delta]$, or two that falsify it.

Example 7. Consider the special case of extension equivalence with respect to a clause (cf. Example 3).

Given any clause $e[\delta]$, and any non-empty set of variables $X \subseteq \text{vars}(\delta)$, any assignment to X will either satisfy one of the corresponding literals specified by δ, or else falsify all of them. If it satisfies at least one of them, then any extension will satisfy the clause, so all such assignments are extension equivalent. If it falsifies all of them, then an extension will satisfy the clauses if and only if it satifies one of the other literals. Hence the equivalence relation $\text{equiv}[e[\delta], X]$ has precisely 2 equivalence classes, one containing the single assignment that falsifies all the literals corresponding to X, and one containing all other assignments.

Definition 12 (Intersecting variables). *Let S be a set of global constraints. We write $\text{iv}(S)$ for the set of variables common to all of their scopes, that is,*

$$\text{iv}(S) = \bigcap_{e[\delta] \in S} \text{vars}(\delta).$$

Definition 13 (Join). *For any set S of global constraints, we define the join of S, denoted $\text{join}(S)$, to be a global constraint $e'[\delta']$ with $\text{vars}(\delta') = \bigcup_{e[\delta] \in S} \text{vars}(\delta)$ such that for any assignment θ to $\text{vars}(\delta')$, we have $\theta \in e'[\delta']$ if and only if for every $e[\delta] \in S$ we have $\theta|_{\text{vars}(\delta)} \in e[\delta]$.*

The join of a set of global constraints may have no simple compact description, and computing its extension may be computationally expensive. However, we introduce this construct simply in order to describe the combined effect of a set of global constraints in terms of a single constraint.

Example 8. Let $V = \{v_1, \ldots, v_n\}$, for some $n \geq 3$, be a set of variables with $D(v_i) = \{a, b, c\}$, and let $S = \{e_1[\delta_1], e_2[\delta_2]\}$ be a set of two global constraints as defined below:

- $e_1[\delta_1]$ is a table constraint with $\mathsf{vars}(\delta_1) = \{v_1, \ldots, v_{n-1}\}$ which enforces *equality*, i.e., $\delta_1 = \{\theta_a, \theta_b, \theta_c\}$, where for each $x \in D(V)$ and $v \in \mathsf{vars}(\delta_1)$, $\theta_x(v) = x$.
- $e_2[\delta_2]$ is a negative constraint with $\mathsf{vars}(\delta_2) = \{v_2, \ldots, v_n\}$ which enforces a *not-all-equal* condition, i.e., $\delta_2 = \{\theta_a, \theta_b, \theta_c\}$, where for each $x \in D(V)$ and $v \in \mathsf{vars}(\delta_2)$, $\theta_x(v) = x$.

We will use substitution notation to write assignments explicitly; thus, an assignment of $\{v, w\}$ that assigns a to both variables is written $\{v/a, w/a\}$.

We have that $\mathsf{iv}(S) = \{v_2, \ldots, v_{n-1}\}$. The equivalence classes of assignments to $\mathsf{iv}(S)$ under $\mathsf{equiv}[\mathsf{join}(S), \mathsf{iv}(S)]$ are $\{\{v_2/a, \ldots, v_{n-1}/a\}\}$, $\{\{v_2/b, \ldots, v_{n-1}/b\}\}$, and $\{\{v_2/c, \ldots, v_{n-1}/c\}\}$, each containing the single assignment shown, as well as (for $n > 3$) a final class containing all other assignments, for which we can choose an arbitrary representative assignment, θ_0, such as $\{v_2/a, v_3/b, \ldots, v_{n-1}/b\}$.

Each assignment in the first 3 classes has just 2 possible extensions that satisfy $\mathsf{join}(S)$, since the value assigned to v_1 must equal the value assigned to $v_2, \ldots, v_{n-1}$, and the value assigned to v_n must be different. The assignment θ_0 has no extensions, since $\mathsf{ext}(\theta_0, e_1[\delta_1]) = \emptyset$.

Hence the number of equivalence classes in $\mathsf{equiv}[\mathsf{join}(S), \mathsf{iv}(S)]$ is at most 4, even though the total number of possible assignments of $\mathsf{iv}(S)$ is 3^{n-2}

Definition 14 (Cooperating constraint catalogue). *We say that a constraint catalogue $\mathcal{C}$ is a cooperating catalogue if for any finite set of global constraints $S \subseteq \mathcal{C}$, we can compute a set of assignments of the variables $\mathsf{iv}(S)$ containing at least one representative of each equivalence class of $\mathsf{equiv}[\mathsf{join}(S), \mathsf{iv}(S)]$ in polynomial time in the size of $\mathsf{iv}(S)$ and the total size of the constraints in S.*

Note that this definition requires two things. First, that the number of equivalence classes in the equivalence relation $\mathsf{equiv}[\mathsf{join}(S), \mathsf{iv}(S)]$ is bounded by some fixed polynomial in the size of $\mathsf{iv}(S)$ and the size of the constraints in S. Secondly, that a suitable set of representatives for these equivalence classes can be computed efficiently from the constraints.

Example 9. Consider a constraint catalogue consisting entirely of clauses (of arbitrary arity). It was shown in Example 7 that for any clause $e[\delta]$ and any non-empty $X \subseteq \mathsf{vars}(\delta)$ the equivalence relation $\mathsf{equiv}[e[\delta], X]$ has precisely 2 equivalence classes.

If we consider some finite set, S, of clauses, then a similar argument shows that the equivalence relation $\mathsf{equiv}[\mathsf{join}(S), \mathsf{iv}(S)]$ has at most $|S| + 1$ classes. These are given by the single assignments of the variables in $\mathsf{iv}(S)$ that falsify the literals corresponding to the variables of $\mathsf{iv}(S)$ in each clause (there are at most $|S|$ of these — they may not all be distinct) together with at most one further equivalence class containing all other assignments (which must satisfy at least one literal in each clause of S).

Hence the total number of equivalence classes in the equivalence relation equiv[join(S), iv(S)] increases at most linearly with the number of clauses in S, and a representative for each class can be easily obtained from the descriptions of these clauses, by projecting the falsifying assignments down to the set of common variables, iv(S), and adding at most one more, arbitrary, assignment.

By same argument, if we consider some finite set, S, of table constraints, then the equivalence relation equiv[join(S), iv(S)] has at most one class for each assignment allowed by each table constraint in S, together with at most one further class containing all other assignments.

In general, arbitrary EGC constraints (cf. Example 2) do not form a cooperating catalogue. However, we will show that if we bound the size of the variable domains, then the resulting EGC constraints do form a cooperating catalogue.

Definition 15 (Counting function). *Let X be a set of variables with domain $D = \bigcup_{x \in X} D(x)$. A counting function for X is any function $K : D \to \mathbb{N}$ such that $\sum_{a \in D} K(a) = |X|$.*

Every assignment θ to X defines a corresponding counting function K_θ given by $K_\theta(a) = |\{x \in X \mid \theta(x) = a\}|$ for every $a \in D$.

It is easy to verify that no EGC constraint can distinguish two assignments with the same counting function; for any EGC constraint, either both assignments satisfy it, or they both falsify it. It follows that two assignments with the same counting function are extension equivalent with respect to EGC constraints.

Definition 16 (Counting constraints). *A global constraint $e[\delta]$ is called a counting constraint if, for any two assignments θ_1, θ_2 of vars(δ) which have the same counting function, either $\theta_1, \theta_2 \in e[\delta]$ or $\theta_1, \theta_2 \notin e[\delta]$.*

EGC constraints are not the only constraint type with this property. Constraints that require the sum (or the product) of the values of all variables in their scope to take a particular value, and constraints that require the minimum (or maximum) value of the variables in their scope to take a certain value, are also counting constraints.

Another example is given by the NValue constraint type, which requires that the number of distinct domain values taken by an assignment is a member of a specified set of acceptable numbers.

Example 10 (NValue constraint type [4,6]). In an NValue constraint, $e[\delta]$, the description δ specifies a finite set of natural numbers $L_\delta \subset \mathbb{N}$. The algorithm e maps an assignment θ to 1 if $|\{\theta(v) \mid v \in \mathsf{vars}(\delta)\}| \in L_\delta$.

The reason for introducing counting functions is the following key property, previously noted by Bulatov and Marx [8].

Property 1. The number of possible counting functions for a set of variables X is at most $\binom{|X|+|D|-1}{|D|-1} = O(|X|^{|D|})$, where $D = \bigcup_{x \in X} D(x)$.

Proof. If every variable $x \in X$ has D as its set of domain elements, that is, $D(x) = D$, then every counting function corresponds to a distinct way of partitioning $|X|$ variables into at most $|D|$ boxes. There are $\binom{|X|+|D|-1}{|D|-1}$ ways of doing so [27, Section 2.3.3]. On the other hand, if there are variables $x \in X$ such that $D(x) \subset D$, then that disallows some counting functions.

Theorem 2. *Any constraint catalogue that contains only counting constraints with bounded domain size, table constraints, and negative constraints, is a cooperating catalogue.*

Proof. Let $\mathcal{C}$ be a constraint catalogue containing only global constraints of the specified types, and let $S \subseteq \mathcal{C}$ be a finite subset of $\mathcal{C}$. Partition S into two subsets: S^C, containing only counting constraints and $S^{\pm}$ containing only table and negative constraints.

Let $\mathcal{K}$ be a set containing assignments of iv(S), such that for every counting function K for iv(S), there is some assignment $\theta_K \in \mathcal{K}$ with $K_\theta = K$. By Property 1, the number of counting functions for iv(S) is bounded by $O(|\text{iv}(S)|^d)$, where d is the bound on the domain size for the counting constraints in $\mathcal{C}$. Hence such a set $\mathcal{K}$ can be computed in polynomial time in the size of iv(S).

For each constraint in $S^{\pm}$ we have that the description is a list of assignments (these are the allowed assignments for the table constraints and the forbidden assignments for the negative constraints, see Example 4).

As we described in Example 9, for each table constraint $e[\delta] \in S$, we can obtain a representative for each equivalence class of equiv$[e[\delta], \text{iv}(S)]$ by taking the projection onto iv(S) of each allowed assignment, which we can denote by $\pi_{\text{iv}(S)}(\delta)$, together with at most one further, arbitrary, assignment, θ_0, that is not in this set. This set of assignments contains at least one representative for each equivalence class of equiv$[e[\delta], \text{iv}(S)]$ (and possibly more than one representative for some of these classes).

Similarly, for each negative constraint $e[\delta] \in S$, we can obtain a representative for each equivalence class of equiv$[e[\delta], \text{iv}(S)]$, by taking the projection onto iv(S) of each forbidden assignment, which we can again denote by $\pi_{\text{iv}(S)}(\delta)$, together with at most one further, arbitrary, assignment, θ_0, that is not in this set.

Now consider the set of assignments $\mathcal{A} = \mathcal{K} \cup \{\theta_0\} \cup \bigcup_{e[\delta] \in S^{\pm}} \pi_{\text{iv}(S)}(\delta)$, where θ_0 is an arbitrary assignment of iv(S) which does not occur in $\pi_{\text{iv}(S)}(\delta)$ for any $e[\delta] \in S$ (if such an assignment exists). We claim that this set of assignments contains at least one representative for each equivalence class of equiv$[\text{join}(S), \text{iv}(S)]$ (and possibly more than one for some classes).

To establish this claim we will show that any assignment θ of iv(S) that is not in $\mathcal{A}$ must be extension equivalent to some member of $\mathcal{A}$. Let θ be an assignment of iv(S) that is not in $\mathcal{A}$ (if such an assignment exists). If $S^{\pm}$ contains any positive constraints, then θ has an empty set of extensions to these constraints, and hence is extension equivalent to θ_0. Otherwise, any extension of θ will satisfy all negative constraints in $S^{\pm}$, so the extensions of θ that satisfy join(S)

are completely determined by the counting function K_θ. In this case θ will be extension equivalent to some element of $\mathcal{K}$.

Moreover, the set of assignments $\mathcal{A}$ can be computed from S in polynomial time in the the the size of $\mathrm{iv}(S)$ and the total size of the descriptions of the constraints in $S^\pm$. Therefore, $\mathcal{C}$ is a cooperating catalogue as described in Definition 14.

Example 11. By Theorem 2, the constraints in Example 1 form a cooperating catalogue.

5 Polynomial-Time Reductions

In this section, we will show that, for any constraint problem over a cooperating catalogue, a set of variables that all occur in exactly the same set of constraint scopes can be replaced by a single new variable with an appropriate domain, to give a polynomial-time reduction to a smaller problem.

Definition 17 (Dual of a hypergraph). *Let $G = \langle V, H \rangle$ be a hypergraph. The dual G^* of G is a hypergraph with vertex set H and a hyperedge $\{h \in H \mid v \in h\}$ for every $v \in V$. For a class $\mathcal{H}$ of hypergraphs, let $\mathcal{H}^* = \{G^* \mid G \in \mathcal{H}\}$.*

Example 12. Consider the hypergraph G in Figure 1. The dual, G^*, of this hypergraph has vertex set $\{C_1, C_2, C_3, C_4, C_5\}$ and five hyperedges $\{C_1, C_2\}$, $\{C_1, C_3\}$, $\{C_2, C_4\}$, $\{C_3, C_4\}$ and $\{C_4, C_5\}$. This transformation is illustrated in Figure 2.

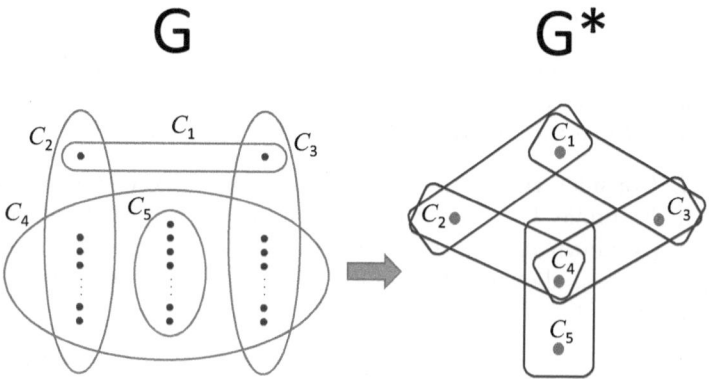

Fig. 2. G and G^* from Example 12

Note that the dual of the dual of a hypergraph is not necessarily the original hypergraph, since we do not allow multiple identical hyperedges.

Example 13. Consider the dual hypergraph G^* defined in Example 12. Taking the dual of this hypergraph yields G^{**}, with vertex set $\{h_1, \ldots, h_5\}$ (corresponding to the 5 hyperedges in G^*) and 5 distinct hyperedges, as shown in Figure 3.

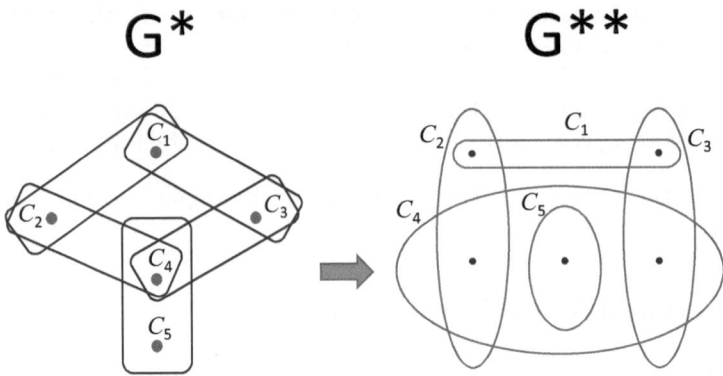

Fig. 3. G^* and G^{**} from Example 13

In the example above, taking the dual of a hypergraph twice had the effect of merging precisely those sets of variables that occur in the same set of hyperedges. It is easy to verify that this is true in general: Taking the dual twice equates precisely those variables that occur in the same set of hyperedges.

Lemma 1. *For any hypergraph G, the hypergraph G^{**} has precisely one vertex corresponding to each maximal subset of vertices of G that occur in the same set of hyperedges.*

Next, we combine the idea of the dual with the usual notion of treewidth to create a new measure of width.

Definition 18 (twDD). *Let G be a hypergraph. The treewidth of the dual of the dual (twDD) of G is* twDD$(G) = $ tw(G^{**}).
For a class of hypergraphs $\mathcal{H}$, we define twDD$(\mathcal{H}) = $ tw$(\mathcal{H}^{**})$.

Example 14. Consider the class $\mathcal{H}$ of hypergraphs of the family of problems described in Example 1. Whatever the value of n, the dual hypergraph, G^* is the same, as shown in Figure 2. Hence for all problems in this family the hypergraph G^{**} is as shown in Figure 3, and can be shown to have treewidth 3. Hence twDD$(\mathcal{H}) = 3$.

When replacing a set of variables in a CSP instance with a single variable, we will use the following definition.

Definition 19 (Quotient of a CSP instance). *Let $P = \langle V, C \rangle$ be a CSP instance and $X \subseteq V$ be a non-empty subset of variables that all occur in the scopes of the same set S of constraints. The quotient of P with respect to X, denoted P^X, is defined as follows.*

- *The variables of P^X are given by $V^X = (V - X) \cup \{v_X\}$, where v_X is a fresh variable, and the domain of v_X is the set of equivalence classes of* equiv[join$(S), X$].

– *The constraints of P^X are unchanged, except that each constraint $e[\delta] \in S$ is replaced by a new constraint $e^X[\delta^X]$, where $\mathsf{vars}(\delta^X) = (\mathsf{vars}(\delta) - X) \cup \{v_X\}$. For any assignment θ of $\mathsf{vars}(\delta^X)$, we define $e^X[\delta^X](\theta)$ to be 1 if and only if $\theta|_{\mathsf{vars}(\delta)-X} \oplus \mu \in e[\delta]$, where μ is a representative of the equivalence class $\theta(v_X)$.*

We note that, by Definition 11, the value of $e^X[\delta^X]$ specified in Definition 19 is well-defined, that is, it does not depend on the specific representative chosen for the equivalence class $\theta(v_X)$, since each representative has the same set of possible extensions.

Lemma 2. *Let $P = \langle V, C \rangle$ be a CSP instance and $X \subseteq V$ be a non-empty subset of variables that all occur in the scopes of the same set of constraints. The instance P^X has a solution if and only if P has a solution.*

Proof (Sketch). Let $P = \langle V, C \rangle$ and X be given, and let $S \subseteq C$ be the set of constraints $e[\delta]$ such that $X \subseteq \mathsf{vars}(\delta)$.

Construct the instance P^X as specified in Definition 19. Any solution to P can be converted into a corresponding solution for P^X, and vice versa. This conversion process just involves replacing the part of the solution assignment that gives values to the variables in the set X with an assignment that gives a suitable value to the new variable v_X.

Theorem 3. *Any CSP instance P can be converted to an instance P' with $\mathsf{hyp}(P') = \mathsf{hyp}(P)^{**}$, such that P' has a solution if and only if P does. Moreover, if P is over a cooperating catalogue, this conversion can be done in polynomial time.*

Proof. Let $P = \langle V, C \rangle$ be a CSP instance. For each variable $v \in V$ we define $S(v) = \{e[\delta] \in C \mid v \in \mathsf{vars}(\delta)\}$ We then partition the vertices of P into subsets $X_1, \ldots, X_k$, where each X_i is a maximal subset of variables v that share the same value for $S(v)$.

We initially set $P_0 = P$. Then, for each X_i in turn, we set $P_i = (P_{i-1})^X$. Finally we set $P' = P_k$. By Lemma 1, $\mathsf{hyp}(P') = \mathsf{hyp}(P)^{**}$, and by Lemma 2, P' has a solution if and only if P has a solution.

Finally, if P is over a cooperating catalogue, then by Definition 14, we can compute the domains of each new variable introduced in polynomial time in the size of each X_i and the total size of the constraints. Hence we can compute P' in polynomial time.

Using Theorem 3, we can immediately get a new tractable CSP class by extending Theorem 1.

Theorem 4. *Let $\mathcal{C}$ be a constraint catalogue and $\mathcal{H}$ a class of hypergraphs. $\mathsf{CSP}(\mathcal{H}, \mathcal{C})$ is tractable if $\mathcal{C}$ is a cooperating catalogue and $\mathsf{twDD}(\mathcal{H}) < \infty$.*

Proof. Let $\mathcal{C}$ be a cooperating catalogue, $\mathcal{H}$ a class of hypergraphs such that $\mathsf{twDD}(\mathcal{H}) < \infty$, and $P \in \mathsf{CSP}(\mathcal{H}, \mathcal{C})$. Reduce P to a CSP instance P' using

Theorem 3. By Definition 18, since $\mathsf{hyp}(P') = \mathsf{hyp}(P)^{**}$, $\mathsf{tw}(\mathsf{hyp}(P')) < \infty$, which means that P' satisfies the conditions of Theorem 1, and hence can be solved in polynomial time.

Recall the family of constraint problems described in Example 1 at the start of this paper. Since the constraints in this problem form a cooperating catalogue (Example 11), and all instances have bounded twDD (Example 14), this family of problems is tractable by Theorem 4.

6   Summary and Future Work

We have identified a novel tractable class of constraint problems with global constraints. In fact, our results generalize several previously studied classes of problems [12]. Moreover, this is the first representation-independent tractability result for constraint problems with global constraints.

Our new class is defined by restricting both the nature of the constraints and the way that they interact. As demonstrated in Example 5, instances with a single global constraint may already be NP-complete [26], so we cannot hope to achieve tractability by structural restrictions alone. In other words, notions such as bounded degree of cyclicity [21] or bounded hypertree width [18] are not sufficient to ensure tractability in the framework of global constraints, where the arity of individual constraints is unbounded. This led us to introduce the notion of a cooperating constraint catalogue, which is sufficiently restricted to ensure that an individual constraint is always tractable.

However, this restriction on the nature of the constraints is still not enough to ensure tractability on any structure: Example 6 demonstrates that not all structures are tractable even with a cooperating constraint catalogue. In fact, a family of problems with acyclic structure (hypertree width one) over a cooperating constraint catalogue can still be NP-complete. This led us to investigate restrictions on the structure that are sufficient to ensure tractability for all instances over a cooperating catalogue. In particular, we have shown that it is sufficient to ensure that the dual of the dual of the hypergraph of the instance has bounded treewidth.

An intriguing open question is whether there are other restrictions on the nature of the constraints or the structure of the instances that are sufficient to ensure tractability in the framework of global constraints. Very little work has been done on this question, apart from the pioneering work of Bulatov and Marx [8], which considered only a single global cardinality constraint, along with arbitrary table constraints, and of Chen and Dalmau [9] on two specific succinct representations. Almost all other previous work on tractable classes has considered only table constraints. This may be one reason why such work has had little practical impact on the design of constraint solvers, which rely heavily on the use of in-built special-purpose global constraints.

We see this paper as a first step in the development of a more robust and applicable theory of tractability for global constraints.

References

1. Aschinger, M., Drescher, C., Friedrich, G., Gottlob, G., Jeavons, P., Ryabokon, A., Thorstensen, E.: Optimization methods for the partner units problem. In: Achterberg, T., Beck, J.C. (eds.) CPAIOR 2011. LNCS, vol. 6697, pp. 4–19. Springer, Heidelberg (2011)
2. Aschinger, M., Drescher, C., Gottlob, G., Jeavons, P., Thorstensen, E.: Structural decomposition methods and what they are good for. In: Schwentick, T., Dürr, C. (eds.) Proc. STACS 2011. LIPIcs, vol. 9, pp. 12–28 (2011)
3. Beeri, C., Fagin, R., Maier, D., Yannakakis, M.: On the desirability of acyclic database schemes. Journal of the ACM 30, 479–513 (1983)
4. Beldiceanu, N.: Pruning for the minimum constraint family and for the number of distinct values constraint family. In: Walsh, T. (ed.) CP 2001. LNCS, vol. 2239, pp. 211–224. Springer, Heidelberg (2001)
5. Bessiere, C., Hebrard, E., Hnich, B., Walsh, T.: The complexity of reasoning with global constraints. Constraints 12(2), 239–259 (2007)
6. Bessiere, C., Katsirelos, G., Narodytska, N., Quimper, C.-G., Walsh, T.: Decomposition of the NValue constraint. In: Cohen, D. (ed.) CP 2010. LNCS, vol. 6308, pp. 114–128. Springer, Heidelberg (2010)
7. Bulatov, A., Jeavons, P., Krokhin, A.: Classifying the complexity of constraints using finite algebras. SIAM Journal on Computing 34(3), 720–742 (2005)
8. Bulatov, A.A., Marx, D.: The complexity of global cardinality constraints. Logical Methods in Computer Science 6(4(4)), 1–27 (2010)
9. Chen, H., Grohe, M.: Constraint satisfaction with succinctly specified relations. Journal of Computer and System Sciences 76(8), 847–860 (2010)
10. Cohen, D.A., Green, M.J., Houghton, C.: Constraint representations and structural tractability. In: Gent, I.P. (ed.) CP 2009. LNCS, vol. 5732, pp. 289–303. Springer, Heidelberg (2009)
11. Cohen, D.A., Jeavons, P., Gyssens, M.: A unified theory of structural tractability for constraint satisfaction problems. Journal of Computer and System Sciences 74(5), 721–743 (2008)
12. Dalmau, V., Kolaitis, P.G., Vardi, M.Y.: Constraint satisfaction, bounded treewidth, and finite-variable logics. In: Van Hentenryck, P. (ed.) CP 2002. LNCS, vol. 2470, pp. 223–254. Springer, Heidelberg (2002)
13. Dechter, R., Pearl, J.: Tree clustering for constraint networks. Artificial Intelligence 38(3), 353–366 (1989)
14. Freuder, E.C.: Complexity of k-tree structured constraint satisfaction problems. In: Proc. AAAI, pp. 4–9. AAAI Press / The MIT Press (1990)
15. Garey, M.R., Johnson, D.S.: Computers and Intractability: A Guide to the Theory of NP-Completeness. W.H. Freeman (1979)
16. Gent, I.P., Jefferson, C., Miguel, I.: MINION: A fast, scalable constraint solver. In: Proc. ECAI 2006, pp. 98–102. IOS Press (2006)
17. Gottlob, G., Leone, N., Scarcello, F.: A comparison of structural CSP decomposition methods. Artificial Intelligence 124(2), 243–282 (2000)
18. Gottlob, G., Leone, N., Scarcello, F.: Hypertree decompositions and tractable queries. Journal of Computer and System Sciences 64(3), 579–627 (2002)
19. Green, M.J., Jefferson, C.: Structural tractability of propagated constraints. In: Stuckey, P.J. (ed.) CP 2008. LNCS, vol. 5202, pp. 372–386. Springer, Heidelberg (2008)

20. Grohe, M.: The complexity of homomorphism and constraint satisfaction problems seen from the other side. Journal of the ACM 54(1), 1–24 (2007)
21. Gyssens, M., Jeavons, P.G., Cohen, D.A.: Decomposing constraint satisfaction problems using database techniques. Artificial Intelligence 66(1), 57–89 (1994)
22. Hermenier, F., Demassey, S., Lorca, X.: Bin repacking scheduling in virtualized datacenters. In: Lee, J. (ed.) CP 2011. LNCS, vol. 6876, pp. 27–41. Springer, Heidelberg (2011)
23. van Hoeve, W.J., Katriel, I.: Global constraints. In: Rossi, F., van Beek, P., Walsh, T. (eds.) Handbook of Constraint Programming, Foundations of Artificial Intelligence, vol. 2, ch. 6, pp. 169–208. Elsevier (2006)
24. Kutz, M., Elbassioni, K., Katriel, I., Mahajan, M.: Simultaneous matchings: Hardness and approximation. Journal of Computer and System Sciences 74(5), 884–897 (2008)
25. Marx, D.: Tractable hypergraph properties for constraint satisfaction and conjunctive queries. In: Proc. STOC 2010, pp. 735–744. ACM (2010)
26. Quimper, C.G., López-Ortiz, A., van Beek, P., Golynski, A.: Improved algorithms for the global cardinality constraint. In: Wallace, M. (ed.) CP 2004. LNCS, vol. 3258, pp. 542–556. Springer, Heidelberg (2004)
27. Rosen, K.H., Michaels, J.G., Gross, J.L., Grossman, J.W., Shier, D.R. (eds.): Handbook of Discrete and Combinatorial Mathematics. Discrete Mathematics and Its Applications. CRC Press (2000)
28. Rossi, F., van Beek, P., Walsh, T. (eds.): The Handbook of Constraint Programming. Elsevier (2006)
29. Samer, M., Szeider, S.: Tractable cases of the extended global cardinality constraint. Constraints 16(1), 1–24 (2011)
30. Wallace, M.: Practical applications of constraint programming. Constraints 1, 139–168 (1996)
31. Wallace, M., Novello, S., Schimpf, J.: ECLiPSe: A platform for constraint logic programming. ICL Systems Journal 12(1), 137–158 (1997)

Postponing Optimization
to Speed Up MAXSAT Solving

Jessica Davies[1] and Fahiem Bacchus[2]

[1] MIAT, UR 875, INRA, F-31326 Castanet-Tolosan, France
jessica.davies@toulouse.inra.fr
[2] Department of Computer Science, University of Toronto,
Toronto, Ontario, Canada, M5S 3H5
fbacchus@cs.toronto.edu

Abstract. MAXSAT is an optimization version of SAT that can represent a wide variety of important optimization problems. A recent approach for solving MAXSAT is to exploit both a SAT solver and a Mixed Integer Programming (MIP) solver in a hybrid approach. Each solver generates information used by the other solver in a series of iterations that terminates when an optimal solution is found. Empirical results indicate that a bottleneck in this process is the time required by the MIP solver, arising from the large number of times it is invoked. In this paper we present a modified approach that postpones the calls to the MIP solver. This involves substituting non-optimal solutions for the optimal ones computed by the MIP solver, whenever possible. We describe the new approach and some different instantiations of it. We perform an extensive empirical evaluation comparing the performance of the resulting solvers with other state-of-the-art MAXSAT solvers. We show that the best performing versions of our approach advance the state-of-the-art in MAXSAT solving.

1 Introduction

MAXSAT, the optimization version of Satisfiability (SAT), is the problem of finding a minimum cost truth assignment for a set of clauses where a cost is incurred for every falsified clause. It is called MAXSAT since in the simplest case where every clause is equally costly to falsify, a solution will satisfy a maximum number of clauses. In the most general version of MAXSAT some clauses are *hard* incurring an infinite cost if they are falsified, while the other clauses are *soft* incurring some integer cost greater than zero. This most general version of MAXSAT is often called **weighted partial MAXSAT** (WPMS) and is what we address in this paper. Many practical problems can be encoded in MAXSAT, so developing effective ways to solve MAXSAT is an important research topic.

There are two standard methods for solving MAXSAT: using Branch and Bound search (e.g. [9,14]), and using a sequence of decision problems, usually encoded as SAT (e.g. [2,3,5,15,10]). In [6] an alternative algorithm for solving MAXSAT, called MAXHS, was presented. MAXHS also solves a sequence of SAT decision problems, but in contrast to existing approaches the SAT problems do not become more difficult for the SAT solver to solve. This is accomplished via a hybrid approach,

C. Schulte (Ed.): CP 2013, LNCS 8124, pp. 247–262, 2013.

whereby a SAT solver and a Mixed Integer Linear Program (MIP) solver are used to cooperatively solve the MAXSAT problem using an approach similar to Bender's Decomposition [11]. The MIP solver is used to find optimal solutions which the SAT solver then tests for feasibility. If the solution is not feasible the SAT solver computes a new constraint to add to the MIP model and the MIP solver is invoked again to find a new optimal solution that additionally satisfies the new constraint.

In this paper we investigate a new technique for improving the performance of this hybrid approach. In [7] it has already been shown that the hybrid approach is one of the state-of-the-art approaches to solving MAXSAT, and thus improving this approach is one way of advancing the state-of-the-art.

Analyzing the performance of the hybrid approach indicates that the main bottleneck is the time spent by the MIP solver. This time mostly accumulates from the number of times that the MIP solver must be called: it is called every time the SAT solver computes a new feasibility constraint, in order to derive a new optimal solution satisfying this additional constraint. Although these calls often take relatively little time, after hundreds of separate calls the total time becomes quite significant.

Inspired by an idea presented by Moreno-Centeno and Karp [16] we developed a method for delaying the calls to the MIP solver for as long as possible. We accomplish this by recognizing situations where non-optimal solutions can be used in place of the optimal solutions produced by the MIP solver without impacting the algorithm's correctness. The SAT solver can use these non-optimal solutions to compute its feasibility constraints and the iterations of feasibility and "optimization" can continue. However, since the optimization phase is now an approximation that can be computed cheaply, each iteration is much more efficient. Eventually, however, an optimal solution must be computed to ensure correctness. So our technique postpones, rather than removes, optimization.

We show that our new technique yields a significant improvement in the performance of the hybrid algorithm and makes it the most robust current approach to solving MAXSAT. One of the reasons why we obtain such a good performance improvement is that the MIP solver is not perfectly incremental. By using non-optimal solutions we collect many feasibility constraints before having to compute their optimal solution. This means that each additional call to the MIP solver involves a model that has been augmented by many feasibility constraints, whereas in the previous approach the model was only augmented by a single feasibility constraint. Although the MIP solver can take advantage of its previous computations when called again, it is not perfectly incremental. That is, its solving time when given k new constraints is typically significantly smaller than the sum of its k solving times when it is given these constraints one at a time and asked to compute an optimal solution each time.

The remainder of the paper is organized as follows. Section 2 provides basic definitions. The MAXHS approach is then reviewed in Section 3 and we show that its main bottleneck is the MIP solving time. Section 4 presents the new algorithm, which addresses this issue by allowing the MIP optimization to be postponed.

An effective additional enhancement, seeding the MIP model with constraints [7], is described in Section 5. The empirical results are reported in Section 6.

2   Background

A MAXSAT instance is specified by a set of propositional clauses $\mathcal{F}$ each having a positive integer or infinite weight $wt(c)$, $c \in \mathcal{F}$. Clauses with infinite weight are called hard clauses and are collectively denoted by $hard(\mathcal{F})$. The other clauses of $\mathcal{F}$ all have finite weight and are called soft clauses, $soft(\mathcal{F})$ ($\mathcal{F} = hard(\mathcal{F}) \cup soft(\mathcal{F})$).

We define the function $cost$ as follows: (a) if H is a set of clauses then $cost(H)$ is the sum of the weights of the clauses in H ($cost(H) = \sum_{c \in H} wt(c)$); and (b) if π is a truth assignment to the variables of $\mathcal{F}$ then $cost(\pi)$ is the sum of the weights of the clauses falsified by π ($\sum_{\{c \mid \pi \not\models c\}} wt(c)$). A solution for $\mathcal{F}$ is a truth assignment π to the variables of $\mathcal{F}$ that satisfies $hard(\mathcal{F})$ and is of minimum cost. We let $mincost(\mathcal{F})$ denote the cost of such a solution. If $hard(\mathcal{F})$ is unsatisfiable, then $\mathcal{F}$ has no solution. In the remainder of the paper, we assume that $hard(\mathcal{F})$ **is satisfiable** (this is easy to test in practice and facilitates clarity). A **core** κ of a MAXSAT formula $\mathcal{F}$ is a subset of $soft(\mathcal{F})$ such that $\kappa \cup hard(\mathcal{F})$ is unsatisfiable. Note that since $hard(\mathcal{F})$ is satisfiable, any solution to $\mathcal{F}$ must falsify at least one clause in κ.

MAXSAT solvers that solve a sequence of decision problems typically insert blocking variables (b-variables) into the soft clauses of the MAXSAT instance.

Definition 1. *If $\mathcal{F}$ is a MAXSAT problem, then its b-**variable relaxation** is a SAT problem $\mathcal{F}^b = \{(c_i \vee b_i) : c_i \in soft(\mathcal{F})\} \cup hard(\mathcal{F})$ where all clause weights are removed. The b-variable b_i appears in the relaxed clause $(c_i \vee b_i)$ and **no** where else in $\mathcal{F}^b$.*

The b-variable relaxation $\mathcal{F}^b$ allows cores of the original MAXSAT formula $\mathcal{F}$ to be computed conveniently, using the **Assumption** mechanism provided by MINISAT [8]. MINISAT can take as input a set of assumptions $\mathcal{A}$, specified as a set of literals, along with a CNF formula $\mathcal{F}$ and then determine if $\mathcal{F} \wedge \mathcal{A}$ is satisfiable. It will return a satisfying truth assignment for $\mathcal{F} \wedge \mathcal{A}$ if one exists. Otherwise it will report unsatisfiability and return a learnt clause c which is a disjunction of negated literals of $\mathcal{A}$. This clause has the property that $\mathcal{F} \models c$. Thus in order to find a core of MAXSAT instance $\mathcal{F}$, we can pass MINISAT the CNF formula $\mathcal{F}^b$ and the set of all negated b-variables as the assumptions. If $\mathcal{F}$ has a core, MINISAT will return UNSAT along with a clause $c = (b_{i_1} \vee \cdots \vee b_{i_k})$ such that $\mathcal{F}^b \models c$ and $\kappa = \{c_{i_1}, ..., c_{i_k}\}$ is a core of $\mathcal{F}$. Any clause over positive b-variables that is entailed by $\mathcal{F}^b$, e.g., $c = (b_{i_1} \vee \cdots \vee b_{i_k})$, is called a **core constraint**.

Besides supporting the computation of cores, the relaxed formula $\mathcal{F}^b$ can also be used to find solutions for the original MAXSAT formula $\mathcal{F}$. To accomplish this we define an objective function $bcost(\pi)$ over truth assignments π to the variables of $\mathcal{F}^b$, equal to the sum of the costs of the clauses whose b-variables are set to *true*: $bcost(\pi) = \sum_{b_i : \pi \models b_i} wt(c_i)$. The minimum *bcost* models of $\mathcal{F}^b$ are MAXSAT solutions.[1]

Proposition 1. $mincost(\mathcal{F}) = \min_{\pi \models \mathcal{F}^b} bcost(\pi)$. *Furthermore, if* $\pi \models \mathcal{F}^b$ *achieves a minimum value of* $bcost(\pi)$, *then* π *restricted to the variables of* $\mathcal{F}$ *is a solution for* $\mathcal{F}$.

However, $\mathcal{F}^b$ has many models whose *bcost* is greater than necessary. For example, if a model π assigns b_i to *true* even though it also satisfies the soft clause c_i, then the *bcost* of π could be reduced by instead assigning b_i to *false*. We can eliminate such models by modifying $\mathcal{F}^b$ so that the *b*-variables are forced to be equivalent to the negation of their corresponding soft clauses.

Definition 2. *Let* $\mathcal{F}$ *be a* MAXSAT *formula. Then*

$$\mathcal{F}^b_{eq} = \mathcal{F}^b \cup \bigcup_{c_i \in soft(\mathcal{F})} \{(\neg b_i \vee \neg \ell) : \ell \in c_i\}$$

is the relaxation of $\mathcal{F}$ *with b-variable equivalences.*

Again, the minimum *bcost* models of $\mathcal{F}^b_{eq}$ are MAXSAT solutions.

Proposition 2. $mincost(\mathcal{F}) = \min_{\pi \models \mathcal{F}^b_{eq}} bcost(\pi)$. *Furthermore, if* $\pi \models \mathcal{F}^b_{eq}$ *achieves a minimum value of* $bcost(\pi)$, *then* π *restricted to the variables of* $\mathcal{F}$ *is a solution for* $\mathcal{F}$.

Propositions 1 and 2 show that we can solve the MAXSAT problem $\mathcal{F}$ by searching for a *bcost* minimal satisfying assignment to $\mathcal{F}^b$ or to $\mathcal{F}^b_{eq}$. Note also that $\mathcal{F}^b_{eq}$ is a stronger theory enabling more inferences than $\mathcal{F}^b$.

3   The MAXHS Approach

In MAXHS [6] a SAT solver is called at every iteration to find a new core of $\mathcal{F}$. A MIP solver is then invoked to find a minimum *bcost* assignment to the *b*-variables that satisfies all of the core constraints found so far. This optimization problem corresponds to a Minimum Cost Hitting Set (MCHS) problem,[2] where the goal is to find a minimum cost set of clauses that hits each of the known cores. The optimal hitting set found by the MIP solver is tested by the SAT solver. If the SAT solver is unable to find another core, it means the MAXSAT solution has been found. Otherwise, the SAT solver returns a new core, the MIP solver re-optimizes, and the iterations continue.

This algorithm is shown in Algorithm 1. The set of cores is initialized on line 2 to the empty set. In the main loop of MAXHS, the MIP solver is invoked to find

---

[1] Recall we assume $hard(\mathcal{F})$ is satisfiable, so $\mathcal{F}^b$ is satisfiable.

[2] An instance of MCHS is given by a universe of weighted elements U and a collection of subsets of these elements, $\mathcal{K} = \{\kappa_1, ..., \kappa_m\}$, $\kappa_i \subseteq U$. The goal is to find a minimum weight set of elements $hs \subseteq U$ such that $hs \cap \kappa_i \neq \emptyset$ for all $\kappa_i \in \mathcal{K}$.

Algorithm 1. The MAXHS algorithm for solving MAXSAT.

---

1 **MaxHS** $(\mathcal{F})$
2 $\mathcal{K} = \emptyset$
3 **while** *true* **do**
4      $hs = \text{FindMinCostHittingSet}(\mathcal{K})$
5      $(\text{sat?},\kappa) = \text{SatSolver}(\mathcal{F} \setminus hs)$
        // If SAT, κ contains the satisfying truth assignment.
        // If UNSAT, κ is a new core.
6   **if** *sat?* **then**
7   **break** // Exit While Loop, κ is a MAXSAT solution.
8      $\mathcal{K} = \mathcal{K} \cup \{\kappa\}$
9 **return** $\big(\kappa,\ cost(\kappa)\big)$

---

a minimum cost hitting set hs of $\mathcal{K}$, on line 4. If removing hs from $\mathcal{F}$ results in a satisfiable formula (tested by the SAT solver on line 5), we break out of the loop on line 7 and return the satisfying assignment as the MAXSAT solution on line 9. Otherwise, the SAT solver will return a new core, κ to add to $\mathcal{K}$ on line 8 and the loop repeats.

The correctness of Algorithm 1 is established by the following theorem.

Theorem 1. *[6] If $\mathcal{K}$ is a set of cores for the MAXSAT problem $\mathcal{F}$, hs is a minimum cost hitting set of $\mathcal{K}$, and π is a truth assignment satisfying $\mathcal{F} \setminus hs$ then $mincost(\mathcal{F}) = cost(\pi) = cost(hs)$.*

This theorem shows that when Algorithm 1 breaks out of its loop, κ is a MAXSAT solution. The argument that the loop must eventually terminate is based on observing that every time the SAT solver returns a core κ, it must be distinct from all previously returned cores (because a hitting set for all previous cores does not hit κ). Since there is a finite number of distinct cores, the SAT solver must eventually be unable to find another new core and the loop will terminate.

3.1 Behaviour of MAXHS

The behaviour of MAXHS in influenced by three potential sources of exponential complexity. These include the time required by the SAT solver to solve $\mathcal{F} \setminus hs$, the time required by the MIP solver to solve the NP-hard MCHS problem, and the number of iterations required. The examples below illustrate that each of these factors can, in the worst case, cause exponential runtime.

Example 1. Let $\mathcal{F}$ be an instance of the Pigeon Hole Principle, where all clauses are considered soft with weight 1. Removing any single clause from $\mathcal{F}$ will make the remaining clauses satisfiable. Therefore, MAXHS will terminate after the first core is found. So only one MCHS problem will be solved, and it is trivial. However, the time spent by the SAT solver to find a single core will be exponential.

Example 2. Let $\mathcal{K}$ be a MCHS instance. We construct a MAXSAT instance $\mathcal{F}$ that is equivalent to $\mathcal{K}$ as follows. For each set $\kappa \in \mathcal{K}$, where $\kappa = \{e_1, ..., e_k\}$, there is a hard clause $(e_1 \vee \cdots \vee e_k)$. Finally, there is a soft clause $(\neg e)$ with weight $wt(e)$ for each element $e \in \bigcup_{\kappa \in \mathcal{K}} \kappa$. A minimal core is a core such that any proper subset is not a core. It is easy to see that the minimal cores of $\mathcal{F}$ correspond to the hard clauses of $\mathcal{F}$ and therefore the total number of minimal cores is equal to $|\mathcal{K}|$. The SAT solver can find each of the minimal cores in polynomial time, by using unit propagation alone. The number of minimal cores required by MAXHS is at most $|\mathcal{K}|$. So the only possible source of exponential runtime on $\mathcal{K}$ is solving the MCHS problems. Assuming that P $\neq$ NP, there must be some MCHS instance $\mathcal{K}$ on which MAXHS will take exponential time and this must arise when MAXHS solves the MCHS problem.

To show that exponential run time can be generated from the number of iterations required we need the following proposition.

Proposition 1. *Let n be an even number and let $E = \{e_1, ..., e_n\}$ be a universe of equally weighted elements. Let $\mathcal{K}_{n,r} = \{\kappa \subset E : |\kappa| = r\}$ be an instance of the MCHS problem where $r = \frac{n}{2}$. Let $\mathcal{K}' = \mathcal{K}_{n,r} \setminus \kappa'$ for some $\kappa' \in \mathcal{K}_{n,r}$. Then the MCHS of $\mathcal{K}'$ is strictly smaller than the MCHS of $\mathcal{K}_{n,r}$.*

Example 3. Let $\mathcal{F}$ be a MAXSAT instance with an even number n of soft unit clauses with weight 1, $(x_1), ..., (x_n)$ and let the hard clauses of $\mathcal{F}$ form a CNF encoding of the cardinality constraint $\Sigma_{i=1}^n x_i < n/2$. On this family of problems, an exponential number of cores will always be required by MAXHS, as we explain next. The solutions to $\mathcal{F}$ are the truth assignments that set as many of the variables to *true* as possible without violating the hard cardinality constraint. Thus a solution to $\mathcal{F}$ will set exactly $\frac{n}{2} - 1$ of the x_i variables to *true* and the rest to *false*, and $\frac{n}{2} + 1$ is the optimal cost. Any subset of the n soft clauses, with size greater than or equal to $\frac{n}{2}$, is a core of $\mathcal{F}$. Therefore, $\mathcal{F}$ has at least $\binom{n}{n/2}$ cores. By Proposition 1, for any number of cores $k < \binom{n}{n/2}$, the cost of their MCHS is less than the optimum. Therefore, MAXHS will require at least $\binom{n}{n/2}$ cores, which is exponential in n.

However, our empirical observations are much more encouraging. In practice, we find that the SAT solving time is typically small.[3] Instead, the performance of MAXHS is most affected by the number of iterations and the time to solve the MCHS problems. Histograms of the percentage of total runtime spent by the SAT solver and the MIP solver are shown in Figure 1 over a set of 4502 Industrial and Crafted instances (the details of the experimental setup are described in Section 6). In order to study the baseline behaviour of Algorithm 1, the improvements presented in prior work [6] are omitted from this implementation. We observe in Figures 1b and 1d that on instances MAXHS failed to solve within

---

[3] If a MAXSAT instance is difficult for a state-of-the-art SAT solver to refute, then any MAXSAT solver that uses a sequence of SAT instance approach will be unable to solve it efficiently.

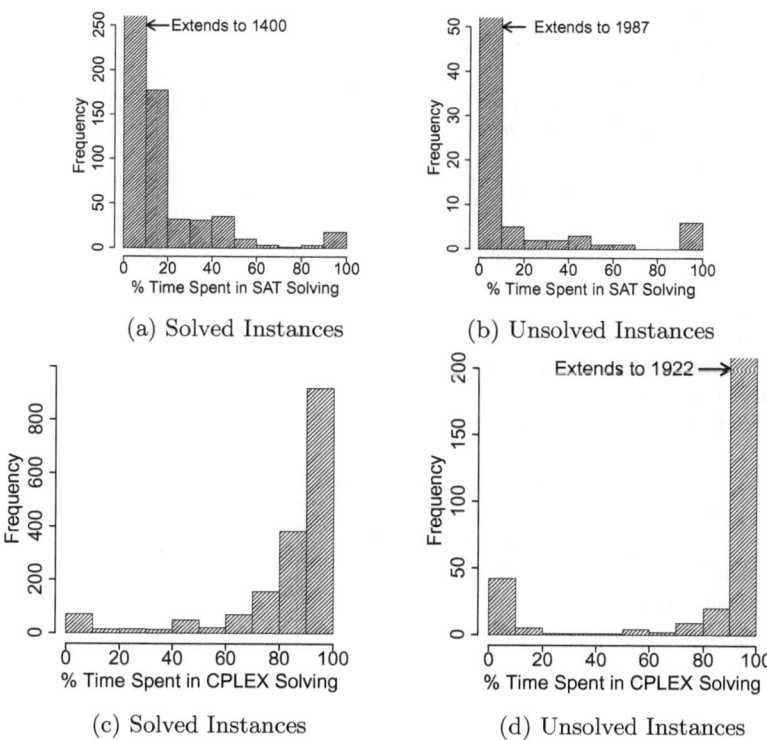

Fig. 1. Histograms over 4502 instances of the percentage of runtime spent in SAT solving and in calls to the MIP solver CPLEX, for Algorithm 1

the resource limits, the time spent by CPLEX is a much larger proportion of the total runtime than the time spent on SAT solving. This is true of the solved instances as well, as shown in Figures 1a and 1c. Thus we are motivated to find ways to reduce the time spent solving the MCHS problems, since this has the greatest potential to reduce the total runtime and thus allow more instances to be solved.

4   Postponing Optimization

We have seen in the previous section that in practice, the execution time of Algorithm 1 is dominated by its multiple calls to the MIP solver. The MIP solver must optimize an NP-hard problem, Minimum Cost Hitting Set, at each iteration. Therefore, in order to improve the performance of MAXHS it is natural to ask if an approximation to MCHS can ever be used instead. In this section we show that by applying a similar approach to [16] the MAXHS algorithm can be modified to use such approximations, in order to postpone the expensive calls to the MIP solver.

Algorithm 2. An algorithm for solving MAXSAT that uses non-optimal hitting sets.

---

1 **MaxHS-nonOPT** $(\mathcal{F})$
2 $\mathcal{K} = \text{DisjointCores}(\mathcal{F})$
3 **while** *true* **do**
4     $hs = \text{FindMinCostHittingSet}(\mathcal{K})$ /* Find **optimal** solution */
5     $(\text{sat?}, \kappa) = \text{SatSolver}(\mathcal{F} \setminus hs)$
        // If SAT, κ contains the satisfying truth assignment.
        // If UNSAT, κ is a new core.
6   **if** *sat?* **then**
7 | **break** // Exit While Loop, κ is a MAXSAT solution.
8     $\kappa = \text{Minimize}(\kappa)$
9     $\mathcal{K} = \mathcal{K} \cup \{\kappa\}$
10     $nonOptLevel = 0$
        // Begin a series of non-optimal solutions
11   **while** *true* **do**
12   **switch** *nonOptLevel* **do**
13   **case** *0*
14 | $hs = \text{FindIncrementalHittingSet}(\mathcal{K}, \kappa, hs)$
15   **case** *1*
16 | $hs = \text{FindGreedyHittingSet}(\mathcal{K})$
17         $(\text{sat?}, \kappa) = \text{SatSolver}(\mathcal{F} \setminus hs)$
18   **if** *sat?* **then**
19   **switch** *nonOptLevel* **do**
20   **case** *0*
21 | $nonOptLevel= 1$
22   **case** *1*
23 | **break**   /* Exit inner while loop */
24   **else**
25             $\kappa = \text{Minimize}(\kappa)$
26             $\mathcal{K} = \mathcal{K} \cup \{\kappa\}$
27             $nonOptLevel= 0$
28 **return** $(\kappa, cost(\kappa))$

The MAXHS algorithm from Algorithm 1 can be modified to use non-optimal hitting set computations, as shown in Algorithm 2. The algorithm operates just like MAXHS in that it terminates at line 7 if $\mathcal{F}$ is satisfiable after removing from it an **optimal** hitting set (computed at line 4). It varies from Algorithm 1 in that if a new core κ is discovered at line 5, it enters an inner loop where non-optimal hitting sets are used in place of optimal ones.

A simpler version of the algorithm uses only one method for computing approximate hitting sets rather than two. This version, which we explain first, is obtained from the version shown, by (1) replacing the **switch** statement on lines 12–16 by only one of lines 14 or 16 (i.e., we perform a single type of approximate hitting set computation); and (2) replacing the **switch** statement on lines 19–23 with a **break** statement. The resultant simpler version repeatedly finds an approximate hitting set and calls the SAT solver again to find a new core.

Eventually, the SAT solver will fail to find any more cores, the inner loop will be terminated, and we will return to line 4 where an optimal hitting set will then be computed. The simple version thus finds as many cores as possible before optimizing.

The more complex version (as shown) uses two levels of approximation: incremental and greedy. It is assumed that the second level (on line 16) computes a better (i.e., smaller) hitting set than the first. The idea here is that we compute a cheap incremental approximate hitting set until the SAT solver can't find any more cores. Then we compute the more expensive greedy approximate hitting set, which because it can be smaller might allow the SAT solver to find a new core. If a new core is found, we continue with the cheap incremental approximation (on line 27 *nonOptLevel* is reset to zero), until we once again fail to find cores with the SAT solver. If the SAT solver fails to find a new core even when using the more expensive greed approximate hitting set (line 23), the inner while loop terminates and we finally return to line 4 to compute an optimal hitting set.[4]

Two other improvements to the basic Algorithm 1 are worth mentioning (originally presented in [7]). First, at line 2, we can use the SAT solver to find a set of disjoint cores. This is accomplished by blocking every clause in the cores found so far (by setting the *b*-variables for the cores' clauses to *true*) and finding another core: the new core will not have any clauses in common with the previous cores. This can only be done at the start before the main loop finds other cores. Second, at lines 8 and 25, after each core is found we use the SAT solver to minimize it. This is accomplished by using a simple minimal unsatisfiable core (MUS) algorithm [17]. This results in a stronger constraint for the MIP solver.[5]

In our implementation we use two different methods for computing approximate hitting sets, both of which are very cheap. **FindIncrementalHittingSet** (line 14), simply adds a clause in the newest core to the current hitting set. The chosen clause can be any clause in κ: we choose the clause that appears most frequently in the set $\mathcal{K}$ of cores found so far. The intuition for this policy is that it takes away clauses that appear in many known cores, so that the next cores found can not use these clauses and thus are more likely to intersect with only a few of the known cores. The second method of computing non-optimal hitting sets, **FindGreedyHittingSet** (line 16), ignores the current hitting set, and instead applies a standard greedy algorithm for the MCHS problem [12].

Theorem 2. *Algorithm 2 returns a solution to the* MAXSAT *problem* $\mathcal{F}$.

Proof. This proof relies on the same argument as the correctness of Algorithm 1. If the algorithm returns on line 28, it must have broken out of the outer while loop at line 7. In this case, κ is a solution of $\mathcal{F} \setminus hs$ (by line 5), where hs is a minimum cost hitting set of $\mathcal{K}$ (by line 4). $\mathcal{K}$ is a collection of cores of $\mathcal{F}$: it is initialized on line 2, and augmented only on lines 9 and 26 with κ, a core of $\mathcal{F} \setminus hs$ (thus

---

[4] As can be seen from the algorithm specification we could add more cases to the **switch** statements if we had multiple approximation algorithms we wished to use.

[5] There are future possibilities for using upper and lower bounds in the algorithm (e.g., at line 17 if we find a satisfying solution its cost is an upper bound).

κ is also a core of $\mathcal{F}$). Therefore, if the algorithm returns, by Theorem 1, κ is a MAXSAT solution. It remains to show that the algorithm eventually terminates. Each time SatSolver is called (line 5 or line 17), hs is a hitting set of all cores in $\mathcal{K}$. So if SatSolver returns a core $\kappa \subseteq \mathcal{F} \setminus hs$ it must be distinct from all cores in $\mathcal{K}$. There is a finite number of cores, so SatSolver can not return $sat? = false$ forever and therefore both while loops must eventually terminate.

5   Additional Enhancements

In previous work we also investigated an alternative approach to reduce the time MAXHS spends in MCHS solving, that was based on using more general, **non-core**, constraints [7]. In [7] it is shown that a very effective technique is to **seed** CPLEX with many non-core constraints as a preprocessing step. In this section we show how seeding can also be applied before Algorithm 2 in order to achieve the same benefit.

Definition 3. *A non-core constraint for* MAXSAT *instance* $\mathcal{F}$ *is a linear inequality constraint over* b-*variables,* c, *such that* $\mathcal{F}_{eq}^b \models c$.

It is sound to add non-core constraints to the MIP model in the MAXHS algorithm. This is easy to see from Proposition 2, which states that a minimum *bcost* solution to $\mathcal{F}_{eq}^b$ corresponds to a MAXSAT solution, and the fact that the non-core constraints are entailed by $\mathcal{F}_{eq}^b$. We obtain a theorem similar to Theorem 1.

Theorem 3. *If* $\mathcal{K}$ *is a set of core and non-core constraints for the* MAXSAT *problem* $\mathcal{F}$, π *is minimum bcost assignment to the* b-*variables that satisfies* $\mathcal{K}$, *and* π' *is a truth assignment extending* π *and satisfying* $\mathcal{F}^b$ *then* $mincost(\mathcal{F}) = bcost(\pi)$ *and* π' *restricted to the variables of* $\mathcal{F}$ *is a* MAXSAT *solution.*

Proof. Since π is a minimum *bcost* assignment to the b-variables that satisfies $\mathcal{K}$, and all constraints in $\mathcal{K}$ are entailed by $\mathcal{F}_{eq}^b$, by Proposition 2 $mincost(\mathcal{F}) \geq bcost(\pi)$. On the other hand, π' extends π to a satisfying assignment of $\mathcal{F}^b$, and since π' sets the same b-variables as π, we have $bcost(\pi') = bcost(\pi)$. So by Proposition 1, $mincost(\mathcal{F}) \leq bcost(\pi') = bcost(\pi)$. Thus $mincost(\mathcal{F}) = bcost(\pi)$. Finally, π' restricted to the variables of $\mathcal{F}$ is a MAXSAT solution by Proposition 1, since π' is a minimum *bcost* satisfying assignment of $\mathcal{F}^b$.

This theorem allows us to modify Algorithm 2 by adding a preprocessing step that identifies a collection of non-core constraints $\mathcal{N}$ as shown in Algorithm 3. Now, when the MIP solver is invoked to find an optimal solution, it no longer solves a pure MCHS problem because it must take into account the seeded non-core constraints $\mathcal{N}$ in addition to the cores $\mathcal{K}$. On line 6, the MIP solver returns an optimal assignment to the b-variables, $\mathcal{A}$, that is then passed as a set of assumptions to the SAT solver on line 7. Note that the SAT solver uses $\mathcal{F}^b$ as input which allows the settings of the b-variables in $\mathcal{A}$ to relax the right set of soft clauses. By Theorem 3, if the SAT solver returns a satisfying assignment it

---

Algorithm 3. An algorithm for solving MAXSAT that uses non-optimal hitting sets and seeded non-core constraints.

---

1 **MaxHS-nonOPT-seed** $(\mathcal{F})$
2 $\mathcal{K} = \text{DisjointCores}(\mathcal{F})$
3 $\mathcal{N} = \text{NonCoreConstraints}(\mathcal{F}^b_{eq})$
4 $obj = wt(c_i) * b_i + \ldots + wt(c_k) * b_k$
5 **while** *true* **do**
6     $\mathcal{A} = \text{Optimize}(\mathcal{K} \cup \mathcal{N}, \; obj)$ /* Find **optimal** solution */
7   (sat?, κ) = AssumptionSatSolver($\mathcal{F}^b, \mathcal{A}$)
    // Subsequent lines identical to Algorithm 2 lines 6–28

---

corresponds to the MAXSAT solution. Otherwise, the SAT solver will return a new **core** constraint κ. Thus once the main loop begins, no more non-core constraints will be derived (this differs from the previous work on non-core constraints) and the rest of the algorithm proceeds as before. The argument that the algorithm terminates remains the same as well.

Theorem 4. *Algorithm 3 returns a solution to the MAXSAT problem $\mathcal{F}$.*

It remains to specify how a collection of non-core constraints are to be found by NonCoreConstraints. We use Eq-Seeding [7] because it was found to be the most effective overall. In Eq-Seeding, we exploit the equivalence between original literals of $\mathcal{F}$ that appear in soft unit clauses, and their b-variables. In $\mathcal{F}^b_{eq}$, $b_i \equiv \neg x$ if there is a unit soft clause $(x) \in \mathcal{F}$. So to generate a collection of constraints, we consider each clause c of $\mathcal{F}^b$, and check whether each literal in c has an equivalent b-literal (or is itself a b-literal). If so, we can derive a new b-variable constraint from c by replacing every original literal by its equivalent b-literal. This constraint is a clause over the b-variables that is entailed by F^b_{eq} and it can be added to $\mathcal{N}$.

6 Experimental Results

We performed an empirical study of ten existing MAXSAT solvers: CPLEX (version 12.2), WPM1 [1], WPM2 (versions 1 and 2 [3]), BINCD [10], WBO [15], MINIMAXSAT [9], SAT4J [5], AKMAXSAT [13], MAXHS-Orig [6], and MAXHS+ [7]. All of these solvers are able to solve MAXSAT in its most general form, i.e., weighted partial MAXSAT, and thus have the widest range of applicability. Our study includes recently developed solvers utilizing a sequence of SAT approach (BINCD, WPM1, WPM2, MAXHS-Orig and MAXHS+), some older solvers (SAT4J and WBO), and two prominent Branch and Bound based solvers (AKMAXSAT and MINIMAXSAT). Also included is the MIP solver CPLEX, which is invoked after applying a standard translation of MAXSAT to MIP [7].

We experiment with three versions of Algorithm 2, that differ by how the non-optimal hitting sets are computed. The first and second versions use the

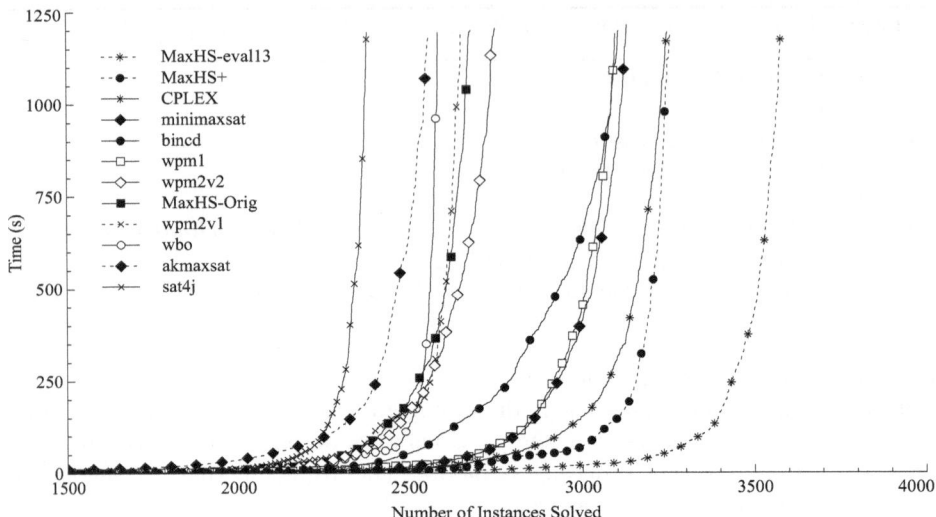

Fig. 2. Performance of solvers on all Crafted and Industrial instances

simple form of Algorithm 2 where only one method for computing approximate hitting sets is used. The first version, **MAXHS-incr**, uses FindIncrementalHittingSet for this computation, and the second version, **MAXHS-greedy**, uses FindGreedyHittingSet. The third version is called **MAXHS-incr-greedy** and it uses the more complex version of Algorithm 2, as specified in the pseudo-code. Finally, we also experiment with the version described in Section 5 that adds Eq-Seeding to MAXHS-incr-greedy. This version uses the same algorithm as the solver submitted to the 2013 MAXSAT Evaluation, and will be called **MAXHS-eval13**.[6]

We obtained **all** problems from the previous seven MAXSAT evaluations [4], discarding all instances in the Random category. After removing duplicate problems (as many as we could find) we ended up with 4502 problems divided into 58 families.[7] The family names and the number of instances in each family are shown in Tables 1 and 2. Our experiments were performed on 2.1 GHz AMD Opteron machines with 98GB RAM shared between 24 cores (about 4GB RAM per core). Each problem was run under a 1200 second timeout and with a memory limit of 2.5GB.

The overall results are shown in Figure 2. We see that the earliest MAXHS-Orig is a reasonable but not distinguished solver. The improvement presented in [7], MAXHS+, is a very good solver being slightly better over all problems than any previous solver. Finally, we see that the best of the versions developed

---

[6] All versions of MAXHS in this study use MINISAT-2.0 and CPLEX version 12.2. The solver submitted to the 2013 Evaluation uses CPLEX version 12.5 and it performs slightly better than the version we report on here.

[7] We include results on 17 families within the Crafted category that were omitted in [7].

Table 1. Results for the Industrial category instances. Shows the number of instances solved by each solver in each benchmark family. The final row gives the total number of instances solved over both the Industrial and Crafted categories.

Family	#	mini	CPLEX	wpm1	bincd	MAXHS Orig	+	Alg. 2 incr	Alg. 2 grdy	Alg. 2 i+g	Alg. 3 eval 13
ms/Safar	112	3	19	**88**	71	75	33	33	34	33	33
ms/circdebug	9	0	1	**9**	7	9	3	4	3	4	3
pms/bcp-fir	59	13	**58**	53	55	16	18	32	18	33	33
pms/bcp-msp	148	108	110	60	117	62	121	90	87	95	**123**
pms/bcp-mtg	215	208	193	**215**	**215**	150	212	**215**	214	**215**	215
pms/bcp-syn	74	27	**71**	40	45	65	**71**	69	69	69	**71**
pms/pb/logic-syn	17	2	**16**	7	7	16	16	16	16	16	16
pms/pbo-rout	15	14	14	**15**	**15**	10	13	12	10	10	13
pms/pseudo/rout	15	14	**15**	**15**	**15**	7	**15**	11	11	10	12
pms/circtracecomp	4	1	0	3	**4**	0	0	0	1	1	1
pms/pb/primes	86	76	78	46	76	59	**80**	77	74	78	**81**
pms/pb-nencdr	128	64	23	69	116	48	104	118	**128**	**128**	127
pms/pb-nlogencdr	128	103	24	88	**128**	78	111	**128**	**128**	**128**	128
pms/aes	7	1	2	0	1	1	2	2	2	**3**	2
pms/hap-asmbly	6	0	2	4	0	**5**	**5**	**5**	**5**	**5**	**5**
pms/bcp-hipp	1183	982	962	1154	**1164**	1125	1142	1141	1138	1137	1140
wpms/haplo-ped	100	0	9	**91**	23	27	28	26	33	25	22
pms/protein-ins	12	**11**	1	1	2	1	1	2	1	2	2
wpms/protein-ins	12	**10**	1	1	2	1	2	2	1	2	1
wpms/timetabling	32	0	0	**13**	12	9	8	7	7	7	6
wpms/upgrade	100	0	**100**	**100**	97	**100**	**100**	**100**	**100**	**100**	100
wpms/up-u98	80	0	**80**	**80**	79	80	80	80	80	80	80
Total	2542	1637	1779	2152	**2251**	1944	2165	2170	2160	2181	2214
Indust + Craft Total	4502	3130	3249	3097	3106	2682	3257	3288	3297	3419	**3578**

in this paper, MAXHS-eval13 achieves a significant performance improvement. Although not shown on the plot, the other versions we develop here all improve over MAXHS+, but are not as good as MAXHS-eval13 (see Tables 1 and 2).

The results are broken down by benchmark family in Tables 1 and 2. Included in the tables are the four competing solvers that performed best overall (as shown in Figure 2): CPLEX, MINIMAXSAT, BINCD and WPM1. Observe that all versions of MAXHS that use non-optimal hitting sets (Alg. 2 and Alg. 3) outperform MAXHS+, which uses non-core constraints more extensively than MAXHS-eval13. Yet when the technique of seeding the MIP model with non-core constraints is added to the use of non-optimal hitting sets, performance is improved (see "Alg. 2 i+g" vs. "Alg. 3 eval 13"). We see that on the industrial problems (Table 1) there is still quite a lot of variance in performance between the different solvers across the different families; that MAXHS-eval13 has fairly robust good performance across the different families; and that the MAXHS approach can be significantly better than just using CPLEX alone, indicating the value of our hybrid approach. On the crafted problems (Table 2) we see that the Branch and Bound approach of MINIMAXSAT is most effective. These problems tend to be smaller than the industrial problems and have

Table 2. Results for the Crafted category instances. Shows the number of instances solved by each solver in each benchmark family. The benchmark families with an asterisk (*) are those we classify as having "random" structure [7]. The final row gives the total number of instances solved over both the Industrial and Crafted categories.

Family	#	bincd	wpm1	CPLEX	mini	MAXHS Orig	+	Alg. 2 incr	grdy	i+g	Alg. 3 eval 13
*ms/spin	20	0	0	19	**20**	0	0	1	4	10	10
*ms/cut/spin	5	1	1	**3**	**3**	0	1	1	2	2	2
*wms/cut/spin	5	0	0	**4**	**4**	0	1	1	2	3	3
*pms/frb	25	0	0	9	5	0	8	5	0	5	**9**
*wms/kexu/frb	35	9	5	**20**	15	10	**20**	16	15	15	**20**
*pms/csp/sprsls	20	**20**	**20**	**20**	**20**	19	11	**20**	**20**	**20**	**20**
*pms/csp/dsls	20	**20**	16	**20**	**20**	5	0	15	15	16	16
*pms/csp/sprstgt	20	**20**	**20**	**20**	**20**	0	0	15	12	**20**	**20**
*pms/csp/dstgt	20	19	19	**20**	**20**	0	0	3	6	**20**	**20**
*ms/ramsey	48	34	34	34	**35**	34	34	34	34	34	34
*wms/ramsey	48	36	34	36	**37**	34	35	34	34	34	34
*pms/clq/rand	96	67	0	**96**	**96**	4	**96**	4	44	59	**96**
*ms/bcut-630	100	0	0	0	**83**	0	0	0	0	0	0
*pms/kbtree	54	15	14	**54**	22	11	15	12	12	12	14
*pms/max1/3sat	80	**80**	71	**80**	**80**	20	**80**	45	44	57	**80**
*ms/cut/rand	40	0	0	4	**40**	0	0	0	0	0	0
*wms/cut/rand	40	0	0	12	**40**	0	0	0	0	0	0
ms/cut/dimacs	62	6	5	20	**48**	4	4	4	4	4	3
wms/cut/dimacs	62	4	5	22	**55**	3	3	4	5	8	9
pms/max1/struc	60	59	30	52	**60**	5	**60**	54	57	**60**	**60**
pms/clique/struc	62	18	8	32	**36**	10	29	12	17	17	34
pms/queens	7	**7**	**7**	**7**	**7**	2	3	5	4	5	5
wpms/QCP	25	**25**	**25**	**25**	20	**25**	**25**	**25**	**25**	**25**	23
pms/pb/gardn	7	5	5	**6**	5	5	**6**	5	**6**	**6**	**6**
wpms/pb/mip	16	**7**	6	6	5	6	**7**	**7**	**7**	**7**	5
wpms/pb/factor	186	**186**	168	**186**	**186**	**186**	**186**	**186**	**186**	**186**	172
wpms/KnotPip	350	0	161	245	117	57	52	**290**	260	**290**	289
wpms/spot5log	21	11	**12**	6	4	6	6	6	6	6	6
wpms/spot5dir	21	11	10	**17**	3	6	6	6	6	6	6
pms/jobshop	4	**4**	3	0	2	**4**	3	**4**	**4**	**4**	**4**
wpms/plan	71	65	64	70	**71**	46	**71**	**71**	**71**	**71**	61
wpms/aucreg	84	6	0	**84**	**84**	34	**84**	4	13	2	76
wpms/aucsch	84	66	**84**	**84**	**84**	82	**84**	76	77	78	75
wpms/aucpath	88	0	52	**88**	**88**	**88**	**88**	**88**	**88**	**88**	78
wpms/ware	18	1	14	**18**	2	1	**18**	9	1	12	**18**
wpms/plan	56	53	52	51	**56**	31	**56**	**56**	**56**	**56**	**56**
Total	1960	855	945	1470	**1493**	738	1092	1118	1137	1238	1364
Indust + Craft Total	4502	3106	3097	3249	3130	2682	3257	3288	3297	3419	**3578**

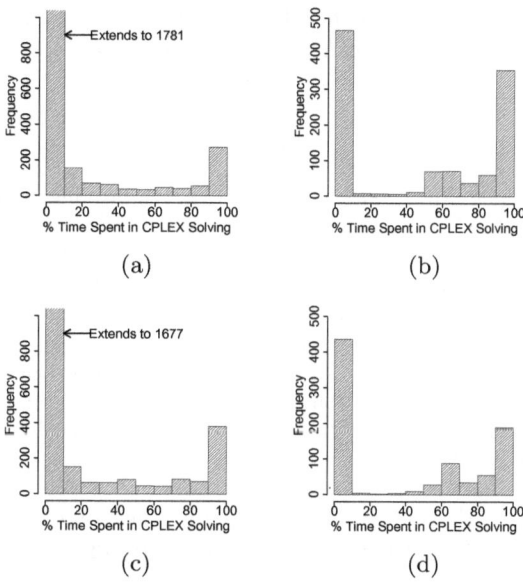

Fig. 3. Histograms over all instances for the percentage of runtime spent in calls to CPLEX for Algorithms 2 and 3. (a) Alg. 2, solved instances; (b) Alg. 2, unsolved instances; (c) Alg. 3, solved instances; (d) Alg. 3, unsolved instances.

tightly interacting variables yielding cores containing a large fraction of the total clauses. The data also shows that although the traditional sequence of SAT solvers BINCD and WPM1 do not perform particularly well on these problems, CPLEX is quite effective, as is the hybrid approach of MAXHS.

The good performance of MAXHS-incr-greedy and MAXHS-eval13, implementing Algorithms 2 and 3 respectively, appears to be due to a significant reduction in the total time spent by CPLEX. In Figure 3 we show the percentage of the total runtime that was spent in calls to CPLEX. Comparing these histograms to those in Figure 1, we observe that the time spent solving the MCHS problems to optimality is now almost always a low percentage of the total runtime. The number of calls to CPLEX generally decreases when we use non-optimal hitting sets, as expected. On average (over all 4502 instances), each run of MAXHS-incr-greedy gave a total of 5419 constraints to CPLEX but solved the optimization problem only 14 times. In contrast, each run of Algorithm 1 gave on average only 972 cores to CPLEX, thus having to solve the MCHS problem 972 times.

In conclusion, we have presented a technique for improving the hybrid MAXHS approach to solving MAXSAT. Our method yields an improvement in the state-of-the-art for MAXSAT solving. Although our method successfully shifts the balance of the runtime away from the MIP solver, a promising avenue for future work is to examine the structure of the constraints given to the MIP solver to see if they could be made more effective.

Acknowledgements. This work has been partly funded by the "Agence nationale de la Recherche", reference ANR-10-BLA-0214.

References

1. Ansótegui, C., Bonet, M.L., Gabàs, J., Levy, J.: Improving SAT-based weighted MaxSAT solvers. In: Milano, M. (ed.) CP 2012. LNCS, vol. 7514, pp. 86–101. Springer, Heidelberg (2012)
2. Ansótegui, C., Bonet, M.L., Levy, J.: Solving (weighted) partial MaxSAT through satisfiability testing. In: Kullmann, O. (ed.) SAT 2009. LNCS, vol. 5584, pp. 427–440. Springer, Heidelberg (2009)
3. Ansótegui, C., Bonet, M.L., Levy, J.: A new algorithm for weighted partial maxsat. In: Proceedings of the AAAI National Conference (AAAI), pp. 3–8 (2010)
4. Argelich, J., Li, C.M., Manyà, F., Planes, J.: The maxSAT evaluations (2007–2011), http://www.maxsat.udl.cat
5. Berre, D.L., Parrain, A.: The sat4j library, release 2.2. JSAT 7(2-3), 59–64 (2010)
6. Davies, J., Bacchus, F.: Solving MAXSAT by solving a sequence of simpler SAT instances. In: Lee, J. (ed.) CP 2011. LNCS, vol. 6876, pp. 225–239. Springer, Heidelberg (2011)
7. Davies, J., Bacchus, F.: Exploiting the power of MIP solvers in MAXSAT. In: Järvisalo, M., Van Gelder, A. (eds.) SAT 2013. LNCS, vol. 7962, pp. 166–181. Springer, Heidelberg (2013)
8. Eén, N., Sörensson, N.: An extensible sat-solver. In: Giunchiglia, E., Tacchella, A. (eds.) SAT 2003. LNCS, vol. 2919, pp. 502–518. Springer, Heidelberg (2004)
9. Heras, F., Larrosa, J., Oliveras, A.: Minimaxsat: An efficient weighted max-sat solver. Journal of Artificial Intelligence Research (JAIR) 31, 1–32 (2008)
10. Heras, F., Morgado, A., Marques-Silva, J.: Core-guided binary search algorithms for maximum satisfiability. In: Proceedings of the AAAI National Conference (AAAI), pp. 36–41 (2011)
11. Hooker, J.N.: Planning and scheduling by logic-based benders decomposition. Operations Research 55(3), 588–602 (2007)
12. Johnson, D.S.: Approximation algorithms for combinatorial problems. In: Symposium on Theory of Computing, pp. 38–49 (1973)
13. Kügel, A.: Improved exact solver for the weighted Max-SAT problem. In: Workshop on the Pragmatics of SAT (2010)
14. Li, C.M., Manyà, F., Mohamedou, N.O., Planes, J.: Resolution-based lower bounds in maxsat. Constraints 15(4), 456–484 (2010)
15. Manquinho, V., Marques-Silva, J., Planes, J.: Algorithms for weighted boolean optimization. In: Kullmann, O. (ed.) SAT 2009. LNCS, vol. 5584, pp. 495–508. Springer, Heidelberg (2009)
16. Moreno-Centeno, E., Karp, R.M.: The implicit hitting set approach to solve combinatorial optimization problems with an application to multigenome alignment. Operations Research 61(2), 453–468 (2013)
17. Marques-Silva, J., Lynce, I.: On improving MUS extraction algorithms. In: Sakallah, K.A., Simon, L. (eds.) SAT 2011. LNCS, vol. 6695, pp. 159–173. Springer, Heidelberg (2011)

Dead-End Elimination for Weighted CSP

Simon de Givry[1,2,*], Steven D. Prestwich[2], and Barry O'Sullivan[2]

[1] MIA-T, UR 875, INRA, F-31320 Castanet Tolosan, France
[2] Cork Constraint Computation Centre, University College Cork, Ireland
degivry@toulouse.inra.fr, {s.prestwich,b.osullivan}@cs.ucc.ie

Abstract. Soft neighborhood substitutability (SNS) is a powerful technique to automatically detect and prune dominated solutions in combinatorial optimization. Recently, it has been shown in [26] that enforcing partial SNS (PSNS^r) during search can be worthwhile in the context of Weighted Constraint Satisfaction Problems (WCSP). However, for some problems, especially with large domains, PSNS^r is still too costly to enforce due to its worst-case time complexity in $O(ned^4)$ for binary WCSP. We present a simplified dominance breaking constraint, called *restricted dead-end elimination* (DEE^r), the worst-case time complexity of which is in $O(ned^2)$. Dead-end elimination was introduced in the context of computational biology as a preprocessing technique to reduce the search space [13, 14, 16, 17, 28, 30]. Our restriction involves testing only one pair of values per variable instead of all the pairs, with the possibility to prune several values at the same time. We further improve the original dead-end elimination criterion, keeping the same time and space complexity as DEE^r. Our results show that maintaining DEE^r during a depth-first branch and bound (DFBB) search is often faster than maintaining PSNS^r and always faster than or similar to DFBB alone.

Keywords: combinatorial optimization, dominance rule, weighted constraint satisfaction problem, soft neighborhood substitutability.

1 Introduction

Pruning by dominance in the context of combinatorial optimization involves reducing the solution space of a problem by adding new constraints to it [19]. We study dominance rules that reduce the domains of variables based on optimality considerations (in relation to the optimization of an objective function). The idea is to automatically detect values in the domain of a variable that are *dominated* by another *dominant* value of the domain such that any solution using the dominant value instead of the dominated ones has a better score. Various dominance rules have been studied recently by the Constraint Programming community [5, 6, 26]. In particular, *soft neighborhood substitutability* (SNS) [3, 26] allows us to detect dominated values in polynomial time under specific conditions for Weighted Constraint Satisfaction Problems (WCSP). In a different community,

---

* This work has been partly funded by the "Agence nationale de la Recherche", reference ANR-10-BLA-0214 and the European Union, reference FP7 *ePolicy* 288147.

C. Schulte (Ed.): CP 2013, LNCS 8124, pp. 263–272, 2013.

similar dominance rules and others, called *dead-end elimination (DEE) crite-ria*, have been studied for many years in the context of *computational protein design* [1, 13, 14, 16, 17, 28, 30]. However to the best of our knowledge, these criteria have never been used during search, possibly due to their high computational cost. Following the work done in [26] showing the interest of maintaining such dominance rule during search, we propose a faster pruning by dominance algorithm combining SNS and DEE in a partial and optimistic way.

2 Weighted Constraint Satisfaction Problems

A Weighted Constraint Satisfaction Problem (WCSP) P is a triplet $P = (X, F, k)$ where X is a set of n variables and F a set of e cost functions. Each variable $x \in X$ has a finite domain, $domain(x)$, of values that can be assigned to it. The maximum domain size is denoted by d. For a set of variables $S \subseteq X$, $l(S)$ denotes the set of all labelings of S, *i.e.*, the Cartesian product of the domain of the variables in S. For a given tuple of values t, $t[S]$ denotes the projection of t over S. A cost function $f_S \in F$, with scope $S \subseteq X$, is a function $f_S : l(S) \mapsto [0, k]$ where k is a maximum integer cost used for forbidden assignments. A cost function over one (resp. zero) variable is called a *unary* (resp. *nullary*, *i.e.*, a constant cost payed by any assignment) cost function, denoted either by $f_{\{x\}}$ or f_x (resp. by $f_\varnothing$). We denote by $\Gamma(x)$ the set of cost functions on variable x, *i.e.*, $\Gamma(x) = \{f_S \in F | \{x\} \subseteq S\}$.

The Weighted Constraint Satisfaction Problem consists in finding a complete assignment t minimizing the combined (*sum*) cost function $\sum_{f_S \in F} f_S(t[S])$. This optimization problem has an associated NP-complete decision problem.

Enforcing a given local consistency property on a problem P involves transforming $P = (X, F, k)$ into a problem $P' = (X, F', k)$ that is equivalent to P (all complete assignments keep the same cost) and that satisfies the considered local consistency property. This enforcing may increase $f_\varnothing$ and provide an improved lower bound on the optimal cost. It is achieved using *Equivalence Preserving Transformations* (EPTs) that move costs between different scopes [8–12, 21, 22, 24, 31]. In particular, node consistency [21] (NC) satisfies $\forall x \in X$, $\min_{a \in domain(x)} f_x(a) = 0, \forall a \in domain(x), f_\varnothing + f_x(a) < k$. Soft arc consistency (AC*) [21, 31] satisfies NC and $\forall f_S \in F, \forall x \in S, \forall a \in domain(x)$, $\min_{t \in l(S \setminus \{x\})} f_S(t \cup \{(x, a)\}) = 0$.

3 Dead-End Elimination

The original dead-end elimination criterion is [14]:

$$\sum_{f_S \in \Gamma(x)} \max_{t \in l(S \setminus \{x\})} f_S(t \cup \{(x, a)\}) \leq \sum_{f_S \in \Gamma(x)} \min_{t \in l(S \setminus \{x\})} f_S(t \cup \{(x, b)\}). \quad (1)$$

This condition implies that value b can be safely removed from the domain of x since the total cost of all the cost functions on x taking their best assignment with x assigned b is still worse than that produced by their worst assignment with x assigned a. This condition was further improved in [17]:

$$\sum_{f_S \in \Gamma(x)} \max_{t \in l(S \setminus \{x\})} f_S(t \cup \{(x,a)\}) - f_S(t \cup \{(x,b)\}) \leq 0. \tag{2}$$

where the best and worst-cases are replaced by the worst difference in costs for any labeling of the remaining variables in the scope of each cost function. It is easy to see that this condition is always stronger than the previous one.

More recently, the authors in [26] reformulated Equation 2 in the specific context of WCSP with bounded cost addition $a \oplus b = \min(k, a + b)$ proving that the reformulated criterion[1] is equivalent to soft neighborhood substitutability when $\Gamma(x)$ is separable (*i.e.*, $\forall f_S, f_{S'} \in \Gamma(x) \times \Gamma(x), S \cap S' = \{x\}$) and $\alpha < k$. In practice, testing Eq. 2 or its reformulation will prune the same values.

They also noticed that if the problem is soft arc consistent then the worst-cost differences are always positive. Equation 1 can be further simplified thanks to soft AC because all the best-case terms are precisely equal to zero:

$$\sum_{f_S \in \Gamma(x)} \max_{t \in l(S \setminus \{x\})} f_S(t \cup \{(x,a)\}) \leq f_x(b). \tag{3}$$

We propose a stronger condition than Eq. 2 or Eq. 3 by discarding forbidden partial assignments with x assigned b when computing the worst-cost difference:

$$\sum_{f_S \in \Gamma(x)} \max_{t \in l(S \setminus \{x\})\ \text{st.}\ C(f_S, t \cup \{(x,b)\}) < k} f_S(t \cup \{(x,a)\}) - f_S(t \cup \{(x,b)\}) \leq 0. \tag{4}$$

where $C(f_S, t) = f_s(t) + \sum_{y \in S, |S| > 1} f_y(t[y]) + f_\emptyset$. This new condition is equivalent to Eq. 2 except that some tuples have been discarded from the max operation. These discarded tuples t are forbidden partial assignments when x is assigned b because the sum of the associated cost function $f_S(t \cup \{(x,b)\})$ plus, if $|S| > 1$, all the unary costs on the variables in S assigned by $t \cup \{(x,b)\}$ plus the current lower bound $f_\emptyset$ is greater than or equal to the current upper bound k. Such tuples t do not need to be considered by the max operation because $t \cup \{(x,b)\}$ does not belong to any optimal solution, whereas $t \cup \{(x,a)\}$ can be.

For CSP (*i.e.*, $k = 1$), Eq. 2 and Eq. 4 are both equivalent to neighborhood substitutability [15]. For Max-SAT, Eq. 3 and Eq. 2 are equivalent if the problem is soft AC, and correspond to the *Dominating 1-clause rule* [29]. In the general case, Eq. 4 is stronger[2] (more domain values can be pruned) than Eq. 2, which is

---

[1] They replace the maximum of cost differences $\alpha - \beta$ by the opposite of the minimum of cost pairs (β, α), ordered by the relation $(\beta, \alpha) \leq (\beta', \alpha') \equiv \beta - \alpha < \beta' - \alpha' \lor (\beta - \alpha = \beta' - \alpha' \land \alpha < \alpha')$. Equation 2 becomes $\sum_{f_S \in \Gamma(x) \cup f_x} \min_{t \in l(S \setminus \{x\})} (f_S(t \cup \{(x,b)\}), f_S(t \cup \{(x,a)\})) \geq 0$ where $(\beta, \alpha) \geq 0$ if $\beta \geq \alpha$.

[2] The definition of soft AC on fair VCSPs [12] makes Eq. 4 and Eq. 2 equivalent.

stronger than Eq. 3. More complex dominance criteria have been defined in the context of protein design (*e.g.*, a value being dominated by a set of values instead of a single one, see [30] for an overview), but they all incur higher computational costs. In the next section, we recall how to enforce Eq. 2 in WCSP, as originally shown in [26]. Then, in Section 5, we present a modified version to partially enforce the two conditions, Eq. 4 and 3, with a lower time complexity.

4   Enforcing Soft Neighborhood Substitutability

Assuming a soft arc consistent WCSP (see *e.g.*, W-AC*2001 algorithm in [24]), enforcing partial[3] soft neighborhood substitutability (PSNS^r) is described by Algorithm 1. For each variable x, all the pairs of values $(a, b) \in domain(x) \times domain(x)$ with $a < b$ are checked by the function DominanceCheck to see if b is dominated by a or, if not, vice versa (line 3). At most one dominated value is added to the value removal queue Δ at each inner loop iteration (line 2). Removing dominated values (line 4) can make the problem arc inconsistent, requiring us to enforce soft arc consistency again. We successively enforce soft AC and PSNS^r until no value removals are made by both enforcing algorithms.

Algorithm 1: Enforce PSNS^r [26]

Procedure PSNS^r(P: AC* consistent WCSP)

    $\Delta := \emptyset$;

1   **foreach** $x \in variables(P)$ **do**

2   **foreach** $(a, b) \in domain(x) \times domain(x)$ such that $a < b$ **do**

            $R :=$ DominanceCheck($x, a \rightarrow b$) ;

3   **if** $R = \emptyset$ **then** $R :=$ DominanceCheck($x, b \rightarrow a$) ;

            $\Delta := \Delta \cup R$;

4   **foreach** $(x, a) \in \Delta$ **do** remove (x, a) from $domain(x)$;

/* Check if value a dominates value b */

Function DominanceCheck($x, a \rightarrow b$): set of dominated values

5   **if** $f_x(a) > f_x(b)$ **then** **return** $\emptyset$;

    $\delta_{a \rightarrow b} := f_x(a)$;

    foreach $f_s \in F$ such that $\{x\} \subset S$ **do**

        $\delta :=$ getDifference($f_s, x, a \rightarrow b$) ;

        $\delta_{a \rightarrow b} := \delta_{a \rightarrow b} + \delta$;

6   **if** $\delta_{a \rightarrow b} > f_x(b)$ **then** **return** $\emptyset$;

    return $\{(x, b)\}$ /* $\delta_{a \rightarrow b} \leq f_x(b)$ */ ;

/* Compute largest difference in costs when using a instead of b */

Function getDifference($f_s, x, a \rightarrow b$): cost

7    $\delta_{a \rightarrow b} := 0$;

    foreach $t \in l(S \setminus \{x\})$ **do**

        $\delta_{a \rightarrow b} := \max(\delta_{a \rightarrow b}, f_s(t \cup \{(x, a)\}) - f_s(t \cup \{(x, b)\}))$;

    return $\delta_{a \rightarrow b}$;

Function DominanceCheck($x, a \rightarrow b$) computes the sum of worst-cost differences as defined by Equation 2 and returns a non-empty set containing value b if Eq. 2 is true, meaning that b is dominated by value a. It exploits early breaks as

---

[3] Enforcing complete soft neighborhood substitutability is co-NP hard as soon as $k \neq +\infty$ (*i.e.*, no restriction on α in the reformulated Equation 2).

soon as Eq. 2 can be falsified (lines 5 and 6). Worst-cost differences are computed by the function getDifference($f_s, x, a \to b$) applied to every cost function related to x. Worst-cost differences are always positive (line 7) due to soft AC.

The worst-case time complexity of getDifference is $O(d^{r-1})$ for WCSP with maximum arity r. DominanceCheck is $O(qd^{r-1})$ where $q = |\Gamma(x)|$. Thus, the time complexity of one iteration of Algorithm 1 (PSNS^r) is $O(nd^2qd^{r-1} + nd) = O(ed^{r+1})$ where $e = nq$. Interleaving PSNS^r and soft AC until a fixed point is reached is done at most nd times, resulting in a worst-case time complexity of PSNS^r in $O(ned^{r+2})$. Its space complexity is $O(nd^2)$ when using *residues* [26].

In the following, we always consider PSNS^r using the better condition given by Equation 4 instead of Eq. 2. This does not change the previous complexities.

5   Enforcing Partial SNS and Dead-End Elimination

In order to reduce the time (and space) complexity of pruning by dominance, we test only one pair of values per variable. The new algorithm is described in Algorithm 2. We select the pair $(a, b) \in domain(x) \times domain(x)$ in an optimistic way such that a is associated with the minimum unary cost and b to the maximum unary cost (lines 8 and 9). Because arc consistency also implies node consistency, we always have $f_x(a) = 0.$[4] When all the unary costs (including the maximum) are null (line 10), we select as b the maximum domain value (or its minimum if this value is already used by a). By doing so, we should favor more pruning on max-closed or submodular subproblems[5].

Instead of checking the new Equation 4 for the pair (a, b) alone, we also check Eq. 3 for all the pairs (a, u) such that $u \in domain(x) \setminus \{a\}$. This is done in the function MultipleDominanceCheck (lines 16 and 17). This function computes at the same time the sum of maximum costs ub_a for value a (lines 12 and 13) and the sum of worst-cost differences $\delta_{a \to b}$ for the pair (a, b). The new function getDifference-Maximum($f_s, x, a \to b$) now returns the worst-cost difference, discarding forbidden assignments with $t \cup \{(x, b)\}$ (line 18), as suggested by Eq. 4, and also the maximum cost in f_S for x assigned a. By construction of the two criteria, we have $\delta_{a \to b} \leq ub_a$, so the stopping condition is unchanged at line 14. When the maximum cost of a value is null for all its cost functions, we can directly remove all the other values in the domain avoiding any extra work (line 15). Finally, if the selected pair (a, b) *prunes* b, then a new pair is checked.

Notice that DEE^r is equivalent to PSNS^r on problems with Boolean variables, such as Weighted Max-SAT. For problems with non-Boolean domains, DEE^r is still able to detect and prune several values per variable. Clearly, its time (resp. space) complexity is $O(ned^r)$ (resp. $O(n)$ using only one residue per variable), reducing by a factor d^2 the time and space complexity compared to PSNS^r.

---

[4] In fact, we set the value a to the *unary support* offered by NC [21] or EDAC [22].

[5] Assuming a problem with two variables x and y having the same domain and a single submodular cost function $f(x, y) = 0$ if $x \leq y$ else $x - y$ or a single max-closed constraint $x < y$, then DEE^r assigns $x = min(domain(x))$ and $y = max(domain(y))$.

Algorithm 2: Enforce DEE^r

Procedure DEE^r(P: AC* consistent WCSP)

$\quad \Delta := \emptyset$;

$\quad$**foreach** $x \in variables(P)$ **do**

$\quad$**8** $\quad\quad a := \arg\min_{u \in domain(x)} f_x(u)$;

$\quad$**9** $\quad\quad b := \arg\max_{u \in domain(x)} f_x(u)$;

$\quad$**10** $\quad\quad$**if** $a = b$ /* $\forall u \in domain(x), f_x(u) = 0$ */ **then**

$\quad\quad\quad\quad$**if** $a = max(domain(x))$ **then**

$\quad\quad\quad\quad\quad | \quad b := min(domain(x))$;

$\quad\quad\quad\quad$**else**

$\quad\quad\quad\quad\quad\quad b := max(domain(x))$;

$\quad\quad\quad R :=$ MultipleDominanceCheck($x, a \rightarrow b$) ;

$\quad$**11** $\quad\quad$**if** $R = \emptyset$ **then** $R :=$ MultipleDominanceCheck($x, b \rightarrow a$) ;

$\quad\quad\quad \Delta := \Delta \cup R$;

$\quad$**foreach** $(x, a) \in \Delta$ **do** remove (x, a) from $domain(x)$;

/* Check if value a dominates value b and possibly other values */

Function MultipleDominanceCheck($x, a \rightarrow b$): set of dominated values

$\quad$**if** $f_x(a) > f_x(b)$ **then return** $\emptyset$;

$\quad \delta_{a \rightarrow b} := f_x(a)$;

12 $\quad ub_a := f_x(a)$;

$\quad$**foreach** $f_s \in F$ such that $\{x\} \subset S$ **do**

$\quad\quad (\delta, ub) :=$ getDifference-Maximum($f_s, x, a \rightarrow b$) ;

$\quad\quad \delta_{a \rightarrow b} := \delta_{a \rightarrow b} + \delta$;

13 $\quad\quad ub_a := ub_a + ub$;

14 $\quad\quad$**if** $\delta_{a \rightarrow b} > f_x(b)$ **then return** $\emptyset$;

15 $\quad$**if** $ub_a = 0$ **then return** $\{(x, u)| u \in domain(x)\} \setminus \{(x, a)\}$;

$\quad R := \{(x, b)\}$ /* $\delta_{a \rightarrow b} \leq f_x(b)$ */ ;

16 $\quad$**foreach** $u \in domain(x)$ such that $u \neq a$ **do**

17 $\quad\quad$**if** $(f_x(u) \geq ub_a)$ **then** $R := R \cup \{(x, u)\}$;

$\quad$**return** R ;

/* Compute largest cost difference and maximum cost for value */

Function getDifference-Maximum($f_s, x, a \rightarrow b$): pair of costs

$\quad \delta_{a \rightarrow b} := 0$;

$\quad ub_a := 0$;

$\quad$**foreach** $t \in l(S \setminus \{x\})$ **do**

18 $\quad\quad$**if** $f_s(t \cup \{(x, b)\}) + f_\emptyset + f_x(b) + \sum_{y \in S \setminus \{x\}} f_y(t[y]) < k$ **then**

$\quad\quad\quad \delta_{a \rightarrow b} := \max(\delta_{a \rightarrow b}, f_s(t \cup \{(x, a)\}) - f_s(t \cup \{(x, b)\}))$;

$\quad\quad ub_a := \max(ub_a, f_s(t \cup \{(x, a)\}))$;

$\quad$**return** $(\delta_{a \rightarrow b}, ub_a)$ /* $\delta_{a \rightarrow b} \leq ub_a$ */ ;

---

6 Experimental Results

We implemented PSNS^r and DEE^r in `toulbar2`[6]. All methods use residues and variable queues with timestamps as in [26]. PSNS^r uses MultipleDominanceCheck and getDifference-Maximum instead of DominanceCheck and getDifference. MultipleDominanceCheck prunes the dominated values directly instead of queuing them into R. It speeds-up further dominance checks without assuming soft AC anymore during the process (soft AC being restored at the next iteration until a fixed point is reached for AC and SNS/DEE). We compared PSNS^r and DEE^r on a collection of binary WCSP benchmarks (http://costfunction.org) (except

---

[6] C++ solver version 0.9.6 mulcyber.toulouse.inra.fr/projects/toulbar2/

for *spot5* using ternary cost functions). The *celar* [4] ($n \le 458, d \le 44$) and *computational protein design* [1] ($n \le 55, d \le 148$) have been selected as they offer good opportunities for neighborhood substitutability, at least in preprocessing as shown in [14, 20]. We added Max SAT *combinatorial auctions* using the CATS generator [27] with 60 goods and a varied number of bids from 70 to 200 (100 to 230 for *regions*) [23]. Other benchmarks were selected by [26] and include: *DIMACS graph coloring* (minimizing edge violations) ($n \le 450, d \le 9$), *optimal planning* [7] ($n \le 1433, d \le 51$), *spot5* ($n \le 1057, d = 4$) [2], and *uncapacitated warehouse location* [22] ($n \le 1100, d \le 300$). Experiments were performed on a cluster of AMD Opteron 2.3 GHz under Linux.

In Table 1, we compared a Depth First Branch and Bound algorithm using EDAC [22] alone (*EDAC* column), EDAC and DEE^r (*EDAC+DEE^r*), EDAC and PSNS^r in preprocessing only (*EDAC+PSNS$^r_{pre}$*), EDAC and PSNS^r in preprocessing and DEE^r during search (*EDAC+PSNS$^r_{pre}$+DEE^r*), EDAC and PSNS^r (*EDAC+PSNS^r*), and no initial upper bound for all. For each benchmark, we report the number of instances, and for each method, the number of instances optimally solved in less than 1,200 seconds. In parentheses, average CPU time over the solved instances (in seconds), average number of nodes, and average number of value removals per search node are reported where appropriate. First, we used a static lexicographic variable ordering and a binary branching scheme (`toolbar2` options -*nopre* -*svo* -*d:*). DEE^r solved always a greater or equal number of instances compared to EDAC alone, and it performed better than PSNS^r on *celar*, *planning*, *protein*, and *warehouse* benchmarks, all having large domains. We also give the results, when available, in terms of the number of solved instances by PSNS^r over the total number of instances solved by at least one method as reported in [26], showing the good performance of our approach. They used the same settings except a cluster of Xeon 3.0 GHz and *max degree* static variable ordering (only identical to our lexicographic ordering for *warehouse*). In addition, we solved the *celar7-sub1* instance with the same *max degree* ordering: *EDAC+DEE^r* solved in (7.7 seconds, 57,584 nodes, 0.96 removals per node), and *EDAC+PSNS^r* in (69.5, 39,346, 7.2), or (86.4, 70,896, 6) as reported in [26]. Secondly, we used a dynamic variable ordering combining Weighted Degree with Last Conflict [25] and an initial Limited Discrepancy Search (LDS) phase [18] with a maximum discrepancy of 2 (option -*l=2*, except for *protein* using also -*sortd* -*d:* as in [1]). This greatly improved the results for all the methods and benchmarks except for *warehouse* where LDS slowed down the methods. DEE^r remained the best method in terms of the number of solved instances; PSNS^r in preprocessing and DEE^r during search being a good alternative, especially on the *protein* benchmark. We compared a subset of our results with the last Max SAT 2012 evaluation (`http://maxsat.ia.udl.cat:81/12`). With roughly the same computation time limit (20 min. with 2.3 GHz instead of 30 min. with AMD Opteron 1.5 GHz), for *auction/paths* and *auction/scheduling*, DEE^r solved 85+82 instances among 170, being in 3rd position among 11 Max SAT solvers.

Table 1. For each method, number of instances optimally solved in less than 1,200 seconds, and in parentheses, average CPU time (in seconds) over the solved instances, average number of search nodes, and average number of value removals per node where appropriate

	#inst.	EDAC	EDAC+DEE^r	EDAC+PSNS$^r_{pre}$	EDAC+PSNS$^r_{pre}$+DEE^r	EDAC+PSNS^r	[26]
Depth First Branch and Bound with static variable ordering							
celar	46	24 (180.6, 954K)	24 (187.2, 877K, 0.80)	24 (188.4, 945K)	24 (187.7, 877K, 0.77)	17 (168.0, 100K, 8.33)	12/16
coloring	40	19 (47.7, 2.4M)	19 (45.7, 2.2M, 0.08)	19 (46.9, 2.2M)	19 (45.6, 2.2M, 0.08)	**20 (103.5, 3.7M, 0.96)**	8/8
planning	76	68 (9.8, 39K)	75 (7.2, 32K, 4.46)	69 (18.3, 127K)	**75 (6.9, 32K, 4.46)**	75 (10.5, 31K, 5.27)	27/27
protein	12	9 (34.4, 70K)	9 (30.9, 42K, 1.50)	9 (26.0, 50K)	**9 (25.7, 40K, 1.32)**	9 (139.0, 31K, 4.37)	
spot5	24	4 (0.1, 68)	7 (93.7, 2.7M, 0.42)	6 (172.7, 3.7M)	7 (93.2, 2.7M, 0.39)	**7 (87.0, 2.5M, 0.42)**	3/3
warehouse	55	**46 (55.6, 709)**	46 (66.1, 542, 34.34)	46 (61.3, 688)	46 (58.6, 542, 34.73)	45 (56.3, 429, 75.00)	29/34
auction/paths	420	138 (225.4, 5.9M)	**148 (212.8, 5.2M, 0.06)**	138 (223.5, 5.9M)	148 (213.7, 5.2M, 0.06)	148 (214.0, 5.2M, 0.06)	
auct./regions	420	364 (137.5, 3.3M)	404 (98.1, 1.9M, 0.03)	373 (131.2, 3.2M)	403 (94.3, 1.9M, 0.03)	**405 (100.2, 2M, 0.03)**	
a./scheduling	420	**392 (115.3, 2.3M)**	392 (118.4, 2.3M, 0.00)	392 (113.3, 2.3M)	391 (115.6, 2.2M, 0.00)	390 (114.1, 2.2M, 0.00)	
total	1513	1064	**1124**	1076	1122	1116	
Depth First Branch and Bound with dynamic variable ordering and initial LDS with maximum discrepancy of 2							
celar	46	40 (22.7, 45K)	40 (24.5, 43K, 1.90)	**40 (19.8, 40K)**	40 (24.9, 38K, 1.80)	38 (114.0, 25K, 10.64)	
coloring	40	23 (6.6, 167K)	24 (39.4, 484K, 0.86)	23 (6.7, 167K)	24 (38.9, 484K, 0.86)	**24 (9.1, 162K, 1.19)**	
planning	76	76 (1.3, 1.5K)	76 (1.2, 1.4K, 3.05)	**76 (0.8, 1.1K)**	76 (1.3, 1.5K, 3.02)	76 (1.2, 1.3K, 3.34)	
protein	12	9 (10.1, 7.7K)	9 (10.5, 8K, 1.77)	9 (9.0, 10K)	**9 (8.5, 8K, 1.33)**	9 (55.0, 11K, 5.67)	
spot5	24	8 (21.7, 669K)	8 (14.1, 418K, 0.13)	8 (27.1, 841K)	8 (16.2, 483K, 0.14)	**8 (12.3, 350K, 0.19)**	
warehouse	55	**45 (67.1, 957)**	43 (30.7, 630, 17.87)	45 (70.8, 949)	43 (30.2, 618, 18.65)	42 (8.7, 411, 31.45)	
auction/paths	420	345 (139.0, 2.5M)	**356 (137.4, 2.4M, 0.16)**	346 (138.5, 2.5M)	356 (137.6, 2.4M, 0.16)	355 (139.0, 2.4M, 0.16)	
auct./regions	420	**420 (2.5, 27K)**	**420 (2.5, 27K, 0.03)**	420 (2.5, 27K)	**420 (2.5, 27K, 0.03)**	**420 (2.5, 27K, 0.03)**	
a./scheduling	420	**413 (54.8, 1.5M)**	413 (57.8, 1.5M, 0.00)	413 (55.5, 1.5M)	413 (57.8, 1.5M, 0.00)	413 (57.8, 1.5M, 0.00)	
total	1513	1379	**1389**	1380	**1389**	1385	

7   Conclusion

We have presented a lightweight algorithm for automatically exploiting a dead-end elimination dominance criterion for WCSPs. Experimental results show that it can lead to significant reductions in search space and run-time on several benchmarks. In future work, we plan to study such dominance criteria applied during search in integer linear programming.

Acknowledgements. We thank the Genotoul Bioinformatic platform for the cluster and Seydou Traoré, Isabelle André, and Sophie Barbe for the protein instances.

References

1. Allouche, D., Traoré, S., André, I., de Givry, S., Katsirelos, G., Barbe, S., Schiex, T.: Computational protein design as a cost function network optimization problem. In: Milano, M. (ed.) CP 2012. LNCS, vol. 7514, pp. 840–849. Springer, Heidelberg (2012)
2. Bensana, E., Lemaître, M., Verfaillie, G.: Earth observation satellite management. Constraints 4(3), 293–299 (1999)
3. Bistarelli, S., Faltings, B.V., Neagu, N.: Interchangeability in Soft CSPs. In: Van Hentenryck, P. (ed.) CP 2002. LNCS, vol. 2470, pp. 31–46. Springer, Heidelberg (2002)
4. Cabon, B., de Givry, S., Lobjois, L., Schiex, T., Warners, J.: Radio link frequency assignment. Constraints Journal 4, 79–89 (1999)
5. Chu, G., Banda, M., Stuckey, P.: Exploiting subproblem dominance in constraint programming. Constraints 17(1), 1–38 (2012)
6. Chu, G., Stuckey, P.J.: A generic method for identifying and exploiting dominance relations. In: Milano, M. (ed.) CP 2012. LNCS, vol. 7514, pp. 6–22. Springer, Heidelberg (2012)
7. Cooper, M., Cussat-Blanc, S., de Roquemaurel, M., Régnier, P.: Soft arc consistency applied to optimal planning. In: Benhamou, F. (ed.) CP 2006. LNCS, vol. 4204, pp. 680–684. Springer, Heidelberg (2006)
8. Cooper, M., de Givry, S., Sanchez, M., Schiex, T., Zytnicki, M., Werner, T.: Soft arc consistency revisited. Artificial Intelligence 174, 449–478 (2010)
9. Cooper, M., de Givry, S., Sanchez, M., Schiex, T., Zytnicki, M.: Virtual arc consistency for weighted CSP. In: Proc. of AAAI 2008, Chicago, IL (2008)
10. Cooper, M.C.: High-order consistency in Valued Constraint Satisfaction. Constraints 10, 283–305 (2005)
11. Cooper, M.C., de Givry, S., Schiex, T.: Optimal soft arc consistency. In: Proc. of IJCAI 2007, Hyderabad, India, pp. 68–73 (January 2007)
12. Cooper, M.C., Schiex, T.: Arc consistency for soft constraints. Artificial Intelligence 154(1-2), 199–227 (2004)
13. Dahiyat, B., Mayo, S.: Protein design automation. Protein Science 5(5), 895–903 (1996)
14. Desmet, J., Maeyer, M., Hazes, B., Lasters, I.: The dead-end elimination theorem and its use in protein side-chain positioning. Nature 356(6369), 539–542 (1992)
15. Freuder, E.C.: Eliminating interchangeable values in constraint satisfaction problems. In: Proc. of AAAI 1991, Anaheim, CA, pp. 227–233 (1991)

16. Georgiev, I., Lilien, R., Donald, B.: Improved pruning algorithms and divide-and-conquer strategies for dead-end elimination, with application to protein design. Bioinformatics 22(14), e174–e183 (2006)
17. Goldstein, R.: Efficient rotamer elimination applied to protein side-chains and related spin glasses. Biophysical Journal 66(5), 1335–1340 (1994)
18. Harvey, W.D., Ginsberg, M.L.: Limited discrepency search. In: Proc. of the 14th IJCAI, Montréal, Canada (1995)
19. Jouglet, A., Carlier, J.: Dominance rules in combinatorial optimization problems. European Journal of Operational Research 212(3), 433–444 (2011)
20. Koster, A.M.C.A.: Frequency assignment: Models and Algorithms. Ph.D. thesis, University of Maastricht, The Netherlands (November 1999), www.zib.de/koster/thesis.html
21. Larrosa, J.: On arc and node consistency in weighted CSP. In: Proc. AAAI 2002, Edmondton (CA), pp. 48–53 (2002)
22. Larrosa, J., de Givry, S., Heras, F., Zytnicki, M.: Existential arc consistency: getting closer to full arc consistency in weighted CSPs. In: Proc. of the 19th IJCAI, Edinburgh, Scotland, pp. 84–89 (August 2005)
23. Larrosa, J., Heras, F., de Givry, S.: A logical approach to efficient max-sat solving. Artif. Intell. 172(2-3), 204–233 (2008)
24. Larrosa, J., Schiex, T.: Solving weighted CSP by maintaining arc consistency. Artif. Intell. 159(1-2), 1–26 (2004)
25. Lecoutre, C., Saïs, L., Tabary, S., Vidal, V.: Reasoning from last conflict(s) in constraint programming. Artificial Intelligence 173, 1592–1614 (2009)
26. Lecoutre, C., Roussel, O., Dehani, D.E.: WCSP Integration of Soft Neighborhood Substitutability. In: Milano, M. (ed.) CP 2012. LNCS, vol. 7514, pp. 406–421. Springer, Heidelberg (2012)
27. Leyton-Brown, K., Pearson, M., Shoham, Y.: Towards a Universal Test Suite for Combinatorial Auction Algorithms. In: ACM E-Commerce, pp. 66–76 (2000)
28. Looger, L., Hellinga, H.: Generalized dead-end elimination algorithms make large-scale protein side-chain structure prediction tractable: implications for protein design and structural genomics. Journal of Molecular Biology 307(1), 429–445 (2001)
29. Niedermeier, R., Rossmanith, P.: New upper bounds for maximum satisfiability. J. Algorithms 36(1), 63–88 (2000)
30. Pierce, N., Spriet, J., Desmet, J., Mayo, S.: Conformational splitting: A more powerful criterion for dead-end elimination. Journal of Computational Chemistry 21(11), 999–1009 (2000)
31. Schiex, T.: Arc consistency for soft constraints. In: Dechter, R. (ed.) CP 2000. LNCS, vol. 1894, pp. 411–424. Springer, Heidelberg (2000)

Solving Weighted CSPs
by Successive Relaxations

Erin Delisle and Fahiem Bacchus

Department of Computer Science, University of Toronto,
Toronto, Ontario, Canada, M5S 3H5
{edelisle,fbacchus}@cs.toronto.edu

Abstract. In this paper we present a new algorithm for solving weighted CSPs (WCSP). This involves first creating an ordinary unweighted CSP, $\mathcal{P}$, by hardening all soft constraints of the WCSP. $\mathcal{P}$ has a solution if and only if the WCSP has a cost zero solution. The algorithm then proceeds by solving relaxations of $\mathcal{P}$ each allowing a particular cost to be incurred. If the relaxation has no solution, a set of its forbidden tuples sufficient to rule out all solutions is computed. From this set of culprit tuples we show how to compute a new relaxation of $\mathcal{P}$ that can again be tested for a solution. If the new relaxation is optimal, incurring a minimum cost, any solution found will also be an solution to the WCSP.

In contrast with traditional branch and bound algorithms our algorithm is a hybrid approach in which a standard CSP solver is used to solve the relaxation and a mixed integer program solver (MIP) is used to compute optimal new relaxations. Our approach is most closely related to unsatisfiable core techniques that have been developed for solving MAXSAT. However by exploiting the fact that at most one tuple in a constraint can be satisfied by any variable assignment we are able to develop a more compact encoding of the optimization problem used to compute the optimal relaxation. We prove that the algorithm is sound, and provide some preliminary empirical results on its performance.

1 Introduction

Many practical problems involve some degree of optimization. That is, typically we are not only interested in finding a solution but in finding low cost solutions, or even optimal solutions, when we can compute them. Weighted CSPs (WCSP) or soft CSPs are a CSP based formalism geared towards representing optimization problems. In this formalism constraints are replaced by cost functions (soft constraints) and instead of aiming to find a solution that satisfies all constraints one aims to find a solution that incurs lowest total cost from the cost functions.

The most prominent methods for solving WCSP employ branch and bound search, e.g., Toulbar [1]. These solvers depend on sophisticated methods for computing lower bounds [2] during search.

A WCSP can be viewed as being the CSP version of weighted MAXSAT. In weighted MAXSAT we have a set of clauses each with a weight and are trying to find a truth assignment that falsifies the lowest total weight of clauses. Research

C. Schulte (Ed.): CP 2013, LNCS 8124, pp. 273–281, 2013.

in MAXSAT has also pursued branch and bound solvers, e.g., [3, 4], however an effective alternative is to solve MAXSAT by solving a sequence of decision problems (typically SAT decision problems), e.g., [5–7]. Empirically, it has been found that on larger problems the sequence of decision problems approach works better. For example in the MAXSAT evaluations [8] branch and bound solvers are not as effective on the larger problems in the industrial category.

In this paper we present a sequence of decision problems approach to solving WCSP. Our approach is based on the ideas presented in [7, 9], but involves some key innovations aimed at better exploiting the additional structure of WCSPs. The approach is a hybrid one in which both a hard-CSP solver and a mixed integer program solver (MIP) solver are used. The idea is similar to the general paradigm of Logic Based Benders Decomposition [10]. In particular, we use the MIP solver to generate candidate optimal solutions and the CSP solver to test their feasibility. If the candidate is feasible we have solved the WCSP.

The novelty of our approach lies in the manner in which we construct the MIP and CSP subproblems: our models are designed to exploit the structure of WCSPs. In the paper we present our approach and prove it to be sound. We close the paper with some preliminary empirical results that indicate the approach has some potential, although more work needs to be done to make it competitive with the far more well developed branch and bound solvers.

2   Background

A weighted CSP (WCSP), $wt\mathcal{P} = (C, V)$ is specified by a set of variables $V = \{v_1, \ldots, v_n\}$, each with an associated domain of values D_i, and a set of soft constraints or cost functions $C = \{c_1, \ldots, c_m\}$. Each $c \in C$ is a function over a subset $scope(c)$ of V, called its scope; c maps tuples of assignments τ over the variables in $scope(c)$ to positive numbers or infinity. If $c(\tau) = \infty$ then τ is **forbidden** by c. Otherwise $c(\tau)$ is the cost incurred by τ from c.

An assignment π is a mapping $v \to d \in D_i$ for all $v \in V$. A partial assignment is a mapping of some subset of the variables. If an assignment π includes all of the variables in $scope(c)$ for some cost function c, then $cost_c(\pi)$ denotes the value of c evaluated on those assignments.

We restrict our attention to cost functions specified extentionally as tables that list all assignment tuples over the function's variables that have non-zero cost (the table also specifies the cost of each such tuple). Tuples that have infinite cost are **hard** tuples. Tuples that specify finite non-zero costs are **soft** tuples. Hard constraints are constraints containing only hard tuples. We may harden tuples, constraints, or weighted CSPs increasing the weight of the corresponding soft tuples to infinity.

The cost of a complete assignment π is the sum of the costs it incurs from cost functions: $cost(\pi) = \sum_{c \in C} cost_c(\pi)$. An assignment π is a **solution** to $wt\mathcal{P}$ if it has **finite cost**. A solution is an **optimal solution** if no other solution has lower cost. Solving a WCSP means finding an optimal solution.

One successful method of solving weighted CSP problems is branch and bound search which uses lower bounds to prune the search space. Soft arc consistency techniques [2] such as EDAC, VAC, and OSAC transform a problem preserving the cost of all solutions, but where as much cost as possible is moved into a special 0-ary cost function $c_\emptyset$ incurred by all assignments. Hence, $c_\emptyset$ provides a lower bound on the optimal cost. Applying these techniques during search provides a lower bound on the solutions extending the current partial assignment. EDAC and VAC used in this manner form the basis of the successful Toulbar2 solver.

In MAXSAT solving a successful alternative solution technique consists of solving a sequence of decision problems. In the sequence of decision problems approach all soft costs are hardened and a proof of unsatisfiability is derived using a SAT solver. The proof of unsatisfiability takes the form of a *core*, a set of culprit soft clauses at least one of which must be falsified. Either new constraints are added to the MAXSAT problem based on the core, e.g, [5, 6], or a relaxation of the hard problem is derived from the set of known cores [9]. This repeats until the decision problem becomes satisfiable.

In adapting the sequence of decision problems approach to weighted CSPs we define a new concept of a core specific to weighted CSPs, as well as a formal definition of a relaxation of the hardened WCSP.

Definition 1. *A **weight vector** for $wt\mathcal{P} = (C, V)$, where $C = \{c_1, \ldots, c_m\}$ is a vector $\langle w_1, \ldots, w_m \rangle$ of numbers where w_i is a cost value that could be assigned by c_i, or zero, or infinity: $w_i \in \{0, \infty\} \cup \{v | \exists \tau \text{ s.t. } c_i(\tau) = v\}$. Note that there are only a finite number of weight vectors for $wt\mathcal{P}$.*

Definition 2. *For a cost function c and weight w, $0 \le w \le \infty$, let $ceiling(c, w)$ be the tightened cost function generated by hardening all tuples τ in c with $c(\tau) \ge w$, i.e., $ceiling(c, w)(\tau) = \infty$ if $c(\tau) \ge w$ and $ceiling(c, w)(\tau) = c(\tau)$ if $c(\tau) < w$. Note that $ceiling(c, \infty) = c$ and $ceiling(c, 0)$ forbids all tuples.*

*A **core** of $wt\mathcal{P} = (C, V)$ is a weight vector $\mathbf{w}$ such that there is no solution (finite cost complete assignment) of the WCSP $(\{ceiling(c_i, w_i) | c_i \in C\}, V)$. That is, if we tighten each cost function c_i by w_i, then all complete assignments have infinite weight.*

This differs from cores in a MAXSAT problem. In MAXSAT each tuple in a cost function corresponds to a soft clause. A core containing many soft clauses corresponds to a collection of hardened tuples sufficient to cause the CSP to have no solution. Such cores might, and typically will, contain many tuples from each cost function. In our formulation, in contrast, there is only one element in the core per cost function. These elements correspond to *sets of tuples* of related weight within the cost function. This more compact representation of a core is possible because the tuples of each cost function are mutually exclusive—no assignment can contain more than one tuple of the cost function.

Definition 3. *A **relaxation** of $wt\mathcal{P} = (C, V)$ is an unweighted CSP $\mathcal{P}_{\mathbf{w}}$ generated by a weight vector $\mathbf{w}$. The CSP $\mathcal{P}_{\mathbf{w}}$ is formed by converting each $c_i \in C$ to a true/false constraint $c_i^{w_i}$ of $\mathcal{P}$ with $c_i^{w_i}(\tau) = true$ iff $c_i(\tau) \le w_i$. The **cost** of a relaxation $\mathcal{P}_{\mathbf{w}}$, $cost(\mathcal{P}_{\mathbf{w}})$, is $\sum_i w_i$.*

Algorithm 1. Algorithm for solving a WCSP $wt\mathcal{P}$

```
1 wtCSP-Solver(wtP)
2 begin
3    │  Cores  ⟵ ∅
4    │  w = 0
5    │  while true do
6    │  │  (solvable?, newcore)  ⟵  relaxAndSolve(wtP, w)
7    │  │  if solvable? then
8    │  │  │  return extractLastSolution()
9    │  │  cores  ⟵  cores ∪ {newcore}
10   │  │  w  ⟵  getOptimalWtVec(cores)
```

The relaxation $\mathcal{P}_0$, in which all weights are zero, admits only assignments incurring zero cost in the WCSP as solutions, and when the weights are all infinite the relaxation admits all assignments as solutions. In general, for weights w, $\mathcal{P}_w$ is a weakening of $\mathcal{P}_0$ that admits as solutions only those assignments that incur a cost $\leq w_i$ from cost function c_i. For convenience we have defined $\mathcal{P}_w$ to be a relaxation of the WCSP, although technically it is a relaxation of the CSP $\mathcal{P}_0$.

Definition 4. *A weight vector w **satisfies a set of cores** $\mathcal{K}$ if for every core (weight vector) $v \in \mathcal{K}$ we have that $w_i \geq v_i$ at some index i. The vector w is **optimal for** $\mathcal{K}$ if it satisfies $\mathcal{K}$ and for all other weight vectors w' that satisfy $\mathcal{K}$ we have $\sum_i w_i \leq \sum_i w'_i$.*

3   Relaxation Based Algorithm for Solving wcsp

To solve a weighted CSP as a sequence of decision problems we rely on two operations, extracting cores from a hard CSP instance with no solution and finding an optimal relaxation of a set of cores. This solve and relax approach has a number of similarities with the MAXHS solver for MAXSAT problems [7, 9]. Starting with the zero cost relaxation $\mathcal{P}_0$ we extract a core from it, calculate an optimal satisfying weight vector w for the current set of cores, form the new relaxation $\mathcal{P}_w$, and repeat until we find a relaxation that has a solution. We will assume for the sake of clarity that the WCSP has at least one solution.

Lemma 1. *Let π be any complete assignment for $wt\mathcal{P}$, and $w_\pi = \langle cost_{c_1}(\pi),$ $\dots cost_{c_m}(\pi)\rangle$, then w_π satisfies all cores of $wt\mathcal{P} = (C, V)$.*

Proof. Let w be any core. By definition of a core π must incur infinite cost on $(\{ceiling(c_i, w_i)|c_i \in C\}, V)$. This means that $cost_{c_i}(\pi) \geq w_i$ for some index i, and thus w_π satisfies w.

Lemma 2. *Let $\mathcal{P}_w$ be a relaxation generated by the weight vector w. If π is a solution of $\mathcal{P}_w$, then $cost(\pi) \leq \sum_i w_i$.*

Proof. For every c_i all tuples of cost greater than w_i are forbidden in $\mathcal{P}_w$. Therefore, $cost_{c_i}(\pi) \leq w_i$ for every c_i.

Lemma 3. *Let $\mathcal{K}$ be a set of cores of $wt\mathcal{P}$ and $\mathbf{w}$ be an optimal weight vector satisfying $\mathcal{K}$. If π is a solution to the relaxation $\mathcal{P}_{\mathbf{w}}$ then $cost(\pi) = \sum_i w_i$.*

Proof. By Lemma 1 the weight vector $\langle cost_{c_1}(\pi), \ldots, cost_{c_m}(\pi)\rangle$ satisfies $\mathcal{K}$. Since $\mathbf{w}$ is optimal for $\mathcal{K}$ we must have $cost(\pi) \geq \sum_i w_i$. By Lemma 2 $cost(\pi) \leq \sum_i w_i$.

Theorem 1. *Let $\mathcal{K}$ be any set of cores of $wt\mathcal{P}$, and $\mathbf{w}$ be an optimal weight vector satisfying $\mathcal{K}$. If π is a solution to the relaxation $\mathcal{P}_{\mathbf{w}}$ then π is an optimal solution for $wt\mathcal{P}$.*

Proof. Any solution for $wt\mathcal{P}$ generates a weight vector satisfying $\mathcal{K}$ by Lemma 1. Therefore, every solution must have cost at least $\sum_i w_i$. By Lemma 3, π has cost equal to $\sum_i w_i$, therefore π has optimal cost.

Each relaxation computed by "getOptimalWtVec" satisfies all previous discovered cores, so the same core cannot be discovered again. Since there are only a finite number of cores (there are only a finite number of weight vectors), the algorithm must terminate. Theorem 1 shows that when the algorithm terminates it has found an optimal solution to $wt\mathcal{P}$.

3.1   Extracting Cores

To extract a core from an unsolvable CSP instance we need to determine a set of culprit tuples sufficient to make the current relaxation $\mathcal{P}_{\mathbf{w}}$ unsatisfiable. The subroutine "relaxAndSolve" in Algorithm 1 uses the clause learning CSP solver MINICSP [11] operating on the strictest relaxation $\mathcal{P}_0$.

In $\mathcal{P}_0$ each cost function is converted to a constraint that forbids all non-zero cost assignments, and MINICSP encodes each of these forbidden tuples as a clause blocking that set of assignments. If the clause C arises from the tuple τ in cost function c_i with $0 < c_i(\tau) < \infty$, then a special "blocking variable" $b_i^{c_i(\tau)}$ is added to C, otherwise C is unchanged. For each cost function c_i and finite non-zero weight w that could be assigned by c_i, we have a single blocking variable b_i^w: all clauses arising from tuples with cost w are "blocked" by the same variable b_i^w.

We use the assumption mechanism of MINICSP in which a set of literals can be given as assumptions. If the CSP is unsatisfiable a subset of these assumptions sufficient to cause unsatisfiablity is returned. In particular, to extract a core from the relaxation $\mathcal{P}_{\mathbf{w}}$ we use the set of assumptions: $\{\neg b_i^v | v > w_i\} \cup \{b_i^v | v \leq w_i\}$.

When b_i^v is true all clauses of weight v from cost function c_i are "blocked". That is, these clauses are immediately satisfied and no longer constrain the theory. When b_i^v is false the clauses of weight v are enforced. Thus any solution to the problem under this set of assumptions must be a solution of $wt\mathcal{P}$ that incurres no more than cost w_i from c_i, for all i.

If there is no solution then MINICSP will return a subset of the assumptions causing unsatisfiablity. The blocking variables appear only positively in the clauses, so no positive b_i^v assumption can contribute to unsatisfiablity. Hence, the subset returned is a set of negated blocking variables asserting that at least one of these variables b_i^v must be made true, i.e., we must incur at least one of these costs in any solution of $wt\mathcal{P}$.

Let U be the set of negated blocking variables causing unsatisfiablity returned by MINICSP. We convert this into the core $\kappa = \{w_i | w_i = \min_v(\{b_i^v \in U\})\}$. That is, the weight vector is determined by the minimum weight tuple of c_i contained in U: to find an optimal cost solution we need only consider U's lower bound on the cost that could be incurred from c_i. If no tuples of c_i contributed to unsatisfiability, we use $w_i = \infty$ in the core.

3.2   Finding Optimal Relaxations

Finding an optimal relaxation of a set of cores is similar to finding a minimum weight hitting set over the cores with some additional constraints. The set of blocking variables, B, used to extract the cores are used again as 0/1 variables in this optimization problem, along with the added constraint that higher cost variables imply the lower cost variables of the same cost function. We use the MIP solver CPLEX to solve this optimization problem.

Assume that the blocking variables within a cost function are sorted in increasing order of cost. Define $\rho(b_i^v)$ for $b_i^v \in B$ to be the next smaller cost blocking variable for c_i or $\emptyset$ if v is the smallest non-zero cost of c_i. ρ will act as a predecessor function, returning the variable which came before it in a cost function. Using ρ we can define δ, a function assigning an adjusted cost to blocking variables. δ compensates for the fact that larger cost variables imply smaller cost variables.

$$\delta(b_i^w) = \begin{cases} cost(b_i^w) & \text{if } \rho(b_i^w) = \emptyset \\ cost(b_i^w) - cost(\rho(b_i^w)) & \text{otherwise} \end{cases} \quad (1)$$

The objective function is the sum of the adjusted costs of all blocking variables. The constraints are that one variable from each core must be selected and larger cost variables imply smaller cost variables within a cost function. For a set of blocking variables B and cores $\mathcal{K}$ we get the following MIP:

$$\text{minimize} \sum_{b_i^w \in B} \delta(b_i^w) \quad (2)$$

Subject to:

$$\rho(b_i^w) - b_i^w \geq 0 \text{ for } b_i^w \in B \text{ such that } \rho(b_i^w) \neq \emptyset \quad (3)$$

$$\forall w \in \mathcal{K}: \sum_{w_i \in w \text{ s.t. } w_i < \infty} b_i^{w_i} \geq 1 \quad (4)$$

The term $\rho(b_i^w) - b_i^w \geq 0$ in Equation 3 is logically equivalent to $b_i^w \Rightarrow \rho(b_i^w)$ since b_i^w and $\rho(b_i^w)$ are binary variables. These constraints ensure that a higher cost variable in a cost function implies all lower cost variables. Equation 2 combined with Equation 3 will cause all $cost(b_i^w)$ terms in $\sum \delta(b_i^w)$ to cancel out except for the largest cost in each cost function. Thus Equation 2 minimizes the sum of the largest cost incurred by each cost function. Equation 4 states that for each core w, a cost greater than or equal to w_i for at least one cost function c_i must be incurred. It can never be optimal to incur cost w_i when $w_i = \infty$ so they can be omitted from the constraint.

The solution S to this optimization problem is a setting of the $0/1$ B variables. From this setting we generate an optimal weight vector for $\mathcal{K}$ $\{w_i | w_i = \max_v \{b_i^v = 1 \in S\}\}$. If in S we have that $b_i^v = 0$ for all v, we set $w_i = 0$: the cores can be satisfied without incurring any cost from c_i.

3.3   Improving Cores

We have found that computing an optimal weight vector for the current set of cores is typically the most time consuming part of our approach. Spending additional effort in the CSP solver to produce stronger cores can often yield better results by reducing the overall effort required in the optimization phase [7].

As described in Sec. 3.1 we compute a core for $\mathcal{P}_w$ by solving $\mathcal{P}_0$ augmented with blocking variables under the assumptions $\{\neg b_i^v | v > w_i\} \cup \{b_i^v | v \leq w_i\}$. MINICSP returns a set U of negated blocking variables from which we compute a core.

To produce stronger cores we sort U so that for each cost function its smaller cost blocking variables appear before its higher cost blocking variables: $\neg b_i^v$ appears before $\neg b_i^w$ if $v < w$. We then test the blocking variables of U in this order to see if they can be removed.

To test if blocking variable b can be removed we solve the CSP again under the new set of assumptions $U - \{b\} \cup \{b_i^w | b_i^w \notin \{U - \{b\}\}\}$. These assumptions block all clauses except those in $U - \{b\}$. If we obtain unsatisfiability we remove b from U and otherwise we retain it. In either case we move on to the next blocking variable of U and test it in the same manner.[1] When we find that b_i^w is necessary for unsatisfiability, we can avoid testing all later variables of the form b_i^v with $v > w$. Intuitively, if we can satisfy the problem by incurring cost w from cost function c_i (i.e., the CSP becomes satisfiable) then we can satisfy the problem by incurring an even higher cost v from c_i.

Once we have updated U by trying to remove each of its variables, we form a core from it exactly as described in Sec. 3.1.

4   Results

The solver described in this paper was implemented and tested on available weighted CSP instances. A few additional optimizations were made. In the MAXHS approach for MAXSAT [7] it was found that computing a new relaxation for every core was time consuming. A technique proposed by Karp [12] was to use a greedy approach for determining the relaxation so as to acquire multiple cores per optimal calculation of the relaxation. Optimal relaxations are generated after a solution is found to a greedy relaxation. Toulbar2 was also used as a preprocessor to provide virtual arc consistency [2].

---

[1] If the CSP solver returns a subset of U we can also reduce U to this subset.

Table 1. Time(s) for selected instances from the Spot5 and Linkage benchmarks

Problem	CSP-Seq	Toulbar2	Problem	CSP-Seq	Toulbar2
Spot5 404	6.14	209.17	Linkage pedigree 18	14.61	119.15
Spot5 503	0.48	-	Linkage pedigree 14	8.92	0.52
Spot5 505	562.75	-	Linkage pedigree 30	27.09	240.71
Linkage pedigree 25	11.64	-	Linkage pedigree 9	63.47	223.85
Linkage pedigree 39	36.14	3.32	Linkage pedigree 20	15.04	0.76
Linkage pedigree 31	-	779.63	Linkage pedigree 44	145.11	-
Linkage pedigree 7	1.52	6.98	Linkage pedigree 13	4.06	0.56
Linkage pedigree 41	270.22	969.88	Linkage pedigree 33	0.76	10.58
Linkage pedigree 51	19.83	-			

Tests were run on the celar, spot5, and linkage benchmarks. The celar benchmark [13] consists of radio frequency link assignment problems and features primarily soft constraints. The spot5 benchmark [14] is about managing satellites and mixes hard and soft constraints. The linkage benchmark problems are about probabilistic inference for genetic linkage. The entries of Table 1 give time in seconds for our solver (CSP-Seq) and Toulbar2 for problems in the spot5 and linkage benchmarks. Excluded are 16 spot5 problems which both solvers time out on, 3 spot5 benchmarks both solvers completed in under 1s, 2 linkage problems both solvers time out on, and 6 linkage problems both solvers completed in under 3s. Tests were run with a 1800s time limit on a AMD Opteron 2435. A table entry of "-" indicates the solver timed out on that problem instance.

Our solver performed poorly on the celar benchmark, with performance dominated by the Toulbar2 solver. We conjecture that this problem set is more well suited to the techniques used in the Toulbar2 solver, and that our technique will be useful for other types of problems, especially those featuring a significant number of hard constraints.

5   Conclusion

We have presented a new method for solving WCSPs based on methods that have been employed in MAXSAT. The main innovation of our approach is to recognize that the notion of a relaxation and a core can be specialized to exploit the specific structure of WCSPs. In particular, in a cost function at most one tuple can be activated by any solution. Thus we can compress the cores down to simply a relaxation weight for each constraint. For future work we are examining a number of methods for improving the performance of the approach. Primary among them is take better advantage of the implication relationships between the b-variables associated with the same constraint. Currently these are encoded directly as implications in the MIP model, but are exploited only via generic methods employed in the MIP solver.

References

1. Toulbar2: WCSP solver, http://mulcyber.toulouse.inra.fr/projects/toulbar2
2. Cooper, M., de Givry, S., Sanchez, M., Schiex, T., Zytnicki, M., Werner, T.: Soft arc consistency revisted. In: Artificial Intelligence, pp. 449–478 (2010)
3. Heras, F., Larrosa, J., Oliveras, A.: Minimaxsat: An efficient weighted max-sat solver. Journal of Artificial Intelligence Research (JAIR) 31, 1–32 (2008)
4. Alsinet, T., Manyà, F., Planès, J.: Improved exact solvers for weighted max-SAT. In: Bacchus, F., Walsh, T. (eds.) SAT 2005. LNCS, vol. 3569, pp. 371–377. Springer, Heidelberg (2005)
5. Heras, F., Morgado, A., Marques-Silva, J.: Core-guided binary search algorithms for maximum satisfiability. In: Proceedings of the AAAI National Conference. AAAI (2011)
6. Ansótegui, C., Bonet, M.L., Gabàs, J., Levy, J.: Improving SAT-based weighted MaxSAT solvers. In: Milano, M. (ed.) CP 2012. LNCS, vol. 7514, pp. 86–101. Springer, Heidelberg (2012)
7. Davies, J., Bacchus, F.: Exploiting the power of MIP solvers in MAXSAT. In: Järvisalo, M., Van Gelder, A. (eds.) SAT 2013. LNCS, vol. 7962, pp. 166–181. Springer, Heidelberg (2013)
8. Argelich, J., Li, C.M., Manyà, F., Planes, J.: The maxsat evaluations (2007–2011), http://www.maxsat.udl.cat
9. Davies, J., Bacchus, F.: Solving MAXSAT by solving a sequence of simpler SAT instances. In: Lee, J. (ed.) CP 2011. LNCS, vol. 6876, pp. 225–239. Springer, Heidelberg (2011)
10. Hooker, J.N.: Planning and scheduling by logic-based benders decomposition. Operations Research 55(3), 588–602 (2007)
11. Katsirelos, G.: MINICSP CSP solver.
    http://www7.inra.fr/mia/T/katsirelos/minicsp.html
12. Moreno-Centeno, E., Karp, R.M.: The implicit hitting set approach to solve combinatorial optimization problems with an application to multigenome alignment. Operations Research 61(2), 453–468 (2013)
13. Cabon, B., de Givry, S., Lobjois, L., Schiex, T., Warners, J.: Radio link frequency assignment. Constraints 4(1), 79–89 (1999)
14. Bensana, E., Lemaitre, M., Verfaillie, G.: Earth observation satellite management. Constraints 4(3), 293–299 (1999)

Constraint-Based Program Reasoning with Heaps and Separation

Gregory J. Duck, Joxan Jaffar, and Nicolas C.H. Koh

Department of Computer Science, National University of Singapore
{gregory,joxan}@comp.nus.edu.sg, kchuenho@dso.org.sg

Abstract. This paper introduces a constraint language $\mathcal{H}$ for *finite partial maps* (a.k.a. *heaps*) that incorporates the notion of *separation* from *Separation Logic*. We use $\mathcal{H}$ to build an extension of *Hoare Logic* for reasoning over *heap manipulating programs* using (constraint-based) *symbolic execution*. We present a sound and complete algorithm for solving quantifier-free (QF) $\mathcal{H}$-formulae based on *heap element propagation*. An implementation of the $\mathcal{H}$-solver has been integrated into a *Satisfiability Modulo Theories* (SMT) framework. We experimentally evaluate the implementation against *Verification Conditions* (VCs) generated from symbolic execution of large (heap manipulating) programs. In particular, we mitigate the *path explosion problem* using subsumption via interpolation – made possible by the constraint-based encoding.

Keywords: Heap Manipulating Programs, Symbolic Execution, Separation Logic, Satisfiability Modulo Theories, Constraint Handling Rules.

1 Introduction

An important part of reasoning over *heap manipulating programs* is the ability to specify properties local to *separate* (i.e. non-overlapping) regions of memory. Most modern formalisms, such as Separation Logic [20], Region Logic [2], and (Implicit) Dynamic Frames [16][22], incorporate some encoding of separation. Separation Logic [20] explicates separation between regions of memory through *separating conjunction* ($*$). For example, the *Separation Logic formula* **list**$(l)*$**tree**(t) represents a program *heap* comprised of two separate sub-heaps: one containing a *linked-list* and the other a *tree* data-structure.

In this paper we explore a reformulation of Separation Logic in terms of a first-order *constraint language* $\mathcal{H}$ over *heaps* (i.e. finite partial maps between pointers and values). Under this approach, separating conjunction ($*$) is re-encoded as a *constraint* $H \simeq H_1 * H_2$ between heaps, indicating that: (1) heaps H_1 and H_2 are *separate* (i.e. disjoint domains) and (2) H is the *heap union* of H_1 and H_2. We can therefore re-encode the above Separation Logic formula as **list**$(l, L) \wedge$ **tree**$(t, T) \wedge \bar{\mathcal{H}} \simeq L * T$ where **list** and **tree** are redefined to be predicates over heaps, and the special variable $\bar{\mathcal{H}}$ represents the *global heap* at the program point where it appears. We can also represent a *singleton heap* as a constraint $\bar{\mathcal{H}} \simeq (p \mapsto v)$.

C. Schulte (Ed.): CP 2013, LNCS 8124, pp. 282–298, 2013.

The motivation behind $\mathcal{H}$ is to lift some of the benefits of Separation Logic to constraint-based reasoning techniques for heap manipulating programs, such as *constraint-based symbolic execution*. Our method is based on an extension of *Hoare Logic* [11] defined in terms of the constraint language $\mathcal{H}$. Whilst Separation Logic guarantees *total correctness* w.r.t. *memory safety* (e.g. no memory errors such as dereferencing dangling pointers, etc.), our reformulation allows for weaker axiomatizations, such as a version that drops the memory-safety requirement. This allows for a *Strongest Post Condition* (SPC) *predicate transformer semantics* [7] to be defined in terms of $\mathcal{H}$, which forms the basis of symbolic execution. The resulting *Verification Conditions* (VCs) can then be discharged using a suitable $\mathcal{H}$-*constraint solver/theorem prover*. This is illustrated with a simple example:

Example 1 (Heap Equivalence). Consider the following Hoare triple:

$$\{H = \bar{\mathcal{H}}\} \; x := \mathbf{alloc}(); \mathbf{free}(x) \; \{H = \bar{\mathcal{H}}\} \tag{1}$$

This triple states that the global heap *before* the code fragment is equal to the heap *after* the fragment, i.e. the global heap is *unchanged*. Here H is a *ghost variable* representing the initial state of the global heap $\bar{\mathcal{H}}$. Symbolic execution of the precondition $P \equiv (H = \bar{\mathcal{H}})$ yields the following $\mathcal{H}$-constraints:

$$Q \;\equiv\; \left(H = H_0 \wedge \underline{H_1 \simeq (x \mapsto \_)*H_0} \wedge \underline{H_1 \simeq (x \mapsto \_)*\bar{\mathcal{H}}}\right)$$

Here H_0 and H_1 represent the initial and intermediate values for $\bar{\mathcal{H}}$ respectively. The underlined $\mathcal{H}$-constraints encode the $\mathbf{alloc}()$ and $\mathbf{free}()$ respectively. Next we can employ an $\mathcal{H}$-constraint solver to prove that the postcondition is implied by Q, i.e. the *Verification Condition* (VC) $Q \rightarrow H = \bar{\mathcal{H}}$ holds, thereby proving the triple (1) valid. □

In order to discharge the VCs generated from symbolic execution we need a solver for the resulting $\mathcal{H}$-formulae. For this we present a simple decision procedure for *Quantifier Free* (QF) $\mathcal{H}$-formulae based on the idea of *heap membership propagation*. We show that the algorithm is both sound and complete, and is readily implementable using *Constraint Handling Rules* (CHR) [10]. We present an implementation of an $\mathcal{H}$-solver that has been integrated into a *Satisfiability Modulo Theories* (SMT) framework using SMCHR [8]. Our decision procedure is related to established algorithms for finite sets.

We use the $\mathcal{H}$-solver as the basis of a simple program verification tool using symbolic execution. In contrast to Separation Logic-based symbolic execution [4], which is based on a set of *rearrangement rules*, our version is based on constraint solving using the $\mathcal{H}$-solver as per Example 1 above. Our encoding allows for some optimization. Namely, we mitigate the *path explosion problem* of symbolic execution by employing *subsumption* via *interpolation* [14][17] techniques.

This paper is organized as follows: Section 2 introduces Hoare and Separation Logic, Section 3 formally introduces the $\mathcal{H}$-language, Section 4 introduces an extension of Hoare Logic based on the $\mathcal{H}$-language, Section 5 presents an $\mathcal{H}$-solver algorithm and implementation, and Section 6 experimentally evaluates the implementation. In summary, the contributions of this paper are the following:

- We define the $\mathcal{H}$-language that encodes separation as a constraint between heaps. We show that satisfiability of quantifier-free $\mathcal{H}$-formulae is decidable, and present a complete algorithm for solving $\mathcal{H}$-formulae.
- We present an extension of Hoare Logic based on the $\mathcal{H}$-language. Our extension is similar to Separation Logic, but allows for strongest post conditions, and is therefore suitable for program reasoning via constraint-based symbolic execution.
- We present an implementation of the $\mathcal{H}$-solver that has been integrated into an SMT framework. We experimentally evaluate the solver against VCs generated from symbolic execution of heap manipulating programs.

2   Preliminaries

This section presents a brief overview of Hoare and Separation Logic.

Hoare Logic [11] is a formal system for reasoning about program correctness. Hoare Logic is defined in terms of axioms over *triples* of the form $\{\phi\}\ C\ \{\varphi\}$, where ϕ is the *pre-condition*, φ is the *post-condition*, and C is some code fragment. Both ϕ and φ are formulae over the *program variables* in C. The meaning of the triple is as follows: for all program states σ_1, σ_2 such that $\sigma_1 \models \phi$ and executing σ_1 through C derives σ_2, then $\sigma_2 \models \varphi$. For example, the triple $\{x < y\}\ x := x+1\ \{x \leq y\}$ is *valid*. Note that under this definition, a triple is automatically valid if C is non-terminating or otherwise has undefined behavior. This is known as *partial correctness*.

Separation Logic [20] is a popular extension of Hoare Logic for reasoning over *heap manipulating programs*. Separation Logic extends predicate calculus with new logical connectives (namely *empty heap* (**emp**), *singleton heap* $(p \mapsto v)$, and *separating conjunction* $(H_1 * H_2)$) such that the structure of assertions reflects the structure of the underlying heap. For example, the pre-condition in the valid Separation Logic triple $\{x \mapsto \_ * y \mapsto 2\}\ [x] := [y]+1\ \{x \mapsto 3 * y \mapsto 2\}$ represents a heap comprised of two *disjoint singleton* heaps, indicating that both x and y are *allocated* and that location y points to the value 2. Here the notation $[p]$ represents pointer dereference. In the post-condition we have that x points to value 3, as expected. Separation Logic also allows *recursively-defined* heaps for reasoning over data-structures, such as **list**(l) and **tree**(t) from Section 1.

Separation Logic triples also have a slightly different meaning versus Hoare triples regarding *memory-safety*. A Separation Logic triple $\{\phi\}\ C\ \{\varphi\}$ additionally guarantees that any state satisfying ϕ will not cause a memory access violation in C. For example, the triple $\{$**emp**$\}\ [x] := 1\ \{x \mapsto 1\}$ is invalid since x is a dangling pointer in any state satisfying the pre-condition.

3   Heaps with Separation

This section formally introduces the syntax and semantics of heaps with separation, which we denote by $\mathcal{H}$, that encodes some of the logical connectives of Separation Logic. We assume as given a countably infinite set Values denoting *values*, e.g. Values $= \mathbb{Z}$. A *heap* is a *finite partial map* between Values, i.e. Heaps $=$ Values $\rightarrow_{\text{fin}}$ Values. This is the same definition as used by Separation Logic. Given a heap $h \in$ Heaps with domain $D = dom(h)$, we sometimes treat h as the set of pairs $\{(p, v) \mid p \in D \land v = h(p)\}$.

The $\mathcal{H}$-language is the first-order language over heaps defined as follows:

Definition 1 (Heap Language). We define the $\mathcal{H}$-signature $\Sigma_{\mathcal{H}}$ as follows:
- *sorts*: Values, Heaps;
- *constants*: (empty heap) $\emptyset$ of sort Heaps;
- *functions*: (singleton heap) $(\_ \mapsto \_)$ of sort Values $\times$ Values $\mapsto$ Heaps.
- *predicates*: (heap constraint) $(\_*\ldots*\_ \mathbin{\triangleq} \_*\ldots*\_)$ of sort Heaps$\times\cdots\times$Heaps $\mapsto$ $\{true, false\}$.

The $\mathcal{H}$-language is the *first-order language* over $\Sigma_{\mathcal{H}}$. $\qquad\square$

Example 1 used heap constraints of the form $H \mathbin{\triangleq} H_1*H_2$, where H, H_1, and H_2 are variables. Throughout this paper we shall use upper-case letters H, I, J, etc., to denote *heap variables*, and lower-case letters p, v, etc., for *value variables*.

A *valuation* s (a.k.a. *variable assignment*) is a function mapping values to Values $\cup$ Heaps. We define the semantics of the $\mathcal{H}$-language as follows:

Definition 2 (Heap Interpretation). Given a valuation s, the $\mathcal{H}$-interpretation $\mathcal{I}$ is a $\Sigma_{\mathcal{H}}$-interpretation such that:
- $\mathcal{I}(v, s) = s(v)$, where v is a variable;
- $\mathcal{I}(\emptyset, s) = \emptyset$ (as a Heap);
- $\mathcal{I}(p \mapsto v, s) = \{(q, w)\}$ where $q = \mathcal{I}(p, s)$ and $w = \mathcal{I}(v, s)$;
- $\mathcal{I}(H_1 * \ldots * H_i \mathbin{\triangleq} H_{i+1} * \ldots * H_n, s) = true$ iff for $h_i = \mathcal{I}(H_i, s)$ we have that:
  1. $dom(h_1) \cap \ldots \cap dom(h_i) = \emptyset$ and $dom(h_{i+1}) \cap \ldots \cap dom(h_n) = \emptyset$; and
  2. $h_1 \cup \ldots \cup h_i = h_{i+1} \cup \ldots \cup h_n$ $\qquad\square$

Note that we treat each configuration of $(*)$ and $(\triangleq)$ as a distinct predicate. Intuitively, a constraint like $H \mathbin{\triangleq} H_1*H_2$ treats $(*)$ in essentially the same way as *separating conjunction* from Separation Logic, except that we give a name H to the conjoined heaps H_1*H_2.

We define $\models_{\mathcal{H}} \ldots [s]$ as the *satisfaction relation* such that $\models_{\mathcal{H}} \phi \, [s]$ holds iff $\mathcal{I}(\phi, s) = true$ for all heap formulae ϕ. We also say that ϕ is *valid* if $\models_{\mathcal{H}} \phi \, [s]$ holds for all s, and *satisfiable* if $\models_{\mathcal{H}} \phi \, [s]$ holds for at least one s.

3.1 Normalization

In the absence of quantifiers, we can restrict consideration of $\mathcal{H}$-formula to a subset in *normal form* defined as follows:

Definition 3 (Normal Form). A quantifier-free (QF) $\mathcal{H}$-formula ϕ is in *normal form* if (1) all heap constraints are restricted to three basic forms:

$$H \mathbin{\triangleq} \emptyset \qquad H \mathbin{\triangleq} (p \mapsto v) \qquad H \mathbin{\triangleq} H_1*H_2$$

where p, v, H, H_1, and H_2 are distinct variables, and (2) there are no negated heap constraints. $\qquad\square$

$$H \simeq E_1 * E_2 * S \longrightarrow H' \simeq E_1 * E_2 \wedge H \simeq H' * S$$
$$H \simeq E_1 * E_2 \longrightarrow H' \simeq E_1 \wedge H \simeq H' * E_2 \qquad (E_1 \text{ non-variable})$$
$$H \simeq H_1 * E_2 \longrightarrow H' \simeq E_2 \wedge H \simeq H_1 * H' \qquad (E_2 \text{ non-variable})$$
$$H_1 \simeq H_2 \longrightarrow H' \simeq \emptyset \wedge H_1 \simeq H_2 * H'$$

$$H \not\simeq E_1 * E_2 * S \longrightarrow \vee \begin{cases} E_1 \simeq (s \mapsto t) * H_1' \wedge E_2 \simeq (s \mapsto u) * H_2' \\ H' \simeq E_1 * E_2 \wedge H \not\simeq H' * S \end{cases}$$

$$H \not\simeq E_1 * E_2 \longrightarrow H' \simeq E_1 \wedge H \not\simeq H' * E_2 \qquad (E_1 \text{ non-variable})$$
$$H \not\simeq H_1 * E_2 \longrightarrow H' \simeq E_2 \wedge H \not\simeq H_1 * H' \qquad (E_2 \text{ non-variable})$$
$$H \not\simeq \emptyset \longrightarrow H \simeq (s \mapsto t) * H'$$

$$H \not\simeq (p \mapsto v) \longrightarrow \vee \begin{cases} H \simeq \emptyset \\ H \simeq (s \mapsto t) * H' \wedge (p \neq s \vee v \neq t) \end{cases}$$

$$H \not\simeq H_1 * H_2 \longrightarrow \vee \begin{cases} H_1 \simeq (s \mapsto t) * H_1' \wedge H_2 \simeq (s \mapsto u) * H_2' \\ H' \simeq H_1 * H_2 \wedge H \not\simeq H' \end{cases}$$

$$H_1 \not\simeq H_2 \longrightarrow \vee \begin{cases} H_1 \simeq (s \mapsto t) * H_1' \wedge H_1 \simeq (s \mapsto u) * H_2' \wedge t \neq u \\ H_1 \simeq I * H_1' \wedge H_2 \simeq I * H_2' \wedge H' \simeq H_1' * H_2' \wedge H' \not\simeq \emptyset \end{cases}$$

Fig. 1. $\mathcal{H}$-formulae normalization rewrite rules

Any given QF $\mathcal{H}$-formula ϕ can be rewritten into *normal form* using the following steps: (1) push negation inwards using De Morgan's laws, and (2) transform the resulting formula using the rewrite rules from Figure 1. Here each rewrite rule is of the form (*head* $\longrightarrow$ *body*), and E_i runs over *heap expressions* (H, $\emptyset$, $(p \mapsto v)$), S runs over $(*)$-sequences of heap expressions (E, $E*E$, etc.), and everything else runs over the variable symbols. A variable that appears in a rule body, but not the rule head, is taken to represent a fresh variable symbol that is introduced each time the rule is applied. For brevity we omit some rules, namely: normalizing the RHS of a ($\simeq$) to a heap variable (as this mirrors the LHS rules), and making variables unique. The main result for normalization is as follows:

Proposition 1 (Normal Form). *For all QF $\mathcal{H}$-formulae ϕ there exists a QF $\mathcal{H}$-formula φ such that (1) φ is in normal form and (2): for all valuations s there exists a valuation s' such that $\models_{\mathcal{H}} \phi\ [s]$ iff $\models_{\mathcal{H}} \varphi\ [s']$ and $s(v) = s'(v)$ for all $v \in vars(\phi)$.*

Proof. (Sketch) By the correctness of, and induction over, the normalization steps from Figure 1. □

Proposition 1 means that, at the expense of an increased formula size, we need only consider a limited subset of the $\mathcal{H}$-language that lacks negation.

3.2   Extensions

We may extend Definitions 1 and 2 to include other kinds of heap constraints, such as:

- *Heap union* $H \simeq H_1 \sqcup H_2$ holds iff there exists a $h \in$ Heaps such that $h = s(H_1) \cup s(H_2)$ as sets and $s(H) = h$.
- *Heap intersection* $H \simeq H_1 \sqcap H_2$ holds iff $s(H) = s(H_1) \cap s(H_2)$ as sets.
- *Heap subset* $H_1 \sqsubseteq H_2$ holds iff $s(H_1) \subseteq s(H_2)$ as sets.

These constraints can similarly be reduced to the normal form from Definition 3.

For some applications we may extend $\mathcal{H}$ with ad hoc *user-defined* heap constraints. For this we can use *Constraint Logic Programming* (CLP) [13] over $\mathcal{H}$, i.e. CLP($\mathcal{H}$). For example, the following CLP($\mathcal{H}$) predicate **list**(l, L) specifies a skeleton *list constraint* under the standard *least model semantics* of CLP:

$$\textbf{list}(0, L) \; :- \; L \simeq \emptyset$$
$$\textbf{list}(l, L) \; :- \; l \neq 0 \wedge L \simeq (l \mapsto n) * L' \wedge \textbf{list}(n, L')$$

We can similarly define predicates for trees and arrays. The inclusion of CLP predicates requires stronger reasoning power in contrast to the base $\mathcal{H}$-language. For this we can employ standard (yet incomplete) methods such as [15].

4   Program Reasoning with $\mathcal{H}$

The core motivation of the $\mathcal{H}$-language is reasoning over heap manipulating programs. For this we consider the following extensions of *Hoare Logic* [11].

4.1   Direct Separation Logic Encoding

Separation Logic [20] is itself an extension of Hoare Logic. Given the similarity in the heap representations, we can re-encode the axioms of Separation Logic directly into Hoare axioms over $\mathcal{H}$-formulae, as shown in Figure 2(B). Each axiom is defined in terms of one of five auxiliary constraints: namely alloced, access, assign, alloc, and free defined in Figure 2(A), which are themselves defined in terms of $\mathcal{H}$-formulae. The alloced(H, x) constraint represents that pointer x is *allocated* in heap H, i.e. $H \simeq (x \mapsto v) * H'$ for some v and H'. The remaining auxiliary constraints encode a *heap manipulation statement* as an $\mathcal{H}$-formula. The statements are:

- *heap access* $(x := [y])$ sets x to be the value pointed to by y;
- *heap assignment* $([x] := y)$ sets the value pointed to by x to be y;
- *heap allocation* $(x :=\textbf{alloc}())$ sets x to point to a freshly allocated heap cell.[1]
- *heap free* $\textbf{free}(x)$ deallocates the cell pointed to by x.

These axioms manipulate the *global heap* that is represented by a *distinguished heap variable* $\bar{\mathcal{H}}$. Under this treatment, $\bar{\mathcal{H}}$ is an *implicit program variable*[2] of type

---

[1] Here we assume the (de)allocation of single heap cells. This can be generalized.

[2] The variable is "implicit" in the sense that it is not explicitly represented in the syntax of the programming language.

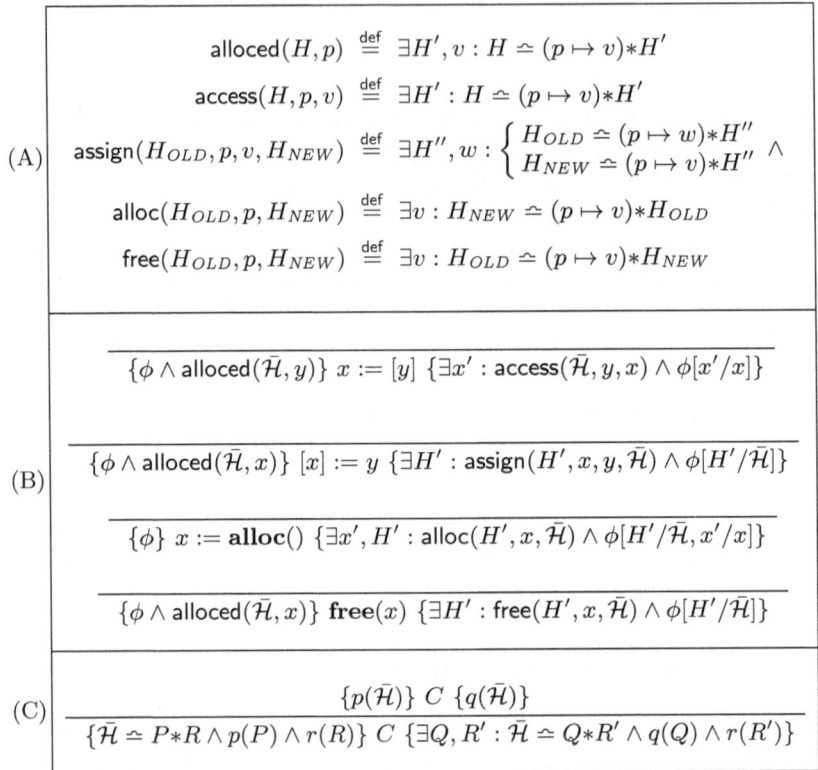

Fig. 2. (A) Auxiliary constraint definitions, (B) basic Hoare inference rules, and (C) the Frame Rule

Heap that is assumed to be threaded throughout the program. Other axioms of Separation Logic, such as the *Frame Rule* [20], can similarly be re-encoded, as shown in Figure 2(C).

It is not surprising that Separation Logic can be re-formulated as Hoare axioms over the $\mathcal{H}$-language. However, there are some important differences to consider. Notably, the $\mathcal{H}$-encoding allows for *explicit heap variables* to express relationships *between* heaps across triples. In Example 1, we use the triple $\{H \simeq \bar{\mathcal{H}}\}\, C\, \{H \simeq \bar{\mathcal{H}}\}$ to express the property that the code fragment C does not change the global heap $\bar{\mathcal{H}}$ through an explicit variable H. Such a global property would require *second order* Separation Logic, e.g., $\forall h : \{h\}\, C\, \{h\}$. Furthermore, with explicit heap variables, we can strengthen the *Frame Rule* by R' for R in the post-condition of Figure 2(C).

The $\mathcal{H}$-based encoding tends to be more verbose compared to Separation Logic, which favors more concise formulae. Whilst not so important for *automated* systems, the $\mathcal{H}$-based encoding is likely less suitable for *manual* proofs of correctness.

$$\frac{}{\{\phi\}\ x := [y]\ \{\exists x' : \mathsf{access}(\bar{\mathcal{H}}, y, x) \wedge \phi[x'/x]\}}$$

$$\frac{}{\{\phi\}\ [x] := y\ \{\exists H' : \mathsf{assign}(H', x, y, \bar{\mathcal{H}}) \wedge \phi[H'/\bar{\mathcal{H}}]\}}$$

$$\frac{}{\{\phi\}\ \mathbf{free}(x)\ \{\exists H' : \mathsf{free}(H', x, \bar{\mathcal{H}}) \wedge \phi[H'/\bar{\mathcal{H}}]\}}$$

Fig. 3. Alternative Hoare inference rules

4.2   Strongest Post-condition Encoding

Separation Logic and the corresponding $\mathcal{H}$-encoding from Figure 2 (B) enforces *total correctness* w.r.t. *memory safety*. That is, a valid triple $\{\phi\}\ C\ \{\varphi\}$ additionally ensures that any state satisfying ϕ will not cause a *memory fault* (e.g. dereferencing a dangling pointer) when executed by C. This is enforced by the *access*, *assignment*, and *free* axioms of Figure 2 (B) by requiring that the pointer x be allocated in the global heap $\mathcal{H}$ in the pre-condition via the $\mathsf{alloced}(\bar{\mathcal{H}}, x)$ constraint.

Memory safety has implications for *forward reasoning* methods such as *symbolic execution*. For example, to symbolically execute a formula ϕ through an assignment $[x] := v$, we must first prove that $\phi \rightarrow \mathsf{alloced}(\bar{\mathcal{H}}, x)$. Such a proof can be arbitrarily difficult in general, e.g. for formulae with quantifiers or recursively-defined $\mathrm{CLP}(\mathcal{H})$ predicates. Furthermore, if memory safety is *not* a property of interest, this extra work is unnecessary.

By decoupling the heap representation ($\mathcal{H}$) from the logic, we can experiment with alternative axiomatizations. One such axiomatization that is *partially correct* modulo memory safety is shown in Figure 3.[3] This version drops the requirement that x be allocated in $\bar{\mathcal{H}}$ in the pre-condition, and therefore treats memory errors the same way as *undefined behavior* (or *non-termination*) in classic Hoare Logic.

There are several advantages to the weaker axiomatization of Figure 3. Firstly, the axioms of Figure 3 specify a *Strongest Post Condition* (SPC) *predicate transformer semantics* and is therefore immediately suitable for automated forward based reasoning techniques such as symbolic execution. This is in contrast to symbolic execution in Separation Logic [4] (or the corresponding axioms from Figure 2), where symbolic execution requires the $\mathsf{alloced}$ condition to be separately proven. The SPC axiomatization allows for weaker, more concise, specifications.

Example 2 (Double List Reverse). For example, consider the following triples in the spirit of Example 1:

---

[3] The axiom for *heap allocation* is the same as Figure 2 (B).

$$\{H \simeq \bar{\mathcal{H}}\}\ l := \mathbf{reverse}(\mathbf{reverse}(l))\ \{H \simeq \bar{\mathcal{H}}\} \qquad (2)$$
$$\{\bar{\mathcal{H}} \simeq L*H' \wedge \mathbf{list}(L,l) \wedge H \simeq \bar{\mathcal{H}}\}\ l := \mathbf{reverse}(\mathbf{reverse}(l))\ \{H \simeq \bar{\mathcal{H}}\} \qquad (3)$$

Both attempt to state the same property: that double *in-place list-reverse* leaves the global heap $\bar{\mathcal{H}}$ unchanged. Suppose that the only property of interest is the heap equivalence (i.e. not memory safety). Triple (2) is valid under the weaker Figure 3 axiomatization, but not the stronger Figure 2 (B) version which requires memory safety. The latter requires a more complex specification, such as Triple (3), where the recursively defined property $\mathbf{list}(L,l)$ ensures l points to a valid allocated list. □

There are also some disadvantages to consider. For obvious reasons, the SPC axiomatization is unsuitable if memory safety *is* a property of interest. Furthermore, the soundness of Separation Logic's *Frame Rule* (or Figure 2 (C)) depends on memory safety, and thus is not valid under the new interpretation. Therefore the SPC axiomatization is not suitable for Separation Logic-style *local reasoning* proofs. In essence, this is a trade-off between local reasoning vs. making symbolic execution "easier", highlighting the flexibility of our overall approach.

5   A Solver for $\mathcal{H}$-Formulae

Automated symbolic execution depends on an $\mathcal{H}$-solver to discharge the generated *Verification Conditions* (VCs). In this section we present a simple, yet sound and complete, algorithm for solving the quantifier-free (QF) fragment of the $\mathcal{H}$-language.

Algorithm. The $\mathcal{H}$-solver algorithm is based on the *propagation* of *heap membership* and *(dis)equality* constraints. Heap membership (a.k.a. heap element) is represented by an auxiliary $in(H, p, v)$ constraint, which is defined as follows:

Definition 4 (Heap Membership). We extend Definitions 1 and 2 to include the *heap membership* constraint $in(H, p, v)$ defined as follows:

$$\models_{\mathcal{H}} in(H, p, v)\ [s] \quad \text{iff} \quad (s(p), s(v)) \in s(H)$$

where H, p, and v are variables. □

Heap element $in(H, p, v)$ is analogous to set membership $x \in S$ from set theory. (Dis)equality is propagated via the usual $x = y$ and $x \neq y$ constraints.

The $\mathcal{H}$-solver operates over conjunctions of *normalized* $\mathcal{H}$-constraints as per Definition 3. Arbitrary QF $\mathcal{H}$-formula ϕ can be normalized to a φ using the rules from Figure 1, such that the solutions to ϕ and φ correspond as per Proposition 1. The arbitrary Boolean structure of φ can be handled using the *Davis-Putnam-Logemann-Loveland* (DPLL) algorithm [6] modulo the $\mathcal{H}$-solver.

$$\mathsf{in}(H, p, v) \wedge \mathsf{in}(H, p, w) \Longrightarrow v = w \tag{1}$$

$$H \simeq \emptyset \wedge \mathsf{in}(H, p, v) \Longrightarrow \mathit{false} \tag{2}$$

$$H \simeq (p \mapsto v) \Longrightarrow \mathsf{in}(H, p, v) \tag{3}$$

$$H \simeq (p \mapsto v) \wedge \mathsf{in}(H, q, w) \Longrightarrow p = q \wedge v = w \tag{4}$$

$$H \simeq H_1 {*} H_2 \wedge \mathsf{in}(H, p, v) \Longrightarrow \mathsf{in}(H_1, p, v) \vee \mathsf{in}(H_2, p, v) \tag{5}$$

$$H \simeq H_1 {*} H_2 \wedge \mathsf{in}(H_1, p, v) \Longrightarrow \mathsf{in}(H, p, v) \tag{6}$$

$$H \simeq H_1 {*} H_2 \wedge \mathsf{in}(H_2, p, v) \Longrightarrow \mathsf{in}(H, p, v) \tag{7}$$

$$H \simeq H_1 {*} H_2 \wedge \mathsf{in}(H_1, p, v) \wedge \mathsf{in}(H_2, q, w) \Longrightarrow p \neq q \tag{8}$$

Fig. 4. $\mathcal{H}$-solver CHR propagation rules

We specify the $\mathcal{H}$-solver as a set of *Constraint Handling Rules* [10] with disjunction (CHR$^\vee$) [1] as shown in Figure 4. Here each rule (*Head* $\Longrightarrow$ *Body*) encodes *constraint propagation*, where the constraints *Body* are added to the store whenever a matching *Head* is found. Rule (1) encodes the functional dependency for finite partial maps; rules (2)–(4) encode propagation for *heap empty* $H = \emptyset$ and *heap singleton* $H \simeq (p \mapsto v)$ constraints; and rules (5)–(8) encode heap membership propagation through *heap separation* $H \simeq H_1 {*} H_2$ constraints. Most of these rules are self-explanatory, e.g., rule (6) states that if $H \simeq H_1 {*} H_2$ and $\mathsf{in}(H_1, p, v)$, then it must be the case that $\mathsf{in}(H, p, v)$, since H_1 is a sub-heap of H. We assume a complete solver for the underlying equality theory ($x = y$, $x \neq y$).

The $\mathcal{H}$-solver employs the standard CHR$^\vee$ execution algorithm with the rules from Figure 4. We shall present a semi-formal summary below. The input is a *constraint store* S defined to be a set[4] of constraints (representing a conjunction). Let Rules be the rules from Figure 4, then the algorithm hsolve(S) is recursively defined as follows:

- (*Propagation Step*) If there exists $R \in$ Rules of the form ($h_1 \wedge \ldots \wedge h_n \Longrightarrow$ *Body*), a subset $\{c_1, \ldots, c_n\} \subseteq S$ of constraints, a subset $E \subseteq S$ of *equality* constraints, and a *matching substitution* θ such that: $E \rightarrow (\theta.h_i = c_i)$ for $i \in 1..n$ then rule R is *applicable* to the store S. We *apply* rule R as follows:
  - If *Body* = *false* then return *false*;
  - If *Body* = $d_1 \wedge \ldots \wedge d_m$ then return hsolve($S \cup \theta.\{d_1, \ldots, d_m\}$); else
  - If *Body* = $d_1 \vee \ldots \vee d_m$ then let $S_i :=$ hsolve($S \cup \theta.\{d_i\}$) for $i \in 1..m$. If there exists an $S_i \neq \mathit{false}$ then return S_i, else return *false*.
- Else if no such R exists, return S.

Propagation proceeds until failure occurs or a fixed point is reached.

Example 3 ($\mathcal{H}$-Solving). Consider the following goal G:

$$H \simeq (p \mapsto v) \wedge H \simeq I {*} J \wedge J \simeq (p \mapsto w) \wedge v \neq w$$

---

[4] We assume a set-based CHR semantics.

$$
\begin{array}{ll}
\{H \simeq (p \mapsto v), \underline{H \simeq I * J}, J \simeq (p \mapsto w), v \neq w\} & (3) \\
\{H \simeq (p \mapsto v), H \simeq I * J, J \simeq (p \mapsto w), v \neq w, \mathsf{in}(H, p, v)\} & (3) \\
\{H \simeq (p \mapsto v), \underline{H \simeq I * J}, \overline{J \simeq (p \mapsto w)}, v \neq w, \mathsf{in}(H, p, v), \mathsf{in}(J, p, w)\} & (7) \\
\{H \simeq (p \mapsto v), H \simeq I * J, J \simeq (p \mapsto w), v \neq w, \mathsf{in}(H, p, v), \overline{\mathsf{in}(J, p, w)}, \mathsf{in}(H, p, w)\} & (1) \\
\{H \simeq (p \mapsto v), H \simeq I * J, J \simeq (p \mapsto w), \underline{v \neq w}, \overline{\mathsf{in}(H, p, v)}, \mathsf{in}(J, p, w), \underline{v = w}\} & (E) \\
false
\end{array}
$$

Fig. 5. $\mathcal{H}$-solving constraint propagation steps

We wish to show that this goal is unsatisfiable using the $\mathcal{H}$-solver from Figure 4. Initially the constraint store contains the initial goal G. Constraint propagation proceeds as shown in Figure 5. Here we apply rules (3), (3), (7), (1), (E) to the underlined constraint(s) in order, where (E) represents an inference made by the underlying equality solver. Propagation leads to failure, and there are no branches – therefore goal G is unsatisfiable. □

Since all the rules from Figure 4 are *propagation rules*, the solving algorithm hsolve(G) will always terminate with some final store S. The $\mathcal{H}$-solver is both sound and complete w.r.t. (un)satisfiability.

Proposition 2 (Soundness). *For all G, S, if hsolve(G) $= S$, then for all valuations s, $\models_{\mathcal{H}} G [s]$ iff $\models_{\mathcal{H}} S [s]$.*

Proof. (Sketch) By the correctness of the rules from Figure 4 w.r.t. Definitions 2 and 4. □

Proposition 3 (Completeness). *For all G, S such that hsolve(G) $= S$, then $\not\models_{\mathcal{H}} G [s]$ for all valuations s (i.e. G is unsatisfiable) iff $S = false$.*

Proof. (Sketch) The "$\Leftarrow$" direction follows from Proposition 2. We consider the "$\Rightarrow$" direction. The rest is proof by contrapositive: assuming $S \neq false$ we show that there exists a valuation s such that $\models_{\mathcal{H}} G [s]$. Let s_E be a valuation for the underlying equality subset of S over integer variables, then let $s(v) = s_E(v)$ for integer variables, and

$$
s(H) = \{(s_E(p), s_E(v)) \mid \mathsf{in}(H, p, v) \in S\} \tag{4}
$$

for all heap variables H. Assume that $\not\models_{\mathcal{H}} S [s]$. By case analysis of Definition 2 we find that a rule must be applicable:
- Case $s(H) \notin$ Heaps: Rule (1);
- Case $H \simeq \emptyset$ and $s(H) \neq \emptyset$: Rule (2);
- Case $H \simeq (p \mapsto v)$ and $s(H) \neq \{(p, v)\}$: Rules (3) or (4);
- Case $H \simeq H_1 * H_2$ and $s(H) \neq s(H_1) \cup s(H_2)$: Rules (5), (6), or (7);
- Case $H \simeq H_1 * H_2$ and $dom(s(H_1)) \cap dom(s(H_2)) \neq \emptyset$: Rule (8)

This contradicts the assumption that S is a final store, therefore if $S \neq false$ then $\models_{\mathcal{H}} S [s]$, and therefore $\models_{\mathcal{H}} G [s]$ by Proposition 2 completes the proof. □

$$H \not\simeq \emptyset \implies \mathsf{in}(H, s, t) \tag{9}$$

$$H \not\simeq (p \mapsto v) \implies \lor \begin{cases} H \simeq \emptyset \\ \mathsf{in}(H, s, t) \land (s \neq p \lor t \neq v) \end{cases} \tag{10}$$

$$H \not\simeq H_1 * H_2 \implies \lor \begin{cases} \mathsf{in}(H, s, t) \land \neg\mathsf{in}(H_1, s, t) \land \neg\mathsf{in}(H_2, s, t) \\ \mathsf{in}(H_1, s, t) \land \neg\mathsf{in}(H, s, t) \\ \mathsf{in}(H_2, s, t) \land \neg\mathsf{in}(H, s, t) \\ \mathsf{in}(H_1, s, t) \land \mathsf{in}(H_2, s, u) \end{cases} \tag{11}$$

$$\mathsf{access}(H, p, v) \iff \mathsf{in}(H, p, v) \tag{12}$$

$$\mathsf{assign}(H_0, p, v, H_1) \implies \mathsf{in}(H_0, p, w) \land \mathsf{in}(H_1, p, v) \tag{13}$$

$$\mathsf{assign}(H_0, p, v, H_1) \land \mathsf{in}(H_0, q, w) \implies p = q \lor \mathsf{in}(H_1, q, w) \tag{14}$$

$$\mathsf{assign}(H_0, p, v, H_1) \land \mathsf{in}(H_1, q, w) \implies p = q \lor \mathsf{in}(H_0, q, w) \tag{15}$$

$$\mathsf{alloc}(H_0, p, H_1) \implies \mathsf{in}(H_1, p, v) \tag{16}$$

$$\mathsf{alloc}(H_0, p, H_1) \land \mathsf{in}(H_0, q, w) \implies p \neq q \land \mathsf{in}(H_1, q, w) \tag{17}$$

$$\mathsf{alloc}(H_0, p, H_1) \land \mathsf{in}(H_1, q, w) \implies p = q \lor \mathsf{in}(H_0, q, w) \tag{18}$$

$$\mathsf{free}(H_0, p, H_1) \iff \mathsf{alloc}(H_1, p, H_0) \tag{19}$$

Fig. 6. Extended $\mathcal{H}$-solver propagation rules

The proof for Proposition 3 is constructive; namely, (4) can be used to construct a *solution* for a satisfiable goal G. Furthermore, we can combine the normalization of Proposition 1 and DPLL(hsolve) to derive a sound and complete algorithm for solving arbitrary QF $\mathcal{H}$-formulae ϕ.

5.1 Extensions

The propagation rules from Figure 4 define a solver for the base $\mathcal{H}$-language. We can use heap membership propagation to define rules for other kinds of $\mathcal{H}$-constraints, as shown in Figure 6.

Rules (9)–(11) handle the negations of the base $\mathcal{H}$-constraints from Definition 3. These rules are an alternative to the decomposition from Figure 1. We can also define rules for directly handling the auxiliary constraints from Figure 2 (A) for program reasoning. For example, rules (13)–(15) handle the $\mathsf{assign}(H_0, p, v, H_1)$ constraint. We similarly provide rules for the other auxiliary constraints. Here, variables appearing in a rule body but not in the rule head are interpreted the same way as with Figure 1.

6 Experiments

In this section we test an implementation of the $\mathcal{H}$-solver against *verification conditions* (VCs) derived from symbolic execution. We compare against Verifast [12] (version 12.12), a program verification system based on Separation Logic. Our motivation for the comparison is: (1) Verifast is based on forward symbolic execution, and (2) Verifast incorporates the notion of separation (via Separation Logic). That said, the Verifast execution algorithm [12] is very different from the $\mathcal{H}$-solver.

We have implemented a version of the $\mathcal{H}$-solver as part of the *Satisfiability Modulo Constraint Handling Rules* (SMCHR) [8] system.[5] SMCHR is a *Satisfiability Modulo Theories* (SMT) framework that supports theory (T) solvers implemented in CHR. The SMCHR system also supports several "built-in" theories, such as a linear arithmetic solver based on [9], that can be combined with the $\mathcal{H}$-solver to handle the underlying (dis)equality constraints. The SMCHR system has also been extended to support *disjunctive propagators* [19] for rules with disjunctive bodies, such as Rule (5).

For these benchmarks we either restrict ourselves to the fragment of Verifast that is fully automatable, or we provide the minimal annotations where appropriate. For the $\mathcal{H}$-solver, we have implemented a prototype symbolic execution tool as a GCC plug-in. Our tool symbolically executes GCC's internal GIMPLE representation to generate *path constraints*. Given a *safety condition* φ, we generate the corresponding *verification condition* $(\exists \bar{x} : \phi) \models \varphi$, which is *valid* iff $\phi \wedge \neg\varphi$ is *unsatisfiable*. Here $\bar{x}$ represents existential variables introduced during symbolic execution. Unsatisfiability is tested for using the $\mathcal{H}$-solver.

A well-known problem with forward symbolic execution is the so-called *path explosion problem*. The number of paths through a (loop-free) program fragment can easily be exponential. We can mitigate this problem using *subsumption via interpolation* [14][17]. The basic idea is as follows: given a VC $\phi_1 \models \varphi$ that holds for path ϕ_1, we generate an *interpolant* ψ_1 for ϕ_1, that, by definition, satisfies $\phi_1 \models \psi_1 \models \varphi$. As symbolic execution continues, we can prune (subsume) all other paths with constraints ϕ_2 such that $\phi_2 \models \psi_1$. The key is that this pruning can occur *early*, as we construct the constraint for each path.

Our interpolation algorithm is based on an improved version of the *constraint deletion* idea from [14]. Given a path constraint $\phi = c_1 \wedge \ldots \wedge c_n$ we find a subset $I \subseteq \{c_1, \ldots, c_n\}$ such that $I \wedge \neg\varphi$ remains *unsatisfiable*. For this we simply re-use the SAT solver's *Unique Implication Point* UIP algorithm over the implication graph formed by the $\mathcal{H}$-solver propagation steps.

We test several programs that exhibit the path explosion problem. These include: subsets_N - sum-of-subsets size N; expr_N - simple virtual machine executing N instructions; stack_N - for all $M \leq N$, do N-pushes, then N-pops; filter_N - filter for TCP/IP packets; sort_N - bubble-sort of length N; search234_N - 234-tree search; insert234_N - 234-tree insert. Most of our examples are derived from unrolling loops of smaller programs.

The results are shown in Figure 7. Here Safety indicates the safety condition (defined below), LOC indicates the number of *lines-of-code*, type indicates heap operations used (with r = read, w = write, and a = allocation/deallocation), #bt is the number of backtracks for our prototype tool, and #forks is the number of

---

[5] SMCHR is available from http://www.comp.nus.edu.sg/~gregory/smchr/

Bench.	Safety	LOC	type	Heaps		Verifast	
				time(s)	#bt	time(s)	#forks
subsets_16	F	50	rw-	0.00	17	10.69	65546
expr_2	F	69	rw-	0.05	124	18.38	136216
stack_80	F	976	rwa	8.66	320	68.20	9963
filter_1	F	192	r--	0.03	80	0.75	8134
filter_2	F	321	r--	0.11	307	–	–
sort_6	F	178	rw-	0.03	54	2.66	35909
search234_3	F	251	r--	0.02	46	0.67	1459
search234_5	F	399	r--	0.05	76	90.65	118099
insert234_5	F	839	rwa	1.19	120	52.87	36885
expr_2	$\sqsubseteq$	69	rw-	0.20	1329	n.a.	n.a.
stack_80	$\sqsubseteq$	976	rwa	8.07	322	n.a.	n.a.
filter_2	OP	321	r--	0.00	2	n.a.	n.a.
stack_80	A	976	rwa	8.90	320	65.68	9801
insert234_5	A	839	rwa	1.50	60	40.64	55423
subsets_16	$\emptyset$	50	rw-	0.00	33	n.a.	n.a.

Fig. 7. Theorem proving and symbolic execution benchmarks

symbolic execution forks for Verifast, and corresponds to the number of paths through the code. All experiments were run on GNU/Linux x86_64 with a Intel® Core™ i5-2500K CPU clocked at 4GHz. A timeout of 10 minutes is indicated by a dash (–). The safety conditions correspond to (some variant of) the following triples:

- Framing (F) with $\{\bar{\mathcal{H}} \simeq (p \mapsto v) * F\} C \{\exists F' : \bar{\mathcal{H}} \simeq (p \mapsto v) * F'\}$ where p is outside the footprint of the code C;
- Operations (OP) where $OP \in \{\sqsubseteq, \sqsupseteq, \simeq\}$ with $\{H \simeq \bar{\mathcal{H}}\} C \{H \ OP \ \bar{\mathcal{H}}\}$;
- Allocation (A) with $\{\ldots\} C \{\exists F', v : \bar{\mathcal{H}} \simeq (p \mapsto v) * F'\}$ for p allocated by C;
- Empty $(\emptyset)$ with $\{\bar{\mathcal{H}} \simeq \emptyset\} C \{false\}$, i.e. C will always fault on memory.

Some safety conditions, namely (OP) and $(\emptyset)$, cannot be encoded directly in Separation Logic or Verifast, and are marked by "n.a.".

Overall our tests exhibit significant search-space pruning thanks to interpolation. In contrast Verifast explores the entire search-space, and thus has exponential runtime behavior. However the *time-per-path* ratio favors Verifast, suggesting that Verifast would perform better on examples that do not have a large search-space, or when interpolation fails to subsume a significant number of branches. Our tool and SMT solver implementation are preliminary and can likely be further optimized.

7   Related Work

Several systems [3][5][12] implement Separation Logic-based symbolic execution, as described in [4]. However, due to the memory-safety requirements of Separation Logic, symbolic execution is limited to formulae over the footprint of the code. Our symbolic execution is based on the SPC Hoare Logic extension and therefore works for arbitrary formulae. This is convenient when memory-safety is not a property of interest, such as Example 2.

Several automatic theorem provers for Separation Logic triples/formulae have been developed, including [4][5][18]. These systems generally rely on a set of *rearrangement rules*, and are usually limited to a subset of all formulae, e.g. those with no non-separating conjunction, etc. In contrast our $\mathcal{H}$-solver uses a different algorithm based on heap-membership propagation, and handles any arbitrary QF $\mathcal{H}$-formulae.

Other formalisms, such as (Implicit) Dynamic Frames [16][22] and Region Logic [2], also encode separation. The underlying approach is to represent the heap H as a (possibly implicit) *total map* over all possible addresses, and to represent access or modification rights as sets of addresses F. Separation is represented as set disjoint-ness, i.e. $F_1 \cap F_2 = \emptyset$. One difficulty is that we must relate H with F, which can make reasoning comparatively more difficult. For example, consider the following VCs:

$$p \notin F \wedge \mathsf{list}(H, F, l) \wedge \mathsf{assign}'(H, p, v, H') \models \mathsf{list}(H', F, l) \quad (5)$$

$$H \simeq L*R \wedge R \simeq (p \mapsto w)*R' \wedge \mathsf{list}(L, l) \wedge \mathsf{assign}(H, p, v, H') \models L \sqsubseteq H' \quad (6)$$

where assign$'$ is a suitable re-encoding of assign for total heaps. Both VCs are natural encodings of the same problem: we wish to prove that l is still a list after writing to a (separate) pointer p. VC (6) holds independently of the recursively defined **list** relation, and can be trivially disposed of using our $\mathcal{H}$-solver. In contrast, VC (5) depends on the recursively-defined **list** predicate as it relates H with F, and is therefore more difficult to prove.

Our $\mathcal{H}$-solving algorithm is related to analogous algorithms for finite sets, such as [23]. Although formalized differently, the basic idea is similar, i.e. based on the propagation of *set membership* $x \in S$ constraints. In [21] this idea was adapted into a decision procedure for Region Logic. Our approach works directly with heaps rather than indirectly via sets.

8   Future Work and Conclusions

In this paper we presented a reformulation of the key ideas behind Separation Logic as a first-order constraint-language $\mathcal{H}$ over heaps. Here we express separation as a constraint between heaps. We present an SPC extension of Hoare Logic based on encoding of heap-manipulating statements in terms of $\mathcal{H}$-formulae. Our extension is suitable for forward reasoning via constraint-based symbolic execution. We present a sound and complete solver for QF $\mathcal{H}$-formulae and have implemented a version as part of an SMT framework. Experimental evaluation yields promising results.

There is significant scope for future work, such as: building theorem provers for recursively-defined properties based on the $\mathcal{H}$-solver, or further developing program verification tools using $\mathcal{H}$-language-based symbolic execution.

References

1. Abdennadher, S., Schütz, H.: CHR $^\vee$: A flexible query language. In: Andreasen, T., Christiansen, H., Larsen, H.L. (eds.) FQAS 1998. LNCS (LNAI), vol. 1495, pp. 1–14. Springer, Heidelberg (1998)
2. Banerjee, A., Naumann, D.A., Rosenberg, S.: Regional logic for local reasoning about global invariants. In: Vitek, J. (ed.) ECOOP 2008. LNCS, vol. 5142, pp. 387–411. Springer, Heidelberg (2008)
3. Berdine, J., Calcagno, C., O'Hearn, P.W.: Smallfoot: Modular automatic assertion checking with separation logic. In: de Boer, F.S., Bonsangue, M.M., Graf, S., de Roever, W.-P. (eds.) FMCO 2005. LNCS, vol. 4111, pp. 115–137. Springer, Heidelberg (2006)
4. Berdine, J., Calcagno, C., O'Hearn, P.W.: Symbolic execution with separation logic. In: Yi, K. (ed.) APLAS 2005. LNCS, vol. 3780, pp. 52–68. Springer, Heidelberg (2005)
5. Botinčan, M., Parkinson, M., Schulte, W.: Separation logic verification of c programs with an SMT solver. Electronic Notes in Theoretical Computer Science 254, 5–23 (2009)
6. Davis, M., Logemann, G., Loveland, D.: A machine program for theorem-proving. Communications of the ACM 5(7), 394–397 (1962)
7. Dijkstra, E.: Guarded commands, nondeterminacy and formal derivation of programs. Communcations of the ACM 18(8), 453–457 (1975)
8. Duck, G.: SMCHR: Satisfiability modulo constraint handling rules. Theory and Practice of Logic Programming 12(4-5), 601–618 (2012); Proceedings of the 28th international conference on Logic Programming
9. Dutertre, B., de Moura, L.: A fast linear-arithmetic solver for DPLL(T). In: Ball, T., Jones, R.B. (eds.) CAV 2006. LNCS, vol. 4144, pp. 81–94. Springer, Heidelberg (2006)
10. Frühwirth, T.: Theory and practice of constraint handling rules. Special Issue on Constraint Logic Programming, Journal of Logic Programming 37 (October 1998)
11. Hoare, C.: An axiomatic basis for computer programming. Communications of the ACM 12(10), 576–580 (1969)
12. Jacobs, B., Smans, J., Philippaerts, P., Vogels, F., Penninckx, W., Piessens, F.: VeriFast: A powerful, sound, predictable, fast verifier for C and Java. In: Bobaru, M., Havelund, K., Holzmann, G.J., Joshi, R. (eds.) NFM 2011. LNCS, vol. 6617, pp. 41–55. Springer, Heidelberg (2011)
13. Jaffar, J., Maher, M.J.: Constraint logic programming: A survey. J. LP 19(20), 503–581 (1994)
14. Jaffar, J., Santosa, A.E., Voicu, R.: An interpolation method for CLP traversal. In: Gent, I.P. (ed.) CP 2009. LNCS, vol. 5732, pp. 454–469. Springer, Heidelberg (2009)
15. Jaffar, J., Santosa, A.E., Voicu, R.: A coinduction rule for entailment of recursively defined properties. In: Stuckey, P.J. (ed.) CP 2008. LNCS, vol. 5202, pp. 493–508. Springer, Heidelberg (2008)
16. Kassios, I.T.: Dynamic frames: Support for framing, dependencies and sharing without restrictions. In: Misra, J., Nipkow, T., Sekerinski, E. (eds.) FM 2006. LNCS, vol. 4085, pp. 268–283. Springer, Heidelberg (2006)
17. McMillan, K.L.: Lazy annotation for program testing and verification. In: Touili, T., Cook, B., Jackson, P. (eds.) CAV 2010. LNCS, vol. 6174, pp. 104–118. Springer, Heidelberg (2010)

18. Nguyen, H.H., David, C., Qin, S., Chin, W.-N.: Automated verification of shape and size properties via separation logic. In: Cook, B., Podelski, A. (eds.) VMCAI 2007. LNCS, vol. 4349, pp. 251–266. Springer, Heidelberg (2007)
19. Ohrimenko, O., Stuckey, P., Codish, M.: Propagation via lazy clause generation. Constraints 14, 357–391 (2009)
20. Reynolds, J.C.: Separation logic: A logic for shared mutable data objects. In: 17th IEEE Symposium on Logic in Computer Science, pp. 55–74. IEEE Computer Society Press (2002)
21. Rosenberg, S., Banerjee, A., Naumann, D.A.: Decision procedures for region logic. In: Kuncak, V., Rybalchenko, A. (eds.) VMCAI 2012. LNCS, vol. 7148, pp. 379–395. Springer, Heidelberg (2012)
22. Smans, J., Jacobs, B., Piessens, F.: Implicit dynamic frames: Combining dynamic frames and separation logic. In: Drossopoulou, S. (ed.) ECOOP 2009. LNCS, vol. 5653, pp. 148–172. Springer, Heidelberg (2009)
23. Zarba, C.G.: Combining sets with elements. In: Dershowitz, N. (ed.) Verification: Theory and Practice. LNCS, vol. 2772, pp. 762–782. Springer, Heidelberg (2004)

Model Combinators for Hybrid Optimization

Daniel Fontaine[1], Laurent Michel[1], and Pascal Van Hentenryck[2]

[1] University of Connecticut, Storrs, CT 06269-2155
[2] NICTA, University of Melbourne

Abstract. In recent years, CML, G12 and SIMPL, have achieved significant progress in automating the generation of hybrid solvers from high-level model specifications. This paper pushes this research direction one step further and introduces the concept of model combinators to provide principled model compositions. These model combinators rely on *runnables* capturing executable models, *runnable signatures* that capture what runnables can produce and consume, and *model hierarchies*, which track relationships among models. These concepts make it possible to enforce the soundness of model compositions and to determine the best model compositions automatically. A prototype of the framework on top of the OBJECTIVE-CP optimization system is presented.

1 Introduction

The *Comet Modeling Language* (CML) [6] demonstrated that the burden of writing a suitable solver for hard industrial problems can be greatly mitigated with high-level language abstractions. With CML, models are specified abstractly in a technology agnostic manner, and then *concretized* into one or more models based on Constraint Programming (CP), Integer Programming (IP) or Constraint-Based Local Search (CBLS) technologies. Concretization typically relied on transformation and reformulation based on rewrite rules [4].

The value of CML primarily resides in the ease with which users can manipulate and combine models into complex hybrid solvers. It is clear that systems such as CML are moving in the direction of providing something akin to *combinators* [8] over models. The ultimate aim is to deliver a clean, well defined and semantically sound collection of operators that allow models to be combined in a multitude of ways into hybrids that expose a clean interface for further composition. CML fell short of this goal due to two shortcomings:

1. First, model composition was based on syntax rather than semantics and the system was not endowed with the ability to track model relationships such as relaxations. In essence, semantics were opaque to end-users and correctness hinged on the modeler's ability to write sound composition expressions.
2. Second, CML did not require the user to specify any property for the inputs or outputs of operators. Instead, most operators relied on implicit assumptions about the models capabilities and functioned by plugging models into static templates.

C. Schulte (Ed.): CP 2013, LNCS 8124, pp. 299–314, 2013.

The purpose of this paper is to push beyond CML's limitations and realize the vision of true *model combinators* with sound semantics and complete compositionality. This introduces two requirements. First, the paper postulates the existence of runnables as first class objects that encapsulate a model and a signature specifying their capabilities. Second, it mandates the introduction of combinators specifying how to derive new runnables from their inputs and what the properties of the derived runnable are. The keystone is based on metadata maintained in the runnables and the leverage it provides to check the soundness of the operators application.

The remainder of the paper is organized as follows. Section 2 first discusses the related work. Section 3 introduces the notion of models and program while Section 4 focuses on runnables. Section 5 turns to combinators with Section 6 discussing implementation and Section 7 presenting empirical results. Section 8 concludes the paper.

2   Related Work

Several systems aim to facilitate solver independent modeling, model reformulation and automated hybrid generation. The Comet Modeling Language (CML) [6] is unique in that it strives to provide a full programming language in which models can be specified and easily manipulated and composed in sophisticated ways without the need for annotations. The work in this paper directly builds off of what was done in CML.

G12 models written with *Zinc* and *mini-Zinc* feature solver independent capabilities and model rewriting, which can be achieved via *Cadmium* [3]. *G12* has been used for column generation and branch-and-price hybrid models [10]. *SIMPL* [14] is a high level modeling language based on search-infer-relax. Users write models at a very high level and rely on constraint-based relaxations and inference rules to assemble a model for a chosen technology. *Essence* [7] is designed for model specification using combinatorial constructs rather than CP specific constructs like global constraints. It has recently been combined with *Conjure* [1] to automate the generation of constraint models from a combinatorial problem specification. The *Z3* SMT [2] engine allows user to specify *tactics* which are used to direct the search procedure. *Tactics* are capable of relaxing parts of the SMT problem and determining whether a particular relaxed subproblem will provide an upper or lower bound for the original problem. These tactics can be queried at runtime to determine how the search should proceed and what to invoke next. Finally, work has already been done in providing a rich language of combinators within the context of search [12] and is revisited in [13].

3   Models and Programs

This section reviews the concepts of models and programs from OBJECTIVE-CP which serves as the foundations for OCPMCL.

Flatten	*Flattening* a model decomposes complex expressions into simpler ones, often adding variables and constraints in the process.
Continuous	Performs a continuous *relaxation* of a model, replacing integer-valued domain constraints with continuous interval domains.
Linear	Creates a linear reformulation to replace global constraints and logical constraints with a set of equivalent linear constraints [11].

Fig. 1. Examples of commonly used model operators

Definition 1. *A model M is of the form $\langle X, C, O \rangle$ where X is the set of model variables, C the model constraints and O the (optional) objective function.*

Definition 2. *A model transformation τ transforms a model $M = \langle X, C, O \rangle$ into another model $\tau(M) = \langle X_o, C_o, O_o \rangle$ satisfying $X \subseteq X_o$.*

Examples of model transformations performed by OBJECTIVE-CP are shown in Figure 1. When models are in flattened form (sufficiently decomposed), they can be concretized in an optimization program.

Definition 3. *A model concretization γ takes a model M in a flattened form and concretizes M into a program $P = \langle M, \gamma \rangle$, where $P = \gamma(M)$. The concretization associates a concrete variable with every model variable, a concrete constraint to every model constraint, and a concrete objective with the model objective.*

To obtain an optimization program P from a model M, OBJECTIVE-CP performs a series of model transformations followed by a concretization, e.g.,

$$P = \gamma(\tau_{k-1}(\cdots \tau_0(M) \cdots)).$$

Model transformations in OBJECTIVE-CP always extend the set of variables, which is convenient both from a semantic and implementation standpoint. In this paper, we ignore that OBJECTIVE-CP can provide a search procedure since it is not relevant for model combinators. Hence, we define a program as a pair (model,concretization).

Definition 4 (Program). *A program is a pair $P = \langle M, \gamma \rangle$, where M is a model in flattened form and γ is a concretization.*

Model transformations impose a natural partial order between models through the concept of relaxation and tightening. These concepts are critical to define sound *combinators* [8].

Definition 5 (Solution Set). *A solution for a model $M = \langle X, C, O \rangle$ is an assignment of all variables in X satisfying C. The set of solutions of model M is denoted by $Sol(M)$.*

Definition 6 (Projection of Solution Sets). *Consider a model $M = \langle X, C, O \rangle$ along with a solution s and $X' \subseteq X$. Then, $Sol|_{X'}(s)$ and $Sol|_{X'}(M)$ denotes the projection of solution s and the solution set of M on the variables in X', respectively.*

We now formalize the concept of relaxation, tightening, and equivalence of transformed models.[1] Without loss of generality, optimization problems are all minimization problems in this paper.

Definition 7 (Relaxations and Tightenings of Satisfaction Problems).
Let $M = \langle X, C \rangle$ and τ be a transformation. The model $\tau(M) = M' = \langle X', C' \rangle$ is a relaxation of M, denoted by $M' \triangle M$, iff $Sol(M) \subseteq Sol|_X(M')$. It is a tightening, denoted by $M' \nabla M$, iff $Sol|_X(M') \subseteq Sol(M)$. M and M' are equivalent, denoted by $M \equiv M'$, if M' is both a relaxation and a tightening of M.

Definition 8 (Relaxations and Tightenings of Minimization Problems).
Let $M = \langle X, C, O \rangle$ and τ be a transformation. The model $\tau(M) = M' = \langle X', C', O' \rangle$ is a relaxation of M, denoted $M' \triangle M$, iff $\langle X', C' \rangle \triangle \langle X, C \rangle$ and

$$\forall s \in Sol(M), s' \in Sol(M') : Sol|_X(s') = s \Rightarrow O'(s') \leq O(s).$$

M' is a tightening of M, denoted by $M' \nabla M$, if $\langle X', C' \rangle \nabla \langle X, C \rangle$ and

$$\forall s \in Sol(M), s' \in Sol(M') : Sol|_{X'}(s) = s' \Rightarrow O'(s') \geq O(s).$$

M and M' are equivalent, denoted by $M \equiv M'$, if M' is both a relaxation and a tightening of M.

The definitions of these concepts are transitive, reflexive and, for equivalences, commutative. We use $\triangle^*$ (resp. ∇^*) to denote the reflexive and transitive closure of $\triangle$ (resp. ∇). We use $\equiv^*$ to denote the reflexive, commutative, and transitive closure of $\equiv$. Our combinators use these relations to enforce pre-conditions and post-conditions on their models.

4   Runnables and Runnable Signatures

This section introduces the concept of a *runnable*. Informaly, a *runnable* can be thought of as a producer/consumer process that uses a *program* to solve an optimization problem, consuming from a number of *input pipes*, and producing into a number of *output pipes* (see Figure 2(a)). The pipes deal with runnable products that are concepts such as solutions and bounds, as well as streams or sets of these products. A runnable is associated with a signature that specifies its inputs and outputs, i.e., the products that it consumes and produces. The implementation creates pipes for each of these inputs and outputs. If a runnable is executed directly, its input and output pipes are not used; the runnable pipes are only useful when it is combined with other runnables through *combinators* (see see Figure 2(b)). Note that stream pipes consume or produce products during the lifetime of a runnable; this is the case when exchanging solutions and bounds during the search.

---

[1] Note that the model M' always has at least the same variables as M since M' is obtained through a transformation of M. Tightenings are only obtained by adding constraints, while relaxations can be obtained by adding variables or removing constraints.

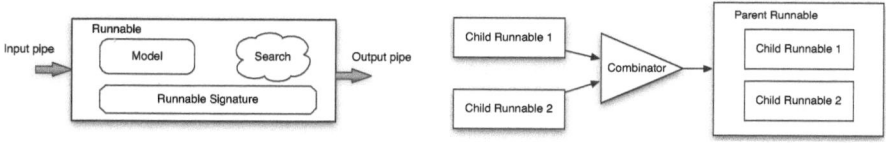

(a) A runnable for solving a *process*   (b) A composite from a combinator.

Fig. 2. Basic and Composite Runnables

Definition 9 (Runnable Products). *A runnable product is specified by the grammar*

⟨*runnable product*⟩ ::= ⟨*basic product*⟩ | [⟨*basic product*⟩] | {⟨*basic product*⟩}

⟨*basic product*⟩ ::= UBD | LBD | COL | CST | SOL

where the basic products UBD, LBD, COL, CST, SOL *represent upper bounds, lower bounds, columns, constraints and solutions,* [p] *represents a stream of products of type* p, *and* {p} *a set of products of type* p.

Definition 10 (Runnable Signature). *A runnable signature is a pair* $S = \langle I, O \rangle$, *where* I *is a set of input runnable products and* O *is a set of output runnable products.*

Definition 11 (Runnable). *A runnable is a pair* $R = \langle P, S \rangle$, *where* P *is an optimization program and* S *is a runnable signature.*

We often abuse language and talk about the model of a runnable to denote the model of its program.

Definition 12 (Pipes of a Runnable). *Let* R *be a runnable* $\langle P, \langle I, O \rangle \rangle$. R *provides the set of input pipes* $\{in(p, R) \mid p \in I\}$ *and the set of output pipes* $\{out(p, R) \mid p \in O\}$.

Our implementation provides a number of primitive runnables. They can be created from a model M, a flattening, and a concretization. For instance, the CPRunnable has a program $\langle \texttt{flatten}(M), \gamma_{CP} \rangle$ and a predefined signature.

5   Model Combinators

This section describes model combinators. We restrict our attention to binary operators for simplicity but it is easy to generalize our results for non-binary combinators. A model combinator $R = C(R_1, R_2)$ combines two runnables R_1 and R_2 to produce another runnable. The combinator requires some properties from its runnables, establishes the links between the pipes of its runnables and its own, and specifies how its model relates to the models of its runnables. Figure 3 illustrates the piping intuitively. More precisely, the specification of a model combinator consists of several parts:

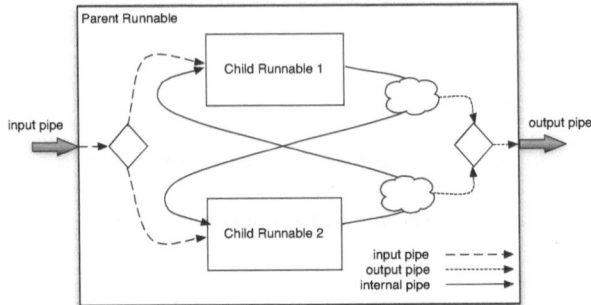

Fig. 3. A Composite Runnable and Its Input, Output and Internal Pipes

1. a precondition that specifies the required relationships between the runnable models and the existence of some input/output products;
2. a set of piping rules for linking the input pipes of the combinators to the input pipes of its runnables;
3. a set of piping rules for linking the output pipes of its runnables to its output pipes;
4. a set of piping rules for linking the pipe of the runnables;
5. a relationship between the model of the combinator and the model of the runnables.

A piping rule is an expression of the form $\pi_1 \to \pi_2$ which specifies that pipe π_1 produces products that are consumed by pipe π_2. For instance, the rule

$$in(\text{SOL},R) \to in(\text{SOL},R_1)$$

specifies that the input pipe for solutions in R produces solutions that are consumed by the input pipe for solution in R_1. If p is a product, an input pipe rule is of the form

$$in(\text{p, R}) \to in(\text{p},R_i)$$

an output pipe rule is of the form

$$out(\text{p},R_i) \to out(\text{p,R})$$

and an internal pipe rule is of the form

$$out(\text{p},R_i) \to in(\text{p},R_j)$$

It is also useful to allow output piping rules with no antecedent, i.e.,

$$\to out(\text{p,R})$$

for situations where the combinator products are not directly taken from the runnables but computed by the combinator itself.

These piping rules have two main purposes: To establish the plumbing inside the combinators and to synthesize the signature of the combinator. It is important to state that the combinator does not have a static signature. Rather OCPMCL synthesizes the most general signature based on the functionalities of its runnables.

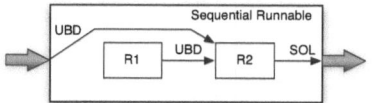

Fig. 4. The Runnable of the Sequential Combinator

Definition 13 (Combinator Specification). *Let R_1 and R_2 be two runnables with signatures $S_i = \langle I_i, O_i \rangle$ and models M_i ($1 \leq i \leq 2$). The specification of a combinator $C(R_1, R_2)$ is a tuple $\langle \mathcal{P}, \mathcal{I}, \mathcal{O}, \mathcal{E}, \mathcal{M} \rangle$, where $\mathcal{P}$ is a precondition on M_i, I_i, and O_i, $\mathcal{I}$, $\mathcal{O}$, $\mathcal{E}$ are sets of input, output, and internal piping rules, and $\mathcal{M}$ specifies the model relationship.*

Obviously, the combinator does not have a model on its own: It combines, sometimes in complex ways, the models of its runnable. Hence, the model relationship specifies the semantics of its products, such as its solutions, its bounds, and streams thereof. For instance, a model relationship $R \triangle R_1$ specifies that the (implicitly defined) combinator model is a relaxation of the model of R_1. The new information is propagated through the transitive closures of $\triangle$ in order to verify preconditions involving R in subsequent combinations. We are now ready to synthesize the combinator signature.

Definition 14 (Combinator Signature). *Let R_1 and R_2 be two runnables with signatures $S_i = \langle I_i, O_i \rangle$ and a combinator $R = C(R_1, R_2)$ with specification $\langle \mathcal{P}, \mathcal{I}, \mathcal{O}, \mathcal{E}, \mathcal{M} \rangle$. The signature of R is $\langle I, O \rangle$ where*

$$I = \{\, p \mid in(p, R) \rightarrow in(p, R_i) \in \mathcal{I} \ \wedge \ p \in I_i \ \wedge \ 1 \leq i \leq 2 \}$$
$$O = \{\, p \mid out(p, R_i) \rightarrow out(p, R) \in \mathcal{O} \ \wedge \ p \in O_i \ \wedge \ 1 \leq i \leq 2 \}.$$

Observe once again that the definition of input/output is dynamic: The piping rules defines what is possible and the actual input/output definitions are derived from the actual input and output functionalities of the combined runnables. If a runnable does not provide a certain product (e.g., streams of lower bounds), this product is not synthesized in the signature, even if a piping rule was specified. We are now ready to present some combinators.

5.1 Sequential Combinator

This section presents a sequential combinator $R = R_1 \triangleright R_2$ which uses R_1 to compute an upper bound which is then passed as an input to R_2. This combinator (see Figure 4) is often used in practice when a heuristic search first finds a high-quality upper bound which is then used to seed a systematic search. The combinator specification is as follows:

$$R = R_1 \triangleright R_2$$

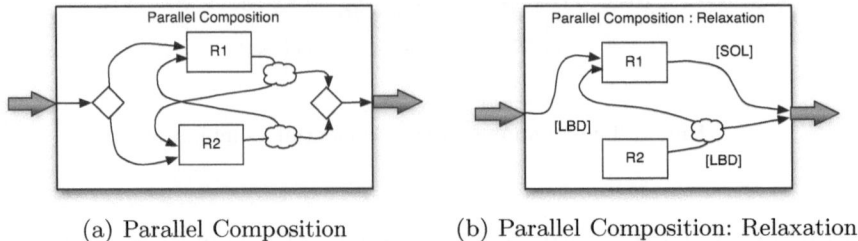

(a) Parallel Composition   (b) Parallel Composition: Relaxation

Fig. 5. Combinators for Parallel Composition

$$
\begin{array}{|l|}
\hline
\mathcal{P}_{\triangleright} = \ R_1 \nabla^* R_2 \wedge UBD \in O_1 \wedge UBD \in I_2 \wedge SOL \in O_2 \\
\hline
\mathcal{I}_{\triangleright} = \ \{in(UBD, R) \rightarrow in(UBD, R_2)\} \\
\hline
\mathcal{O}_{\triangleright} = \ \{out(SOL, R_2) \rightarrow out(SOL, R)\} \\
\hline
\mathcal{E}_{\triangleright} = \ \{out(UBD, R_1) \rightarrow in(UBD, R_2)\} \\
\hline
\mathcal{M}_{\triangleright} = R \equiv R_2 \\
\hline
\end{array}
$$

The precondition requires that M_1 be a tightening of M_2 to ensure that the upper bound of M_1 is indeed an upper bound for M_2. The input piping rule allows the upper bound of R to be consumed by R_2; It cannot be passed to R_1 since R_1 is a tightening of M_2. The output piping rule allows the solution of R_2 to be produced as a solution to R. The internal piping rule specifies that the upper bound produced by R_1 can be consumed by R_2. The model relationship specifies that the resulting combinator is equivalent to R_2.

The signature that is synthesized here is trivial, since the piping rules are only concerned with required inputs and outputs. It will simply be $\langle\{UBD\}, \{SOL\}\rangle$. If the output piping rule

$$
out(LBD, R_2) \rightarrow out(LBD, R)
$$

had been present, and LBD would belong to O_2, the synthesized signature would have been $\langle\{UBD\}, \{SOL, LBD\}\rangle$.

5.2 Parallel Combinator

We now turn to the parallel composition of two equivalent runnables exchanging solutions (see Figure 5(a)). Its specification is

$$
R = R_1 \parallel R_2
$$

$\mathcal{P}_{\parallel} =$	$R_1 \equiv^* R_2 \wedge [SOL] \in I_1 \wedge [SOL] \in O_1 \wedge [SOL] \in I_2 \wedge [SOL] \in O_2$
$\mathcal{I}_{\parallel} =$	$\{in([SOL], R) \rightarrow in([SOL], R_1), in([SOL], R) \rightarrow in([SOL], R_2),$
	$\quad in([UBD], R) \rightarrow in([UBD], R_1), in([UBD], R) \rightarrow in([UBD], R_2),$
	$\quad in([LBD], R) \rightarrow in([LBD], R_1), in([LBD], R) \rightarrow in([LBD], R_2)\}$
$\mathcal{O}_{\parallel} =$	$\{out([SOL], R_1) \rightarrow out([SOL], R), out([SOL], R_2) \rightarrow out([SOL], R),$
	$\quad out([UBD], R_1) \rightarrow out([UBD], R), out([UBD], R_2) \rightarrow out([UBD], R),$
	$\quad out([LBD], R_1) \rightarrow in([LBD], R), out([LBD], R_2) \rightarrow out([LBD], R)\}$
$\mathcal{E}_{\parallel} =$	$\{out([SOL], R_1) \rightarrow in([SOL], R_2), out([SOL], R_2) \rightarrow in([SOL], R_1),$
	$\quad out([UBD], R_1) \rightarrow in([UBD], R_2), out([UBD], R_2) \rightarrow in([UBD], R_1),$
	$\quad out([LBD], R_1) \rightarrow in([LBD], R_2), out([LBD], R_2) \rightarrow in([LBD], R_1)\}$
$\mathcal{M}_{\parallel} = R \equiv R_1$	

The precondition $\mathcal{P}_{\parallel}$ ensures that the two runnables are equivalent and the input and output of both runnables include a stream of solutions. The piping rules are very explicit this time and allow for the exchanges of upper and lower bounds as well. In particular, if the runnables provide lower bounds, the implementation will ensure that the internal piping provides that functionality. Similarly, the input and output piping will synthesize streams of upper and lower bounds if the combined runnables provide these products.

Observe that this combinator can be used for composing three runnables: It suffices to use $(R_1 \parallel R_2) \parallel R_3$, since the parallel combinator will satisfy its own precondition. Also, Figure 5(a) shows the flow of solutions within this parallel runnable using black arrows, The small clouds waiting at the outputs of the child runnables represents small blocks of code used by the parent to intercept output solutions coming from the children.

5.3   Relaxed Parallel Combinator

The relaxed parallel combinator runs two optimization programs concurrently but one of the models is a relaxation of the other and streams lower bounds. The specification of the combinator is

$$R = R_1 \dashv R_2$$

$\mathcal{P}_{\dashv} =$	$R_2 \triangle^* R_1 \wedge [SOL] \in O_1 \wedge [LBD] \in I_1 \wedge [LBD] \in O_2$
$\mathcal{I}_{\dashv} =$	$\{in([LBD], R) \rightarrow in([LBD], R_1), in([UBD], R) \rightarrow in([UBD], R_1),$
	$\quad in([SOL], R) \rightarrow in([SOL], R_1)\}$
$\mathcal{O}_{\dashv} =$	$\{out([SOL], R_1) \rightarrow out([SOL], R), out([UBD], R_1) \rightarrow out([UBD], R),$
	$\quad out([LBD], R_1) \rightarrow in([LBD], R)\}$
$\mathcal{E}_{\dashv} =$	$\{out([LBD], R_2) \rightarrow in([LBD], R_1)\}$
$\mathcal{M}_{\dashv} = R \equiv R_1$	

The precondition $\mathcal{P}_{\dashv}$ ensures that R_2 is a relaxation of R_1, R_1 produces a stream of solutions, and R_2 produce lower bounds. The input piping rules $\mathcal{I}_{\dashv}$

states that streams of lower bounds, upper bounds or solutions consumed by $R_\dashv$ can be consumed by R_1. The output piping rules $\mathcal{O}_\dashv$ state that $R_\dashv$ produces the output streams produced by R_1. The internal piping rules ensure that the stream of lower bounds produced by R_2 can be consumed by R_1. The combinator then produces a model equivalent to R_1. Figure 5(b) illustrates the flow of runnable products through this runnable assuming the children meet only the minimum preconditions for simplicity.

5.4   Column-Generation Combinator

Automating column-generation solvers has been done previously in systems such as CML [6] and the G12 Project [9], but the use of runnables allows for a cleaner expression of the semantics as well as a much more compositional interface. The column-generation combines a master problem and a slave problem. The master runnable consumes columns and generates solutions, while the slave runnable consumes solutions and produces columns. The column-generation combinator produces a solution and a stream of upper bounds on its own. An implementation of the combinator copies the master runnable, before starting the column-generation process, in order to allow the master runnable to be reused in other combinators. As a result, the combinator does not use the output of its runnables but generates products on its own. The common terminology for the resulting model is the Restricted Master Problem (RMP), but this paper will refer to it as a *relaxation* of the master model, since column generation adds new columns (i.e., variables). Note that the column-generation combinator is very general: It does not impose how the slave process uses the solution (though the dual values are captured in the solution). As a consequence, it can implement a traditional column-generation algorithm or use a heuristic approach to generate columns based on the problem structure.

$$R = MA \bowtie SL \text{ with } MA = \langle I_m, O_m \rangle \text{ and } SL = \langle I_s, O_s \rangle$$

$\mathcal{P}_\bowtie =$	$COL \subseteq I_m \wedge SOL \subseteq O_m \wedge SOL \subseteq I_s \wedge COL \subseteq O_s$
$\mathcal{I}_\bowtie =$	$\{\}$
$\mathcal{O}_\bowtie =$	$\{\rightarrow out(SOL, R), \rightarrow out([UBD], R)\}$
$\mathcal{E}_\bowtie =$	$\{out(\{COL\}, SL) \rightarrow in(\{COL\}, MA), out(SOL, MA) \rightarrow in(SOL, SL)\}$
$\mathcal{M}_\bowtie = R\triangle MA$	

Figure 6 illustrates the combinator.

5.5   Logical Benders Decomposition

Consider now a combinator for implementing Logical Benders decomposition. Logical Benders decomposition was not supported in CML but this section highlights that it is in fact the dual of the column-generation combinator and is easily supported in OCPMCL. Informally speaking, in its simplest form, a Benders decomposition features a master that relaxes some of the constraints of an original

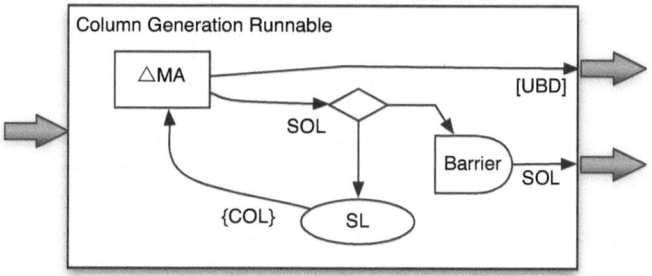

Fig. 6. The Column-Generation Combinator

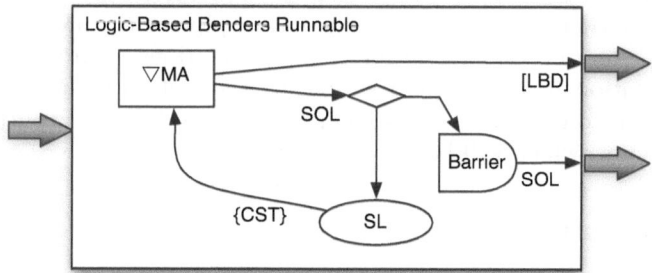

Fig. 7. A Combinator for Logical Benders Decomposition

model and a slave that checks if the solution produces by this master are feasible for the relaxed constraints. If these constraints are infeasible, the slave generates new constraints that are added to the master. The process is repeated until a feasible (and optimal) solution is found. Once again, the combinator receives a master and a slave runnable. The master runnable is copied and the combinator implementation keeps adding constraints to the master until an optimal solution is found. The slave receives the solutions to the master and generates new constraints. The combinator produces a stream of lower bounds and a final solution. The model specification closely mirrors the combinator for column generation, with upper bounds being replaced by lower bounds. Moreover, the combinator is now a tightening of the master program since the Benders decomposition adds new constraints. Figure 7 illustrates the combinator.

$$R = MA \otimes SL \text{ with } MA = \langle I_m, O_m \rangle \text{ and } SL = \langle I_s, O_s \rangle$$

$\mathcal{P}_\otimes =$	$CST \subseteq I_m \wedge SOL \subseteq O_m \wedge SOL \subseteq I_s \wedge CST \subseteq O_s$
$\mathcal{I}_\otimes =$	$\{\}$
$\mathcal{O}_\otimes =$	$\{\rightarrow out(SOL, R), \rightarrow out([LBD], R)\}$
$\mathcal{E}_\otimes =$	$\{out(\{CST\}, SL) \rightarrow in(\{CST\}, MA), out(SOL, MA) \rightarrow in(SOL, SL)\}$
$\mathcal{M}_\otimes =$	$R \nabla MA$

```
1 id<ORModel> root = ... // Def. of AP Model
2 id<ORModel> L = [ORFactory linearizeModel:   root];
3 id<ORModel> C = [ORFactory continuousRelax:  root];
4 id<ORRunnable> r0 = [ORFactory CPRunnable: root];
5 id<ORRunnable> r1 = [ORFactory IPRunnable: L];
6 if<ORRunnable> r2 = [ORFactory LPRunnable: C];
7 id<ORRunnable> cmp = [ORCombinator parallel: cp0 with: ip1];
8 id<ORRunnable> rlx = [ORCombinator relaxedParallel: complete with: 1p2];
9 [relaxed run];
```

Fig. 8. A parallel hybrid for the Asymmetric Traveling Salesman Problem

The structure of the benders runnable is shown in figure 7. The precondition $\mathcal{P}_\otimes$ checks that the *master* accepts a pool of constraints and generates a solution. The input relations $\mathcal{I}_\otimes$ states that the master can receive a stream of constraints. The output relations $\mathcal{O}_\otimes$ states that the output pipe of $R_\otimes$ produces a solution taken from the master. Finally, the internal pipe relations ensures that the master outputs a solution to the slave closure. The slave closure will use the solved master problem to generate and run a slave problem before outputting a set of constraints (cuts) which will be injected into the master.

6   Implementation

OCPCML offers a combinator model library built atop OBJECTIVE-CP. The library provides protocols for all the concepts including *abstract models* (ORModel), *runnables* (ORRunnable), *runnable signatures* (ORSignature), *runnable specification* (ORSpecification) and *combinators* (ORCombinator). The delicate part of the implementation is focused on interpreting the signatures to synthesize the pipes connecting components within a combinator.

Example. Consider as an example the simple Assignment Problem (AP) given as an abstract model (ORModel) that should be solved with the parallel combination of three models: a complete CP model, a complete IP model and a linear relaxation. This can be achieved with the OBJECTIVE-CP code in Figure 8. Line 2 linearizes the root model into L while line 3 stores in C a continuous relaxation. Lines 4–6 concretize the three models with a CP solver (OBJECTIVE-CP), an integer-programming solver (using GUROBI) and a linear-programming solver (using GUROBI too). Lines 7–8 combines the first two models with the parallel combinator which is fed to the **relaxedParallel** combinator alongside runnable r2 which holds the linear relaxation. Line 10 executes the resulting code. The excerpt shows that, with only a few lines of code, three models are easily composed in parallel, producing and consuming solutions, upper bounds and lower bounds. The combinators automatically account for the fact that one model is a relaxation of the other two. (see Figure 11).

Precondition and Signature. Figure 9 shows the OCPCML code for the precondition of the relaxed parallel combinator. It is simply a function working on the signatures of the runnables and verifying that the required properties hold.

```
1 BOOL pre(id<ORRunnable> r1, id<ORRunnable> r2) {
2  return [r2 isRelaxationOf: r1] && [r1.sig acceptsLowerBoundsStream] &&
3         [r1.sig producesSolutionStream]&&[r2.sig producesLowerBoundStream];
4 }
```

Fig. 9. Precondition closure for `ORRelaxedParallelCombinator`

```
1 void internal(id<ORRunnable> parent,id<ORRunnable> r1, id<ORRunnable> r2) {
2   if([[r1 sig] providesSolutionStream] && [[r2 sig] acceptsSolutionStream])
3     [[r1 outSolutionStream] wheneverNotifiedDo: ^void(id<ORSolution> s) {
4        [[r2 inSolutionStream] notifyWith: s];
5     }];
6   ...
7   if([[r1 sig] providesSolutionStream])
8     [[r1 outSolutionStream] wheneverNotifiedDo: ^void(id<ORSolution> s) {
9        [[parent outSolutionStream] notifyWith: s];
10    }];
11 }
```

Fig. 10. Internal pipe closure for ORCompleteParallelCombinator

The Piping Infrastructure. Communication relies heavily on an event infrastructure provided by OBJECTIVE-CP and is similar to COMET events. In OBJECTIVE-CP, an `ORInformer` object embodies a thread-safe event delivery mechanism and meshes with Objective-C closures. Two operations are available on an informer. One can **notify** the occurrence of the event it represents and one can register a closure to *listen and respond* to an event occurrence. The arguments of the closure offer a simple way to hand-off data. In OCPCML, runnables provide informers to represent the input and output pipes of their signature. Combinators then generate the right producer/consumer glue to transmit the solutions.

Figure 10 highlights fragments responsible for setting up the plumbing infrastructure for a parallel combinator. Lines 3–5 setup the pipelining for the output solution pipe from r_1 into the input solution pipe of r_2. This is realized with a simple closure relaying solutions from one informer to the other. The plumbing code is only generated when needed (line 2). Naturally, the other plumbing code (from r_2 to r_1) is generated similarly. Lines 7–10 install the output pipe connecting r_1's output solution stream to the output solution stream of the parent. This is realized with a simple closure that listens to inbound solutions from r_1 and relays them. The other connections are similar but not shown for brevity.

7   Empirical Results

This section presents benchmark results to assess the practicality of OCPMCL. The goal is not to give comprehensive results on a wide variety of benchmarks but to give preliminary evidence that OCPMCL is a promising approach to ease the building of hybrid optimization algorithms.

The first benchmark is the Location-Allocation Problem implemented using the logical Benders approach in [5]. It will allow us to compare the efficiency of an OCPMCL model with a hand-crafted implementation. The experiments feature 6

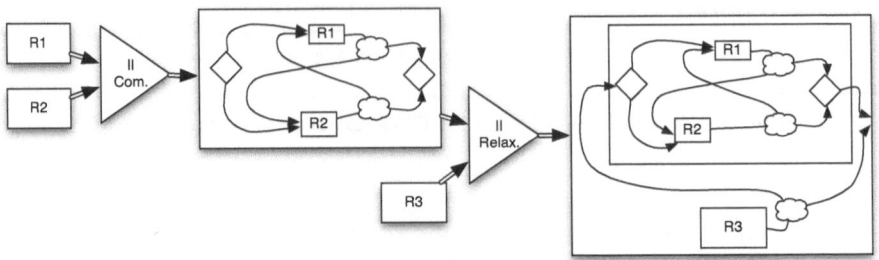

Fig. 11. Nested Parallel composition of two equivalent models with a linear relaxation

Instance #	min	max	avg
1	19.69	21.23	20.23
2	22.53	26.66	23.82
3	11.26	13.30	11.92
4	5.81	7.61	6.37
5	94.81	110.13	99.31
6	67.28	79.73	70.91
overall avg			38.76

(a) Logic-Based Bender's

Inst	min	max	avg	min ‖	max ‖	avg ‖
8 × 20	2.2	2.5	2.3	1.8	2.3	2.1
8 × 30	6.1	6.5	6.3	2.5	4.3	3.5
8 × 40	7.9	16.2	9.7	4.5	9.5	5.7
9 × 20	72.0	89.8	75.3	48.6	62.3	55.5
9 × 30	27.6	30.2	28.3	21.8	27.9	24.1
9 × 40	155.2	174.5	165.6	72.6	89.5	80.2

(b) Assignment Problem

Fig. 12. Benchmarks for OBJECTIVE-CP runnables

instances from the original paper, namely, the *uncorrelated* 20 × 10 (20 clients, 10 facilities) instances[2]. The authors reported an average running time of 33 seconds for these instances. The OCPCML results are based on 20 runs of each instance and are given in Table 12(a). The Bender's runnable runs in about 39 seconds on average which is remarkably close to the results for the hand-written model. The experiment was carried out on a 2.13 Ghz Intel Core 2 Duo with 4 GB of RAM running Mac OS X (10.8) which is comparable to the machines in the original paper (Duo Core AMD 270 CPU, 4 GB Ram, Red Hat Linux). Table 13 reports the results of an instrumentation of code to measure the time spent in the master, in the slave, and otherwise, considering that the remainder of time was attributed to the combinator. This is an overestimate of the true combinator cost as any other overhead is attributed to the combinator. The Master columns report the total time spent in the master. The % column report the fraction of the total that this represents. The same applies for the Slave and Combinator columns. For any row, the percentages add up to 100%. Overall, the combinator overhead never exceeds $\frac{1}{2}$% of the runtime and demonstrates that the approach is competitive. This should be contrasted with the brief OCPMCL-based implementation which weighs in at 90 lines of Objective-C code (without data reading) to create the models and setup the Bender's combinator.

The second benchmark is a simple Assignment Problem (AP) in which we run a standard CP implementation [6] in parallel with a CP linear reformulation using the Complete Parallel Combinator. Note that there are better approaches to solving the AP, we only aim to show the benefit of using OCPMCL combinators

---

[2] A detailed description of the problem can be found in [5].

Instance#	Master			Slave			Combinator		
	μ	σ	%	μ	σ	%	μ	σ	%
1	20.12	0.40	99.41	0.09	0.02	0.42	0.12	0.40	0.17
2	23.72	0.90	99.57	0.06	0.02	0.25	0.10	0.90	0.18
3	11.84	0.46	99.37	0.04	0.01	0.31	0.08	0.47	0.32
4	6.31	0.57	99.10	0.03	0.01	0.43	0.05	0.57	0.47
5	99.14	3.68	99.82	0.14	0.05	0.14	0.18	3.71	0.04
6	70.80	3.10	99.84	0.07	0.03	0.12	0.11	3.10	0.04

Fig. 13. Time allocation between Master/Slave/Combinator

to generate a parallel runnable. The linear reformulation is substantially slower (particularly as a CP model). The table gives the running time of solving the linear model alone and within the parallel runnable. Results are based on random instances with sizes $n \times m$ where n is the number of agents/tasks and m is the maximum allowed cost (cost range $\in [1, m]$). Columns *min, max, avg* in Table 12(b) represent the minimum maximum and average running time (in secs) of the standalone linear CP problem, while *min $\|$, max $\|$, avg $\|$* refer to the parallel runnable.

8   Conclusion

This paper proposes the concept of model combinators, and its implementation in OCPMCL, to simplify the design of hybrid optimization algorithms and provide a foundation for combining complex models. In earlier work, such as in CML, semantic ambiguity prevented models from being truly composible, as they lacked systematic mechanisms for synthesizing input/output interfaces and verifying preconditions on models. To address these shortcomings, this paper introduces a number of concepts:

1. The definition of relationships between models, tightenings and relaxations, which are derived through model transformations. The transitive closures of these relationships enables OCPMCL to verify preconditions on the models. Combinators can also specify their relationship with the underlying models.
2. The concept of runnable and runnable signatures that specify the functionalities supported by an optimization program and its model.
3. The concept of model specifications, including input/output/internal pipes and pipe rules, that enables the synthesis of the signature of the combinators from the signature of their components.

These concepts were used to specify a number of model combinators, including sequential and parallel composition, column generation, and Benders decomposition. These high-level concepts are implemented in OCPMCL using thread-safe informers, an event mechanism provided by OBJECTIVE-CP. Preliminary experimental results on logical Benders decomposition algorithms presented in the literature and artificial benchmarks indicates that the approach promises to be practical.

References

1. Akgun, O., Miguel, I., Jefferson, C., Frisch, A., Hnich, B.: Extensible automated constraint modelling (2011)
2. De Moura, L., Bjørner, N.: Satisfiability modulo theories: introduction and applications. Commun. ACM 54(9), 69–77 (2011)
3. Duck, G.J., De Koninck, L., Stuckey, P.J.: Cadmium: An implementation of ACD term rewriting. In: Garcia de la Banda, M., Pontelli, E. (eds.) ICLP 2008. LNCS, vol. 5366, pp. 531–545. Springer, Heidelberg (2008)
4. Duck, G., Stuckey, P., Brand, S.: Acd term rewriting. In: Etalle, S., Truszczyński, M. (eds.) ICLP 2006. LNCS, vol. 4079, pp. 117–131. Springer, Heidelberg (2006)
5. Fazel-Zarandi, M.M., Beck, J.C.: Solving a Location-Allocation Problem with Logic-Based Benders' Decomposition. In: Gent, I.P. (ed.) CP 2009. LNCS, vol. 5732, pp. 344–351. Springer, Heidelberg (2009)
6. Fontaine, D., Michel, L.: A high level language for solver independent model manipulation and generation of hybrid solvers. In: Beldiceanu, N., Jussien, N., Pinson, É. (eds.) CPAIOR 2012. LNCS, vol. 7298, pp. 180–194. Springer, Heidelberg (2012)
7. Frisch, A., Harvey, W., Jefferson, C., Martínez-Hernández, B., Miguel, I.: Essence: A constraint language for specifying combinatorial problems. Constraints 13, 268–306 (2008)
8. Seldin, J.P., Roger Hindley, J.: Lambda-Calculus and Combinators An Introduction, 2nd edn. Cambridge University Press (2008)
9. Puchinger, J., Stuckey, P.J., Wallace, M., Brand, S.: From high-level model to branch-and-price solution in g12 (2008)
10. Puchinger, J., Stuckey, P.J., Wallace, M.G., Brand, S.: Dantzig-wolfe decomposition and branch-and-price solving in g12. Constraints 16(1), 77–99 (2011)
11. Refalo, P.: Linear Formulation of Constraint Programming Models and Hybrid Solvers. In: Dechter, R. (ed.) CP 2000. LNCS, vol. 1894, pp. 369–383. Springer, Heidelberg (2000)
12. Schrijvers, T., Tack, G., Wuille, P., Samulowitz, H., Stuckey, P.J.: Search combinators. In: Lee, J. (ed.) CP 2011. LNCS, vol. 6876, pp. 774–788. Springer, Heidelberg (2011)
13. Hentenryck, P.V., Michel, L.: Search = Continuations + Controllers. In: Schulte, C. (ed.) CP 2013. LNCS, vol. 8124, Springer, Heidelberg (2013)
14. Yunes, T., Aron, I.D., Hooker, J.N.: An integrated solver for optimization problems. Oper. Res. 58(2), 342–356 (2010)

Modelling Destructive Assignments

Kathryn Francis[1,2], Jorge Navas[2], and Peter J. Stuckey[1,2]

[1] National ICT Australia, Victoria Research Laboratory
[2] The University of Melbourne, Victoria 3010, Australia

Abstract. Translating procedural object oriented code into constraints is required for many processes that reason about the execution of this code. The most obvious is for *symbolic execution* of the code, where the code is executed without necessarily knowing the concrete values. In this paper, we discuss translations from procedural object oriented code to constraints in the context of solving optimisation problems defined via simulation. A key difficulty arising in the translation is the modelling of state changes. We introduce a new technique for modelling destructive assignments that outperforms previous approaches. Our results show that the optimisation models generated by our technique can be as efficient as equivalent hand written models.

1 Introduction

Symbolic reasoning has been the crux of many software applications such as verifiers, test-case generation tools, and bug finders since the seminal papers of Floyd and Hoare [6,8] in program verification and King [10] in symbolic execution for testing. Common to these applications is their translation of the program or some abstraction of it into equivalent *constraints* which are then fed into a *constraint solver* to be checked for (un)satisfiability.

The principal challenge for this translation is effective handling of destructive state changes. These both influence and depend on the flow of control, making it necessary to reason disjunctively across possible execution paths. In object oriented languages with field assignments, the disjunctive nature of the problem is further compounded by potential aliasing between object variables.

In this paper we introduce a new, *demand-driven* technique for modelling destructive assignments, designed specifically to be effective for the difficult case of field assignments. The key idea is to view the value stored in a variable not as a function of the current state, but as a function of the *relevant assignment statements*. This allows us to avoid maintaining a representation of the entire program state, instead only producing constraints for expressions which are actually required.

The particular application we consider for our new technique is the tool introduced in [7], which aims to provide Java programmers with more convenient access to optimisation technology. The tool allows an optimisation problem to be expressed in simulation form, as Java code which computes the objective value given the decisions. This code could be used directly to solve the optimisation problem by searching over possible combinations of decisions and comparing the computed results, but this is likely to be very inefficient. Instead, the tool in [7]

C. Schulte (Ed.): CP 2013, LNCS 8124, pp. 315–330, 2013.
© Springer-Verlag Berlin Heidelberg 2013

translates the simulation code into constraints, and then uses an *off-the-shelf* CP solver to find a set of decisions resulting in the optimal return value.

Experimental results using examples from this tool demonstrate that our new technique for modelling destructive assignments is superior to previous approaches, and can produce optimisation models comparable in efficiency to a simple hand written model for the same problem.

1.1   Running Example

As a running example throughout the paper we consider a smartphone app for group pizza ordering. Each member of the group nominates a number of slices and some ingredient preferences. The app automatically generates a joint order of minimum cost which provides sufficient pizza for the group, assuming that a person will only eat pizza with at least one ingredient they like and no ingredients they dislike. After approval from the user, the order is placed electronically.

Our focus is on the optimisation aspect of the application: finding the cheapest acceptable order. We assume that for each type of pizza both a price per pizza and a price per slice is specified. The order may include surplus pizza if it is cheaper to buy a whole pizza than the required number of individual slices.

Figure 1 shows a Java method defining this optimisation problem, called build-Order. The problem parameters are the contents of the people list and the details stored in the menu object when buildOrder is called. Each call to the method choiceMaker.chooseFrom indicates a decision to be made, where the possible options are the OrderItem objects included in the list pizzas (the Order constructor creates an OrderItem for each pizza on the menu, all initially for 0 slices). The objective is to minimise the return value, which is the total cost of the order.

1.2   Translating Code into Constraints

To evaluate different possible translations from procedural code to constraints we use examples from the tool in [7]. This tool actually performs the translation on demand at run-time (not as a compile time operation), which complicates the translation process somewhat. For the purpose of this paper we will ignore such implementation details, using the following abstraction to simplify the description of the different translations.

We consider the translation to be split into two phases. In the first phase the code is flattened into a linear sequence of assignment statements, each of which has some conditions attached. We describe this transformation briefly in Section 2. In the second phase, which is the main focus of the paper, the flattened sequence of assignments is translated into constraints.

2   Flattening

In the Java programming language only the assignment statement changes the state of the program. All other constructs simply influence which other statements will be executed. It is therefore possible to emulate the effect of a piece of Java code using a sequence of assignment statements, each with an attached set

```
int buildOrder() {                              class OrderItem
  order = new Order(menu);                       {
  for(Person person : people) {                    int pizzaPrice;
    // Narrow down acceptable pizzas               int slicePrice;
    pizzas.clear();                                int fullPizzas = 0;
    for(OrderItem item : order.items)             int numSlices = 0;
      if(person.willEat(item))
        pizzas.add(item);                          void addSlice() {
    // Choose from these for each slice              numSlices = numSlices + 1;
    for(int i = 0; i < person.slices; i++) {        if(numSlices == slicesPerPizza) {
      OrderItem pizza =                                numSlices = 0;
        choiceMaker.chooseFrom(pizzas);                fullPizzas = fullPizzas + 1;
      pizza.addSlice();                            } }
  } }
  return order.totalCost();                        int getCost() {
}                                                    int cost = fullPizzas * pizzaPrice;
                                                     if(numSlices > 0) {
class Order {                                          int slicesCost =
  List<OrderItem> items;                                numSlices * slicePrice;
  int totalCost() {                                   if(slicesCost > pizzaPrice)
    int totalcost = 0;                                  slicesCost = pizzaPrice;
    for(OrderItem item : items)                        cost = cost + slicesCost;
      totalcost += item.getCost();                 } }
    return totalcost;                            }
  }
}
```

Fig. 1. A Java simulation of a pizza ordering optimisation problem

of conditions controlling whether or not it should be executed. The conditions reflect the circumstances under which this statement would be reached during the execution of the original code.

The flattening process involves unrolling loops[1], substituting method bodies for method calls, and removing control flow statements after adding appropriate execution conditions for the child statements. As an example, consider the method getCost shown in Figure 1. To flatten an if statement we simply add the if condition to the execution conditions of every statement within the then part. The body of getCost can be flattened into the following sequence of conditional assignment statements.

Conditions	Variable	Assigned Value
1.	cost :=	fullPizzas × pizzaPrice
2. (numSlices > 0)	: slicesCost :=	numSlices × slicePrice
3. (numSlices > 0, slicesCost>pizzaPrice)	: slicesCost :=	pizzaPrice
4. (numSlices > 0)	: cost :=	cost + slicesCost

---

[1] The tool in [7] only supports loops with exit conditions unaffected by the decisions, or iteration over bounded collections. This means the number of loop iterations is always bounded. For unbounded loops partial unrolling can be performed, and the final model will be an under-approximation of the behaviour of the program.

Note that each assignment statement applies to a specific variable. This may be a local variable identified by name (as above), or an object field o.f where o is a variable storing an object, and f is a field identifier. We call an assignment to an object field a *field assignment*. The value of the object variable o may depend on the decisions, so the concrete object whose field is updated by a field assignment is not necessarily known.

An important optimisation is to consider the declaration scope of variables. For example, if a variable is declared inside the then part of an if statement (as is the case for the slicesCost variable above), assignments to that variable need not depend on the if condition. In any execution of the original code where the if condition does not hold, this variable would not be created, and therefore its value is irrelevant. This means assignments 2 and 3 above do not need the condition numSlices > 0.

We also need to record the initial program state. For variables which exist outside the scope of the code being analysed, we add an unconditional assignment at the beginning of the list setting the variable to its initial value. We call this an *initialising assignment*. For object fields we add an initialising assignment for each concrete object.

Figure 2 shows the sequence of assignments produced by flattening our example function buildOrder for an instance with two people and three pizza types. Note that calls to ChoiceMaker methods are left untouched (these represent the creation of new decision variables), and expressions which do not depend on the decisions are calculated upfront. For example, the code used to find acceptable order items for each person does not depend on any decisions, so rather than including assignments originating in this part of the code in the flattened sequence, we simply calculate these lists and then use them as constants. Where these expressions are used as if conditions or loop exit conditions we exclude from the translation any unreachable code.

In the following sections, we assume our input is this flattened list of conditional assignment statements. We also use the notation $Dom(v, i)$ to refer to the set of possible values for variable v at (just before) assignment i. This is easily calculated from the list of assignments. A conditional assignment adds values to the domain of the assigned-to variable, while an unconditional assignment replaces the domain.

3 Modelling Assignments: Existing Techniques

Using the flattening transformation described above and a straightforward translation of mathematical and logical expressions, we reduce the problem of representing Java code by constraints to that of modelling (conditional) assignment statements. In this section we describe two existing approaches to this, while in the next section we introduce a new proposed approach.

3.1 Typical CP Approach

One obvious technique for modelling assignments, and that used in [7,2,4], is to create a new version of the assigned-to variable for each assignment, and then

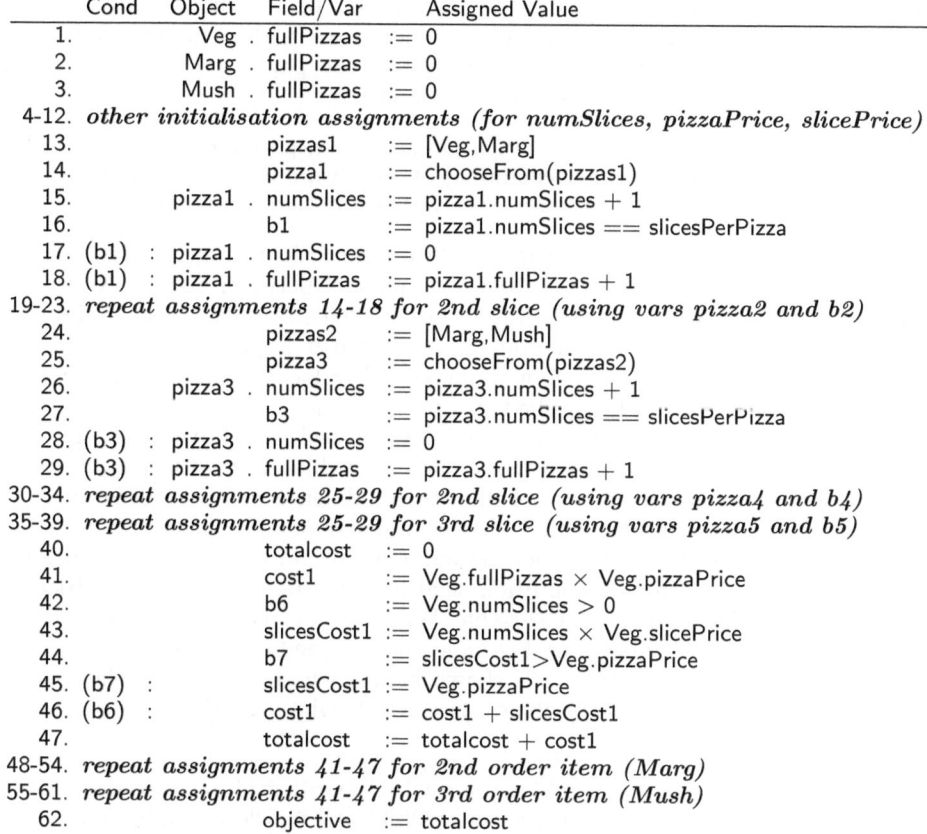

	Cond	Object	Field/Var	Assigned Value
1.		Veg .	fullPizzas	:= 0
2.		Marg .	fullPizzas	:= 0
3.		Mush .	fullPizzas	:= 0
4-12.		*other initialisation assignments (for numSlices, pizzaPrice, slicePrice)*		
13.			pizzas1	:= [Veg,Marg]
14.			pizza1	:= chooseFrom(pizzas1)
15.		pizza1 .	numSlices	:= pizza1.numSlices + 1
16.			b1	:= pizza1.numSlices == slicesPerPizza
17.	(b1) :	pizza1 .	numSlices	:= 0
18.	(b1) :	pizza1 .	fullPizzas	:= pizza1.fullPizzas + 1
19-23.		*repeat assignments 14-18 for 2nd slice (using vars pizza2 and b2)*		
24.			pizzas2	:= [Marg,Mush]
25.			pizza3	:= chooseFrom(pizzas2)
26.		pizza3 .	numSlices	:= pizza3.numSlices + 1
27.			b3	:= pizza3.numSlices == slicesPerPizza
28.	(b3) :	pizza3 .	numSlices	:= 0
29.	(b3) :	pizza3 .	fullPizzas	:= pizza3.fullPizzas + 1
30-34.		*repeat assignments 25-29 for 2nd slice (using vars pizza4 and b4)*		
35-39.		*repeat assignments 25-29 for 3rd slice (using vars pizza5 and b5)*		
40.			totalcost	:= 0
41.			cost1	:= Veg.fullPizzas × Veg.pizzaPrice
42.			b6	:= Veg.numSlices > 0
43.			slicesCost1	:= Veg.numSlices × Veg.slicePrice
44.			b7	:= slicesCost1>Veg.pizzaPrice
45.	(b7) :		slicesCost1	:= Veg.pizzaPrice
46.	(b6) :		cost1	:= cost1 + slicesCost1
47.			totalcost	:= totalcost + cost1
48-54.		*repeat assignments 41-47 for 2nd order item (Marg)*		
55-61.		*repeat assignments 41-47 for 3rd order item (Mush)*		
62.			objective	:= totalcost

Fig. 2. Flattened version of buildOrder method. We assume an instance where the menu lists three different types of pizza (vegetarian, margharita and mushroom), meaning the order will contain three OrderItems [Veg, Marg, Mush], and where the people list contains two Person objects, the first willing to eat vegetarian or margharita and requiring two slices, and the second willing to eat margharita or mushroom and requiring three slices. The b variables have been introduced to store branching conditions. Variables from methods called more than once and those used as the iteration variable in a loop are numbered to distinguish between the different versions.

use the latest version whenever a variable is referred to as part of an expression. If the assignment has some conditions, the new version of the variable can be constrained to equal either the assigned value or the previous version, depending on whether or not the conditions hold. This is easily achieved using a pair of implications, or alternatively using an element constraint with the condition as the index. The element constraint has the advantage that some propagation is possible before the condition is fixed, so we will use this translation.

The constraint arising from a local variable assignment is shown below, where localvar0 is the latest version of localvar before the assignment, and localvar1 is the new variable that results from the assignment, which will become the new latest version of localvar. Note that we assume arrays in element constraints are indexed from 1. For convenience, in the rest of the paper we use a simplified syntax for these constraints (also shown below).

assignment: condition : localvar := expression
constraint: element(bool2int(condition)+1, [localvar0, expression], localvar1)
simple syntax: localvar1 = [localvar0, expression][condition]

This translation is only correct for local variables. Field assignments are more difficult to handle due to the possibility of aliasing between objects. However, if the set of concrete objects which may be referred to by an object variable is finite (which is the case for our application), then it is possible to convert all field assignments into equivalent assignments over local variables, after which the translation above can be applied.

For each concrete object, a local variable is created to hold the value of each of its fields. In the following we name these variables using the object name and the field name separated by an underscore. Then every field assignment is replaced by a sequence of local variable assignments, one for each of the possibly affected concrete objects. These new assignments retain the original conditions, and each also has one further condition: that its corresponding concrete object is the one referred to by the object variable. Where necessary to avoid duplication, an intermediate variable is created to hold the assigned expression.

An example of this conversion is shown below, where we assume the assignment is on line n and $Dom(\text{objectvar}, n) = \{\text{Obj1, Obj2, Obj3}\}$.

field assignment: condition : objectvar.field := expression
assignments: condition $\land$ (objectvar = Obj1) : Obj1_field := expression
                    condition $\land$ (objectvar = Obj2) : Obj2_field := expression
                    condition $\land$ (objectvar = Obj3) : Obj3_field := expression

The final requirement is to handle references to object fields. We need to look up the field value for the concrete object corresponding to the current value of the object variable. To achieve this we use a pair of element constraints sharing an index as shown below, where fieldrefvar is an intermediate variable representing the retrieved value. We assume the same domain for objectvar.

field reference: objectvar.field
constraints: element(indexvar, [Obj1,Obj2,Obj3], objectvar)
                   element(indexvar, [Obj1_field,Obj2_field,Obj3_field], fieldrefvar)

In summary, this approach involves two steps. First the list of assignments is modified to replace field assignments with equivalent local variable assignments, introducing new variables as required. Then the new list (now containing only local variable assignments) is translated into constraints, with special handling for field references. This approach is quite simple, but can result in a very large model if fields are used extensively. To see the result of applying this translation to a portion of our running example, see Figure 3(a).

3.2   Typical SMT Approach

One of the main reasons for the significant advances in program symbolic reasoning (e.g. verification and testing) during the last decade has been the remarkable progress in modern SMT solvers (we refer the reader to [1] for details).

When using SMT, local variable assignments can be translated in the same way as for the CP approach (adding a new version of the variable for each assignment), but using an if-then-else construct (ite below) instead of an element constraint.

assignment: condition : localvar := expression
formula: localvar1 = ite(condition, expression, localvar0)

For field assignments, it is more convenient to use the theory of arrays. This theory extends the theory of uninterpreted functions with two interpreted functions *read* and *write*. McCarthy proposed [11] the main axiom for arrays:

$$\forall a, i, j, x \text{ (where } a \text{ is an array, } i \text{ and } j \text{ are indices and } x \text{ is a value)}$$
$$i = j \to read(write(a, i, x), j) = x$$
$$i \neq j \to read(write(a, i, x), j) = read(a, j)$$

Note that since we are not interested in equalities between arrays we only focus on the non-extensional fragment.

Following the key idea of Burstall [3] and using the theory of arrays, we define one array variable for each object field. Conceptually, this array contains the value of the field for every object, indexed by object. Note however that there are no explicit variables for the elements.

An assignment to a field is modelled as a write to the array for that field, using the object variable as the index. The result is a new array variable representing the new state of the field for all objects. This is much more concise and efficient than creating an explicit new variable for each concrete object.

We still need to handle assignments with conditions. If the condition does not hold all field values should remain the same, so we can simply use an ite to ensure that in this case the new array variable is equal to the previous version.

field assignment: cond : objectvar.field := expression
formula: field1 = ite(cond, *write*(field0, objectvar, expression), field0)

A reference to an object field is represented as a read of the latest version of the field array, using the object variable as the lookup index.

field reference: objectvar.field
formula: *read*(field0, objectvar)

For a more complete example, see Figure 3(b). This example clearly demonstrates that the SMT formula can be much more concise than the CP model arising from the translation discussed in the previous section. Its weakness is its inability to reason over disjunction (compared to *element* in the CP approach). The approaches are compared further in Section 4.3.

pizza1 in {Veg, Marg}
element(index1, [Veg,Marg], pizza1)
element(index1, [Veg_numSlices0,Marg_numSlices0], pizza1_numSlices0)
temp1 = pizza1_numSlices0 + 1
Marg_numSlices1 = [Marg_numSlices0,temp1][pizza1 = Marg]
Veg_numSlices1 = [Veg_numSlices0,temp1][pizza1 = Veg]
element(index2, [Veg,Marg], pizza1)
element(index2, [Veg_numSlices1,Marg_numSlices1], pizza1_numSlices1)
b1 = (pizza1_numSlices1 == slicesPerPizza)
Marg_numSlices2 = [Marg_numSlices1,0][b1 ∧ pizza1 = Marg]
Veg_numSlices2 = [Veg_numSlices1,0][b1 ∧ pizza1 = Veg]

(a) CP Translation: Constraints

(pizza1 = Marg) ∨ (pizza1 = Veg)
numSlicesArray1 = write(numSlicesArray0, pizza1, read(numSlicesArray0,pizza1)+1)
b1 = (read(numSlicesArray1,pizza1) = slicesPerPizza)
numSlicesArray2 = ite(b1, write(numSlicesArray1,pizza1,0), numSlicesArray1)

(b) SMT Translation: Formula

Fig. 3. Translation of assignments 14-17 of the running example (Figure 2) using (a) the obvious CP approach, and (b) the SMT approach

4   A New Approach to Modelling Assignments

The main problem with the CP approach presented earlier is the excessive number of variables created to store new field values for every object possibly affected by a field assignment. Essentially we maintain a representation of the complete state of the program after each execution step.

This is not actually necessary. Our only real requirement is to ensure that the values retrieved by variable references are correctly determined by the assignment statements. Maintaining the entire state is a very inefficient way of achieving this, since we may make several assignments to a field using different object variables before ever referring to the value of that field for a particular concrete object. To take advantage of this observation, we move away from the state-based representation, instead simply creating a variable for each field reference, and constraining this to be consistent with the relevant assignments.

4.1   The General Case

We first need to define which assignment statements are *relevant* (i.e. may affect the retrieved value) for a given variable reference. Let a_i be the assignment on line i of the flattened list, and o_i, f_i and c_i be the object, field identifier and set of conditions for this assignment. For a reference to variable obj.field occurring on line n, assignment a_j is *relevant* iff the following conditions hold.

$j < n$ and f_j=field   (occurs before the reference, uses the correct field)
$Dom(\text{obj}, n) \cap Dom(o_j, j) \neq \emptyset$ (assigns to an object which may equal obj)
$\nexists u : o_u$=obj,f_u=field,$c_u=\emptyset$,$j<u<n$ (not overwritten by an unconditional assignment)

As an example, consider the reference to Veg.fullPizzas on line 41 in Figure 2. Of the eight assignments to the fullPizzas field (all of which occur before this reference), the following three are relevant. The others cannot affect the retrieved value as they use object variables (e.g. pizza3) whose domains do not include Veg.

```
 1.              Veg . fullPizzas := 0
18.   (b1) : pizza1 . fullPizzas := pizza1.fullPizzas + 1
23.   (b2) : pizza2 . fullPizzas := pizza2.fullPizzas + 1
```

For a correct model we need constraints ensuring that the retrieved value (Veg_fullPizzas) corresponds to the most recent assignment which updated the read variable. To achieve this we introduce a new integer variable indexvar whose value indicates which of the relevant assignments this is. We use three element constraints to ensure that the selected assignment applies to the correct object, has true execution conditions, and assigns a value equal to the result.

```
element(indexvar, [Veg,pizza1,pizza2], Veg)
element(indexvar, [true,b1,b2], true)
element(indexvar, [0, pizza1_fullPizzas + 1, pizza2_fullPizzas + 1], Veg_fullPizzas)
```

Note that pizza1_fullPizzas and pizza2_fullPizzas are the variables introduced for the field references used as part of the assigned values. These would be constrained using their own list of relevant assignments.

The only remaining requirement is that we must choose the *latest* applicable assignment. Using the natural order for the arrays this corresponds to the greatest index. We therefore add constraints stating that if at index i the object variables are equal and the execution condition is true, then the selected index must be no less than i.

$$(b1 \wedge pizza1 = Veg) \rightarrow indexvar \geq 2$$
$$(b2 \wedge pizza2 = Veg) \rightarrow indexvar \geq 3$$

The general form of the constraints used for field references is shown below. References to local variables are treated as field references where the object variable is the same as that used for all relevant assignments. When this is the case the first element constraint is not required (as it is trivially satisfied), and the implications can be simplified. Other obvious simplifications are also applied.

field reference: queryobj.field
relevant assignments: cond1 : obj1.field := expr1
$$...$$
condn : objn.field := exprn
constraints: element(indexvar, [obj1, ..., objn], queryobj)
element(indexvar, [cond1, ..., condn], true)
element(indexvar, [expr1, ..., exprn], queryobj_field)
$$(cond2 \wedge queryobj = obj2) \rightarrow indexvar \geq 2$$
$$...$$
$$(condn \wedge queryobj = objn) \rightarrow indexvar \geq n$$

As an optimisation, when the code contains more than one reference to some variable v, we insert an unconditional assignment to v at the time of the earlier read, using the read result as the assigned value. This will become the earliest relevant assignment for the later read, which helps to avoid duplication of expressions and constraints.

There is some similarity between our extraction of relevant assignments and the dynamic slicing technique proposed in [9]. Note however that slicing is used only to reduce the number of statements to be translated into constraints. The actual translation is still based on the standard CP approach.

4.2   Special Cases

In some cases, we can detect a pattern to the relevant assignments which allows us to use a more specialised constraint. The three special cases we look for are Boolean variables, sequences of assignments representing a sum calculation, and sequences of assignments representing a maximum or minimum calculation. The translation of assignments automatically detects the cases described below and uses the more efficient translation.

Boolean Variables. When the referenced variable is of type bool, we can define a Boolean expression for the retrieved value rather than using element constraints and implications. The expression (shown below) is true if some assignment with a true value applies and no later assignment with a false value applies.

field reference: q.field
assignments: $c_i : o_i$.field $:= e_i$ $i \in 1..n$

expression: $$\bigvee_{i \in 1..n} \left(c_i \wedge e_i \wedge (o_i = q) \wedge \bigwedge_{j \in i+1..n} (e_j \vee \neg c_j \vee (o_j \neq q)) \right)$$

Local variables are handled in the same way except without the object equalities. Using this constraint instead of the generic constraint eliminates the need to introduce an index variable, and allows simplifications to be performed when objects are known to be equal, an assigned value is fixed, or an assignment is unconditional.

Sum Calculations. Computations often involve taking the sum of a set of numbers. In a procedural language, sums are commonly calculated by iteratively adding each number to a variable representing the total. This coding pattern results in a sequence of writes where each written value is an addition of the previous value of this variable and some other number. When this pattern is detected, we can replace the usual constraints with a sum constraint.

relevant assignments: total := 0
                         total := total + value1
                         total := total + value2
                         total := total + value3
constraint: finaltotal = sum([0,value1,value2,value3])

In the example above all assignments were unconditional and used exactly the same variable (the local variable total). It is also possible to use a general form of this constraint for sequences of assignments which do not represent a pure sum but a related calculation, such as counting the objects which satisfy some condition. Consider again the relevant writes for the reference to Veg.fullPizzas discussed earlier. A better constraint for Veg_fullPizzas is shown below.

assignments: Veg . fullPizzas := 0
    (b1) : pizza1 . fullPizzas := pizza1.fullPizzas + 1
    (b2) : pizza2 . fullPizzas := pizza2.fullPizzas + 1
constraint: Veg_fullPizzas =
    sum([0, bool2int(b1 $\wedge$ pizza1=Veg), bool2int(b2 $\wedge$ pizza2=Veg)])

A major advantage of this constraint is that it removes the need to create and constrain the variables pizza1_fullPizzas and pizza2_fullPizzas. These can be excluded from the model entirely as they were only used to re-assign to the same variable, and are now not required to define the values retrieved from this field.

The general form of the alternative constraint used for sums is given below. The first assignment gives the initial value of qobj.field. If not already present it is created from the initialisation assignments for objects in the domain of qobj.

field reference: qobj.field
assignments: qobj.field := init
  cond_i : obj_i.field := obj_i.field + expr_i $i \in 1..n$
constraint: q_field = sum([init, $\text{expr}_i \times$ bool2int($\text{cond}_i \wedge \text{obj}_i = $ qobj) $|i \in 1..n$])

Max/Min Calculations. Another common coding pattern is to calculate a maximum or minimum by iterating through a list of values overwriting a variable each time a smaller/larger value is found. As with sum, we can detect this pattern when building the constraints for the final read of the variable. This time in order for the alternative constraint to apply, every non-initialisation assignment must have a condition which compares the current value of the variable with the assigned value. When there are no other assignment conditions, and the variable is a local variable (or the assigned-to object is known to equal the read object for all relevant assignments), we can use a max/min constraint as shown below.

assignments: max := init
    (value1 > max) : max := value1
    (value2 > max) : max := value2
constraint: finalmax = max([init, value1, value2])

We can again extend this to apply to field assignments and assignments with additional conditions. When extra conditions are present, we are calculating the maximum or minimum value for which these additional conditions hold. For a maximum, we constrain the result to be no less than any value for which the extra conditions hold, and to equal one of the values for which the conditions hold. Minimum is handled equivalently.

field reference: queryobj.field
assignments: $(\text{cond}_i \wedge \text{value}_i > \text{obj}_i.\text{field})$: $\text{obj}_i.\text{field} := \text{value}_i$ $i \in 1..n$
constraints: $\bigvee_{i \in 1..n} \text{cond}_i \wedge (\text{obj}_i = \text{queryobj}) \wedge (\text{queryobj\_field} = \text{value}_i)$
  $\bigwedge_{i \in 1..n} (\text{cond}_i \wedge \text{obj}_i = \text{queryobj}) \rightarrow \text{queryobj\_field} \geq \text{value}_i$

Table 1. Comparing three approaches to modelling destructive assignments

Problem		smt	Time (secs) orig	orig+	new	new+	hand	Failures (000s) orig	orig+	new	new+	hand
proj1	200	2.2	23.0	0.1	12.1	0.1	0.1	56	0	34	0	0
	225	2.4	3.2	0.1	1.5	0.1	0.1	9	0	4	0	0
	250	1.6	61.9_3	0.1	61.7_3	0.1	0.1	99	0	127	0	0
proj2	22	115.6_2	84.8_1	42.7_1	51.7_2	23.7_1	7.6	39	31	110	35	22
	24	221.1_7	286.9_9	167.6_5	170.9_6	129.2_4	92.2_4	92	89	368	239	280
	26	262.7_8	376.2_{16}	293.3_{10}	255.9_{11}	137.9_6	128.9_5	120	144	583	251	452
pizza	3	56.0	37.4_1	25.1	7.0	3.1	2.0	175	118	30	14	0
	4	226.4_8	180.9_7	175.7_7	138.0_4	79.3_2	2.1	544	541	377	252	1
	5	480.9_{22}	411.8_{18}	407.5_{18}	343.4_{13}	298.3_{12}	2.2	1170	1216	865	945	7

4.3 Comparison with Earlier Approaches

We compared the three presented translation techniques experimentally, using the pizza ordering example plus two benchmarks used in [7] (the other benchmarks require support for collection operations, as discussed in the next section). We used 30 instances for each of several different sizes to evaluate scaling behaviour. For the original and new CP approaches we show the effect of adding special cases (orig+ and new+). Special cases can be detected in the original method, but only for local variables. Using the new translation makes these cases also recognisable for fields. As a reference we also include a fairly naive hand written model for each problem. The Java code defining the problems and all compared constraint models are available online at www.cs.mu.oz.au/~pjs/optmodel.

The CP models were solved using the lazy clause generation solver Chuffed. The SMT versions were solved using Z3 [5]. Z3, like most SMT solvers, does not have built-in support for optimisation. We used a technique similar to [12] to perform optimisation using SMT: in incremental mode, we repeatedly ask for a solution with a better objective value until a not satisfiable result is returned.

Table 1 shows average time to solve and failures for the different models. The small number next to the time indicates the number of timeouts (> 600s). These were included in the average calculations. The results show that while the SMT approach does compete with the original approach, with special case treatment it does not. The new approach is quite superior and in fact has a synergy with special cases (since more of them are visible). new+ competes with hand except for pizza where it appears that the treatment of the relationship between slices and full pizzas used in hand is massively more efficient than the iterative approach in the simulation.

5 Collection Operations

The code for the pizza ordering example makes use of collection classes from the Java Standard Library: Set, List and Map. In this case no special handling is required as all collection operations are independent of the decisions, but often it is more natural to write code where that is not the case. For example, say we wished to extend our application to choose between several possible pizza outlets, each with a different menu. We could do this by adding one extra line at the beginning of the buildOrder function.

menu = chooseFrom(availableMenus);

This change means the contents of the OrderItem list in the Order class will depend on the decisions, so the for loop iterating over this list (in buildOrder) will perform an unknown (though bounded) number of iterations, and the result of any query operation on this list will also depend on the decisions.

In [7], collection operations were supported by introducing appropriately constrained variables representing the state of each collection after each update operation (e.g. List.add). Query operations (e.g. List.get) were represented as a function of the current state of the relevant collections. This is analogous to the way field assignments were handled, with the same drawbacks.

Fortunately our new technique can also be extended to apply to collection operations, resulting in a much more efficient representation. Where previously the flattened list of state changing operations contained only assignments, we now also include collection update operations. Then every query operation on a collection is treated analogously to a field reference. That is, a new variable is created to hold the returned value, and constraints are added to ensure that this value is consistent with the relevant update operations.

Below we provide details of the constraints used for List operations. Set and Map operations are treated similarly; a detailed description is omitted for brevity. We then give experimental results using collection-related benchmarks from [7].

5.1   Example: List

For the List class we support update operations add (at end of list) and replace (item at index), and query operations get (item at index) and size.

A code snippet containing one of each operation type is shown below. Also shown are the assumed possible variable values and initial list contents, and the flattened list of collection update operations. Each update operation has an associated condition, list, index and item. For the add operation, the index is a variable size1 holding the current size of list1. The first three operations in the table reflect the original contents of the lists.

```
if(cond) {
  list1.add(A);
  list1.replace(0, item);
}
if(list2.size() > ind)
  item = list2.get(ind);
```

(a) Code

list1 ∈ {L1,L2,L3}
ind ∈ {0,1,2}
list2 ∈ {L1,L2,L3}
item ∈ {A,B,C}
cond ∈ {true,false}
L1:[A,B], L2:[C], L3:[]

(b) Variables

Cond	List	Index		Item	
	L1	[0]	:=	A	(add)
	L1	[1]	:=	B	(add)
	L2	[0]	:=	C	(add)
cond :	list1	[size1]	:=	A	(add)
cond :	list1	[0]	:=	item	(repl)

(c) Update Operations

With our limited set of supported update operations (which is nevertheless sufficient to cover all code used in the benchmarks from [7]), the size of a list is

Table 2. Comparison on examples which use variable collections

Benchmark		Time (secs)				Failures (000s)			
		orig+	new	new+	hand	orig+	new	new+	hand
bins	12	2.6	5.3	1.1	1.2	8.1	32.5	6.2	13.8
	14	82.8 ₁	129.6 ₃	7.6	18.0	95.4	612.9	75.1	169.9
	16	327.2 ₁₅	391.6 ₁₅	84.8	141.6 ₅	315.1	1617.6	749.5	1355.0
golf	4,3	0.7	0.2	0.2	21.3	0.7	0.8	0.7	159.7
	4,4	3.4	2.0	0.3	0.1	0.8	6.7	0.0	0.0
	5,2	2.4	0.8	0.3	1.5	0.4	0.3	0.0	12.3
golomb	8	1.3	1.2	1.2	1.2	10.8	10.4	10.4	24.0
	9	14.0	12.9	12.9	13.7	55.4	51.9	51.9	149.4
	10	161.5	144.1	151.8	178.8	281.1	284.5	284.5	1211.0
knap1	70	2.1	8.3	2.8	1.8	33.3	2.2	1.9	33.3
	80	7.5	18.4	7.1	6.8	95.7	3.5	3.5	95.7
	90	14.2	31.9	12.7	13.9	180.2	4.5	4.5	180.2
knap2	70	20.9	23.2	22.4	34.7	247.8	245.4	245.4	425.8
	80	88.4 ₂	87.7 ₂	93.9 ₂	117.5 ₃	935.2	901.8	915.0	1253.1
	90	223.6 ₅	229.9 ₅	230.9 ₅	207.0 ₅	2263.9	2182.7	2199.3	2085.5
knap3	40	26.2	0.9	0.3	0.2	14.3	0.5	0.4	1.3
	50	81.1	2.2	1.3	0.1	25.0	0.8	0.6	2.4
	60	295.2 ₆	4.2	1.8	0.4	58.7	1.4	1.2	10.2
proj3	10	153.9 ₅	2.3	2.4	0.1	289.3	9.3	11.6	0.1
	12	509.4 ₂₄	28.0	20.7	0.1	778.5	83.5	92.1	0.2
	14	600.0 ₃₀	133.9 ₂	102.9 ₁	0.1	807.3	299.5	394.5	0.5
route	5	34.2	1.7	1.7	0.2	34.0	6.3	6.3	2.3
	6	338.3 ₃	43.7	43.1	0.8	195.8	57.5	57.5	7.6
	7	600.0 ₃₀	536.9 ₂₀	502.5 ₁₇	2.7	263.2	286.9	333.1	19.2
talent	3,8	11.1	3.4	0.9	0.8	25.9	17.2	5.0	8.9
	4,9	170.8	42.7	8.8	7.3	159.7	127.3	31.1	52.4
	4,10	545.5 ₂₂	223.0 ₁	77.9	54.6	459.7	510.8	178.1	212.5

simply the number of preceding add operations applying to this list and having true execution conditions. Note that the replace operation is not relevant to size.

query: sizeresult := list2.size()
constraint: sizeresult = sum([bool2int(list2=L1), bool2int(list2=L1),
                          bool2int(list2=L2), bool2int(list2=list1∧cond)])

A get query is treated almost exactly like a field reference. The value returned must correspond to the most recent update operation with true execution condition which applied to the correct list and index. There is however one extra complication to be considered. Constraining the get result to correspond to an update operation has the effect of forcing the index to be less than the size of the list. This is only valid if the get query is actually executed.

In the constraints shown below, the final element of each array has been added to leave the index unconstrained and assign an arbitrary value A to our result variable when the get would not be executed (sizeresult>ind is false). Without this the constraints would force ind to correspond to an operation on list2 regardless of whether or not the get query is actually executed, incorrectly causing failure when list2 is empty. We fix the result rather than leaving it unconstrained to avoid searching over its possible values.

query: getresult := list2.get(ind)
constraints: element(indexvar, [L1,L1,L2,list1,list1,list2], list2)
                element(indexvar, [0,1,0,size1,0,ind], ind)
                element(indexvar, [true,true,true,cond,cond,¬(sizeresult>ind)], true)
                element(indexvar, [A,B,C,A,item,A], getresult)
                (list2=L1) ∧ (ind=1)   → indexvar ≥ 2
                (list2=L2) ∧ (ind=0)   → indexvar ≥ 3
                (list2=list1) ∧ (ind=size1) ∧ cond → indexvar ≥ 4
                (list2=list1) ∧ (ind=0) ∧ cond   → indexvar ≥ 5
                ¬(sizeresult>ind)   → indexvar ≥ 6

5.2 Comparison on Benchmarks with Collections

Table 2 compares the various translation approaches (excluding smt and orig which were shown to be not competitive in Table 1) and hand written models, using problems involving collections from [7]. It is clear that the new translation substantially improves on the old in most cases, and is never very much worse (bins,knap1). With the addition of special case treatment the new translation is often comparable to the hand written model, though certainly not always (proj3,route). In a few instances it is superior (bins,golf), this may be because it uses a sequential search based on the order decisions are made in the Java code, or indeed that the intermediate variables it generates give more scope for reusable nogood learning.

6 Conclusion

Effective modelling of destructive assignment is essential for any form of reasoning about procedural code. We have developed a new encoding of assignment and state that gives effective propagation of state-related information. We demonstrate the effectiveness of this encoding for the automatic generation of optimisation models from simulation code, showing that the resulting model can be comparable in efficiency to a hand-written optimization model.

In the future we will investigate the use of this encoding for applications such as test generation. The main difference is the lack of a known initial state. This will require the creation of variables to represent unknown initial field values, with constraints ensuring that if a pair of object variables are equal then their corresponding initial field variables are also equal. Uncertainty about the initial state will also affect the number of relevant assignments for field references. For a query object with unbounded domain all assignments to the same field occurring prior to the read are relevant, unless one of these is an unconditional assignment using this exact variable. These differences may mean that redundant constraints relating reads to each other (which we have not discussed due to their lack of impact for our application) become more important for effective propagation.

330 K. Francis, J. Navas, and P.J. Stuckey

Acknowledgments. NICTA is funded by the Australian Government as represented by the Department of Broadband, Communications and the Digital Economy and the Australian Research Council.

References

1. Biere, A., Heule, M.J.H., van Maaren, H., Walsh, T.: Handbook of Satisfiability. Frontiers in Artificial Intelligence and Applications, vol. 185. IOS Press (February 2009)
2. Brodsky, A., Nash, H.: CoJava: Optimization modeling by nondeterministic simulation. In: Benhamou, F. (ed.) CP 2006. LNCS, vol. 4204, pp. 91–106. Springer, Heidelberg (2006)
3. Burstall, R.: Some techniques for proving correctness of programs which alter data structures. Machine Intelligence 7, 23–50 (1972)
4. Collavizza, H., Rueher, M., Van Hentenryck, P.: CPBPV: a constraint-programming framework for bounded program verification. Constraints 15(2), 238–264 (2010)
5. de Moura, L., Bjørner, N.S.: Z3: An efficient SMT solver. In: Ramakrishnan, C.R., Rehof, J. (eds.) TACAS 2008. LNCS, vol. 4963, pp. 337–340. Springer, Heidelberg (2008)
6. Floyd, R.W.: Assigning meanings to programs. In: Proceedings of the American Mathematical Society Symposia on Applied Mathematics, vol. 19, pp. 19–31 (1967)
7. Francis, K., Brand, S., Stuckey, P.J.: Optimisation modelling for software developers. In: Milano, M. (ed.) CP 2012. LNCS, vol. 7514, pp. 274–289. Springer, Heidelberg (2012)
8. Hoare, C.A.R.: An axiomatic basis for computer programming. Communications of the ACM 12(10), 576–580 (1969)
9. Hofer, B., Wotawa, F.: Combining slicing and constraint solving for better debugging: The CONBAS approach. Advances in Software Engineering 2012, Article ID 628571 (2012)
10. King, J.C.: Symbolic Execution and Program Testing. Com. ACM, 385–394 (1976)
11. McCarthy, J.: Towards a mathematical science of computation. In: IFIP Congress, pp. 21–28 (1962)
12. Sebastiani, R., Tomasi, S.: Optimization in SMT with $LA(Q)$ cost functions. In: IJCAR, pp. 484–498 (2012)

An Improved Search Algorithm for Min-Perturbation

Alex Fukunaga

The University of Tokyo

Abstract. In many scheduling and resource assignment problems, it is necessary to find a solution which is as similar as possible to a given, initial assignment. We propose a new algorithm for this minimal perturbation problem which searches a space of variable commitments and uses a lower bound function based on the minimal vertex covering of a constraint violation graph. An empirical evaluation on random CSPs show that our algorithm significantly outperforms previous algorithms, including the recent two-phased, hybrid algorithm proposed by Zivan, Grubshtein, and Meisels.

1 Introduction

In many CP applications it is necessary to find solutions that are as similar as possible to a given, initial assignment of values to variables. For example, in a meeting scheduling problem or resource scheduling problem, constraints can change unexpectedly after a solution has been generated. This is a type of dynamic constraint satisfaction problem. Similarly, there are situations where there is an "ideal" (but possibly infeasible) assignment of values to variables for a CSP, and the goal is to find an assignment which differs as little as possible from the target. Another scenario where a solution similar to a given initial state is desired occurs in staff scheduling. Employees express preferences regarding when they want to work, but their preferences must be balanced against the staffing demands and constraints of the business, requiring a schedule that satisfies staffing requirements while deviating minimally from employee preferences.

This paper considers search algorithms for this class of *minimal perturbation problem* (MPP) for CSPs, where we seek a solution that minimizes the number of variables whose values differ from a target assignment, or equivalently, the minimal number of variable changes that are necessary to a CSP solution when some of the constraints change unexpectedly. In particular, we focus on minimal perturbation for binary CSPs. Previously, Ran et al. proposed an iterated deepening algorithm for the MPP that searches the space of variable assignments that differ from the target/initial assignment by at most d assignments, where d is the iterative deepening bound [9]. More recently, Zivan, Grubshtein, and Meisels proposed a two-phased algorithm that interleaves the problem of bounding the number of necessary perturbations from the initial assignment, and the problem of testing if such an assignment is possible [11].

We propose a new search algorithm for the MPP, where the main features are (1) a search space where nodes represent a set of *committed* variable assignments, (2) a lower bound based on the minimal vertex covering of the current set of violated constraints, which dominates the lower bound by Zivan et al. This generalizes an earlier, domain-specific MPP algorithm proposed in [4].

C. Schulte (Ed.): CP 2013, LNCS 8124, pp. 331–339, 2013.

2   Problem Definition and Preliminaries

The *Minimal Perturbation Problem* (MPP) is defined as follows: Let $C = (V, D, C)$ be a CSP, where $V = v_1, ..., v_n$ is a set of variables, $D = D_1, ..., D_n$ is a set of domains where D_i is a finite discrete set of possible values for variable v_i, and $C = c_1, ..., c_m$ is a set of constraints which restricts the set of values that the variables can be simultaneously assigned.

Let I be a complete assignment for C. The objective of the MPP is to find an assignment A such that all of the constraints are satisfied, and the *number* of variables in A whose value differs from I is minimized. Following [11], the value of variable v in the original assignment is called the *Starting Variable Assignment* of v, or its SVA.

While previous work [9,11] defined the MPP more generally, i.e., a general distance function, and a partial initial assignment, the lower bound functions used in the previous work assume the definition above, and the actual experimental evaluations of the previous algorithms were performed on binary CSPs based on this definition.

3   Previous Algorithms for the MPP

The first algorithm which specifically addressed the MPP defined in Sec 2 was the *Repair-Based algorithm with Arc-Consistency (RB-AC)*, by Ran et al. [9]. Given an initial variable assignment $I = \{x_1 = v_1, ..., x_n = v_n\}$, let D_i be the set of states which have exactly i variables whose value are different from that of the initial state I. We call the set $D = D_1 \cup ...D_n$ the *difference space*, or *D-space*. The root node of this search space is I. Nodes at depth d of the search tree contain variable assignments which differ by d assignments from I. Each edge in the tree changes the value of one variable which has not yet been changed by any ancestor. RB-AC searches D-space using a depth-first iterative deepening strategy, IDA* [7]. The d-th iteration of IDA* explores the subset of the depth-first branch-and-bound D-space search tree where at each node, the sum $f = g + h \le d$, where g is the number of differences from the initial state in the current solution, and h is the lower bound on the additional number of differences required to find a conflict-free solution. RB-AC uses a simple lower bound, L_1, which is the number of variables that do not have the SVA in its domain.

Zivan, Grubshtein, and Meisels proposed HS_MPP, a "hybrid" search algorithm for the MPP [11]. Their algorithm consists of two, interleaved phases: The first phase performs branch-and-bound on a binary search tree where each node represents a variable, and the branches correspond to a decision regarding whether to assign the variable to the same value as in the initial assignment. At each node, HS_MPP-Phase1 computes a lower bound on the number of perturbations, and prunes the search if this exceeds or equals the current upper bound. Then, v, variable such that $SVA(v) \in dom(v)$ is selected. If there is no such variable (i.e., all remaining variables must be perturbed), then HS_PP-Phase2, described below, is called to test for feasibility. Otherwise, HS_MPP branches: The left branch assigns v its SVA and recursively searches the remaining variables; the right branch of the binary search tree, HS_MPP eliminates the SVA from the domain of v, and recursively searches the remaining variables.

The HS_MPP algorithm uses a lower bound, which we denote L_Z, to prune the branch-and-bound tree in HS_MPP-Phase1. This bound improves upon L_1 by exploiting the fact that if there is a pair of variables which have the SVA in the domain, but the SVAs conflict with each other, then one of these variables must be assigned a non-SVA, so the bound can be increased relative to L_1 by accounting for such pairs (see [11]).

After each decision in Phase 1, the following, limited filtering function is applied: For each remaining variable v, $SVA(v)$ is removed from $domain(v)$ if $SVA(v)$ is inconsistent with the current assignment of SVAs.

HS_MPP-Phase2 applies a standard MAC (maintaining-arc-consistency) algorithm to the remaining variables (i.e., variables which do not have the SVA in the domain and must be perturbed). If the MAC algorithm finds a satisfying assignment of values to v_r, then this is a solution to the MPP.

Finally, a third previous approach is by Hebrard, O'Sullivan and Walsh, who proposed a GAC for distance constraints [6]. Zivan et al compared HS_MPP to this GAC method and showed that HS_MPP performed significantly better on random binary CSPs (30-40 variables) and meeting rescheduling problems.

Related Work

Other previous work has addressed problems that are related to (but different from) the MPP formulation treated in this paper. A Dynamic CSP is a sequence of constraint satisfaction problems where each instance is derived from the previous instance by modifying some constraints [2]. Verfaillie and Schiex solved Dynamic CSPs by repairing the solution to the previous CSP instance [10]. They proposed a depth-first backtracking algorithm in D-space. Since the goal is to solve the Dynamic CSP instance, there is no mechanism to guarantee minimal perturbation, although they incorporate variable ordering heuristics that tend to bias the search towards a minimal perturbation solution. El Sakkout and Wallace [3] investigated a minimal cost repair problem for scheduling. They consider difference functions that can be expressed linearly (our MPP difference count objective is nonlinear). Their probe backtracking algorithm does not explicitly consider the initial schedule, and reschedules from scratch [3]. Barták et al. investigated overconstrained CSPs for which there is likely to be no feasible solution without violated constraints [1], and studied methods to seek a maximal assignment of consistent variables which also differs minimally from an initial state. They also studied an iterative repair (local search) algorithm biased to seek minimal perturbation solutions for course timetabling [8].

4   A Commitment-Space Search Algorithm for the MPP

We now describe our algorithm for the MPP. Unlike RB-AC, which searches D-space, and HS_MPP, which searches a 2-phase search in the space of variable assignments, our algorithm searches the space of variable commitments.

In a *commitment-based search space* (C-space) for the MPP, each node in the search tree represents a complete assignment of values to variables, where some subset of the variables are committed to their current value. Edges in the search tree represent a decision to commit a variable to some value. For each variable, we represent its current value, as well as whether a commitment has been made to the value. The root node of this search space is the initial assignment I. We say that a variable x is *committed* to value v at node N if x is assigned to v at N and every descendant of N, and *uncommitted* otherwise.

Each node represents the result of committing some variable to a particular value. Thus, this search space has a branching factor of d, the domain size, and a maximum depth of n, the number of variables. We originally proposed C-space for minimal perturbation in [4]. However that previous work focused on a specific type of MPP (bin packing constraint repair e.g., virtual machine reassignment in data centers), and C-space has not been evaluated for standard, domain-independent binary CSPs. C-space has a narrower structure (smaller branching factor) compared to D-space, at the cost of some redundancy. See [4] for an analysis, as well as a figure illustrating example search trees.

We evaluated both a standard depth-first branch-and-bound strategy, as well as an iterative deepening (IDA*) strategy [7] for C-space. Although iterative deepening can repeatedly visit the same state, in cases where the minimal perturbation solution is close to the initial solution, the IDA* search strategy would be expected to be faster than depth-first branch-and-bound.

For both of these strategies, a standard, most-constrained variable ordering is used, and a min-conflicts (with respect to the original values in the initial assignment) value ordering is used. At each node, arc consistency (AC-3) is applied for filtering. The depth-first branch-and-bound version is shown in Algorithm 1.

Lower Bound

The new lower bound function is based on a constrained vertex covering of a constraint violation graph. At every node in the search tree, there is a non-empty set of violated constraints. Given the set of all violated constraints, we construct a *constraint violation graph* G where each variable corresponds to a vertex in G, and there is an edge between vertex v_i and v_j if a constraint between variables x_i and x_j is violated. A vertex cover (VC) of a graph is a subset $vc \subset V$ of the vertices such that for every edge $e = (v_a, v_b)$ in G, either $v_a \in vc$ or $v_b \in vc$. A minimal vertex cover of G is a covering of G which has minimal cardinality.

The minimal VC of a constraint violation graph is clearly a relaxation of the MPP. The minimal VC identifies a subset of variables that could possibly eliminate all constraint violations, without identifying the actual values that must be assigned. The covering has one additional constraint: variables which no longer have the SVA in the domain are forced to be included in the covering. Thus, the cardinality of the (constrained) minimal VC is a lower bound on the number of perturbations required to result in a conflict-free assignment of values to variables. It is easy to see that this bound dominates the L_Z bound [11].

Although computing a minimal vertex cover is NP-complete [5], computing the minimal VC of a graph is much easier than solving the MPP (the search space is a binary tree with depth $= \#vars$, as opposed to a tree with branch factor $|Domain|$ for the MPP C-space search), so the minimal VC can be used as the basis for a lower bound. Our current implementation performs a straightforward branch-and-bound search where each node determines whether a variable is included or excluded from the cover. A simple filtering/pruning rule is used: for every edge (v_a, v_b), if v_a is excluded from the covering, then v_b must be included; conversely, if v_b is excluded, then v_a must be included. No other lower-bounding techniques or optimizations are used in the minimal VC computation, but as shown below, this simple implementation suffices in practice.

Algorithm 1. C-space Search Algorithm

---

mpp_search(uncommittedVars,committedVars,numChanges)

if get_conflicts(uncommittedVars,committedVars)==∅ **then**
    if count_num_perturbations(committedVars) < minimalChanges **then**
        minimalChanges = count_num_perturbations(commitedVars) {replace best-so-far solution}
    return s uccess
if lowerbound(uncommitedVars,committedVars) > minimalChanges **then**
    return f ailure {pruning based on lower bound}
V = select(uncommittedVars)
for all val in Order(domain(V)) **do**
    commit(V,val) {commitment also applies filtering (arc-consistency)}
    r = mpp_search(uncommittedVars \ V, committedVars ∪ V)
    if r==success **then**
        return s uccess
return f ailure

---

5 Experimental Evaluation

We evaluated the performance of the MPP algorithms using problems derived from standard, randomly generated binary CSPs. The classes of MPPs used in the experiments are defined by 5 parameters (n, k, p_1, p_2, δ). The first 4 parameters are used to first generate a random, uniform binary CSP, C, where n is the number of variables, k is domain size of all variables, p_1 is the constraint density (probability that any 2 variables have a constraint), and p_2 is the tightness (probability that any 2 values in a pair of constraint variables are a nogood). Then, C is solved using a standard CSP solver. If C is unsatisfiable, then it is discarded. If C is satisfiable, then the solution that is found is used as I, the initial assignment for the MPP. Then, C is perturbed by replacing some fraction δ of the constraints, resulting in a perturbed CSP C', and the MPP instance is (C', I). For n=30 and 40 variables, we generated 30 candidate binary CSPs each for all combinations of p_1, p_2, δ, where $p_1 \in 0.3, 0.4, 0.5, 0.6, 0.7$, $p_2 \in 0.3, 0.4, 0.5, 0.6, 0.7$, and $\delta \in 0.05, 0.10, 0.25, 0.50, 0.75, 1.00$. All of these were tested for solvability using a standard CSP solver (i.e., whether there is any satisfying assignment, regardless of distance from the initial configuration I). Of these, 2676 of the 30-variable instances and 1578 of the 40-variable instances were satisfiable MPPs. Similarly, for n=50 variables, we generated 30 candidate MPPs for all combinations of p_1, p_2 taken from $p_1 \in 0.3, 0.4, 0.5$, $p_2 \in 0.3, 0.4, 0.5$, $\delta \in 0.05, 0.10, 0.25, 0.50, 0.75, 1.00$. and 936 were satisfiable.

In the experiments below, we compare the algorithms in such a way that only these solvable instances matter, i.e., comparisons of the time to find solutions, with a time limit of 900 seconds. This is because unsolvable instances can be detected by running a standard CSP solver much more quickly than any of the MPP algorithms (clearly, checking satisfiability is a simpler problem than seeking a minimal perturbation). Our new algorithm is at least as fast as the previous algorithms in detecting unsatisfiability. In practice, the best strategy would be to first run a standard CSP to check for satisfiability, then run a dedicated MPP solver to minimize the perturbations.

We compared the following algorithms:

- C-space/L_{vc} - our C-space search algorithm using the new L_{vc} lower bound and depth-first branch-and-bound.
- C-space/L_{vc}/ID - Iterative Deepening C-space search algorithm using the L_{vc} lower bound.
- HS_MPP - The hybrid algorithm by Zivan et al [11].
- RB-AC/L_{vc} - A modified version of RB-AC algorithm by Ran et al [9], which uses use our L_{VC} lower bound instead of the L_1 bound [9] and searches D-space using iterative-deepening.
- C-space/L_Z - C-space search algorithm using the L_Z lower bound [11]. This comparison isolates the effect of the lower bound function L_{vc} compared to L_z.

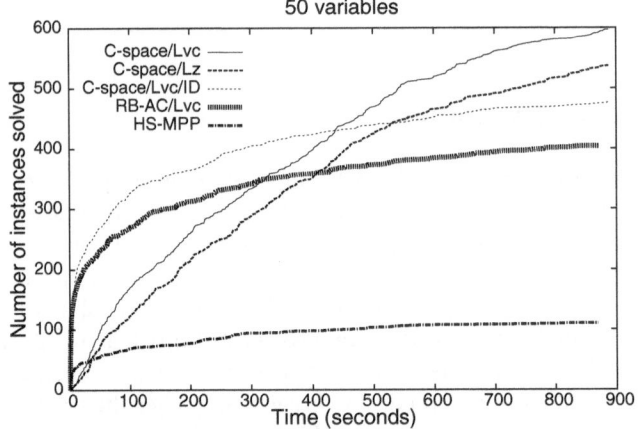

Fig. 1. n=50, cumulative number of problems that can be solved after a given time

Each algorithm was executed on each of the 30, 40, and 50-variable random binary MPP instances, with a 900 second time limit per run. Note that although we focus on runtime due to space restrictions, comparisons of the number of backtracks and constraint checks are qualitatively similar to the runtime results.

Figure 1 shows an overall comparison of the MPP algorithms, and plots the cumulative number of problems solved (y-axis) as the amount of time increases (x-axis) by each algorithm for the 50-variable problems. For example, the C-space/L_{vc} algorithm solved around 500 instances within 500 seconds. Overall, C-space/L_{vc} performed best on the hardest instances (which require > 400 seconds), while C-space/L_{vc}/ID performed best on problems requiring less than 400 seconds. Another interesting result is that RB-AC with the L_{vc} bound performs significantly better than HS_MPP, suggesting that the success of HS_MPP compared to the original RB-AC algorithm was due much more to the lower bound than to the hybrid search strategy.

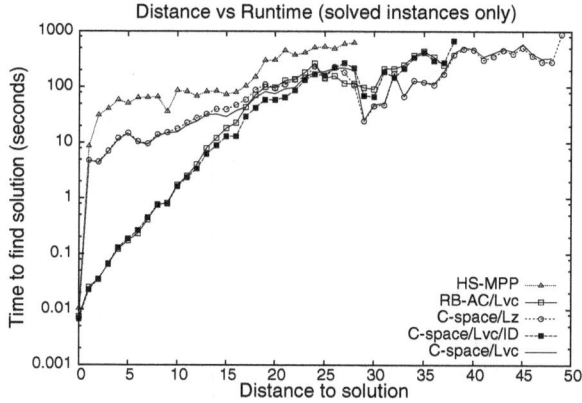

Fig. 2. Effect of Distance from Initial/Target State

While results for 30 and 40 variable problems are not shown due to space, they look similar, except that C-space/L_{vc}/ID performs relatively better with fewer variables.

Figure 2 plots average runtime required to solve instances as a function of the distance of the solution found to the initial assignment. This only includes successful runs and excludes runs that timed out, so some of the lines appear truncated (otherwise, for the less successful algorithms, it is difficult to see the impact of distance because there were so many failed runs). Overall, if the distance to a solution is within 10-15 variable assignment changes (i.e., the amount of repair required is small), the faster algorithms such as C-space/L_{vc} can solve the problems within 10 seconds (if at all).

Figure 3 compares key pairs of MPP algorithms on all of the 30, 40, and 50 variable problems. Each figure plots the runtimes for all instances on a pair of algorithms A_1, A_2, where the x-coordinate is the runtime of A_1 on the instance, and the y-coordinate is the runtime of A_2. An x or y value of 900 indicates failure to solve the instance. The straight diagonal line is (x=y), i.e., points above the line indicate that C-space/L_{vc} performed better, while points below the line indicate that the other algorithm performed better.

Figure 3a shows that C-space/L_{vc} clearly outperforms HS_MPP, the previous state-of-the-art algorithm. The average ratio of runtimes for HS_MPP vs C-space/L_{vc} is 86.48 for all problems that were solved by at least 1 of these solvers, and 160.35 for problems that took more than 60 seconds for the faster solver on each instance.

Figure 3d compares C-space/L_{vc}/ID and RB-AC/L_{vc}. These two algorithms, which both use iterative deepening search and the same lower bound (L_{vc}) differ mainly in the choice of search space (C-space and D-space, respectively). Figure 3d shows that C-space/L_{vc}/ID clearly outperforms RB-AC/L_{vc} on almost every problem instance, suggesting that C-space is better structured for search than D-space. However, the advantage of C-space over D-space seems to be less pronounced for this class of benchmarks compared to the virtual machine reassignment problem in [4].

Figure 3b compares C-space/L_{vc} and C-space/L_Z. Combined with Figures. 1, and 2 the results show that iterative deepening is a good strategy for quickly solving relatively easy problems (problems where the distance from I to a solution is small); however,

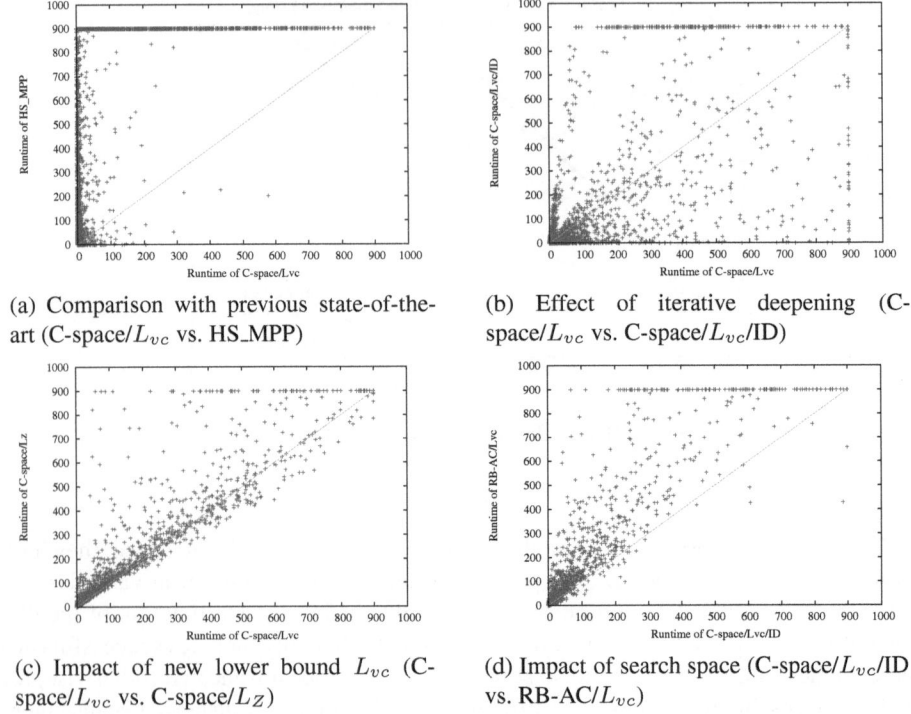

(a) Comparison with previous state-of-the-art (C-space/L_{vc} vs. HS_MPP)

(b) Effect of iterative deepening (C-space/L_{vc} vs. C-space/L_{vc}/ID)

(c) Impact of new lower bound L_{vc} (C-space/L_{vc} vs. C-space/L_Z)

(d) Impact of search space (C-space/L_{vc}/ID vs. RB-AC/L_{vc})

Fig. 3. Pairwise comparison of MPP algorithms (includes all 30,40, and 50 variable problems)

for harder problems (where the distance from I to a solution is large), straightforward depth-first branch-and-bound seems to be a more robust choice.

Figure 3c compares C-space/L_{vc} and C-space/L_Z. The results show that the new vertex-cover based lower bound L_{vc} clearly outperforms the previous lower bound L_Z by Zivan et al [11]. The average ratio of runtimes using lower bound L_Z vs L_{vc} is 1.73 for all instances solved by at least one solver, and 2.10 for instances that required 60 seconds or more for the faster solver.

6 Discussion and Conclusions

We proposed a search algorithm for optimal solutions to the min-perturbation problem. Our main contributions are: (1) We showed that our new CSpace/L_{vc} algorithm significantly improves upon the previous state of the art (HS_MPP) for random binary CSPs generated with a wide range of parameters. (2) We showed that both L_{vc}, the new lower bound for the MPP based on vertex covering of the constraint graph, as well as the C-space search space contribute significantly to the performance of the new algorithm (Fig. 3). Future work includes evaluation on applications such as employee shift rescheduling and meeting rescheduling.

References

1. Barták, R., Müller, T., Rudová, H.: A new approach to modeling and solving minimal perturbation problems. In: Apt, K.R., Fages, F., Rossi, F., Szeredi, P., Váncza, J. (eds.) CSCLP 2003. LNCS (LNAI), vol. 3010, pp. 233–249. Springer, Heidelberg (2004)
2. Dechter, R., Dechter, A.: Belief maintenance in dynamic constraint networks. In: Proc. AAAI, pp. 37–42 (1988)
3. El-Sakkout, H., Wallace, M.: Probe backtrack search for minimal perturbation in dynamic scheduling. Constraints 5, 359–388 (2000)
4. Fukunaga, A.S.: Search spaces for min-perturbation repair. In: Gent, I.P. (ed.) CP 2009. LNCS, vol. 5732, pp. 383–390. Springer, Heidelberg (2009)
5. Garey, M., Johnson, D.: Computers and Intractability: A Guide to the Theory of NP-Completeness. W.H. Freeman and Company (1979)
6. Hebrard, E., O'Sullivan, B., Walsh, T.: Distance constraints in constraint satisfaction. In: Proc. IJCAI, pp. 106–111 (2007)
7. Korf, R.: Depth-first iterative-deepening: an optimal admissible tree search. Artificial Intelligence 27(1), 97–109 (1985)
8. Müller, T., Rudová, H., Barták, R.: Minimal perturbation problem in course timetabling. In: Burke, E.K., Trick, M.A. (eds.) PATAT 2004. LNCS, vol. 3616, pp. 126–146. Springer, Heidelberg (2005)
9. Ran, Y., Roos, N., van den Herik, H.: Approaches to find a near-minimal change solution for dynamic CSPs. In: Proc. CP-AI-OR, pp. 378–387 (2002)
10. Verfaillie, G., Schiex, T.: Solution reuse in dynamic constraint satisfaction problems. In: Proc. AAAI, Seattle, Washington, pp. 307–312 (1994)
11. Zivan, R., Grubshtein, A., Meisels, A.: Hybrid search for minimal perturbation in dynamic CSPs. Constraints 16, 228–249 (2011)

Explaining Propagators
for Edge-Valued Decision Diagrams

Graeme Gange[1], Peter J. Stuckey[1,2], and Pascal Van Hentenryck[1,2]

[1] National ICT Australia, Victoria Laboratory
[2] Department of Computer Science and Software Engineering
The University of Melbourne, Vic. 3010, Australia
ggange@csse.unimelb.edu.au,
{peter.stuckey,pvh}@nicta.com.au

Abstract. Propagators that combine reasoning about satisfiability and reasoning about the cost of a solution, such as weighted all-different, or global cardinality with costs, can be much more effective than reasoning separately about satisfiability and cost. The COST-MDD constraint is a generic propagator for reasoning about reachability in a multi-decision diagram with costs attached to edges (a generalization of COST-REGULAR). Previous work has demonstrated that adding nogood learning for MDD propagators substantially increases the size and complexity of problems that can be handled by state-of-the-art solvers. In this paper we show how to add explanation to the COST-MDD propagator. We demonstrate on scheduling benchmarks the advantages of a learning COST-MDD global propagator, over both decompositions of COST-MDD and MDD with a separate objective constraint using learning.

1   Introduction

Optimization constraints merge the checking of feasibility and optimization conditions into a single propagator. A propagator for an optimization constraint filters decisions for variables which cannot take part in a solution which is better than the best known solution. They also propagate the bounds on the cost variable to keep track of its lower bound, and hence allow fathoming of the search, when no better solution can be found. There is a significant body of work on optimization constraints including: *weighted alldifferent* [1] and *global cardinality with costs* [2]. In this paper we examine the COST-MDD optimization constraint which is a generalization of the COST-REGULAR [3] constraint.

Previous work has explored the use of *Boolean Decision Diagrams* (BDDs) [4,5] and *Multi-valued Decision Diagrams* (MDDs) [6] for automatically constructing efficient global propagators. But these propagators do not handle costs. And adding a separate objective function constraint to encode the costs, leads to significantly weaker propagation.

COST-MDD is a generic constraint that can be used to encode many problems where the feasibility of a sequence of decisions is represented by an MDD, and the costs of the sequence of decisions is given by the sum of the weights on the

C. Schulte (Ed.): CP 2013, LNCS 8124, pp. 340–355, 2013.

edges taken in this MDD. COST-REGULAR [3] is encoded as a particular form of COST-MDD where the set of states at each level is uniform, and the transition from one level to another is uniform. The WEIGHTED-GRAMMAR constraint [7] is a similar optimization constraint which permits a more concise encoding of some constraints than COST-MDD, but is less convenient to construct and manipulate.

In this paper we investigate how to incorporate COST-MDD global propagators into a lazy clause generation [8] based constraint solver. The principle challenge is to be able to explain propagations as concisely as possible, in order that the nogoods learnt are as reusable as possible. We give experimental evidence that explaining COST-MDD propagators outperform both decompositions of COST-MDD and previous MDD-based propagators.

2   Preliminaries

Constraint programming solves constraint satisfaction problems by interleaving propagation, which remove impossible values of variables from the domain, with search, which guesses values. All propagators are repeatedly executed until no change in domain is possible, then a new search decision is made. If propagation determines there is no solution then search undoes the last decision and replaces it with the opposite choice. If all variables are fixed then the system has found a solution to the problem. For more details see e.g. [9].

We assume we are solving a constraint satisfaction problem over set of variables $x \in \mathcal{V}$, each of which takes values from a given initial finite set of values or *domain* $D_{init}(x)$. The domain D keeps track of the current set of possible values $D(x)$ for a variable x. Define $D \sqsubseteq D'$ iff $D(x) \subseteq D'(x), \forall x \in \mathcal{V}$. We let $\mathsf{lb}_D(x) = \min D(x)$ and $\mathsf{ub}_D(x) = \max D(x)$, and will omit the D subscript when D is clear from the context. The constraints of the problem are represented by propagators f which are functions from domains to domains which are monotonically decreasing $f(D) \sqsubseteq f(D')$ whenever $D \sqsubseteq D'$, and contracting $f(D) \sqsubseteq D$.

We make use of constraint programming with learning using the lazy clause generation [8] approach. Learning keeps track of what caused changes in domain to occur, and on failure computes a *nogood* which records the reason for failure. The nogood prevents search making the same incorrect set of decisions later.

In a lazy clause generation solver integer domains are also represented using Boolean variables. Each variable x with initial domain $D_{init}(x) = [l..u]$ is represented by two sets of Boolean variables $[\![x = d]\!], l \le d \le u$ and $[\![x \le d]\!], l \le d < u$ which define which values are in $D(x)$. We use $[\![x \ne d]\!]$ as shorthand for $\neg [\![x = d]\!]$, and $[\![x \ge d]\!]$ as shorthand for $\neg [\![x \le d - 1]\!]$. A lazy clause generation solver keeps the two representations of the domain in sync. For example if variable x has initial domain $[0..5]$ and at some later stage $D(x) = \{1, 3\}$ then the literals $[\![x \le 3]\!], [\![x \le 4]\!], \neg [\![x \le 0]\!], \neg [\![x = 0]\!], \neg [\![x = 2]\!], \neg [\![x = 4]\!], \neg [\![x = 5]\!]$ will hold. Explanations are defined by clauses over this Boolean representation of the variables.

Example 1. Consider a simple constraint satisfaction problem with constraints $b \leftrightarrow x + y \leq 2$, $x + y \leq 2$, $b' \leftrightarrow x \leq 1$, $b \rightarrow b'$, with initial domains $D_{init}(b) = D_{init}(b') = \{0,1\}$, and $D_{init}(x) = D_{init}(y) = \{0,1,2\}$. There is no initial propagation. Setting $x = 2$ makes the third constraint propagate $D(b') = \{0\}$ with explanation $[\![x = 2]\!] \rightarrow [\![b' = 0]\!]$, this makes the last constraint propagate $D(b) = \{0\}$ with explanation $[\![b' = 0]\!] \rightarrow [\![b = 0]\!]$. The first constraint propagates that $D(y) = \{1,2\}$ with explanation $[\![b = 0]\!] \rightarrow [\![y \geq 1]\!]$ and the second constraint determines failure with explanation $[\![x = 2]\!] \wedge [\![y \geq 1]\!] \rightarrow false$. The graph of the implications is

$$[\![b' = 0]\!] \longrightarrow [\![b = 0]\!] \longrightarrow [\![y \geq 1]\!]$$

$$[\![x = 2]\!] \longrightarrow false$$

Any cut separating the decision $[\![x = 2]\!]$ from *false* gives a nogood. The simplest one is $[\![x = 2]\!] \rightarrow false$ or equivalently $[\![x \neq 2]\!]$.   □

2.1 Edge-Valued Decision Diagrams

A *Multi-valued Decision Diagram (MDD)* encodes a propositional formula as a directed acyclic graph with a single terminal $\mathcal{T}$ representing *true* (the *false* terminal is typically omitted for MDDs). In an MDD G, each internal node $n = node(x_i, [(v_1, n_1), (v_2, n_2), \ldots, (v_k, n_k)])$ is labelled with a variable x_i, and outgoing edges consisting of a value v_j and destination node n_j. Each node represents the formula

$$\langle n \rangle \Leftrightarrow \bigvee_{j=1}^{k} (x = v_j \wedge \langle n_j \rangle)$$

where $\langle n \rangle$ is a Boolean representing the reachability of node n, and $\langle \mathcal{T} \rangle = true$. The MDD constraint enforces $\langle G.root \rangle = true$ where $G.root$ is the root of the MDD.

In this paper we restrict ourselves to *layered* MDDs. In a layered MDD G each node n is assigned to a layer k and all its child nodes must be at layer $k+1$. Each node at layer k is labelled with the *same* variable x_k, and the root node $G.root$ is at layer 1. This encodes an ordered MDD with no *long edges*, which typically propagate faster than MDDs with long edges [6]. Each assignment satisfying the constraint represented by G corresponds to a path from the root $G.root$ to the terminal $\mathcal{T}$. If, at the i-th layer, the path follows an edge with value v_j, the corresponding assignment has $x_i = v_j$.

An *Edge-valued MDD (EVMDD)* G is a (layered) MDD with a weight attached to each edge. Hence nodes are of the form

$$n = node(x_i, [(v_1, w_1, n_1), (v_2, w_2, n_2), \ldots, (v_k, w_k n_k)]),$$

where w_j is the weight of the jth outgoing edge. The cost of a solution $\theta = [x_1 = d_1, x_2 = d_2, \ldots, x_n = d_n]$ which defines a path from the root of G to $\mathcal{T}$ is given

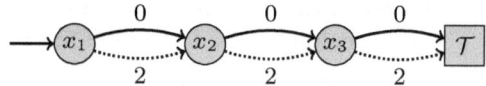

Fig. 1. A simple EVMDD with only paths of even cost

by the sum of the weights along the corresponding path in the EVMDD. Each node n enforces the constraint:

$$\langle\!\langle n \rangle\!\rangle = \begin{cases} 0 & n = \mathcal{T} \\ \min\{w_j + \langle\!\langle n_j \rangle\!\rangle \mid j = 1, .., k \wedge x_i = v_j\} & \text{otherwise} \end{cases}$$

where $\langle\!\langle n \rangle\!\rangle$ holds the cost of the minimal weight path from n to $\mathcal{T}$.

For convenience, we denote edges by 4-tuples $(n, x_i = v_j, w_j, n_j)$, representing the edge with source n (in layer i), destination n_j (in layer $i + 1$) and weight w_j corresponding to the value v_j. We will refer to the components as $(e.begin, e.var = e.val, e.weight, e.end)$.

We use $s.out\_edges$ to refer to all the edges of the form $(s, \_, \_, \_)$, i.e. those leaving node s, and $d.in\_edges$ to refer to edges of the form $(\_, \_, \_, d)$, i.e. those entering node d. We use $G.edges(x_i, v_j)$ to record the set of edges of the form $(\_, x_i = v_j, \_, \_)$ in EVMDD G.

The COST-MDD constraint COST-MDD$(G, [x_1, \ldots, x_n], \bowtie, C)$ requires that $\langle\!\langle G.root \rangle\!\rangle \bowtie C$ where $\bowtie \in \{\leq, =, \geq\}$. Note that the constraint (except the $\geq$ incarnation) enforces satisfiability, i.e., that there is a path from $G.root$ to $\mathcal{T}$, since otherwise $\langle\!\langle G.root \rangle\!\rangle = \infty$. The COST-MDD constraint can represent COST-REGULAR as well as other constraints representable by automata with counters.

Our definition of EVMDDs differs from the standard treatment of edge-valued BDDs [10], apart from the extension from Boolean variables to finite-domain variables. We do not require the graph to be deterministic; a single node may have multiple edges annotated with the same value. Also, we do not require the edge weights to be normalized; normalization may reduce the size of the graph by inducing additional sharing, but does not affect propagation or explanation.

3   EVMDD Propagation

An incremental algorithm for propagating COST-REGULAR constraints was described in [3]. This algorithm essentially converts the COST-REGULAR constraint into a COST-MDD constraint where $\bowtie$ is $=$, then performs propagation on this transformed representation. This algorithm operates by incrementally maintaining the distance of the shortest $up[n]$ and longest $lup[n]$ path from the root to each node n, and the distance of the shortest $dn[n]$ and longest $ldn[n]$ path from each node n to $\mathcal{T}$. Given a constraint COST-MDD$(G, [x_1, \ldots, x_n], =, C)$, an edge e may be used to build a path from $G.root$ to $\mathcal{T}$ only if $up[e.start] + e.weight + dn[e.end] \leq \mathsf{ub}(C)$ and $lup[e.start] + e.weight + ldn[e.end] \geq \mathsf{lb}(C)$.

The description in [3] does not mention how changes to the bounds of C are handled. When the upper bound of C is reduced, the lengths of all shortest paths remain the same; however, the domains of variables x_i may change, if the shortest path through $x_i = v_j$ is longer than the updated bound.

Example 2. Consider the EVMDD (EVBDD) G shown in Figure 1 where edges for value 0 are shown dotted, and edges for value 1 are shown full. The constraint COST-MDD$(G, [x_1, x_2, x_3], =, C)$ encodes the equation $2x_1 + 2x_2 + 2x_3 = C$. If we initially have $D(C) = [0..2]$, no values may be eliminated, as every edge can occur on a path of cost at most 2. However, if we reduce ub(C) to 1, we must eliminate $x_i = 1$ from the domain of each variable.

The authors claim that their propagation algorithm enforces domain consistency on the x variables in a COST-MDD constraint. This statement is not correct.

Example 3. Consider again the EVMDD G shown in Figure 1. The algorithm of [3] makes no propagation for the constraint COST-MDD$(G, [x_1, x_2, x_3], =, C)$ when $D(C) = \{3\}$. This is because every edge can take part in a path which is both longer (length 4) or shorter (length 2) than the bounds of C. But there is no support for any value of x_i since there is no path of length exactly 3.   □

In fact even bounds propagation is NP-hard for COST-MDD where $\bowtie$ is $=$, using any applicable definition of bounds consistency [11].

Theorem 1. *Domain propagation, bounds($\mathcal{Z}$) or bounds($\mathcal{D}$) consistent propagation for* COST-MDD$(G, [x_1, \ldots, x_n], =, C)$ *is NP-hard*

Proof. We map SUBSETSUM to COST-MDD *propagation. Given a set $S = \{s_1, \ldots, s_m\}$ of numbers and target T we build an EVBDD with m 0-1 variables $x_1, \ldots, x_m$ and m nodes $n_1, \ldots, n_m$ ($n_{m+1} = T$) with $2m$ edges $(n_i, x_i = 0, 0, n_{i+1})$ and $(n_i, x_i = 1, s_i, n_{i+1})$. Enforcing domain (or equivalently in this case bounds($\mathcal{Z}$) or bounds($\mathcal{D}$)) consistency on* COST-MDD$(G, [x_1, \ldots, x_n], =, C)$ *with $D(C) = \{T\}$ generates a false domain unless the SUBSETSUM holds.*   □

In this paper we restrict consideration to the COST-MDD constraint of the form COST-MDD$(G, [x_1, \ldots, x_n], \leq, C)$. This is the critical form of the constraint when we are trying to minimize costs. Treatment of COST-MDD$(G, [x_1, \ldots, x_n], \geq, C)$ is identical by negating each edge weight and the cost variable; the treatment of COST-MDD$(G, [x_1, \ldots, x_n], =, C)$ in [3] is effectively combining propagators for each of COST-MDD$(G, [x_1, \ldots, x_n], \leq, C)$ and COST-MDD$(G, [x_1, \ldots, x_n], \geq, C)$.

We give a non-incremental propagation algorithm for the constraint COST-MDD$(G, [x_1, \ldots, x_n], \leq, C)$ in Figure 2.[1] evmdd_prop first records the shortest path (given the current domain D) from each node n to $\mathcal{T}$ in $dn[n]$ using mark_paths. It returns the shortest path from $G.root$ to $\mathcal{T}$. It then visits using infer all the edges reachable from $G.root$ that appear on paths of length less than ub(C). Initially the negation of all edge labels are placed in *inferences*. When an edge that appears on a path of length less than or equal to ub(C) is discovered, the negation of its label is removed from *inferences*. The algorithm returns the a lower bound of C (which may not be new) and any new inferences on x_i variables.

---

[1] This is not novel with respect to [3] but they don't formally define their algorithm.

Example 4. Consider the propagation that occurs with the EVMDD of Figure 1 with $C \leq 2$ when we set $x_1 \neq 1$ ($x_1 = 0$) and $x_2 \neq 0$ ($x_2 = 1$). mark_paths sets $dn[\mathcal{T}] = 0$, $dn[x_3] = 0$ (using the variable name for the node name), $dn[x_2] = 2$ and $dn[x_1] = 2$ and returns 2. infer initially starts with *inferences* $= \{\llbracket x_1 \neq 0 \rrbracket, \llbracket x_1 \neq 1 \rrbracket, \llbracket x_3 \neq 0 \rrbracket, \llbracket x_3 \neq 1 \rrbracket\}$. It sets $up[x_1] = 0$ then removes $\llbracket x_1 \neq 0 \rrbracket$ from *inferences* setting $up[x_2] = 0$. It then removes $\llbracket x_2 \neq 2 \rrbracket$ from *inferences* setting $up[x_3] = 2$. It removes $\llbracket x_3 \neq 0 \rrbracket$ from *inferences*, but then when examining the full edge from x_3 the distance test fails. Hence it returns $\{\llbracket x_3 \neq 1 \rrbracket\}$. The final inferences are $\{\llbracket C \geq 2 \rrbracket, \llbracket x_3 \neq 1 \rrbracket\}$.

Proposition 1. *evmdd_prop maintains domain consistency for* COST-MDD$(G, [x_1, \ldots, x_n], \leq, C)$.

Proof. After **evmdd_prop** *finishes if* $v_j \in D(x_i)$ *then there is an edge* $(s, x_i = v_j, w, d)$ *in* G *where* $up[s] + w + dn[d] \leq \mathsf{ub}(C)$. *Hence there is a path of edges from* $G.root$ *to* s *of length* $up[s]$ *and a path of edges from* d *to* $\mathcal{T}$ *of length* $dn[d]$. *If we set each variable to the value given on this path and* $C = \mathsf{ub}(C)$ *we have constructed a solution supporting* $x_i = v_j$. *Similarly, given* $l = \mathsf{lb}(C)$ *then after* **evmdd_prop** *finishes there is a path from* $G.root$ *to* $\mathcal{T}$ *of length* l. *If we set each variable to the value given on this path, and* C *to any value* $d \in D(C)$ *domain we have constructed a solution supporting* $C = d$. $\square$

It is straightforward to make the above algorithm incremental in changes in x variables. A removed edge $e = (s, x = v, w, d)$ forces the recalculation of $dn[s]$ which may propagates upward, and $up[s]$ which may propagate downwards. If a change reaches $G.root$ or $\mathcal{T}$ then the lower bound on C may change. When the upper bound of C changes, we simply scan the edges for each value until we find one that is still feasible (infeasible edges are not checked on later calls).

4 Explaining EVMDD Propagation

A nogood learning solver, upon reaching a conflict, analyses the inference graph to determine some subset of assignments that results in a conflict. This subset is then added to the solver as a *nogood* constraint, preventing the solver from making the same set of assignments again, and reducing the search space. In order to be incorporated in a nogood learning solver, the EVMDD propagator must be able to explain its inferences.

4.1 Minimal Explanation

The explanation algorithm is similar in concept to that used for BDDs and MDDs. To explain $\llbracket x \neq v \rrbracket$ we assume $\llbracket x = v \rrbracket$ and hence make the EVMDD unsatisfiable. A correct explanation is (the negation of) all the values for other variables which are currently false. We then progressively remove assignments (unfix literals) from this explanation while ensuring the constraint as a whole remains unsatisfiable. We are guaranteed to create *a minimal explanation* (but

```
evmdd_prop(G, [x₁, ..., xₙ], C, D)
    ĉ := mark_paths(G, D)
    L := infer(G, [x₁, ..., xₙ], D, ub(C))
    return {[[C ≥ ĉ]]} ∪ L

mark_paths(G, D)
    for(n ∈ G.nodes) dn[n] := ∞
    dn[T] := 0; queue := {T}
    while(queue ≠ ∅)
        nqueue := {} % Record nodes of interest on the next level.
        for(node in queue)
            for(e in node.in_edges)
                if(e.val ∈ D(e.var))
                    dn[e.begin] := min(dn[s.begin], e.weight + dn[node])
                    nqueue ∪= {e.begin}
        queue := nqueue
    return dn[G.root]

infer(G, [x₁, ..., xₙ], D, u)
    inferences := {[[xᵢ ≠ vⱼ]] | 1 ≤ i ≤ n, vⱼ ∈ D(xᵢ)}
    for(n ∈ G.nodes) up[n] := ∞
    up[G.root] := 0; queue := {G.root}
    while(queue ≠ ∅)
        nqueue := {} % Record nodes of interest on the next level.
        for(node in queue)
            for(e ∈ node.out_edges)
                if(e.val ∈ D(e.var))
                    if(up[node] + e.weight + dn[e.end] ≤ u)
                        inferences := inferences − {[[e.var ≠ e.val]]}
                        up[e.end] := min(up[e.end], e.weight + up[node])
                        nqueue ∪= {e.end}
        queue := nqueue
    return inferences
```

Fig. 2. Algorithm for inferring newly propagated literals

not the smallest minimal explanation) $\bigwedge_{l \in expln} l \rightarrow [[x \neq v]]$ since removing any literal l' from the *expln* would mean COST-MDD$(G, [x_1, \ldots, x_n], \leq, C) \wedge$ $\bigwedge_{l \in expln-\{l'\}} l \wedge [[x = v]]$ is satisfiable. Constructing a smallest minimal explanation for an EVMDD is NP-hard just as for BDDs [12].

We adapt the minimal MDD explanation algorithm used in [6] to COST-MDD constraints. The propagator conflicts when the shortest path from $G.root$ to T (under the current domain) is longer than ub(C). To construct a minimal explanation, we begin with the set of values that have been removed from variable domains, and progressively restore any values which would not re-introduce a path of length $\leq$ ub(C).

The minimal explanation algorithm is illustrated in Figure 3. To explain $[[x \neq v]]$ under current domain D, we first create the domain D' where $D'(x) =$

---

evmdd_explain($G, C, D, [\![x \neq v]\!]$)
  $D' := D$ with $D(x)$ replaced by $D'(x) = \{v\}$
  $\hat{c} :=$ mark_paths(G, D')
  if($\hat{c} < \infty$ **or** *choice*) $u :=$ ub(C) $+ 1$
  else $u := \infty$
  return $[\![x \neq v]\!] \leftarrow$ collect_expln(G, C, x, v, u)

evmdd_explain_lb($G, C, D, [\![C \geq l]\!]$)
  mark_paths(G, D) % unnecessary if just run evmdd_prop
  return $[\![C \geq l]\!] \leftarrow$ collect_expln($G, C, \bot, \bot, l$) $- \{[\![C \leq l - 1]\!]\}$

collect_expln(G, C, x, v, u)
  queue $:= \{G.root\}$; $up[G.root] := 0$; $s := \infty$
  while(*queue* $\neq \emptyset$)
    for(*node* **in** *queue*)
      for($e \in node.out\_edges$)
        $up[e.end] := \infty$
        if($e.var \neq x$ **and** $up[node] + e.weight + dn[e.end] < u$)
          $explanation \cup\!= [\![e.var \neq e.val]\!]$
        else $s := \min(s, up[node] + e.weight + dn[e.end])$
    nqueue $:= \{\}$ % Record nodes of interest on the next level.
    for(*node* **in** *queue*)
      for($e \in node.out\_edges$)
        if (($e.var = x$ **and** $e.val = v$)
            or ($e.var \neq x$ **and** $[\![e.var \neq e.val]\!] \notin explanation$))
          nqueue $\cup\!=\{e.end\}$
          $up[e.end] := \min(up[e.end], up[node] + e.weight)$
    queue $:=$ *nqueue*
  return *explanation* $\cup$ $[\![C \leq s - 1]\!]$

Fig. 3. Algorithms for computing a minimal explanation

$\{v\}$ and otherwise D' agrees with D. With this domain the constraint is unsatisfiable. We use mark_paths to compute the shortest path from each node n to $\mathcal{T}$ and store this in $dn[n]$. It returns the shortest path $\hat{c}$ from root to $\mathcal{T}$. If $\hat{c}$ is finite, or we choose to (by setting global *choice* true) we use an upper bound of C in the explanation, by setting $u \neq \infty$. collect_expln traverses the EVMDD from the root, building an explanation of literals which if not true would cause a path of length $< u$ to be created in the EVMDD. The algorithm examines all reachable nodes on a level (initially just the root) and if adding an edge would create a path shorter than u then the (negation of) the label on the edge is added to the explanation, if not then we update s which records the shortest path found from root to $\mathcal{T}$ with length $\geq u$. The algorithm then adds all the nodes of the next level which are still reachable, and updates the shortest path from the root to each such node n storing this in $up[n]$. This continues while there are still some reachable nodes. At the end the algorithm returns the collected explanation, plus the relaxed upper bound literal $[\![C \leq s - 1]\!]$, which ensures that none of the paths found from root to $\mathcal{T}$ can be traversed.

Since the procedures mark_paths and collect_expln perform one and two breadth-first traversals of the graph, respectively, the explanation requires $O(|G|)$ time.

Proposition 2. *evmdd_explain$(G, C, D, [\![x \neq v]\!]$ returns a correct minimal explanation for $[\![x \neq v]\!]$.*

Proof. (Sketch) The algorithm implicitly maintains the invariant that there is no path in G through an edge labelled $x = v$ of length less than or equal to $lb(C)$ which does not make use of an edge in DE. Initially DE is the set of edges e where $e.var \neq x$ and $e.val \notin D(e.var)$. The base case holds using the correctness of evmdd_prop. During collect_expl we remove processed edges from this implicit set DE, except those kept in explanation. Whenever we remove an edge from DE the shortest path through the edge that uses an edge labeled $x = v$ and none of the edges in DE is $> ub(C)$. This demonstrates the correctness of the algorithm, since the explanation literals force that $[\![x \neq v]\!]$ holds since there is no feasible path through any edges labelled $x = v$.

For minimality we can reason that if we remove any literal from the explanation, then we would have added a path that was too short passing through an edge labelled $x = v$. The minimality of the bound constraint $[\![C \leq s - 1]\!]$ follows since if we relax it we will allow a path of length s through $x = v$. □

Explaining a new lower bound l for C is similarly defined by evmdd_explain_lb. We compute $dn[n]$ for each reachable node using mark_paths with the current domain D, then choose a set of literals to ensure no shorter paths are allowed. In this case collect_expln will always return $[\![C \leq l - 1]\!]$ in the explanation which we can safely omit. Explaining failure of the whole constraint is identical to explaining why $C \geq \infty$.

Example 5. Consider the constraint defined by the EVMDD shown in Figure 4, which encodes a simple scheduling constraint requiring shifts to be of even length.

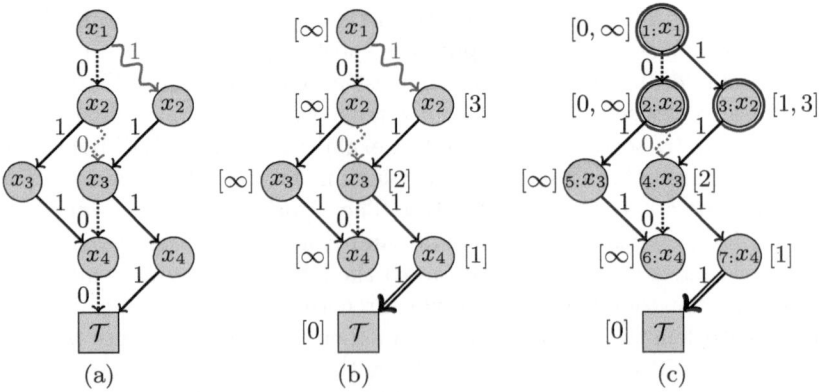

(a)   (b)   (c)

Fig. 4. (a) An EVMDD which requires shifts to be assigned in blocks of two. (b) We compute the shortest path from each node to $\mathcal{T}$. (c) Enqueued nodes are shown circled, and have been annotated with the shortest path from n_1 under the current assignment.

Assume the solver first propagates $C \leq 2$, then fixes $[\![x_1 \neq 1]\!]$ and $[\![x_2 \neq 0]\!]$. The only satisfying assignment is then $[x_1, x_2, x_3, x_4] = [0, 1, 1, 0]$.

If we are asked to explain the inference $x_4 \neq 1$, we first compute the shortest paths from each node to $\mathcal{T}$ through $x_4 = 1$, using mark_paths. This is shown in Figure 4(b). Notice that the cost at the root node is ∞. This indicates that, even without a cost bound, there is no feasible path through $x_4 = 1$. We have the choice of either omitting the cost bound (obtaining an explanation not dependent on C) or including it and possibly obtaining a smaller explanation.

Whether or not we include a bound on C, we proceed by sweeping down level-by-level from the r_1. Assuming we include the bound $[\![C \leq 2]\!]$, so $u = 3$, we first check if any of the outgoing edges would introduce a path of length less than 3. We find that the edge from n_1 to n_3 can safely be restored, since $up[n_1] + 1 + dn[n_3] = 4 \geq u = 3$. We update $s = 4$. As no edges introduce a feasible path, we update up for both n_2 and n_3, and add them to the queue for the next level.

At the second level, we discover that restoring the edge from n_2 to n_4 would introduce a feasible path, as $up[n_2] + 0 + dn[n_4] = 2 < u = 3$. The literal $[\![x_2 \neq 0]\!]$ must then be added to the explanation. Since n_4 is still reachable via n_3, both n_4 and n_5 are added to the queue for the next level; however, $up[n_4]$ is only updated by the edge from n_3, and not from n_2. This process continues until no further nodes remain. At the end $s = 4$ so we didn't need the bound on paths to be $\mathsf{ub}(C)$ it could have been looser. Hence we add $[\![C \leq 3]\!]$ to the explanation. The explanation returned is $[\![x_4 \neq 1]\!] \leftarrow [\![C \leq 3]\!] \wedge [\![x_2 \neq 0]\!]$. This is a minimal explanation.

If we omit the cost bound, then we cannot restore the edge from n_1 to n_3; so we construct the alternate explanation $[\![x_4 \neq 1]\!] \leftarrow [\![x_1 \neq 1]\!] \wedge [\![x_2 \neq 0]\!]$ which is also minimal. Note we omit the redundant literal $[\![C \leq \infty - 1]\!]$ created by collect_expl. □

4.2 Incremental Explanation

Example 6. Unfortunately, on large EVMDDs, constructing a minimal explanation can be expensive since explaining each inference may involve exploring the entire EVMDD. For these cases, we present a greedy algorithm for constructing valid, but not necessarily minimal, explanations in an incremental manner, often only examining a small part of the EVMDD.

We adapt the incremental MDD algorithm of [6] to COST-MDD. As in the MDD case, we explain $[\![x \neq v]\!]$ beginning from the set of edges corresponding to $x = v$. For all such edges $e = (s, x = v, w, d)$, we know that $up[s] + w + dn[d] > \mathsf{ub}(C)$. If we have $up[s] + w + dn[d] = \mathsf{ub}(C) + 1$, then there is no flexibility in the bounds; we must select an explanation which ensures the shortest path from $G.root$ to s has cost $up[s]$, and the shortest path from e to $\mathcal{T}$ has cost $dn[d]$. We record the amount of cost that needs to be explained on all paths to s; this is denoted by $upe[s]$. We then sweep upwards, level-by-level, collecting an explanation which guarantees this minimum cost. At each level, we maintain the set of edges which need to be explained. If for some edge we have $up[s] + w < upe[d]$, then $[\![x \neq v]\!]$

```
mdd_inc_explain(G, x, v, u)
    for(n ∈ G.nodes) upe[n] := ∞
    for(n ∈ G.nodes) dne[n] := ∞
    kfa := {} % edges killed from above
    kfb := {} % edges killed from below
    for(e in G.edges(x, v))
        % Split possible supports
        Assign p_up, p_dn subject to:
            p_up + e.weight + p_dn ≥ u ∧ p_up ≤ up[e.begin] ∧ p_dn ≤ dn[e.end]
        if(p_up > up_0[e.begin])
            kfa ∪={edge}
            upe[e.begin] := max(upe[e.begin], p_up)
        if(p_dn > dn_0[e.end])
            kfb ∪={edge}
            dne[e.end] := max(dne[e.end], p_dn)
    % Explain all those killed from below
    return explain_down(kfb)
    % And all those killed from above
        ∪ explain_up(kfa)
```

Fig. 5. Top-level wrapper for incremental explanation

must be added to the current explanation; otherwise, a feasible path would be introduced. We perform an initial pass over the edges at the current level to determine which values must be included in the explanation; during the second pass, we update upe for the source node of each edge that hasn't been excluded, and enqueue the set of incoming edges to be processed at the next level. If at any point we have $upe[s]$ is no greater than $up_0[s]$ (the shortest path to s under the initial variable domains), then we don't need to enqueue the incoming edges, as an empty explanation is sufficient.

If $up[s] + w + dn[d] > \mathsf{ub}(C) + 1$, then we can potentially relax the generated explanation. Obviously, the amount by which we relax $up[s]$ affects the amount of slack available to $dn[d]$. To relax the bounds as far as possible, we would initially allocate as much slack as possible to $up[s]$, and collect the corresponding explanation. Before performing the downward pass, we would then propagate the newly reduced path lengths back to the current layer, to determine how much slack remains for the explanation of d.

Instead, we determine *a priori* how the slack is allocated in the explanation. If either $up[s]$ or $dn[d]$ is ∞, then we build the explanation in only that direction (if both, we arbitrarily explain upwards). Otherwise, we explain as much as possible in the upward pass, and allocate all possible slack to the downwards pass. Alternative strategies for relaxing the bounds is interesting future work.

Consider again the case described in Example 5. During incremental propagation, we maintain up and dn for each node. These are shown in Figure 7(a). To explain $[\![x_4 \neq 1]\!]$, we need to eliminate some set of values which ensures that $up[n_7] + 1 + dn[\mathcal{T}] \geq 3$.

```
explain_down(kfb)
   reason = {}
   % Traverse the MDD downwards, breadth first
   while(¬is_empty(kfb))
      % Scan the current level for edges that will need explaining.
      pending = {}
      for(e in kfb)
         % For each edge requiring explanation
         if(e.val ∉ D(e.var) and
            e.weight + dn[e.end] < dne[e.begin])
            % There is no later explanation,
            % so add ⟦e.var ≠ e.val⟧ to the reason.
            reason ∪={⟦e.var ≠ e.val⟧}
         else
            pending ∪={e}
      next = {}
      % Collect the edges that haven't been explained at this level.
      for(e in pending)
         if(⟦e.var ≠ e.val⟧ ∉ reason and e.weight + dn₀[e.end] < dne[e.begin])
            % If e is not explained already collect its outgoing edges
            next ∪= e.end.out_edges
            dne[e.end] := max(dne[e.end], dne[e.begin] − e.weight)
      % Continue with the next layer of edges.
      kfb = next
   return reason
```

Fig. 6. Pseudo-code for incremental explanation of EVMDDs. explain_up acts in exactly the same fashion as *explain_down*, but in the opposite direction.

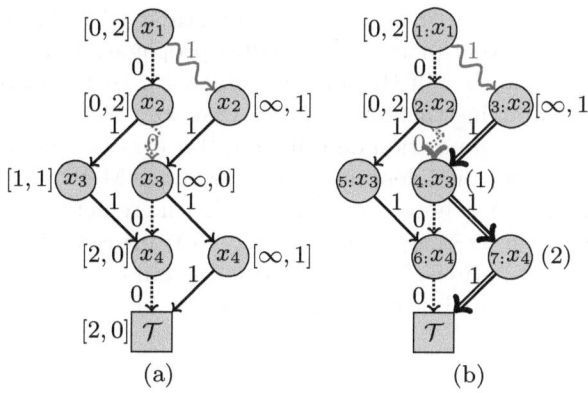

(a)   (b)

Fig. 7. The EVMDD from Example 5. (a) Values of $[up, dn]$ for each node. (b) Edges enqueued while explaining $⟦x_4 ≠ 1⟧$.

Under the current assignment, $up[n_7] = \infty$. However, as our current cost bound is 2, we only need to ensure $up[n_7] + 1 \geq 3$. We set $upe[n_7] = 2$, the amount of cost that must be guaranteed from above, and add n_7 to the queue. Expanding n_7, we find that it has only one parent, which is the edge from n_4 with weight 1. This edge cannot be eliminated, so we set $upe[n_4] = upe[n_7] - 1 = 1$, and enqueue n_4.

n_4 has 2 incoming edges, so we first check both edges to determine if any values must be added to the explanation. Examining the edge from n_2 to n_4, we have $upe[n_4] - 0 = 1 > up[n_2]$. This indicates that, if the edge from n_2 to n_4 were restored, a path of length 2 would be introduced. $[\![x_2 \neq 0]\!]$ is therefore added to the explanation. The edge from n_3 to n_4 is safe, as $up[n_3] = \infty \geq upe[n_3] - 1$.

We then make a second pass through the edges, to determine which new nodes must be enqueued. As $[\![x_2 \neq 0]\!]$ is in the explanation, we don't need to expand the node from n_2 to n_4. The edge from n_3 to n_4 is traversable, so we update $upe[n_3]$. However, since $upe[n_3] = 0$, and the base cost to reach n_3 is 1 (that is, $upe[n_3] \leq upo[n_3]$), we don't need to enqueue n_3, since the cost to n_3 will always be at least 0. Since we have no nodes enqueued, the upwards pass is finished. Since there are no nodes which must be propagated downwards, this yields the final explanation $[\![x_4 \neq 1]\!] \leftarrow [\![C \leq 2]\!] \wedge [\![x_2 \neq 0]\!]$.

Observe that this is not minimal, since the explanation is still valid if we replace $[\![C \leq 2]\!]$ with $[\![C \leq 3]\!]$.   □

5   Experimental Results

Experiments were conducted on a 3.00GHz Core2 Duo with 4 Gb of RAM running Ubuntu GNU/Linux 10.04. The propagators were implemented in `chuffed`, a state-of-the-art lazy-clause generation [8] based constraint solver. All experiments were run with a 10 minute time limit. For the minimal explanation algorithm, we always selected to use upper bounds in the explanation if possible.

We evaluate the COST-MDD constraints on a standard set of shift scheduling benchmarks. For the experiments, `dec` denotes propagation using a decomposition of COST-MDD like that of [13] but introducing a cost variable per layer of the EVMDD and summing them to compute cost, and `dec`$_{mdd}$ uses the domain-consistent Boolean decomposition described in [14] (or equivalently in [6]) and a separate cost constraint. `mdd` denotes using a separate MDD propagator [6] and cost constraint, `ev-mdd` denotes COST-MDD using incremental propagation and minimal explanations, `ev-mdd`$_I$ denotes COST-MDD using incremental propagation and greedy explanations. We also tried a domain consistent decomposition of COST-MDD based on [7] but it failed to solve any of the shown instances, and is omitted.

5.1   Shift Scheduling

Shift scheduling, a problem introduced in [3], allocates n workers to shifts such that (a) each of k activities has a minimum number of workers scheduled at

Table 1. Comparison of different methods on shift scheduling problems

Inst.	dec		dec_{mdd}		mdd		ev-mdd		ev-mdd$_I$	
	time	fails	time	fails	time	fails	time	fails	time	fails
1,2,4	—	—	14.51	39700	3.49	21888	0.20	607	**0.17**	635
1,3,6	—	—	11.25	40675	19.00	76348	**0.87**	4045	0.91	4156
1,4,6	36.48	86762	2.62	7582	0.69	3518	**0.11**	350	0.27	1077
1,5,5	5.64	32817	0.41	1585	0.52	3955	0.07	239	**0.06**	238
1,6,6	7.32	35064	0.40	1412	0.21	1161	**0.08**	249	0.11	413
1,7,8	27.58	77757	4.03	13149	2.43	12046	**0.73**	3838	0.83	4279
1,8,3	67.74	126779	0.85	5002	0.39	3606	**0.06**	219	0.07	262
1,10,9	321.44	441884	17.55	44222	19.77	68688	**1.23**	5046	1.31	7419
2,1,5	1.29	12520	0.14	691	0.24	1490	0.02	78	**0.01**	45
2,2,10	—	—	—	—	131.29	286747	**43.62**	99583	49.05	100958
2,3,6	—	—	188.77	187760	144.99	289568	**2.39**	6443	5.94	13695
2,4,11	—	—	—	—	391.59	918438	42.38	111567	92.89	220568
2,5,4	—	—	25.85	59635	12.18	50340	0.65	1545	**0.48**	1541
2,6,5	—	—	83.78	104911	30.27	80046	**6.18**	12100	7.63	16074
2,8,5	—	—	90.28	153331	34.69	110917	**4.99**	15507	10.02	26565
2,9,3	—	—	6.10	20472	9.17	42105	0.86	1898	**0.47**	1593
2,10,8	—	—	349.88	303227	95.61	168720	**8.85**	26331	17.22	37356
Total	—	—	—	—	896.53	2139581	**113.29**	289645	187.44	436874
Mean	—	—	—	—	52.74	125857.71	**6.66**	17037.94	11.03	25698.47
Geom.	—	—	—	—	7.92	31465.82	**0.86**	2667.65	1.03	3428.35

any given time, and (b) the overall cost of the schedule is minimised, without violating any of the additional constraints:

- An employee must work on a task (A_i) for at least one hour, and cannot switch tasks without a break (b).
- A part-time employee (P) must work between 3 and 5.75 hours, plus a 15 minute break.
- A full-time employee (F) must work between 6 and 8 hours, plus 1 hour for lunch (L), and 15 minute breaks before and after.
- An employee can only be rostered while the business is open.

These constraints can be formulated as a **grammar** constraint as follows:

$$S \rightarrow RP^{[13,24]}R \mid RF^{[30,38]}R$$

$$F \rightarrow PLP \qquad P \rightarrow WbW$$
$$W \rightarrow A_i^{[4,\dots]} \qquad A_i \rightarrow a_i A_i \mid a_i$$
$$L \rightarrow llll \qquad R \rightarrow rR \mid r$$

We convert the **grammar** constraint into a Boolean formula, as described in [13]; however, we convert the formula directly into an MDD, rather than a **s-DNNF** circuit; the MDD and cost-MDD propagators, as well as the decompositions, are all constructed from this MDD. This process is similar to the reformulation described in [15]. Note that some of the productions for P, F and A_i are annotated with restricted intervals – while this is no longer strictly context-free, it can be integrated into the graph construction with no additional cost.

The coverage constraints and objective function are implemented using the monotone BDD decomposition described in [16].

The model using MDD is substantially better than the COST-MDD decomposition, and also superior to the MDD decomposition. It already improves upon the best published CP/SAT models for these problems[2] in [15]. The results for ev-mdd show that modelling the problem using COST-MDD is substantially better than separately modelling cost and an MDD constraint. Incremental greedy explanation can improve on minimal explanations, but the results demonstrate that minimal explanations are preferable. This contrasts with results for explaining MDD [6] where greedy incremental explanations were almost always superior. This may be because the presence of path costs in EVMDDs means that decisions higher in the graph have a greater impact on explanations further down (whereas for MDDs, the explanation only changes if a node is rendered completely unreachable).

6   Conclusion

In this paper we have defined how to explain the propagation of an EVMDD. Interestingly we have a trade-off between using cost bounds or literals on x to explain the same propagation. We define non-incremental minimal and incremental non-minimal explanation algorithms for EVMDDs. Using EVMDD with explanation to define a COST-MDD constraint, we are able to substantially improve on other modelling approaches for solving problems with COST-MDD with explanation.

Acknowledgments. NICTA is funded by the Australian Government as represented by the Department of Broadband, Communications and the Digital Economy and the Australian Research Council.

References

1. Focacci, F., Lodi, A., Milano, M.: Cost-based domain filtering. In: Jaffar, J. (ed.) CP 1999. LNCS, vol. 1713, pp. 189–203. Springer, Heidelberg (1999)
2. Régin, J.-C.: Arc consistency for global cardinality constraints with costs. In: Jaffar, J. (ed.) CP 1999. LNCS, vol. 1713, pp. 390–404. Springer, Heidelberg (1999)
3. Demassey, S., Pesant, G., Rousseau, L.-M.: A cost-regular based hybrid column generation approach. Constraints 11(4), 315–333 (2006)
4. Cheng, K.C.K., Yap, R.H.C.: Maintaining generalized arc consistency on ad hoc r-ary constraints. In: Stuckey, P.J. (ed.) CP 2008. LNCS, vol. 5202, pp. 509–523. Springer, Heidelberg (2008)

---

[2] The best results for these problems use dynamic programming as a column generator in a branch-and-price solution [17].

5. Gange, G., Stuckey, P., Lagoon, V.: Fast set bounds propagation using a BDD-SAT hybrid. Journal of Artificial Intelligence Research 38, 307–338 (2010)
6. Gange, G., Stuckey, P.J., Szymanek, R.: MDD propagators with explanation. Constraints 16(4), 407–429 (2011)
7. Katsirelos, G., Narodytska, N., Walsh, T.: The weighted grammar constraint. Annals OR 184(1), 179–207 (2011)
8. Ohrimenko, O., Stuckey, P., Codish, M.: Propagation via lazy clause generation. Constraints 14(3), 357–391 (2009)
9. Schulte, C., Stuckey, P.: Efficient constraint propagation engines. ACM Transactions on Programming Languages and Systems 31(1), Article No. 2 (2008)
10. Vrudhula, S.B., Pedram, M., Lai, Y.-T.: Edge valued binary decision diagrams. In: Representations of Discrete Functions, pp. 109–132. Springer (1996)
11. Choi, C.W., Harvey, W., Lee, J.H.M., Stuckey, P.J.: Finite domain bounds consistency revisited. In: Sattar, A., Kang, B.-H. (eds.) AI 2006. LNCS (LNAI), vol. 4304, pp. 49–58. Springer, Heidelberg (2006)
12. Subbarayan, S.: Efficient reasoning for nogoods in constraint solvers with bDDs. In: Hudak, P., Warren, D.S. (eds.) PADL 2008. LNCS, vol. 4902, pp. 53–67. Springer, Heidelberg (2008)
13. Quimper, C.G., Walsh, T.: Global grammar constraints. In: Benhamou, F. (ed.) CP 2006. LNCS, vol. 4204, pp. 751–755. Springer, Heidelberg (2006)
14. Jung, J.C., Barahona, P., Katsirelos, G., Walsh, T.: Two encodings of DNNF theories. In: ECAI Workshop on Inference Methods Based on Graphical Structures of Knowledge (2008)
15. Katsirelos, G., Narodytska, N., Walsh, T.: Reformulating global grammar constraints. In: van Hoeve, W.-J., Hooker, J.N. (eds.) CPAIOR 2009. LNCS, vol. 5547, pp. 132–147. Springer, Heidelberg (2009)
16. Abío, I., Nieuwenhuis, R., Oliveras, A., Rodríguez-Carbonell, E.: BDDs for pseudo-boolean constraints – revisited. In: Sakallah, K.A., Simon, L. (eds.) SAT 2011. LNCS, vol. 6695, pp. 61–75. Springer, Heidelberg (2011)
17. Côté, M.C., Gendron, B., Rousseau, L.M.: Grammar-based integer programming models for multiactivity shift scheduling. Management Science 57(1), 151–163 (2011)

A Simple and Effective Decomposition for the Multidimensional Binpacking Constraint

Stefano Gualandi[1] and Michele Lombardi[2]

[1] Università di Pavia, Dipartimento di Matematica
[2] Università di Bologna, Dipartimento di Informatica: Scienza ed Ingegneria
stefano.gualandi@unipv.it, michele.lombardi2@unibo.it

Abstract. The `multibin_packing` constraint captures a fundamental substructure of many assignment problems, where a set of items, each with a fixed number of dimensions, must be assigned to a number of bins with limited capacities. In this work we propose a simple decomposition for `multibin_packing` that uses a `bin_packing` constraint for each dimension, a set of `all_different` constraints automatically derived from a conflict graph, plus two alternative symmetry breaking approaches. Despite its simplicity, the proposed decomposition is very effective on a number of instances recently proposed in the literature.

1 Introduction

Given a set $I = \{1, \ldots, n\}$ of items and a set $K = \{1, \ldots, k\}$ of dimensions, where each item i has a weight $w_{i,l}$ for every dimension l, and given a set $B = \{1, \ldots, m\}$ of bins j with a capacity $c_{j,l}$ for every dimension l, the `multibin_pa-cking` constraint states that every item must be packed into a single bin while the sum of weights for each bin and for each dimension cannot exceed the corresponding bin capacity. In particular, we extend the formulation of the `bin_packing` constraint from [10] by using the following signature:

$$\texttt{multibin\_packing}([y_{j,l}], [x_i], [w_{i,l}]) \tag{1}$$

where $x_i = j$ if item i is assigned to bin j and $y_{j,l}$ is a load variable ranging in $[0, c_{j,l}]$ that represents the total weight packed on bin j for dimension l. Indeed, the semantic of constraint (1) is equivalent to the following relations:

$$\sum_{i \in I: x_i = j} w_{i,l} = y_{j,l} \leq c_{j,l} \qquad \forall j \in B, \forall l \in K. \tag{2}$$

The case of a single dimension $k = 1$ reduces (1) to the well-known `bin_packing` constraint. Therefore, a natural decomposition of (1) is to use a single `bin_packing` for each dimension as follows:

$$\texttt{bin\_packing}([y_{1,l}, \ldots, y_{m,l}], [x_i], [w_{i,l}]) \qquad \forall l \in K. \tag{3}$$

C. Schulte (Ed.): CP 2013, LNCS 8124, pp. 356–364, 2013.

This decomposition is used for instance in [6] to formulate the Machine Reassignment problem proposed in the Roadef Google Challenge 2012[1]. In the Machine Reassignment problem, each item represents an application to be (re)assigned to a server, and each dimension represents a resource consumed by an application, such as, for instance, CPU time, memory, and bandwidth. Since every server has a limited capacity for each resource, the `multibin_packing` constraint captures a fundamental substructure of the Machine Reassignment problem.

In [2], the authors propose a constraint based on Multivalued Decisions Diagrams (MDD) for multidimensional bin packing problems and they show that the approach is very effective on a set of randomly generated instances. The MDD approach definitely outperforms a basic CP model based on decomposition (3) and solved with a basic first-unassigned min-value branching strategy. In several cases, the model with the MDD constraint outperforms a Mixed Integer Programming approach as well. In particular, the MDD approach is attractive for hard instances on the transition phases from infeasible to feasible, likely due to the ability of MDD to handle symmetries. While their approach is very interesting, it is hard to embed into existing CP solvers.

The first filtering algorithm for the `bin_packing` constraint was presented by Shaw in [10], where the author combined the algorithm for the Knapsack constraint introduced in [11] with a well-known lower bound on the minimum bin packing problem (e.g., see Chapter 10 in [5]). Shaw's algorithm is implemented in several CP solvers, it is listed in the global constraint catalog [1], and it is useful for several industrial applications (e.g., see [9]).

The contribution of this paper is to propose a simple decomposition for the `multibin_packing` constraint that, in addition to constraints (3), uses a collection of `all_different` constraints automatically derived from a conflict graph, and that posts symmetry breaking constraints. The benefits of the `all_different` constraints are twofold: they perform additional filtering and they help the branching to identify "most-conflicting" variables. We show experimentally that our approach is very effective on the instances recently presented in [2]. Since the proposed `multibin_packing` decomposition is based on existing constraints, it can be easily implemented in any existing CP solver.

The outline of the paper is as follows: Section 2 introduces the problem decomposition, while Section 3 presents two alternative symmetry breaking strategies. Section 4 reports computational results and concludes the paper.

2   Constraint Decomposition

The `multibin_packing` constraint naturally decomposes into k independent `bin_packing` constraints. However, this basic decomposition does not account for the different dimensions of each item: while two items may fit in the same bin while considering a given dimension l_1, they may be in conflict while considering a different dimension l_2, since the sum of their weights might exceed the

---

[1] http://challenge.roadef.org/2012/en/, last visited April, 24-th, 2013.

bin capacity for l_2. However, the interdependencies among the item dimensions across the bins can be exploited systematically via a conflict graph.

Given an instance of the `multibin_packing` constraint, we build an undirected *conflict* graph $G = (V, E)$ by looking for pairs of conflicting items. The conflict graph G is constructed as follows. First, for each item i in I, we add a vertex to V: we have a one-to-one mapping between items and vertices of G. Second, we add an edge $\{i_1, i_2\}$ to E for each pair of items with $i_1, i_2 \in I$ and $i_1 < i_2$, and such that the following relation holds:

$$\forall j \in B. \quad \exists l \in K: \quad w_{i_1,l} + w_{i_2,l} > c_{j,l} \tag{4}$$

Indeed, two items i_1 and i_2 are in conflict if for each bin there exists at least a dimension l such that the sum of the two item weights exceed the bin capacity for dimension l.

We show next how we use the conflict graph. Recall that a *clique* is complete subgraph of G, and that a clique is *maximal* if it is not a subset of any other clique. Given a subset of items $J \subseteq I$, we denote by x_J the subset of the $[x_i]$ variables corresponding to the items in J.

Proposition 1. *Given an instance of `multibin_packing`, the cardinality $\omega(G)$ of the maximum clique, denoted by C^*, of the conflict graph gives a lower bound on the number of bins necessary in any feasible assignment of items to bins.*

Clearly, every pair of items corresponding to vertices in C^* must be assigned to different bins, therefore we need at least $\omega(G)$ bins to have a feasible assignment. However, we do not really need a maximum clique in order to detect infeasibility, as shown next.

Proposition 2. *Given an instance of `multibin_packing`, if the conflict graph G contains a clique C of cardinality strictly greater than the number of available bins, that is $|C| > m$, then the `multibin_packing` constraint cannot hold.*

Therefore, to declare infeasibility of a given instance of `multibin_packing`, we are interested in a finding clique of cardinality equal to $m + 1$. Despite finding a clique of a given size is an NP-complete problem [3], we have that $m \ll n$ and that the conflict graph is sparse (otherwise the `multibin_packing` instance would be likely infeasible). In practice, exact algorithms (e.g., see [7]) are extremely efficient in checking if a clique of size $m + 1$ exists in the `multibin_packing` instances taken from the literature. For easy instances, the set of edges E may even be empty.

It is possible to exploit the cliques in G for more than pure consistency checking. Specifically, since for any clique C with $1 < |C| \leq m$, it is possible to post an `all_different` constraint on the item variables corresponding to the vertices in the clique. More formally:

Proposition 3. *Given an instance of `multibin_packing`, every subset $J \subset I$ of items corresponding to vertices of any (maximal) clique C of G, with $1 < |C| \leq m$, must be assigned to different bins, that is the corresponding subset of item variables x_J must take a different value: `all_different`($[x_J]$) must hold.*

The proposition holds for any clique, but it is better to consider maximal cliques only, since they are fewer in number.

While the number of maximal cliques of G is, in theory, exponential in the number of vertices, in many practical cases their number is definitely reasonable. In addition, we are not forced to list every maximal cliques, and it is easy to devise heuristics that limit the number of cliques considered. For instance, it is possible to look for every vertex i of G a maximal clique that contains vertex i. Simple greedy heuristics for finding maximal cliques have worst-case time complexity of $O(n^2)$.

3   Symmetry Breaking

In case the bins have identical capacities for every dimension, that is whenever $c_{jl} = c_l$ for all j in B, and in case there is not other special constraint nor costs on the assignment of items to bins, then the multibin_packing constraint admits several symmetric assignments, since every possible permutation of bin assignment gives equivalent solutions. For instance, given a feasible assignment, if we exchange all the items assigned to the first bin with all the items assigned to the second bin, we get another feasible solution. This kind of situation seems to occur quite often in practice.

3.1   Symmetry Breaking by Variable Fixing

In order to partially break symmetries, we can again exploit the conflict graph G. In practice, we can take the largest clique C with cardinality smaller than or equal to m (otherwise the constraint is infeasible) and we can fix every item variable corresponding to vertices in C to a different value from its own domain. Let $dom(x_i)$ be the domain of variable x_i; initially, every variables has the same domain. We take the first variable in C, and we assign it the minimum value in its domain; then we take the second variable in C and we assign it the second value in its domain; and so on. With a small abuse of notation, we write:

$$x_i \leftarrow \min\{dom(x_i)\} \qquad \text{for } i \in C. \tag{5}$$

where after each assignment, propagation reduces the other variable domains.

Note that whenever it is possible to use this variable fixing technique, then, once the variables have been fixed and all other problem constraints have propagated, it is possible to construct a new conflict graph G using the residual bin capacity obtained by considering the fixing in (5). The new conflict graph can again be used for consistency checking and, likely, to post additional all_different constraints.

3.2   Symmetry Breaking Constraints

If the assignment of items to bins is symmetrical, as discussed previously, a different but standard way to break symmetries during search consists in posting an additional constraint that invalidates equivalent feasible assignments (e.g., see Chapter 10 in [8]).

If we consider a single dimension l, for instance the first dimension $l = 1$, we can post an ordering constraint on the corresponding subset of load variables:

$$y_{j,1} \geq y_{(j+1),1} \qquad \forall j \in \{1, \ldots, m-1\}. \tag{6}$$

However, no ordering constraint can be posted on the other load variables.

Note that constraint (6) reduces the number of equivalent feasible assignments of items to bins, and therefore on *easy* instances of multibin_packing it might have a negative effect when looking for a single feasible solution. We will discuss this issue in the computational results section.

The symmetry breaking constraints based on (6) are only applicable when every bin has the same capacity for every dimension. If we perform an item-bin variable fixing for statically break some symmetries, as described in Section 3.1, we are in practice modifying the problem and, as a consequence, the bins have no longer an identical capacity for each dimension. For this reason, the two symmetry breaking strategies are incompatible.

4   Computational Results

In order to evaluate our approach, we have implemented the proposed decomposition of multibin_packing within the Gecode constraint system v.3.7.3 [4]. Note that Gecode has a very efficient implementation of the Shaw's Bin Packing constraint [10]. In order to list every maximal cliques of the conflict graph we used cliquer-1.21 that is the state-of-the-art exact clique finder for sparse graphs [7]. Everything was compiled using the gnu-gcc v4.7.2 compiler. All the tests were run on standard computer with Linux as operating system, with 4GB of RAM and an AMD Opteron 2.4GHz CPU, but using a single thread and limiting the process memory size to 1GB. Our implementation of the constraint decomposition is available online[2].

We have run experiments using combinations of the following decomposition:

$$\text{multibin\_packing}([y_{j,l}], [x_i], [w_{i,l}]) =$$

$$\text{bin\_packing}([y_{1,l}, \ldots, y_{m,l}], [x_i], [w_{i,l}]) \qquad \forall l \in K \qquad (7)$$

$$\text{all\_different}([x_C]) \qquad \forall C \in \mathcal{C} \qquad (8)$$

$$y_{j,1} \geq y_{(j+1),1} \qquad \forall j \in \{1, \ldots, m-1\} \qquad (9)$$

where $\mathcal{C}$ is the collection of all maximal cliques of the conflict graph defined for the given instance of multibin_packing.

We denote by:

(A) the decomposition using only constraints (7)
(B) the decomposition using constraints (7) and (8)

---

[2] http://github.com/stegua/binpacking/tree/master/release

(C) the decomposition using constraints (7) and (9)
(D) the decomposition using constraints (7), (8), and (9)
(E) the decomposition using constraints (7), (8), and (5)

The rationale for studying decompositions (B) and (C) is to assess the impact of the additional constraints (7) and (8), and for decompositions (D) and (E) is to compare the two alternative symmetry breaking strategies.

In order to make a fair comparison between the different decomposition strategies, in our computational results, we did not update the conflict graph for strategy (E). In practice, constraints (8) are posted on the same collection of maximal clique for method (B), (D), and (E).

We consider a set of instances taken from [2] that corresponds to randomly generated instances with 18 items, 6 dimensions, and 6 bins. In order to have a range of instances from "easy infeasible" to "easy feasible" while passing through a "hard" phase transition, the instances differ on the tightness of the bin capacities (i.e. the bin slacks), according to a parameter β (for the details on the instances generation see [2]). The values of β range in $\{0, \ldots 35\}$, and for each value of β there are 52 instances, for a total of 1872. The phase transition happens for $\beta \in \{16, \ldots, 24\}$.

Figures (1.a)–(1.d) show the average computation times in seconds (vertical axis) in function of the percentage of bin slacks, that is, for value of β from 0 to 35. Plot (1.a) clearly shows that the hard instances correspond to values of β ranging from 15 to 25, while the plots (1.b)–(1.d) show that:

- For the hard instances (1.b) and the easy infeasible instances (1.c) the decomposition (E) that exploits both the all_different constraints and the variable fixing on the maximum clique, outperforms the other decompositions by a large margin: the average computation time is below 1 second also for the instances with $\beta = 20$, and, hence, it is a *simple* and *effective* alternative to the multibin_packing constraint based on Multivalues Decision Diagrams proposed in [2].
- For the easy infeasible (1.c) and the easy feasible instances (1.d), the symmetry breaking constraint (9) does not always pay off, while it plays an important role on the hard instances (1.b), since decomposition (C) and (D) are always more efficient than (A) and (B).
- As expected, for the easy feasible instances (1.d), the simplest decomposition (A) is quite efficient, since in this case the filtering algorithms play a minor role.

Figures (2.a) and (2.b) give a different view of the same results by showing the empirical cumulative distributions of the fraction of instances solved as a function of the run times. Figure (2.a) are the distributions for all the 1872 instances proposed in [2], while Figure (2.b) considers only the instances with $\beta = 20$, i.e. among the hardest instances. Again decomposition (E) outperforms all the others by a large margin. The decomposition (B) based on the all_different constraints outperforms the simple decomposition (A) for a large number of instances, but not for all of them. The decomposition (C) and (D) that exploit

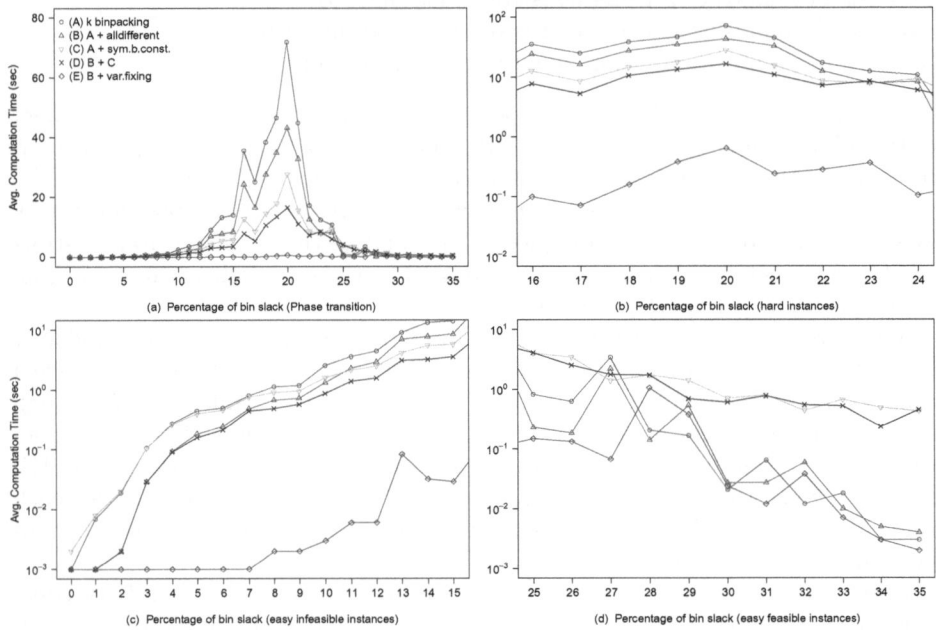

Fig. 1. Average Computation Time vs. Percentage of Bin Slack

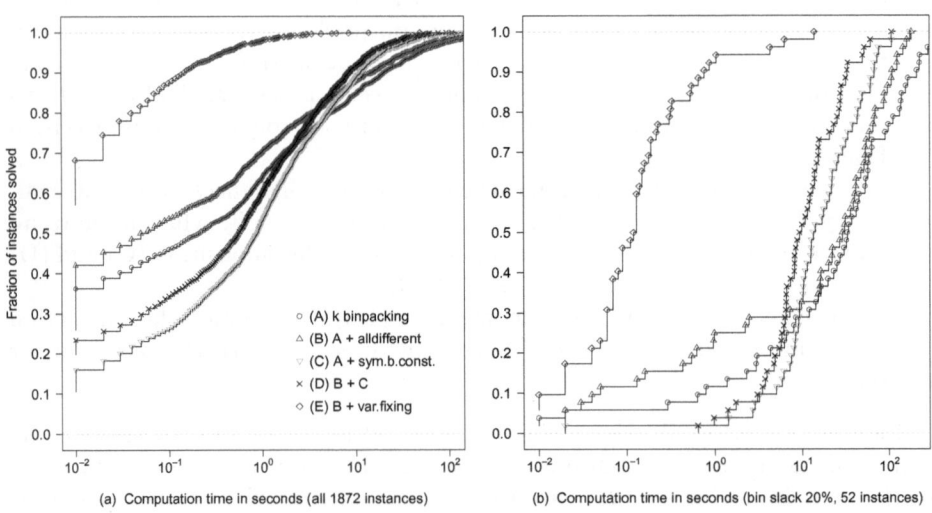

Fig. 2. Empricial Cumulative Distribution of the fraction of instances solved as a function of computation times

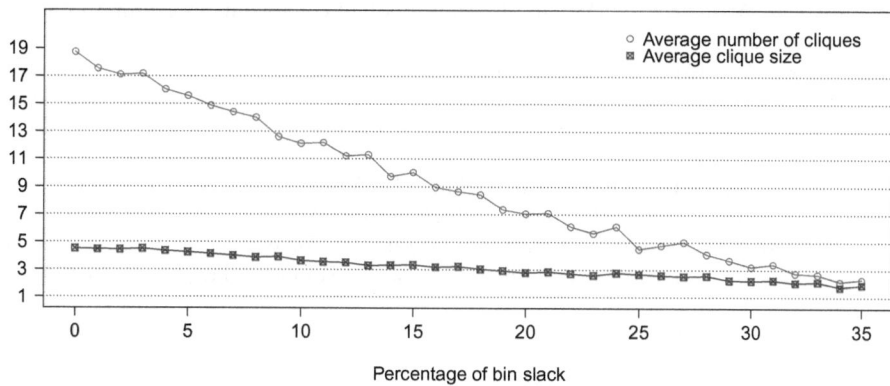

Fig. 3. Average number of cliques and average clique size vs. bin slacks

the symmetry breaking constraints are useful on the hard instances, as shown in Figure (2.b), but in some cases the overhead of constraints (9) does not pay off.

Finally, since the main contribution of our approach is based on detecting maximal cliques in the conflict graph and on posting an `all_different` constraint for every clique found, Figure 3 characterizes the instances in terms of average clique number and average clique size as a function of the bin slack. Note that for the easy infeasible instances (i.e. $\beta < 15$), there are many and large cliques, while for the easy feasible instances (i.e. $\beta > 25$) the cliques are few and rather small (sometimes are not even present). Clearly, our decomposition approach is better suited for instances with several large cliques.

References

1. Beldiceanu, N., Carlsson, M., Rampon, J.X.: Global constraint catalog. SICS Research Report (2005)
2. Kell, B., van Hoeve, W.-J.: An MDD approach to multidimensional bin packing. In: Gomes, C., Sellmann, M. (eds.) CPAIOR 2013. LNCS, vol. 7874, pp. 128–143. Springer, Heidelberg (2013)
3. Garey, M.R., Johnson, D.S.: Computers and intractability. Freeman, New York (1979)
4. Gecode Team. Gecode: Generic constraint development environment (2013), http://www.gecode.org
5. Martello, S., Toth, P.: Knapsack problems: algorithms and computer implementations. John Wiley & Sons, Inc. (1990)
6. Mehta, D., O'Sullivan, B., Simonis, H.: Comparing solution methods for the machine reassignment problem. In: Milano, M. (ed.) CP 2012. LNCS, vol. 7514, pp. 782–797. Springer, Heidelberg (2012)
7. Östergård, P.R.J.: A fast algorithm for the maximum clique problem. Discrete Applied Mathematics 120(1), 197–207 (2002)

8. Rossi, F., Van Beek, P., Walsh, T.: Handbook of constraint programming. Elsevier Science (2006)
9. Schaus, P., Régin, J.-C., Van Schaeren, R., Dullaert, W., Raa, B.: Cardinality reasoning for bin-packing constraint: Application to a tank allocation problem. In: Milano, M. (ed.) CP 2012. LNCS, vol. 7514, pp. 815–822. Springer, Heidelberg (2012)
10. Shaw, P.: A constraint for bin packing. In: Wallace, M. (ed.) CP 2004. LNCS, vol. 3258, pp. 648–662. Springer, Heidelberg (2004)
11. Trick, M.A.: A dynamic programming approach for consistency and propagation for knapsack constraints. Annals of Operations Research 118(1), 73–84 (2003)

Maintaining Soft Arc Consistencies in BnB-ADOPT$^+$ during Search*

Patricia Gutierrez[1], Jimmy H.M. Lee[2], Ka Man Lei[2],
Terrence W.K. Mak[3], and Pedro Meseguer[1]

[1] IIIA - CSIC, Universitat Autònoma de Barcelona, 08193 Bellaterra Spain
{patricia,pedro}@iiia.csic.es
[2] Department of Computer Science and Engineering,
The Chinese University of Hong Kong, Shatin, N.T., Hong Kong
{jlee,kmlei}@cse.cuhk.edu.hk
[3] NICTA Victoria Laboratory & University of Melbourne, VIC 3010, Australia
Terrence.Mak@nicta.com.au

Abstract. Gutierrez and Meseguer show how to enforce consistency in BnB-ADOPT$^+$ for distributed constraint optimization, but they consider unconditional deletions only. However, during search, more values can be pruned conditionally according to variable instantiations that define subproblems. Enforcing consistency in these subproblems can cause further search space reduction. We introduce efficient methods to maintain soft arc consistencies in every subproblem during search, a non trivial task due to asynchronicity and induced overheads. Experimental results show substantial benefits on three different benchmarks.

1 Introduction

Distributed Constraint Optimization Problems (DCOPs) have been applied in modeling and solving a substantial number of multiagent coordination problems, such as meeting scheduling [1], sensor networks [2] and traffic control [3]. Several distributed algorithms for optimal DCOP solving have been proposed: ADOPT [4], DPOP [5], BnB-ADOPT [6], NCBB [7] and others.

BnB-ADOPT$^+$-AC/FDAC [8] incorporate consistency enforcement during search into BnB-ADOPT$^+$ [9], obtaining substantial efficiency improvements. Enforcing consistency allows to prune some values, making the search space smaller. This previous work considers unconditional deletions only so as to avoid overhead in handling assignments and backtracking. However, values that could be deleted conditioned to some assignments will not be pruned with this strategy, so that search space reduction opportunities are missed. In this paper, we propose an efficient way to maintain soft

---

* We are grateful to the anonymous referees for their constructive comments. The work of Lei and Lee was generously supported by grants CUHK413808, CUHK413710 and CUHK413713 from the Research Grants Council of Hong Kong SAR. The work of Gutierrez and Meseguer was partially supported by the Spanish project TIN2009-13591-C02-02 and Generalitat de Catalunya 2009-SGR-1434. The work of Gutierrez, Lee and Meseguer was also jointly supported by the CSIC/RGC Joint Research Scheme grants S-HK003/12 and 2011HK0017. The work of Mak was performed while he was at CUHK.

C. Schulte (Ed.): CP 2013, LNCS 8124, pp. 365–380, 2013.

arc consistencies, considering any kind of deletions resulting from enforcing consistency in asynchronous distributed constraint solving, something that—to the best of our knowledge—has not been explored before.

A search-based constraint solving algorithm forms subproblems of the original problem by assignments. We maintain soft arc consistencies in each subproblem, so that variable assignments during search are also considered in consistency enforcement. As a result, we can explore more value pruning opportunities and thus further reduce the search space. Gutierrez and Meseguer introduce an extra copy of cost functions in each agent, so that search and consistency enforcement are done asynchronously. Our contribution goes further maintaining soft arc consistencies in each subproblem during search, so that (i) search and consistency enforcement are done asynchronously, introducing some extra copies of cost functions; (ii) the induced overhead caused by backtracking and undoing assignments and deletions is minimized. The asynchronicity requirement and different cost measurements require us to introduce novel techniques over those used in centralized CP. Experimentally, we show the benefits of our proposal on benchmarks usually unamenable to solvers without consistency.

2   Preliminaries

DCOP. A DCOP is defined by $\langle \mathcal{X}, \mathcal{D}, \mathcal{C}, \mathcal{A}, \alpha \rangle$, where $\mathcal{X} = \{x_1, \ldots, x_n\}$ is a set of variables; $\mathcal{D} = \{D_1, \ldots, D_n\}$ is a set of finite domains for $\mathcal{X}$; $\mathcal{C}$ is a set of cost functions; $\mathcal{A} = \{1, ..., n\}$ is a set of n agents and $\alpha : \mathcal{X} \rightarrow \mathcal{A}$ maps each variable to one agent. We use binary and unary cost functions only, which produce non-negative costs. The cost of a complete assignment is the sum of all unary and binary cost functions evaluated on it. An *optimal solution* is a complete assignment with minimum cost. Each agent holds exactly one variable, so that variables and agents can be used interchangeably. Agents communicate through messages, which are never lost and delivered in the order they were sent, for any agent pair.

DCOPs can be arranged in a *pseudo-tree*, where nodes correspond to variables and edges correspond to binary cost functions. There is a subset of edges, called tree-edges, that form a rooted tree. The remaining edges are called back-edges. Variables involved in the same cost function appear in the same branch. Tree edges connect parent-child nodes. Back-edges connect a node with its pseudo-parents and pseudo-children.

BnB-ADOPT and BnB-ADOPT$^+$. BnB-ADOPT [6] is an algorithm for optimal DCOP solving. It uses the communication framework of ADOPT [4] (agents are arranged in a pseudo-tree), but it changes the search strategy to depth first branch-and-bound. It shows improvements over ADOPT. Each agent holds a context, as a set of assignments involving some of the agent's ancestors that is updated with message exchanges. Message types are: VALUE, COST and TERMINATE. A BnB-ADOPT agent executes this loop: it reads and processes all incoming messages and assigns its value. Then, it sends a VALUE to each child or pseudochild and a COST to its parent. BnB-ADOPT$^+$ [9] is a version of BnB-ADOPT that prevents from sending most redundant messages, keeping optimality and termination. It substantially reduces communication.

Soft Arc Consistency. Let (i, a) represents x_i taking value a, $\top$ is the lowest unacceptable cost, C_{ij} is the binary cost function between x_i and x_j, C_i is the unary cost function on x_i values, C_ϕ is a zero-ary cost function (lower bound of the cost of any solution). We consider the following local consistencies [10,11]:

- *Node Consistency* (NC): (i, a) is NC if $C_\phi + C_i(a) < \top$; x_i is NC if all its values are NC and $\exists b \in D_i$ s.t. $C_i(b) = 0$. P is NC if every variable is NC.
- *Arc Consistency* (AC): (i, a) is AC w.r.t. C_{ij} if $\exists b \in D_j$ s.t. $C_{ij}(a, b) = 0$; b is a *support* of a; x_i is AC if all its values are AC w.r.t. every binary cost function involving x_i; P is AC if every variable is AC and NC.
- *Directional Arc Consistency* (DAC): (i, a) is DAC w.r.t. $C_{ij}, j > i$, if $\exists b \in D_j$ s.t. $C_{ij}(a, b) + C_j(b) = 0$; b is a *full support* of a; x_i is DAC if all its values are DAC w.r.t. every C_{ij}; P is DAC if every variable is DAC and NC.
- *Full DAC* (FDAC): P is FDAC if it is DAC and AC.

AC/DAC can be reached by forcing supports/full supports to NC values and pruning values that are not NC. Supports can be forced by projecting the minimum cost from its binary cost functions to its unary costs, and then projecting the minimum unary cost into C_ϕ. Full supports can be forced in the same way, but first it is needed to extend from the unary costs of neighbors to the binary cost functions the minimum cost required to perform in the next step the projection over the value. The systematic application of *projection* and *extension* does not change the optimum cost [10,11]. When we prune a value from x_i, we need to recheck AC/DAC on every variable that x_i is constrained with, since the deleted value could be the support/full support of a value of a neighbor variable. So, a deleted value in one variable might cause further deletions in others. The AC/DAC check must be done until no further values are deleted.

BnB-ADOPT$^+$ and Soft Arc Consistencies. BnB-ADOPT$^+$ has been combined with AC and FDAC [8]. Search is based on BnB-ADOPT$^+$, maintaining the same data and communication structure. Soft arc consistencies are enforced on a copy of the original cost functions, limited to unconditional deletions. This combination has caused a number of modifications in the original algorithm, both in messages and in computation.

Regarding messages, (i) COST messages include *subtreeContr* that aggregates the costs of unary projections to C_ϕ made on every agent; (ii) VALUE messages include $\top$ and C_ϕ; (iii) a new DEL message is added to inform of value deletions; when received, neighbors recheck AC/FDAC, which may lead to further deletions; (iv) a new UCO message is added when FDAC is enforced, to inform the unary costs needed for enforcing DAC; when received, agents enforce DAC with any other higher constrained agents and recheck FDAC, which may lead to further deletions.

Regarding computation, each agent holds one copy of constrained agents' domains and related binary cost functions for consistency enforcement. Handling value deletions require some extra effort. Only the agent owner of a variable can modify its domain.

3   Maintaining Soft Arc Consistencies

We enforce AC and FDAC asynchronously in all subproblems during search by utilizing additional copies of variable domains and cost functions in each agent. To explain our *Maintaining AC* (MAC) and *Maintaining FDAC* (MFDAC) algorithms, we first outline

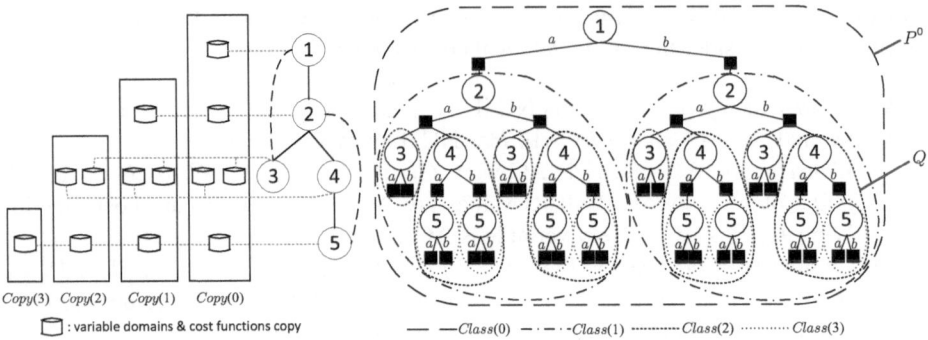

Fig. 1. Left: The pseudo-tree of a DCOP with five variables, the variable domains and cost functions copies they maintain. Right: Search tree (a/b domains), subproblems and classes of subproblems. Subproblems at the same depth belong to same class.

an agent classing scheme based on the position of an agent in the problem structure. The scheme governs the required number of copies of variable domains and cost functions. Second, we provide the information of messages in our methods and an overview of the changes in the overall message handling mechanism after adopting our new methods in BnB-ADOPT$^+$. Third, we provide methods for reinitializing variable domains and cost functions copies in an agent when the context of a subproblem changes. Such reinitialization is needed since conditional deletions are no longer valid. Thus, consistency enforcement has to start from scratch again using the new context. Fourth, we propose a new message type and the handling mechanism for backtracking, when an agent arrives at the empty domain within a subproblem. This means that the assignments of some ancestor agents cannot lead to the optimal solution and should be pruned. Fifth, we reduce costs by transferring deletions from subproblems to inner subproblems. Sixth, we present an ordering scheme and asynchronous messaging mechanism to ensure that the two separate copies of the same cost function stored in the two constrained agents are identical even in the presence of simultaneous consistency operations. Finally, we describe how we ensure optimality and termination after introducing the new methods.

3.1 Classes of Subproblems

In BnB-ADOPT$^+$ [9], all agents are organized in a pseudo-tree (Fig. 1 Left). The variable ordering of the corresponding AND-OR search tree [6] (Fig. 1 Right) follows the (partial) order defined in the pseudo-tree. When an agent is assigned a value, the descendant agents together with the current assignments form a subproblem. Notations: P^0 is the original DCOP; P is a subproblem of P^0; T^0 is a pseudo-tree that defines the variable ordering in P^0; d_j is the *depth* of agent j in T^0 as the distance from the root node to j excluding back-edges; $vars(P)$ is the set of variables of P; $depth(P)$ is the smallest depth among all variables in $vars(P)$; $ancestors(P)$ is the set of *ancestor variables* satisfying (1) they are in $vars(P^0)$ but not in $vars(P)$, (2) they have depths smaller than $depth(P)$, and (3) they are constrained with at least one variable in $vars(P)$; $context(P)$ is the variable assignments of $ancestors(P)$; $context_j$, the *context* of agent j, is the set $\{(i,a,t)|i = j$ or i is an ancestor of j, a is agent i value which is assigned to i at timestamp $t\}$. Two contexts are *compatible* if no agent takes on different values in the two contexts. Every *subproblem* P of P^0 is uniquely identified by $(depth(P), ancestors(P), context(P), vars(P))$.

Fig. 1 Right illustrates the search tree and subproblems of a DCOP with 5 agents. Each circular node is the root node of a subproblem and there are 19 such subproblems (including the original problem) in the example. The original problem P^0 is $(0, \emptyset, \emptyset, \{1,2,3,4,5\})$ and $(2, \{1,2\}, \{(1,b),(2,b)\}, \{4,5\})$ (labeled Q in the figure) is the subproblem of P^0 after instantiating agent 1 and agent 2 to value b. We define a class of subproblems as follows. A subproblem P of P^0 is of *Class d* if $depth(P) = d$. We further define $Class(d) = \{P|depth(P) = d\}$.

Fig. 1 Right also illustrates the classes of subproblems of the DCOP. There are four classes of subproblems: $Class(0)$ involves the original problem only. $Class(1)$ includes two subproblems $(0, \{1\}, \{(1,a)\}, \{2,3,4,5\})$ and $(0, \{1\}, \{(1,b)\}, \{2,3,4,5\})$, $Class(2)$ includes eight subproblems in which four are rooted at node 3 and the other four are rooted at node 4. All $Class(2)$ subproblems hold the assignment information of agents 1 and 2 (their context). $Class(3)$ includes eight subproblems which are all rooted at node 5 and hold assignment information of agents 1, 2 and 4.

In BnB-ADOPT$^+$-AC/FDAC [8], search and consistency enforcement are done asynchronously: an extra copy of each cost function is used for consistency enforcement and they do not interfere with the original copy used for search. We use the same idea for MAC and MFDAC: we include extra copies of variable domains and cost functions for enforcing consistency in different subproblems, but not a copy for each subproblem. Each agent i of depth d_i will hold one copy $Copy(d)$ for each class $Class(d)$ of subproblems where $d \leq d_i$. For instance, in Fig. 1 Left, agents keep the following copies of cost functions and domains: agent 1 one copy, agent 2 two copies, agent 3 and 4 three copies, and agent 5 four copies. Then, each agent i will hold $d_i + 1$ copies of variable domains and cost functions and the space complexity of each agent is $O(dhm^2)$ where d is the agent's depth, h is the pseudo-tree's height and m is the maximum domain size of agents. These copies will play a key role in reinitializing domains and cost functions when conditional deletions are no longer valid in a context change.

3.2 Maintaining Consistencies in All Subproblems: An Overview

To maintain soft arc consistencies in every subproblem, extra operations and information exchanges are needed. The major additional operations include (1) reinitialization, (2) backtracking to the culprit when an empty domain is detected and (3) transferring

Table 1. Messages of AC, FDAC, MAC and MFDAC. New fields are underlined. DEL messages contain ACC or DACC depending on the AC or FDAC consistency level enforced.

AC/FDAC	MAC/MFDAC
VALUE(*src,dest,value,threshold*,⊤,C_ϕ)	VALUE(*src,dest,value,threshold*,⊤,C_ϕ[],*context*)
COST(*src,dest,lb,ub,reducedContext,subtreeContr*)	COST(*src,dest,lb,ub,context,subtreeContr*[])
DEL(*src,dest,value,ACC|DACC*)	DEL(*src,dest,depth,values*[],*context,ACC*[]|*DACC*[])
UCO*(*src,dest,vectorOfExtensions,ACC*)	UCO**(*src,dest,depth,vectorOfExtensions,context,ACC*)
	BTK(*src,dest,targetDepth,context*)
* Only in FDAC	** Only in MFDAC

procedure ProcessVALUE(msg)
 do the work as in BnB-ADOPT$^+$
 Reinitialize($msg.src$,$msg.context$)
 update $\top$ and C_ϕ if applicable

procedure ProcessCOST(msg)
 do the work as in BnB-ADOPT$^+$
 Reinitialize($msg.src$,$msg.context$)
 aggregate C_ϕ from $msg.subtreeContr$
 update C_ϕ if applicable

procedure ProcessDEL(msg)
 Reinitialize($msg.src$,$msg.context$)
 $d \leftarrow msg.depth$
 $vars \leftarrow$ set of variables i in $context_{self}$ where $d_i \in [0, msg.depth - 1]$
 if values of $vars$ in $msg.context$ are compatible with those in $context_{self}$ **then**
  for $d' = d \rightarrow d_{self}$ **do**
   delete $msg.values[]$ from $msg.src$'s domain in $Copy(d')$
   undo disordered operations in $Copy(d')$ if necessary
   perform projection in $Copy(d')$
   update ACC counter if necessary

procedure ProcessUCO(msg)
 Reinitialize($msg.src$,$msg.context$)
 $d \leftarrow msg.depth$
 $vars \leftarrow$ set of variables i in $context_{self}$ where $d_i \in [0, msg.depth - 1]$
 if values of $vars$ in $msg.context$ are compatible with those in $context_{self}$ **then**
  if $ACC_{self \rightarrow msg.src} = msg.ACC$ **then**
   perform extension in $Copy(d)$
   update DACC counter if necessary

Fig. 2. Pseudocode for handling VALUE, COST, DEL and UCO messages

deletions to subproblems. Reinitialization is needed for ensuring the correctness of the algorithm. Backtracking to the culprit and transferring deletions to subproblems are not necessary for correctness but they can improve performance. Besides, to ensure the agents maintain the same cost functions in each copy, Gutierrez and Meseguer [12] proposed to include two new messages to synchronize deletions. However, these messages introduce an extra overhead and slow down the consistency enforcement. We propose a new method to allow agents to undo and reorder some of their operations in order to ensure identical cost functions copies.

Consistency enforcement in each subproblem is similar to that of Gutierrez and Meseguer [8], in which consistency is only enforced in the copy for the original problem. In our case, consistency is enforced in the copy for every class of subproblems at the same time. Extra information is embedded in the existing messages (TERMINATE message same as the one in BnB-ADOPT$^+$ and UCO only in FDAC and MFDAC) and only one new message type (BTK) is added. Table 1 summarizes the information per type. Fig. 2 shows the pseudocode for handling these messages (pseudocode for BTK appears in Section 3.4).

When an agent receives a VALUE or COST message, it first performs the BnB-ADOPT$^+$ process, and then it checks for reinitialization. When an agent receives a DEL message, it does the following steps (1) reinitialization checking, (2) compatibility checking, (3) value deletions, (4) maintaining identical cost function copies, (5) projections and (6) update projection counter. Similarly, when an agent receives an UCO message, it checks for reinitialization first and then performs the extension and extension counter update. After an agent i has processed a VALUE, COST, DEL or UCO message, AC/DAC may be re-enforced in $Copy(d)$ where $d \in [1, d_i]$ if (1) $Copy(d)$ is reinitialized, (2) a better $\top$ or C_ϕ is found in $Copy(d)$, (3) some values are deleted in $Copy(d)$, and (4) some unary costs are increased in $Copy(d)$ (only apply for DAC).

3.3   Reinitialization

When enforcing consistencies in a subproblem P (excluding the original problem), the conditional deletions generated depend on the variable assignments information ($context(P)$) that P holds. These conditional deletions may not occur in other subproblems in $Class(depth(P))$, when the variable assignments have changed. Therefore, conditionally deleted values have to be recovered when values of ancestor agents change. When an agent $i \in ancestors(P)$ of a subproblem P in $Class(d)$ changes its value, the context no longer matches that of P. Search should be now switched to another $P' \in Class(d)$ such that $context(P')$ matches the new value of agent i and other existing assignments. In addition, the copies of cost functions owned by the agents in $vars(P')$ should be reset using the corresponding copies from upper classes and updated with the variable assignments in $context(P')$. Otherwise, the search algorithm will search for solution based on obsolete value pruning information and may result in suboptimal solution. This rationale justifies our next rule.

Rule 1. When an agent i changes its value, all agents $j \in vars(P)$ where $P \in \{P' | i \in ancestors(P')\}$ should reinitialize $Copy(d)$ where $d_i < d \leq d_j$ to be the corresponding subproblem based on the updated context. The reinitialization in j is done in a top-down sequence as follows. For $d = d_i + 1$ to d_j: (1) $Copy(d) = Copy(d - 1)$; (2) Transform each binary cost function C_{jk} where $k \in ancestors(P')$ to unary cost functions C_j by assigning each k to value a where $(k, a) \in context(P')$.

Next we describe how to implement Rule 1 in BnB-ADOPT$^+$ with MAC and MF-DAC. Rule 1 affects an agent when there is a context change. We use VALUE, COST, DEL and UCO messages to carry the context information.

Receiving a VALUE message always signifies a context change in i's parent or pseudo-parent. Thus, agent i always performs reinitialization before deciding whether to change its own value. Receiving a COST message from any of its children may cause a context change. Agent i should always first compare the timestamps of the child's context and i's own context. If the child's context is older than or equal to i's context, i performs nothing. Otherwise, there is a context change and i will perform reinitialization before performing other BnB-ADOPT$^+$ operations. When receiving a DEL or UCO message, an agent performs similar checking before proceeding to consistency enforcement operations—if any—. Strictly speaking, reinitialization in an agent is needed only when handling VALUE and COST messages to ensure correctness of the solving result; skipping the reinitialization step for DEL and UCO messages will only miss pruning opportunities, and thus losing efficiency.

Agents need to have the context of *all* ancestors to check for context changes and to do reinitialization. In our VALUE, COST, DEL and UCO messages we include the context of the sender agent, instead of the agent's reduced context as used in BnB-ADOPT$^+$, which does not necessarily contain the information of all ancestors.

Fig. 3 shows how to reinitialize, and `Reinitialize`($src,context_{src}$) is called whenever an agent $self$ receives a VALUE, COST, DEL or UCO message from src

```
procedure Reinitialize(src,context_src)
    mindepth = ∞
    for d = d_src → 0 do
        if d >= d_self then continue
        var ← variable of depth d in context_self
        if Time(context_src, var) > Time(context_self, var) then
            context_self(var) ← context_src(var)
            mindepth ← d + 1
    if mindepth ≠ ∞ then
        for d' = mindepth → d_self do
            Copy(d') ← Copy(d' - 1)
            var' ← variable of depth (d' - 1) in context_self
            TransformBinaryToUnary(Copy(d'), var', context_self)

function Timestamp(context, var) return t where (var, a, t) ∈ context

procedure TransformBinaryToUnary(Copy, var, context)
    if self is constrained with var and (var, a, t) ∈ context then
        for each b ∈ D_self do
            Copy.C_self(b) ← Copy.C_self(b) + Copy.C_{self,var}(b, a)
```

Fig. 3. Pseudocode for performing reinitialization

(as shown in Fig. 2 and 3). When $self$ receives $context_{src}$, it first checks whether the variable assignments that src holds are the latest information by comparing the timestamps of each variable assignment in $context_{src}$ and $context_{self}$. If the information in $context_{src}$ is more updated, $self$ updates $context_{self}$ according to $context_{src}$. If $self$'s context is updated, it has to perform reinitialization starting from the class of subproblems $Class(mindepth)$ where $mindepth$ is the smallest depth d_i and agent i's context has been changed in $context_{self}$. The operations for reinitializing $Copy(d)$ are described in Rule 1. We do not reinitialize the subproblem from the original problem but by duplicating from $Copy(d - 1)$. Thus the works done in the current subproblem of $Class(d - 1)$ will not have to be repeated in $Class(d)$.

3.4   Backtracking

Enforcing consistencies in a subproblem P can lead to an empty domain in some agent of P. In this case, $context(P)$ is inconsistent and it should be changed. Upon backtracking, the current assigned value a to the parent, say j, of the root of P should be changed: value a is removed from D_j, and agent j can then pick another value from D_j. This justifies our next rule.

Rule 2. If an agent i obtains an empty domain in the subproblem P during consistency enforcement, the agent $j \in ancestors(P)$ with $d_j = depth(P) - 1$ can delete its value a from its domain in $Copy(d_j)$, where $(j, a) \in context(P)$, provided that $context_j$ is compatible with $context(P)$.

We add a new message BTK to notify backtrackings. When agent i obtains an empty domain in P, i sends a message BTK($i,k,depth(P) - 1, context(P)$) to its parent k. The BTK message is sent to the parent agent for propagation because agents can only communicate with constrained agents but the targeted agent may not be a constrained agent. Therefore, this message is propagated up the pseudo-tree until it reaches agent $j \in ancestors(P)$ where $d_j = depth(P) - 1$.

Fig. 4 shows how to handle an incoming BTK message. When an agent other than j receives a BTK message, it forwards the message to its parent. When j receives that

procedure ProcessBTK(msg)
   if $d_{self} \neq msg.targetDepth$ **then**
      sendMsg:$(BTK, self, parent, msg.targetDepth, msg.context)$
   else
      if $msg.context$ is compatible with $context_{self}$ **then**
         DeleteValue($Copy(msg.targetDepth), a$) where $(self, a, t) \in msg.context$

Fig. 4. Pseudocode for handling the BTK message

message, j checks whether the attached context is compatible with its own context. If yes, it knows that its current assignment $(j, a) \in context(P)$ will not lead to an optimal solution and it deletes a from $Copy(d_j)$. Otherwise, j ignores the message.

3.5 Transferring Deletions to Subproblems

Redundant deletions may appear in embedded subproblems. It is easy to see that if P' is a subproblem of P, the values deleted in P can also be deleted in P'. We can transfer the deletions in P to P' and no need to send out redundant information of these deletions for P'. Transferring deletions to subproblems not just avoid redundant DEL messages, it may also increase the chance of reducing more search space. Since the consistency enforcement in different subproblem is different, the suboptimal values found in P may not be found in P'. If we transfer these suboptimal values from P to P', more pruning opportunities may be found in P'.

When an agent deletes values in subproblem P, $depth(P) = d$, it can also apply the deletions to subproblems P' where $depth(P') > d$. A DEL message is labeled by d. When other agents receive that message, they apply the deletions to all the subproblems P' s.t. $depth(P') \geq d$. The pseudocode of transferring deletions to subproblems when receiving a DEL message is covered in the ProcessDEL() procedure in Fig. 2.

3.6 Keeping Cost Functions Copies Identical

Each of the two agents constrained by a cost function holds a separate copy of the cost function for consistency enforcement. It is thus of paramount importance to ensure the two copies being identical but this task is made difficult by the asynchronous nature of the search algorithm. Fig. 5 gives a simple example of simultaneous deletions [12] in constrained agents i and j, which cause projections from C_{ij} to C_i in agent j and C_j in agent i respectively. The asynchronous nature of message exchanges can result in the projections/extensions performed in different order and thus different C_{ij} copies in agents i and j respectively.

Gutierrez and Meseguer [12] propose to include two new messages to synchronize deletions but the overhead is high. by allowing one of the two agents to undo and reorder the operations. With this *Undo Mechanism* we keep the asynchronicity and avoid extra messages. We give preference to one of the two agents. The operations will be done in the order of the preferred agent, while the non-preferred one must undo the operations that do not follow that order.

Let us consider two constrained agents i and j, and the cost function between them C_{ij}; i and j each holds a copy of it, denoted by C_{ij}^i and C_{ij}^j respectively. Both agents maintain AC. The projection from C_{ij} to C_i has to be done on both C_{ij}^i and C_{ij}^j. $C_{ij}^i \Rightarrow$

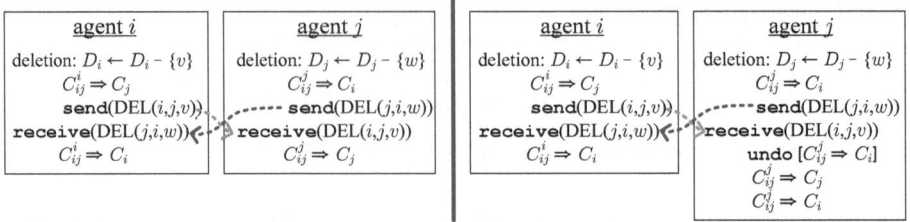

Fig. 5. Left: issue by two simultaneous deletions Right: proposed solution when maintaining AC

C_i represents i performs projection from C_{ij} to C_i on i's copy and $C_{ij}^j \Rightarrow C_i$ represents j performs projection from C_{ij} to C_i on j's copy. If value v is deleted from D_i and value w from D_j simultaneously, both i and j will process these deletions (which imply each agent projecting from C_{ij} to each other) and they will send DEL messages to each other (Fig. 5 Left). If i is the preferred agent, upon receipt of the DEL message from j, it performs $C_{ij}^i \Rightarrow C_i$ and updates C_{ij}^i. However, when j receives the DEL message from i, if j realizes that it has done more projections $C_{ij}^j \Rightarrow C_i$ than the agent i, then it has to undo some of these projections, until both have done the same number of projections. The proposed solution appear in Fig. 5 Right. The same ordering of operations in both agents is achieved as follows. Agent i keeps a counter $ACC_{j\rightarrow i}$ to record the number of projections $C_{ij}^i \Rightarrow C_i$ (and $DACC_{j\rightarrow i}$ to record the number of extensions from agent j to i in FDAC/MFDAC cases). These counter and stack are stored in the copy of each class of subproblems. Agent j keeps a stack $P_{j\rightarrow i}^j$ that records each projection operation $C_{ij}^j \Rightarrow C_i$. The operations of the Undo Mechanism on C_{ij} between agents i and j for AC and MAC are:

Agent i:

- When there is a value deletion, perform projection $C_{ij}^i \Rightarrow C_j$. Attach $ACC_{j\rightarrow i}^i$ in a DEL message and send it to j. Then, reset $ACC_{j\rightarrow i}^i$ to zero.
- When i receives a DEL message from j, perform projection $C_{ij}^i \Rightarrow C_i$ and increment $ACC_{j\rightarrow i}^i$ by 1.

Agent j:

- When there is a value deletion, perform projection $C_{ij}^j \Rightarrow C_i$. Push this projection in the stack $P_{j\rightarrow i}^j$. Send the DEL message to i.
- When j receives a DEL message from i, pop and undo $|P_{j\rightarrow i}^j| - ACC_{j\rightarrow i}^i$ number of projection records from the stack $P_{j\rightarrow i}^j$, where $|P_{j\rightarrow i}^j|$ is the size of the stack $P_{j\rightarrow i}^j$, and clear the stack. Then, the DEL message is processed, projecting $C_{ij}^j \Rightarrow C_j$. If there is at least one pop/undo performed, then perform projection $C_{ij}^j \Rightarrow C_i$.

To maintain FDAC between two constrained agents i and j, DAC is maintained in one direction (e.g. j to i) and AC in the other (e.g. i to j). In FDAC, preference should be given to agent i if AC is enforced from C_{ij} to C_j since the enforcement of DAC from j to i is ensured under the assumption that i is AC w.r.t. C_{ij} [10] (in AC, any agent i or j may be preferred). Due to space limits, we skip the details for FDAC and MFDAC.

3.7   Optimality and Termination

Enforcing MAC and MFDAC during BnB-ADOPT$^+$ search maintains the optimality and termination properties of BnB-ADOPT$^+$, as we see next.

Projections and extensions to maintain MAC and MFDAC are done on a copy of the cost functions. In this way, the search process is based on the unmodified original copy of the cost functions. The only changes with respect to the BnB-ADOPT$^+$ operations come from the fact that inconsistent values discovered by local consistency enforcement are removed from the domain of agents.

Termination is justified as follows. BnB-ADOPT$^+$ always terminates [6,9] and the only change that BnB-ADOPT$^+$-MAC introduces is AC enforcement after variable assignments. AC enfocement terminates, because the number of agents involved is finite and their domains are also finite. When enforcing AC in a particular subproblem, after a finite amount of time all subproblem variables become AC (possibly after some value deletions) reaching a fixpoint.

Optimality is justified as follows. In the case of unconditional deletions, deleted values are suboptimal values which will not be present in the optimal solution, so it is completely legal to remove them. In the case of conditional deletions, deleted values are values proved inconsistent conditioned to the current assignment of ancestor agents. They are properly restored using a reinitialization mechanism when the assignments of ancestors change. Operation is as follows. An agent may change its assigned value, selecting another one from its domain, only after it receives a VALUE or COST message. Reinitialization is done whenever an agent receives a VALUE or COST message and there is context change. Thus, reinitialization is guaranteed to be performed before any agent changes its value, so that no obsolete value deletions will be considered. Then, in both cases all solutions potentially optimal are visited. Next we detail these operations, showing they do not affect optimality and termination.

In MAC (both unconditional and conditional deletions), we perform projections over the cost functions (projections from binary to unary cost functions, and from unary to C_ϕ). Projection is an equivalence preserving transformation [11]. Its application maintains the optimum cost and the set of optimal solutions. In our approach (distributed context), we assure identical copies of any binary cost function in the two involved agents: cost projections are performed in the same order in the two agents (Section 3.6). Therefore, costs cannot be duplicated when projections are performed inside each agent (equivalence is preserved) or when costs are propagated to other agents. Since each agent contributes to C_ϕ projecting on its unary cost functions, we can conclude that projections of different agents into C_ϕ does not duplicate costs. Proving that a value a of variable x_i is not NC involves its unary cost $C_i(a)$ and C_ϕ. Since we have seen that, neither $C_i(a)$ nor C_ϕ contains duplicated costs, the NC detection is correct and a's deletion is legal. Because of the NC definition, the first found optimal solution can never be pruned, since the cost of their values will never reach $\top$.

In the case of conditional deletions, the reinitialization mechanism (Section 3.3) ensures the correctness of values deletions in different copies. For each copy, projections and deletions are performed conditioned to the ancestor assignments. For example, in $Copy(0)$, projections are performed contemplating no previous assignments, and only unconditional deletions are detected; in $Copy(1)$, inconsistent values are discovered

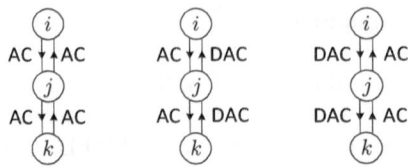

Fig. 6. Directions of enforcing AC/FDAC consistencies

and deleted conditioned to the first-level ancestor's assignment; in $Copy(2)$, inconsistent values are deleted conditioned to the first and second level agent's assignment, and so on. Each time an agent of depth d changes its variable assignment, the $Copy(d)$ of descendant agents are restored to $Copy(d-1)$. In this way, all modifications that were performed according to the previous variable assignment are undone.

Regarding values pruned by backtracking messages, the justification of its correctness is as follows. When an empty domain is found in $Copy(d)$ in one agent, we have discovered that the current assignment of the ancestor at depth d is inconsistent, and so it must be removed. This is implemented by sending a BTK message to that ancestor. Note that only BTK messages containing a compatible context are accepted in the ancestor. In this way, it is assured that the ancestor agent changes its value if the empty domain of the descendant agent was generated considering a compatible context. Otherwise, either the descendant or the ancestor is missing one or several messages that will properly update their contexts. Upon receipt of these messages, proper actions, depending on the missing messages, will be taken by the ancestor/descendant.

Regarding MFDAC, in addition to projections, we have to take into account extensions, another equivalence preserving transformation [10]. Our approach (distributed case) is correct, since each agent can extend its own unary costs only. So no cost duplication may occur. The process is done in such a way that the copies of any binary cost function are kept identical in the two involved agents. From this point on, only projections are done, and arguments from previous paragraphs apply.

4   Experimental Results

We evaluate the efficiency of BnB-ADOPT$^+$-MAC/MFDAC (abbrev as MAC/MFDAC) by comparing to BnB-ADOPT$^+$-AC/FDAC (abbrev as AC/FDAC). For AC and MAC algorithms, AC is enforced in both directions of each binary cost function. The direction of DAC enforcement matters in FDAC and MFDAC algorithms. Fig. 6 shows the direction of AC and DAC enforcement between agents, where i (j) is the parent or pseudo-parent of j (k). For FDAC algorithm, we use the direction as shown in Fig. 6 Middle. DAC is enforced bottom-up so that the unary costs are pushed upward so as to hopefully increase the opportunities of pruning more values in upper agents (pruning values in upper agents is more preferred because BnB-ADOPT$^+$ is a depth-first search algorithm). For MFDAC, we evaluate both directions: MFDAC1 uses the direction shown in Fig. 6 Middle and MFDAC2 uses the direction in Fig. 6 Right. We evaluate both because of the possible tradeoff between backtracking and direct pruning in upper agents. With MFDAC1, unary costs will float upward and increase the opportunities of pruning values directly in upper agents. However, MFDAC2 pushes the unary costs downward and increases the opportunities of reaching empty domains in lower agents, which can possibly increase the pruning opportunities in upper agents.

Our simulator runs in cycles, during which every agent reads its incoming messages, performs computation and sends its outcoming messages. Without delays, a message sent in a cycle is delivered in the next cycle. To make a more realistic evaluation, a random delay of $[0, 50]$ cycles is introduced to each message in our experiments. Besides, we have an extensive number of instances over three benchmarks. Since AC is too slow to generate results for hard or large-scale problems within a reasonable time, we set a 2×10^8 NCCCs limit in our simulator. One can expect that setting the NCCCs limit is to our *disadvantage* since MAC/MFDAC can improve even more on harder or larger-scale problems (normally taking bigger effort to solve but some of these problems are skipped because of the NCCCs limit). Three measures of performance are thus compared: (1) the number of messages to evaluate the communication cost, (2) the number of non-concurrent constraint checks (NCCCs) to evaluate the computation effort, and (3) the number of instances that can be solved within the 2×10^8 NCCCs limit to evaluate the general efficiency of each algorithm. In addition, we assume that each randomly delayed cycle costs 100 NCCCs and it is counted in the total NCCCs accordingly.

We test our algorithms on three sets of benchmarks: binary random DCOPs [8], Soft Graph Coloring Problems (SGCP) and Radio Link Frequency Assignment Problem (RLFAP) [13]. We run 50 instances for each parameter setting. Results are reported in Tables 2, 3 and 4. The columns show (from left to right) the problem, algorithm, the number of instances that can be solved within limit, the number of commonly solved instances (the number of messages and NCCCs are averaged over this number), total number of messages, number of VALUE, COST, DEL, BTK and UCO messages, and NCCCs. The best results for each measure are highlighted in bold.

Binary random DCOPs [8] are characterized by $\langle n, d, p \rangle$, where n is the number of variables, d is the domain size and p is the network connectivity. We have generated random DCOP instances: $\langle n = 10, d = 10, p \in \{0.3, 0.4, 0.5, 0.6\} \rangle$. Costs are selected from a uniform cost distribution. Following Guiterrez and Meseguer [8], two types of binary cost functions are used, small and large. Small cost functions randomly extract costs from the set $\{0, ..., 10\}$ while large ones randomly extract costs from the set $\{0, ..., 1000\}$. The proportion of large cost functions is 1/4 of the total number of cost functions. Results are reported in Table 2.

Soft Graph Coloring Problems are the softened version of graph coloring problems by allowing the inequalities to return costs from the violation measure $M^2 - |v_i - v_j|^2$, where M is the maximum domain size, v_i and v_j are the values of agent i and j respectively. Each SGCP is also characterized by $\langle n, d, p \rangle$, where n is the number of variables, d is the domain size and p is the network connectivity. We evaluate four sets of instances: $\langle n \in \{6, 7, 8, 9\}, d = 8, p = 0.4 \rangle$. Results are shown in Table 3.

We generate the Radio Link Frequency Assignment Problems according to two small but hard CELAR sub-instances [13], which are extracted from CELAR6. All instances are generated with parameters $\langle i, n, d \rangle$, where i is the index of the CELAR sub-instances, n is an even number of links, and d is an even number of allowed frequencies. For each instance, we randomly extract a sequence of n links from the corresponding CELAR sub-instance and fix a domain of d frequencies. If two links are restricted not to take

Table 2. Random DCOPs

p	Algorithm	#instances solved within NCCCs limit	Avg. over (common instances)	#Msgs	#VALUE	#COST	#DEL	#BTK	#UCO	NCCCs
	AC	50		6,802	1,619	5,099	59	0	0	5,622,762
	FDAC	50		4,645	1,062	3,389	117	0	53	3,857,078
0.3	MAC	50	50	5,610	1,124	3,569	760	134	0	4,203,119
	MFDAC1	50		**3,656**	726	2,346	338	13	184	**2,738,511**
	MFDAC2	50		5,036	923	2,911	495	249	435	3,511,191
	AC	47		56,632	11,581	44,946	79	0	0	42,210,453
	FDAC	48		39,560	8,043	31,188	195	0	105	29,477,148
0.4	MAC	50	47	36,309	6,692	25,564	2,399	1,628	0	24,845,040
	MFDAC1	50		**28,493**	5,271	20,541	1,430	236	967	**19,236,541**
	MFDAC2	50		29,814	5,116	20,255	1,523	1,441	1,451	19,434,413
	AC	35		106,194	20,796	85,260	106	0	0	78,603,224
	FDAC	38		75,074	14,412	60,231	247	0	152	55,129,851
0.5	MAC	43	34	63,571	11,238	46,279	2,694	3,329	0	43,949,687
	MFDAC1	44		**54,564**	9,490	39,791	2,926	286	2,018	**36,699,194**
	MFDAC2	**46**		57,150	9,497	39,535	2,245	3,651	2,191	37,488,828
	AC	9		124,222	26,839	97,268	86	0	0	91,145,921
	FDAC	16		90,850	14,867	55,465	277	0	211	51,437,525
0.6	MAC	20	9	47,586	8,973	35,153	2,143	1,288	0	34,059,166
	MFDAC1	**24**		**37,697**	6,883	27,900	1,141	463	1,255	**27,122,566**
	MFDAC2	20		45,988	8,093	31,699	2,011	2,047	2,109	31,074,814

Table 3. Soft Graph Coloring Problems

n	Algorithm	#instances solved within NCCCs limit	Avg. over (common instances)	#Msgs	#VALUE	#COST	#DEL	#BTK	#UCO	NCCCs
	AC	50		459	123	321	8	0	0	572,082
	FDAC	50		376	91	240	29	0	7	438,002
6	MAC	50	50	358	81	190	67	11	0	361,607
	MFDAC1	50		**287**	51	127	54	9	30	**248,106**
	MFDAC2	50		367	71	161	70	17	40	333,807
	AC	50		1,349	370	961	9	0	0	1,534,451
	FDAC	50		875	225	594	37	0	8	974,678
7	MAC	50	50	888	213	507	143	14	0	841,000
	MFDAC1	50		**628**	127	314	95	20	51	**521,659**
	MFDAC2	50		883	185	437	143	27	81	733,084
	AC	50		8,611	2,072	6,523	5	0	0	8,562,373
	FDAC	50		5,764	1,359	4,354	29	0	11	5,727,394
8	MAC	50	50	4,955	1,044	3,166	625	109	0	4,261,463
	MFDAC1	50		**4,359**	905	2,942	287	61	138	3,799,575
	MFDAC2	50		4,695	857	2,615	613	163	437	**3,553,383**
	AC	46		39,199	8,659	30,525	3	0	0	32,353,604
	FDAC	46		30,189	6,580	23,559	23	0	14	24,858,245
9	MAC	47	46	23,164	4,554	15,882	2,545	170	0	17,448,119
	MFDAC1	47		25,738	5,265	19,124	795	69	453	19,829,021
	MFDAC2	47		**20,219**	3,547	12,624	2,081	493	1,461	**13,863,427**

frequencies f_i and f_j with distance less than t, we measure the costs of interference using a binary cost function $\max(0, t - |f_i - f_j|)$. Results of evaluating three sets of instances, $A\langle 0, 10, 12\rangle$, $B\langle 1, 6, 6\rangle$, and $C\langle 1, 6, 8\rangle$, are reported in Table 4.

As we see in Tables 2, 3 and 4, MAC, MFDAC1 and MFDAC2 substantially further reduce the total number of messages and NCCCs, and be able to solve the same number or more instances within the NCCCs limit over all three benchmarks. Moreover, MAC outperforms FDAC in almost all cases even when MAC is maintaining a weaker form of consistency than FDAC. Although our methods introduce overhead, i.e., increase in the number of DEL, BTK and UCO messages, the reduction in the number of VALUE

Table 4. Radio Link Frequency Assignment Problems

	Algorithm	#instances solved within NCCCs limit	Avg. over (common instances)	#Msgs	#VALUE	#COST	#DEL	#BTK	#UCO	NCCCs
A	AC	50		28,837	5,064	23,751	0	0	0	24,522,945
	FDAC	50		28,894	5,069	23,790	0	0	13	24,621,897
	MAC	50	50	22,840	,3802	16,540	1,954	522	0	17,447,513
	MFDAC1	50		**18,054**	2,937	12,000	1,606	378	1,090	13,051,121
	MFDAC2	50		19,233	2,888	11,711	1,861	1,250	1,501	**12,845,773**
B	AC	21		56,943	10,466	46,455	11	0	0	67,658,716
	FDAC	21		57,964	10,635	47,267	39	0	9	69,091,598
	MAC	50	21	29,120	4,930	21,521	1,061	1,596	0	37,861,737
	MFDAC1	50		**18,080**	3,228	13,900	403	433	100	**20,881,354**
	MFDAC2	50		25,430	4,490	19,489	541	702	197	2,937,7041
C	AC	18		29,385	5,505	23,853	0	0	0	34,158,516
	FDAC	18		31,133	5,814	25,250	47	0	9	36,259,302
	MAC	50	18	13,914	2,464	10,787	297	356	0	15,890,040
	MFDAC1	50		**11,964**	2,183	9,394	177	123	71	**14,067,760**
	MFDAC2	50		13,454	2,431	10,496	220	207	89	15,731,062

and COST messages (and thus search space) outweighs the overhead. Therefore, we conclude that maintaining soft arc consistencies during search is beneficial.

We also observe that the improvement of MFDAC over AC and FDAC in random DCOPs increases as constraint density increases. More constraints in the problem implies more pruning opportunities and thus substantial smaller search space. Similar observations cannot be concluded for Soft Graph Coloring and Radio Link Frequency Assignment Problems since these problems have particular problem structures affecting the efficiency and power of consistency enforcement.

To compare the different directions of DAC enforcement, we can see MFDAC1 outperforms MFDAC2 in some instances while MFDAC2 outperforms MFDAC1 in others. For random DCOPs and Radio Link Frequency Assignment Problem, MFDAC1 performs the best in almost all instances. However, for Soft Graph Coloring Problem, MFDAC2 performs better for instances with $n = 9$ and MFDAC1 performs better on another three sets of instances. From these results we can see that the directions of DAC enforcement can affect the efficiency and the effects are problem-specific.

5   Conclusion

In this paper, we propose methods to maintaining soft arc consistencies in every subproblem during search. In order to preserve the asynchronicities of search and consistency enforcement, we propose to include extra copies (a small number) of variable domains and cost functions. Besides, we minimize the induced overhead caused by backtracking and undoing assignments and deletions by attaching information in the existing messages rather than creating new ones. We present the issues and solutions for maintaining consistencies in subproblems and ensure their correctness: (i) reinitializing variables' domains and cost functions after context changes in subproblems to ensure the search algorithm would not search on values using obsolete value pruning information, (ii) backtracking when an agent arrives at the empty domain within a subproblem so as to prune the value in upper agents which could not lead to an optimal solution, (iii) transferring deletions from subproblems to further subproblems to avoid redundant messages, and (iv) asynchronous methods to ensure identical cost functions copies in different agents by ensuring the ordering of consistency operations between every two agents. Our experimental results show that our methods can substantially further reduce the communication and computation efforts compared to BnB-ADOPT$^+$-AC/FDAC, which only consider unconditional deletions. These results allow us to consider the proposed methods as important steps to maintain consistencies in every subproblems asynchronously during search and improve the efficiency of optimal DCOP solving. As a future work, we may go further to maintain the even stronger Existential Directional Arc Consistency (EDAC) [14] during distributed and asynchronous search, but preserving privacy is a concern [15]. The study of how DAC enforcement directions affect efficiency and the possible heuristics for such ordering is a worthwhile direction.

References

1. Maheswaran, R.T., Tambe, M., Bowring, E., Pearce, J.P., Varakantham, P.: Taking DCOP to the real world: Efficient complete solutions for distributed multi-event scheduling. In: Proc. AAMAS 2004, pp. 310–317 (2004)
2. Jain, M., Taylor, M., Tambe, M., Yokoo, M.: DCOPs meet the realworld: exploring unknown reward matrices with applications to mobile sensor networks. In: Proc. IJCAI 2009, pp. 181–186 (2009)
3. Junges, R., Bazzan, A.L.C.: Evaluating the performance of DCOP algorithms in a real world, dynamic problem. In: Proc. AAMAS 2008, pp. 599–606 (2008)
4. Modi, P.J., Shen, W.M., Tambe, M., Yokoo, M.: ADOPT: Asynchronous distributed constraint optimization with quality guarantees. Artificial Intelligence 161, 149–180 (2005)
5. Petcu, A., Faltings, B.: A scalable method for multiagent constraint optimization. In: Proc. IJCAI 2005, pp. 266–271 (2005)
6. Yeoh, W., Felner, A., Koenig, S.: BnB-ADOPT: an asynchronous branch-and-bound DCOP algorithm. JAIR 38, 85–133 (2010)
7. Chechetka, A., Sycara, K.: No-commitment branch and bound search for distributed constraint optimization. In: Proc. AAMAS 2006, pp. 1427–1429 (2006)
8. Gutierrez, P., Meseguer, P.: BnB-ADOPT$^+$ with several soft arc consistency levels. In: Proc. ECAI 2010, pp. 67–72 (2010)
9. Gutierrez, P., Meseguer, P.: Saving redundant messages in BnB-ADOPT. In: Proc. AAAI 2010, pp. 1259–1260 (2010)
10. Larrosa, J., Schiex, T.: In the quest of the best form of local consistency for weighted CSP. In: Proc. IJCAI 2003, pp. 239–244 (2003)
11. Larrosa, J., Schiex, T.: Solving weighted csp by maintaining arc consistency. Artificial Intelligence 159(1), 1–26 (2004)
12. Gutierrez, P., Meseguer, P.: Improving BnB-ADOPT$^+$-AC. In: Proc. AAMAS 2012, pp. 273–280 (2012)
13. Cabon, B., Givry, S.D., Lobjois, L., Fcabon, L.L., Schiex, T.: Radio link frequency assignment. Constraints 4, 79–89 (1999)
14. de Givry, S., Heras, F., Larrosa, J., Zytnicki, M.: Existential arc consistency: getting closer to full arc consistency in weighted CSPs. In: Proc. IJCAI 2005, pp. 84–89 (2005)
15. Gutierrez, P., Meseguer, P.: Enforcing soft local consistency on multiple representations for DCOP solving. In: CP 2010, Workshop: Preferences and Soft Constraints, pp. 98–113 (2010)

Solving String Constraints:
The Case for Constraint Programming

Jun He[1,2], Pierre Flener[1], Justin Pearson[1], and Wei Ming Zhang[2]

[1] Uppsala University, Department of Information Technology, Uppsala, Sweden
[2] National University of Defense Technology,
School of Information System and Management, Changsha, Hunan, China
{Jun.He,Pierre.Flener,Justin.Pearson}@it.uu.se, wmzhang@nudt.edu.cn

Abstract. We improve an existing propagator for the context-free grammar constraint and demonstrate experimentally the practicality of the resulting propagator. The underlying technique could be applied to other existing propagators for this constraint. We argue that constraint programming solvers are more suitable than existing solvers for verification tools that have to solve string constraints, as they have a rich tradition of constraints for membership in formal languages.

1 Introduction

For constraint programming (CP) languages, user-level extensibility has been an important goal for over a decade. Global constraints for formal languages are promising for this purpose. The REGULAR constraint [16] requires a sequence of decision variables to belong to a regular language, specified by a deterministic finite automaton (DFA) or a regular expression; the AUTOMATON constraint [2] takes a DFA with counters. The CFG constraint [17,20] requires a sequence of decision variables to belong to a context-free language, specified by a context-free grammar (CFG). For many applications, the length n of a sequence constrained to belong to some formal language is known in advance. Since every fixed-size language is finite and hence regular, the need for a CFG constraint in such applications depends on the grammar and the complexities of the propagators. It takes $O(n\,|A|)$ time to achieve generalised arc consistency (GAC) for a REGULAR constraint with an automaton A, but $O(n^3\,|G|)$ time for a CFG constraint with a grammar G. In [12], the authors introduce a reformulation of a grammar into an automaton for a fixed length n, and show that this reformulation is preferable if the resulting automaton is not huge. However, their reformulation itself needs a CFG propagator to achieve domain consistency at the root of the search tree so that the resulting automaton is smaller. In [7], the authors introduce a forklift scheduling problem, where there is no tractable reformulation of a grammar into an automaton as the size of the resulting automaton is exponential in n. Hence, a CFG propagator is necessary in this case. To the best of our knowledge, no CP solver includes the CFG constraint.

In the analysis, testing, and verification of string-manipulating programs, constraints on sequences (strings) of decision variables arise. Kieżun et al. [14] argue

C. Schulte (Ed.): CP 2013, LNCS 8124, pp. 381–397, 2013.

that custom string solvers should not be designed any more, for sustainability reasons, since powerful off-the-shelf solvers are available: their tool, HAMPI, translates a REGULAR or CFG constraint on a fixed-size string into bit-vector constraints so as to solve them using the SMT solver STP [6], much more efficiently than three custom tools and even up to three orders of magnitude faster than the SAT-based CFGAnalyzer tool [1]. The solver KALUZA [19] handles constraints over multiple string variables, unlike the restriction of HAMPI to one such variable, and it also generates bit-vector constraints that are passed to STP. Fu *et al.* [5] argue that it is important to model regular replacement operations, which are not supported by HAMPI and KALUZA, and introduce the custom string solver SUSHI, which models string constraints via automata instead of a bit-vector encoding. So the question arises whether the formal language constraints of CP are competitive with HAMPI, KALUZA, and SUSHI.

In this paper, we revisit the CFG constraint and make the following *contributions*:

- We improve the CFG propagator of [11], which improves the one of [20], by exploiting an idea of [14] for reformulating a grammar into a regular expression for a fixed string length. We conjecture that this idea also applies to the CFG propagators of [7,13,17,18]. (Section 3)
- We implement our CFG propagator for the GECODE [8] open-source CP solver, and demonstrate experimentally its practicality. (Sections 4.1 to 4.3)
- We show that the CP solver GECODE with our CFG propagator (or even its ancestor [11]) systematically beats HAMPI and KALUZA, by up to four orders of magnitude, on HAMPI's benchmark (Section 4.3). We show that GECODE with the built-in REGULAR propagator systematically beats KALUZA and SUSHI, by a factor up to 130, on SUSHI's benchmark (Section 4.4).

2   Background

We first give some background material on grammars (e.g., see [10]).

2.1   Context-Free Grammars

A *CFG* is a tuple $\langle \Sigma, N, P, S \rangle$, where Σ is the alphabet and any value $v \in \Sigma$ is called a terminal, N is the finite set of non-terminals, $P \subseteq N \times (\Sigma \cup N)^*$ is the finite set of productions, and $S \in N$ is the start non-terminal. A CFG is said to be in *Chomsky normal form* (CNF) iff $P \subseteq N \times (\Sigma \cup N^2)$. Every CFG can be converted into an equivalent grammar in CNF.

Example 1. Consider the CFG $G_B = \langle \Sigma, N, P, S \rangle$, where $\Sigma = \{\ell, r\}$, $N = \{S\}$, and $P = \{S \rightarrow \ell r,\ S \rightarrow SS,\ S \rightarrow \ell Sr\}$. It defines a language of correctly bracketed expressions (e.g., $\ell r \ell r$ and $\ell \ell r r$), with 'ℓ' denoting the left bracket and 'r' the right one. Its CNF is $G'_B = \langle \Sigma, N', P', S \rangle$, where $N' = \{L, M, R, S\}$ and $P' = \{S \rightarrow LR,\ S \rightarrow SS,\ S \rightarrow MR,\ M \rightarrow LS,\ L \rightarrow \ell,\ R \rightarrow r\}$.

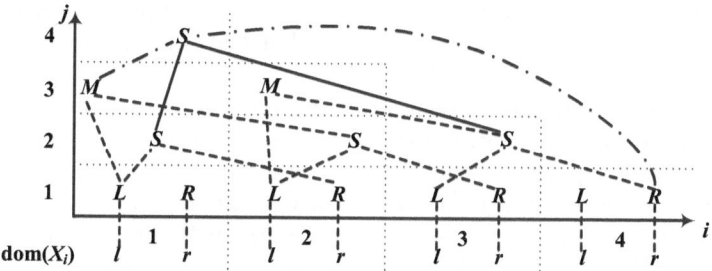

Fig. 1. The CYK-based propagator parses a sequence $\langle X_1, \ldots, X_4 \rangle$ of $n = 4$ decision variables with the same domain $\{\ell, r\}$ under the CFG G'_B of Example 1

The Cocke-Younger-Kasami (CYK) algorithm is a parser for CFGs in CNF. We describe it for a sequence of decision variables instead of values. Given a CFG $\langle \Sigma, N, P, S \rangle$ in CNF and a sequence $\langle X_1, \ldots, X_n \rangle$ of n decision variables, the CYK parser computes a table V, where $V_{i,j}$ (with $1 \leq j \leq n$ and $1 \leq i \leq n+1-j$) is the set of non-terminals (or at most the start non-terminal S for $i = 1$ and $j = n$) that can be parsed using a sequence of j values in the domains of X_i to X_{i+j-1} respectively, using dynamic programming:

$$
V_{i,j} = \begin{cases} \{W \mid (W \to b) \in P \;\wedge\; b \in \mathrm{dom}\,(X_i)\} & \text{if } j = 1 \\[2mm] \displaystyle\bigcup_{k=1}^{j-1} \left\{ W \;\middle|\; \begin{array}{l} (W \to YZ) \in P \;\wedge\; (j < n \;\vee\; W = S) \\ \wedge\; Y \in V_{i,k} \;\wedge\; Z \in V_{i+k,j-k} \end{array} \right\} & \text{otherwise} \end{cases}
$$

For example, Figure 1 gives the CYK table V when parsing a sequence X of 4 decision variables with the same domain $\{\ell, r\}$ under the grammar G'_B of Example 1. We have $V_{1,1} = \{L, R\}$ and $V_{1,4} = \{S\}$. Note that we use $\mathrm{dom}\,(X_i)$ to denote the domain of the decision variable X_i.

Given a word $w \in \Sigma^n$, let w_i (with $1 \leq i \leq n$) denote the letter at position i of w. If all decision variables X_i have $\mathrm{dom}\,(X_i) = \{w_i\}$, then w is *accepted* by G iff $V_{1,n} = \{S\}$.

2.2   The CFG Constraint

The CFG constraint is defined as $\mathrm{CFG}(X, G)$, where X is a sequence of decision variables and G is a grammar. An assignment w to X is a solution iff w is a word accepted by G.

Given a CFG $G = \langle \Sigma, N, P, S \rangle$ in CNF and a sequence X of n variables, let $|G| = \sum_{p \in P} |p|$ be the size of G, and $|p|$ the number of (non-)terminals in the production p. The propagator of [11] achieves GAC for the $\mathrm{CFG}(X, G)$ constraint in $O\left(n^3 \,|G|\right)$ time with $O\left(n^2 \,|G|\right)$ space, which is better than the propositional satisfiability (SAT) based propagator of [18], which decomposes and achieves GAC for the CFG constraint in $O\left(n^3 \,|G|\right)$ time and space. More recently, another SAT-based propagator is introduced in [7], which works similarly to the propagator of [11] and outperforms the propagator of [18].

In this paper, we use the propagator of [11] *as an example* to show how to improve a CFG propagator. We conjecture that the same idea can be used to improve the propagators of [7,13,17,18].

To describe elegantly the propagator of [11] and ours (given in Section 3), we first introduce a novel concept. Informally, given a non-terminal W in $V_{i,j}$ of the CYK table, a *low support* for this W, namely $(W \to YZ, k)$, denotes that two non-terminals lower down in V, namely Y in $V_{i,k}$ and Z in $V_{i+k,j-k}$, support the existence of W in $V_{i,j}$; and this low support corresponds to two *high supports*, namely $(W \to YZ, j)$ of Y in $V_{i,k}$ and Z in $V_{i+k,j-k}$. Formally:

Definition 1 (Support). *For any $1 < j \le n$, $1 \le i \le n + 1 - j$, and non-terminal W in $V_{i,j}$ of the CYK table, the set $\overline{\mathrm{LS}}_{i,j}(W) = \{(W \to YZ, k) \mid (W \to YZ) \in P \wedge 0 < k < j\}$ is called the* candidate low-support set *for W in $V_{i,j}$. The set $\mathrm{LS}_{i,j}(W) = \{(W \to YZ, k) \in \overline{\mathrm{LS}}_{i,j}(W) \mid Y \in V_{i,k} \wedge Z \in V_{i+k,j-k}\}$ is called the* low-support set *for W in $V_{i,j}$. For $j = 1$ and any $1 \le i \le n$ and non-terminal W in $V_{i,1}$, we define $\overline{\mathrm{LS}}_{i,1}(W) = \{(W \to b) \in P\}$ and $\mathrm{LS}_{i,1}(W) = \{(W \to b) \in \overline{\mathrm{LS}}_{i,1}(W) \mid b \in \mathrm{dom}(X_i)\}$.*

For any $1 \le j < n$, $1 \le i \le n + 1 - j$, and non-terminal W in $V_{i,j}$ of the CYK table, the set $\overline{\mathrm{HS}}_{i,j}(W) = \{(Y \to QZ, k) \mid (Y \to QZ) \in P \wedge (W = Q \vee W = Z) \wedge j < k \le n\}$ is called the candidate high-support set *of W in $V_{i,j}$. The set $\mathrm{HS}_{i,j}(W) = \{(Y \to QZ, k) \in \overline{\mathrm{HS}}_{i,j}(W) \mid (W = Q \wedge Y \in V_{i,k} \wedge Z \in V_{i+j,k-j}) \vee (W = Z \wedge Y \in V_{i-j,k} \wedge Q \in V_{i-j,k-j})\}$ is called the* high-support set *of W in $V_{i,j}$. For any $1 \le i \le n$ and value b in $\mathrm{dom}(X_i)$, we define $\overline{\mathrm{HS}}_i(b) = \{(W \to b) \in P\}$ and $\mathrm{HS}_i(b) = \{(W \to b) \in \overline{\mathrm{HS}}_i(b) \mid b \in \mathrm{dom}(X_i)\}$.* □

For example, in the CYK table V of Figure 1, $\overline{\mathrm{LS}}_{1,4}(S) = \{S \to LR, S \to SS, S \to MR\} \times \{1, 2, 3\}$ has 9 candidate low supports; only 2 thereof are low supports for non-terminal S in $V_{1,4}$, namely $(S \to SS, 2)$ (depicted by the solid arcs), and $(S \to MR, 3)$ (depicted by the dash-dotted arcs). The low support $(S \to SS, 2)$ for S in $V_{1,4}$ denotes that it is supported by S in $V_{1,2}$ and $V_{3,2}$, hence the low support corresponds to 2 high supports, namely $(S \to SS, 4)$ of S in $V_{1,2}$ and $V_{3,2}$.

The propagator of [11] achieves GAC for the $\mathrm{CFG}(X, G)$ constraint as follows: (1) The CYK parser computes the table V. (2) A bottom-up process finds the *first* low support in every $\overline{\mathrm{LS}}_{i,j}(W)$. A top-down process finds the *first* high support in every $\overline{\mathrm{HS}}_{i,j}(W)$. All non-terminals W with no support are removed from V. (3) The *first* high support in every $\overline{\mathrm{HS}}_i(b)$ is found, and all values b in any $\mathrm{dom}(X_i)$ with no high support are removed from $\mathrm{dom}(X_i)$. When a support is found in steps 2 and 3, its position in the candidate support set is recorded. When a support is lost as the domains shrink, the next support is to be found starting *after* the previous support in the candidate support set. The propagator is incremental, and explores all candidate supports at most once.

3   An Improved Propagator

Inspired by [14], we present, verify, and analyse an improved version of the propagator of [11] for the CFG constraint.

3.1 Motivation and Theoretical Foundation

There are two dependent opportunities for improving the propagator of [11].

Encoding the Support Sets Space-Efficiently. The propagator of [11] explores *all* candidate supports once in the worst case, hence its time complexity is bounded by

$$|\overline{\text{LS}}| + |\overline{\text{HS}}| = \sum_{j=1}^{n} \sum_{i=1}^{n+1-j} \sum_{W \in V_{i,j}} |\overline{\text{LS}}_{i,j}(W)| + |\overline{\text{HS}}_{i,j}(W)| = O\left(n^3 |G|\right). \text{ If we can}$$

make the propagator run on the small support sets instead of the large candidate support sets, then the propagator probably runs faster. Consider that $\overline{\text{LS}}_{i,j}(W) \supseteq \text{LS}_{i,j}(W)$ and $\overline{\text{HS}}_{i,j}(W) \supseteq \text{HS}_{i,j}(W)$ (from Definition 1), and that the gaps may be huge. For example in Figure 1, $\text{LS}_{1,4}(S) = \{(S \rightarrow SS, 2), (S \rightarrow MR, 3)\}$ is of size 2, while $\overline{\text{LS}}_{1,4}(S) = \{S \rightarrow LR, S \rightarrow SS, S \rightarrow MR\} \times \{1,2,3\}$ is of size 9; $\text{HS}_{2,1}(R) = \{(S \rightarrow LR, 2)\}$ is of size 1, while $\overline{\text{HS}}_{2,1}(R) = \{S \rightarrow LR, S \rightarrow MR\} \times \{2,3,4\}$ is of size 6. However, the challenge is to avoid having to pay with space what we save in time.

Given a CFG $G = \langle \Sigma, N, P, S \rangle$ in CNF and n decision variables, Kadıoğlu and Sellmann [11] claim that storing all support sets takes $O\left(n^3 |G|\right)$ space, which is expensive. Their propagator thus runs on the large *candidate* support sets, which can be encoded very space-efficiently. Two sets $\text{Out}(W) = \{(W \rightarrow YZ) \in P\}$ and $\text{In}(W) = \{(Y \rightarrow QZ) \in P \mid W = Q \vee W = Z\}$ are computed for any $W \in N$, so that $\overline{\text{LS}}_{i,j}(W) = \text{Out}(W) \times \{1, \ldots, j-1\}$ and $\overline{\text{HS}}_{i,j}(W) = \text{In}(W) \times \{j+1, \ldots, n\}$. For any j, the sets $\{1, \ldots, j-1\}$ and $\{j+1, \ldots, n\}$ need not be stored. Hence encoding all candidate support sets only takes $O(|G|)$ space by storing all $\text{Out}(W)$ and $\text{In}(W)$. As it takes $O\left(n^2 |G|\right)$ space to store the CYK table V, the overall space complexity is $O\left(n^2 |G|\right)$.

However, we *can* decrease the space requirement for encoding all low-support sets and a superset of all high-support sets (given in Theorem 2 below) from $O\left(n^3 |G|\right)$ to $O\left(n^2 |G|\right)$, which is the *same* as the one needed to store the CYK table V, by using an idea of [14] for reformulating a grammar into a regular expression for a fixed string length n. In that reformulation, a regular expression is obtained by using the *same* domains: $\text{dom}(X_i) = \Sigma$ for *all* $1 \leq i \leq n$. A regular expression $E_{1,j}$ for the sub-sequence $\langle X_1, \ldots, X_j \rangle$ is computed and stored as a template for every $1 \leq j \leq n$, and then the regular expression $E_{i,j}$ for the sub-sequence $\langle X_i, \ldots, X_{i+j-1} \rangle$ turns out to be equal to $E_{1,j}$ for every $1 < i \leq n+1-j$. Similarly, in Figure 1, we find that $V_{i,j} = V_{1,j}$ and every nonterminal in $V_{i,j}$ has the the same low supports as in $V_{1,j}$. For example, $V_{3,2} = V_{2,2} = V_{1,2} = \{S\}$ and $\text{LS}_{3,2}(S) = \text{LS}_{2,2}(S) = \text{LS}_{1,2}(S) = \{(S \rightarrow LR, 1)\}$. Based on this observation, we give the following theorem (we show in Section 3.2 how to lift the same-domain restriction):

Theorem 1. *Given a CFG $G = \langle \Sigma, N, P, S \rangle$ in CNF and a sequence $\langle X_1, \ldots, X_n \rangle$ of n decision variables, if all X_i have the same domain, then for any $1 \leq j \leq n$ and $1 < i \leq n+1-j$:*

1. $V_{i,j} = V_{1,j}$
2. $\forall W \in V_{i,j} :\ \mathrm{LS}_{i,j}(W) = \mathrm{LS}_{1,j}(W)$

Proof: We prove claim 1 by complete induction on j.

(Base: $j = 1$) For any non-terminal W, we have $W \in V_{i,1}$ iff there exists a production $(W \to b) \in P$ such that $b \in \mathrm{dom}(X_i)$. As $\mathrm{dom}(X_i) = \mathrm{dom}(X_1)$, we have $W \in V_{i,1}$ iff $W \in V_{1,1}$.

(Step: $1 < j \leq n$) For any $1 \leq j' < j$, the induction hypothesis is $V_{i,j'} = V_{1,j'}$ for any $1 < i$. We want to prove $V_{i,j} = V_{1,j}$ for any $1 < i \leq n + 1 - j$. For any non-terminal W, we have $W \in V_{i,j}$ iff there exists a production $(W \to YZ) \in P$ and $1 \leq k < j$ such that $Y \in V_{i,k}$ and $Z \in V_{i+k,j-k}$. As $V_{i,k} = V_{1,k}$ and $V_{i+k,j-k} = V_{1,j-k} = V_{1+k,j-k}$, we have $W \in V_{i,j}$ iff $W \in V_{1,j}$.

Using this, claim 2 follows from Definition 1. $\qquad\square$

The next theorem enables a space-efficient encoding of the support sets (again, we show in Section 3.2 how to lift the same-domain restriction).

Theorem 2. *Given a CFG $G = \langle \Sigma, N, P, S \rangle$ in CNF and a sequence $\langle X_1, \ldots, X_n \rangle$ of n decision variables, if all X_i have the same domain, then it takes $O(n^2 |G|)$ space to encode the CYK table V and all support sets.*

Proof: For any $1 \leq j \leq n$ and $1 < i \leq n + 1 - j$:

By Theorem 1, we have $V_{i,j} = V_{1,j}$. Hence we obtain the whole CYK table V by storing all $V_{1,j}$ in $\sum_{j=1}^{n} |V_{1,j}| = O(n|N|) = O(n|G|)$ space, as $|G| = \sum_{p \in P} |p| > |N|$.

By Theorem 1, we have $\mathrm{LS}_{i,j}(W) = \mathrm{LS}_{1,j}(W)$. Hence we obtain all low supports by storing all $\mathrm{LS}_{1,j}(W)$ in $\displaystyle\sum_{j=1}^{n} \sum_{W \in V_{1,j}} |\mathrm{LS}_{1,j}(W)| \leq \sum_{j=1}^{n} |P \times \{k \mid 1 \leq k < j\}|$

$= \displaystyle\sum_{j=1}^{n} O(n|G|) = O(n^2|G|)$ space, as $|G| = \sum_{p \in P} |p| > |P|$ and each low support takes constant space.

Considering the high-support set $\mathrm{HS}_{i,j}(W)$, it takes $O(n^3|G|)$ space to store all $\mathrm{HS}_{i,j}(W)$ as $\mathrm{HS}_{i,j}(W) = \mathrm{HS}_{1,j}(W)$ is not true for all $1 \leq j \leq n$ and $i > 1$. For example in Figure 1, we have $\mathrm{HS}_{2,1}(R) = \{(S \to LR,\ 2)\}$, while $\mathrm{HS}_{1,1}(R) = \emptyset$. To save space, we compute the set $\mathrm{HS}'_{i,j}(W) = \bigcup_{k=1}^{n+1-j} \mathrm{HS}_{k,j}(W)$ instead of $\mathrm{HS}_{i,j}(W)$, as we can encode $\mathrm{HS}'_{i,j}(W)$ efficiently. Note that we still have $\mathrm{HS}'_{i,j}(W) \subseteq \overline{\mathrm{HS}}_{i,j}(W)$ as $\overline{\mathrm{HS}}_{i,j}(W) = \overline{\mathrm{HS}}_{1,j}(W)$ (its formulation in Definition 1 is independent of i) and $\mathrm{HS}'_{i,j}(W) = \bigcup_{k=1}^{n+1-j} \mathrm{HS}_{k,j}(W) \subseteq \bigcup_{k=1}^{n+1-j} \overline{\mathrm{HS}}_{k,j}(W) = \overline{\mathrm{HS}}_{1,j}(W)$. Hence we obtain all $\mathrm{HS}'_{i,j}(W)$ by computing and storing all $\mathrm{HS}'_{1,j}(W)$ in $O(n^2|G|)$ space, as $\mathrm{HS}'_{i,j}(W) = \mathrm{HS}'_{1,j}(W)$ and $\displaystyle\sum_{j=1}^{n} \sum_{W \in V_{1,j}} |\mathrm{HS}'_{1,j}(W)| \leq$

$2 \displaystyle\sum_{j=1}^{n} \sum_{W \in V_{1,j}} |\mathrm{LS}_{1,j}(W)| = O(n^2|G|)$ (the definition of $\mathrm{HS}'_{i,j}(W)$ is independent of i and one low support corresponds to at most two high supports).

Hence we can encode the CYK table V, all $\mathrm{LS}_{i,j}(W)$, and all $\mathrm{HS}'_{i,j}(W)$ in $O\left(n^2 |G|\right)$ space. $\square$

Using Theorem 2, it *is* practical to make the propagator run on $\mathrm{LS}_{i,j}(W)$ and $\mathrm{HS}'_{i,j}(W)$, which are subsets of the candidate support sets, with $O\left(n^2 |G|\right)$ space. Although Theorem 2 requires all $\mathrm{dom}(X_i)$ to be the *same*, this is not an obstacle in practice, as shown in Section 3.2 below. Note that $|\mathrm{LS}| + |\mathrm{HS}'|$ and $|\overline{\mathrm{LS}}| + |\overline{\mathrm{HS}}|$ are asymptotically the same (as shown in Section 3.3 below), hence we cannot improve the propagator of [11] asymptotically.

Counting the Supports. For each non-terminal W in the CYK table, the propagator of [11], which is based on the arc-consistency (AC) algorithm AC-6 [3], decides whether W has low and high supports by exhibiting two actual supports (one low and one high). However, this is not necessary. We can simply *count* the supports for W as in AC-4 [15], and then just decrease the counter by one when a support is lost. Although Bessière [3] shows that AC-4 is worse than AC-6 for *binary* CSPs given *extensionally* because initialising the counters is expensive, in our case initialisation is much cheaper because we have $|\mathrm{LS}_{i,j}(W)| = |\mathrm{LS}_{1,j}(W)|$ initially when using our efficient encoding of the support sets. However, by using counters, we do not need complex data structures and operations to trace which non-terminal in the CYK table is currently supporting and supported by which non-terminal(s), as in [11]. Indeed, our experiments (omitted for space reasons, see Appendix C of [9]) show that counting with our efficient encoding of the support sets works better (up to 12 times) than using *only* the latter, which already works better (up to 20 times) than the propagator of [11].

3.2   Description and Proof of Our Propagator

Consider a CFG $G = \langle \Sigma, N, P, S \rangle$ in CNF and a sequence $X = \langle X_1, \ldots, X_n \rangle$ of n decision variables. We introduce a propagator for the $\mathrm{CFG}(X, G)$ constraint using the AC-4 framework, which computes all supports and counts them when *posting* the constraint (see Algorithm 1), and then *only* decreases the support counters during propagation (see Algorithm 2), without changing the support sets. Hence, to satisfy the condition of Theorem 2, we *only* need to make all decision variables *temporarily* take the same domain when *posting* the constraint. Our propagator has *no* limitation on the initial domains of the decision variables, as we will show how our propagator lifts the temporary restriction at no asymptotic overhead.

Let $C_{i,j}^{\mathrm{LS}}(W)$ (or $C_{i,j}^{\mathrm{HS}}(W)$) denote the number of low (or high) supports for (or of) a non-terminal W in $V_{i,j}$ of the CYK table during propagation. Similarly, let $C_i^{\mathrm{LS}}(b)$ (or $C_i^{\mathrm{HS}}(b)$) denote the number of low (or high) supports for (or of) a terminal b in $\mathrm{dom}(X_i)$. Note that every (non-)terminal has two counters and there is no sharing of counters between any two (non-)terminals, as the counters will be changed *independently* during propagation. Using Theorem 2, Algorithm 1 posts the $\mathrm{CFG}(X, G)$ constraint, encodes the CYK table

and support sets, counts the supports, and achieves GAC. Given all propagator state variables, which are also shared by Algorithm 2, initialised so that
$V_{1,j} = \mathrm{LS}_{1,j}(W) = \mathrm{HS}'_{1,j}(W) = \emptyset$ and $C^{\mathrm{HS}}_{i,j}(W) = 0$ (lines 2 to 4), Algorithm 1
works as follows. First, it constructs a virtual domain $\mathrm{Dom}' = \bigcup_{i=1}^{n} \mathrm{dom}(X_i)$
(line 5), and uses it to post the $\mathrm{CFG}(X, G)$ constraint, hence the condition of
Theorem 2 is satisfied as all domains are now the same. Using the virtual domain may introduce extra solutions, and we show in the last step how to avoid
this. Second, it uses a bottom-up process (lines 6 to 17) based on the CYK
parser to compute all $V_{1,j}$, $\mathrm{LS}_{1,j}(W)$, $\mathrm{HS}'_{1,j}(W)$, and $C^{\mathrm{LS}}_{i,j}(W)$. Note that we
only need to compute $V_{1,j}$ by Theorem 1, and any reference to $V_{i,j}$ is replaced
by $V_{1,j}$. The same holds for $\mathrm{LS}_{i,j}(W)$, and $\mathrm{HS}'_{i,j}(W)$ (by its definition independently of i in Theorem 2). If the start non-terminal S is not in $V_{1,n}$, then
it fails (line 18; no word from the current domains is accepted by G, hence no
solution exists). Third, it uses a top-down process (lines 19 to 25) to compute all
$C^{\mathrm{HS}}_{i,j}(W)$. Fourth, it removes all values with no high support from the domains
(lines 26 to 28). Finally, it constructs a set Δ of all variable-value pairs that
are not in the domains of X but in the virtual domain (line 29), and calls the
function filterFromUpdate (in Algorithm 2, discussed next) to re-establish GAC
after removing all such variable-value pairs (line 30). Hence the side effect of using the virtual domain is lifted; we show in Section 3.3 that calling the function
filterFromUpdate does not increase the asymptotic complexity of Algorithm 1.

Given a set Δ of all recently filtered variable-value pairs by other propagators
or a branching of the search tree, the function filterFromUpdate in Algorithm 2
incrementally re-establishes GAC for the $\mathrm{CFG}(X, G)$ constraint as follows. First,
it creates two arrays Q_{LS} and Q_{HS} of initially empty queues (line 2), with $Q_{\mathrm{LS}}[j]$
(or $Q_{\mathrm{HS}}[j]$) storing all non-terminals W in the j-th row of the CYK table with
no low (or high) supports due to the domain changes Δ. Second, it iterates over
all removed values in Δ, decreasing the counter $C^{\mathrm{LS}}_{i,1}(W)$ for all non-terminals W
in the bottom row supported by a removed value, and adding all W with no low
support to the queue $Q_{\mathrm{LS}}[1]$ (lines 3 to 7). Third, a bottom-up process (lines 8
to 11) calls the procedure rmNoLS handling all W in the queue $Q_{\mathrm{LS}}[j]$. Given
a non-terminal W with no low support, rmNoLS iterates over each high support of W, decreasing the three counters related with this lost high support,
and enqueuing $Q_{\mathrm{LS}}[j]$ (or $Q_{\mathrm{HS}}[j]$) whenever a low (or high) support counter is
zero (lines 22 to 33). Fourth, a top-down process (lines 12 to 14) calls the procedure rmNoHS (omitted for space reasons, see Appendix C of [9]), which works
similarly to rmNoLS, handling all W in the queue $Q_{\mathrm{HS}}[j]$. Finally, it removes inconsistent values (with no high support) from the domains of X (lines 15 to 20),
and reaches a fixpoint (line 21). Note that Algorithm 2 is a direct usage of the
AC-4 framework. Once Algorithm 1 initialises the support sets and counters
correctly, the correctness of Algorithm 2 is guaranteed by the AC-4 framework.

Theorem 3. *Our propagator achieves GAC for* $\mathrm{CFG}(X, G)$.

Proof: A value is removed by our propagator from the domains of X iff it has
no high supports, as with the propagator of [11]. Hence the two propagators are
equivalent. The result follows from Theorem 2 on page 132 of [11].   □

Algorithm 1. An improved propagator for the $\text{CFG}(X, G)$ constraint, where $X = \langle X_1, \ldots, X_n \rangle$ is a sequence of n decision variables and $G = \langle \Sigma, N, P, S \rangle$ is a CFG in CNF

1: **function** $\text{post}(\text{CFG}(X, G))$
2: **for all** $W \in N$ and $j \leftarrow 1$ **to** n **do**
3:     $V_{1,j} \leftarrow \text{LS}_{1,j}(W) \leftarrow \text{HS}'_{1,j}(W) \leftarrow \emptyset$
4:         **for all** $i \leftarrow 1$ **to** $n+1-j$ **do** $C^{\text{HS}}_{i,j}(W) \leftarrow 0$
5: $\text{Dom}' \leftarrow \bigcup_{i=1}^{n} \text{dom}(X_i)$
6: $V_{1,1} \leftarrow \{W \mid (W \to b) \in P \ \wedge \ b \in \text{Dom}'\}$
7: $\text{LS}_{1,1}(W) \leftarrow \{W \to b \mid (W \to b) \in P \ \wedge \ b \in \text{Dom}'\}$
8: $\text{HS}'_1(b) \leftarrow \{W \to b \mid (W \to b) \in P \ \wedge \ b \in \text{Dom}'\}$
9: **for all** $j \leftarrow 2$ **to** n **do**
10:     **for all** $(W \to YZ) \in P$ and $k \leftarrow 1$ **to** $j-1$ **do**
11:         **if** $Y \in V_{1,k} \wedge Z \in V_{1,j-k} \wedge (j < n \vee W = S)$ **then**
12:             $V_{1,j} \leftarrow V_{1,j} \cup \{W\}$
13:             $\text{LS}_{1,j}(W) \leftarrow \text{LS}_{1,j}(W) \cup \{(W \to YZ, \ k)\}$
14:             $\text{HS}'_{1,k}(Y) \leftarrow \text{HS}'_{1,k}(Y) \cup \{(W \to YZ, \ j)\}$
15:             $\text{HS}'_{1,j-k}(Z) \leftarrow \text{HS}'_{1,j-k}(Z) \cup \{(W \to YZ, \ j)\}$
16: **for all** $j \leftarrow 1$ **to** n and $W \in V_{1,j}$ **do**
17:     **for all** $i \leftarrow 1$ **to** $n+1-j$ **do** $C^{\text{LS}}_{i,j}(W) \leftarrow |\text{LS}_{1,j}(W)|$
18: **if** $S \notin V_{1,n}$ **then return** *failed*
19: **for all** $j \leftarrow n$ **to** 2, $W \in V_{1,j}$, and $i \leftarrow 1$ **to** $n+1-j$ **do**
20:     **if** $C^{\text{HS}}_{i,j}(W) > 0 \vee j = n$ **then**
21:         **for all** $(W \to YZ, \ k) \in \text{LS}_{1,j}(W)$ **do**
22:             $C^{\text{HS}}_{i,k}(Y) \ {++}; \ C^{\text{HS}}_{i+k,j-k}(Z) \ {++}$
23: **for all** $W \in V_{1,1}$ and $i \leftarrow 1$ **to** n **do**
24:     **if** $C^{\text{HS}}_{i,1}(W) > 0$ **then**
25:         **for all** $(W \to b) \in \text{LS}_{1,1}(W)$ **do** $C^{\text{HS}}_i(b) \ {++}$
26: **for all** $i \leftarrow 1$ **to** n and $b \in \text{dom}(X_i)$ **do**
27:     **if** $C^{\text{HS}}_i(b) = 0$ **then** $\text{dom}(X_i) \leftarrow \text{dom}(X_i) \setminus \{b\}$
28:     **if** $\text{dom}(X_i) = \emptyset$ **then return** *failed*
29: $\Delta \leftarrow \{(X_i, b) \mid X_i \in X \wedge b \in \text{Dom}' \setminus \text{dom}(X_i)\}$
30: **return** $\text{filterFromUpdate}(\text{CFG}(X, G), \Delta)$

3.3 Complexity Analysis

We first investigate the worst-case *time* complexity of our propagator for the $\text{CFG}(X, G)$ constraint. In Algorithm 1, the time complexity of lines 2 to 29 is dominated by lines 19 to 25, which explore at most all low-support sets $\text{LS}_{i,j}(W)$ (referenced as $\text{LS}_{1,j}(W)$) once in $\sum_{j=1}^{n} \sum_{i=1}^{n+1-j} \sum_{W \in V_{1,j}} |\text{LS}_{1,j}(W)| <$

$n \sum_{j=1}^{n} \sum_{W \in V_{1,j}} |\text{LS}_{1,j}(W)| = O\left(n^3 |G|\right)$ time, by Theorem 2; line 30 calls the function filterFromUpdate in Algorithm 2, which explores once all $\text{LS}_{i,j}(W)$

Algorithm 2. Given a set Δ of domain changes, the function filterFromUpdate incrementally re-establishes GAC for the $\mathrm{CFG}(X, G)$ constraint on a sequence $X = \langle X_1, \ldots, X_n \rangle$ of n decision variables.

1: **function** filterFromUpdate($\mathrm{CFG}(X, G), \Delta$)
2: **for all** $j \leftarrow 1$ **to** n **do** $Q_{\mathrm{LS}}[j] \leftarrow [\,]; \ Q_{\mathrm{HS}}[j] \leftarrow [\,]$
3: **for all** $(X_i, b) \in \Delta$ **do**
4:    $C_i^{\mathrm{HS}}(b) \leftarrow 0$
5:    **for all** $(W \rightarrow b) \in \mathrm{HS}_1'(b)$ **do**
6:      **if** $C_{i,1}^{\mathrm{LS}}(W) > 0$ **then**
7:        **if** $--C_{i,1}^{\mathrm{LS}}(W) = 0$ **then** $Q_{\mathrm{LS}}[1].\mathbf{enqueue}((W, i))$
8: **for all** $j \leftarrow 1$ **to** n **do**
9:    **while** $Q_{\mathrm{LS}}[j] \neq [\,]$ **do**
10:      **if** $j = n$ **then return** *failed* as $S_{1,n}$ has no low support
11:      $(W, i) \leftarrow Q_{\mathrm{LS}}[j].\mathbf{dequeue}(); \ \mathrm{rmNoLS}(W, i, j, Q_{\mathrm{LS}}, Q_{\mathrm{HS}})$
12: **for all** $j \leftarrow n - 1$ **to** 2 **do**
13:    **while** $Q_{\mathrm{HS}}[j] \neq [\,]$ **do**
14:      $(W, i) \leftarrow Q_{\mathrm{HS}}[j].\mathbf{dequeue}(); \ \mathrm{rmNoHS}(W, i, j, Q_{\mathrm{LS}}, Q_{\mathrm{HS}})$
15: **while** $Q_{\mathrm{HS}}[1] \neq [\,]$ **do**
16:    $(W, i) \leftarrow Q_{\mathrm{HS}}[1].\mathbf{dequeue}()$
17:    **for all** $(W \rightarrow b) \in \mathrm{LS}_{1,1}(W)$ **do**
18:      **if** $C_i^{\mathrm{HS}}(b) > 0$ **then**
19:        **if** $--C_i^{\mathrm{HS}}(b) = 0$ **then** $\mathrm{dom}(X_i) \leftarrow \mathrm{dom}(X_i) \setminus \{b\}$
20:        **if** $\mathrm{dom}(X_i) = \emptyset$ **then return** *failed*
21: **return** *at-fixpoint*

22: **procedure** $\mathrm{rmNoLS}(W, i, j, Q_{\mathrm{LS}}, Q_{\mathrm{HS}})$
23: **if** $C_{i,j}^{\mathrm{HS}}(W) > 0$ **then**
24:    **for all** $(F \rightarrow YZ, \ k) \in \mathrm{HS}_{1,j}'(W)$ **do**
25:      **if** $W = Y \wedge F \in V_{i,k} \wedge Z \in V_{i+j,k-j}$ **then**
26:        $(i_F, j_F, B, i_B, j_B) \leftarrow (i, k, Z, i + j, k - j)$
27:      **else if** $W = Z \wedge F \in V_{i-j,k} \wedge Y \in V_{i-j,k-j}$ **then**
28:        $(i_F, j_F, B, i_B, j_B) \leftarrow (i - j, k, Y, i - j, k - j)$
29:      **else** skip lines 30 to 33
30:      **if** $C_{i_F, j_F}^{\mathrm{LS}}(F) > 0 \wedge C_{i_B, j_B}^{\mathrm{HS}}(B) > 0 \wedge C_{i,j}^{\mathrm{HS}}(W) > 0$ **then**
31:        **if** $--C_{i_F, j_F}^{\mathrm{LS}}(F) = 0$ **then** $Q_{\mathrm{LS}}[j_F].\mathbf{enqueue}((F, i_F))$
32:        **if** $--C_{i_B, j_B}^{\mathrm{HS}}(B) = 0$ **then** $Q_{\mathrm{HS}}[j_B].\mathbf{enqueue}((B, i_B))$
33:        **if** $--C_{i,j}^{\mathrm{HS}}(W) = 0$ **then** $Q_{\mathrm{HS}}[j].\mathbf{enqueue}((W, i))$; **return**

---

and $\mathrm{HS}_{i,j}'(W)$ in the worst case, hence takes $\sum_{j=1}^{n} \sum_{i=1}^{n+1-j} \sum_{W \in V_{i,j}} |\mathrm{LS}_{1,j}(W)| + |\mathrm{HS}_{1,j}'(W)| = O\left(n^3 |G|\right)$ time, for similar reasons. Hence there is no asymptotic overhead by line 30, and the overall time complexity is $O\left(n^3 |G|\right)$.

Consider now the worst-case *space* complexity of our propagator. By Theorem 2, encoding the CYK table V, all $\mathrm{LS}_{i,j}(W)$, and all $\mathrm{HS}_{i,j}'(W)$ takes $O\left(n^2 |G|\right)$ space. There are $\sum_{j=1}^{n} \sum_{i=1}^{n+1-j} |V_{i,j}| = \sum_{j=1}^{n} \sum_{i=1}^{n+1-j} |V_{1,j}| = O\left(n^2 |N|\right) =$

$O\left(n^2 |G|\right)$ non-terminals in V, hence storing the support counters for all non-terminals takes $O(n^2 |G|)$ space. There are $n |\Sigma|$ terminals in the domains, hence storing the support counters for all terminals takes $O(n |G|)$ space. The two arrays Q_{LS} and Q_{HS} of queues contain at most all non-terminals in V, hence take $O(n^2 |G|)$ space. The overall space complexity is thus $O\left(n^2 |G|\right)$.

Although our propagator has the same *worst*-case time and space complexity as the one of [11], which is probably optimal anyway, our experiments below show that our propagator systematically beats it in practice (by up to two orders of magnitude), which might be confirmed by an *average*-case complexity analysis.

4   Experimental Evaluation

We now demonstrate the speed-up of our CFG propagator over its ancestor [11]. We implemented our propagator and the one of [11] in GECODE [8]. Katsirelos *et al.* [12] show how to reformulate a CFG into a DFA for a fixed length, as propagation for the REGULAR constraint is much cheaper than for CFG. This reformulation needs a propagator for the CFG constraint to shrink the initial domains of all decision variables to achieve GAC for all constraints at the root of the search tree, so that the obtained DFA is smaller. Hence this reformulation also benefits from a more efficient propagator for the CFG constraint.

Note that Sections 4.3 and 4.4 demonstrate that CP outperforms some state-of-the-art solvers from the verification literature by orders of magnitude on their own benchmarks. Our experimental results show that those benchmarks are trivial, *but these benchmarks were not known to be trivial* before this paper, and we have neither discarded any non-trivial benchmarks (of HAMPI and SUSHI) nor included the benchmarks that were in the meantime known to be trivial.

We use the GECODE built-in REGULAR propagator. We ran the experiments of Sections 4.1, 4.2, and 4.3 under GECODE 3.7.3, HAMPI 20120213, and Ubuntu Linux 11.10 on 1.8 GHz Intel Core 2 Duo with 3GB RAM; and we ran the experiment of Section 4.4 under GECODE 3.7.3, KALUZA, SUSHI 2.0, and Ubuntu Linux 10.04 with 1GB RAM in Oracle VirtualBox 4.2.4 (recommended by the SUSHI developers) on the same hardware. As our chosen search heuristics do not randomise, all instances of Sections 4.1, 4.2, and 4.3 were run once. However, for Section 4.4, we ran each instance 10 times and recorded the average runtime, as the performance of the virtual machine might vary significantly.

4.1   A Shift Scheduling Problem

Demassey *et al.* [4] introduce a real-life shift scheduling problem for staff in a retail store. Let w be the number of workers, p the number of periods of the scheduling horizon, and a the number of work activities. The aim is to construct a $w \times p$ matrix of values in $[1, \ldots, a + 3]$ (there are 3 non-work activities, namely break, lunch, and rest) to satisfy work regulation constraints, which can be modelled with a CFG constraint for each worker over the p periods and some global cardinality constraints (GCC).

Katsirelos *et al.* [12] model this problem as an optimisation problem, so that the reformulation of the grammar into a DFA takes only a tiny part of the runtime; they show that this optimisation problem is extremely difficult for CP-based CFG and REGULAR propagators. We are here, like [11], primarily interested in the first solution to the satisfaction version of this problem. We use the search heuristic of [11], namely selecting the second-largest value from the first decision variable with the minimum domain size in the last period with unassigned variables. HAMPI cannot handle multiple variables, while HAMPI, KALUZA, and SUSHI cannot model GCC, so we do not compare with them.

Table 1 gives our results: each row gives the instance, the search tree size, the DFA size after the reformulation of [12] of CFG into REGULAR, and the runtimes of four methods in seconds, namely our propagator (denoted by G++), the one of [11] (denoted by G), and the reformulation, using the two CFG propagators respectively (denoted by DFA$_{G++}$ and DFA$_G$). We find that G++ always works much better (up to 18 times) than G; DFA$_{G++}$ always works much better (up to 10 times) than DFA$_G$, as the reformulation of [12] itself needs a CFG propagator to shrink the initial domains at the root of the search tree (the reformulation, which is *instance-dependent*, is here taken on-line and takes about 85% of the total runtime) and as G++ works better than G; overall, G++ wins on 15 instances, and DFA$_{G++}$ wins on the other 2 instances. When solving for all or best solutions, DFA$_{G++}$ gradually takes over as the best method, as predicted by [12], but G++ continues to dominate G, and DFA$_{G++}$ decreasingly dominates DFA$_G$, as instances get harder.

4.2   A Forklift Scheduling Problem

Gange and Stuckey [7] introduce a forklift scheduling problem. Let s be the number of stations, i the number of items, and n the length of the scheduling horizon. There is a unique forklift and a shipping list giving the initial and final stations of each item. The aim is to construct an array of n actions, where an action can move the forklift from a station to any other station with a cost of 3, load an item from the current station onto the top of the forklift tray with a cost of 1, unload the item from the top of the forklift tray at the current station with a cost of 1, or do nothing with a cost of 0, so that the shipping list is accomplished with a minimised cost under forklift behaviour constraints, which can be modelled with one CFG constraint and i REGULAR constraints. We use the first-fail search heuristic, namely selecting the smallest value from the first decision variable with the minimum domain size, to solve this optimisation problem. Since HAMPI, KALUZA, and SUSHI cannot solve optimisation problems, we do not compare with them.

Table 2 gives our results over the instances solvable in one CPU hour: each row specifies the instance and gives the runtimes of two methods in seconds, namely our propagator (denoted by G++) and the one of [11] (denoted by G). We find that G++ always works better (up to 5 times) than G. The reformulation of [12] of the CFG constraint into the REGULAR constraint is not suitable for this problem, as the resulting automaton is of size exponential in n.

Table 1. Runtimes for the shift scheduling problem

benchmark ($p = 96$)			search tree size			DFA	runtimes of four methods in seconds			
instance	a	w	#nodes	#propagations	#fails	$|A|$	G++	DFA$_{G++}$	DFA$_G$	G
1_1	1	1	11	438	1	446	**0.24**	0.49	4.26	3.93
1_2	1	3	133	2123	33	998	**0.90**	3.78	15.38	12.87
1_3	1	4	349	5790	137	998	**1.68**	4.10	19.48	19.49
1_4	1	5	95	1836	7	814	**1.18**	2.41	21.99	20.53
1_5	1	4	71	1332	3	722	**0.92**	1.75	16.95	16.32
1_6	1	5	76	1567	3	722	**1.17**	2.01	21.16	20.17
1_7	1	6	3623	56635	1773	814	7.87	**2.97**	25.56	47.48
1_8	1	2	57	1005	10	998	**0.52**	3.59	10.76	8.47
1_9	1	1	19	460	1	630	**0.22**	0.80	4.41	3.94
1_10	1	7	12699	209988	6305	814	23.31	**4.02**	30.14	100.95
2_1	2	2	46	1414	8	984	**0.93**	1.69	16.76	15.97
2_5	2	4	83	2208	20	1209	**1.02**	3.15	18.51	16.41
2_6	2	5	89	1801	12	1207	**1 35**	2.94	23.03	21.57
2_7	2	6	258	5847	104	944	**1.97**	2.63	32.22	32.03
2_8	2	2	1046	28691	500	1774	**2.86**	7.75	23.09	24.09
2_9	2	1	35	1249	8	1460	**0.63**	4.11	14.21	11.03
2_10	2	7	4690	100007	2302	1506	**7.64**	7.82	43.24	53.90

Table 2. Runtimes for the forklift scheduling problem

instance			runtimes in seconds		instance			runtimes in seconds	
s	i	n	G++	G	s	i	n	G++	G
3	4	15	**4.35**	20.02	3	4	16	**22.64**	103.75
3	4	17	**20.98**	100.48	3	4	18	**76.77**	382.31
3	4	19	**72.66**	338.69	3	4	20	**197.98**	1013.78
3	5	16	**67.54**	297.55	3	5	17	**81.67**	368.65
3	5	18	**200.91**	1058.17	3	6	18	**1134.58**	5008.90
4	5	17	**388.92**	1631.94	4	5	18	**819.82**	3876.87

4.3 Intersection of Two Context-Free Languages

HAMPI [14] selects a subset of 100 CFG pairs (from the benchmark of CFGAnalyzer [1]), where a string of length $1 \le n \le 50$ accepted by both CFGs in each pair is to be found (8 instances are satisfiable and 92 are unsatisfiable; disjointness of two context-free languages is undecidable). The CFGs of this benchmark have 10 to 600 productions in CNF and up to 18 alphabet symbols. This problem can also be solved using tools from automata theory. On this benchmark, HAMPI beats CFGAnalyzer by a large margin. HAMPI also beats other ad hoc solvers on other benchmarks, which are too easy (HAMPI solves them in one second), hence any improvements might be subject to runtime measurement errors.

Instead of running each CFG pair 50 times with the n-th run to find a string of length n accepted by both CFGs, we search once, namely for the first solution string of length *up to* 50 for each pair. Given a CFG $G = \langle \Sigma, N, P, S \rangle$, we create a new CFG $G' = \langle \Sigma', N', P', S' \rangle$ with $\Sigma' = \Sigma \cup \{\#\}$ (let $\# \notin \Sigma$ denote a dummy symbol), $N' = N \cup \{S'\}$, and $P' = P \cup \{S' \to S \mid S'\#\}$. If a string s' of length

n is accepted by G', then the string s obtained by removing all '#' at the end of s' has a length up to n and is accepted by G.

Given a CFG pair (G_1, G_2), our model is $\mathrm{CFG}(X, G_1') \wedge \mathrm{CFG}(X, G_2')$, where X is a sequence of n decision variables with $\mathrm{dom}\,(X_i) = \Sigma_1' \cup \Sigma_2'$. Our search heuristic is to select the first value from the last unassigned variable. Figure 2 gives the runtimes of HAMPI and the two CFG propagators for the 55 instances where HAMPI takes at least one second. Each '×' (or '+') denotes the comparison between our propagator (or the one of [11]) and HAMPI; each 'Δ' denotes the solving time of the bit-vector solver STP. For all 100 instances, the two propagators always work much better (up to 9000 times) than HAMPI, and even always work much better than STP when the fixed-sizing of the grammar into a regular expression and the transformation into bit-vector constraints are taken off-line; our propagator always works much better (up to 250 times) than the one of [11]. As 97 instances turn out to be solvable at the root of the search tree, the reformulation of [12] of the CFG constraint into the REGULAR constraint has similar results; for the other 3 instances, our CFG propagator is 3 to 5 orders of magnitude faster (details omitted for space reasons, see Appendix C of [9]). The two CFG propagators always beat HAMPI for all $n < 50$ (up to 380 times even with $n = 10$), and whether run on the CFG pair (G_1', G_2') or the original pair (G_1, G_2). We get similar speed-ups (details omitted for space reasons, see Appendix C of [9]) over 99% of the CFG pairs even with the first-fail search heuristic. Note that KALUZA uses HAMPI's functionality to solve the CFG constraint, hence KALUZA has the same performance as HAMPI on this benchmark.

4.4 Solving String Equations

Fu *et al.* [5] introduce just one benchmark of 5 string equations with a parameter $1 \leq n \leq 37$ to demonstrate the practicality of their string solver SUSHI against KALUZA. SUSHI handles string variables of *unbounded* length. Like KALUZA, we expect a user-given parameter $\bar{n}$ and look for the first solution string of up to $\bar{n}$ symbols. Unlike KALUZA, which tries all lengths until $\bar{n}$, we allow strings to end with dummy symbols '#' (as in Section 4.3) and add length constraints. For a sequence $X = \langle X_1, \ldots, X_{\bar{n}} \rangle$, let decision variable n_X with $\mathrm{dom}\,(n_X) = \{0, \ldots, \bar{n}\}$ denote the index of the right-most non-dummy symbol in X. The length constraint is $\forall 1 \leq i \leq \bar{n} : X_i = \# \Leftrightarrow n_X < i$. String concatenation $X = Y + Z$ is modelled as $n_X = n_Y + n_Z \wedge \langle X_1, \ldots, X_{n_X} \rangle = \langle Y_1, \ldots, Y_{n_Y}, Z_1, \ldots, Z_{n_Z} \rangle$ with reification constraints. Regular language membership $X \in L(R)$, where $L(R)$ denotes the language accepted by the regular expression R, is modelled as $\mathrm{REGULAR}(X, R\#^*)$. We use the first-fail search heuristic. Table 3 gives the runtimes of GECODE, SUSHI, and KALUZA for equations 1 to 3 with the *hardest* setting $n = 37$ and the KALUZA models (for a fair comparison). As KALUZA solves the equations for some $n \leq \bar{n} < 3n$, we *pessimistically* set $\bar{n} = 4n$ for GECODE, and GECODE *still* beats SUSHI and KALUZA, by up to 130 times. GECODE solves our better models than the KALUZA ones of equations 4 and 5 within 0.10 seconds, beating SUSHI and KALUZA by up to 3000 times.

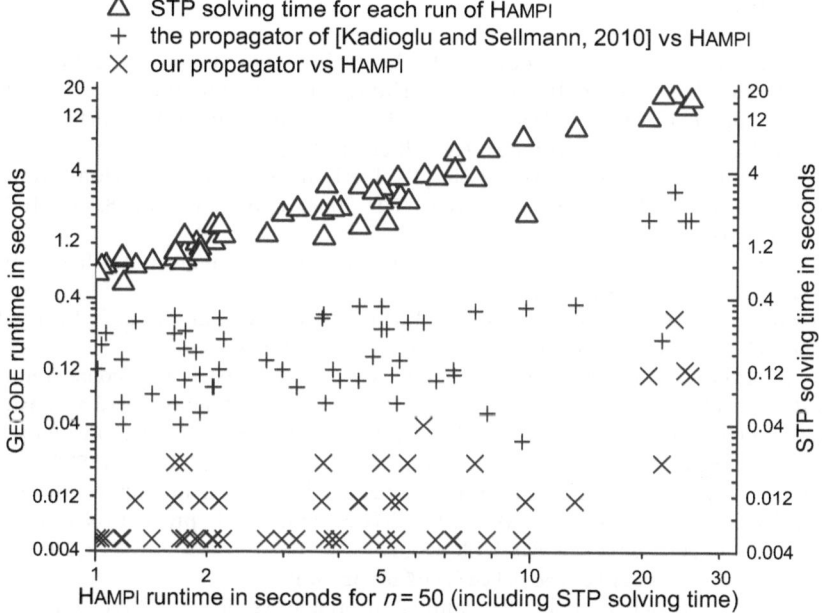

Fig. 2. Runtimes for the CFG-intersection problem

Table 3. Runtimes (in seconds) for solving string equations

	eq1: 3 string variables			eq2: 2 string variables			eq3: 4 string variables		
n	GECODE	SUSHI	KALUZA	GECODE	SUSHI	KALUZA	GECODE	SUSHI	KALUZA
37	**0.15**	1.34	10.40	**0.05**	1.82	3.94	**0.07**	2.52	5.71

5 Conclusion

We argue that CP solvers are more suitable than existing solvers for verification tools that solve string constraints. Indeed, CP has a rich tradition of constraints for membership in formal languages: their propagators run directly on descriptions, such as automata and grammars, of these languages. Apparently tricky features, such as string equality or multiple string variables (with shared characters), pose no problem to CP. Future work includes designing propagators for string constraints over strings of (un)bounded length.

Acknowledgements. The first three authors are supported by grants 2007-6445 and 2011-6133 of the Swedish Research Council (VR), and Jun He is also supported by grant 2008-611010 of China Scholarship Council and the National University of Defence Technology of China. Many thanks to Xiang Fu, Serdar Kadıoğlu, George Katsirelos, Adam Kieżun, Prateek Saxena, and Guido Tack for useful discussions during the preparation of this work.

References

1. Axelsson, R., Heljanko, K., Lange, M.: Analyzing context-free grammars using an incremental SAT solver. In: Aceto, L., Damgård, I., Goldberg, L.A., Halldórsson, M.M., Ingólfsdóttir, A., Walukiewicz, I. (eds.) ICALP 2008, Part II. LNCS, vol. 5126, pp. 410–422. Springer, Heidelberg (2008)
2. Beldiceanu, N., Carlsson, M., Petit, T.: Deriving filtering algorithms from constraint checkers. In: Wallace, M. (ed.) CP 2004. LNCS, vol. 3258, pp. 107–122. Springer, Heidelberg (2004)
3. Bessière, C.: Arc-consistency and arc-consistency again. Artificial Intelligence 65(1), 179–190 (1994)
4. Demassey, S., Pesant, G., Rousseau, L.-M.: A cost-regular based hybrid column generation approach. Constraints 11(4), 315–333 (2006)
5. Fu, X., Powell, M.C., Bantegui, M., Li, C.-C.: Simple linear string constraints. Formal Aspects of Computing (2012), Published on-line in January 2012 and available from http://dx.doi.org/10.1007/s00165-011-0214-3. SUSHI is available from http://people.hofstra.edu/Xiang_Fu/XiangFu/projects/SAFELI/SUSHI.php
6. Ganesh, V., Dill, D.L.: A decision procedure for bit-vectors and arrays. In: Damm, W., Hermanns, H. (eds.) CAV 2007. LNCS, vol. 4590, pp. 519–531. Springer, Heidelberg (2007), STP is available from https://sites.google.com/site/stpfastprover/
7. Gange, G., Stuckey, P.J.: Explaining propagators for s-DNNF circuits. In: Beldiceanu, N., Jussien, N., Pinson, É. (eds.) CPAIOR 2012. LNCS, vol. 7298, pp. 195–210. Springer, Heidelberg (2012)
8. Gecode Team. Gecode: A generic constraint development environment (2006), http://www.gecode.org/
9. He, J.: Constraints for Membership in Formal Languages under Systematic Search and Stochastic Local Search. PhD thesis, Uppsala University, Sweden (2013), http://urn.kb.se/resolve?urn=urn:nbn:se:uu:diva-196347
10. Hopcroft, J.E., Ullman, J.D.: Introduction to Automata Theory, Languages, and Computation. Addison Wesley, New York (1979)
11. Kadioğlu, S., Sellmann, M.: Grammar constraints. Constraints 15(1), 117–144 (2008); An early version is published in the Proceedings of the 23rd AAAI Conference on Artificial Intelligence in 2008
12. Katsirelos, G., Narodytska, N., Walsh, T.: Reformulating global grammar constraints. In: van Hoeve, W.-J., Hooker, J.N. (eds.) CPAIOR 2009. LNCS, vol. 5547, pp. 132–147. Springer, Heidelberg (2009)
13. Katsirelos, G., Narodytska, N., Walsh, T.: The weighted GRAMMAR constraint. Annals of Operations Research 184(1), 179–207 (2011), An early version is published in the Proceedings of the 5th International Conference on Integration of AI and OR Techniques in Constraint Programming for Combinatorial Optimization Problems in 2008
14. Kieżun, A., Ganesh, V., Guo, P.J., Hooimeijer, P., Ernst, M.D.: HAMPI: A solver for string constraints. In: Proceedings of the 18th International Symposium on Software Testing and Analysis, Chicago, USA, July 2009, pp. 105–116. ACM Press (2009), HAMPI is available from http://people.csail.mit.edu/akiezun/hampi/
15. Mohr, R., Henderson, T.C.: Arc and path consistency revisited. Artificial Intelligence 28(2), 225–233 (1986)
16. Pesant, G.: A regular language membership constraint for finite sequences of variables. In: Wallace, M. (ed.) CP 2004. LNCS, vol. 3258, pp. 482–495. Springer, Heidelberg (2004)

17. Quimper, C.-G., Walsh, T.: Global grammar constraints. In: Benhamou, F. (ed.) CP 2006. LNCS, vol. 4204, pp. 751–755. Springer, Heidelberg (2006)
18. Quimper, C.-G., Walsh, T.: Decomposing global grammar constraints. In: Bessière, C. (ed.) CP 2007. LNCS, vol. 4741, pp. 590–604. Springer, Heidelberg (2007)
19. Saxena, P., Akhawe, D., Hanna, S., Mao, F., McCamant, S., Song, D.: A symbolic execution framework for javascript. In: Proceedings of the 31st IEEE Symposium on Security and Privacy, California, USA, pp. 513–528. IEEE Press (May 2010), Kaluza is available from `http://webblaze.cs.berkeley.edu/2010/kaluza/`
20. Sellmann, M.: The theory of grammar constraints. In: Benhamou, F. (ed.) CP 2006. LNCS, vol. 4204, pp. 530–544. Springer, Heidelberg (2006)

Blowing Holes in Various Aspects
of Computational Problems, with Applications
to Constraint Satisfaction

Peter Jonsson, Victor Lagerkvist, and Gustav Nordh

Department of Computer and Information Science, Linköping University, Sweden
{peter.jonsson,victor.lagerkvist,gustav.nordh}@liu.se

Abstract. We consider methods for constructing NP-intermediate problems under the assumption that P $\neq$ NP. We generalize Ladner's original method for obtaining NP-intermediate problems by using parameters with various characteristics. In particular, this generalization allows us to obtain new insights concerning the complexity of CSP problems. We begin by fully characterizing the problems that admit NP-intermediate subproblems for a broad and natural class of parameterizations, and extend the result further such that structural CSP restrictions based on parameters that are hard to compute (such as tree-width) are covered. Hereby we generalize a result by Grohe on width parameters and NP-intermediate problems. For studying certain classes of problems, including CSPs parameterized by constraint languages, we consider more powerful parameterizations. First, we identify a new method for obtaining constraint languages Γ such that CSP(Γ) are NP-intermediate. The sets Γ can have very different properties compared to previous constructions (by, for instance, Bodirsky & Grohe) and provides insights into the algebraic approach for studying the complexity of infinite-domain CSPs. Second, we prove that the propositional abduction problem parameterized by constraint languages admits NP-intermediate problems. This settles an open question posed by Nordh & Zanuttini.

1   Introduction

Ladner [20] explicitly constructed NP-intermediate problems (under the assumption P $\neq$ NP) by removing strings of certain lengths from NP-complete languages via a diagonalization technique that is colloquially known as *blowing holes in problems*. The languages constructed via blowing are unfortunately famous for being highly artificial: Arora and Barak [1] write the following.

> We do not know of a natural decision problem that, assuming NP $\neq$ P, is proven to be in NP \ P but not NP-complete, and there are remarkably few candidates for such languages

More natural examples are known under other complexity-theoretic assumptions. For instance, LOGCLIQUE (the problem of deciding whether an n-vertex

C. Schulte (Ed.): CP 2013, LNCS 8124, pp. 398–414, 2013.

graph contains a clique of size $\log n$) is NP-intermediate under the exponential-time hypothesis (ETH). The lack of natural NP-intermediate computational problems makes it important to investigate new classes of NP-intermediate problems and, hopefully, increase our understanding of the borderline between P and NP.

We begin (in Section 3) by presenting a diagonalization method for obtaining NP-intermediate problems, based on parameterizing decision problems in different ways. In our framework, a parameter, or a *measure function*, is simply a function ρ from the instances of some decision problem X to the non-empty subsets of $\mathbb{N}$. We say that such a function is *single-valued* if $\rho(I)$ is a singleton set for every instance of X, and *multi-valued* otherwise. Depending on the parameter one obtains problems with different characteristics. Simple applications of our method include the connection between the complexity class XP and NP-intermediate problems observed by Chen et al. [9]. Even though our method is still based on diagonalization we claim that the intermediate problems obtained are qualitatively different from the ones obtained by Ladner's original method, and that they can be used for gaining new insights into the complexity of computational problems. We demonstrate this on different CSP problems in the following sections.

In Section 4, we analyze the applicability of the diagonalization method for single-valued measure functions. Under mild additional assumptions, we obtain a full understanding of when NP-intermediate problems arise when the measure function is single-valued and polynomial-time computable. Unfortunately, CSPs under structural restrictions (i.e. when considering instances with bounded width parameters) are not captured by this result since width parameters are typically not polynomial-time computable. To remedy this, we present a fairly general method for obtaining NP-intermediate problems based on structurally restricted CSPs in Section 4.2. This is a generalization of a result by Grohe [15] who has shown that, under the assumption that FPT $\neq$ W[1], NP-intermediate CSP problems can be obtained by restricting the tree-width of their corresponding primal graphs. Our result imply that this holds also under the weaker assumption that P $\neq$ NP and for many different width parameters. NP-intermediate problems based on structural restrictions have also been identified by Bodirsky & Grohe [4].

Multi-valued measure functions are apparently much harder to study and a full understanding appears difficult to obtain. Despite this, multi-valued measure functions have highly useful properties and we exploit them for studying constraint satisfaction problems parameterized by constraint languages. Our first result is inspired by Bodirsky & Grohe [4] who have proved that there exists an infinite constraint language Γ such that CSP(Γ) is NP-intermediate. We extend this and prove that whenever an infinite language Γ does not satisfy the so called *local-global* property, i.e. when CSP(Γ) $\notin$ P but CSP(Γ') $\in$ P for all finite $\Gamma' \subset \Gamma$, then there exists a language closely related to Γ such that the resulting CSP problem is NP-intermediate. The only requirement is that Γ can be extended by certain operators $\langle \cdot \rangle$. We then provide two very different extension operators. The first operator $\langle \cdot \rangle_{pow}$ works for languages over both finite and infinite domains but gives relations of arbitrarily high arity. The second operator $\langle \cdot \rangle_+$ is limited to idempotent languages over infinite domains but does have the advantage that the arity of any relation is only increased by a small constant factor. Together with the language Γ° from Jonsson & Lööw [18] which does not satisfy the local-global property we are thus able to identify a concrete language $\langle \Gamma^\circ \rangle_+$ such that $\text{CSP}(\langle \Gamma^\circ \rangle_+)$ is NP-complete, $\text{CSP}(\Gamma') \in \text{P}$ for any finite $\Gamma' \subset \langle \Gamma^\circ \rangle_+$, and there exists a $\Gamma'' \subset \langle \Gamma^\circ \rangle_+$ such that $\text{CSP}(\Gamma'')$ is NP-intermediate. The so-called *algebraic approach* [3,6] has been very successful in studying the computational complexity of both finite- and infinite-domain CSPs. However, this approach is, to a large extent, limited to constraint languages that are finite. If one only considers tractable finite subsets of $\langle \Gamma^\circ \rangle_+$, we miss that there are both NP-intermediate and NP-complete problems within $\text{CSP}(\langle \Gamma^\circ \rangle_+)$. Hence the constraint language $\langle \Gamma^\circ \rangle_+$ clearly shows the algebraic approach in its present shape is not able to give a full understanding of $\text{CSP}(\langle \Gamma^\circ \rangle_+)$ and its subclasses.

Our second result (which is presented in Section 5.3) is the propositional *abduction* problem $\text{ABD}(\Gamma)$. This problem can be viewed as a non-monotonic extension of propositional logic and it has numerous important applications ranging from automated diagnosis, text interpretation to planning. The complexity of propositional abduction has been intensively studied from a complexity-theoretic point of view (cf. [13,23]) and the computational complexity is known for every finite Boolean constraint language Γ and many infinite languages [23]. In Nordh & Zanuttini [23], the question of whether such a classification is possible to obtain for infinite languages was left open. Since the abduction problem can loosely be described as a combination of the SAT and UNSAT problems, it might be expected that it, like the parameterized SAT($\cdot$) problem, does not contain any NP-intermediate problems. By exploiting our diagonalization method, we present a constraint language Γ such that $\text{ABD}(\Gamma)$ is NP-intermediate.

2   Preliminaries

Let Γ denote a (possibly infinite) set of finitary relations over some (possibly infinite) set D. We call Γ a *constraint language*. Given a relation $R \subseteq D^k$, we let $ar(R) = k$. The reader should note that we will sometimes express Boolean relations as conjunctions of Boolean clauses. The *constraint satisfaction problem* over Γ (abbreviated as $\text{CSP}(\Gamma)$) is defined as follows.

INSTANCE: A set V of variables and a set C of constraint applications $R(v_1, \ldots, v_k)$ where $R \in \Gamma$, $k = ar(R)$, and $v_1, \ldots, v_k \in V$.
QUESTION: Is there a total function $f : V \to D$ such that $(f(v_1), \ldots, f(v_k)) \in R$ for each constraint $R(v_1, \ldots, v_k)$ in C?

For an arbitrary decision problem X, we let $I(X)$ denote its set of instances, and $||I||$ to denote the number of bits needed for representing $I \in I(X)$. By a *polynomial-time reduction* from problem X to problem X', we mean a Turing reduction from X to X' that runs in time $O(p(||I||))$ for some polynomial p.

Definition 1. *Let X be a decision problem. A total and computable function $\rho : I(X) \rightarrow 2^{\mathbb{N}} \setminus \{\emptyset\}$ is said to be a* measure function.

If $\rho(I)$ is a singleton set for every $I \in I(X)$, then we say that ρ is *single-valued*, and otherwise that it is *multi-valued*. We abuse notation in the first case and simply assume that $\rho : I(X) \rightarrow \mathbb{N}$. The measure function ρ combined with a decision problem X yields a problem $X_\rho(S)$ parameterized by $S \subseteq \mathbb{N}$.

INSTANCE. Instance I of X such that $\rho(I) \subseteq S$.
QUESTION. Is I a yes-instance?

For examples of both single- and multi-valued measure functions we refer the reader to Section 3.2. Finally, we prove a simple lemma regarding single-valued measure functions that will be important later on.

Lemma 2. *Let ρ be a single-valued and polynomial-time computable measure function. Let $S \subseteq \mathbb{N}$ and let T be a non-empty subset of S such that $S \setminus T = \{s_1, \ldots, s_k\}$. If $X_\rho(\{s_i\})$, $1 \leq i \leq k$, is in P, then there is a polynomial-time reduction from $X_\rho(S)$ to $X_\rho(T)$.*

Proof. Let I be an arbitrary instance of $X_\rho(S)$. Compute (in polynomial time) $\rho(I)$. If $\rho(I) \in \{s_1, \ldots, s_k\}$, then we can compute the correct answer in polynomial time. Otherwise, I is an instance of $X_\rho(T)$ and the reduction is trivial. $\square$

3   Generation of NP-Intermediate Problems

We will now extend Ladner's method to parameterized problems. Section 3.1 contains the main result and Section 3.2 contains some examples.

3.1   Diagonalization Method

Theorem 3. *Let $X_\rho(\cdot)$ be a computational decision problem with a measure function ρ. Assume that $X_\rho(\cdot)$ and $S \subseteq \mathbb{N}$ satisfies the following properties:*

P0: $I(X)$ is recursively enumerable.
P1: $X_\rho(S)$ is NP-complete.
P2: $X_\rho(T)$ is in P whenever T is a finite subset of S.
P3: $X_\rho(S)$ is polynomial-time reducible to $X_\rho(T)$ whenever $T \subseteq S$ and $S \setminus T$ is finite.

Then, if $P \neq NP$, there exists a set $S' \subset S$ such that $X_\rho(S')$ is in $NP \setminus P$ and $X_\rho(S)$ is not polynomial-time reducible to $X_\rho(S')$.

Before the proof, we make some observations that will be used without explicit references. If ρ is single-valued and polynomial-time computable, then P2 implies P3 by Lemma 2. In many examples, $S = \mathbb{N}$ which means that P1 can be restated as NP-completeness of X. If P1 holds, then property P3 simply states that $X_\rho(T)$ is NP-complete for every cofinite $T \subseteq S$. Finally, we remind the reader that the polynomial-time bounds may depend on the choice of S in the definitions of P2 and P3.

The proof is an adaption of Papadimitriou's [24] proof where we use the abstract properties P0 – P3 instead of focusing on the size of instances. Papadimitriou's proof is, in turn, based on Ladner's original proof [20]. It may also be illuminating to compare with Schöning [25] and Bodirsky & Grohe [4].

In the sequel, we let $X_\rho(\cdot)$ be a computational decision problem that together with $S \subseteq \mathbb{N}$ satisfies properties P0 – P3. Let A_X be an algorithm for $X_\rho(S)$, let $M_1, M_2, \ldots$ be an enumeration of all polynomial-time bounded deterministic Turing machines, and let $R_1, R_2, \ldots$ be an enumeration of all polynomial-time Turing reductions. Such enumerations are known to exist, cf. Papadimitriou [24].

We define a function $f : \mathbb{N} \to \mathbb{N}$ that is computed by a Turing machine F and the input n is given to F in unary representation. We let $f(0) = f(1) = 0$. The computation of $f(n)$ starts with the computation of $f(0), f(1), f(2), \ldots$, until the total number of steps F has used in computing this sequence exceeds n. This is possible since F has access to its own description by Kleene's fixed point theorem. Let i be the largest value for which F was able to completely compute $f(i)$ (during these n steps) and let $k = f(i)$.

In the final phase of the execution of the machine F we have two cases depending on whether k is even or odd. In both cases, if this phase requires F to run for more than n computation steps, F stops and returns k (i.e., $f(n) = k$).

The first case is when k is even: here, F enumerates all instances I of $X_\rho(S)$ — this is possible by property P0. For each instance I, F simulates $M_{k/2}$ on the encoding of I, determines whether $A_X(I)$ is accepted, and finally, F computes f for all $x \in \rho(I)$. If $M_{k/2}$ rejects and $A_X(I)$ was accepted, and $f(x)$ is even for all $x \in \rho(I)$, then F returns $k + 1$ (i.e., $f(n) = k + 1$). F also returns $k + 1$ if $M_{k/2}$ accepts and I is not accepted by A_X and $f(x)$ is even for all $x \in \rho(I)$.

The second case is when k is odd. Again, F enumerates all instances I of $X_\rho(S)$. Let $E = \emptyset$. Now, for each instance I, F begins simulating $R_{\lfloor k/2 \rfloor}$ on the encoding of I with an oracle for A_X. Whenever the simulation notices that $R_{\lfloor k/2 \rfloor}$ enters an oracle state, we calculate $\rho(I') = E'$ (where I' is the $X_\rho(S)$ instance corresponding to the input of the oracle tape), and add the members of E' to E. When the simulation is finished we first calculate $f(x)$ for every $x \in E$. If the result of any $f(x)$ operation is odd we return $k + 1$. We then compare the result of the reduction with $A_X(I)$. If the results do not match, i.e. if one is accepted or rejected while the other is not, we return $k + 1$. This completes the definition of f. Note that f can be computed in polynomial time (regardless of the time complexity of computing ρ and A_X) since the input is given in unary.

We now show that f is increasing, i.e. for all $n \geq 0$, $f(n) \leq f(n + 1)$ and $\{f(n) \mid n \in \mathbb{N}\}$ is an unbounded set, unless P = NP. To see this, we first prove by induction that $f(n) \leq f(n + 1)$ for all $n \geq 0$. This obviously holds for $n = 0$ and $n = 1$. Assume that this holds for an arbitrary number $i > 1$. By definition $f(i + 1)$ cannot return a smaller number than $f(i)$ in the first phase of the computation, since the Turing machine F simulates $f(i')$ for all $i' < i$, and returns the largest k for which $f(i')$ was successfully computed within the allotted time. In the second phase, the argument to f is used to determine the total amount of computation steps, and since f will either return the k from the first phase, or $k+1$, there is no possibility that $f(i) > f(i+1)$.

Let $S_e = \{x \mid x \in S \text{ and } f(x) \text{ is even}\}$. We continue by showing that there is no n_0 such that $f(n) = k_0$ for all $n > n_0$ unless P = NP. If there is such a n_0, then there is also a n_1 such that for all $n > n_1$ the value k computed in the phase where F computes $f(1), f(2), \ldots$ (in n steps) is k_0. If k_0 is even, then on all inputs $n > n_1$ the machine $M_{k_0/2}$ correctly decides $X_\rho(S_e)$ and thus $X_\rho(S_e)$ is in P. But since $f(n) = k_0$ for all $n > n_1$, we have that $S \setminus S_e$ is finite, and thus $X_\rho(S)$ is polynomial-time reducible to $X_\rho(S_e)$ by Property P3, which is a contradiction since $X_\rho(S)$ is NP-complete by Property P1. Similarly if k_0 is odd, then on all inputs $n > n_1$ the function $R_{\lfloor k_0/2 \rfloor}$ is a valid reduction from $X_\rho(S)$ to $X_\rho(S_e)$ and thus $X_\rho(S_e) \notin$ P. But since $f(n) = k_0$ for all $n > n_1$, we have that S_e is finite, and we conclude that $X_\rho(S_e)$ is in P by Property P2, which is a contradiction since $X_\rho(S)$ is NP-complete by Property P1.

We conclude the proof by showing that $X_\rho(S_e)$ is neither in P, nor is $X_\rho(S)$ polynomial-time reducible to $X_\rho(S_e)$, unless P = NP. By Property P1, $X_\rho(S_e)$ is in NP since $S_e \subseteq S$. Assume now that $X_\rho(S_e)$ is in P. Then there is an i such that M_i solves $X_\rho(S_e)$. Thus, by the definition of f, there is an n_1 such that for all $n > n_1$ we have $f(n) = 2i$; this contradicts that f is increasing. Similarly, assume that $X_\rho(S)$ is polynomial-time reducible to $X_\rho(S_e)$. Then, there is an i such that R_i is a polynomial-time reduction from $X_\rho(S)$ to $X_\rho(S_e)$. It follows from the definition of f that there is an n_1 such that $f(n) = 2i - 1$ for all $n > n_1$, and this contradicts that f is increasing. $\qquad\square$

If the measure function is polynomially bounded (e.g. $\rho(I) \leq p(||I||)$ for some polynomial p), then checking whether an integer x written in binary is in S_e or not can be decided in polynomial time. This follows from the fact that x written in binary can be converted to x written in unary in polynomial time. Another useful observation is the following: it follows from the proof that property P1 (i.e. the NP-hardness of the original problem) can be replaced by hardness for other complexity classes within NP. By noting that $X_\rho(S_e)$ is recursively enumerable, this implies that we can construct infinite chains of problems $X_\rho(T_1), X_\rho(T_2), \ldots$ such that $S_e = T_1 \supset T_2 \supset \ldots$, there does not exist any polynomial-time reductions from $X_\rho(T_i)$ to $X_\rho(T_{i+1})$, and $X_\rho(T_i)$ is not in P for any $i \geq 1$.

3.2   Examples

Ladner's result is now a straightforward consequence of Theorem 3. Let X be an arbitrary NP-complete problem such that $I(X)$ is recursively enumerable. For an arbitrary instance $I \in I(X)$, we let the single-valued measure function ρ be defined such that $\rho(I) = ||I||$. We verify that $X_\rho(\mathbb{N})$ satisfies properties P0 – P3 and conclude that there exists a set $T \subseteq \mathbb{N}$ such that $X_\rho(T)$ is NP-intermediate. Properties P0 and P1 hold by assumption and property P2 holds since $X_\rho(U)$ can be solved in constant time whenever U is finite. If $U \subseteq \mathbb{N}$ and $\mathbb{N} \setminus U = \{x_1, \ldots, x_k\}$, then $X_\rho(\{x_i\})$, $1 \leq i \leq k$, is solvable in constant time and we can apply Lemma 2(2). Thus, property P3 holds, too.

Another straightforward application of single-valued measure functions is the following: Chen et al. [9] have discovered a striking connection between NP-intermediate problems and the parameterized complexity class XP (XP denotes the class of decision problems X that are solvable in time $O(||I||^{f(k)})$ for some polynomial-time computable parameter k and some computable function f).

Proposition 4. *Let X be a decision problem and ρ a polynomial-time computable single-valued measure function such that $X_\rho(\cdot)$ satisfies conditions P0 and P1, and $X_\rho \in$ XP. Then there exists a $T \subseteq \mathbb{N}$ such that $X_\rho(T)$ is NP-intermediate.*

Proof. We note that $X_\rho(S)$ is in P whenever S is a finite subset of $\mathbb{N}$. Hence, X_ρ satisfies P2 and consequently P3. The result follows from Theorem 3.   □

To illustrate multi-valued measure functions, we turn our attention to the SUBSET-SUM problem [19].

INSTANCE: A finite set $Y \subseteq \mathbb{N}$ and a number $k \in \mathbb{N}$.
QUESTION: Is there a $Y' \subseteq Y$ such that $\sum Y' = k$?

We define a multi-valued measure function by letting $\rho((Y,k)) = Y$. Once again, properties P0 and P1 hold by assumption so it is sufficient to prove that SUBSET-SUM$_\rho(\mathbb{N})$ satisfies P2 and P3. Property P2: instances of SUBSET-SUM can be solved in time $O(\text{poly}(||I||) \cdot c(I))$, where $c(I)$ denotes the difference between the largest and smallest number in Y [14]. This difference is finite whenever we consider instances of SUBSET-SUM$_\rho(S)$ where $S \subseteq \mathbb{N}$ is finite. Property P3: arbitrarily choose $S \subseteq \mathbb{N}$ such that that $\mathbb{N} \setminus S$ is finite. We present a polynomial-time Turing reduction from SUBSET-SUM$_\rho(\mathbb{N})$ to SUBSET-SUM$_\rho(S)$. Let $I = (Y,k)$ be an instance of SUBSET-SUM$_\rho(\mathbb{N})$. Let $T = Y \setminus S$, i.e. the elements of the instance which are not members of the smaller set S. Since $\mathbb{N} \setminus S$ is finite, T is a finite set, too. Let $Z = Y \cap S$. For every subset $T'_i = \{x_1, \ldots, x_{im}\}$ of T, we let $I'_i = (Z, k'_i)$, where $k'_i = k - (x_1 + \ldots + x_{im})$. Then, it is easy to see that I is a yes-instance if and only if at least one I'_i is a yes-instance. Finally, we note that the reduction runs in time $O(\text{poly}(||I||) \cdot 2^c)$, where $c = |\mathbb{N} \setminus S|$, and this is consequently a polynomial-time reduction for every fixed S.

4   Single-Valued Measure Functions

This section is divided into two parts: Section 4.1 is concerned with polynomial-time computable single-valued measure functions and Section 4.2 is concerned with structurally restricted CSPs.

4.1   Polynomial-Time Computable Measure Functions

By Theorem 3, we know that properties P0 – P3 are sufficient to assure the existence of NP-intermediate problems. A related question is to what degree the properties are also necessary. Here, we investigate the scenario when P2 and P3 do not necessarily hold.

Theorem 5. *Assume X is a decision problem and ρ is a single-valued measure function such that $X_\rho(\mathbb{N})$ satisfies P0 and P1. Let $S_P = \{s \in \mathbb{N} \mid X_\rho(\{s\}) \in$ P$\}$ and assume membership in S_P is a decidable problem. Then, at least one of the following holds: (1) there exists a set $T \subseteq S_P$ such that $X_\rho(T)$ is NP-intermediate, (2) there exists a $t \in \mathbb{N}$ such that $X_\rho(\{t\})$ is NP-intermediate, or (3) X_ρ admits no NP-intermediate subproblems.*

Proof. If $X_\rho(\{s\})$ is NP-complete for every $s \in \mathbb{N}$, then we are in case (3) so we assume this is not the case. If there exists $s \in \mathbb{N}$ such that $X_\rho(\{s\})$ is NP-intermediate, then we are in case (2) so we assume this does not hold either. Thus, we may henceforth assume that there exists $s \in \mathbb{N}$ such that $X_\rho(\{s\}) \in P$ and that $X_\rho(\{u\})$ is NP-complete whenever $u \in \mathbb{N} \setminus S_P$. This implies that S_P is non-empty. Once again, we single out two straightforward cases: if $X_\rho(S_P)$ is NP-intermediate, then we are in case (1), and if $X_\rho(S_P)$ is in P, then we are in case (3) (since $X_\rho(\{u\})$ is NP-complete whenever $u \notin S_P$). Hence, we may assume that $X_\rho(S_P)$ is NP-complete (note that $X_\rho(S_P) \in$ NP since $X_\rho(\mathbb{N}) \in$ NP by P1), i.e. $X_\rho(S_P)$ satisfies P1. Furthermore, $X_\rho(S_P)$ satisfies P0 since S_P is a decidable set and the instances of X are recursively enumerable. To generate the instances of $X_\rho(S_P)$, we generate the instances of X one after another and output instance I if and only if $\rho(I)$ is in S_P.

We finally show that $X_\rho(S_P)$ satisfies P2 and P3. It is sufficient to prove that $X_\rho(S_P)$ satisfies P2 since ρ is single-valued. Assume there exists a finite set $K \subseteq S_P$ such that $X_\rho(K) \notin P$. Let $\emptyset \subset K' \subseteq K$ be a subset such that $X_\rho(K')$ is a member of P; such a set exists since $K \subseteq S_P$. For every $k' \in K'$, we know that $X_\rho(\{k'\}) \in P$. Hence, we can apply Lemma 2 and deduce that there exists a polynomial-time reduction from $X_\rho(K)$ to $X_\rho(K')$. This contradicts the fact that $X_\rho(K)$ is not a polynomial-time solvable problem. We can now apply Theorem 3 and conclude that there exists a set $T \subseteq S_P$ such that $X_\rho(T)$ is NP-intermediate, i.e. we are in case (1). □

Problems parameterized by multi-valued measure functions are apparently very different from those parameterized by single-valued functions. For instance, Lemma 2 breaks down which indicates that the proof strategy used in Theorem 5 is far from sufficient to attack the multi-valued case.

4.2   Structurally Restricted CSPs

When identifying tractable (i.e. polynomial-time solvable) fragments of constraint satisfaction problems and similar problems, two main types of results have been considered in the literature. The first one is to identify constraint languages Γ such that $\mathrm{CSP}(\Gamma) \in P$, and the second one is to restrict the structure induced by the constraints on the variables. The second case is often concerned with associating some structure with each instance and then identifying sets of structures that yield tractable problems. The classical example of this approach is to study the *primal graph* or *hypergraph* of CSP instances. Given a CSP instance I with variable set V, we define its primal graph $G = (V, E)$ such that $(v_i, v_j) \in E$ if and only if variables v_i, v_j occur simultaneously in some constraint, and we define the hypergraph $\mathcal{H} = (V, \mathcal{E})$ such that the hyperedge $\{v_{i_1}, ..., v_{i_k}\} \in \mathcal{E}$ if and only if there is a constraint $R(v_{i_1}, \ldots, v_{i_k})$ in I.

When it comes to defining structurally restricted problems that are tractable, one is typically interested in certain parameters of these (hyper)graphs such as *tree-width, fractional hypertree width* [16], or *submodular width* [22]. It is, for instance, known that any finite-domain CSP instance I with primal graph $G = (V, E)$ can be solved in $||I||^{O(\text{tw}(G))}$ time [11] where $\text{tw}(G)$ denotes the tree-width of G, and it can be solved in $||I||^{O(\text{fhw}(\mathcal{H}))}$ time [16] where $\text{fhw}(\mathcal{H})$ denotes the fractional hypertree width of $\mathcal{H}$. Since these results rely on the domains being finite, we restrict ourselves to finite-domain CSPs throughout this section. Now note that if given a finite constraint language Γ, then the instances of $\text{CSP}(\Gamma)$ are recursively enumerable and $\text{CSP}(\Gamma)$ is in NP. If Γ is infinite, then this is not so evident and it may, in fact, depend on the representation of relations. We adopt a simplistic approach and represent a relation by listing its tuples. Under this assumption, the instances of $\text{CSP}(\Gamma)$ are recursively enumerable and $\text{CSP}(\Gamma)$ is in NP.

By restricting the CSP problem to instances with tree-width or fractional hypertree width $\leq k$ (for some constant k), it is known that the resulting problem is solvable in polynomial time. This immediately implies that problems like CSP_{tw} and CSP_{fhw}[1] have property P2. If the width parameter under consideration is polynomial-time computable, then we have property P3 (via Lemma 2), too, and conclude that NP-intermediate fragments exist. Unfortunately, this is typically not the case. It is for instance NP-complete to determine whether a given graph G has treewidth at most k or not [2] if k is part of the input. This is a common feature that holds for, or is suspected to hold for, many different width parameters. Hence, width parameters are a natural source of single-valued measure functions that are not polynomial-time computable. Such measure functions are problematic since we cannot prove the existence of NP-intermediate subproblems by using simplifying results like Proposition 4 or Theorem 5. By a few additional assumptions we can however still prove the applicability of Theorem 3. Note that if k is fixed, and thus not part of the input, then the graphs with tree-width $\leq k$ can be recognized in linear time [5]. This is not uncommon when studying width parameters — determining the width exactly is computationally hard but it can be computed or estimated in polynomial time under additional assumptions. We arrive at the following result.

Proposition 6. *Assume that X is a decision problem and ρ is a single-valued measure function such that $X_\rho(\cdot)$ satisfies conditions P0 and P1. Furthermore suppose that for each set $\{0, \ldots, k\}$ there exists a promise algorithm A_k for $X_\rho(\{0, \ldots, k\})$ with the following properties:*

- *if $\rho(I) \leq k$, then A_k returns the correct answer in $p_k(||I||)$ steps where p_k is a polynomial only depending on k, and*

---

[1] We slightly abuse notation since tw and fhw are not directly defined on problem instances.

 – if $\rho(I) > k$, then A_k either return a correct answer or do not answer at all.

Then there exists a set $S \subset \mathbb{N}$ such that $X_\rho(S)$ is NP-intermediate.

Proof. Let X^k denote the computational problem X restricted to instances $I \in I(X)$ such that $\rho(I) \geq k$. Assume there exists a k such that $X^k \in P$ and let B be an algorithm for this problem running in time $q(||I||)$ for some polynomial q. For $X_\rho(\{0, \dots, k-1\})$, we have algorithm A_{k-1} described above. Given an arbitrary instance I of X, we may not be able to compute $\rho(I)$ and choose which algorithm to run. Do as follows: run algorithm A_{k-1} for $p_{k-1}(||I||)$ steps on input I. If A_{k-1} produces an answer, then this is correct. If A_{k-1} does not produce an answer, then we know that $\rho(I) > k-1$ and we can apply algorithm B. All in all, this takes $O(p_{k-1}(||I||) + q(||I||))$ time so $X \in P$ which leads to a contradiction.

If X^k is in NPI for some k, then we simply let $S = \{k, k+1, \dots\}$. We can henceforth assume that X^k is NP-complete for all k. Obviously, $X_\rho(\mathbb{N})$ satisfies property P2 since algorithm A_k, $k \geq 0$, runs in polynomial time. We show that it satisfies property P3, too. Let $T \subseteq \mathbb{N}$ be a finite set and let $m = \max T$. We know that X^{m+1} is NP-complete. Hence, there exists a polynomial-time reduction from the NP-complete problem $X_\rho(\mathbb{N})$ to X^{m+1} which, in turn, admits a trivial polynomial-time reduction to $X_\rho(\mathbb{N} \setminus T)$ since $\{m+1, m+2, \dots\} \subseteq \mathbb{N} \setminus T$. We can now apply Theorem 3 and obtain the set S.   □

We apply this result to CSP_{tw} and CSP_{fhw}, respectively. Clearly, both these problems satisfy properties P0 and P1 due to the assumptions that we have made. For CSP_{tw}, we let A_k work as follows: given a CSP instance I, check whether I has treewidth $\leq k$ using Bodlaender's [5] algorithm. If the algorithm answers "no", then go into an infinite loop. Otherwise, decide whether I has a solution or not in $||I||^{O(k)}$ time. Proposition 6 implies that there exists a set $T \subseteq \mathbb{N}$ such that $CSP_{tw}(T)$ is NP-intermediate. We observe that Grohe [15] has shown a similar result under the assumption that FPT $\neq$ W[1] instead of P $\neq$ NP. Many other width parameters can also be used for obtaining NP-intermediate problems. One example is CSP_{fhw} for which the proof is very similar but is instead based on Theorem 4.1 in Marx [21].

5   Multi-valued Measure Functions

In this section we turn our attention to multi-valued measure functions and apply them to constraint problems. Throughout this section we assume that P $\neq$ NP. Here, we want to associate the complexity of CSPs with constraint languages and multi-valued measure functions are convenient for this purpose. Given a constraint satisfaction problem parameterized with a constraint language Γ, let ρ denote the single-valued measure function defined to return the highest arity of any constraint in a given instance: $\rho((V, C)) = \max\{k \mid R(v_1, \dots, v_k) \in C\}$. Let $CSP_\rho^*(X)$ denote the $CSP(\Gamma)$ problem restricted to instances I such that $\rho(I) \in X$, and assume there exists a set $X \subset \mathbb{N}$ such that $CSP_\rho^*(X)$ is NP-intermediate. Can we from this conclude that there exists a constraint language

$\Gamma' \subset \Gamma$ such that $\mathrm{CSP}(\Gamma')$ is NP-intermediate? In general, the answer is no since the set of valid instances of $\mathrm{CSP}_\rho^*(X)$ are not in a one-to-one correspondence with any constraint language restriction. Note that $\mathrm{CSP}_\rho^*(X)$ is not the same problem as $\mathrm{CSP}(\{R \in \Gamma \mid ar(R) \in X\})$. If we on the other hand define the multi-valued measure function $\sigma((V,C)) = \{k \mid R(v_1,\ldots,v_k) \in C\}$, then for every $X \subset \mathbb{N}$ the problem $\mathrm{CSP}_\sigma^*(X)$ is equivalent to $\mathrm{CSP}(\{R \in \Gamma \mid ar(R) \in X\})$.

5.1 Constraint Satisfaction Problems and the Local-Global Conjecture

A constraint language Γ is said to have the *local-global* property [4] if $\mathrm{CSP}(\Gamma') \in$ P for every finite set $\Gamma' \subset \Gamma$ implies $\mathrm{CSP}(\Gamma) \in$ P. The non-existence of languages not having the local-global property is known as the *local-global conjecture*. In Bodirsky & Grohe [4] it is proven that if Γ is a constraint language over a finite domain D that does not exhibit the local-global property, then there exists a constraint language Γ' over D such that $\mathrm{CSP}(\Gamma')$ is NP-intermediate. In this section we prove a more general result not restricted to finite domains based on the notion of *extension operators*. If R is a k-ary relation and Γ a constraint language over a domain D we say that R has a *primitive positive* (p.p.) definition in Γ if $R(x_1,\ldots,x_k) \equiv \exists y_1,\ldots,y_l . R_1(\mathbf{x_1}) \wedge \ldots R_i(\mathbf{x_i})$, where each $R_j \in \Gamma \cup \{=\}$ and each $\mathbf{x_i}$ is a vector over $x_1,\ldots,x_k, y_1,\ldots,y_l$.

Definition 7. *Let Γ be a recursively enumerable constraint language (with a suitable representation of relations in Γ). We say that $\langle \cdot \rangle$ is an extension operator if (1) $\langle \Gamma \rangle$ is a recursively enumerable set of p.p. definable relations over Γ and (2) whenever $\Delta \subset \langle \Gamma \rangle$ and $\langle \Gamma \rangle \setminus \Delta$ is finite, then every $R \in \langle \Gamma \rangle \setminus \Delta$ is p.p. definable in Δ.*

Another way of viewing this is that the expressive power of $\langle \Gamma \rangle$ does not change when removing finitely many relations. Since Γ and $\langle \Gamma \rangle$ are recursively enumerable we can enumerate relations in Γ or $\langle \Gamma \rangle$ as $R_1, R_2, \ldots$, and it is not hard to see that this implies that instances of $\mathrm{CSP}(\Gamma)$ and $\mathrm{CSP}(\langle \Gamma \rangle)$ are also recursively enumerable. Given an instance I of $\mathrm{CSP}(\Gamma)$ containing the relations $R_{i_1},\ldots,R_{i_k}$, we let $\rho(I) = \{i_1,\ldots,i_k\}$. Let $\mathrm{CSP}_\rho^*(S)$ denote the $\mathrm{CSP}(\Gamma)$ problem over instances I such that $\rho(I) \subseteq S$. Define the measure function ρ' analogous to ρ but for instances over $\mathrm{CSP}(\langle \Gamma \rangle)$, and let $\mathrm{CSP}_{\rho'}^\times(S)$ be the $\mathrm{CSP}(\langle \Gamma \rangle)$ problem restricted to instances I such that $\rho'(I) \subseteq S$.

Theorem 8. *Assume Γ is a constraint language such that $\mathrm{CSP}_\rho^*(\mathbb{N})$ satisfies property P0 – P2. Let $\langle \cdot \rangle$ be an extension operator such that $\mathrm{CSP}_{\rho'}^\times(\langle \Gamma \rangle)$ satisfies property P0 – P1. Then there exists a $\Gamma' \subset \langle \Gamma \rangle$ such that $\mathrm{CSP}(\Gamma')$ is NP-intermediate.*

Proof. We prove that $\mathrm{CSP}_{\rho'}^\times(\mathbb{N})$ satisfies property P0 – P3. The first two properties are trivial by assumption. For property P2 let $T = \{i_1,\ldots,i_k\}$ be an arbitrary finite subset of $\mathbb{N}$ and let $\Theta = \{R_{i_1},\ldots,R_{i_k}\}$. Note that Θ might contain relations which are not included in Γ. For every such relation $R \in \Theta$ we can however replace it by its p.p. definition in Γ. Let the resulting set of relations be Θ' and let $S = \{i \mid R_i \in \Theta'\}$. Then $\text{CSP}^\times_{\rho'}(T)$ and $\text{CSP}^*_\rho(S)$ are polynomial-time equivalent since T is a finite set. Since $\text{CSP}^*_\rho(S)$ is solvable in polynomial time by assumption, $\text{CSP}^\times_{\rho'}(T)$ is polynomial-time solvable too.

For property P3 let $T \subset \mathbb{N}$ such that $\mathbb{N} \setminus T = \{t_1, \ldots, t_k\}$. To see that there exists a polynomial-time reduction from $\text{CSP}^\times_{\rho'}(\mathbb{N})$ to $\text{CSP}^\times_{\rho'}(T)$, we let I be an arbitrary instance of $\text{CSP}^\times_{\rho'}(\mathbb{N})$. Assume I contains the constraint $R_i(x_1, \ldots, x_m)$, $i \in \mathbb{N} \setminus T$. Since $\langle \cdot \rangle$ is an extension operator the relation R_i is p.p. definable in $\langle \Gamma \rangle \setminus \Delta$ where $\Delta = \{R_i \mid i \in \mathbb{N} \setminus T\}$. Thus, we can replace $R_i(x_1, \ldots, x_m)$ with its p.p. definition in $\langle \Gamma \rangle \setminus \Delta$, and by doing this for all constraints that are not allowed by T, we end up with an instance I' of $\text{CSP}^\times_{\rho'}(T)$ that is satisfiable if and only if I is satisfiable. This is a polynomial-time reduction since $\mathbb{N} \setminus T$ is a finite set.

By applying Theorem 3, we can now identify a set $S \subset \mathbb{N}$ such that $\text{CSP}^\times_{\rho'}(S)$ is NP-intermediate. This implies that $\text{CSP}(\Gamma')$ is NP-intermediate when $\Gamma' = \{R_i \in \langle \Gamma \rangle \mid i \in S\}$. $\qquad \square$

Our first extension operator is based on the idea of extending a relation into a relation with higher arity. For any relation $R \subseteq D^n$, we define the kth power of R to be the relation $R^k(x_0, \ldots, x_{k \cdot n - 1}) \equiv R(x_0, \ldots, x_{n-1}) \wedge R(x_n, \ldots, x_{n+n-1}) \wedge R(x_{2n}, \ldots, x_{2n+n-1}) \wedge \ldots \wedge R(x_{(k-1)n}, \ldots, x_{(k-1)n+n-1})$. Given a constraint language Γ, let $\langle \Gamma \rangle_{pow} = \{R^k \mid R \in \Gamma \text{ and } k \in \mathbb{N}\}$. We represent each relation in $\langle \Gamma \rangle_{pow}$ as a pair (R, k). It is easy to see that $\text{CSP}(\langle \Gamma \rangle_{pow}) \in \text{NP}$ if $\text{CSP}(\Gamma) \in \text{NP}$ from which it follows that $\text{CSP}(\langle \Gamma \rangle_{pow})$ is NP-complete. Now assume that $\Delta \subset \langle \Gamma \rangle_{pow}$ and that $\langle \Gamma \rangle_{pow} \setminus \Delta$ is finite. First, for every $R^k \in \langle \Gamma \rangle_{pow} \setminus \Delta$ we can p.p. define R^k in Δ as $R(x_1, \ldots, x_n) \equiv \exists x_{n+1}, \ldots, x_{k' \cdot n + n - 1}.R^{k'+1}(x_1, \ldots, x_n, x_{n+1}, \ldots, x_{k' \cdot n + n - 1})$, where $k' > k$. Such a k' must exist since we have only removed finitely many relations from $\langle \Gamma \rangle_{pow}$. Hence $\langle \cdot \rangle_{pow}$ is an extension operator. Extension operators are not uncommon in the literature. Well studied examples (provided relations can be suitably represented) include closure under p.p. definitions (known as *co-clones*) and closure under p.p. definitions without existential quantification (known as *partial co-clones*). These are indeed extension operators since $\langle \Gamma \rangle_{pow}$ is always a subset of the partial co-clone of Γ and hence also of the co-clone of Γ. For a general introduction to the field of clone theory we refer the reader to Lau [26].

Let $R_{a,b,c,U} = \{(x, y) \in \mathbb{Z}^2 \mid ax - by \leq c, 0 \leq x, y \leq U\}$ for arbitrary $a, b, U \in \mathbb{N}$ and $c \in \mathbb{Z}$. Furthermore let $\Gamma'_U = \{R_{a,b,c,U} \mid a, b \in \mathbb{N}, c \in \mathbb{Z}\}$ for any $U \in \mathbb{N}$ and the language Γ° be defined as $\Gamma^\circ = \bigcup_{i=0}^\infty \Gamma'_i$. Note that we can represent each relation in Γ° compactly by four integers written in binary. Due to Jonsson & Lööw [18] it is known that Γ° does not satisfy the local-global property. By combining the language Γ° and the extension operator $\langle \cdot \rangle_{pow}$ with Theorem 8 we thus obtain the following result.

Theorem 9. *There exists a $\Gamma' \subset \langle \Gamma^\circ \rangle_{pow}$ such that CSP(Γ') is NP-intermediate.*

Due to the work of Bodirsky & Grohe [4] we already know that the CSP problem over infinite domains is non-dichotomizable. Their result is however based on reducing an already known NP-intermediate problem to a CSP problem while our language $\Gamma' \subset \langle \Gamma^\circ \rangle_{pow}$ is an explicit example of a locally tractable language obtained via blowing holes.

5.2   Locally Tractable Languages with Bounded Arity

The downside of the $\langle \cdot \rangle_{pow}$ operator is that the construction creates relations of arbitrary high arity even if the language only contain relations of bounded arity. In this section we show that simpler extensions are sometimes applicable for constraint languages over infinite domains. For any k-ary relation R we define the $(k+1)$-ary relation R_a as $R_a(x_1, \ldots, x_n, y) \equiv R(x_1, \ldots, x_n) \wedge (y = a)$, where $a \in D$ and $(y = a)$ is the constraint application of the relation $\{(a)\}$. Let $\langle \Gamma \rangle_+ = \{R_a \mid R \in \Gamma, a \in D\}$. If we represent each relation in $\langle \Gamma \rangle_+$ as a tuple (R, a) then obviously $\langle \Gamma \rangle_+$ is recursively enumerable if Γ is recursively enumerable. Now assume that Γ is an infinite constraint language and that $\langle \Gamma \rangle_+ \setminus \Delta$ is finite. For any relation $R_a \in \langle \Gamma \rangle_+ \setminus \Delta$ we first determine a b such that $R_b \in \Delta$. By construction there exists such a b since $\langle \Gamma \rangle_+ \setminus \Delta$ is finite. Then, since Γ is infinite, there exists an m-ary relation $R' \in \Gamma$ such that $R'_a \in \Delta$. Hence we can implement R_a as $R_a(x_1, \ldots, x_n, y) \equiv \exists y', x'_1, \ldots, x'_m.R_b(x_1, \ldots, x_n, y') \wedge R'_a(x'_1, \ldots, x'_m, y)$, by which it follows that $\langle \cdot \rangle_+$ is an extension operator.

Say that a language Γ is *idempotent* if for all $a \in D$ it holds that $\{(a)\}$ is p.p. definable in Γ. We assume that we can find the p.p. definition of $\{(a)\}$) in Γ in polynomial time.

Theorem 10. *Let Γ be an idempotent language over an infinite domain such that Γ does not satisfy the local-global property. Then there exists a constraint language Γ' such that (1) CSP(Γ') is NP-intermediate and (2) Γ' contains only relations of arity $k + 1$, where k is the highest arity of a relation in Γ.*

Proof. Let $R_1, R_2, \ldots$ be an enumeration of Γ and define the measure function ρ over an instance I containing the relations $R_{i_1}, \ldots, R_{i_k}$ as $\rho(I) = \{i_1, \ldots, i_k\}$. We note that Γ must be infinite since it does not satisfy the local-global property. Let $CSP^*_\rho(S)$ denote the CSP(Γ) problem over instances I such that $\rho(I) \subseteq S$. Then $CSP^*_\rho(\mathbb{N})$ obviously satisfies property P0–P2, and since $\langle \cdot \rangle_+$ is an extension operator, we only need to prove that $CSP(\langle \Gamma \rangle_+)$ is NP-complete. NP-hardness is easy since CSP(Γ) is trivially polynomial-time reducible to $CSP(\langle \Gamma \rangle_+)$. For membership in NP we give a polynomial-time reduction from $CSP(\langle \Gamma \rangle_+)$ to CSP(Γ). Let I be an arbitrary instance of $CSP(\langle \Gamma \rangle_+)$. For any constraint $R_a(x_1, \ldots, x_n, y)$ we replace it by $R(x_1, \ldots, x_n) \wedge \phi(x'_1, \ldots, x'_m, y)$, where $\exists x'_1, \ldots, x'_m.\phi$ is the p.p. definition of $y = a$, which is computable in polynomial time by assumption. If we repeat the procedure for all R_a in I we get an instance I' of CSP(Γ) which is satisfiable if and only if I is satisfiable. Hence there exists a $\Gamma' \subset \langle \Gamma \rangle_+$ such that CSP(Γ') is NP-intermediate by Theorem 8. Let k denote the highest arity of a relation in Γ. By definition every relation in $\langle \Gamma \rangle_+$ then has its arity bounded by $k + 1$, which trivially also holds for Γ'. $\square$

It is not hard to see that for the constraint language Γ° defined in the previous section any constant relation is p.p. definable in polynomial time. For any $a \in \mathbb{N}$ we simply let $(y = a) \equiv \exists x.R_{0,1,a,a}(x,y)$, i.e. the relation $0 \cdot x - 1 \cdot y \leq a \wedge 0 \leq x, y \leq a$. By Theorem 10 and the fact that Γ° only contains relations of arity 2 we therefore obtain the following.

Theorem 11. *There exists a $\Gamma' \subset \langle \Gamma^\circ \rangle_+$ such that (1) $CSP(\Gamma')$ is NP-intermediate and (2) Γ' contains only relations of arity 3.*

5.3 Propositional Abduction

Abduction is a fundamental form of nonmonotonic reasoning whose computational complexity has been thoroughly investigated [10,13,23]. It is known that the abduction problem parameterized with a finite constraint language is always in P, NP-complete, coNP-complete or Σ_2^P-complete. For infinite languages the situation differs and the question of whether it is possible to obtain a similar classification was left open in [23]. We will show that there exists an infinite constraint language such that the resulting abduction problem is NP-intermediate.

Let Γ denote a constraint language and define the propositional abduction problem $\textsc{Abd}(\Gamma)$ as follows.

INSTANCE. An instance I of $\textsc{Abd}(\Gamma)$ consists of a tuple (V, H, M, KB), where V is a set of Boolean variables, H is a set of literals over V (known as the *set of hypotheses*), M is a literal over V (known as the the *manifestation*), and KB is a set of constraint applications $C_1(\mathbf{x}_1) \wedge \ldots \wedge C_k(\mathbf{x}_k)$ where C_i denotes an application of some relation in Γ and $\mathbf{x_i}$, $1 \leq i \leq k$, is a vector of variables in V (*KB* is known as the *knowledge base*).

QUESTION. Does there exist an *explanation* for I, i.e., a set $E \subseteq H$ such that $KB \wedge \bigwedge E$ is satisfiable and $KB \wedge \bigwedge E \models M$, i.e. $KB \wedge \bigwedge E \wedge \neg M$ is not satisfiable.

Let Γ_{IHSB-} be the infinite constraint language consisting of the relations expressed by the clauses $(x), (\neg x \vee y)$ and all negative clauses, i.e., $\{(\neg x_1 \vee \cdots \vee \neg x_n) \mid n \geq 1\}$. We may represent each relation is Γ_{IHSB-} with a natural number in the obvious way. Let the finite constraint language $\Gamma_{IHSB-/k}$ be the subset of Γ_{IHSB-} that contains all clauses C such that $ar(C) = k$. In light of this we define the multi-valued measure function $\rho(I) = \{ar(C) \mid C$ is a negative clause of KB in $I\}$. With the chosen representation of relations, ρ is obviously polynomial-time computable. We define the corresponding parameterized abduction problem $\textsc{Abd}_\rho^*(\Gamma)$ such that $I(\textsc{Abd}^*)$ is the set of abduction instances over Γ_{IHSB-}. We now verify that $\textsc{Abd}_\rho^*(\mathbb{N})$ fulfills property P0 – P3.

Property P0 holds trivially while property P1 follows from [23]. For property P2, we note that if T is an arbitrary finite subset of $\mathbb{N}$, then there exists a $k \in T$ such that the clauses of every $\textsc{Abd}_\rho^*(T)$ instance is bounded by k. By [23], we know that $\textsc{Abd}(\Gamma_{IHSB-/k})$ is in P for every k, and hence that $\textsc{Abd}_\rho^*(T)$ is in P for every finite subset of S. To show property P3, we present a polynomial-time reduction from $\textsc{Abd}_\rho^*(\mathbb{N})$ to $\textsc{Abd}_\rho^*(T)$ when $\mathbb{N} \setminus T$ is finite. Let $k = \max(\mathbb{N} \setminus T)$. Arbitrarily choose an instance $I = (V, H, M, KB)$ of $\textsc{Abd}_\rho^*(\mathbb{N})$. Then, for every clause $C = (\neg x_1 \vee \ldots \vee \neg x_l) \in KB$ such that $l \in S \setminus T$, replace C by the logically equivalent clause $C' = (\neg x_1 \vee \ldots \vee \neg x_{l-1} \vee \neg x_l \vee \underbrace{\neg x_l \ldots \vee \neg x_l}_{k+1-l \ \neg x_l\text{'s}})$ of length $k+1$.

If we let the resulting knowledge base be KB' then $I' = (V, H, M, KB')$ is an instance of $\text{ABD}_\rho^*(T)$ which has a solution if and only if I has a solution.

From this and Theorem 3 it follows that that there exists a $S' \subset \mathbb{N}$ such that $\text{ABD}_\rho^*(S')$ is NP-intermediate. Hence we conclude the following.

Theorem 12. *There exists a constraint language $\Gamma'_{IHSB-} \subset \Gamma_{IHSB-}$ such that $\text{ABD}(\Gamma'_{IHSB-})$ is NP-intermediate.*

6   Future Work

One way of obtaining genuinely new NP-intermediate problems is to consider other complexity-theoretic assumptions than $P \neq NP$. We have pointed out that the LOGCLIQUE problem is NP-intermediate under the ETH, and that the main difficulty is to provide a lower bound, i.e. proving that LOGCLIQUE $\notin P$. One may suspect that providing lower bounds is the main difficulty also when considering other problems. We have seen that CSP problems constitute a rich source of NP-intermediate problems via different kinds of parameterization, Hence, it appears feasible that methods for studying the complexity of parameterized problems will become highly relevant. In particular, *linear fpt-reductions* [7,8] have been used for proving particularly strong lower bounds which may be used for linking together NP-intermediate problems, parameterized problems, and lower bound assumptions. Another way is to adapt and use recent methods for studying the time complexity of Boolean CSP problems [17]. These methods aim at obtaining reductions that provide a fine-grained picture of time complexity and this may be useful when studying NP-intermediate problems. Additionally, recent results by Dell and van Melkebeek [12] can be used for proving the non-existence of such reductions.

We have shown that the propositional abduction problem has NP-intermediate fragments. One may view abduction as a problem that is closely related to Boolean CSPs. However, there is an important difference: the $CSP(\Gamma)$ problem is either a member of P or NP-complete for *all* choices of Boolean Γ. Hence, it would be interesting to determine which finite-domain CSP-related problems can be used for obtaining NP-intermediate problems and which of them have the local-global property. Inspired by our result on the abduction problem, we view other forms of non-monotonic reasoning such as circumscription and default logic as potential candidates. Unfortunately, many problems of this type are polynomial-time solvable only in very restricted cases, which makes it hard to find a candidate language resulting in a problem not having the local-global property. Thus, more powerful methods than blowing may be needed for identifying NP-intermediate problems in this and similar cases.

References

1. Arora, S., Barak, B.: Computational Complexity: A Modern Approach, 1st edn. Cambridge University Press, New York (2009)
2. Arnborg, S., Corneil, D., Proskurowski, A.: Complexity of finding embeddings in a k-tree. SIAM Journal on Matrix Analysis and Applications 8(2), 277–284 (1987)
3. Bodirsky, M.: Complexity Classification in Infinite-Domain Constraint Satisfaction. Habilitation thesis. Univ. Paris 7 (2012)
4. Bodirsky, M., Grohe, M.: Non-dichotomies in constraint satisfaction complexity. In: Aceto, L., Damgård, I., Goldberg, L.A., Halldórsson, M.M., Ingólfsdóttir, A., Walukiewicz, I. (eds.) ICALP 2008, Part II. LNCS, vol. 5126, pp. 184–196. Springer, Heidelberg (2008)
5. Bodlaender, H.: A linear-time algorithm for finding tree-decompositions of small treewidth. SIAM Journal on Computing 25(6), 1305–1317 (1996)
6. Bulatov, A., Jeavons, P., Krokhin, A.: Classifying the computational complexity of constraints using finite algebras. SIAM Journal on Computing 34(3), 720–742 (2005)
7. Chen, J., Chor, B., Fellows, M., Huang, X., Juedes, D., Kanj, I., Xia, G.: Tight lower bounds for certain parameterized np-hard problems. In: Proc. IEEE Conference on Computational Complexity (CCC 2004), pp. 150–160 (2004)
8. Chen, J., Huang, X., Kanj, I., Xia, G.: Linear fpt reductions and computational lower bounds. In: Proc. 36th ACM Symposium on Theory of Computing (STOC-2004), pp. 212–221 (2004)
9. Chen, Y., Thurley, M., Weyer, M.: Understanding the complexity of induced subgraph isomorphisms. In: Aceto, L., Damgård, I., Goldberg, L.A., Halldórsson, M.M., Ingólfsdóttir, A., Walukiewicz, I. (eds.) ICALP 2008, Part I. LNCS, vol. 5125, pp. 587–596. Springer, Heidelberg (2008)
10. Creignou, N., Schmidt, J., Thomas, M.: Complexity of propositional abduction for restricted sets of boolean functions. In: Proc. 12th International Conference on the Principles of Knowledge Representation and Reasoning, KR 2010 (2010)
11. Dechter, R.: Constraint Processing. Elsevier Morgan Kaufmann (2003)
12. Dell, H., van Melkebeek, D.: Satisfiability allows no nontrivial sparsification unless the polynomial-time hierarchy collapses. In: Proc. 42nd ACM Symposium on Theory of Computing (STOC 2010), pp. 251–260 (2010)
13. Eiter, T., Gottlob, G.: The complexity of logic-based abduction. Journal of the ACM 42(1), 3–42 (1995)
14. Garey, M., Johnson, D.: "Strong" NP-completeness results: motivation, examples and implications. Journal of the ACM 25(3), 499–508 (1978)
15. Grohe, M.: The complexity of homomorphism and constraint satisfaction problems seen from the other side. Journal of the ACM 54(1), article 1 (2007)
16. Grohe, M., Marx, D.: Constraint solving via fractional edge covers. In: Proc. 17th Annual ACM-SIAM Symposium on Discrete Algorithms (SODA-2006), pp. 289–298 (2006)
17. Jonsson, P., Lagerkvist, V., Nordh, G., Zanuttini, B.: Complexity of SAT problems, clone theory and the exponential time hypothesis. In: Proc. the Twenty-Fourth Annual ACM-SIAM Symposium on Discrete Algorithms, SODA 2013 (2013)
18. Jonsson, P., Lööw, T.: Computational complexity of linear constraints over the integers. Artificial Intelligence 195, 44–62 (2013)
19. Karp, R.M.: Reducibility Among Combinatorial Problems. In: Miller, R.E., Thatcher, J.W. (eds.) Complexity of Computer Computations, pp. 85–103. Plenum Press (1972)

20. Ladner, R.: On the structure of polynomial time reducibility. Journal of the ACM 22, 155–171 (1975)
21. Marx, D.: Approximating fractional hypertree width. ACM Transactions on Algorithms 6(2) (2010)
22. Marx, D.: Tractable hypergraph properties for constraint satisfaction and conjunctive queries. In: Proc. 42nd ACM Symposium on Theory of Computing (STOC 2010), pp. 735–744 (2010)
23. Nordh, G., Zanuttini, B.: What makes propositional abduction tractable. Artificial Intelligence 172, 1245–1284 (2008)
24. Papadimitriou, C.H.: Computational Complexity. Addison-Wesley (1994)
25. Schöning, U.: A uniform approach to obtain diagonal sets in complexity classes. Theoretical Computer Science 18, 95–103 (1982)
26. Lau, D.: Function Algebras on Finite Sets: Basic Course on Many-Valued Logic and Clone Theory. Springer Monographs in Mathematics. Springer-Verlag New York, Inc., Secaucus (2006)

Solving QBF with Free Variables

William Klieber[1], Mikoláš Janota[2], Joao Marques-Silva[2,3], and Edmund Clarke[1]

[1] Carnegie Mellon University, Pittsburgh, PA, USA
[2] IST/INESC-ID, Lisbon, Portugal
[3] University College Dublin, Ireland

Abstract. An *open* quantified boolean formula (QBF) is a QBF that contains free (unquantified) variables. A solution to such a QBF is a quantifier-free formula that is logically equivalent to the given QBF. Although most recent QBF research has focused on closed QBF, there are a number of interesting applications that require one to consider formulas with free variables. This article shows how clause/cube learning for DPLL-based closed-QBF solvers can be extended to solve QBFs with free variables. We do this by introducing *sequents* that generalize clauses and cubes and allow learning facts of the form "under a certain class of assignments, the input formula is logically equivalent to a certain quantifier-free formula".

1   Introduction

In recent years, significant effort has been invested in developing efficient solvers for Quantified Boolean Formulas (QBFs). So far this effort has been almost exclusively directed at solving closed formulas — formulas where each variable is either existentially or universally quantified. However, in a number of interesting applications (such as symbolic model checking and automatic synthesis of a boolean reactive system from a formal specification), one needs to consider *open* formulas, i.e., formulas with free (unquantified) variables. A solution to such a QBF is a formula equivalent to the given one but containing no quantifiers and using only those variables that appear free in the given formula. For example, a solution to the open QBF formula $\exists x.\ (x \wedge y) \vee z$ is the formula $y \vee z$.

This article shows how DPLL-based closed-QBF solvers can be extended to solve QBFs with free variables. In [14], it was shown how clause/cube learning for DPLL-based QBF solvers can be reformulated in terms of *sequents* and extended to non-CNF, non-prenex formulas. This technique uses *ghost variables* to handle non-CNF formulas in a manner that is symmetric between the existential and universal quantifiers. We show that this sequent-based technique can be naturally extended to handle QBFs with free variables.

A naïve way to recursively solve an open QBF Φ is shown in Figure 1. Roughly, we Shannon-expand on the free variables until we're left with only closed-QBF problems, which are then handed to a closed-QBF solver. As an example, consider the formula $(\exists x.\ x \wedge y)$, with one free variable, y. Substituting y with true in Φ yields $(\exists x.\ x)$; this formula is given to a closed-QBF solver, which yields

C. Schulte (Ed.): CP 2013, LNCS 8124, pp. 415–431, 2013.
© Springer-Verlag Berlin Heidelberg 2013

```
function solve(Φ) {
    if (Φ has no free variables) {return closed_qbf_solve(Φ);}
    x := (a free variable in Φ);
    return ite(x, solve(Φ with x substituted with True),
                  solve(Φ with x substituted with False));
}
```

Fig. 1. Naive algorithm. The notation "ite(x, ϕ_1, ϕ_2)" denotes a formula with an *if-then-else* construct that is logically equivalent to $(x \wedge \phi_1) \vee (\neg x \wedge \phi_2)$.

true. Substituting y with false in Φ immediately yields false. So, our final answer is the formula (y ? true : false), which simplifies to y. In general, if the free variables are always branched on in the same order, then the algorithm effectively builds an ordered binary decision diagram (OBDD) [7], assuming that the ite function is memoized and performs appropriate simplification.

The above-described naïve algorithm suffers from many inefficiencies. In terms of branching behavior, it is similar to the DPLL algorithm, but it lacks non-chronological bracktracking and an equivalent of clause learning. The main contribution of this paper is to show how an existing closed-QBF algorithm can be modified to directly handle formulas with free variables by extending the existing techniques for non-chronological backtracking and clause/cube/sequent learning.

2   Preliminaries

Grammar. We consider prenex formulas of the form $Q_1 X_1 ... Q_n X_n . \phi$, where $Q_i \in \{\exists, \forall\}$ and ϕ is quantifier-free and represented as a DAG. The logical connectives allowed in ϕ are conjunction, disjunction, and negation. We say that $Q_1 X_1 ... Q_n X_n$ is the **quantifier prefix** and that ϕ is the **matrix**.

Assignments. Let π be a partial assignment of boolean values to variables. For convenience, we identify π with the set of literals made true by π. For example, we identify the assignment $\{(e_1, \text{true}), (u_2, \text{false})\}$ with the set $\{e_1, \neg u_2\}$. We write "vars(π)" to denote the set of variables assigned by π.

Quantifier Order. In a formula such as $\forall x . \exists y . \phi$, where the quantifier of y occurs inside the scope of the quantifier of x, and the quantifier type of x is different from the quantifier type of y, we say that y is **downstream** of x. Likewise, we say that x is **upstream** of y. All quantified variables in a formula are considered downstream of all free variables in the formula. In the context of an assignment π, we say that a variable is an **outermost** unassigned variable iff it is not downstream of any variables unassigned by π.

QBF as a Game. A closed QBF formula Φ can be viewed as a game between an existential player (Player $\exists$) and a universal player (Player $\forall$):

- Existentially quantified variables are *owned* by Player $\exists$.
- Universally quantified variables are *owned* by Player $\forall$.
- Players assign variables in quantification order (starting with outermost).
- The *goal* of Player $\exists$ is to make Φ be true.
- The *goal* of Player $\forall$ is to make Φ be false.
- A player *owns* a literal ℓ if the player owns $var(\ell)$.

If both players make the best moves possible, then the existential player will win iff the formula is true, and the universal player will win if the formula is false.

Substitution. Given a partial assignment π, we define "$\Phi|\pi$" to be the result of the following: For every assigned variable x, we replace all occurrences of x in Φ with the assigned value of x (and delete the quantifier of x, if any).

Gate Variables. We label each conjunction and disjunction with a *gate variable*. If a formula ϕ is labelled by a gate variable g, then $\neg\phi$ is labelled by $\neg g$. The variables originally in the formulas are called "input variables", in distinction to gate variables.

2.1   Tseitin Transformation's Undesired Effects in QBF

The Tseitin transformation [20] is the usual way of converting a formula into CNF. In the Tseitin transformation, all the gate variables (i.e., Tseitin variables) are existentially quantified in the innermost quantification block and clauses are added to equate each gate variable with the subformula that it represents. For example, consider the following formula:

$$\Phi_{\text{in}} \;:=\; \exists e.\, \forall u.\; \underbrace{(e \wedge u)}_{g_1} \vee \underbrace{(\neg e \wedge \neg u)}_{g_2}$$

This formula is converted to:

$$\Phi'_{\text{in}} \;=\; \exists e.\, \forall u.\, \exists \boldsymbol{g}.\; (g_1 \vee g_2) \wedge (g_1 \Leftrightarrow (e \wedge u)) \wedge (g_2 \Leftrightarrow (\neg e \wedge \neg u)) \qquad (1)$$

The biconditionals defining the gate variables are converted to clauses as follows:

$$(g_1 \Leftrightarrow (e \wedge u)) \;=\; (\neg e \vee \neg u \vee g_1) \wedge (\neg g_1 \vee e) \wedge (\neg g_1 \vee u)$$

Note that the Tseitin transformation is asymmetric between the existential and universal players: In the resulting CNF formula, the gate variables are existentially quantified, so the existential player (but not the universal player) loses if a gate variable is assigned inconsistently with the subformula that it represents. For example, in Equation 1, if $e|\pi = $ false and $g_1|\pi = $ true, then the existential player loses $\Phi'_{\text{in}}|\pi$. This asymmetry can be harmful to QBF solvers. For example, consider the QBF

$$\forall \boldsymbol{x}.\, \exists y.\; y \vee \underbrace{\psi(\boldsymbol{x})}_{g_1} \qquad (2)$$

This formula is trivially true. A winning move for the existential player is to make y be true, which immediately makes the matrix of the formula true, regardless of ψ. Under the Tseitin transformation, Equation 2 becomes:

$$\forall \boldsymbol{x}.\, \exists y.\, \exists \boldsymbol{g}.\, (y \vee g_1) \wedge (\text{clauses equating gate variables})$$

Setting y to be true no longer immediately makes the matrix true. Instead, for each assignment of universal variables $\boldsymbol{x}$, the QBF solver must actually find a satisfying assignment to the gate variables. This makes it much harder to detect when the existential player has won. Experimental results [1,22] indicate that purely CNF-based QBF solvers would, in the worst case, require time exponential in the size of ψ to solve the CNF formula, even though the original problem (before translation to CNF) is trivial.

3   Ghost Variables and Sequents

We employ *ghost variables* to provide a modification of the Tseitin transformation that is symmetric between the two players. The idea of using a symmetric transformation was first explored in [22], which performed the Tseitin transformation twice: once on the input formula, and once on its negation. Similar ideas have been used to handle non-prenex formulas in [14] and to handle "don't care" propagation in [12].

For each gate variable g, we introduce two *ghost variables*: an existentially quantified variable $g^{\exists}$ and a universally quantified variable $g^{\forall}$. We say that $g^{\exists}$ and $g^{\forall}$ *represent* the formula labeled by g. Ghost variables are considered to be downstream of all input variables.

We now introduce a semantics with ghost variables for the game formulation of QBF. As in the Tseitin transformation, the existential player should lose if an existential ghost variable $g^{\exists}$ is assigned a different value than the subformula that it represents. Additionally, the universal player should lose if an universal ghost variable $g^{\forall}$ is assigned a different value than the subformula that it represents.

In this paper, we never consider formulas (other than single literals) in which ghost variables occur as actual variables. In particular, if Φ is the input formula to the QBF solver, then in a substitution $\Phi|\pi$, ghost variables in π have no effect.

Definition 1 (Consistent assignment to ghost literal). Given a quantifier type $Q \in \{\exists, \forall\}$ and an assignment π, we say that a ghost literal g^Q is assigned **consistently** under π iff $g^Q|\pi = $ (the formula represented by g^Q)$|\pi$.

Definition 2 (Winning under a total assignment). Given a formula Φ, a quantifier type $Q \in \{\exists, \forall\}$, and an assignment π to all the input variables and a subset of the ghost variables, we say "Player Q **wins** Φ under π" iff:

- $\Phi|\pi = \mathsf{true}$ if Q is $\exists$, and
- $\Phi|\pi = \mathsf{false}$ if Q is $\forall$, and
- Every ghost variable owned by Q in vars(π) is assigned consistently.
  (Intuitively, a winning player's ghost variables must "respect the encoding").

For example, if $\Phi = \exists e.\forall u.\,(e \wedge u)$ and g labels $(e \wedge u)$ then neither player wins Φ under $\{\neg e, u, g^\forall, \neg g^\exists\}$. The existential player loses because $\Phi|\pi = \mathsf{false}$, and the universal player loses because $g^\forall|\pi \neq$ (the formula represented by $g^\forall)|\pi$.

Definition 3 (Losing under a total assignment). Given a formula Φ and an assignment π that assigns all the input variables, we say "Player Q **loses** Φ under π" iff Player Q does not win Φ under π.

Definition 4 (Losing under a partial assignment). Given a formula Φ, an assignment π, and an outermost unassigned input variable x, we say "Player Q **loses** Φ under π" iff either:

- Player Q loses Φ under both $\pi \cup \{(x, \mathsf{true})\}$ and $\pi \cup \{(x, \mathsf{false})\}$, or
- Q's opponent owns x and Player Q loses Φ under either $\pi \cup \{(x, \mathsf{true})\}$ or $\pi \cup \{(x, \mathsf{false})\}$.

For example, consider a formula $\Phi = \exists e.\,x \wedge e$, where x is a free variable. The existential player loses Φ under $\{\neg x\}$ and under $\{\neg e\}$. Neither player can be said to lose Φ under the empty assignment, because the value of Φ depends on the free variable x. Now let us make a few general observations about when a player loses under an arbitrary partial assignment.

Observation 1. If $\Phi|\pi = \mathsf{true}$, then Player $\forall$ loses Φ under π.

Observation 2. If $\Phi|\pi = \mathsf{false}$, then Player $\exists$ loses Φ under π.

Observation 3. If a ghost variable owned by Q in $\mathrm{vars}(\pi)$ is assigned inconsistently under π, then Player Q loses Φ under π.

Observation 4. If the opponent of Q owns a literal ℓ that is unassigned under π, and Q loses Φ under $\pi \cup \{\ell\}$, then Q loses Φ under π.

Definition 5 (Game-State Specifier, Match). A **game-state specifier** is a pair $\langle L^{\mathrm{now}}, L^{\mathrm{fut}} \rangle$ consisting of two sets of literals, L^{now} and L^{fut}. We say that $\langle L^{\mathrm{now}}, L^{\mathrm{fut}} \rangle$ **matches** an assignment π iff:

1. for every literal ℓ in L^{now}, $\ell|\pi = \mathsf{true}$, and
2. for every literal ℓ in L^{fut}, either $\ell|\pi = \mathsf{true}$ or $\ell \notin \mathrm{vars}(\pi)$.

For example, $\langle \{u\}, \{e\} \rangle$ matches the assignments $\{u\}$ and $\{u, e\}$, but does not match $\{\}$ or $\{u, \neg e\}$. Note that, for any literal ℓ, if $\{\ell, \neg\ell\} \subseteq L^{\mathrm{fut}}$, then $\langle L^{\mathrm{now}}, L^{\mathrm{fut}} \rangle$ matches an assignment π only if π doesn't assign ℓ. The intuition behind the names "L^{now}" and "L^{fut}" is as follows: Under the game formulation of QBF, the assignment π can be thought of as a state of the game, and π matches $\langle L^{\mathrm{now}}, L^{\mathrm{fut}} \rangle$ iff every literal in L^{now} is already true in the game and, for every literal ℓ in L^{fut}, it is possible that ℓ can be true in a future state of the game.

Definition 6 (Game Sequent). The sequent "$\langle L^{\mathrm{now}}, L^{\mathrm{fut}} \rangle \models (Q \text{ loses } \Phi)$" means "Player Q loses Φ under all assignments that match $\langle L^{\mathrm{now}}, L^{\mathrm{fut}} \rangle$."

As an example, let Φ be the following formula:

$$\forall u.\, \exists e.\, (e \vee \neg u) \wedge (u \vee \neg e) \wedge \overbrace{(x_1 \vee e)}^{g_3}$$

Note that sequent $\langle \{u\}, \{e\} \rangle \models (\forall \text{ loses } \Phi)$ holds true: in any assignment π that matches it, $\Phi|\pi = \text{true}$. However, $\langle \{u\}, \varnothing \rangle \models (\forall \text{ loses } \Phi)$ does not hold true: it matches the assignment $\{u, \neg e\}$, under which Player $\forall$ does not lose Φ. Finally, $\langle \{g_3^{\vee}\}, \{e, \neg e\} \rangle \models (\forall \text{ loses } \Phi)$ holds true. Let us consider why Player $\forall$ loses Φ under the assignment $\{g_3^{\vee}\}$. The free variable x_1 is the outermost unassigned variable, so under Definition 4, Player $\forall$ loses under $\{g_3^{\vee}\}$ iff Player $\forall$ loses under both $\{g_3^{\vee}, x_1\}$ and $\{g_3^{\vee}, \neg x_1\}$. Under $\{g_3^{\vee}, x_1\}$, Player $\forall$ loses because $\Phi|\{g_3^{\vee}, x_1\}$ evaluates to true. Under $\{g_3^{\vee}, \neg x_1\}$, Player $\forall$ loses because e is owned by the opponent of Player $\forall$ and $g_3^{\vee}$ is assigned inconsistently under $\{g_3^{\vee}, \neg x_1, \neg e\}$.

Note that a clause $(\ell_1 \vee ... \vee \ell_n)$ in a CNF formula Φ_{in} is equivalent to the sequent $\langle \{\neg \ell_1, ..., \neg \ell_n\}, \varnothing \rangle \models (\exists \text{ loses } \Phi_{\text{in}})$. (Sequents in this form can also be considered similar to *nogoods* [19].) Likewise, a cube $(\ell_1 \wedge ... \wedge \ell_n)$ in a DNF formula Φ_{in} is equivalent to the sequent $\langle \{\ell_1, ..., \ell_n\}, \varnothing \rangle \models (\forall \text{ loses } \Phi_{\text{in}})$.

3.1   Sequents with Free Variables

Above, we introduced sequents that indicate if a player loses a formula Φ. Now, we will generalize sequents so that they can indicate that Φ evaluates to a quantifier-free formula involving the free variables. To do this, we first introduce a logical semantics for QBF with ghost variables. Given a formula Φ and an assignment π that assigns all the input variables, we want the semantic evaluation $[\![\Phi]\!]\pi$ to have the following properties:

1. $[\![\Phi]\!]\pi = \text{true}$ iff the existential player wins Φ under π.
2. $[\![\Phi]\!]\pi = \text{false}$ iff the universal player wins Φ under π.

Note that the above properties cannot be satisfied in a two-valued logic if both players lose Φ under π. So, we use a three-valued logic with a third value dontcare. We call it "don't care" because we are interested in the outcome of the game when both players make the best possible moves, but if both players fail to win, then clearly at least one of the players failed to make the best possible moves. In our three-valued logic, a conjunction of boolean values evaluates to false if any conjunct is false, and otherwise it evaluates to dontcare if any conjunct is dontcare. Disjunction is defined analogously. The negation of dontcare is dontcare. In a truth table:

x	y	$x \wedge y$	$x \vee y$
true	dontcare	dontcare	true
false	dontcare	false	dontcare

Definition 7. Given an assignment π to all the input variables and a subset of the ghost variables, we define $[\![\Phi]\!]_\pi$ as follows:

$$[\![\Phi]\!]_\pi := \begin{cases} \text{true} & \text{if Player } \exists \text{ wins } \Phi \text{ under } \pi \\ \text{false} & \text{if Player } \forall \text{ wins } \Phi \text{ under } \pi \\ \text{dontcare} & \text{if both players lose } \Phi \text{ under } \pi \end{cases}$$

For convenience in defining $[\![\Phi]\!]_\pi$ for a partial assignment π, we assume that the formula is prepended with a dummy "quantifier" block for free variables. For example, $(\exists e.\, e \wedge z)$ becomes $(\mathcal{F}z.\, \exists e.\, e \wedge z)$, where $\mathcal{F}$ denotes the dummy block for free variables. If Φ contains free variables unassigned by π then $[\![\Phi]\!]_\pi$ is a formula in terms of these free variables. We define $[\![\Phi]\!]_\pi$ as follows for a partial assignment π that assigns only a proper subset of the input variables:

$$[\![Qx.\, \Phi]\!]_\pi = [\![\Phi]\!]_\pi \quad \text{if } x \in \text{vars}(\pi)$$
$$[\![\exists x.\, \Phi]\!]_\pi = [\![\Phi]\!]_{(\pi \cup \{x\})} \vee [\![\Phi]\!]_{(\pi \cup \{\neg x\})} \quad \text{if } x \notin \text{vars}(\pi)$$
$$[\![\forall x.\, \Phi]\!]_\pi = [\![\Phi]\!]_{(\pi \cup \{x\})} \wedge [\![\Phi]\!]_{(\pi \cup \{\neg x\})} \quad \text{if } x \notin \text{vars}(\pi)$$
$$[\![\mathcal{F}x.\, \Phi]\!]_\pi = x \ ? \ [\![\Phi]\!]_{(\pi \cup \{x\})} : [\![\Phi]\!]_{(\pi \cup \{\neg x\})} \quad \text{if } x \notin \text{vars}(\pi)$$

The notation "$x \ ? \ \phi_1 : \phi_2$" denotes a formula with an *if-then-else* construct that is logically equivalent to $(x \wedge \phi_1) \vee (\neg x \wedge \phi_2)$. Note that the branching on the free variables here is similar to the Shannon expansion [17].

Remark. Do we really need to add the dummy blocks for free variables and have the rule for $[\![\mathcal{F}x.\, \Phi]\!]_\pi$ in Definition 7? Yes, because if π contains a ghost literal g^Q that represents a formula containing variables free in Φ, then it doesn't make sense to ask if g^Q is assigned consistently under π unless all the variables in the formula represented by g^Q are assigned by π.

Definition 8 (Sometimes-Dontcare). A formula ϕ is said to be **sometimes-dontcare** iff there is an assignment π under which ϕ evaluates to dontcare. For example, $(x \vee \text{dontcare})$ is sometimes-dontcare, while $(x \vee (x \wedge \text{dontcare}))$ is not sometimes-dontcare (because it evaluates to true if x is true and evaluates to false if x is false).

Definition 9 (Free Sequent). The sequent "$\langle L^{\text{now}}, L^{\text{fut}} \rangle \models \Phi \Leftrightarrow \psi$" means "for all assignments π that match $\langle L^{\text{now}}, L^{\text{fut}} \rangle$, if $[\![\Phi]\!]_\pi$ is not sometimes-dontcare, then $[\![\Phi]\!]_\pi$ is logically equivalent to $\psi|\pi$".

Remark. The sequent definitions in Definitions 9 and 6 are related as follows:
- "$\langle L^{\text{now}}, L^{\text{fut}} \rangle \models (\exists \text{ loses } \Phi)$" means the same as "$\langle L^{\text{now}}, L^{\text{fut}} \rangle \models (\Phi \Leftrightarrow \text{false})$".
- "$\langle L^{\text{now}}, L^{\text{fut}} \rangle \models (\forall \text{ loses } \Phi)$" means the same as "$\langle L^{\text{now}}, L^{\text{fut}} \rangle \models (\Phi \Leftrightarrow \text{true})$".

We treat a game sequent as interchangeable with the corresponding free sequent.

Sequents of the form $\langle L^{\text{now}}, L^{\text{fut}} \rangle \models \Phi \Leftrightarrow \psi$ extend clause/cube learning by allowing ψ to be a formula (in terms of the variables free in Φ) in addition to the constants true and false. This enables handling of formulas with free variables.

4   Algorithm

The top-level algorithm, shown in Figure 2, is based on the well-known DPLL algorithm, except that sequents are used instead of clauses. Similar to how SAT solvers maintain a *clause database* (i.e., a set of clauses whose conjunction is equisatisfiable with the original input formula Φ_{in}), our solver maintains a *sequent database*. A SAT solver's clause database is initialized to contain exactly the set of clauses produced by the Tseitin transformation of the input formula Φ_{in} into CNF. Likewise, our sequent database is initialized (§ 4.1) to contain a set of sequents analogous to the clauses produced by the Tseitin transformation.

In the loop on lines 4–7, the solver chooses an outermost unassigned literal, adds it to π_{cur}, and performs boolean constraint propagation (BCP). BCP may add further literals to π_{cur}, as described in detail in § 4.4; such literals are referred to as *forced literals*, in distinction to the literals added by DecideLit, which are referred to as *decision literals*. The stopping condition for the loop is when the current assignment matches a sequent already in the database. (The analogous stopping condition for a SAT solver would be when a clause is falsified.) When this stopping condition is met, the solver performs an analysis similar to that of *clause learning* [18] to learn a new sequent (line 8). If the L^{now} component of the learned sequent is empty, then the solver has reached the final answer, which it returns (line 9). Otherwise, the solver backtracks to the earliest decision level at which the newly learned sequent will trigger a forced literal in BCP. (The learning algorithm guarantees that this is possible.) The solver then performs BCP (line 11) and returns to the inner loop at line 4.

The intuition behind BCP for quantified variables is fairly straightforward; a literal owned by Q is forced by a sequent if the sequent indicates that Q need to make ℓ true to avoid losing. For free variables, the intuition is slightly different. Free variables are forced to prevent the solver from re-exploring parts of the

---

```
1.   initialize_sequent_database();
2.   π_cur := ∅; Propagate();

3.   while (true) {
4.      while (π_cur doesn't match any database sequent) {
5.         DecideLit();
6.         Propagate();
7.      }
8.      Learn();
9.      if (learned seq has form ⟨∅, L^fut⟩ ⊨ (Φ_in ⇔ ψ)) return ψ;
10.     Backtrack();
11.     Propagate();
12.  }
```

Fig. 2. Top-Level Algorithm. Details have been omitted for sake of clarity.

search space that it has already seen, so that the solver is continuously making progress in exploring the search space, thereby guaranteeing it would eventually terminate (given enough time and memory). (Actually, this intuition also applies to quantified variables.)

The solver maintains a list of assigned literals in the order in which they were assigned; this list is referred to as the *trail* [9]. Given a decision literal ℓ_d, we say that all literals that appear in the trail after ℓ_d but before any other decision literal belong to the same *decision level* as ℓ_d.

For prenex formulas without free variables, the algorithm described here is operationally very similar to standard DPLL QBF solvers, except that L^{now} and L^{fut} do not need to be explicitly separated, since L^{now} always consists exactly of all the loser's literals. However, for formulas with free variables, it is necessary to explicitly record which literals belong in L^{now} and which in L^{fut}.

4.1   Initial Sequents

We initialize the sequent database to contain a set of *initial sequents*, which correspond to the clauses produced by the Tseitin transformation of the input formula Φ_{in}. The set of initial sequents must be sufficient to ensure the loop on line 4–6 of Figure 2 (which adds unassigned literals to the current assignment until it matches a sequent in the database) operates properly. That is, for every possible total assignment π, there must be at least one sequent that matches π.

First, let us consider a total assignment π in which both players assign all their ghost variables consistently (Definition 1). In order to handle this case, we generate the following two initial sequents, where g_{in} is the label of the input formula Φ_{in}: $\langle \{\neg g_{\text{in}}^{\exists}\}, \varnothing \rangle \models (\exists \text{ loses } \Phi_{\text{in}})$ and $\langle \{g_{\text{in}}^{\vee}\}, \varnothing \rangle \models (\forall \text{ loses } \Phi_{\text{in}})$.

Since all ghost variables are assigned consistently in π, it follows that, for each gate g, $g^{\exists}|\pi$ must equal $g^{\vee}|\pi$, since both $g^{\exists}$ and $g^{\vee}$ must each be assigned the same value as the formula that g labels. In particular, $g_{\text{in}}^{\exists}|\pi$ must be equal to $g_{\text{in}}^{\vee}|\pi$, so π must match exactly one of the two above initial sequents.

Now let us consider a total assignment π in which at least one player assigns a ghost variable inconsistently. In order to handle this case, we generate a set of initial sequents for every conjunction and disjunction in Φ_{in}. Let g_* be the label of an arbitrary conjunction in Φ_{in} of the form

$$\left(x_1 \wedge ... \wedge x_n \wedge \underbrace{\phi_1}_{g_1} \wedge ... \wedge \underbrace{\phi_m}_{g_m} \right)$$

where x_1 through x_n are input literals. The following initial sequents are produced from this conjunction for each $Q \in \{\exists, \forall\}$:

1. $\langle \{g_*^Q, \neg x_i\}, \varnothing \rangle \models (Q \text{ loses } \Phi_{\text{in}})$ for $i \in \{1, ..., n\}$
2. $\langle \{g_*^Q, \neg g_i^Q\}, \varnothing \rangle \models (Q \text{ loses } \Phi_{\text{in}})$ for $i \in \{1, ..., m\}$
3. $\langle \{\neg g_*^Q, x_1, ..., x_n, g_1^Q, ..., g_m^Q\}, \varnothing \rangle \models (Q \text{ loses } \Phi_{\text{in}})$

Now let g_*^Q denote a ghost literal such that (1) g_*^Q is inconsistently assigned under π and (2) no proper subformula of the formula represented by g_*^Q is labelled by a inconsistently-assigned ghost variable. Then π must match one of the above-listed initials sequents.

4.2   Normalization of Initial Sequents

Note that all the initial sequents have the form $\langle L^{\mathrm{now}}, L^{\mathrm{fut}} \rangle \models (Q \text{ loses } \Phi)$ where $L^{\mathrm{fut}} = \varnothing$. We normalize these sequents by moving all literals owned by Q's opponent from L^{now} to L^{fut}, in accordance with the following inference rule:

$$\frac{\text{The opponent of } Q \text{ owns } \ell, \text{ and } \neg\ell \notin L^{\mathrm{fut}} \qquad \langle L^{\mathrm{now}} \cup \{\ell\}, L^{\mathrm{fut}} \rangle \models (Q \text{ loses } \Phi)}{\langle L^{\mathrm{now}}, L^{\mathrm{fut}} \cup \{\ell\} \rangle \models (Q \text{ loses } \Phi)}$$

To prove the above inference rule, we consider an arbitrary assignment π that matches $\langle L^{\mathrm{now}}, L^{\mathrm{fut}} \cup \{\ell\} \rangle$, assume that the premises of inference rule hold true, and prove that Player Q loses under π:

1. π matches $\langle L^{\mathrm{now}}, L^{\mathrm{fut}} \cup \{\ell\} \rangle$ (by assumption).
2. $\pi \cup \{\ell\}$ matches $\langle L^{\mathrm{now}} \cup \{\ell\}, L^{\mathrm{fut}} \rangle$ (using the premise that $\neg\ell \notin L^{\mathrm{fut}}$).
3. Q loses Φ under $\pi \cup \{\ell\}$ (by the premise $\langle L^{\mathrm{now}} \cup \{\ell\}, L^{\mathrm{fut}} \rangle \models (Q \text{ loses } \Phi)$).
4. Q loses Φ under π (by Observation 4 on page 419).

4.3   Properties of Sequents in Database

After the initial sequents have been normalized (as described in § 4.2), the solver maintains the following invariants for all sequents in the sequent database, including sequents added to the database as a result of learning (§ 4.5):

1. In a sequent of the form $\langle L^{\mathrm{now}}, L^{\mathrm{fut}} \rangle \models (Q \text{ loses } \Phi_{\mathrm{in}})$:
   (a) Every literal in L^{now} either is owned by Q or is free in Φ_{in}.
   (b) Every literal in L^{fut} is owned by the opponent of Q.
2. In a sequent of the form $\langle L^{\mathrm{now}}, L^{\mathrm{fut}} \rangle \models (\Phi_{\mathrm{in}} \Leftrightarrow \psi)$, every variable in ψ appears both positively and negatively in L^{fut} (i.e., if r occurs in ψ, then $\{r, \neg r\} \subseteq L^{\mathrm{fut}}$). This is guaranteed by the learning algorithm in § 4.5.

4.4   Propagation

The `Propagate` procedure is similar to that of closed-QBF solvers. Consider a sequent $\langle L^{\mathrm{now}}, L^{\mathrm{fut}} \rangle \models (\Phi_{\mathrm{in}} \Leftrightarrow \psi)$ in the sequent database. If, under π_{cur},

1. there is exactly one unassigned literal ℓ in L^{now}, and
2. no literals in $L^{\mathrm{now}} \cup L^{\mathrm{fut}}$ are assigned false, and
3. ℓ is not downstream of any unassigned literals in L^{fut},

then $\neg\ell$ is *forced* — it is added to the current assignment π_{cur}. In regard to the 3rd condition, if an unassigned literal r in L^{fut} is upstream of ℓ, then r should get assigned before ℓ, and if r gets assigned false, then ℓ shouldn't get forced at all by the sequent. Propagation ensures that the solver never re-explores areas of the search space for which it already knows the answer, ensuring continuous progress and eventual termination. It is instructive to consider how the propagation rule applies in light of the properties of sequents discussed in § 4.3:

1. A sequent of the form $\langle L^{\text{now}}, L^{\text{fut}} \rangle \models (Q \text{ loses } \Phi_{\text{in}})$ can force a literal that is either owned by Q or free in Φ_{in}; it cannot force a literal owned by Q's opponent. If ℓ is owned by Q, then the reason for forcing $\neg\ell$ is intuitive: the only way for Q to avoid losing is to add $\neg\ell$ to the current assignment. If ℓ is free in Φ_{in}, then $\neg\ell$ is forced because the value of $[\![\Phi_{\text{in}}]\!]_{\pi_{\text{cur}}} \cup \{\ell\}$ is already known and the solver shouldn't re-explore that same area of the search space.

2. A sequent of the form $\langle L^{\text{now}}, L^{\text{fut}} \rangle \models (\Phi_{\text{in}} \Leftrightarrow \psi)$, where ψ contains free variables, can only force a literal that is free in Φ_{in}. Although L^{now} can contain literals owned by Player $\exists$ and Player $\forall$, such literals cannot be forced by the sequent. To prove this, we consider two cases: either there exists a variable v that occurs in ψ and is assigned by π_{cur}, or all variables that occur ψ are left unassigned by π_{cur}. If there is variable v in ψ that is assigned by π_{cur}, then π_{cur} cannot match $\langle L^{\text{now}}, L^{\text{fut}} \rangle \models (\Phi_{\text{in}} \Leftrightarrow \psi)$, since $\{v, \neg v\} \subseteq L^{\text{fut}}$. If there is a variable v in ψ that is left unassigned by π_{cur}, then $\langle L^{\text{now}}, L^{\text{fut}} \rangle \models (\Phi_{\text{in}} \Leftrightarrow \psi)$ cannot force any quantified variable, since v occurs in L^{fut} and all quantified variables are downstream of free variable v.

We employ a variant of the watched-literals rule designed for SAT solvers [16] and adapted for QBF solvers [10]. For each sequent $\langle L^{\text{now}}, L^{\text{fut}} \rangle \models (\Phi \Leftrightarrow \psi)$, we watch two literals in L^{now} and one literal in L^{fut}.

4.5   Learning

In the top-level algorithm in Figure 2, the solver performs *learning* (line 8) after the current assignment π_{cur} matches a sequent in the database. The learning procedure is based on the clause learning introduced for SAT in [18] and adapted for QBF in [24]. We use inference rules shown in Figure 4 to add new sequents to the sequent database. These rules, in their L^{now} components, resemble the *resolution* rule used in SAT (i.e., from $(A \vee r) \wedge (\neg r \vee B)$ infer $A \vee B$). The learning algorithm ensures that the solver remembers the parts of the search space for which it has already found an answer. This, together with propagation, ensures that solver eventually covers all the necessary search space and terminates.

The learning procedure, shown in Figure 3, works as follows. Let seq be the database sequent that matches the current assignment π_{cur}. Let r be the literal in the L^{now} component of seq that was most recently added to π_{cur} (i.e., the latest one in the *trail*). Note that r must be a forced literal (as opposed to a decision literal), because only an outermost unassigned literal can be picked as a decision literal, but if r was outermost immediately before it added to π_{cur},

```
func Learn() {
    seq := (the database sequent that matches π_cur);
    do {
        r := (the most recently assigned literal in seq.Lⁿᵒʷ)
        seq := Resolve(seq, antecedent[r]);
    } until (seq.Lⁿᵒʷ = ∅ or has_good_UIP(seq));
    return seq;
}
```

Fig. 3. Procedure for learning new sequents

Resolving on a literal r owned by Player Q (case 1):

The quantifier type of r in Φ is Q

$\langle L_1^{\text{now}} \cup \{r\}, L_1^{\text{fut}} \rangle \models (Q \text{ loses } \Phi_{\text{in}})$

$\langle L_2^{\text{now}} \cup \{\neg r\}, L_2^{\text{fut}} \rangle \models (Q \text{ loses } \Phi_{\text{in}})$

r is not downstream of any ℓ such that $\ell \in L_1^{\text{fut}}$ and $\neg \ell \in (L_1^{\text{fut}} \cup L_2^{\text{fut}})$

$\overline{\langle L_1^{\text{now}} \cup L_2^{\text{now}}, L_1^{\text{fut}} \cup L_2^{\text{fut}} \rangle \models (Q \text{ loses } \Phi_{\text{in}})}$

Resolving on a literal r owned by Player Q (case 2):

The quantifier type of r in Φ is Q

$\langle L_1^{\text{now}} \cup \{r\}, L_1^{\text{fut}} \rangle \models (Q \text{ loses } \Phi_{\text{in}})$

$\langle L_2^{\text{now}} \cup \{\neg r\}, L_2^{\text{fut}} \rangle \models (\Phi_{\text{in}} \Leftrightarrow \psi)$

r is not downstream of any ℓ such that $\ell \in L_1^{\text{fut}}$ and $\neg \ell \in (L_1^{\text{fut}} \cup L_2^{\text{fut}})$

$\overline{\langle L_1^{\text{now}} \cup L_2^{\text{now}}, L_1^{\text{fut}} \cup L_2^{\text{fut}} \cup \{\neg r\} \rangle \models (\Phi_{\text{in}} \Leftrightarrow \psi)}$

Resolving on a variable r that is free in Φ_{in}:

Literal r is free

$\langle L_1^{\text{now}} \cup \{r\}, L_1^{\text{fut}} \rangle \models (\Phi_{\text{in}} \Leftrightarrow \psi_1)$

$\langle L_2^{\text{now}} \cup \{\neg r\}, L_2^{\text{fut}} \rangle \models (\Phi_{\text{in}} \Leftrightarrow \psi_2)$

$\overline{\langle L_1^{\text{now}} \cup L_2^{\text{now}}, L_1^{\text{fut}} \cup L_2^{\text{fut}} \cup \{r, \neg r\} \rangle \models (\Phi_{\text{in}} \Leftrightarrow (r \text{ ? } \psi_1 : \psi_2))}$

Fig. 4. Resolution-like inference rules

then no unassigned literal in the L^{fut} component of *seq* was upstream of r, so *seq* would have forced $\neg r$ in accordance with §4.4. We use the inference rules in Figure 4 to infer a new sequent from *seq* and the *antecedent* of r (i.e., the sequent that forced r). This is referred to as *resolving* due to the similarity of the inference rules to the clause resolution rule. We stop and return the newly inferred sequent if it has a "good" unique implication point (UIP) [24], i.e., if there is a literal ℓ in the L^{now} component such that

1. Every literal in $(L^{\text{now}} \setminus \{\ell\})$ belongs to an earlier decision level than ℓ,
2. Every literal in L^{fut} upstream of ℓ belongs to a decision level earlier than ℓ.
3. If *seq* has the form $\langle L^{\text{now}}, L^{\text{fut}} \rangle \models (Q \text{ loses } \Phi_{\text{in}})$, then the decision variable of the decision level of ℓ is not owned by the opponent of Q.

Otherwise, we resolve the sequent with the antecedent of the most recently assigned literal in its L^{now} component, and continue this process until the stopping conditions above are met or L^{now} is empty. Note that if the most recently assigned literal in L^{now} is a decision literal, then it is a good UIP.

Note that in the resolution rule for resolving on a free variable r, we add both r and $\neg r$ to L^{fut}. This is not necessary for soundness of the resolution itself. Rather, it is to ensure that the properties in §4.3 hold true. Without these properties, a quantified variable could be forced by a sequent that is not equivalent to a clause or a cube.

Example. Below, we give several applications of the resolution rules. For brevity, we omit free variables from the L^{fut} component.

$$\exists e_3. \underbrace{(i_1 \wedge e_3)}_{g_5} \vee \underbrace{(i_2 \wedge \neg e_3)}_{g_4}$$

1. Start: $\langle \{\neg i_1, \neg i_2\}, \{\} \rangle \models (\Phi_{\text{in}} \Leftrightarrow \text{false})$

2. Resolve $\neg i_1$ via $\langle \{i_1, \neg g_5^{\vee}\}, \{e_3\} \rangle \models (\Phi_{\text{in}} \Leftrightarrow \text{true})$
   Result: $\langle \{\neg i_2, \neg g_5^{\vee}\}, \{e_3\} \rangle \models (\Phi_{\text{in}} \Leftrightarrow i_1)$

3. Resolve $\neg i_2$ via $\langle \{i_2, \neg g_4^{\vee}\}, \{\neg e_3\} \rangle \models (\Phi_{\text{in}} \Leftrightarrow \text{true})$
   Result: $\langle \{\neg g_5^{\vee}, \neg g_4^{\vee}\}, \{e_3, \neg e_3\} \rangle \models (\Phi_{\text{in}} \Leftrightarrow (i_1 \vee i_2))$

4. Resolve $\neg g_4^{\vee}$ via $\langle \{g_4^{\vee}\}, \{\} \rangle \models (\Phi_{\text{in}} \Leftrightarrow \text{true})$
   Result: $\langle \{\neg g_5^{\vee}\}, \{e_3, \neg e_3, \neg g_4^{\vee}\} \rangle \models (\Phi_{\text{in}} \Leftrightarrow (i_1 \vee i_2))$

5. Resolve $\neg g_5^{\vee}$ via $\langle \{g_5^{\vee}\}, \{\} \rangle \models (\Phi_{\text{in}} \Leftrightarrow \text{true})$
   Result: $\langle \{\}, \{e_3, \neg e_3, \neg g_4^{\vee}, \neg g_5^{\vee}\} \rangle \models (\Phi_{\text{in}} \Leftrightarrow (i_1 \vee i_2))$

4.6   Justification of Inference Rules

The first inference rule in Figure 4 is analogous to long-distance resolution [23] and can be proved by similar methods (e.g., [2]). Intuitively, if the current

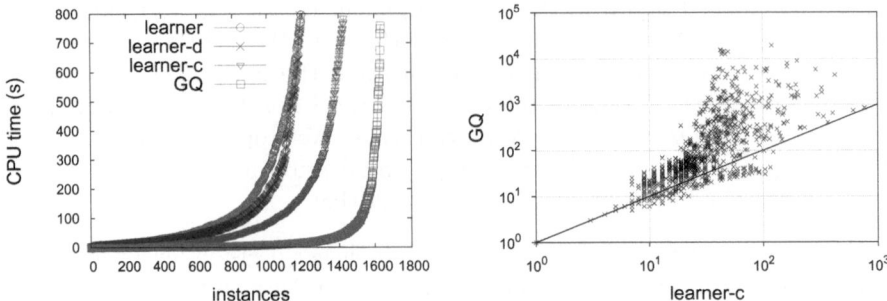

Fig. 5. Time and size comparisons, instances solved by all solvers in less than 10 s are not included in the time comparison

assignment matches $\langle L_1^{\mathrm{now}} \cup L_2^{\mathrm{now}},\ L_1^{\mathrm{fut}} \cup L_2^{\mathrm{fut}} \rangle$, then the opponent of Q can make Q lose Φ_{in} by assigning true to all the literals in L_1^{fut} that are upstream of r. This forces Q to assign $r = \mathsf{false}$ to avoid matching the first sequent in the premise of the inference rule, but assigning $r = \mathsf{false}$ makes the current assignment match the second sequent in the premise.

If the current assignment π_{cur} matches the sequent in the conclusion of the second inference rule, there are two possibilities. For simplicity, assume that π_{cur} assigns all free variables and that neither L_1^{fut} nor L_2^{fut} contains any free literals (since, as mentioned earlier, free literals can be removed from L^{fut} without affecting soundness of the sequent). If Q loses ψ under π_{cur}, then the situation is similar to first inference rule. If the opponent of Q loses ψ under π_{cur}, then Q can make his opponent lose Φ_{in} by assigning $r = \mathsf{false}$, thereby making the current assignment match the second sequent of the premise.

For the third inference rule, we don't need a condition about r not being downstream of other literals, since no free variable is downstream of any variable.

5   Experimental Results

We extended the existing closed-QBF solver GhostQ [14] to implement the techniques described in this paper. For comparison, we used the solvers and load-balancer benchmarks from [3].[1] The benchmarks contain multiple alternations of quantifiers and are derived from problems involving the automatic synthesis of a reactive system from a formal specification. The experimental results were obtained on Intel Xeon 5160 3-GHz machines with 4 GB of memory. The time limit was 800 seconds and the memory limit to 2 GB.

---

[1] The results do not exactly match the results reported in [3] because we did not preprocess the QDIMACS input files. We found that sometimes the output of the preprocessor was not logically equivalent to its input. With the unpreprocessed inputs, the output formulas produced by the learner family of solvers were always logically equivalent to the output formulas of GhostQ.

There are three solvers from [3], each with a different form of the output: CDNF (a conjunction of DNFs), CNF, and DNF. We will refer to these solvers as "Learner" (CNDF), "Learner-C" (CNF), and "Learner-D" (DNF). Figure 5 compares these three solvers with GhostQ on the "hard" benchmarks (those that not all four solvers could solve within 10 seconds). As can be seen on the figure, GhostQ solved about 1600 of these benchmarks, Learner-C solved about 1400, and Learner-D and Learner each solved about 1200. GhostQ solved 223 instances that Learner-C couldn't solve, while Learner-C solved 16 instances that GhostQ couldn't solve. GhostQ solved 375 instances that neither Learner-DNF nor Learner could solver, while there were only 2 instances that either Learner-DNF or Learner could solve but GhostQ couldn't solve.

Figure 5 shows a comparison of the size of the output formulas for GhostQ and Learner-C, indicating that the GhostQ formulas are often significantly larger. The size is computed as 1 plus the number of edges in the DAG representation of the formula, not counting negations, and after certain simplifications. E.g., the size of x is 1, the size of $\neg x$ is also 1, and the size of $x \wedge y$ is 3.

6   Related Work

Ken McMillan [15] proposed a method to use SAT solvers to perform quantifier elimination on formulas of the form $\exists x.\,\phi$, generating CNF output. This problem (i.e, given a formula $\exists x.\,\phi$, return a logically equivalent quantifier-free CNF formula) has received attention recently. Brauer, King, and Kriener [6] designed an algorithm that combines model enumeration with prime implicant generation. Goldberg and Manolios [11] developed a method based on *dependency sequents*; experimental results show that it works very well on forward and backward reachability on the Hardware Model Checking Competition benchmarks. For QBFs with arbitrary quantifier prefixes, the only other work of which we are aware is that of Becker, Ehlers, Lewis, and Marin [3], which uses computational learning to generate CNF, DNF, or CDNF formulas, and that of Benedetti and Mangassarian [5], which adapts sKizzo [4] for open QBF. The use of SAT solvers to build unordered BDDs [21] and OBDDs [13] has also been investigated.

7   Conclusion

This paper has shown how a DPLL-based closed-QBF solver can be extended to handle free variables. The main novelty of this work consists of generalizing clauses/cubes (and the methods involving them), yielding sequents that can include a formula in terms of the free variables. Our extended solver GhostQ produces unordered BDDs, which have several favorable properties [8]. However, in practice, the formulas tended to fairly large in comparison to equivalent CNF representations. Unordered BDDs can often be larger than equivalent OBDDs, since logically equivalent subformulas can have multiple distinct representations in an unordered BDD, unlike in an OBDD. Although our BDDs are necessarily unordered due to unit propagation, in future work it may be desirable to investigate techniques aimed at reducing the size of the output formula.

References

1. Ansótegui, C., Gomes, C.P., Selman, B.: The Achilles' Heel of QBF. In: AAAI 2005 (2005)
2. Balabanov, V., Jiang, J.-H.R.: Unified QBF certification and its applications. Formal Methods in System Design 41(1), 45–65 (2012)
3. Becker, B., Ehlers, R., Lewis, M., Marin, P.: ALLQBF Solving by Computational Learning. In: Chakraborty, S., Mukund, M. (eds.) ATVA 2012. LNCS, vol. 7561, pp. 370–384. Springer, Heidelberg (2012)
4. Benedetti, M.: sKizzo: A Suite to Evaluate and Certify QBFs. In: Nieuwenhuis, R. (ed.) CADE 2005. LNCS (LNAI), vol. 3632, pp. 369–376. Springer, Heidelberg (2005)
5. Benedetti, M., Mangassarian, H.: QBF-Based Formal Verification: Experience and Perspectives. In: JSAT (2008)
6. Brauer, J., King, A., Kriener, J.: Existential Quantification as Incremental SAT. In: Gopalakrishnan, G., Qadeer, S. (eds.) CAV 2011. LNCS, vol. 6806, pp. 191–207. Springer, Heidelberg (2011)
7. Bryant, R.E.: Graph-based algorithms for boolean function manipulation. IEEE Transactions on Computers 100(8), 677–691 (1986)
8. Darwiche, A., Marquis, P.: A Knowledge Compilation Map. J. Artif. Intell. Res (JAIR) 17, 229–264 (2002)
9. Eén, N., Sörensson, N.: An Extensible SAT-solver. In: Giunchiglia, E., Tacchella, A. (eds.) SAT 2003. LNCS, vol. 2919, pp. 502–518. Springer, Heidelberg (2004)
10. Gent, I.P., Giunchiglia, E., Narizzano, M., Rowley, A.G.D., Tacchella, A.: Watched Data Structures for QBF Solvers. In: Giunchiglia, E., Tacchella, A. (eds.) SAT 2003. LNCS, vol. 2919, pp. 25–36. Springer, Heidelberg (2004)
11. Goldberg, E., Manolios, P.: Quantifier elimination by Dependency Sequents. In: FMCAD, pp. 34–43. IEEE (2012)
12. Goultiaeva, A., Bacchus, F.: Exploiting QBF Duality on a Circuit Representation. In: AAAI (2010)
13. Huang, J., Darwiche, A.: Using DPLL for Efficient OBDD Construction. In: H. Hoos, H., Mitchell, D.G. (eds.) SAT 2004. LNCS, vol. 3542, pp. 157–172. Springer, Heidelberg (2005)
14. Klieber, W., Sapra, S., Gao, S., Clarke, E.: A Non-prenex, Non-clausal QBF Solver with Game-State Learning. In: Strichman, O., Szeider, S. (eds.) SAT 2010. LNCS, vol. 6175, pp. 128–142. Springer, Heidelberg (2010)
15. McMillan, K.L.: Applying SAT Methods in Unbounded Symbolic Model Checking. In: Brinksma, E., Larsen, K.G. (eds.) CAV 2002. LNCS, vol. 2404, pp. 250–264. Springer, Heidelberg (2002)
16. Moskewicz, M.W., Madigan, C.F., Zhao, Y., Zhang, L., Malik, S.: Chaff: Engineering an Efficient SAT Solver. In: DAC 2001 (2001)
17. Shannon, C.E.: The Synthesis of Two Terminal Switching Circuits. Bell System Technical Journal 28, 59–98 (1949)
18. Silva, J.P.M., Sakallah, K.A.: GRASP - a new search algorithm for satisfiability. In: ICCAD, pp. 220–227 (1996)
19. Stallman, R.M., Sussman, G.J.: Forward Reasoning and Dependency-Directed Backtracking in a System for Computer-Aided Circuit Analysis. Artif. Intell. 9(2), 135–196 (1977)
20. Tseitin, G.S.: On the complexity of derivation in propositional calculus. Studies in Constructive Mathematics and Mathematical Logic 2(115-125), 10–13 (1968)

21. Wille, R., Fey, G., Drechsler, R.: Building free binary decision diagrams using SAT solvers. Facta Universitatis-series: Electronics and Energetics (2007)
22. Zhang, L.: Solving QBF by Combining Conjunctive and Disjunctive Normal Forms. In: AAAI 2006 (2006)
23. Zhang, L., Malik, S.: Conflict Driven Learning in a Quantified Boolean Satisfiability Solver. In: ICCAD 2002 (2002)
24. Zhang, L., Malik, S.: Towards a Symmetric Treatment of Satisfaction and Conflicts in Quantified Boolean Formula Evaluation. In: Van Hentenryck, P. (ed.) CP 2002. LNCS, vol. 2470, pp. 200–215. Springer, Heidelberg (2002)

Globalizing Constraint Models*

Kevin Leo[1], Christopher Mears[1], Guido Tack[1,2], and Maria Garcia de la Banda[1,2]

[1] Faculty of IT, Monash University, Australia
[2] National ICT Australia (NICTA), Victoria Laboratory
{kevin.leo,chris.mears,guido.tack,maria.garciadelabanda}@monash.edu

Abstract. We present a method that, given a constraint model, suggests global constraints to replace parts of it. This helps non-expert users to write higher-level models that are easier to reason about and may result in better solving performance. Our method exploits the *structure* of the model by considering combinations of the constraints, collections of variables, parameters and loops already present in the model, as well as parameter data from several data files. We assign a score to a candidate global constraint by comparing a sample of its solution space with that of the part of the model it is intended to replace. The top-scoring global constraints are presented to the user through an interactive display, which shows how they could be incorporated into the model. The *MiniZinc Globalizer*, our implementation of the method for the MiniZinc modelling language, is available on the web.

1 Introduction

Constraint problems can usually be modelled in many different ways, and the choice of model can have a significant impact on the effectiveness of the resulting constraint program. Developing good models is often a very challenging iterative process that requires considerable levels of expertise and consumes significant amounts of resources. This paper introduces a method that supports users through this iterative process: given a constraint problem model and a few input data files, the method suggests global constraints as possible replacements for certain sets of constraints in the model.

Replacing simpler constraints by global constraints — *"globalizing" the model* — has three significant advantages. First, many solvers implement specialised algorithms for global constraints. Therefore, having the global constraint in the model can improve the efficiency of the solving process considerably. Second, more information is made available regarding the underlying structure of the model. The additional information can help, for example, to detect symmetries, which can then be broken either by adding symmetry breaking constraints or by modifying the search. As another example, even if the chosen solver does not yet support the inferred global constraint, its presence in the model can be used to select better decompositions than the ones originally used by the modeller. And third, the higher-level model obtained by the globalization may improve the modeller's understanding of the problem and even make it more readable.

---

* NICTA is funded by the Australian Government as represented by the Department of Broadband, Communications and the Digital Economy and the Australian Research Council. This research was partly sponsored by the Australian Research Council grant DP110102258.

C. Schulte (Ed.): CP 2013, LNCS 8124, pp. 432–447, 2013.

Our method is based on *splitting* a constraint model into submodels, *generating candidate* global constraints for each submodel, and *ranking and filtering* these candidates to produce the output returned to the user. Critically, each of these steps makes extensive use of the existing structure in the model, such as loops and collections of variables, as well as the provided instance data. Note that the correctness of replacing constraints in the model by the candidate global constraints needs to be determined by the user. This approach is similar to that successfully used for symmetry detection [10], which analyses several small instances of a model (i.e., several combinations of model with input data) to obtain candidate symmetries, and then lifts this information from the instances to the model itself.

Our method has many novel characteristics when compared to other automatic model transformation methods (e.g., [6,8,9,7,4,5,2,1]; see Section 6 for a detailed discussion). First, other methods focus on directly inferring a combination of constraints for the entire model, rather than on splitting it into submodels. Splitting allows us to directly associate the candidate global constraint with the group of constraints it replaces (those in the submodel). Second, the generation of arguments for the candidate global constraints uses the variables, parameters and collections of variables appearing in the associated submodel. This allows us to generate likely constraint arguments efficiently. Further, it means the candidate global constraints are defined at the model level rather than at the instance level. This is important not only for the user, but also for our third novel characteristic: our method uses the solutions from different instances (rather than from a single one) to generate, rank and filter the candidates. This increases its accuracy considerably (as shown experimentally in Section 5).

We have implemented the method for the MiniZinc modelling language [11]. The resulting tool – the *MiniZinc Globalizer* – can be accessed through a web interface at `http://www.minizinc.org/globalizer/`. The presented techniques are however not specific to MiniZinc and apply to any representation of a constraint model.

2   Background

We distinguish between *constraint problems*, *models*, and *instances*. A constraint (satisfaction or optimization) problem is the abstract problem we want to solve, e.g., the Graph-colouring problem. A model is a concrete specification of the problem in terms of variables, domains, constraints, and parameters. For the Graph-colouring problem, a model could have variables representing the nodes, domains representing the colours, parameters for the graph and number of colours used, and constraints stating that no two connected nodes can have the same colour. A model together with one concrete set of input data – such as a concrete graph and set of colours – is an instance.

All models used herein are written in MiniZinc. A MiniZinc model consists of a list of variable declarations, parameter declarations, and constraints, as well as a *solve item* that may specify an objective function. The subset of MiniZinc used in this paper should be mostly self-explanatory.

```
1    int: p; int: nh; int: ng;
2    set of int : HostBoats = 1..nh;
3    set of int : GuestCrews = 1..ng;
4    set of int : Time = 1..p;
5    array [GuestCrews] of int : crew;
6    array [HostBoats]  of int : capacity;
7
8    array [GuestCrews, Time] of var HostBoats : hostedBy;
9    array [GuestCrews, HostBoats, Time] of var 0..1 : visits;
10   constraint forall (g in GuestCrews, h in HostBoats, t in Time)
11     (visits[g,h,t] = 1 <-> hostedBy[g,t]=h);              % channel
12
13   constraint forall (h in HostBoats)
14   ( forall (g in GuestCrews)
15       (sum (t in Time) (visits[g,h,t]) <= 1)              % distinct_visits
16   /\ forall (t in Time)
17       (sum (g in GuestCrews) (crew[g]*visits[g,h,t]) <= capacity[h]));
18                                                           % capacity
19
20   array [GuestCrews, GuestCrews, Time] of var 0..1 : meet;
21   constraint forall (k, l in GuestCrews where k<l) (
22       forall (t in Time)
23         (hostedBy[k,t] = hostedBy[l,t] -> meet[k,l,t] = 1)  % will_meet
24   /\  sum (t in Time) (meet[k,l,t]) <= 1 );               % meet_once
```

Fig. 1. A Progressive Party model in MiniZinc

Running Example: The Progressive Party Problem

Throughout the paper, we will use a version of the *Progressive Party Problem* as a running example. This problem can be described as follows: to organise a party at a yacht club, certain boats are designated as hosts, while the crews of the remaining boats in turn visit the host boats for several successive fixed-time periods. Every boat has a given maximum capacity for hosting guests, a guest crew cannot revisit a host, and guest crews cannot meet more than once.

As shown in [12], the first known model for this problem was a zero-one integer program by the University of Southampton. Since this model introduced a huge number of constraints, an alternative one was given [12] which found a 13-host solution.

A MiniZinc version of the second model is shown in Fig. 1. Lines 1–6 introduce the parameters: number of time periods, host boats, and guest crews, as well as the sets of designated host boats and guest crews. The main decision variables (line 8) express that at time t, guest crew g is hosted by boat hostedBy[g,t]. Lines 9–11 introduce auxiliary zero-one variables visits[g,h,t] that are 1 if and only if hostedBy[g,t]=h. These variables are used in line 15 to express that each guest crew visits each host boat at most once; and in line 17 to model the capacity constraints. Finally, lines 20–24 model that guest crews can meet at most once.

The expert modeller can immediately see that line 15 expresses an *alldifferent* constraint on the hostedBy variables for each g in GuestCrews. The fact that line 17 can be expressed using a set of *bin packing* constraints is slightly less obvious.

To simplify the discussion of our running example, the remaining sections will use the following shorthand notation to express the main structure of the above model: $(\forall GHT : channel) \land (\forall H : (\forall G : distinct\_visits) \land (\forall T : capacity)) \land (\forall GG : (\forall T : will\_meet) \land meet\_once)$, where *channel* denotes the constraint appearing in lines 10–11, *distinct_visits* that in line 15, *capacity* that in line 17, *will_meet* that in line 23

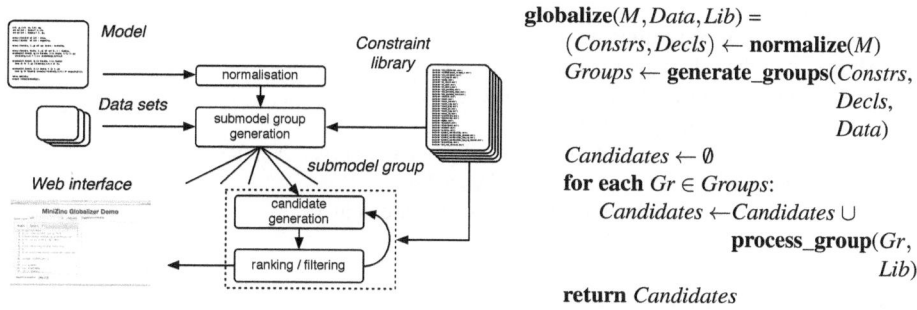

$$\text{globalize}(M, Data, Lib) =$$
$$(Constrs, Decls) \leftarrow \text{normalize}(M)$$
$$Groups \leftarrow \text{generate\_groups}(Constrs,$$
$$Decls,$$
$$Data)$$
$$Candidates \leftarrow \emptyset$$
$$\textbf{for each } Gr \in Groups:$$
$$Candidates \leftarrow Candidates \cup$$
$$\textbf{process\_group}(Gr,$$
$$Lib)$$
$$\textbf{return } Candidates$$

Fig. 2. An overview of model globalization

and *meet_once* that in line 24. Universal quantifications over G, H and T correspond to loops over the sets GuestCrews, HostBoats, and Time, respectively. We call G, H and T the *index sets* of their loops. For simplicity, we always write nested forall loops using a single quantifier and disregard the order of their index sets, e.g., $\forall GHT$ is equivalent to $\forall T \forall G \forall H$, to $\forall G \forall T \forall H$, and so on.

3 Globalization

Figure 2 provides a graphical and algorithmic view of the main steps of our method, which are as follows. First, the input model file M together with one or more data files *Data* are read. Then, M is *normalized* by splitting conjoined constraints and separating the constraints *Constrs* from the variable and parameter declarations *Decls*. After normalization, several *submodel instance groups* are generated, where each such group $Gr \in Groups$ corresponds to the instantiation of a single submodel with each of the data files in *Data*. A submodel is formed by the combination of *Decls* with a subset of *Constrs*. For each group Gr, the method generates a set of *candidate global constraints* from those present in the constraint library *Lib*, where each global constraint in this set is a candidate for equivalence to the submodel associated to Gr. The generated candidates are then *scored* according to how well their solution space matches that of the submodel, and are *filtered out* if their score is below a given threshold. Finally, the resulting candidates are shown to the user by means of an interactive GUI. The following sections discuss each of these steps in detail.

3.1 Generating Submodel Instance Groups

The algorithms for normalizing a model M and generating its submodel instance groups *Groups* are shown in Fig. 3. The **normalize** procedure partitions M into two sets: the set *Constrs* of *normalized* constraints and the set *Decls* of original variable and parameter declarations. Constraints are normalized by exhaustively applying two rewriting rules that (a) turn top-level conjunctions into individual constraints and (b) split forall loops that contain conjunctions into individual forall loops. For example, after normalizing the Progressive Party model, *Constrs* will contain the following five

normalize(*M*)
    Constrs ← set of constraints in *M*
    while one of the following rules applies:
        if there is a $c \in$ *Constrs* of the form $(c_1 \wedge \ldots \wedge c_n)$:
            replace c with c_i for each i
        if there is a $c \in$ *Constrs* of the form $(\forall A_1 \ldots \forall A_n : c_1 \wedge \ldots \wedge c_m)$:
            replace c with $(\forall A_1 \ldots \forall A_n : c_i)$ for each i
    Decls ← set of all variable and parameter declarations in *M*
    return (*Constrs*,*Decls*)

generate_groups(*Constrs*,*Decls*,*Data*)
    Groups ← ∅
    for each $SC \subseteq$ *Constrs* such that *SC* is connected:
        Groups ← *Groups* ∪ { **instantiate**(∅, *SC*, *Decls*, *Data*) }
                         ∪ **unroll_loops**(∅, *SC*, *Decls*, *Data*)
    return *Groups*

unroll_loops(*Fix*, *SC*, *Decls*, *Data*)
    Groups ← ∅
    for each *A* such that $(\forall \ldots A \ldots : d)$ is in all $c \in SC$:
        for each combination *L* of loops ∀*A*, one for each $c \in SC$:
            Groups ← *Groups* ∪ { **instantiate**($\{L\} \cup Fix, SC \setminus L, Decls, Data$) }
                         ∪ **unroll_loops**($\{L\} \cup Fix, SC \setminus L, Decls, Data$)
    return *Groups*

instantiate(*Fix*, *SC*, *Decls*, *Data*)
    Gr ← ∅
    for all combinations of min, max for all $L \in Fix$ and all $D \in Data$
        create submodel instance *SI* from submodel ($SC \cup Decls$)
        Gr ← *Gr* ∪ {*SI*}
    return *Gr*

Fig. 3. Splitting a model *M* into groups of submodel instances

constraints $\forall GHT$: *channel*, $\forall GH$: *distinct_visits*, $\forall HT$: *capacity*, $\forall GGT$: *will_meet* and $\forall GG$: *meet_once*.

    Normalization is vital for discovering global constraints that describe parts of a top level constraint. For example, we said that the combination of constraint $\forall GH$: *distinct_visits* with channelling constraint $\forall GHT$: *channel* in the Progressive Party is equivalent to a conjunction of `alldifferents`. To discover this, we need to consider each component constraint separately, so that we can combine them appropriately.

    Once normalization is complete, **generate_groups** produces every *connected* subset of constraints in *Constrs*.[1] Two constraints are connected if they share at least one variable. A set of constraints is connected if for each pair of constraints c, c' in the set, a path of constraints can be found starting with c and ending with c', such that consecutive constraints on the path are connected. For example, the subset of normalized constraints

---

[1] Our implementation has a parameter to limit the maximum size of the generated subsets (the default is 3 as our experiments have not found globals from larger conjunctions of constraints).

$SC = \{\forall GHT : channel, \forall GG : meet\_once\}$ in the Progressive Party is not connected, since their sets of variables are $\{\texttt{visits}, \texttt{hostedBy}\}$ and $\{\texttt{meet}\}$, respectively. Since the set of global constraints inferred for a non-connected SC would at best be identical to the union of those found for each of its constraints separately, we can discard SC.

Every connected subset $SC \in Constrs$ is passed to **instantiate**, together with the set of declarations $Decls$ and the input data files $Data$, to create a group of related submodel instances: one per combination of the submodel $SC \cup Decls$ with a data file D in $Data$.

Considering all connected subsets of the normalized top-level constraints is, however, not enough. This can be illustrated with the Progressive Party model: since no global constraint describes a conjunction of $\texttt{alldifferent}$ constraints, the normalized constraints $\forall GH : distinct\_visits$ and $\forall GHT : channel$ must be combined, unrolled and instantiated in such a way as to make them discoverable. Loop unrolling achieves this by recursively combining and instantiating the loops in each SC subset of $Constrs$ as follows. For each index set A that appears in every constraint of SC, it computes each *combination L* of $\texttt{forall}$ loops over A, choosing one from each constraint $c \in SC$. Let us explain how such a combination is computed for subset $SC = \{\forall GHT : channel, \forall GH : distinct\_visits, \forall GGT : will\_meet, \forall GG : meet\_once\}$ and loop G (which appears in every constraint of SC). We first label the individual loops to be able to identify them: $\forall G_1 HT : channel$, $\forall G_2 H : distinct\_visits$, $\forall G_3 G_4 T : will\_meet$, $\forall G_5 G_6 : meet\_once$. We then compute all combinations of G that have one index set from each constraint, obtaining 4 combinations: $L_1 = \{\forall G_1, \forall G_2, \forall G_3, \forall G_5\}$, $L_2 = \{\forall G_1, \forall G_2, \forall G_3, \forall G_6\}$, $L_3 = \{\forall G_1, \forall G_2, \forall G_4, \forall G_5\}$, and $L_4 = \{\forall G_1, \forall G_2, \forall G_4, \forall G_6\}$.

For each combination L, loop unrolling calls **instantiate** with L added to the set of combinations Fix to be *fixed*, and removed from SC. Fixing a loop means instantiating its index variable to a particular value from its index set. Our algorithm fixes indices to two values: the minima and maxima of their index set. Consider, for example, index set G and $SC = \{\forall G_1 HT : channel, \forall G_2 H : distinct\_visits\}$. The only combination L for G is $L = \{\forall G_1, \forall G_2\}$. Thus, **instantiate** will fix the index variable g of G_1 and G_2 to the same value resulting in the following two submodels:

$$(g = \min(G)) \wedge (\forall HT : channel) \wedge (\forall H : distinct\_visits)$$
$$(g = \max(G)) \wedge (\forall HT : channel) \wedge (\forall H : distinct\_visits)$$

each describing an $\texttt{alldifferent}$ constraint over the variables $\texttt{HostedBy[g,h]}$, for a fixed value of $\texttt{g}$. Both submodels are then instantiated with any provided data, resulting in submodel instances that are added to the same group. Each will also be submitted to further loop unrolling leading to the following submodels:

$$(g = \min(G) \wedge h = \min(H)) \wedge (\forall T : channel) \wedge (distinct\_visits)$$
$$(g = \min(G) \wedge h = \max(H)) \wedge (\forall T : channel) \wedge (distinct\_visits)$$
$$(g = \max(G) \wedge h = \min(H)) \wedge (\forall T : channel) \wedge (distinct\_visits)$$
$$(g = \max(G) \wedge h = \max(H)) \wedge (\forall T : channel) \wedge (distinct\_visits)$$

This process is repeated recursively until all loops have been unrolled.

A special case to be considered is what the algorithm should do upon fixing all the loops in a constraint when the original $\texttt{forall}$ loop had a $\texttt{where}$ clause. During unrolling, and as long as there is one $\texttt{forall}$ the algorithm leaves the $\texttt{where}$ clause

generate_candidates(*SI*,*Decls*, *Lib*)=
    Candidates ← ∅
    Solutions ← random sample of solutions of submodel instance *SI*
    Template ← (*SI* \ *Decls*) ∪ *Solutions*
    Base_arguments ←
           (variable and parameter collections in *SI*) ∪
           (variable and parameter sub-collections in constraints of *SI*) ∪
    Arguments ← *Base_arguments* ∪
           (array accesses of elements of *Base_arguments*) ∪
           { constant 0 } ∪
           { blank symbol }
    for each constraint cons in *Lib*:
        for each tuple args that can be built from *Arguments*:
        Replace blank symbols in args by their value
        Instance ← *Template* ∪ (constraints for cons(args))
        if *Instance* is satisfiable
            add cons(args) to *Candidates*
    return *Candidates*

Fig. 4. Generating candidate constraints for a submodel instance

as part of that `forall`, but when there is no `forall` left, the resulting constraint gets wrapped in an `if-then-else` expression to avoid creating incorrect submodels.

3.2   Candidate Generation

As explained in the previous section, once **generate_groups** finishes, *Groups* has all submodel instance groups, where each group contains different instances of the same submodel. Recall that each instance has different parameter values due either to different data files given by the user, or to the different minima and maxima values chosen during loop unrolling. Each group *Gr* in *Groups* is then processed to generate a set of candidate constraints, which are added to the final set *Candidates*. The algorithm for generating candidate constraints for each submodel instance $SI \in Gr$, given the set *Decls* of declarations of the original model *M*, and the library of global constraints *Lib*, is shown in Fig. 4. Note that each constraint entry in *Lib* has a name, arity, and type of arguments. In addition, arguments can have associated information indicating whether they are functionally dependent on other arguments, and stating conditions that must be met for the argument to be used. See Section 4.2 for details on the particular *Lib* used by our implementation.

The algorithm proceeds as follows. After finding a random sample of solutions of *SI*, we build a template model by replacing the parameters and variables in *Decls* by the sample solutions. The template includes *all* the sample solutions, as we want the candidate global constraint to satisfy all of these sample solutions: a single sample solution that violates the constraint is sufficient evidence to discard the constraint. This template model is trivially satisfiable. Intuitively, a global constraint will be considered as a candidate if it is satisfied by the sample solutions of *SI* — that is, if after adding the candidate, the template model remains satisfiable.

Candidate constraints are obtained by combining each constraint cons in *Lib* with an *A*-tuple args of arguments, where *A* is the arity of the constraint. The arguments are drawn from the identifiers that appear in *SI*. These include the variable and parameter collections whose identifiers appear in *SI*, those same collections restricted to their subsets that are actually used in the constraints of *SI*, array access expressions composed from the two previous groups (referred to as base_arguments), the constant zero, and a special blank symbol. This blank symbol is used as a place-holder for arguments known to be functionally defined by the others. Once all non-blank arguments are selected, the blank symbol is replaced by its corresponding value. Note that this value must be the same for all sample solutions. If the constraint is not functional, or if the sample solutions disagree on what the value should be, args is discarded. Functionally-defined arguments are further required to take the same value *across all instances*. This is however not a significant issue, as they are only used when no named parameter is found.

Finally, the candidate global constraint cons(args) is added to the template model and the resulting model is evaluated. If the constraint holds, cons(args) is added to the list of candidate global constraints for *SI*.

Let us illustrate this process with the subinstance *SI* formed by combining the constraint $(g = \min(G)) \wedge (\forall HT : channel)$ of the Progressive Party model (where $\min(G) = 1$) with the variable and parameter declarations in the model, and some data file $D \in Data$. The base arguments for *SI* include the variable collections hostedBy and visits, the variable sub-collections hostedBy[1,t] and visits[1,h,t], and the parameter collections HostBoats, GuestCrews, p, nh, ng, Time, and the index g itself (with value 1). The arguments are the constant 0, the blank symbol, the base arguments, and array accesses formed by combining an array with a parameter, e.g., crew[nh] and hostedBy[p,capacity]. After considering all constraints in *Lib* with these arguments, the following candidate constraints are generated (among many others):

- lex2(hostedBy)
- alldifferent([hostedBy[g,1],hostedBy[g,2],. . .,hostedBy[g,p]])
- sliding_sum(0, p, g, [hostedBy[g,1],hostedBy[g,2],. . .,hostedBy[g,p]])

Note that in the first constraint the entire hostedBy array is used as an argument, while in the second and third constraints only that subset of the array that participates in the constraints of the submodel – where g is fixed to 1 – is used as an argument.

An alternative to this form of candidate generation is to syntactically match groups of constraints to known (correct) reformulations. We do not take this approach as it would be much too restrictive, requiring the modeler to have implemented exactly the constraints we are looking for.

3.3   Ranking and Filtering

As shown in Figure 5, submodel instances are processed in groups, where each group *Gr* contains submodel instances of the same model. This allows us to accurately determine the candidate constraints for *Gr* by taking the intersection of the candidate constraints found for each submodel instance *SI* in *Gr*. For the first *SI* being processed, the full set of constraints and argument tuples (written as the special symbol *Universe*)

process_group(*Gr,Decls,Lib*)=
   Candidates ← Universe
   for each submodel instance *SI* in group *Gr*:
     Candidates ← Candidates ∩ **generate_candidates**(*SI,Decls,Lib*)
     for each candidate constraint cons(args) in *Candidates*:
       SolutionsC ← random sample of solutions of cons(args)
       SolutionsM ← subset of *SolutionsC* that are also solutions of *SI*
       if |*SolutionsM*| ÷ |*SolutionsC*| < *equivalenceThreshold*
         Delete cons(args) from *Candidates*
         for each possible context *B*:
           SolutionsC ← random sample of solutions of cons(args) ∧ *B*
           SolutionsM ← subset of *SolutionsC* that are also solutions of *SI*
           if |*SolutionsM*| ÷ |*SolutionsC*| ≥ *equivalenceThreshold*
             Add cons(args) with context *B* to *Candidates*
   return *Candidates*

Fig. 5. Ranking and filtering constraints

is considered, so that the intersection is the candidate set generated for this *SI*. Due to filtering the set decreases for subsequent instances in the group and, after processing the final one, the remaining candidates are exactly the intersection we seek to compute.

For each candidate constraint cons(args) inferred for a given *SI*, **process_group** measures how closely it matches *SI*. To achieve this, we collect a random sample of solutions of cons(args), and compute the fraction of these solutions that are also solutions to *SI*. If the constraint is equivalent to the submodel of *SI*, this fraction must be 1; if the constraint is a poor match, the fraction should be close to 0. We filter the candidates by keeping only those constraints whose matching fraction is greater than a given threshold. A threshold of 0.5 has been shown experimentally to be sufficient to eliminate imperfect matches, and we use that value in our implementation.

In some cases a constraint in a model is equivalent to a candidate global constraint only in the context of another constraint. Consider a submodel instance *SI* containing constraints *A* and *B*, and a candidate global constraint cons(args), where *A* is not equivalent to cons(args), but the conjunction $A \wedge B$ is equivalent to the conjunction cons(args)$\wedge B$. We call *B* the *context* in which *A* is equivalent to cons(args). For example, let *SI* have the constraints $(t = \min(1..p)) \wedge \forall GH : channel \wedge \forall H : capacity$ from the Progressive Party. The global constraint bin_packing_capa(capacity,hostedBy[1..ng,t], crew) is equivalent to $\forall H : capacity$, but only in the context of $\forall GH : channel$. In general, a contextually-equivalent constraint cons(args) will be implied by *SI* but appear weaker than the instance and, thus, will score badly during ranking. In this case, we try using one of *SI*'s constraints as the context constraint *B*, and test via sampling whether cons(args)$\wedge B$ implies the submodel instance. If this scores well, we say that *A* is equivalent to cons(args) under the context of *B*, and add this to the list of candidates.

For the purpose of scoring and filtering, a candidate constraint should now be considered a pair of the cons(args) and its context. Note that the context may be empty. This means that for a context-dependent constraint to pass the filtering tests, it must pass with the same context in all instances of the group.

Table 1. Library of global constraints

alldifferent	circuit	global_cardinality	nvalue
alldifferent_except_0	count	increasing	sliding_sum
all_equal	cumulative	inverse	sort
atleast	decreasing	lex_less	strict_lex2
atmost	diffn	lex_lesseq	subcircuit
bin_packing	distribute	lex2	unary
bin_packing_capa	element	maximum	value_precede
bin_packing_load	exactly	minimum	
channel	gcc	member	

4 Implementation

We have implemented the globalization system for MiniZinc models. We use the libmzn C++ library for parsing and manipulating MiniZinc model and data files. The model evaluator is written in Haskell, and uses bindings to call libmzn.

4.1 Checking versus Solving

As shown before, when generating the candidates of a given submodel instance SI of group Gr, our method first solves SI to find a random sample of its solutions. To do this, our implementation uses the standard MiniZinc tool mzn2fzn to flatten SI, and the Gecode constraint solver to find 30 random solutions for it. Theses are found with a search that selects values in random order, and restarts from scratch whenever a solution is found. If the search is not complete within 60 seconds, SI is discarded.

Later in the process our method checks the satisfiability of the instance resulting from adding to the template the possible candidate constraints. Since this template has no variables, and the added constraints are simply evaluated, no search is required. Such checks are performed very often, and the expense of flattening the instance and calling a full constraint solver is crippling. To avoid this, we have implemented a simple evaluator of MiniZinc instances known not to have variables. In practice, this optimization is crucial, as the number of evaluations is usually in the hundreds of thousands.

4.2 Library of Global Constraints

Table 1 lists the global constraints in our implementation of *Lib*, which is used for candidate generation. These are all the global constraints defined in MiniZinc's standard library (version 1.6) over integer arguments (sets are not handled yet by our prototype implementation), with the addition of the following constraints:

- channel(x,a): channels an integer variable x to an array of 0-1 variables a.
- gcc(x,counts): a special case of global_cardinality where the "cover" argument, which specifies a map from indices to values, is fixed to the identity map.
- unary(s,d): a special case of cumulative where the resource capacity and the usage for each task are fixed to 1, implementing a unary resource constraint.

We are able to evaluate most of the constraints using the default decomposition given in the MiniZinc library. However, some decompositions introduce variables which are not handled by our simple satisfiability check evaluator (recall that our satisfiability check does not perform search). Thus, in these cases we evaluate the constraint directly.

As mentioned in Section 3.3, each constraint is annotated with conditions for its use to prevent the constraint being considered as a candidate when it is trivially true or otherwise useless. For example, the alldifferent constraint specifies that its argument must be an array of variables with arity greater than one since, otherwise, the constraint is trivially true or nonsensical. As another example, the sliding_sum(l,u,n,x) constraint specifies that every n-length subsequence of x must sum to a value between l and u. When generating candidates, we ensure that $l < u$, $1 < n < length(x)$, and $l > n \times$ lb_array(x) $\vee$ u $< n \times$ ub_array(x). The first two conditions ensure that the parameters make sense, while the third one ensures that the constraint is tighter than what is already imposed by the domains of the variables in x. This last condition is added for efficiency reasons, as the ranking and filtering process would have taken care of it.

4.3   Web Interface

The MiniZinc Globalizer is implemented as an asynchronous web server that queues the requests made by clients and can execute several requests in parallel. Requests can be cancelled by the user and are automatically cancelled when the session is terminated, for instance, when a user closes the browser window. The Globalizer is publicly available at http://www.minizinc.org/globalizer/.

Figure 6 shows a screen shot of the web interface. Users can enter their model and instance data through an embedded editor (seen on the left). Clicking ANALYZE launches the request, initiating a progress bar that provides the user periodic progress updates. When the analysis finishes, the right hand side of the window displays the results. Clicking on any candidate constraint highlights in yellow the part of the original model that the constraint could replace, and highlights in orange the candidate's context, if any. The interface allows the user to select parts of the model and restrict the analysis to the selected parts by selecting "Only selection" in the lower left corner of the window. The analysis will then only use the selected constraints. This is useful when the analysis is taking a long time, or the user wants to focus on a particular part of the model.

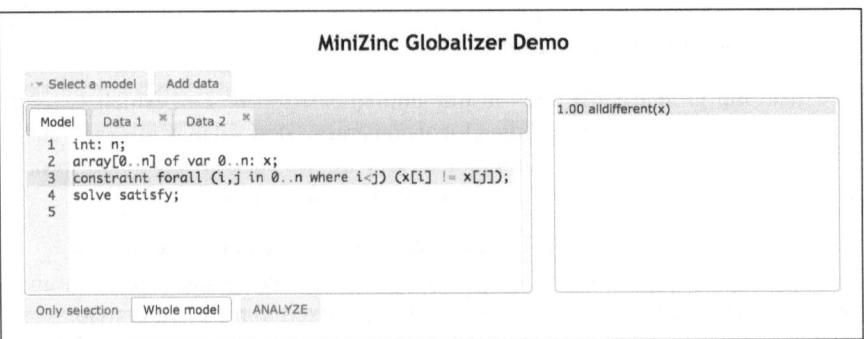

Fig. 6. The web interface to the MiniZinc Globalizer

Table 2. Experimental Results

problem	time	*Groups*	calls	evals	top candidates
Cars	215	12	2910	31335	**gcc(step_class,cars_in_class)** count(step_class,c,cars_in_class[c]) **sliding_sum(sliding_sum(0,option_max_per_block[p],** **option_block_size[p],step_option_use[1..10,*]))**
Jobshop	23	16	410	3620	**unary(s[1..n,*], d[1..n,*])**
Party	691	48	4214	36429	**bin_packing_capa(spareCapacity,hostedBy[i..n,*],crew)** **alldifferent(hostedBy[*,1..4])** **channel(hostedBy[*,*], visits[*,1..4,*])** unary(hostedBy[*,1..4],visits[*,*,1..4])
Packing	1659	29	32174	114597	**diffn(x,y,pack_s,pack_s)** **diffn(y,x,pack_s,pack_s)**
Schedule	174	13	3077	22835	**gcc(x, [1,0,1,0,1,0,1])**
Sudoku 1	3996	166	6305	24465	**alldifferent(p[*,1..9])** gcc(p[*,1..9], [1,1,1,1,1,1,1,1,1]) **alldifferent(p[1..9,*])** gcc(p[1..9,*], [1,1,1,1,1,1,1,1,1])
Sudoku 2	335	7	338	2347	-
Warehouses	245	24	3853	69871	**gcc(supplier,use)**

5 Experiments

This section evaluates the accuracy and practicality of the prototype implementation of our MiniZinc Globalizer. The evaluation is performed over a set of constraint problems, each with a number of different data sets. The MiniZinc models used for these problems are available at the MiniZinc Globalizer website.

The results are shown in Table 2, where for each problem model we show: the name of the problem (**problem**), the time in seconds to run the MiniZinc Globalizer (**time**), the number of submodel instance groups obtained (|*Groups*|), the number of calls to Gecode to obtain sample solutions of either submodel instances or global constraint candidates (**calls**), the number of satisfiability tests performed (**evals**), and the global constraints proposed as candidates with score 1 (**top candidates**), where high quality candidates appear in bold. We have manually simplified the output, and excluded some duplicate constraints where the system was unable to distinguish two parameters that appear different, but actually refer to the same value. The set of problems used in the table is as follows.

Cars is a version of the car sequencing problem (CSPLib 001) as implemented in the MiniZinc distribution. It uses simple arithmetic and counting constraints to express the capacity and sequence restrictions of the problem. The Globalizer finds the corresponding sliding_sum and global_cardinality constraints.

Jobshop is a simple job-shop scheduling problem taken from the MiniZinc distribution. It implements the non-overlapping of two tasks on a unary resource using simple reified constraints. Globalization finds the unary scheduling constraint.

Party is our running example from Figure 1. As discussed earlier, the Globalizer finds the bin packing and alldifferent constraints. It also finds the channel global constraint.

Packing packs n squares into a rectangle. The source code was taken from the MiniZinc distribution. The Globalizer finds diffn constraints that express the non-overlapping of rectangles.

Schedule is a contrived scheduling example from [4]. The schedule is constrained in a way that one task needs to start on every even time point, which implies a global cardinality constraint with argument $[1, 0, 1, 0, ...]$. Our analysis can find this constraint as long as all instances have the same schedule length, as otherwise the arguments differ in length between instances and are thus discarded. The generalization of such sequences is left to future work.

Sudoku 1 and 2 are different models for the Sudoku puzzle. The first one uses a zero-one integer linear programming formulation, and the Globalizer finds some `alldifferent` constraints. The second model posts binary not-equal constraints on variables that are organised by row and column, in a complicated set of nested `forall` loops. Here, the loop unrolling is not strong enough to generate candidates that correspond to individual rows, columns, or blocks. As a result, Globalizer cannot find any replacement global constraints.

Warehouses is a warehouse allocation problem (CSPLib 034), whose source code was taken from the MiniZinc distribution. The Globalizer finds that a loop containing counting constraints can be aggregated into one `global cardinality` constraint.

Discussion

The models discussed here are taken either from the literature or from the example suite that comes with MiniZinc. In most cases, the Globalizer has been able to find the global constraints that an expert modeller would have used.

The current prototype is not optimized for performance. The time to analyse a reasonably complex model with 3-4 data sets is in the range of minutes up to an hour, depending mainly on the number of candidates that need to be checked for satisfiability (a number that grows considerably with the number of possible arguments). There is still great potential for improving performance by both avoiding and parallelizing unnecessary candidate checks and parameter instantiations which, as indicated, make up for the bulk of the run time.

The number of generated groups is relatively small (usually less than 50), which means that the problem splitting algorithm achieves a good level of pruning. Generating only a small number of groups and top scoring constraints is important since the results are meant to be presented to a human user.

Looking at the number of satisfiability tests, which can reach hundreds of thousands, it becomes clear that each check needs to be very efficient. This justifies the introduction of a dedicated constraint evaluator as discussed in Section 4.1.

It is interesting to note the effect of using more than one data file for a given problem. For example, analysing the **packing** problem with a single data file results in 54 candidate global constraints with a score of 1. Adding the additional data files increases the discriminative power of the system by reducing the candidates with a score of 1 to the two shown in Table 2. Similarly, analysing **Warehouses** with only one data file results in 6 candidate global constraints with a score of 1, as opposed to one as in the table above. For the **Schedule** and **Jobshop** problems, however, a single data file was enough to narrow the candidates down to a single constraint with a score of 1.

6 Related Work

There are two main lines of research related to this work: *constraint acquisition* and *automatic model transformation*.

In the *acquisition* line of work, Constraint Seeker [1] infers global constraints from positive and negative examples of solutions, and Model Seeker [2] infers an entire model (i.e., conjunctions of constraints) from complete solutions to a constraint problem. This differs from the method presented in this paper both in motivation and methodology. Our motivation is to identify parts of a given model that can be replaced by global constraints. Having access to an initial model significantly affects our methodology, as it allows us to make extensive use of the information contained in the model. In particular, it allows us to (a) focus on submodels that are equivalent to a single global constraint, as opposed to a conjunction of them, (b) significantly reduce the search for possible combinations of global constraint arguments, while increasing the likelihood of obtaining meaningful ones, and (c) consider not only the solution variables, but any other intermediate variables in the model and its input data. Having the input data also affects our methodology, as it allows us to (a) better generate candidates and (b) automatically generate as many solutions as we require for our rankings. For example, the input data enables us to derive bin packing constraints for the Progressive Party problem, while Model Seeker cannot infer these from just the solutions.

Note that we could use Model Seeker to generate more complex candidates for each submodel, and Constraint Seeker to infer and rank candidate constraints. We would like to experimentally evaluate and compare these approaches to our own submodel and constraint generators when the two tools become publicly available.

Our method is also related to the CGRASS system [6,8], which among other model transformations, includes a specialised component to detect alldifferent global constraints for instances of the problem. The main differences are that our Globalizer aims at inferring any of a set of global constraint using a general (rather than specialised) method, and does so for a model, rather than for each of its instances.

Other acquisition approaches focus on the automatic generation of implied constraints. A general method is described in [5], where machine learning is used to induce constraints for the solutions for small problems, and a theorem prover is then used to show the constraints hold for the model. The generality of the method results in applicability restrictions: the model data can only be a single integer, and the model needs to be expressible in first order logic. In our case the constraints are already pre-determined (the list of global constraints considered) and, thus, the data can be as complex as necessary. Further, we do not attempt to prove the correctness of the constraints as this reduces to proving the equivalence of two models, which is undecidable.

Another related method is that of CONACQ [3] which, given examples of solutions and non-solutions for a target problem and a library of constraints, *acquires* constraint networks, that is, conjunctions of constraints in the library that are consistent with the given solutions and non-solutions. CONACQ uses SAT-based *version space algorithm*, where the version space is the set of all constraint networks defined from the library that are consistent with the examples. While general and powerful, it considers instances of models, rather than models themselves. Further, it relies on the library of constraints being relatively small. This is not the case for our approach. As far as we know, CONACQ currently only handles binary constraints and is not publicly available. Finally, [4] describes how implied parametric constraints can be learned by adding a large disjunction of constraints with different parameters that together are guaranteed to be implied, and then successively pruning that disjunction by checking if certain sets of parameters can be removed without changing the solution space. This method goes further than what we attempt in that it infers parameters from solutions, while we only try to match parameters that are already given in the model. For functional dependencies, however, we can infer parameter sets, as in the Schedule example in Section 5.

In the area of model transformations, the work on Essence [9,7] is somewhat related. These systems transform a model specified in a highly abstract manner into a more concrete one. Our method moves in the opposite direction: we detect parts of a concrete model that are instances of a more generic model pattern. While currently this generic pattern is restricted to global constraints, it is straightforward to extend the method to use any other useful constraint pattern. In fact, globalization and automatic transformation are complementary: starting from a low-level model, globalization yields a high-level model that is then amenable to automatic transformation.

7   Conclusion

This paper has introduced a method for *globalizing* constraint models. Given a constraint model, the method proposes global constraints to replace parts of it. This helps users improve their models, since global constraints capture the inherent structure of a model and can thus help obtain a better translation to the underlying solving technology and faster solving using specialised algorithms.

The inference process is based on splitting a model into submodels that correspond to subsets of its constraints, potentially unrolling loops, and instantiating each of the resulting submodels with different data sets into a group of submodel instances. From these groups of instances, candidate constraints are generated by sampling the solution space of both the group and the candidate constraints. The candidates are ranked and filtered based on how well their search spaces match.

We have presented experimental evidence that the method is both practical and accurate. Our implementation, the MiniZinc Globalizer, is available as a web-based tool.

Regarding future work, while the system already provides useful results, there are some improvements we are planning to explore. First, we would like to incorporate ranking techniques from Constraint Seeker, such as using known implications between constraints to eliminate more imperfect candidates. Second, in order to make contexts more useful, we need to detect global constraints on alternative viewpoints by automatically introducing channeling constraints (Model Seeker follows a similar approach). Third, we would like to generalise argument sequences (such as the [1,0,1,0...] in the simple scheduling example from Section 5) and to detect more complex expressions as constraint arguments. Fourth, the confidence in the suggested constraints may be improved by using theorem proving or other techniques to prove equivalence in cases where it is possible. Finally, we would like to integrate the system into an IDE that lets users refactor models automatically using the suggestions generated by the Globalizer.

References

1. Beldiceanu, N., Simonis, H.: A constraint seeker: Finding and ranking global constraints from examples. In: Lee, J. (ed.) CP 2011. LNCS, vol. 6876, pp. 12–26. Springer, Heidelberg (2011)
2. Beldiceanu, N., Simonis, H.: A model seeker: Extracting global constraint models from positive examples. In: Milano, M. (ed.) CP 2012. LNCS, vol. 7514, pp. 141–157. Springer, Heidelberg (2012)
3. Bessiere, C., Coletta, R., Koriche, F., O'Sullivan, B.: A SAT-based version space algorithm for acquiring constraint satisfaction problems. In: Gama, J., Camacho, R., Brazdil, P.B., Jorge, A.M., Torgo, L. (eds.) ECML 2005. LNCS (LNAI), vol. 3720, pp. 23–34. Springer, Heidelberg (2005)
4. Bessiere, C., Coletta, R., Petit, T.: Learning implied global constraints. In: Veloso, M.M. (ed.) IJCAI, pp. 44–49 (2007)
5. Charnley, J., Colton, S., Miguel, I.: Automatic generation of implied constraints. In: European Conference on Artificial Intelligence, ECAI, vol. 141, pp. 73–77. IOS Press (2006)
6. Frisch, A., Miguel, I., Walsh, T.: Extensions to proof planning for generating implied constraints. In: Calculemus 2001 (2001)
7. Frisch, A.M., Jefferson, C., Martínez-Hernández, B., Miguel, I.: The rules of constraint modelling. In: International Joint Conference on Artificial Intelligence, vol. 19, pp. 109–116. Lawrence Erlbaum Associates LTD. (2005)
8. Frisch, A.M., Miguel, I., Walsh, T.: CGRASS: A system for transforming constraint satisfaction problems. In: O'Sullivan, B. (ed.) CologNet 2002. LNCS (LNAI), vol. 2627, pp. 15–30. Springer, Heidelberg (2003)
9. Gent, I.P., Miguel, I., Rendl, A.: Tailoring solver-independent constraint models: A case study with ESSENCE/ and MINION. In: Miguel, I., Ruml, W. (eds.) SARA 2007. LNCS (LNAI), vol. 4612, pp. 184–199. Springer, Heidelberg (2007)
10. Mears, C., Garcia de la Banda, M., Wallace, M., Demoen, B.: A novel approach for detecting symmetries in CSP models. In: Perron, L., Trick, M.A. (eds.) CPAIOR 2008. LNCS, vol. 5015, pp. 158–172. Springer, Heidelberg (2008)
11. Nethercote, N., Stuckey, P.J., Becket, R., Brand, S., Duck, G.J., Tack, G.: MiniZinc: Towards a standard CP modelling language. In: Bessiere, C. (ed.) CP 2007. LNCS, vol. 4741, pp. 529–543. Springer, Heidelberg (2007)
12. Smith, B.M., Brailsford, S.C., Hubbard, P.M., Williams, H.P.: The progressive party problem: Integer linear programming and constraint programming compared. Constraints 1(1), 119–138 (1996)

A New Propagator for Two-Layer Neural Networks in Empirical Model Learning

Michele Lombardi[1] and Stefano Gualandi[2]

[1] DISI, Università of Bologna, Dipartimento di Informatica: Scienza ed Ingegneria
michele.lombardi2@unibo.it
[2] University of Pavia, Dipartimento di Matematica
stefano.gualandi@unipv.it

Abstract. This paper proposes a new propagator for a set of Neuron Constraints representing a two-layer network. Neuron Constraints are employed in the context of the Empirical Model Learning technique, that enables optimal decision making over complex systems, beyond the reach of most conventional optimization techniques. The approach is based on embedding a Machine Learning-extracted model into a combinatorial model. Specifically, a Neural Network can be embedded in a Constraint Model by simply encoding each neuron as a Neuron Constraint, which is then propagated individually. The price for such simplicity is the lack of a global view of the network, which may lead to weak bounds. To overcome this issue, we propose a new network-level propagator based on a Lagrangian relaxation, that is solved with a subgradient algorithm. The approach is tested on a thermal-aware dispatching problem on multicore CPUs, and it leads to a massive reduction of the size of the search tree, which is only partially countered by an increased propagation time.

1 Introduction

Pushed by research advancements in the last decades, Combinatorial Optimization techniques have been successfully applied to a large number of industrial problems. Yet, many real-world domains are still out-of-reach for such approaches. To a large extent, this is due to difficulties in the formulation of an accurate declarative model for the system to be optimized.

The Empirical Model Learning technique (EML), introduced in [1], has been designed to enable optimal decisions making over complex systems considered beyond the reach of traditional combinatorial approaches. In EML, an approximate model of the target system is extracted via Machine Learning. Such *empirical model* captures the effect of the user decisions on one or more observables of interest (e.g. a cost measure or a constrained parameter). Then, the empirical model is encoded using a combinatorial technology and embedded into a combinatorial model to perform optimization.

Currently, the EML approach has been instantiated using Artificial Neural Networks (ANN) and Constraint Programming, respectively as Machine Learning and Combinatorial Optimization technologies. Specifically, in [1] an ANN is

C. Schulte (Ed.): CP 2013, LNCS 8124, pp. 448–463, 2013.

employed to learn the effect of task mapping decisions on the temperature of a quad-core CPU. In [2], the authors tackle a workload dispatching problem on a 48-core system with thermal controllers: in this case, bad mapping decisions may lead to overheating, which may cause a loss of efficiency when the controllers slow down the cores to decrease their temperature. ANNs are employed to predict the mapping-dependent efficiency loss, i.e the combined effect of the thermal physics and the action of the on-line controllers.

The use of automatically extracted models for cost computation has been previously employed in the context of metaheuristic methods. EML stands out from those approaches for two main reasons: 1) because it makes the empirical model a *component*, easy to integrate with traditional constraints; 2) because it makes the empirical model *active*, rather than an simple function evaluator. In [1] an [2], this is achieved by encoding each neuron in the ANN as a *Neuron Constraint*, which is then propagated to narrow the search space.

Using individual constraints for the neurons is simple, but the loss of the network global view may degrade the propagation effectiveness. To address this issue, we propose a new propagator for the most common ANN structure in practice, i.e. a two-layer, feed forward network. We assume to have sigmoid neurons in the hidden layer, since they are a common choice [15], but the method easily extends to any differentiable activation function. The new propagator does not replace the use of multiple Neuron Constraints, but provides tighter bounds (hence stronger filtering) on the network output variables. The bounds are obtained via a Lagrangian relaxation, with the Lagrangian multipliers being optimized via a subgradient method. We test the approach on a simplified version of the thermal-aware dispatching problem from [2]: the new propagator leads to a substantial (sometimes massive) reduction of the search tree size, in particular for larger instances. This is however partially countered by an increased propagation time. Fortunately, on the basis of a rather strong conjecture that we give at the end of the paper, we believe a complexity reduction is possible.

The paper is structured as follows: Section 2 provides background information. Section 3 describes our Lagrangian relaxation and its solution method, while Section 4 explains how the Lagrangian multipliers are optimized. Section 5 provides our experimental results and Section 6 the concluding remarks.

2   Background and Related Works

Artificial Neural Neworks: An ANN is a system emulating the behavior of a biological network of neurons. Each ANN unit (artificial neuron) corresponds to the following function:

$$z = f\left(b + \sum_{i=0}^{n-1} w_i x_i\right) \tag{1}$$

where x_i are the neuron inputs and w_i are their weights, b is called the bias and z is the neuron output. All the terms are $\in \mathbb{R}$. Besides, $f : \mathbb{R} \to \mathbb{R}$ is

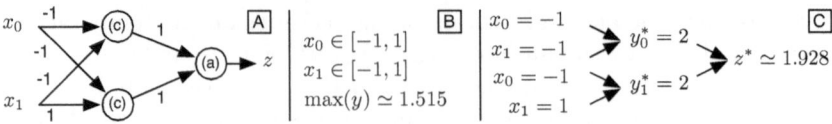

Fig. 1. $\boxed{\text{A}}$ A two-layer, feed forward ANN. $\boxed{\text{B}}$ Domains for the ANN inputs and actual output maximum. $\boxed{\text{C}}$ Output bound computed by the existing propagators.

called *activation function* and is monotone non-decreasing. Some examples of activation functions follow:

$$(a)\ f(y) = y \quad (b)\ f(y) = \begin{cases} 1 \text{ if } y \geq 0 \\ -1 \text{ otherwise} \end{cases} \quad (c)\ f(y) = \frac{2}{1 + e^{-2y}} - 1 \quad (2)$$

Case (a), (b) and (c) respectively correspond to a linear, step and sigmoid neuron. Function (c) is called *tansig* and it is an accurate, faster to compute, approximation of $tanh(y)$. The neurons are connected in a network structure, the most common being an acyclic (i.e. feed-forward), two-layer graph. The first layer is referred as *hidden*, the second is called the *output layer* (because it provides network output). Typically, sigmoid neurons are employed in the hidden layer. Figure 1A shows an example of such a network. Each node represents a neuron, the weights are reported as label on the arcs, x_i are the network input and z is the network output. The weights of an ANN can be assigned automatically by minimizing the average square error on a known set of examples: there are many specifically designed, readily available, algorithms for this purpose [13,3,11].

Neuron Constraints: A Neuron Constraint is a constraint that encodes and propagates Equation (1) and is equivalent to the following pair of constraints:

$$(c_0) \quad z = f(y) \qquad\qquad (c_1) \quad y = b + \sum_{i=0}^{n-1} w_i x_i \qquad\qquad (3)$$

where z, y and x_i are real-valued decision variables[1] with interval domain, i.e. $z \in [\underline{z}, \overline{z}]$, $y \in [\underline{y}, \overline{y}]$ and $x_i \in [\underline{x}_i, \overline{x}_i]$. The term y is called the neuron *activity*. It is possible to embed an ANN in a CP model by building a Neuron Constraint for each node in the network and by introducing decision variables to represent the output of each hidden neuron. Each Neuron Constraint is implemented either as a single entity or as an actual pair of constraints. In the second case, we must explicitly introduce a decision variable to model each neuron activity.

Motivating Example: The propagator for a Neuron Constraint enforces bound consistency on (c_0) and (c_1). For a single neuron, this leads to the tightest possible bounds on all the variables. However, this approach is much less effective once more complex networks are taken into account. Consider the two layer

---

[1] Note that real-valued variables with fixed precision can be modeled via integer variables (e.g. a number in $[0, 1]$ with precision 0.01 corresponds to a number $\in \{0..100\}$).

network from Figure 1A, having *tansig* neurons in the hidden layer and a single linear neuron connected to the output.

Assuming both x_0 and x_1 range in $[-1, 1]$, the maximum possible value for the output z is $\simeq 1.515$ (see Figure 1B). Now, let y_j denote the activity of the j-th hidden neuron. The upper bound on z computed by the output neuron (i.e. z^*), is obtained by fixing both y_0 and y_1 to their maximum possible values (i.e. y_0^* and y_1^*). We have $y_0^* = 2$, obtained by fixing both x_0 and x_1 to -1. We have also $y_1^* = 2$, corresponding to $x_0 = -1$ and $x_1 = 1$. Therefore, z^* is $\simeq 1.928$. The loose bound is obtained since each neuron is propagated separately, thus allowing the network inputs to take incompatible values. This issue can be overcome by employing a global, network-level propagator, which is what this paper is about.

Related Works: Neural Networks have been used as cheap-to-compute cost function evaluators in the context of metaheuristics: in [4] a Genetic Algorithm exploits an ANN to estimate the performance of an absorption chiller. The work [14] proposes a custom heuristic for workload dispatching in a data center and uses an ANN for temperature estimation. In Control Theory, ANNs are employed on-line as predictors (i.e. dynamic system models) and their parameters are continuously adjusted according to the prediction error [6]. This is a specific case of *system identification* [12], which is the process of learning a (typically linear) system model to be used for on-line control, mostly at a local scale. A few works, such as [10], have employed ANNs for solution checking. Others have used ANNs as a surrogate system model for the back-computation of hidden parameters: in [8], this is done to estimate the condition of road pavement layers. Finally, in the OptQuest metaheuristic system [7], a neural network is trained during search with the aim to avoid trivially bad solutions.

As a common trait, in all the mentioned approaches the ANN is exploited in a rather limited fashion, namely as black-box function evaluator. Conversely, Empirical Model Learning has the ability to actively employ the extracted model to improve the performance of the optimization process.

3   Computing Bounds for the Network Output

In this work, we design a new propagator for computing bounds to the output of a two-layer, feed-forward network. Without loss of generality, we consider the problem of finding an upper bound for a single output variable, i.e. on solving:

$$\mathbf{P0}: \quad \max z = \hat{b} + \sum_{j=0}^{m-1} \hat{w}_j f(y_j) \tag{4}$$

$$\text{s.t. } y_j = b_j + \sum_{i=0}^{n-1} w_{j,i} x_i \qquad \forall j = 0..m-1 \tag{5}$$

$$x_i \in [\underline{x}_i, \overline{x}_i] \qquad \forall i = 0..n-1 \tag{6}$$

where x_i are the network inputs (n in total), y_j are the activities of the hidden layer neurons (m in total). The term $w_{j,i}$ is the weight of the i-th input in activity of the j-th hidden neuron and b_j is the bias for the j-th hidden neuron. The term $\hat{w}_j$ is the weight of the output of the j-th hidden neuron in the activity of the output neuron and $\hat{b}$ is the bias for the output neuron. The z variable represents the activity of the output neuron: since all activation functions are monotone non-decreasing, an upper bound on z corresponds to an upper bound on the network output.

Problem Relaxation: Problem **P0** is non-linear, non-convex and cannot be solved in polynomial time in general. Therefore, we resort to a relaxation in order to obtain a scalable solution approach. Specifically, we employ a Lagrangian relaxation for Constraints (5), obtaining:

$$\textbf{LP0}(\lambda): \quad \max_{x,y} \; z(\lambda) = \hat{b} + \sum_{j=0}^{m-1} \hat{w}_j f(y_j) + \tag{7}$$

$$+ \sum_{j=0}^{m-1} \lambda_j \left(b_j + \sum_{i=0}^{n-1} w_{j,i} x_i - y_j \right) \tag{8}$$

$$x_i \in [\underline{x}_i, \overline{x}_i] \qquad \forall i = 0..n-1 \tag{9}$$

$$y_j \in [\underline{y}_j, \overline{y}_j] \qquad \forall j = 0..m-1 \tag{10}$$

where λ is the vector of Lagrangian multipliers λ_j, acting as parameters for the relaxation. The notations x and y refer to the vectors of the x_i and y_j variables. Constraints (10) have been added to prevent **LP0**(λ) from becoming unbounded. The values $\underline{y}_j$ and $\overline{y}_j$ are chosen so that Constraints (10) are redundant in the original problem. In particular:

$$\underline{y}_j = b_j + \sum_i \begin{cases} w_{j,i} \underline{x}_i \text{ if } w_{j,i} \geq 0 \\ w_{j,i} \overline{x}_i \text{ otherwise} \end{cases} \tag{11}$$

and the value $\overline{y}_j$ is computed similarly. Now, since problem **LP0**(λ) is a relaxation, its feasible space includes that of **P0**. Additionally, for all points where Constraints (5) are satisfied, we have $z = z(\lambda)$, for every possible λ. Therefore, the set of solutions of **LP0**(λ) contains all the solutions of **P0**, with the same objective value. Hence the optimal solution $z^*(\lambda)$ of **LP0**(λ) is always a valid bound on the optimal solution z^* of **P0**.

Solving the Relaxation: Problem **LP0**(λ) can be decomposed into two independent subproblems **LP1**(λ) and **LP2**(λ) such that:

$$z^*(\lambda) = \hat{b} + \sum_{j=0}^{m-1} \lambda_j b_j + z^*_{LP1}(\lambda) + z^*_{LP2}(\lambda) \tag{12}$$

with:

$$z^*_{LP1}(\lambda) = \max_x z_{LP1}(\lambda) = \sum_{i=0}^{n-1} \left(\sum_{j=0}^{m-1} \lambda_j w_{j,i} \right) x_i \qquad \textbf{LP1}(\lambda) \qquad (13)$$

$$\text{s.t.} \quad x_i \in [\underline{x}_i, \overline{x}_i] \qquad \forall i = 0..n-1 \qquad (14)$$

$$z^*_{LP2}(\lambda) = \max_y z_{LP2}(\lambda) = \sum_{j=0}^{m-1} (\hat{w}_j f(y_j) - \lambda_j y_j) \qquad \textbf{LP2}(\lambda) \qquad (15)$$

$$\text{s.t.} \quad y_j \in [\underline{y}_j, \overline{y}_j] \qquad \forall j = 0..m-1 \qquad (16)$$

The two subproblems can be addressed separately.

Solving LP1(λ): Problem **LP1(λ)** can be solved by assigning each x_i either to $\underline{x}_i$ or to $\overline{x}_i$, depending on the sign of the reduced weight $\tilde{w}_i(\lambda) = \sum_{j=0}^{m-1} \lambda_j w_{j,i}$. In detail:

$$z^*_{LP1}(\lambda) = \sum_{i=0}^{n-1} \begin{cases} \tilde{w}_i(\lambda) \overline{x}_i \text{ if } \tilde{w}_i(\lambda) \geq 0 \\ \tilde{w}_i(\lambda) \underline{x}_i \text{ otherwise} \end{cases} \qquad (17)$$

The process requires nm steps to compute the reduced weights and n steps to obtain the final solution, for a worst case time complexity of $O(nm)$.

Solving LP2(λ): Problem **LP2(λ)** can be further decomposed into a sum of maximization problems of non-linear, non-convex, monovariate functions with box constraints. Each of the subproblems is in the form:

$$\max_{y_j} g_j(y_j, \lambda) = \hat{w}_j f(y_j) - \lambda_j y_j \qquad (18)$$

$$\text{s.t. } y_j \in [\underline{y}_j, \overline{y}_j] \qquad (19)$$

Each subproblem can be solved analytically, in case f is differentiable, which is a very realistic assumption given that in most practical applications f is a sigmoid. In such case, the objective function from Equation (18) will have a shape similar to the one depicted in Figure 2A. Hence the maximum can be found by comparing the value of $g_j(y_j, \lambda)$ on $\underline{y}_j$, on $\overline{y}_j$ (depending on the value of $\hat{w}_j$ and λ_j) or on the y_j value corresponding to a local maximum. The presence of at most one local maximum is guaranteed by the fact that both $f(y_j)$ and $\lambda_j y_j$ are mononote. Now, for the local minimum and maximum the derivative of $g_j(y_j, \lambda)$ will be null, i.e. $\hat{w}_j f'(y_j) - \lambda_j = 0$. Assuming a *tansig* activation function, this means that:

$$\hat{w}_j \frac{4e^{-2y_j}}{(1 + e^{-2y_j})^2} - \lambda_j = 0 \qquad (20)$$

By substituting $u = e^{-2y_j}$ in Equation (20), we get:

$$4\hat{w}_j u - \lambda_j \left(1 + 2u + u^2\right) = 0 \qquad (21)$$

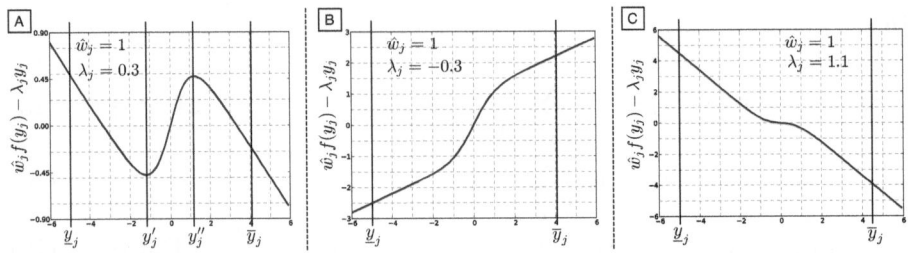

Fig. 2. Shapes for the cost function in the decomposition of **LP2**(λ)

Note that if $\lambda_j = 0$, then no local maximum exists. The same holds if $\hat{w}_j = 0$. Hence it is safe to assume $\lambda_j, \hat{w}_j \neq 0$ and we can get:

$$u^2 + \left(2 - 4\frac{\hat{w}_j}{\lambda_j}\right) u + 1 = 0 \tag{22}$$

Which can be solved via the classic quadratic formula for second degree equations, yielding two solutions u' and u''. The solutions are non-complex iff:

$$\left(2 - 4\frac{\hat{w}_j}{\lambda_j}\right)^2 - 4 \geq 0 \quad \Leftrightarrow \quad \frac{\hat{w}_j}{\lambda_j}\left(\frac{\hat{w}_j}{\lambda_j} - 1\right) \geq 0 \tag{23}$$

I.e. if $\hat{w}_j$ and λ_j are equal in sign and if $|\hat{w}_j| \geq \lambda_j$ (or equivalently if $\hat{w}_j/\lambda_j > 1$). In this case, the y_j values corresponding to the local minimum and maximum are given by:

$$y'_j = -\frac{1}{2}\log u', \quad y''_j = -\frac{1}{2}\log u'' \tag{24}$$

It can be shown that u' and u'' are guaranteed to be positive. If the conditions to have a real-valued solution for Equation (22) do not hold, then the maximum corresponds to either $\underline{y}_j$ or $\overline{y}_j$: in detail, if $sign(\hat{w}_j) \neq sign(\lambda_j)$, then this happens because both $f(y_j)$ and $-\lambda_j y_j$ are non-decreasing or non-increasing (see Figure 2B). If $|\hat{w}_j| < \lambda_j$ no local maximum exists (see Figure 2C), because the derivative of the *tansig* is always ≤ 1.

Hence, the solution of each subproblem in the decomposition of **LP2**(λ) can be found by solving Equation (20) (if the conditions are met) and by comparing the value of $g_j(y_j\lambda)$ for at most four y_j values. The process takes constant time. Solving **LP2**(λ) requires m such computations, plus nm iterations to obtain all $\underline{y}_j$ and $\overline{y}_j$. The overall worst case time complexity is $O(nm)$.

Summary: For a fixed value of all the Lagrangian multipliers λ, the relaxed subproblem **LP0**(λ) can therefore be solved via the process described in Algorithm 1, to yield an upper bound on z. In the algorithm, $z^*(\lambda)$ denotes the bound, while $x_i^*(\lambda)$ and $y_j^*(\lambda)$ are the values of x_i and y_j in the corresponding solution. We recall that a feasible solution for **LP0**(λ) may be infeasible for **P0**. The algorithm to compute a lower bound is analogous.

Algorithm 1. (Computing an upper bound on z by solving **LP0**(λ))

---

initialize $z^*(\lambda) = \hat{b} + \sum_{j=0}^{m-1} \lambda_j b_j$
for $i = 0..n-1$ do
   set $x_i^*(\lambda) = \begin{cases} \overline{x}_i \text{ if } \tilde{w}_i(\lambda) \geq 0 \\ \underline{x}_i \text{ otherwise} \end{cases}$
   update $z^*(\lambda) = z^*(\lambda) + \tilde{w}_i(\lambda)x_i^*(\lambda)$
for $j = 0..m-1$ do
   compute $\underline{y}_j$ and $\overline{y}_j$
   set $y_j^*(\lambda) = \text{argmax}\{g_j(\underline{y}_j, \lambda), g_j(\overline{y}_j, \lambda)\}$
   if $\lambda_j \neq 0$ and $\tilde{w}_j/\lambda_j > 1$ then
     set $y_j', y_j'' = -\frac{1}{2}\log\left(\frac{-\beta \pm \sqrt{\beta^2 - 4}}{2}\right)$, with $\beta = 2 - 4\frac{\tilde{w}_j}{\lambda_j}$
     if $y_j' \in]\underline{y}_j, \overline{y}_j[$ and $g_j(y_j', \lambda_j) > g_j(y_j^*(\lambda), \lambda_j)$ then $y_j^*(\lambda) = y_j'$
     if $y_j'' \in]\underline{y}_j, \overline{y}_j[$ and $g_j(y_j'', \lambda_j) > g_j(y_j^*(\lambda), \lambda_j)$ then $y_j^*(\lambda) = y_j''$
   set $z^*(\lambda) = z^*(\lambda) + g_j(y_j^*(\lambda), \lambda)$
return $z^*(\lambda)$

---

4 Optimizing the Lagrangian Multipliers

Any assignment of the multipliers λ yields a valid bound on the output variable z. Hence it is possible to improve the bound quality by optimizing the multiplier values, i.e. by solving the following unconstrained minimization problem:

$$\textbf{L0}: \quad \min_{\lambda} z^*(\lambda) \tag{25}$$

Where $z^*(\lambda)$ is here a function that denotes the optimal solution of **LP0**(λ). Problem **L0** is convex in λ and hence has a unique minimum. This is true even if **LP0**(λ) is non-convex: in fact, the two problems are defined on different variables (i.e. λ versus x and y). The minimum point can therefore be found via a descent method. Now, let λ' be an assignment of λ such that the corresponding solution of **LP0**(λ) does not change for very small variations of the multipliers, i.e. $x^*(\lambda') = x^*(\lambda'')$ and $y^*(\lambda') = y^*(\lambda'')$, with $\|\lambda' - \lambda''\| \to 0$. Then $z^*(\lambda)$ is differentiable in λ' and in particular:

$$\frac{\partial z^*(\lambda')}{\partial \lambda_j} = s_j = b_j + \sum_{i=0}^{n-1} w_{j,i}x_i^*(\lambda') - y_j^*(\lambda') \tag{26}$$

Equation (26) is obtained by differentiating the objective of **LP0**(λ) under the above mentioned assumptions. When such assumptions do not hold, the s_j values provide a valid *subgradient*. The optimum value of **L0** can therefore be found via a subgradient method, by starting from an assignment $\lambda^{(0)}$ and iteratively applying the update rule:

$$\lambda^{(k+1)} = \lambda^{(k)} - \sigma^{(k)}s^{(k)} \tag{27}$$

where $\lambda^{(k)}$ denotes the multipliers for the k-th step, $s^{(k)}$ is the vector of all s_j (i.e. the subgradient) and $\sigma^{(k)}$ is a scalar, representing a step length.

Step Update Policy: We have chosen to employ the corrected Polyak step size policy with non-vanishing threshold from [5]. This guarantees the convergence to the optimal multipliers (given infinitely many iterations), with bounded error. Other policies from the literature are more accurate, but have a slower convergence rate, which is in our case *the* critical parameter (since we will run the subgradient method within a propagator). In detail, we have:

$$\sigma^{(k)} = \beta \frac{z^*(\lambda^{(k)}) - (z^{best} - \delta^{(k)})}{\|s^{(k)}\|^2} \tag{28}$$

where β is a scalar value in $]0,2[$. The term $z^{best} - \delta^{(k)}$ is an estimate of the **L0** optimum: it is computed as the difference between the best (lowest) bound found so far z^{best}, and a scalar $\delta^{(k)}$ dynamically adjusted during search. Hence, the step size is directly proportional to the distance of the current bound from the estimated optimal one, i.e. $z^*(\lambda^{(k)}) - (z^{best} - \delta^{(k)})$. The larger $\delta^{(k)}$, the larger the estimated gap w.r.t the best bound and the larger the step size.

The value of $\delta^{(k)}$ is non-vanishing, which means it is constrained to be larger than a threshold δ^*. This ensures to have $\sigma^{(k)} > 0$ and prevents the subgradient optimization from getting stuck. We determine the δ^* value when the propagator is first executed at the root of the search tree. Specifically, we choose $\delta^* = \gamma z^*(\lambda^{(0)})$, with γ being a small positive value. During search, we compute $\delta^{(k)}$ according to the following rules:

$$\delta^{(k+1)} = \begin{cases} \max(\delta^*, \nu\delta^{(k)}) & \text{if } z^*(\lambda^{(k)}) > z^{best} - \delta^{(k)} \\ \max(\delta^*, \mu z^*(\lambda^{(k)})) & \text{otherwise} \end{cases} \tag{29}$$

where $\nu, \mu \in]0,1[$. In practice, if the last computed bound $z^*(\lambda^{(k)})$ does not improve over the estimated optimum $z^{best} - \delta^{(k)}$, then we reduce the current $\delta^{(k)}$ value, i.e. we make the estimated optimum closer to z^{best}. Conversely, when an improvement is obtained, we "reset" $\delta^{(k)}$, i.e. we assume that the estimated optimum is $\mu\%$ lower than z^{best}.

Deflection: Subgradient methods are known to exhibit a zig-zag behavior when close to an area where the cost function is non-differentiable. In this situation the convergence rate can be improved via deflection techniques. In its most basic form (the one we adopt), a deflection technique consists in replacing the subgradient in Equation (27) and (28) with the following vector (see [5]):

$$d^{(k)} = \alpha s^{(k)} + (1 - \alpha)d^{(k-1)} \tag{30}$$

where $d^{(k)}$ is called search direction and α is a scalar in $]0,1]$, meaning that $d^{(k)}$ is a convex combination of the last search direction and the current subgradient.

The components s_j having alternating sign in consecutive gradients (such that $s_j^{(k)} s_j^{(k-1)} < 0$) tend to cancel one each other in the deflected search direction.

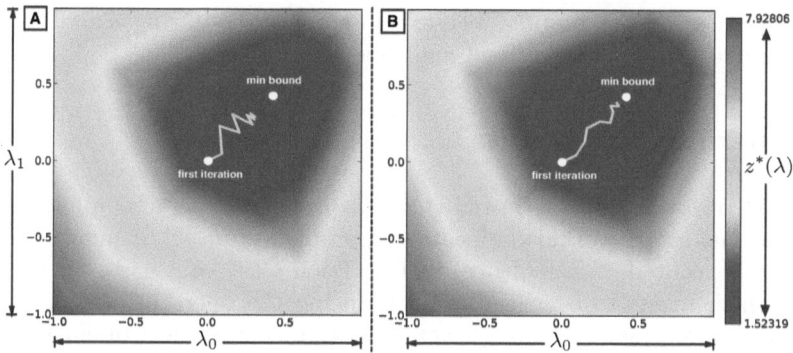

Fig. 3. $\boxed{\text{A}}$ Subgradient optimization trace (10 iterations, no deflection). $\boxed{\text{B}}$ Subgradient optimization trace (10 iterations, with deflection).

This behavior can be observed in Figure 3. This depicts the bound value as a function of λ for the network from Figure 1, together with the trace of the first 10 subgradient iterations. The use of the deflection allows to get considerably closer to the best possible bound ($\simeq 1.523$ in this case). Note that bound is not tight (the actual network maximum is $\simeq 1.515$), but it remarkably better than the value obtained from the propagation of individual neuron constraints ($\simeq 1.928$). When using the deflection technique, the value β from Equation (28) must be $\leq \alpha$ for the method to converge.

Propagator Configuration: We stop the subgradient optimization after a fixed number of steps. At the end of the process, we keep the best multipliers λ^* we have found and the corresponding bound $z^*(\lambda^*)$. We compute both an upper and a lower bound on the network output. The bound computation algorithm does not replace the propagation of individual neuron constraints, that we implement as pair of separated constraints as from Equation (3). We rely on individual neuron constraints to perform propagation on the network inputs, and for computing the bounds $\underline{y}_j$, $\overline{y}_j$ on the activity of the hidden neurons.

The new propagator is scheduled with the lowest possible priority in the target Constraint Solver. When the constraint is propagated for the first time, we perform 100 subgradient iterations, starting with all-zero multipliers ($\lambda_j^{(0)} = 0 \; \forall j = 0..m-1$). After that, when the constraint is triggered we perform only 3 iterations, starting from the best multipliers λ^* from the last activation. We keep the multipliers also when branching from a node of the search tree to one of its children, as a simple (but important) form of incremental computation.

We always use $\alpha = 0.5$ for the deflection and we keep $\beta = \alpha$. We re-initialize δ^* every time the constraint is triggered, using $\gamma = 0.01$. Therefore, the correction factor $\delta^{(k)}$ is always at least 1% of the computed bound computed at the first subgradient iteration. The attenuation factor ν for $\delta^{(k)}$, used when no improvement is obtained, is fixed to 0.75. The μ factor, used to reset $\delta^{(k)}$ when the estimated bound is improved, is 0.25 for the first constraint propagation and 0.05 for all the following ones. This choice is done on the basis that small updates of the network inputs (such as those occurring during search) result in small modifications of the optimal multipliers.

5   Experimental Results

Target Problem: We have tested the new propagator on a simplified version of the thermal-aware workload dispatching problem from [2]. A number of tasks need to be executed on a multi-core CPU. Each CPU core has a thermal controller, which reacts to overheating by reducing the operating frequency until the temperature is safe. The frequency reduction causes a loss of efficiency that depends on the workload of the core, on that of the neighboring cores, on the thermal physics, and on the controller policy itself. An ANN is used to obtain an approximate model of the efficiency of each core, as a function of the workload and the room temperature. We target a synthetic quad core CPUs, simulated via an internally developed tool based on the popular Hotspot system [9]. A training set has been generated by mapping workloads at random on the platform and then obtaining the corresponding core efficiencies via the simulator. We have then trained a two-layer ANN for each core, with *tansig* neurons in the hidden layer and a single linear neuron in the output layer.

Each task i is characterized by a value cpi_i, measuring the degree of its CPU usage: lower cpi_i values correspond to more computation intensive (and heat generating) tasks. An equal number of tasks must be mapped on each core. The input of the ANN is the average cpi_i of each core and the room temperature t. The goal is to find a task-to-core mapping such that no efficiency is below a minimum threshold θ. We use the vector of integer variables p to model the task mapping, with $p_i = k$ iff task i is mapped to core k. Our model is as follows:

$$gcc\left(p, [0..n_c - 1], {}^{n_t}/n_c\right) \tag{31}$$

$$acpi_k = \frac{n_c}{n_t} \sum_{i=0}^{n_t-1} cpi_i(p_i = k) \qquad \forall k = 0..n_c - 1 \tag{32}$$

$$e_k = \hat{b}_k + \sum_{j=0}^{n_h-1} \hat{w}_{k,j} y_{k,j} \qquad \forall k = 0..n_c - 1 \tag{33}$$

$$y_{k,j} = tansig\left(b_{k,j} + \sum_{h=0}^{n_c-1} w_{k,j,h} acpi_h + w_{k,j,n_c} t\right) \qquad \begin{array}{l} \forall k = 0..n_c - 1, \\ \forall j = 0..n_h - 1 \end{array} \tag{34}$$

$$e_k \geq \theta \qquad \forall k = 0..n_c - 1 \tag{35}$$

$$p_i \in \{0..n_c - 1\} \qquad \forall i = 0..n_t \tag{36}$$

where n_t is the number of tasks and n_c is the number of cores (4 in our case). In (31) we use the *gcc* global constraint to have exactly ${}^{n_t}/n_c$ tasks per core. For simplicity, we assume n_t is a multiple of n_c. Constraints (32) are used to obtain the average cpi_i per core (i.e. the $acpi_k$ variables). Constraints (33) and (34) define the ANN structure and are implemented using Neuron Constraints. The

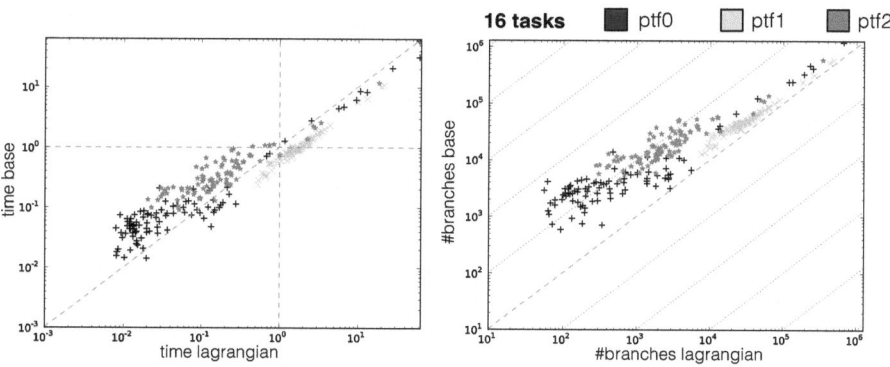

Fig. 4. Results for the 16 task workloads, on platforms 0-2

value n_h is the number of hidden neurons per ANN ($n_h = 5$ in our case), $\hat{b}_k$ is the bias of the output neuron in the ANN for core k, $\hat{w}_{k,j}$ are the neuron weights. Similarly, $b_{k,j}$ is bias of the hidden neuron j in the neural network for core k, while $w_{k,j,h}$, w_{k,j,n_c} are the weights. The value t is the room temperature, which is fixed for each problem instance. Each e_k variable represents the efficiency of the core k and is forced to be higher than θ by Constraints (35).

Experimental Setup: We tested two variants of the above model, where the new Lagragian propagator is respectively used (*lag*) and not used (*base*). A comparison with an alternative approach (e.g. a meta-hueristic using the ANN as a black-box), although very interesting, is outside the scope of this paper, which is focused on improving a filtering algorithm. We solve the problem via depth-first search by using a static search heuristic, namely by selecting for branching the first unbounded variable and always assigning the minimum value in the domain. The choice of a static heuristic allows a fair comparison of different propagators: pruning a value at a search node has the effect of skipping the corresponding sub-tree, but does not affect the branching decisions in an unpredictable fashion. As an adverse side effect, static heuristics are not well suited to solve this specific problem. Therefore, we limit ourselves to relatively small instances with either 16 or 20 tasks, which are nevertheless be sufficient to provide a sound evaluation. We consider 100 task sets for each size value. We performed experiments on 6 synthetic quad-core platforms, effectively testing $4 \times 6 = 24$ networks. For each combination of task set size and platform, we have empirically determined an efficiency threshold θ such that finding a feasible solution is non-trivial in most cases. Each experiment is run with a 60 seconds time limit. This is usually enough to find a solution, but it is never sufficient for proving infeasibility (which appears to take a very long time, mainly due to the chosen search heuristic). We have implemented everything on top of the Google or-tools solver. All the tests are run on a 2.8 GHz Intel Core i7.

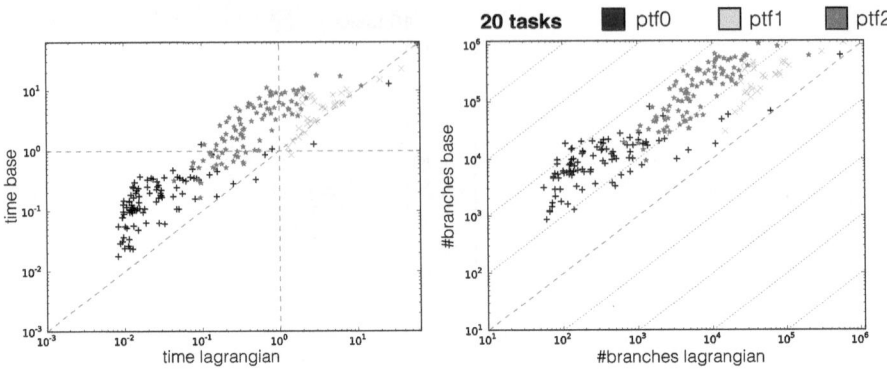

Fig. 5. Results for the 20 task workloads, on platforms 0-2

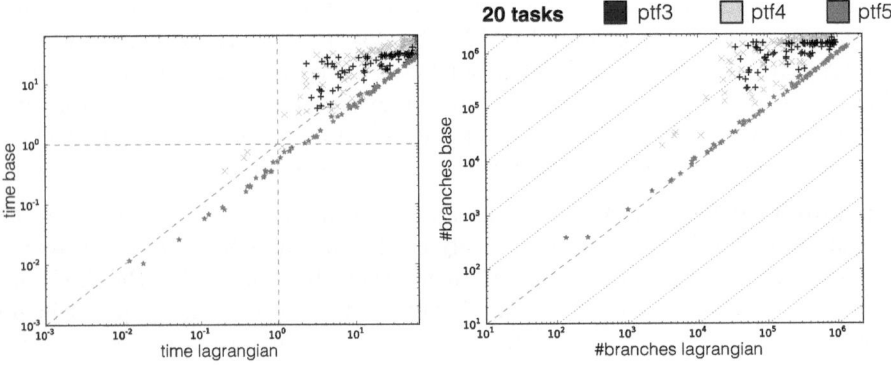

Fig. 6. Results for the 16 task workloads, on platforms 3-5

Results: The results of our experimentation are reported in Figures 4, 5, 6, 7. Each of them refers to 100 instances (with either 16 or 20 tasks) tested on three different platforms, and contains two scatter plots in log scale. The left-most diagram reports the solution times, with *lag* on the x axis and *base* on the y axis. Each instance is represented by a point and different colors and markers are used to distinguish between different platforms. Points above the diagonal represent instances where an improvement was obtained. A horizontal and a vertical line highlight the position of 1-second run times. The right-most plot is similar, except that it shows the number of branches and refers only to the instances for which a solution was found by both approaches. Each of the dotted diagonal lines represents a one-order-of-magnitude improvement.

The dramatic good news here is that the novel propagator achieves an impressive reduction in the number of branches, in a significant number cases. The gain may be as large as 2-3 orders or magnitude. This is an important result, pointing out that the bound improvement provided by the Lagrangian relaxation is far from negligible. Interestingly, the benefits tend to be higher for larger instances: a reasonable explanation for this behavior is that additional propagation is

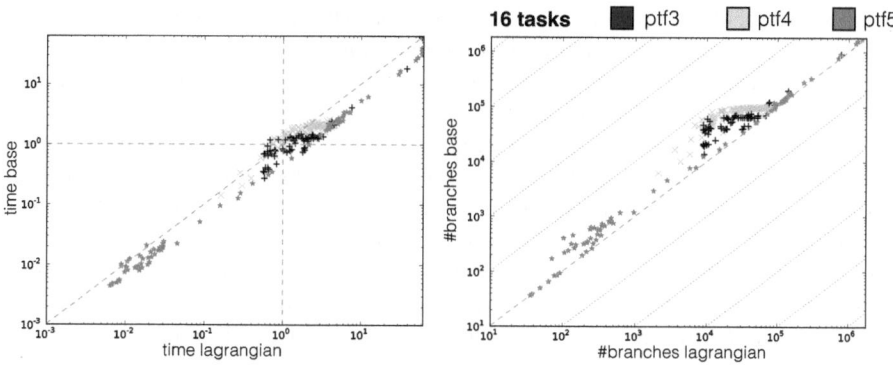

Fig. 7. Results for the 20 task workloads, on platforms 3-5

performed relatively high in the search tree, thus pruning larger subtrees as the instance size grows.

On the flip side, the new propagator comes with a considerable computational burden at each search node. As a general trend, on the 16 task instances this is sufficient to counter the benefits of the smaller number of branches: the *lag* approach therefore tends to be slower than the *base* one, although not much slower. For the 20 task workloads, there is a significant gain in solution time on platform 2 and 1, and a slight improvement on platform 0. The novel propagator behaves nicely on platform 3 and 4 as well, solving more instances than *base* respectively for the first 28 and 47 seconds. The method reports however a larger number of time-outs at the end of the 60 seconds. The *base* approach considerably outperforms the *lag* one on platform 5.

In general, the effectiveness of the Lagrangian propagator is non-uniform across different platforms: the reduction in the number of branches is much larger for platforms 0, 2, 3 and 4 than it is for platforms 1 and 5. This rises interest in investigating techniques to identify the network weight configurations that are more likely to benefit from the new propagator. The results seem to be much more consistent for different workloads on a single platform, although this may be due in part to the way our task sets are generated.

Finally, it is worth noting that the higher scalability (on the time side) of the Lagrangian approach is in part due to the use of subgradient optimization. We recall from Section 4 that for each constraint activation (except for the first one) we perform only 3 subgradient iterations. Since such number is fixed regardless of the number of tasks, the computational cost of the new propagator grows proportionally slower as the instances become larger.

6 Concluding Remarks

Summary: We have introduced a novel propagator for two-layer, feed forward ANNs, to be used in Empirical Model Learning. The new propagator is based on a Lagrangian relaxation, which is solved for a fixed assignment of the multipliers via a fast, dedicated, approach. The multipliers themselves are optimized via a subgradient method. The current implementation works for *tansig* sigmoids in the hidden layer, but targeting other activation functions should be easy enough, provided they are differentiable.

The novel propagation does not replace the existing ones, but allows the computation of tighter bound on the ANN output variables. The approach manages to obtain a substantial reduction of the number of branches (up to 2-3 orders of magnitude) in our test set. The method seems to work best for comparatively larger instances. On the other side, the new propagation is computationally expensive, countering in part the benefits of the smaller search tree. Nevertheless, a gain in terms of solution time is obtained in a significant number of cases.

Future Work: A natural direction for future research is devising a way to filter the x_i variables, based on the Lagrangian relaxation. Second, the highest priority for future developments is achieving a reduction in the computation time, in order to fully exploit the reduction in the number of branches. This goal can be pursued (1) via the application of additional incremental techniques or (2) by improvements in the multiplier optimization routine. The computation of $\underline{y}_j$, $\overline{y}_j$ can be easily be made incremental, since they are linear expression. The incremental update of the **LP0**(λ) solution upon changes in λ is trickier, since all the multipliers tend to change after every subgradient iterations. We believe however that the convergence of the multiplier optimization routine offers large room for improvements, on the basis of the following conjecture.

The Conjecture: Let us assume that the relaxed problem $z^*(\lambda)$ from Section 4 is differentiable for the optimal multipliers λ^*. As a consequence, it must hold $\frac{\partial z^*(\lambda^*)}{\partial \lambda_j} = 0$ for every λ_j. Now, the partial derivatives are given by Expressions (26), which also represents the violation degree of Constraints (5). Therefore, if $z^*(\lambda)$ is differentiable in λ^*, then the relaxation solution $x^*(\lambda^*)$, $y^*(\lambda^*)$ is feasible for the original problem and the bound is tight. This means that the original problem can be solved via convex optimization.

Since we know problem **P0** is non-convex and hard to solve in general, we expect the above situation to be symptomatic of tractable subclasses, which can be probably identified by an analysis of the network weights. For example we know that, if the products $w_{j,i} \hat{w}_j$ have constant sign $\forall j$, then propagating the individual Neuron Constraints is sufficient to compute tight bounds on z.

Therefore, we expect that non-trivial Lagrangian bounds correspond to non-differentiable points of $z^*(\lambda)$. Such non-differentiable areas are given in our case by a set of hyperplanes in $\mathbb{R}^m$ (i.e. on the space of the multipliers), with the coefficients of the hyperplanes being easy to compute. This information can be exploited to focus the search for the optimal λ to a much smaller space, improving the rate of convergence and decreasing the overall computation time.

References

1. Bartolini, A., Lombardi, M., Milano, M., Benini, L.: Neuron Constraints to Model Complex Real-World Problems. In: Lee, J. (ed.) CP 2011. LNCS, vol. 6876, pp. 115–129. Springer, Heidelberg (2011)
2. Bartolini, A., Lombardi, M., Milano, M., Benini, L.: Optimization and Controlled Systems: A Case Study on Thermal Aware Workload Dispatching. Proc. of AAAI (2012)
3. Belew, R.K., McInerney, J., Schraudolph, N.N.: Evolving networks: Using the genetic algorithm with connectionist learning. In: Proc. of Second Conference on Artificial Life, pp. 511–547 (1991)
4. Chow, T.T., Zhang, G.Q., Lin, Z., Song, C.L.: Global optimization of absorption chiller system by genetic algorithm and neural network. Energy and Buildings 34(1), 103–109 (2002)
5. Frangioni, A., D'Antonio, G.: Deflected Conditional Approximate Subgradient Methods (Tech. Rep. TR-07-20). Technical report, University of Pisa (2007)
6. Ge, S.S., Hang, C.C., Lee, T.H., Zhang, T.: Stable adaptive neural network control. Springer Publishing Company, Incorporated (2010)
7. Glover, F., Kelly, J.P., Laguna, M.: New Advances for Wedding optimization and simulation. In: Proc. of WSC, pp. 255–260 (1999)
8. Gopalakrishnan, K., Ph, D., Asce, A.M.: Neural Network Swarm Intelligence Hybrid Nonlinear Optimization Algorithm for Pavement Moduli Back-Calculation. Journal of Transportation Engineering 136(6), 528–536 (2009)
9. Huang, W., Ghosh, S., Velusamy, S.: HotSpot: A compact thermal modeling methodology for early-stage VLSI design. IEEE Trans. on VLSI 14(5), 501–513 (2006)
10. Jayaseelan, R., Mitra, T.: A hybrid local-global approach for multi-core thermal management. In: Proc. of ICCAD, pp. 314–320. ACM Press, New York (2009)
11. Kiranyaz, S., Ince, T., Yildirim, A., Gabbouj, M.: Evolutionary artificial neural networks by multi-dimensional particle swarm optimization. Neural Networks 22(10), 1448–1462 (2009)
12. Ljung, L.: System identification. Wiley Online Library (1999)
13. Montana, D.J., Davis, L.: Training feedforward neural networks using genetic algorithms. In: Proc. of IJCAI, pp. 762–767 (1989)
14. Moore, J., Chase, J.S., Ranganathan, P.: Weatherman: Automated, Online and Predictive Thermal Mapping and Management for Data Centers. In: Proc. of IEEE ICAC, pp. 155–164. IEEE (2006)
15. Zhang, G., Patuwo, B.E., Hu, M.Y.: Forecasting with artificial neural networks: The state of the art. International Journal of Forecasting 14(1), 35–62 (1998)

Bandit-Based Search
for Constraint Programming

Manuel Loth[1], Michèle Sebag[2], Youssef Hamadi[3], and Marc Schoenauer[2]

[1] Microsoft Research – INRIA joint centre, Palaiseau, France
[2] TAO, INRIA – CNRS – LRI, Université Paris-Sud, Orsay, France
[3] Microsoft Research, Cambridge, United Kingdom

Abstract. Constraint Programming (CP) solvers classically explore the solution space using tree-search based heuristics. Monte-Carlo Tree Search (MCTS), aimed at optimal sequential decision making under uncertainty, gradually grows a search tree to explore the most promising regions according to a specified reward function. At the crossroad of CP and MCTS, this paper presents the Bandit Search for Constraint Programming (BaSCoP) algorithm, adapting MCTS to the specifics of the CP search. This contribution relies on i) a generic reward function suited to CP and compatible with a multiple restart strategy; ii) the use of depth-first search as roll-out procedure in MCTS. BaSCoP, on the top of the Gecode constraint solver, is shown to significantly improve on depth-first search on some CP benchmark suites, demonstrating its relevance as a generic yet robust CP search method.

Keywords: adaptive search, value selection, bandit, UCB, MCTS.

1 Introduction

A variety of algorithms and heuristics have been designed in constraint programming (CP), determining which (variable, value) assignment must be selected at each step, how to backtrack on failures, and how to restart the search [1]. The selection of the algorithm or heuristics most appropriate to a given problem instance, intensively investigated since the late 70s [2], most often relies on supervised machine learning (ML) [3–7].

This paper advocates the use of another ML approach, namely reinforcement learning (RL) [8], to support the CP search. Taking inspiration from earlier work [9–12], the paper contribution is to extend the Monte-Carlo Tree Search (MCTS) algorithm to control the exploration of the CP search tree.

Formally, MCTS upgrades the multi-armed bandit framework [13, 14] to sequential decision making [15], leading to breakthroughs in the domains of e.g. games [16, 17] or automated planning [18]. MCTS proceeds by growing a search tree through consecutive tree walks, gradually biasing the search toward the most promising regions of the search space. Each tree walk, starting from the root, iteratively selects a child node depending on its empirical reward estimate and the confidence thereof, enforcing a trade-off between the exploitation of the best results found so far, and the exploration of the search space (more in section 2.3). The use of MCTS within the CP search faces two main difficulties. The first one is to define an appropriate reward attached to a tree node (that is, a partial assignment of the variables). The second difficulty is due to the fact that the CP search frequently involves multiple restarts [19]. In each restart, the current search tree is erased and a brand new search tree is built based on a new variable ordering (reflecting the variable criticality after e.g. their weighted degree, impact or activity). As the rewards attached to all nodes cannot be maintained over multiple restarts for tractability reasons, MCTS cannot be used as is.

A first contribution of the presented algorithm, named *Bandit-based Search for Constraint Programming* (BASCOP), is to associate to each (variable, value) assignment its average *relative failure depth*. This average can be maintained over the successive restarts, and used as a reward to guide the search. A second contribution is to combine BASCOP with a depth-first search, enforcing the search completeness in the no-restart case. A proof of principle of the approach is given by implementing BASCOP on the top of the Gecode constraint solver [20]. Its experimental validation on three benchmark suites, respectively concerned with the job-shop (JSP) [21], the balanced incomplete block design (BIBD) [22], and the car-sequencing problems, comparatively demonstrates the merits of the approach.

The paper is organized as follows. Section 2 discusses the respective relevance of supervised learning and reinforcement learning with regard to the CP search control, and describes the Monte Carlo Tree Search. Section 3 gives an overview of the BASCOP algorithm, hybridizing MCTS with CP search. Section 4 presents the experimental setting for the empirical validation of BASCOP and discusses the empirical results. The paper concludes with some perspectives for further research.

2   Machine Learning for Constraint Programming

This section briefly discusses the use of supervised machine learning and reinforcement learning for the control of CP search algorithms. For the sake of completeness, the Monte-Carlo Tree Search algorithm is last described.

2.1   Supervised Machine Learning

Most approaches to the control of search algorithms exploit a dataset that records, for a set of benchmark problem instances, i) the description of each problem instance after appropriate static and dynamic features [3, 23]; ii) the associated target result, e.g. the runtime of a solver. Supervised machine learning is applied on the dataset to extract a model of the target result based on the descriptive features of the problem instances. In SATzilla [3], a regression model predicting the runtime of each solver on a problem instance is built from the known instances, and used on unknown instances to select the solver with minimal expected run-time. Note that this approach can be extended to accommodate several restart strategies [24]. CPHydra [4] uses a similarity-based approach (case-based reasoning) and builds a switching policy based on the most efficient solvers for the problem instance at hand. In [5], ML is likewise applied to adjust the CP heuristics online. The *Adaptive Constraint Engine* [25] can be viewed as an ensemble learning approach, where each heuristic votes for a possible (variable,value) assignment to solve a CSP. The methods *Combining Multiple Heuristics Online* [6] and *Portfolios with Deadlines* [26] are designed to build a scheduler policy in order to switch the execution of black-box solvers during the resolution process. Finally, optimal hyper-parameter tuning [7, 27] is tackled by optimizing the estimate of the runtime associated to parameter settings depending on the current problem instance.

2.2   Reinforcement Learning

A main difference between supervised learning and reinforcement learning is that the former focuses on taking a single decision, while the latter is interested in sequences of decisions. Reinforcement learning classically considers a Markov decision process framework $(\mathcal{S}, \mathcal{A}, p, r)$, where $\mathcal{S}$ and $\mathcal{A}$ respectively denote the state and the action spaces, p is the transition model ($p(s, a, s')$ being the probability of being in state s' after selecting action a in state s in a probabilistic setting; in a deterministic setting, $tr(s, a)$ is the node s' reached by selecting action a in state s) and $r : \mathcal{S} \mapsto \mathbb{R}$ is a bounded reward function. A policy $\pi : \mathcal{S} \mapsto \mathcal{A}$, starting in some initial state until arriving in a terminal state or reaching a time horizon, gathers a sum of rewards. The RL goal is to find an optimal policy, maximizing the expected cumulative reward.

RL is relevant to CP along two frameworks, referred to as offline and online frameworks. The offline framework aims at finding an optimal policy w.r.t. a family of problem instances. In this framework, the set of states describes the search status of any problem instance, described after static and dynamic feature values; the set of actions corresponds e.g. to the CP heuristics to be applied for a given lapse of time. An optimal policy associates an action to each state, in such a way that, over the family of problem instances (e.g., on average), the policy reaches optimal performances (finds a solution in the satisfiability setting, or reaches the optimal solution in an optimization setting) as fast as possible.

The online framework is interested in solving a single problem instance. In this framework, the set of states corresponds to a partial assignment of the variables and the set of admissible actions corresponds to the (variable, value) assignments consistent with the current state. An optimal policy is one which finds as fast as possible a solution (or, the optimal solution) for the problem instance at hand.

In the remainder of the paper, only the online framework will be considered; *states* and *nodes* will be used interchangeably. This online framework defines a specific RL landscape. Firstly, the transition model is known and deterministic; the next state $s' = tr(s, a)$ reached from a state s upon the (variable,value) assignment action a, is the conjunction of s and the (variable,value) assignment. Secondly, and most importantly, there is no clearly defined reward to be attached to intermediate states: e.g. in the satisfiability context, intrinsic rewards (satisfiability or unsatisfiability) can only be attached to terminal states. Furthermore, such intrinsic rewards are hardly informative (e.g. all but a negligible fraction of the terminal states are unsatisfiable; and the problem is solved in general after a single satisfiable assignment is found).

The online framework thus makes it challenging for mainstream RL approaches to adjust the Exploration *vs* Exploitation trade-off at the core of RL. For this reason, the Monte-Carlo Tree Search approach is considered.

2.3   Monte Carlo Tree Search

The best known MCTS algorithm, referred to as Upper Confidence Tree (UCT) [15], extends the Upper Confidence Bound algorithm [14] to tree-structured spaces. UCT simultaneously explores and builds a search tree, initially restricted to its root node, along N tree-walks. Each tree-walk involves three phases:

The **bandit phase** starts from the root node (initial state) and iteratively selects a child node (action) until arriving in a leaf node of the MCTS tree. Action selection is handled as a multi-armed bandit problem. The set $\mathcal{A}_s$ of admissible actions a in node s defines the child nodes (s, a) of s; the selected action a^* maximizes the Upper Confidence Bound:

$$\bar{r}_{s,a} + C\sqrt{\log(n_s)/n_{s,a}} \tag{1}$$

over a ranging in $\mathcal{A}_s$, where n_s stands for the number of times node s has been visited, $n_{s,a}$ denotes the number of times a has been selected in node s, and $\bar{r}_{s,a}$ is the average cumulative reward collected when selecting action a from node s. The first (respectively the second) term in Eq. (1) corresponds to the exploitation (resp. exploration) term, and the exploration vs exploitation trade-off is controlled by parameter C. In a deterministic setting, the selection of the child node (s, a) yields a single next state $tr(s, a)$, which replaces s as current node.

The **tree building phase** takes place upon arriving in a leaf node s; some action a is (randomly or heuristically) selected and $tr(s, a)$ is added as child node of s. The growth rate of the MCTS tree can be controlled through an *expand rate* parameter k, by adding a child node after the leaf node has been visited k times. Accordingly, the number of nodes in the tree is N/k, where N is the number of tree-walks.

The **roll-out phase** starts from the leaf node $tr(s, a)$ and iteratively (randomly or heuristically) selects an action until arriving in a terminal state u; at this point the reward r_u of the whole tree-walk is computed and used to update the cumulative reward estimates in all nodes (s, a) visited during the tree-walk:

$$\begin{aligned} n_{s,a} &\leftarrow n_{s,a} + 1; \quad n_s \leftarrow n_s + 1 \\ \bar{r}_{s,a} &\leftarrow \bar{r}_{s,a} + (r_u - \bar{r}_{s,a})/n_{s,a} \end{aligned} \tag{2}$$

Additional heuristics have been considered, chiefly to prevent over-exploration when the number of admissible arms is large w.r.t the number of simulations (the so-called many-armed bandit issue [28]). Notably, the *Rapid Action Value*

Estimate (RAVE) heuristics is used to guide the exploration of the search space and the tree-building phase [16] when node rewards are based on few samples (tree-walks) and are thus subject to a high variance. In its simplest version, $RAVE(a)$ is set to the average reward taken over all tree-walks involving action a. The action selection is based on a weighted sum of the RAVE and the Upper Confidence Bound (Eq. (1)), where the RAVE weight decreases with the number n_s of visits to the current node [16].

A few work have pioneered the use of MCTS to explore a tree-structured assignment search space, in order to solve satisfiability or combinatorial optimization problem instances. In [9], MCTS is applied to boolean satisfiability; the node reward is set to the ratio of clauses satisfied by the current assignment, tentatively estimating how far this assignment goes toward finding a solution. In [11], MCTS is applied to Mixed Integer Programming, and used to control the selection of the top nodes in the CPLEX solver; the node reward is set to the maximal value of solutions built on this node. In [10], MCTS is applied to Job Shop Scheduling problems; it is viewed as an alternative to Pilot or roll-out methods, featuring an integrated and smart look-ahead strategy. Likewise, the node reward is set to the optimal makespan of the solutions built on this node.

3   The BaSCoP Algorithm

This section presents the BaSCoP algorithm (Algorithm 1), defining the proposed reward function and describing how the reward estimates are exploited to guide the search. Only binary variables will be considered in this section for the sake of simplicity; the extension to n-ary variables is straightforward, and will be considered in the experimental validation of BaSCoP (section 4). Before describing the structure of the BaSCoP search tree, let us first introduce the main two ideas behind the proposed hybridization of MCTS and the CP search.

Among the principles guiding the CP search [29], a first one is to select variables such that an eventual failure occurs as soon as possible (*First Fail* principle). A second principle is to select values that maximize the number of possible assignments. The *First Fail* principle is implemented by hybridizing MCTS with a mainstream variable-ordering heuristics (wdeg is used in the experiments). The latter principle will guide the definition of the proposed reward (section 3.2).

A second issue regards the search strategy used in the MCTS roll-out phase. The use of random search is not desirable, among other reasons as it does not enforce the search completeness in the no-restart context. Accordingly, the roll-out strategy used in BaSCoP implements a complete strategy, the depth first search.

3.1   Overview

The overall structure of the BaSCoP search space is displayed in Fig. 1. BaSCoP grows a search tree referred to as *top-tree* (the filled nodes in Fig. 1), which is a subtree of the full search tree. Each node is a partial assignment s (after

Algorithm 1. BaScoP

input	: number N of tree-walks, restart schedule, selection rule SR, expand rate k.
data structure:	a node stores

 - a *state* : partial assignment as handled by the solver,
 - the *variable* to be assigned next,
 - children nodes corresponding to its admissible values,
 - a *top* flag marking it as subject to SR or DFS,
 - statistics: number n of visits, average failure depth avg.

Every time a new node must be created (first visit), its state is computed in the solver by adding the appropriate literal, and its variable is fetched from the solver.

All numeric variables are initialized to zero.

main loop :

> search tree $\mathcal{T} \leftarrow$ **new Node**(*empty state*)
> **for** N *tree-walks* **do**
>> **if** *restart* **then** $\mathcal{T} \leftarrow$ **new Node**(*empty state*)
>>
>> **if** Tree-walk($\mathcal{T}$).*state.success* **then**
>> process returned solution

function Tree-walk(node) **returns** (depth, state) :

> **if** node.state *is terminal (failure,success)* **then**
>> *close the node, and its ancestors if necessary*
>> **return** $(0, \text{node.state})$
>
> **if** node.top $=$ false **then**
>> **once every** k, node.top $\leftarrow$ true
>> **otherwise, return** DFS(node)
>
> node.n $\leftarrow$ node.n $+ 1$
> *Use SR to select* value *among admissible ones*
> $(\text{d}, \text{s}) =$ Tree-walk(node's *child associated to* value)
> node.avg $\leftarrow$ node.avg $+ (\text{d} - \text{node.avg})/\text{node.n}$
> **if** d $>$ node.avg **then** reward $= 1$
> **else** reward $= 0$
> **let** $\ell = (\text{node.variable}, \text{value})$:
>> $n_\ell \leftarrow n_\ell + 1$
>> $RAVE_\ell \leftarrow RAVE_\ell + (\text{reward} - RAVE_\ell)/n_\ell$
>
> **return** $(\text{d} + 1, \text{s})$

function DFS(node) **returns** (depth, state) :

> **if** node.state *is terminal (failure,success)* **then**
>> *close the node, and its ancestors if necessary*
>> **return** $(0, \text{node.state})$
>
> $(\text{d}, \text{s}) =$ DFS(*leftmost admissible child*)
> **return** $(\text{d} + 1, \text{s})$

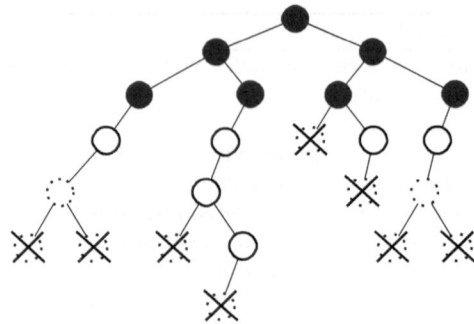

Fig. 1. Overview of the BASCOP search space. The top tree (filled nodes) is explored and extended along the MCTS framework. The bottom tree involves the tree paths under the top-tree leaves, iteratively updated by depth-first search. The status of a bottom-node is open (unfilled) or closed (dotted).

the constraint propagation achieved by the CP solver). The possible actions in s are to assign a fixed variable X (fetched from the variable-ordering heuristics) to value *true* or *false*, respectively represented as ℓ_X and $\ell_{\bar{X}}$ literals. Each child node of s (noted $s \wedge \ell$ with $\ell = \ell_X$ or $\ell_{\bar{X}}$) is associated a status: *closed* if the sub-tree associated to $s \wedge \ell$ has been fully explored; *open* if the sub-tree is being explored; *to-be-opened* if the node has not yet been visited. The value assigned to X is selected depending on the reward of the child nodes (section 3.2) and the selection rule (section 3.3).

BASCOP simultaneously explores and extends the top-tree along the MCTS framework, following successive tree-walks from the root until reaching a leaf node of the top-tree. The growth of the top-tree is controlled through the *expand rate* parameter k (section 2.3), where a child node is added below a leaf node s after s has been visited k times.

Upon reaching a leaf node of the top-tree, the BASCOP roll-out phase is launched until reaching a terminal state (failure or complete assignment). The roll-out phase uses the depth-first-search (DFS) strategy. DFS only requires to maintain a tree path below each leaf node; specifically, it requires to maintain the status of every node in these tree paths, referred to as bottom nodes (depicted as unfilled nodes in Fig. 1). By construction, DFS proceeds by selecting the left child node unless it is closed. Thereby, BASCOP enables a systematic exploration of the subtrees below its leaf nodes, thus enforcing a complete search in the no-restart setting.

3.2 Relative Failure Depth Reward

In the MCTS spirit, the choice among two child nodes must be guided by the average performance or reward attached to these child nodes, and the confidence thereof. Defining a reward attached to a partial assignment however raises several difficulties, as discussed in section 2. Firstly, the performance attached to the terminal states below a node might be poorly informative, e.g. in the satisfiability context. Secondly and most importantly, a heuristics commonly involved in the CP search is that of *multiple restarts*. Upon each restart, the current CP search tree is erased; the memory of the search is only reflected through some indicators (e.g. weighted degree, weighted dom-degree, impact, activity or nogoods) maintained over the restarts. When rebuilding the CP search tree from scratch, a new variable ordering is computed from these indicators, expectedly resulting in more efficient and shorter tree-paths. Naturally, BASCOP must accommodate multiple restarts in order to define a generic CP search strategy. For tractability reasons however, BASCOP can hardly maintain all top-trees built along multiple restarts, or the rewards attached to all nodes in these top-trees. On the other hand, estimating node rewards from scratch after each restart is poorly informative too, as the rewards are estimated from insufficiently many tree-walks.

Taking inspiration from the RAVE heuristics (section 2.3), it thus comes to associate a reward to each ℓ_X and $\ell_{\bar{X}}$ literals, where X ranges over the variables of the problem. The proposed reward measures the impact of the literal on the depth of the failure, as follows. Formally, let s denote a current node with ℓ_X and $\ell_{\bar{X}}$ as possible actions. Let $\bar{d}_{s,f}$ denote the average depth of the failures occurring below s. Literal ℓ (with $\ell = \ell_X$ or $\ell_{\bar{X}}$) receives an instant reward 1 (respectively, 0) if the failure of the current tree-path occurs at depth $d > \bar{d}_{s,f}$ (resp., $d < \bar{d}_{s,f}$). The rationale for this reward definition is twofold. On the one hand, the values to be assigned to a variable only need to be assessed relatively to each other (recall that the variable ordering is fixed and external to BASCOP). On the other hand, everything else being equal, the failure due to a (variable, value) assignment should occur later than sooner: intuitively, a shorter tree-walk likely contains more bad literals than a longer tree-walk, everything else being equal.

Overall, the BASCOP reward associated to each literal ℓ, noted $r(\ell)$, averages the instant rewards gathered in all tree-paths where ℓ is selected in a top-tree node s. Indicator $n(\ell)$ counts the number of times literal ℓ is selected in a top tree node. As desired, reward $r(\ell)$ and indicator $n(\ell)$ can be maintained over multiple restarts, and thus based on sufficient evidence. Their main weakness is to aggregate the information from different contexts due to dynamic variable ordering (in particular the top-tree nodes s where literal ℓ is selected might be situated at different tree-depths) and due to multiple restarts. The aggregation might blur the estimate of the literal impact; however, the blurring effect is mitigated as the aggregation affects both ℓ_X and $\ell_{\bar{X}}$ literals in the same way.

3.3   Selection Rules

Let s and X respectively denote the current node and the variable to be assigned. BASCOP uses different rules in order to select among the possible assignments of X (literals ℓ_X and $\ell_{\bar{X}}$) depending on whether the current node s belongs to the top or the bottom tree.

In the bottom tree, the depth-first-search applies, always selecting the left child node unless its status is *closed*. Note that DFS easily accommodates value ordering: in particular, the local neighborhood search [21] biased toward the neighborhood of the last found solution (see section 4.2) can be enforced by setting the left literal to the one among ℓ_X and $\ell_{\bar{X}}$ which is satisfied by this last solution.

In the top-tree, several selection rules have been investigated:

- **Balanced SR** alternatively selects ℓ_X and $\ell_{\bar{X}}$;
- ϵ-**left SR** selects ℓ_X with probability $1 - \epsilon$ and $\ell_{\bar{X}}$ otherwise, thus implementing a stochastic variant of the limited discrepancy search [30];
- **UCB SR** selects the literal with maximal reward upper-bound (Eq. (1))

$$\text{select} \ \underset{\ell \in \{\ell_X, \ell_{\bar{X}}\}}{\arg \max} \ r(\ell) + C \sqrt{\frac{\log(n(\ell_X) + n(\ell_{\bar{X}}))}{n(\ell)}}$$

- **UCB-Left SR**: same as UCB SR, with the difference that different exploration constants are attached to literals ℓ_X and $\ell_{\bar{X}}$ in order to bias the exploration toward the left branch. Formally, $C_{\text{left}} = \rho C_{\text{right}}$ with $\rho > 1$ the strength of the left bias.

Note that balanced and ϵ-left selection rules are not adaptive; they are considered to comparatively assess the merits of the adaptive UCB and UCB-Left selection rules.

3.4   Computational Complexity

BaSCoP undergoes a time complexity overhead compared to DFS, due to the use of tree-walks instead of the optimized backtrack procedure, directly jumping to a parent or ancestor node. A tree-walk involves: i) the selection of a literal in each top-node; ii) the creation of a new node every k visits to a leaf node; iii) the update of the reward values for each literal. The tree-walk overhead thus amounts to h arithmetic computations, where h is the average height of the top-tree.

However, in most cases these computations are dominated by the cost of creating a new node, which involves constraint propagation upon the considered assignment.

With regard to its space complexity, BaSCoP includes N/k top nodes after N tree-walks, where k is the expand rate; it also maintains the DFS tree-paths behind each top leaf node, with complexity $\mathcal{O}(Nh'/k)$, where h' is the average height of the full tree. The overall space complexity is thus increased by a multiplicative factor N/k; however no scalability issue was encountered in the experiments.

4   Experimental Validation

This section reports on the empirical validation of BaSCoP on three binary and n-ary CP problems: job shop scheduling problems (JSP) [31], balance incomplete block design (BIBD) and car sequencing (the last two problems respectively correspond to problems 28 and 1 in [32]).

4.1   Experimental Setting

BaSCoP is implemented on the top of the state-of-the-art Gecode framework [20]. The goal of the experiments is twofold. On the one hand, the adaptive exploration *vs* exploitation MCTS scheme is assessed comparatively to the depth-first-search baseline. On the other hand, the relevance of the relative-depth-failure reward (section 3.2) is assessed by comparing the adaptive selection rules to the fixed balanced and ϵ-left selection rules (section 3.3).

The BaSCoP expand rate parameter k is set to 5, after a few preliminary experiments showing good performances in a range of values around 5. The performances (depending on the problem family) are reported versus the number of tree-walks, averaged over 11 independent runs unless otherwise specified. The computational time is similar for all considered approaches, being granted that the DFS baseline uses the same tree-walk implementation as BaSCoP[1]. The comparison of the runtimes is deemed to be fair as most of BaSCoP computational effort is spent in the tree-walk part, and will thus take advantage of an optimized implementation in further work.

4.2   Job Shop Scheduling

Job shop scheduling, aimed at minimizing the schedule makespan, is modelled as a binary CP problem [21]. Upon its discovery, a new solution is used to i) update the model (requiring further solutions to improve on the current one); ii) bias the search toward the neighborhood of this solution along a local neighborhood search strategy. The search is initialized using the solutions of randomized Werner schedules, that is, using the insertion algorithm of [33] with randomized flips in the duration-based ranking of operations. The variable ordering heuristics is based on wdeg-max [34]. Multiple restarts are scheduled along a Luby sequence with factor 64.

The performance indicator is the *mean relative error* (MRE), that is the relative distance to the best known makespan m^* $((makespan - m^*)/m^*)$, averaged over the runs and problem instances of a series. MRE is monitored over 50 000 BaSCoP tree-walks, comparing the following selection rules: *none*, which corresponds to DFS standalone; *balanced*, which corresponds to a uniform exploration of the top nodes; *ϵ-left*, where the exploration is biased towards the left child nodes, and the strength of the bias is controlled from parameter ϵ; *UCB-left*, where the exploration-exploitation trade-off based on the relative-depth-failure reward is controlled from parameter C, and the bias toward the left is controlled from parameter ρ. The results on the first four series of Taillard instances are

---

[1] This implementation is circa twice longer than the optimized tree-walk Gecode implementation − which did not allow however the solution-guided search procedure used for the JSP and car sequencing problems at the time of the experiments.

Table 1. BASCOP experimental validation on the Taillard job shop problems: mean relative error w.r.t. the best known makespan, averaged on 11 runs (50 000 tree walks)

Selection rule			Results on instance sets			
			1-10	11-20	21-30	31-40
None (DFS)			0.51	2.07	2.31	13.55
Balanced			0.39	1.76	2.00	3.29
ε-left	ϵ					
	0.05		0.57	1.58	1.58	2.56
	0.1		0.45	1.65	1.74	2.24
	0.15		0.58	1.46	1.63	2.37
	0.2		0.46	1.67	1.88	2.55
	average		0.51	1.59	1.71	2.43
UCB	ρ	C				
	1	0.05	0.35	1.61	1.59	2.24
	1	0.1	0.39	1.53	1.51	2.34
	1	0.2	0.41	1.52	1.65	2.57
	1	0.5	0.42	**1.39**	1.71	2.37
	2	0.05	0.32	1.51	1.47	2.22
	2	0.1	0.40	1.57	1.49	2.16
	2	0.2	0.43	1.48	1.48	2.37
	2	0.5	0.55	1.77	1.67	2.38
	4	0.05	0.34	1.57	1.60	2.19
	4	0.1	0.43	1.55	1.68	2.33
	4	0.2	0.44	1.53	1.63	2.39
	4	0.5	0.40	1.40	**1.42**	2.46
	8	0.05	0.36	1.51	1.62	**2.04**
	8	0.1	0.45	1.52	1.59	2.33
	8	0.2	0.46	1.51	1.62	2.39
	8	0.5	**0.29**	1.51	1.65	2.55
	average		0.40	1.53	1.59	2.33

Table 2. Best makespans obtained out of 11 runs of 200 000 tree-walks on the 11-20 series of Taillard instances, comparing DFS and BASCOP with UCB-Left selection rule with parameters $C = 0.05, \rho = 2$. Bold numbers indicate best known results so far.

	Ta11	Ta12	Ta13	Ta14	Ta15	Ta16	Ta17	Ta18	Ta19	Ta20
DFS	1365	1367	1343	**1345**	1350	1360	1463	1397	1352	1350
BASCOP	**1357**	1370	**1342**	1345	**1339**	1365	**1462**	1407	**1332**	1356

reported in Table 1, showing that BASCOP robustly outperforms DFS for a wide range of parameter values. Furthermore, the adaptive UCB-based search improves on average on all fixed strategies, except for the 1-10 series.

Complementary experiments displayed in Table 2, show that BASCOP discovers some of the current best-known makespans, previously established using dedicated CP and local search heuristics [35], at similar computational cost (circa one hour on Intel Xeon E5345, 2.33GHz for 200 000 tree-walks).

4.3   Balance Incomplete Block Design (BIBD)

BIBD is a family of challenging Boolean satisfaction problems, known for their many symmetries. We considered instances from [22], characterized from their $v, k,$ and λ parameters. A simple Gecode model with lexicographic order of the rows and columns is used. Instances for which no solution could be discovered by any method within 50 000 tree-walks are discarded. Two goals are tackled: finding a single solution; finding them all.

Table 3. BASCoP experimental validation on BIBD: number of tree-walks needed to find the first solution. Best results are indicated in bold; '-' indicates that no solution was found after 50 000 tree-walks.

v	k	λ	DFS	bal.	BASCoP C 0.05	C 0.1	C 0.2	C 0.5	C 1
9	3	2	49	49	49	49	49	49	49
9	4	3	45	45	45	45	45	45	45
10	3	2	63	63	63	63	63	63	63
10	4	2	45	45	45	45	45	45	45
10	5	4	333	669	357	355	355	**256**	509
11	5	2	45	45	45	45	45	45	45
13	3	1	**161**	331	176	176	176	243	265
13	4	1	40	40	40	40	40	40	40
13	4	2	**202**	935	216	216	216	499	463
15	3	1	131	131	131	131	131	131	131
15	7	3	567	1579	**233**	**233**	**233**	451	370
16	4	1	164	166	164	164	164	164	164
16	4	2	**639**	12583	1297	1279	1282	1324	2492
16	6	2	503	821	315	315	315	**314**	407
16	6	3	7880	-	3200	3198	**2559**	2594	4394
19	3	1	671	-	**493**	**493**	**493**	709	3541
19	9	4	-	-	26251	25310	25383	**2004**	-
21	3	1	-	-	**779**	**779**	**779**	1183	6272
21	5	1	261	634	**217**	**217**	**217**	**217**	277
25	5	1	3425	11168	636	636	636	643	**541**
25	9	3	-	-	-	35940	-	**30131**	-
31	6	1	13889	36797	**882**	**882**	**882**	953	893

After preliminary experiments, neither variable ordering nor value ordering (e.g. based on the local neighborhood search) heuristics were found to be effective. Accordingly, BASCoP with UCB selection rule is assessed comparatively to the DFS standalone and BASCoP with balanced selection rule.

Table 3 reports the number of iterations needed to find the first solution; a single run is considered. Satisfactory results are obtained for low values of the trade-off parameter C. On-going experiments consider lower C values.

The All-solution setting is considered to investigate the search efficiency of BASCoP. On easy problems where all solutions can be found after 50 000 tree-walks, same number of tree-walks is needed to find all solutions. The search

Table 4. BASCoP experimental validation on BIBD: Number of tree-walks needed to find 50% of the solutions when all solutions are found in 50 000 tree-walks

v	k	λ	DFS	bal.	BASCoP				
					C 0.05	C 0.1	C 0.2	C 0.5	C 1
9	3	2	8654	8000	8862	8860	7473	7317	**7264**
9	4	3	13291	15144	12821	12824	**12794**	13524	13753
10	4	2	156	215	**153**	**153**	**153**	**153**	181
11	5	2	45	45	45	45	45	45	45
13	4	1	40	40	40	40	40	40	40
15	7	3	5007	5254	**1877**	1878	**1877**	1961	2773
16	4	1	**322**	394	377	379	378	392	340
16	6	2	1677	1947	**1130**	1131	1133	1139	1270
21	5	1	507	799	**484**	**484**	**484**	495	537
average			3300	3538	2865	2866	**2709**	2785	2911

Table 5. BASCoP experimental validation on BIBD: Number of solutions found in 50 000 tree-walks

v	k	λ	DFS	bal.	BASCoP				
					C 0.05	C 0.1	C 0.2	C 0.5	C 1
10	3	2	19925	11136	17145	17172	17031	18309	**22672**
10	5	4	1454	1517	1552	1554	1550	1556	**1558**
13	4	2	824	1457	16597	**16654**	16596	2063	1898
15	3	1	21884	2443	22496	22505	22497	**23142**	15273
16	4	2	190	6	4726	**4727**	4725	247	392
16	6	3	180	-	416	416	**425**	306	64
19	3	1	18912	-	**19952**	**19952**	**19952**	15794	10190
19	9	4	-	-	18	18	18	**36**	-
21	3	1	-	-	16307	16289	**16329**	14764	9058
25	5	1	416	260	**460**	**460**	**460**	**460**	420
25	9	3	-	-	-	**12**	-	8	-
31	6	1	253	34	**347**	342	**347**	**347**	342
average			7388	3279	**9173**	8473	**9166**	6684	6516

efficiency is therefore assessed from the number of tree-walks needed to find 50% of the solutions, displayed in Table 4. Likewise, there exists a plateau of good results for low values of parameter C.

For more complex problems, the number of solutions found after 50 000 tree-walks is displayed in Table 5.

Overall, BASCoP consistently outperforms DFS, particularly so for low values of the exploration constant C, while DFS consistently outperforms the non-adaptive balanced strategy. For all methods, the computational cost is ca 2 minutes on Intel Xeon E5345, 2.33GHz for 50 000 tree-walks).

4.4   Car Sequencing

Car sequencing is a CP problem involving circa 200 n-ary variables, with n ranging over $[20, 30]$. As mentioned, the UCB decision rule straightforwardly

Table 6. BASCoP experimental validation on car-sequencing: top line: violation after 10 000 tree-walks, averaged over 70 problem instances. bottom line: significance of the improvement over DFS after Wilcoxon signed-rank test.

	DFS	bal.	BASCoP			
			C 0.05	C 0.1	C 0.2	C 0.5
average gap	17.1	17.1	16.6	16.7	16.6	16.5
p-value	-	0	10^{-3}	$5\,10^{-3}$	10^{-3}	10^{-3}

extends beyond the binary case. After preliminary experiments, multiple restart strategies were not considered as they did not bring any improvements. Variable ordering based on *activity* [36] was used together with a static value ordering. 70 instances (ranging in 60-01 to 90-10 from [32]) are considered; the algorithm performance is the violation of the capacity constraint (number of extra stalls) averaged over the solutions found after 10 000 tree-walks.

The experimental results (Table 6) show that CP solvers are far from reaching state-of-the-art performance on these problems, especially when using the classical relaxation of the capacity constraint [37]. Still, while DFS and balanced exploration yield same results, BASCoP with UCB selection rule significantly improves on DFS after a Wilcoxon signed-rank test; the improvement is robust over a range of parameter settings, with C ranging in [.05, .5].

5   Discussion and Perspectives

The generic BASCoP scheme presented in this paper achieves the adaptive control of the variable-value assignment in the CP search along the Monte-Carlo Tree Search ideas. The implementation of BASCoP on the top of the Gecode solver and its comparative validation on three families of CP problems establish, as a proof of principle that cues about the relevance of some (variable,value) assignments can be efficiently extracted and exploited online.

A main contribution of the proposed scheme is the proposed (variable,value) assignment reward, enforcing the BASCoP compatibility with multiple restart strategies. Importantly, BASCoP can (and should) be hybridized with CP heuristics, such as dynamic variable ordering or local neighborhood search; the use of the depth-first search strategy as roll-out policy is a key issue commanding the completeness of the BASCoP search, and its efficiency.

This work opens several perspectives for further research. Focussing on the no-restart CP context, a first perspective is to apply the proposed relative failure depth reward to partial assignments. Another extension concerns the use of progressive-widening [38] or X-armed bandits [39] to deal with respectively many-valued or continuous variables.

A mid-term perspective concerns the parallelization of BASCoP, e.g. through adapting the parallel MCTS approaches developed in the context of games [40]. In particular, parallel BASCoP could be hybridized with the parallel CP approaches based on work stealing [41], and contribute to the collective identification of the most promising parts of the search tree.

Acknowledgments. The authors warmly thank Christian Schulte for his help and many insightful suggestions about the integration of MCTS within the Gecode solver.

References

1. van Beek, P.: Backtracking Search Algorithms. In: Handbook of Constraint Programming (Foundations of Artificial Intelligence), pp. 85–134. Elsevier Science Inc., New York (2006)
2. Rice, J.: The algorithm selection problem. In: Advances in Computers, pp. 65–118 (1976)
3. Xu, L., Hutter, F., Hoos, H., Leyton-Brown, K.: Satzilla: Portfolio-based algorithm selection for SAT. JAIR 32, 565–606 (2008)
4. O'Mahony, E., Hebrard, E., Holland, A., Nugent, C., O'Sullivan, B.: Using case-based reasoning in an algorithm portfolio for constraint solving. In: AICS (2008)
5. Samulowitz, H., Memisevic, R.: Learning to solve QBF. In: AAAI, 255–260 (2007)
6. Streeter, M., Golovin, D., Smith, S.: Combining multiple heuristics online. In: AAAI, pp. 1197–1203 (2007)
7. Hutter, F., Hoos, H.H., Leyton-Brown, K., Stützle, T.: Paramils: An automatic algorithm configuration framework. J. Artif. Intell. Res (JAIR) 36, 267–306 (2009)
8. Sutton, R., Barto, A.: Reinforcement Learning: an introduction. MIT Press (1998)
9. Previti, A., Ramanujan, R., Schaerf, M., Selman, B.: Monte-carlo style UCT search for boolean satisfiability. In: Pirrone, R., Sorbello, F. (eds.) AI*IA 2011. LNCS, vol. 6934, pp. 177–188. Springer, Heidelberg (2011)
10. Runarsson, T.P., Schoenauer, M., Sebag, M.: Pilot, Rollout and Monte Carlo Tree Search Methods for Job Shop Scheduling. In: Hamadi, Y., Schoenauer, M. (eds.) LION 2012. LNCS, vol. 7219, pp. 160–174. Springer, Heidelberg (2012)
11. Sabharwal, A., Samulowitz, H., Reddy, C.: Guiding combinatorial optimization with UCT. In: Beldiceanu, N., Jussien, N., Pinson, É. (eds.) CPAIOR 2012. LNCS, vol. 7298, pp. 356–361. Springer, Heidelberg (2012)
12. Loth, M.: Hybridizing constraint programming and Monte-Carlo Tree Search: Application to the job shop problem. In: Nicosia, G., Pardalos, P. (eds.) Learning and Intelligent Optimization Conference (LION 7), Springer, Heidelberg (2013)
13. Lai, T., Robbins, H.: Asymptotically efficient adaptive allocation rules. Advances in Applied Mathematics 6, 4–22 (1985)
14. Auer, P., Cesa-Bianchi, N., Fischer, P.: Finite-time analysis of the multiarmed bandit problem. Machine Learning 47(2-3), 235–256 (2002)
15. Kocsis, L., Szepesvári, C.: Bandit based monte-carlo planning. In: Fürnkranz, J., Scheffer, T., Spiliopoulou, M. (eds.) ECML 2006. LNCS (LNAI), vol. 4212, pp. 282–293. Springer, Heidelberg (2006)
16. Gelly, S., Silver, D.: Combining online and offline knowledge in UCT. In: International Conference on Machine Learning, pp. 273–280. ACM (2007)
17. Ciancarini, P., Favini, G.: Monte-Carlo Tree Search techniques in the game of Kriegspiel. In: International Joint Conference on Artificial Intelligence, pp. 474–479 (2009)

18. Nakhost, H., Müller, M.: Monte-Carlo exploration for deterministic planning. In: Boutilier, C. (ed.) International Joint Conference on Artificial Intelligence, pp. 1766–1771 (2009)
19. Luby, M., Sinclair, A., Zuckerman, D.: Optimal speedup of las vegas algorithms. Information Processing Letters 47(4), 173–180 (1993)
20. Gecode Team: Gecode: Generic constraint development environment (2012), www.gecode.org
21. Beck, J.C.: Solution-guided multi-point constructive search for job shop scheduling. Journal of Artificial Intelligence Research 29, 49–77 (2007)
22. Mathon, R., Rosa, A.: Tables of parameters for BIBD's with $r \leq 41$ including existence, enumeration, and resolvability results. Ann. Discrete Math. 26, 275–308 (1985)
23. Hutter, F., Hamadi, Y., Hoos, H., Leyton-Brown, K.: Performance prediction and automated tuning of randomized and parametric algorithms. In: Benhamou, F. (ed.) CP 2006. LNCS, vol. 4204, pp. 213–228. Springer, Heidelberg (2006)
24. Haim, S., Walsh, T.: Restart strategy selection using machine learning techniques. In: Kullmann, O. (ed.) SAT 2009. LNCS, vol. 5584, pp. 312–325. Springer, Heidelberg (2009)
25. Epstein, S., Freuder, E., Wallace, R., Morozov, A., Samuels, B.: The adaptive constraint engine. In: Van Hentenryck, P. (ed.) CP 2002. LNCS, vol. 2470, pp. 525–540. Springer, Heidelberg (2002)
26. Wu, H., Van Beek, P.: Portfolios with deadlines for backtracking search. In: IJAIT, vol. 17, pp. 835–856 (2008)
27. Schneider, M., Hoos, H.H.: Quantifying homogeneity of instance sets for algorithm configuration. In: Hamadi, Y., Schoenauer, M. (eds.) LION 2012. LNCS, vol. 7219, pp. 190–204. Springer, Heidelberg (2012)
28. Wang, Y., Audibert, J., Munos, R.: Algorithms for infinitely many-armed bandits. In: Advances in Neural Information Processing Systems, pp. 1–8 (2008)
29. Refalo, P.: Impact-based search strategies for constraint programming. In: Wallace, M. (ed.) CP 2004. LNCS, vol. 3258, pp. 557–571. Springer, Heidelberg (2004)
30. Harvey, W., Ginsberg, M.: Limited discrepancy search. In: International Joint Conference on Artificial Intelligence, pp. 607–615 (1995)
31. Taillard, E.: Benchmarks for basic scheduling problems. European Journal of Operational Research 64(2), 278–285 (1993)
32. Gent, I., Walsh, T.: Csplib: A benchmark library for constraints. In: Jaffar, J. (ed.) CP 1999. LNCS, vol. 1713, pp. 480–481. Springer, Heidelberg (1999)
33. Werner, F., Winkler, A.: Insertion techniques for the heuristic solution of the job shop problem. Discrete Applied Mathematics 58(2), 191–211 (1995)
34. Boussemart, F., Hemery, F., Lecoutre, C., Sais, L.: Boosting systematic search by weighting constraints. In: ECAI, pp. 146–150 (2004)
35. Beck, J., Feng, T., Watson, J.P.: Combining constraint programming and local search for job-shop scheduling. INFORMS Journal on Computing 23(1), 1–14 (2011)
36. Michel, L., Van Hentenryck, P.: Activity-based search for black-box constraint programming solvers. In: Beldiceanu, N., Jussien, N., Pinson, É. (eds.) CPAIOR 2012. LNCS, vol. 7298, pp. 228–243. Springer, Heidelberg (2012)
37. Perron, L., Shaw, P.: Combining forces to solve the car sequencing problem. In: Régin, J.-C., Rueher, M. (eds.) CPAIOR 2004. LNCS, vol. 3011, pp. 225–239. Springer, Heidelberg (2004)

38. Coulom, R.: Efficient Selectivity and Backup Operators in Monte-Carlo Tree Search. In: van den Herik, H.J., Ciancarini, P., Donkers, H.H.L.M(J.) (eds.) CG 2006. LNCS, vol. 4630, pp. 72–83. Springer, Heidelberg (2007)
39. Bubeck, S., Munos, R., Stoltz, G., Szepesvári, C.: X-armed bandits. Journal of Machine Learning Research 12, 1655–1695 (2011)
40. Chaslot, G.M.J.-B., Winands, M.H.M., van den Herik, H.J.: Parallel Monte-Carlo Tree Search. In: van den Herik, H.J., Xu, X., Ma, Z., Winands, M.H.M. (eds.) CG 2008. LNCS, vol. 5131, pp. 60–71. Springer, Heidelberg (2008)
41. Chu, G., Schulte, C., Stuckey, P.: Confidence-based work stealing in parallel constraint programming. In: Gent, I.P. (ed.) CP 2009. LNCS, vol. 5732, pp. 226–241. Springer, Heidelberg (2009)

Focused Random Walk with Configuration Checking and Break Minimum for Satisfiability

Chuan Luo[1,*], Shaowei Cai[2], Wei Wu[1,3], and Kaile Su[3,4]

[1] Key Laboratory of High Confidence Software Technologies,
Peking University, Beijing, China
[2] Queensland Research Laboratory, National ICT Australia, QLD, Australia
[3] College of Mathematics, Physics and Information Engineering,
Zhejiang Normal University, Jinhua, China
[4] Institute for Integrated and Intelligent Systems,
Griffith University, Brisbane, Australia
{chuanluosaber,shaoweicai.cs,william.third.wu}@gmail.com,
k.su@griffith.edu.au

Abstract. Stochastic local search (SLS) algorithms, especially those adopting the focused random walk (FRW) framework, have exhibited great effectiveness in solving satisfiable random 3-satisfiability (3-SAT) instances. However, they are still unsatisfactory in dealing with huge instances, and are usually sensitive to the clause-to-variable ratio of the instance. In this paper, we present a new FRW algorithm dubbed FrwCB, which behaves more satisfying in the above two aspects. The main idea is a new heuristic called CCBM, which combines a recent diversification strategy named configuration checking (CC) with the common break minimum (BM) variable-picking strategy. By combining CC and BM in a subtle way, CCBM significantly improves the performance of FrwCB, making FrwCB achieve state-of-the-art performance on a wide range of benchmarks. The experiments show that FrwCB significantly outperforms state-of-the-art SLS solvers on random 3-SAT instances, and competes well on random 5-SAT, random 7-SAT and structured instances.

1   Introduction

The satisfiability problem (SAT) is a prototypical NP-complete problem, and has been widely studied due to its significant importance in both theories and applications. Given a propositional formula in conjunctive normal form (CNF), the SAT problem consists in finding an assignment to the variables such that all clauses are satisfied.

Algorithms for solving SAT can be mainly categorized into two classes: complete algorithms and stochastic local search (SLS) algorithms. Although SLS algorithms are incomplete in that they cannot prove an instance to be unsatisfiable, they are very efficient in solving satisfiable instances. The basic schema of an SLS algorithm for SAT works as follows. After initializing a random (complete) assignment, the algorithm flips a variable in each step according to a heuristic for selecting the flipping variable, until it seeks out a satisfiable assignment or timeout.

---

* Corresponding author.

C. Schulte (Ed.): CP 2013, LNCS 8124, pp. 481–496, 2013.
© Springer-Verlag Berlin Heidelberg 2013

There has been much interest in studying the performance of SLS algorithms on uniform random k-SAT instances, especially 3-SAT ones. The random 3-SAT instances, at the phase transition region, have been cited as the hardest group of SAT problems [16]. The random 3-SAT problem is an important special case of SAT, and is also a classic problem in combinatorics, at the heart of computational complexity studies [1]. Random 3-SAT instances have been widely used as a testing ground in the literature [22,6,18,2,3,8,20], as well as in SAT competitions[1].

In the past two decades, there have been numerous works devoted to improving SLS algorithms, especially for random 3-SAT instances. Heuristics in SLS algorithms for SAT can be divided into three categories: GSAT [24,18], focused random walk (FRW) [23,12,3] and dynamic local search (DLS) [14]. Recent solvers usually combine these three kinds of heuristics, such as the winners of SAT Competition 2011 and SAT Challenge 2012 namely Sparrow [2] and CCASat [8].

FRW algorithms conduct the search by always selecting a variable to flip from an unsatisfied clause chosen randomly in each step [21]. On solving random 3-SAT instances, the FRW framework performs better than others. WalkSAT [23], as the first practical FRW algorithm and one of the most influential representatives, still shows state-of-the-art performance in solving random 3-SAT instances. The recent FRW algorithm probSAT [3] makes progress in this field and, to the best of our knowledge, is the current best SLS solver for solving random 3-SAT. However, the performance of probSAT is still not satisfactory on huge instances with more than one million variables and the ones with different clause-to-variable ratios near the phase transition.

This work is devoted to improving the effectiveness and robustness of FRW algorithms. We propose a new heuristic, called CCBM, which combines the configuration checking (CC) strategy [7] and the break minimum (BM) strategy effectively in a subtle way. The BM strategy prefers to pick the variable which brings fewest number of clauses from satisfied to unsatisfied, and is a commonly used strategy in FRW algorithms, such as WalkSAT. Originally proposed in [10], the CC strategy reduces the cycling problem by checking the circumstance information. It has been successfully used in non-FRW algorithms, leading to several state-of-the-art SLS solvers such as CCASat. However, the direct application of CC in the FRW framework does not work. This work combines CC and BM in a novel way to improve FRW algorithms.

We utilize the CCBM heuristic to develop a new algorithm named FrwCB (focused random walk with configuration checking and break minimum). We compare FrwCB against five state-of-the-art solvers, namely WalkSAT, probSAT, CCASat, Swqcc [20] and Sattime [19] on a broad range of instances. The experiments illustrate that FrwCB significantly outperforms its competitors on huge random 3-SAT instances with up to 4 million variables. Also, FrwCB demonstrates a satisfactory robustness by performing best on the benchmark consisting of 3-SAT instances with different clause-to-variable ratios near the phase transition from SAT Challenge 2012. Additionally, FrwCB can cooperate well with the survey propagation (SP) algorithm [5], and their combination can push forward state of the art in solving huge random 3-SAT (with 10^7 variables).

The robustness of FrwCB is further demonstrated by its good performance on other kinds of instances, including random 5-SAT instances, random 7-SAT instances and

---

[1] http://www.satcompetition.org

structured instances. On these instances, FrwCB is highly competitive with state-of-the-art solvers.

The remainder of this paper is structured as follows. In the following, we provide necessary definitions and the clause states based CC strategy. Then, we present the CCBM heuristic combining focused random walk with configuration checking. After that, we use the CCBM heuristic to develop an SLS algorithm called FrwCB. Experiments demonstrating the performance of FrwCB and some discussions about CCBM are presented next. Finally we conclude the paper and give some future work.

2 Preliminaries

In this section, we first give some basic definitions and notations in local search for SAT. Then we introduce the details of clause states based configuration checking.

2.1 Definitions and Notations

Given a set of n Boolean variables $V = \{x_1, x_2, \cdots, x_n\}$ and also the set of literals $L = \{x_1, \neg x_1, x_2, \neg x_2, \cdots, x_n, \neg x_n\}$ corresponding to these variables, a *clause* is a disjunction of literals. Using clauses and the logical operation AND ($\wedge$), we can construct a CNF formula, i.e., $F = c_1 \wedge \cdots \wedge c_m$, where the number of clauses in F is denoted as m, and $r = m/n$ is its clause-to-variable ratio. A formula can be described as a set of clauses. A k-SAT formula is a formula in which each clause has exactly k literals. We use $V(F)$ to denote the set of all variables appearing in formula F. Two variables are neighbors when they appear in at least one clause, and $N(x) = \{y \mid y \in V(F),\ y \text{ and } x \text{ are neighbors}\}$ is the set of all neighbors of variable x. We also denote $CL(x) = \{c \mid c \text{ is a clause which } x \text{ appears in}\}$.

A mapping $\alpha : V(F) \rightarrow \{True, False\}$ is called an *assignment*. If α maps all variables to a Boolean value, it is *complete*. For local search algorithms for SAT, a candidate solution is a complete assignment. Given a complete assignment α, each clause has two possible *states*: *satisfied* or *unsatisfied*: A clause is satisfied if at least one literal in that clause is true under α; otherwise, it is unsatisfied. An assignment α satisfies a formula F if α satisfies all clauses in F. Given a CNF formula F, the SAT problem is to find an assignment that satisfies all clauses in F.

The method of selecting the flipping variable in each step is usually guided by a scoring function. In each step, the flipping variable is usually selected based on its properties, such as *make*, *break* and *score*. For a variable x, the property $make(x)$ is defined as the number of clauses that would become satisfied if the variable is flipped; the property $break(x)$ is the number of clauses that would become unsatisfied if the variable is flipped; the property $score(x)$ is the increment in the number of satisfied clauses if the variable is flipped, and can be understood as $make(x) - break(x)$. The heuristic in FrwCB utilizes *break* and *score* to select the flipping variable.

2.2 Clause States Based Configuration Checking

Configuration checking (CC) techniques have proven successful in SLS algorithms [7,8,20,9]. The main idea of configuration checking is to forbid flipping any variable whose circumstance information has not been changed since its last flip. For each variable, the circumstance information is formally defined as the concept of *configuration*.

In the context of SAT, the first definition of configuration was introduced in [7], where the configuration of a variable x refers to a vector consisting of Boolean values of $N(x)$ (x's all neighboring variables). This original CC strategy has been successfully used in non-FRW algorithms [7,8]. However, when applied to FRW algorithms, this variable based strategy makes almost all candidate variables *configuration changed* during the search process, and thus loses its power.

An alternative CC strategy was proposed in [20], where the configuration of a variable x refers to a vector consisting of Boolean values of $CL(x)$ (the clauses x appears in). This paper also adopts the clause states based configuration.

Definition 1. *Given a CNF formula F and a complete assignment α to $V(F)$, the configuration of a variable $x \in V(F)$ is a vector configuration(x) consisting of the states of all clauses in $CL(x)$ under assignment α.*

For a variable x, a change on any bit of *configuration*(x) is considered as a change on the whole *configuration*(x) vector. For a variable $x \in V(F)$, if the *configuration* of x has not been changed since x's last flip, then x should not be flipped.

An implementation of the clause states based CC strategy is to employ an integer array *ConfTimes* for variables. For each variable x, *ConfTimes*(x) measures the frequency (i.e., the number of times) that *configuration*(x) has been changed since x's last flip. The array *ConfTimes* is maintained as follows.

- Rule 1: In the beginning, for each variable $x \in V(F)$, *ConfTimes*(x) is set to 1.
- Rule 2: Whenever a variable x is flipped, *ConfTimes*(x) is reset to 0. Then each clause $c \in CL(x)$ is checked whether its state is changed by flipping x. If this is the case, for each variable y ($y \neq x$) in c, *ConfTimes*(y) is increased by 1.

Apparently, a variable x's configuration has been changed since its last flip if and only if *ConfTimes*$(x) > 0$. An important notion is the concept of configuration changed decreasing (CCD) variables, which is defined as follows.

Definition 2. *Given a CNF formula F and a complete assignment α to $V(F)$, a variable x is configuration changed decreasing (CCD) if and only if score$(x) > 0$ and ConfTimes$(x) > 0$.*

This work uses *CCDVars(c)* to denote the set of all CCD variables in clause c.

3   The CCBM Heuristic and The FrwCB Algorithm

In this section, we utilize the clause states based CC strategy in a novel way, so that the CC strategy cooperates well with FRW algorithms. Especially, we combine the CC strategy with the common break minimum variable-picking strategy, resulting in the CCBM (configuration checking with break minimum) heuristic. We then utilize CCBM to develop an FRW algorithm called FrwCB for SAT. Finally, we discuss the differences between the FrwCB algorithm and the Swqcc algorithm [20], which also employs a clause states based CC strategy.

3.1 The CCBM Heuristic

In this subsection, we propose a new heuristic which combines focused random walk with configuration checking effectively, and is referred to as CCBM. We first give the definition of the BM variable of a clause, and it is an important concept in CCBM.

Definition 3. *Given a CNF formula F and a complete assignment α to $V(F)$, for each clause c, a variable x is the break minimum (BM) variable of clause c if and only if $break(x) = \min\{break(y) \mid y \text{ appears in } c\}$.*

In this work, we use *BMVars(c)* to denote the set of all BM variables of clause c.

The main idea of the CCBM heuristic is to prefer to flip CCD variables and BM variables from a random unsatisfied clause. Flipping a CCD variable brings down the number of unsatisfied clauses, and at the same time prevents the algorithm from revisiting the scenario the algorithm recently faced. Although previous works such as [7,8] also prefer to flip CCD variables, they survey CCD variables globally, i.e., searching CCD variables from all the variables. In contrast, the CCBM heuristic picks a CCD variable from an unsatisfied clause. Whenever no CCD variable is present, CCBM prefers to pick a BM variable of a random unsatisfied clause to flip, leading the algorithm to search deeply.

In more detail, the CCBM heuristic works as follows. After selecting an unsatisfied clause c, it switches between two levels, namely the CCD level and the probability (PROB) level, depending on whether *CCDVars(c)* is empty or not. If *CCDVars(c)* is not empty, CCBM works in the CCD level; otherwise it works in the PROB level. In the CCD level, CCBM does a gradient decreasing walk, i.e., selecting the variable with the greatest *score* in *CCDVars(c)* to flip. In the PROB level, with a probability p, CCBM chooses the variable with the greatest *ConfTimes* in *BMVars(c)*; in the remaining case, it employs a diversification strategy to pick a variable in c. In this work, this is accomplished by selecting the one with the greatest *ConfTimes* from clause c.

3.2 The FrwCB Algorithm

In this subsection, we use the CCBM heuristic to develop a new focused random walk algorithm named FrwCB (Focused Random Walk with Configuration Checking and Break Minimum). The FrwCB algorithm is outlined in Algorithm 1, as described below.

At the beginning of the algorithm, a complete assignment α is generated randomly, and *ConfTimes(x)* is initialized as 1 for each variable x.

After the initialization, the algorithm executes search steps iteratively until it seeks out a satisfiable assignment or the number of search steps exceeds *maxSteps*, which is the step limit. In each search step, FrwCB first picks an unsatisfied clause c randomly, and then it employs the CCBM heuristic to select a variable to flip from c as follows.

The CCD Level: If *CCDVars(c)* is not empty, the FrwCB algorithm selects the variable x with the greatest *score(x)* appearing in *CCDVars(c)* to flip, breaking ties by preferring the one with the greatest *ConfTimes(x)* (lines 7-8).

The PROB Level: If *CCDVars(c)* is empty, with a fixed probability p, FrwCB selects the variable x with greatest *ConfTimes(x)* in *BMVars(c)*, breaking ties by preferring the least recently flipped one; otherwise FrwCB diversifies the search by selecting the variable x in clause c with the greatest $ConfTimes(x)$ to flip, breaking ties by preferring the least recently flipped one (lines 9-12).

After picking the flipping variable, the algorithm flips the chosen variable. FrwCB repeats picking and flipping a variable and updating $ConfTimes$ until it seeks out a satisfiable assignment or reaches the step limit. If the algorithm finds a satisfiable assignment, it outputs the satisfiable assignment; otherwise it reports $Unknown$.

Algorithm 1. FrwCB

Input: CNF-formula F, $maxSteps$
Output: A satisfiable assignment α of F or $Unknown$

1 **begin**
2   generate a random assignment α;
3   initialize $ConfTimes(x)$ as 1 for each variable x;
4   **for** $step \leftarrow 1$ **to** $maxSteps$ **do**
5   **if** α *satisfies* F **then return** α;
6          $c \leftarrow$ an unsatisfied clause chosen randomly;
7   **if** *CCDVars(c) is not empty* **then**
8              $v \leftarrow x$ with the greatest $score(x)$ in *CCDVars(c)*, breaking ties by preferring the one with the greatest $ConfTimes(x)$;
9   **else if** *with the fixed **probability** p* **then**
10              $v \leftarrow x$ with the greatest $ConfTimes(x)$ in *BMVars(c)*, breaking ties by preferring the least recently flipped one;
11   **else**
12              $v \leftarrow x$ with the greatest $ConfTimes(x)$ in clause c, breaking ties by preferring the least recently flipped one;
13   flip v and update $ConfTimes$;
14   **return** $Unknown$;
15 **end**

3.3 Discussion on Differences between Swqcc and FrwCB

The most related work is the Swqcc algorithm [20], which adopts a clause states based configuration checking heuristic named QCC. In the following, we discuss the differences between the Swqcc algorithm and the FrwCB algorithm.

The most important difference is that Swqcc and FrwCB adopt different local search paradigms. While Swqcc is a two-mode (GSAT-like + random walk) SLS algorithm [20], FrwCB is a single-mode (focused random walk) one.

The two algorithms employ different heuristics to pick a variable to flip. Swqcc employs the QCC heuristic: if there exist candidate variables (described in [20]) for the greedy mode, QCC selects the one with greatest $score$; otherwise QCC always picks a variable with the greatest $ConfTimes$ in a random unsatisfied clause. In contrast, FrwCB uses the CCBM heuristic: after picking an unsatisfied clause c randomly, if there exist candidate variables (i.e., CCD variables) in c, CCBM selects the one with greatest *score*; otherwise CCBM selects either a variable with the greatest *ConfTimes* in *c* or a variable with the minimum *break* in *c*.

Also, we conduct a direct comparison between Swqcc and FrwCB, referring to Section 4. Additionally, we compare the underlying heuristic in Swqcc, namely QCC, with the CCBM heuristic in this work, and the experimental analysis can be found in Section 5.2.

4 Experimental Results

In this section, we first introduce the benchmarks and some preliminaries about our experiments. Then we divide the experiments into four parts. Part 1 is to compare FrwCB with its competitors on random 3-SAT instances. Part 2 is to compare FrwCB with its competitors on 5-SAT and 7-SAT instances. Part 3 is to compare FrwCB against its competitors on structured instances. In part 4, we combine FrwCB with the SP algorithm, resulting in a new solver called SP+FrwCB, and then we investigate the performance of different solvers combining SP with different SLS solvers on random 3-SAT instances with 10 million variables.

4.1 The Benchmarks

We evaluate FrwCB on random instances as well as structured ones. Random 3-SAT instances are the best studied random instances and thus we have four different sets of random 3-SAT instances. Specifically, we adopt the following benchmarks:

1. all 100 large satisfiable 3-SAT instances in the random category of SAT Competition 2011[2] ($r = 4.2, 2500 \leqslant \#var \leqslant 50000$, 10 instances each size);
2. all 120 satisfiable random 3-SAT instances in SAT Challenge 2012[3] ($4.2 \leqslant r \leqslant 4.267, 2000 \leqslant \#var \leqslant 40000$, 12 instances each ratio);
3. 200 huge satisfiable random 3-SAT instances ($r = 4.2, 0.1M \leqslant \#var \leqslant 4.0M$ where $1.0M = 10^6$, 20 instances each size), generated according to the fixed clause length random model (no duplicate clauses, no duplicate literals in a clause);
4. 20 extremely huge satisfiable random 3-SAT instances ($r = 4.2, \#var = 10.0M = 10^7$), generated according to the fixed clause length random model.

The medium-sized satisfiable random 3-SAT instances in SAT Competition 2011 are too easy for modern SLS solvers, and thus are not included in our experiments.

For random 5-SAT and 7-SAT instances, we adopt the testing benchmark used in [25,3]. The benchmark contains 250 satisfiable random 5-SAT instances[4] ($r = 20, \#var = 500$) and 250 satisfiable random 7-SAT instances[5] ($r = 85, \#var = 90$).

---

[2] http://www.cril.univ-artois.fr/SAT11/bench/
SAT11-Competition-SelectedBenchmarks.tar

[3] http://baldur.iti.kit.edu/SAT-Challenge-2012/
downloads/sc2012-random.tar

[4] http://people.cs.ubc.ca/~davet/captain-jack/
5sat500.test.tar.gz

[5] http://people.cs.ubc.ca/~davet/captain-jack/7sat90.test.tar.gz

For structured instances, we adopt satisfiable crafted instances from SATLIB[6], including the ais, blocksworld, gcp, jnh, logistics, par8, par8-c, par16 and par16-c classes, which have been widely tested in the literature [13,27,26], as well as the largest and thus the most difficult frb instances[7] (frb50-23, frb53-24, frb56-25 and frb59-26). Note that these frb instances are generated randomly in the phase transition area according to the Model RB [28], and are very difficult to solve by current techniques in spite of their relative small size. These frb instances have been extensively used in the SAT competitions and MAX-SAT evaluations, and in the literature [17,15,11].

4.2   Experimental Preliminaries

The FrwCB algorithm is implemented in programming language C and statically compiled by gcc with the '-O3' option. We set the parameter probability p to 0.6 for 3-SAT with $r < 4.26$, 0.63 for 3-SAT with $r \geqslant 4.26$, 0.65 when FrwCB cooperates with SP, 0.8 for 5-SAT, 0.9 for 7-SAT and 0.95 for structured instances.

We compare FrwCB with five state-of-the-art SLS solvers, including WalkSAT, probSAT, CCASat, Swqcc and Sattime. WalkSAT is the most famous FRW solver, and is still highly competitive with the state of the art on random 3-SAT instances. The probSAT solver is the current best SLS solver for random 3-SAT instances especially the huge ones. CCASat is the winner of the random track in SAT Challenge 2012 and the current best solver using the CC strategy. Swqcc is the other solver that adopts the clause states based configuration checking strategy. The Sattime solver is the current best SLS solver for solving crafted instances.

For WalkSAT, we adopt the latest version (Version 50) from its author's website[8], and we set the noise parameter to 0.567 for 3-SAT, 0.25 for 5-SAT and 0.1 for both 7-SAT and structured instances, as reported in [3]. The binary of probSAT is provided by its author, and that of CCASat is downloaded online[9]. The parameters of Swqcc are identical to those reported in [20]. The Sattime solver we adopt is the one submitted to SAT Competition 2011[10]. We get the source code of SP from its author's website[11].

All the experiments are carried out on a machine with Intel Core i7 2.7GHz CPU and 7.8GB RAM under Linux. We report the number of successful runs ('#suc') as well as averaged run time ('avg time') for each solver on each instance class, as most works on SLS for SAT do.

4.3   Results on Random 3-SAT Instances

On the Instances from SAT Competition 2011: Table 1 shows experimental results on the large random 3-SAT instances from the SAT Competition 2011, where each

---

[6] http://www.satlib.org/

[7] http://www.nlsde.buaa.edu.cn/~kexu/benchmarks/benchmarks.htm

[8] http://www.cs.rochester.edu/~kautz/walksat/Walksat_v50.zip

[9] http://shaoweicai.net/research.html

[10] http://www.cril.univ-artois.fr/SAT11/solvers/
SAT2011-sources.tar.gz

[11] http://www.ictp.trieste.it/~zecchina/SP/sp-1.4b.tgz

Table 1. Comparative results on large random 3-SAT instances from SAT Competition 2011. Each solver is performed 100 times on each instance class.

Instance Class	WalkSAT		probSAT		CCASat		Swqcc		FrwCB	
	#suc	avg time	#suc	avg time	#suc	avg time	#suc	avg time	#suc	avg time
3SAT-v2500	95	152	99	88	100	9	100	6	100	37
3SAT-v5000	100	31	100	13	100	11	100	13	100	12
3SAT-v10000	100	19	100	21	100	19	100	37	100	10
3SAT-v15000	100	24	100	24	100	29	100	73	100	13
3SAT-v20000	100	35	100	37	100	44	100	118	100	26
3SAT-v25000	100	54	100	56	100	73	100	172	100	36
3SAT-v30000	100	56	100	63	100	92	100	186	100	42
3SAT-v35000	100	122	100	108	100	147	100	279	100	61
3SAT-v40000	100	114	100	84	100	125	100	240	100	56
3SAT-v50000	99	206	100	145	100	250	99	403	100	99

Table 2. Comparative results on all the random 3-SAT instances from the SAT Challenge 2012

# Total Runs	WalkSAT		probSAT		CCASat		Swqcc		FrwCB	
	#suc	avg time	#suc	avg time	#suc	avg time	#suc	avg time	#suc	avg time
1200	964	658	1003	598	967	693	986	731	**1043**	**499**

solver is performed 10 runs for each instance with a cutoff time of 2000 seconds. On the instances with $\#var = 2500$, CCASat and Swqcc outperform all FRW algorithms, among which FrwCB performs best. On the instances with $\#var = 5000$, FrwCB outperforms its competitors (but CCASat). On the other hand, FrwCB significantly outperforms its competitors on the instances with $\#var > 5000$. Especially, on the largest sized instances (with $\#var = 50000$), FrwCB performs about 4 times as fast as Swqcc does, about 2.5 times as fast as CCASat does, about 2 times as fast as WalkSAT does, and about 1.5 times as fast as probSAT does. Indeed, FrwCB is so efficient that it solves all large random 3-SAT instance classes in SAT Competition 2011 with an averaged time less than 100 seconds.

On the Instances from SAT Challenge 2012: Table 2 reports experimental results on all random 3-SAT instances from SAT Challenge 2012, whose clause-to-variable ratios range from 4.2 to 4.267. Each solver is performed 10 runs for each instance with a cutoff time of 2000 seconds. FrwCB outperforms its competitors in terms of both success rate and averaged run time on this benchmark. FrwCB succeeds in 1043 (out of 1200) runs, 40 more than the second best solver probSAT does. Moreover, the overall averaged run time of FrwCB on this benchmark is only 499 seconds, while this number is 598 for probSAT, 658 for WalkSAT, 693 for CCASat and 731 for Swqcc. The excellent performance of FrwCB on these instances with various clause-to-variable ratios indicates its good robustness on random 3-SAT instances.

On the Huge Instances: To evaluate the genuine solving ability on random 3-SAT, we compare FrwCB with its competitors on the huge instances. The experimental results are reported in Table 3 and summarized in Figure 1, where each solver is performed

Table 3. Comparative results on huge random 3-SAT instances. Each solver is performed 20 times for each instance class. Swqcc fails to solve any instance in this benchmark, so we do not report its results.

Instance Class	WalkSAT		probSAT		CCASat		FrwCB	
$(1.0M = 10^6)$	#suc	avg time	#suc	avg time	#suc	avg time	#suc	avg time
3SAT-v0.1M	20	375	20	266	20	955	20	**227**
3SAT-v0.3M	20	920	20	934	10	9064	20	**393**
3SAT-v0.5M	20	2150	20	1905	0	>10000	20	**789**
3SAT-v1.0M	20	4691	20	4358	0	>10000	20	**1865**
3SAT-v1.5M	20	7696	20	6838	0	>10000	20	**3248**
3SAT-v2.0M	3	9964	15	9360	0	>10000	**20**	**4197**
3SAT-v2.5M	0	>10000	0	>10000	0	>10000	**20**	**5045**
3SAT-v3.0M	0	>10000	0	>10000	0	>10000	**20**	**6463**
3SAT-v3.5M	0	>10000	0	>10000	0	>10000	**20**	**7797**
3SAT-v4.0M	0	>10000	0	>10000	0	>10000	**15**	**9530**

one run for each instance with a cutoff time of 10000 seconds (less than 3 hours). We would like to note that Swqcc fails to solve any instance in this benchmark, so we do not report its results in both Table 3 and Figure 1. FrwCB stands out as the best solver and dramatically outperforms others. While other solvers all fail to solve any instance with $\#var \geqslant 2.5M$, FrwCB consistently solves all instances with up to $3.5M$ variables, and the results clearly show the superiority of FrwCB to other solvers. Furthermore, FrwCB remains effective on the instances with $4.0M$ variables. This is, to our best knowledge, the first time that such huge random 3-SAT instances are solved by an SLS algorithm in reasonable time. We conclude this section by remarking that the averaged run time of FrwCB seems to scale linearly (or close to that) in the number of variables of the instance.

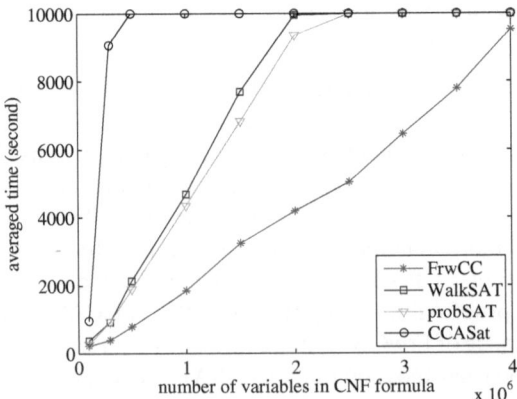

Fig. 1. Averaged time of FrwCB and other competitors on the huge random 3-SAT instances. Swqcc fails to solve any instance in this benchmark, so we do not report its results.

4.4   Results on Random 5-SAT and 7-SAT Instances

Table 4 reports experimental results on the benchmark of random 5-SAT and 7-SAT instances, where each solver is performed one run for each instance with a cutoff time of 2000 seconds. On the 5-SAT and 7-SAT instances, FrwCB is competitive with state-of-the-art solvers for random SAT. In a conclusion, FrwCB shows promising performance on random k-SAT instances with $k > 3$. We believe that the performance of FrwCB on these instances can be improved by tuning the parameter p.

Table 4. Comparative results on 5-SAT and 7-SAT instances. The number of total runs on each instance class is 250.

Instance Class	WalkSAT		probSAT		CCASat		Swqcc		FrwCB	
	#suc	avg time	#suc	avg time	#suc	avg time	#suc	avg time	#suc	avg time
5SAT-v500	250	17.7	250	9.0	250	**7.0**	250	37.8	250	13.1
7SAT-v90	250	28.7	250	37.4	250	43.2	250	**14.4**	250	25.8

4.5   Results on Structured Instances

We compare FrwCB with WalkSAT, probSAT, CCASat, Swqcc and Sattime on a broad range of structured instances. Table 5 illustrates the results on structured instances, where each solver is performed 10 runs for each instance with a cutoff time of 2000 seconds. On these structured instances, although FrwCB performs worse than Sattime on structured instances, it does show improvement over WalkSAT, probSAT, CCASat and Swqcc, especially on frb instances, indicating that the CCBM heuristic does improve the FRW algorithms on structured instances.

We also note that Sattime performs pre-process before local search, which is helpful for solving structured instances. To investigate the influence of pre-process on par16 instances, we run the pre-processor in lingeling[12] [4] to simplify the instances. It turns out that the simplified par16 instances can be solved by WalkSAT, CCASat, Swqcc and FrwCB (probSAT is able to solve 4 instances), compared to the fact that all solvers but Sattime cannot solve the original instances. Specially, the averaged time of FrwCB on solving the simplified par16 instances is 13.2 seconds, while those of WalkSAT, probSAT, CCASat, Swqcc and Sattime are 102.6 seconds, 1294.4 seconds, 90.8 seconds, 18.9 seconds and 2.3 seconds, respectively.

4.6   Results of SP+FrwCB on Random 3-SAT Instances with 10^7 Variables

Although SP exhibits the best performance on random 3-SAT instances, it needs to call an SLS solver (such as WalkSAT) to solve the sub-formula after it simplifies the original formula. We also perform some experiments to show the good cooperation of SP and FrwCB. As reported in [5], SP calls WalkSAT to solve the simplified formula, and we refer to this hybrid solver as SP+WalkSAT. We replace WalkSAT with probSAT, CCASat, Swqcc and FrwCB, and then obtain four new hybrid solvers which are referred to as

---

[12] http://fmv.jku.at/lingeling/
lingeling-ala-b02aa1a-121013.tar.gz

Table 5. Comparative performance results on the structured instances

Instance Class	#inst.	WalkSAT #suc avg time	probSAT #suc avg time	CCASat #suc avg time	Swqcc #suc avg time	Sattime #suc avg time	FrwCB #suc avg time
ais	4	40 0.3	40 2.6	40 <0.1	40 <0.1	40 0.3	40 0.3
blocksworld	7	70 2.8	70 1.2	70 0.6	63 214.7	70 0.5	70 1.4
gcp	4	40 2.7	21 984.6	40 61.5	32 483.5	40 2.2	40 1.3
jnh	16	160 0.11	160 <0.1	160 <0.1	160 <0.1	160 <0.1	160 <0.1
logistics	4	40 0.3	40 0.1	40 <0.1	40 <0.1	40 <0.1	40 0.2
par8	5	50 8.0	50 23.5	50 16.8	50 4.6	50 <0.1	50 2.1
par8-c	5	50 <0.1	50 <0.1	50 <0.1	50 <0.1	50 <0.1	50 <0.1
par16	5	0 >2000	0 >2000	0 >2000	0 >2000	50 23.7	0 >2000
par16-c	5	6 1898.4	50 264.9	45 310.9	50 161.6	50 16.0	50 99.4
frb50-23	5	42 553.6	38 765.0	40 571.6	39 744.2	49 248.7	47 332.8
frb53-24	5	35 880.4	19 1598.2	34 990.6	20 1550.5	47 539.7	46 624.0
frb56-25	5	34 992.6	24 1364.3	30 1137.8	15 1555.8	43 656.7	40 739.1
frb59-26	5	17 1555.5	10 1697.8	15 1627.6	11 1771.6	34 1094.9	27 1317.2

Table 6. Comparative results on the extremely huge random 3-SAT instances ($r = 4.2$, $\#var = 10^7$). 'a.t.sls' means the averaged time of the SLS component in SP+SLS hybrid solver.

# Total Runs	SP+WalkSAT #suc a.t.sls		SP+probSAT #suc a.t.sls		SP+CCASat #suc a.t.sls		SP+Swqcc #succ a.t.sls		SP+FrwCB #suc a.t.sls	
20	20	810.5	20	685.2	0	>2000	0	>2000	20	**251.7**

SP+probSAT, SP+CCASat, SP+Swqcc and SP+FrwCB, respectively. The results of all the five SP+SLS hybrid solvers on 20 extremely huge random 3-SAT instances ($r = 4.2$, $\#var = 10^7$) are summarized in Table 6, where each solver is performed one run for each of the 20 instances. Each SLS solver is given 2000 seconds to solve each sub-formula simplified by SP using the parameter $f = 5\%$ (with 6320 seconds on average).

To solve the simplified instances, the averaged time of FrwCB is 251.7 seconds, while those of WalkSAT and probSAT are 810.5 and 685.2 seconds, respectively. Also, CCASat and Swqcc fail to solve all these simplified instances within the cutoff time.

The results show that FrwCB cooperates well with SP, and SP+FrwCB is the best solver for huge random 3-SAT instances (with $r = 4.2$) to the best of our knowledge.

5   Discussions

In this section, we analyze effectiveness of each component adopted in the CCBM heuristic; we also discuss the differences between the CCBM heuristic and the QCC heuristic in Swqcc, both of which are based on clause states.

5.1   Effectiveness of Each Component in CCBM

There are three components in CCBM: the CCD component corresponding to the CCD level, the BM component choosing a BM variable of a random unsatisfied clause in the PROB level, and the diversification component based on *ConfTimes*. We perform three additional experiments to analyze the effectiveness of these components. Our analysis suggests that the good performance of FrwCB is mainly due to the first two components.

Effectiveness of the CCD Component: Recalling that FrwCB works in two levels, i.e., the CCD level and the PROB level. We modify FrwCB to obtain an alternative algorithm Frw1 which works without the CCD level, that is, deleting lines 7 and 8 in Algorithm 1. The probability p is set to 0.86 for Frw1 according to some tunings. We run Frw1 on the first benchmark, one time for each instance within 2000 seconds, and it fails to solve any instance when $\#var \geqslant 15000$, indicating the importance of the CCD component.

Effectiveness of the BM Component: By removing the BM component (deleting lines 9 and 10 in Algorithm 1), we obtain another degenerating algorithm Frw2, which differs from FrwCB only in the PROB level. Specifically, in the PROB level of Frw2, it always selects the variable with the greatest *ConfTimes* in the chosen unsatisfied clause to flip. We test Frw2 on the first benchmark, one time for each instance within 2000 seconds. Our experimental results show that Frw2 fails to solve any instance from the first benchmark, demonstrating the significance of the BM component.

Effectiveness of the Diversification Component Based on *ConfTimes*: We replace the diversification component based on *ConfTimes* in FrwCB with the one based on age, resulting in the third alternative algorithm Frw3. Frw3 differs from FrwCB in lines 8, 10 and 12 in Algorithm 1: in line 8, Frw3 selects the variable x with the greatest $score(x)$ appearing in *CCDVars(c)* to flip, breaking ties by preferring the least recently flipped one; in line 10, Frw3 selects the least recently flipped variable in *BMVars(c)*; in line 12, Frw3 selects the least recently flipped variable in clause c. We set the parameter p to 0.6 for Frw3 according to some tunings. We test Frw3 on the first benchmark, one time for each instance within 2000 seconds. The results show that Frw3 succeeds in all runs, and the averaged time of Frw3 is 69 seconds. These observations suggest that the performance of Frw3 is comparable with that of FrwCB (at least on the large random 3-SAT benchmark from SAT Competition 2011). This also indicates that the CCBM heuristic is a general heuristic and can work well with different diversification strategies, presenting its good extendibility.

5.2 Experimental Analysis of QCC and CCBM on Focused Random Walk

In this subsection, we perform some experiments to analyze the performance of the QCC and CCBM heuristics on FRW algorithms.

If we apply the QCC heuristic to design an FRW algorithm, we would obtain the Frw2 algorithm. As mentioned in the preceding subsection, Frw2 works as follows. It first picks an unsatisfied clause c. If there exist CCD variables in c, it flips the one with the greatest *score*, corresponding to the greedy mode; otherwise, it picks the variable with the greatest *ConfTimes* in c, corresponding to the random mode.

Now we can see the differences between QCC and CCBM heuristics on FRW algorithms concerning about their behaviors when no CCD variable exists. In this situation, QCC simply selects the variable with the greatest *ConfTimes* in an unsatisfied clause, whilst CCBM adopts a hybrid strategy which selects the variable with the minimum *break* or the one with the greatest *ConfTimes* in the unsatisfied clause. While picking the variable with the greatest *ConfTimes* contributes to diversification, picking the one with the minimum *break* is quite greedy.

This hybrid strategy plays a key role in CCBM, as the CC strategy based on clause states is too strict for FRW algorithms. We test FrwCB on 20 random 3-SAT instances with 0.1 million variables. For each instance, FrwCB is performed 5 runs within 2000 seconds. The experiments are summarized in Table 7 and show that for FrwCB, only in approximately 20% steps exist CCD variables. Therefore, if the algorithm always picks the variable with the greatest *ConfTimes* when no CCD variable exists, it would bias too much towards diversification, which is not reasonable. By employing the hybrid strategy, FrwCB strikes a good balance between diversification and intensification.

As mentioned in the preceding subsection, FrwCB performs significantly better than its alternative algorithm Frw2, which directly uses the QCC heuristic. Actually, Frw2 fails to solve any random 3-SAT instance from SAT Competition 2011. To the best of our knowledge, the CCBM heuristic, which is carefully designed for FRW algorithms, is the only CC heuristic that can be used to improve FRW algorithms.

Table 7. Frequencies of each type of search steps in FrwCB

Instance Class	CCD		BM		Diversification	
$(1.0M = 10^6)$	avg step	freq	avg step	freq	avg step	freq
3SAT-v0.1M	23251674	19.6%	57479605	48.3%	38203681	32.1%

6 Conclusions and Future Work

We proposed an effective heuristic named CCBM which combines two strategies namely configuration checking and break minimum to improve FRW algorithms. We utilized CCBM to develop a new SLS algorithm called FrwCB. It is the first time to apply CC in the FRW framework. According to the experimental results, FrwCB significantly outperforms state-of-the-art solvers including WalkSAT, probSAT, CCASat and Swqcc on huge random 3-SAT instances with up to 4 million variables. Our experiments indicate that the run time of FrwCB seems to scale linearly in the number of variables of an instance. Also, FrwCB cooperates well with SP on random 3-SAT instances with 10 million variables.

Moreover, the experiments show that FrwCB performs better than its competitors on random 3-SAT instances with different clause-to-variable ratios from SAT Challenge 2012, indicating its robustness. The robustness of FrwCB is further confirmed by the experimental results on structured instances, where FrwCB performs better than WalkSAT, probSAT, CCASat and Swqcc, and is competitive with Sattime on structured instances.

Some preliminary experiments also suggest that CCBM cooperates well with other diversification strategies, such as *age*. Our algorithm FrwCB cannot handle industrial instances well, and we would like to improve the performance of FrwCB on industrial instances.

Acknowledgements. This work is partially supported by 973 Program 2010CB328103, ARC Future Fellowship FT0991785, National Natural Science Foundation of China (61073033, 61003056 and 60903054), and Fundamental Research Funds for the Central Universities of China (21612414). The authors would like to thank the anonymous referees for their helpful comments in this paper.

References

1. Aurell, E., Gordon, U., Kirkpatrick, S.: Comparing beliefs, surveys and random walks. Advances in Neural Information Processing Systems 17, 49–56 (2005)
2. Balint, A., Fröhlich, A.: Improving stochastic local search for SAT with a new probability distribution. In: Strichman, O., Szeider, S. (eds.) SAT 2010. LNCS, vol. 6175, pp. 10–15. Springer, Heidelberg (2010)
3. Balint, A., Schöning, U.: Choosing probability distributions for stochastic local search and the role of make versus break. In: Cimatti, A., Sebastiani, R. (eds.) SAT 2012. LNCS, vol. 7317, pp. 16–29. Springer, Heidelberg (2012)
4. Biere, A.: Lingeling and friends at the SAT competition 2011. Technical Report FMV Reports Series, Institute for Formal Models and Verification, Johannes Kepler University (2011), http://fmv.jku.at/papers/Biere-FMV-TR-11-1.pdf
5. Braunstein, A., Mézard, M., Zecchina, R.: Survey propagation: An algorithm for satisfiability. Random Struct. Algorithms 27(2), 201–226 (2005)
6. Brüggemann, T., Kern, W.: An improved deterministic local search algorithm for 3-SAT. Theor. Comput. Sci. 329(1-3), 303–313 (2004)
7. Cai, S., Su, K.: Local search with configuration checking for SAT. In: Proc. of ICTAI 2011, pp. 59–66 (2011)
8. Cai, S., Su, K.: Configuration checking with aspiration in local search for SAT. In: Proc. of AAAI 2012, pp. 434–440 (2012)
9. Cai, S., Su, K., Luo, C., Sattar, A.: NuMVC: An efficient local search algorithm for minimum vertex cover. J. Artif. Intell. Res (JAIR) 46, 687–716 (2013)
10. Cai, S., Su, K., Sattar, A.: Local search with edge weighting and configuration checking heuristics for minimum vertex cover. Artif. Intell. 175(9-10), 1672–1696 (2011)
11. Guo, W., Yang, G., Hung, W.N.N., Song, X.: Complete boolean satisfiability solving algorithms based on local search. J. Comput. Sci. Technol. 28(2), 247–254 (2013)
12. Hoos, H.H.: An adaptive noise mechanism for WalkSAT. In: Proc. of AAAI 2002, pp. 655–660 (2002)

13. Hoos, H.H., Stützle, T.: Systematic vs. local search for SAT. In: Proc. of KI 1999, pp. 289–293 (1999)
14. Hutter, F., Tompkins, D.A.D., Hoos, H.H.: Scaling and probabilistic smoothing: Efficient dynamic local search for SAT. In: Van Hentenryck, P. (ed.) CP 2002. LNCS, vol. 2470, pp. 233–248. Springer, Heidelberg (2002)
15. Järvisalo, M., Biere, A., Heule, M.: Simulating circuit-level simplifications on CNF. J. Autom. Reasoning 49(4), 583–619 (2012)
16. Kirkpatrick, S., Selman, B.: Critical behavior in the satisfiability of random boolean formulae. Science 264, 1297–1301 (1994)
17. Lewis, M.D.T., Schubert, T., Becker, B.W.: Speedup techniques utilized in modern SAT solvers. In: Bacchus, F., Walsh, T. (eds.) SAT 2005. LNCS, vol. 3569, pp. 437–443. Springer, Heidelberg (2005)
18. Li, C.-M., Huang, W.Q.: Diversification and determinism in local search for satisfiability. In: Bacchus, F., Walsh, T. (eds.) SAT 2005. LNCS, vol. 3569, pp. 158–172. Springer, Heidelberg (2005)
19. Li, C.M., Li, Y.: Satisfying versus falsifying in local search for satisfiability. In: Cimatti, A., Sebastiani, R. (eds.) SAT 2012. LNCS, vol. 7317, pp. 477–478. Springer, Heidelberg (2012)
20. Luo, C., Su, K., Cai, S.: Improving local search for random 3-SAT using quantitative configuration checking. In: Proc. of ECAI 2012, pp. 570–575 (2012)
21. Papadimitriou, C.H.: On selecting a satisfying truth assignment. In: Proc. of FOCS 1991, pp. 163–169 (1991)
22. Parkes, A.J.: Scaling properties of pure random walk on random 3-SAT. In: Van Hentenryck, P. (ed.) CP 2002. LNCS, vol. 2470, pp. 708–713. Springer, Heidelberg (2002)
23. Selman, B., Kautz, H.A., Cohen, B.: Noise strategies for improving local search. In: Proc. of AAAI 1994, pp. 337–343 (1994)
24. Selman, B., Levesque, H.J., Mitchell, D.G.: A new method for solving hard satisfiability problems. In: Proc. of AAAI 1992, pp. 440–446 (1992)
25. Tompkins, D.A.D., Balint, A., Hoos, H.H.: Captain jack: New variable selection heuristics in local search for SAT. In: Sakallah, K.A., Simon, L. (eds.) SAT 2011. LNCS, vol. 6695, pp. 302–316. Springer, Heidelberg (2011)
26. Wei, W., Li, C.-M., Zhang, H.: Switching among non-weighting, clause weighting, and variable weighting in local search for SAT. In: Stuckey, P.J. (ed.) CP 2008. LNCS, vol. 5202, pp. 313–326. Springer, Heidelberg (2008)
27. Wu, Z., Wah, B.W.: An efficient global-search strategy in discrete lagrangian methods for solving hard satisfiability problems. In: Proc. of AAAI 2000, pp. 310–315 (2000)
28. Xu, K., Boussemart, F., Hemery, F., Lecoutre, C.: A simple model to generate hard satisfiable instances. In: Proc. of IJCAI 2005, pp. 337–342 (2005)

Multi-Objective Constraint Optimization with Tradeoffs

Radu Marinescu[1], Abdul Razak[2], and Nic Wilson[2]

[1] IBM Research, Ireland
radu.marinescu@ie.ibm.com
[2] Cork Constraint Computation Centre
University College Cork, Ireland
{a.razak,n.wilson}@4c.ucc.ie

Abstract. In this paper, we consider the extension of multi-objective constraint optimization algorithms to the case where there are additional tradeoffs, reducing the number of optimal solutions. We focus especially on branch-and-bound algorithms which use a mini-buckets algorithm for generating the upper bound at each node (in the context of maximizing values of objectives). Since the main bottleneck of these algorithms is the very large size of the guiding upper bound sets we introduce efficient methods for reducing these sets, yet still maintaining the upper bound property. We also propose much faster dominance checks with respect to the preference relation induced by the tradeoffs. Furthermore, we show that our tradeoffs approach which is based on a preference inference technique can also be given an alternative semantics based on the well known Multi-Attribute Utility Theory. Our comprehensive experimental results on common multi-objective constraint optimization benchmarks demonstrate that the proposed enhancements allow the algorithms to scale up to much larger problems than before.

1 Introduction

Multi-objective Constraint Optimization (MOCOP) is a general framework that can be used to model many real-world problems involving multiple, conflicting and sometimes non-commensurate objectives that need to be optimized simultaneously. The solution space of these problems is typically only partially ordered and can contain many non-inferior or undominated solutions which must be considered equally good in the absence of information concerning the relevance of each objective relative to the others.

Solutions are compared on more than one (real-valued) objective, so that each complete assignment to the decision variables has an associated multi-objective utility value, represented by a vector in $\mathbb{R}^p$, where p is the number of objectives. The utility vectors associated to solutions can be compared using the Pareto ordering. However, the Pareto ordering is rather weak, which can lead to the Pareto-undominated set becoming too large for the decision maker (DM) to handle. At the other extreme, it can be undesirable to force the decision maker to define precise tradeoffs between objectives, since they may have no clear idea about them, and it may lead to somewhat arbitrary decisions.

The approach we take, based on that of [1], is to allow as input a number of preferences between utility vectors, which may come e.g., from a brief elicitation procedure. These input preferences are used to strengthen the preference relation over $\mathbb{R}^p$; even a small number of such tradeoffs can greatly reduce the size of the undominated set.

C. Schulte (Ed.): CP 2013, LNCS 8124, pp. 497–512, 2013.
© Springer-Verlag Berlin Heidelberg 2013

In this paper, we extend MOCOP algorithms for the case with tradeoffs, including both branch-and-bound and variable elimination algorithms. The branch-and-bound algorithms perform a depth-first traversal of an AND/OR search tree that captures the underlying structure of the problem, and use a mini-buckets algorithm for generating an upper bound, which is a set of utility vectors, at each node of the search tree. The main contributions of the paper are as follows. First, the guiding upper bound sets can become quite large and therefore can have a dramatic impact on the performance of the search algorithms. To remedy this issue, we propose efficient methods for reducing these sets, yet still ensuring the upper bound property for both the Pareto and tradeoffs case. Second, the MOCOP algorithms need to make many dominance checks with respect to the preference relation induced by the tradeoffs. For computational efficiency we compile this dominance check by use of a matrix and show that in practice it can achieve almost an order of magnitude speed up over the current approach of [1] which is based on solving a linear program. Third, we show that our approach for handling tradeoffs can be given an alternate semantics based on the well known Multi-Attribute Utility Theory. Fourth, we show empirically on a variety of MOCOP benchmarks that our improved algorithms outperform the current state-of-the-art solvers by a significant margin and therefore they can scale up to much larger problems than before.

Following background on MOCOPs and on AND/OR search spaces for MOCOPs (Section 2), Section 3 defines the formalism for tradeoffs and shows how the induced notion of preference domination can be computed. Section 4 describes our proposed methods for reducing the size of the upper bound sets used by branch-and-bound search algorithms. Section 5 presents our empirical evaluation. Section 6 overviews related work, while Section 7 provides a summary and concluding remarks.

2 Background

2.1 Multi-objective Constraint Optimization

Consider an optimization problem with p objectives. A *utility vector* $\boldsymbol{u} = (u_1, \ldots, u_p)$ is a vector with p components where each component $u_i \in \mathbb{R}$ represents the utility (or value) with respect to objective $i \in \{1, \ldots, p\}$. We assume the standard pointwise arithmetic operations, namely $\boldsymbol{u} + \boldsymbol{v} = (u_1 + v_1, \ldots, u_p + v_p)$ and $q \times \boldsymbol{u} = (q \times u_1, \ldots, q \times u_p)$, where $q \in \mathbb{R}$ is a real-valued scalar.

A *Multi-objective Constraint Optimization Problem* (MOCOP) is a tuple $\mathcal{M} = \langle \mathbf{X}, \mathbf{D}, \mathbf{F} \rangle$, where $\mathbf{X} = \{X_1, \ldots, X_n\}$ is a set of decision variables having finite domains of values $\mathbf{D} = \{D_1, \ldots, D_n\}$ and $\mathbf{F} = \{f_1, \ldots, f_r\}$ is a set of utility functions.[1] A utility function $f_i(Y) \in \mathbf{F}$ is defined over a subset of variables $Y \subseteq \mathbf{X}$, called its *scope*, and associates a utility vector to each assignment of Y. The objective function is $\mathcal{F}(\mathbf{X}) = \sum_{i=1}^{r} f_i(Y_i)$. A *solution* is a complete assignment $\bar{x} = (x_1, \ldots, x_n)$ and is characterized by a utility vector $\boldsymbol{u} = \mathcal{F}(\bar{x})$. Therefore, the comparison of solutions reduces to that of their corresponding p-dimensional vectors.

We are interested in partial orders $\succcurlyeq$ on $\mathbb{R}^p$ satisfying the following two monotonicity properties, where $\boldsymbol{u}, \boldsymbol{v}, \boldsymbol{w} \in \mathbb{R}^p$ are arbitrary vectors:

---

[1] Since we are expressing the optimization in terms of maximizing rather than minimizing, we use the terminology *utility function/vector* as opposed to *cost function/vector*.

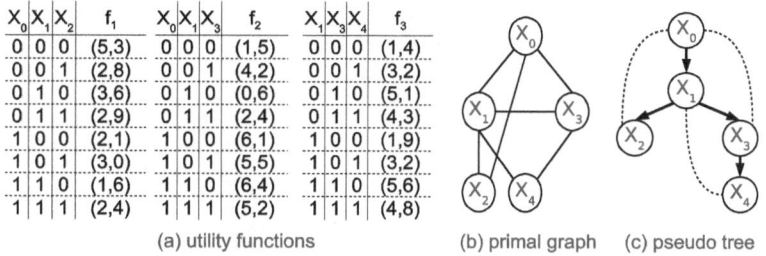

X_0	X_1	X_2	f_1
0	0	0	(5,3)
0	0	1	(2,8)
0	1	0	(3,6)
0	1	1	(2,9)
1	0	0	(2,1)
1	0	1	(3,0)
1	1	0	(1,6)
1	1	1	(2,4)

X_0	X_1	X_3	f_2
0	0	0	(1,5)
0	0	1	(4,2)
0	1	0	(0,6)
0	1	1	(2,4)
1	0	0	(6,1)
1	0	1	(5,5)
1	1	0	(6,4)
1	1	1	(5,2)

X_1	X_3	X_4	f_3
0	0	0	(1,4)
0	0	1	(3,2)
0	1	0	(5,1)
0	1	1	(4,3)
1	0	0	(1,9)
1	0	1	(3,2)
1	1	0	(5,6)
1	1	1	(4,8)

(a) utility functions   (b) primal graph   (c) pseudo tree

Fig. 1. A MOCOP instance with 2 objectives

Independence: if $u \succcurlyeq v$ then $u + w \succcurlyeq v + w$;
Scale-Invariance: if $u \succcurlyeq v$ and $q \in \mathbb{R}$, $q \geq 0$ then $q \times u \succcurlyeq q \times v$.

An important example of such a partial order is the weak Pareto order, defined as follows.

Definition 1 (weak Pareto order). *Let $u, v \in \mathbb{R}^p$ such that $u = (u_1, \ldots, u_p)$ and $v = (v_1, \ldots, v_p)$. We define the binary relation $\geq$ on $\mathbb{R}^p$ by $u \geq v \iff \forall i \in \{1, \ldots, p\}, u_i \geq v_i$.*

Given $u, v \in \mathbb{R}^p$, if $u \succcurlyeq v$ then we say that u *dominates* v. As usual, the symbol $\succ$ refers to the asymmetric part of $\succcurlyeq$, namely $u \succ v$ if and only if $u \succcurlyeq v$ and it is not the case that $v \succcurlyeq u$. Given finite sets $U, V \subseteq \mathbb{R}^p$, we say that U *dominates* V, denoted $U \succcurlyeq V$, if $\forall v \in V \exists u \in U$ such that $u \succcurlyeq v$. In particular, relation $\geq$ (resp. $>$) is also called *weak Pareto dominance* (resp. *Pareto dominance*).

Definition 2 (maximal/Pareto set). *Given a partial order $\succcurlyeq$ and a finite set of vectors $U \subseteq \mathbb{R}^p$, we define the maximal set, denoted by $\max_{\succcurlyeq}(U)$, to be the set consisting of the undominated elements in U, i.e., $\max_{\succcurlyeq}(U) = \{u \in U \mid \nexists v \in U, v \succ u\}$. When $\succcurlyeq$ is the weak Pareto ordering $\geq$, we call $\max_{\succcurlyeq}(U)$ the Pareto set (or Pareto frontier).*

Solving a MOCOP instance means finding the set of optimal solutions that generate maximal utility vectors, namely values in the set $\max_{\succcurlyeq}\{\mathcal{F}(\bar{x}) \mid \text{solution } \bar{x}\}$.

Given a MOCOP instance $\mathcal{M} = \langle \mathbf{X}, \mathbf{D}, \mathbf{F} \rangle$, the scopes of the utility functions in $\mathbf{F}$ imply a *primal graph* G (nodes correspond to the variables and edges connect any two nodes whose variables belong to the same function) with certain *induced width* [2].

Example 1. Figure 1 shows a MOCOP instance with 5 bi-valued variables $\{X_0, X_1, X_2, X_3, X_4\}$ and 3 ternary utility functions $f_1(X_0, X_1, X_2)$, $f_2(X_0, X_1, X_3)$, and $f_3(X_1, X_3, X_4)$, respectively. Its corresponding primal graph is depicted in Figure 1(b). The Pareto set of the problem contains 8 solutions with undominated utility vectors: (3,24), (8,21), (9,19), (10,16), (11,14), (12,12), (13,8) and (14,6), respectively.

2.2   AND/OR Search Spaces for MOCOPs

The concept of AND/OR search spaces for graphical models [3] has been extended recently to multi-objective constraint optimization to better capture the problem structure

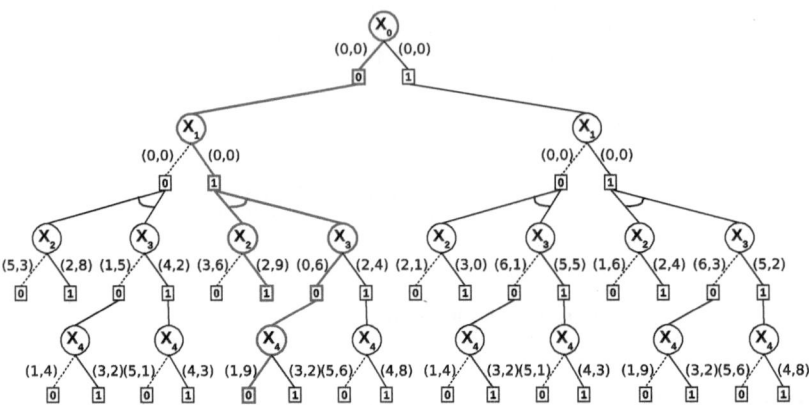

Fig. 2. Weighted AND/OR search tree

during search [4]. The search space is defined using a *pseudo tree* of the primal graph which captures problem decomposition as follows.

Definition 3 (pseudo tree). *A* pseudo tree *of an undirected graph $G = (V, E)$ is a directed rooted tree $\mathcal{T} = (V, E')$, such that every arc of G not included in E' is a back-arc in $\mathcal{T}$, namely it connects a node in $\mathcal{T}$ to an ancestor in $\mathcal{T}$. The arcs in E' may not all be included in E.*

Weighted AND/OR Search Tree. Given a MOCOP instance $\mathcal{M} = \langle \mathbf{X}, \mathbf{D}, \mathbf{F} \rangle$, its primal graph G and a pseudo tree $\mathcal{T}$ of G, the associated AND/OR search tree $S_{\mathcal{T}}$ has alternating levels of OR and AND nodes. Its structure is based on the underlying pseudo tree. The root node of $S_{\mathcal{T}}$ is an OR node labeled by the root of $\mathcal{T}$. The children of an OR node $\langle X_i \rangle$ are AND nodes labeled with value assignments $\langle X_i, x_j \rangle$; the children of an AND node $\langle X_i, x_j \rangle$ are OR nodes labeled with the children of X_i in $\mathcal{T}$, representing conditionally independent subproblems. The OR-to-AND arcs in $S_{\mathcal{T}}$ are annotated by *weights* derived from the input utility functions, while each node $n \in S_{\mathcal{T}}$ is associated with a *value* $v(n)$, defined as the set of utility vectors corresponding to the optimal solutions of the conditioned subproblem rooted at n. The node values can be computed recursively based on the values of their successors, as shown in [4].

The size of the AND/OR search tree associated with a MOCOP instance with pseudo tree of depth d is $O(n \cdot k^d)$, where n is the number of variables and k bounds the domain size [3]. Figure 2 shows the AND/OR search tree of the MOCOP instance from Figure 1, relative to the pseudo tree given in Figure 1(c). The utility vectors displayed on the OR-to-AND arcs are the weights corresponding to the input utility functions. An optimal solution tree corresponding to the assignment $(X_0 = 0, X_1 = 1, X_2 = 1, X_3 = 0, X_4 = 0)$ with utility vector $(3,24)$ is highlighted.

Multi-objective AND/OR Branch-and-Bound. One of the most effective methods for solving MOCOPs is the multi-objective AND/OR Branch-and-Bound (MOAOBB) introduced recently by [4]. For completeness, we next review the algorithm. As usual, each node n in the search tree is associated with a heuristic estimate $h(n)$ of $v(n)$.

Algorithm 1. MOAOBB

Data: $\mathcal{M} = \langle \mathbf{X}, \mathbf{D}, \mathbf{F} \rangle$, pseudo tree $\mathcal{T}$, heuristic h.
Result: Maximal set of $\mathcal{M}$, $\max_{\succcurlyeq}\{\mathcal{F}(\bar{x}) \mid \text{solution } \bar{x}\}$.

1 create an OR node s labeled by the root of $\mathcal{T}$
2 $OPEN \leftarrow \{s\}; CLOSED \leftarrow \emptyset$
3 **while** $OPEN \neq \emptyset$ **do**
4 $\quad$ move top node n from OPEN to CLOSED
5 $\quad$ expand n by creating its successors $succ(n)$
6 $\quad$ **foreach** $n' \in succ(n)$ **do**
7 $\quad\quad$ evaluate $h(n')$ and add n' on top of OPEN
8 $\quad\quad$ let T' be the current partial solution tree with tip n'
9 $\quad\quad$ **if** $v(s) \succcurlyeq f(T')$ **then**
10 $\quad\quad\quad$ remove n' from OPEN and $succ(n)$
11 $\quad$ **while** $\exists n \in CLOSED$ s.t. $succ(n) = \emptyset$ **do**
12 $\quad\quad$ remove n from CLOSED and let p be n's parent
13 $\quad\quad$ **if** p *is AND* **then** $v(p) \leftarrow v(p) + v(n)$
14 $\quad\quad$ **else** $v(p) \leftarrow \max_{\succcurlyeq}\{v(p) \cup \{w(p, n) + v(n)\}\}$
15 $\quad\quad$ remove n from $succ(p)$
16 **return** $v(s)$

MOAOBB is described by Algorithm 1. Assuming maximization of the objective values, MOAOBB traverses the weighted AND/OR search tree in a depth-first manner while maintaining at the root node s of the search tree the set $v(s)$ of best solution vectors found so far. During node expansion, the algorithm uses the $h(n)$ values to compute an upper bound set $f(T')$ on the set of optimal solutions extending the current partial solution tree T', and prunes the subproblem below the current tip node n' if $f(T')$ is dominated by $v(s)$ (i.e., all utility vectors in $f(T')$ are dominated by at least one element in $v(s)$). The node values are updated recursively in a bottom-up manner, starting from the terminal nodes in the search tree - AND nodes by summation, OR nodes by maximization (undominated closure wrt $\succcurlyeq$).

The time complexity of MOAOBB is bounded by $O(n \cdot k^d)$, the size of the weighted AND/OR search tree. Since the utility vectors are in $\mathbb{R}^p$, it is not easy to predict the size of the maximal set $\max_{\succcurlyeq}\{\mathcal{F}(\bar{x}) \mid \text{solution } \bar{x}\}$, and therefore MOAOBB may use prohibitively large amounts of memory to store it.

Mini-Bucket Heuristics. The heuristic function $h(n)$ that we used in our experiments is the *multi-objective mini-bucket heuristic* [4, 5]. It was shown that the intermediate functions generated by the mini-bucket approximation of multi-objective variable elimination can be used to derive a heuristic function that is *admissible*, namely in a maximization context it generates a set of utility vectors (upper bound set) such that the utility vectors of any optimal solution below node n in the search tree is dominated by some element in $h(n)$. A control parameter, called i-bound, allows a tradeoff between accuracy of the heuristic and its time-space requirements.

3   Handling Imprecise Tradeoffs

In this section, we present our approach for handling tradeoffs between the objectives of a MOCOP. We first introduce some notation. For $W \subseteq \mathbb{R}^p$, define $\mathbf{C}(W)$, *the positive convex cone generated by* W, to be the set consisting of all vectors $\boldsymbol{u}$ such that there exists $k \geq 0$ and non-negative real scalars $q_1, \ldots, q_k$ and $\boldsymbol{w}_j \in W$ with $\boldsymbol{u} \geq \sum_{j=1}^{k} q_j \boldsymbol{w}_j$, where $\geq$ is the weak Pareto relation (and an empty summation is taken to be equal to $\mathbf{0}$, the zero vector $(0, \ldots, 0)$ in $\mathbb{R}^p$). $\mathbf{C}(W)$ is the set of vectors that weakly-Pareto dominate some (finite) positive linear combination of elements of W.

For $W \subseteq \mathbb{R}^p$, define W^* to be the set of vectors $\boldsymbol{u} \in \mathbb{R}^p$ such that $\boldsymbol{u} \cdot \boldsymbol{v} \geq 0$ for all $\boldsymbol{v} \in W$, where $\boldsymbol{u} \cdot \boldsymbol{v}$ means $\sum_{i=1}^{p} u_i v_i$. A standard result for finitely generated convex cones (see e.g., [6]) states that for finite $W \subseteq \mathbb{R}^p$, W^{**} (i.e., $(W^*)^*$) equals $\mathbf{C}(W)$.

3.1   Deducing Preferences from Additional Inputs

We assume that we have learned some preferences of the decision maker (DM), i.e., a set Θ of pairs of the form $(\boldsymbol{u}, \boldsymbol{v})$ meaning that the decision maker prefers $\boldsymbol{u}$ to $\boldsymbol{v}$. We will use this input information to deduce further preferences, in two different ways, the first taken from [1], the second based on a multi-attribute utility theory model.

Partial Order-Based Inference: We will use the input preferences Θ to infer a preference relation $\succcurlyeq_\Theta$ extending Θ. Here we assume that the DM has some unknown partial order $\succcurlyeq$ over $\mathbb{R}^p$, representing their preferences. We further assume that $\succcurlyeq$ extends Pareto (i.e., extends the weak Pareto order), and satisfies Independence and Scale-Invariance. We are given the information that the DM's preference relation includes pairs Θ, i.e., $\succcurlyeq$ extends Θ. This naturally leads to the following definitions:

- We say that Θ *is consistent* if there exists some partial order $\succcurlyeq$ (on $\mathbb{R}^p$) that extends Θ, extends Pareto, and satisfies Scale-Invariance and Independence.
- For consistent Θ, we define the induced preference relation $\succcurlyeq_\Theta$ on $\mathbb{R}^p$ by $\boldsymbol{u} \succcurlyeq_\Theta \boldsymbol{v}$ $\iff \boldsymbol{u} \succcurlyeq \boldsymbol{v}$ for all partial orders $\succcurlyeq$ such that $\succcurlyeq$ extends Pareto and Θ, and satisfies Independence and Scale-Invariance.

MAUT-Based Inference: Here we take a different approach: we assume that the DM uses a weighted sum of the objectives to compare objective vectors, as in the additive form of the Multi-attribute Utility Theory (MAUT) model [7]. We say that pre-order $\succcurlyeq$ on $\mathbb{R}^p$ is MAUT-based if there exists some vector $\boldsymbol{w} \in \mathbb{R}^p$ with only non-negative values, such that, for all $\boldsymbol{u}, \boldsymbol{v} \in \mathbb{R}^p$, $\boldsymbol{u} \succcurlyeq \boldsymbol{v} \iff \sum_{i=1}^{p} u_i w_i \geq \sum_{i=1}^{p} v_i w_i$. If we knew the weights vector $\boldsymbol{w}$, the problem would reduce to a single-objective problem. However, all we know is that the induced preference relation satisfies the pairs Θ. This leads to the induced preference relation $\succcurlyeq_\Theta^2$ defined as follows:

$$\boldsymbol{u} \succcurlyeq_\Theta^2 \boldsymbol{v} \text{ if and only if } \boldsymbol{u} \succcurlyeq \boldsymbol{v} \text{ holds for all MAUT-based } \succcurlyeq \text{ extending } \Theta.$$

Thus $\boldsymbol{u}$ is preferred to $\boldsymbol{v}$ if and only if the preference holds for every compatible MAUT ordering.

Let $W_\Theta = \{\boldsymbol{u} - \boldsymbol{v} : (\boldsymbol{u}, \boldsymbol{v}) \in \Theta\}$. The following result from [1] gives a characterization of the first induced preference relation $\succcurlyeq_\Theta$.

Proposition 1. *Let Θ be a consistent set of pairs of vectors in $\mathbb{R}^p$. Then $u \succeq_\Theta v$ if and only if $u - v \in C(W_\Theta)$.*

In fact, for finite consistent Θ, the two induced preference relations are equal:

Proposition 2. *For finite Θ, we have $u \succ_\Theta^2 v \iff u - v \in C(W_\Theta)$. Therefore, if Θ is consistent then the relations $\succeq_\Theta$ and $\succ_\Theta^2$ are equal.*

Proof. Any MAUT-based order $\succeq$ has an associated vector w, so we write $\succeq$ as $\succeq_w$. We then have $u \succ_w v \iff (u - v) \cdot w \geq 0$. Relation $\succeq_w$ extends Θ if and only if, for all $u \in W_\Theta$, $w \cdot u \geq 0$, i.e., iff $w \in (W_\Theta)^*$. Thus $u \succ_\Theta^2 v \iff (u - v) \cdot w \geq 0$ for all $w \in (W_\Theta)^*$, i.e., iff $u - v \in (W_\Theta)^{**}$. Since $(W_\Theta)^{**} = C(W_\Theta)$, the result follows.

Example 2. Suppose that the decision maker has told us that she prefers $(0, 1)$ to $(1, 0)$, so that a unit of the second objective is considered more valuable than a unit of the first objective. Θ is then equal to $\{((0, 1), (1, 0))\}$. The Independence property implies $(0, 0) \succ_\Theta (1, -1)$, and also, for example, $(8, 21) \succ_\Theta (9, 20)$, and thus $(8, 21) \succeq_\Theta (9, 19)$ since $\succeq_\Theta$ extends Pareto. In Example 1 the additional preference implies that $(3, 24)$ and $(8, 21)$ are the only $\succ_\Theta$-undominated solutions, illustrating that even a single tradeoff can greatly reduce the number of undominated solutions.

3.2 Implementing Dominance Tests

For multi-objective constraint optimization, we will need to make many dominance checks with respect to the induced preference relation $\succeq_\Theta$. In [1], a dominance check is performed using a linear programming (LP) solver, based on Proposition 1. For computational efficiency we compile this dominance check by use of a matrix. Specifically, we generate a matrix $\mathbf{A}$ that represents $\succeq_\Theta$ in the sense that $u \succeq_\Theta v$ if and only if $\mathbf{A}(u - v) \geq \mathbf{0}$ (which is if and only if $\mathbf{A}u \geq \mathbf{A}v$). Lemma 1 below shows that we can construct such a matrix $\mathbf{A}$ by finding a generating set of the dual cone $(C(W_\Theta))^*$, where $(C(W_\Theta))^*$ is the set of vectors u in $\mathbb{R}^p$ such that $\sum_{i=1}^p u_i v_i \geq 0$ for all $v \in C(W_\Theta)$. We use the approach from [8] for this task.

Lemma 1. *Matrix A represents $\succeq_\Theta$ if and only if the dual cone $(C(W_\Theta))^*$ is equal to the cone generated by the rows of A, i.e., every row of A is in $(C(W_\Theta))^*$ and every element of $(C(W_\Theta))^*$ is a positive convex combination of rows of A.*

Proof. Let R be the set of rows of matrix $\mathbf{A}$. We have $u \in R^*$ if and only if $\mathbf{A}u \geq 0$. Abbreviate $C(W_\Theta)$ to $\mathbf{C}$. It easily follows from Proposition 1, that matrix $\mathbf{A}$ represents $\succeq_\Theta$ if and only if the following equivalence holds for all vectors $u \in \mathbb{R}^p$: $u \in \mathbf{C} \iff \mathbf{A}u \geq 0$. Thus $\mathbf{A}$ represents $\succeq_\Theta$ if and only $\mathbf{C} = R^*$. Now, $\mathbf{C} = R^*$ implies that $\mathbf{C}^* = R^{**} = \mathbf{C}(R)$. Conversely, if $\mathbf{C}^* = \mathbf{C}(R)$ then $\mathbf{C} = \mathbf{C}^{**} = (\mathbf{C}(R))^* = R^*$. Therefore, $\mathbf{A}$ represents $\succeq_\Theta$ if and only if $\mathbf{C}^* = \mathbf{C}(R)$.

Figure 3 compares the proposed matrix based dominance checks against the LP based ones on random multi-objective influence diagrams from [1]. These problems have 5 decisions, an increasing number of chance variables, and involve 3 and 5 objectives, respectively. Each data point represents an average over the number of solved instances by both methods out of 100 instances generated for the respective problem size. We can see that the former method clearly dominates the latter across all reported problem sizes, and in some cases it can achieve almost one order of magnitude speedup.

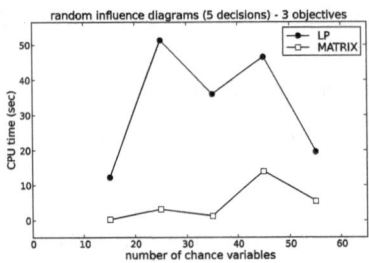
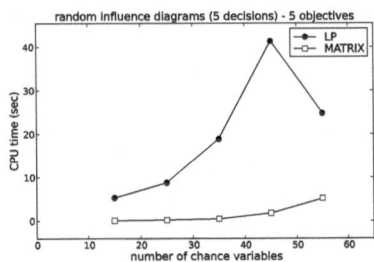

Fig. 3. Comparing matrix based versus LP based dominance checks on problems with 3 and 5 objectives. CPU time in seconds as a function of problem size. Time limit 20 minutes.

4   Reducing the Upper Bound Sets

Branch-and-bound algorithms such as MOAOBB involve use of an upper bound set during search, i.e., a set of utility vectors in $\mathbb{R}^p$ such that the utility vector for every assignment below the current node is weakly dominated by some element in the set. The upper bound sets can grow quite large and thus their manipulation during search may become computationally very expensive. In this section, we will require the upper bound set to have restricted cardinality, at most B (≥ 1). We thus need a method for taking a larger upper bound set $\mathcal{U}$, with $|\mathcal{U}| > B$, and reducing it to have cardinality at most B, whilst maintaining its property of being an upper bound set at the node.

We do this by iteratively choosing a selection $v_1, \dots, v_k$ of elements and producing an element u that is an upper bound of all of them. Elements $v_1, \dots, v_k$ are then removed from the upper bound set, along with the elements that u dominates, and u is added. The new set is still an upper bound set. This gets repeated until we have $|\mathcal{U}| \leq B$. (For the last iteration we reduce the cluster size k to being $|\mathcal{U}| - B + 1$ to avoid excessive "overshooting", so as to achieve a final upper bound set with cardinality closer to B.) For the case when $B = 1$, we use a single iteration with $k = |\mathcal{U}|$.

We go into more detail for the Pareto and tradeoffs cases below. Both make use of the Pareto least upper bound v of vectors $v_1, \dots, v_k$, given by $v = \max_{j=1}^k v_j$, where the max is applied point-wise.

4.1   Pareto (No Tradeoffs) Case

We remove only two elements from $\mathcal{U}$ in each iteration. We choose randomly $v \in \mathcal{U}$, and find $w \in \mathcal{U}$ that minimizes the Manhattan distance from v, i.e., that minimizes $\sum_{i=1}^p |v_i - w_i|$, where v_i and w_i are the ith components of v and w, respectively. We add u, the Pareto least upper bound of v and w, to $\mathcal{U}$, and remove v and w from $\mathcal{U}$ and the elements that are dominated by u; this procedure is iterated until $|\mathcal{U}| \leq B$.

We also implemented a number of variations of this method, which seemed to perform slightly less well, including replacing every two elements of $\mathcal{U}$ with their Pareto least upper bound that are randomly selected, minimizing the Manhattan distance, maximizing the dot product value and maximizing the dot product value of the normalized vectors.

4.2   Tradeoffs Case

We assume a consistent set of inputs Θ, leading to a preference relation $\succeq_\Theta$ which we abbreviate to $\succeq$. As described above, we make use of matrix $\mathbf{A}$ to represent $\succeq$. The key step is to choose a selection of k elements in $\mathcal{U}$ and to generate an upper bound of them. We make use of the following result:

Proposition 3. *Let $v_1, \ldots, v_k$ be vectors in $\mathbb{R}^p$, and let $v = \max_{j=1}^k v_j$ be their Pareto least upper bound, and let $w = \max_{j=1}^k \mathbf{A} v_j$, (with $\max$ being applied pointwise in both cases). Then,*

(i) $v \succeq v_1, \ldots, v_k$, i.e., the Pareto least upper bound is an upper bound with respect to $\succeq$;

(ii) for $u \in \mathbb{R}^p$, $u \succeq v_1, \ldots, v_k \iff \mathbf{A} u \geq w$; and

(iii) $\mathbf{A} v \geq w$.

Proof. (i): Let j be an arbitrary element of $\{1, \ldots, k\}$. By definition, $v \geq v_j$. Since $\succeq$ extends Pareto, we have $v \succeq v_j$.
(ii): $u \succeq v_1, \ldots, v_k \iff$ for all $j = 1, \ldots, k$, $u \succeq v_j$, $\iff$ for all $j = 1, \ldots, k$, $\mathbf{A} u \geq \mathbf{A} v_j \iff \mathbf{A} u \geq \max_{j=1}^k \mathbf{A} v_j$.
(iii) follows immediately from (i) and (ii).

Our first approach for generating an upper bound u of vectors $v_1, \ldots, v_k$ is to just use $u = v$, the Pareto least upper bound, which is an upper bound (w.r.t. $\succeq$) by Proposition 3(i). However, especially if $\succeq$ is much stronger than the Pareto ordering, we may be able to obtain a much tighter upper bound, leading to potentially much stronger pruning. To obtain an upper bound, we minimize objective function $\min \sum_i u_i$ (where u_i is the ith value of vector u), subject to $\mathbf{A} u \geq w$, plus the extra constraints that $u \leq \max_{j=1}^k v_j$, i.e., that u is weakly Pareto dominated by the Pareto least upper bound v. Proposition 3(iii) shows that the constraints are satisfiable, since v is a solution, and Proposition 3(ii) shows that the solution will be an upper bound.

Example 3. Continuing Example 2, suppose that we are wanting to replace the pair of utility vectors $\{(21, 3), (3, 15)\}$ by an upper bound. We have $(9, 15) \succeq_\Theta (21, 3), (3, 15)$ so that $(9, 15)$ is a much tighter upper bound than the Pareto least upper bound $(21, 15)$.

The cluster $v_1, \ldots, v_k$ is chosen randomly. We again tried a number of variations of this approach. This included (1) replacing an element and its $k - 1$ nearest neighbors (with respect to Manhattan distance) minimizing the sum of Manhattan distances (between the element and its neighbors) with their upper bound generated using the linear programming approach; (2) we set k to 2 and iteratively replace pairs of elements with their upper bound generated using the linear programming approach.

5   Experiments

In this section, we evaluate empirically the performance of the proposed improvements to branch-and-bound algorithms on problem instances derived from three classes of

MOCOP benchmarks: random networks, combinatorial auctions and vertex covering problems. All algorithms were implemented in C++ (32 bit) and the experiments were run on a 2.6GHz quad-core processor with 4 GB of RAM.

For our purpose, we consider the following random problem generators:

- **Random Networks:** Our random networks are characterized by parameters $\langle n, c \rangle$, where n is the number of variables and c is the number of binary utility functions. For consistency, we used similar parameters to [4] and generated random instances with $n \in [10, 160]$ and $c = 1.6n$ having 2, 3, 4 and 5 objectives. The components of the utility vectors were uniformly distributed between 0 and 10. The induced width of these problems ranged between 5 and 14, respectively.
- **Vertex Coverings:** Given a graph $G = (V, E)$, the task is to find a *vertex covering* $S \subseteq V$ such that $\forall (u, v) \in E$, either $u \in S$ or $v \in S$, and $F(S) = \sum_{v \in S} \boldsymbol{w}(v)$ is maximized, where $\boldsymbol{w}(v) = (w_1, \ldots, w_p)$ is a p-dimensional utility vector corresponding to vertex $v \in V$. Following [4], we generated random graphs with $|V| \in [10, 180]$ vertices, $|E| = 1.6|V|$ edges and having 2, 3, 4 and 5 objectives. The components of the utility vectors were generated randomly between -10 and 0. The induced width of these problems ranged between 9 and 25, respectively.
- **Combinatorial Auctions:** In our multi-objective combinatorial auctions each bid is associated with the price, the probability of failing the payment upon acceptance, and the quality of service measure. The task is to determine the subset of winning bids that simultaneously maximize the profit, minimize the risk of not getting the full revenue and maximize the overall quality of the services represented by the selected bids. We generated auctions with 30 goods and increasing number of bids from the *paths* distribution of the CATS suite [9] and randomly added failure probabilities to the bids in the range (0,0.3) while the quality of service associated with each bid was set uniformly at random between 1 and 10. The induced width of these problems ranged between 6 and 61, respectively.
- **Tradeoffs:** Given a problem instance we generated consistent random tradeoffs between its objectives, using the generator from [1] with parameters (K, T) where K and T are the number of pairwise and 3-way tradeoffs, respectively. For a pair (i, j) of objectives picked randomly out of p objectives we generate two tradeoffs $ae_i - be_j$ and $be_j - ace_i$, where e_i and e_j are the i-th and j-th unit vectors. Intuitively, one of the tradeoffs indicates how much of objective i one sacrifices to gain a unit of objective j, and the other is vice versa. We generate a 3-way tradeoff between three objectives (i, j, k) picked randomly as well in the form of the tradeoff vector $ae_i + be_j - ce_k$, where $a, b, c \in [0.1, 1)$.

We consider the following solving alternatives:

- **MOAOBB(i)** – the multi-objective AND/OR Branch-and-Bound from Section 2, where parameter i is the mini-bucket i-bound and controls the accuracy of the heuristic. Larger values of i typically yield more accurate estimates but they are more expensive to compute.
- **B=b (PLUB)** – the extension of MOAOBB(i) that uses the Pareto least upper bound based method described in Section 4.1 to reduce the upper bound set to at most b (≥ 1) utility vectors, for both the Pareto and tradeoffs cases.

Table 1. CPU time in seconds as a function of the upper bound set cardinality (B) for vertex covering problems with 3 and 5 objectives. Mini-bucket i-bound is 10. Time limit 20 minutes.

	B=1	B=2	B=4	B=10	B=50	B=100
	vertex coverings (110 vars) - 3 obj - pareto					
PLUB	**59**	67	70	95	188	238
	vertex coverings (110 vars) - 5 obj - pareto					
PLUB	**270**	581	598	646	894	983
vertex coverings (160 vars) - 3 obj - $(K=2, T=1)$ tradeoffs						
LP	496	243	**227**	269	340	341
PLUB	552	**94**	150	268	340	340
vertex coverings (160 vars) - 5 obj - $(K=5, T=2)$ tradeoffs						
LP	629	370	**346**	376	485	493
PLUB	972	**166**	211	345	487	495

- **B=b (LP)** – the extension of MOAOBB(i) that uses the LP based method from Section 4.2 to compute the upper bound sets, for the tradeoffs case only. The cluster size was set to 30.
- **VE** – the variable elimination algorithm introduced recently by [1] for evaluating multi-objective influence diagrams, which we adapted here to solve MOCOPs. Unlike the branch-and-bound algorithms which can operate in linear space (ignoring the optimal solution sets), VE is time and space exponential in the induced width of the problem instance.

All competing algorithms were restricted to a static variable ordering obtained as a depth-first traversal of the guiding pseudo tree which was computed using a min-fill heuristic [3, 4]. The AND/OR search algorithms order the subproblems rooted at each node in the search tree in lexicographic order. In all our experiments we report the average CPU time in seconds and the number of problem instances solved (we omit the nodes expanded for space reasons). The best performance points are highlighted.

Impact of the Upper Bound Set Size. Table 1 displays the average CPU time as a function of the upper bound set size (B) for vertex covering problems with 3 and 5 objectives, respectively. We consider both the Pareto (top two rows) and tradeoffs (bottom two rows) cases. For each problem size we generated 10 random instances and for each instance we generated 10 random sets of tradeoffs using the (K, T) parameters shown in the table. The mini-bucket i-bound was set to 10. We can see clearly that using a singleton upper bound set which has a reduced computational overhead is best for the Pareto case. For example, on problems with 110 variables and 5 objectives, MOAOBB(10) manipulates very large upper bound sets with more than 5000 elements and consequently is able to solve only one instance within the 20 minute time limit. In contrast, algorithm B=1 (PLUB) solves all problem instances in less than 5 minutes on average. When looking at the tradeoffs case, we can see a different picture. Namely, the best option is to use an upper bound set with small cardinality (up to 5 elements) for both the Pareto and the LP based bounds. The singleton upper bound set, although less expensive to compute, is highly inaccurate and causes the algorithms to explore a much larger search space thus deteriorating their performance considerably. The results

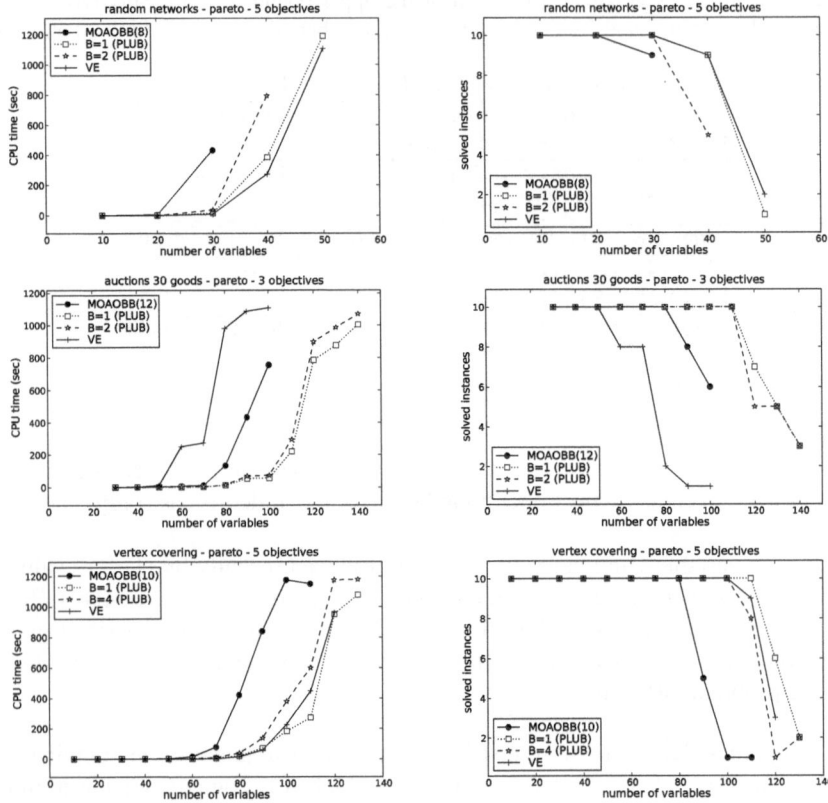

Fig. 4. CPU time in seconds (left) and number of problem instances solved (right) for random networks with 5 objectives, combinatorial auctions with 3 objectives and vertex covering problems with 5 objectives, respectively. Using the Pareto ordering. Time limit 20 minutes.

obtained on the other problem classes displayed a similar pattern and therefore were omitted for space reasons.

Comparison with State-of-the-Art Approaches. Figure 4 shows the results obtained for random networks with 5 objectives, combinatorial auctions with 3 objectives and vertex covering problems with 5 objectives, respectively, using the Pareto ordering. Each data point represents an average over 10 random instances of the corresponding problem size (number of variables). The mini-bucket i-bounds were chosen as follows: 8 for random networks, 10 for vertex covering and 12 for combinatorial auctions, respectively. We can see that algorithm B=1 (PLUB) using a singleton upper bound set outperforms the state-of-the-art MOAOBB by a significant margin and thus offers the overall best performance. The second best algorithm across all reported problem sizes is B=2 (PLUB) which is slightly slower than B=1 (PLUB) due to computational overhead issues. For example, both B=1 (PLUB) and B=2 (PLUB) scale up to auctions with 140 bids, whereas MOAOBB runs out of time beyond problems with 100 bids. VE is

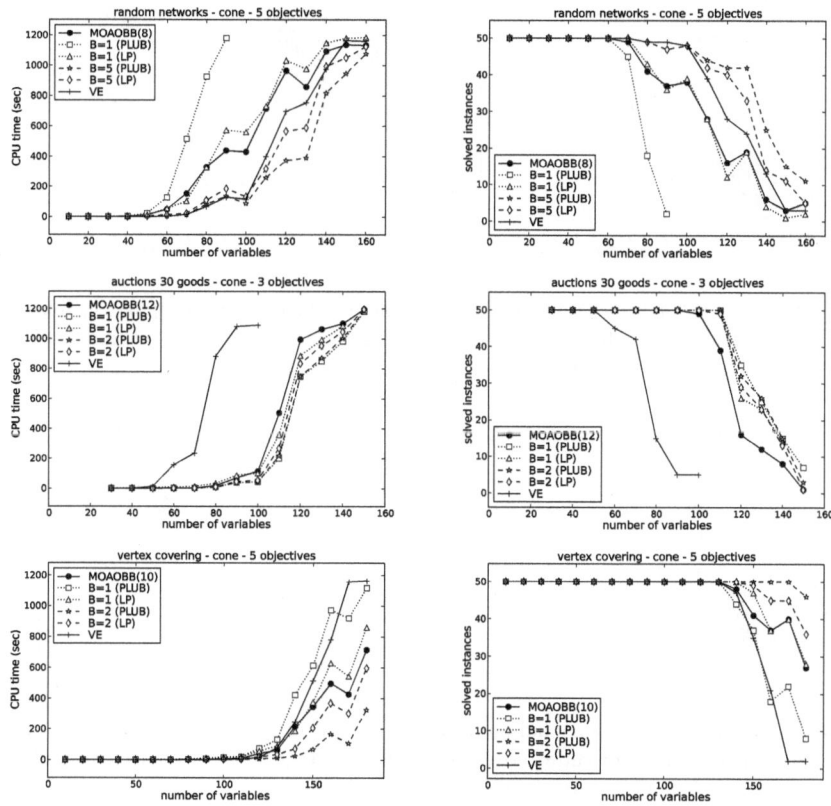

Fig. 5. CPU time in seconds (left) and number of problem instances solved (right) for random networks with 5 objectives, combinatorial auctions with 3 objectives and vertex covering problems with 5 objectives, respectively. Tradeoffs generated with parameters $(K = 6, T = 3)$ for random networks, $(K = 2, T = 1)$ for combinatorial auctions and $(K = 5, T = 2)$ for vertex covering. Time limit 20 minutes.

competitive only for medium size problems and quickly runs out of memory on larger problems (e.g., combinatorial auctions) because of larger induced widths.

In Figure 5, we summarize the results for the same problem classes using tradeoffs generated with parameters (K, T) shown in the caption. The singleton upper bounds are very weak in this case and consequently algorithms B=1 (PLUB) and B=1 (LP) perform very poorly. The best performance is offered by the algorithms using upper bound sets with 5 (random networks) and 2 (auctions and vertex covering) elements, respectively. In summary, the algorithms using upper bound sets of relatively small cardinality, which are much less expensive to compute and manipulate during search, are superior to the current state-of-the-art solvers over a wide range of problem instances, and in many cases they scale up to much larger problems.

Impact of the Number of Tradeoffs. Figure 6 plots the CPU time as a function of the number of pairwise tradeoffs K for vertex covering problems with 160 variables and

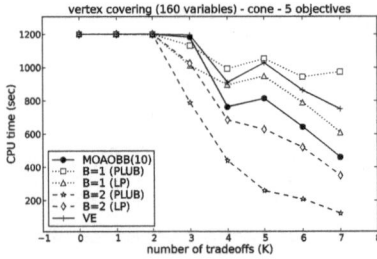
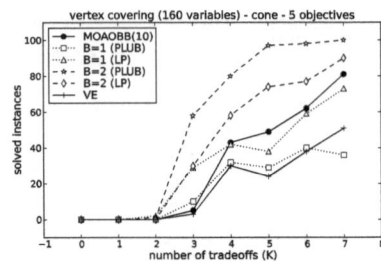

Fig. 6. CPU time in seconds (left) and number of problem instances solved (right) as a function of the number of pairwise tradeoffs (K) for vertex covering problems with $n = 160$ and 5 objectives ($T = 1$). The mini-bucket i-bound is 10. Time limit 20 minutes.

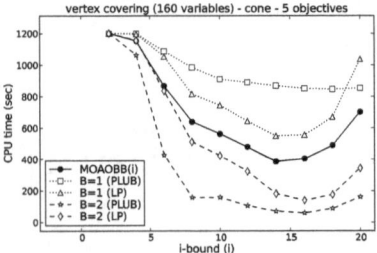
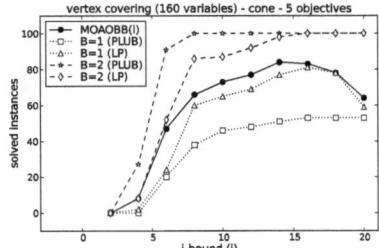

Fig. 7. CPU time in seconds (left) and number of problem instances solved (right) as function of the mini-bucket i-bound for vertex covering problems with $n = 160$, 5 objectives and ($K = 5, T = 2$) tradeoffs. Time limit 20 minutes.

5 objectives for fixed number of 3-way tradeoffs ($T = 1$). As more tradeoffs become available, the running time of the algorithms decreases substantially because the $\succeq_\Theta$-dominance gets stronger and therefore it can prune the search space more effectively. The singleton upper bounds are quite loose and therefore algorithms B=1 (PLUB) and B=1 (LP) have a relatively flat performance across the different values of K. We observed a similar behavior for fixed $K = 1$ and increasing number of 3-way tradeoffs T (results omitted for lack of space).

Impact of the Heuristic Information. Figure 7 plots the CPU time (left) and number of problem instances solved (right) as a function of the mini-bucket i-bound for vertex covering problems with 160 variables and 5 objectives. We notice the U-shaped curve characteristic of search algorithms using mini-bucket heuristics. As the i-bound increases, the total time decreases because the heuristics get stronger and prune the search space more effectively. But then as i increases further, the heuristic strength does not outweigh its computational overhead and the time starts to increase again.

For the tradeoffs case, we observed that the LP-based upper bounds were typically tighter than the corresponding Pareto-based ones across all benchmark problems, but they incurred a much higher computational overhead. Therefore, the pruning power of the former did not outweigh their overhead, except for the case when the upper bound set was restricted to a single element ($B = 1$).

6   Related Work

The optimization algorithms we use are built on the approach of [4]. The use of upper bound sets in the context of mini-buckets for branch-and-bound was also developed in [5, 10–12], and bound sets have been used in the approaches described in [13–16]. Constraint programming approaches for multi-criteria optimization include [17–19].

As mentioned in Section 3, the formalism for tradeoffs derives from that described in [1], and relates to convex cone-based approaches for multi-objective preferences such as [8, 20–22]. The variable elimination technique for the tradeoffs case derives from that in [1] and the correctness follows from the results in [23]; it can also be related with the general algorithmic approach described in [24].

7   Summary and Conclusion

We extended multi-objective constraint optimization algorithms – including a variable elimination algorithm and variants of a branch-and-bound algorithm – to the case where there are additional tradeoffs. The tradeoffs approach is based on a preference inference technique. We show that the inference technique from [1] can be given an alternative semantics based on Multi-Attribute Utility Theory, where it is assumed that the decision maker compares utility vectors by a weighted sum of the individual values.

The branch-and-bound algorithms use a mini-buckets procedure for generating the upper bound set at each node. Because the upper bound set can get large we consider different methods for reducing its size. This is achieved by incrementally replacing a selection of the elements by an upper bound of them. In almost all our experimental results for the Pareto (no tradeoffs) case, we found that using a singleton upper bound set is best, and this can considerably improve the current state-of-the-art. Although using a larger upper bound set pruned slightly more, it was not sufficient to make up for the additional overhead. For the tradeoffs case, our results suggest that it is usually best to use a non-singleton upper bound set, but which has quite small cardinality; even allowing a 2-element upper bound set can improve dramatically the efficiency of the algorithm because of the extra pruning power.

Acknowledgments. Abdul Razak is funded by IRCSET and IBM through the IRCSET Enterprise Partnership Scheme. This work was also supported in part by the Science Foundation Ireland under grant no. 08/PI/I1912.

References

1. Marinescu, R., Razak, A., Wilson, N.: Multi-objective influence diagrams. In: Proceedings of the 28th Conference on Uncertainty in Artificial Intelligence (UAI), pp. 574–583 (2012)
2. Dechter, R., Rish, I.: Mini-buckets: A general scheme of approximating inference. Journal of ACM 50(2), 107–153 (2003)
3. Dechter, R., Mateescu, R.: AND/OR search spaces for graphical models. Artificial Intelligence 171(2-3), 73–106 (2007)
4. Marinescu, R.: Exploiting problem decomposition in multi-objective constraint optimization. In: Gent, I.P. (ed.) CP 2009. LNCS, vol. 5732, pp. 592–607. Springer, Heidelberg (2009)

5. Rollon, E.: Multi-Objective Optimization for Graphical Models. PhD thesis, Universitat Politècnica de Catalunya, Barcelona, Spain (2008)
6. Nering, D.E.: Linear Algebra and Matrix Theory, 2nd edn. John Wiley and Sons (1970)
7. Figueira, J., Greco, S., Ehrgott, M.: Multiple Criteria Decision Analysis—State of the Art Surveys. Springer International Series in Operations Research and Management Science, vol. 76. Springer (2005)
8. Tamura, K.: A method for constructing the polar cone of a polyhedral cone, with applications to linear multicriteria decision problems. Journal of Optimization Theory and Applications 19(4), 547–564 (1976)
9. Leyton-Brown, K., Pearson, M., Shoham, Y.: Towards a universal test suite for combinatorial auction algorithms. In: Proceedings of the 2nd ACM Conference on Electronic Commerce (EC), pp. 66–76 (2000)
10. Rollon, E., Larrosa, J.: Bucket elimination for multi-objective optimization problems. Journal of Heuristics 12, 307–328 (2006)
11. Rollon, E., Larrosa, J.: Constraint optimization techniques for exact multiobjective optimization. In: Proceedings of the 7th International Conference on Multi-Objective Programming and Goal Programming (2006)
12. Wilson, N., Fargier, H.: Branch-and-bound for soft constraints based on partially ordered degrees of preference. In: Proceedings of the ECAI Workshop on Inference Methods Based on Graphical Structures of Knowledge, WIGSK 2008 (2008)
13. Villarreal, B., Karwan, M.: Multicriteria integer programming: A (hybrid) dynamic programming recursive approach. Mathematical Programming 21, 204–223 (1981)
14. Ehrgott, M., Gandibleux, X.: Bounds and bound sets for biobjective combinatorial optimization problems. Notes in Economics and Mathematical Systems 507, 241–253 (2001)
15. Sourd, F., Spanjaard, O.: A multiobjective branch-and-bound framework: Application to the biobjective spanning tree problem. INFORMS Journal on Computing 20(3), 472–484 (2008)
16. Delort, C., Spanjaard, O.: Using bound sets in multiobjective optimization: Application to the biobjective binary knapsack problem. In: Festa, P. (ed.) SEA 2010. LNCS, vol. 6049, pp. 253–265. Springer, Heidelberg (2010)
17. Junker, U.: Preference-based search and multi-criteria optimization. Annals of Operations Research 130, 75–115 (2004)
18. Gavanelli, M.: Partially ordered constraint optimization problems. In: Walsh, T. (ed.) CP 2001. LNCS, vol. 2239, p. 763. Springer, Heidelberg (2001)
19. Gavanelli, M.: An implementation of Pareto optimality in CLP(FD). In: Proceedings of the 4th International Workshop on Integration of AI and OR Techniques in Constraint Programming for Combinatorial Optimization Problems (CPAIOR), pp. 49–63 (2002)
20. Yu, P.: Cone convexity, cone extreme points, and nondominated solutions in decision problems with multiobjectives. Journal of Optimization Theory and Applications 14(3), 319–377 (1974)
21. Wiecek, M.: Advances in cone-based preference modeling for decision making with multiple criteria. Decision Making in Manufacturing and Services 1(1-2), 153–173 (2007)
22. Hunt, B., Wiecek, M., Hughes, C.: Relative importance of criteria in multiobjective programming: A cone-based approach. European Journal of Operational Research 207(2), 936–945 (2010)
23. Wilson, N., Marinescu, R.: An axiomatic framework for influence diagram computation with partially ordered utilities. In: Proceedings of the 13th International Conference on Principles of Knowledge Representation and Reasoning (KR), pp. 210–220 (2012)
24. Fargier, H., Rollon, E., Wilson, N.: Enabling local computation for partially ordered preferences. Constraints 15(4), 516–539 (2010)

Multidimensional Bin Packing Revisited

Michael D. Moffitt

IBM Corp.
mdmoffitt@us.ibm.com

Abstract. *Multidimensional bin packing* is a challenging combinatorial problem with applications to cloud computing, virtualized datacenters, and machine reassignment. In contrast to the classical bin packing model, item sizes and bin capacities both span a vector of values, requiring that feasible assignments honor capacity constraints across all dimensions. Recent work has yielded significant improvements over traditional CP and MIP encodings by incorporating *multivalued decision diagrams* (MDDs) into a heuristic-driven CSP-based search. In this paper, we consider a radically different approach to multidimensional bin packing, in which the complete contents of bins are considered sequentially and independently. Our algorithm remains depth-first, yet adopts a powerful *least commitment* strategy for items when their exclusion from a bin is attempted. We abandon the use of MDDs, and instead aggregate capacity over incomplete bins to establish significantly stronger bounds on the solution quality of a partial assignment. Empirical results demonstrate that our approach outperforms the state-of-the-art by up to *four orders of magnitude*, and can even solve some previously intractable problems within a fraction of a second.

1   Introduction

Within the scope of combinatorial optimization, applications and extensions of *bin packing* have motivated considerable progress in the constraint programming and operations research communities [20,30,6,16,7]. In the classical formulation, a finite set of *items* (and their corresponding sizes) must be distributed across a minimal number of *bins* with fixed capacities. Since the relative order of items within a bin is irrelevant, a solution merely maps items to disjoint subsets such that the sum of sizes within each subset respects the appropriate capacity constraints.

Inspired (in part) by the recent Google ROADEF/EURO challenge[1], a more expressive variant of bin packing has recently been introduced by Kell and van Hoeve [18] in which both item sizes and bin capacities are augmented to provide multidimensional support. This enhanced encoding allows the modeling of independent resource requirements (e.g., CPU time, RAM, etc.) that cannot adequately be captured by simple scalar values. The authors develop a novel heuristic-driven CSP-based search – utilizing *multivalued decision diagrams*

---

[1] Online at http://challenge.roadef.org/2012/en/

C. Schulte (Ed.): CP 2013, LNCS 8124, pp. 513–528, 2013.

(MDDs) – that is shown to yield significant improvements over conventional CP and MIP approaches.

A key advantage that decision diagrams offer in the context of combinatorial search is their ability to model and manipulate the equivalence of partial solutions. By merging isomorphic assignments, MDDs collapse symmetric and/or dominated solutions into shared paths to prevent the creation and exploration of redundant search nodes. However, this flexibility often comes at a high price, since the exponentially-many variable combinations are continually maintained in memory and can impose a substantial footprint. Even if approximation schemes are employed, the resource requirements for a moderately-sized instance can easily eclipse the capabilities of a modern high-end workstation. Furthermore, the use of decision diagrams alone does not necessarily serve as a substitute for the pruning power offered by strong inference procedures; this issue is of particular importance in a heavily quantitative domain such as bin packing, where the ability to establish tight lower bounds can have a profound effect on the efficiency of search. Despite a simplification made in recent works that fixes the number of bins to a constant, existing techniques for inference in the multidimensional case remain relatively weak.

In this paper, we consider a radically different approach to multidimensional bin packing. In contrast to previous schemes that branch on the assignment of individual items, our technique branches on the complete contents of individual bins, resurrecting a set-CSP reformulation originally proposed for one-dimensional bin packing more than a decade ago [14]. The resulting algorithm remains depth-first, yet adopts a powerful *least commitment* strategy for individual items when their exclusion from a bin is attempted. We abandon the use of MDDs altogether, and instead aggregate capacity over incomplete bins to establish significantly stronger bounds on the solution quality of partial assignments. Our algorithm is simple and compact, yet empirical results demonstrate that it outperforms the state-of-the-art by up to *four orders of magnitude*, and can even solve some previously intractable problems within a fraction of a second.

The remainder of the paper is organized as follows. In Section 2 we formally define the multidimensional bin packing problem and review prior work on this topic. In Section 3, we explore the limitations of conventional search and the challenges required to perform inference on partial assignments in the CSP formulation. Our new approach is presented in Section 4, where we describe our reformulation and several techniques to increase efficiency in this alternative search space. Experimental results are provided in Section 5, in which we compare the performance of our algorithm to the previous state-of-the-art. We briefly review related work in Section 6, and finally conclude in Section 7.

2   Background

We consider the formulation proposed in [18] which specifies n item sizes $(s_1, \ldots, s_n)$ and m bin capacities $(c_1, \ldots, c_m)$. Each item size s_i is a d-tuple of nonnegative integers $(s_{i,1}, \ldots, s_{i,d})$, and likewise for each bin capacity $c_j = (c_{j,1}, \ldots, c_{j,d})$.

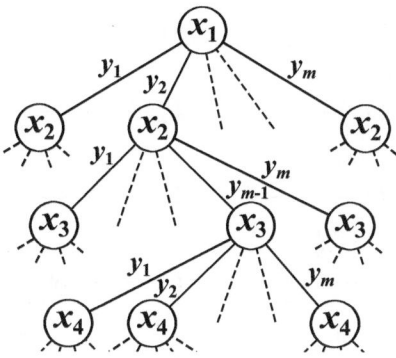

Fig. 1. Solving multidimensional bin packing as a CSP. Nodes correspond to items, branches correspond to bins.

The objective is to produce an assignment S such that each of the n items is assigned to exactly one of the m bins without exceeding the capacity of any bin/dimension pair:

$$\sum_{S(i)=j} s_{i,k} \leq c_{j,k} \quad \forall j \in [1, m], k \in [1, d]$$

Since a constant number of bins is specified in advance, this formulation is a strict decision problem, as opposed to the optimization variant that has been more commonly studied in the literature on one-dimensional bin packing.

Multidimensional bin packing formulations are far from new, with some dating back even to the late 1970's. Kou and Markowsky [24] considered a traditional limitation that constrains each $s_{i,k}$ to the range $[0, 1]$, and adapted popular heuristics from the one-dimensional case (e.g., first fit, best fit, etc.) to accommodate all dimensions. Several studies have considered geometric varieties [25,26,1,2] in which orthogonal dimensions are tightly coupled; in such models, two- and three- dimensional items respectively correspond to rectangles and boxes in Euclidean space. Chekuri and Khanna [9] obtained a variety of approximability and inapproximability results for vector scheduling and vector bin packing.

More recently, Kell and van Hoeve [18] transform multidimensional bin packing into a constraint satisfaction problem (CSP) where a variable x_i is created for each item whose domain $D_i = \{y_1, ..., y_m\}$ corresponds to the set of available bins. The CSP has $m \times d$ constraints over the subsets that compose each bin.

Definition: A *partial assignment* P at depth p in the CSP formulation of multidimensional bin packing is a mapping $(x_1, ..., x_p) \rightarrow (y_{P(1)}, ..., y_{P(p)})$ of a subset of items to their respective bins such that $\sum_{P(x_i)=y_j} s_{i,k} \leq c_{j,k}$ for all $j \in [1, m]$ and $k \in [1, d]$. A *complete assignment* is any P where $|P| = n$.   □

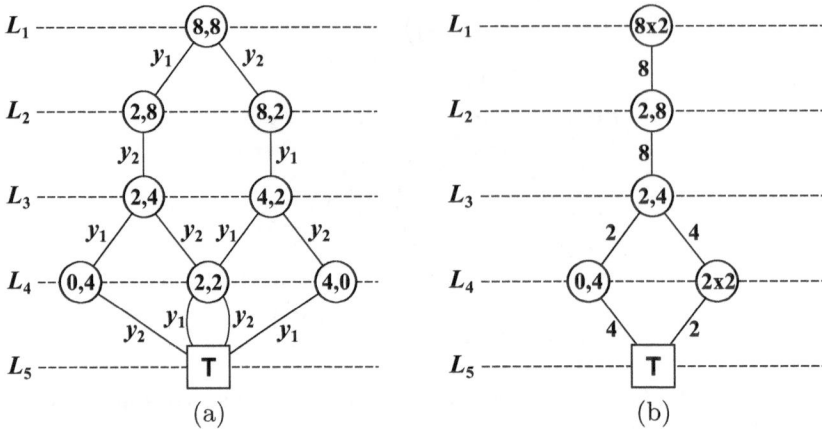

Fig. 2. MDD representations of bin packing with item sizes $\{6, 4, 2, 1\}$ and two bins with capacity 8. (a) A Direct MDD, where nodes are labeled by ullage vectors and edges are labeled by bin indices. (b) A Ullage MDD, where nodes are labeled by ullage multiplicities and edges are labeled by ullage references.

2.1   An MDD Approach

Constraint satisfaction problems are traditionally solved in a depth-first search manner (see Figure 1) in which variables are selected sequentially and each value leads to a separate branch of search. The search state is encoded *implicitly* in the call stack, with each recursive call forming a node along a partial path. When applied to the domain of one-dimensional bin packing, a generic CSP approach mirrors the combinatorial search formulation originally championed by Martello and Toth [27].

For the CSP formulation of multidimensional bin packing, Kell and van Hoeve take a different approach, in which the various combinations of item-bin assignments are stored *explicitly* in a *multivalued decision diagram* (MDD) [17,3]. For a problem with n variables, the MDD is comprised of $n + 1$ layers denoted $L_1, \ldots, L_{n+1}$. Each layer contains a set of nodes, and each edge connects nodes in adjacent layers L_i and L_{i+1} reflecting an assignment to the variable x_i. The final layer L_{n+1} contains a single node (the sink) representing feasibility. Any path in the MDD from root to sink corresponds to a complete solution, whose values can be derived from the labels of edges along the path. Partial paths, of course, correspond only to partial solutions that may or may not extend to feasibility.

In a direct MDD representation, nodes in each layer are labeled with *states* that map each bin to its remaining multidimensional capacities (its "ullage" vector). Edges are labeled with the index of the bin y_j corresponding to the appropriate assignment for x_i. Hence, the ullage vectors of two nodes u and v at respective layers L_i and L_{i+1} that are connected by an edge y_j differ only at position j (i.e., $v_j = u_j - s_i$). Note that if two nodes at the same layer were to share identical ullage vectors across all bins, the set of feasible completions beneath these nodes would be identical. To prevent the construction and expansion of duplicate structures, the MDD stores only one copy of each node at a layer, effectively merging the paths of isomorphic partial assignments. This not only decreases the MDD's overall size, but also reduces the effort required to find a complete path from root to sink.

> **Example:** In Figure 2(a) the node $(2, 2)$ at layer L_4 reflects a merger of two partial assignments: the extension of $(2, 4)$ at layer L_3 with $x_3 \leftarrow y_2$, and the extension of $(4, 2)$ at layer L_3 with $x_3 \leftarrow y_1$. $\qquad\square$

Kell and van Hoeve explore a variety of MDD representations, including *approximate* MDDs that represent a relaxation to the original problem instance, and also *ullage* MDDs like the one in Figure 2(b) that exploit bin symmetry by collapsing nodes with identical ullage multiplicities.

In order to create the MDD, the algorithm maintains a set of nodes whose children have yet to be constructed, akin to an *open set* in traditional breadth-first search. As each node is processed, all possible extensions for each subsequent child node are enqueued. A *unique table* serves to prevent equivalent (and therefore redundant) nodes from being regenerated. This process continues until all nodes have been expanded, at which point an optimal solution may be extracted from the MDD. To enable exploratory construction, the ordering of node expansion is not required to populate complete layers one-by-one. Instead, a heuristic estimate that resembles nested monte-carlo search [8] is used to guide search toward the early consideration of favorable solutions: remaining items are randomly assigned to bins with sufficient available capacity, and nodes that minimize the total number of unplaced items are given higher priority. Because the layers of the MDD are tightly bound to specific items, a static variable ordering strategy is employed, interleaving the position of large and small items.

Empirical results have shown that the performance of the MDD-based approach is quite competitive with traditional methods; orders of magnitude improvement were observed over an industrial constraint solver, and moderate (but nevertheless significant) gains were produced in comparison to mixed-integer programming formulations using CPLEX.

3   Limitations of Conventional and MDD-Based Search

A fundamental challenge that arises in any practical implementation of multidimensional bin packing is the computation of strong lower bounds, i.e., estimating the minimum number of additional bins ultimately needed to extend a partial assignment. Even in the decision variant of the problem where precisely m bins are available, such bounding is critical in determining wasted space and pruning nodes for which any complete extension is incapable of remaining within the available resource envelope.

> **Example:** Consider a d-dimensional bin packing instance with m bins and $n = pm + 2$ items for some $p \geq 1$. For all $1 \leq i \leq n - 2$ and $1 \leq k \leq d$, we set $s_{i,k}$ to m. For the final two items (s_{n-1} and s_n),

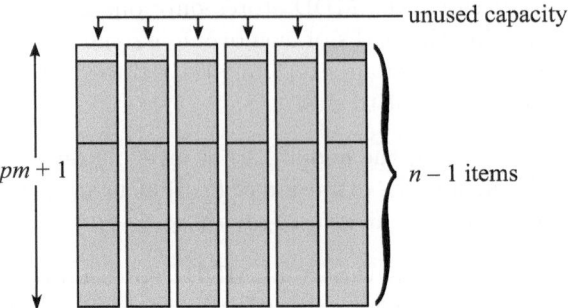

Fig. 3. An example for which only $n - 1$ items can be successfully packed into m bins

we set their sizes across all dimensions to $m - 1$ and 1 respectively. Bin capacities $c_{j,k}$ are set to $pm + 1$ for all $1 \leq j \leq m$ and $1 \leq k \leq d$. By construction, the combined capacity for this instance appears sufficient to accommodate all items; for any dimension k, the quantity $\sum_j c_{j,k}$ is equal to $m(pm + 1) = pm^2 + m$, which is equivalent to the combined demand across all $pm + 2$ items. However, each bin can accommodate at most p items of size m if $m > 1$, requiring at least one unit of wasted space in all but a single bin.   □

The above example is clearly contrived to be infeasible, as demonstrated in Figure 3 for the case where $m = 6$ and $p = 3$. In any partial assignment, only 19 of the 20 items can be packed into the bins without overflow.

In a *conventional* approach using direct combinatorial search – that is, one where each of the m branches for an item are issued consecutively (including the aforementioned CSP and MDD frameworks) – effective lower bounds can be difficult to compute. Only at leaf nodes in search are all items bound to individual bins; until this point, the provable amount of wasted space for any single bin typically cannot be determined. Our pathological example illustrates an extreme case of this limitation, as the number of partial assignments that successfully place all but the last two items grows exponentially with p and m:

$$\binom{pm}{p} \times \binom{p(m - 1)}{p} \times \binom{p(m - 2)}{p} \times \cdots \times 1$$

Even if slightly improved inference rules are adopted, these often depend heavily on structural properties of the instance. For example, if it were not for our unit-sized item s_n, modulo arithmetic of the kind suggested by Gent and Walsh [13] could quickly detect that the maximum capacity utilized by any bin is at most pm. However, in typical bin packing benchmarks produced by random generators [7], it is unlikely that item sizes will share a large common denominator, and even less probable when considering the compounding effect of multiple dimensions. Kell and van Hoeve remove the unused capacity of *dead bins* if and only if none of the remaining items will fit; our example demonstrates that this technique is

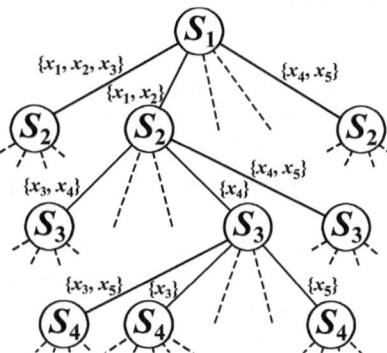

Fig. 4. Solving multidimensional bin packing as a set-CSP. Nodes correspond to bins, branches correspond to item subsets.

of little use if small items exist whose assignment is deferred until the tail end of search.

4   An Alternative Approach

Since precise calculation of wasted space is difficult for conventional forms of search, we consider an entirely different solution space for multidimensional bin packing. Conceptually, our alternative model may be viewed as a variant of a *set-CSP* [4,29,5] in which each set variable S_j corresponds to the contents of bin j, and each value $v \in D_j$ corresponds to a complete subset of items ($D_j \subseteq 2^n$). We impose a constraint $C_{j,j'}$ between every pair of set variables S_j and S'_j to ensure that subset contents are disjoint (i.e., $S_j \cap S_{j'} = \varnothing$). Finally, we enforce a constraint over all set variables to ensure that every item is assigned to some bin in any complete assignment: $\bigcup S_j = \{x_1, \ldots, x_n\}$.

> **Definition:** A *partial assignment* P at depth p in our alternative reformulation of multidimensional bin packing is a sequence of item subsets $(S_1, ..., S_p)$ such that $\sum_{x_i \in S(j)} s_{i,k} \leq c_{j,k}$ for all $j \in [1, p]$ and $k \in [1, d]$, and $S_j \cap S_{j'} = \varnothing$ for any $j \neq j'$. A *complete assignment* is any such P where $|P| = m$ and $\bigcup S_j = \{x_1, \ldots, x_n\}$.   □

In contrast to the original CSP where the contents of each bin are subject to change until leaves of search, our reformulation commits to specific complete subsets as each branch is explored (see Figure 4). The above transformation to set-CSPs should come as little surprise, since one-dimensional bin packing constituted one of several key applications motivating early work on the subject [14]. A similar technique known as *bin completion* [20,11] remains one of the best known algorithms for one-dimensional packing and covering problems, motivating its use in the broader multidimensional case.

4.1   Dynamic Domain Generation

If the domains of our set variables were to be represented explicitly (and subsequently modeled by a generic set-CSP solver[2]), as many as 2^n values reflecting each possible subset of items would be required for a multidimensional bin packing problem containing n items. Although it may be possible to enumerate all such subsets in advance if n is relatively small, this process clearly becomes intractable for sufficiently large n.

One common method to avoid this exhaustive encoding is to represent the domain of a set variable S by an interval $[L(S), U(S)]$ expressing its lower and upper bounds [32]. This is the approach taken in the original Conjunto solver [14]. However, note that our reformulation embodies a special case of a set-CSP where in any feasible solution, the set variables' assignments form a complete *partition* over the n items [33,4]. Since bin contents must be mutually exclusive and collectively exhaustive (therefore covering each item x_i exactly once), this enables a specialized implementation that dynamically populates each value for a set variable S_j on-the-fly through nested recursion. To achieve this, our search procedure maintains a global set U of as-yet unassigned items (e.g., items not consumed by previously instantiated set variables) and branches on subsets selected from this set. Specifically, each item $x_i \in U$ invokes a disjunction:

$$include(x_i, S_j) \vee exclude(x_i, S_j)$$

that is resolved by exploring an inclusion / exclusion tree. Every leaf node of this tree corresponds to a fully instantiated value for S_j. Partial combinations of items are tested for feasibility by incrementally subtracting the demand of each item from an available capacity vector that begins at $\langle c_{j,1}, \cdots, c_{j,d} \rangle$. Subsets are pruned whenever the remaining capacity along any dimension becomes negative, ensuring that only a fraction of the full $2^{|U|}$ subsets are considered. In this way, we implicitly ensure that set variables will ultimately assume disjoint values, and avoid the potential overhead that would otherwise be required to propagate constraints across their respective domains.

It is useful to compare the order in which partial assignments are expanded in our reformulation versus the original CSP. Once an item x_i is excluded from subset S_j in the conventional approach, it is immediately committed to inclusion in some other specific subset $S_{j'}$. In contrast, our implementation effectively defers the ultimate assignment of x_i, taking a *least commitment* approach and instead selecting some replacement $x_{i'}$ for inclusion in S_j. As will be shown in subsequent sections, this strategy allows substantial inference to be performed on partial assignments.

---

[2] Support for set variables in modern CP solvers has waned in recent years; for instance, they are no longer modeled in IBM's ILOG CP Optimizer [31], despite earlier support in the ILOG Solver (its predecessor).

4.2   Aggregating Capacities to Improve Inference

The nodes explored while searching our alternative CSP roughly correspond to the same nodes explored in the CSP formulation (albeit in a different order): in both cases, combinations of items to bins are considered exhaustively with basic pruning rules serving to prohibit bin overflow along any partial path.

However, since each partial assignment $P = (S_1, ..., S_p)$ in our search space corresponds to a sequence of *complete* subsets, our search procedure is prohibited from retroactively inserting items into previously assigned bins, mandating that all remaining items must be distributed across only the remaining (uninstantiated) set variables $\{S_{p+1}, ..., S_m\}$. This knowledge can be leveraged to strengthen the inference performed at intermediate nodes.

> **Theorem:** Consider a partial assignment $P = (S_1, ..., S_p)$ that leaves $U - \{x_1, ..., x_n\} - \bigcup S_j$ items unassigned. For any complete (and satisfying) assignment P' that descends from P, it must be the case that $\sum_{x_i \in U} s_{i,k} \leq \sum_{j \in [p+1,m]} c_{j,k}$ for all $k \in [1, d]$. $\square$

The proof of this theorem is trivial: if there exists any dimension for which the total amount of remaining demand exceeds the total amount of remaining capacity, no feasible assignment of items to bins can be achieved that is capable of respecting all capacity constraints. We use this rule as the basis for a stronger pruning criterion that aggregates capacities of uninstantiated bins and abandons search if the cumulative sum of deferred item sizes exceeds this calculation.

Returning to our previous example, we observe that is impossible for search to populate more than two bins without consuming a nonzero amount of wasted space, regardless of the ordering of items. In Figure 5(a), the first bin is only partially occupied, and because the total bin capacity is precisely equal to cumulative item size, this assignment will be pruned immediately. If the unit item is included in the initial bin, search may descend one additional level to reach the solution shown in Figure 5(b), but can proceed no further if the limited capacity of remaining bins are taken into account.

4.3   Pruning Dominated Solutions

As is the case with the classical CSP formulation, our search space prohibits bin contents from becoming oversubscribed by pruning any intermediate node that violates the capacity constraints in one or more bins. However, the potential still remains for a bin to be assigned too *few* elements. Solutions that needlessly defer the assignment of viable items are often fruitless, and therefore it is helpful to explicitly prevent wasted space from being carelessly accumulated.

> **Example:** Consider the trace shown in Figure 4; following the assignment $S_1 \leftarrow \{x_1, x_2, x_3\}$, a (weaker) assignment $S_1 \leftarrow \{x_1, x_2\}$ is subsequently attempted. Provided that both solutions honor all capacity constraints, the latter partial assignment is clearly dominated by the former; while it may indeed lead to a feasible solution, any such solution

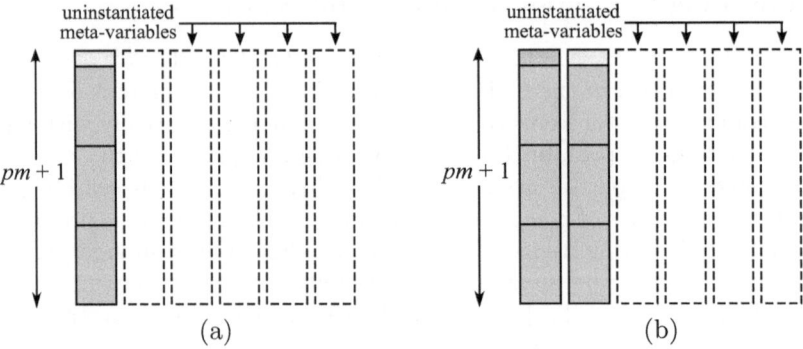

Fig. 5. Terminal partial assignments in the reformulation of our toy example. When the capacity of uninstantiated set variables are aggregated and used for pruning, search will abort immediately after wasted space is consumed by any bin.

would have been found already by exploring extensions to the assignment for which S_1 also included x_3.   □

In response, we avoid expanding any search node $(S_1, ..., S_p)$ whenever there exists an item x_i excluded from S_j such that $\sum_{i' \in S_j \cup \{i\}} s_{i',k} \leq c_{p,k}$ for all $k \in [1, d]$. If such an item exists, the most recently instantiated bin is undersubscribed, and is therefore dominated by partial assignments considered earlier in search.

4.4   Exploiting Symmetry

The multidimensional bin packing formulation considered in this work makes no explicit assumption regarding the similarity of bin capacities. However, in one-dimensional bin packing, it is common to assume that all bins are identical. Indeed, the benchmarks produced by Kell and van Hoeve [18] assign the same capacity profile to all m bins, and heavily exploit this property in their MDD representation by merging nodes with identical ullage multiplicities.

Within our alternative encoding, symmetry can be exploited in a similar manner by forcing one specific unassigned item into the bin currently under consideration. For instance, at the topmost level in search in Figure 4, the inclusion of item x_1 in S_1 can be imposed upon all partial assignments, pruning cases such as $S_1 \leftarrow \{x_4, x_5\}$ whose extensions are isomorphic to assignments previously considered.

4.5   The Complete Algorithm

In Algorithm 1, we present the complete pseudocode for our approach to multidimensional bin packing. The recursive function *Solve* accepts j as the index of the bin whose contents are being considered, U as the remaining items to assign, S_j as the items to be included in bin j, and $\overline{S_j}$ as the items to be

---

Algorithm 1: $Solve(j, U, S_j, \overline{S_j})$

Data: j (bin index), U (remaining items to assign), S_j (items included in bin j), $\overline{S_j}$ (items excluded from bin j)

Result: SAT or $UNSAT$

1 **begin**
2   **if** $\exists_k(\sum_{i \in S_j} s_{i,k} > c_{j,k})$ **then**
3          $\lfloor$ return $UNSAT$
4   **if** $\exists_k(\sum_{i \in \overline{S_j}} s_{i,k} > \sum_{j'>j} c_{j',k})$ **then**   // Section 4.2
5          $\lfloor$ return $UNSAT$
6   **if** $U = \varnothing$ **then**
7   **if** $j = m$ **then**
8              $\lfloor$ return SAT
9   **if** $\exists_{i \in \overline{S_j}} \sum_{i' \in S_j \cup \{i\}} s_{i',k} \leq c_{j,k}$ **then**   // Section 4.3
10             $\lfloor$ return $UNSAT$
11         $\lfloor$ return $Solve(j+1, \overline{S_j}, \varnothing, \varnothing)$
12     $i \leftarrow Select(U)$ // Section 4.1
13   **if** $Solve(j, U - \{i\}, S_j \cup \{i\}, \overline{S_j})$ **then**
14         $\lfloor$ return SAT
15   **if** $S_j \neq \varnothing$ **then**   // Section 4.4
16   **if** $Solve(j, U - \{i\}, S_j, \overline{S_j} \cup \{i\})$ **then**
17             $\lfloor$ return SAT
18   return $UNSAT$

---

excluded from bin j. The top-level call to *Solve* is invoked with the parameters $\langle j, U, S_j, \overline{S_j} \rangle = \langle 1, \{x_1, ..., x_n\}, \varnothing, \varnothing \rangle$. Lines 2-3 and 4-5 check capacity constraints for the current and remaining bins, respectively. Lines 6-11 handle the case where all numbers have been included in (or excluded from) bin j; if there are no additional bins to complete, a solution has been found and the algorithm terminates. Otherwise, dominance detection is tested before invoking a recursive call to establish the contents of subset $j + 1$.

If there are items in U to be assigned, one is selected (line 12) and recursively explored in the inclusion branch (lines 13-14) and exclusion branch $\overline{S_j}$ (lines 15-17). In our implementation, we choose the item whose average size across all dimensions is largest, as its placement has the largest impact on the bounding criteria used for pruning. To exploit symmetry, this latter branch is taken only if S_j is non-empty. If neither of these attempts leads to a satisfying assignment, the failed partial assignment is abandoned (line 18).

We note that in contrast to the MDD approach, our algorithm avoids the creation of large or complex data structures, and therefore is not constrained by memory limitations. Since we consider bin contents dynamically and independently, our approach also extends easily to optimization problems with variable numbers of bins.

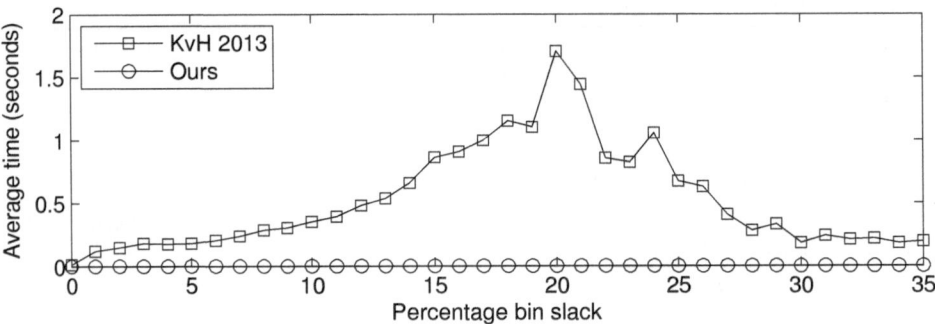

Fig. 6. Hardness profiles for instances having 6 dimensions, 18 items, and 6 bins

5   Experimental Results

In order to evaluate the efficacy of our approach, we compare our algorithm against the most recently published results for multidimensional bin packing as reported by Kell and van Hoeve [18]. Our solver is implemented in C++, and all experiments were executed on a 64-bit 3GHz AMD Opteron Processor. We use the same test case generator in prior work that accepts parameters $\langle d, n, m, \beta \rangle$, where d is the number of dimensions, n is the number of items, m is the number of bins, and β is the so-called *bin slack percentage*. For every item s_i, sizes along each dimension $s_{i,k}$ are randomly and uniformly chosen from the range $[0, 1000]$. All bins are assigned identical capacity vectors such that:

$$c_{j,k} = \lceil (1 + \beta/100) \times \sum_{i=1}^{n} s_{i,k}/m \rceil$$

We begin with the original benchmarks where $\langle d, n, m \rangle = \langle 6, 18, 6 \rangle$.[3] These benchmarks assign values to β in the range of 0 to 35, and include 52 instances for each setting. As shown in Figure 6, our approach dramatically outperforms the MDD-based solver across all benchmarks. The latter requires more than one second per instance on average for the hardest problems, whereas our solver consistently consumes negligible runtime (on the order of milliseconds) regardless of problem difficulty.

Since average runtime provides only limited insight into the performance distribution exhibited by these algorithms, we produce a runtime profile for both solvers over all 52 instances at the hardness peak where $\beta = 20\%$. These profiles are shown in Figure 7, which relates the fraction of instances solved with cumulative solver runtime (note the logarithmic scale). Our approach successfully completes the *entire suite of problems* in less time than is required for the

---

[3] Available for download from
http://www.math.cmu.edu/~bkell/6-18-6-instances.txt

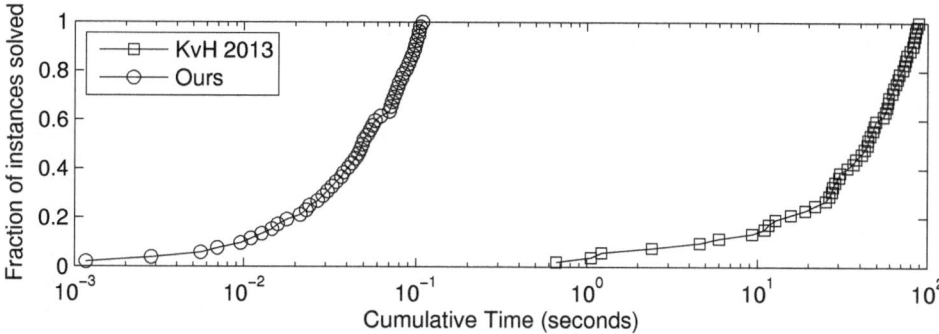

Fig. 7. Performance profile on the subset of instances having 20% bin slack

MDD-based approach to solve *just the first instance*. Since our combined runtime over all problems remains well under one second, these benchmarks should be classified as trivial, and are unlikely to provide meaningful comparisons for future studies of multidimensional bin packing.

To obtain a more complete analysis of the runtime behavior of these solvers, we significantly increased the spectrum of problem sizes by varying n in the range $\{18, \ldots, 40\}$ and m in the range $\{6, \ldots, 9\}$. Through empirical observation, we devised the following formula for the bin-slack percentage β that appears to maintain a relatively consistent phase transition:

$$\beta_{n,m} = 4.5^{3-n/18} * 2^{(m-6)/3}$$

Results of these experiments are shown in Table 1. The runtime of the MDD-based solver reaches over a minute for the setting $\langle n, m \rangle = \langle 22, 6 \rangle$. This is over **10000×** slower than our approach, which requires only ten milliseconds on average. Problems with larger n are intractable for the MDD solver, which tends to eclipse its self-imposed 512 MB memory limit after several minutes and is therefore unable to produce a definitive result for most cases. In contrast, our algorithm scales to as many as 33 items without exceeding an average runtime of one second, and nearly doubles the size of problems that can be solved in under a minute from 21 to 40. Similar behavior is observed for larger values of m, although the highest achievable n for the MDD solver increases slightly.

6   Related Work

A number of subset-oriented exploration techniques have been developed in prior work for various incarnations of partitioning [22,23,28], knapsack [10], and one-dimensional bin packing problems [14,20,21,11]. A recent approach to the steel mill slab problem [12] enumerates all possible slab designs in a preprocessing phase, and subsequently encodes these candidates as 0/1 variables in an integer linear program [15]. The approach bears some resemblance to our formulation, although a principal reason for its scalability is the presence of *color constraints* that severely constrain the space of feasible slab combinations.

Table 1. Average runtime (in seconds) across problems with varying n and m

n	$m = 6$			$m = 7$			$m = 8$			$m = 9$		
	Kv'13	Ours	Ratio	Kv'13	Ours	Ratio	Kv'13	Ours	Ratio	Kv'13	Ours	Ratio
18	1.457	0.002	**904×**	0.798	0.002	**510×**	0.579	0.001	**412×**	0.368	0.001	**292×**
19	3.107	0.002	**1750×**	1.247	0.002	**772×**	1.000	0.002	**439×**	0.620	0.002	**378×**
20	7.323	0.003	**2798×**	2.783	0.002	**1348×**	1.915	0.005	**349×**	0.796	0.002	**483×**
21	42.75	0.006	**7771×**	8.750	0.004	**1996×**	3.317	0.006	**597×**	2.071	0.003	**739×**
22	106.6	0.010	**10879×**	22.91	0.005	**4712×**	7.635	0.006	**1207×**	4.005	0.009	**439×**
23	—	0.008		136.5	0.012	**11561×**	42.67	0.015	**2852×**	8.883	0.014	**625×**
24	—	0.016		—	0.051		133.4	0.017	**8061×**	45.89	0.035	**1305×**
25	—	0.033		—	0.046		—	0.046		81.69	0.025	**3306×**
26	—	0.046		—	0.094		—	0.102		—	0.085	
27	—	0.079		—	0.140		—	0.357		—	0.107	
28	—	0.065		—	0.256		—	0.337		—	0.185	
29	—	0.168		—	0.297		—	0.609		—	0.794	
30	—	0.146		—	0.761		—	1.008		—	1.192	
31	—	0.351		—	1.120		—	2.085		—	2.749	
32	—	0.715		—	1.081		—	2.160		—	3.458	
33	—	0.951		—	3.789		—	4.892		—	6.195	
34	—	1.511		—	4.628		—	8.638		—	8.074	
35	—	2.346		—	6.266		—	16.96		—	13.49	
36	—	3.785		—	13.75		—	21.25		—	15.46	
37	—	5.798		—	18.30		—	28.80		—	23.95	
38	—	14.78		—	30.03		—	26.62		—	51.66	
39	—	24.12		—	39.02		—	32.32		—	54.26	
40	—	42.68		—	67.14		—	53.91		—	71.28	

Several aspects of our implementation can be viewed as employing a variant of *set branching*, originally proposed by Kitching and Bacchus [19] for general-purpose constraint optimization problems. Values of each variable's domain are clustered into sets, thereby allowing branch-and-bound to branch on assignments to these sets rather than on the original individual values. The structure of bin packing allows our algorithm to adopt a special case of this policy, where all variables in the classical CSP formulation (i.e., the items) share an identical domain space (i.e., the bins), and all bins aside from the one currently being populated are collectively grouped into a single exclusion set.

7 Conclusions

In this paper, we have considered a radically different approach to multidimensional bin packing. In contrast to previous approaches that branch on the assignment of individual items, our technique instantiates the contents of bins sequentially and independently. Our algorithm remains depth-first, yet adopts a powerful *least commitment* strategy for individual items when their exclusion from a bin is attempted. We abandon the use of MDDs, and instead aggregate capacity over incomplete bins to establish significantly stronger bounds on the solution quality of a partial assignment. Empirical results have shown that our approach outperforms the state-of-the-art by several orders of magnitude.

Acknowledgments. The author wishes to thank Brian Kell for supplying the MDD solver and benchmark generator source codes. We also thank the anonymous reviewers for their thoughtful and constructive feedback, which helped greatly to improve the quality of our manuscript.

References

1. Bansal, N., Caprara, A., Sviridenko, M.: Improved approximation algorithms for multidimensional bin packing problems. In: Proceedings of the 47th Annual IEEE Symposium on Foundations of Computer Science (FOCS 2006), pp. 697–708 (2006)
2. Bansal, N., Caprara, A., Sviridenko, M.: A new approximation method for set covering problems, with applications to multidimensional bin packing. SIAM J. Comput. 39(4), 1256–1278 (2009)
3. Bergman, D., van Hoeve, W.-J., Hooker, J.N.: Manipulating MDD relaxations for combinatorial optimization. In: Achterberg, T., Beck, J.C. (eds.) CPAIOR 2011. LNCS, vol. 6697, pp. 20–35. Springer, Heidelberg (2011)
4. Bessiere, C., Hebrard, E., Hnich, B., Walsh, T.: Disjoint, partition and intersection constraints for set and multiset variables. In: Wallace, M. (ed.) CP 2004. LNCS, vol. 3258, pp. 138–152. Springer, Heidelberg (2004)
5. Bodirsky, M., Hils, M., Krimkevitch, A.: Tractable set constraints. In: Proceedings of the 22nd International Joint Conference on Artificial Intelligence (IJCAI 2011), pp. 510–515 (2011)
6. Cambazard, H., O'Sullivan, B.: Propagating the bin packing constraint using linear programming. In: Cohen, D. (ed.) CP 2010. LNCS, vol. 6308, pp. 129–136. Springer, Heidelberg (2010)
7. Castiñeiras, I., De Cauwer, M., O'Sullivan, B.: Weibull-based benchmarks for bin packing. In: Milano, M. (ed.) CP 2012. LNCS, vol. 7514, pp. 207–222. Springer, Heidelberg (2012)
8. Cazenave, T.: Nested monte-carlo search. In: Proceedings of the 21st International Joint Conference on Artificial Intelligence (IJCAI 2009), pp. 456–461 (2009)
9. Chekuri, C., Khanna, S.: On multidimensional packing problems. SIAM J. Comput. 33(4), 837–851 (2004)
10. Forrest, J.J.H., Kalagnanam, J., Ladányi, L.: A column-generation approach to the multiple knapsack problem with color constraints. INFORMS Journal on Computing 18(1), 129–134 (2006)
11. Fukunaga, A.S., Korf, R.E.: Bin completion algorithms for multicontainer packing, knapsack, and covering problems. J. Artif. Intell. Res (JAIR) 28, 393–429 (2007)
12. Gargani, A., Refalo, P.: An efficient model and strategy for the steel mill slab design problem. In: Bessiere, C. (ed.) CP 2007. LNCS, vol. 4741, pp. 77–89. Springer, Heidelberg (2007)
13. Gent, I.P., Walsh, T.: From approximate to optimal solutions: Constructing pruning and propagation rules. In: Proceedings of the 15th International Joint Conference on Artificial Intelligence (IJCAI 1997), pp. 1396–1401 (1997)
14. Gervet, C.: Interval propagation to reason about sets: Definition and implementation of a practical language. Constraints 1(3), 191–244 (1997)
15. Heinz, S., Schlechte, T., Stephan, R., Winkler, M.: Solving steel mill slab design problems. Constraints 17(1), 39–50 (2012)
16. Hermenier, F., Demassey, S., Lorca, X.: Bin repacking scheduling in virtualized datacenters. In: Lee, J. (ed.) CP 2011. LNCS, vol. 6876, pp. 27–41. Springer, Heidelberg (2011)

17. Hoda, S., van Hoeve, W.-J., Hooker, J.N.: A systematic approach to MDD-based constraint programming. In: Cohen, D. (ed.) CP 2010. LNCS, vol. 6308, pp. 266–280. Springer, Heidelberg (2010)
18. Kell, B., van Hoeve, W.-J.: An MDD approach to multidimensional bin packing. In: Gomes, C., Sellmann, M. (eds.) CPAIOR 2013. LNCS, vol. 7874, pp. 128–143. Springer, Heidelberg (2013)
19. Kitching, M., Bacchus, F.: Set branching in constraint optimization. In: Proceedings of the 21st International Joint Conference on Artificial Intelligence (IJCAI 2009), pp. 532–537 (2009)
20. Korf, R.E.: A new algorithm for optimal bin packing. In: Proceedings of the 18th National Conference on Artificial Intelligence (AAAI 2002), pp. 731–736 (2002)
21. Korf, R.E.: An improved algorithm for optimal bin packing. In: Proceedings of the 18th International Joint Conference on Artificial Intelligence (IJCAI 2003), pp. 1252–1258 (2003)
22. Korf, R.E.: Multi-way number partitioning. In: Proceedings of the 21st International Joint Conference on Artificial Intelligence (IJCAI 2009), pp. 538–543 (2009)
23. Korf, R.E.: A hybrid recursive multi-way number partitioning algorithm. In: Proceedings of the 22nd International Joint Conference on Artificial Intelligence (IJCAI 2011), pp. 591–596 (2011)
24. Kou, L.T., Markowsky, G.: Multidimensional bin packing algorithms. IBM Journal of Research and Development 21(5), 443–448 (1977)
25. Lodi, A., Martello, S., Monaci, M.: Two-dimensional packing problems: A survey. European Journal of Operational Research 141(2), 241–252 (2002)
26. Lodi, A., Martello, S., Vigo, D.: Heuristic algorithms for the three-dimensional bin packing problem. European Journal of Operational Research 141(2), 410–420 (2002)
27. Martello, S., Toth, P.: Lower bounds and reduction procedures for the bin packing problem. Discrete Applied Mathematics 28(1), 59–70 (1990)
28. Moffitt, M.D.: Search strategies for optimal multi-way number partitioning. In: Proceedings of the 23rd International Joint Conference on Artificial Intelligence, IJCAI 2013, pp. 623–629 (2013)
29. Sellmann, M., Hentenryck, P.V.: Structural symmetry breaking. In: Proceedings of the 19th International Joint Conference on Artificial Intelligence (IJCAI 2005), pp. 298–303 (2005)
30. Shaw, P.: A constraint for bin packing. In: Wallace, M. (ed.) CP 2004. LNCS, vol. 3258, pp. 648–662. Springer, Heidelberg (2004)
31. Shaw, P.: IBM ILOG CP Optimizer, CPAIOR, Masterclass (2009)
32. van Hoeve, W.-J., Sabharwal, A.: Filtering atmost1 on pairs of set variables. In: Trick, M.A. (ed.) CPAIOR 2008. LNCS, vol. 5015, pp. 382–386. Springer, Heidelberg (2008)
33. Walsh, T.: Consistency and propagation with multiset constraints: A formal viewpoint. In: Rossi, F. (ed.) CP 2003. LNCS, vol. 2833, pp. 724–738. Springer, Heidelberg (2003)

A Parametric Propagator
for Discretely Convex Pairs of Sum Constraints[*]

Jean-Noël Monette[1], Nicolas Beldiceanu[2], Pierre Flener[1], and Justin Pearson[1]

[1] Uppsala University, Dept. of Information Technology, 751 05 Uppsala, Sweden
FirstName.LastName@it.uu.se
[2] Mines de Nantes, TASC Team (CNRS/INRIA), 44307 Nantes, France
Nicolas.Beldiceanu@Mines-Nantes.fr

Abstract. We introduce a propagator for abstract pairs of Sum constraints, where the expressions in the sums respect a form of convexity. This propagator is parametric and can be instantiated for various concrete pairs, including Deviation, Spread, and the conjunction of Sum and Count. We show that despite its generality, our propagator is competitive in theory and practice with state-of-the-art propagators.

1 Introduction

Many constraint problems involve a Sum constraint, along with other constraints. It is however well-known that a Sum constraint taken in isolation is not able to perform a lot of pruning since the estimation of the minimum or maximum of a sum does not take other constraints into account. Several authors have studied how to include other constraints (sharing some variables) in the propagator for Sum, either in particular cases (e.g., Spread [9], IncreasingSum [11], and Sum with cliques [12]), or in general (e.g., ObjectiveSum [15]).

In the present work, we focus on a parametric problem, which can be cast as

$$\sum_{i \in [1,n]} f_i(x_i) \leq \overline{f} \tag{1}$$

$$\underline{g} \leq \sum_{i \in [1,n]} g_i(x_i) \leq \overline{g} \tag{2}$$

for any $n \geq 1$. The f_i and g_i are functions from integers to integers and the f_i (resp. g_i) can differ for each i. In this work, $\overline{f}$, $\underline{g}$, and $\overline{g}$ are constants, but Section 5 shows how to use variables instead. In Section 5, we also consider a lower bound $\underline{f}$ on the first sum.

Finding a solution to the conjunction of (1) and (2) is in general NP-complete as it includes as a special case the knapsack problem. There is however a large class of f_i and g_i functions for which either domain consistency or bounds($\mathbb{Z}$) consistency (see, e.g., [19] for definitions) can be achieved in polynomial time.

[*] This work is supported by grants 2011-6133 and 2012-4908 of the Swedish Research Council (VR). We thank the reviewers for their constructive comments.

C. Schulte (Ed.): CP 2013, LNCS 8124, pp. 529–544, 2013.

In this paper, we present a *parametric propagator* for this class of functions and show how to instantiate it for various functions f_i and g_i. We show that the considered class of problems includes among others the (bounds($\mathbb{Z}$) consistent) constraints DEVIATION [17], SPREAD [9], and WEIGHTEDAVERAGE [3] (with variable weights and constant values) and the (domain consistent) conjunction of LINEAR and COUNT [13]. In several cases, we match the theoretical complexity and practical efficiency of previously published specialised propagators.

Our approach for propagating the conjunction of (1) and (2) contains two parts. First (as discussed in Section 2), we compute a sharp lower bound on $\sum_{i \in [1,n]} f_i(x_i)$ under constraint (2), together with a witnessing assignment. The conjunction is feasible if this lower bound, which we call the *feasibility bound*, is at most $\overline{f}$. To compute this feasibility bound, we introduce new functions derived from the f_i and g_i. We show that the feasibility bound can be greedily computed if the newly introduced functions are *discretely convex*.

In the second part of the propagator (discussed in Section 3), the domain of each variable x_j is filtered by computing for each value u in its domain a sharp lower bound on $\sum_{i \in [1,n]} f_i(x_i)$ under constraint (2) when x_j is assigned u. If this lower bound is larger than $\overline{f}$, then u is removed from the domain of x_j. The lower bound for each pair (x_j, u) is computed *incrementally* from the witnessing assignment for the feasibility bound thanks to the discrete convexity property. We also present an improved propagator for an additional property of f_j and g_j.

The resulting propagator is parametric, depending on the f_i and g_i. The time complexity and the achieved level of consistency depend on the shape of the f_i and g_i and on the values given to the parameters. We study the complexity in Section 4 and give some implementation notes. Afterwards, we present in Section 5 several instantiations of the propagator, including a case study of DEVIATION. Finally, Section 6 presents some experimental results showing that the genericity of our approach is not detrimental to performance.

2   Feasibility Test

Given a variable x, let D_x denote the current domain of that variable. For a function f and value v, we write $f^{-1}(v)$ for the *set* of values $\{u \mid f(u) = v\}$. For a function f and set S, we write $f(S)$ for $\{f(u) \mid u \in S\}$. We use x_i, v_i, f_i to represent single variables, values, and functions, while $\mathbf{x}, \mathbf{v}, \mathbf{f}$ represent the respective vectors of all variables, values, and functions (e.g., $\mathbf{x} = \langle x_1, x_2, \ldots, x_n \rangle$).

The conjunction of (1) and (2) is satisfiable if and only if the *cost* (i.e., the value of the objective function) of an optimal solution to the following problem is at most $\overline{f}$:

$$\text{minimise} \quad \sum_{i \in [1,n]} f_i(x_i)$$

$$\text{such that} \quad \underline{g} \le \sum_{i \in [1,n]} g_i(x_i) \le \overline{g} \tag{3}$$

$$x_i \in D_{x_i}, \quad \forall i \in [1, n]$$

We gradually show in the next sub-sections how to compute greedily this cost, called the *feasibility bound*, together with a witnessing assignment.

2.1 Problem Reformulation

We reformulate problem (3) in two steps. The first step introduces for each i a new function h_i that captures the relation between f_i and g_i. The second step splits the resulting reformulated problem into two subproblems.

First Step. After introducing new variables y_i, so that $y_i = g_i(x_i)$ for each i, we propose the following new problem:

$$\text{minimise} \quad \sum_{i \in [1,n]} h_i(y_i)$$

$$\text{such that} \quad \underline{g} \leq \sum_{i \in [1,n]} y_i \leq \overline{g} \tag{4}$$

$$y_i \in g_i(\mathbf{D}_{x_i}), \quad \forall i \in [1, n]$$

where we introduce a new function $h_i \colon g_i(\mathbf{D}_{x_i}) \to f_i(\mathbf{D}_{x_i})$ for each i. This function is defined as $h_i(v) = \min f_i(g_i^{-1}(v)) = \min\{f_i(u) \mid u \in \mathbf{D}_{x_i} \land g_i(u) = v\}$, that is $h_i(v)$ is the smallest value of $f_i(x_i)$ that can be attained when $g_i(x_i)$ is equal to v. Note that the definition of h_i depends on the current domain of x_i. We now prove that the feasibility bound can also be computed from problem (4).

Lemma 1. *All optimal solutions to problems (3) and (4) have the same cost.*

Proof. Let $\mathbf{v}$ denote a vector of values for the vector $\mathbf{y}$ of variables. For each value v_i, we choose an arbitrary value u_i in $\mathbf{D}_{x_i}$ such that $g_i(u_i) = v_i$ and $f_i(u_i) = h_i(v_i)$. Such a value u_i always exists, by the definition of h_i. Then the vector $\mathbf{u}$ is a feasible solution to problem (3) if and only if $\mathbf{v}$ is a feasible solution to problem (4), and they have the same cost. In addition, any other assignment $\mathbf{u}'$ such that $g_i(u_i') = v_i$ for each i has a cost larger than or equal to the cost of $\mathbf{u}$ and $\mathbf{v}$, by the definition of h_i. Hence $\mathbf{u}$ is optimal if and only if $\mathbf{v}$ is optimal. □

Second Step. We define a new function, called H, from integers to integers:

$$H(b) = \min \left\{ \sum_{i \in [1,n]} h_i(y_i) \ \middle| \ \sum_{i \in [1,n]} y_i = b \land \forall i \in [1, n] : y_i \in g_i(\mathbf{D}_{x_i}) \right\} \tag{5}$$

That is, $H(b)$ is the minimum of the sum of the $h_i(y_i)$ when the sum of the y_i is equal to b. For a given b, we define $\mathbf{w}^b$ to be an assignment of $\mathbf{y}$ such that $b = \sum_{i \in [1,n]} w_i^b$ and $H(b) = \sum_{i \in [1,n]} h_i(w_i^b)$, i.e., an optimal solution to (5). We call $\mathbf{w}^b$ a *witnessing assignment* of b. We propose the following new problem:

$$\text{minimise} \quad H(z)$$

$$\text{such that} \quad \underline{g} \leq z \leq \overline{g} \tag{6}$$

Table 1. Several instantiations of f_i and g_i, and the corresponding h_i. The notation [*cond*] uses the Iverson bracket and is defined to be 1 if *cond* is true, and 0 otherwise.

Common Name	$f_i(u)$	$g_i(u)$	$h_i(v)$
LINEAR	$a_i \cdot u$	0	$\begin{cases} a_i \cdot \min D_{x_i} & \text{if } a_i > 0 \\ a_i \cdot \max D_{x_i} & \text{if } a_i \leq 0 \end{cases}$
WEIGHTEDAVERAGE [3]	$a_i \cdot u$	u	$a_i \cdot v$
DEVIATION [17]	$\|n \cdot u - n \cdot \mu\|$	u	$\|n \cdot v - n \cdot \mu\|$
SPREAD [9]	$(n \cdot u - n \cdot \mu)^2$	u	$(n \cdot v - n \cdot \mu)^2$
L_p-NORM, $0 < p < +\infty$	$\|n \cdot u - n \cdot \mu\|^p$	u	$\|n \cdot v - n \cdot \mu\|^p$
LINEAR and COUNT [13]	$a_i \cdot u$	$[u \in \mathcal{V}]$	$\begin{cases} a_i \cdot \min (D_{x_i} \setminus \mathcal{V}) & \text{if } v = 0 \wedge a_i > 0 \\ a_i \cdot \max (D_{x_i} \setminus \mathcal{V}) & \text{if } v = 0 \wedge a_i \leq 0 \\ a_i \cdot \min (D_{x_i} \cap \mathcal{V}) & \text{if } v = 1 \wedge a_i > 0 \\ a_i \cdot \max (D_{x_i} \cap \mathcal{V}) & \text{if } v = 1 \wedge a_i \leq 0 \end{cases}$
LINEAR and MAXIMUM	$a_i \cdot u$	$[u \geq m]$	(omitted, similar to previous pair)
MODANDDIV ($a_i > 0$)	$u - a_i \cdot \lfloor u/a_i \rfloor$	$\lfloor u/a_i \rfloor$	$\max (0, \min D_{x_i} - a_i \cdot v)$

where z is a fresh variable. The feasibility bound can also be computed from problem (6), as the latter has the same optimal cost as problem (4), and thus as problem (3): this is shown by replacing $H(z)$ by its definition (5) in the formulation of problem (6). Problems (4) and (6) are more interesting than problem (3) in three respects. First, it is simpler to reason with only one function per variable (namely h_i) instead of two (namely f_i and g_i). Second, the domain D_{y_i}, which is equal to $g_i(D_{x_i})$, might be much smaller than D_{x_i}. Third, introducing H allows us to compute the feasibility bound in two steps: (i) construct H from the h_i, and (ii) find an optimal solution to (6). This can be done greedily if all h_i are discretely convex.

Definition 1. *A function $f\colon A \to B$, where $A, B \subseteq \mathbb{Z}$, is discretely convex if*
1. *A is an interval, and*
2. *$\forall v \in A\colon (v - 1) \in A \wedge (v + 1) \in A \Rightarrow 2 \cdot f(v) \leq f(v - 1) + f(v + 1)$.*

The notion of discrete convexity is an adaptation of the usual convexity from the reals to the integers. This notion has been studied in depth, for instance in [7]. It is also related to the notion of submodular functions on sets [4].

The first condition in Definition 1 restricts in some cases the application of our approach to domains with no holes. This is discussed further in Section 5.1.

Table 1 presents the f_i, g_i, and h_i for several pairs. The h_i are convex for all those examples. Before providing algorithms, we need to introduce some notions.

2.2 Deltas, Segments, Slopes, Breakpoints, Reasoning on Infinity

Let $f\colon A \to B$ be a function with $A, B \subseteq \mathbb{Z}$. Given some value v in A, we call *right delta* (resp. *left delta*) the increase of f when v increases (resp. decreases) by 1. Formally: $\Delta^+(f, v) = f(v + 1) - f(v)$ and $\Delta^-(f, v) = f(v - 1) - f(v)$; the value of $\Delta^+(f, v)$ (resp. $\Delta^-(f, v)$) is $+\infty$ when $v + 1$ (resp. $v - 1$) is not in A.

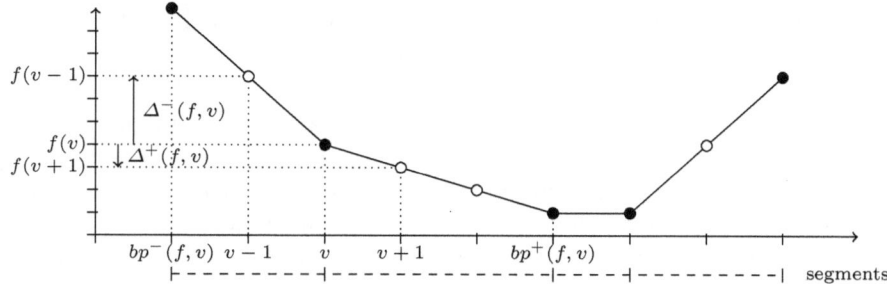

Fig. 1. Illustration of the notions of Section 2.2. Filled points are at breakpoints.

A *segment* of f is a maximal interval $[\ell, u]$ of its domain where the (right or left) delta is constant. Formally: $\Delta^+(f, v) = \Delta^+(f, v+1)$ for all $v \in [\ell, u-1]$, with $\ell \leq u$, $\Delta^+(f, \ell-1) \neq \Delta^+(f, \ell)$, and $\Delta^+(f, u-1) \neq \Delta^+(f, u)$. The endpoints ℓ and u of a segment $[\ell, u]$ of f are called *breakpoints* of f. The *length* of a segment $[\ell, u]$ is $u - \ell$. The *slope* of a segment $[\ell, u]$ is $\Delta^+(f, \ell)$. Hence the slope of a function is constant inside any of its segments and changes at its breakpoints.

The domain of f can be uniquely partitioned into its segments, and each value of the domain belongs to one or two segments. For a value v, the *breakpoint on the right* of v, denoted by $\mathrm{bp}^+(f, v)$, is u if v is in some segment $[\ell, u]$ with $u \neq v$, and otherwise undefined, denoted by $+\infty$. Similarly, $\mathrm{bp}^-(f, v)$ denotes the *breakpoint on the left* of v, if any, otherwise $-\infty$.

Let f be a discretely convex function. For any two contiguous segments, the slope of the former is smaller than the slope of the latter, hence no two segments have the same slope. Also, $\Delta^+(f, v) = +\infty$ only for the largest value v in A, as A is an interval, and $\Delta^-(f, v) = +\infty$ only for the smallest value v in A.

Figure 1 illustrates these notions on a discretely convex function.

The basic properties of $+\infty$ and $-\infty$ used in our algorithms are, for any $v \in \mathbb{Z}$: $-\infty < v < +\infty$, $v + (+\infty) = +\infty$, $v + (-\infty) = -\infty$, $v - (-\infty) = +\infty$, $v - (+\infty) = -\infty$, $\min(v, +\infty) = v$, and $v/+\infty = 0$.

2.3   Characterisation of the H Function

When the h_i are discretely convex, problem (6) is easy to solve by greedy search, because H is then also discretely convex and can be calculated efficiently.

Before proving those claims, we need to study the relationship between $H(b)$, $H(b+1)$, and $H(b-1)$, and their respective witnessing assignments. For any j and $k \neq j$, the sum $w_1^b + \cdots + (w_j^b + 1) + \cdots + (w_k^b - 1) + \cdots + w_n^b$ equals b, and hence by definition of H (since w_i^b are the values that minimise $H(b)$), we have $H(b) \leq h_1(w_1^b) + \cdots + h_j(w_j^b + 1) + \cdots + h_k(w_k^b - 1) + \cdots + h_n(w_n^b)$.

Rearranging and cancelling out common terms gives

$$h_k(w_k^b) - h_k(w_k^b - 1) \leq h_j(w_j^b + 1) - h_j(w_j^b) \tag{7}$$

If h_j is discretely convex, then we have that $h_k(w_k^b) - h_k(w_k^b - 1) \leq h_j(w_j^b + 1) - h_j(w_j^b) \leq h_j(w_j^b + 2) - h_j(w_j^b + 1)$. Thus $h_1(w_1^b) + \cdots + h_j(w_j^b + 1) + \cdots + h_k(w_k^b) + \cdots + h_n(w_n^b) \leq h_1(w_1^b) + \cdots + h_j(w_j^b + 2) + \cdots + h_k(w_k^b - 1) + \cdots + h_n(w_n^b)$, so that adding two to any single w_j^b and reducing another w_k^b by one to arrive at the sum $b + 1$ will have a higher cost than simply adding one to a single w_j^b. Because each h_i is discretely convex, this is true for any increment larger than one. Hence it is possible to find a witnessing assignment $\mathbf{w}^{b+1}$ for $b + 1$ from a witnessing assignment $\mathbf{w}^b$ for b by increasing any suitable w_i^b by one. Similarly it is possible to find a $\mathbf{w}^{b-1}$ by subtracting one from any suitable w_i^b.

Lemma 2. *H is discretely convex whenever each h_i is discretely convex.*

Proof. The domain of each h_i is an interval $[\ell_i, u_i]$, so that the domain of H is the interval $\left[\sum_{i \in [1,n]} \ell_i, \sum_{i \in [1,n]} u_i\right]$. We need to show that $H(b) - H(b-1) \leq H(b+1) - H(b)$. If w_i^b is a witnessing assignment for some b then by the discussion above there are some k and j such that $H(b-1) = h_1(w_1^b) + \cdots + h_k(w_k^b - 1) + \cdots + h_n(w_n^b)$ and $H(b+1) = h_1(w_1^b) + \cdots + h_j(w_j^b + 1) + \cdots + h_n(w_n^b)$. Therefore $H(b) - H(b-1) = h_k(w_k^b) - h_k(w_k^b - 1)$ and $H(b+1) - H(b) = h_j(w_j^b + 1) - h_j(w_j^b)$ and by (7) $H(b) - H(b-1) \leq H(b+1) - H(b)$. Hence H is discretely convex. $\square$

We now show how to calculate H efficiently by giving a characterisation of its minimum and segments. Here, for any set S and function f, the expression $\operatorname{argmin}_{i \in S} f(i)$ returns *one* (arbitrary) value $i \in S$ that minimises $f(i)$.

Lemma 3. *A witnessing assignment $\mathbf{w}^{b^*}$ of a value b^* that minimises H is such that $w_i^{b^*} = \operatorname{argmin}_{v_i \in g_i(D_{x_i})} h_i(v_i)$.*

Proof. If $\mathbf{w}^{b^*}$ is a witnessing assignment of b^*, then b^* is equal to $\sum_{i \in [1,n]} w_i^{b^*}$ and $H(b^*) = \sum_{i \in [1,n]} h_i(w_i^{b^*})$. Since each $w_i^{b^*} = \operatorname{argmin}_{y_i \in g_i(D_{x_i})} h_i(y_i)$ corresponds to the minimum value obtainable by h_i, it is not possible to reduce the value $\sum_{i \in [1,n]} h_i(w_i^{b^*})$ by picking a different value for any $w_i^{b^*}$. $\square$

There exist potentially several $\mathbf{w}^{b^*}$ that minimise H. The correctness of our approach does not depend on a particular choice for those values.

We now characterise the segments of H.

Lemma 4. *If $\mathbf{w}^b$ is a witnessing assignment for b, then $\Delta^+(h_i, w_i^b) \geq \Delta^+(H, b)$ and $\Delta^-(h_i, w_i^b) \geq \Delta^-(H, b)$ for all $i \in [1, n]$.*

Proof. If b is increased by one, then one of the w_i^b must be increased by one as discussed previously. To reach the minimum value for $b+1$, one needs to increase the value of a variable y_k that has the smallest $\Delta^+(h_k, w_k^b)$. So the increase of H, namely $\Delta^+(H, b)$, is equal to $\Delta^+(h_k, w_k^b)$, which is smaller than or equal to $\Delta^+(h_i, w_i^b)$ for any other i. A similar argument is used for a decrease of b. $\square$

Lemma 5. *The length of each segment of H is equal to the sum of the lengths of the segments in the h_i functions with the same slope.*

Proof. As in the proof of Lemma 4, $\Delta^+(H, b)$ is equal to a minimal $\Delta^+(h_k, w_k^b)$. If one wants to increase b by more than one, the increase per unit stays constant as long as there is at least one variable with slope equal to $\Delta^+(H, b)$. This defines a segment of slope $\Delta^+(H, b)$, whose length is equal to the sum of the lengths of the segments of all h_i functions with the same slope.   □

We can use Lemmas 3 and 5 to construct H efficiently. Section 4 presents two ways to implement this construction in practice.

2.4   Computing the Feasibility Bound and a Witnessing Assignment

We can now show a case when problem (6) can be solved in a greedy way.

Theorem 1. *Problem (6) can be solved greedily if each h_i is discretely convex.*

Proof. If each function h_i is discretely convex, then the function H is also discretely convex (by Lemma 2) and can be constructed from the h_i (by Lemmas 3 and 5). Finding the minimum of a discretely convex function under some bound constraints can be done greedily, as a local minimum of a discretely convex function is also a global minimum (see, e.g., Theorem 2.2 in [7]).   □

Given the function H, problem (6) can be solved by first finding b^* minimising H, and then greedily increasing or decreasing b^* if b^* is not in $[\underline{g}, \overline{g}]$. In addition, it is useful for the filtering to compute the witnessing assignment $\mathbf{w}^{b^*}$ of b^*.

Thanks to Lemma 4, this can be achieved as in Algorithm 1. From now on, we simply write $\mathbf{w}$ to refer to $\mathbf{w}^{b^*}$. An assignment $\mathbf{w}$ that minimises the value of H without considering the bounds of b is initially constructed (lines 2–4). If b is in $[\underline{g}, \overline{g}]$, then the initial assignment is the final one. Otherwise the assignment is iteratively modified in order to satisfy the bounds of b. We assume $b < \underline{g}$ in line 5 (the case $b > \overline{g}$ is symmetrical and not shown). Then some w_i must be increased until b is equal to $\underline{g}$. This is done in two steps. In lines 6–10, the segment of H where $\underline{g}$ lies is found. Its slope is stored in $\Delta^{\max}$, and the distance between $\mathrm{bp}^-(H, \underline{g})$ and $\underline{g}$ is stored in *slack*. Those two values allow us then to modify each w_i separately (lines 11–17). For each i, first w_i is moved from breakpoint to breakpoint of h_i while the slope of the segment is smaller than $\Delta^{\max}$. Next, if the slope of the segment on the right of w_i is equal to $\Delta^{\max}$, then w_i is moved further on this segment, without exceeding the remaining slack (line 15).

The algorithm returns the witnessing assignment $\mathbf{w}$ (line 20), or "null" if the constraint is unsatisfiable (line 8), which triggers propagator failure and happens if there exists no value in the domains of the h_i such that $b \in [\underline{g}, \overline{g}]$.

3   Domain Filtering

To filter the domain of a variable, we extend the reasoning presented in Section 2.1. Indeed, variable x_j can take the value u if the cost of an optimal solution to the following problem is smaller than or equal to $\overline{f}$:

Algorithm 1. Greedy algorithm to compute a witnessing assignment

1: **function** GETWITNESSLOWERBOUND($\mathbf{h}, H, \underline{g}, \overline{g}$)
2:     **for all** $i \in [1, n]$ **do**
3:         $w_i := \text{argmin}_{v \in g_i(D_{x_i})} h_i(v)$ } initial bound
4:     $b := \sum_{i \in [1,n]} w_i$
5:     **if** $b < \underline{g}$ **then**
6:         **while** $\Delta^+(H, b) < +\infty$ **and** $\text{bp}^+(H, b) < \underline{g}$ **do**
7:             $b := \text{bp}^+(H, b)$
8:         **if** $\Delta^+(H, b) = +\infty \wedge b < \underline{g}$ **then return** null   } sharp bound
9:         $\Delta^{\max} := \Delta^+(H, b)$
10:        $slack := \underline{g} - b$
11:        **for all** $i \in [1, n]$ **do**
12:            **while** $\Delta^+(h_i, w_i) < \Delta^{\max}$ **do**
13:                $w_i := \text{bp}^+(h_i, w_i)$
14:            **if** $\Delta^+(h_i, w_i) = \Delta^{\max}$ **and** $slack > 0$ **then**   } modifying $\mathbf{w}$
15:                $w' := \min\left(\text{bp}^+(h_i, w_i), w_i + slack\right)$
16:                $slack := slack - w_i + w'$
17:                $w_i := w'$
18:     **else if** $b > \overline{g}$ **then**
19:         [analogous algorithm]
20:     **return w**

$$\text{minimise} \quad f_j(u) + \sum_{i \neq j \in [1,n]} f_i(x_i)$$

$$\text{such that} \quad \underline{g} \leq g_j(u) + \sum_{i \neq j \in [1,n]} g_i(x_i) \leq \overline{g} \tag{8}$$

$$x_i \in D_{x_i}, \quad \forall i \neq j \in [1, n]$$

Problem (8) resembles problem (3) but x_j is fixed to u. Hence we can use the same reformulation as in Section 2.1. We introduce the following new function:

$$H_j(b) = \min \left\{ \sum_{i \neq j \in [1,n]} h_i(y_i) \;\middle|\; \sum_{i \neq j \in [1,n]} y_i = b \wedge \forall i \neq j \in [1, n] : y_i \in g_i(D_{x_i}) \right\}$$

That is, $H_j(b)$ is similar to $H(b)$ in (5) but it only uses the functions h_i for i different from j. The optimal cost of problem (8) is the optimal cost of the following new problem:

$$\text{minimise} \quad f_j(u) + H_j(z)$$
$$\text{such that} \quad \underline{g} \leq g_j(u) + z \leq \overline{g} \tag{9}$$

where value u is given and z is the only variable. The result of the following lemma can be used to compute H_j.

Lemma 6. *The function H_j is discretely convex if all h_i are convex. The value b_j^* that minimises H_j is equal to the value b^* that minimises H minus the value v^* that minimises h_j. The length of each segment of H_j is equal to the length of the linear segment of H of the same slope minus the length of the linear segment of h_j of the same slope (if any).*

The proof (omitted for space reasons) of this lemma uses similar arguments to the ones of Lemmas 2 to 5. We show hereafter two ways to use H_j to filter the domains. The first way is applicable in general (provided H_j is discretely convex). The second way makes use of an additional property of f_j and g_j.

3.1   Filtering in the General Case

As several values u of x_j can have the same image v through g_j, the set of values in D_{x_j} that are consistent with constraints (1) and (2) can be partitioned as:

$$\bigcup_{v \in g_j(D_{x_j})} \left\{ u \;\middle|\; g_j(u) = v \wedge f_j(u) \leq \overline{f} - \min_{\underline{g} \leq z + v \leq \overline{g}} H_j(z) \right\}$$

That is, for each v, we have the set of values u in $g_j^{-1}(v)$ such that the optimal cost of problem (9) is no larger than $\overline{f}$, hence which are consistent. The domain of x_j can be made domain consistent by filtering the following unary constraint for each value $v \in g_j(D_{x_j})$:

$$g_j(x_j) = v \Rightarrow f_j(x_j) \leq \overline{f} - \min_{\underline{g} \leq z + v \leq \overline{g}} H_j(z) \qquad (10)$$

The function H_j being discretely convex, one can compute $\min_{\underline{g} \leq z + v \leq \overline{g}} H_j(z)$ (which is independent from a particular u) incrementally from a value v to $v+1$. In addition, if v is equal to w_j, the value of y_j in the witnessing assignment $\mathbf{w}$ computed in Section 2.4, then $H_j(\sum_{i \neq j \in [1,n]} w_i) + h_j(w_j) = H(\sum_{i \in [1,n]} w_i)$. This leads to Algorithm 2, which is used to filter the domain of x_j for the values v larger than w_j. This algorithm traverses h_j and H_j. The only complication is that in some cases (captured by the Boolean variable dec_b defined in lines 6 and 11) reaching an optimal solution to $\min_{\underline{g} \leq z + v \leq \overline{g}} H_j(z)$ involves decrementing b, which is the current value of z (line 9). Domain filtering according to constraint (10) takes place in lines 5 and 10. The algorithm ends when the optimal cost of problem (9) for $v + 1$ is larger than $\overline{f}$ (line 7). A complementary algorithm is used for the values smaller than w_j. Algorithm 2 achieves domain consistency provided the h_i are discretely convex. Section 5.1 discusses more precisely the link between the shape of the h_i and the consistency level.

3.2   Filtering in a Special Case

We now present a special case to avoid useless computation. Let us define $k_j(v) = \max f_j(g_j^{-1}(v))$, that is $k_j(v)$ is the largest value $f_j(u)$ for u such that $g_j(u) = v$. The function k_j is similar to h_j but the 'max' operator replaces the 'min' one.

Algorithm 2. Filtering algorithm for values larger than w_j (general case)

1: **function** FORWARDFILTER($j, \mathbf{h}, \mathbf{w}, H, \overline{f}$)
2:     $H_j := \text{COMPUTEHJ}(H, h_j)$
3:     $b := \sum_{i \in [1,n]} w_i - w_j$
4:     $v := w_j$
5:     FILTER($g_j(x_j) = v \Rightarrow f_j(x_j) \leq \overline{f} - H_j(b)$)
6:     $dec_b := b + v \geq \overline{g} \vee \Delta^-(H_j, b) < 0$
7:     **while** $H_j(b) + h_j(v) + (\textbf{if } dec_b \textbf{ then } \Delta^-(H_j, b) \textbf{ else } 0) + \Delta^+(h_j, v) \leq \overline{f}$ **do**
8:         $v := v + 1$
9:         **if** dec_b **then** $b := b - 1$
10:        FILTER($g_j(x_j) = v \Rightarrow f_j(x_j) \leq \overline{f} - H_j(b)$)
11:        $dec_b := b + v \geq \overline{g} \vee \Delta^-(H_j, b) < 0$
12:    FILTER($g_j(x_j) \leq v$)

If $h_j(v) \geq k_j(v - 1)$ for any value v larger than $v^* = \text{argmin}_{u \in g_j(D_{x_j})} h_j(u)$ and $h_j(v) \geq k_j(v + 1)$ for any v smaller than v^*, then there exists a value $v^{\max}$ such that for all values $v \in g_j(D_{x_j})$ smaller than $v^{\max}$ (but larger than or equal to w_j), all values $u \in g_j^{-1}(v)$ are consistent, and for all v larger than $v^{\max}$, there is no consistent u. We then need not consider all values but only find $v^{\max}$ and filter according to the two constraints $g_j(x_j) \leq v^{\max}$ and $g_j(x_j) = v^{\max} \Rightarrow f_j(x_j) \leq \overline{f} - \min_{\underline{g} \leq z + v^{\max} \leq \overline{g}} H_j(z)$. A similar argument holds for a $v^{\min}$.

Finding $v^{\max}$ amounts to computing the largest value v such that $h_j(v) + \min_{\underline{g} \leq z + v \leq \overline{g}} H_j(z) \leq \overline{f}$. As h_j and H_j are both convex, this problem can be solved by incrementally increasing v until the bound is reached. Algorithm 3 presents the steps to find $v^{\max}$. This algorithm is very similar to Algorithm 2, but it does not need to iterate over all the values v, only over the ones that are at a breakpoint of h_j or H_j. The increment is stored in ℓ (lines 6, 11, and 12).

An example of the special case is when g_j is the identity function. Then g_j is injective. Hence $h_j = k_j$ and, by convexity, h_j is non-decreasing right of v^* and non-increasing left of v^*.

4   A Parametric Propagator and Its Complexity

Our propagator is generic in the sense that it works correctly for any functions f_i and g_i that respect the condition of Theorem 1. However, we call it a *parametric* propagator, because rather than resorting to a fully generic implementation, we use hook functions and procedures that need to be provided. This allows us to get a lower time complexity. The parameters to provide for an instantiation are shown in Table 2: they are used in Algorithms 1 to 3. We now study the time and space complexity of our propagator, based on a few implementation notes.

Feasibility Test. We implement the H function as a linked list of segments, plus two integers for the values b^* and $H(b^*)$. The value of $H(b)$ is never queried for arbitrary values of b, but only for b^* and for incrementally modified values of b,

---

Algorithm 3. Filtering algorithm for values larger than w_j (special case)

1: **function** FORWARDFILTER$(j, \mathbf{h}, \mathbf{w}, H, \overline{f})$
2: $\quad H_j := \text{COMPUTEHJ}(H, h_j)$
3: $\quad b := \sum_{i \in [1,n]} w_i - w_j$
4: $\quad v := w_j$
5: $\quad dec_b := b + v \geq \overline{g} \vee \Delta^-(H_j, b) < 0$
6: $\quad \ell := \min\left\{b - \text{bp}^-(H_j, b),\ \text{bp}^+(h_j, v) - v,\ \text{if } dec_b \text{ then } +\infty \text{ else } \overline{g} - b - v\right\}$
7: $\quad$**while** $H_j(b) + h_j(v) + \ell \cdot ((\text{if } dec_b \text{ then } \Delta^-(H_j, b) \text{ else } 0) + \Delta^+(h_j, v)) \leq \overline{f}$ **do**
8: $\quad\quad v := v + \ell$
9: $\quad\quad$**if** dec_b **then** $b := b - \ell$
10: $\quad\quad dec_b := b + v \geq \overline{g} \vee \Delta^-(H_j, b) < 0$
11: $\quad\quad \ell := \min\{b - \text{bp}^-(H_j, b),\ \text{bp}^+(h_j, v) - v,\ \text{if } dec_b \text{ then } +\infty \text{ else } \overline{g} - b - v\}$
12: $\quad \ell := (\overline{f} - H_j(b) - h_j(v))/(\Delta^+(h_j, v) + (\text{if } dec_b \text{ then } \Delta^-(H_j, b) \text{ else } 0))$
13: $\quad v := v + \ell$
14: $\quad$FILTER$(g_j(x_j) \leq v)$
15: $\quad$FILTER$(g_j(x_j) = v \Rightarrow f_j(x_j) \leq \overline{f} - H_j(b))$

Table 2. Parameters to instantiate

Functions	Procedures
$\text{argmin}_{v \in g_i(\mathrm{D}_{x_i})} h_i(v)$	FILTER$(g_i(x_i) \leq v)$
$\Delta^+(h_i, v)$	FILTER$(g_i(x_i) \geq v)$
$\Delta^-(h_i, v)$	FILTER$(g_i(x_i) = v \Rightarrow f_i(x_i) \leq u)$
$\text{bp}^+(h_i, v)$	
$\text{bp}^-(h_i, v)$	

so that $H(b)$ can also be computed incrementally. This is also true for h_i, and is reflected by the absence of $h_i(u)$ from the parameters in Table 2. Using that linked list and some bookkeeping, the computation of $H(b)$, $\Delta^+(H, b)$, $\Delta^-(H, b)$, $\text{bp}^+(H, b)$, and $\text{bp}^-(H, b)$ can be performed in constant time for all values of b used in the algorithms.

Constructing the linked list of H can be done in various ways. A first way is to traverse each function h_i in turn and to build H incrementally by traversing the linked list in parallel. This takes $\mathcal{O}(n \cdot (s(h) \cdot p + s(H)))$ time, where $s(h)$ is the maximum number of segments among the h_i functions, $s(H)$ is the number of segments of H, and p is the highest complexity of the parametric functions. A second way is to collect all the segments from all the functions in a list, to sort this list, and to construct H by traversing the list. This takes $\mathcal{O}(n \cdot s(h) \cdot (p + \log(n \cdot s(h))))$ time and is asymptotically better than the first way when $s(H) > s(h) \cdot \log(n \cdot s(h))$.

Algorithm 1 computes a witnessing assignment in $\mathcal{O}(s(H) + n \cdot s(h))$ time. This is dominated by the prior construction of H, as $s(H) \leq n \cdot s(h)$.

Filtering. We implement Algorithm 2 to run in $\mathcal{O}(r(h) \cdot c)$ time, where $r(h) = |g_j(\mathrm{D}_{x_j})|$ and c is the highest complexity of the procedures in Table 2. The segments of H_j are computed on the fly from h_j and H. The sum in line 3 of

Table 3. Time complexity of the different versions of the propagator

Propagator	Time complexity
Traversing, general case	$\mathcal{O}(n \cdot (s(h) \cdot p + s(H) + r(h) \cdot c)$
Sorting, general case	$\mathcal{O}(n \cdot (s(h) \cdot p + s(h) \cdot \log(n \cdot s(h)) + r(h) \cdot c))$
Traversing, special case	$\mathcal{O}(n \cdot (s(h) \cdot p + s(H) + c))$
Sorting, special case	$\mathcal{O}(n \cdot (s(h) \cdot p + s(h) \cdot \log(n \cdot s(h)) + s(H) + c))$

Algorithm 2 is actually provided by our implementation of H, so it need not be recomputed each time. Algorithm 3 takes $\mathcal{O}(s(h) + s(H) + c)$ time.

The Whole Propagator. The time complexity of our propagator is obtained by multiplying the filtering complexity by n (the number of variables) and adding the complexity of computing H. Table 3 summarises this for the different versions of the propagator. Note that $s(h) \leq r(h) \leq |\mathrm{D}_x|$ and $s(H) \leq n \cdot s(h)$.

The space complexity of our propagator is $\mathcal{O}(n + s(H))$, as we need to store a constant amount of information (namely w_i) for each variable and the whole function H (which amounts to a constant amount for each of its segments). The functions h_i and H_j are not stored explicitly.

5   Instantiating the Parametric Propagator

We now show how our propagator can be used for particular pairs of constraints.

Note that if h_i is a linear function, then $-h_i$ is also discretely convex. This means that one can put a lower bound $\underline{f}$ on $\sum_{i \in [1,n]} f_i(x_i)$ and run the propagator twice, first with constraint (1) being $\sum_{i \in [1,n]} f_i(x_i) \leq \overline{f}$, then with constraint (1) being $-\sum_{i \in [1,n]} f_i(x_i) \leq -\underline{f}$.

Our propagator can also be extended to handle variables as the upper and lower bounds of the constraints. In such a case, the largest values in the domains of $\overline{f}$ and $\overline{g}$, and the smallest values in the domains of $\underline{f}$ and $\underline{g}$ are used in the propagator. In addition, the other bound of each variable can be constrained by the H function. Only bounds($\mathbb{Z}$) consistency can be achieved on those variables.

5.1   Instantiations and Consistency

We now discuss for which functions f_i and g_i our propagator can be used and how it affects the consistency of the propagator. The required discrete convexity of the h_i functions puts a strong restriction on the shape of the g_i. Recall that $g_i(\mathrm{D}_{x_i})$ must be an interval by the first condition in Definition 1. Note that the discrete convexity must be respected for *all* D_{x_i} that arise during the search.

If D_{x_i} can be any set of integers, then the only instantiations of g_i satisfying the first condition of Definition 1 are those whose image contains only two values, which must be consecutive. We call these *characteristic functions*. In such a case, the second condition of Definition 1 is always respected and the f_i can be any (integer) functions.

If D_{x_i} can only be an interval, then the class of g_i functions satisfying the first condition of Definition 1 is more general, namely all functions where

$$|g_i(u) - g_i(u+1)| \leq 1 \quad \forall u, u+1 \in D_{x_i} \tag{11}$$

If there are holes in a domain D_{x_i}, then D_{x_i} can be relaxed to the smallest containing interval without losing the correctness of the approach. Some propagation may be lost, but this compromise is often acceptable for global constraints. In particular, we do not achieve domain consistency, but bounds($\mathbb{Z}$) consistency.

Among others, the identity function respects equation (11). If g_i is the identity function, then f_i must be discretely convex, because $h_i = f_i$. For other instantiations of g_i satisfying (11), the restrictions on f_i are varying.

5.2 Example Instantiations

We now show that many existing (pairs of) constraints fit our parametric problem, optionally extended with a lower bound $\underline{f}$ and with variable bounds. Table 1 presents several instantiations of f_i and g_i, together with the derived h_i. We discuss below various constraints and their time complexity. The concrete complexities are derived from the complexities in Table 3 by replacing $s(h)$, $s(H)$, $r(h)$, p, and c by suitable values derived from the h_i.

If $g_i(u) = 0$ for all i, then the second constraint vanishes and we can use our propagator for a single SUM constraint, e.g., a linear inequation. Our parametric propagator is however too general for this simple case, as it runs in $\mathcal{O}(n \cdot \log n)$ time, while a dedicated bounds($\mathbb{Z}$) consistent propagator runs in $\mathcal{O}(n)$ time [6].

The case $g_i(u) = u$ covers many interesting constraints already presented in the literature. In particular, it covers the bounds($\mathbb{Z}$) consistent propagators for the statistical constraints DEVIATION and SPREAD with a fixed rational mean. Interestingly, it can be generalised to any L_p-norm, with $p > 0$ (except $L_{+\infty}$). One can also give a different penalty for deviations over and under the average. The time complexity of our propagator is $\mathcal{O}(n)$ for DEVIATION, which matches the best published propagator [17]. For SPREAD (and higher norms), the time complexity of our propagator is $\mathcal{O}(n \cdot d)$, with $d = \left| \cup_{i \in [1,n]} D_{x_i} \right|$. This is incomparable to the complexity $\mathcal{O}(n \cdot \log n)$ of the best published propagator [9]. Note that our propagator achieves bounds($\mathbb{Z}$) consistency, which has only been achieved very recently in the case of SPREAD [18].

As an example, we show in Table 4 the instantiation of the parameters for DEVIATION (symmetric parameters are omitted). For DEVIATION, h_i has (up to) three segments, joining at the breakpoints $\lfloor \mu \rfloor$ and $\lceil \mu \rceil$.

The case $g_i(u) = u$ and $f_i(u) = a_i \cdot u$ can be used to model a restricted version of the WEIGHTEDAVERAGE constraint [3], where the weight are variables, the values are constants, and the average must take an integer value. The time complexity of our bounds($\mathbb{Z}$) consistent propagator is $\mathcal{O}(n \cdot \log n)$, though the dedicated propagator runs in $\mathcal{O}(n)$ time.

If g_i is a characteristic function, then f_i can be any function. A characteristic function may be used to count, as is the case of the COUNT family of constraints

Table 4. Expressions for instantiating a propagator for DEVIATION. The conditions are not always mutually exclusive and are to be evaluated in top-down order.

Parameter	Instantiation
$\operatorname{argmin}_{v \in g_i(\mathrm{D}_{x_i})} h_i(v)$	$\begin{cases} \lceil \mu \rceil & \text{if } \min \mathrm{D}_{x_i} \leq \mu \leq \max \mathrm{D}_{x_i} \wedge \lceil \mu \rceil - \mu < \mu - \lfloor \mu \rfloor \\ \lfloor \mu \rfloor & \text{if } \min \mathrm{D}_{x_i} \leq \mu \leq \max \mathrm{D}_{x_i} \wedge \lceil \mu \rceil - \mu \geq \mu - \lfloor \mu \rfloor \\ \min \mathrm{D}_{x_i} & \text{if } \mu < \min \mathrm{D}_{x_i} \\ \max \mathrm{D}_{x_i} & \text{if } \mu > \max \mathrm{D}_{x_i} \end{cases}$
$\Delta^+(h_i, v)$	$\begin{cases} +\infty & \text{if } v = \max \mathrm{D}_{x_i} \\ -n & \text{if } v < \lfloor \mu \rfloor \\ n \cdot (\lceil \mu \rceil + \lfloor \mu \rfloor) - 2 \cdot n \cdot \mu & \text{if } v = \lfloor \mu \rfloor \wedge \lfloor \mu \rfloor \neq \lceil \mu \rceil \\ n & \text{if } v \geq \lceil \mu \rceil \end{cases}$
$\mathrm{bp}^+(h_i, v)$	$\begin{cases} +\infty & \text{if } v = \max \mathrm{D}_{x_i} \\ \min(\max \mathrm{D}_{x_i}, \lfloor \mu \rfloor) & \text{if } v < \lfloor \mu \rfloor \\ \lceil \mu \rceil & \text{if } v = \lfloor \mu \rfloor \wedge \lfloor \mu \rfloor \neq \lceil \mu \rceil \\ \max \mathrm{D}_{x_i} & \text{if } v \geq \lceil \mu \rceil \end{cases}$
FILTER($g_i(x_i) \leq v$)	FILTER($x_i \leq v$)
FILTER($g_i(x_i) = v \Rightarrow f_i(x_i) \leq u$)	FILTER($|n \cdot v - n \cdot \mu| > u \Rightarrow x_i \neq v$)

(e.g., AMONG [1,2]). But characteristic functions can also be used to represent the MAXIMUM constraint. Indeed, the constraint $m = \max_{i \in [1,n]} x_i$ can be decomposed as $\forall i \in [1,n] : m \geq x_i \wedge \sum_{1 \in [1,n]}(\text{if } x_i \geq m \text{ then } 1 \text{ else } 0) \geq 1$. Table 1 gives the definition of h_i for LINEAR and EXACTLY, in which case our propagator is domain consistent and runs in $\mathcal{O}(n \cdot (\log n + p + c))$ time, as does the dedicated propagator presented in [13].

Many other pairs can be instantiated. Note that the f_i or g_i functions can differ for each i, i.e., one can mix in the same sum terms of different forms (e.g., some linear and some quadratic), as long as each function h_i is discretely convex.

6   Experimental Evaluation

To show that the genericity of our propagator is not detrimental not only to asymptotic complexity (as seen in Section 5) but also to performance, we propose a small experiment to compare custom propagators with instantiations of our parametric propagator. We selected the DEVIATION [17] and SPREAD [18] constraints as their bounds($\mathbb{Z}$)-consistent propagators are freely available in the distribution of OscaR [8]. We performed the comparison on the 100 instances of the Balanced Academic Curriculum Problem (BACP) that were introduced in [16],[1] modelled as in the OscaR distribution (we only slightly modified the search heuristic to make it deterministic, so that the search trees are the same).

---

[1] They are available from http://becool.info.ucl.ac.be/resources/bacp

For DEVIATION, we used the 44 instances that are solved to optimality in more than 1 second (to avoid measurement errors) but less than 12 hours (3 instances timed out). When using our parametric propagator, the time to solve an instance is on average only 7% longer than when using the custom propagator (with a standard deviation of 5%). The numbers of nodes in the search tree and calls to the propagator are *exactly* the same for both propagators due to their common level of consistency and the deterministic search procedure.

For SPREAD, we used the 33 instances that are solved to optimality in more than 1 second but less than 12 hours (2 instances timed out). When using our parametric propagator, the time to solve an instance is on average 28% *shorter* than when using the custom propagator (with a standard deviation of 10%). Again, the numbers of nodes in the search tree and calls to the propagator are *exactly* the same for both propagators. This improvement is explained by a different algorithmic approach, which is in our favour when the domains of the variables are small, as is the case for the BACP instances.

Our Java implementation is available at `http://www.it.uu.se/research/group/astra/software/convexpairs` and a package for replication at `http://recomputation.org` [5].

7 Conclusion, Related Work, and Future Work

We have studied how to propagate pairs of SUM constraints that respect a discrete convexity condition. From this condition, we have derived a parametric propagator, which can be instantiated to be competitive with previously published propagators, often matching their time complexity, despite its generality.

Our approach of first computing a feasibility bound and then incrementally adapting it is not new and has been used in the design of several propagators. Among others, this is the case for the constraints covered by our own propagator. However, the novelty of our work is that for the first time we abstract from the details of each constraint to focus on their common properties. This is close in spirit to what has been done with SEQBIN [10] for another class of constraints.

When the g_i are characteristic functions, our conjunction of sum constraints can be represented using COSTGCC [14]. However, this requires the explicit representation of all variable-value pairs and induces a larger time complexity than our propagator. On the other hand, COSTGCC can handle more than one counting constraint in one propagator.

There are a number of open questions we plan to address in the future. Can we *automatically* generate the instantiation of the parameters from the definitions of the f_i and g_i? Can we make an *incremental* propagator that has a better time complexity along a branch of the search tree? Can we extend the approach to functions that take more than one argument, say $f_i(x_i, y_i)$ for variables y_i distinct from each other, or $f_i(x_i, y)$ for a shared variable y? Can we deal with more than two sum constraints in one propagator? Beside when there are holes in the domains, when is it correct and useful to use a relaxation of h_i when this function is not discretely convex?

References

1. Beldiceanu, N., Contejean, E.: Introducing global constraints in CHIP. Mathematical and Computer Modelling 20(12), 97–123 (1994)
2. Bessière, C., Hebrard, E., Hnich, B., Kiziltan, Z., Walsh, T.: Among, common and disjoint constraints. In: Hnich, B., Carlsson, M., Fages, F., Rossi, F. (eds.) CSCLP 2005. LNCS (LNAI), vol. 3978, pp. 29–43. Springer, Heidelberg (2006)
3. Bonfietti, A., Lombardi, M.: The weighted average constraint. In: Milano, M. (ed.) CP 2012. LNCS, vol. 7514, pp. 191–206. Springer, Heidelberg (2012)
4. Fujishige, S.: Submodular Functions and Optimization. In: Annals of Discrete Mathematics, 2nd edn., Elsevier (2005)
5. Gent, I.P.: The recomputation manifesto. CoRR, abs/1304.3674 (2013)
6. Harvey, W., Schimpf, J.: Bounds consistency techniques for long linear constraints. In: Proceedings of TRICS 2002, the Workshop on Techniques foR Implementing Constraint programming Systems, pp. 39–46 (2002)
7. Murota, K.: Recent developments in discrete convex analysis. In: Cook, W., Lovász, L., Vygen, J. (eds.) Research Trends in Combinatorial Optimization, pp. 219–260. Springer (2009)
8. OscaR Team. OscaR: Scala in OR (2012), https://bitbucket.org/oscarlib/oscar
9. Pesant, G., Régin, J.-C.: SPREAD: A balancing constraint based on statistics. In: van Beek, P. (ed.) CP 2005. LNCS, vol. 3709, pp. 460–474. Springer, Heidelberg (2005)
10. Petit, T., Beldiceanu, N., Lorca, X.: A generalized arc-consistency algorithm for a class of counting constraints. In: IJCAI 2011, pp. 643–648. AAAI Press (2011), revised edition available at http://arxiv.org/abs/1110.4719
11. Petit, T., Régin, J.-C., Beldiceanu, N.: A $\Theta(n)$ bound-consistency algorithm for the increasing sum constraint. In: Lee, J. (ed.) CP 2011. LNCS, vol. 6876, pp. 721–728. Springer, Heidelberg (2011)
12. Puget, J.-F.: Improved bound computation in presence of several clique constraints. In: Wallace, M. (ed.) CP 2004. LNCS, vol. 3258, pp. 527–541. Springer, Heidelberg (2004)
13. Razakarison, N., Beldiceanu, N., Carlsson, M., Simonis, H.: GAC for a linear inequality and an atleast constraint with an application to learning simple polynomials. In: SoCS 2013, AAAI Press (2013)
14. Régin, J.-C.: Cost-based arc consistency for global cardinality constraints. Constraints 7(3-4), 387–405 (2002)
15. Régin, J.-C., Petit, T.: The objective sum constraint. In: Achterberg, T., Beck, J.C. (eds.) CPAIOR 2011. LNCS, vol. 6697, pp. 190–195. Springer, Heidelberg (2011)
16. Schaus, P.: Solving balancing and bin-packing problems with constraint programming, PhD Thesis, Université catholique de Louvain, Belgium (2009)
17. Schaus, P., Deville, Y., Dupont, P.: Bound-consistent deviation constraint. In: Bessière, C. (ed.) CP 2007. LNCS, vol. 4741, pp. 620–634. Springer, Heidelberg (2007)
18. Schaus, P., Régin, J.-C.: Bound-consistent spread constraint, application to load balancing in nurse to patient assignments (submitted)
19. Schulte, C., Stuckey, P.J.: When do bounds and domain propagation lead to the same search space? ACM Transactions on Programming Languages and Systems 27(3), 388–425 (2005)

Breaking Symmetry with Different Orderings

Nina Narodytska and Toby Walsh

NICTA and UNSW,
Sydney, Australia
{nina.narodytska,toby.walsh}@nicta.com.au

Abstract. We can break symmetry by eliminating solutions within each symmetry class. For instance, the Lex-Leader method eliminates all but the smallest solution in the lexicographical ordering. Unfortunately, the Lex-Leader method is intractable in general. We prove that, under modest assumptions, we cannot reduce the worst case complexity of breaking symmetry by using other orderings on solutions. We also prove that a common type of symmetry, where rows and columns in a matrix of decision variables are interchangeable, is intractable to break when we use two promising alternatives to the lexicographical ordering: the Gray code ordering (which uses a different ordering on solutions), and the Snake-Lex ordering (which is a variant of the lexicographical ordering that re-orders the variables). Nevertheless, we show experimentally that using other orderings like the Gray code to break symmetry can be beneficial in practice as they may better align with the objective function and branching heuristic.

1 Introduction

Symmetry occurs in many combinatorial problems. For example, when coloring a graph, we can permute the colors in any proper coloring. Symmetry can also be introduced by modelling decisions (e.g. using a set of finite domain variables to model a set of objects will introduce the symmetries that permute these variables). A common method to deal with symmetry is to add constraints which eliminate symmetric solutions (e.g. [1–13]). Unfortunately, breaking symmetry by adding constraints to eliminate symmetric solutions is intractable in general [2]. More specifically, deciding if an assignment is the smallest in its symmetry class for a matrix with row and column symmetries is NP-hard, supposing rows are appended together and compared lexicographically. There is, however, nothing special about appending rows together or comparing solutions lexicographically. We could use *any* total ordering over assignments. For example, we could break symmetry with the Gray code ordering. That is, we add constraints that eliminate symmetric solutions within each symmetry class that are not smallest in the Gray code ordering. This is a total ordering over assignments used in error correcting codes. Such an ordering may pick out different solutions in each symmetry class, reducing the conflict between symmetry breaking, problem constraints, objective function and the branching heuristic. The Gray code ordering has some properties that may be useful for symmetry breaking. In particular, neighbouring assignments in the ordering only differ at one position, and flipping one bit reverses the ordering of the subsequent bits.

C. Schulte (Ed.): CP 2013, LNCS 8124, pp. 545–561, 2013.

As a second example, we can break row and column symmetry with the Snake-Lex ordering [14]. This orders assignments by lexicographically comparing vectors constructed by appending the variables in the matrix in a "snake like" manner. The first row is appended to the reverse of the second row, and this is then appended to the third row, and then the reverse of the fourth row and so on. As a third example, we can break row and column symmetry by ordering the rows lexicographically and the columns with a multiset ordering [15]. This is incomparable to the Lex-Leader method.

We will argue theoretically that breaking symmetry with a different ordering over assignments cannot improve the worst case complexity. However, we also show that other orderings can be useful in practice as they pick out different solutions in each symmetry class. Our argument has two parts. We first argue that, under modest assumptions which are satisfied by the Gray code and Snake-Lex orderings, we cannot reduce the computational complexity from that of breaking symmetry with the lexicographical ordering which considers variables in a matrix row-wise. We then prove that for the particular case of row and column symmetries, breaking symmetry with the Gray code or Snake-Lex ordering is intractable (as it was with the lexicographical ordering). Many dynamic methods for dealing with symmetry are equivalent to posting symmetry breaking constraints "on the fly" (e.g. [16–24]).

Hence, our results have implications for such dynamic methods too.

2   Background

A symmetry of a set of constraints S is a bijection σ on complete assignments that maps solutions of S onto other solutions of S. Many of our results apply to the more restrictive definition of symmetry which considers just those bijections which map individual variable-value pairs [25]. However, this more general definition captures also conditional symmetries [26]. In addition, a few of our results require this more general definition. In particular, Theorem 3 only holds for this more general definition[1]. The set of symmetries form a group under composition. Given a symmetry group Σ, a subset Π generates Σ iff any $\sigma \in \Sigma$ is a composition of elements from Π. A symmetry group Σ partitions the solutions into symmetry classes (or orbits). We write $[A]_\Sigma$ for the symmetry class of solutions symmetric to the solution A. Where Σ is clear from the context, we write $[A]$. A set of symmetry breaking constraints is *sound* iff it leaves at least one solution in each symmetry class, and *complete* iff it leaves at most one solution in each symmetry class.

We will study what happens to symmetries when problems are reformulated onto equivalent problems. For example, we might consider the Boolean form of a problem in which $X_i = j$ maps onto $Z_{ij} = 1$. Two sets of constraints, S and T over possibly different variables are *equivalent* iff there is a bijection between their solutions. Suppose U_i and V_i for $i \in [1, k]$ are partitions of the sets U and V into k subsets. Then the two partitions are *isomorphic* iff there are bijections $\pi : U \mapsto V$ and $\tau : [1, k] \mapsto [1, k]$ such that $\pi(U_i) = V_{\tau(i)}$ for $i \in [1, k]$ where $\pi(U_i) = \{\pi(u) \mid u \in U_i\}$. Two symmetry groups Σ and Π of constraints S and T respectively are *isomorphic* iff S and T are equivalent, and their symmetry classes of solutions are isomorphic.

---

[1] We thank an anonymous reviewer for pointing this out.

3   Using Other Orderings

The Lex-Leader method [2] picks out the lexicographically smallest solution in each symmetry class. For every symmetry σ, it posts a lexicographical ordering constraint: $\langle X_1, \ldots, X_n \rangle \leq_{\text{lex}} \sigma(\langle X_1, \ldots, X_n \rangle)$ where X_1 to X_n is some ordering on the variables in the problem. Many static symmetry breaking constraints can be derived from such Lex-Leader constraints. For example, DOUBLELEX constraints to break row and column symmetry can be derived from them [27]. As a second example, PRECEDENCE constraints to break the symmetry due to interchangeable values can also be derived from them [5, 8]. Efficient algorithms exist to propagate such lexicographical constraints (e.g. [28–30]).

We could, however, break symmetry by using another ordering on assignments like the Gray code ordering. We define the Gray code ordering on Boolean variables. For each symmetry σ, we could post an ordering constraint:

$$\langle X_1, \ldots, X_n \rangle \leq_{\text{Gray}} \sigma(\langle X_1, \ldots, X_n \rangle)$$

Where the k-bit Gray code ordering is defined recursively as follows: 0 is before 1, and to construct the $k+1$-bit ordering, we append 0 to the front of the k-bit ordering, and concatenate it with the reversed k-bit ordering with 1 appended to the front. For instance, the 4-bit Gray code orders assignments as follows:

$$0000, 0001, 0011, 0010, 0110, 0111, 0101, 0100,$$
$$1100, 1101, 1111, 1110, 1010, 1011, 1001, 1000$$

The Gray code ordering is well founded. Hence, every set of complete assignments will have a smallest member under this ordering. This is the unique complete assignment in each symmetry class selected by posting such Gray code ordering constraints. Thus breaking symmetry with Gray code ordering constraints is sound and complete.

Proposition 1. *Breaking symmetry with Gray code ordering constraints is sound and complete.*

In Section 6, we propose a propagator for the Gray code ordering constraint. We cannot enforce the Gray code ordering by ordering variables and values, and using a lexicographical ordering constraint. For example, we cannot map the 2-bit Gray code onto the lexicographical ordering by simply re-ordering variables and values. To put it another way, no reversal and/or inversion of the bits in the 2-bit Gray code will map it onto the lexicographical ordering. The 2-bit Gray code orders 00, 01, 11 and then 10. We can invert the first bit to give: 10, 11, 01 and then 00. Or we can invert the second bit to give: 01, 00, 10, and then 11. Or we can invert both bits to give: 11, 10, 00, and then 01. We can also reverse the bits to give: 00, 10, 11, and then 01. And we can then invert one or both bits to give: 10, 00, 01, and then 10; or 01, 11, 10, and then 00; or 11, 01, 00, and then 10. Note that none of these re-orderings and inversions is the 2-bit lexicographical ordering: 00, 01, 10, and then 11.

4   Complexity of Symmetry Breaking

We will show that, under some modest assumptions, we cannot make breaking symmetry computationally easier by using a new ordering like the Gray code ordering. Our argument breaks into two parts. First, we observe how the symmetry of a problem changes when we reformulate onto an equivalent problem. Second, we argue that we can map onto an equivalent problem on which symmetry breaking is easier.

Proposition 2. *If a set of constraints S has a symmetry group Σ, S and T are equivalent sets of constraints, π is any bijection between solutions of S and T, and $\Pi \subseteq \Sigma$ then:*

(a) $\pi\Sigma\pi^{-1}$ is a symmetry group of T;
(b) Σ and $\pi\Sigma\pi^{-1}$ are isomorphic symmetry groups;
(c) if Π generates Σ then $\pi\Pi\pi^{-1}$ generates $\pi\Sigma\pi^{-1}$.

We will use this proposition to argue that symmetry breaking with any ordering besides the lexicographical ordering is intractable. We consider only *simple* orderings. In a simple ordering, we can compute the position of any assignment in the ordering in polynomial time, and given any position in the ordering we can compute the assignment at this position in polynomial time. We now give our main result.

Proposition 3. *Given any simple ordering $\preceq$, there exists a symmetry group such that deciding if an assignment is smallest in its symmetry class according to $\preceq$ is NP-hard.*

Proof. Deciding if an assignment is smallest in its symmetry class according to $\leq_{\mathrm{lex}}$ is NP-hard [2]. Since $\preceq$ and $\leq_{\mathrm{lex}}$ are both simple orderings, there exist polynomial functions f to map assignments onto positions in the $\leq_{\mathrm{lex}}$ ordering, and g to map positions in the $\preceq$ ordering to assignments. Consider the mapping π defined by $\pi(A) = g(f(A))$. Now π is a permutation that is polynomial to compute which maps the total ordering of assignments of $\leq_{\mathrm{lex}}$ onto that for $\preceq$. Similarly, π^{-1} is a permutation that is polynomial to compute which maps the total ordering of assignments of $\preceq$ onto that for $\leq_{\mathrm{lex}}$. Let Σ_{rc} be the row and column symmetry group. By Theorem 2, the problem of finding the lexicographical least element of each symmetry class for Σ_{rc} is equivalent to the problem of finding the least element of each symmetry class according to $\preceq$ for $\pi\Sigma_{rc}\pi^{-1}$. Thus, for the symmetry group $\pi\Sigma_{rc}\pi^{-1}$ deciding if an assignment is smallest in its symmetry class according to $\preceq$ is NP-hard. $\square$

It follows that there exists an infinite family of symmetry groups such that checking a constraint which is only satisfied by the smallest member of each symmetry class is NP-hard. Note that the Gray code and Snake-Lex orderings are simple. Hence, breaking symmetry with either ordering is NP-hard for some symmetry groups. Note that we are not claiming that deciding if an assignment is smallest in its symmetry class is NP-complete. First, we would need to worry about the size of the input (since we are considering the much larger class of symmetries that act on complete assignments rather than on literals). Second, to decide that an assignment is the smallest, we are also answering a complement problem (there is *no* smaller symmetric assignment). This will take us to DP-completeness or above.

5   Breaking Matrix Symmetry

We next consider a common type of symmetry. In many models, we have a matrix of decision variables in which the rows and columns are interchangeable [31–33]. We will show that breaking row and column symmetry specifically is intractable with the Gray code and the Snake-Lex orderings, as it is with the lexicographical ordering that considers the variables in a row-wise order.

Proposition 4. *Finding the smallest solution up to row and column symmetry for the Snake-Lex ordering is NP-hard.*

Proof. We reduce from the problem of finding the Lex-Leader solution of a matrix B. Let B be an $n \times m$ matrix of Boolean values. W.l.o.g. we assume B does not contain a row of only ones since any such row can be placed at the bottom of the matrix. We embed B in the matrix M such that finding $\sigma(M)$, denoted M', the smallest row and column symmetry of M in the Snake-Lex ordering is equivalent to finding the Lex-Leader of B. We ensure that even rows in the Snake-Lex smallest symmetric solution of M are taken by dummy identical rows. Then in odd rows, where Snake-Lex moves from the left to the right along a row like Lex does, we embed the Lex-Leader solution of B.

Let z be the maximum number of zeros in any row of B. We construct M with $2n+1$ rows and $(z+2) + (z+1) + m$ columns so that it contains three sets of rows. The first set consists of a single row that contains $z+2$ zeros followed by $(z+1) + m$ ones. The second set contains n identical rows with $z+2+m$ ones followed by $z+1$ zeros in each row. The third set of rows contains n rows such that at the ith row the first $(z+2)$ positions are ones, the next m positions are the ith row from B and the last $z+1$ positions are ones again. Schematic representation of M is shown at Figure 1(a).

We determine positions of rows and columns that must be fixed in M' up to permutation of identical rows and columns. The first row of M has to be the first row of M' as no other row contains $z+2$ zeros. Note that this also fixes the position of columns from 1 to $z+2$ in M to be the first columns in the M'. Note also that these columns are identical and each of them contains the zero in the first row only.

One of the rows in the second set has to be the second row of M', as none of the rows that embed rows from B contains $z+1$ zeros. As we move from the right to the left on even rows, this also makes sure that last $z+1$ columns from M must be the last

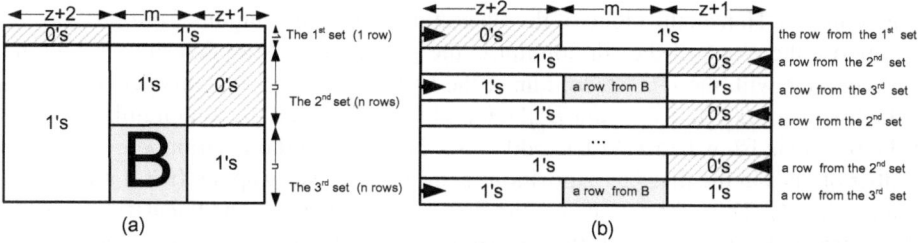

(a)   (b)

Fig. 1. (a) Construction of M (b) Partial construction of M'. The first and all even rows are fixed.

columns in M'. We summarise that at this point the first two rows are fixed and the first $z + 2$ columns and the last $z + 1$ columns in M' must be equal to a permutation of the first $z + 2$ identical columns and the last $z + 1$ identical columns in M, respectively.

By assumption, B does not contain rows with all ones. Moreover, only rows that embed rows from B can have the value zero at columns from $(z+2)+1$ to $(z+2)+m$ in M'. Hence, a row from the third set that embeds a row from B has to be the third row in M'. We do not specify which row it is at this point. The fourth row has to be again a row from the second set as any of remaining rows from the second set has $z + 1$ zeros in the last $z + 1$ columns in M' while any row that embeds B has at most z zeros. We can repeat this argument for the remaining rows. A schematic representation of the positions of rows from the first and second sets are shown in Figure 1(b). Note that the first and all even rows in M' are fixed. The only part of M' not yet specified is the ordering of odd rows of m columns from $(z + 2) + 1$ to $(z + 2) + m$. These are exactly all rows from B. Hence, finding M' is reduced to ordering of this set of rows and columns that embed B. Now, all columns from $(z + 2) + 1$ to $(z + 2) + m$ are interchangeable, all odd rows except the first are interchangeable, and all elements of M' except elements of B are fixed by construction. As the Snake-Lex ordering goes from the left to the right on odd rows like the Lex ordering, finding M' is equivalent to finding the Lex-Leader of B □

To show that finding the smallest row and column symmetry in the Gray code ordering is NP-hard, we need a technical lemma about cloning columns in a matrix. We use rowwise ordering in a matrix. Suppose we clone each column in a $n \times m$ Boolean matrix B to give the matrix B^c. Let B_{gl}^c be the smallest row and column symmetry of B^c in the Gray code ordering.

Lemma 1. *Any original column of B is followed by its clone in B_{gl}^c ignoring permutation of identical original columns.*

Proof. By contradiction. Suppose there exists an element $B_{gl}^c[j, i + 1]$ such that the original column i and the next column $i + 1$ are different at the jth row. We denote by k the $[j, i + 1]$ element of B_{gl}^c in its row-wise linearization. We ignore the rows from $j + 1$ to n at this point as they are not relevant to this discrepancy.

Each pair of columns coincide on the first j rows for the first $i - 1$ columns and on the first $j - 1$ rows for the columns from i to m. We conclude that (1) i is odd and $i + 1$ is even; (2) the number of ones between the first and the $(k - 2)$th positions in the linearization of B_{gl}^c is even as each value is duplicated; (3) the clone of the ith column cannot be among the first $i - 1$ columns as each such column is followed by its clone by assumption. Hence, the clone of the ith column is among columns from $i + 2$ to m.

Suppose the clone of the ith column is the pth column. Note that the pth column must coincide with the $i + 1$th column at the first $j - 1$ rows. We consider two cases. In the first case, $B_{gl}^c[j, i] = 1$ and $B_{gl}^c[j, i + 1] = 0$. Note that the total number of ones at the positions from 1 to $k - 1$ is odd as we have one in the position $k - 1$ and the number of ones in the first $k - 2$ positions is even. Next we swap the $(i + 1)$th and pth columns in B_{gl}^c. This will not change the first $k - 1$ elements in the linearization as the pth column must coincide with the $i + 1$th column at the first $j - 1$ rows. Moreover, this swap puts 1 in position k. As the number of ones up to the $(k - 1)$th position is odd then 1 goes before 0 at position k in the Gray code ordering. Hence, by swapping the $(i+1)$th and pth columns we obtain a matrix that is smaller than B_{gl}^c in the Gray code ordering. This is a contradiction. In the second case, $B_{gl}^c[j,i] = 0$ and $B_{gl}^c[j,i+1] = 1$. Note that the total number of ones at positions 1 to $k-1$ in the linearization is even as we have zero at the position $k-1$ and the number of ones in the first $k-2$ positions is even. Therefore, 0 precedes 1 at position k in the Gray code ordering. By swapping the $(i+1)$th and pth columns we obtain a matrix that is smaller than B_{gl}^c in the Gray code ordering as 0 appears at the position k instead of 1. This is a contradiction.   □

Proposition 5. *Finding the smallest solution up to row and column symmetry for the Gray code ordering is NP-hard.*

Proof. We again reduce from the problem of finding the Lex-Leader solution of a matrix B. We clone every column of B and obtain a new matrix B^c. Let B_{gl}^c be the smallest row and column symmetry of B^c in the Gray code ordering. Lemma 1 shows that each original column is followed by its clone in B_{gl}^c. Next we delete all clones by removing every second column. We call the resulting matrix B_l. We prove that B_l is the Lex-Leader of B by contradiction. Suppose there exists a matrix M which is the Lex-Leader of B that is different from B_l. Hence, M is also the Lex-Leader of B_l. We find the first element $M[j,i]$ where $B_l[j,i] \neq M[j,i]$ in the row-wise linearization of M and B_l, so that $B_l[j,i] = 1$ and $M[j,i] = 0$. We denote by k the position of the $[j,i]$ element of M in its row-wise linearization. We clone each column of M once and put each cloned column right after its original column. We obtain a new matrix M^c. We show that M^c is smaller than B_{gl}^c in the Gray code ordering to obtain a contradiction.

As $B_l[j,i] = 1$ and $M[j,i] = 0$ then $B_{gl}^c[j,2i-1] = 1$ and $M^c[j,2i-1] = 0$ because the matrices B_{gl}^c and M^c are obtained from B_l and M by cloning each column and putting each clone right after its original column. As B_l and M coincide on the first $k-1$ positions then B_{gl}^c and M^c coincide in the first $2k-2$ positions. By transforming B_l and M to B_{gl}^c and M^c, we duplicated each value in positions from 1 to $k-1$. Hence, the total number of ones in positions from 1 to $2k-2$ in $B_{gl}^c[j,i]$ and $M^c[j,i]$ is even. Therefore, the value zero precedes the value one at position $2k-1$ in the Gray code ordering. By assumption, the value in the position $2k-1$ in B_{gl}^c, which is $B_{gl}^c[j,2i-1]$, is 1, and the position $2k-1$ in M^c, which is $M^c[j,2i-1]$, is 0. Hence, M^c is smaller than B_{gl}^c in the Gray code ordering.   □

We conjecture that row and column symmetry will be intractable to break for other simple orderings. However, each such ordering may require a new proof.

6   Other Symmetry Breaking Constraints

Despite these negative theoretical results, there is still the possibility for other orderings on assignments to be useful when breaking symmetry in practice. It is interesting therefore to develop propagation algorithms for different orderings. Propagation algorithms are used to prune the search space by enforcing properties like domain consistency. A constraint is *domain consistent* (DC) iff when a variable is assigned any value in its domain, there exist compatible values in the domains of the other variables.

6.1 Gray Code Constraint

We give an efficient encoding for the new global constraint $Gray([X_1,\ldots,X_n],[Y_1,\ldots,Y_n])$ that ensures $\langle X_1,\ldots,X_n \rangle$ is before or equal in position to $\langle Y_1,\ldots,Y_n \rangle$ in the Gray code ordering where X_i and Y_j are 0/1 variables. We encode the transition relation of an automaton with 0/1/-1 state variables, Q_1 to Q_{n+1} that reads a sequence $\langle X_1, Y_1, \ldots, X_n, Y_n \rangle$ and ensures that the two sequences are ordered appropriately. We consider the following decomposition where $1 \leq i \leq n$:

$$Q_1 = 1, \ Q_i \neq 1 \vee X_i \leq Y_i, \ Q_i \neq -1 \vee X_i \geq Y_i,$$
$$X_i = Y_i \vee Q_{i+1} = 0, \ X_i = 0 \vee Y_i = 0 \vee Q_{i+1} = -Q_i.$$

We can show that this decomposition not only preserves the semantics of the constraint but also does not hinder propagation.

Proposition 6. *Unit propagation on this decomposition enforces domain consistency on* $Gray([X_1,\ldots,X_n],[Y_1,\ldots,Y_n])$ *in* $O(n)$ *time.*

Proof. (Correctness) $Q_i = 0$ as soon as the two vectors are ordered correctly. $Q_i = 1$ iff X_i and Y_i are ordered in the Gray code ordering with 0 before 1. $Q_i = -1$ iff the ith bits, X_i and Y_i are ordered in the Gray code ordering with 1 before 0. Q_{i+1} stays the same polarity as Q_i iff $X_i = Y_i = 0$ and flips polarity iff $X_i = Y_i = 1$.

(Completeness) This follows from the completeness of CNF encoding of the corresponding automaton [34] and the fact that unit propagation on this set of constraints enforces DC on a table constraint that encodes the transition relation.

(Complexity) There are $O(n)$ disjuncts in the decomposition. Hence unit propagation takes $O(n)$ time. In fact, it is possible to show that the *total* time to enforce DC down a branch of the search tree is $O(n)$.   □

Note that this decomposition can be used to break symmetry with the Gray code ordering in a SAT solver.

6.2 Snake-Lex Constraint

For row and column symmetry, we can break symmetry with the DOUBLELEX constraint that lexicographically orders rows and columns, or the SNAKELEX constraint. This is based on the smallest row and column permutation of the matrix according to an ordering on assignments that linearizes the matrix in a snake-like manner [14]. The (columnwise) SNAKELEX constraint can be enforced by a conjunction of $2m - 1$ lexicographical ordering constraints on pairs of columns and $n - 1$ lexicographical constraints on pairs of intertwined rows. To obtain the rowwise SNAKELEX constraint, we transpose the matrix and then order as in the columnwise SNAKELEX. Note that DOUBLELEX and SNAKELEX *only break a subset* of the row and colum symmetries. However, they are very useful in practice. It was shown in [12], that enforcing DC on the DOUBLELEX constraint is NP-hard. Hence we typically decompose it into separate row and column constraints. Here, we show that enforcing DC on the SNAKELEX constraint is also NP-hard. It is therefore also reasonable to propagate SNAKELEX by decomposition.

Proposition 7. *Enforcing DC on the* SNAKELEX *constraint is NP-hard.*

Proof. (Sketch) A full proof is in the online technical report. Let X be a n by m matrix of Boolean variables. The main idea is to embed X in to a specially constructed matrix in such a way that enforcing DC on the DOUBLELEX constraint on X (which we already know is NP-hard) is equivalent to enforcing DC on the SNAKELEX constraint on this larger matrix. □

7   Experimental Results

We tested two hypotheses that provide advice to the modeller when breaking symmetry.

1. other orderings besides the lexicographical ordered can be effective when breaking symmetry in practice;
2. symmetry breaking should align with the branching heuristic, and with the objective function.

All our experiments report the time to find an optimal solution *and* prove it optimal. We believe that optimisation is often a more realistic setting in which to illustrate the practical benefits of symmetry breaking, than satisfaction experiments which either find one or all solutions. Breaking symmetry in optimisation problems is important as we must traverse the whole search space when proving optimality. All our experiments used the BProlog 7.7 constraint solver. This solver took second place in the ASP 2011 solver competition. The three sets of experiments took around one CPU month on a MacBook Pro with an Intel Core i5 2 core 2.53 GHz processor, with 4GB of memory. The three domains were chosen as representative of optimisation problems previously studied in symmetry breaking. We observed similar results in these as well as other domains.

7.1   Maximum Density Still Life Problem

This is prob032 in CSPLib [35]. This problem arises in Conway's Game of Life, and was popularized by Martin Gardner. Given a n by n submatrix of the infinite plane, we want to find and prove optimal the pattern of maximum density which does not change from generation to generation. For example, an optimal solution for $n = 3$ is:

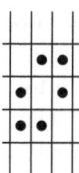

This is a still life as every live square has between 2 and 3 live neighbours, and every dead square does not have 3 live neighbours. We use the simple 0/1 constraint model from [36]. This problem has the 8 symmetries of the square as we can rotate or reflect any still life to obtain a new one. Bosch and Trick argued that *"... The symmetry embedded in this problem is very strong, leading both to algorithmic insights and algorithmic difficulties... "*.

Our first experiment used the default search strategy to find and prove optimal the still life of maximum density for a given n. The default strategy instantiates variables row-wise across the matrix. Our goal here is to compare the different symmetry breaking methods with an "out of the box" solver. We then compare the impact of the branching heuristic on symmetry breaking. We broke symmetry with either the lexicographical or Gray code orderings, finding the smallest (lex, gray) or largest (anti-lex, anti-gray) solution in each symmetry class. In addition, we linearized the matrix either row-wise (row), column-wise (col), snake-wise along rows (snake), snake-wise along columns (col-snake), or in a clockwise spiral (spiral). Table 1 gives results for the 20 different symmetry breaking methods constructed by using 1 of the 4 possible solution orderings and the 5 different linearizations, as well with no symmetry breaking (none).

Table 1. Backtracks required to find and prove optimal the maximum density still life of size n by n using the default branching heuristic. Column winner is in *emphasis*.

Symmetry breaking	$n = 4$	5	6	7	8
none	176	1,166	12,205	231,408	5,867,694
gray row	91	446	5,702	123,238	2,507,747
anti-lex row	84	424	5,473	120,112	2,416,266
anti-gray col-snake	68	500	5,770	72,691	2,332,085
gray spiral	86	541	6,290	120,051	2,311,854
gray snake	80	477	5,595	120,601	2,264,184
anti-lex col-snake	79	660	4,735	66,371	2,254,325
anti-lex spiral	81	507	6,174	119,262	2,241,660
anti-lex col	74	718	3,980	68,330	2,215,936
anti-lex snake	68	457	5,379	117,479	2,206,189
lex spiral	48	434	4,025	90,289	2,028,624
lex col-snake	77	359	5,502	76,400	2,003,505
lex col	80	560	4,499	83,995	2,017,935
lex row	33	406	2,853	87,781	1,982,698
lex snake	35	407	2,965	86,331	1,980,498
anti-gray col	70	522	5,666	75,930	1,925,613
gray col	65	739	3,907	87,350	1,899,887
gray col-snake	62	693	3,833	82,736	1,880,506
anti-gray row	*26*	269	2,288	38,476	1,073,659
anti-gray spiral	27	279	2,404	40,224	1,081,006
anti-gray snake	28	*262*	*2,203*	*38,383*	*1,059,704*

We make some observations about these results. First, the Lex-Leader method (lex row) is beaten by many methods. For example, the top three methods all use the anti-Gray code ordering. Second, lex tends to work better than anti-lex, but anti-gray better than gray. We conjecture this is because anti-gray tends to align better with the maximization objective than gray, but anti-lex is too aggressive as the maximum density still life can have more dead cells than alive cells. Third, although we eliminate all 7 non-identity symmetries, the best method is only about a factor 6 faster than not breaking symmetry at all.

To explore the interaction between symmetry breaking and the branching heuristic, we report results in Table 2 using branching heuristics besides the default row-wise variable ordering. We used the best symmetry breaking method for the default row-wise branching heuristic (anti-gray snake), the worst symmetry breaking method for the default branching heuristic (gray row), a standard method (lex row), as well as no symmetry breaking (none). We compared the default branching heuristic (row heuristic)

Table 2. Backtracks required to find the 8 by 8 still life of maximum density and prove optimality for different branching heuristics and symmetry breaking constraints. Overall winner is in **bold**.

Branching/SymBreak	none	gray row	lex row	anti-gray snake
spiral-out heuristic	196,906,862	24,762,297	194,019,848	222,659,696
spiral-in heuristic	65,034,993	18,787,751	12,662,207	9,292,164
constr heuristic	*5,080,541*	2,816,355	3,952,445	8,590,077
degree heuristic	6,568,195	2,024,955	6,528,018	7,053,908
col-snake heuristic	5,903,851	*1,895,920*	1,849,702	2,127,122
col heuristic	5,867,694	2,212,104	*1,634,016*	1,987,864
snake heuristic	5,903,851	1,868,303	2,043,473	1,371,200
row heuristic	5,867,694	2,507,747	1,982,698	**1,059,704**

with branching heuristics that instantiate variables column-wise (col heuristic), snake-wise along rows (snake heuristic), snake-wise along columns (col-snake heuristic), in a clockwise spiral from top left towards the middle (spiral-in heuristic), in an anti-clockwise spiral from the middle out to the top left (spiral-out heuristic), by order of degree (degree heuristic), and by order of the number of attached constraints (constr heuristic). Note that there is no value in reporting results for domain ordering heuristics like fail-first as domains sizes are all binary.

We make some observations about these results. First, the symmetry breaking method with the best overall performance (anti-gray snake + row heuristic) had the worst performance with a different branching heuristic (anti-gray snake + spiral-out heuristic). Second, we observed good performance when the branching heuristic aligned with the symmetry breaking (e.g. anti-gray snake + snake heuristic). Third, a bad combination of branching heuristic and symmetry breaking constraints (e.g. anti-gray snake + spiral-out heuristic) was worse than all of the branching heuristics with no symmetry breaking constraints. Fourth, the default row heuristic was competitive. It was best or not far from best in every column.

7.2 Low Autocorrelation Binary Sequences

This is prob005 in CSPLib [35]. The goal is to find the binary sequence of length n with the lowest autocorrelation. We used a standard model from one of the first studies into symmetry breaking [19]. This model contains a triangular matrix of 0/1 decision variables, in which the sum of the kth row equals the kth autocorrelation. Table 3 reports results to find the sequence of lowest autocorrelation and prove it optimal. We used the default variable ordering heuristic (left2right) that instantiates variables left to right from the beginning of the sequence to the end. The model has 7 non-identity symmetries which leave the autocorrelation unchanged. We can reverse the sequence, we can invert the bits, we can invert just the even bits, or we can do some combination of these operations. We broke all 7 symmetries by posting the constraints that, within its symmetry class, the sequence is smallest in the lexicographical or Gray code orderings (lex, gray) or largest (anti-lex, anti-gray). In addition, we also considered symmetry breaking constraints that took the variables in reverse order from right to left (rev), alternated the variables from both ends inwards to the middle (outside-in), and from the middle out to both ends (inside-out).

We make some observations about these results. First, the best two symmetry breaking methods both look at variables starting from the middle and moving outwards to

Table 3. Backtracks required to find the n bit binary sequence of lowest autocorrelation and prove optimality with the default branching heuristic

Symmetry breaking	$n = 12$	14	16	18	20	22	24
none	2,434	9,487	36,248	126,057	474,915	1,725,076	7,447,186
anti-gray outside-in	2,209	6,177	18,881	92,239	310,473	1,223,155	4,966,068
gray outside-in	1,351	5,040	19,152	68,272	350,790	903,441	4,526,114
lex outside-in	869	3,057	11,838	43,669	262,935	557,790	3,330,931
gray	704	2,400	10,158	36,854	158,080	468,317	3,048,723
lex	707	2,408	10,178	36,885	158,132	468,390	3,047,241
gray rev	699	1,790	9,892	25,551	147,911	329,897	2,706,466
anti-lex outside-in	1,262	2,704	14,059	67,848	179,219	544,116	2,579,981
anti-gray	1,036	2,226	9,889	45,375	167,916	606,977	2,436,236
anti-lex	1,522	3,087	10,380	51,162	281,789	920,543	2,415,736
lex rev	634	1,751	7,601	23,218	127,438	299,877	2,160,463
anti-lex rev	549	1,707	9,398	32,638	117,367	398,822	2,092,787
gray inside-out	662	1,582	6,557	25,237	89,365	248,135	1,667,262
lex inside-out	640	1,549	6,478	25,049	88,978	247,558	1,665,054
anti-gray rev	1,007	1,661	6,894	29,689	86,198	312,038	1,422,693
anti-gray inside-out	*412*	1,412	5,934	22,942	*82,673*	*245,259*	1,271,986
anti-lex inside-out	629	*1,320*	*4,558*	*19,811*	138,337	291,050	*927,321*

Table 4. Backtracks required to find the 22 bit sequence of lowest autocorrelation and prove optimality with different branching heuristics and symmetry breaking constraints

Branching/SymBreak	none	anti-gray outside-in	gray	lex	anti-gray inside-out	anti-lex inside-out
left2right heuristic	*1,725,076*	1,223,155	468,317	468,390	245,259	291,050
right2left heuristic	*1,725,076*	*322,291*	329,897	299,877	**224,540**	269,628
degree heuristic	2,024,484	603,857	329,897	400,228	500,415	*268,173*
constr heuristic	2,024,484	1,624,765	349,025	313,817	1,097,303	297,616
inside-out heuristic	1,786,741	2,787,164	1,406,831	1,055,918	326,938	268,206
outside-in heuristic	2,053,179	364,469	*284,417*	*284,526*	2,044,042	2,767,059

both ends (inside-out). By comparison, symmetry breaking constraints that reverse this ordering of variables (outside-in) perform poorly. We conjecture this is because the middle bits in the sequence are more constrained, appearing in more autocorrelations, and so are more important to decide early in search. Second, although we only eliminate 7 symmetries, the best method offers a factor of 8 improvement in search over not breaking symmetry.

To explore the interaction between symmetry breaking and branching heuristics, we report results in Table 4 to find the optimal solution and prove optimality using different branching heuristics. We used the best two symmetry breaking methods for the default left to right branching heuristic (anti-gray inside-out, and anti-lex inside-out), the worst symmetry breaking method for the default branching heuristic (anti-gray outside-in), a standard symmetry breaking method (lex), the Gray code alternative (gray), as well as no symmetry breaking (none). We compared the default branching heuristic (left2right heuristic) with branching heuristics that instantiate variables right to left (right2left heuristic), alternating from both ends inwards to the middle (outside-in heuristic), from the middle alternating outwards to both ends (inside-out heuristic), by order of degree (degree heuristic), and by order of the number of attached constraints (constr heuristic). Note that all domains are binary so there is again no value for a heuristic like ff that considers domain size.

We make some observations about these results. First, the best overall performance is observed when we break symmetry with the anti-Gray code ordering (anti-gray inside-out + right2left heuristic). Second, we observe better performance when the symmetry breaking constraint aligns with the branching heuristic than when it goes against it (e.g. anti-gray outside-in + outside-in heuristic is much better than anti-gray outside-in + inside-out heuristic). Third, the default heuristic (left2right) is again competitive.

7.3   Peaceable Armies of Queens

The goal of this optimisation problem is to place the largest possible equal-sized armies of white and black queens on a chess board so that no white queen attacks a black queen or vice versa [37]. We used a simple model from an earlier study of symmetry breaking [38]. The model has a matrix of 0/1/2 decision variables, in which $X_{ij} = 2$ iff a black queen goes on square (i, j), $X_{ij} = 1$ iff a white queen goes on square (i, j), and 0 otherwise. Note that our model is now ternary, unlike the binary models considered in the two previous examples. However, the Gray code ordering extends from binary to ternary codes in a straight forward. Similarly, we can extend the decomposition to propagate Gray code ordering constraints on ternary codes.

Table 5 reports results to find the optimal solution and prove optimality for peaceable armies of queens. This model has 15 non-identity symmetries, consisting of any combination of the symmetries of the square and the symmetry that swaps white queens for black queens. We broke all 15 symmetries by posting constraints to ensure that we only find the smallest solution in each symmetry class according to the lexicographical or Gray code orderings (lex, gray), or the largest solution in each symmetry class according to the two orders (anti-lex, anti-gray). We also considered symmetry breaking constraints that take the variables in row-wise order (row), in column-wise order (col), in a snake order along the rows (snake), in a snake order along the columns (col-snake), or in a clockwise spiral (spiral). We again used the default variable ordering that instantiates variables in the lexicographical row-wise order.

We make some observations about these results. First, finding the largest solution in each symmetry class (anti-gray and anti-lex) is always better than finding the smallest (gray and lex). We conjecture that this is because symmetry breaking lines up better with the objective of maximizing the number of queens on the board. Second, symmetry breaking in a "conventional" way (lex, row) is beaten by half of the symmetry breaking methods. In particular, all 10 methods which find the largest solution up to symmetry in the Gray order (anti-gray) or lexicographical ordering (anti-lex) beat the "conventional" method (lex row). Third, ordering the variables row-wise in the symmetry breaking constraint is best for lex, but for every other ordering (anti-lex, gray, anti-gray) ordering variables row-wise is never best. In particular, anti-lex spiral beats anti-lex row and all other anti-lex methods, gray snake beats gray row and all other gray methods, and anti-gray col beats anti-gray row and all other anti-gray methods. Fourth, a good symmetry breaking method (e.g. anti-gray col) offers up to a 12-fold improvement over not breaking the 15 non-identity symmetries.

To explore the interaction between symmetry breaking and branching heuristics, we report results in Table 6 using different branching heuristics. We used the best symmetry breaking method for the default row-wise branching heuristic (anti-gray col), the worst

Table 5. Backtracks required to solve the n by n peaceable armies of queens problem to optimality with the default branching heuristic

Symmetry breaking	$n = 3$	4	5	6	7	8
none	19	194	2,588	37,434	679,771	19,597,858
lex col-snake	13	98	1,014	8,638	199,964	5,299,787
lex col	23	87	1,042	10,792	198,032	5,197,013
gray col	26	101	1,118	9,763	214,391	5,008,279
gray col-snake	13	100	1,059	8,973	205,453	4,877,014
gray spiral	18	104	913	10,795	169,725	4,690,071
lex spiral	18	93	887	10,694	169,293	4,674,458
gray row	19	73	680	6,975	116,725	3,705,591
gray snake	19	81	685	7,070	117,489	3,683,558
lex snake	19	80	661	7,043	117,590	3,682,438
lex row	19	73	679	6,880	115,999	3,652,269
anti-gray spiral	8	43	466	4,381	108,214	2,402,049
anti-gray snake	8	47	472	4,333	106,317	2,367,290
anti-gray row	8	44	452	4,326	105,837	2,357,024
anti-lex col-snake	18	59	560	4,513	70,950	2,346,875
anti-lex col	18	57	485	4,373	69,484	2,291,512
anti-lex row	9	29	315	3,417	101,530	2,037,336
anti-lex snake	9	34	314	3,366	100,472	2,010,354
anti-lex spiral	9	30	326	3,432	105,717	2,007,586
anti-gray col-snake	19	40	471	4,061	71,079	1,709,744
anti-gray col	19	40	385	4,317	70,632	1,698,492

symmetry breaking method for the default branching heuristic (lex col-snake), a standard method (lex row), the Gray code alternative (gray row), as well as no symmetry breaking (none). We compared the same branching heuristics as with the maximum density still life problem. As domains are now not necessarily binary, we also included the ff heuristic that order variables by their domain size tie-breaking with the row heuristic (ff heuristic). Given the good performance of the spiral and ff heuristics individually, we also tried a novel heuristic that combines them together, branching on variables by order of the domain size and tie-breaking with the spiral-in heuristic (ff-spiral heuristic).

Table 6. Backtracks required to solve the 8 by 8 peaceable armies of queens problem to optimality for different branching heuristics and symmetry breaking constraints

Branching/SymBreak	none	lex col-snake	gray row	lex row	anti-gray col
col-snake heuristic	20,209,357	4,270,637	6,372,404	5,836,975	7,363,488
col heuristic	19,597,858	4,384,086	6,338,413	5,775,781	6,811,345
spiral-out heuristic	8,196,693	4,894,264	5,099,899	5,126,074	6,478,506
degree heuristic	19,597,858	3,129,599	4,216,463	4,343,792	6,351,547
snake heuristic	20,209,357	5,261,095	4,258,903	4,221,336	1,946,556
constr heuristic	7,305,061	2,757,360	2,650,590	2,645,054	1,789,444
row heuristic	19,597,858	5,299,787	3,705,591	3,652,269	1,698,492
ff heuristic	12,826,856	3,371,419	2,495,788	2,521,351	1,309,529
ff-spiral heuristic	13,400,485	2,447,867	3,147,237	2,162,657	1,222,607
spiral-in heuristic	15,577,982	1,787,653	2,387,067	2,430,499	**1,193,988**

We make some observations about these results. First, the best symmetry breaking constraint with the default branching heuristic (anti-gray col + row heuristic) was either very good or very bad with the other branching heuristics. It offers the best overall performance in this experiment (viz. anti-gray col + spiral-in heuristic), and is the best of all the symmetry breaking methods for 5 other heuristics. However, it also the worst of all the symmetry breaking methods with 4 other heuristics. Second, aligning the branching heuristic with the symmetry breaking constraint at best offers middle of the road performance (e.g. lex row + row heuristic) but can also be counter-productive (e.g. anti-gray col + col heuristic). Third, the spiral-in heuristic offer some of the best performance. This heuristic provided the best overall result, and was always in the top 2 for every symmetry breaking method. Recall that the spiral-in heuristic was one of the worst heuristics on the maximum density still life problem. We conjecture that this is because it delays constraint propagation on the still life problem constraints but not on the constraints in the peaceable armies of queens problem. Fourth, a bad combination of branching heuristic and symmetry breaking constraints is worse than not breaking symmetry if we have a good branching heuristic (e.g. none + constr heuristic beats anti-gray col + col-snake heuristic).

These results support both our hypotheses. Other orderings besides the simple lexicographical ordering can be effective for breaking symmetry, and symmetry breaking should align with both the branching heuristic and the objective function. Unfortunately, as the last example demonstrated, the interaction between problem constraints, symmetry breaking and branching heuristic can be complex and difficult to predict. Overall, the Gray code ordering appears useful. Whilst it is conceptually similar to the lexicographical ordering, it looks at more than one bit at a time. This is reflected in the automaton for the Gray code ordering which has more states than that required for the lexicographical ordering.

8   Conclusions

We have argued that in general breaking symmetry with a different ordering over assignments than the usual lexicographical ordering does not improve the computational complexity of breaking with symmetry. Our argument had two parts. First, we argued that under modest assumptions we cannot reduce the worst case complexity from that of breaking symmetry with a lexicographical ordering. These assumptions are satisfied by the Gray code and Snake-Lex orderings. Second, we proved that for the particular case of row and column symmetries, breaking symmetry with the Gray code or Snake-Lex ordering is intractable (as it was with the lexicographical ordering). We then explored algorithms to break symmetry with other orderings. In particular, we gave a linear time propagator for the Gray code ordering constraint, and proved that enforcing domain consistency on the SNAKELEX constraint, like on the DOUBLELEX constraint, is NP-hard. Finally, we demonstrated that other orderings have promise in practice. We ran experiments on three standard benchmark domains where breaking symmetry with the Gray code ordering was often better than with the Lex-Leader or Snake-Lex methods.

References

1. Puget, J.F.: On the satisfiability of symmetrical constrained satisfaction problems. In: Komorowski, J., Raś, Z.W. (eds.) ISMIS 1993. LNCS (LNAI), vol. 689, pp. 350–361. Springer, Heidelberg (1993)

2. Crawford, J., Luks, G., Ginsberg, M., Roy, A.: Symmetry breaking predicates for search problems. In: Proceedings of the 5th International Conference on Knowledge Representation and Reasoning (KR 1996), pp. 148–159 (1996)

3. Shlyakhter, I.: Generating effective symmetry-breaking predicates for search problems. In: Proceedings of LICS Workshop on Theory and Applications of Satisfiability Testing, SAT 2001 (2001)

4. Aloul, F., Sakallah, K., Markov, I.: Efficient symmetry breaking for Boolean satisfiability. In: Proceedings of the 18th International Joint Conference on AI, International Joint Conference on Artificial Intelligence, pp. 271–276 (2003)

5. Law, Y.C., Lee, J.H.M.: Global constraints for integer and set value precedence. In: Wallace, M. (ed.) CP 2004. LNCS, vol. 3258, pp. 362–376. Springer, Heidelberg (2004)

6. Puget, J.F.: Breaking symmetries in all different problems. In: Proceedings of 19th IJCAI, International Joint Conference on Artificial Intelligence, pp. 272–277 (2005)

7. Law, Y., Lee, J.: Symmetry Breaking Constraints for Value Symmetries in Constraint Satisfaction. Constraints 11(2-3), 221–267 (2006)

8. Walsh, T.: Symmetry breaking using value precedence. In: Proc. of the 17th European Conference on Artificial Intelligence (ECAI 2006), European Conference on Artificial Intelligence. IOS Press (2006)

9. Walsh, T.: General symmetry breaking constraints. In: Benhamou, F. (ed.) CP 2006. LNCS, vol. 4204, pp. 650–664. Springer, Heidelberg (2006)

10. Law, Y.C., Lee, J.H.M., Walsh, T., Yip, J.Y.K.: Breaking symmetry of interchangeable variables and values. In: Bessière, C. (ed.) CP 2007. LNCS, vol. 4741, pp. 423–437. Springer, Heidelberg (2007)

11. Walsh, T.: Breaking value symmetry. In: Fox, D., Gomes, C. (eds.) Proceedings of the 23rd National Conference on AI, Association for Advancement of Artificial Intelligence, pp. 1585–1588 (2008)

12. Katsirelos, G., Narodytska, N., Walsh, T.: On the complexity and completeness of static constraints for breaking row and column symmetry. In: Cohen, D. (ed.) CP 2010. LNCS, vol. 6308, pp. 305–320. Springer, Heidelberg (2010)

13. Rossi, F., van Beek, P., Walsh, T. (eds.): Handbook of Constraint Programming. Foundations of Artificial Intelligence. Elsevier (2006)

14. Grayland, A., Miguel, I., Roney-Dougal, C.M.: Snake lex: An alternative to double lex. In: Gent, I.P. (ed.) CP 2009. LNCS, vol. 5732, pp. 391–399. Springer, Heidelberg (2009)

15. Frisch, A., Hnich, B., Kiziltan, Z., Miguel, I., Walsh, T.: Multiset ordering constraints. In: Proceedings of the 18th International Joint Conference on Artificial Intelligence (IJCAI-2003), International Joint Conference on Artificial Intelligence (2003)

16. Benhamou, B., Sais, L.: Theoretical study of symmetries in propositional calculus and applications. In: Kapur, D. (ed.) CADE 1992. LNCS, vol. 607, pp. 281–294. Springer, Heidelberg (1992)

17. Benhamou, B., Sais, L.: Tractability through symmetries in propositional calculus. Journal of Automated Reasoning 12(1), 89–102 (1994)

18. Backofen, R., Will, S.: Excluding symmetries in constraint-based search. In: Jaffar, J. (ed.) CP 1999. LNCS, vol. 1713, pp. 73–87. Springer, Heidelberg (1999)

19. Gent, I., Smith, B.: Symmetry breaking in constraint programming. In: Horn, W. (ed.) Proceedings of ECAI 2000, pp. 599–603. IOS Press (2000)

20. Fahle, T., Schamberger, S., Sellmann, M.: Symmetry breaking. In: Walsh, T. (ed.) CP 2001. LNCS, vol. 2239, pp. 93–107. Springer, Heidelberg (2001)

21. Sellmann, M., Hentenryck, P.V.: Structural symmetry breaking. In: Proceedings of 19th Internatinal Joint Conference on AI (IJCAI 2005), International Joint Conference on Artificial Intelligence, pp. 298–303 (2005)

22. Puget, J.-F.: Dynamic lex constraints. In: Benhamou, F. (ed.) CP 2006. LNCS, vol. 4204, pp. 453–467. Springer, Heidelberg (2006)
23. Katsirelos, G., Walsh, T.: Symmetries of symmetry breaking constraints. In: Proc. of the 19th European Conference on Artificial Intelligence (ECAI 2010), European Conference on Artificial Intelligence. IOS Press (2010)
24. Narodytska, N., Walsh, T.: An adaptive model restarts heuristic. In: Gomes, C., Sellmann, M. (eds.) CPAIOR 2013. LNCS, vol. 7874, pp. 369–377. Springer, Heidelberg (2013)
25. Cohen, D., Jeavons, P., Jefferson, C., Petrie, K., Smith, B.: Symmetry definitions for constraint satisfaction problems. Constraints 11(2-3), 115–137 (2006)
26. Gent, I.P., Kelsey, T., Linton, S.A., McDonald, I., Miguel, I., Smith, B.M.: Conditional symmetry breaking. In: van Beek, P. (ed.) CP 2005. LNCS, vol. 3709, pp. 256–270. Springer, Heidelberg (2005)
27. Flener, P., Frisch, A., Hnich, B., Kizil tan, Z., Miguel, I., Pearson, J., Walsh, T.: Symmetry in matrix models. In: Proceedings of the CP 2001 Workshop on Symmetry in Constraints (SymCon 2001), Held alongside CP, Also APES-30-2001 technical report (2001)
28. Frisch, A., Hnich, B., Kiziltan, Z., Miguel, I., Walsh, T.: Global constraints for lexicographic orderings. In: Van Hentenryck, P. (ed.) CP 2002. LNCS, vol. 2470, pp. 93–108. Springer, Heidelberg (2002)
29. Frisch, A., Hnich, B., Kiziltan, Z., Miguel, I., Walsh, T.: Propagation algorithms for lexicographic ordering constraints. Artificial Intelligence 170(10), 803–908 (2006)
30. Katsirelos, G., Narodytska, N., Walsh, T.: Combining symmetry breaking and global constraints. In: Oddi, A., Fages, F., Rossi, F. (eds.) CSCLP 2008. LNCS, vol. 5655, pp. 84–98. Springer, Heidelberg (2009)
31. Flener, P., Frisch, A., Hnich, B., Kiziltan, Z., Miguel, I., Walsh, T.: Matrix modelling. In: Proceedings of the CP 2001 Workshop on Modelling and Problem Formulation, Held alongside CP 2001 (2001)
32. Flener, P., Frisch, A., Hnich, B., Kiziltan, Z., Miguel, I., Walsh, T.: Matrix modelling: Exploiting common patterns in constraint programming. In: Proceedings of the International Workshop on Reformulating Constraint Satisfaction Problems, Held alongside CP 2002 (2002)
33. Flener, P., Frisch, A.M., Hnich, B., Kiziltan, Z., Miguel, I., Pearson, J., Walsh, T.: Breaking row and column symmetries in matrix models. In: Van Hentenryck, P. (ed.) CP 2002. LNCS, vol. 2470, pp. 462–472. Springer, Heidelberg (2002)
34. Quimper, C.-G., Walsh, T.: Decomposing global grammar constraints. In: Bessière, C. (ed.) CP 2007. LNCS, vol. 4741, pp. 590–604. Springer, Heidelberg (2007)
35. Gent, I., Walsh, T.: CSPLib: a benchmark library for constraints. Technical report, Technical report APES-09-1999, A shorter version appears in the Proceedings of the 5th International Conference on Principles and Practices of Constraint Programming, CP-99 (1999)
36. Bosch, R., Trick, M.: Constraint programming and hybrid formulations for three life designs. Annals OR 130(1-4), 41–56 (2004)
37. Bosch, R.: Peaceably coexisting armies of Queens. Optima (Newsletter of the Mathematical Programming Society) 62, 6–9 (1999)
38. Smith, B., Petrie, K., Gent, I.: Models and symmetry breaking for peacable armies of queens. In: Régin, J.-C., Rueher, M. (eds.) CPAIOR 2004. LNCS, vol. 3011, pp. 271–286. Springer, Heidelberg (2004)

Time-Table Extended-Edge-Finding
for the Cumulative Constraint

Pierre Ouellet and Claude-Guy Quimper

Université Laval, Québec, Canada

Abstract. We propose a new filtering algorithm for the cumulative constraint. It applies the Edge-Finding, the Extended-Edge-Finding and the Time-Tabling rules in $O(kn \log n)$ where k is the number of distinct task heights. By a proper use of tasks decomposition, it enforces the Time-Tabling rule and the Time-Table Extended-Edge-Finding rule. Thus our algorithm improves upon the best known Extended-Edge-Finding propagator by a factor of $O(\log n)$ while achieving a much stronger filtering.

1 Introduction

Scheduling problems consist of deciding when a task should start and which resource should execute it. Many side constraints can enrich the problem definition. For instance, a precedence constraint can force a task to complete before another can start. The need to cope with side constraints makes constraint programming a very attractive tool since it is handy to specify extra requirements in the problem without tweaking the scheduling algorithms provided by the constraint solver.

The CUMULATIVE constraint encodes a large variety of scheduling problems. It allows the tasks to request a portion of a cumulative resource. Tasks can execute concurrently as long as the workload is below the capacity of the resource.

There exist multiple techniques to filter the cumulative constraint. Most of these techniques are filtering rules that reason over a time interval and that deduce the relative positions between the tasks or the position relative to a given time point. Among the popular rules, there are the not-first/not-last [1], the Time-Tabling [2,3,4,5], the Edge-Finding [6,7], the Extended-Edge-Finding [8], and the Energetic Reasoning rule [9]. The later rule dominates them all except for the not-first/not-last. Vilím [10] proposes to combine the Edge-Finding rule to the Time-Tabling rule to obtain a level of filtering greater than what is obtained by individually applying the Edge-Finding and Time-Tabling. He calls this new technique the *Timetable Edge Finding*. Schutt et al. [11] combines the technique with the use of nogoods and obtain impressive results.

We propose an algorithm that performs both Edge-Finding and Extended-Edge-Finding filtering. It is largely inspired by Vilím's Edge-Finder [6] and is mostly an extension of it. We also propose an algorithm that performs Time-Tabling and Time-Table Extended-Edge-Finding, using the pruning rules from Vilím [10]. However, our algorithm differs from [10] in three points: 1) the algorithm we present performs Time-Tabling as well as Time-Table Extended Edge-Finding; 2) when the number of distinct task heights is constant, the new algorithm runs in time $O(n \log n)$; 3) both algorithms

C. Schulte (Ed.): CP 2013, LNCS 8124, pp. 562–577, 2013.

are non-idempotent but the new algorithm guaranties to perform some filtering on all tasks for which the Edge-Finding, Extended-Edge Finding, Time-Tabling, and Time-Table Extended-Edge-Finding rules apply.

2   Preliminaries

We consider a set $\mathcal{I}$ of n non-preemptive tasks. A task $i \in \mathcal{I}$ is specified by its *earliest starting time* (est$_i$), its *latest completion time* (lct$_i$), its *processing time* (p_i), and its *height* (h_i). From the previous attributes, one can compute the *earliest completion time* (ect$_i$) of a task i with the relation ect$_i$ = est$_i$ +p_i and its *latest starting time* (lst$_i$) with the relation lst$_i$ = lct$_i$ −p_i. The *energy* (e_i) of a task i is the amount of consumption of the resource during its execution and satisfies $e_i = p_i h_i$. We extend this notation to a subset of tasks $\Omega \subseteq \mathcal{I}$ as follows.

$$\text{est}_\Omega = \min_{i \in \Omega} \text{est}_i \qquad \text{lct}_\Omega = \max_{i \in \Omega} \text{lct}_i \qquad e_\Omega = \sum_{i \in \Omega} e_i \qquad (1)$$

A cumulative resource is characterized by its capacity C. A task i starts at time s_i and executes during p_i units of time. The task consumes h_i units of the cumulative resource over the time period $[s_i, s_i + p_i)$. Solving a cumulative scheduling problem consists of finding, for each task $i \in \mathcal{I}$, the starting times s_i such that est$_i \leq s_i \leq$ lst$_i$ and such that at any time t, the cumulative usage of the resource does not exceed C.

$$\sum_{i \in \mathcal{I} | t \in [s_i, s_i + p_i)} h_i \leq C \qquad \qquad \forall t \in \mathbb{Z} \qquad (2)$$

Deciding whether there exists a solution to the cumulative scheduling problem is NP-Complete, even for the disjunctive case where $C = 1$.

The CUMULATIVE constraint encodes the cumulative scheduling problem (CuSP). This constraint restrains the starting times to satisfy Equation (2). It takes as parameter the vector of starting time variables, the vector of processing times, the vector of heights, and the resource capacity. The earliest starting times and latest completion times are encoded in the domains of the starting time variables by setting dom(S_i) = [est$_i$, lst$_i$].

$$\text{CUMULATIVE}([S_1, \ldots, S_n], [p_1, \ldots, p_n], [h_1, \ldots, h_n], C) \qquad (3)$$

2.1   Slack, E-Feasibility and Energy Envelope

For a given time interval $[a, b)$, let $\Omega = \{i \in \mathcal{I} \mid a \leq est_i \wedge \text{lct}_i \leq b\}$, the slack ($Sl_\Omega$) is the remaining energy of the resource within the interval once all tasks in Ω are processed.

$$Sl_\Omega = C(b - a) - e_\Omega \qquad (4)$$

A CuSP is said to be energy-feasible (*E-Feasible*) if it has no interval of negative slack.

The *envelope* or *energy envelope* of a task i (Env$_i$), is a measure of the potential consumed energy of the resource up to the completion of i. It takes into account the full resource capacity prior to the starting time of task i regardless of its effective usage. We extend the definition of the envelope to a subset of tasks $\Omega \subseteq \mathcal{I}$.

$$\text{Env}_i = Cest_i + e_i \qquad\qquad \text{Env}_\Omega = \max_{\Theta \subseteq \Omega}(Cest_\Theta + e_\Theta) \qquad (5)$$

2.2 Edge-Finding

Edge-Finding aims at finding necessary orderings within the tasks and deducing related time-bound adjustments. The filtering usually occurs in two steps. The first step detects a relation of precedence $\Omega \lessdot i$ where $\Omega \subset \mathcal{I}$ and $i \in \mathcal{I} \setminus \Omega$. Such a precedence implies that task i finishes after all tasks in Ω are completed and is detected when the task i cannot be scheduled in the interval $[\text{est}_{\Omega \cup \{i\}}, \text{lct}_\Omega]$ along with the other tasks in Ω.

$$C(\text{lct}_\Omega - \text{est}_{\Omega \cup \{i\}}) < e_{\Omega \cup \{i\}} \Rightarrow \Omega \lessdot i \qquad (6)$$

The second step consists in pruning the domain of the task i based on the detected precedence $\Omega \lessdot i$. Although several techniques exist, the goal is to deduce the availability of the resource for the task i within the interval. Nuijten [12] uses the following method. Given a set $\Theta \subseteq \Omega$, she divides and assigns the energy in e_Θ into two blocks. The first block of $(C - h_i)(\text{lct}_\Theta - \text{est}_\Theta)$ units of energy evenly consumes $C - h_i$ units of the resource over the time interval $[\text{est}_\Theta, \text{lct}_\Theta)$. The second block of energy is scheduled at its earliest time within the interval $[\text{est}_\Theta, \text{lct}_\Theta)$ using the remaining h_i units of resource. When this second block completes, the task i can start its execution.

$$\Omega \lessdot i \Rightarrow \text{est}'_i = \max_{\Theta \subseteq \Omega} \text{est}_\Theta + \left\lceil \frac{e_\Theta - (C - h_i)(\text{lct}_\Theta - \text{est}_\Theta)}{h_i} \right\rceil \qquad (7)$$

Vilím [6] detects all precedences in $O(n \log n)$ and shows how to perform the optimal pruning in $O(kn \log n)$ where $k = |\{h_i \mid i \in \mathcal{I}\}|$ is the number of distinct task heights. By comparing tasks with *minimum slack intervals*, Kameugne et al. [7] produce a single-step quadratic algorithm. It finds all tasks that need to be adjusted according to the Edge-Finder rule and prunes them. Although their algorithm does not always deduce the best adjustment (7) on the first detection, multiple executions of their algorithm converge to the same fixed point.

2.3 Extended-Edge-Finding

The Extended-Edge-Finding rule stipulates that if the task i, when starting at its earliest time, overlaps the time interval $[\text{est}_\Omega, \text{lct}_\Omega)$ and that the energy of task i over this interval plus the energy e_Ω overloads the resource, then i must finish after all tasks in Ω have completed.

$$\text{est}_\Omega \in [\text{est}_i, \text{ect}_i) \wedge e_\Omega + h_i(\text{ect}_i - \text{est}_\Omega) > C(\text{lct}_\Omega - \text{est}_\Omega) \Rightarrow \Omega \lessdot i \qquad (8)$$

Mercier and Van-Hentenryck [8] detect and prune the precedences in time $O(kn^2)$ where $k = |\{h_i \mid i \in \mathcal{I}\}|$ is the number of distinct task heights.

2.4   Time Tabling

Time Tabling consists of finding the necessary usage of the resource over a time interval. For a task that satisfies $\text{lst}_i < \text{ect}_i$, the interval $[\text{lst}_i, \text{ect}_i)$ determines the fixed part of the task. Let $f(\Omega, t)$ be the aggregate of the fixed parts that spans over time t by the tasks in Ω and let $f(\Omega, [a, b))$ be the aggregate of the fixed parts over the time interval $[a, b)$ by the tasks in Ω.

$$f(\Omega, t) = \sum_{i \in \Omega | t \in [\text{lst}_i, \text{ect}_i)} h_i \qquad f(\Omega, [a, b)) = \sum_{t \in [a, b)} f(\Omega, t) \qquad (9)$$

If a task i cannot complete before time t and $h_i + f(\mathcal{I} \setminus \{i\}, t) > C$, then the task i must start after time t.

$$\text{ect}_i > t \wedge C < h_i + f(\mathcal{I} \setminus \{i\}, t) \Rightarrow \text{est}'_i > t \qquad (10)$$

Figure 1 depicts the Time-Tabling rule. Letort et al. [5] introduce a *sweep* technique that iterates over time and gradually enlarges the aggregate while pruning the tasks. Their method is later improved [13] and copes with very large sets of tasks. Beldiceanu et al. [4] propose an original technique reasoning over slack using a relation with the problem of rectangles placement.

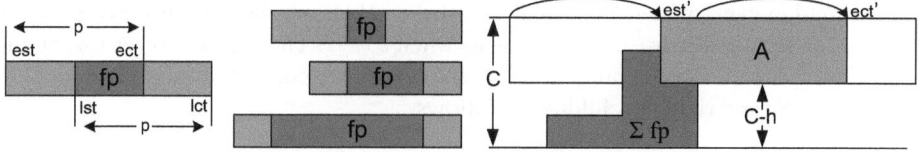

Fig. 1. A task with a fixed part, all tasks with a fixed part, the aggregate of the fixed parts and the Time-Tabling rule applied to task A

2.5   Time-Table Extended-Edge-Finding

Recent efforts [10,11] enhanced the Edge-Finding and Extended-Edge-Finding rules by taking into account the necessary usage of the resource due to fixed parts. The Time-Table Extended Edge-Finding combines the techniques of the Time-Tabling, the Edge-Finding, and the Extended-Edge-Finding. Let e^f_Ω be the energy of the tasks in Ω plus the fixed energy of the tasks in $\mathcal{I} \setminus \Omega$ spent within the interval $[\text{est}_\Omega, \text{lct}_\Omega)$.

$$e^f_\Omega = e_\Omega + f(\mathcal{I} \setminus \Omega, [\text{est}_\Omega, \text{lct}_\Omega)) \qquad (11)$$

Substituting e_Ω by e^f_Ω in (6) and (8) leads to the Time-Table Extended-Edge-Finding rules. Substituting e_Θ by e^f_Θ in (7) gives the new adjustment rule.

2.6   The Cumulative Tree and the Overload Checking Test

The algorithm we propose utilizes a cumulative tree similar to those introduced by Vilím [6,14,15]. The cumulative tree is an essentially complete binary tree with n leaves. Its main purpose is to compute the time interval $[a, b)$ that optimizes functions of the form $f(a, b, \Omega, \Lambda)$. Its leaves from left to right are associated to the tasks sorted in non-decreasing order of earliest starting time (est). The leaf of task i is labeled $\{i\}$ and their association holds throughout the execution of the algorithm. When the algorithm moves a task from a set to another, values in its associated leaf are re-initialized accordingly and the functions are updated from the leaf up to the root in $O(\log n)$. This data structure has proven very effective in particular for the Overload Checking that tests the E-feasibility. We illustrate in the following.

The function to optimize is the envelope of the subset $\Omega \subseteq \mathcal{I}$. From Equation (5).

$$\mathrm{Env}_\Omega = \max_{\Theta \in \Omega}(Cest_\Theta + e_\Theta)$$

The algorithm initializes all tasks as member of Ω. It iterates over every task j in decreasing order of lct_j. The algorithm ends an iteration by moving task j from Ω to Λ triggering a sequence of updates from its associated leaf. It results in the root holding the maximum envelope value of all intervals $[est_\Theta, \mathrm{lct}_\Theta) \subseteq [est_\Omega, \mathrm{lct}_\Omega)$ where $\mathrm{lct}_\Theta = \mathrm{lct}_\Omega = \mathrm{lct}_j$ at the beginning of any iteration. If $\mathrm{Env}_\Omega > C\,\mathrm{lct}_j$, the algorithm detects an overload.

To achieve the computation (see Figure 2), an envelope value and an energy value are required in every nodes. For a leaf $\{i\}$, these values are those of its corresponding task $e_{\{i\}} = p_i h_i$ and $\mathrm{Env}_{\{i\}} = Cest_i + e_i$ when $i \in \Omega$. They are set to zero when the task is moved to Λ. For the inner nodes v, the values are computed from the ones held by their left (l) and right (r) children as follows.

$$e_v = e_l + e_r \qquad\qquad \mathrm{Env}_v = \max\{\mathrm{Env}_l + e_r , \mathrm{Env}_r\}$$

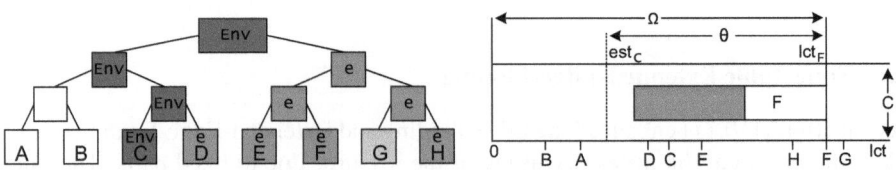

Fig. 2. A cumulative tree with its leaves sorted in increasing order of *est* (left) and a schematic representation of the cumulative resource with the time axis labeled with the lct (right). The algorithm moved task G to Λ at the previous iteration. It now iterates over task F. At this point, all tested intervals are upper bounded by lct_F. In this instance, the maximum envelope value is induced by the leaf associated to task C. The right part of the figure shows the optimal interval $[est_\Theta, \mathrm{lct}_\Theta) = [est_C, \mathrm{lct}_F)$. It is composed by the set of tasks $\{C, D, E, F, H\}$. The left part of the figure shows all the values that are cumulated up to the root resulting in the optimal value.

3   New Filtering Rules

The algorithm we present enforces the Edge-Finding and Extended-Edge-Finding rules to filter the lower bound of the starting time variables. A symmetric algorithm can filter the upper bounds. It proceeds by detecting any surplus of energy within a time interval $[\text{est}_\Omega, \text{lct}_\Omega)$ should a task i start at its *earliest starting time* est_i. If the surplus is positive, the algorithm detects that task i cannot start at time est_i and performs the exact adjustment to the lower bound of task i that erases the surplus.

We consider two cases where the Edge-Finding rule applies. The *weak case* occurs when the Edge-Finding rule (6) applies and $\text{ect}_i < \text{lct}_\Omega$ holds. We denote this rule EF^w. The *strong case* occurs when $\text{ect}_i \geq \text{lct}_\Omega$ and leads to the strong Edge-Finding rule denoted EF^s. The weak and strong cases also apply to the Extended-Edge-Finding rule (8) and leads to the two detection rules EEF^w and EEF^{s1}. In all cases, the Edge-Finding rules apply when $\text{est}_\Omega < \text{est}_i$ and the Extended-Edge-Finding rules apply when $\text{est}_\Omega \geq \text{est}_i$.

When one of these four rules detects a *surplus*, we denote by $\sigma_{EF^w}(i, \Omega)$, $\sigma_{EEF^w}(i, \Omega)$, $\sigma_{EF^s}(i, \Omega)$, and $\sigma_{EEF^s}(i, \Omega)$ the extra energy requirement in the time interval $[\text{est}_\Omega, \text{lct}_\Omega)$ should task i start at time est_i.

$$\sigma_{EF^w}(i, \Omega) = e_{\Omega \cup \{i\}} - C(\text{lct}_\Omega - \text{est}_\Omega) \tag{12}$$

$$\sigma_{EEF^w}(i, \Omega) = e_\Omega + h_i(\text{ect}_i - \text{est}_\Omega) - C(\text{lct}_\Omega - \text{est}_\Omega) \tag{13}$$

$$\sigma_{EF^s}(i, \Omega) = e_\Omega + h_i(\text{lct}_\Omega - \text{est}_i) - C(\text{lct}_\Omega - \text{est}_\Omega) \tag{14}$$

$$\sigma_{EEF^s}(i, \Omega) = e_\Omega - (C - h_i)(\text{lct}_\Omega - \text{est}_\Omega) \tag{15}$$

These quantities are used to combine the detection and the adjustment rules into a single rule that adjusts the earliest starting time of task i. In the weak case ($\text{ect}_i < \text{lct}_\Omega$), we obtain these two rules.

$$EF^w : \text{est}_i \geq \text{est}_\Omega \wedge \sigma_{EF^w}(i, \Omega) > 0 \Rightarrow \text{est}'_i = \text{lct}_\Omega - p_i + \left\lceil \frac{\sigma_{EF^w}(i, \Omega)}{h_i} \right\rceil$$

$$EEF^w : \text{est}_i < \text{est}_\Omega \wedge \sigma_{EEF^w}(i, \Omega) > 0 \Rightarrow$$
$$\text{est}'_i = \text{lct}_\Omega - (\text{ect}_i - \text{est}_\Omega) + \left\lceil \frac{\sigma_{EEF^w}(i, \Omega)}{h_i} \right\rceil$$

In the strong case ($\text{ect}_i \geq \text{lct}_\Omega$), we have this adjustment rule for the Edge-Finding

$$EF^s : \text{est}_i \geq \text{est}_\Omega \wedge \sigma_{EF^s}(i, \Omega) > 0 \Rightarrow \text{est}'_i = \text{est}_i + \left\lceil \frac{\sigma_{EF^s}(i, \Omega)}{h_i} \right\rceil$$

and the following one for the Extended-Edge-Finding

$$EEF^s : \text{est}_i < \text{est}_\Omega \wedge \sigma_{EEF^s}(i, \Omega) > 0 \Rightarrow \text{est}'_i = \text{est}_\Omega + \left\lceil \frac{\sigma_{EEF^s}(i, \Omega)}{h_i} \right\rceil$$

---

[1] The rules EF^w, EEF^w, EF^s, and EEF^s respectively represents the cases *inside*, *left*, *right*, and *through* in [10].

We show that these new adjustment rules are identical to the adjustment rule (7) when the relation $\Theta = \Omega$ holds. The case when $\Theta \subset \Omega$ is handled later.

Lemma 1. *The rules EF^w, EEF^w, EF^s, and EEF^s are equivalent to the adjustment rule (7) when $\Theta = \Omega$.*

Proof. The adjustment for the rule EF^s is

$$\text{est}'_i = \text{est}_i + \left\lceil \frac{e_\Omega + h_i(\text{lct}_\Omega - \text{est}_i) - C(\text{lct}_\Omega - \text{est}_\Omega)}{h_i} \right\rceil$$
$$= \text{est}_\Omega + \left\lceil \frac{e_\Omega - (C - h_i)(\text{lct}_\Omega - \text{est}_\Omega)}{h_i} \right\rceil$$

which is equivalent to rule (7) when $\Theta = \Omega$. The adjustment for the rule EEF^s is

$$\text{est}'_i = \text{est}_\Omega + \left\lceil \frac{e_\Omega - (C - h_i)(\text{lct}_\Omega - \text{est}_\Omega)}{h_i} \right\rceil$$

which is equivalent to rule (7) when $\Theta = \Omega$. The adjustment for the rule EF^w is

$$\text{est}'_i = \text{lct}_\Omega - p_i + \left\lceil \frac{e_\Omega + e_i - C(\text{lct}_\Omega - \text{est}_\Omega)}{h_i} \right\rceil = \text{lct}_\Omega + \left\lceil \frac{e_\Omega - C(\text{lct}_\Omega - \text{est}_\Omega)}{h_i} \right\rceil$$
$$= \text{est}_\Omega + \left\lceil \frac{e_\Omega - (C - h_i)(\text{lct}_\Omega - \text{est}_\Omega)}{h_i} \right\rceil$$

which is equivalent to rule (7) when $\Theta = \Omega$. The adjustment for the rule EEF^w is

$$\text{est}'_i = \text{lct}_\Omega - (\text{ect}_i - \text{est}_\Omega) + \left\lceil \frac{e_\Omega + h_i(\text{ect}_i - \text{est}_\Omega) - C(\text{lct}_\Omega - \text{est}_\Omega)}{h_i} \right\rceil$$
$$= \text{lct}_\Omega + \left\lceil \frac{e_\Omega - C(\text{lct}_\Omega - \text{est}_\Omega)}{h_i} \right\rceil$$

This form was already proved equivalent to rule (7) when $\Theta = \Omega$.   □

We show that successively applying, in no particular order, the rules EF^w, EEF^w, EF^s, and EEF^s leads to the same fixed point as the adjustment rule (7).

Lemma 2. *After applying the rules EF^w and EEF^w, the inequality $\text{ect}'_i \geq \text{lct}_\Omega$ holds, where ect'_i is the new earliest completion time of task i.*

Proof. After applying the rule EF^w, we obtain $\text{ect}'_i = \text{lct}_\Omega + \left\lceil \frac{\sigma_{EF^w}(i,\Omega)}{h_i} \right\rceil$. Since $\sigma_{EF^w}(i, \Omega) > 0$, we have $\text{ect}'_i > \text{lct}_\Omega$. The same applies for the rule EEF^w.   □

Lemma 2 ensures that when there are tasks for which the weak rules EF^w and EEF^w apply, after the adjustment of the rules, only the strong rules EF^s and EEF^s can apply.

Lemma 3. *Successively applying the adjustment rules EF^w, EEF^w, EF^s, and EEF^s leads to the same fixed point obtained by using the adjustment rule (7).*

Proof. Let Θ be the set that maximizes the expression in (7). Since Lemma 1 covers the case where $\Theta = \Omega$, we suppose that $\Theta \subset \Omega$. In the strong case, we have the inequalities $\text{lct}_\Theta \leq \text{lct}_\Omega \leq \text{ect}_i$. In the weak case, Lemma 2 ensures that these inequalities also hold after applying the rules EF^w or EEF^w. Therefore, we only need to check whether the rules EF^s and EEF^s can be applied with the set of tasks Θ. Since the set Θ leads to an adjustment, the numerator in (7) is positive which implies $e_\Theta > (C - h_i)(\text{lct}_\Theta - \text{est}_\Theta)$. If $\text{est}_i < \text{est}_\Theta$ then the rule EEF^s applies and leads to the same filtering as rule (7).

Suppose that $\text{est}_i \geq \text{est}_\Theta$ and that the adjustment rule (7) prunes the earliest starting time est_i further. Then this inequality holds.

$$\text{est}_i < \text{est}_\Theta + \frac{e_\Theta - (C - h_i)(\text{lct}_\Theta - \text{est}_\Theta)}{h_i} \tag{16}$$

This is equivalent to $0 < e_\Theta + h_i(\text{lct}_\Theta - \text{est}_i) - C(\text{lct}_\Theta - \text{est}_\Theta)$. Therefore, $\sigma_{EF^s}(i, \Theta) > 0$ and the rule EF^s prunes the est at the same position as rule (7) does. Consequently, after adjusting est_i, either $\Theta = \Omega$ and the adjustment is equivalent to the rule (7) or $\Theta \subset \Omega$ and the rules EF^s and EEF^s can still be applied in a future iteration. □

4 A New Extended-Edge-Finding Algorithm

We present a new algorithm that performs the Extended-Edge-Finding. Algorithm 1 is largely based on Vilím's algorithm [6] for the Edge-Finding of the cumulative constraint and its cumulative tree data structure. We broaden the scope of the cumulative tree with two more sets, Ψ and Γ and substitute Ω for Θ to go along our notation. Therefore, the algorithm uses a cumulative $\Omega, \Lambda, \Psi, \Gamma$ tree. These four sets are different status of the tasks during the execution of the algorithm and serve computational purposes. The mechanic of the cumulative tree is illustrated in Section 2.6.

An essentially complete binary tree of $|\mathcal{I}|$ leaves is built, with leaves from left to right associated to the tasks sorted in non-decreasing order of est, breaking ties on the smallest lct. The algorithm iterates on heights in $\{h_i \mid i \in \mathcal{I} \wedge \text{ect}_i < \text{lct}_i\}$ in arbitrary order, with h being the current height. These operations occur within an iteration.

The cumulative tree is initialized with all its tasks in Ω. It iterates through the tasks in non-increasing order of latest completion time (lct). We say that j is the current task. Thus, lct_j is the upper bound of all optimized intervals at the current iteration.

The algorithm partitions the tasks $\mathcal{I}$ into four sets: Γ is the set of excluded tasks, $\Omega = \{i \in \mathcal{I} \setminus \Gamma \mid \text{lct}_i \leq \text{lct}_j\}$ is the set of unprocessed tasks, $\Lambda = \{i \in \mathcal{I} \setminus (\Omega \cup \Gamma) \mid h_i = h, \text{ect}_i < \text{lct}_j\}$ is the set of processed tasks of height h with earliest completion time smaller than lct_j, and $\Psi = \{i \in \mathcal{I} \setminus (\Omega \cup \Gamma) \mid h_i = h, \text{ect}_i \geq \text{lct}_j\}$ is the set of processed tasks of height h with earliest completion time greater than or equal to lct_j. As it iterates through the tasks, the current latest completion time lct_j changes and might result in moving tasks from Λ to Ψ. At any time, a task can move from Λ and Ψ to the set of excluded tasks. Those are tasks that are ignored for the rest of the iteration. At the end of the iteration, the task j is removed from Ω and added to Λ if $h_j = h \wedge \text{ect}_j < \text{lct}_j$, otherwise, the task cannot be further filtered and is added to Γ.

The algorithm utilizes the cumulative tree to optimize the surplus functions (12) to (15) and performs an overload check. Whenever a detection applies, the corresponding task is pruned according to the adjustment rule and then moved to Γ. Then, the algorithm updates the nodes from the leaf associated to the pruned task up to the root and checks for an other detection. To efficiently compute the functions, eleven values are held in the nodes. Some of these values are function of the horizon $\text{Hor} = \max_{i \in \mathcal{I}} \text{lct}_i$, i.e. the latest time when a task can complete. For a leaf node v, these values are.

$$
e_v = \begin{cases} e_i & \text{if } i \in \Omega \\ 0 & otherwise \end{cases}
\qquad
\text{Env}_v = \begin{cases} C\,\text{est}_i + e_i & \text{if } i \in \Omega \\ -\infty & \text{otherwise} \end{cases}
\tag{17}
$$

$$
\text{Env}_v^h = \begin{cases} (C-h)\,\text{est}_i + e_i & \text{if } i \in \Omega \\ -\infty & \text{otherwise} \end{cases}
\qquad
e_v^\Lambda = \begin{cases} e_i & \text{if } i \in \Lambda \\ -\infty & \text{otherwise} \end{cases}
\tag{18}
$$

$$
\text{Env}^\Lambda = \begin{cases} C\,\text{est}_i + e_i & \text{if } i \in \Lambda \\ -\infty & \text{otherwise} \end{cases}
\qquad
\text{ex}_v^\Lambda = \begin{cases} h\,\text{ect}_i & \text{if } i \in \Lambda \\ -\infty & \text{otherwise} \end{cases}
\tag{19}
$$

$$
e_v^\Psi = \begin{cases} h(\text{Hor} - \text{est}_i) & \text{if } i \in \Psi \\ -\infty & \text{otherwise} \end{cases}
\qquad
\text{Env}_v^\Psi = \begin{cases} C\,\text{est}_i + e^\Psi & \text{if } i \in \Psi \\ -\infty & \text{otherwise} \end{cases}
\tag{20}
$$

$$
\text{ex}_v^\Psi = \begin{cases} h_i\,\text{Hor} & \text{if } i \in \Psi \\ -\infty & \text{otherwise} \end{cases}
\tag{21}
$$

$$
\text{Envx}_v^\Lambda = -\infty
\qquad\qquad
\text{Envx}_v^\Psi = -\infty
\tag{22}
$$

For an inner node v, its left child and right child are denoted $\text{left}(v)$ and $\text{right}(v)$. These values are computed recursively as follows.

$$
e_v = e_{\text{left}(v)} + e_{\text{right}(v)}
\tag{23}
$$

$$
\text{Env}_v = \max(\text{Env}_{\text{left}(v)} + e_{\text{right}(v)}, \text{Env}_{\text{right}(v)})
\tag{24}
$$

$$
\text{Env}_v^h = \max(\text{Env}_{\text{left}(v)}^h + e_{\text{right}(v)}, \text{Env}_{\text{right}(v)}^h)
\tag{25}
$$

$$
e_v^\Lambda = \max(e_{\text{left}(v)}^\Lambda + e_{\text{right}(v)}, e_{\text{left}(v)} + e_{\text{right}(v)}^\Lambda)
\tag{26}
$$

$$
\text{Env}_v^\Lambda = \max(\text{Env}_{\text{left}(v)}^\Lambda + e_{\text{right}(v)}, \text{Env}_{\text{left}(v)} + e_{\text{right}(v)}^\Lambda, \text{Env}_{\text{right}(v)}^\Lambda)
\tag{27}
$$

$$
\text{ex}_v^\Lambda = \max(\text{ex}_{\text{left}(v)}^\Lambda, \text{ex}_{\text{right}(v)}^\Lambda)
\tag{28}
$$

$$
e_v^\Psi = \max(e_{\text{left}(v)}^\Psi + e_{\text{right}(v)}, e_{\text{left}(v)} + e_{\text{right}(v)}^\Psi)
\tag{29}
$$

$$
\text{Env}_v^\Psi = \max(\text{Env}_{\text{left}(v)}^\Psi + e_{\text{right}(v)}, \text{Env}_{\text{left}(v)} + e_{\text{right}(v)}^\Psi, \text{Env}_{\text{right}(v)}^\Psi)
\tag{30}
$$

$$
\text{ex}_v^\Psi = \max(\text{ex}_{\text{left}(v)}^\Psi, \text{ex}_{\text{right}(v)}^\Psi)
\tag{31}
$$

$$
\text{Envx}_v^\Lambda = \max(\text{Envx}_{\text{left}(v)}^\Lambda + e_{\text{right}(v)}, \text{ex}_{\text{left}(v)}^\Lambda + \text{Env}_{\text{right}(v)}^h, \text{Envx}_{\text{right}(v)}^\Lambda)
\tag{32}
$$

$$
\text{Envx}_v^\Psi = \max(\text{Envx}_{\text{left}(v)}^\Psi + e_{\text{right}(v)}, \text{ex}_{\text{left}(v)}^\Psi + \text{Env}_{\text{right}(v)}^h, \text{Envx}_{\text{right}(v)}^\Psi)
\tag{33}
$$

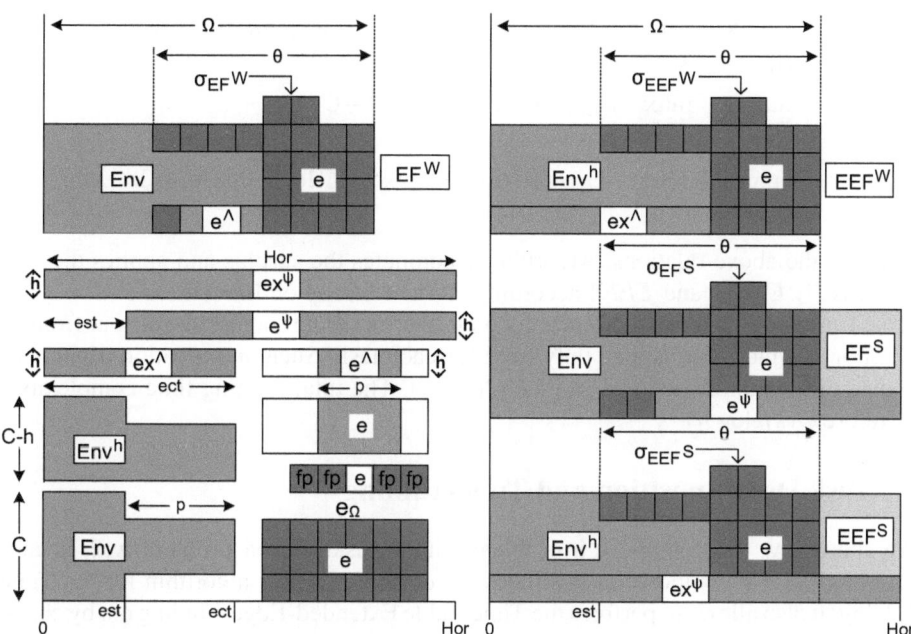

Fig. 3. Geometric illustration of the values cumulated by the tree, the four filtering rules and their detected surplus. The blue squares depict the cumulated energy of all tasks in Ω. The figure shows the optimal interval $[\text{est}_\Theta, \text{lct}_\Theta)$ within $[\text{est}_\Omega, \text{lct}_\Omega)$. All four rules are a combination of the energy of a task $i \notin \Omega$ and an optimal envelope. In this figure, each rule detects a surplus of 2 units of energy.

At the root of the tree, four values are particularly important and have the following equivalences. We use these relations to rewrite the conditions of the Edge-Finding and Extended-Edge-Finding rules.

$$\text{Env}^{\Lambda} = \max_{\substack{\Theta \subseteq \Omega \\ \text{lct}_\Theta = \text{lct}_\Omega}} \max_{\substack{i \in \Lambda \\ \text{est}_\Theta \leq \text{est}_i}} C \,\text{est}_\Theta + e_\Theta + e_i \tag{34}$$

$$\text{Env}^{\Psi} = \max_{\substack{\Theta \subseteq \Omega \\ \text{lct}_\Theta = \text{lct}_\Omega}} \max_{\substack{i \in \Psi \\ \text{est}_\Theta \leq \text{est}_i}} C \,\text{est}_\Theta + e_\Theta + h(\text{Hor} - \text{est}_i) \tag{35}$$

$$\text{Envx}^{\Lambda} = \max_{\substack{\Theta \subseteq \Omega \\ \text{lct}_\Theta = \text{lct}_\Omega}} \max_{\substack{i \in \Lambda \\ \text{est}_i \leq \text{est}_\Theta}} (C - h) \,\text{est}_\Theta + e_\Theta + h \,\text{ect}_i \tag{36}$$

$$\text{Envx}^{\Psi} = \max_{\substack{\Theta \subseteq \Omega \\ \text{lct}_\Theta = \text{lct}_\Omega}} \max_{\substack{i \in \Psi \\ \text{est}_i \leq \text{est}_\Theta}} (C - h) \,\text{est}_\Theta + e_\Theta + h \,\text{Hor} \tag{37}$$

The functions $\sigma_{EF^w}(i, \Omega)$, $\sigma_{EEF^w}(i, \Omega)$, $\sigma_{EF^s}(i, \Omega)$, and $\sigma_{EEF^s}(i, \Omega)$ can be optimized using the functions above.

$$\max_{\substack{\Theta \subseteq \Omega \\ \text{lct}_\Theta = \text{lct}_\Omega}} \max_{\substack{i \in \Lambda \\ \text{est}_\Theta \leq \text{est}_i}} \sigma_{EF^w}(i, \Theta) = \text{Env}^{\Lambda} - C \,\text{lct}_j \tag{38}$$

$$\max_{\substack{\Theta \subseteq \Omega \\ \mathrm{lct}_\Theta = \mathrm{lct}_\Omega \; \mathrm{est}_\Theta \leq \mathrm{est}_i}} \max_{i \in \Psi} \; \sigma_{EF^s}(i, \Theta) = \mathrm{Env}^\Psi - C \, \mathrm{lct}_j - h(\mathrm{Hor} - \mathrm{lct}_j) \tag{39}$$

$$\max_{\substack{\Theta \subseteq \Omega \\ \mathrm{lct}_\Theta = \mathrm{lct}_\Omega \; \mathrm{est}_i \leq \mathrm{est}_\Theta}} \max_{i \in \Lambda} \; \sigma_{EEF^w}(i, \Theta) = \mathrm{Envx}^\Lambda - C \, \mathrm{lct}_j \tag{40}$$

$$\max_{\substack{\Theta \subseteq \Omega \\ \mathrm{lct}_\Theta = \mathrm{lct}_\Omega \; \mathrm{est}_i \leq \mathrm{est}_\Theta}} \max_{i \in \Psi} \; \sigma_{EEF^s}(i, \Theta) = \mathrm{Envx}^\Psi - C \, \mathrm{lct}_j - h(\mathrm{Hor} - \mathrm{lct}_j) \tag{41}$$

Using the above relations, Algorithm 1 computes the surplus and applies the rules EF^w, EF^s, EEF^w, and EEF^s accordingly. The for loop on line 1 iterates $k = |\{h_i \mid i \in \mathcal{I}\}|$ times. Each time the repeat loop on line 2 executes, a task moves out from the set Λ or Ψ which can happen only once for each task. Such an operation triggers the update of the cumulative tree in time $\Theta(\log n)$. The total running time complexity is therefore $O(kn \log n)$.

5   Task Decomposition and Time-Tabling

We show how to decompose a problem with n tasks into a problem with at most $5n$ tasks. This decomposition facilitates the design of a new algorithm for the Time-Tabling. It also allows to perform the Time-Table Extended-Edge-Finding not by changing the Algorithm 1, but rather by changing its input. Task decomposition is a technique also used by Schutt et al. [1] and Vilím [10].

The tasks in $\mathcal{I}$ are decomposed into two sets of tasks: the depleted tasks $\mathcal{T}$ and the fixed tasks $\mathcal{F}$. For every task i such that $\mathrm{lst}_i < \mathrm{ect}_i$, there is a fixed energy of height h_i in the interval $[\mathrm{lst}_i, \mathrm{ect}_i)$. We replace the task $i \in \mathcal{I}$ by the task $i' \in \mathcal{T}$ with $\mathrm{est}_{i'} = \mathrm{est}_i$, $\mathrm{lct}_{i'} = \mathrm{lct}_i$, $p_{i'} = p_i - \mathrm{ect}_i + \mathrm{lst}_i$, and $h_{i'} = h_i$. If $\mathrm{lst}_i \geq \mathrm{ect}_i$, we create a task $i' \in \mathcal{T}$ that is a copy of the original task i. Let Z be the set of all time points est_i, lst_i, ect_i, and lct_i. We consider two consecutive time points a and b in Z with positive fixed energy, i.e. $f(\mathcal{I}, [a, b)) > 0$. We create a *fixed task* $f \in \mathcal{F}$ with $\mathrm{est}_f = a$, $\mathrm{lct}_f = b$, $p_f = b - a$, $h_f = f(\mathcal{I}, a)$. This task has no choice but to execute at its earliest starting time.

Since $|Z| \leq 4n$, there are fewer than $4n$ fixed tasks and the decomposition has fewer than $5n$ tasks. Two distinct tasks $f_1, f_2 \in \mathcal{F}$ produce two disjoint intervals $[\mathrm{est}_{f_1}, \mathrm{lct}_{f_1})$ and $[\mathrm{est}_{f_2}, \mathrm{lct}_{f_2})$. Figure 4 depicts this transformation.

5.1   Task Decomposition Algorithm

Algorithm 2 takes as input the set of original tasks $\mathcal{I}$ and returns the set of depleted tasks $\mathcal{T}$ and the set of fixed tasks $\mathcal{F}$. The algorithm has a running time complexity of $O(n \log n)$. Indeed, the dimension of vector r is at most $4n$ and requires $O(n \log n)$ to sort. The function IndexOf can be implemented with a binary search with time complexity $O(\log n)$ and is called at most n times. The first and second for loop have a time complexity of $O(n \log n)$ and $O(n)$ for a total of $O(n \log n)$.

5.2   Time-Tabling Algorithm

Algorithm 3 sorts the tasks $\mathcal{T}$ in non-decreasing heights and the fixed tasks $\mathcal{F}$ in non-increasing heights. It maintains, using an AVL tree, a set S of time intervals in which

---

Algorithm 1. ExtendedEdgeFinder($\mathcal{I}$)

Hor $\leftarrow \max_{i \in \mathcal{I}} \text{lct}_i$;

1 **for** $h \in \{h_i \mid i \in \mathcal{I} \wedge \text{ect}_i < \text{lct}_i\}$ **do**
$\quad \Omega \leftarrow \mathcal{I}$;
$\quad \Lambda \leftarrow \emptyset$;
$\quad \Psi \leftarrow \emptyset$;
$\quad$ **for** $j \in \mathcal{I}$ *in non-increasing order of* lct_j **do**
$\quad\quad$ **if** $\text{Env} > C \, \text{lct}_j$ **then** Fail;
$\quad\quad \Delta \leftarrow \{i \in \Lambda \mid \text{ect}_i \geq \text{lct}_j\}$;
$\quad\quad \Lambda \leftarrow \Lambda \setminus \Delta$;
$\quad\quad \Psi \leftarrow \Psi \cup \Delta \setminus \{i \in \Psi \mid \text{est}_i \geq \text{lct}_j\}$;
2 $\quad\quad$ **repeat**
$\quad\quad\quad \sigma(EF^w) \leftarrow \text{Env}^\Lambda - C \, \text{lct}_j$;
$\quad\quad\quad \sigma(EEF^w) \leftarrow \text{Envx}^\Lambda - C \, \text{lct}_j$;
$\quad\quad\quad \sigma(EF^s) \leftarrow \text{Env}^\Psi - C \, \text{lct}_j - h(\text{Hor} - \text{lct}_j)$;
$\quad\quad\quad \sigma(EEF^s) \leftarrow \text{Envx}^\Psi - C \, \text{lct}_j - h(\text{Hor} - \text{lct}_j)$;
$\quad\quad\quad m \leftarrow \max\{\sigma(EEF^w), \sigma(EEF^s), \sigma(EF^w), \sigma(EF^s), \}$;
$\quad\quad\quad$ **if** $\sigma(EEF^w) = m > 0$ **then**
$\quad\quad\quad\quad$ Let $i \in \Lambda$ be the unique task whose value ex^Λ is used for the computation of Envx^Λ;
$\quad\quad\quad\quad$ Let $k \in \Omega$ be the unique task whose value est_k is used for the computation of Env^h;
$\quad\quad\quad\quad \text{est}'_i \leftarrow \text{lct}_j - (\text{ect}_i - \text{est}_k) + \left\lceil \frac{\sigma(EEF^w)}{h_i} \right\rceil$;
$\quad\quad\quad\quad \Lambda \leftarrow \Lambda \setminus \{i\}$;
$\quad\quad\quad$ **else if** $\sigma(EEF^s) = m > 0$ **then**
$\quad\quad\quad\quad$ Let $i \in \Psi$ be the task with smallest est whose value ex^Ψ is used for the computation of Envx^Ψ;
$\quad\quad\quad\quad$ Let $k \in \Omega$ be the unique task whose value est_k is used for the computation of Env^Ψ;
$\quad\quad\quad\quad \text{est}'_i \leftarrow \text{est}_k + \left\lceil \frac{\sigma(EEF^s)}{h_i} \right\rceil$;
$\quad\quad\quad\quad \Psi \leftarrow \Psi \setminus \{i\}$;
$\quad\quad\quad$ **else if** $\sigma(EF^w) = m > 0$ **then**
$\quad\quad\quad\quad$ Let $i \in \Lambda$ be the unique task whose value e_v^Λ is used for the computation of Env^Λ;
$\quad\quad\quad\quad \text{est}'_i \leftarrow \text{lct}_j - p_i + \left\lceil \frac{\sigma(EF^w)}{h_i} \right\rceil$;
$\quad\quad\quad\quad \Lambda \leftarrow \Lambda \setminus \{i\}$;
$\quad\quad\quad$ **else if** $\sigma(EF^s) = m > 0$ **then**
$\quad\quad\quad\quad$ Let $i \in \Psi$ be the unique task whose value e^Ψ is used for the computation of Env^Ψ;
$\quad\quad\quad\quad \text{est}'_i \leftarrow \text{est}_i + \left\lceil \frac{\sigma(EF^s)}{h_i} \right\rceil$;
$\quad\quad\quad\quad \Psi \leftarrow \Psi \setminus \{i\}$;
$\quad\quad$ **until** $m \leq 0$;
$\quad\quad$ **if** $h_j = h \wedge \text{ect}_j < \text{lct}_j$ **then** $\Lambda \leftarrow \Lambda \cup \{j\}$;
$\quad\quad \Omega \leftarrow \Omega \setminus \{j\}$;

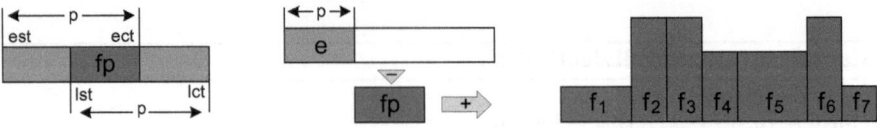

Fig. 4. A task with a fixed part, the same task after depletion of its fixed energy, and an energy aggregate turned into a set of fixed tasks $\mathcal{F}$

Algorithm 2. TimeTableTaskDecomposition($\mathcal{I}$)

Create the sorted vector $r = \{\mathrm{est}_i, \mathrm{ect}_i, \mathrm{lst}_i, \mathrm{lct}_i\}$ for all $i \in \mathcal{I}$ without duplicates;
Create the null vector c of dimension $|r|$;
$\mathcal{T} \leftarrow \emptyset, \mathcal{F} \leftarrow \emptyset$;
for $i \in \mathcal{I}$ **do**
    if $\mathrm{ect}_i > \mathrm{lst}_i$ **then**
        $a \leftarrow \texttt{IndexOf}(\mathrm{lst}_i, r)$;
        $b \leftarrow \texttt{IndexOf}(\mathrm{ect}_i, r)$;
        $c[a] \leftarrow c[a] + h_i$;
        $c[b] \leftarrow c[b] - h_i$;
        $\mathcal{T} \leftarrow \mathcal{T} \cup \{\texttt{Task}(\mathrm{est} = \mathrm{est}_i, \mathrm{lct} = \mathrm{lct}_i, h = h_i, p = p_i - \mathrm{ect}_i + \mathrm{lst}_i)\}$;
    else
        $\mathcal{T} \leftarrow \mathcal{T} \cup \{\texttt{Task}(\mathrm{est} = \mathrm{est}_i, \mathrm{lct} = \mathrm{lct}_i, h = h_i, p = p_i)\}$;
for $l = 1..|r| - 1$ **do**
    $c[l] \leftarrow c[l] + c[l - 1]$;
    if $c[l - 1] > C$ **then** Failure;
    if $c[l - 1] > 0$ **then**
        $\mathcal{F} \leftarrow \mathcal{F} \cup \{\texttt{Task}(\mathrm{est} = r[l - 1], \mathrm{lct} = r[l], h = c[l - 1], p = r[l] - r[l - 1])\}$;
return $(\mathcal{T}, \mathcal{F})$

---

the unprocessed tasks in $\mathcal{T}$ cannot execute concurrently with the fixed tasks. The set S grows as the algorithm iterates through $\mathcal{T}$. While processing the task $i' \in \mathcal{T}$, if there exists an interval $[a, b) \subseteq S$ such that $\mathrm{est}_{i'} < b$ and $\mathrm{est}_{i'} + p_{i'} > a$ then the algorithm retrieves the original task $i \in \mathcal{I}$ associated to i' and performs the pruning $\mathrm{est}_i \leftarrow \min(\mathrm{lst}_i, b)$. When $\mathrm{lst}_i < b$, the earliest starting time is set to lst_i to force the task to start at the beginning of its fixed part. The AVL tree finds the interval $[a, b)$ in $O(\log |\mathcal{F}|)$. Sorting the tasks require $O(|\mathcal{T}| \log |\mathcal{T}|)$ and $O(|\mathcal{F}| \log |\mathcal{F}|)$. Since $|\mathcal{T}|, |\mathcal{F}| \in O(n)$, the overall complexity is $O(n \log n)$.

5.3 Time-Table Extended-Edge-Finding

We use the decomposition to perform Time-Table Extended-Edge-Finding. After reaching a fixed point with Algorithm 2 and 3, we pass the tasks $\mathcal{T} \cup \mathcal{F}$ as input to Algorithm 1. Since the fixed tasks will not be filtered, the for loop on line 1 can restrict the iterations over the heights of the tasks in $\mathcal{T}$. When the earliest starting time of task $i' \in \mathcal{T}$ is filtered to time t, we filter the est of the original task $i \in \mathcal{I}$

Algorithm 3. FilterTimeTabling($\mathcal{T}, \mathcal{F}$)

Sort the fixed tasks $\mathcal{F}$ in non-increasing order of heights;
$S \leftarrow \{\infty\}; j \leftarrow 0;$
for $i' \in \mathcal{T}$ *in non-decreasing order of height* **do**
    while $j < |\mathcal{F}| \wedge h_{\mathcal{F}[j]} > C - h_{i'}$ **do**
        $S \leftarrow S \cup [\text{est}_{\mathcal{F}[j]}, \text{lct}_{\mathcal{F}[j]});$
        $j \leftarrow j + 1;$
    $b \leftarrow \min\{b \notin S \mid b - 1 \in S \wedge \text{est}_{i'} < b\};$
    $a \leftarrow \min\{a \in S \mid [a, b) \subseteq S\};$
    if $\text{est}_{i'} + p_{i'} > a$ **then**
        if $\text{lst}_i \geq \text{ect}_i$ **then** $\text{est}_i \leftarrow b;$
        else $\text{est}_i \leftarrow \min(\text{lst}_i, b);$

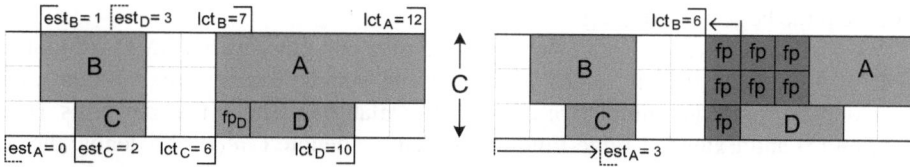

Fig. 5. The left part depicts a CuSP with 4 tasks. The upper and lower parts of the time axis indicates the earliest starting times and the latest completion times. The grid determines the energy units. The processing times and heights are to scale. By not taking into account the fixed part of task D, neither the Time-Tabling rule nor the Extended-Edge-Finding rule can deduct a pruning. A decomposition of task D leads to two consecutive updates. The rule EEF^w updates the lower bound of Task A to 3 which creates 6 new units of fixed energy. Then, the Time-Tabling rule adjusts the upper bound of task B to 6. The right part depicts the resulting CuSP.

to time $\text{est}_i \leftarrow \min(t, \text{lst}_i)$. This ensures to perform Time-Table Edge-Finding in time $O(kn \log n)$. Figure 5 shows an example where a task is filtered by Time-Table Extended-Edge-Finding.

6   Experiments

We tested the different versions of the algorithm with the PSLIB benchmark (Projection Scheduling Problem Library) [16]. More precisely, we solved instances of the single-mode resource-constrained project scheduling problem (SMRCPSP). Those instances are based on series of tasks that can be completed before a given horizon limit. A number of resources is given with varying capacities of production. Each task has a duration and an amount of a specific resources used during its execution. Each task also has a list of other tasks, its successors, that can be started only after this task is completed.

    The model is based on two constraints. We use a precedence constraint to ensure the order of the successors is respected and we use a cumulative constraint for each resource that ensures the execution of the tasks does not overload the resources. We set the makespan to the best known value reported for the benchmark. We use a binary

Table 1. Experimental results. Section *Benchmark* reports the number of tasks n, the number of instances, and the time out (in seconds) used for the experiment. For each filtering algorithm, we report the number of instances solved (*solved*). We report the cumulative number of backtracks (*bt*) and the cumulative time (*time*) required to solve all instances that are commonly solved by the three algorithms.

Benchmark			Choco			EEF+TT			TTEEF		
n	#instances	time out	solved	bt	time	solved	bt	time	solved	bt	time
30	480	10	364	8757	223	377	8757	50	377	8379	54
60	480	20	332	3074	1527	340	3074	269	341	2861	291
90	480	50	321	5024	5522	327	5024	857	329	4635	913

variable to enforce a precedence between each relevant pair of tasks. We branch on the precedence constraints that involve the tasks with the most similar resource consumptions and the largest processing times.

We used the CP solver Choco version 2.1.5 on a computer with a AMD Athlon(tm) II P340 Dual-Core running at 2.20GHz. We ran simultaneously 2 experiments, one per core. We used the cumulative constraint available in Choco that performs Time-Tabling [5] and Extended-Edge-Finding [8] that we denote *Choco*. We denote the Algorithm 1 combined with the Algorithm 3 *EEF+TT* and the Time-Table Extended-Edge-Finding *TTEEF*. Table 1 reports the results.

Choco and EEF+TT produce the same number of backtracks since they offer the same filtering. However, EEF+TT is significantly faster than Choco and solves more instances. TTEEF is slightly slower in time than EEF+TT but solves few more instances in fewer backtracks.

7   Conclusion

We presented three new algorithms that filter the CUMULATIVE constraint. The first algorithm is an Extended-Edge-Finder with a time complexity of $O(kn \log n)$. The second filtering algorithm performs Time-Tabling in time $O(n \log n)$. The third algorithm performs Time-Table Extended-Edge-Finding in time $O(kn \log n)$. These new algorithms proved to be very efficient in practice offering a fast and strong filtering.

References

1. Schutt, A., Wolf, A.: A new $\wr(n^2 \log n)$ not-first/Not-last pruning algorithm for cumulative resource constraints. In: Cohen, D. (ed.) CP 2010. LNCS, vol. 6308, pp. 445–459. Springer, Heidelberg (2010)
2. Aggoun, A., Beldiceanu, N.: Extending chip in order to solve complex scheduling and placement problems. Mathematical and Computer Modelling 17(7) (1993)
3. Baptiste, P., Pape, C.L.: Constraint propagation techniques for disjunctive scheduling: The preemptive case. In: Proceedings of the 12th European Conference on Artificial Intelligence, ECAI 1996 (1996)
4. Beldiceanu, N., Carlsson, M., Poder, E.: New filtering for the **[Equation image]** constraint in the context of non-overlapping rectangles. In: Trick, M.A. (ed.) CPAIOR 2008. LNCS, vol. 5015, pp. 21–35. Springer, Heidelberg (2008)

5. Beldiceanu, N., Carlsson, M.: A new multi-resource *cumulatives* constraint with negative heights. In: Van Hentenryck, P. (ed.) CP 2002. LNCS, vol. 2470, pp. 63–79. Springer, Heidelberg (2002)
6. Vilím, P.: Edge finding filtering algorithm for discrete cumulative resources in $\iota(kn\log n)$. In: Gent, I.P. (ed.) CP 2009. LNCS, vol. 5732, pp. 802–816. Springer, Heidelberg (2009)
7. Kameugne, R., Fotso, L.P., Scott, J., Ngo-Kateu, Y.: A quadratic edge-finding filtering algorithm for cumulative resource constraints. In: Lee, J. (ed.) CP 2011. LNCS, vol. 6876, pp. 478–492. Springer, Heidelberg (2011)
8. Mercier, L., Van Hentenryck, P.: Edge finding for cumulative scheduling. INFORMS Journal on Computing 20(1), 143–153 (2008)
9. Baptiste, P., Pape, C.L., Nuijten, W.: Constraint-Based Scheduling. Kluwer Academic Publishers (2001)
10. Vilím, P.: Timetable edge finding filtering algorithm for discrete cumulative resources. In: Achterberg, T., Beck, J.C. (eds.) CPAIOR 2011. LNCS, vol. 6697, pp. 230–245. Springer, Heidelberg (2011)
11. Schutt, A., Feydy, T., Stuckey, P.J.: Explaining time-table-edge-finding propagation for the cumulative resource constraint. In: Gomes, C., Sellmann, M. (eds.) CPAIOR 2013. LNCS, vol. 7874, pp. 234–250. Springer, Heidelberg (2013)
12. Nuijten, W.: Time and Resource Constrained Scheduling. PhD thesis, Eindhoven University of Technology (1994)
13. Letort, A., Beldiceanu, N., Carlsson, M.: A scalable sweep algorithm for the *cumulative* constraint. In: Milano, M. (ed.) CP 2012. LNCS, vol. 7514, pp. 439–454. Springer, Heidelberg (2012)
14. Vilím, P.: $O(n \log n)$ filtering algorithms for unary reource constraint. In: Régin, J.-C., Rueher, M. (eds.) CPAIOR 2004. LNCS, vol. 3011, pp. 335–347. Springer, Heidelberg (2004)
15. Vilím, P.: Max energy filtering algorithm for discrete cumulative resources. In: van Hoeve, W.-J., Hooker, J.N. (eds.) CPAIOR 2009. LNCS, vol. 5547, pp. 294–308. Springer, Heidelberg (2009)
16. Kolisch, R., Sprecher, A.: Psplib - a project scheduling library. European Journal of Operational Research 96, 205–216 (1996), http://webserver.wi.tum.de/psplib/

Revisiting the Cardinality Reasoning
for BinPacking Constraint

François Pelsser[1], Pierre Schaus[1], and Jean-Charles Régin[2]

[1] UCLouvain, ICTEAM,
Place Sainte-Barbe 2,
1348 Louvain-la-Neuve, Belgium
pierre.schaus@uclouvain.be
[2] University of Nice-Sophia Antipolis,
I3S UMR 6070, CNRS, France
jcregin@gmail.com

Abstract. In a previous work, we introduced a filtering for the Bin-Packing constraint based on a cardinality reasoning for each bin combined with a global cardinality constraint. We improve this filtering with an algorithm providing tighter bounds on the cardinality variables. We experiment it on the Balanced Academic Curriculum Problems demonstrating the benefits of the cardinality reasoning for such bin-packing problems.

Keywords: Constraint Programming, Global Constraints, Bin-Packing.

1 Introduction

The BinPacking($[X_1, ..., X_n], [w_1, ..., w_n], [L_1, ..., L_m]$) global constraint captures the situation of allocating n indivisible weighted items to m capacitated bins:

- X_i is an integer variable representing the bin where item i, with strictly positive integer weight w_i, is placed. Every item must be placed *i.e.* $Dom(X_i) \subseteq [1..m]$.
- L_j is an integer variable representing the sum of items weights placed into that bin.

The constraint enforces the following relations:

$$\forall j \in [1..m] : \sum_{i|X_i=j} w_i = L_j$$

The initial filtering algorithm proposed for this constraint in [8] essentially filters the domains of the X_i using a knapsack-like reasoning to detect if forcing an item into a particular bin j would make it impossible to reach a load L_j for that bin. This procedure is very efficient but can say that an item is OK for a particular bin while it is not. A failure detection algorithm was also introduced in [8] computing a lower bound on the number of bins necessary to complete the

C. Schulte (Ed.): CP 2013, LNCS 8124, pp. 578–586, 2013.

partial solution. This last consistency check has been extended in [2]. Cambazard and O'Sullivan [1] propose to filter the domains using an LP arc-flow formulation.

In classical bin-packing problems, the capacity of the bins $\overline{L_j}$ are constrained while the lower bounds $\underline{L_j}$ are usually set to 0 in the model. This is why existing filtering algorithms use the upper bounds of the load variables $\overline{L_j}$ (*i.e.* capacity of the bins) and do not focus much on the lower bounds of these variables $\underline{L_j}$.

Recently [7] introduced an additional cardinality based filtering counting the number of items in each bin. We can view this extension as a generalization $\texttt{BinPacking}([X_1, ..., X_n], [w_1, ..., w_n], [L_1, ..., L_m], [C_1, ..., C_m])$ of the constraint where C_j are counting variables, that is defined by $\forall j \in [1..m] : C_j = |\{i | X_i = j\}|$. This formulation for the $\texttt{BinPacking}$ constraint is well suited when

- the lower bounds on load variables are also constrained initially $\underline{L_j} > 0$,
- the items to be placed are approximately equivalent in weight (the bin-packing is dominated by an assignment problem), or
- there are cardinality constraints on the number of items in each bin.

The idea of [7] is to introduce a redundant global cardinality constraint [5]:

$$\begin{aligned} \texttt{BinPacking}([X_1, ..., X_n], [w_1, ..., w_n], [L_1, ..., L_m], [C_1, ..., C_m]) &\equiv \\ \texttt{BinPacking}([X_1, ..., X_n], [w_1, ..., w_n], [L_1, ..., L_m]) &\wedge \qquad (1) \\ \texttt{GCC}([X_1, ..., X_n], [C_1, ..., C_m]) & \end{aligned}$$

with a specialized algorithm used to adjust the upper and lower bounds of the C_j variables when the bounds of the L_j's and/or the domains of the X_i's change. Naturally the tighter are the bounds computed on the cardinality variables, the stronger will be the filtering induced by the GCC constraint.

We first introduce some definitions, then we recall the greedy algorithm introduced in [7] to update the cardinality variables.

Definition 1. *We denote by $pack_j$ the set of items already packed in bin j : $pack_j = \{i | Dom(X_i) = \{j\}\}$ and by $cand_j$ the candidate items available to go in bin j: $cand_j = \{i | j \in Dom(X_i) \wedge |Dom(X_i)| > 1\}$. The sum of the weights of a set of items S is $sum(S) = \sum_{i \in S} w_i$.*

As explained in [7], a lower bound on the number of items that can be additionally packed into bin j can be obtained by finding the size of the smallest cardinality set $A_j \subseteq cand_j$ such as $sum(A_j) \geq \underline{L_j} - sum(pack_j)$. Then we have $C_j \geq |pack_j| + |A_j|$. Thus we can filter the lower bound of the cardinality $\underline{C_j}$ as follows:

$$\underline{C_j} \leftarrow \max(\underline{C_j}, |pack_j| + |A_j|).$$

This set A_j is obtained in [7] by scanning greedily elements in $cand_j$ with decreasing weights until an accumulated weight of $\underline{L_j} - sum(pack_j)$ is reached. It can be done in linear time assuming the items are sorted initially by weight.

Example 1. Five items with weights $3, 3, 4, 5, 7$ can be placed into bin 1 having a possible load $L_1 \in [20..22]$. Two other items are already packed into that bin with weights 3 and 7 ($|pack_1| = 2$ and $l_1 = 10$). Clearly we have that $|A_1| = 2$ obtained with weights $5, 7$. The minimum value of the domain of the cardinality variable C_1 is thus set to 4.

A similar reasoning can be used to filter the upper bound of the cardinality variable $\overline{C_j}$.

This paper further improves the cardinality based filtering, introducing

1. In Section 2, an algorithm computing tighter lower/upper bounds on the cardinality variables C_j of each bin j, and
2. In Section 3, an algorithm to update the load variables L_j based on the cardinality information.

The new filtering is experimented on the Balanced Academic Curriculum Problem in Section 4.

2   Filtering the Cardinality Variables

The lower (upper) bound computation on the cardinality C_j introduced in [7] only considers the possible items $cand_j$ and the minimum (maximum) load value to reach i.e. $\underline{L_j}$ ($\overline{L_j}$). Stronger bounds can possibly be computed by also considering the cardinality variables of other bins. Indeed, an item which is used for reaching the minimum cardinality or minimum load for a bin j, may not be usable again for computing the minimum cardinality of another bin k as illustrated on next example:

Example 2. A bin j can accept items having weights $3, 3, 3$ with a minimum load of 6 and thus a minimum cardinality of 2 items. A bin k with a minimum load of 5 can accept the same items plus two items of weight 1. Clearly, the bin k can not take more than one item with weight 3 for computing its minimum cardinality because it would prevent the bin j to reach its minimum cardinality of 2. Thus the minimum cardinality of bin k should be 3 and not 2 as would be computed with the lower bound of [7].

Minimum Cardinality of bin j Algorithm 1 computes a stronger lower bound also taking into account the cardinality variables of other bins $\underline{C_k}$ $\forall k \neq j$. The intuition is that it prevents to reuse again an item if it is required for reaching a minimum cardinality in another bin. This is achieved by maintaining for every other bin k the number of items this bin is ready to give without preventing it to fulfill its own minimum cardinality requirement $\underline{C_k}$.

Clearly if a bin k must pack at least $\underline{C_k}$ items and has already packed $|pack_k|$ items, this bin can not give more than $|cand_k| - (\underline{C_k} - |pack_k|)$ items to bin j. This information is maintained into the variables $availableForOtherBins_k$ initialized at line 5.

Example 3. Continuing on Example 2, bin j will have $availableForOtherBins_j = 3 - (2 - 0) = 1$ because this bin can give at most one of its item to another bin.

Since items are iterated in decreasing weight order at line 7, the other bins accept to give first their "heaviest" candidate items. This is an optimistic situation from the point of view of bin j, justifying why the algorithm computes a valid lower bound on the cardinality variable C_j. Each time an item is used by bin j, the other bins (where this item was candidate) reduce their quantities $availableForOtherBins_k$ since they "consume" their flexibility to give items. If at least one other bin k absolutely needs the current item i to fulfill its own minimum cardinality (detected at line 13), `available` is set to `false` meaning that this item can not be used in the computation of the cardinality of bin j to reach the minimum load.

On the other hand, if the current item can be used (`available=true`), then other bins which agreed to give this item have one item less available. The $availableForOtherBins_k$ numbers are decremented at line 22.

Finally notice that the algorithm may detect unfeasible situations when it is not able to reach the minimum load at line 28.

Maximum Cardinality The algorithm to compute the maximum cardinality is similar. The changes to bring to Algorithm 1 are:

1. The variable $binMinCard$ should be named $binMaxCard$
2. The items are considered in increasing weight order at line 7, and
3. The stopping criteria at line 8 becomes $binLoad + w_i > \overline{L_j}$.
4. There is no feasibility test at lines 27 - 29.

Complexity Assuming the items are sorted initially in decreasing weights, this algorithm runs in $O(n \cdot m)$ with n the number of items and m the number of bins. Hence adjusting the cardinality of every bins takes $O(n \cdot m^2)$. This algorithm has no guarantee to be idempotent. Indeed the bin j may consider an item i as available, but the later adjustment of the minimum cardinality of another bin k may cause this item to be unavailable if bin j is considered again.

Example 4. The instance considered - depicted in Figure 1 (a) - is the following:

$$\text{BinPacking}([X_1, \ldots, X_4], [w_1, \ldots, w_4], [L_1, \ldots, L_3])$$
$$X_1 \in \{1, 2\}, X_2 \in \{1, 2\}, X_3 \in \{2, 3\}, X_4 \in \{2, 3\},$$
$$w_1 = 1, w_2 = 1, w_3 = 3, w_4 = 3 \tag{2}$$
$$L_1 \in \{1, 2\}, L_2 \in \{2, 3\}, L_3 \in \{2, 4\}$$

We consider first the computation of the cardinality of bin 2. This bin must have at least one item to reach its minimum load. We now consider the maximum cardinality of this bin. Items 1 and 2 can both be packed into bin 2 but doing so would prevent bin 1 to achieve its minimum load requirement of 1. Hence only one of these items can be used during the computation of the maximum cardinality for bin 2. Assuming that item 1 is used, the next item considered is item 3 having a weight of 3. But Adding this item together with item 1 would exceed the maximum load $(4 > 3)$ (stopping criteria for the maximum

---

Algorithm 1. Computes a lower bound on the cardinality of bin j

---

Data: j a bin index
Result: $binMinCard$ a lower bound on the min cardinality for the bin j
1 $binLoad \leftarrow sum(pack_j)$;
2 $binMinCard \leftarrow |pack_j|$;
3 $othersBins \leftarrow \{1, \dots, m\} \setminus j$;
4 **foreach** $k \in otherBins$ **do**
5 $\quad$ $availableForOtherBins_k \leftarrow |cand_k| - (\underline{C_k} - |pack_k|)$;
6 **end**
7 **foreach** $i \in cand_j$ in decreasing weight order **do**
8 $\quad$ **if** $binLoad \geq \underline{L_j}$ **then**
9 $\quad\quad$ break ;
10 $\quad$ **end**
11 $\quad$ $available \leftarrow$ true;
12 $\quad$ **for** $k \in othersBins$ **do**
13 $\quad\quad$ **if** $k \in Dom(X_i) \land availableForOtherBins_k = 0$ **then**
14 $\quad\quad\quad$ $available \leftarrow$ false ;
15 $\quad\quad$ **end**
16 $\quad$ **end**
17 $\quad$ **if** $available$ **then**
18 $\quad\quad$ $binLoad \leftarrow binLoad + w_i$;
19 $\quad\quad$ $binMinCard \leftarrow binMinCard + 1$;
20 $\quad\quad$ **for** $k \in othersBins$ **do**
21 $\quad\quad\quad$ **if** $k \in Dom(X_i)$ **then**
22 $\quad\quad\quad\quad$ $availableForOtherBins_k \leftarrow availableForOtherBins_k - 1$;
23 $\quad\quad\quad$ **end**
24 $\quad\quad$ **end**
25 $\quad$ **end**
26 **end**
27 **if** $binLoad < \underline{L_j}$ **then**
28 $\quad$ The constraint is unfeasible ;
29 **end**

---

cardinality computation). Hence the final maximum cardinality for bin 2 is one. The cardinality reasoning also deduces that bin 1 must have between one and two items and bin 3 must have exactly one item. Based on these cardinalities, the global cardinality constraint (GCC) is able to deduce that item 1 and 2 must be packed into bin 1. This filtering is illustrated on Figure 1 (b).

The algorithm from [7] deduces that bin 2 must have between one and two items (not exactly one as the new filtering). The upper bound of two items is obtained with the two lightest items 1 and 2. As for the new algorithm, it deduces that bin 1 must have between one and two items and bin 3 must have exactly one item. Unfortunately, the GCC is not able to remove any bin from the item's domains based on these cardinality bounds. Thus, this algorithm is less powerful than the new one.

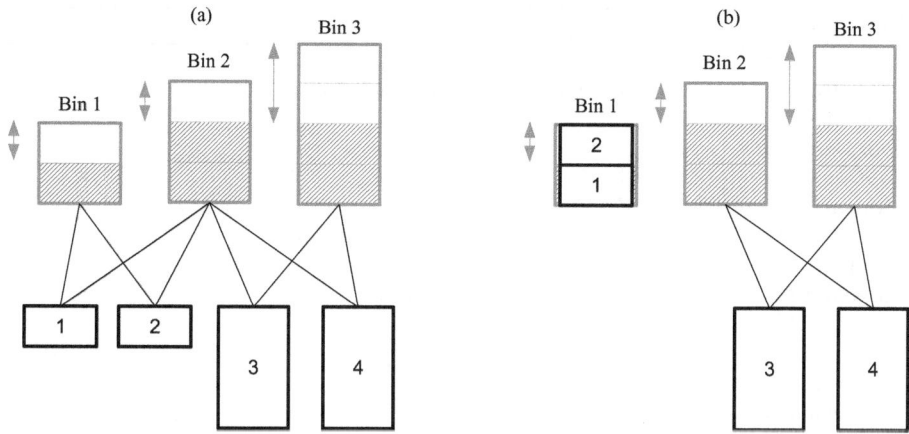

Fig. 1. (a) BinPacking instance with 3 bins and 4 items. The arcs represent for each item, the possible bins. (b) Domains resulting from the filtering induced with the tighter computation of the cardinalities. The grey in a bin stands for the minimum level to reach.

3 Filtering the Load Variables

We introduce a filtering of the load variable taking the cardinality information into account. No such filtering was proposed in [7]. Algorithm 2 is similar to Algorithm 1 except that we try to reach the minimum cardinality requirements by choosing first the "lightest" items until the minimum cardinality $\underline{C_j}$ is reached (line 8). Again a similar reasoning can be done to compute an upper bound on the maximum load.

4 Experiments

The Balanced Academic Curriculum Problem (BACP) is recurrent in Universities. The goal is to schedule the courses that a student must follow in order to respect the prerequisite constraints between courses and to balance as much as possible the workload of each period. Each period also has a minimum and maximum number of courses. The largest of the three instances available on CSPLIB (http://www.csplib.org) with 12 periods, 66 courses having a weight between 1 and 5 (credits) and 65 prerequisites relations, was modified in [6] to generate 100 new instances[1] by giving each course a random weight between 1 and 5 and by randomly keeping 50 out of the 65 prerequisites. Each period must have between 5 and 7 courses. As shown in [3], a better balance property is obtained by minimizing the variance instead of the maximum load. For each instance, we test three different filtering configurations for bin-packing:

---

[1] Available at `http://becool.info.ucl.ac.be/resources/bacp`

Algorithm 2. Computes a lower bound on load of bin j

Data: j a bin index

Result: $binMinLoad$ a lower bound on the load of bin j

1   $binCard \leftarrow |pack_j|$;

2   $binMinLoad \leftarrow sum(pack_j)$;

3   $othersBins \leftarrow \{1, \ldots, m\} \setminus j$;

4   **foreach** $k \in otherBins$ **do**

5      $\mid$ $availableForOtherBins_k \leftarrow |cand_k| - (\underline{C_k} - |pack_k|)$;

6   **end**

7   **foreach** $i \in cand_j$ *in increasing weight order* **do**

8   **if** $binCard \geq (\overline{C_j})$ **then**

9          $\mid$ break ;

10   **end**

11     $available \leftarrow$ true;

12   **for** $k \in othersBins$ **do**

13   **if** $k \in Dom(X_i) \wedge availableForOtherBins_k = 0$ **then**

14          $\mid$ $available \leftarrow$ false ;

15   **end**

16   **end**

17   **if** $available$ **then**

18        $binMinLoad \leftarrow binLoad + w_i$;

19        $binCard \leftarrow binCard + 1$;

20   **for** $k \in othersBins$ **do**

21   **if** $k \in Dom(X_i)$ **then**

22             $\mid$ $availableForOtherBins_k \leftarrow availableForOtherBins_k - 1$;

23   **end**

24   **end**

25   **end**

26   **end**

27   **if** $binCard < \underline{C_j}$ **then**

28   The constraint is unfeasible ;

29   **end**

Table 1. Number of instances for which is was possible to prove optimality within the time limit

limit(s)	A	B	C
15	13	27	**41**
30	18	34	**46**
60	21	37	**51**
120	25	43	**57**
1800	37	62	**69**

- A: The `BinPacking` constraint from [8] + a `GCC` constraint,
- B: A + the cardinality filtering from [7],
- C: A + the cardinality filtering introduced in this paper.

The experiments were conducted on a Macbook Pro 2.3 Ghz, I7. The solver used is OscaR [4] running on JVM 1.7 of Oracle and implemented with Scala 2.10. The source code of the constraint is available on OscaR repository.

Table 2. Detailed statistics obtained on some significant instances

instance	time (ms)			best bound			number of failures		
	A	B	C	A	B	C	A	B	C
inst2.txt	timeout	timeout	**679**	3243	3247	**3237**	835459	1064862	**829**
inst14.txt	timeout	45625	**6925**	3107	**3105**	**3105**	1043251	228294	**8530**
inst22.txt	timeout	13971	**281**	3045	**3041**	**3041**	811852	48482	**353**
inst30.txt	timeout	118964	**192**	3416	**3402**	**3402**	795913	707487	**129**
inst36.txt	timeout	timeout	**337**	2685	2685	**2671**	847641	915849	**364**
inst47.txt	timeout	timeout	**112**	3309	3309	**3303**	2561038	3812512	**269**
inst65.txt	timeout	timeout	**222**	3416	3414	**3402**	921694	1091396	**168**
inst70.txt	timeout	timeout	**101060**	3043	3043	**3041**	1917729	1516627	**125270**
inst87.txt	16275	15089	**251**	**3643**	**3643**	**3643**	109173	65493	**207**
inst98.txt	timeout	timeout	**48**	2987	2987	**2979**	7023383	8261509	**261**

Table 1 gives the number of solved instances for increasing timeout values. Table 2 illustrates the detailed numbers (time, best bound, number of failures) for some instances with a 30 minutes timeout. As can be seen, the new filtering allows to solve more instances sometimes cutting the number of failures by several order of magnitudes.

5   Conclusion

We introduced stronger cardinality bounds on the BinPacking constraint by also integrating the cardinality requirements of other bins during the computation. These stronger bounds have a direct impact on the filtering of placement variables through the GCC constraint. The improved filtering was experimented on the BACP allowing to solve more instances and reducing drastically the number of failures on some instances.

References

1. Cambazard, H., O'Sullivan, B.: Propagating the bin packing constraint using linear programming. In: Cohen, D. (ed.) CP 2010. LNCS, vol. 6308, pp. 129–136. Springer, Heidelberg (2010)
2. Dupuis, J., Schaus, P., Deville, Y.: Consistency Check for the Bin Packing Constraint Revisited. In: Lodi, A., Milano, M., Toth, P. (eds.) CPAIOR 2010. LNCS, vol. 6140, pp. 117–122. Springer, Heidelberg (2010)
3. Monette, J.-N., Schaus, P., Zampelli, S., Deville, Y., Dupont, P.: A CP approach to the balanced academic curriculum problem. In: Seventh International Workshop on Symmetry and Constraint Satisfaction Problems, vol. 7 (2007)
4. OscaR Team. OscaR: Scala in OR (2012), https://bitbucket.org/oscarlib/oscar
5. Régin, J.-C.: Generalized arc consistency for global cardinality constraint. In: Proceedings of the Thirteenth National Conference on Artificial Intelligence, vol. 1, pp. 209–215. AAAI Press (1996)

6. Pierre Schaus, et al.: Solving balancing and bin-packing problems with constraint programming. PhD thesis, PhD thesis, Universit catholique de Louvain Louvain-la-Neuve (2009)
7. Schaus, P., Régin, J.-C., Van Schaeren, R., Dullaert, W., Raa, B.: Cardinality reasoning for bin-packing constraint: Application to a tank allocation problem. In: Milano, M. (ed.) CP 2012. LNCS, vol. 7514, pp. 815–822. Springer, Heidelberg (2012)
8. Shaw, P.: A constraint for bin packing. In: Wallace, M. (ed.) CP 2004. LNCS, vol. 3258, pp. 648–662. Springer, Heidelberg (2004)

Value Interchangeability in Scenario Generation

Steven D. Prestwich[1], Marco Laumanns[2], and Ban Kawas[3]

[1] Cork Constraint Computation Centre, Department of Computer Science,
University College Cork, Ireland
[2] IBM Research – Zurich, 8803 Rueschlikon, Switzerland
[3] IBM Thomas J. Watson Research Center, NY, USA
s.prestwich@cs.ucc.ie, mlm@zurich.ibm.com, bkawas@us.ibm.com

Abstract. Several types of symmetry have been identified and exploited in Constraint Programming, leading to large reductions in search time. We present a novel application of one such form of symmetry: detecting dynamic value interchangeability in the random variables of a 2-stage stochastic problem. We use a real-world problem from the literature: finding an optimal investment plan to strengthen a transportation network, given that a future earthquake probabilistically destroys links in the network. Detecting interchangeabilities enables us to bundle together many equivalent scenarios, drastically reducing the size of the problem and allowing the exact solution of cases previously considered intractable and solved only approximately.

1 Introduction

Constraint Programming (CP) and Mixed Integer Programming (MIP) usually address *deterministic* problems, in which a solution is simply assignments to a set of decision variables. However, many real-world problems are inherently *stochastic*: they contain aspects outside our control, which are often represented as random variables in Stochastic Programming (SP) and Stochastic Constraint Programming (SCP). We assume a basic knowledge of SP and/or SCP, and refer readers unfamiliar with these fields to [4] and [26] respectively.

Much SP and SCP research is devoted to *single-stage* problems in which a solution is simply a value for each decision variable. This solution is then evaluated by examining the scenarios generated by assigning values to the random variables. In *multi-stage* problems we must set the values of the stage-1 decision variables, then explore alternative assignments to the stage-1 random variables, then move on to stage 2, and so on. Multi-stage problems are particularly hard to solve exactly, because of the intractable number of scenarios that must often be considered. Problems with many scenarios have motivated *scenario sampling* techniques, which allow us to work with a manageable subset of the scenarios but lose exactness.

In this paper we apply CP symmetry breaking methods to scenario generation. In (non-stochastic) CP several symmetry breaking methods have been devised, and they can lead to spectacular reduction in search times. The MIP literature also contains work on symmetry (a recent survey is given in [18]), but we restrict our attention to the CP literature which turns out to contain exactly the type of symmetry needed for the problem under consideration. Symmetry breaking on the decision variables of a stochastic

C. Schulte (Ed.): CP 2013, LNCS 8124, pp. 587–595, 2013.

problem is not essentially different to symmetry breaking on a deterministic problem. However, symmetry breaking on the random variables could reduce the number of scenarios needed to evaluate a solution, possibly leading to an exact solution instead of an inexact one found by scenario sampling. As far as we know this connection between symmetry breaking and stochastic problems is unexplored.

We test the idea on a problem in the literature: finding an optimal investment plan for a transportation network, given that a future disaster such as an earthquake will probabilistically destroy links in the network. This can be modelled as a 2-stage stochastic program, but the case we consider is challenging as it has over a billion scenarios. For this reason it has previously been solved only by an approximation technique, and only for a small scenario sample. We exploit symmetries between scenarios to reduce greatly the size of the problem, allowing us to find exact solutions. Section 2 presents the problem, Section 3 describes our new method and gives experimental results, Section 4 discusses related work, and Section 5 concludes the paper and outlines future work.

2   A Pre-disaster Planning Problem

The problem was first described by Peeta *et al.* [21] who cite evidence that the probability of a major earthquake occurring in the next few decades with its epicentre in Istanbul has been estimated as $62.6 \pm 15\%$; that this is likely to cause tens of billions of dollars worth of damage; that the Turkish government plans to invest \$400 million to strengthen infrastructure for earthquake resistance; and that a key element of this plan is to retrofit selected highways to maximise accessibility after an earthquake.

The Istanbul road network is represented by an undirected graph $G = (V, E)$ with 25 nodes V and 30 edges or links E. Each link represents a highway and may fail with some given probability, while each node represents a junction. The failure probability of a link can be reduced by investing money in it, but there is a budget limiting the total investment. To maximise post-quake accessibility, an interesting objective is to minimise the expected shortest path between a specified origin and destination node in the network, by investing in carefully-chosen links. In fact the actual objective is to minimise a weighted sum of shortest path lengths between several origin-destination (O-D) pairs, the choice of which is based on likely earthquake scenarios in the Japan International Cooperation Agency Report of 2002.

We now sketch the stochastic model. For each link $e \in E$ define a binary decision variable y_e which is 1 if we invest in that link and 0 otherwise. Define a binary random variable r_e which is 1 if link e survives and 0 if it fails. Denote the survival (non-failure) probability of link e by p_e without investment and q_e with, the investment required for link e by c_e, the length of link e by t_e (the units used in [21] are not specified but are proportional to the actual distances), and the budget by B. If the O-D pair are unconnected then the path length is taken to be a fixed number M representing (for example) the cost of using a helicopter. Actually, if they are only connected by long paths then they are considered to be unconnected, as in practice rescuers would resort to alternatives such as rescue by helicopter or sea. So Peeta *et al.* only consider a few (4–6) shortest paths for each O-D pair, and we shall refer to these as the *allowed paths*. In each case M is chosen to be the smallest integer that is greater than the longest allowed path length. They also consider a larger value of $M = 120$ that places a greater importance on connectivity, though using the same paths as with the smaller M values. To distinguish between these two usages we replace M by M_a (the length below which a path is allowed) and M_p (the penalty imposed when no allowed path exists). We fix M_a to the smaller values (not 120) for each O-D pair, and generate two sets of instances using $M_p = M_a$ and $M_p = 120$. All q_e values are set to 1 based on feedback from structural engineers. Three budget levels B_1, B_2, B_3 are considered, corresponding to 10%, 20% and 30% of the total cost of investing in all links. All problem parameters can be found in Peeta *et al.* and are based on the 2003 Master Earthquake Plan of the Istanbul municipality.

The earthquake problem is a 2-stage problem. In the first stage we decide which links to invest in by assigning values to the y_e, then link failures occur randomly with probabilities depending on the y_e, causing values to be assigned to the r_e. In the second stage we choose a shortest path between the O-D pair, given the surviving links. If they are no longer connected by an allowed path then the value M_p is used instead of a path length. For a given O-D pair the expected length is computed over all scenarios, and minimising this value is the objective. This is a challenging problem because each of the 30 links is independently affected by an earthquake, giving 2^{30} scenarios. Though optimisation time is not critical in pre-disaster planning, a billion scenarios is intractable. Instead Peeta *et al.* sample a million scenarios, and approximate the objective function by a monotonic multilinear function. They show that their method gives optimal or near-optimal results on smaller instances, and present results on the full-scale problem.

3   Scenario Bundling

This section describes our new method. First we provide background on the CP symmetry breaking ideas on which it is based. An early form of symmetry that has received considerable attention is *(value) interchangeability* [9]:

Definition. *A value a for variable v is fully interchangeable with value b if and only if every solution in which v = a remains a solution when b is substituted for a and vice-versa.*

If two values are interchangeable then one of them can be removed from the domain, reducing the size of the problem; alternatively they can be replaced by a single meta-value, and thus collected together in a Cartesian product representation of the search space. Both approaches avoid revisiting equivalent solutions. Several variants of interchangeability were defined in [9] and subsequent work in this area is surveyed in [14]. The relevant variant here is called *dynamic interchangeability*:

Definition. *A value a for variable v is dynamically interchangeable for b with respect to a set A of variable assignments if and only if they are fully interchangeable in the subproblem induced by A.*

Values may become interchangeable during backtrack search after some variables have been assigned values, so even a problem with no interchangeable values may exhibit dynamic interchangeability under some search strategy. This is an example of the more general concept of *conditional symmetry* [10] in which symmetry occurs at certain nodes in a search tree.

Interchangeable values can be exploited to group similar solutions together in *bundles*, a term used in [5,11,15] and other work. Bundles are Cartesian products of sets of values, which have been used in CP to represent related solutions compactly in *solution bundles* [11], *cross product representations* [13], *maximal consistent decisions* [16], *solution clusters* [20] and the SAT *maximal encoding* [23]. A drawback with interchangeability is that it does not seem to occur in many real applications [6,19,27] so it has recieved less attention than (for example) variable and value symmetries. Properties related to dynamic interchangeability were also investigated in [2,22] but otherwise little or no work has been done on it. One of the contributions of this paper is to demonstrate the usefulness of dynamic interchangeability in a stochastic problem.

We shall detect and exploit dynamic interchangeability in the random variables of the earthquake problem. As an illustration consider the simple example in Figure 1 with links $e \in \{1, \ldots, 4\}$. We set $t_e = 1$, $p_e = 0.8$, $q_e = 1$, $c_e = 1$ ($\forall e$), $B = 1$ and $M_a = M_p = 3.5$ so that both possible paths between nodes 1 and 4 are allowed. We must choose 1 link to invest in, to minimise the expected shortest path length between nodes 1–4. There are 16 scenarios and the optimal policy is to invest in link 1, giving an expected shortest path length of 2.4888. This is computed as $\sum_{i=1}^{16} p_i \ell_i$ where p_i is the probability and ℓ_i the path length in scenario i.

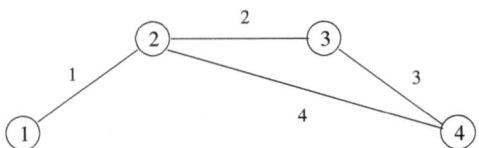

Fig. 1. A small network example

Some scenarios can be considered together instead of separately. For example consider the four scenarios 1001, 1011, 1101 and 1111, where the numbers indicate the survival (1) or failure (0) of links 1–4. Survival has probability 0.8 and failure 0.2 so these scenarios have probabilities 0.0256, 0.1024, 0.1024 and 0.4096 respectively. As links 1 and 4 survive in all four scenarios, it is irrelevant whether or not links 2 and 3 survive because they cannot be part of a shortest path: the path containing links 1 and 4 is shorter. We can therefore merge these four scenarios into a single expression 1**1 where the meta-value * denotes interchangeability: the values 0 and 1 for links 2 and 3 are interchangeable. The expression represents the Cartesian product $\{1\} \times \{0,1\} \times \{0,1\} \times \{1\}$ of scenarios. The probability associated with this product of scenarios is $0.8 \times (0.8 + 0.2) \times (0.8 + 0.2) \times 0.8 = 0.64$, which is equal to the sum of the 4 scenario probabilities.

Table 1. Two scenario bundle sets for the small example

links 3 2 4 1	p
0 * 0 *	0.0400
0 * 1 0	0.0320
0 * 1 1	0.1280
1 0 0 *	0.0320
1 0 1 0	0.0256
1 0 1 1	0.1024
1 1 0 0	0.0256
1 1 0 1	0.1024
1 1 1 0	0.1024
1 1 1 1	0.4096

links 1 4 2 3	p
0 * * *	0.2000
1 0 0 *	0.0320
1 0 1 0	0.0256
1 0 1 1	0.1024
1 1 * *	0.6400

We shall call a product such as 1**1 a *scenario bundle* by analogy with solution bundles in CP. Note that this usage is distinct from *bundle methods* in SP [24], which are quite different and apply to the class of nonsmooth convex programming problems.

Bundling scenarios together may lead to faster solution of some stochastic problems. However, for the earthquake problem it is impractical to enumerate a billion scenarios then look for ways of bundling some of them together, as we did in the above example. Instead we enumerate scenarios by tree search on the random variables (the *scenario tree*) and apply symmetry breaking as we search.

Consider a node in the scenario tree at which links $1 \ldots i-1$ have been realised, so that random variables $r_1 \ldots r_{i-1}$ have been assigned values, and we are about to assign a value to r_i corresponding to link i. Denote by S_i the shortest O-D path length including i, under the assumption that all unrealised links survive; and denote by F_i the shortest O-D path length not including i, under the assumption that all unrealised links fail (using M_p when no path exists). So S_i is the minimum shortest path length including i in all scenarios below this scenario tree node, while F_i is the maximum shortest path length not including i in the same scenarios. They can be computed by temporarily assigning $r_i \ldots r_{|E|}$ to 1 or 0 respectively, and applying a shortest path algorithm. Now if $S_i \geq F_i$ then the value assigned to r_i is irrelevant: the shortest path length in each scenario under this tree node is independent of the value of r_i, so the values are interchangeable. This observation is the core of our method.

The order in which we assign the s variables affects the cardinality of the bundle set. Two bundle sets for the example are shown in Table 1 along with their link permutations, where p is the bundle probability. Note that once we have obtained a bundle set we can discard the permutation used to derive it. We can also replace the symbol * by any domain value (we choose 0) and treat each bundle as an ordinary scenario. For example the bundle 11** under link permutation (1,4,2,3) can be replaced by the scenario 1001 under link permutation (1,2,3,4), with the same associated probability.

The problem of finding the smallest cardinality scenario bundle set corresponds exactly to the problem of finding a variable permutation that minimizes the number of paths in a binary decision tree. This is known to be NP-complete [28] so we shall apply heuristic search to the problem. First we use a greedy heuristic to quickly find a good permutation. We assign a score $\lambda_{oa}\lambda_{da} + \lambda_{ob}\lambda_{db}$ to each link (a, b) given O-D

Table 2. Bundle set sizes for the earthquake problem

instance	O-D pair	M_a	bundles
1	14–20	31	67
2	14–7	31	45
3	12–18	28	79
4	9–7	19	26
5	4–8	35	124

pair (o, d), where λ_{xy} denotes the shortest path length between nodes x and y. Then we sort the links into ascending order of score. The motivation is to realise links closest to the O-D pair first, and in experiments this led to good results. We further improve the permutation by a limited amount of hill-climbing: apply some number of 3-exchange moves, accepting moves that improve or leave unchanged the number of bundle sets.

We now apply scenario bundling to the earthquake problem, using 1000 hill-climbing moves to improve the bundle sets. The method is implemented in the Eclipse [1] constraint logic programming system (which provides a library of graph algorithms) and executed on a 2.8 GHz Pentium 4 with 512 MB RAM. The results are given in Table 2 for each O-D pair considered separately, and took approximately 1 minute each to compute. The table shows the instances numbered 1–5, the O-D pairs, the constant M_a, and the size of the corresponding bundle set. For each O-D pair the bundle sets are remarkably small, representing scenario reduction of several orders of magnitude.

However, Peeta et al. do not use a single O-D pair. Instead they minimise the expected weighted sum $\mathbb{E}\left\{\sum_{i=1}^{5} w_i \lambda_i\right\}$ of shortest path lengths λ_i between several O-D pairs for weights w_i, which are all set to 1 [25]. Unfortunately, there is likely to be little interchangeability in this problem, especially if (as we would expect) the O-D pairs are chosen to cover most of the network: for a given link to be irrelevant to the lengths of several paths is much less likely than for one path. But we can avoid this drawback by exploiting linearity of expectation and rewriting the objective as $\sum_{i=1}^{5} w_i \mathbb{E}\{\lambda_i\}$ so that each expected path length can be computed separately using its own bundle set.

We have replaced 1 billion scenarios by a total of 341 bundles, so on average each bundle replaces approximately 3 million scenarios. This reduction allows us to find exact solutions to the problem using a MIP model (to be described in an extended version of this paper). Solution times range between 14 and 26 seconds on a 2.4GHz Intel Core i5-520M with 4GB RAM using IBM ILOG CPLEX Optimizer Version 12.3[1] so our total solution times are dominated by the scenario bundling phase. The total times for both our method and that of Peeta et al. are a few minutes.

Table 3 show the approximate results of Peeta et al. and our exact results, including our exact evaluation of the objective function values of their approximate solutions. The results validate the method of Peeta et al. as their solutions are of good quality. However, the exact solutions are roughly 1–10% better than the approximate solutions, so the improvement is worthwhile.

---

[1] IBM, ILOG, and CPLEX are trademarks of International Business Machines Corporation, registered in many jurisdictions worldwide. Other product and service names might be trademarks of IBM or other companies.

Table 3. Approximate and exact solutions

B	link investment plan	objective
	approximate solutions (low M_p)	
B_1	20 21 22 23	86.7168
B_2	10 17 20 21 22 23 25	70.0352
B_3	10 13 16 17 20 21 22 25	59.5317
	exact solutions (low M_p)	
B_1	10 17 21 22 23 25	83.0801
B_2	4 10 12 17 20 21 22 25	66.1877
B_3	3 4 10 16 17 20 21 22 25	57.6802
	approximate solutions (high M_p)	
B_1	9 10 12 15 21 22 23 25	215.67
B_2	4 9 10 17 20 21 22 23 25	121.818
B_3	4 5 7 9 10 12 13 15 17 20 21 22 23 25	87.9268
	exact solutions (high M_p)	
B_1	10 17 21 22 23 25	212.413
B_2	3 4 10 12 17 20 21 22 25	120.08
B_3	4 10 16 17 20 21 22 23 25	78.4017

Peeta *et al.* remark that links 10, 20, 21, 22, 23 and 25 are invested in under most of their plans, and the same is true of ours. However, in some cases our plans look quite different to theirs. For example with B_1 and low M_p we invest in more links than they do, while with B_3 and high M_p the reverse is true. It is not obvious in either case why one solution is better than another, illustrating the impracticality of finding good solutions manually.

Further experiments on random road networks indicate that our method scales up well to larger instances. On networks with up to 77 links and up to 5 allowed paths it reduced the number of scenarios by up to 20 orders of magnitude. However, allowing more paths causes the bundle sets to grow rapidly, which is a limitation of our method.

4   Related Work

Scenario bundling has connections to other work. One way of viewing symmetry among random variables is as *stochastic dominance* [17], a concept from Decision Theory: the objective function associated with one choice (0 or 1) is at least as good as with another choice (1 or 0). Because this holds in every scenario, it is the simplest form of stochastic dominance: *statewise* (or *zeroth order*) *dominance*. However, this is usually defined as a strict dominance by adding an extra condition: that one choice is strictly better than the other in at least one state (or scenario). In our case neither value is better so this is a *weak dominance*. If both alternatives weakly dominate each other then they are *indifferent*, and the indifference relation is of course a symmetric relation. There does not seem to be an accepted term such as *stochastic symmetry* for this phenomenon, so we propose using this term to describe symmetries between scenarios.

There has been considerable work on scenario reduction methods for convex SP problems [8]. But these often start with a large set of scenarios then try to reduce it, rather than try to construct a reduced set from scratch. They also approximate the optimal solution, unlike our method. Sampling methods such as Monte Carlo sampling and Latin hypercube sampling have been used in both SP [4] and Artificial Intelligence approaches such as SCP [12] but these also approximate the optimal solution. The Network Reliability literature [7] describes methods for evaluating and approximating the reliability of a network. These include ways of pruning irrelevant parts of a network and have similarities to our method, but they are usually concerned with connectivity rather than path length. The literature on pre-disaster planning and robust networks is too large to review here, but a survey is given in [21].

5   Conclusion

We showed that a type of symmetry from Constraint Programming called dynamic interchangeability occurs in the random variables of a 2-stage stochastic program, and can be exploited by a method we call *scenario bundling*. Though this form of symmetry does not appear to occur significantly in constraint satisfaction problems, bundling can reduce the number of scenarios in a stochastic program by many orders of magnitude. This enables us to find exact solutions to a real-world pre-disaster planning problem that was previously considered intractable, and solved only approximately.

Scenario bundling can potentially be developed in several directions. (1) We expect that it will be useful for other stochastic problems, in particular those involving stochastic shortest paths. (2) It can be generalised so that, instead of performing tree search on a permutation of the random variables, it uses a dynamic branching heuristic. This should detect more interchangeability. (3) It could speed up the fitness computation in metaheuristics for stochastic problems [3]. (4) It establishes a new link between SP/SCP scenario reduction and CP symmetry breaking, and further links might emerge. We might call such a collection of techniques *stochastic symmetry breaking*,

Acknowledgments. This work was partly funded by the IBM/IDA-funded Risk Collaboratory project.

References

1. Apt, K.R., Wallace, M.: Constraint Logic Programming using Eclipse. Cambridge University Press (2007)
2. Beckwith, A.M., Choueiry, B.Y.: On the Dynamic Detection of Interchangeability in Finite Constraint Satisfaction Problems. In: Walsh, T. (ed.) CP 2001. LNCS, vol. 2239, p. 760. Springer, Heidelberg (2001)
3. Bianchi, L., Dorigo, M., Gambardella, L.M., Gutjahr, W.J.: A Survey on Metaheuristics for Stochastic Combinatorial Optimization. Natural Computing 8(2), 239–287 (2009)
4. Birge, J., Louveaux, F.: Introduction to Stochastic Programming. Springer Series in Operations Research (1997)
5. Choueiry, B.Y., Davis, A.M.: Dynamic Bundling: Less Effort for More Solutions. In: Koenig, S., Holte, R. (eds.) SARA 2002. LNCS (LNAI), vol. 2371, pp. 64–82. Springer, Heidelberg (2002)
6. Choueiry, B.Y., Noubir, G.: On the Computation of Local Interchangeability in Discrete Constraint Satisfaction Problems. In: 15th National Conference on Artificial Intelligence and 10th Innovative Applications of Artificial Intelligence Conference, pp. 326–333 (1998)

7. Colbourn, C.J.: Concepts of Network Reliability. Wiley Encyclopedia of Operations Research and Management Science. John Wiley & Sons, Inc. (2010)
8. Dupăcová, J., Gröwe-Kuska, N., Römisch, W.: Scenario Reduction in Stochastic Programming: an Approach Using Probability Metrics. Mathematical Programming Series A 95, 493–511 (2003)
9. Freuder, E.C.: Eliminating Interchangeable Values in Constraint Satisfaction Problems. In: National Conference on Artificial Intelligence, pp. 227–233 (1991)
10. Gent, I.P., Kelsey, T., Linton, S.A., Pearson, J., Roney-Dougal, C.M.: Groupoids and Conditional Symmetry. In: Bessière, C. (ed.) CP 2007. LNCS, vol. 4741, pp. 823–830. Springer, Heidelberg (2007)
11. Haselböck, A.: Exploiting Interchangeabilities in Constraint Satisfaction Problems. In: 13th International Joint Conference on Artificial Intelligence, pp. 282–287 (1993)
12. Hnich, B., Rossi, R., Tarim, S.A., Prestwich, S.: A Survey on CP-AI-OR Hybrids for Decision Making under Uncertainty. In: Milano, M., Van Hentenryck, P. (eds.) Hybrid Optimization: the 10 Years of CP-AI-OR. Springer Optimization and its Applications 45, 227–270 (2011)
13. Hubbe, P.D., Freuder, E.C.: An Efficient Cross Product Representation of the Constraint Satisfaction Problem Search Space. In: 10th National Conference on Artificial Intelligence, San Jose, California, USA, pp. 421–427 (1992)
14. Karakashian, S., Woodward, R., Choueiry, B.Y., Prestwich, S.D., Freuder, E.C.: A Partial Taxonomy of Substitutability and Interchangeability. In: 10th International Workshop on Symmetry in Constraint Satisfaction Problems (2010) (Journal paper in preparation)
15. Lal, A., Choueiry, B.Y., Freuder, E.C.: Neighborhood Interchangeability and Dynamic Bundling for Non-Binary Finite CSPs. In: 10th National Conference on Artificial Intelligence and 17th Innovative Applications of Artificial Intelligence Conference, pp. 397–404 (2005)
16. Lesaint, D.: Maximal Sets of Solutions for Constraint Satisfaction Problems. In: 11th European Conference on Artificial Intelligence, pp. 110–114 (1994)
17. Levy, H.: Stochastic Dominance and Expected Utility: Survey and Analysis. Management Science 38, 555–593 (1992)
18. Margot, F.: Symmetry in Integer Linear Programming. 50 Years of Integer Programming 1958–2008, pp. 647–686 (2010)
19. Neagu, N.: Studying Interchangeability in Constraint Satisfaction Problems. In: Van Hentenryck, P. (ed.) CP 2002. LNCS, vol. 2470, pp. 787–788. Springer, Heidelberg (2002)
20. Parkes, A.J.: Exploiting Solution Clusters for Coarse-Grained Distributed Search. In: IJCAI Workshop on Distributed Constraint Reasoning (2001)
21. Peeta, S., Salman, F.S., Gunnec, D., Viswanath, K.: Pre-Disaster Investment Decisions for Strengthening a Highway Network. Computers & Operations Research 37, 1708–1719 (2010)
22. Prestwich, S.D.: Full Dynamic Interchangeability with Forward Checking and Arc Consistency. In: ECAI Workshop on Modeling and Solving Problems With Constraints (2004)
23. Prestwich, S.D.: Full Dynamic Substitutability by SAT Encoding. In: Wallace, M. (ed.) CP 2004. LNCS, vol. 3258, pp. 512–526. Springer, Heidelberg (2004)
24. Ruszczyński, A.: Decomposition Methods in Stochastic Programming. Mathematical Programming 79, 333–353 (1997)
25. Salman, S.: Personal communication
26. Walsh, T.: Stochastic Constraint Programming. In: 15th European Conference on Artificial Intelligence, pp. 111–115 (2002)
27. Weigel, R., Faltings, B.V., Choueiry, B.Y.: Context in Discrete Constraint Satisfaction Problems. In: 12th European Conference on Artificial Intelligence, pp. 205–209 (1996)
28. Zantema, H., Bodlaender, H.L.: Sizes of Ordered Decision Trees. International Journal of Foundations of Computer Science 13(3), 445–458 (2002)

Embarrassingly Parallel Search*

Jean-Charles Régin, Mohamed Rezgui, and Arnaud Malapert

Université Nice-Sophia Antipolis, I3S UMR 6070, CNRS, France
jcregin@gmail.com, rezgui@i3s.unice.fr, arnaud.malapert@unice.fr

Abstract. We propose the Embarrassingly Parallel Search, a simple and efficient method for solving constraint programming problems in parallel. We split the initial problem into a huge number of independent subproblems and solve them with available workers, for instance cores of machines. The decomposition into subproblems is computed by selecting a subset of variables and by enumerating the combinations of values of these variables that are not detected inconsistent by the propagation mechanism of a CP Solver. The experiments on satisfaction problems and optimization problems suggest that generating between thirty and one hundred subproblems per worker leads to a good scalability. We show that our method is quite competitive with the work stealing approach and able to solve some classical problems at the maximum capacity of the multi-core machines. Thanks to it, a user can parallelize the resolution of its problem without modifying the solver or writing any parallel source code and can easily replay the resolution of a problem.

1 Introduction

There are two mainly possible ways for parallelizing a constraint programming solver. On one hand, the filtering algorithms (or the propagation) are parallelized or distributed. The most representative work on this topic has been carried out by Y. Hamadi [5]. On the other hand, the search process is parallelized. We will focus on this method. For a more complete description of the methods that have been tried for using a CP solver in parallel, the reader can refer to the survey of Gent et al. [4].

When we want to use k machines for solving a problem, we can split the initial problem into k disjoint subproblems and give one subproblem to each machine. Then, we gather the different intermediate results in order to produce the results corresponding to the whole problem. We will call this method: simple static decomposition method. The advantage of this method is its simplicity. Unfortunately, it suffers from several drawbacks that arise frequently in practice: the times spent to solve subproblems are rarely well balanced and the communication of the objective value is not good when solving an optimization problem (the workers are independent). In order to balance the subproblems that have to be solved some works have been done about the decomposition of the search tree based on its size [8,3,7]. However, the tree size is only approximated and is not strictly correlated with the resolution time. Thus, as mentioned by Bordeaux et al. [1], it is quite difficult to ensure that each worker will receive the same amount

---

* This work was partially supported by the Agence Nationale de la Recherche (Aeolus ANR-2010-SEGI-013-01 and Vacsim ANR-11-INSE-004) and OSEO (Pajero).

C. Schulte (Ed.): CP 2013, LNCS 8124, pp. 596–610, 2013.

of work. Hence, this method lacks scalability, because the resolution time is the maximum of the resolution time of each worker. In order to remedy for these issues, another approach has been proposed and is currently more popular: the work stealing idea.

The work stealing idea is quite simple: workers are solving parts of the problem and when a worker is starving, it "steals" some work from another worker. Usually, it is implemented as follows: when a worker W has no longer any work, it asks another worker V if it has some work to give it. If the answer is positive, then the worker V splits its current problem into two subproblems and gives one of them to the starving worker W. If the answer is negative then W asks another worker U, until it gets some work to do or all the workers have been considered.

This method has been implemented in a lot of solvers (Comet [10] or ILOG Solver [12] for instance), and into several ways [14,6,18,2] depending on whether the work to be done is centralized or not, on the way the search tree is split (into one or several parts), or on the communication method between workers.

The work stealing approach partly resolves the balancing issue of the simple static decomposition method, mainly because the decomposition is dynamic. Therefore, it does not need to be able to split a problem into well balanced parts at the beginning. However, when a worker is starving it has to avoid stealing too many easy problems, because in this case, it have to ask for another work almost immediately. This happens frequently at the end of the search when a lot of workers are starving and ask all the time for work. This complicates and slows down the termination of the whole search by increasing the communication time between workers. Thus, we generally observe that the method scales well for a small number of workers whereas it is difficult to maintain a linear gain when the number of workers becomes larger, even thought some methods have been developed to try to remedy for this issue [16,10].

In this paper, we propose another approach: the embarrassingly parallel search (EPS) which is based on the embarrassingly parallel computations [15].

When we have k workers, instead of trying to split the problem into k equivalent subparts, we propose to split the problem into a huge number of subproblems, for instance $30k$ subproblems, and then we give successively and dynamically these subproblems to the workers when they need work. Instead of expecting to have equivalent subproblems, we expect that *for each worker the sum of the resolution time of its subproblems will be equivalent*. Thus, the idea is not to decompose a priory the initial problem into a set of equivalent problems, but to decompose the initial problem into a set of subproblems whose resolution time can be shared in an equivalent way by a set of workers. Note that we do not know in advance the subproblems that will be solved by a worker, because this is dynamically determined. *All the subproblems are put in a queue and a worker takes one when it needs some work.*

The decomposition into subproblems must be carefully done. We must avoid subproblems that would have been eliminated by the propagation mechanism of the solver in a sequential search. Thus, *we consider only problems that are not detected inconsistent by the solver*.

The paper is organized as follows. First, we recall some principles about embarrassingly parallel computations. Next, we introduce our method for decomposing the initial problems. Then, we give some experimental results. At last, we make a conclusion.

2   Preliminaries

2.1   A Precondition

Our approach relies on the assumption that the resolution time of disjoint subproblems is equivalent to the resolution time of the union of these subproblems. If this condition is not met, then the parallelization of the search of a solver (not necessarily a CP Solver) based on any decomposition method, like simple static decomposition, work stealing or embarrassingly parallel methods may be unfavorably impacted.

This assumption does not seem too strong because the experiments we performed do not show such a poor behavior with a CP Solver. However, we have observed it in some cases with a MIP Solver.

2.2   Embarrassingly Parallel Computation

A computation that can be divided into completely independent parts, each of which can be executed on a separate process(or), is called *embarrassingly parallel* [15]. For the sake of clarity, we will use the notion of *worker* instead of process or processor.

An embarrassingly parallel computation requires none or very little communication. This means that workers can execute their task, i.e. any communication that is without any interaction with other workers. Some well-known applications are based on embarrassingly parallel computations, like Folding@home project, Low level image processing, Mandelbrot set (a.k.a. Fractals) or Monte Carlo Calculations [15].

Two steps must be defined: the definition of the tasks (TaskDefinition) and the task assignment to the workers (TaskAssignment). The first step depends on the application, whereas the second step is more general. We can either use a static task assignment or a dynamic one. With a static task assignment, each worker does a fixed part of the problem which is known a priori. And with a dynamic task assignment, a work-pool is maintained that workers consult to get more work. The work pool holds a collection of tasks to be performed. Workers ask for new tasks as soon as they finish previously assigned task. In more complex work pool problems, workers may even generate new tasks to be added to the work pool.

In this paper, we propose to see the search space as a set of independent tasks and to use a dynamic task assignment procedure. Since our goal is to compute one solution, all solutions or to find the optimal solution of a problem, we introduce another operation which aims at gathering solutions and/or objective values: TaskResultGathering. In this step, the answers to all the sub-problems are collected and combined in some way to form the output (i.e. the answer to the initial problem).

For convenience, we create a master (i.e. a coordinator process) which is in charge of these operations: it creates the subproblems (TaskDefinition), holds the work-pool and assigns tasks to workers (TaskAssignment) and fetches the computations made by the workers (TaskResultGathering).

In the next sections, we will see how the three operations can be defined in order to be able to run the search in parallel and in an efficient way.

3 Problem Decomposition

3.1 Principles

We have seen that decomposing the initial problem into the same number of subproblems as workers may cause unbalanced resolution time for each worker. Thus, our idea is to strongly increase the number of considered subproblems, in order to define an embarrassingly parallel computation leading to good performance.

Before going into further details on the implementation, we would like to establish a property. While solving a problem, we will call:

- *active time of a worker* the sum of the resolution times of a worker (the decomposition time is excluded).
- *inactive time of a worker* the difference between the elapsed time for solving all the subproblems (the decomposition time is excluded) and the active time of the worker.

Our approach is mainly based on the following remark:

Remark 1. *The active time of all the workers may be well balanced even if the resolution time of each subproblem is not well balanced*

The main challenge of a static decomposition is not to define equivalent problems, it is to avoid some workers without work whereas some others are running. We do not need to know in advance the resolution time of each subproblem. We just expect that the workers will have equivalent activity time. In order to reach that goal we propose to decompose the initial problem into a lot of subproblems. This increases our chance to obtain well balanced activity times for the workers, because we increase our chance to be able to obtain a combination of resolution times leading to the same activity time for each worker.

For instance, when the search space tends to be not balanced, we will have subproblems that will take a longer time to be solved. By having a lot of subproblems we increase our chance to split these subproblems into several parts having comparable resolution time and so to obtain a well balanced load of the workers at the end. It also reduces the relative importance of each subproblem with respect to the resolution of the whole problem.

Here is an example of the advantage of using a lot of subproblems. Consider a problem which requires 140s to be solved and that we have 4 workers. If we split the problem into 4 subproblems then we have the following resolution times: $20, 80, 20, 20$. We will need 80s to solve these subproblems in parallel. Thus, we gain a factor of $140/80 = 1.75$. Now if we split again each subproblem into 4 subproblems we could obtain the following subproblems represented by their resolution time: $((5, 5, 5, 5), (20, 10, 10, 40), (2, 5, 10, 3), (2, 2, 8, 8))$. In this case, we could have the following assignment: worker1 $: 5+20+2+8 = 35$; worker2 $: 5+10+2+10 = 27$; worker3 $: 5+10+5+3+2+8 = 33$ and worker4 $: 5 + 40 = 45$. The elapsed time is now 45s and we gain a factor of $140/45 = 3.1$. By splitting again the subproblems, we will reduce the average resolution time of the subproblems and expect to break the 40s subproblem. Note that decomposing more a subproblem does not increase the risk of increasing the elapsed time.

Property 1. *Let P be an optimization problem, or a satisfaction problem for which we search for all solutions. If P is split into subproblems whose maximum resolution time is $tmax$, then*

(i) *the minimum resolution time of the whole problem is $tmax$*

(ii) *the maximum inactivity time of a worker is less than or equal to $tmax$.*

Suppose that a worker W has an inactivity time which is greater than $tmax$. Consider the moment where W started to wait after its activity time. At this time, there is no more available subproblems to solve, otherwise W would have been active. All active workers are then finishing their last task, whose resolution is bounded by $tmax$. Thus, the remaining resolution time of each of these other workers is less than $tmax$. Hence a contradiction.

3.2 Subproblems Generation

Suppose we want to split a problem into q disjoint subproblems. Then, we can use several methods.

A Simple Method. We can proceed as follows:

1. We consider any ordering of the variables $x_1,...x_n$.
2. We define by A_k the Cartesian product $D(x_1) \times ... \times D(x_k)$.
3. We compute the value k such that $|A_{k-1}| < q \leq |A_k|$.

Each assignment of A_k defines a subproblem and so A_k is the sought decomposition.

This method works well for some problems like the n-queen or the Golomb ruler, but it is really bad for some other problems, because a lot of assignments of A may be trivially not consistent. Consider for instance that x_1, x_2 and x_3 have the three values $\{a, b, c\}$ in their domains and that there is an alldiff constraint involving these three variables. The Cartesian product of the domains of these variables contains 27 tuples. Among them only 6 ((a, b, c), (a, c, b), (b, a, c),(b, c, a),(c, a, b), (c, b, a)) are not inconsistent with the alldiff constraint. That is, only $6/27 = 2/9$ of the generated problems are not trivially inconsistent. It is important to note that most of these inconsistent problems would never be considered by a sequential search. For some problems we have observed more than 99% of the generated problems were detected inconsistent by running the propagation. Thus, we present another method to avoid this issue.

Not Detected Inconsistent (NDI) Subproblems. We propose to *generate only subproblems that are not detected inconsistent by the propagation.* The generation of q such subproblems becomes more complex because the number of NDI subproblems may be not related to the Cartesian product of some domains. A simple algorithm could be to perform a Breadth First Search (BFS) in the search tree, until the desired number of NDI subproblems is reached. Unfortunately, it is not easy to perform efficiently a BFS mainly because a BFS is not an incremental algorithm like a Depth First Search (DFS). Therefore, we propose to use a process similar to an iterative deepening depth-first search [9]: we repeat a Depth-bounded Depth First Search (DBDFS), in other words a DFS which never visits nodes located at a depth greater than a given value, increasing the bound until generating the right number of subproblems. Each branch of a search tree computed by this search defines an assignment. We will denote by NDI_k the set of assignments computed for the depth k. For generating q subproblems, we repeat the DBDFS until we reach a level k such that $|\text{NDI}_{k-1}| < q \leq |\text{NDI}_k|$. For convenience and simplicity, we use a static ordering of the variables.

We improve this method in three ways:

1. We try to estimate some good values for k in order to avoid repeating too many DBDFS. For instance, if for a given depth u we produce only $q/1000$ subproblems and if the size of the domains of the three next non assigned variables is 10, then we can deduce that we need to go at least to the depth $u + 3$.

2. In order to avoid repeating the same DFS for the first variables while repeating DBDFS, we store into a table constraint the previous computed assignments. More precisely, if we have computed NDI_k then we use a table constraint containing all these assignments when we look for NDI_l with $l > k$.

3. We parallelize our decomposition algorithm in a simple way. Consider we have w workers. We search for w NDI subproblems. Then, each worker receives one of these subproblems and decomposes it into q/w NDI subproblems by using our algorithm. The master gathers all computed subproblems. If a worker is not able to generate q/w subproblems because it solves its root NDI problem by decomposing it, the master asks the workers to continue to decompose their subproblems into smaller ones until reaching the right number of subproblems. Note that the load balancing of the decomposition is not really important because once a worker has finished its decomposition work it begins to solve the available subproblems.

Large Domains. Our method can be adapted to large domains. A new step must be introduced in the algorithm in the latest iteration. If the domain of the latest considered variable, denoted by lx, is large then we cannot consider each of its values individually. We need to split its domain into a fix number of parts and use each part as a value. Then, either the desired number of subproblems is generated or we have not been able to reach that number. In this latter case, we need to split again the domain of lx, for instance by splitting each part into two new parts (this multiplies by at most 2 the number of generated subproblems) and we check if the generated number of subproblems is fine or not. This process is repeated until the right number of subproblems is generated or the domain of lx is totally decomposed, that is each part corresponds to a value. In this latter case, we continue the algorithm by selecting a new variable.

3.3   Implementation

Satisfaction Problems

- The TaskDefinition operation consists of computing a partition of the initial problem P into a set S of subproblems.
- The TaskAssignment operation is implemented by using a FIFO data structure (i.e. a queue). Each time a subproblem is defined it is added to the back of the queue. When a worker needs some work it takes a subproblem from the queue.
- The TaskResultGathering operation is quite simple : when searching for a solution it stops the search when one is found; when searching for all solutions, it just gathers the solutions returned by the workers.

Optimization Problems

In case of optimization problems we have to manage the best value of the objective function computed so far. Thus, the operations are slightly modified.

- The TaskDefinition operation consists of computing a partition of the initial problem P into a set S of subproblems.
- The TaskAssignment operation is implemented by using a queue. Each time a subproblem is defined it is added to the back of the queue. The queue is also associated with the best objective value computed so far. When a worker needs some work, the master gives it a subproblem from the queue. It also gives it the best objective value computed so far.
- The TaskResultGathering operation manages the optimal value found by the worker and the associated solution.

Note that there is no other communication, that is when a worker finds a better solution, the other workers that are running cannot use it for improving their current resolution. So, if the absence of communication may increase our performance, this aspect may also lead to a decrease of performance. Fortunately, we do not observe this bad behavior in practice. We can see here another argument for having a lot of subproblems in case of optimization problems: the resolution of a subproblem should be short for improving the transmission of a better objective value and for avoiding performing some work that could have been ignored with a better objective value.

3.4   Size of the Partition

One important question is: how many subproblems do we generate? This is mainly an experimental question. However, we can notice that if we want to have a good scalability then this number should be defined in relation to the number of workers that are involved. More precisely, it is more consistent to have q subproblems per worker than a total of q subproblems.

3.5   Replay

One interesting advantage of our method in practice is that we can simply replay a resolution in parallel by saving the order in which the subproblems have been executed. This costs almost nothing and helps a lot the debugging of applications.

4   Related Work

The decomposition of some hard parts of the problem into several subproblems in order to fill a work-pool has been proposed by [10] in conjunction with the work-stealing approach.

Yun and Epstein proposed to parallelize a sequential solver in order to find one solution for a satisfaction problem [17]. Their approach strongly relies on a weight learning mechanism and on the use of a restart strategy. A first portfolio phase allows to initialize the weights as well as solving easy problems. In the splitting phase, the manager distributes subproblems to the worker with a given search limit. If the worker is not able to solve the problem within the limit, it returns the problem to the manager for further partitioning by iterative bisection partitioning.

We can already notice three major differences with our approach. First, we partition statically the problems at the beginning of the search whereas they use an on-demand dynamic partitioning. Second, there is much more communication between the manager and the workers since the workers have to notify the manager of encountered search limit. Last, the same part of the search tree can be explored several times since the worker do not learn clauses form the unsuccessful runs. Therefore, it really complicated to adapt their approach for solution enumeration whereas it is straightforward with ours.

5   Experiments

Machines. All the experiments have been made on a Dell machine having four E7-4870 Intel processors, each having 10 cores with 256 GB of memory and running under Scientific Linux. Our experiments can be reproduced by downloading the program EPSearch for Linux from [13].

Solvers. We implemented our method on the top of two CP solvers: or-tools rev2555 by Google and Gecode 4.0.0 (http://www.gecode.org/).

Experimental Protocol. Our code just performs three operations:

1. Read a FlatZinc model or create directly the model with the solver API.
2. Create the threads and define an instance of the solver for each thread
3. Compute the subproblems, feed the threads with them and gather the results.

For each problem, we will either search for all solutions of satisfaction problems or solve the whole optimization problem (i.e. find the best solution and prove its optimality). The resolution times represent the elapsed time to solve the whole problem that is they include the decomposition time and the times needed by the workers to solve subproblems.

Note that testing the performance of a parallel method is more complex with an optimization problem because the chance may play a role. It can advantage or disadvantage us. However, in the real life, optimization problems are quite common therefore it is important to test our method on them.

The means that are given are geometric means.

Selected Benchmarks. We selected a wide range of problems that are representative of the types of problems solved in CP. Some are coming from the CSP lib and have been modeled by Hakan Kjellerstrandk (http://www.hakank.org/) and some are coming from the MiniZinc distribution (1.5 see [11]). We selected instances that are not too easy (more than 20s) and that are not too long to be solved (less than 3600s) with the Gecode solver.

The examples coming from Hakan Kjellerstrandk are: golombruler-13 ; magicsequence-40000 ; sportsleague-10 ; warehouses (number of warehouses = 10, number of stores = 20 and a fixed cost = 180) ; setcovering (Placing of fire stations with 80 cities and a min distance fixed at 24) ; allinterval-15 (the model of Regin and Puget of the CSP Lib is used).

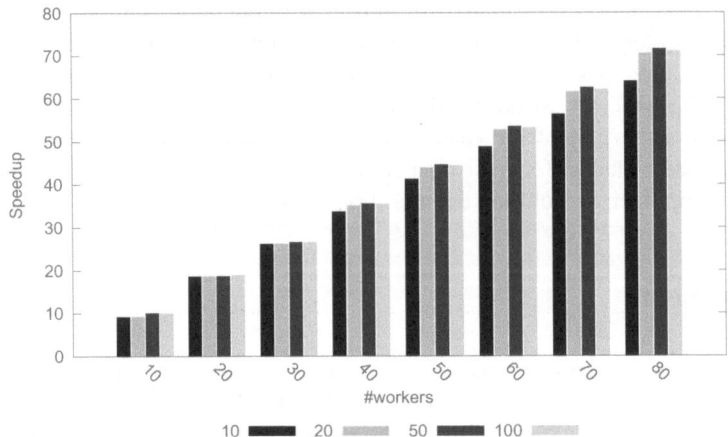

Fig. 1. 17-queens: performance as a function of #sppw (10,20,50 and 100). We no longer observe the limit of a gain factor of 29

The Flatzinc instances coming from the MiniZinc distribution are: 2DLevelPacking (5-20-6), depotPlacement (att48-5; rat99-5), fastfood (58), openStacks (01-problem-15-15;01-wbp-30-15-1), sugiyama (2-g5-7-7-7-7-2), patternSetMining (k1-german-credit), sb-sb (13-13-6-4), quasigroup7-10, non-non-fast-6, radiation-03, bacp-7, talent-scheduling-alt-film116.

Tests. Let #sppw denote the number of subproblems per worker. We will study the following aspects of our method:

5.1 the scalability compared to other static decompositions
5.2 the inactivity time of the workers as a function of the value of #sppw
5.3 the difficulty of the subproblems when dealing with a huge number of them
5.4 the advantage of parallelizing the decomposition
5.5 the influence of the value of #sppw on the factor of improvements
5.6 its performance compared to the work-stealing approach
5.7 the influence of the CP solver that is used

5.1 Comparison with a Simple Static Decomposition

We consider the famous n-queens problem, because it is a classical benchmark and because some proposed methods [6,1] were not able to observe a factor greater than 29 with a simple static decomposition of the problems even when using 64 workers. Figure 1 shows that our method scales very well when #sppw is greater than 20. The limit of the scalability (a maximum ratio of 29) described in [1] clearly disappeared. Note that we used the same search strategy as in [1] and two 40-cores Dell machines for this experiment.

5.2 Ratio of NDI Subproblems

Figure 2 shows the percentage of NDI problems generated by the simple method of decomposition for all problems. The geometric mean is a bold line and the dashed lines

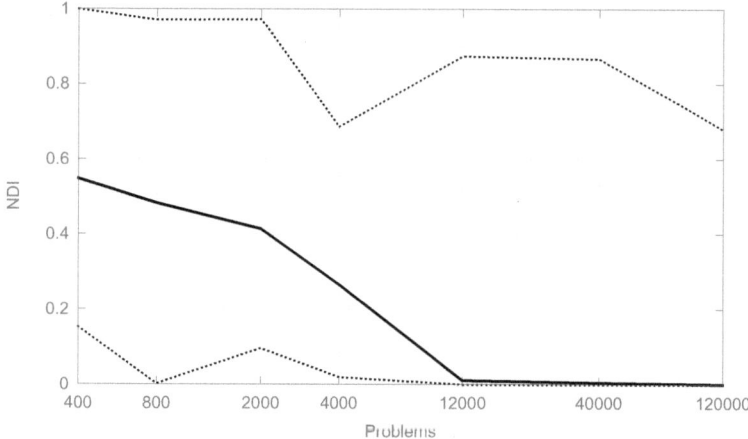

Fig. 2. Percentage of NDI problems generated by the simple decomposition method

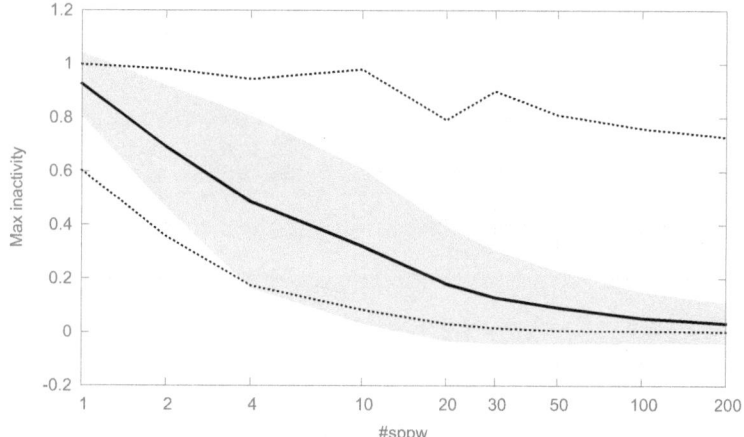

Fig. 3. Percentage of maximum inactivity time of the workers (geometric mean)

represent the minimum and maximum values. We can see that this number depends a lot on the considered instances. For some instances, the number is close to 100% whereas for some others it can be really close to 0% which indicates a decomposition issue. The mean starts at 55% and decreases according to the number of subproblems to end at 1%. Most of the inconsistent problems generated by the simple decomposition method would not have been considered by a sequential search. Therefore, for some instances this method should not be used. This is the reason why we do not generate any non NDI problems.

5.3   The Inactivity Time as a Function of #sppw

Figure 3 shows that the percentage of the maximum inactivity time of the workers decreases when the number of subproblems per worker is increased. The geometric

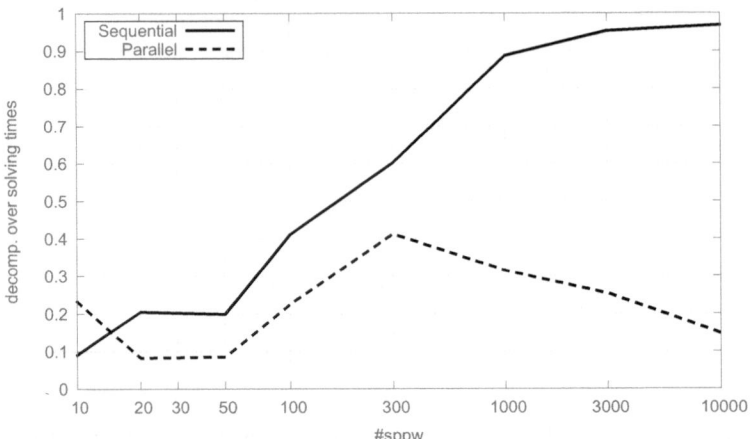

Fig. 4. Percentage of the total time spent in the decomposition

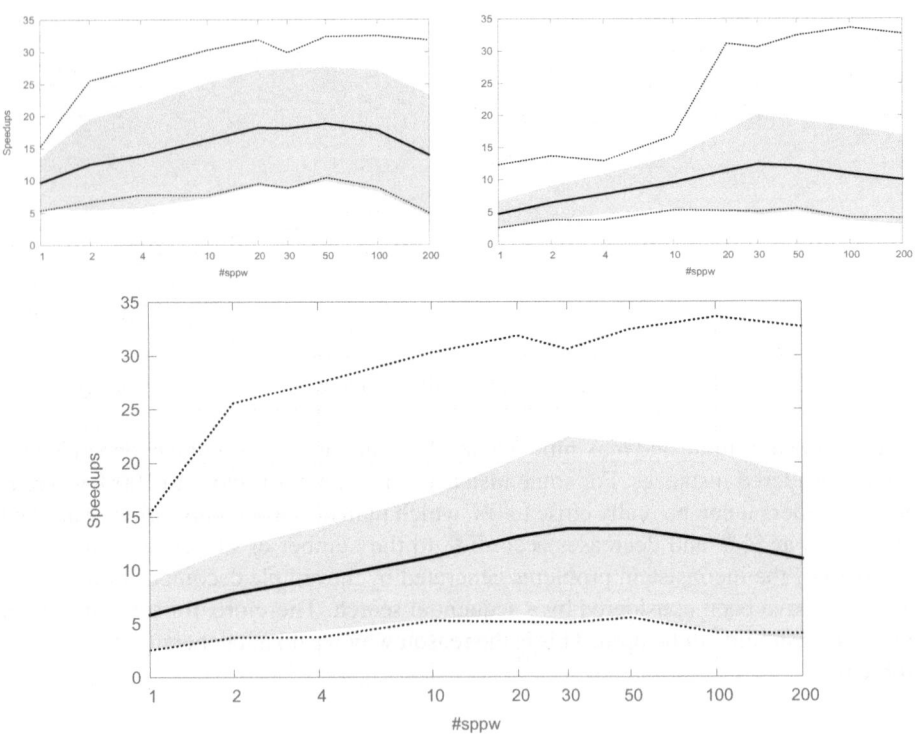

Fig. 5. Speed up as a function of the number of subproblems for finding all solutions of satisfaction problems (top left), for finding and proving the optimality of optimization problems (top right) and for all the problems (bottom).

mean is a bold line, and the dashed lines represent the minimum and maximum values, and the standard deviation is indcated by a gray area. From 20 subproblems per worker, we observe that in average the maximum inactivity time represents less than 20% of the resolution time

5.4   Parallelism of the Decomposition

Figure 4 compares the percentage of total time needed to decompose the problem when only the master performs this operation or when the workers are also involved. We clearly observe that the parallelization of the decomposition saves some time, especially for a large number of subproblems per worker.

5.5   Influence of the Number of Considered Subproblems

Figure 5 describes the speed up obtained by the Embarrassignly Parallel Search (EPS) as a function of the number of subproblems for Gecode solver. The best results are obtained with a number subproblems per worker between 30 and 100. In other words, we propose to start the decomposition with $q = 30w$, where w is the number of workers.

It is interesting to note that a value of #sppw in [30,100] is good for all the considered problems and seems independent from them.

The reduction of the performance when increasing the value of #sppw comes from the fact that the decomposition process solves an increasing part of the problem; and this process is slower than a resolution procedure.

Note that with our method, only 10% of the resolution time is lost if we use a sequential decomposition instead of a parallel one.

5.6   Comparison with the Work Stealing Approach

Table 1 presents a comparison between the EPS and the work stealing method available in Gecode. The column t gives the solving time in seconds and the column s give the speed-up. The last row shows the sum of the resolution times or the geometric mean of the speed-ups. The geometric average gain factor with the work stealing method is 7.7 (7.8 for satisfaction problems and 7.6 for optimization problems) whereas with the EPS it is 13.8 (18.0 for satisfaction problems and 12.3 for optimization problems). Our method improves the work stealing approach in all cases but one.

5.7   Influence of the CP Solver

We also performed some experiments by using or-tools solver. With or-tools the speed up of the EPS are increased (See Table 2). We obtain a geometric average gain factor of 13.8 for the Gecode solver and 21.3 for or-tools.

Table 1. Resolution with 40 workers, and #sspw=30 using Gecode 4.0.0

Instance	Seq.	Work stealing		EPS	
	t	t	s	t	s
allinterval_15	262.5	9.7	27.0	8.8	**29.9**
magicsequence_40000	328.2	592.6	0.6	37.3	**8.8**
sportsleague_10	172.4	7.6	22.5	6.8	**25.4**
sb_sb_13_13_6_4	135.7	9.2	14.7	7.8	**17.5**
quasigroup7_10	292.6	14.5	20.1	10.5	**27.8**
non_non_fast_6	602.2	271.3	2.2	56.8	**10.6**
golombruler_13	1355.2	54.9	24.7	44.3	**30.6**
warehouses	148.0	25.9	5.7	21.1	**7.0**
setcovering	94.4	16.1	5.9	11.1	**8.5**
2DLevelPacking_Class5_20_6	22.6	13.8	1.6	0.7	**30.2**
depot_placement_att48_5	125.2	19.1	6.6	10.2	**12.3**
depot_placement_rat99_5	21.6	6.4	3.4	2.6	**8.3**
fastfood_ff58	23.1	4.5	5.1	3.8	**6.0**
open_stacks_01_problem_15_15	102.8	6.1	16.9	5.8	**17.8**
open_stacks_01_wbp_30_15_1	185.7	15.4	12.1	11.2	**16.6**
sugiyama2_g5_7_7_7_7_2	286.5	22.8	12.6	10.8	**26.6**
pattern_set_mining_k1_german-credit	113.7	22.3	5.1	13.8	**8.3**
radiation_03	129.1	33.5	3.9	25.6	**5.0**
bacp-7	227.2	15.6	14.5	9.5	**23.9**
talent_scheduling_alt_film116	254.3	13.5	**18.8**	35.6	**7.1**
total (t) or geometric mean (s)	488.2	1174.8	7.7	334.2	**13.8**

Table 2. Resolution with 40 workers, and #sspw=30 using or-tools (revision 2555)

Instance	Seq.	EPS	
	t	t	s
allinterval_15	2169.7	67.7	32.1
magicsequence_40000	−	−	−
sportsleague_10	−	−	−
sb_sb_13_13_6_4	227.6	18.1	12.5
quasigroup7_10	−	−	−
non_non_fast_6	2676.3	310.0	8.6
golombruler_13	16210.2	573.6	28.3
warehouses	−	−	−
setcovering	501.7	33.6	14.9
2DLevelPacking_Class5_20_6	56.2	3.6	15.5
depot_placement_att48_5	664.9	13.7	48.4
depot_placement_rat99_5	67.0	2.8	23.7
fastfood_ff58	452.4	25.1	18.0
open_stacks_01_problem_15_15	164.7	7.1	23.2
open_stacks_01_wbp_30_15_1	164.9	6.3	26.0
sugiyama2_g5_7_7_7_7_2	298.8	20.5	14.6
pattern_set_mining_k1_german-credit	270.7	12.8	21.1
radiation_03	416.6	23.5	17.7
bacp-7	759.7	23.8	32.0
talent_scheduling_alt_film116	575.7	15.7	36.7
total (t) or geometric mean (s)	25677.2	1158.1	21.3

6   Conclusion

In this paper we have presented the Embarrassingly Parallel Search (EPS) a simple method for solving CP problems in parallel. It proposes to decompose the initial problem into a set of k subproblems that are not detected inconsistent and then to send them to workers in order to be solved. After some experiments, it appears that splitting the initial problem into 30 such subproblems per worker gives an average factor of gain equals to 21.3 with or-tools and 13.8 with Gecode while searching for all the solutions or while finding and proving the optimality, on a machine having 40 cores. This is competitive with the work stealing approach.

Acknowledgments. We would like to thank very much Laurent Perron and Claude Michel for their comments which helped improve the paper.

References

1. Bordeaux, L., Hamadi, Y., Samulowitz, H.: Experiments with Massively Parallel Constraint Solving. In: Boutilier, C. (ed.) IJCAI, pp. 443–448 (2009)
2. Chu, G., Schulte, C., Stuckey, P.J.: Confidence-Based Work Stealing in Parallel Constraint Programming. In: Gent, I.P. (ed.) CP 2009. LNCS, vol. 5732, pp. 226–241. Springer, Heidelberg (2009)

3. Cornuéjols, G., Karamanov, M., Li, Y.: Early Estimates of the Size of Branch-and-Bound Trees. INFORMS Journal on Computing 18(1), 86–96 (2006)
4. Gent, I.P., Jefferson, C., Miguel, I., Moore, N.C.A., Nightingale, P., Prosser, P., Unsworth, C.: A Preliminary Review of Literature on Parallel Constraint Solving. In: Proceedings PMCS 2011 Workshop on Parallel Methods for Constraint Solving (2011)
5. Hamadi, Y.: Optimal Distributed Arc-Consistency. Constraints 7, 367–385 (2002)
6. Jaffar, J., Santosa, A.E., Yap, R.H.C., Zhu, K.Q.: Scalable Distributed Depth-First Search with Greedy Work Stealing. In: ICTAI, pp. 98–103. IEEE Computer Society (2004)
7. Kilby, P., Slaney, J.K., Thiébaux, S., Walsh, T.: Estimating Search Tree Size. In: AAAI, pp. 1014–1019 (2006)
8. Knuth, D.E.: Estimating the efficiency of backtrack programs. Mathematics of Computation 29, 121–136 (1975)
9. Korf, R.: Depth-first Iterative-Deepening: An Optimal Admissible Tree Search. Artificial Intelligence 27, 97–109 (1985)
10. Michel, L., See, A., Van Hentenryck, P.: Transparent Parallelization of Constraint Programming. INFORMS Journal on Computing 21(3), 363–382 (2009)
11. MiniZinc (2012), http://www.g12.csse.unimelb.edu.au/minizinc/
12. Perron, L.: Search Procedures and Parallelism in Constraint Programming. In: Jaffar, J. (ed.) CP 1999. LNCS, vol. 1713, pp. 346–361. Springer, Heidelberg (1999)
13. Régin, J.-C.: (2013), http://www.constraint-programming.com/people/regin/papers
14. Schulte, C.: Parallel Search Made Simple. In: Beldiceanu, N., Harvey, W., Henz, M., Laburthe, F., Monfroy, E., Müller, T., Perron, L., Schulte, C. (eds) Proceedings of TRICS: Techniques for Implementing Constraint programming Systems, a Post-Conference Workshop of CP 2000, Singapore (September 2000)
15. Wilkinson, B., Allen, M.: Parallel Programming: Techniques and Application Using Networked Workstations and Parallel Computers, 2nd edn. Prentice-Hall Inc. (2005)
16. Xie, F., Davenport, A.: Massively Parallel Constraint Programming for Supercomputers: Challenges and Initial Results. In: Lodi, A., Milano, M., Toth, P. (eds.) CPAIOR 2010. LNCS, vol. 6140, pp. 334–338. Springer, Heidelberg (2010)
17. Yun, X., Epstein, S.L.: A Hybrid Paradigm for Adaptive Parallel Search. In: Milano, M. (ed.) CP 2012. LNCS, vol. 7514, pp. 720–734. Springer, Heidelberg (2012)
18. Zoeteweij, P., Arbab, F.: A Component-Based Parallel Constraint Solver. In: De Nicola, R., Ferrari, G.-L., Meredith, G. (eds.) COORDINATION 2004. LNCS, vol. 2949, pp. 307–322. Springer, Heidelberg (2004)

Multi-Objective Large Neighborhood Search

Pierre Schaus and Renaud Hartert

UCLouvain, ICTEAM,
Place sainte barbe 2,
1348 Louvain-la-Neuve, Belgium
{pierre.schaus,renaud.hartert}@uclouvain.be

Abstract. Large neighborhood search (LNS) [25] is a framework that combines the expressiveness of constraint programming with the efficiency of local search to solve combinatorial optimization problems. This paper introduces an extension of LNS, called multi-objective LNS (MO-LNS), to solve multi-objective combinatorial optimization problems ubiquitous in practice. The idea of MO-LNS is to maintain a set of nondominated solutions rather than just one best-so-far solution. At each iteration, one of these solutions is selected, relaxed and optimized in order to strictly improve the hypervolume of the maintained set of nondominated solutions. We introduce modeling abstractions into the OscaR solver for MO-LNS and show experimentally the efficiency of this approach on various multi-objective combinatorial optimization problems.

Keywords: Constraint Programming, Multi-Objective Combinatorial Optimization, Large Neighborhood Search.

Multi-Objective Combinatorial Optimization (MOCO) problems are ubiquitous in real-world applications. Decision makers often face the problem of dealing with several objectives *e.g.* the cost and the risk. In this situation, people are mostly interested to see a set of solutions representing the optimal compromises between objectives instead of one solution resulting from an *a priori* preference between these objectives.

Not surprisingly, the last decades have seen a growth of interest in the theory and the methodology for MOCO problems (see [7,26] for a review). Currently, hybridized-meta-heuristics between Evolutionary Algorithm (EA) and Local Search (LS) obtain state-of-the-art results[1] on most standard MOCO problems such as the traveling salesman, the binary knapsack, and the quadratic assignment problems (see [1] for a review of these methods). However – despite the implementation facilities offered by libraries such as ParadisEO [4] and jMetal [6] – these approaches are quite far from "model and run" ones. Indeed, users still have to provide several implementation blocks (for crossover, mutations, moves and neighborhood, etc.) requiring a great knowledge and expertise on the problems and the used algorithms. Furthermore, meta-heuristic methods for MOCO problems are more and more specific and strongly related to the optimization problem to solve [8]. This tendency increases the difficulty to design a single universal method or solver.

---

[1] The LS and EA communities are probably the most active ones on the domain of MOCO.

C. Schulte (Ed.): CP 2013, LNCS 8124, pp. 611–627, 2013.
© Springer-Verlag Berlin Heidelberg 2013

Conversely, Constraint Programming (CP) offers a high level declarative language and has shown to be a competitive approach for solving single-objective constrained optimization problems (COP). In particular, the LNS (Large Neighborhood Search) framework [25] – which combines the efficiency of LS with the expressiveness of CP – allowed to solve large scale problems such as vehicle routing [3,25], scheduling [14,21], and assignment/bin-packing problems [18,23] successfully.

We believe that the expressiveness of CP can have a real added value to tackle some MOCO problems by reducing the amount of work required from the modeler.[2] This work is one step in the direction of extending the LNS framework in the multi-objective context (MO-LNS). The goal of MO-LNS is to quickly discover good nondominated sets of solutions for large scale MOCO problems while keeping a declarative CP model.

This paper introduces the MO-LNS framework. We demonstrate experimentally its flexibility on standard MOCO problems as well as on a real-world bi-objective version of the Tank Allocation Problem (TAP) [24]. We also introduce modeling abstractions, explaining in depth an MO-LNS model implemented with the OscaR open source library [20].

Outline.   Section 1 gives definitions related to constraint programming and multi-objective optimization. Section 2 reviews the related work of existing CP approaches to solve MOCO problems. Section 3 introduces MO-LNS. Section 4 details an MO-LNS model for the quadratic assignment problem in the OscaR [20] solver. Section 5 experiments the MO-LNS approach on various MOCO problems. Section 6 gives perspectives and concludes.

1   Definitions

The typical MOCO problem we want to solve has m integer objective variables to minimize while satisfying some constraints:

$$\text{Minimize} \quad obj = (obj_1, obj_2, \ldots, obj_m)$$
$$\text{Subject to} \quad constraints \tag{1}$$

Solutions of this problem are defined as follows:

Definition 1 (Solution). *Let $\mathcal{P}$ be a MOCO problem, a solution of the problem $\mathcal{P}$ is an assignment of the decision variables and objective variables of $\mathcal{P}$ that satisfies all the constraint of this problem. In the following, $sol(x)$ denotes the value assigned to the variable x in the solution sol.*

The conflicting nature of the objectives usually prevents the existence of a unique solution sol^* that is optimal in all objectives. Hence, one is usually interested in the set of all the optimal compromises known as *Pareto optimal* solutions.

---

[2] The lack of hybridization with CP approaches for solving MOCO problems was recently underlined by Ehrgott in [8].

Definition 2 (Pareto dominance). *Let sol and sol' be two solutions of a MOCO problem* $\mathcal{P}$. *We say that sol dominates sol', denoted* $sol \prec sol'$, *if and only if:*

$$\forall j \in [1..m] : sol(obj_j) \leq sol'(obj_j)$$
$$\wedge \quad \exists j \in [1..m] : sol(obj_j) < sol'(obj_j) \tag{2}$$

Besides, we say that sol weakly-dominates sol', denoted $sol \preceq sol'$, *if and only if the first part of Equation 2 holds.*[3]

Definition 3 (Pareto optimality). *Let* $sols(\mathcal{P})$ *denotes all the feasible solutions of a MOCO problem* $\mathcal{P}$. *A solution* sol^* *is Pareto optimal if and only if there is no solution* sol' *in* $sols(\mathcal{P})$ *that dominates* sol^*:

$$\nexists sol' \in sols(\mathcal{P}) : sol' \prec sol^* \tag{3}$$

In other words, a solution is said to be Pareto optimal if it is impossible to improve the value of one objective without degrading the value of at least one other objective.

The set of all the Pareto optimal solutions is known as the *Pareto set* and is defined as follows:

Definition 4 (Pareto set). *The Pareto set of a MOCO problem* $\mathcal{P}$ *is the set of all the Pareto optimal solutions of this problem:*

$$\{sol \in sols(\mathcal{P}) \mid \nexists sol' \in sols(\mathcal{P}) : sol' \prec sol\} \tag{4}$$

Definition 5 (Pareto front). *The Pareto front of a MOCO problem* $\mathcal{P}$ *is the projection of its Pareto set in the objective space.*

Unfortunately, discovering the exact Pareto set may be impracticable on difficult MOCO problems. We are thus interested in finding an approximation of this set, also known as the *archive*.

Definition 6 (Archive). *An archive* $\mathcal{A}$ *is a set of solutions such that there is no solution in the archive that dominates an other solution in the archive. This property is known as the domination-free property:*

$$\forall sol \in \mathcal{A}, \nexists sol' \in \mathcal{A} : sol' \prec sol \tag{5}$$

As illustrated in Fig. 1, an archive can be used to partition the objective space into three subspaces:

- The *dominated subspace* consists of all the solutions that are dominated by at least one solution in the archive (see Fig. 1a);
- The *diversification subspace* consists of all the solutions that neither dominate nor are dominated by any solution in the archive (see Fig. 1b);
- The *intensification subspace* consists of all the solutions that dominate at least one solution in the archive (see Fig. 1c).

Clearly the archive quality can only be improved by adding new solutions from:

---

[3] In the remainder of this paper, we abuse of these notations to compare solutions with vectors.

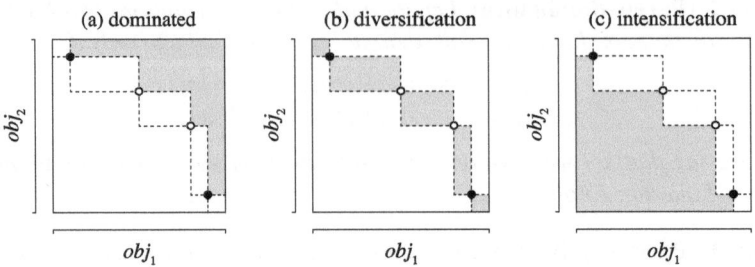

Fig. 1. An archive partitions the objective space into three subspaces: (a) the dominated subspace, (b) the diversification subspace, and (c) the intensification subspace

- the intensification subspace where a new solution replaces at least one solution in the archive;
- the diversification subspace where a new solution is added into the archive without replacing any other solutions.

In the following, we suppose that an archive maintains its domination-free property by removing the solutions that are dominated by a new solution from the intensification space. Therefore, adding new solutions into the archive increases the size of the dominated subspace. The size of the dominated subspace is a common indicator used to measure the quality of an archive known as the hypervolume indicator $\mathcal{H}$ [30]:

Definition 7 (Hyper-volume indicator). *The hypervolume $\mathcal{H}$ is an unary quality indicator (to be maximized) which measures the volume of the objective subspace dominated by a given archive.*

The hypervolume indicator is mostly used for bi-objective problems since its computation increases exponentially with the number of objectives.

Every solution of the Pareto front is not equally difficult to discover. Supported solutions can be discovered using a single objective optimization approach by minimizing a linear aggregation of the objectives while non supported ones cannot [7]:

Definition 8 (Supported Pareto optimal solutions). *A supported Pareto optimal solution is an extreme point on the convex hull of the Pareto front.*

The Pareto front has no guarantee to be convex, justifying the need for more advanced techniques to tackle MOCO problems.

2   Related Work

While multi-objective combinatorial optimization problems have gained a lot of traction over last decade in the Local Search and Evolutionary Search communities (with algorithms such as NSGA-II [5] and SPEA-II [29]), not so many methods have been proposed for CP.

One approach detailed in Section 2.1 has been initially proposed to solve bi-objective problems by solving a sequence of problems. Another approach detailed in Section 2.2 allows to solve arbitrary multi-objective problems in one search using an adaptation of Branch and Bound (BnB) search with a special global constraint to filter the objective variables.

2.1 Bi-Objective Optimization

In bi-objective optimization problems, improving the value of the first objective of a Pareto optimal solution cannot be done without degrading the value of the second objective. The approach proposed by van Wassenhove and Gelders [27] exploits this property in order to find the exact Pareto optimal set of solutions of bi-objective optimization problems. The idea is as follows:[4]

1. Find the Pareto optimal solution with the best value for the first objective;
2. If this solution exists, the search is restarted with an additional constraint enforcing the value of the second objective to be strictly better than its value in the previous solution.

2.2 Multiple-Objective Optimization with CP (MO-CP)

In [9], Gavanelli suggested a framework to solve multi-objective optimization with CP allowing to find all the Pareto optimal solutions in a single search. This framework is presented as a specialized BnB search making use of no-goods recording, corresponding to nondominated solutions. Although not presented this way in [9], we view this approach as the introduction of a new global constraint defined on the objective variables and an archive $\mathcal{A}$ that is domination-free:

$$\texttt{Pareto}(obj_1, \ldots, obj_m, \mathcal{A} = \{sol_1, \ldots, sol_n\}) \qquad (6)$$

where sol_i is a solution to Problem (1). The $\texttt{Pareto}$ constraint ensures that the next discovered solution is nondominated w.r.t. $\mathcal{A}$:

$$\nexists sol \in \mathcal{A} : sol \preceq (obj_1, \ldots, obj_m) \qquad (7)$$

Let $obj_i^{\min}$ and $obj_i^{\max}$ denote the lower and upper bounds of the objective variable obj_i. The filtering of $obj_i^{\max}$ achieved in [9] considers first the *dominated point* DP_i that is defined as follows:

$$DP_i = (obj_1^{\min}, \ldots, obj_{i-1}^{\min}, obj_i^{\max}, obj_{i+1}^{\min}, \ldots, obj_m^{\min}) \qquad (8)$$

Then it finds a solution $sol^* \in \mathcal{A}$ dominating the dominated point *i.e.* such that $sol^* \preceq DP_i$. If such a solution exists, $sol^*(obj_i) - 1$ is an upper bound for obj_i that can be used to filter its domain:

$$obj_i^{\max} \leftarrow sol^*(obj_i) - 1. \qquad (9)$$

---

[4] The approach of van Wassenhove and Gelders can be seen as a particular instance of the ϵ-constraint method [10].

Since we are interested in finding the tightest upper bound for objective i, the idempotent filtering rule is:

$$obj_i^{\max} \leftarrow \min(\{obj_i^{\max}\} \cup \{sol(obj_i) - 1 \mid sol \in \mathcal{A} \wedge sol \preceq DP_i\}) \qquad (10)$$

In this scheme, each time a new solution is found, it is added into $\mathcal{A}$ possibly filtering out dominated solutions to maintain its domination-free property.

It has been demonstrated in [9] that MO-CP, although more general, is also more efficient than the approach of van Wassenhove and Gelders to solve bi-objective knapsack problems.[5]

Example 1. Consider Pareto($obj_1, obj_2, obj_3, \mathcal{A}$) with domains $D(obj_1) = [3..5]$, $D(obj_2) = [2..5]$, $D(obj_3) = [2..5]$ and $\mathcal{A} = \{(1, 4, 2), (4, 2, 3), (2, 3, 1), (2, 1, 4)\}$. No filtering for $obj_1^{\max}$ is possible because $(obj_1^{\max} = 5, obj_2^{\min} = 2, obj_3^{\min} = 2)$ is not dominated by any point in $\mathcal{A}$. For obj_2 some filtering is possible since $(obj_1^{\min} = 3, obj_2^{\max} = 5, obj_3^{\min} = 2)$ is dominated by $(1, 4, 2)$ and $(2, 3, 1)$. We can set $obj_2^{\max} \leftarrow \min(4 - 1, 3 - 1) = 2$. The domain of obj_3 can also be filtered since $(obj_1^{\min} = 2, obj_2^{\min} = 2, obj_3^{\max} = 5)$ is dominated by $(2, 1, 4)$. We can thus set $obj_3^{\max} \leftarrow 4 - 1 = 3$.

3   Multi-Objective LNS

Large Neighborhood Search (LNS) [25] is an hybridization between CP and LS. At each iteration (called *restart* in the LNS context), a best-so-far solution is considered for improvement by exploration of a neighborhood using CP. This solution is relaxed and optimized again with CP, replacing the best-so-far solution on each improvement. This process is repeated until a stopping criterion is met (for instance a maximum number of restarts). LNS has the main advantage that the neighborhood to explore at each restart is potentially very large, permitting to escape local minima most of the time. Every CP optimization model can be turned easily into an LNS by providing the following information/implementation to the solver:

- A *relaxation procedure*. This procedure (also called fragment selection) defines the neighborhood to explore. It adds some constraints to the problem coming from the structure of the best-so-far solution while allowing some flexibility for re-optimization. This relaxation procedure generally includes some randomness.
- A *search limit*. This limit, although optional, prevents the search from spending too much time in the exploration of the neighborhood. It can for instance be a time limit, or a limit on the number of backtracks.

Finding the right relaxation procedure, relaxation size and search limit is a challenging problem (see [15,16,22] for attempts to automatize LNS parameters). This work proposes to adapt the LNS scheme in a multi-objective context.

---

[5] This is probably due to the fact that the approach of Gavanelli does not need to restart the search at each discovered solution. Besides, the already discovered solutions provide some supports to prune dominated branches of the search tree.

3.1 Restarting from a Nondominated Solution

Instead of a unique best-so-far solution, the MO-LNS framework maintains a best-so-far approximation $\mathcal{A}$ of the Pareto set *i.e.* an *archive*. The `Pareto` constraint (using the set $\mathcal{A}$) is added to the model ensuring that only new nondominated solutions w.r.t. $\mathcal{A}$ can be discovered.

Any solution in $\mathcal{A}$ can be used as restarting point. We distinguish two kinds of improvements of the archive:

– finding a new point in the diversification subspace. We call this a *diversification* of the archive. The resulting archive has one more element;
– finding a new point in the intensification subspace. We call this an *intensification* of the archive. The resulting archive is not larger after this insertion since some points may disappear from $\mathcal{A}$.

Notice that both improvements strictly increase the hypervolume (see Definition 7) and both improvements are allowed by the `Pareto` constraint which guarantees that only nondominated solutions w.r.t. the archive can be discovered.

3.2 Guiding Diversification-Intensification

The discovery of a new solution may contribute to diversify the archive, or it may improve existing solutions. A good strategy in terms of filtering for the `Pareto` constraint could consist in finding quickly a limited number of solutions very close to the Pareto set. On the contrary, it would be less efficient to quickly discover a large number of nondominated solutions while being far from the Pareto set. Those two situations are illustrated in Fig. 2.

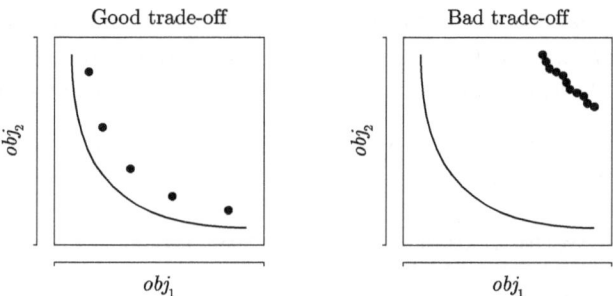

Fig. 2. Situations resulting from (left) a good diversification/intensification trade-off (right) too much diversification at the early MO-LNS iterations. The exact Pareto front is represented with the plain curve.

An archive with many solutions quite far from the Pareto set is the consequence of a too large number of diversification at the early iterations of MO-LNS. It is thus important to have a good trade-off between the number of diversification and intensification[6].

---

[6] Beck [2] also proposes to control diversification and intensification of a pool of elite solutions for single objective problems. We owe this observation to an anonymous referee. Thanks!

A first idea to control the ratio of diversification/intensification is to adapt the search heuristic dynamically. One could for instance have two different search heuristics *e.g.* one that favors intensification and the other one favoring diversification. This approach has the main disadvantage of requiring a good knowledge of the problem and an additional implementation work by the modeler. A better approach forces diversification or intensification at each restart, based on a *dynamic* change of the filtering behavior of the different objectives. Each objective can be set into three different filtering modes during the BnB search:

1. *No-Filtering*: it means that the filtering of the objective is deactivated, having no impact at all.
2. *Weak-Filtering*: each time a new solution is discovered during the search, the upper bound of the objective is updated such that the next discovered solution has a lower or equal upper bound for this objective.
3. *Strong-Filtering*: each time a new solution is discovered during the search, the upper bound of the objective is updated such that the next discovered solution strictly improves the upper bound of this objective.

We propose to use this idea to control the diversification/intensification rates along the restarts.

Intensification. The goal of intensification restarting from a solution *sol* is to discover new solutions dominating it. We propose two different ways to guarantee that the next discovered solution dominates *sol* by adjusting all the objective's upper bounds to their value in *sol* and setting the objectives in one of both following configurations:

– *Strong Intensification*. All the objectives are set in *Strong-Filtering* mode;
– *Driven Intensification*. All the objectives are set in *Weak-Filtering* mode except one that is set into *Strong-Filtering* mode. This objective drives the intensification.

Both configuration are illustrated in Fig. 3 where a possible sequence of successive discovered solutions is given.

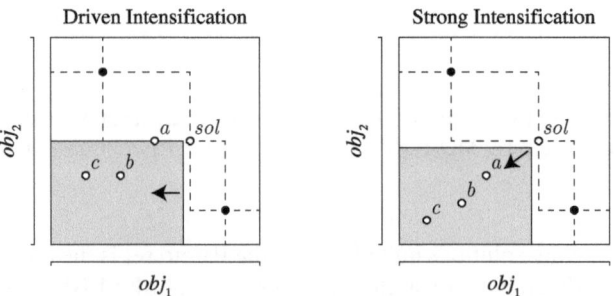

Fig. 3. Intensification. (left) obj_1 is set in *Strong-Filtering* mode and obj_2 is set in *Weak-Filtering* mode. (right) obj_1 and obj_2 are both set in *Strong-Filtering* mode. For both configurations, a possible sequence of successive discovered solutions is given.

Diversification. The diversification mode attempts to find new nondominated solutions without necessarily trying to dominate existing ones. To achieve this, we set all the objectives in *No-Filtering* mode and we let the `Pareto` constraint force the discovery of new nondominated solutions.

Fig. 4 illustrates the benefit of including intensification along the restarts on a bi-objective knapsack (maximization) problem with 100 items from MOCOLib [28]. In the first setting, only diversification restarts are used. In the second setting, 50% are diversification, the others are intensification restarts. One can see on the left, that after 5 seconds, the quality of the nondominated solutions is clearly superior when using intensification restarts. On the right the evolution of the hypervolume (averaged on 10 runs) is depicted. As expected, the hypervolume grows faster when including intensification restarts.

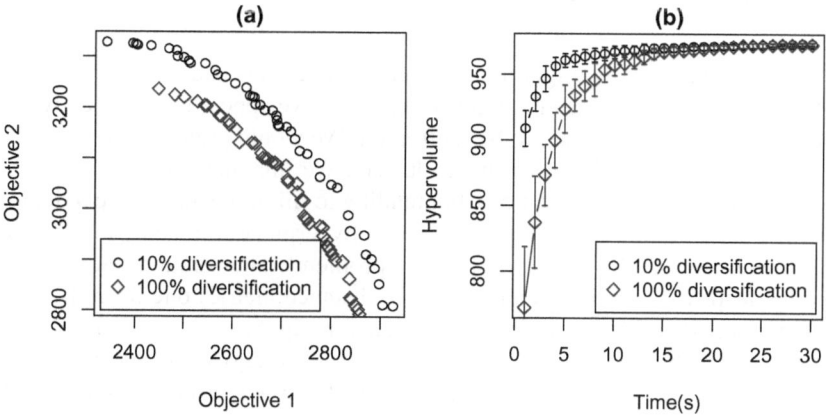

Fig. 4. Impact of the diversification/intensification ratio on a 100 items bi-objective knapsack problem. (left) Nondominated solutions obtained after 5 seconds. (right) Evolution of the hypervolume.

To summarize, the actions that must be taken at each MO-LNS restart are:

– select a solution *sol* from the set of nondominated solutions;
– relax *sol*;
– configure all objectives either in intensification or in diversification mode.

The question of selecting the nondominated solution *sol* is addressed next.

3.3   Selection of the Restarting Solution

Choosing the next solution to restart from can have a strong impact on the quality of the archive. Intuitively, a relaxed solution has a higher chance to generate new solutions close to this one in the objective space when doing diversification. We call this the *locality effect*. Having a final set of nondominated solutions spreading over the frontier is a desired property supported by many researchers [17]. A very simple idea, quite effective in practice, is to select randomly and uniformly the solution to restart from.

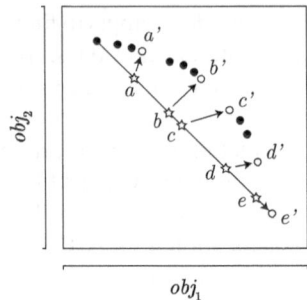

Fig. 5. Selection of solutions according to the nearest neighbor strategy. The straight line represents the hyperplane defined by the extremities of the archive. The stars a, b, c, d, and e correspond to possible points randomly generated on the hyperplane. The solutions a', b', c', d', and e' are the solutions that would be selected for each of the random points.

Unfortunately, this strategy might have negative side effects caused by the locality effect. If at some point, clusters of solutions in the archive appear in the objective space, those clusters have high chances to be reinforced. We would prefer a selection strategy helping to fill in the gaps between those clusters. We imagined another strategy, also randomized (to ensure diversification), but tending to fill in the gaps more quickly. The idea is to select an uniform random point on the hyperplane formed by the extremities of the archive (*i.e.* on a line for a bi-objective problem). The solution selected to relax is then the nearest (according to an Euclidean distance metric) one from this random point. This *nearest neighbor strategy* is illustrated in Fig. 5.

Fig. 6 presents the benefits of the nearest neighbor strategy over the purely randomized selection strategy on a 200 items bi-objective knapsack problem from MOCOLib [28]. We have initially added 6 Pareto optimal solutions in the archive, then 20 diversification restarts were executed with both strategies. While the pure randomized strategy (right) quickly focuses on a particular region of the objective space, the nearest neighbor strategy (left) diversifies better the objective space trying to discover solutions between the gaps on the frontier. The reason is that with the randomized nearest neighbor strategy, solutions close to the gaps are selected more frequently.

4   Modeling an MO-Quadratic Assignment Problem with MO-LNS

This section introduces the MO-LNS modeling of the Multi-Objective Quadratic Assignment Problem (MOQAP) in OscaR [20] and provides some implementation details.

In this problem a set of n facilities must be assigned to n different locations. For each pair of locations, a distance is specified and for each pair of facilities a weight or flow is specified (*e.g.* , the amount of supplies transported between the two facilities). The problem is to assign all facilities to different locations with the goal of minimizing the sum of the distances multiplied by the corresponding weights. More formally, if $x(i)$ represents the location assigned to facility i, the objective is to minimize the weighted sum $\sum_{i,j\in[1..n]} w(i,j)\cdot d(x(i),x(j))$ with w and d respectively the weight and distance matrices.

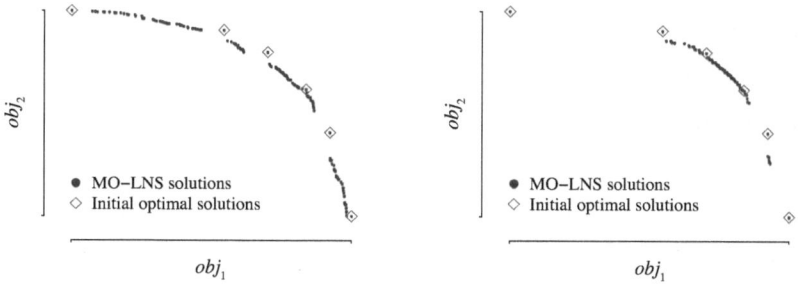

Fig. 6. Impact of selection strategy after 20 restarts on a 200 items bi-objective knapsack problem using 100% of diversification starting from 6 initial Pareto optimal solutions. (left) Using the randomized nearest neighbor strategy. (right) Using a pure randomized strategy.

The multi-objective QAP with multiple weight matrices naturally models any facility layout problem where we are concerned with the flow of more than one type of item or agent [12]. The OscaR model for a bi-objective QAP is given in Statement 1.

The data declaration is specified in lines 1 - 6 and should be self-explanatory. The distance matrix and the two weight matrices are declared. Then comes the CP model. A solver object is created in line 8. An array x of n decision variables is created at line 10 representing the location of each facility. The distance variables between any two facility are initialized at line 11 using 2D element constraints. The two objective functions are initialized in lines 12 - 13 multiplying each distance entry by the corresponding weight and summing them all.

Notice that `paretoMinimize`, `subjectTo` and `exploration` are methods of the `CPSolver` class each returning the `CPSolver` caller instance. This allows to chain directly the calls. The `paretoMinimize` method call at line 15 implicitly adds the `Pareto` global constraint (6) to the model. The search in the exploration block is a nondeterministic search [11]. Although hidden from the user point of view, all the discovered solutions are added into the archive $\mathcal{A}$ used by the `Pareto` constraint. The `run` method takes two optional arguments: a limit on the number of solutions and a limit on the number of backtracks. Both are set to infinity by default. The search to find the first feasible solution is started at line 23.

Lines 20 and 21 are iterated until all variables are bound and each iteration nondeterministically assigns a facility `x(i)` to a location `v` computed by the variable value heuristic introduced in [19]. Notice that this heuristic receives a weight matrix in argument. In the non-deterministic search `exploration` block, the weight matrix is randomly chosen between `w1` and `w2` at line 20.

The MO-LNS procedure is implemented at lines 30 - 40, after that the first feasible solution is found. The search executes 1000 LNS restarts. Each restart has a limit of 200 failures and use the search defined in the exploration block. On each restart a solution is selected from the current archive according to the nearest neighbor strategy (line 32). Then, the objectives are configured into intensification or diversification mode w.r.t. to a user defined probability. The `runSubjectTo` method is similar to the `run` method except that all the constraints added in its block are temporary constraints

```
1    // DATA AND CONSTANTS
2    val N = 0 until n                // number of locations
3    var w1: Array[Array[Int]] = ... // weight matrix 1
4    var w2: Array[Array[Int]] = ... // weight matrix 2
5    var d:  Array[Array[Int]] = ... // distance matrix
6    val rand = Random(0)             // random number generator
7    // CP MODEL
8    val cp = CPSolver()
9    // the location chosen for each facility
10   val x    = Array.fill(n){CPVarInt(cp, N)}
11   val dist = Array.tabulate(n, n){(i, j) => d(x(i))(x(j))}
12   val obj1 = sum(n, n){(i, j) => dist(i)(j) * w1(i)(j)}
13   val obj2 = sum(n, n){(i, j) => dist(i)(j) * w2(i)(j)}
14   // CONSTRAINT AND EXPLORATION
15   cp.paretoMinimize(obj1, obj2) subjectTo {
16     cp.add(allDifferent(x), Strong)
17   } exploration {
18     // compute variable, value heuristic randomly on w1 or w2
19     while (!allBounds(x)) {
20       val (i, v) = heuristic(if (rand.nextBoolean) w1 else w2)
21       cp.branch(cp.post(x(i) == v))(cp.post(x(i) != v))
22     }
23   } run(nbSolution = 1)      // only search for an initial solution
24   // MO LNS PARAMETERS
25   val maxRestarts    = 1000 // number of restarts
26   val maxFailures    = 200  // max number of failures at each restart
27   val relaxSize      = 5    // number of relaxed variables at each restart
28   val probaIntensify = 30   // probability (%) of intensification
29   // MO LNS FRAMEWORK
30   for (restart <- 1 to maxRestarts) {
31     // next solution to restart from
32     val sol = nearestNeibhborSol()
33     // random selection between intensification or diversification
34     if (rand.nextInt(100) < probaIntensify) cp.objective.intensify(sol)
35     else cp.objective.diversify()
36     // search
37     cp.runSubjectTo(failureLimit = maxFailures) {
38       relaxRandomly(x, sol, relaxSize)
39     }
40   }
```

Statement 1. Model of the multi-objective QAP in OscaR/Scala.

that will be removed before the next restart. Some constraints are added through the `relaxRandomly` at line 38 to restore the assignments on x from the selected solution `sol` except for 5 randomly chosen variables.

5   Experiments

This section compares the performances of MO-LNS over MO-CP on bi-objective problems.[7] The tested problems are: 1) the multi-objective QAP, 2) the multi-objective binary knapsack and 3) a bi-objective tank allocation problem. Although multi-objective

---

[7] Instances and optimal fronts available at
http://becool.info.ucl.ac.be/resources/mo-lns

Table 1. Results on MO-QAP instances from [13] with MO-LNS and MO-CP

Instance	$\lvert S^* \rvert$	$\mathcal{H}_{S^*}(10^8)$	$\mathcal{H}_S(10^8)$		$\lvert S \rvert$		$\lvert S \cap S^* \rvert$	
			MO-LNS	MO-CP	MO-LNS	MO-CP	MO-LNS	MO-CP
KC10-2fl-1uni	13	117.38	**115.43**	99.09	9	15	**6.6**	0
KC10-2fl-2uni	1	91.56	**87.11**	76.99	1.4	2	**0.6**	0
KC10-2fl-3uni	130	90.50	**87.42**	78.79	84.4	65	**30.6**	0
KC10-2fl-1rl	58	606005.79	**604932.58**	598146.60	54	41	**50.4**	13
KC10-2fl-2rl	15	604864.40	**604864.40**	604864.40	15	15	**15**	15
KC10-2fl-3rl	55	623898.50	**623076.50**	565772.58	47	37	**43**	0
KC10-2fl-4rl	53	732716.05	**732716.05**	732716.05	53	53	**53**	53
KC10-2fl-5rl	49	1819669.88	**1819669.88**	1819669.88	49	49	**49**	49

heuristics are the methods of choice to tackle the two first problems, those are interesting standard benchmarks to study, with known exact Pareto front. Problem 3 however, is more constrained and probably more suited for constraint programming. All experiments were conducted with the OscaR open-source solver [20] on an Intel® Core i7 ™ 2.6GHz CPU.

5.1 Multi-Objective Quadratic Assignment Problem

We experiment the MO-LNS model introduced in Section 4 on instances.[8] with 10 facilities from [13] Table 1 reports the results obtained with a 30 seconds timeout, averaged over 10 runs for MO-LNS. The size and the hypervolume of the exact Pareto fronts are given in columns 2 and 3. The hypervolumes obtained with MO-LNS and MO-CP ($\mathcal{H}_S$) are given in columns 4 and 5. The sizes of the archives obtained with MO-LNS and MO-CP ($\lvert S \rvert$) are given in columns 6 and 7. The number of optimal solutions obtained with each approach ($\lvert S \cap S^* \rvert$) is presented in columns 8 and 9.

As can be seen, 3 instances are optimally solved with MO-CP. MO-LNS is also able to solve these instances optimally and obtain strictly better results on the other instances. The hypervolume values reached by MO-LNS are very close to the optimal ones.

5.2 Multi-Objective 0/1 Knapsack Problem

The multi-objective 0/1 knapsack problem is defined as follows:

$$\text{Maximize} \quad obj = (obj_1, \dots, obj_m) \quad \text{with} \quad obj_j = \sum_{i=1}^{n} x_i \cdot p_{ij}$$

$$\text{Subject to} \quad \sum_{i=1}^{n} x_i \cdot w_i \leq C \tag{11}$$

with p_{ij} the profit of item i according to objective j, w_i the weight of item i, and C the capacity of the knapsack. The binary variables x_i represent the selection status of each item i.

MO-LNS Settings. The next solution *sol* to restart from is chosen in the archive with the nearest neighbor strategy. The idea of the relaxation procedure is to keep fixed some *good items* (w.r.t. to one objective) already selected in *sol* :

---

[8] Only the ones for which the optimal Pareto front is available.

Table 2. Comparison of MO-CP and MO-LNS on standard instances of the Bi-Objective Knapsack Problem

Instance	$|S^*|$	$\mathcal{H}_{S^*}(10^8)$	$\mathcal{H}_S(10^8)$		$|S|$		$|S \cap S^*|$	
			MO-LNS	MO-CP	MO-LNS	MO-CP	MO-LNS	MO-CP
2KP100A	172	15,59	**15,59**	15,05	**172**	128	**172**	112
2KP100B	174	15,12	**15,12**	14,62	**170,6**	124	**164,7**	93
2KP100C	64	16,68	**16,68**	16,68	**64**	64	**64**	64
2KP100D	76	16.31	**16.31**	16.28	**76**	73	**76**	73
2KP150A	244	39.66	**39.66**	35.42	**226.2**	88	**187**	31
2KP150B	348	41.46	**41.46**	36.31	**303.3**	91	**192.4**	51
2KP150C	166	34.17	**34.17**	33.15	**155.6**	83	**127**	34
2KP150D	207	36.04	**36.04**	32.88	**199.8**	88	**155.6**	62
2KP200A	439	64.34	**64.34**	57.43	**361**	86	**178.2**	34
2KP200B	397	65.78	**65.77**	58.82	**345.2**	108	**232.8**	54
2KP200C	328	57.48	**57.46**	48.30	**297.8**	60	**187.1**	18
2KP200D	361	73.42	**73.40**	62.71	**304.1**	64	**176.9**	29
2KP250A	629	94.37	**94.34**	78.68	**433.5**	70	**95.5**	12
2KP250B	629	89.67	**89.65**	74.81	**410.9**	90	**107.9**	44
2KP250C	528	91.25	**91.24**	75.30	**383.3**	72	**108.9**	22
2KP250D	424	66.56	**66.55**	56.98	**303.4**	51	**142.8**	26

1. select randomly one objective index $j \in [1..m]$,
2. select randomly 90% of the items in the set $\{i \in [1..n] \mid sol(x_i) = 1\}$ according to a probability function proportional to p_{ij}/w_i,
3. keep fixed the items selected at step 2. *i.e.* force them to be in the knapsack,
4. chose randomly to diversify or intensify with equal probability.

The variable-value heuristic used when re-optimizing after the relaxation is dependent of the selected objective j in the relaxation. The heuristic selects first the unbound variable x(i) with the largest ratio p_{ij}/w_i, selecting this item on the left and removing it on the right branch. Each restart is given a limit of 1000 backtracks.

Results. We compare the MO-CP and MO-LNS approaches on instances from MO-COLib [28] ranging from 50 to 250 items. Table 2 reports the results obtained with a 60 seconds timeout, averaged over 10 runs for MO-LNS. The size and the hypervolume of the exact Pareto fronts[9] are given in columns 2 and 3. The hypervolumes obtained with MO-LNS and MO-CP ($\mathcal{H}_S$) are given in columns 4 - 5. The sizes of the archives obtained with MO-LNS and MO-CP ($|S|$) are given in columns 6 - 7. The number of optimal solutions obtained with each approach ($|S \cap S^*|$) is presented in columns 8 - 9. As can be seen, MO-LNS consistently obtains better or equivalent results compared to the MO-CP approach. The hypervolume values reached by MO-LNS are very close to the optimal ones.

5.3   Tank Allocation Problem

The tank allocation problem involves the assignment of different cargoes (volumes of chemical products to be shipped by the vessel) to the available tanks of the vessel [24] while satisfying hard segregation constraints *e.g.* to avoid placing dangerous cargoes

---

[9] The optimal fronts were provided by the creator of MOCOLib [28].

Table 3. Comparison of MO-CP and MO-LNS on real-life instances of the Bi-Objective Tank Allocation Problem. MO-LNS finds all the exact Pareto fronts.

Instance	$|S^*|$	$\mathcal{H}_{S^*}(10^8)$	$\mathcal{H}_S(10^8)$		$|S|$		$|S \cap S^*|$	
			MO-LNS	MO-CP	MO-LNS	MO-CP	MO-LNS	MO-CP
chemicalA	6	1848	**1848**	1022	**6**	4	**6**	1
chemicalB	7	2976	**2976**	1010	**7**	3	**7**	0
chemicalC	8	3597	**3597**	852	**8**	2	**8**	0
chemicalD	8	5358	**5358**	555	**8**	1	**8**	0

in adjacent tanks. An ideal loading plan should maximize the total volume of unused tanks (*i.e.* free space) to minimize cleaning costs (objective 1). Minimizing the number of used tanks (objective 2) is also desirable in order to maximize the chances of accommodating other cargoes in next visited ports. The LNS model used in our experiment is the same as the one introduced in [24] except that it is now bi-objective. The MO-LNS parameters are:

- The relaxed solution is chosen randomly from the current archive.[10]
- Chose randomly to diversify or intensify with equal probability.

Results are given in Table 5.3. The columns have the same meaning as in previous result tables. The exact Pareto fronts were generated with the van Wassenhove and Gelders Algorithm using a MIP solver (Gurobi 5.02) that took about 3 minutes to run on each instance. The MO-LNS framework finds the exact Pareto front of each instance within a timeout of 60 seconds. We also indicate for the same model the results obtained for a MO-CP approach with a timeout of 300 seconds. This MO-CP approach is only able to discover one solution of the exact Pareto front of the first instance (chemicalA). The hypervolume indicator shows that other nondominated solutions discovered with MO-CP remain quite far from optimal ones.

6   Future Works and Conclusion

This paper introduced the MO-LNS framework; an extension of LNS to efficiently solve MOCO problems with CP. MO-LNS uses the `Pareto` constraint to maintain a best-so-far archive that is iteratively improved by diversification and intensification. Modeling abstractions were presented into the OscaR solver to select at each restart a solution in the archive, and to diversify or intensify the front MO-LNS was experimented on various MOCO problems showing its superiority over MO-CP to get close to optimal hypervolumes. The Scala source-code of our implementation as well as some complete MO-LNS examples are available on OscaR repository [20].

As future work, we plan to study adaptive diversification/intensification strategies. We would like to explore the parallelization of MO-LNS. Finally we want to tackle more complex problems with MO-LNS, such as multi-objective scheduling or vehicle routing problems.

---

[10] No real impact since the optimal front is very small.

626 P. Schaus and R. Hartert

References

1. Abraham, A., Jain, L.: Evolutionary multiobjective optimization. Evolutionary Multiobjective Optimization (2005)
2. Christopher Beck, J.: Solution-guided multi-point constructive search for job shop scheduling. J. Artif. Int. Res. 29(1), 49–77 (2007)
3. Bent, R., Hentenryck, P.V.: A two-stage hybrid algorithm for pickup and delivery vehicle routing problems with time windows. Computers & Operations Research 33(4), 875–893 (2006)
4. Cahon, S., Melab, N., Talbi, E.-G.: Paradiseo: A framework for the reusable design of parallel and distributed metaheuristics. Journal of Heuristics 10(3), 357–380 (2004)
5. Deb, K., Pratap, A., Agarwal, S., Meyarivan, T.: A fast and elitist multiobjective genetic algorithm: Nsga-2. IEEE Transactions on Evolutionary Computation 6(2), 182–197 (2002)
6. Durillo, J.J., Nebro, A.J.: jmetal: A java framework for multi-objective optimization. Advances in Engineering Software 42, 760–771 (2011)
7. Ehrgott, M.: Multicriteria optimization, vol. 2. Springer, Berlin (2005)
8. Ehrgott, M., Gandibleux, X.: Hybrid metaheuristics for multi-objective combinatorial optimization. In: Blum, C., Aguilera, M.J.B., Roli, A., Sampels, M. (eds.) Hybrid metaheuristics. SCI, vol. 114, pp. 221–259. Springer, Heidelberg (2008)
9. Gavanelli, M.: An algorithm for multi-criteria optimization in csps. ECAI 2, 136–140 (2002)
10. Haimes, Y.Y., Lasdon, L.S., Wismer, D.A.: On a bicriterion formulation of the problems of integrated system identification and system optimization. IEEE Transactions on Systems, Man, and Cybernetics 1(3), 296–297 (1971)
11. Van, P., Van Hentenryck, P., Michel, L.: Nondeterministic control for hybrid search. Michel 11(4), 353–373 (2006)
12. Knowles, J.: Towards landscape analyses to inform the design of a hybrid local search for the multiobjective quadratic assignment problem. Soft Computing Systems: Design, Management and Applications 2002, 271–279 (2002)
13. Knowles, J., Corne, D.: Instance generators and test suites for the multiobjective quadratic assignment problem. In: Fonseca, C.M., Fleming, P.J., Zitzler, E., Deb, K., Thiele, L. (eds.) EMO 2003. LNCS, vol. 2632, pp. 295–310. Springer, Heidelberg (2003)
14. Laborie, P., Godard, D.: Self-adapting large neighborhood search: Application to single-mode scheduling problems. In: Proceedings MISTA 2007, Paris, pp. 276–284 (2007)
15. Mairy, J.-B., Deville, Y., Van Hentenryck, P.: Reinforced adaptive large neighborhood search. In: 8th Workshop on Local Search techniques in Constraint Satisfaction (LSCS 2011). A Satellite Workshop of CP, Perugia, Italy (2011)
16. Mairy, J.-B., Schaus, P., Deville, Y.: Generic adaptive heuristics for large neighborhood search. In: Seventh International Workshop on Local Search Techniques in Constraint Satisfaction (LSCS 2010). A Satellite Workshop of CP (2010)
17. Timothy Marler, R., Arora, J.S.: Survey of multi-objective optimization methods for engineering. Structural and Multidisciplinary Optimization 26(6), 369–395 (2004)
18. Mehta, D., O'Sullivan, B., Simonis, H.: Comparing solution methods for the machine reassignment problem. In: Milano, M. (ed.) CP 2012. LNCS, vol. 7514, pp. 782–797. Springer, Heidelberg (2012)
19. Michel, L., Shvartsman, A., Sonderegger, E., Van Hentenryck, P.: Optimal deployment of eventually-serializable data services. In: Integration of AI and OR Techniques in Constraint Programming for Combinatorial Optimization Problems, pp. 188–202 (2008)
20. OscaR Team. OscaR: Scala in OR (2012), https://bitbucket.org/oscarlib/oscar.

21. Pacino, D., Van Hentenryck, P.: Large neighborhood search and adaptive randomized decompositions for flexible jobshop scheduling. In: Proceedings of the Twenty-Second International Joint Conference on Artificial Intelligence, vol. 3, pp. 1997–2002. AAAI Press (2011)
22. Perron, L., Shaw, P., Furnon, V.: Propagation guided large neighborhood search. In: Wallace, M. (ed.) CP 2004. LNCS, vol. 3258, pp. 468–481. Springer, Heidelberg (2004)
23. Schaus, P., Van Hentenryck, P., Monette, J.N., Coffrin, C., Michel, L., Deville, Y.: Solving steel mill slab problems with constraint-based techniques: Cp, lns, and cbls. Constraints 16(2), 125–147 (2011)
24. Schaus, P., Régin, J.-C., Van Schaeren, R., Dullaert, W., Raa, B.: Cardinality reasoning for bin-packing constraint: Application to a tank allocation problem. In: Milano, M. (ed.) CP 2012. LNCS, vol. 7514, pp. 815–822. Springer, Heidelberg (2012)
25. Shaw, P.: Using constraint programming and local search methods to solve vehicle routing problems. In: Maher, M.J., Puget, J.-F. (eds.) CP 1998. LNCS, vol. 1520, pp. 417–431. Springer, Heidelberg (1998)
26. Ulungu, E.L., Teghem, J.: Multi-objective combinatorial optimization problems: A survey. Journal of Multi-Criteria Decision Analysis 3(2), 83–104 (1994)
27. Van Wassenhove, L.N., Gelders, L.F.: Solving a bicriterion scheduling problem. European Journal of Operations Research 4, 42–48 (1980)
28. Gandibleux, X.: A collection of test instances for multiobjective combinatorial optimization problems (2013), http://xgandibleux.free.fr/MOCOlib/
29. Zitzler, E., Laumanns, M., Thiele, L.: Spea2: Improving the strength pareto evolutionary algorithm (2001)
30. Zitzler, E., Thiele, L., Laumanns, M., Fonseca, C.M., da Fonseca, V.G.: Performance assessment of multiobjective optimizers: An analysis and review. IEEE Transactions on Evolutionary Computation 7(2), 117–132 (2003)

Scheduling Optional Tasks with Explanation

Andreas Schutt[1,2], Thibaut Feydy[1,2], and Peter J. Stuckey[1,2]

[1] Optimisation Research Group, National ICT Australia
[2] Department of Computing and Information Systems,
The University of Melbourne, Victoria 3010, Australia
{andreas.schutt,thibaut.feydy,peter.stuckey}@nicta.com.au

Abstract. Many scheduling problems involve reasoning about tasks which may or may not actually occur, so called optional tasks. The state-of-the-art approach to modelling and solving such problems makes use of interval variables which allow a start time of $\bot$ indicating the task does not run. In this paper we show we can model interval variables in a lazy clause generation solver, and create explaining propagators for scheduling constraints using these interval variables. Given the success of lazy clause generation on many scheduling problems, this combination appears to give a powerful new solving approach to scheduling problems with optional tasks. We demonstrate the new solving technology on well-studied flexible job-shop scheduling problems where we are able to close 36 open problems.

1 Introduction

Many resource-constrained scheduling problems involve reasoning about tasks which may or may not actually occur, so called optional tasks. The state-of-the-art approach in Constraint Programming (CP) to modelling and solving such problems makes use of so-called *interval variables* [12] which represent a start time, end time, and duration of a task, or $\bot$ indicating the task does not run. Propagation algorithms can update the possible start and end times of a task, without knowing whether the task actually runs or not.

In 2008, Laborie and Rogerie [12] introduce interval variables for resource-constrained scheduling to IBM ILOG CP Optimizer [11] as a "first-class citizen" variable type for CP systems. In that work and later follow up work [13,14], they show how to handle these variables in the context of planning and scheduling. The benefits of interval variables are not only in giving a neat conceptual model for representing optional tasks, but also the additional propagation obtained that is possible by reasoning on start and end times even without knowing whether a task executes. However, interval variables do not come for free, they may introduce additional variables into a model, and their propagation is more complex.

Standard CP systems that do not support interval variables are still able to model and solve problems with optional tasks, but suffer from the weaker propagation. For example, each optional task can be associated with a Boolean

C. Schulte (Ed.): CP 2013, LNCS 8124, pp. 628–644, 2013.

variable representing whether it executes or not [1,2] or a non-optional task composed of exclusive optional tasks can be associated with an index variable representing which one of the optional tasks runs [16]. In order to strengthen the propagation, special global constraints have been introduced (see, *e.g.*, [1,2,16]).

For CP systems that support interval variables, propagation algorithms have been proposed for the resource constraints disjunctive and cumulative with optional tasks (see, *e.g.*, [26,30,29,28]). These algorithms record tentative start and end times of optional task and once the optional task is known to execute these become the actual start and end time variables.

In this paper, we not only show how to mimic interval variables with integer variables, but also how propagators defined for constraints on optional tasks can be extended to explain their propagation, which is required for CP solvers with learning. One of those solvers is a lazy clause generation (LCG) (LCG) [19] solver which has proven to be remarkably effective on many scheduling problems defining the state-of-the-art in RCPSP [24,23], RCPSP/max [25], and RCPSPDC [22] problems. We implement the handling of optional tasks in the re-engineered LCG solver [7], and then demonstrate the combination on the well-studied flexible job shop scheduling problem, where we are able to close a number of open instances.

2 Preliminaries

At first, we introduce lazy clause generation and then scheduling with optional tasks.

2.1 Lazy Clause Generation

CP solves constraint satisfaction problems by interleaving propagation, which remove impossible values of variables from the domain, with search, which guesses values. All propagators are repeatedly executed until no change in domain is possible, then a new search decision is made. If propagation determines there is no solution then search undoes the last decision and replaces it with the opposite choice. If all variables are fixed then the system has found a solution to the problem. For more details see, *e.g.*, [21].

We assume we are solving a constraint satisfaction problem over set of variables $x \in \mathcal{V}$, each of which takes values from a given initial finite set of values or *domain* $D^0(x)$. The domain D keeps track of the current set of possible values $D(x)$ for a variable x. Define $D \sqsubseteq D'$ iff $D(x) \subseteq D'(x), \forall x \in \mathcal{V}$. The constraints of the problem are represented by propagators f which are functions from domains to domains which are monotonically decreasing $f(D) \sqsubseteq f(D')$ whenever $D \sqsubseteq D'$, and contracting $f(D) \sqsubseteq D$. If all values are removed from one domain of a variable x, *i.e.*, $D(x) = \emptyset$ then the constraints cannot be satisfied with the search decisions made and a failure is triggered. Given a domain D then $lb_D(x) = \min D(x)$ and $ub_D(x) = \max D(x)$. We will omit the subscript D when the domain is clear from the context.

We make use of CP with learning using the Lazy Clause Generation (LCG) [19] approach. Learning keeps track of what caused changes in domain to occur, and on failure computes a *nogood* which records the reason for failure. The nogood prevents search making the same incorrect set of decisions again.

In an LCG solver integer domains are also represented using Boolean variables. Each variable x with initial domain $D^0(x) = [l..u]$ is represented by two sets of Boolean variables $[\![x = d]\!], l \le d \le u$ and $[\![x \le d]\!], l \le d < u$ which define which values are in $D(x)$. We use $[\![x \ne d]\!]$ as shorthand for $\neg[\![x = d]\!]$, and $[\![d \le x]\!]$ as shorthand for $\neg[\![x \le d - 1]\!]$. An LCG solver keeps the two representations of the domain in sync. For example if variable x has initial domain $[0..5]$ and at some later stage $D(x) = \{1,3\}$ then the literals $[\![x \le 3]\!], [\![x \le 4]\!], \neg[\![x \le 0]\!], \neg[\![x = 0]\!], \neg[\![x = 2]\!], \neg[\![x = 4]\!], \neg[\![x = 5]\!]$ will hold. Explanations are defined by clauses over this Boolean representation of the variables.

Example 1. Consider a simple constraint satisfaction problem with constraints $b \to x + 3 \le y, \neg b \to y + 3 \le x, b' \to y \le 3, \neg b' \to x \le 3$, with initial domains $D^0(b) = D^0(b') = \{0,1\}$, and $D^0(x) = D^0(y) = \{0,1,2,3,4,5,6\}$. There is no initial propagation. Setting $[\![y = 2]\!]$ makes the first constraint propagate $D(b) = \{0\}$ with explanation $[\![y = 2]\!] \to \neg b$, then the second constraint propagates $D(x) = \{5,6\}$ with explanation $\neg b \wedge [\![y = 2]\!] \to [\![5 \le x]\!]$. The third constraint propagates $D(b') = \{0\}$ with explanation $[\![y = 2]\!] \to \neg b'$ and the last constraint sets $D(x) = \emptyset$, with explanation $[\![5 \le x]\!] \wedge \neg b' \to false$. The graph of the implications is

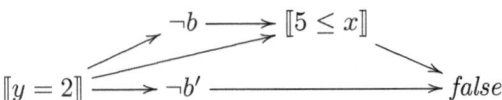

Any cut separating the decision $[\![y = 2]\!]$ from *false* gives a nogood. The simplest one is $[\![y = 2]\!] \to false$. □

2.2   Scheduling and Optional Tasks

Scheduling applications deal with non-optional and optional tasks. A typical task is specified by a *start time* variable S_i and a *processing time/duration* d_i (which may also be variable). For simplicity we assume durations are fixed, it is easy to extend the results of the paper to variable durations. Given a task and current domain D we define the *earliest start time* $ect_i = lb_D(S_i)$, *earliest completion time* $ect_i = lb_D(S_i) + d_i$, *latest start time* $lst_i = ub_D(S_i)$, and *latest completion time* $lst_i = ub_D(S_i) + d_i$.

Some tasks need resources, such as *e.g.*, labour, space, or particular machinery, from a limited pool for their execution. A schedule of those tasks must ensure that the demand on a resource does not exceed the resource capacity in any time period. In this work, we consider *renewable resources* characterised by the constant *resource capacity* R over time. Such a resource can be modelled by the *cumulative constraint*

$$\texttt{cumulative}([S_1, \ldots, S_n], [d_1, \ldots, d_n], [r_1, \ldots, r_n], R)$$
$$\equiv \left(\forall \tau : \sum_{i=1}^{n} runs_{i\tau} \cdot r_i \leq R \right),$$

where τ is a time period, r_i is the *resource usage* of task i, and $runs_{i\tau}$ expresses whether task i runs at time period τ.

The disjunctive constraint $\texttt{disjunctive}$, requiring that no two tasks are executing at the same time, encodes the special case of cumulative when the resource capacity is 1, and the resource usage for each task is 1.

$$\texttt{disjunctive}([S_1, \ldots, S_n], [d_1, \ldots, d_n])$$
$$\equiv \texttt{cumulative}([S_1, \ldots, S_n], [d_1, \ldots, d_n], [1, \ldots, 1], 1)$$

Specialised propagation algorithms [26,30] are available for the $\texttt{disjunctive}$ constraint.

Laborie and Rogerie [12] introduce interval variables to represent optional tasks. The domain of an interval variable ranges over $\perp \cup \{[s, e) \mid s, e \in \mathbb{Z}, e \geq s\}$. A fixed interval variable represents either an *absent* interval $\perp$ or a *present* interval $[s, e)$. Accordingly, an optional task is absent or present if its interval is absent or present respectively. If the interval $[s, e)$ is present then s and e respectively represent its start and end time and $e - s$ its length and it must be that $s \leq e$. If a is an interval variable then let s^a, e^a, and x^a denote the start time, end time, and presence state, respectively.

A task 0 can be composed of other tasks $1, \ldots, n$ and modelled with interval variables a_0 and $a_1, \ldots, a_n$, respectively. Then the relation between the tasks is described via a *span* constraint [14]:

$$span(a_0, \{a_1, \ldots, a_n\}) \equiv \begin{cases} (x_0^a \leftrightarrow \bigvee_{i=1}^n x_i^a) \\ \wedge \ (s_0^a = \min_{1 \leq i \leq n : x_i^a} s_i^a) \\ \wedge \ (e_0^a = \max_{1 \leq i \leq n : x_i^a} e_i^a) \end{cases} \tag{1}$$

That is, task 0 starts when the earliest task in $\{1, \ldots, n\}$ that is present starts, and ends when the latest task that is present ends. It is present iff at least one of tasks $1, \ldots, n$ is present.

An important specialisation of the span constraint is the *alternative* constraint [14] which allows only one task $1, \ldots, n$ to be present (thus representing a choice for task 0).

$$alternative(a_0, \{a_1, \ldots, a_n\}) \equiv \begin{cases} \sum_{i=1}^n x_i^a \leq 1 \\ \wedge \ span(a_0, \{a_1, \ldots, a_n\}) \end{cases} \tag{2}$$

Note if the task 0 is present then exactly one task in $\{1, \ldots, n\}$ is present too; otherwise all are absent.

3   Modelling Optional Tasks

The crucial requirement for effective modelling of optional tasks is to be able to reason about finite domain integer variables which have an additional value $\bot$, which we will call $\text{int}_\bot$ variables. These variables can then represent start times of optional tasks. They can also be useful for other reasoning, for example reasoning about databases with null values. In this section we show how to model $\text{int}_\bot$ variables using integer and Boolean variables. We then discuss how to model compositional constraints such as *span* and *alternative*. Finally we discuss how tracking implications between presence of tasks can be modelled, to help improve propagation.

3.1   Integers with Bottom

LCG solvers do not currently support $\text{int}_\bot$ variables. But we can make use of existing integer and Boolean variables to model an $\text{int}_\bot$ variable and thus interval variables.

We model an $\text{int}_\bot$ variable S with initial domain $D^0(S) = [l^S..u^S]$ and $\bot$ as $S = (\underline{S}, \overline{S}, x^S)$ using two integer variables $\underline{S}$, $\overline{S}$, and a Boolean variable x^S: $\underline{S}$ holds the lower bound of the $\text{int}_\bot$ variable S; while $\overline{S}$ holds the upper bound of the $\text{int}_\bot$ variable S; and x^S holds the presence state of the $\text{int}_\bot$ variable. The initial domains are $D^0(\underline{S}) = l^S..u^S + 1$ are $D^0(\overline{S}) = l^S - 1..u^S$. The representatives $(\underline{S}, \overline{S}, x^S)$ are constrained by

$$intbot(S) \quad \equiv \quad x^S \leftrightarrow (\underline{S} = \overline{S}) \quad \wedge \quad \neg x^S \leftrightarrow \underline{S} > u^S \quad \wedge \quad \neg x^S \leftrightarrow \overline{S} < l^S$$

Thus if the $\text{int}_\bot$ variable S is present, *i.e.*, $S \neq \bot$, the lower and upper bounds are identical. If the lower and upper bound are not compatible, *i.e.*, $\underline{S} > \overline{S}$, then the $\text{int}_\bot$ variable must be absent, *i.e.*, $S = \bot$, and if the $\text{int}_\bot$ variable is absent we set the lower bound to $u^S + 1$ and the upper bound to $l^S - 1$. Note $\underline{S} < \overline{S}$ never holds.

The constraint $\underline{S} \geq v$ represents that $S \geq v \vee S = \bot$. The constraint $\overline{S} \leq v$ represents that $S \leq v \vee S = \bot$.

Propagation on the $\text{int}_\bot$ variable is enforced using the appropriate bound. Hence a new (tentative) lower bound $S \geq v$ is enforced by $\underline{S} \geq v$, and a new (tentative) upper bound $S \leq v$ is enforced as $\overline{S} \leq v$. Asserting that $S \neq v$ is enforced by $\underline{S} \neq v \wedge \overline{S} \neq v$. Asserting $S = v$ if S is present is enforced by $\underline{S} \geq v \wedge \overline{S} \leq v$. Two integer variables are required to model an $\text{int}_\bot$ variable so that if the bounds cross we do not get a domain wipe-out, which would incorrectly trigger a failure.

Care must be taken in using the tripartite representation of $\text{int}_\bot$ variables, because of the special role taken by the *sentinel values* $u^S + 1$ for $\underline{S}$ and $l^S - 1$ for $\overline{S}$. If a propagator ever tries to set $\underline{S} \geq k$ where $k > u^S + 1$, this should be replaced by setting $\underline{S} \geq u^S + 1$. Similarly if a propagator ever tries to set $\overline{S} \leq k$ where $k < l^S - 1$, we should instead set $\overline{S} \leq l^S - 1$. Since propagators are aware that they are dealing with $\text{int}_\bot$ variables, they can be modified to act accordingly, without changing the integer variables used to represent $\underline{S}$ and $\overline{S}$.

Given that we have $\texttt{int}_\bot$ variables, we can model an interval variable a as a pair (S, d) of an $\texttt{int}_\bot$ variable $S = (\underline{S}, \overline{S}, x^S)$ and an integer d by $x^a = x^S$, $s^a = lb(\underline{S})$, and $e^a = ub(\overline{S}) + d$. Note that [12,13,14] introduce interval variables as an abstract type for tasks and here we consider tasks with fixed duration, thus an end time variable is not required.

3.2 Compositional Constraints

The *span* constraint can be modelled using $\texttt{int}_\bot$ variables and constraints supported by most CP solvers as follows:

$span((S_0, d_0), [(S_1, d_1), \ldots, (S_n, d_n)])$
$$\equiv \begin{cases} & \underline{S}_0 \geq \min\{\underline{S}_i + (1 - x_i^S)(u_0^S - u_i^S) \mid 1 \leq i \leq n\} \cup \{u_0^S + 1\} \\ \wedge & \overline{S}_0 \leq \max\{\overline{S}_i + (1 - x_i^S)(l_0^S - l_i^S) \mid 1 \leq i \leq n\} \cup \{l_0^S - 1\} \\ \wedge & \underline{d}_0 \geq \min\{\underline{S}_i + d_i - \overline{S}_0 + (1 - x_i^S)(u_0^d + 1 - d_i) \mid 1 \leq i \leq n\} \cup \{u_0^d + 1\} \\ \wedge & \overline{d}_0 \leq \max\{\overline{S}_i + d_i - \underline{S}_0 + (1 - x_i^S)(l_0^d - 1 - d_i) \mid 1 \leq i \leq n\} \cup \{l_0^d - 1\} \\ \wedge & x_0^S \geq \sum_{i=1}^n x_i^S , \end{cases}$$

The interval S_0 is constrained to be lie around the S_i that are present. The duration interval d_0 is constrained to be large enough to reach the minimal end time of tasks that is present, and small enough not to reach beyond the last possible end time of a task which is present. Note the last element in each line ensures that none of the upper or lower bound variables is every bound too strongly to remove the sentinel value.

The *alternative* constraint can be modelled similarly. It propagates more strongly if it is modelled directly rather than making use of *span*. The model is:

$alternative((S_0, d_0), [(S_1, d_1), \ldots, (S_n, d_n)])$
$$\equiv \begin{cases} & \underline{S}_0 \geq \min\{\underline{S}_i + (1 - x_i^S)(u_0^S - u_i^S) \mid 1 \leq i \leq n\} \cup \{u_0^S + 1\} \\ \wedge & \overline{S}_0 \leq \max\{\overline{S}_i + (1 - x_i^S)(l_0^S - l_i^S) \mid 1 \leq i \leq n\} \cup \{l_0^S - 1\} \\ \wedge & \underline{d}_0 \geq \min\{d_i + (1 - x_i^S)(u_0^d + 1 - d_i) \mid 1 \leq i \leq n\} \cup \{u_0^d + 1\} \\ \wedge & \overline{d}_0 \leq \max\{d_i + (1 - x_i^S)(l_0^d - 1 - d_i) \mid 1 \leq i \leq n\} \cup \{l_0^d - 1\} \\ \wedge & x_0^S = \sum_{i=1}^n x_i^S , \end{cases}$$

The duration d_0 is easier to model since it must be one of the durations of the alternatives. The last constraint enforces that exactly one optional task is actually present if the task 0 is present.

3.3 Presence Implications

Laborie and Rogerie [12] illustrate how reasoning about the presence of optional tasks can substantially improve propagation. The key knowledge is, given two tasks, i and j, does the presence of i imply the presence of j, i.e., $x_i^S \to x_j^S$.

Such knowledge allows one to perform propagation on i using the information of j even when the presence of both tasks is still unknown. This relationship might initially be available in the modelling stage or might dynamically become available during the solving stage.

Define $impl(i, j)$ as the representation of $x_i^S \rightarrow x_j^S$ we shall use in explanation. For models where there is no information about relative presence we just use $impl(i, j) = x_j^S$. If presence implications can be statically determined from the model we can define the representation statically, hence $impl(i, j) = true$ if task i is present then so must be j, and x_j^S otherwise. We also add the constraint $x_i^S \rightarrow x_j^S$ to enforce the presence relationship.

For models where the relative execution information is dynamically determined we introduce new Boolean variables $I_{i,j}$ to represent the information and let $impl(i, j) = I_{i,j}$. We also add a transitivity constraint $transitive(I, [x_1^S, \ldots, x_n^S])$ which ensures that $I_{i,j} \wedge I_{j,k} \rightarrow I_{i,k}$ and $I_{i,j} \leftrightarrow (\neg x_i^S \vee x_j^S)$. In practice the Boolean variables $I_{i,j}$ can be created as required during the execution, they do not all need to be created initially. Our use of *transitive* corresponds to the logical network of [12].

Example 2. Suppose we have a model with tasks i, j, and k and variable *sum* where we know that $x_i^S \rightarrow x_j^S$, and if $sum \geq 0$ then $x_i^S \rightarrow x_k^S$, but nothing else about presence implications. For this model we have that $impl(i, j) = true$, $impl(i, k) = I_{i,k}$ where $sum \geq 0 \rightarrow I_{i,k}$ and $I_{i,k} \leftrightarrow (\neg x_i^S \vee x_k^S)$. Since we can never determine any presence implications between j and k, $impl(j, k) = x_k^S$, and similarly $impl(k, j) = x_j^S$, $impl(k, i) = impl(j, i) = x_i^S$.   □

4   Explanations for Propagation with Optional Tasks

Propagation with optional tasks requires the generation of explanations for use in a CP solver with nogood learning. Here, we present explanations for pruning on lower bounds of the start time variables making use of generalised precedences, detectable precedences, and time-table, and energetic reasoning propagation. Pruning on corresponding upper bounds is symmetric and thus omitted. These explanations are extensions of the explanation presented in [24,23] and the same generalisation steps apply for optional tasks for creating a strongest explanation as possible. However, we omit consideration of generalisation here, since it works equivalently to the non-optional tasks case.

For the remainder of this paper, we only consider optional tasks. A non-optional task with a start time variable S and duration d can be represented as an optional task with start time $\underline{S} = \overline{S} = S$ and $x^S = true$ and duration d. While we only consider fixed durations, the explanations can all be extended to use variable durations by replacing d with $lb(d)$ and adding literals $[\![lb(d) \leq d]\!]$ to explanations.

We assume a given domain D, for which we are defining explanations. We lift the definitions of lst_i and ect_i to optional tasks.

$$lst_i := ub(\overline{S}_i) \quad ect_i := lb(\underline{S}_i) + d_i \quad lct_i := ub(\overline{S}_i) + d_i \quad est_i := lb(\underline{S}_i)$$

If $lst_i < ect_i$ then we say the task i has a *compulsory part* $[lst_i, ect_i)$.

Generalised Precedences. Given the constraint $S_j + v \leq S_i$ where S_i and S_j are $int_\perp$ variables and v is an integer, then we can propagate on the lower bound of S_i if $impl(i,j)$ is currently known to be *true*. The lower bound is $est_j + v$. In order to prevent the wipe out of all values in $\underline{S}_i$ if the new bound is greater than $u_i^S + 1$ we reduce it to this. Consequently, only an update to $\min(est_j + v, u_i^S + 1)$ is permissible. The corresponding explanation is

$$impl(i,j) \wedge [\![est_j \leq \underline{S}_j]\!] \rightarrow [\![\min(est_j + v, u_i^S + 1) \leq \underline{S}_i]\!]$$

Note that the explanation holds regardless of whether i or j executes.

We can extend this reasoning to half-reified [6] precedences of the form $b \rightarrow S_j + v \leq S_i$ by simply adding b to the left hand of the explanation.

Example 3. Suppose that $S_k + 3 \leq S_i$ for the tasks described in Example 2. Suppose $I_{i,k}$ is currently true, and $D(\underline{S}_i) = [2..5]$ and $D(\underline{S}_k) = [6..10]$. The we propagate $[\![9 \leq \underline{S}_i]\!]$ assuming $u_i^S \geq 8$ with an explanation $I_{i,k} + [\![3 \leq \underline{S}_k]\!] \rightarrow [\![9 \leq \underline{S}_i]\!]$. Suppose instead that $u_i^S = 7$, then we propagate with explanation $I_{i,k} + [\![3 \leq \underline{S}_k]\!] \rightarrow [\![8 \leq \underline{S}_i]\!]$ which will cause $x_i^S = false$.   □

Detectable Precedences. Given the constraint $\mathtt{disjunctive}([S_1, \ldots, S_n], [d_1, \ldots, d_n])$ over n tasks with start time $int_\perp$ variables S_i and fixed duration d_i, $1 \leq i \leq n$. Then two tasks i, j can not be run concurrently if $lst_j < ect_i$ and we can conclude that j must finish before i ($j \ll i$) if they are both present. If we detect that currently $lst_j < ect_i$ holds and also $impl(i,j)$ then we can propagate as in the case above. The new bound is $\min(ect_j, u_i^S + 1)$ with explanation:

$$impl(i,j) \wedge [\![t + 1 - d_i \leq \underline{S}_i]\!] \wedge [\![\overline{S}_j \leq t]\!] \rightarrow [\![\min(ect_j, u_i^S + 1) \leq \underline{S}_i]\!]$$

where t can be any integer in $[lst_j, ect_i)$.

Time-Table Propagation. Given n tasks which are competing for a resource with capacity R. Then $\mathtt{cumulative}([S_1, \ldots, S_n], [d_1, \ldots, d_n], [r_1, \ldots, r_n], R)$ must hold. Let i be a task for which we want to propagate the lower bound and Ω be subset of tasks $\{j \mid 1 \leq j \neq i \leq n\}$ which are known to be present if i is present, i.e., $impl(i,j), j \in \Omega$ are known to be true currently. If the tasks $j \in \Omega$ create a compulsory part overlapping the interval $[begin, end)$, i.e., $lst_j \leq begin$ and $end \leq ect_j$, and it holds that $begin < ect_i$ and $r_i + \sum_{j \in \Omega} r_i > R$ then the lower bound of S_i can be updated to $\min(end, u_i^S + 1)$. If $ect_i < end$ then LCG solvers break down the propagation in several steps, so that $ect_i \geq end$ holds for the interval considered (see [24] for details). Then, the point-wise explanation [24] is

$$[\![end - d_i \leq \underline{S}_j]\!] \wedge \bigwedge_{j \in \Omega} impl(i,j) \wedge [\![end - d_j \leq \underline{S}_j]\!] \wedge [\![\overline{S}_j \leq end - 1]\!]$$

$$\rightarrow [\![\min(end, u_i^S + 1) \leq \underline{S}_i]\!]$$

Explaining conditional task overload requires a set of tasks $\Omega \subseteq \{1, \ldots, n\}$ that are all either present together or none present, that is all of $impl(i,j)$ currently hold for $\{i,j\} \in \Omega$, and all have a compulsory part overlapping $[begin, end)$ where $\sum_{i \in \Omega} r_i > R$. Then none of the tasks in Ω can be present, which can be explained as:

$$\bigwedge_{\{i,j\} \in \Omega} impl(i,j) \wedge \bigwedge_{j \in \Omega} [\![t - d_j \leq \underline{S_j}]\!] \wedge [\![\overline{S}_j \leq t - 1]\!] \rightarrow \bigwedge_{j \in \Omega} \neg x_j^S \ ,$$

where t can be any value in $[begin, end)$. Note that this explanation creates $|\Omega|$ clauses due to the conjunction on the right hand side.

Energetic Reasoning Propagation. Given n tasks which are competing for a resource with capacity R. Then $\texttt{cumulative}([S_1, \ldots, S_n], [d_1, \ldots, d_n], [r_1, \ldots, r_n], R)$ must hold. Let i be a task for which we want to propagate the lower bound and Ω be subset of tasks $\{j \mid 1 \leq j \neq i \leq n\}$ which are known to be present if i is present, *i.e.*, $impl(i,j), j \in \Omega$ are known to be true currently. If the tasks $j \in \Omega$ are partially processed in the interval $[begin, end)$, *i.e.*, $begin < ect_j$ and $lst_j < end$ for $j \in \Omega$, then the lower bound of S_i can be updated to $min(begin + \lceil rest/r_i \rceil, u_i^S + 1)$ if $begin < ect_i$, $rest > 0$, and $min(d_i, end - begin) + \sum_{j \in \Omega} r_j \cdot p_j(begin, end) > R \cdot (end - begin)$ where

$$rest = \sum_{j \in \Omega} r_j \cdot p_j(begin, end) - (R - d_i) \cdot (end - begin) \qquad \text{and}$$

$$p_j(begin, end) = \max(0, \min(ect_j - begin, end - lst_j, end - begin)) \quad j \in \Omega \ .$$

Thus, the explanation is as follows with $t = min(begin + \lceil rest/r_i \rceil, u_i)$.

$$[\![begin - d_i < \underline{S_i}]\!] \wedge \bigwedge_{j \in \Omega} impl(i,j) \wedge [\![\overline{S}_j \leq end - p_j(begin, end)]\!] \wedge$$

$$\bigwedge_{j \in \Omega} [\![begin + p_j(begin, end) - d_j \leq \underline{S_j}]\!] \rightarrow [\![t \leq \underline{S_i}]\!]$$

Note that t might not be the largest lower bound for this update, but just as for time-table propagation, for LCG solvers using energetic reasoning it is preferable to perform a step-wise update (see [25] for details). Moreover, since energetic reasoning generalises (extended) edge-finding and time-tabling edge-finding propagation, the explanation presented covers these cases too.

5   Experiments on Flexible Job Shop Scheduling

Experiments were carried out on challenging flexible job-shop scheduling problems (FJSP) [5] where we seek a minimal makespan. FJSP consists of a set of jobs J to be executed on a set of machines M. Each job $j \in J$ is made up of a sequence of tasks $T_{j1}, \ldots T_{jn_j}$, and the tasks can be executed on different machines which may cause them to have different duration. Executing a task T_{jk} on machine $m \in M$ requires d_{jkm} time. The aim is to complete all the tasks in the minimum amount of time.

5.1   Model

For FJSP instance, we model each task T_{jk} using a integer start time variable S_{jk} and duration variable d_{jk} (if the processing time of the task differs on different machines), as well as $\text{int}_\perp$ start time variables S_{jkm} and fixed durations d_{jkm} for the optional task of execution task T_{jk} on machine m. The constraints of the model are

$$\bigwedge_{m \in M} \cdot \texttt{disjunctive}([S_{jkm} \mid j \in J, k \in [1..n_j]], [d_{jkm} \mid j \in J, k \subset [1..n_j]]) \wedge$$
$$\bigwedge_{j \in J, k \in [1..n_j]} \cdot alternative(S_{jk}, d_{jk}, [S_{jkm} \mid m \in M], [d_{jkm} \mid m \in M]) \wedge$$
$$\bigwedge_{j \in J, k \in [1..n_j - 1]} \cdot S_{jkm} + d_{jkm} \leq S_{jk+1m} \wedge$$
$$\bigwedge_{j \in J, k \in [1..n_j], m \in M} intbot(S_{jkm})$$

We can add a redundant $\texttt{cumulative}$ constraint to improve propagation

$$\texttt{cumulative}([S_{jk} \mid j \in J, k \in [1..n_j]], [d_{jk} \mid j \in J, k \in [1..n_j]],$$
$$[1 \mid j \in J, k \in [1..n_j]], |M|) \ .$$

In this model there are no presence implications and $impl(a_i, a_j) = x_j^a$ and similarly for b.

Example 4. Consider a FJSP problem with 2 machines (a, b) and 5 jobs each made up of a single task where the durations (d_a, d_b) of each task if it is executed on machine a,b respectively are given by $(12,9), (5,11), (6,7), (9,6), (7,8)$. We aim to schedule the tasks on the two machines with no two tasks on the same machine overlapping within a makespan of at most 22. This is modelled with 5 (non-optional) tasks with start times S_1, S_2, S_3, S_4, S_5 and (variable) durations $d_1 \in [9..12], d_2 \in 5..11, d_3 \in 6..7, d_4 \in 6..9, d_5 \in 7..8$. And 5 optional tasks with time-intervals a_1, a_2, a_3, a_4, a_5 and fixed durations $da = [12, 5, 6, 9, 7]$ representing that task i runs on machine a. And 5 optional tasks with time-intervals b_1, b_2, b_3, b_4, b_5 with fixed durations $db = [9, 11, 7, 6, 8]$ representing that task i runs on machine b. This constraints of the model are:

$$\texttt{disjunctive}([a_1, a_2, a_3, a_4, a_5], [12, 5, 6, 9, 6])$$
$$\wedge \quad \texttt{disjunctive}([b_1, b_2, b_3, b_4, b_5], [9, 11, 7, 6, 8])$$
$$\wedge \quad \texttt{cumulative}([S_1, S_2, S_3, S_4, S_5], [d_1, d_2, d_3, d_4, d_5], [1, 1, 1, 1, 1], 2)$$
$$\wedge \quad \bigwedge_{i=1}^{5} alternative(S_i, d_i, [a_i, b_i], [da_i, db_i])$$
$$\wedge \quad \bigwedge_{i=1}^{5} intbot(a_i) \quad \wedge \quad \bigwedge_{i=1}^{5} intbot(b_i) \quad \wedge \quad \bigwedge_{i=1}^{5} S_i + d_i \leq 22$$

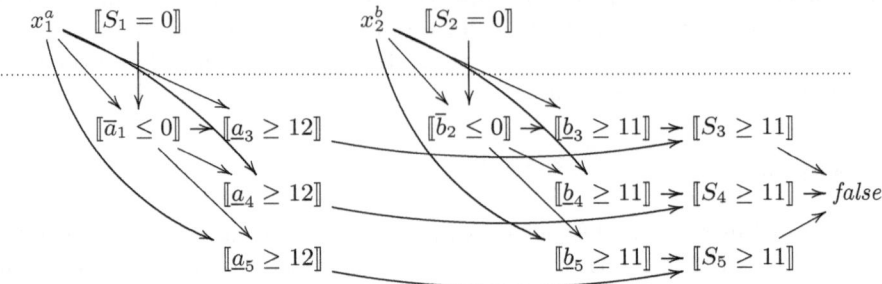

Fig. 1. Implication graph for the search of Example 4. Literals above the dotted lines are decisions.

The first `disjunctive` ensures that no tasks that run on machine a overlap, while the second ensures the same for machine b. The `cumulative` is a redundant constraint that ensures that at most two tasks run at any time. The *alternative* constraints model the relationship between each (non-optional) task and its two alternatives running on machines a and b. Finally the *intbot* constraints ensure that the interval variables are accurately modelled by triples.

Suppose search first schedules task 1 on machine a, setting $x_1^a = true$ and $S_1 = 0$. This forces $\underline{a}_1 = \overline{a}_1 = 0$. The first `disjunctive` constraint then imposes that $\underline{a}_2 \geq 12$, $\underline{a}_3 \geq 12$, $\underline{a}_4 \geq 12$, $\underline{a}_5 \geq 12$. Suppose search next schedules task 2 on machine b, setting $x_2^b = true$ and $S_2 = 0$. This forces $\underline{b}_2 = \overline{b}_2 = 0$. The second `disjunctive` constraint then imposes that $\underline{b}_3 \geq 11$, $\underline{b}_3 \geq 11$, $\underline{b}_3 \geq 11$. The *alternative* constraints enforce that $S_3 \geq 11$, $S_4 \geq 11$ and $S_5 \geq 11$. The `cumulative` constraint discovers that task 3 has a compulsory part in $[16, 17)$, task 4 has a compulsory part in $[16, 17)$ and task 5 has a compulsory part in $[15, 18)$. This leads to a resource overload at time 16 and failure is detected. The (relevant part) of the implication graph is shown in Figure 1. The 1UIP nogood is: $[\![\underline{a}_3 \geq 12]\!] \wedge [\![\underline{a}_4 \geq 12]\!] \wedge [\![\underline{a}_5 \geq 12]\!] \wedge x_2^b \wedge [\![S_2 = 0]\!] \rightarrow false$. Note how the interval variables play an important role in propagation and in the final nogood.   □

5.2   Experiments

The experiments were run on a X86-64 architecture running GNU/Linux and an Intel(R) Core(TM) i7 CPU processor at 2.8 GHz. The code was written using the G12 Constraint Programming Platform [27]. The model was written in MiniZinc [18] and executed by `mzn-g12lazy`, the LCG solver described in [7]. The `disjunctive` propagator in the LCG solver performs the time-table and edge-finding consistency check before filtering the bounds on the start times via edge-finding (denoted disjEF). We also ran the experiments with filtering via detectable precedences and edge-finding, but the results were very similar to disjEF. Thus, we present only the results of disjEF. We compare our results with the current best known lower and upper bounds of the makespan.

Table 1. Overview of the benchmark suites used

suite	sub-suite	#inst	#mach	#jobs	#task	#o-task
CB		21	11–18	10–15	100–225	110–270
Bᴍ [4]		10	4–15	10–20	55–240	115–716
Hᴜ [9]	edata	43	5–10	6–20	36–225	42–341
	rdata	43	5–10	6–20	36–225	74–592

We used different benchmark suites for which a brief overview is given in Table 1 where #inst is the number of instances considered, #mach the range of the number of machines, #jobs the range of the number of jobs, #task the range of the number of task per job, and #o-task the range of the total number of optional tasks.

5.3 Upper Bounds Computations

Upper bound computations approach the optimal solution by generating feasible solutions, potentially sub-optimal, and then restricting the objective correspondingly before continuing the search. Many methods (see, *e.g.*, [15,20,32]) have been proposed for finding feasible solutions. Most of them are incomplete, *i.e.*, they have no guarantee for finding the optimal solution and proving its optimality, but they are fast.

We use branch and bound for minimising the makespan and an activity-based search (an adaption of Vsɪᴅs [17]) with restart. A geometric restart policy [31] on the number of node failures was used with a factor of 2.0 and a base of 256. The upper bound on the makespan was initialised to the rounded up value of the average makespan computed by [15], because [15] provides a method that quickly finds high quality solutions. For each instance, a second run was executed where the initial makespan was made looser by 5%. These are indicated by UB^0+5%.

Tables 2–4 are organised as follow: the column Inst provides the instance names; the column LB-UB the best known lower and upper bound with respect to [10,15,20,3,32]; the column Initial Sol presents the rounded up average UB obtained by [15] over several runs and its average run time in seconds;[1] the column disjEF shows the best obtained UB and the run time in seconds in which a bold UB indicates that disjEF could improve the best known bound or closed the instance, and an asterisk after UB indicates that the disjEF was able to find the optimal solution and prove it. An entry n/a in UB indicates that the LCG solver was not able to find a solution with the given initial UB within the run time limit of 10 minutes. An entry t/o in time indicates that run was aborted after hitting the run time limit.

---

[1] The numbers were taken from the appendix of [15] provided at
http://www.idsia.ch/~monaldo/fjsp.html

Table 2. Results on CB with initial UB from [15]

Inst	#o-task	LB-UB	Initial Sol UB	time	disjEF UB	time	disjEF (UB⁰+5%) UB	time
mt10c1	110	655-927	928	2.33s	**927***	4.47s	**927***	7.02s
mt10cc	120	655-908	910	10.04s	**908***	3.66s	**908***	4.45s
mt10x	110	655-918	918	4.31s	**918***	2.45s	**918***	4.59s
mt10xx	120	655-918	918	1.73s	**918***	2.21s	**918***	6.86s
mt10xxx	130	655-918	918	1.10s	**918***	2.87s	**918***	7.16s
mt10xy	120	655-905	906	4.02s	**905***	4.41s	**905***	5.66s
mt10xyz	130	655-847	851	5.50s	**847***	2.98s	**847***	16.33s
setb4c9	165	857-914	920	14.02s	**914***	12.45s	**914***	20.26s
setb4cc	180	857-907	912	12.95s	**907***	8.60s	**907***	7.08s
setb4x	165	846-925	925	7.45s	**925***	12.86s	**925***	29.72s
setb4xx	180	846-925	927	14.87s	**925***	14.31s	**925***	22.15s
setb4xxx	195	846-925	925	7.99s	**925***	15.02s	**925***	23.13s
setb4xy	180	845-910	916	3.15s	**910***	8.40s	**910***	14.72s
setb4xyz	195	838-903	909	7.35s	**902***	6.77s	**902***	16.52s
seti5c12	240	1027-1171	1175	19.49s	**1169***	54.68s	**1169***	209.77s
seti5cc	255	955-1136	1137	11.91s	**1135***	95.27s	**1135***	175.97s
seti5x	240	955-1198	1204	15.85s	**1198***	28.34s	**1198***	35.1s
seti5xx	255	955-1197	1201	23.64s	**1194***	12.43s	**1194***	14.84s
seti5xxx	270	955-1197	1199	23.51s	**1194***	8.84s	**1194***	28.39s
seti5xy	255	955-1136	1137	11.91s	**1135***	95.20s	**1135***	175.95s
seti5xyz	270	955-1125	1127	17.13s	**1125***	337.98s	**1125***	486.92s

Table 3. Results on BM with initial UB from [15]

Inst	#o-task	LB-UB	Initial Sol UB	time	disjEF UB	time	disjEF (UB⁰+5%) UB	time
Mk01	115	40	40	0.01s	40*	0.25s	40*	0.25s
Mk02	238	24-26	26	0.73s	n/a	t/o	n/a	t/o
Mk03	451	204	204	0.01s	204*	2.10s	204*	7.99s
Mk04	172	48-60	60	0.08s	**60***	0.45s	**60***	0.57s
Mk05	181	168-172	173	0.96s	173	t/o	175	t/o
Mk06	490	33-57	59	3.26s	59	t/o	60	t/o
Mk07	283	133-139	147	8.91s	n/a	t/o	n/a	t/o
Mk08	322	523	523	0.02s	523*	4.95s	523*	5.7s
Mk09	606	307	307	0.15s	307*	9.69s	307*	143.75s
Mk10	716	165-196	200	7.69s	n/a	t/o	n/a	t/o

Our method performed exceptionally well on instances from CB (see Table 2), all instances could be solved within the time limit given, even with the looser initial UB. In contrast, the instances from BM were harder to solve for our method (see Table 3), although the instance Mk04 was closed. Only five instances could be solved and a solution could be found for only one of the remaining five instances.

Table 4. Results on HU with initial UB from [15]

Inst	#o-task	LB-UB	Initial Sol		disjEF		disjEF (UB⁰+5%)	
			UB	time	UB	time	UB	time
edata/la11	113	1087-1103	1103	1.91s	**1103***	0.45s	**1103***	5.92s
edata/la21	173	895-1009	1024	2.83s	1013	t/o	1014	t/o
edata/la22	173	832-880	883	4.29s	**880***	6.21s	**880***	12.28s
edata/la23	171	950	950	2.97s	950*	17.26s	950*	230.93s
edata/la24	174	881-908	912	3.88s	**908***	80.21s	**908***	83.92s
edata/la25	174	894-936	945	1.76s	**936***	21.89s	**936***	33.98s
edata/la26	227	1089-1107	1127	5.48s	1127	t/o	1147	t/o
edata/la27	227	1181	1189	9.25s	1189	t/o	1221	t/o
edata/la28	226	1116-1142	1149	3.44s	1144	t/o	1163	t/o
edata/la29	227	1058-1111	1121	5.47s	1121	t/o	1133	t/o
edata/la30	227	1147-1195	1214	9.22s	1208	t/o	1227	t/o
edata/la31	341	1523-1533	1541	9.58s	1538	t/o	1583	t/o
edata/la32	341	1698	1698	1.85s	1698*	101.44s	1762	t/o
edata/la33	339	1547	1547	1.40s	1547*	27.53s	1564	t/o
edata/la34	339	1592-1599	1600	9.35s	**1599***	52.03s	1620	t/o
edata/la35	339	1736	1736	0.41s	1736*	3.37s	1736*	105.45s
edata/la36	258	1006-1160	1164	8.08s	**1160***	26.01s	**1160***	88.71s
edata/la37	258	1397	1397	3.48s	1397*	1.59s	1397*	6.41s
edata/la38	257	1019-1143	1147	6.90s	**1141***	436.15s	**1141***	485.95s
edata/la39	257	1151-1184	1186	8.68s	**1184***	13.12s	**1184***	26.96s
edata/la40	258	1034-1144	1152	7.78s	**1144***	473.22s	**1144***	505.94s
rdata/la02	94	529-530	531	1.31s	**529***	431.20s	**529***	464.95s
rdata/la19	196	647-700	702	1.90s	**700***	1.35s	**700***	4.42s
rdata/la21	301	808-833	841	7.81s	n/a	t/o	869	t/o
rdata/la22	306	737-758	764	5.14s	764	t/o	774	t/o
rdata/la23	306	816-832	846	6.50s	n/a	t/o	877	t/o
rdata/la24	297	775-801	814	4.06s	n/a	t/o	830	t/o
rdata/la25	302	752-785	795	3.38s	793	t/o	801	t/o
rdata/la26	391	1056-1061	1064	7.69s	n/a	t/o	1114	t/o
rdata/la27	392	1085-1090	1093	7.47s	n/a	t/o	1145	t/o
rdata/la28	402	1075-1080	1082	7.54s	n/a	t/o	1133	t/o
rdata/la29	399	993-997	999	4.03s	n/a	t/o	1029	t/o
rdata/la30	392	1068-1078	1081	7.78s	n/a	t/o	1126	t/o
rdata/la31	576	1520-1521	1522	8.61s	n/a	t/o	1578	t/o
rdata/la32	585	1657-1659	1660	12.67s	n/a	t/o	1735	t/o
rdata/la33	581	1497-1498	1500	11.48s	n/a	t/o	n/a	t/o
rdata/la34	584	1535-1536	1537	7.28s	n/a	t/o	1612	t/o
rdata/la35	592	1549-1550	1551	15.28s	n/a	t/o	n/a	t/o
rdata/la36	439	1016-1028	1032	4.90s	**1023***	38.98s	**1023***	73.6s
rdata/la37	437	989-1066	1081	9.52s	1077	t/o	1091	t/o
rdata/la38	444	943-960	968	9.32s	**954***	44.19s	**954***	168.92s
rdata/la39	436	966-1018	1034	2.78s	**1011***	539.80s	**1016**	t/o
rdata/la40	441	955-956	974	6.11s	968	t/o	978	t/o

Table 4 show the result on a subset of instances from HU. Due to space limits, we omit the instances mt06, mt10, mt20, and la01–la20 from the sub-suites edata and rdata since they are easily solvable. We also omit the entire sub-suite vdata, because no new results could be obtained. The LCG solver disjEF solves all instances of the sub-suite edata except 7 and closes 9 of them. For the sub-suite rdata, the LCG solver closes 5 instances, but for 13 instances it could not find a solution within the time limit. If disjEF was started with the looser initial UB then it finds a solution for each instance in edata and rdata except two and could close only two instances less than before.

Overall we close 36 open instances and improve the best known upper bounds of 11 instances. Our approach is strongest on examples without too many tasks, we plan to investigate the combination with large neighbourhood search (as in [20]) to improve results on larger problems.

6   Conclusion and Outlook

Scheduling with optional tasks generalises the case of scheduling with tasks that must always execute. It provides considerable expressiveness for defining complex scheduling problems. In this paper we show how to extend LCG solvers to support scheduling with optional tasks. The resulting system combines the advantages of scheduling with optional tasks with learning. We demonstrate the power of the combination on hard flexible job-shop scheduling problems. In the future we plan to extend our implementation for optional tasks to more propagators, and probably to implement a native $int_\perp$ variable in our LCG solvers (although we do not expect the native implementation to be much more efficient than the tripartite model we use here).

Acknowledgements. NICTA is funded by the Australian Government as represented by the Department of Broadband, Communications and the Digital Economy and the Australian Research Council through the ICT Centre of Excellence program. This work was partially supported by Asian Office of Aerospace Research and Development (AOARD) grant FA2386-12-1-4056.

References

1. Barták, R., Čepek, O.: Temporal networks with alternatives: Complexity and model. In: Proceedings of the Twentieth International Florida AI Research Society Conference (FLAIRS), pp. 641–646 (2007)
2. Beck, J.C., Fox, M.S.: Scheduling alternative activities. In: Proceedings of the National Conference on Artificial Intelligence, pp. 680–687. John Wiley & Sons, Ltd. (1999)
3. Behnke, D., Geiger, M.J.: Test instances for the flexible job shop scheduling problem with work centers. Research Report RR-12-01-01, Helmut-Schmidt-Universität, Hamburg, Germany (2012)
4. Brandimarte, P.: Routing and scheduling in a flexible job shop by tabu search. Annals of Operations Research 41(3), 157–183 (1993)

5. Brucker, P., Schlie, R.: Job-shop scheduling with multi-purpose machines. Computing 45(4), 369–375 (1990)
6. Feydy, T., Somogyi, Z., Stuckey, P.J.: Half reification and flattening. In: Lee, J. (ed.) CP 2011. LNCS, vol. 6876, pp. 286–301. Springer, Heidelberg (2011)
7. Feydy, T., Stuckey, P.J.: Lazy clause generation reengineered. In: Gent [8], pp. 352–366
8. Gent, I.P. (ed.): CP 2009. LNCS, vol. 5732. Springer, Heidelberg (2009)
9. Hurink, J., Jurisch, B., Thole, M.: Tabu search for the job-shop scheduling problem with multi-purpose machines. Operations-Research-Spektrum 15(4), 205–215 (1994)
10. Jurisch, B.: Scheduling jobs in shops with multi-purpose machines. Ph.D. thesis, Universität Osnabrück (1992)
11. Laborie, P.: IBM ILOG CP Optimizer for detailed scheduling illustrated on three problems. In: van Hoeve, W.-J., Hooker, J.N. (eds.) CPAIOR 2009. LNCS, vol. 5547, pp. 148–162. Springer, Heidelberg (2009)
12. Laborie, P., Rogerie, J.: Reasoning with conditional time-intervals. In: Wilson, D.C., Lane, H.C. (eds.) Proceedings of the Twenty-First International Florida Artificial Intelligence Research Society Conference, pp. 555–560. AAAI Press (2008)
13. Laborie, P., Rogerie, J., Shaw, P., Vilím, P.: Reasoning with conditional time-intervals part II: An algebraic model for resources. In: Lane, H.C., Guesgen, H.W. (eds.) Proceedings of the Twenty-First International Florida Artificial Intelligence Research Society Conference, pp. 201–206. AAAI Press (2009)
14. Laborie, P., Rogerie, J., Shaw, P., Vilím, P., Katai, F.: Interval-based language for modeling scheduling problems: An extension to constraint programming. In: Kallrath, J. (ed.) Algebraic Modeling Systems. Applied Optimization, vol. 104, pp. 111–143. Springer, Heidelberg (2012)
15. Mastrolilli, M., Gambardella, L.M.: Effective neighbourhood functions for the flexible job shop problem. Journal of Scheduling 3(1), 3–20 (2000)
16. Moffitt, M.D., Peintner, B., Pollack, M.E.: Augmenting disjunctive temporal problems with finite-domain constraints. In: Proceedings of the National Conference on Artificial Intelligence, pp. 1187–1192. AAAI Press (2005)
17. Moskewicz, M.W., Madigan, C.F., Zhao, Y., Zhang, L., Malik, S.: Chaff: Engineering an efficient SAT solver. In: Proceedings of Design Automation Conference – DAC 2001, pp. 530–535. ACM, New York (2001)
18. Nethercote, N., Stuckey, P.J., Becket, R., Brand, S., Duck, G.J., Tack, G.R.: MiniZinc: Towards a standard CP modelling language. In: Bessière, C. (ed.) CP 2007. LNCS, vol. 4741, pp. 529–543. Springer, Heidelberg (2007)
19. Ohrimenko, O., Stuckey, P.J., Codish, M.: Propagation via lazy clause generation. Constraints 14(3), 357–391 (2009)
20. Pacino, D., Van Hentenryck, P.: Large neighborhood search and adaptive randomized decompositions for flexible jobshop scheduling. In: Proceedings of the Twenty-Second International Joint Conference on Artificial Intelligence, IJCAI 2011, pp. 1997–2002. AAAI Press (2011)
21. Schulte, C., Stuckey, P.J.: Efficient constraint propagation engines. ACM Transactions on Programming Languages and Systems 31(1), Article No. 2 (2008)
22. Schutt, A., Chu, G., Stuckey, P.J., Wallace, M.G.: Maximising the net present value for resource-constrained project scheduling. In: Beldiceanu, N., Jussien, N., Pinson, É. (eds.) CPAIOR 2012. LNCS, vol. 7298, pp. 362–378. Springer, Heidelberg (2012)
23. Schutt, A., Feydy, T., Stuckey, P.J.: Explaining time-table-edge-finding propagation for the cumulative resource constraint. In: Gomes, C., Sellmann, M. (eds.) CPAIOR 2013. LNCS, vol. 7874, pp. 234–250. Springer, Heidelberg (2013)

24. Schutt, A., Feydy, T., Stuckey, P.J., Wallace, M.G.: Explaining the cumulative propagator. Constraints 16(3), 250–282 (2011)
25. Schutt, A., Feydy, T., Stuckey, P.J., Wallace, M.G.: Solving RCPSP/max by lazy clause generation. Journal of Scheduling 16(3), 273–289 (2013)
26. Seipel, D., Hanus, M., Wolf, A. (eds.): INAP 2007. LNCS, vol. 5437. Springer, Heidelberg (2009)
27. Stuckey, P.J., de la Banda, M.G., Maher, M.J., Marriott, K., Slaney, J.K., Somogyi, Z., Wallace, M., Walsh, T.: The G12 project: Mapping solver independent models to efficient solutions. In: Gabbrielli, M., Gupta, G. (eds.) ICLP 2005. LNCS, vol. 3668, pp. 9–13. Springer, Heidelberg (2005)
28. Vilím, P.: Edge finding filtering algorithm for discrete cumulative resources in $\mathcal{O}(kn \log n)$. In: Gent [8], pp. 802–816
29. Vilím, P.: Max energy filtering algorithm for discrete cumulative resources. In: van Hoeve, W.-J., Hooker, J.N. (eds.) CPAIOR 2009. LNCS, vol. 5547, pp. 294–308. Springer, Heidelberg (2009)
30. Vilím, P., Barták, R., Čepek, O.: Extension of $O(n \log n)$ filtering algorithms for the unary resource constraint to optional activities. Constraints 10(4), 403–425 (2005)
31. Walsh, T.: Search in a small world. In: Proceedings of Artificial intelligence – IJCAI 1999, pp. 1172–1177. Morgan Kaufmann (1999)
32. Yuan, Y., Xu, H.: Flexible job shop scheduling using hybrid differential evolution algorithms. Computers & Industrial Engineering 65(2), 246–260 (2013)

Residential Demand Response
under Uncertainty

Paul Scott[1,3], Sylvie Thiébaux[1,3],
Menkes van den Briel[1,3], and Pascal Van Hentenryck[2,3]

[1] Australian National University, ACT, Australia
[2] University of Melbourne, VIC, Australia
[3] NICTA, Australia
{first.lastname}@nicta.com.au

Abstract. This paper considers a residential market with real-time electricity pricing and flexible electricity consumption profiles for customers. Such a market raises an optimisation problem for home automation systems where they need to schedule consumption activities to reduce costs, whilst maintaining a base level of comfort and convenience. This optimisation problem faces uncertainty in real-time prices, weather conditions, and occupant behaviour. The paper presents two online stochastic combinatorial optimisation algorithms that produce fast, high-quality solutions to this problem. These algorithms are compared with reactive control strategies and a clairvoyant controller. Our results demonstrate the value of stochastic information and online stochastic optimisation in residential demand response.

1   Introduction

Electricity consumption in residential markets will undergo fundamental changes in the next decade due to the availability of solar panels and novel pricing mechanisms, progress in batteries and electric cars, and the emergence of smart appliances and home automation. These technologies provide residential customers with the ability to actively participate in smart grid activities such as demand response where loads are shifted to times favourable for the network as a whole.

Having an intelligent Home Automation System (HAS) within each home is a key component in this vision. The HAS receives information about device operating characteristics, usage requests and network signals, and can send control requests back to smart devices. The HAS provides occupants with feedback on their consumption habits and, more importantly, can make control decisions for itself. This control can be used to meet one or more of the following objectives:

1. Improve occupant comfort,
2. Reduce overall electricity consumption,
3. Perform demand response for network.

These objectives are often conflicting, so occupants need to indicate how they value comfort against cost savings in order to get the right balance for them

C. Schulte (Ed.): CP 2013, LNCS 8124, pp. 645–660, 2013.
© Springer-Verlag Berlin Heidelberg 2013

overall. The task of the HAS is to decide on a series of control actions to take over time, which produces an optimal solution to the combined objectives. The HAS can implement simple policies to attempt to meet these conflicting objectives. Or, more interestingly, it can use sophisticated stochastic optimisation technology which exploits forecasts and observed patterns in prices, weather, residential activities and smart device usage.

This paper aims to determine the benefits of online stochastic optimisation for a HAS that is exposed to Real-Time Pricing (RTP) as a demand response mechanism. A number of research projects have started examining this very issue (see the related work section) but they often give an incomplete picture of the benefits of optimisation and the value of stochastic information. These projects often consider simpler uncertainty models, which give a partial understanding of the true benefits that optimisation can bring to this setting. In contrast, this paper makes two primary contributions: one conceptual and one algorithmic.

At the conceptual level, the paper presents a compositional architecture for HAS optimisation, where each device can be modelled independently in terms of a collection of functions that encapsulate its behaviour. These devices are then assembled into a model of a home, from which optimisation problems for the HAS can derive.

At the algorithmic level, the paper presents a comprehensive study of the value of HAS optimisation in the presence of uncertainty about future prices, occupant behaviour and environmental conditions. Our formulation uses models representative of physical devices and stochastic models trained on real weather and network demand data. These device and stochastic models are used in two online stochastic optimisation algorithms which are compared to simple control systems based on reactive policies.

The experimental results not only show the value of stochastic information, but also that stochastic optimisation provides solutions that are close to the clairvoyant solutions which have perfect knowledge of the future. The online stochastic algorithms using MILP technology are fast and produce significantly better solutions than the reactive controllers. Also of interest is the comparison between the two online stochastic algorithms, and an experiment that investigates the optimal rolling horizon duration.

The rest of the paper presents the deterministic HAS optimisation problem, its stochastic version, the stochastic models and finally the experimental results.

2   Deterministic HAS Optimisation

A house contains a collection of controllable devices which influence the amount of power consumed in the house and the level of comfort that residents experience. We consider the operation of these devices over discrete time steps[1]: $\forall i \in \mathbb{Z} : t_i \in \mathbb{R}$ where $t_i > t_{i-1}$ and $\forall i \in \mathbb{Z} : t_i^{stp} = t_i - t_{i-1}$.

Given a real time price for electricity and other input parameters (e.g., external temperatures and device requests), optimal operation of these devices is

---

[1] Variable time step sizes will be used to focus computational time where most needed.

achieved by minimising the sum of monetary and comfort costs. The optimisation decision variables are the device actions at each time step, which are constrained by device characteristics and total power limits on the house.

2.1 Formal Definition

We start with a new formal definition of a device, which is a collection of functions that govern the device operation. These include functions for permissible device actions, state updates, electrical power consumption/generation, and any non-power-related operation costs. The operation costs are always positive and may include any occupant comfort costs, fuel consumption or wear and tear on the equipment. By convention power consumed by the device is negative and power generated (e.g., by a rooftop photovoltaic system) is positive.

Definition 1 (Device). *A device is a tuple* $d = (A_d, S_d, R_d, q_d, g_d, f_d, l_d)$, *where:*

- $A_d \subseteq \mathbb{R}^{m_d} \times \mathbb{Z}^{m'_d}$ *is the set of device actions*
- $S_d \subseteq \mathbb{R}^{k_d} \times \mathbb{Z}^{k'_d}$ *is the set of device states*
- $R_d \subseteq \mathbb{R}^{w_d} \times \mathbb{Z}^{w'_d}$ *is the set of device input parameters*
- $q_d : S_d \times R_d \longrightarrow \mathcal{P}(A_d)$ *is the permissible action function*
- $g_d : A_d \times S_d \times R_d \longrightarrow S_d$ *is the state update function*
- $f_d : A_d \times S_d \longrightarrow \mathbb{R}$ *is the electrical power function*
- $l_d : A_d \times S_d \times R_d \times \mathbb{R} \longrightarrow \mathbb{R}$ *is the operational cost function*

A house is simply a set of devices, together with bounds on the instantaneous amount of power the house can transfer to or from the grid:

Definition 2 (House). *A house is a tuple* $h = (D_h, \underline{p}_h, \bar{p}_h)$, *where:*

- D_h *is the set of devices*
- $\underline{p}_h, \bar{p}_h \in \mathbb{R}$ *are the lower and upper power limits*

We now turn to the deterministic formulation of the HAS optimisation problem which will be later used as a building block for our stochastic formulation. The deterministic formulation assumes that the input parameters are known over a horizon of n time steps. The objective is to choose device actions to reduce the total cost over the horizon, which includes device operational costs and monetary costs from trading power with the network.[2] Inputs include the device initial states, the RTP, the house background power usage[3] and the device input parameters at each time step. The variables at each time step include the device actions and states, and the device and house power consumptions and costs.

We use the following notation: $(a)_+ = |a|$ if $a > 0$ and 0 otherwise, and similarly $(a)_- = |a|$ if $a < 0$ and 0 otherwise, where $a \in \mathbb{R}$.

---

[2] The RTP for net consumption or generation can be different.
[3] This aggregates uncontrollable electrical consumption, e.g., lighting, entertainment and cooking.

Definition 3 (Deterministic HAS Optimisation Problem)
For a house $h = (D_h, \underline{p}_h, \bar{p}_h)$, the HAS optimisation problem over a horizon of $n \in \mathbb{N}^*$ time steps is the following:

Inputs
    for each device $d = (A_d, S_d, R_d, q_d, g_d, f_d, l_d) \in D_h$
    – $s_{d,0} \in S_d$ *is the device initial state*
    for each device $d \in D_h$ and time step $i \in \{1 \ldots n\}$
    – $r_{d,i} \in R_d$ *are the device input parameters*
    for each time step $i \in \{1 \ldots n\}$
    – $p_{h,i}^b \in \mathbb{R}_-$ *is the house background power*
    – $v_i \in \mathbb{R}^2$ *is the real-time price (buying, selling)*

Decision variables
    for each device $d \in D_h$ and time step $i \in \{1 \ldots n\}$
    – $a_{d,i} \in A_d$ *are the device action variables*

Other variables
    for each device $d \in D_h$ and time step $i \in \{1 \ldots n\}$
    – $s_{d,i} \in S_d$ *are the device state variables*
    – $p_{d,i} \in \mathbb{R}$ *is the device power*
    – $c_{d,i} \in \mathbb{R}_+$ *is the device operation cost*
    for each time step $i \in \{1 \ldots n\}$
    – $p_{h,i} \in [\underline{p}_h, \bar{p}_h]$ *is the total power*
    – $c_{h,i} \in \mathbb{R}$ *is the total cost*

Constraints
    for each device $d \in D_h$ and time step $i \in \{1 \ldots n\}$
    – $a_{d,i} \in q_d(s_{d,i-1}, r_{d,i})$ *is the action permissibility constraint*
    – $s_{d,i} = g_d(a_{d,i}, s_{d,i-1}, r_{d,i})$ *is the state update constraint*
    – $p_{d,i} = f_d(a_{d,i}, s_{d,i})$ *is the device power constraint*
    – $c_{d,i} = l_d(a_{d,i}, s_{d,i}, r_{d,i}, t_i^{stp})$ *is the device cost constraint*
    for each time step $i \in \{1 \ldots n\}$
    – $p_{h,i} = \sum_{d \in D_h} p_{d,i} + p_{h,i}^b$ *is the house power constraint*
    – $\underline{p}_h \leq p_{h,i} \leq \bar{p}_h$ *is the house power limits constraint*
    – $c_{h,i} = \sum_{d \in D_h} c_{d,i} + t_i^{stp} v_{i,1}(p_{h,i})_- - t_i^{stp} v_{i,2}(p_{h,i})_+$ *is the house cost constraint*

Objective
    $\min \sum_{i=1}^n c_{h,i}$

2.2 Modelled Devices

In our experiments we consider a modern house with electrical HVAC, hot water heating, solar panels, a washing machine, a clothes dryer and a dish washer. We also include two devices that are expected to become popular within the next decades: an electric vehicle (EV) and a dedicated battery for storing electrical energy. Descriptions of these devices are given in this section. Some liberty has been used in these descriptions to aid understanding, however note that with slight reformulation they all fit into the rigorous device definition of the previous section. Device electrical powers and operational costs are consistently represented by the variables p_i and c_i, and power consumed by a device takes on a negative number.

The physical behaviour of devices has been approximated by linearising their physical equations and discretising time. Only significant steps of this process are mentioned in the device descriptions. For the experiments parameters were selected to be representative of typical devices. For example, the EV battery capacity is equivalent to that of a Nissan Leaf, and the house floor area for heating purposes is typical of an average-sized house. Some parameters were difficult to source so had to be estimated such as the efficiency of the EV battery.

Battery. A battery has a stored energy state $E \in [0, \bar{E}]$ and a charge/discharge power $p \in [\underline{p}, \bar{p}]$ action variable. The battery has a fixed efficiency $\eta \in [0, 1]$. The stored energy state update function is given by:

$$E_i = E_{i-1} + t_i^{stp} \left(\eta(p_i)_- - (p_i)_+ \right) \tag{1}$$

A battery lifetime cost c is associated with power that is discharged from the battery through a lifetime price v: $c_i = v(p_i)_+$.

Electric Vehicle. An electric vehicle (EV) is like the battery above, but with a few additional constraints. Firstly the input parameter $x^h \in \{0, 1\}$ indicates whether the EV is home, and the battery can only be charged/discharged when this is the case:

$$x_i^h = 0 \implies p_i = 0 \tag{2}$$

The input parameter $p^d \in \mathbb{R}_+$ represents the power drawn from the battery whilst driving. This modifies the state update function as follows:

$$E_i = E_{i-1} + t_i^{stp} \left(\eta(p_i)_- - (p_i)_+ - p_i^d \right) \tag{3}$$

The final constraint is on the amount of energy stored in the battery. The house occupants provide an input parameter $E^m \in [0, \bar{E}]$ that represents the minimum energy that the EV battery should have in it at each time. This value represents how much energy the occupant expects to need if they drive away in the car at a given time. This is not a hard constraint as the draw from driving can bring the battery charge below this limit, but it ensures that if the battery power does fall below, then it charges back up as fast as possible.

$$x_i^h = 1 \implies E_i \geq \min \left[E_{i-1} + t_i^{stp} \left(-\eta\underline{p} - p_i^d \right), E_i^m \right] \tag{4}$$

Hot Water Heating. The hot water system is made up of a storage tank and an electric heating element. We ignore the details of the interaction between hot and cold water in the tank and consider the state of the tank as being the amount of energy $E \in [0, \bar{E}]$ it contains above the inlet cold water temperature. The tank is considered empty of hot water when this value is zero. The action variable is the power setting of the electric heater $p \in [\underline{p}, 0]$ at each time step. The input parameter $p^d \in \mathbb{R}_+$ is the amount of power drawn from the tank to meet occupant demand. The energy state update function is given by:

$$E_i = E_{i-1} + t_i^{stp} \left(-p_i - p_i^d - p_i^l + p_i^u\right) \tag{5}$$

The variable $p^l \in \mathbb{R}_+$ represents thermal losses from the tank to the outdoor environment. The rate of loss depends on how full the tank is and the difference in temperature between the water set point $T^s \in \mathbb{R}$ and the outdoor temperature $T^o \in \mathbb{R}$ through a resistivity $R \in \mathbb{R}_+$:

$$p_i^l = \frac{1}{R} \frac{E_i}{\bar{E}} (T^s - T_i^o) \tag{6}$$

The variable $p^u \in \mathbb{R}_+$ is a recourse variable that is used to indicate the amount of hot water demand which goes unmet, i.e. water drawn from the tank when it is empty. This is heavily penalised as a cost c through an unmet demand price v: $c_i = v p_i^u$.

The hot water system has a minimum stored energy level $E^m \in [0, \bar{E}]$, much like the electric vehicle. If drawn water brings the energy level of the tank below this value then the heater must work as hard as possible to bring the energy back up. Occupants can adjust this input parameter to reduce the likelihood of running out of hot water.

$$E_i \geq \min \left[E_{i-1} + t_i^{stp} \left(-\underline{p} - p_i^d - p_i^l + p_i^u\right), E^m\right] \tag{7}$$

Under-Floor Heating/Cooling. The house temperature is controlled by a heat pump that heats/cools water, which is then pumped through piping embedded in the floor of the house. The temperatures of the floor and the air in the room $T^f, T^a \in \mathbb{R}$ are the device state variables. The action variable is the amount of thermal energy that is supplied to the floor of the house $p^t \in \mathbb{R}$. This is limited by the heat pump electrical power consumption $p \in [\underline{p}, 0]$ through heating and cooling Coefficients of Performance (COP) $\eta^h \in [\underline{\eta}^h, \bar{\eta}^h], \eta^c \in [\underline{\eta}^c, \bar{\eta}^c]$:

$$p_i = -\frac{1}{\eta_i^h} (p_i^t)_+ - \frac{1}{\eta_i^c} (p_i^t)_- \tag{8}$$

The COPs depend on the temperatures of the two thermal wells between which the heat pump is operating. We assume that the internal thermal well is at a constant temperature and that the external well is at the outdoor temperature $T^o \in \mathbb{R}$. We approximate the COPs as linear functions of T^o for some constants $a^h, a^c \in \mathbb{R}_+$ and $b^h, b^c \in \mathbb{R}$, with hard upper and lower limits:

$$\eta_i^h = \min \left[\max \left[a^h T_i^o + b^h, \underline{\eta}^h\right], \bar{\eta}^h\right], \quad \eta_i^c = \min \left[\max \left[-a^c T_i^o + b^c, \underline{\eta}^c\right], \bar{\eta}^c\right] \tag{9}$$

Heat can transfer between the floor and the outdoor environment $p^{fo} \in \mathbb{R}$, the floor and the air in the room $p^{fa} \in \mathbb{R}$, and the air in the room and the outdoor environment $p^{ao} \in \mathbb{R}$. We use simple lumped thermal resistivities $R^{fo}, R^{fa}, R^{ao} \in \mathbb{R}_+$ to govern these heat flows:

$$p_i^{fo} = \frac{1}{R^{fo}}(T_i^f - T_i^o), \quad p_i^{fa} = \frac{1}{R^{fa}}(T_i^f - T_i^a), \quad p_i^{ao} = \frac{1}{R^{ao}}(T_i^a - T_i^o) \quad (10)$$

The temperature state update functions are given by:

$$T_i^f = T_{i-1}^f + \frac{t_i^{stp}}{m^f \kappa^f}\left(p_i^t - p_i^{fo} - p_i^{fa} + A^f I_i\right) \quad (11)$$

$$T_i^a = T_{i-1}^a + \frac{t_i^{stp}}{m^a \kappa^a}\left(p_i^{fa} - p_i^{ao} + p_i^g\right) \quad (12)$$

where $m^f, m^a, \kappa^f, \kappa^a \in \mathbb{R}_+$ are the floor and air, mass and specific heat capacity coefficients respectively. Sunlight enters through the windows at an irradiance $I \in \mathbb{R}_+$ and lands on a floor area $A^f \in \mathbb{R}_+$. The input $p^g \in \mathbb{R}_+$ is thermal power generated by occupant metabolisms and background electric appliances which contributes to heating the air in the room.

The final relation we have is for the comfort cost c which depends on the distance of the air temperature from an occupant specified set point temperature $T^s \in \mathbb{R}$. The occupants also specify two time-varying comfort prices v^a, v^b, one of which is only included after a threshold temperature difference ΔT^{th}:

$$c_i = \begin{cases} v_i^a |T_i^a - T_i^s| & \text{if } |T_i^a - T_i^s| < \Delta T^{th} \\ (v_i^a + v_i^b)|T_i^a - T_i^s| & \text{otherwise} \end{cases} \quad (13)$$

Shiftable Loads. Shiftable loads are devices that need to run once within a time window. An occupant sets two input parameters: a start time i^s and a last allowed start time i^l, between which the controller must schedule the device to run. Examples of this kind of device include washing machines, clothes dryers and dish washers. We model non-preemptive shiftable loads which can have time varying power consumptions.

The start of run indicators $x \in \{0, 1\}$ act as both the device action and state variables. A shiftable load has a cumulative energy consumption function $\psi : \mathbb{R}_+ \longrightarrow \mathbb{R}_+$ which takes a run duration and returns the cumulative amount of energy that the device has consumed up to that duration. Constraints on the run indicator variables and the device power $p \in \mathbb{R}_-$ are given by:

$$\sum_{k=i^s}^{i^l} x_k = 1, \quad p_i = -\sum_{k=i^s}^{i} x_k \frac{\psi(t_i - t_{k-1}) - \psi(t_{i-1} - t_{k-1})}{t_i^{stp}} \quad (14)$$

Photovoltaics. The photovoltaic (PV) panels have no action variables, the amount of electricity they generate is purely determined by the solar irradiance input parameter. We model a PV system ignoring temperature and shading effects and by assuming the panels lay on a horizontal surface. The generated electric power $p \in \mathbb{R}_+$ is then a simple function of the panel area $A \in \mathbb{R}_+$, efficiency $\eta \in [0, 1]$ and global irradiance input parameter $I \in \mathbb{R}_+$: $p_i = \eta A I_i$.

3   Stochastic HAS Optimisation

So far we have considered the deterministic HAS formulation that requires perfect foresight about what will happen over the time horizon. In practice, almost all the input parameters are uncertain and their uncertainty is only revealed in real time (e.g., outdoor temperature) or in some cases a few time steps in advance (e.g., RTP). This motivates the use of online stochastic optimisation [1], which exploits statistical models of the uncertain parameters in order to make the best decisions on average.

3.1   The Stochastic Model

In the stochastic HAS problem the RTP v_i, background house power $p_{h,i}^b$ and device input parameters $r_{d,i}$ are random variables. We denote their real-world realisations (i.e. their values when the uncertainty is revealed) with the symbol $*$. For instance, T_i^{o*} denotes the real outdoor temperature at time step i. For notational convenience, all inputs are combined into one vector

$$z_i = (v_i, p_{i,h}^b, r_{d_1,i}, r_{d_2,i}, \ldots)^T \tag{15}$$

where we index elements with a k (e.g., $z_{i,k}$). Random variables at time step i may be dependent on each other and on the variables at previous time steps. Therefore the joint distribution for random variables up to time step i is given by:

$$P(z_i, z_{i-1}, \ldots) \tag{16}$$

Let t^* represent the current real world time. Each input $z_{i,k}$ is revealed a fixed amount of time $\Delta t_k^{rev} \in \mathbb{R}_+$ in advance (or in real time if $\Delta t_k^{rev} = 0$). This means, that for a given t^* an input $z_{i,k}$ is known to be $z_{i,k}^*$ if $t_i \leq t^* + \Delta t_k^{rev}$, otherwise it is a random variable. Given i and t^* we use $K_{i,t^*} = \{k | t_i \leq t^* + \Delta t_k^{rev}\}$ to denote the set of known input vector indices.

3.2   Online Stochastic Optimisation

In online stochastic optimisation decisions are made one step at a time using stochastic information about future events. After each time step the uncertainty and the effect of all actions is revealed, updating the state of the system. Decisions for the next period are computed and the process is repeated. It has been used successfully on a wide variety of problems (e.g., [2,1]).

Our algorithms use a rolling finite horizon as illustrated in Fig. 1, where the time steps $1, \ldots, n$ are aligned to each horizon with $t_0 = t^*$. Optimisation is performed for each horizon using stochastic information for any unrevealed inputs and then the actions for the first time step are executed in the real world.

It might not be possible to execute actions produced by the optimisation if the real world input parameters z_1^* differ from what the optimisation anticipated. For example, if the optimisation decides to run the hot water heater at full power and the tank unexpectedly reaches its capacity (due to less demand for hot water

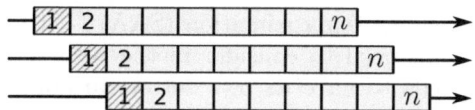

Fig. 1. Rolling horizon for 3 consecutive iterations

than expected), then the power of the heating action will need to be reduced so as to remain within the tank's capacity. Our HAS handles this automatically in the execution step, by using very simple executives for each device which select the closest feasible action.

The following sections introduce two approaches to solving the stochastic optimisation problem within each horizon: the expectation and the 2-stage algorithms.

3.3 Expectation Formulation

The expectation online stochastic algorithm takes the conditional expected value of any unrevealed inputs in the optimisation horizon, and solves the deterministic version of the problem given in Definition 3. We use the term expected value loosely because in truth we calculate the expected value only where it makes sense, which is typically for continuous inputs. For the rest of the inputs the most likely value is calculated instead. For example, expected value is used for outdoor temperatures and most likely value for the washing machine requests. Both of these calculations are performed using the joint distribution for inputs in the horizon, conditioned on any known inputs in and prior to the horizon:

$$P(z_n, z_{n-1}, \dots, z_1 | (z_{n,k}^*, \forall k \in K_{n,t_0}), \dots, (z_{1,k}^*, \forall k \in K_{1,t_0}), z_0^*, \dots) \quad (17)$$

3.4 2-Stage Formulation

In this algorithm 2-stage stochastic programming is used within each horizon. This provides an approximation to a full multi-stage stochastic program which are, in general, known to be extremely challenging computationally [3]. The first stage includes time step 1, and the second stage time steps $2, \dots, n$. Traditionally, in 2-stage stochastic programming there is no uncertainty in the first stage [4]. However, in our problem we are required to make decisions before all inputs in the first stage are revealed. To resolve this, first stage inputs are set to their real values if revealed, otherwise their conditional expected value is taken (as described in Section 3.3).

The second stage uses sampled scenarios to represent the uncertainty in the input parameters. We define a second stage scenario s as being a sample from the joint distribution of random variables in the second stage, conditioned on any revealed inputs in the second stage, and inputs in and prior to the first stage:

$$s \sim P(z_n, z_{n-1}, \dots, z_2 | (z_{n,k}^*, \forall k \in K_{n,t_0}), \dots, (z_{2,k}^*, \forall k \in K_{2,t_0}),$$
$$z_1, z_0^*, \dots) \quad (18)$$

We use the Sample Average Approximation (SAA) [4] to limit the number of scenarios $S \in \mathbb{N}$ that we need to consider in the second stage. Each scenario in the second stage needs to have its own set of variables in the optimisation problem. For example we denote the power of device d at time step i in scenario s by $p_{d,i,s}$. The 2-stage objective function is given by:

$$\min \left[c_1 + \frac{1}{S} \sum_{s \in \{s_1, \ldots, s_S\}} \sum_{i=2}^{n} c_{i,s} \right] \tag{19}$$

3.5 Stochastic Inputs

Stochastic inputs include the real-time pricing (RTP), outdoor temperature, solar irradiance, background power, internal heat generation, hot water demand, EV usage and shiftable load requests. Accurately modelling any of these random processes is a significant undertaking in itself. The models we developed, while not the most sophisticated, suit the purposes of our experiments by capturing the fundamental nature of these stochastic processes. We investigated a number of different model types before settling on Generalised Additive Models (GAM) [5] for the continuous variables like temperature, and Markov Models for the more discrete occupant driven behaviours such as shiftable device requests.

Generalised Additive Models. In order to predict future values, the GAMs take advantage of weather forecasts that can be readily obtained from national weather services. These forecast values include daily maximum and minimum temperatures, as well as morning and afternoon cloud cover and wind speeds. The models also take in the value from the previous time step and temporal information. The models were trained on data obtained from the Bureau of Meteorology[4] and Australian Energy Market Operator[5] relevant to the states of NSW and the ACT in Australia.

The best way of implementing RTP in retail markets is still an open question and so is worth particular mention. It is unlikely that it will be a simple replication of the wholesale spot market price due to its high volatility. More likely it will be set by retailers, but it will have a shape that is representative of the wholesale market. We designed our RTP to be a quadratic function[6] of the amount of power that fossil fuel sources must supply to meet total network load. This is the total network demand minus the generation from renewable sources such as wind and solar. We used a GAM for the total network demand and the generation from renewables is a function of wind speed and solar irradiance. The RTP is only revealed to a house 30 minutes in advance.

Markov Models. Semi-Markov models were used to capture the behaviour of four occupants of a specific house in the ACT. These models provide the consumption patterns and profiles for input parameters such as hot water demand,

---

[4] Bureau of Meteorology (BOM), www.bom.gov.au
[5] Australian Energy Market Operator (AEMO), www.aemo.com.au
[6] The quadratic is representative of an increasing marginal supply price [6].

shiftable load requests and EV usage. Each model identifies the key activities of an occupant (e.g., sleeping, taking a shower and leaving for work), and specifies the probabilities of transiting from one activity to the next within certain time windows. Each activity is associated with a series of actions (e.g., watching TV and requesting the dish washer) that trigger changes in input parameters. Conditional sampling through these models is used to generate scenarios.

Whilst this scheme was convenient for our experiments, other more data-driven options are possible: we could simply gather and use a database of raw scenarios, or learn model parameters from disaggregated demand data [7,8].

4   Experiments

We implemented the expectation and 2-stage online algorithms using Gurobi as a backend to solve the MILP within each horizon. The devices in Section 2.2 were implemented and included in the experimental house, and conditional samplers were created for the uncertain input parameters in Section 3.5. We created a simple simulator that uses the same physical equations as the optimisation to simulate the execution of actions in the real world. We compare the performance of the expectation and 2-stage controllers with naive and smart reactive controllers, and a controller that has perfect information.

The *Naive* reactive controller represents a household that either has no ability or no desire to respond to a RTP. It starts shiftable devices as soon as a request is received, fills up the hot water tank in off-peak hours, charges the EV only if it is below the requested minimum level, maintains the room at the set point temperature and never uses the battery bank.

The *Smart* reactive controller uses simple device action policies to decide how to respond to changes in RTP. It delays running a shiftable device until it reaches either a cheap price or the last available start time; uses thresholds about a moving average of the RTP to decide when to charge or discharge energy from the batteries, EV and hot water system; and maintains the room at the set point temperature like the naive controller.

The *Perfect* controller has perfect foresight about what will happen in the future. It optimises the deterministic problem in Definition 3 over the whole experiment duration with full knowledge of z_i^*. This controller (which is infeasible in practice) is used to give a lower bound on the objective that can be achieved by the other controllers.

4.1   Controller Comparison

A total of 9 sets of input parameters typical for the month of February were generated. These were used in 9 separate experimental runs, each with a duration of 7 days. The online algorithms had 16 hour optimisation horizons, with 15 minutes for the first two time steps and 30 minute time steps for the remainder of the horizon.[7] The reactive and perfect controllers had 15 minute time steps.

---

[7] By using larger time steps for more uncertain values further into the future we reduce the computational burden with only a minor reduction to solution quality.

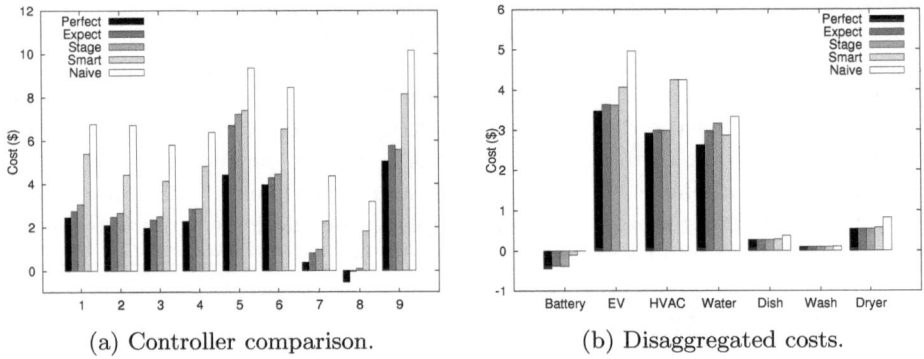

(a) Controller comparison.   (b) Disaggregated costs.

Fig. 2. Costs for each experimental run and device costs averaged over runs

The 2-stage algorithm sampled 60 scenarios for each second stage. The amount of time spent optimising in Gurobi per day was on average 1 second for the expectation algorithm and 4 minutes for the 2-stage algorithm (using a single core of an Intel i7-2600 3.4GHz CPU). Whilst the 2-stage is much slower, its computational time can still be considered small when spread out over a day.

The controller costs are plotted in Fig. 2a for each experimental run. These results are adjusted to account for any energy that remains in the battery, EV, or hot water system at the end of an experimental run. This is done by valuing the left-over energy at the average RTP for the last 24 hours. Without this adjustment it would not be a fair comparison since any controller that anticipates the need to store energy for a future purpose would perform poorly if it does so just before the experiment ends. This is an artefact of the finite length of our experimental runs; with very long durations this problem goes away as the left-over energy costs become insignificant.

We see that the expectation and 2-stage algorithms get quite close to the performance of the controller with perfect foresight and they achieve significant cost reductions over the two reactive controllers ($\sim$ 35% less than smart reactive controller). The expectation controller outperforms the 2-stage controller on average and in all runs except 9. This appears to suggest that the expectation controller is superior as it requires less computation and achieves lower costs. There are however a few subtleties to this that are worth discussing.

Fig. 2b shows the average split in costs between devices, ignoring the PV. The costs for the expectation and 2-stage controllers are essentially identical except when it comes to the hot water heater. The hot water heater is different from the other devices because it has a recourse variable for unmet hot water demand which has a very high cost associated with it. This recourse variable takes on a non-zero value in run 5 for the expectation and 2-stage controllers, and run 9 for the expectation controller, in all cases because the controllers fail to anticipate large spikes in hot water demand until it is too late. The reactive controllers have to be quite conservative with the amount of hot water they keep stored, so they never encounter a demand that they cannot meet (at the expense of higher prices paid for heating the water and greater thermal losses over time).

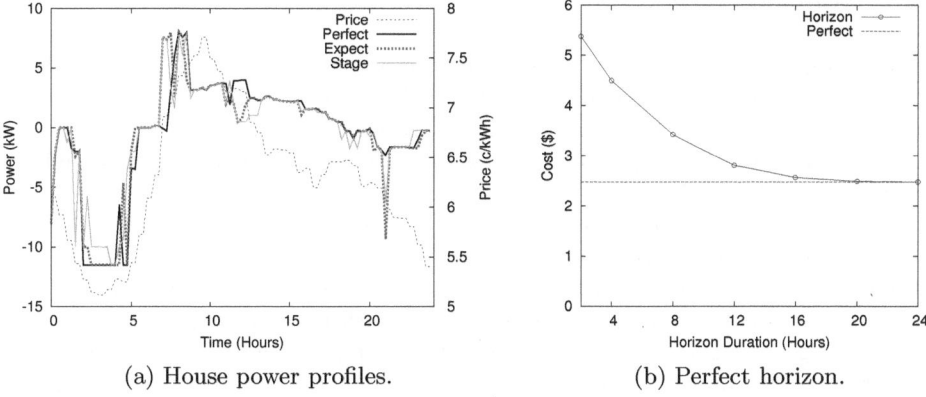

(a) House power profiles.   (b) Perfect horizon.

Fig. 3. House power profiles for one day and experimenting with horizon lengths

We in general expect the 2-stage algorithm to be more conservative than the expectation algorithm and to avoid having any unmet demand because through sampling it can identify upcoming peaks. This is the case in all runs except 5. With further investigation we found that in run 5 the 2-stage algorithm failed to generate the scenario with high demand when it was needed. The reason for this occurring was not due to too few second stage scenarios, but because of the way that second stage scenarios are conditionally sampled from the first stage which can take on expected values. In this instance, at a critical point in time, the first stage expected values precluded the high demand scenario from being able to occur in the second stage even though it was still physically possible.

What these results show is that the expectation algorithm typically does a very good job, but there are certain types of devices and random processes for which it performs poorly. We initially thought that the online 2-stage algorithm presented in this paper would be able to overcome these limitations, but there appears to be some problems associated with taking the expected value of first stage variables.

Fig. 3a gives an example of the power exchanged between the house and the grid for one day, along with the RTP. As expected, most consumption occurs when the price is low and when the price is high power is sold back to the grid from the battery, EV and PV. The expectation and 2-stage controllers follow the general trend of the perfect controller with some small divergences.

Fig. 3b shows the results of an experiment where we investigated how performance changes with the horizon duration. This plot shows the performance of the perfect controller running as an online algorithm where it is restricted to only having perfect foresight a certain distance into the future. The experiment is performed on run 1 for a number of different horizon durations and the results are compared to the original perfect controller that can see the full 7 days. The results show that there is little to be gained by looking any further into the future than 20 hours.

5   Related Work

Much of the existing literature on residential demand response focuses on deterministic formulations over fixed horizons where the scheduler has perfect foresight [6,9]. Those that have considered uncertainty in the problem typically focus on just one aspect (e.g., real-time pricing) [10], or use very simple models for random variables [11].

Model-predictive control has been used to account for the uncertainty of estimated device model parameters and measurement noise [12], but not the uncertainty of the type we model. In general, model-predictive control is best suited to unconstrained, purely continuous settings with limited uncertainty.

Dynamic programming [11,13] and Q-learning [14] have been used in conjunction with Markov Decision Process (MDP) formulations of the residential load scheduling problem, to generate policies that allocate power to each device. MDP approaches suffer from severe scalability issues, especially since the state space needs to be discretised. Moreover, MDPs seem somewhat excessive for our problem, given that uncertainty does not depend on the decisions taken. Our stochastic programming approach which uses scenario sampling is more scalable and more natural in the presence of exogenous uncertainty.

One paper [11] found that acting on the basis of the optimal dynamic programming solution did not provide any benefit over acting on expectations. For the most part we found this to be true, but as discussed in the experimental section we identified circumstances where the greedy nature of an expectation algorithm can lead to poor results. Our use of more complex uncertainty models and different devices is likely the reason why we had this extra observation.

The paper closest to ours compares two-stage stochastic programming and robust optimisation techniques for scheduling residential loads [15]. Uncertainty is restricted to the RTP which is known for the first stage but becomes uncertain thereafter. The objective includes minimising expected price and the probability mass of "risky" scenarios whose price exceeds a certain threshold. Comfort is handled by imposing hard constraints under which appliances must run, rather than by inclusion into the objective. In this setting, two-stage stochastic programming was observed to provide benefits over robust scheduling.

The scope of our analysis goes significantly beyond these results, by exploring uncertainty from a large range of sources and by identifying the value of stochastic information. We enable richer sources of uncertainty to be considered in our framework, by allowing inputs to be revealed at arbitrary points in time.

Commercially available residential DR solutions[8] typically focus on direct load control or simple reactive policies. Such systems could experience more optimal DR performance and greater residential customer satisfaction by using our algorithms.

---

[8] E.g., comverge: www.comverge.com, nest: www.nest.com and Cooper Power Systems: www.cooperindustries.com

6   Conclusion and Future Work

This paper contributes to the growing body of work on residential control of loads and storage under real-time pricing, by developing a framework that accounts for uncertainty. To our knowledge, it is the first work that provides a scalable and accurate solution in the presence of uncertainty about future prices, occupant behaviour and environmental conditions. Using models representative of physical devices and random processes, we have shown the monetary and comfort cost savings that can be achieved by using online stochastic algorithms over reactive control, and the comparison of performance between a 2-stage approach and acting on expectations. Studies such as the one in this paper are import for rallying industry and customers towards more effective energy management schemes.

Further research is needed to investigate how closely reality can be modelled with random processes, and if in turn they are suitable for online learning. We also need to further investigate how time step sizes and the number of second stage scenarios influence performance, and to conduct more experiments for different months of the year. The experimental set up we have developed can be used to experiment with and compare different pricing schemes. For example, time of use pricing and RTP where the price for generation is different from that for consumption. We also plan on investigating how multiple houses react to a RTP and what sort of emergent behaviour develops when they are all learning their statistical models online.

Acknowledgements. This work is supported by NICTA's Optimisation Research Group as part of the Future Energy Systems project. We thank our project members and reviewers for useful discussions and helpful suggestions. NICTA is funded by the Australian Government as represented by the Department of Broadband, Communications and the Digital Economy and the Australian Research Council through the ICT Centre of Excellence program.

References

1. Van Hentenryck, P., Bent, R.: Online Stochastic Combinatorial Optimization. The MIT Press, Cambridge (2006)
2. Powell, W., Simao, H., Bouzaiene-Ayari, B.: Approximate dynamic programming in transportation and logistics: a unified framework. EURO Journal on Transportation and Logistics 1, 237–284 (2012)
3. Shapiro, A.: On Complexity of Multistage Stochastic Programs. Operations Research Letters 34, 1–8 (2006)
4. Shapiro, A., Dentcheva, D., Ruszczyński, A.: Lectures on Stochastic Programming: Modeling and Theory. MPS-SIAM series on optimization. Society for Industrial and Applied Mathematics (SIAM, 3600 Market Street, Floor 6, Philadelphia, PA 19104) (2009)
5. Hastie, T., Tibshirani, R.: Generalized Additive Models. In: Monographs on Statistics and Applied Probability Series. Chapman & Hall, CRC Press (1990)

6. Ramchurn, S.D., Vytelingum, P., Rogers, A., Jennings, N.R.: Agent-based control for decentralised demand side management in the smart grid. In: Tumer, Y., Sonenberg, S. (eds.) Proc. of 10th Int. Conf. on Autonomous Agents and Multi-agent Systems - Innovative Applications Track (AAMAS 2011), Taipei, Taiwan, pp. 330–331 (2011)

7. Kolter, J.Z., Batra, S., Ng, A.Y.: Energy disaggregation via discriminative sparse coding. In: 24th Annual Conference on Neural Information Processing Systems (NIPS), pp. 1153–1161 (2010)

8. Parson, O., Ghosh, S., Weal, M., Rogers, A.: Non-intrusive load monitoring using prior models of general appliance types. In: Proceedings of Twenty-Sixth Conference on Artificial Intelligence (AAAI 2012), Toronto, CA (2012)

9. Gatsis, N., Giannakis, G.B.: Residential load control: Distributed scheduling and convergence with lost ami messages. IEEE Transactions on Smart Grid PP, 1–17 (2012)

10. Mohsenian-Rad, A.H., Leon-Garcia, A.: Optimal residential load control with price prediction in real-time electricity pricing environments. IEEE Transactions on Smart Grid 1, 120–133 (2010)

11. Tischer, H., Verbic, G.: Towards a smart home energy management system - a dynamic programming approach. In: Innovative Smart Grid Technologies Asia (ISGT), pp. 1–7. IEEE PES (2011)

12. Yu, Z., McLaughlin, L., Jia, L., Murphy-Hoye, M.C., Pratt, A., Tong, L.: Modeling and stochastic control for home energy management. In: 2012 Power and Energy Society General Meeting (2012)

13. Kim, T., Poor, H.: Scheduling power consumption with price uncertainty. IEEE Transactions on Smart Grid 2, 519–527 (2011)

14. Levorato, M., Goldsmith, A., Mitra, U.: Residential demand response using reinforcement learning. In: 2010 First IEEE International Conference on Smart Grid Communications (SmartGridComm), pp. 409–414 (2010)

15. Chen, Z., Wu, L., Fu, Y.: Real-time price-based demand response management for residential appliances via stochastic optimization and robust optimization. IEEE Transactions on Smart Grid 3, 1822–1831 (2012)

Lifting Structural Tractability
to CSP with Global Constraints

Evgenij Thorstensen*

Department of Computer Science, University of Oxford, UK
evgenij.thorstensen@cs.ox.ac.uk

Abstract. A wide range of problems can be modelled as constraint satisfaction problems (CSPs), that is, a set of constraints that must be
satisfied simultaneously. Constraints can either be represented extensionally, by explicitly listing allowed combinations of values, or implicitly, by
special-purpose algorithms provided by a solver. Such implicitly represented constraints, known as global constraints, are widely used; indeed,
they are one of the key reasons for the success of constraint programming
in solving real-world problems.

In recent years, a variety of restrictions on the structure of CSP instances that yield tractable classes have been identified. However, many
such restrictions fail to guarantee tractability for CSPs with global constraints. In this paper, we investigate the properties of extensionally represented constraints that these restrictions exploit to achieve tractability,
and show that there are large classes of global constraints that also possess these properties. This allows us to lift these restrictions to the global
case, and identify new tractable classes of CSPs with global constraints.

1 Introduction

Constraint programming (CP) is widely used to solve a variety of practical problems such as planning and scheduling [22,30], and industrial configuration [2,21].
Constraints can either be represented explicitly, by a table of allowed assignments, or implicitly, by specialized algorithms provided by the constraint solver.
These algorithms may take as a parameter a *description* that specifies exactly
which kinds of assignments a particular instance of a constraint should allow.
Such implicitly represented constraints are known as global constraints, and a
lot of the success of CP in practice has been attributed to solvers providing
them [15, 28, 31].

The theoretical properties of constraint problems, in particular the computational complexity of different types of problem, have been extensively studied and
quite a lot is known about what restrictions on the general *constraint satisfaction problem* are sufficient to make it tractable [3, 6, 8, 16, 19, 25]. In particular,
many structural restrictions, that is, restrictions on how the constraints in a
problem interact, have been identified and shown to yield tractable classes of

---

* Work supported by EPSRC grant EP/G055114/1.

C. Schulte (Ed.): CP 2013, LNCS 8124, pp. 661–677, 2013.

CSP instances [17, 20, 25]. However, much of this theoretical work has focused on problems where each constraint is explicitly represented, and most known structural restrictions fail to yield tractable classes for problems with global constraints, even when the global constraints are fairly simple [23].

Theoretical work on global constraints has to a large extent focused on developing efficient algorithms to achieve various kinds of local *consistency* for individual constraints. This is generally done by pruning from the domains of variables those values that cannot lead to a satisfying assignment [4,29]. Another strand of research has explored conditions that allow global constraints to be replaced by collections of explicitly represented constraints [5]. These techniques allow faster implementations of algorithms for *individual constraints*, but do not shed much light on the complexity of problems with multiple *overlapping* global constraints, which is something that practical problems frequently require.

As such, in this paper we investigate what properties of explicitly represented constraints structural restrictions rely on to guarantee tractability. Identifying such properties will allow us to find global constraints that also possess them, and lift well-known structural restrictions to instances with such constraints.

As discussed in [7], when the constraints in a family of problems have unbounded arity, the way that the constraints are *represented* can significantly affect the complexity. Previous work in this area has assumed that the global constraints have specific representations, such as propagators [18], negative constraints [9], or GDNF/decision diagrams [7], and exploited properties particular to that representation. In contrast, we will use a definition of global constraints that allows us to discuss different representations in a uniform manner. Furthermore, as the results we obtain will rely on a relationship between the size of a global constraint and the number of its satisfying assignments, we do not need to reference any specific representation.

As a running example, we will use the connected graph partition problem (CGP) [13, p. 209], defined below. The CGP is the problem of partitioning the vertices of a graph into bags of a given size while minimizing the number of edges that span bags. The vertices of the graph could represent components to be placed on circuit boards while minimizing the number of inter-board connections.

Problem 1 (Connected graph partition (CGP)). We are given an undirected and connected graph $\langle V, E \rangle$, as well as $\alpha, \beta \in \mathbb{N}$. Can V be partitioned into disjoint sets $V_1, \ldots, V_m$ with $|V_i| \leq \alpha$ such that the set of broken edges $E' = \{\{u, v\} \in E \mid u \in V_i, v \in V_j, i \neq j\}$ has cardinality β or less?

This problem is NP-complete [13, p. 209], even for fixed $\alpha \geq 3$. We are going to use the results in this paper to show a new result, namely that the CGP is tractable for every fixed β.

2   Global Constraints

In this section, we define the basic concepts that we will use throughout the paper. In particular, we give a precise definition of global constraints, and illustrate it with a few examples.

Definition 1 (Variables and assignments). *Let V be a set of variables, each with an associated set of domain elements. We denote the set of domain elements (the domain) of a variable v by $D(v)$. We extend this notation to arbitrary subsets of variables, W, by setting $D(W) = \bigcup_{v \in W} D(v)$.*

An assignment *of a set of variables V is a function $\theta : V \to D(V)$ that maps every $v \in V$ to an element $\theta(v) \in D(v)$. We denote the restriction of θ to a set of variables $W \subseteq V$ by $\theta|_W$. We also allow the special assignment $\perp$ of the empty set of variables. In particular, for every assignment θ, we have $\theta|_\emptyset = \perp$.*

Definition 2 (Projection). *Let Θ be a set of assignments of a set of variables V. The* projection *of Θ onto a set of variables $X \subseteq V$ is the set of assignments $\pi_X(\Theta) = \{\theta|_X \mid \theta \in \Theta\}$.*

Note that when $\Theta = \emptyset$ we have $\pi_X(\Theta) = \emptyset$, but when $X = \emptyset$ and $\Theta \neq \emptyset$, we have $\pi_X(\Theta) = \{\perp\}$.

Definition 3 (Disjoint union of assignments). *Let θ_1 and θ_2 be two assignments of disjoint sets of variables V_1 and V_2, respectively. The* disjoint union *of θ_1 and θ_2, denoted $\theta_1 \oplus \theta_2$, is the assignment of $V_1 \cup V_2$ such that $(\theta_1 \oplus \theta_2)(v) = \theta_1(v)$ for all $v \in V_1$, and $(\theta_1 \oplus \theta_2)(v) = \theta_2(v)$ for all $v \in V_2$.*

Global constraints have traditionally been defined, somewhat vaguely, as constraints without a fixed arity, possibly also with a compact representation of the constraint relation. For example, in [22] a global constraint is defined as "a constraint that captures a relation between a non-fixed number of variables".

Below, we offer a precise definition similar to the one in [4], where the authors define global constraints for a domain D over a list of variables σ as being given intensionally by a function $D^{|\sigma|} \to \{0, 1\}$ computable in polynomial time. Our definition differs from this one in that we separate the general *algorithm* of a global constraint (which we call its *type*) from the specific description. This separation allows us a better way of measuring the size of a global constraint, which in turn helps us to establish new complexity results.

Definition 4 (Global constraints). *A* global constraint type *is a parameterized polynomial-time algorithm that determines the acceptability of an assignment of a given set of variables.*

Each global constraint type, e, has an associated set of descriptions, *$\Delta(e)$. Each description $\delta \in \Delta(e)$ specifies appropriate parameter values for the algorithm e. In particular, each $\delta \in \Delta(e)$ specifies a set of variables, denoted by $\mathcal{V}(\delta)$.*

A global constraint $e[\delta]$, *where $\delta \in \Delta(e)$, is a function that maps assignments of $\mathcal{V}(\delta)$ to the set $\{0, 1\}$. Each assignment that is allowed by $e[\delta]$ is mapped to 1, and each disallowed assignment is mapped to 0. The* extension *or* constraint relation *of $e[\delta]$ is the set of assignments, θ, of $\mathcal{V}(\delta)$ such that $e[\delta](\theta) = 1$. We also say that such assignments* satisfy *the constraint, while all other assignments* falsify *it.*

When we are only interested in describing the set of assignments that satisfy a constraint, and not in the complexity of determining membership in this set, we will sometimes abuse notation by writing $\theta \in e[\delta]$ to mean $e[\delta](\theta) = 1$.

As can be seen from the definition above, a global constraint is not usually explicitly represented by listing all the assignments that satisfy it. Instead, it is represented by some description δ and some algorithm e that allows us to check whether the constraint relation of $e[\delta]$ includes a given assignment. To stay within the complexity class NP, this algorithm is required to run in polynomial time. As the algorithms for many common global constraints are built into modern constraint solvers, we measure the *size* of a global constraint's representation by the size of its description.

Example 1 (EGC). A very general global constraint type is the *extended global cardinality* constraint type [29]. This form of global constraint is defined by specifying for every domain element a a finite set of natural numbers $K(a)$, called the cardinality set of a. The constraint requires that the number of variables which are assigned the value a is in the set $K(a)$, for each possible domain element a.

Using our notation, the description δ of an EGC global constraint specifies a function $K_\delta : D(\mathcal{V}(\delta)) \to \mathcal{P}(\mathbb{N})$ that maps each domain element to a set of natural numbers. The algorithm for the EGC constraint then maps an assignment θ to 1 if and only if, for every domain element $a \in D(\mathcal{V}(\delta))$, we have that $|\{v \in \mathcal{V}(\delta) \mid \theta(v) = a\}| \in K_\delta(a)$.

Example 2 (Table and negative constraints). A rather degenerate example of a a global constraint type is the *table* constraint.

In this case the description δ is simply a list of assignments of some fixed set of variables, $\mathcal{V}(\delta)$. The algorithm for a table constraint then decides, for any assignment of $\mathcal{V}(\delta)$, whether it is included in δ. This can be done in a time which is linear in the size of δ and so meets the polynomial time requirement.

Negative constraints are complementary to table constraints, in that they are described by listing *forbidden* assignments. The algorithm for a negative constraint $e[\delta]$ decides, for any assignment of $\mathcal{V}(\delta)$, whether whether it is *not* included in δ. Observe that disjunctive clauses, used to define propositional satisfiability problems, are a special case of the negative constraint type, as they have exactly one forbidden assignment.

We observe that any global constraint can be rewritten as a table or negative constraint. However, this rewriting will, in general, incur an exponential increase in the size of the description.

As can be seen from the definition above, a table global constraint is explicitly represented, and thus equivalent to the usual notion of an explicitly represented constraint.

Definition 5 (CSP instance). *An instance of the constraint satisfaction problem (CSP) is a pair $\langle V, C \rangle$ where V is a finite set of variables, and C is a set of global constraints such that for every $e[\delta] \in C$, $\mathcal{V}(\delta) \subseteq V$. In a CSP instance, we call $\mathcal{V}(\delta)$ the scope of the constraint $e[\delta]$.*

A classic *CSP instance is one where every constraint is a table constraint.*

A solution *to a CSP instance* $P = \langle V, C \rangle$ *is an assignment* θ *of* V *which satisfies every global constraint, i.e., for every* $e[\delta] \in C$ *we have* $\theta|_{V(\delta)} \in e[\delta]$. *We denote the set of solutions to* P *by* $\mathsf{sol}(P)$.

The size *of a CSP instance* $P = \langle V, C \rangle$ *is* $|P| = |V| + \sum_{v \in V} |D(v)| + \sum_{e[\delta] \in C} |\delta|.$

Example 3 (The CGP encoded with global constraints). Given a connected graph $G = \langle V, E \rangle$, α, and β, we build a CSP instance $\langle A \cup B, C \rangle$ as follows. The set A will have a variable v for every $v \in V$ with domain $D(v) = \{1, \ldots, |V|\}$, while the set B will have a boolean variable e for every edge in E.

The set of constraints C will have an EGC constraint C^α on A with $K(i) = \{0, \ldots, \alpha\}$ for every $1 \leq i \leq |V|$. Likewise, C will have an EGC constraint C^β on B with $K(0) = \{0, \ldots, |E|\}$ and $K(1) = \{1, \ldots, \beta\}$.

Finally, to connect A and B, the set C will have for every edge $\{u, v\} \in E$, with corresponding variable $e \in B$, a table constraint on $\{u, v, e\}$ requiring $u \neq v \rightarrow e = 1$.

As an example, Figure 1 shows this encoding for the CGP on the graph C_5, that is, a simple cycle on five vertices.

This encoding follows the definition of Problem 1 quite closely, and can be done in polynomial time.

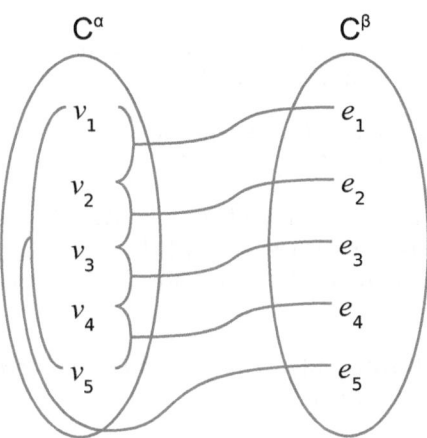

Fig. 1. CSP encoding of the CGP on the graph C_5

3   Structural Restrictions

In recent years, there has been a flurry of research into identifying tractable classes of classic CSP instances based on restrictions on the hypergraphs of CSP instances, known as structural restrictions. Below, we present and discuss a few representative examples. To present the various structural restrictions, we will use the framework of width functions, introduced by Adler [1].

Definition 6 (Hypergraph). *A hypergraph* $\langle V, H \rangle$ *is a set of vertices* V *together with a set of hyperedges* $H \subseteq \mathcal{P}(V)$.

Given a CSP instance $P = \langle V, C \rangle$, *the hypergraph of* P, *denoted* $\mathsf{hyp}(P)$, *has vertex set* V *together with a hyperedge* $\mathcal{V}(\delta)$ *for every* $e[\delta] \in C$.

Definition 7 (Tree decomposition). *A* tree decomposition *of a hypergraph* $\langle V, H \rangle$ *is a pair* $\langle T, \lambda \rangle$ *where* T *is a tree and* λ *is a labelling function from nodes of* T *to subsets of* V, *such that*

1. *for every* $v \in V$, *there exists a node* t *of* T *such that* $v \in \lambda(t)$,
2. *for every hyperedge* $h \in H$, *there exists a node* t *of* T *such that* $h \subseteq \lambda(t)$, *and*
3. *for every* $v \in V$, *the set of nodes* $\{t \mid v \in \lambda(t)\}$ *induces a connected subtree of* T.

Definition 8 (Width function). *Let* $G = \langle V, H \rangle$ *be a hypergraph. A* width function *on* G *is a function* $f : \mathcal{P}(V) \to \mathbb{R}^+$ *that assigns a positive real number to every nonempty subset of vertices of* G. *A width function* f *is* monotone *if* $f(X) \leq f(Y)$ *whenever* $X \subseteq Y$.

Let $\langle T, \lambda \rangle$ *be a tree decomposition of* G, *and* f *a width function on* G. *The* f-width *of* $\langle T, \lambda \rangle$ *is* $\max(\{f(\lambda(t)) \mid t \text{ node of } T\})$. *The* f-width *of* G *is the minimal* f-width *over all its tree decompositions.*

In other words, a width function on a hypergraph G tells us how to assign weights to nodes of tree decompositions of G.

Definition 9 (Treewidth). *Let* $f(X) = |X| - 1$. *The* treewidth $\mathsf{tw}(G)$ *of a hypergraph* G *is the* f-width *of* G.

Let $G = \langle V, H \rangle$ be a hypergraph, and $X \subseteq V$. An edge cover for X is any set of hyperedges $H' \subseteq H$ that satisfies $X \subseteq \bigcup H'$. The edge cover number $\rho(X)$ of X is the size of the smallest edge cover for X. It is clear that ρ is a width function.

Definition 10 ([1, Chapter 2]). *The* generalized hypertree width $\mathsf{hw}(G)$ *of a hypergraph* G *is the* ρ-width *of* G.

Next, we define a relaxation of hypertree width known as fractional hypertree width, introduced by Grohe and Marx [20].

Definition 11 (Fractional edge cover). *Let* $G = \langle V, H \rangle$ *be a hypergraph, and* $X \subseteq V$. *A* fractional edge cover *for* X *is a function* $\gamma : H \to [0, 1]$ *such that* $\sum_{v \in h \in H} \gamma(h) \geq 1$ *for every* $v \in X$. *We call* $\sum_{h \in H} \gamma(h)$ *the* weight *of* γ. *The* fractional edge cover number $\rho^*(X)$ *of* X *is the minimum weight over all fractional edge covers for* X. *It is known that this minimum is always rational [20].*

Definition 12. *The* fractional hypertree width *fhw*(G) *of a hypergraph* G *is the* ρ^**-width of* G.

For a class of hypergraphs $\mathcal{H}$ and a notion of width α, we write $\alpha(\mathcal{H})$ for the maximal α-width over the hypergraphs in $\mathcal{H}$. If this is unbounded we write $\alpha(\mathcal{H}) = \infty$; otherwise $\alpha(\mathcal{H}) < \infty$.

All the above restrictions can be used to guarantee tractability for classes of CSP instances where all constraints are table constraints.

Theorem 1 ([10, 17, 20]). *Let* $\mathcal{H}$ *be a class of hypergraphs. For every* $\alpha \in \{hw, fhw\}$, *any class of classic CSP instances whose hypergraphs are in* $\mathcal{H}$ *is tractable if* $\alpha(\mathcal{H}) < \infty$.

To go beyond fractional hypertree width, Marx [24,25] recently introduced the concept of submodular width. This concept uses a set of width functions satisfying a condition (submodularity), and considers the f-width of a hypergraph for every such function f.

Definition 13 (Submodular width function). *Let* $G = \langle V, H \rangle$ *be a hypergraph. A width function* f *on* G *is* submodular *if for every set* $X, Y \subseteq V$, *we have* $f(X) + f(Y) \geq f(X \cap Y) + f(X \cup Y)$.

Definition 14 (Submodular width). *Let* G *be a hypergraph. The* submodular width *subw*(G) *of* G *is the maximum* f*-width of* G *taken over all monotone submodular width functions* f *on* G.

For a class of hypergraphs $\mathcal{H}$, *we write* subw$(\mathcal{H})$ *for the maximal submodular width over the hypergraphs in* $\mathcal{H}$. *If this is unbounded we write* subw$(\mathcal{H}) = \infty$; *otherwise* subw$(\mathcal{H}) < \infty$.

Unlike for fractional hypertree width and every other structural restriction discussed so far, the running time of the algorithm given by Marx for classic CSP instances with bounded submodular width has an exponential dependence on the number of vertices in the hypergraph of the instance. The class of classic CSP instances with bounded submodular width is therefore not tractable. However, this class is what is called fixed-parameter tractable [11, 12].

Definition 15 (Fixed-parameter tractable). *A* parameterized problem instance *is a pair* $\langle k, P \rangle$, *where* P *is a problem instance, such as a CSP instance, and* $k \in \mathbb{N}$ *a parameter.*

Let S *be a class of parameterized problem instances. We say that* S *is* fixed-parameter tractable *(in* FPT*) if there is a function* f *of one argument, as well as a constant* c, *such that every problem* $\langle k, P \rangle \in S$ *can be solved in time* $O(f(k) \times |P|^c)$.

The function f can be arbitrary, but must only depend on the parameter k. For CSP instances, a natural parameterization is by the size of the hypergraph of an instance, measured by the number of vertices. Since the hypergraph of an instance has a vertex for every variable, for every CSP instance $P = \langle V, C \rangle$ we consider the parameterized instance $\langle |V|, P \rangle$.

Theorem 2 ([24]). *Let $\mathcal{H}$ be a class of hypergraphs. If* $\mathsf{subw}(\mathcal{H}) < \infty$, *then a class of classic CSP instances whose hypergraphs are in $\mathcal{H}$ is in* FPT.

The three structural restrictions that we have just presented form a hierarchy [20, 24]: For every hypergraph G, $\mathsf{subw}(G) \leq \mathsf{fhw}(G) \leq \mathsf{hw}(G)$.

As the example below demonstrates, Theorem 1 does not hold for CSP instances with arbitrary global constraints, even if we have a fixed, finite domain.

Example 4. The NP-complete problem of 3-colourability [13] is to decide, given a graph $\langle V, E \rangle$, whether the vertices V can be coloured with three colours such that no two adjacent vertices have the same colour.

We may reduce this problem to a CSP with EGC constraints (cf. Example 1) as follows: Let V be the set of variables for our CSP instance, each with domain $\{r, g, b\}$. For every edge $\langle v, w \rangle \in E$, we post an EGC constraint with scope $\{v, w\}$, parameterized by the function K such that $K(r) = K(g) = K(b) = \{0, 1\}$. Finally, we make the hypergraph of this CSP instance have low width by adding an EGC constraint with scope V parameterized by the function K' such that $K'(r) = K'(g) = K'(b) = \{0, \ldots, |V|\}$. This reduction clearly takes polynomial time, and the hypergraph G of the resulting instance has $\mathsf{hw}(G) = \mathsf{fhw}(G) = \mathsf{subw}(G) = 1$.

As the constraint with scope V allows all possible assignments, any solution to this CSP is also a solution to the 3-colourability problem, and vice versa.

Likewise, Theorem 2 does not hold for CSP instances with arbitrary global constraints if we allow the variables unbounded domain size, that is, change the above example to k-colourability. With that in mind, in the rest of the paper we will identify properties of extensionally represented constraints that these structural restrictions exploit to guarantee tractability. Then, we are going to look for restricted classes of global constraints that possess these properties. To do so, we will use the following definitions.

Definition 16 (Constraint catalogue). *A constraint catalogue is a set of global constraints. A CSP instance $\langle V, C \rangle$ is said to be over a constraint catalogue Γ if for every $e[\delta] \in C$ we have $e[\delta] \in \Gamma$.*

Definition 17 (Restricted CSP class). *Let Γ be a constraint catalogue, and let $\mathcal{H}$ be a class of hypergraphs. We define* $\mathrm{CSP}(\mathcal{H}, \Gamma)$ *to be the class of CSP instances over Γ whose hypergraphs are in $\mathcal{H}$.*

Definition 17 allows us to discuss classic CSP instances alongside instances with global constraints. Let **Ext** be the constraint catalogue containing all table global constraints. The classic CSP instances are then precisely those that are over **Ext**. In particular, we can now restate Theorems 1 and 2 as follows.

Theorem 3. *Let $\mathcal{H}$ be a class of hypergraphs. For every $\alpha \in \{\mathsf{hw}, \mathsf{fhw}\}$, the class of CSP instances $\mathrm{CSP}(\mathcal{H}, \mathbf{Ext})$ is tractable if $\alpha(\mathcal{H}) < \infty$. Furthermore, if $\mathsf{subw}(\mathcal{H}) < \infty$ then $\mathrm{CSP}(\mathcal{H}, \mathbf{Ext})$ is in* FPT.

4   Properties of Extensional Representation

We are going to start our investigation by considering fractional hypertree width in more detail. To obtain tractability for classic CSP instances of bounded fractional hypertree width, Grohe and Marx [20] use a bound on the number of solutions to a classic CSP instance, and show that this bound is preserved when we consider parts of a CSP instance. The following definition formalizes what we mean by "parts", and is required to state the algorithm that Grohe and Marx use in their paper.

Definition 18 (Constraint projection). *Let $e[\delta]$ be a constraint. The projection of $e[\delta]$ onto a set of variables $X \subseteq \mathcal{V}(\delta)$ is the constraint $\mathsf{pj}_X(e[\delta])$ such that $\mu \in \mathsf{pj}_X(e[\delta])$ if and only if there exists $\theta \in e[\delta]$ with $\theta|_X = \mu$.*

For a CSP instance $P = \langle V, C \rangle$ and $X \subseteq V$ we define $\mathsf{pj}_X(P) = \langle X, C' \rangle$, where C' is the least set containing for every $e[\delta] \in C$ such that $X \cap \mathcal{V}(\delta) \neq \emptyset$ the constraint $\mathsf{pj}_{X \cap \mathcal{V}(\delta)}(e[\delta])$.

Their algorithm is given as Algorithm 1, and is essentially the usual recursive search algorithm for finding all solutions to a CSP instance by considering smaller and smaller sub-instances using constraint projections.

Algorithm 1. Enumerate all solutions of a CSP instance

---

    procedure ENUMSOLUTIONS(CSP instance $P = \langle V, C \rangle$)   ▷ Returns sol(P)
        Solutions ← ∅
        if $V = \emptyset$ **then**
            return $\{\bot\}$ ▷ The empty assignment
        else
            $w \leftarrow$ chooseVar(V)   ▷ Pick a variable from V
            $\Theta =$ EnumSolutions($\mathsf{pj}_{V-\{w\}}(P)$)
            for $\theta \in \Theta$ **do**
                for $a \in D(w)$ **do**
                    if $\theta \cup \langle w, a \rangle$ is a solution to P **then**
                        Solutions.add($\theta \cup \langle w, a \rangle$)
                  end if
                end for
            end for
        end if
        return Solutions
    end procedure

To show that Algorithm 1 does indeed find all solutions, we will use the following property of constraint projections.

Lemma 1. *Let $P = \langle V, C \rangle$ be a CSP instance. For every $X \subseteq V$, we have $\mathsf{sol}(\mathsf{pj}_X(P)) \supseteq \pi_X(\mathsf{sol}(P))$.*

Proof. Given $P = \langle V, C \rangle$, let $X \subseteq V$ be arbitrary, and let $C' = \{e[\delta] \in C \mid X \cap \mathcal{V}(\delta) \neq \emptyset\}$. For every $\theta \in \mathsf{sol}(P)$ and constraint $e[\delta] \in C'$ we have that $\theta|_{\mathcal{V}(\delta)} \in e[\delta]$ since θ is a solution to P. By Definition 18, it follows that for every $e[\delta] \in C'$, $\theta|_{X \cap \mathcal{V}(\delta)} \in \mathsf{pj}_{X \cap \mathcal{V}(\delta)}(e[\delta])$. Since the set of constraints of $\mathsf{pj}_X(P)$ is the least set containing for each $e[\delta] \in C'$ the constraint $\mathsf{pj}_{X \cap \mathcal{V}(\delta)}(e[\delta])$, we have $\theta|_X \in \mathsf{sol}(\mathsf{pj}_X(P))$, and hence $\mathsf{sol}(\mathsf{pj}_X(P)) \supseteq \pi_X(\mathsf{sol}(P))$. Since X was arbitrary, the claim follows.

Theorem 4 (Correctness of Algorithm 1). *Let P be a CSP instance. We have that* $\mathrm{EnumSolutions}(P) = \mathsf{sol}(P)$.

Proof. The proof is by induction on the set of variables V in P. For the base case, if $V = \emptyset$, the empty assignment is the only solution.

Otherwise, choose a variable $w \in V$, and let $X = V - \{w\}$. By induction, we can assume that $\mathrm{EnumSolutions}(\mathsf{pj}_X(P)) = \mathsf{sol}(\mathsf{pj}_X(P))$. Since for every $\theta \in \mathsf{sol}(P)$ there exists $a \in D(w)$ such that $\theta = \theta|_X \cup \langle w, a \rangle$, and furthermore $\theta|_X \in \pi_X(\mathsf{sol}(P))$, it follows by Lemma 1 that $\theta|_X \in \mathsf{sol}(\mathsf{pj}_X(P))$. Since Algorithm 1 checks every assignment of the form $\mu \cup \langle w, a \rangle$ for every $\mu \in \mathsf{sol}(\mathsf{pj}_X(P))$ and $a \in D(w)$, it follows that $\mathrm{EnumSolutions}(P) = \mathsf{sol}(P)$.

The time required for this algorithm depends on three key factors, which we are going to enumerate and discuss below. Let

1. $s(P)$ be the maximum of the number of solutions to each of the instances $\mathsf{pj}_{V-\{w\}}(P)$,
2. $c(P)$ be the maximum time required to check whether an assignment is a solution to P, and
3. $b(P)$ be the maximum time required to construct any instance $\mathsf{pj}_{V-\{w\}}(P)$.

There are $|V|$ calls to EnumSolutions. For each call, we need $b(P)$ time to construct the projection, while the double loop takes at most $s(P) \times |D(w)| \times c(P)$ time. Therefore, letting $d = \max(\{|D(w)| \mid w \in V\})$, the running time of Algorithm 1 is bounded by $O(|V| \times (s(P) \times d \times c(P) + b(P)))$.

Since constructing the projection of a classic CSP instance can be done in polynomial time, and likewise checking that an assignment is a solution, the whole algorithm runs in polynomial time if $s(P)$ is a polynomial in the size of P. For fractional hypertree width, Grohe and Marx show the following.

Lemma 2 ([20]). *A classic CSP instance P has at most $|P|^{\mathsf{fhw}(\mathsf{hyp}(P))}$ solutions.*

Since fractional hypertree width is a monotone width function, it follows that for any instance $P = \langle V, C \rangle$ and $X \subseteq V$, $\mathsf{fhw}(\mathsf{hyp}(\mathsf{pj}_X(P))) \leq \mathsf{fhw}(\mathsf{hyp}(P))$. Therefore, for classic CSP instances of bounded fractional hypertree width $s(P)$ is indeed polynomial in $|P|$.

5   CSP Instances with Few Solutions in Key Places

Having few solutions for every projection of a CSP instance is thus a property that makes fractional hypertree width yield tractable classes of classic CSP instances. More importantly, we have shown that this property allows us to find all solutions to a CSP instance P, even with global constraints, if we can build arbitrary projections of P in polynomial time. In other words, with these two conditions we should be able to reduce instances with global constraints to classic instances in polynomial time.

However, on reflection there is no reason why we should need few solutions for *every* projection. Instead, consider the following reduction.

Definition 19 (Partial assignment checking). *A global constraint catalogue Γ allows partial assignment checking if for any constraint $e[\delta] \in \Gamma$ we can decide in polynomial time whether a given assignment θ to a set of variables $W \subseteq \mathcal{V}(\delta)$ is contained in an assignment that satisfies $e[\delta]$, i.e. whether there exists $\mu \in e[\delta]$ such that $\theta = \mu|_W$.*

As an example, a catalogue that contains arbitrary EGC constraints (cf. Example 1) does not satisfy Definition 19, since checking whether an arbitrary EGC constraint has a satisfying assignment is NP-hard [26]. On the other hand, a catalogue that contains only EGC constraints whose cardinality sets are intervals does satisfy Definition 19 [27].

If a catalogue Γ satisfies Definition 19, we can for any constraint $e[\delta] \in \Gamma$ build arbitrary projections of it, that is, construct the global constraint $\mathsf{pj}_X(e[\delta])$ for any $X \subseteq \mathcal{V}(\delta)$, in polynomial time.

Definition 20 (Intersection variables). *Let $\langle V, C \rangle$ be a CSP instance. The set of intersection variables of any constraint $e[\delta] \in P$ is $\mathsf{iv}(\delta) = \bigcup\{\mathcal{V}(\delta) \cap \mathcal{V}(\delta') \mid e'[\delta'] \in C - \{e[\delta]\}\}$.*

Definition 21 (Table constraint induced by a global constraint). *Let $P = \langle V, C \rangle$ be a CSP instance. For every $e[\delta] \in C$, let μ^* be the assignment to $\mathcal{V}(\delta) - \mathsf{iv}(\delta)$ that assigns a special value $*$ to every variable. The table constraint induced by $e[\delta]$ is $\mathsf{ic}(e[\delta]) = e'[\delta']$, where $\mathcal{V}(\delta') = \mathcal{V}(\delta)$, and δ' contains for every assignment $\theta \in \mathsf{sol}(\mathsf{pj}_{\mathsf{iv}(\delta)}(P))$ the assignment $\theta \oplus \mu^*$.*

If every constraint in a CSP instance $P = \langle V, C \rangle$ allows partial assignment checking, then building $\mathsf{ic}(e[\delta])$ for any $e[\delta] \in C$ can be done in polynomial time when $|\mathsf{sol}(\mathsf{pj}_X(P))|$ is itself polynomial in the size of P for every subset X of $\mathsf{iv}(\delta)$. To do so, we can invoke Algorithm 1 on the instance $\mathsf{pj}_{\mathsf{iv}(\delta)}(P)$. The definition below expresses this idea.

Definition 22 (Sparse intersections). *A class of CSP instances $\mathcal{P}$ has sparse intersections if there exists a constant c such that for every constraint $e[\delta]$ in any instance $P \in \mathcal{P}$, we have that for every $X \subseteq \mathsf{iv}(\delta)$, $|\mathsf{sol}(\mathsf{pj}_X(P))| \leq |P|^c$.*

If a class of instances $\mathcal{P}$ has sparse intersections, and the instances are all over a constraint catalogue that allows partial assignment checking, then we can for every constraint $e[\delta]$ of any instance from $\mathcal{P}$ construct $\mathsf{ic}(e[\delta])$ in polynomial time. While this definition considers the instance as a whole, one special case of it is the case where every constraint has few solutions in the size of its description, that is, there is a constant c and the constraints are drawn from a catalogue Γ such that for every $e[\delta] \in \Gamma$, we have that $|\{\mu \mid \mu \in e[\delta]\}| \leq |\delta|^c$.

Theorem 5. *Let $\mathcal{P}$ be a class of CSP instances over a catalogue that allows partial assignment checking. If $\mathcal{P}$ has sparse intersections, then we can in polynomial time reduce any instance $P \in \mathcal{P}$ to a classic CSP instance P_{CL} with $\mathsf{hyp}(P) = \mathsf{hyp}(P_{CL})$, such that P_{CL} has a solution if and only if P does.*

Proof. Let $P = \langle V, C \rangle$ be an instance from such a class $\mathcal{P}$. For each $e[\delta] \in C$, P_{CL} will contain the table constraint $\mathsf{ic}(e[\delta])$ from Definition 21. Since P is over a catalogue that allows partial assignment checking, and $\mathcal{P}$ has sparse intersections, computing $\mathsf{ic}(e[\delta])$ can be done in polynomial time by invoking Algorithm 1 on $\mathsf{pj}_{\mathsf{iv}(\delta)}(P)$.

It is clear that $\mathsf{hyp}(P) = \mathsf{hyp}(P_{CL})$. All that is left to show is that P_{CL} has a solution if and only if P does. Let θ be a solution to $P = \langle V, C \rangle$. For every $e[\delta] \in C$, we have that $\theta|_{\mathsf{iv}(\delta)} \in \mathsf{pj}_{\mathsf{iv}(\delta)}(P)$ by Definitions 18 and 20, and the assignment μ that assigns the value $\theta(v)$ to each $v \in \bigcup_{e[\delta] \in C} \mathsf{iv}(\delta)$, and $*$ to every other variable is therefore a solution to P_{CL}.

In the other direction, if θ is a solution to P_{CL}, then θ satisfies $\mathsf{ic}(e[\delta])$ for every $e[\delta] \in C$. By Definition 21, this means that $\theta|_{\mathsf{iv}(\delta)} \in \mathsf{sol}(\mathsf{pj}_{\mathsf{iv}(\delta)}(P))$, and by Definition 18, there exists an assignment $\mu^{e[\delta]}$ with $\mu^{e[\delta]}|_{\mathsf{iv}(\delta)} = \theta|_{\mathsf{iv}(\delta)}$ that satisfies $e[\delta]$. By Definition 20, the variables not in $\mathsf{iv}(\delta)$ do not occur in any other constraint in P, so we can combine all the assignments $\mu^{e[\delta]}$ to form a solution μ to P such that for $e[\delta] \in C$ and $v \in \mathcal{V}(\delta)$ we have $\mu(v) = \mu^{e[\delta]}(v)$.

From Theorem 5, we get tractable and fixed-parameter tractable classes of CSP instances with global constraints.

Corollary 1. *Let $\mathcal{H}$ be a class of hypergraphs, and Γ a catalogue that allows partial assignment checking. If $\mathrm{CSP}(\mathcal{H}, \Gamma)$ has sparse intersections, then $\mathrm{CSP}(\mathcal{H}, \Gamma)$ is tractable or in FPT if $\mathrm{CSP}(\mathcal{H}, \mathbf{Ext})$ is.*

Proof. Let $\mathcal{H}$ and Γ be given. By Theorem 5, we can reduce any $P \in \mathrm{CSP}(\mathcal{H}, \Gamma)$ to an instance $P_{CL} \in \mathrm{CSP}(\mathcal{H}, \mathbf{Ext})$ in polynomial time. Since P_{CL} has a solution if and only if P does, tractability or fixed-parameter tractability of $\mathrm{CSP}(\mathcal{H}, \mathbf{Ext})$ implies the same for $\mathrm{CSP}(\mathcal{H}, \Gamma)$.

5.1   Applying Corollary 1 to the CGP

Recall the connected graph partition problem (Problem 1): Given a connected graph G, as well as natural numbers α and β, can the vertices of G be partitioned into bags of size at most α, such that no more than β edges are broken. Using the CSP encoding we gave in Example 3, as well as Corollary 1, we will show a new result, that this problem is tractable if β is fixed. To simplify the analysis, we assume without loss of generality that $\alpha < |V|$, which means that any solution has at least one broken edge.

We claim that if β is fixed, then the constraint $C^\beta = e^\beta[\delta^\beta]$ allows partial assignment checking, and has only a polynomial number of satisfying assignments. The latter implies that for any instance P of the CGP, $|\mathsf{sol}(\mathsf{pj}_{\mathsf{iv}(\delta^\beta)}(P))|$ is polynomial in the size of P for every subset of $\mathsf{iv}(\delta^\beta)$. Furthermore, we will show that for the constraint $C^\alpha = e^\alpha[\delta^\alpha]$, we also have that $|\mathsf{sol}(\mathsf{pj}_{\mathsf{iv}(\delta^\alpha)}(P))|$ is polynomial in the size of P. That C^α allows partial assignment checking follows from a result by Régin [27], since the cardinality sets of C^α are intervals.

First, we show that the number of satisfying assignments to C^β is limited. Since C^β limits the number of ones in any solution to β or fewer, the number of satisfying assignments to this constraint is the number of ways to choose up to β variables to be assigned one. This is bounded by $\sum_{i=1}^{\beta} \binom{|E|}{i} \le (|E|+1)^\beta$, and so we can generate them all in polynomial time.

Now, let θ be such a solution. How many solutions to P contain θ? Well, every constraint on $\{u, v, e\}$ with $e = 1$ allows at most $|V|^2$ assignments, and there are at most β such constraints. So far we therefore have at most $(|E|+1)^\beta \times |V|^{2\beta}$ assignments.

On the other hand, a ternary constraint with $e = 0$ requires $u = v$. Consider the graph G_0 containing for every constraint on $\{u, v, e\}$ with $e = 0$ the vertices u and v as well as the edge $\{u, v\}$. Since the original graph was connected, every connected component of G_0 contains at least one vertex which is in the scope of some constraint with $e = 1$. Therefore, since equality is transitive, each connected component of G_0 allows at most one assignment for each of the $(|E|+1)^\beta \times |V|^{2\beta}$ assignments to the other variables of P. We therefore get a total bound of $(|E|+1)^\beta \times |V|^{2\beta}$ on the total number of solutions to P, and hence to $\mathsf{pj}_{\mathsf{iv}(\delta^\alpha)}(P)$.

The hypergraph of any CSP instance P encoding the CGP has two hyperedges covering the whole problem, so the hypertree width of this hypergraph is two. Therefore, we may apply Corollary 1 and Theorem 1 to obtain tractability when β is fixed. As this problem is NP-complete for fixed $\alpha \ge 3$ [13, p. 209], β is a natural parameter to try and use.

As it happens, in this problem we can drop the requirement of partial assignment checking for the constraint C^α. All its variables are intersection variables, and the instance has few solutions even if we disregard C^α. Thus, we need only check whether any of those solutions satisfy C^α, and checking whether an assignment to the whole scope of a constraint satisfies it can always be done in polynomial time by Definition 4. In the next section, we turn this observation into a general result.

6   Back Doors

If a class of CSP instances includes constraints from a catalogue that is not known to allow partial assignment checking, we may still obtain tractability in some cases by applying the notion of a back door set. A (strong) back door set [14,32] is a set of variables in a CSP instance that, when assigned, make the instance easy to solve. Below, we are going to adapt this notion to individual constraints.

Definition 23 (Back door). *Let Γ be a global constraint catalogue. A back door for a constraint $e[\delta] \in \Gamma$ is any set of variables $W \subseteq \mathcal{V}(\delta)$ (called a back door set) such that we can decide in polynomial time whether a given assignment θ to a set of variables $\mathcal{V}(\theta) \supseteq W$ is contained in an assignment that satisfies $e[\delta]$, i.e. whether there exists $\mu \in e[\delta]$ such that $\mu|_{\mathcal{V}(\theta)} = \theta$.*

Trivially, for every constraint $e[\delta]$ the set of variables $\mathcal{V}(\delta)$ is a back door set, since by Definition 4 we can always check in polynomial time if an assignment to $\mathcal{V}(\delta)$ satisfies the constraint $e[\delta]$.

The key point about back doors is that given a catalogue Γ, adding to each $e[\delta] \in \Gamma$ with back door set W an arbitrary set of assignments to W produces a catalogue Γ' that allows partial assignment checking. Adding a set of assignments Θ means to add Θ to the description, and modify the algorithm e to only accept an assignment if it contains a member of Θ in addition to previous requirements. Furthermore, given a CSP instance P containing $e[\delta]$, as long as $\Theta \supseteq \pi_W(\text{sol}(P))$, adding Θ to $e[\delta]$ produces an instance that has exactly the same solutions. This point leads to the following definition.

Definition 24 (Sparse back door cover). *Let Γ_{PAC} be a catalogue that allows partial assignment checking and Γ_{BD} a catalogue. For every instance $P = \langle V, C \rangle$ over $\Gamma_{PAC} \cup \Gamma_{BD}$, let $P \cap \Gamma_{PAC}$ be the instance with constraint set $C' = C \cap \Gamma_{PAC}$ and set of variables $\bigcup \{V \cap \mathcal{V}(\delta) \mid e[\delta] \in C'\}$.*

A class of CSP instances $\mathcal{P}$ over $\Gamma_{PAC} \cup \Gamma_{BD}$ has sparse back door cover *if there exists a constant c such that for every instance $P = \langle V, C \rangle \in \mathcal{P}$ and constraint $e[\delta] \in C$, if $e[\delta] \notin \Gamma_{PAC}$, then there exists a back door set W for $e[\delta]$ with $|\text{sol}(\text{pj}_X(P \cap \Gamma_{PAC}))| \leq |P|^c$ for every $X \subseteq W$.*

Sparse back door cover means that for each constraint that is not from a catalogue that allows partial assignment checking, we can in polynomial time get a set of assignments Θ for its back door set using Algorithm 1, and so turn this constraint into one that does allow partial assignment checking. This operation preserves the solutions of the instance that contains this constraint.

Theorem 6. *If a class of CSP instance $\mathcal{P}$ has sparse back door cover, then we can in polynomial time reduce any instance $P \in \mathcal{P}$ to an instance P' such that $\text{hyp}(P) = \text{hyp}(P')$ and $\text{sol}(P) = \text{sol}(P')$. Furthermore, the class of instances $\{P' \mid P \in \mathcal{P}\}$ is over a catalogue that allows partial assignment checking.*

Proof. Let $P = \langle V, C \rangle \in \mathcal{P}$. We construct P' by adding to every $e[\delta] \in C$ such that $e[\delta] \notin \Gamma_{PAC}$, with back door set W, the set of assignments $\mathsf{sol}(\mathsf{pj}_W(P \cap \Gamma_{PAC}))$, which we can obtain using Algorithm 1. By Definition 24, we have for every $X \subseteq W$ that $|\mathsf{sol}(\mathsf{pj}_W(P \cap \Gamma_{PAC}))| \leq |P|^c$, so Algorithm 1 takes polynomial time since Γ_{PAC} does allow partial assignment checking.

It is clear that $\mathsf{hyp}(P') = \mathsf{hyp}(P)$, and since $\mathsf{sol}(\mathsf{pj}_W(P \cap \Gamma_{PAC})) \supseteq \pi_W(\mathsf{sol}(P))$, the set of solutions stays the same, i.e. $\mathsf{sol}(P') = \mathsf{sol}(P)$. Finally, since we have replaced each constraint $e[\delta]$ in P that was not in Γ_{PAC} by a constraint that does allow partial assignment checking, it follows that P' is over a catalogue that allows partial assignment checking.

One consequence of Theorem 6 is that we can sometimes apply Theorem 5 to a CSP instance that contains a constraint for which checking if a partial assignment can be extended to a satisfying one is hard. We can do so when the variables of that constraint are covered by the variables of other constraints that do allow partial assignment checking — but only if the instance given by those constraints has few solutions.

As a concrete example of this, consider again the encoding of the CGP that we gave in Example 3. The variables of constraint C^α are entirely covered by the instance P' obtained by removing C^α. As the entire set of variables of a constraint is a back door set for it, and the instance P' has few solutions (cf. Section 5.1), this class of instances has sparse back door cover. As such, the constraint C^α could, in fact, be arbitrary without affecting the tractability of this problem. In particular, the requirement that C^α allows partial assignment checking can be dropped.

7   Summary and Future Work

In this paper, we have investigated properties that many structural restrictions rely on to yield tractable classes of CSP instances with explicitly represented constraints. In particular, we identify a relationship between the number of solutions and the size of a CSP instance as being one such property. Using this insight, we show that known structural restrictions yield tractability for any class of CSP instances with global constraints that satisfies this property. In particular, the above implies that the structural restrictions we consider yield tractability for classes of instances where every global constraint has few satisfying assignments relative to its size.

To illustrate our result, we apply it to a known problem, the connected graph partition problem, and use it to identify a new tractable case of this problem. We also demonstrate how the concept of back doors, subsets of variables that make a problem easy to solve once assigned, can be used to relax the conditions of our result in some cases.

As for future work, one obvious research direction to pursue is to find a complete characterization of tractable classes of CSP instances with sparse intersections. Another avenue of research would be to apply the results in this paper to various kinds of valued CSP.

References

1. Adler, I.: Width Functions for Hypertree Decompositions. Doctoral dissertation, Albert-Ludwigs-Universität Freiburg (2006)
2. Aschinger, M., Drescher, C., Friedrich, G., Gottlob, G., Jeavons, P., Ryabokon, A., Thorstensen, E.: Optimization methods for the partner units problem. In: Achterberg, T., Beck, J.C. (eds.) CPAIOR 2011. LNCS, vol. 6697, pp. 4–19. Springer, Heidelberg (2011)
3. Aschinger, M., Drescher, C., Gottlob, G., Jeavons, P., Thorstensen, E.: Structural decomposition methods and what they are good for. In: Schwentick, T., Dürr, C. (eds.) Proc. STACS 2011. LIPIcs, vol. 9, pp. 12–28 (2011)
4. Bessiere, C., Hebrard, E., Hnich, B., Walsh, T.: The complexity of reasoning with global constraints. Constraints 12(2), 239–259 (2007)
5. Bessiere, C., Katsirelos, G., Narodytska, N., Quimper, C.-G., Walsh, T.: Decomposition of the NVALUE constraint. In: Cohen, D. (ed.) CP 2010. LNCS, vol. 6308, pp. 114–128. Springer, Heidelberg (2010)
6. Bulatov, A., Jeavons, P., Krokhin, A.: Classifying the complexity of constraints using finite algebras. SIAM Journal on Computing 34(3), 720–742 (2005)
7. Chen, H., Grohe, M.: Constraint satisfaction with succinctly specified relations. Journal of Computer and System Sciences 76(8), 847–860 (2010)
8. Cohen, D., Jeavons, P., Gyssens, M.: A unified theory of structural tractability for constraint satisfaction problems. Journal of Computer and System Sciences 74(5), 721–743 (2008)
9. Cohen, D.A., Green, M.J., Houghton, C.: Constraint representations and structural tractability. In: Gent, I.P. (ed.) CP 2009. LNCS, vol. 5732, pp. 289–303. Springer, Heidelberg (2009)
10. Dalmau, V., Kolaitis, P.G., Vardi, M.Y.: Constraint satisfaction, bounded treewidth, and finite-variable logics. In: Van Hentenryck, P. (ed.) CP 2002. LNCS, vol. 2470, pp. 310–326. Springer, Heidelberg (2002)
11. Downey, R.G., Fellows, M.R.: Parameterized Complexity. Monographs in Computer Science. Springer (1999)
12. Flum, J., Grohe, M.: Parameterized Complexity Theory. Texts in Theoretical Computer Science. Springer (2006)
13. Garey, M.R., Johnson, D.S.: Computers and Intractability: A Guide to the Theory of NP-Completeness. W. H. Freeman (1979)
14. Gaspers, S., Szeider, S.: Backdoors to satisfaction. In: Bodlaender, H.L., Downey, R., Fomin, F.V., Marx, D. (eds.) Fellows Festschrift. LNCS, vol. 7370, pp. 287–317. Springer, Heidelberg (2012)
15. Gent, I.P., Jefferson, C., Miguel, I.: MINION: A fast, scalable constraint solver. In: Proc. ECAI 2006, pp. 98–102. IOS Press (2006)
16. Gottlob, G., Leone, N., Scarcello, F.: A comparison of structural CSP decomposition methods. Artificial Intelligence 124(2), 243–282 (2000)
17. Gottlob, G., Leone, N., Scarcello, F.: Hypertree decompositions and tractable queries. Journal of Computer and System Sciences 64(3), 579–627 (2002)
18. Green, M.J., Jefferson, C.: Structural tractability of propagated constraints. In: Stuckey, P.J. (ed.) CP 2008. LNCS, vol. 5202, pp. 372–386. Springer, Heidelberg (2008)
19. Grohe, M.: The complexity of homomorphism and constraint satisfaction problems seen from the other side. Journal of the ACM 54(1), 1–24 (2007)

20. Grohe, M., Marx, D.: Constraint solving via fractional edge covers. In: Proc. SODA 2006, pp. 289–298. ACM (2006)
21. Hermenier, F., Demassey, S., Lorca, X.: Bin repacking scheduling in virtualized datacenters. In: Lee, J. (ed.) CP 2011. LNCS, vol. 6876, pp. 27–41. Springer, Heidelberg (2011)
22. van Hoeve, W.J., Katriel, I.: Global constraints. In: Rossi, F., van Beek, P., Walsh, T. (eds.) Handbook of Constraint Programming, Foundations of Artificial Intelligence, vol. 2, pp. 169–208. Elsevier (2006)
23. Kutz, M., Elbassioni, K., Katriel, I., Mahajan, M.: Simultaneous matchings: Hardness and approximation. Journal of Computer and System Sciences 74(5), 884–897 (2008)
24. Marx, D.: Tractable hypergraph properties for constraint satisfaction and conjunctive queries. CoRR abs/0911.0801 (2009)
25. Marx, D.: Tractable hypergraph properties for constraint satisfaction and conjunctive queries. In: Proc. STOC 2010, pp. 735–744. ACM (2010)
26. Quimper, C.-G., López-Ortiz, A., van Beek, P., Golynski, A.: Improved algorithms for the global cardinality constraint. In: Wallace, M. (ed.) CP 2004. LNCS, vol. 3258, pp. 542–556. Springer, Heidelberg (2004)
27. Régin, J.C.: Generalized Arc Consistency for Global Cardinality Constraint. In: Proc. AAAI 1996, pp. 209–215. AAAI Press (1996)
28. Rossi, F., van Beek, P., Walsh, T. (eds.): The Handbook of Constraint Programming. Elsevier (2006)
29. Samer, M., Szeider, S.: Tractable cases of the extended global cardinality constraint. Constraints 16(1), 1–24 (2011)
30. Wallace, M.: Practical applications of constraint programming. Constraints 1, 139–168 (1996)
31. Wallace, M., Novello, S., Schimpf, J.: ECLiPSe: A platform for constraint logic programming. ICL Systems Journal 12(1), 137–158 (1997)
32. Williams, R., Gomes, C.P., Selman, B.: Backdoors to typical case complexity. In: Proc. IJCAI 2003, pp. 1173–1178 (2003)

Empirical Study of the Behavior
of Conflict Analysis in CDCL Solvers

Djamal Habet and Donia Toumi

Aix-Marseille University
LSIS UMR CNRS 7296
Avenue Escadrille Normandie Niemen
13397 Marseille Cedex 20, France
{djamal.habet,donia.toumi}@lsis.org

Abstract. The Conflict Driven Clause Learning (CDCL) Boolean Satisfiability (SAT) solvers are very effective in solving large and numerous crafted and industrial instances. Paradoxically, we do not know much about the reasons for their effectiveness and their running is hard to trace. This paper participates in the quest to understand the CDCL solvers. Specifically, we empirically study the behavior of one of their essential components which is the conflict analysis module. We show that this module returns generally a relevant backjump level whatever the analyzed clause. We also classify the falsified clauses according to their capacity to produce pertinent learned clauses. We use this classification to induce the apparition of specific clauses in the implication graph by ordering the list of clauses watched by the propagated literals. Finally, we advance some explanations on the effectiveness of CDCL solvers.

Keywords: CDCL Solvers, Conflict Analysis, Empirical Study.

1   Introduction

The satisfiability problem (SAT) consists in deciding whether a Boolean formula in Conjunctive Normal Form (CNF) is satisfiable. In recent decades, one of the major advances in SAT is the great success of the so-called modern solvers, based on the CDCL (Conflict Driven Clause Learning) scheme [6,11,12]. Indeed, in the previous international SAT competitions[1], such solvers showed their ability to handle instances, mainly industrial and crafted ones, with many thousands even millions of variables and clauses. However and despite the CDCL solver improvements, only few works (for instance [2,10]) tried to understand why these solvers are so efficient on most industrial instances.

Unlike lookahead SAT solvers, CDCL based algorithms are difficult to analyze and their behavior is hardly predictable. Also, these solvers are the result of a subtle mix of many components (clause learning [12,13], VSIDS heuristic [11], restarts policies [8,9], etc.) but this combination remains sensitive to even slight changes.

---

[1] www.satcompetition.org

C. Schulte (Ed.): CP 2013, LNCS 8124, pp. 678–693, 2013.

One of the main components of CDCL solvers is the conflict analysis which is launched when a conflict is detected during the propagation process. This analysis aims to explain the failure by detecting the responsible literals and it consequently produces an assertive clause (nogood) which is added to the set of clauses of the instance. Also, one of the key successes of a CDCL solver is the management of the set of the learned clauses to avoid its exponential increase and to keep only useful clauses.

Usually, the conflict analysis is launched at the first empty clause encountered. But is there any change in the solver behavior if we continue propagation, even if a conflict is reached, and stop propagation according to the properties of the reached falsified clause? Maybe analyzing the reason for falsifying the original clauses is more suitable, which seems to be intuitive while the aim is to check the satisfiability of the original formula[2]. Hence, in this paper we address the question of the relevance of the falsified clause to analyze. To the best of our knowledge, this is the first study on this question.

In this aim, we conduct several experimental studies on the *glucose* solver [2] which is efficient, especially to solve industrial instances. Moreover, it incorporates several features that will be useful in our study. Firstly, we consider the relationship between the falsified clause and the properties of the learned clause generated by analyzing the conflict above. This learned clause is crucial in the solving process because it is used to calculate the backjump level and its properties (for instance, its size) determine its relevance to the search. One of the major results of this first step of experiments will concern the relevance of the backjump levels returned by the analysis of the falsified clauses, at each reached conflict. Secondly and according to the observations above, we try to predict the relevant falsified clause to analyze. One main finding of this phase is the establishment of a classification of the falsified clauses according to their ability to produce short clauses. Thirdly, we use this prediction to introduce an ordering on the clauses watched by the literals to propagate. By this sorting, we favor the apparition in the implication graph of the clauses which could be more suitable for the conflict analysis. To the best of our knowledge, this is the first attempt to impact the quality of the learned clauses by such ordering. We also hope that this work is a progress in the understanding of the CDCL solvers.

We organize our paper by giving, in Section 2, the necessary background for the rest of the paper. In Section 3, we present the empirical study on the conflict analysis to identify the relationships between the analyzed clause and the properties of the learned clause. According to the observed relationships, we try to determine the most useful clauses to analyze when a conflict is reached. In Section 4, we attempt to exploit the results obtained above by re-ordering the clauses which are watched by the propagated literals. We finish Section 4 by observing the behavior of the *glucose* solver when it is always forced to learn a particular kind of clauses. Finally, we conclude in Section 5.

---

[2] In the CDCL solver implementations, the propagation is usually started on the original clauses of the formula.

2   Definitions, Notations and Background

An instance $\mathcal{F}$ of the satisfiability problem (SAT) is defined by the pair $\mathcal{F} = (\mathcal{X}, \mathcal{C})$, such that $\mathcal{X} = \{x_1, x_2, \cdots x_n\}$ is a set of Boolean variables (their values belong to the set $\{true, \ false\}$) and $\mathcal{C} = \{c_1, c_2, \cdots c_m\}$ is a set of clauses. A clause $c_i \in \mathcal{C}$ is a finite disjunction of literals and a literal is either a variable (x_i) or its negation $(\neg x_i)$. A clause is represented by the set of its literals. The empty clause is denoted by $\square$ and it is always false.

An interpretation (or an assignment) $\mathcal{I}$ of the variables of $\mathcal{F}$ is defined by a set of literals. If all the variables are assigned in $\mathcal{I}$ then $\mathcal{I}$ is complete, otherwise it is partial. $\mathcal{F}_{|l}$ denotes the formula $\mathcal{F}$ simplified by the assignment of the literal l to $true$. Similarly, if $\mathcal{I} = \{l_1 \ldots l_i\}$, $\mathcal{F}_{|\mathcal{I}}$ denotes the formula $\mathcal{F}$ simplified by the assignment of the literals $l_1 \ldots l_i$ to $true$. The clause $c_j \in \mathcal{C}$ is satisfied by the interpretation $\mathcal{I}$ iff it contains at least one satisfied literal and c_j is falsified if all its literals are falsified by $\mathcal{I}$. A model of $\mathcal{F}$ is an interpretation that satisfies all its clauses. Finally, the satisfiability problem (SAT) consists in deciding whether $\mathcal{F}$ has a model. If this is the case then $\mathcal{F}$ is said to be satisfiable, otherwise $\mathcal{F}$ is unsatisfiable.

The CDCL (Conflict Driven Clauses Learning) solvers are an extension of the complete search method DPLL [4]. Furthermore, they involve a number of additional key techniques such as clause learning mechanisms [12], activity based heuristic VSIDS [11], lazy data structures (watched literals) [13] and restart techniques [8,9]. The general structure of a CDCL solver is given in Algorithm 1.

Algorithm 1. CDCL Solver

Input: $\mathcal{F}$ CNF formula, $maxconflits$.
Output: SAT($\mathcal{F}$ is satisfiable) or UNSAT ($\mathcal{F}$ is unsatisfiable).
while $true$ **do**
  $\mathcal{L} \leftarrow \emptyset$;
  $\mathcal{I} \leftarrow \emptyset$;
  $dl \leftarrow 0$;
  $nbconflits \leftarrow 0$;
  while $(nbconflits < maxconflits)$ **do**
    $confl \leftarrow$ Unitpropagation($\mathcal{F}',\mathcal{I}$); $//\mathcal{F}' = (\mathcal{X}, \mathcal{C} \cup \mathcal{L})$
    if $(confl \neq null)$ **then**
      $nbconflits + +$;
      if $(dl = 0)$ **then** return UNSAT;
      $(\alpha, bj) \leftarrow$ ConflictAnalysis($confl$);
      $\mathcal{L} \leftarrow \mathcal{L} \cup \{\alpha\}$;
      $dl \leftarrow bj$;
      CancelUntil(dl);
    end
    else
      if $(|\mathcal{I}| = |\mathcal{V}|)$ **then** return SAT;
      $dl + +$;
      $l \leftarrow$ PickBranchLit($\mathcal{V},\{0,1\}$);
      $\mathcal{I} \leftarrow \mathcal{I} \cup \{l\}$;
    end
  end
  //restart
end

At each step of the search, the algorithm selects, through the *PickBranchLit()* function, a decision literal l (via the VSIDS heuristic). This literal is added to the current interpretation $\mathcal{I}$ along with the literals involved by applying the function *Unitpropagation()* to $\mathcal{F}'_{|\mathcal{I}}$, where $\mathcal{F}' = (\mathcal{X}, \mathcal{C} \cup \mathcal{L})$ and $\mathcal{L}$ is the set of the clauses learned during the search. As soon as a clause c_i from $\mathcal{F}'$ is falsified, the function *Unitpropagation()* is immediately interrupted and returns this clause in $confl$. To explain the reasons of this failure, Algorithm 1 calls the *ConflictAnalysis(confl)* function which generates an assertive clause (a nogood) α and defines a backjump level bj which is assigned to dl. The unsatisfiability of $\mathcal{F}$ is proven if a conflict occurs at the root of the tree search ($dl = 0$), in which case Algorithm 1 returns UNSAT. Otherwise, the new learned clause α is added to $\mathcal{L}$ and the function *CancelUntil(dl)* backjumps to the decision level dl. When a maximum limit of conflicts is reached, Algorithm 1 performs a restart. If all the variables are assigned without falsifying any clause then the algorithm returns SAT, which means that $\mathcal{F}$ is satisfiable. Finally, Algorithm 1 iterates this process until the satisfiability or the unsatisfiability of the formula is proven.

The mechanism of conflict analysis and clause learning requires the implementation of a strategy of management of the learned clauses to identify and to remove the clauses assumed useless for the search. Otherwise, the set $\mathcal{L}$ can increase exponentially. In this context, different approaches to reduce $\mathcal{L}$ have been studied. Particularly in [2], the authors propose a relevant management of the quality of learned clauses based on the notion of LBD (Literals Blocks Distance). A LBD value is assigned to each learned clause, corresponding to the number of different levels of literals belonging to this clause. The lower the LBD of a clause, the more useful this clause is judged. Finally, a part of the clauses with high LBD values are deleted from $\mathcal{L}$.

3   Empirical Study of Conflict Analysis

As explained in the previous section, a CDLC solver starts the conflict analysis process on the first falsified clause (which we will denote by c_{f_0}) reached during the propagation phase of the enqueued literals. Classically, this analysis is done according to the first implication point [13] by applying a sequence of resolution steps between the clauses involved in the conflict. The clause c_{f_0} is the first to be used in this sequence. Also, for combinatorial reasons, keeping all learned clauses during the search is shown to be unsuccessful. Hence, some learned clauses are kept and others are deleted, according to some parameters (clause activities, LBD values, etc.).

However, what about the relevance of learned clauses regarding conflict analysis? Is it relevant to accomplish the analysis on the first reached falsified clause? Is there any difference in the behavior of a CDCL solver if we restrict the analysis on the basis of a particular falsified clause? To our knowledge, there is no work in the state of the art of CDCL solvers which addresses such a study.

As it is well-known, the behavior of a CDCL solver is so variable that it is hard to study formally its components. For this reason, we will attempt to answer the question above by the way of a large empirical study. The solver used in our experiments is *glucose 2.1* [2,3] with the *SatElite* preprocessing [5]. *glucose* is a minisat-like solver [6] with a particular learned clause set reduction based on the LBD (Literals Blocks Distance) values. Indeed, *glucose* considers that a learned clause with a small LBD (≤ 2) is of a better quality than a clause having a higher LBD value. Accordingly, *glucose* deletes from $\mathcal{L}$ some learned clauses having a size > 2 and a LBD value > 2 (for more details, a reader can refer to [2,3]). Finally, like any CDCL solver, *glucose* propagates the enqueued literals and stops the propagations at the first falsified clause.

3.1 Impact of the Analyzed Clause on the Properties of the Learned Clause

In this Section, we study empirically the impact of the empty clause to analyze on the features of the generated assertive clause, regarding three criteria: its size, its LBD value and finally the backjump level it produces. Indeed, we know that the lower the LBD and the smaller the size of a learned clause, the more useful this clause could be for the search. Also, the higher the backjump level in the search tree (small backjump level values), the more interesting it is for the search [1].

For instance, suppose that the assignment of the literal x adds the literals x_1, x_2 and x_3 to the list of the enqueued literals. Now, let us consider what happens when a conflict is reached by the satisfaction of x_1. Such failure is obtained because a clause, say $\neg x_1 \lor \neg y_1 \lor \neg y_2$, has been falsified under the current partial assignment (due to the assignment of the literal x_1 to *true*). In a classical CDCL solver, the *UnitPropagation* function is interrupted and the clause $\neg x_1 \lor \neg y_1 \lor \neg y_2$ is returned as a conflict. This clause is the first to be falsified by the current assignment. However, other clauses would be falsified if we continue the propagation of the two remaining enqueued literals x_2 and x_3. Nevertheless, this is not done and these clauses are ignored, mainly for performance considerations.

In our study and in the case of a failure, we propose to continue the propagation of all enqueued literals and determine the set of the falsified clauses, which we denote by $\mathcal{C}_f$. Hence, at a given conflict k, we construct the set of the falsified clauses $\mathcal{C}_f(k)$. In order to examine the impact of the empty clause to analyze on the features of the generated assertive clause, we apply the *ConflictAnalysis()* function to each falsified clause $cf_i \in \mathcal{C}_f(k)$ and we record the backjump level $bj_i(k)$ generated by this analysis, as well as the LBD ($lbd_i(k)$) and the size ($size_i(k)$) of the learned clause obtained. When $i = 0$, $cf_0(k)$ corresponds to the first falsified clause which is reached and usually analyzed by *glucose*.

Study 1. At a given conflict k, we want to know whether we can improve the properties of the generated assertive clause (namely its size, its LBD and the backjump level that it returns) by varying the clause $cf_i(k)$ to analyze. Thus, for all falsified clauses $cf_i(k)$ ($i = \{1 \ldots |\mathcal{C}_f(k)|\}$), we observe whether there is a backjump level $bj_i(k)$ such that $bj_i(k) < bj_0(k)$ i.e. $bj_i(k)$ improves $bj_0(k)$.

If a such bj_i exists then we increment $count_{bj}$ (initialized to 0), which counts the number of times when bj_0 is improved during the search. Finally, we calculate $A_{bj}^+ = 100 \times (count_{bj} / \#conflicts)\%$ the improvement rate of the backjump levels overall $\#conflicts$ conflicts encountered since the starting of the search.

We proceed in the same way to calculate A_{lbd}^+ (the improvement rate of the LBD values) and A_{size}^+ (the improvement rate of the clause sizes). We run *glucose* on 300 crafted and industrial instances and on 30 random instances issued from the SAT 2011 competition. We limit the search to the first 10^6 conflicts for the industrial and crafted instances and to 1800 seconds for random ones.

Table 1 summarizes the results obtained: the column $\#inst.$ gives the number of instances in a series and the columns b^+, l^+ and s^+ correspond to the average of A_{bj}^+, A_{lbd}^+ and A_{size}^+ respectively. Table 1 clearly shows that b^+ is generally low: it does not exceed 10% for the crafted and the random instances. Although b^+ is slightly higher for some industrial instances but does not exceed 19.50%. The values of s^+ and l^+ are strongly related to the series of the instances: l^+ is between 13.56% and 45.40% while s^+ ranges from 16.67% tp 78.37%. For some series such as anton, jarvisalo, kullman, spence/sat and leberre, the high values of s^+ could suggest a possible improvement in the size of the learned clauses.

Study 2. In the previous study, we have observed the possible improvements of the results of the analysis of cf_0. We did not look for the best improvement. For the backjump level and at the kth conflict, this improvement could be given by a comparison to $bj_{min}(k) = min\{bj_i(k)\}$, $i = \{1 \dots |\mathcal{C}_f(k)|\}$.

Table 1. Rate improvements of the properties of the first learned clause (i.e. bj_0, lbd_0 and $size_0$) obtained by the analysis of the other falsified clauses at the same conflict

Series	#inst.	b^+	l^+	s^+
Industrial Instances				
fuhs/AProVE11	10	9.90%	16.60%	22.10%
fuhs/bottom	15	8.60%	31.13%	48.07%
fuhs/top	9	16.44%	38.56%	54.00%
jarvisalo	47	19.50%	40.50%	65.83%
kullmann/128	4	6.50%	20.75%	23.50%
kullmann/32	5	2.40%	22.20%	42.00%
kullmann/64	4	4.25%	30.00%	39.25%
leberre/	17	16.89%	38.44%	60.44%
manthey/	9	7.22%	30.00%	58.11%
Crafted Instances				
anton	28	9.74%	43.58%	78.37%
kullmann/G*	10	6.88%	40.75%	58.13%
kullman/V*	26	6.92%	34.00%	65.85%
mosoi/sat	24	4.29%	24.14%	20.57%
mosoi/unsat	6	2.83%	17.67%	16.83%
skvortsov/automata	12	6.67%	23.08%	44.08%
skvortsov/battleship	24	8.08%	40.50%	54.58%
spence/sat	10	8.00%	45.40%	68.80%
spence/unsat	9	1.22%	13.56%	16.67%
Random Instances				
3SAT/UNKNOWN/360	10	8.60%	42.70%	53.00%
5SAT/UNKNOWN/100	10	7.20%	41.90%	49.80%
7SAT/UNKNOWN/60	10	6.00%	36.30%	44.40%

In the following, we want to know whether $bj_{min}(k)$ is often reached by the analysis of the first falsified clause. Also, we want to know whether the bj_{min} is often reached by the analysis of the other falsified clauses. For this purpose, we run *glucose* on the same series of instances of Table 1. At each conflict k, we increment the counter $count_{bj0}$ by 1 if $bj_0(k) = bj_{min}(k)$ and the counter $count_{bj\_}$ by the number of the other clauses that produce a backjump level $bj_i(k) = bj_{min}(k)$, $i \in \{1 \ldots |\mathcal{C}_f(k)|\}$. Finally, we calculate $E^+_{bj0} = 100 \times (count_{bj0}/\#conf)\%$ the success rate of cf_0 to produce bj_{min} and $E^+_{bj} = 100 \times (count_{bj\_}/\#conf)\%$ the success rate of the other falsified clauses to produce also bj_{min}. We proceed in the same way to calculate E^+_{lbd0} (resp. E^+_{size0}) which is the success rate of cf_0 to reach lbd_{min} (resp. $size_{min}$) and E^+_{lbd} (resp. E^+_{size}) corresponding to the success rate of the other falsified clauses to reach lbd_{min} (resp. $size_{min}$). The results are given in Tables 2 and 3, where bj^+_0, bj^+, lbd^+_0, lbd^+, $size^+_0$ and $size^+$ give respectively the average of E^+_{bj0}, E^+_{bj}, E^+_{lbd0}, E^+_{lbd}, E^+_{size0} and E^+_{size} over the instances in the same series. We limit the search to the first 10^6 conflicts for the crafted and industrial instances and to 1800 seconds for the random ones.

Table 2 shows that bj^+_0 is between 65% and 90% which means that bj_{min} is often reached by the analysis of cf_0. The values of lbd^+_0 is between 53% and 70% for the industrial instances, while it is between 44% and 57% for the crafted instances, except the series Skvortsov/automata (70%) and Kullman/V* (29%). We note that lbd^+_0 is lower for the random instances with an average of 30%. $size^+_0$ varies between 33% and 55% except in the case of anton instances (19.67%) and Kullmann/128 instances (62%).

Table 3 shows that the bj^+ values are generally high with an average of 86.87%. Hence, bj_{min} is also often reached by the analysis of the other falsified clauses. lbd^+ is between 24% and 47% for the industrial instances and between 20% and 53% for the crafted ones. We note that lbd^+ is higher for random instances with an average of 58.13%. The values of $size^+$ vary between 20% (kullman/32) and 50.20% except for some series in which $size^+$ is very low (for example anton, spence/sat and kullman/64 series).

Study 3. At this point of the discussion, we have showed that analyzing other falsified clauses allows the features of the learned clauses (especially their size and their LBD) to be improved, for some instances.

However, is it really a strong improvement or just a slight effect? To answer this question, we focus on the cases where the values of lbd^+_0 and $size^+_0$ are high. According to this last criterion, we run *glucose* on a selection of 20 industrial and 53 crafted ones. For each instance and at each conflict k, we firstly store the results $(bj_i(k), lbd_i(k)$ and $size_i(k)$, $i = 1 \cdots |\mathcal{C}_f|)$ returned by *ConflictAnalysis(cf_i)*, $cf_i \in \mathcal{C}_f(k)$. Secondly, we calculate $bj_{min}(k)$, $bj_{max}(k)$, $bj_{avg}(k)$ and $bj_\sigma(k)$ which correspond respectively to the minimum, the maximum, the average and the standard deviation of the backjump levels $bj_i(k)$, $i = 1 \cdots |\mathcal{C}_f(k)|$. We proceed in the same manner for the LBD and the size of the assertive clauses and we compute: $lbd_{min}(k)$, $size_{min}(k)$, $lbd_{max}(k)$, $size_{max}(k)$,

Table 2. Success rate of the first analysis to produce a relevant assertive clause regarding its size and its LBD and the backjump level defined by this clause

Series	#inst.	bj_0^+	lbd_0^+	$size_0^+$
Industrial Instances				
fuhs/AProVE11	10	65.60%	59.10%	53.80%
fuhs/bottom*	15	82.73%	60.73%	44.13%
fuhs/top*	9	75.44%	53.78%	38.67%
jarvisalo	47	72.10%	53.60%	33.20%
kullmann/128/	4	80.25%	66.00%	62.75%
kullmann/32	5	89.80%	70.40%	51.40%
kullmann/64	4	90.75%	65.00%	55.75%
leberre/	17	76.33%	55.00%	33.33%
manthey/	9	88.00%	65.22%	37.56%
Crafted Instances				
anton	28	88.14%	54.19%	19.67%
kullmann/G*	10	90.00%	56.50%	39.25%
kullman/V*	26	81.92%	29.73%	33.92%
mosoi/sat/	24	66.50%	52.00%	55.25%
mosoi/unsat/	6	65.40%	44.93%	46.13%
skvortsov/automata	12	86.75%	70.92%	50.42%
skvortsov/battleship	24	69.05%	45.05%	48.50%
spence/sat	10	65.40%	46.80%	47.70%
spence/unsat	9	68.00%	57.00%	53.78%
Random Instances				
3SAT/UNKNOWN/360	10	83.90%	28.10%	28.90%
5SAT/UNKNOWN/100	10	85.70%	30.30%	38.50%
7SAT/UNKNOWN/60	10	87.00%	37.50%	50.90%

$lbd_{avg}(k)$, $size_{avg}(k)$, $lbd_\sigma(k)$ and $size_\sigma(k)$. Finally, let n be the number of conflicts, we compute the following parameters concerning the backjump levels:

$$- \delta 1_{bj} = \frac{1}{n} \sum_{k=1\cdots n} bj_0(k) - bj_{min}(k) \qquad - \delta 2_{bj} = \frac{1}{n} \sum_{k=1\cdots n} bj_0(k) - bj_{max}(k)$$

$$- \delta 3_{bj} = \frac{1}{n} \sum_{k=1\cdots n} |bj_0(k) - bj_{avg}(k)| \qquad - bj_\sigma = \frac{1}{n} \sum_{k=1\cdots n} bj_\sigma(k)$$

Concerning the LBD of the assertive clauses, the computed parameters are:

$$- \delta 1_{lbd} = \frac{1}{n} \sum_{k=1\cdots n} lbd_0(k) - lbd_{min}(k) \qquad - \delta 2_{lbd} = \frac{1}{n} \sum_{k=1\cdots n} lbd_0(k) - lbd_{max}(k)$$

$$- \delta 3_{lbd} = \frac{1}{n} \sum_{k=1\cdots n} |lbd_0(k) - lbd_{avg}(k)| \qquad - lbd_\sigma = \frac{1}{n} \sum_{k=1\cdots n} lbd_\sigma(k)$$

For their sizes, the parameters are:

$$- \delta 1_{size} = \frac{1}{n} \sum_{k=1\cdots n} size_0(k) - size_{min}(k) \qquad - \delta 2_{size} = \frac{1}{n} \sum_{k=1\cdots n} size_0(k) - size_{max}(k)$$

$$- \delta 3_{size} = \frac{1}{n} \sum_{k=1\cdots n} |size_0(k) - size_{avg}(k)| \qquad - size_\sigma = \frac{1}{n} \sum_{k=1\cdots n} size_\sigma(k)$$

The results are given in Table 4. $\delta 1_{bj}^m$ (respectively $\delta 2_{bj}^m$, $\delta 3_{bj}^m$, bj_σ^m, $\delta 1_{lbd}^m$, $\delta 2_{lbd}^m$, $\delta 3_{lbd}^m$, lbd_σ^m, $\delta 1_{size}^m$, $\delta 2_{size}^m$, $\delta 3_{size}^m$, $size_\sigma^m$) indicates the average of $\delta 1_{bj}$

Table 3. Success rate of the analysis of the other falsified clauses to produce a relevant assertive clause regarding its size and its LBD and the backjump level defined by this clause

Series	#inst.	bj^+	lbd^+	$size^+$
Instances Industrielles				
fuhs/AProVE11	10	62.90%	47.40%	38.90%
fuhs/bottom	15	77.40%	39.87%	25.33%
fuhs/top	9	71.11%	33.11%	20.89%
jarvisalo	46	64.88%	36.94%	22.94%
kullmann/128	4	76.75%	38.25%	34.75%
kullmann/32	5	87.00%	43.20%	34.80%
kullmann/64	4	86.25%	24.50%	15.75%
leberre/	17	67.44%	33.22%	20.78%
manthey/	9	81.33%	44.00%	21.89%
Instances Crafted				
anton	28	80.28%	32.48%	8.16%
kullmann/G*	10	80.00%	20.00%	20.33%
kullman/V*	26	81.92%	29.73%	33.92%
mosoi/sat/	24	65.08%	45.83%	47.67%
mosoi/unsat/	6	64.47%	39.87%	40.13%
skvortsov/automata	12	83.67%	47.08%	38.92%
skvortsov/battleship	24	77.54%	33.08%	22.21%
spence/sat	10	83.00%	24.40%	8.00%
spence/unsat	9	67.44%	53.22%	49.33%
Random Instances				
3SAT/UNKNOWN/360	10	90.00%	56.40%	39.50%
5SAT/UNKNOWN/100	10	91.00%	57.00%	45.30%
7SAT/UNKNOWN/60	10	91.00%	61.00%	50.20%

Table 4. Measures of the dispersion of the backjump levels (resp. LBDs and sizes of the learned clauses) obtained by the analysis of all the falsified clauses

Series	#inst.	$\delta 1_{bj}^m$	$\delta 2_{bj}^m$	bj_σ^m	$\delta 3_{bj}^m$	$\delta 1_{lbd}^m$	$\delta 2_{lbd}^m$	lbd_σ^m	$\delta 3_{lbd}^m$	$\delta 1_{size}^m$	$\delta 2_{size}^m$	$size_\sigma^m$	$\delta 3_{size}^m$
Industrial Instances													
fuhs/bottom*	15	0.71	-0.51	0.43	1.82	1.15	-3.41	1.62	2.07	3.14	-11.72	5.15	4.92
fuhs/top*	9	1.76	-1.32	1.03	2.82	1.45	-4.57	2.13	2.4	3.24	-12.35	5.15	4.76
javisalo	4	2.27	-1.41	1.27	3.85	1.73	-5.75	2.45	2.46	17.01	-90.50	33.28	18.71
leberre	7	1.46	-1.31	0.84	1.38	2.71	-7.34	3.21	2.90	11.45	-48.85	18.50	12.96
Crafted Instances													
anton	28	0.27	-0.34	0.14	0.24	6.42	-10.24	3.65	3.60	39.52	-260.08	67.09	40.72
kullman/G*	10	0.19	-0.25	0.07	0.09	6.83	-79.3	16.09	5.41	10.64	-105.01	21.69	4.41
kullman/V*	26	0.11	-0.21	0.08	0.13	0.98	-3.71	1.04	0.45	8.17	-11.87	5.06	7.13
skvortsov	8	0.22	-0.17	0.11	0.36	6.37	-12.68	5.45	5.32	12.43	-27.64	10.74	8.66
spence	7	0.15	-0.16	0.09	0.22	2.23	-5.54	2.11	1.36	5.54	-17.24	6.159	2.92

(respectively $\delta 2_{bj}, \delta 3_{bj}, bj_\sigma, \delta 1_{lbd}, \delta 2_{lbd}, \delta 3_{lbd}, lbd_\sigma, \delta 1_{size}, \delta 2_{size}, \delta 3_{size}, size_\sigma$) over the instances in the same series.

The low values of $\delta 1_{bj}, \delta 2_{bj}, \delta 3_{bj}$ and bj_σ indicate that the backjump level values are close. Also, the values of $\delta 1_{lbd}, \delta 2_{lbd}, \delta 3_{lbd}$ and lbd_σ are slightly higher. However, the values of $\delta 1_{size}, \delta 2_{size}, \delta 3_{size}$ and $size_\sigma$ are generally higher. These results confirm our findings from the previous studies.

Partial Conclusion. These empirical results could be interpreted as follows: first, accomplishing the analysis on the first reached empty clause is generally sufficient to reach a relevant backjump level. Indeed, the backjump level provided by the first emptied clause is poorly improved. This result is interesting and may be one explanation of the efficiency of the CDCL solvers. Secondly, the choice of the conflicting clause cf_i may improve the LBD of the learned clause. Finally, the size of the learned clause is more dependent on the analyzed clause and may also be improved. This raises the question of which falsified clause is the most suitable to analyze? We will try to respond to this question in Section 3.2

3.2   Identifying Good Clauses to Analyze

We have showed that the choice of the falsified clause to analyze does not significantly affect the backjump level, but may improve the LBD and the size of the learned clause. In this section, our goal is to try to identify which clause is the most suitable to analyze. More precisely, we want to know whether some classes of clauses are more interesting than others.

In this order, we conduct the following experiment: we run *glucose* on the instances selected in the previous sections. We limit the search to the first 10^6 conflicts. For each instance and at the kth conflict, we determine $\mathcal{C}_f(k)$, $lbd_{min}(k)$ and $size_{min}(k)$ as explained in Section 3.1. Then, we classify the clauses of $\mathcal{C}_f(k)$ as described below:

- $\mathcal{C}_{f_I}(k) = \{cf_i \in \mathcal{C}_f(k) \cap \mathcal{C}\}$
- $\mathcal{C}_{f_{L2}}(k) = \{cf_i \in \mathcal{C}_f(k) \cap \mathcal{L}$ and $lbd_i(k) \leq 2\}$
- $\mathcal{C}_{f_{L>2}}(k) = \{cf_i \in \mathcal{C}_f(k) \cap \mathcal{L}$ and $lbd_i(k) > 2\}$
- $\mathcal{C}_{f_{S2}}(k) = \{cf_i \in \mathcal{C}_f(k)$ and $size_i(k) = 2\}$
- $\mathcal{C}_{f_{S3}}(k) = \{cf_i \in \mathcal{C}_f(k)$ and $size_i(k) = 3\}$
- $\mathcal{C}_{f_{S>3}}(k) = \{cf_i \in \mathcal{C}_f(k)$ and $size_i(k) > 3\}$

By this classification, we want to compare the results obtained by the analysis of the initial clauses $\mathcal{C}_{f_I}$ to the ones obtained by the analysis of the learned clauses. Similarly, we compare the results obtained by the analysis of short clauses $\mathcal{C}_{f_{S2}}$ and $\mathcal{C}_{f_{S3}}$ to the ones obtained by the analysis of large ones $\mathcal{C}_{f_{S>3}}$. Also, we compare the results obtained by the analysis of learned clauses with LBD ≤ 2 $\mathcal{C}_{f_{L2}}$ (because of the importance of such clauses [2]), to the ones obtained by the analysis of learned clauses with LBD > 2, $\mathcal{C}_{f_{L>2}}$. For each class of clauses $\mathcal{C}_{f_i}$, $i \in \{I, L2, L > 2, S2, S3, S > 3\}$ and at each conflict k, we measure:

- $r_{lbd}(\mathcal{C}_{f_i}(k))$: the number of clauses in $\mathcal{C}_{f_i}(k)$ which produce a learned clause whose LBD is equal to $lbd_{min}(k)$, divided by the number of falsified clauses in the same class.
- $r_{size}(\mathcal{C}_{f_i}(k))$: the number of clauses in $\mathcal{C}_{f_i}(k)$ which produce a learned clause whose size is equal to $size_{min}(k)$, divided by the number of falsified clauses in the same class.

Table 5. An estimate of good classes over $\mathcal{C}_{f_i}$, $i \in \{I, L2, L > 2, S2, S3, S > 3\}$

Series	#inst.	L^+_{L2}	$L^+_{L>2}$	L^+_I	L^+_{S2}	L^+_{S3}	$L^+_{S>3}$	S^+_{L2}	$S^+_{L>2}$	S^+_I	S^+_{S2}	S^+_{S3}	$S^+_{S>3}$
		Industrial instances											
fuhs/AProVE11	10	69.6	63.2	75.0	63.0	72.0	65.0	58.9	54.3	67.7	52.7	63.9	56.2
fuhs/bottom*	15	43.2	42.4	55.5	53.3	48.9	44.8	26.2	28.5	41.9	39.3	34.3	29.6
fuhs/top*	9	36.3	35.5	52.3	48.5	41.4	41.1	22.9	23.8	41.4	37.1	31.1	28.8
jarvisalo	46	27.1	25.3	36.9	30.9	29.1	31.7	12.4	12.1	22.9	17.8	16.0	17.5
kullmann	13	43.0	43.6	47.2	35.1	39.7	46.0	35.0	35.4	41.5	29.3	33.6	39.3
leberre	17	33.0	32.0	40.1	36.5	35.4	33.7	20.0	19.9	28.4	25.7	24.8	21.5
manthey	9	45.9	45.8	48.3	48.6	46.8	46.2	24.8	26.5	27.7	26.8	26.2	25.3
		Crafted instances											
anton	28	37.4	38.7	35.7	36.2	59.7	36.0	13.0	16.3	10.8	11.8	30.6	11.5
kullman/G*	10	22.4	23.3	23.4	-	26.5	23.1	21.4	21.4	22.1	-	23.8	22.0
kullman/V_3k*	10	26.0	32.3	28.0	-	28.0	28.2	31.6	29.5	30.3	-	29.4	30.2
kullman/V_pd_3k	16	39.5	39.9	32.3	37.3	32.1	32.7	35.5	36.5	38.2	33.0	36.7	38.1
mosoi/sat	24	-	55.9	75.3	58.1	-	74.0	-	55.5	77.2	54.3	-	76.1
mosoi/unsat	6	-	49.7	80.1	41.1	-	77.6	-	49.5	80.6	39.8	-	78.3
skvortsov	36	53.4	45.3	46.6	47.1	53.1	43.2	40.2	34.3	38.1	38.6	40.0	34.3
spence	19	48.3	46.2	58.8	28.5	86.4	48.1	36.9	36.8	49.7	13.7	81.9	38.0

For example at conflict number 500, consider that $size_{min}(500) = 3$ and the number of initial falsified clauses is $|\mathcal{C}_{f_I}(k)| = 5$. If after the analysis of these five clauses, only two have produced an assertive clause of size $= 3$ then $r_{size}(\mathcal{C}_{f_I}(500))$ is equal to 2/5.

At the end of the search, let n be the total number of conflicts, we consider:
$L_i = \frac{1}{n} \times \sum_{k=0\cdots n} r_{lbd}(\mathcal{C}_{f_i}(k))$ and $S_i = \frac{1}{n} \times \sum_{k=0\cdots n} r_{size}(\mathcal{C}_{f_i}(k))$, $i \in \{I, L2, L > 2, S2, S3, S > 3\}$. These two parameters give an estimate of the good class of clauses for the conflict analysis. For instance, the higher the S_i value is, the more interesting this class of clauses is. Indeed, this suggests that the majority of the clauses of this class probably produces a learned clause with the smallest size. This is also applicable to L_i.

The results are summarized in Table 5. The symbol '-' indicates that, for a given instance, the class $\mathcal{C}_{f_i} = \emptyset$, $i \in \{I, L2, L > 2, S2, S3, S > 3\}$. The results concerning the LBD of the learned clause are contrasting and they differ according to the category of instances (industrial or crafted). However, concerning the size of the learned clauses for the industrial instances, the initial clauses $\mathcal{C}_{f_I}$ are the most useful to produce clauses with $size_{min}$. We rank the rest of the classes according to their ability to produce clauses with $size_{min}$ as follows: $\mathcal{C}_{f_{S2}}$ (except the case of fuhs/AProVE11 and Kullman series) then $\mathcal{C}_{f_{S3}}$ (except the case of Kullman and leberre series) and finally $\mathcal{C}_{f_{S>3}}$. For the crafted instances, the results are mixed and differ according to the series.

4   Impact of the Empirical Results on a CDCL Solver

In this section, we attempt to take advantage of the results obtained in Section 3.2, where we have classified the clauses according to their originality, their size and also their LBD (for the learned clauses). Therefore, we propose an approach that makes it possible to prefer the analysis of clauses which satisfy certain characteristics. During the propagation of a literal l, a CDCL solver checks in all clauses where the literal $\neg l$ appears as a watched literal, if the assignment of the literal l to true ($\neg l$ to false) falsifies or propagates other clauses. Let $\mathcal{W}$ be the set of the clauses in which $\neg l$ appears as a watched literal. We classify the clauses in $\mathcal{W}$ according to their originality, their size and their LBD as follows:

- $\mathcal{W}_I = \{c_i \in \mathcal{W} \text{ and } c_i \in \mathcal{C}\}$
- $\mathcal{W}_A = \{c_i \in \mathcal{W} \text{ and } c_i \in \mathcal{L}\}$
- $\mathcal{W}_{L2} = \{c_i \in \mathcal{W} \cap \mathcal{L} \text{ and } lbd_i \leq 2\}$
- $\mathcal{W}_{S2} = \{c_i \in \mathcal{W} \text{ and } size_i = 2\}$
- $\mathcal{W}_{S3} = \{c_i \in \mathcal{W} \text{ and } size_i = 3\}$
- $\mathcal{W}_{S>3} = \{c_i \in \mathcal{W} \text{ and } size_i > 3\}$

Accordingly, we propose to impose a certain order on the handling of this set $\mathcal{W}$ in order to propagate firstly the clauses that we consider the most useful for the search. For example, we first propagate all the original clauses $\mathcal{W}_I$ (while a conflict is not reached) then we propagate the learned clauses $\mathcal{W}_A$. In this way, we favor and stimulate the apparition in the implication graph of certain clauses which could be more useful for the search without imposing additional computing costs. We consider six variants of *glucose* which differ in the starting criteria of the propagation process. The tested solvers are run on blade servers with the GNU/Linux operating system. Each node (blade) has 2 Intel Xeon 2.4 Ghz processors and 24 GB of RAM. Each processor includes 4 physical cores. The time limit is fixed to 5400 seconds. The *glucose* variants are:

- *glucose_o*: while a conflict is not reached, propagate firstly the initial clauses $\mathcal{W}_I$ and secondly the learned clauses $\mathcal{W}_A$.
- *glucose_a*: while a conflict is not reached, propagate firstly the assertive clauses $\mathcal{W}_A$ and secondly the initial clauses $\mathcal{W}_I$.
- *glucose_ol2*: while a conflict is not reached, propagate (1) initial clauses, (2) assertive clauses with LBD ≤ 2 ($\mathcal{W}_{L2}$) and (3) the rest of the clauses in $\mathcal{W}$.
- *glucose_s3l2o*: while a conflict is not reached, propagate (1) clauses with size ≤ 3 ($\mathcal{W}_{S3}$), (2) assertive clauses with LBD ≤ 2 ($\mathcal{W}_{L2}$), (3) initial clauses and (4) the rest of the clauses in $\mathcal{W}$.
- *glucose_s3ol2*: while a conflict is not reached, propagate (1) clauses with size ≤ 3 ($\mathcal{W}_{S3}$), (2) initial clauses, (3) assertive clauses with LBD ≤ 2 ($\mathcal{W}_{L2}$) and (4) propagate the rest of the clauses in $\mathcal{W}$.
- *glucose_l2s3o*: while a conflict is not reached, propagate (1) assertive clauses with LBD ≤ 2 ($\mathcal{W}_{L2}$), (2) clauses with size ≤ 3 ($\mathcal{W}_{S3}$), (3) initial clauses and (4) the rest of the clauses in $\mathcal{W}$.

Note that *glucose* starts always the propagations under the binary clauses $\mathcal{W}_{S2}$. We keep this treatment on all its variants presented above.

The following cactus plots (Figures 1 and 2) give the results obtained by all the solvers. We can remark, for both industrial and crafted instances, that the curves are similar and close. Also, *glucose* and *glucose_s3ol2* have approximatively the same performance. For the other solvers, the results are close.

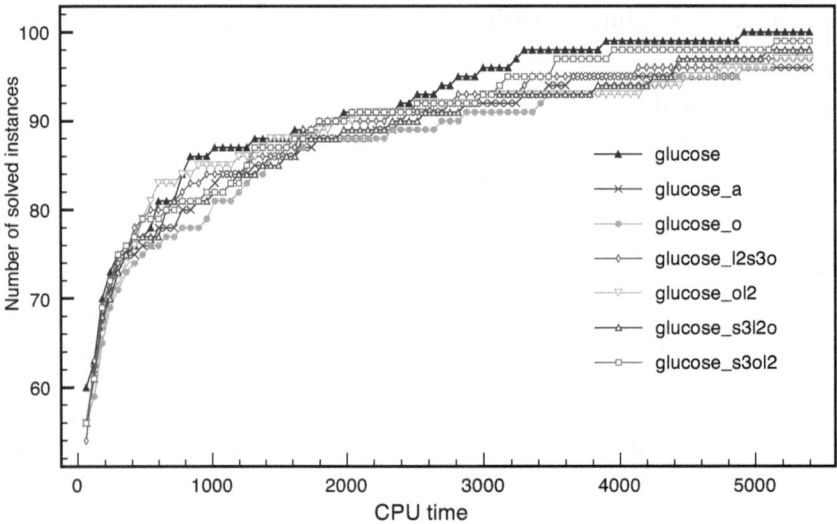

Fig. 1. Variants of *glucose* which differ in the order of propagation of clauses: results on industrial instances

In this study, we have attempted to produce interesting clauses regarding their size and their LBD. In this context, we have tried to favor the apparition of certain types of clauses in the implication graph. Consequently, such clauses are falsified first and analyzed first in the case of a conflict. The results show that such preferences modify slightly the performances of *glucose*, without being able to improve them significantly. This leads us to suppose that focusing on analyzing or producing such clauses is not critical in a CDCL solver. Of course, such interrogation must be handled gingerly. Indeed, it is implicitly assumed that learning short clauses and/or clauses with small LBD could improve significantly the performance of CDCL solvers. However, although this seems reasonable, it is not strictly proven.

We examine this issue as follows: at each conflict, we firstly generate the set of all the assertive clauses and secondly, we retain the shortest learned clause or the one whose LBD value is the smallest. In this aim, we run *glucose* on 300 instances used in the previous studies and, at conflict k, we continue the propagation of all enqueued literals to determine the set of the falsified clauses $\mathcal{C}_f(k)$. Then, we apply the *ConflictAnalysis()* function to each falsified clause $cf_i \in \mathcal{C}_f(k)$ and calculate $size_{min}(k)$ and $lbd_{min}(k)$. Accordingly, we define the variant *glucose_S* (resp. *glucose_L*) of *glucose* which learns, at each conflict k, the assertive clause

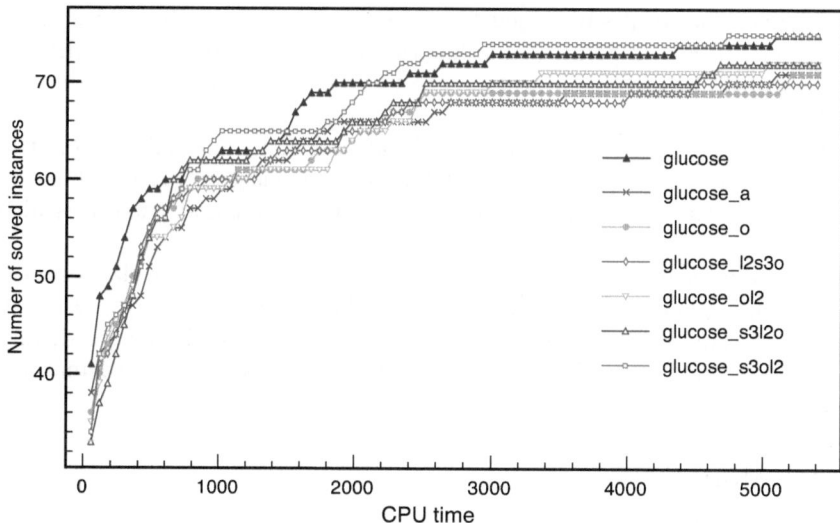

Fig. 2. Variants of *glucose* which differ in the order of propagation of clauses: results on crafted instances

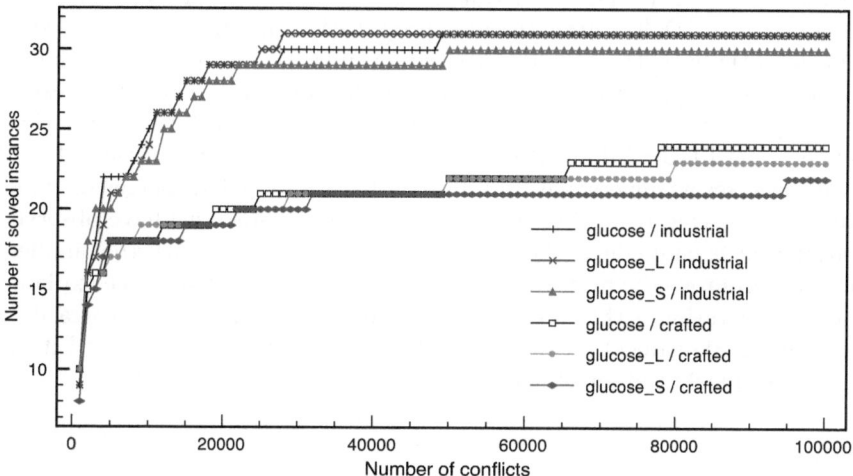

Fig. 3. At each conflict, *glucose_S* (resp. *glucose_L*) learns the clause with the minimum size (resp. the clause with the minimum LBD)

with the minimum size, $size_{min}(k)$ (resp. the minimum LBD, $lbd_{min}(k)$)[3]. We compare the number of the solved instances by *glucose_S*, *glucose_L* and *glucose* regarding the necessary number of the conflicts.

---

[3] The version of zChaff used in SAT 2004 Competition generates and analyzes all the conflicting clauses and keeps the shortest assertive clause [7].

Figure 3 gives the results obtained by these three solvers with a cutoff time of 1800 seconds. It can be pointed out that the maximum number of conflicts is 10^5. This is caused by the time consumed to learn the clauses with the smallest size/LBD.

For the industrial instances, *glucose_L* and *glucose* have approximatively the same performance. *glucose_S* is the worst but remains close to *glucose*. For the crafted instances, the performance gaps are not important and *glucose* is also the best solver.

These results indicate that learning short clauses (*glucose_S*) or clauses with the smallest LBD (*glucose_L*) does not improve the performance of *glucose*. We could conclude that, in a CDCL solver and when a conflict is analyzed, any clause can be learned regardless of its quality and the solver has to manage them during the search.

5   Conclusion

One of the goals of this paper is to contribute to understanding the behavior of CDCL solvers. We have focused our study on one of their components which is the conflict analysis. Our experiments showed that the choice of the falsified clause to analyze does not affect significantly the backjump level defined by the analysis and the conflict analysis module returns generally a relevant backjump level. In a classical CDCL solver, the unit propagations are firstly accomplished on the initial (original) clauses of the treated formula. We saw that these clauses are the most able to produce learned clauses with short sizes. These two results can be an explanation of the powerful of CDCL solvers. Concerning the relationship between the LBD of the learned clause and the properties of the falsified clause which causes its generation, it is difficult to draw a global conclusion while the results are varying regarding the treated instances. Also, we have observed that learning, at each conflict, the shortest clauses or clauses with the smallest LBD does not improve the performance of the CDCL solver. A possible explanation of this observation is that a CDLC solver learns relevant clauses (regarding their sizes and LBD values) enough early. The irrelevant ones are deleted if the solver manages nicely the set of learned clauses.

References

1. Audemard, G., Bordeaux, L., Hamadi, Y., Jabbour, S., Sais, L.: A Generalized Framework for Conflict Analysis. In: Kleine Büning, H., Zhao, X. (eds.) SAT 2008. LNCS, vol. 4996, pp. 21–27. Springer, Heidelberg (2008)
2. Audemard, G., Simon, L.: Predicting Learnt Clauses Quality in Modern sat Solvers. In: Proceedings of the 21st International Joint Conference on Artificial Intelligence (IJCAI 2009), pp. 399–404 (2009)
3. Audemard, G., Simon, L.: Refining Restarts Strategies for SAT and UNSAT. In: Milano, M. (ed.) CP 2012. LNCS, vol. 7514, pp. 118–126. Springer, Heidelberg (2012)

4. Davis, M., Logemann, G., Loveland, D.: A machine program for theorem-proving. Commun. ACM 5(7), 394–397 (1962)
5. Eén, N., Biere, A.: Effective Preprocessing in SAT Through Variable and Clause Elimination. In: Bacchus, F., Walsh, T. (eds.) SAT 2005. LNCS, vol. 3569, pp. 61–75. Springer, Heidelberg (2005)
6. Eén, N., Sörensson, N.: An Extensible SAT-solver. In: Giunchiglia, E., Tacchella, A. (eds.) SAT 2003. LNCS, vol. 2919, pp. 502–518. Springer, Heidelberg (2004)
7. Z. Fu, Y. Mahajan, S. Malik. New Features of the SAT'04 Version of zChaff. In SAT 2004 Competition. Solver Descriptions (2004)
8. Gomes, C.P., Selman, B., Kautz, H.: Boosting Combinatorial Search Through Randomization. In: Proceedings of the Fifteenth National Conference on Artificial Intelligence (AAAI 1998), pp. 431–437. AAAI Press (1998)
9. Huang, J.: The Effect of Restarts on the Efficiency of Clause Learning. In: Proceedings of the 20th International Joint Conference on Artificial Intelligence (IJCAI 2007), pp. 2318–2323 (2007)
10. Katsirelos, G., Simon, L.: Eigenvector Centrality in Industrial SAT Instances. In: Milano, M. (ed.) CP 2012. LNCS, vol. 7514, pp. 348–356. Springer, Heidelberg (2012)
11. Moskewicz, M.W., Madigan, C.F., Zhao, Y., Zhang, L., Malik, S.: Chaff: Engineering an Efficient SAT Solver. In: Annual ACM IEEE Design Automation Conference, pp. 530–535. ACM (2001)
12. Silva, J.P.M., Sakallah, K.A.: GRASP: A Search Algorithm for Propositional Satisfiability. IEEE Trans. Computers 48(5), 506–521 (1999)
13. Zhang, L., Madigan, C.F., Moskewicz, M.H., Malik, S.: Efficient Conflict Driven Learning in a Boolean Satisfiability Solver. In: Proceedings of the 2001 IEEE/ACM International Conference on Computer-Aided Design (ICCAD 2001), pp. 279–285. IEEE Press (2001)

Primal and Dual Encoding from Applications into Quantified Boolean Formulas

Allen Van Gelder

University of California
http://www.cse.ucsc.edu/~avg

Abstract. Quantified Boolean Formulas (QBF) provide a good language for modeling many complex questions about deterministic systems, especially questions involving control of such systems and optimizing choices. However, translators typically have one set way to encode the description of a system and a property, or question about the system, often without distinguishing between the two. In many cases there are choices about encoding methods, and one method will be much easier for a particular solver. This paper shows how to encode a large class of problems into primal and dual versions with opposite answers, while avoiding the blow-up associated with a simple negation of the first encoding.

The main point is that these encodings require knowledge of the underlying application; they are not automatic translations on a QBF. Trying to divide an arbitrary QBF into a system part and a property part is actually co-NP-hard.

For proof of concept, primal and dual encodings were implemented for several problem families in QBFLIB. (The primal encodings were already in QBFLIB.) For leading publicly available QBF solvers, solving times often differed by factors over 100, between the primal and dual encoding of the same underlying problem. Therefore, running both encodings in parallel and stopping when one encoding is solved is an attractive strategy, even though CPU time on both processors is charged. For some families and some solvers, this strategy was significantly faster than running only the primal on all problems, or running only the dual on all problems. Implications for certificates are also discussed briefly.

1 Introduction

Quantified Boolean Formulas (QBF) provide a good language for modeling many complex questions about deterministic systems, especially questions involving control of such systems and optimizing choices. Recent progress in the strength of QBF solvers is beginning to make this a method of choice for analyzing such systems, in much the same way that progress in SAT solvers over the last decade has made translation to SAT the method of choice for many hardware and software design and verification problems. For example, Benedetti and Mangassarian exploit QBF with free variables for model checking [3], while Marin *et al.* use QBF to model designs with incomplete specifications [19]. However, there is a tendency for research to focus on one area or another of the overall problem of

C. Schulte (Ed.): CP 2013, LNCS 8124, pp. 694–707, 2013.

using QBF effectively in applications. Many papers focus on improving solvers [7,6,23,5,8,17,15], some investigate certificates [2,11,20,1,10], while others focus on using QBF in applications [12,25,18,3,16]. Narizzano *et al.* give an extensive survey and additional bibliography [20]. The purpose of this paper is to step back a little and show how considering all aspects of the problem suggests a better way for the parts to work together.

First, we show that many applications can encode both a primal and dual version of their problem. The benefit of having both encodings is the extensive experience with search-based solvers such as Qube and depqbf that valid QBFs are usually more difficult to handle than invalid QBFs. For a given problem, the primal and dual QBFs have opposite values.

In terms of certificates, invalid QBFs have a well understood proof system called Q-resolution [13]. Methods to extract winning strategies from Q-resolution refutations have been recently reported [10,1]. Jussila *et al.* report difficulties in producing adequate certificates for valid formulas [11]. By having both the primal and dual encoding to work with, every problem has a Q-resolution refutation to serve as a certificate. Steps in this direction have appeared for non-prenex, non-CNF solvers [15,10]. Our methodology is essentially independent of the encoding used, and allows certificate methodologies for either valid or invalid formulas to be brought to bear.

The basic idea of using the dual of a QBF is not new. The main novelty in our approach is that the overall problem is partitioned into a deterministic system and a property being queried. The deterministic system, which may be most of the formula, is encoded identically in the primal and dual; only the property is negated in the dual. In instances patterned after the widely used qdimacs format, we propose "d" as a supplement to "a" and "e" to declare deterministic functional variables. Sabharwal *et al.* consider encoding certain two-player games, but their duality is simply logical negation and the methodology consists of a specific encoding that mixes CNF and DNF in a single QBF [24]. Goultiaeva and Bacchus encode a QBF simultaneously as a primal circuit and a dual circuit [8]. However, their duality is also simply the logical negation of the entire formula and exists only as internal data structures in their solver. In more recent work, the same authors use heuristics to attempt to identify a deterministic system within the encoding [9].

We show in Section 5 that, in general, automated post-processing cannot detect the deterministic system embedded in the encoding and construct the dual, at least not efficiently. Therefore, both encodings must be designed with knowledge of the application domain. Construction of the dual proposed in Section 3 involves negating a boolean formula and flipping some quantifiers, which is not technically challenging. At the modeling stage the primal formula is not even CNF. Therefore we believe this approach does not impose a major burden on the developers of software that models their applications as QBF problems.

Empirical evaluations tentatively support the conjecture that having dual encodings available should benefit many applications. In the future, we expect system modelers to incorporate both primal and dual encodings for a wide variety of applications.

The second, and possibly more immediately fruitful, result of having both a primal and a dual encoding for the same problem is that major CPU resource savings might be realized by attempting to solve both in tandem (in parallel, if multiple cores are available). One encoding might be unsolvable within any reasonable time, while the other is quite tractable. Preliminary experimental results, reported in Section 6 (see Table 2), show surprisingly high ratios of solution times for the primal and dual encodings of the same problem. The ratios are so large that, besides saving time on the clock, total CPU resources usually are reduced by attempting to solve both encodings in tandem, then killing the other process when one succeeds. Benchmark instances and generators for the reported experiments are publicly available.[1]

2   QBF and Two-Player Game Strategies

A closed QBF formula $\Psi = \vec{Q} . \mathcal{F}$ can be interpreted as a game with two players, E (for $\exists$, or *existential*) and A (for $\forall$, or *universal*), whose "moves" involve setting their variables. The variables are set in quantifier-prefix order, from outer to inner scopes. This section briefly reviews this well-known interpretation.

The paper uses mostly standard logical operator notation, with "$\wedge$" for "and," "$\vee$" for "or", "$\neg$" for "not", "$\rightarrow$" for "implies", and occasionally "$\oplus$" for "exclusive or." However, "$\equiv$" in some contexts means "is defined as."

Besides the standard existential ($\exists x$) and universal ($\forall x$) quantifiers, we informally introduce *functional quantifiers*, denoted by $* x$, explained in Section 3. Functional quantifiers will become existential or universal, depending on the encoding strategy. The notation $* x$ conveys the information that the value for x is functionally determined by the assignments to the variables that precede x in the quantifier prefix. Part of the quantifier-free formula $\mathcal{F}$ defines the function that determines x.

The quantifier prefix can be partitioned into *quantifier blocks* (*blocks*, for short) of differing quantifier type. The *quantifier depth* of a block begins at 1 for the outermost block and increases by 1 with each alternation. It is known that variable order within a quantifier block is immaterial in determining the value of a QBF formula, so there is no loss of generality, and considerable simplification, in assuming that each player sets all the variables in one block in a single turn. Thus turns are taken in order of increasing quantifier depth, and the quantifier type of the outermost unset block determines whose turn it is to play. By the time all variables have been set, $\mathcal{F}$, the body of the QBF, will be reduced to a formula that evaluates either to true, in which case E is the winner, or to false, in which case A is the winner, for a particular "play" of the game.

Every closed QBF formula is either *valid* (i.e., it evaluates to true) or invalid (evaluates to false). If it is valid, then there exists a *winning strategy* for player E, such that player E can win every play of the game, no matter how player A plays. If it is invalid, then a winning strategy exists for player A. A strategy for player E is a set of boolean functions, one for each existential variable in the QBF.

---

[1] See http://www.cse.ucsc.edu/~avg/QBFdual/

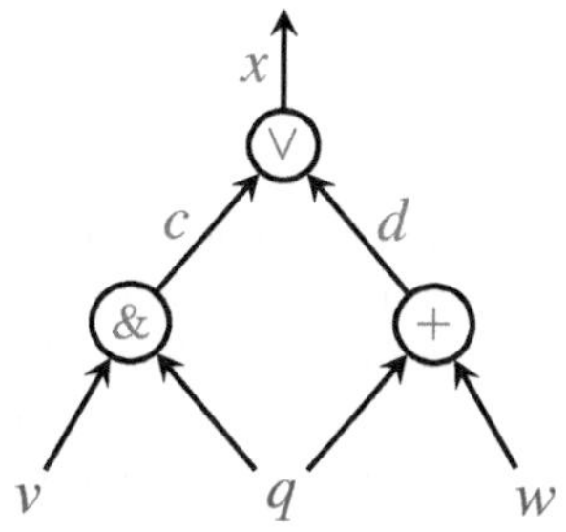

$$G \equiv (c = (q \wedge v)) \wedge (d = (q \oplus w)) \wedge (x = (c \vee d))$$

$$\Psi_1 \equiv \exists p \exists q \forall v \forall w \exists x \exists c \exists d \, (G \wedge (x = p))$$
$$\Psi_2 \equiv \exists p \exists q \forall v \forall w \exists x \exists c \exists d \, (G \wedge ((v = w) \rightarrow (x = p)))$$

Fig. 1. A circuit represented by the quantifier-free formula G, and two QBFs encoding questions about it

All existential variables in a single block have as parameters the variables in blocks of lesser quantifier depth. Suppose $\mathbf{A}$ is an assignment to all variables of lesser quantifier depth. The function for an existential variable e_i in the outermost unassigned block, denoted by $e_i(\mathbf{A})$, specifies what assignment player E should make for e_i at this stage of the game. Strategies for player A are similar except that they are functions that specify assignments for universal variables in the outermost unassigned block. A *winning strategy* is a strategy that guarantees that its player wins. In general terms, a winning strategy for player E tries to satisfy *all* clauses, while a winning strategy for player A tries to falsify *some* clause.

3 QBF Modeling and Duals

QBF is a good language for modeling complex questions about deterministic systems. We use the circuit shown in Figure 1 as a first example. A more substantial example is developed in Section 4. In this example, $G(q, v, w, c, d, x)$ is a quantifier-free formula that encodes the behavior of a deterministic system, the circuit in Figure 1. G must encode a boolean function from the inputs (q, v, w) to the remaining state variables (c, d, x). That is, for each assignment to (q, v, w) there must be exactly one assignment to (c, d, x) for which G evaluates to true. We are not concerned with the details of this encoding, at this point.

Properties, or yes-no questions, about the deterministic system, can be encoded in QBF. We illustrate this idea with the following two questions about the circuit:

1. Is there a value for q such that the output x has the same value (call it p) for all settings of the remaining inputs, v and w? That is, property P_1 is $(x = p)$. Ψ_1 in Figure 1 encodes this question.
2. Same question, with the further restriction that the remaining inputs are ganged (forced to the same value). That is, property P_2 is $((v = w) \rightarrow (x = p))$. Ψ_2 in Figure 1 encodes this question.

The general pattern is ($\equiv$ should be read "is defined as"):

$$\Psi \equiv \overrightarrow{Q}_P \exists_G (G \wedge P). \tag{1}$$

The quantifier prefix $\overrightarrow{Q}_P$ depends on the question or property, and includes inputs to the deterministic system and variables local to P. The quantifier prefix $\exists_G$ denotes existential quantification of the non-input state variables of the deterministic system; if G is thought of as a function, these are its outputs. In Section 4 we begin using the "functional quantifier" $*_G$ instead of $\exists_G$ for these situations.

The key observation is that there is an alternative QBF representation of the same problem:

$$\Psi_{alt} \equiv \overrightarrow{Q}_P \forall_G (G \rightarrow P). \tag{2}$$

Theorem 3.1. If G has the stated functional property, then Ψ_{alt} has the same truth value as Ψ.

Proof. (Sketch) This is most easily realized by considering the associated two-person games. In Ψ_{alt} the A player has control of the state variables, whereas in Ψ the E player has control. Rewriting the body of Ψ_{alt} as $(\neg(G) \vee P)$, we see that the A player tries to make $\neg(G)$ false in Ψ_{alt}, while the E player tries to make G true in Ψ. Given any assignment to the system inputs, we see that both players will choose the unique setting of the state variables that allows G to evaluate to true. The remaining parts of Ψ and Ψ_{alt} are identical, so the logical equivalence is shown. ∎

Now suppose we want to consider the negation of Ψ. For many QBF solvers, the natural form of this negation is very cumbersome [8], so instead we define Ψ_{dual} to be the negation of Ψ_{alt}. We use the notation that $inverse(\overrightarrow{Q}_P)$ flips $\forall$ and $\exists$ keeping variables in the same order as $\overrightarrow{Q}_P$.

$$\Psi_{dual} \equiv inverse(\overrightarrow{Q}_P) \exists_G (G \wedge \neg(P)). \tag{3}$$

Notice that this expression differs from (1) in that property P is negated and the non-G quantifiers are inverted. However the deterministic system is represented the same as in Ψ.

Let us see the duals for our running example. Note that the non-G variables whose quantifiers get flipped are the circuit inputs, q, v, w, and the property p.

$$\Psi_{1,dual} \equiv \forall p \, \forall q \, \exists v \, \exists w \, \exists x \, \exists c \, \exists d \, (G \wedge (x \neq p))$$
$$\Psi_{2,dual} \equiv \forall p \, \forall q \, \exists v \, \exists w \, \exists x \, \exists c \, \exists d \, (G \wedge ((v = w) \wedge (x \neq p)))$$

We anticipate that applications with tools that encode domain problems into QBF will benefit by distinguishing between the deterministic system G and property P. They should generate both a primal and a dual formula, whatever the underlying deterministic system is. One formula will be valid and the other will

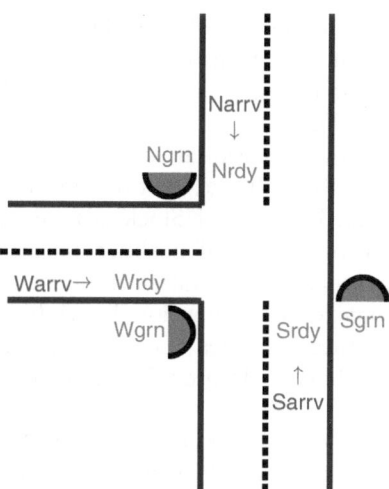

Fig. 2. A traffic controller for a three-way intersection. Inputs are Narrv, Sarrv, Warrv, as well as Wright, not shown. The main state variables are Nrdy, Srdy, Wrdy, Ngrn, Sgrn, Wgrn.

be invalid. Extracting additional information from the solution may proceed generally in the same way, whatever tools are available to the application, because exactly one of the primal and dual is valid, and exactly one is invalid.

Experience indicates that most solvers handle one type of QBF (i.e., valid or invalid) better than the other type. A possible additional benefit is that a solver might be able to solve one encoding (i.e., primal or dual) much more easily than another. One of the main contributions of this paper is the conjecture that running one solver on the primal and dual in tandem will shorten the overall time on a suite of valid and invalid formulas. Experimental evidence supporting this conjecture is given in Section 6.

4   Example: Modeling a Traffic Controller

We illustrate the modeling process with a simple traffic controller for a three-way intersection. This example is inspired by a tutorial by E. Nurvitadhi [21], but is different in several ways from that tutorial. Many technical details are omitted due to similarity to that tutorial, but the code to generate instances and the QBF files that were tested are publicly available at the URL given in Section 1. The setup is shown in Figure 2.

The variables are interpreted as follows. Nrdy is a state variable that means a car is ready to enter the intersection from the north. Narrv is an input variable that means Nrdy will be true at the next time step. Ngrn is a state variable that is true when the north-facing light is green, meaning that a car from the north passes the intersection by the next time step; in this case, Nrdy becomes false at the next time step unless Narrv is also true. The "south" and "west" directions are similar. Wright is an input that enables "right turn on red" by a car at Wrdy if there is no car at Nrdy, an exception to the rule that a green light is required to proceed. This is an input because the car might wish to turn left, in which case it remains there until Wgrn is true.

The traffic controller follows functional rules, depending on the inputs at the current and preceding time steps. It uses a state variable NSlock to assist in the computation. The intention is that if NSlock is true, then Wgrn is false. The named state variables are persistent: if nothing occurs to change the value of one of these variables, it keeps that value at the next time step. The encoding into CNF introduces about 20 other functional variables per time step that are not consistent unless the state variables obey the state transition functions.

The safety condition is that, if Ngrn or Sgrn is true, then Wgrn must *not* be true in the same time step.

In the primal encoding, the A player controls (N,S,W)arrv, while the E player controls Wright. The E player wins if the safety condition *holds*, and the formula is valid in these cases.

The logic of the traffic controller has an intentional bug, in that it is possible that NSlock is false in a configuration that allows both Wgrn and Ngrn to be true at the next time step. To activate the bug requires a specific sequence of arrivals over four time steps. It would be very easy for the A player to expose the bug if the arrv inputs were unrestricted. Therefore, an additional functional system limits the time steps at which Narrv, Sarrv, and Warrv are enabled. (When an input is enabled it is still the A player's choice whether to make it true.) To permit variety, a set of initial conditions controls which arrv variables are enabled at which time steps.

The system is simulated for a fixed number of time steps, n, which is chosen so that about 50 percent of the various initial conditions allow the bug to be manifested. Our generator is able to generate suites of problems with varying size parameters. Within a suite, all problems are about the same size, and approximately half are valid and half are invalid, for the primal encoding.

In the reported tests, $n = 21$. There are 42 primal problems and 42 dual problems in the suite. Each encoding uses 747 quantified variables; the number of clauses is 2023 for the primals, 2043 for the duals.

Let us now step through the process of forming the primal and dual encodings. Recall that we use $*x$ to denote that x is a system variable that is functionally determined by the input variables that precede x in the quantifier prefix. The p_k below constitute initial conditions, so there are no preceding input variables; they will be forced to take on certain constant values to satisfy the subformula system$_0$. At a high level the primal quantifier prefix takes the form shown in Figure 3. The primal quantifier-free formula encodes

$$\text{system}_0 \wedge \left(\bigwedge_{i=1}^{n} \text{system}_i \right) \wedge \left(\bigwedge_{j=1}^{n} \text{safe}_j \right)$$

Here, system$_i$ is a quantifier-free formula that encodes a function from parameters (N,S,W)arrv$_i$, Wright$_i$, and earlier input variables, to output state variables

$* p_k$
$\forall$ (N,S,W)arrv$_1$ $\exists$ Wright$_1$ $*$ (N,S,W)rdy,(N,S,W)grn,NSlock$_1$
$\forall$ (N,S,W)arrv$_2$ $\exists$ Wright$_2$ $*$ (N,S,W)rdy,(N,S,W)grn,NSlock$_2$
$\cdots$
$\forall$ (N,S,W)arrv$_n$ $\exists$ Wright$_n$ $*$ (N,S,W)rdy,(N,S,W)grn,NSlock$_n$

Fig. 3. Traffic Controller primal quantifier prefix

$* p_k$
$\exists$ (N,S,W)arrv$_1$ $\forall$ Wright$_1$ $*$ (N,S,W)rdy,(N,S,W)grn,NSlock$_1$
$\exists$ (N,S,W)arrv$_2$ $\forall$ Wright$_2$ $*$ (N,S,W)rdy,(N,S,W)grn,NSlock$_2$
$\cdots$
$\exists$ (N,S,W)arrv$_n$ $\forall$ Wright$_n$ $*$ (N,S,W)rdy,(N,S,W)grn,NSlock$_n$

Fig. 4. Traffic Controller dual quantifier prefix

(N,S,W)rdy$_i$, (N,S,W)grn$_i$, NSlock$_i$. That is, for each possible assignment to the parameter variables, exactly one assignment to the output variables satisfies **system**$_i$.

The formula **safe**$_i$ encodes the mutual exclusion $(\neg \text{Wgrn}_i \vee (\neg \text{Ngrn}_i \wedge \neg \text{Sgrn}_i))$.

In the standard primal encoding, the functionally quantified variables are assigned to the E player. That is, the "$*$" symbols become "$\exists$". Of course, the E player has no real choice about these assignments because he or she loses immediately if any part of **system**$_i$ is false.

The alternate primal encoding has the same quantifier prefix with functional quantifiers included, but the quantifier-free formula encodes

$$\left(\text{system}_0 \wedge \bigwedge_{i=1}^{n} \text{system}_i \right) \rightarrow \left(\bigwedge_{j=1}^{n} \text{safe}_j \right)$$

In the alternate primal encoding, the functionally quantified variables are assigned to the A player. That is, the "$*$" symbols become "$\forall$". As with the standard primal, the A player has no real choice about these assignments because he or she loses immediately if any part of **system**$_i$ is false. As stated in Theorem 3.1, the alternate primal encoding is logically equivalent to the standard primal encoding.

Finally, the dual encoding is the negation of the *alternate* primal encoding, so the functional quantifiers are back to existential and the other quantifiers flip, giving the prefix shown in Figure 4. The negated formula becomes

$$\text{system}_0 \wedge \left(\bigwedge_{i=1}^{n} \text{system}_i \right) \wedge \left(\bigvee_{j=1}^{n} \neg \text{safe}_j \right)$$

That is, only the safety property needs to be negated. Note that negating the "and" creates an n-way "or".

5   Dual Encoding Requires Domain Knowledge

A natural question is whether the dual encoding can be deduced automatically from the primal. Both `cirqit2` [8] and `ghostq` [15] make some attempt in this direction, but do not have a way to distinguish the functional description of the deterministic system from the property being queried. Partial duality, proposed in [9] but for which code is not publicly available, also cannot make this distinction. We now argue that it is not feasible to detect this distinction automatically.

The traffic controller example illustrated the difficulty. In the primal encoding Wright and Wrdy are both existential, but in the dual, only one of them becomes universal. How does an automatic translator know this? In limited cases, the translator can pattern match clauses to typical encodings of logic gates, and perhaps deduce that Wright is an input, while Wrdy is a gate output. Theory and experience both tell us this does not always succeed.

Essentially, to know that a certain set of clauses (or any quantifier-free subformula) encodes a boolean function, the translator must be able to tell that certain clauses have exactly one satisfying assignment. This problem is at least as hard as satisfiability itself [14, Ch. 3.2, Problem 31], even if only one set of clauses needs to be considered.

In practice, the translator would need to consider *many* subsets of the entire set of clauses to try to identify those that define a deterministic system. We cannot expect a translator to do this within reasonable time constraints.

Therefore, the people with domain knowledge of the application need to be involved to identify the functional variables. We proposed the "*" quantifier as mathematical notation for this purpose. In an instance patterned after the widely used `qdimacs` format, we propose "d" as a supplement to "a" and "e." It is up to the designer to ensure that declared functional variables really only can take on one value to satisfy the defining clauses, once the relevant inputs are assigned.

6   Experimental Results

To evaluate the conjecture that dual encodings are beneficial to CPU resources, we tested several prominent QBF solvers on QBF families that had both primal and dual encodings of the underlying application problem. The hardware was a 48-core AMD64 processor with 2.0 GHz clock and about 190 GB of memory. The system managed the load so that the number of processes never exceeded the number of cores.

The code to generate duals and the QBF files that were tested are publicly available at the URL given in Section 1. For the families reported in Tables 3 and 4, `awk` and `csh` scripts identify the "input" variables and a single "property" variable by an *ad hoc* technique. To convert a primal instance into its dual, the quantifier types of the "input" variables were inverted, and the "property" variable in inverted in its unit clause.

For calibration, Table 1 summarizes how four solvers performed on the 568 benchmarks used in QBFEVAL-10 [22]. Preprocessors are emerging as being

Table 1. Solver profiles for 568 instances in QBFEVAL-10

Solver	Number solved total (valid,invalid)	avg. secs.	total seconds
qube7.2(2)	436 (205,231)	135	296484
depqbf0.1(1)	315 (156,159)	150	503340
cirqit2(1)	251 (106,145)	30	578940
ghostq(1)	172 (74, 98)	29	716788
ghostq	150 (60, 90)	101	767557
bloqqer(3)	148 (62, 86)	5.5	3117

NOTES FOR ALL TABLES:
(1) includes `bloqqer` preprocessor.
(2) includes built-in preprocessor.
(3) incomplete preprocessor, never timed out.
avg. secs. includes solved only.
total seconds includes 1800 (the timeout) for unsolved.
PAR10 hours includes 18000 secs. (10 × timeout) for unsolved.

critical to the overall performance of QBF solvers [4], so tests were conducted with a preprocessor included. `Qube` and `depqbf` are competition leaders, while `cirqit2` [8] and `ghostq` [15] are solvers designed to exploit the duality of QBF. The latter solvers are also designed for non-CNF inputs, and are not always at their best with the prevalent CNF format. The report on `ghostq` indicates that it analyzes the gate structure of the input file, so we also tested it *without* a preprocessor. In limited preliminary tests, `ghostq` performed better without a general-purpose preprocessor on the circuit-oriented instances in Tables 2 and 3, so its results are presented that way.

Although `cirqit` and `ghostq` attempt to exploit the duality of QBF, they consider a dual that essentially negates the entire formula, whereas the duals that we encoded negate only the "property" part of the formula.

The ***Dual-Tandem*** solving strategy starts two processes using the same solver on a multi-core machine. One process attempts the primal, and the other process attempts the dual, of the same underlying problem. The timeout is cut in half, compared to single-encoding strategies. As soon as one process gets a solution, it kills the partner process.

Results for the traffic-controller suite of 42 problems, with the primal encoding and dual encoding each for each problem, are summarized in Table 2. These problems were homogeneous enough that all solvers solved all problems, so there are no time-outs to confuse the data. The primal encoding had 25 valid and 17 invalid instances. The *Dual-Tandem* strategy is compared to just attempting the primal, and just attempting the dual. The *Dual-Tandem* CPU time is the sum of the times used by both processes. The others are single-process times. This convention is followed for all tests.

The traffic controller was too easy, as most instances were solved by the pre-processor, and the solver behavior was not well tested. However, it shows that

Table 2. Total seconds for traffic controller suite 021 (42 problems, primal valid on 25, invalid on 17), for four solvers and three encoding strategies. No time limits; otherwise see notes to Table 1.

Solver	Primal	Dual	Dual-Tandem
qube7.2(2)	50509.20	0.59	1.18
depqbf0.1(1)	0.02	3150.26	0.04
cirqit2(1)	0.02	12.15	0.04
ghostq	3.51	0.69	1.36

the encoding can make a big difference in solving difficulty. Dual-Tandem offers insurance against guessing wrong about which encoding to use.

Results for the tipfixpoint family in QBFLIB, consisting of 446 problems, with the primal encoding and dual encoding each for each problem, are summarized in Table 3. Not all problems were solved. A timeout of 1800 was used (900 each for Dual-Tandem). The family tipdiam in QBFLIB (203 problems, two encodings each) produced a similar pattern, and is shown in Table 4.

The tipfixpoint and tipdiam families encode genuine bounded model checking problems. It is important to note that generation of the duals required domain knowledge and could *not* be done simply by referring to the (primal) benchmarks in QBFLIB. Published documentation [12], the SMV specifications [21], and publicly available software contained enough information about the underlying problems for the authors to develop a translator into the dual encoding for each family. Most families encoded in QBFLIB do not have sufficient documentation, or require domain expertise beyond that of the authors, to undertake independent encodings.

The tipfixpoint and tipdiam families are two of the families that ghostq was reported as doing very well on [15], but they only tested on the parts of the families that were selected for the QBFEVAL-08 event. The tables show that ghostq is the clear leader for these entire families, and that dual encoding is not of benefit to it.

The other three solvers benefited greatly from the dual encoding and the Dual-Tandem strategy. More problems were solved and total CPU resources were reduced. The PAR10 measure (*penalized average runtime*, with failures charged 10 times the CPU limit) combines both factors.

7   Conclusion

We have argued for an overall approach to using QBF for applications that includes primal and dual encoding. We showed that, in general, automated postprocessing cannot detect the functional variables and construct the dual, at least not efficiently. Therefore, both encodings must be designed with knowledge of the application domain. In instances patterned after the widely used qdimacs format, we propose "d" as a supplement to "a" and "e" to declare determin-

Table 3. Total seconds for `tipfixpoint` family in QBFLIB (446 problems), for four solvers and three encoding strategies

Solver	Primal			Dual			Dual-Tandem		
	Number solved total (valid,invalid)	total seconds	PAR10 hours	Number solved total (valid,invalid)	total seconds	PAR10 hours	Number solved	total seconds	PAR10 hours
qube7.2(2)	265 (196,69)	359863	914	266 (46,220)	346832	906	313	261173	671
depqbf0.1(1)	145 (74,71)	545117	1506	180 (48,132)	483306	1331	211	426447	1176
cirqit2(1)	129 (63,66)	575197	1586	134 (39,95)	574776	1564	172	504375	1373
ghostq	379 (307,72)	125032	336	378 (72,306)	126372	341	379	128934	337

(1) , (2) See notes to Table 1.

Table 4. Total seconds for `tipdiam` family in QBFLIB (203 problems), for four solvers and three encoding strategies

Solver	Primal			Dual			Dual-Tandem		
	Number solved total (valid,invalid)	total seconds	PAR10 hours	Number solved total (valid,invalid)	total seconds	PAR10 hours	Number solved	total seconds	PAR10 hours
qube7.2(2)	164 (145,19)	78999	197	146 (8,138)	108185	287	166	73314	187
depqbf0.1(1)	128 (118,10)	136972	376	132 (15,117)	128258	355	146	103393	285
cirqit2(1)	105 (100,5)	181743	491	124 (9,115)	144380	396	132	128291	355
ghostq	175 (155,20)	51321	140	175 (20,155)	51247	140	175	51992	140

(1) , (2) See notes to Table 1.

istic functional variables. Empirical evaluations tentatively support the conjecture that having dual encodings available should benefit many applications. In the future, we expect system modelers to incorporate both primal and dual encodings for a wide variety of applications.

Acknowledgment. We thank Fahiem Bacchus and Alexandra Goultiaeva for helpful discussions. We thank Armin Biere for providing data and programs that were constructed as part of Jussila and Biere [12].

References

1. Balabanov, V., Jiang, J.R.: Unified QBF certification and its applications. Formal Methods in System Design 41, 45–65 (2012)
2. Benedetti, M.: Extracting certificates from quantified boolean formulas. In: Proceedings of the International Joint Conference on Artificial Intelligence (IJCAI) (2005)
3. Benedetti, M., Mangassarian, H.: QBF-based formal verification: Experience and perspectives. JSAT 5, 133–191 (2008)
4. Biere, A., Lonsing, F., Seidl, M.: Blocked clause elimination for QBF. In: Bjørner, N., Sofronie-Stokkermans, V. (eds.) CADE 2011. LNCS, vol. 6803, pp. 101–115. Springer, Heidelberg (2011)
5. Biere, A.: Resolve and expand. In: Hoos, H.H., Mitchell, D.G. (eds.) SAT 2004. LNCS, vol. 3542, pp. 59–70. Springer, Heidelberg (2005)
6. Giunchiglia, E., Narizzano, M., Tacchella, A.: Clause/term resolution and learning in the evaluation of quantified boolean formulas. JAIR 26, 371–416 (2006)
7. Giunchiglia, E., Narizzano, M., Tacchella, A.: QBF reasoning on real-world instances. In: Hoos, H.H., Mitchell, D.G. (eds.) SAT 2004. LNCS, vol. 3542, pp. 105–121. Springer, Heidelberg (2005)
8. Goultiaeva, A., Bacchus, F.: Exploiting QBF duality on a circuit representation. In: AAAI (2010)
9. Goultiaeva, A., Bacchus, F.: Recovering and utilizing partial duality in QBF. In: Järvisalo, M., Van Gelder, A. (eds.) SAT 2013. LNCS, vol. 7962, pp. 83–99. Springer, Heidelberg (2013)
10. Goultiaeva, A., Van Gelder, A., Bacchus, F.: A uniform approach for generating proofs and strategies for both true and false QBF formulas. In: Proc. IJCAI (2011)
11. Jussila, T., Biere, A., Sinz, C., Kroning, D., Wintersteiger, C.M.: A first step towards a unified proof checker for QBF. In: Marques-Silva, J., Sakallah, K.A. (eds.) SAT 2007. LNCS, vol. 4501, pp. 201–214. Springer, Heidelberg (2007)
12. Jussila, T., Biere, A.: Compressing BMC encodings with QBF. Electr. Notes Theor. Comput. Sci. 174(3), 45–56 (2007)
13. Kleine Büning, H., Karpinski, M., Flögel, A.: Resolution for quantified boolean formulas. Information and Computation 117, 12–18 (1995)
14. Kleine Büning, H., Lettmann, T.: Propositional Logic: Deduction and Algorithms. Cambridge University Press (1999)
15. Klieber, W., Sapra, S., Gao, S., Clarke, E.: A non-prenex, non-clausal QBF solver with game-state learning. In: Strichman, O., Szeider, S. (eds.) SAT 2010. LNCS, vol. 6175, pp. 128–142. Springer, Heidelberg (2010)

16. Kontchakov, R., Pulina, L., Sattler, U., Schneider, T., Selmer, P., Wolter, F., Zakharyaschev, M.: Minimal module extraction from DL-lite ontologies using QBF solvers. In: Proceedings of the International Joint Conference on Artificial Intelligence (IJCAI), pp. 836–841 (2009)

17. Lonsing, F., Biere, A.: Integrating dependency schemes in search-based QBF solvers. In: Strichman, O., Szeider, S. (eds.) SAT 2010. LNCS, vol. 6175, pp. 158–171. Springer, Heidelberg (2010)

18. Mangassarian, H., Veneris, A.G., Safarpour, S., Benedetti, M., Smith, D.: A performance-driven QBF-based iterative logic array representation with applications to verification, debug and test. In: International Conference on Computer-Aided Design. pp. 240–245 (2007)

19. Marin, P., Miller, C., Lewis, M., Becker, B.: Verification of partial designs using incremental QBF solving. In: Proc. DATE (2012)

20. Narizzano, M., Peschiera, C., Pulina, L., Tacchella, A.: Evaluating and certifying QBFs: A comparison of state-of-the-art tools. AI Commun. 22(4), 191–210 (2009)

21. Nurvitadhi, E.: SMV tutorial—part I (2009), http://www.cs.cmu.edu/-~emc/15817-f09/nurvitadhi.SMV-tutorial-part1.pdf

22. Peschiera, C., Pulina, L., Tacchella, A., Bubeck, U., Kullmann, O., Lynce, I.: The seventh QBF solvers evaluation (QBFEVAL'10). In: Strichman, O., Szeider, S. (eds.) SAT 2010. LNCS, vol. 6175, pp. 237–250. Springer, Heidelberg (2010)

23. Pulina, L., Tacchella, A.: A structural approach to reasoning with quantified boolean formulas. In: IJCAI, pp. 596–602 (2009)

24. Sabharwal, A., Ansótegui, C., Gomes, C.P., Hart, J.W., Selman, B.: QBF modeling: Exploiting player symmetry for simplicity and efficiency. In: Biere, A., Gomes, C.P. (eds.) SAT 2006. LNCS, vol. 4121, pp. 382–395. Springer, Heidelberg (2006)

25. Staber, S., Bloem, R.: Fault localization and correction with QBF. In: Marques-Silva, J., Sakallah, K.A. (eds.) SAT 2007. LNCS, vol. 4501, pp. 355–368. Springer, Heidelberg (2007)

Asynchronous Forward Bounding Revisited

Mohamed Wahbi[1], Redouane Ezzahir[2], and Christian Bessiere[3]

[1] TASC (INRIA/CNRS), Mines Nantes, France
[2] ENSA Agadir, University Ibn Zohr, Morroco
[3] University of Montpellier, France
mohamed.wahbi@emn.fr, red.ezzahir@gmail.com,
bessiere@lirmm.fr

Abstract. The Distributed Constraint Optimization Problem (DCOP) is a powerful framework for modeling and solving applications in multi-agent coordination. Asynchronous Forward Bounding (AFB_BJ) is one of the best algorithms to solve DCOPs. We propose AFB_BJ$^+$, a revisited version of AFB_BJ in which we refine the lower bound computations. We also propose to compute lower bounds for the whole domain of the last assigned agent instead of only doing this for its current assignment. This reduces both the number of messages needed and the time future agents remain idle. In addition, these lower bounds can be used as a value ordering heuristic in AFB_BJ$^+$. The experimental evaluation on standard benchmark problems shows the efficiency of AFB_BJ$^+$ compared to other algorithms for DCOPs.

1 Introduction

Distributed Constraint Optimization Problem (DCOP) is a powerful framework to model a wide range of applications in multi-agent coordination such as distributed scheduling [14], distributed planning [4], distributed resource allocation [17], target tracking in sensor networks [15] distributed vehicle routing [12], etc. A DCOP consists of a group of autonomous agents, where each agent has an independent computing power. Each agent owns a local constraint network. Variables owned by different agents are connected by constraints. These constraints specify a non-negative constraint cost for combinations of values assigned to the variables they connect. In general, constraints or value assignments may be strategic information or private choice that should not be revealed or delegated to other agents. Thus, each agent only has control on its variables and only knows constraints that involve them. DCOP addresses problems in which agents must, *in a distributed manner*, assign values to their variables such that the sum of the constraint costs of all constraints is minimized.

Several complete algorithms for solving DCOPs have been proposed in the last decade. The pioneer complete asynchronous algorithm is Adopt [15]. Later on, the closely related BnB-Adopt [20] was presented. BnB-Adopt changes the nature of the search from Adopt best-first search to a depth-first branch-and-bound strategy, obtaining better performance. Gutierrez and Meseguer show that some of the messages exchanged by Adopt and BnB-Adopt turned out to be redundant [9]. By removing

C. Schulte (Ed.): CP 2013, LNCS 8124, pp. 708–723, 2013.

these redundant messages they obtain more efficient algorithms: Adopt$^+$ and BnB-Adopt$^+$. The algorithms mentioned so far perform assignments concurrently and asynchronously. Thereby, the perception of agents on the variable assignments of other agents is in general inconsistent.

Another category of algorithms for solving DCOPs is that of algorithms performing assignments sequentially and synchronously. The synchronous branch and bound (SyncBB) [10] is the basic systematic search algorithm in this category. In SyncBB, only the agent holding the token is allowed to perform an assignment while the other agents remain idle. Once it assigns its variables, it passes on the token and then remains idle. Thus, SyncBB does not make any use of concurrent computation. No-Commitment Branch and Bound (NCBB) is another synchronous polynomial-space search algorithm for solving DCOPs [5]. To capture independent sub-problems, NCBB arranges agents in constraint tree ordering. NCBB incorporates, in a synchronous search, a concurrent computation of lower bounds in non-intersecting areas of the search space based on the constraint tree structure.

Another attempt to incorporate a concurrent computation in a synchronous search was applied in Asynchronous Forward Bounding (AFB) [6]. AFB can be seen as an improvement of SyncBB where agents extend a partial assignment as long as the lower bound on its cost does not exceed the global upper bound (i.e., the cost of the best solution found so far). In AFB, the lower bounds are computed concurrently by unassigned agents. Thus, each synchronous extension of the current partial assignment is followed by an asynchronous forward bounding phase. Forward bounding propagates the bounds on the cost of the partial assignment by sending to all unassigned agents copies of the extended partial assignment. When the lower bound of all assignments of an agent exceeds the upper bound, it performs a simple backtrack to the previous assigned agent. Later, the AFB has been enhanced by the addition of a backjumping mechanism, resulting in the AFB_BJ algorithm [7]. The authors report that AFB_BJ, especially combined with the minimal local cost value ordering heuristic performs significantly better than other DCOP algorithms.

In this paper, we propose AFB_BJ$^+$, a revisited version of AFB_BJ in which we refine the lower bound computations. We also propose to compute lower bounds for the whole domain of the last assigned agent instead of only doing this for its current assignment. Thus, an unassigned agent computes the lower bound for each value in the domain of the agent requesting it. This reduces both the number of messages needed and the time future agents remain idle. Hence, we take all possible advantage from the asynchronicity of the system. In addition, these lower bounds are used as a value ordering heuristic in AFB_BJ$^+$. Thus, an agent assigns first values with minimal lower bound.

This paper is structured as follows. Section 2 gives the necessary background on DCOP and a short description of the AFB algorithms. We present the AFB_BJ$^+$ algorithm in Section 3. Correctness proofs are given in Section 4. We report experimental results in Section 5. Finally, we conclude in Section 6.

2   Background

2.1   Basic Definitions and Notations

The *Distributed Constraint Optimization Problem* (DCOP) has been formalized in [15] as a quadruple $(\mathcal{A}, \mathcal{X}, \mathcal{D}, \mathcal{C})$, where $\mathcal{A}$ is a set of p agents $\{A_1, \ldots, A_p\}$, $\mathcal{X}$ is a set of n variables $\{x_1, \ldots, x_n\}$, where each variable x_j is controlled by one agent in $\mathcal{A}$. $\mathcal{D} = \{D_1, \ldots, D_n\}$ is a set of n domains, where D_j is the set of possible values to which variable x_j may be assigned. Only the agent controlling a variable can assign a value to it and has knowledge of its domain. $\mathcal{C} = \{c_{ij} : D_i \times D_j \to \mathbb{R}^+\}$ is a set of binary utility constraints (i.e., soft constraints). Each utility constraint $c_{ij} \in \mathcal{C}$ is defined over the pair of variables $\{x_i, x_j\} \subseteq \mathcal{X}$. We say that x_i and x_j are *neighbors*.

For simplicity purposes, we consider a restricted version of DCOP where each agent holds exactly one variable ($p = n$). Thus, we use the terms agent (A_j) and variable (x_j) interchangeably and we identify the agent ID with its variable index (j). Furthermore, all agents store a unique total order $\prec$ on agents. Agents appearing before an agent $A_j \in \mathcal{A}$ in the total order are the higher priority agents and those appearing after A_j are the lower priority agents. The order $\prec$ divides the set $\Gamma(x_j)$ of neighbors of A_j into higher priority neighbors $\Gamma^-(x_j)$, and lower priority neighbors $\Gamma^+(x_j)$. For sake of clarity, we assume that the total order is the lexicographic ordering $[A_1, A_2, \ldots, A_n]$. In the rest of the paper, we consider a generic agent $A_j \in \mathcal{A}$. Thus, j is the level of agent A_j.

An *assignment* for agent A_j is a tuple (x_j, v_j), where v_j is a value from D_j. A_j maintains a counter t_j and increments it whenever it changes its value. The value of the counter *tags* each generated assignment. When comparing two assignments for the same agent, the most up to date is the one with the greatest tag. A *current partial assignment* (CPA) is an ordered set of assignments, e.g., $Y = [(x_1, v_1), \ldots, (x_j, v_j)]$ s.t. $x_1 \prec \ldots \prec x_j$. The set of all variables assigned in Y is denoted by $var(Y) = \{x_1, \ldots, x_j\}$. A *time-stamp* associated to a CPA Y is an ordered list of counters $[t_1, t_2, \ldots, t_j]$ where t_i is the tag of the variable x_i s.t. $x_i \in var(Y)$ [16,19]. When comparing two CPAs, the *strongest* one is that associated with the lexicographically greater time-stamp. That is, the CPA with greatest value on the first counter on which they differ, if any, otherwise the longest one. Let $Y = [(x_1, v_1), \ldots, (x_i, v_i), \ldots, (x_j, v_j)]$ be a CPA, the subset of Y including all variables down to x_i is denoted by $Y^i = [(x_1, v_1), \ldots, (x_i, v_i)]$.

The guaranteed cost of a CPA Y, denoted by $gc(Y)$, is the sum of all utility constraints c_{ij} s.t. x_i and x_j are assigned in Y (Eq. 1).

$$gc(Y) = \sum_{c_{ij} \in \mathcal{C}} c_{ij}(v_i, v_j) \mid (x_i, v_i), (x_j, v_j) \in Y. \tag{1}$$

A full assignment Y is a CPA that involves all variables of the problem, i.e., $var(Y) = \mathcal{X}$. The goal of a DCOP solver is to distributively find a full assignment Y^* with minimal cost, that is, $Y^* = \arg \min_Y \{gc(Y) \mid var(Y) = \mathcal{X}\}$.

In the following, we will present standard AFB algorithms. We will use a nomenclature of messages and data structures different from those used in the original paper [7] in order to be closer to those used in our approach.

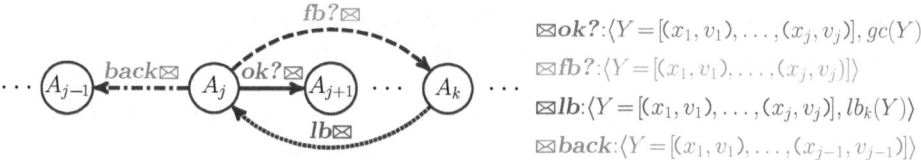

2.2   Asynchronous Forward Bounding (AFB) Algorithm

In Asynchronous Forward Bounding (AFB) [6], agents assign their variables sequentially and unassigned agents asynchronously try to compute lower bounds on the CPA, say Y. Agents perform assignments of their variables only when they hold the current partial assignments Y (i.e., Y is the token). Each extension of the CPA Y, is followed by a Forward Bounding (FB) phase. The FB phase is performed by sending forward copies of Y to all unassigned agents. In the FB phase, it is required from unassigned agents to compute a lower bound on the cost increment caused by an assignment of their variables on Y. Once computed, the lower bounds are sent back to the agent that sent the request, i.e., the last assigned agent in Y. Due to the asynchronous nature of the FB phase, multiple CPAs may be present at a given moment in time. However, the time-stamp mechanism is used by agents to discard obsolete ones.

The lower bounds collected from unassigned agents are used to compute a lower bound on the CPA. When the computed lower bound becomes larger than the current upper bound (i.e., the cost of the best full CPA found so far), the CPA is pruned. Concretely, whenever agent A_j receives a valid lower bound from an unassigned agent, it adds it to that received from other agents and checks if the cumulative lower bound exceeds the upper bound. In such a case, A_j tries to assign an alternative value to its variable. If such value is not available, it needs to backtrack. When agent A_j takes the decision to backtrack, it sends the CPA (Y^{j-1}) backwards to the last agent assigned on it (i.e., A_{j-1}). However, if A_j is the first agent in the ordering, it ends the search process after claiming this to other agents. The AFB algorithm then reports that the optimal solution is the best full CPA found so far.

Fig. 1 shows the messages exchanged by the AFB algorithm.[1] AFB agents exchange the following types of messages:

ok?: a message which contains the CPA Y with its cost $gc(Y)$. When A_j assigns its variable, it sends this message to the next agent in the ordering (A_{j+1}).

back: a message which contains an inconsistent CPA. It is sent back to agent A_{j-1} requiring it to change its assignment.

fb?: a message which contains a copy of an **ok?** message. It is sent by A_j to unassigned agents to compute a lower bound on the CPA it carries.

lb: a message which contains a lower bound on the current partial assignment. It is sent as response to a **fb?** message.

---

[1] The names of the message types here are closer to that used in our approach and then different from that used in the original AFB paper [6].

The computation of lower bounds on AFB is performed as follows. In a preprocessing step, A_j computes the minimal future cost estimation for each possible value in its domain incurred by every lower priority agent A_k (Eq. 2). $fc_j(v_j)$ is a lower bound on the cost of constraints involving the assignment (x_j, v_j) and all its lower priority neighbors. Agents compute these estimations only once and store them.

$$fc_j(v_j) = \sum_{x_k \in \Gamma^+(x_j)} \min_{v_k \in D_k} \{c_{jk}(v_j, v_k)\}. \tag{2}$$

Given a current partial assignment Y^i, and an unassigned agent A_j, the local cost of assigning a value v_j to A_j is the sum of the constraints costs of this value with all assignments in Y^i s.t. $i < j$ (Eq. 3).

$$lc_j(Y^i, v_j) = \sum_{(x_h, v_h) \in Y^i \ s.t. \ h \leq i < j} c_{hj}(v_h, v_j) \tag{3}$$

Summing the local cost of an assignment (x_j, v_j) and the minimal future cost incurred by lower priority neighbors provides a future cost of an eventual extension of Y^i with (x_j, v_j). The lower bound of Y^i on a unassigned agent (A_j) is the minimal future cost over all its values (Eq. 4). Thus, whenever a higher priority agent A_i requires from A_j to compute a lower bound on a CPA Y^i, it responds by sending back the minimal lower bound of Y^i over all values in its domain, i.e., $lb_j(Y^i)$.

$$lb_j(Y^i) = \min_{v_j \in D_j} \{lc_j(Y^i, v_j) + fc_j(v_j)\}. \tag{4}$$

By collecting lower bounds from lower priority agents, agent A_j can compute a lower bound on its CPA Y^j. The lower bounds on a CPA Y^j reported by lower priority agents are accumulated and summed up with the guaranteed cost of Y^j to provide a lower bound on the cost of a complete extension of Y^j (Eq. 5).

$$lb(Y^j) = gc(Y^j) + \sum_{A_k \succ A_j} lb_k(Y^j) \tag{5}$$

If the computed lower bound $lb(Y^j)$ exceeds the current known upper bound (UB_j), A_j needs to change its current value v_j on Y^j by a new value v'_j generating a stronger CPA. Then, search continues with the generated CPA. If A_j has already tested all possible values for its variable, it backtracks, asking the previous agent in the ordering (A_{j-1}) to assign a new value to its variable through a **back** message.

2.3   Asynchronous Forward Bounding with CBJ (AFB_BJ)

The Asynchronous Forward Bounding with backjumping (AFB_BJ) was obtained by adding a backjumping mechanism to standard AFB [7]. When the lower bounds of all values exceed the upper bound, instead of backtracking to the most recently assigned variable, AFB_BJ tries to jump to the last assigned agent such that its re-assignment could possibly lead to a solution. To this end, agents in AFB_BJ use some maintained data structures that we introduce in the following. Another feature of AFB_BJ is that the agent that performs an assignment, uses the minimal local cost (Eq. 3) as value ordering heuristic.

When an agent A_j assigns its variable, it sends an **ok?** message to the next agent in the ordering. In AFB_BJ, the **ok?** message contains the current partial assignment Y^j and an array of guaranteed costs, one for each level. $gc(Y^j)[j]$ is the cost of $Y^j = [(x_1, v_1), \ldots, (x_j, v_j)]$ where $gc(Y^j)[0] = 0$. For each $i \in 1..j-1$, $gc(Y^j)[i]$ equals that received from A_{j-1}, i.e., $gc(Y^{j-1})[i]$.

$$gc(Y^j)[j] = gc(Y^j) = gc(Y^{j-1}) + lc_j(Y^{j-1}, v_j) \tag{6}$$

In order to perform backjumping, AFB_BJ agents compute a lower bound for each level on the CPA. Concretely, instead of computing the lower bound for the whole received CPA Y^i, A_j computes it for each Y^h where $h \le i < j$. To this end, A_j first computes the local cost for each level h (Eq. 7).

$$lc_j(Y^i, v_j)[h] = lc_j(Y^h, v_j) \tag{7}$$

The lower bound at level h (Eq. 8) is then the minimal lower bound of Y^h over all values in D_j. When a higher agent A_i requests from A_j to compute its lower bound on a CPA Y^i, it answers by sending an array of lower bounds, one for each level h where $h \le i < j$.

$$lb_j(Y^i)[h] = \min_{v_j \in D_j} \{lc_j(Y^i, v_j)[h] + fc_j(v_j)\}. \tag{8}$$

When A_j successfully assigns a value v_j to its variable, it sends forward copies of the extended CPA, Y^j, to each unassigned agent A_k and awaits for receiving from them the array of lower bounds. The lower bounds denoted by $lb_k(Y^j)$, is an array in which the ith element ($1 \le i \le j$) contains a lower bound on the cost of assigning a value to A_k with respect to the assignment on Y^i (Eq. 8). Once A_j receives lower bounds arrays, it computes a lower bound on the cost of any full assignment (Eqs. 9).

$$lb(Y^j)[i] = gc(Y^j)[i] + \sum_{A_k \succ A_j} lb_k(Y^j)[i]. \tag{9}$$

These lower bounds are used by the AFB_BJ to determinate the backjumping level. For more details about the way in which the level of backjumping is calculated we refer the reader to [7].

3   Asynchronous Forward Bounding Revisited

AFB_BJ$^+$ is a revisited version of AFB_BJ in which we propose a refinement of the lower bound computations. We also propose to compute lower bounds for the whole domain of the last assigned agent instead of only doing this for its current assignment. Thus, an unassigned agent computes the lower bound for each value in the domain of the agent requesting it. In addition, these lower bounds are used as a value ordering heuristic.

3.1   Lower Bound Refinement

When an agent A_i successfully assigns a value v_i to its variable, it sends forward copies of the extended CPA, Y^i, to each unassigned agent and awaits for receiving from them the array of lower bounds. When agent A_j receives this CPA Y^i (through a **fb?** message), it computes the lower bound for each level h where $h \le i < j$ (Eq. 8). When computing the lower bound of level h only assignments on Y^h are considered. We also add the cost of assigning v_i to x_i (i.e., (x_i, v_i)) to this lower bound. Moreover, we also add the minimal cost of constraints with variables (x_m) between x_h and x_i. Thus, instead of being a lower bound on a possible extension of Y^h by a possible assignment of A_j, it will be a lower bound on a possible extension of Y^h by both A_i and A_j and agents between x_h and x_i. Hence, we revise Eq. 8 to get Eq. 10 where the first and the last terms remain as in the original equation.

$$
lb_j(Y^i)[h] = \min_{v_j \in D_j} \left\{ lc_j(Y^i, v_j)[h] + \sum_{m=h+1}^{i-1} \min_{v_m \in D_m} \{c_{mj}(v_m, v_j)\} \right.
$$
$$
\left. + c_{ij}(v_i, v_j) + fc_j(v_j) \right\} \tag{10}
$$

At first glance, it seems that this will require more computational effort from unassigned agents, however it is not the case. One can simply compute the array of lower bounds, as is already done in AFB_BJ, and at the end it adds to each level the cost with variable x_i. We obtain the addition of the third term, i.e., $c_{ij}(v_i, v_j)$. To get the quantity to be added by the second term (i.e., $\sum_{m=h+1}^{i-1} \min_{v_m \in D_m} \{c_{mj}(v_j)\}$), we use the same principle used in Eq. 2. Agents compute for each value the estimations of each level only once and store them.

The refinement of the lower bounds computation allows agents to get more accurate lower bounds on their assignments. Thus, the accumulated lower bound at each level is increased. This mechanism allows earlier detection of CPAs with lower bound larger than the upper bound. In addition, by doing this, the **back** message will be sent as high as possible in the agent ordering, thus saving unnecessary search effort.

3.2 Lower Bounds for the Whole Domain

In AFB_BJ, the forward bounding phase is very expensive in term of communication load. FB requires for each value in D_j, $2 \times (n - j)$ messages (a **fb?** and a **lb** message for each lower agent). Thus, FB needs, for each CPA Y^{j-1}, $2 \times |D_j| \times (n-j)$ messages.

In AFB_BJ$^+$, we propose to compute lower bounds for the whole domain of the last assigned agent instead of only computing this for its current assignment. When an agent receives a **fb?** message it answers by sending back a two-dimensional array, an array for each value in the domain of the receiver agent. Hence, the forward bounding phase will need, for each CPA Y^{j-1}, only $2(n - j)$ messages, 2 messages for each lower agent. When agent A_j receives a **fb?** from agent A_i, instead of computing $lb_j(Y^i)[h]$ only for the current value of x_i, A_j computes it for each value v_i in D_i, $lb_j(Y^{i-1})[h][v_i]$ using Eq. 11.[2]

$$
\forall h \in 1..i-1, \ \forall v_i \in D_i, \ lb_j(Y^{i-1})[h][v_i] = lb_j(Y^{i-1} \cup (x_i, v_i))[h] \tag{11}
$$

---

[2] If x_i and x_j are not neighbors, a simple array is sufficient since the lower bound is the same for all values in D_i. Moreover, x_j does not known D_i.

3.3   Avoiding Redundancy

Another feature of the AFB_BJ$^+$ algorithm is that agents retain and maintain the received lower bounds to avoid redundant messages. When an agent A_j receives a *fb?* message from agent A_i, instead of clearing all information it stores (namely the collected lower bounds and the computed ones with their local costs), it clears only irrelevant information w.r.t the received CPA Y^i. Concretely, A_j compares the time-stamp of the received CPA with its CPA. If its CPA is stronger than the received one, the message is discarded. Otherwise, A_j gets the index $h \le i$ of the first counter on which they differ. All local costs and lower bounds on the current partial assignment Y^{h-1} remain valid. Thus, agent A_j will not re-compute lower bounds for this part.

The same thing is done for *ok?* messages. Whenever agent A_j receives a CPA Y^{j-1}, it updates all stored information by only removing parts that are not compatible with Y^{j-1}. When A_j succeeds in assigning its variable, it sends forward copy of the extended CPA Y^j in *fb?* messages to its lower agents. However, some of these messages are redundant. To avoid this, each agent A_j stores, for each lower priority agent A_k, the agent which is the closest to A_j in the neighbors of A_k higher than A_j. As long as the assignment of such agent or agents higher than him were not updated, there is no need to send *fb?* message to A_k. Thus, redundant messages and computations are saved. Moreover, the agent assigning its variable has more accurate lower bounds for all values in its domain. As long as the new complete array of lower bounds has not yet been received, the remaining valid part can be used as a lower bound estimation for each value in the current domain using Eq. 12. h is the lowest valid level for lower bound received from A_k.

$$lb(Y^j) = gc(Y^j) + \sum_{k \succ j} lb_k(Y^{j-1})[h][v_j] \ \ s.t. \ (x_j, v_j) \in Y^j \qquad (12)$$

3.4   Promising Value Ordering Heuristic

Unlike AFB_BJ that uses minimal local cost as value ordering heuristic, an AFB_BJ$^+$ agent uses a different strategy for reordering values in its current domain. All computations performed so far by unassigned agents to calculate lower bounds are used to reorder values in the current domain. Thus, when receiving an *ok?* message, A_j computes the lower bounds for all values in its domain using Eq. 12. A_j chooses to assign first values with minimal lower bound. Then, instead of considering only costs with past variable, both costs with past variables and estimations of costs on future variables are considered. We mimic an informed memory-bounded version of A^*, instead of simulating an uninformed memory-bounded version of A^*.

3.5   AFB_BJ$^+$ Description

Fig. 2 presents the pseudo-code of AFB_BJ$^+$ executed by every agent A_j. Agent A_j maintains a variable UB_j that stores the current upper-bound (the cost of the best solution found so far) initialized to $+\infty$, v_j^* that stores the value of A_j on the solution, Y that stores the strongest received CPA, GC an array of size $j - 1$ that stores the

procedure AFB-BJ^+ ( )

01. $UB_j \leftarrow +\infty;\ v_j^* \leftarrow empty;\ Y \leftarrow \{\};\ GC[1..j-1] \leftarrow [0,\dots,0];$

02. $mustSendFB \leftarrow True;$

03.   **foreach** ($A_k \succ A_j$) **do**

04.   **foreach** ($v_j \succ D_j$) **do** $lb_k(Y)[0][v_j] \leftarrow \min\limits_{v_k \in D_k} \{c_{jk}(v_j, v_k)\}$;

05.   **if** ($A_j = A_1$) **then** ExtendCPA ( ) ;

06.   **while** ($\neg end$) **do**

07. $msg \leftarrow$ getMsg ( ) ;

08.   **if** ($msg.UB < UB_j$) **then** $UB_j \leftarrow msg.UB;\quad v_j^* \leftarrow v_j$;

09.   **if** ($msg.Y$ is stronger than Y) **then**

10. $Y \leftarrow msg.Y;\ GC \leftarrow msg.GC$;

11.   clear irrelevant lower bounds ;

12.   **switch** ($msg.type$) **do**

13.   **ok?** : $mustSendFB \leftarrow true$; ExtendCPA ( ) ;

14.   **back** : $Y \leftarrow Y^{j-1}$; ExtendCPA ( ) ;

15.   **fb?** : sendMsg : **lb** $\langle lb_j(Y^i)[], msg.Y \rangle$ **to** A_i ;   /* A_i is msg sender */

16.   **lb** : ProcessLB (msg) ;

17.   **stp** : $end \leftarrow true$;

procedure ExtendCPA ( )

18. $v_j \leftarrow \arg \min\limits_{v_j' \in D_j} \{lb(Y \cup (x_j, v_j'))\}$;   /* Eq. 12 */

19.   **if** ($lb(Y \cup (x_i, v_i)) \geq UB_j$) **then** Backtrack ( ) ;

20.   **else**

21. $Y \leftarrow \{Y \cup (x_j, v_j)\};\quad t_j \leftarrow t_j + 1;$

22.   **if** ($var(Y) = \mathcal{X}$) **then**

23. $UB_j \leftarrow gc(Y)$;   /* $A_j = A_n$ */

24. $v_j^* \leftarrow v_j$;

25. $Y \leftarrow Y^{j-1};$

26.   ExtendCPA ( ) ;

27.   **else**

28.   sendMsg : **ok?** $\langle Y, GC, UB_j \rangle$ **to** A_{j+1} ;

29.   **if** ($mustSendFB$) **then**

30.   **foreach** ($A_k \succ A_j$) **do** sendMsg : **fb?** $\langle Y, GC, UB_j \rangle$ **to** A_k ;

31. $mustSendFB \leftarrow false$;

procedure Backtrack ( )

32.   **for** ($i \leftarrow j\text{-}1$ **dowTo** 1) **do**

33.   **if** ($lb(Y)[i-1] < UB_j$) **then**

34.   sendMsg : **back** $\langle Y^i, UB_j \rangle$ **to** A_i ; **return**;

35.   broadcastMsg : **stp** $\langle UB_j \rangle$;

36. $end \leftarrow true$;

procedure ProcessLB (msg)

37. $lb_k(Y^j) \leftarrow msg.lb$;   /* A_k is the sender of msg */

38.   **if** ($lb(Y^j) \geq UB_j$) **then** ExtendCPA ( ) ;

Fig. 2. The AFB_BJ^+ algorithm running on agent A_j

guaranteed costs where $GC[i] = gc(Y^i)$, and $lb_k(Y)[]$ that stores the lower bounds received from a lower agent A_k. Since $lb_k(Y)[0][v_j]$ depends only on the assignments of x_j and x_k, it is initialized to $\min_{v_k \in D_k} \{c_{jk}(v_j, v_k)\}$. Thus, it is a valid lower bound for all CPAs that contains (x_j, v_j). Eq. 2 is obtained by summing $lb_k(Y)[0][v_j]$ for each lower agent A_k.

AFB_BJ$^+$ starts by initializing the local data structures of A_j (lines 1-4). A_j then enters in the waiting and processing message loop (line 6). Each received message holds a CPA $msg.Y$ and its corresponding guaranteed costs $msg.GC$. Due to the asynchronous nature of the algorithm, some messages may be obsolete. A_j uses the time-stamping mechanism to discard those messages (line 9). If the received CPA ($msg.Y$) is stronger than Y, A_j updates Y and GC by the received ones (line 10). Then, A_j clears all irrelevant lower bounds computed or received so far (line 11). Agent A_j attaches to each message it sends its UB_j. The upper bound UB_j and v_j^* are updated when a received message carries a new upper bound smaller than the stored one (line 8).

Upon receiving an **ok?** message, A_j marks that it must send **fb?** messages by setting $mustSendFB$ to $true$. Next, it attempts to extend the received CPA by calling procedure ExtendCPA() (line 13).

When calling ExtendCPA(), A_j tries to find a value with the minimum lower bound (Eq. 12) without exceeding UB_j (lines 18-19). If such value does not exist, A_j backtracks (Backtrack() call, line 19). Otherwise, A_j extends the CPA by adding its new assignment and increments its counter t_j. If the resulting CPA includes assignments of all agents (line 22), a solution is found and then the upper bound is updated. Instead of broadcasting the new solution and its associated upper-bound, A_j calls ExtendCPA() to continue the search (line 26). Since UB_j is always attached to the exchanged messages, other agents will be informed of this new upper bound when continuing the search. At the end of the search, the best assignment of A_j is stored in v_j^*. If A_j is not the last agent on the ordering, it sends the extended CPA to the next agent (line 28). Afterwards, A_j sends **fb?** messages to all lower priority agents (lines 30-31).[3]

When A_j receives a **fb?** message, it computes for each value from the domain of the sender a lower bound on the cost increment caused by adding an assignment to its variable using Eq. 10. These lower bounds are sent back to the agent who sent the **fb?** message through a **lb** message.

When A_j receives a valid **lb** message, it saves the attached lower bounds (line 37). It checks if this new information causes the current partial assignment to exceed the upper-bound. In such a case, A_j calls ExtendCPA() in order to change its assignment (line 38).

Agent A_j calls procedure Backtrack() whenever the lower bounds of all its values exceed the upper-bound. When this occurs, A_j computes to which agent the CPA Y should be sent to (the backtracking target). A_j goes over all candidates, from $j - 1$ down to 1, looking for the first agent it finds that its reassignment could lead to a full assignment with a cost lower than UB_j. This agent is the latest assigned agent A_i where $lb(Y)[i-1] < UB_j$ (line 33). If such an agent exists, A_j sends him a **back** message

---

[3] In our implementation the **fb?** messages are sent under certain conditions to avoid redundancy, see Section 3.3.

(line 34). Otherwise, A_j reports this to other agents through **stp** messages (line 35) and terminates its execution.

4   Correctness Proofs

Lemma 1. *AFB_BJ$^+$ is guaranteed to terminate.*

Proof. (Sketch) The proof is close to the one given in [19]. It can easily be obtained by induction on the agent ordering that there will be a finite number of new generated CPAs (at most d^n, where n is the number of variables and d is the maximum domain size), and that agents can never fall into an infinite loop for a given CPA.   □

To prove that AFB_BJ$^+$ is correct, we need to prove that the correctness inherent to AFB_BJ is not violated by the lower bound refinements and the non-broadcasting of solution messages.

(Sketch) Assuming the correctness of AFB_BJ, the lower bounds without refinement terms are consistent. It is thus enough to prove that the costs included in the refinement terms are not redundant. All constraints considered in the calculation of the second and third terms of Eq. 10 have not been included in the first and fourth terms. Moreover, these constraints are not included in the lower bounds computed by other lower priority agents. Therefore, costs added by refinement terms are not redundant and then Eq. 12 is a lower bound on Y^j.

Lemma 2. *By the end of AFB_BJ$^+$, each agent stores in its UB_j the cost of the optimal solution Y^* and in v_j^* its value on Y^*.*

Proof. (Sketch) In agent A_n, $lb(Y^n)$ equals $gc(Y^n)$ because it does not have lower priority agents (Eq. 12). A_n updates its UB_n and v_n^* only when it generates a full CPA (Y^n) with $gc(Y^n)$ smaller than its current upper bound (lines 23-24). Thus, UB_n only decreases. Let σ be the smallest generated UB_n, i.e., σ is the cost of the latest generated full CPA Y^n. In AFB_BJ$^+$, (i) each agent A_j attaches to each message it sends its UB_j. Agent A_j only updates its UB_j and its v_j^* when the upper bound carried in a received message is smaller than the stored one (line 8, Fig. 2). (ii) All agents will receive at least one message after the generation of σ (at least they will receive **stp** messages before they stop their execution). (iii) Messages are only sent after receiving and processing other messages. A_n attaches σ to each message it sends. Hence, all messages that follow the generation of σ will contain it. Because σ is the smallest generated upper bound and following (i), (ii) and (iii), when the search is ended, UB_j of each agent A_j equals σ and its v_j^* equals that assigned to x_j in Y^n. Now, one needs to prove that σ is the cost of the optimal solution Y^* (i.e., σ equals $gc(Y^*)$ and Y^n equals Y^*). To prove that σ equals $gc(Y^*)$, it is enough to demonstrate that during search no CPA that can lead to a solution of lower cost than σ is discarded. In AFB_BJ$^+$, the CPAs are discarded only in three places (line 19, procedure ExtendCPA ( ) , line 38, procedure ProcessLB ( ) , and line 33, procedure Backtrack ( )). In all cases above, we are ensured that the lower bound of the discarded CPAs exceeds UB_j. Thus, they cannot lead to a solution with a cost smaller than UB_j. Now, since $\sigma \leq UB_j$ when discarding those CPAs, we are ensured that they have a cost larger than σ. Thus, σ is the cost of the optimal solution Y^* and then Y^n equals Y^*. Therefore, when AFB_BJ$^+$ terminates, v_j^* is the assignment of x_j on Y^*. Then, the lemma is proved. This also completes the correctness proof of the AFB_BJ$^+$ algorithm.   □

Corollary 1. *AFB_BJ$^+$ is sound, complete, and terminates.*

5   Experiments

In this section we experimentally compare AFB_BJ$^+$ to AFB_BJ [7], BnB-Adopt$^+$ [8], and BnB-Adopt-DP2$^+$ (BnB-Adopt$^+$ combined with DP2 value ordering heuristic [1]). Algorithms are evaluated on four commonly used benchmarks: binary random Max-DisCSPs, binary random DCOPs, meeting scheduling and sensor networks. All experiments were performed on the DisChoco 2.0 platform[4] [18], in which agents are simulated by Java threads that communicate only through message passing. We evaluate the performance of the algorithms by communication load and computation effort. Communication load is measured by the total number of exchanged messages among agents during algorithm execution ($\#msg$) [13]. Computation effort is measured by the number of non-concurrent constraint checks ($\#ncccs$) [22]. $\#ncccs$ is the metric used in distributed constraint solving to simulate the computation time.

We simulate two scenarios of communication: fast communication (where message delay is null), and slow communication with uniform random message delay, where the delay costs between 0 and 100 $\#ncccs$ for each message. On slow communication, the trends are similar to those observed for fast communication, so the results are not reported here.

5.1   The Benchmark Settings

Uniform binary random Max-DisCSPs are characterized by $\langle n, d, p_1, p_2 \rangle$, where n is the number of agents/variables, d is the number of values per variable, p_1 is the network connectivity defined as the ratio of existing binary constraints, and p_2 is the constraint tightness defined as the ratio of forbidden value pairs (with a cost of 1). We solved instances of two classes of constraint graphs: sparse graphs $\langle 10, 10, .4, p_2 \rangle$ and dense graphs $\langle 10, 10, .7, p_2 \rangle$. We varied the tightness from 0.6 to 0.9 by steps of 0.1 and from 0.9 to 0.98 by steps of 0.02. For each pair of fixed density and tightness (p_1, p_2) we report average over 50 instances.

Binary random DCOPs are characterized by $\langle n, d, p_1 \rangle$, where n, d and p_1 are as in Max-DisCSPs [8]. For each value combination a cost is selected randomly from the set $\{0, \dots, 100\}$. For each $p_1 = 0.4, \dots, 0.8$, we have generated 50 instances in the class $\langle n = 10, d = 10, p_1 \rangle$.

The meeting scheduling consists of a set of agents, each having a personal private calendar and a set of meetings each taking place in a specified location. The meeting scheduling is encoded as follows. Variables/agents represent meetings. Each meeting/variable has as domain the time slots possible for it. There are constraints between

---

[4] http://dischoco.sourceforge.net/

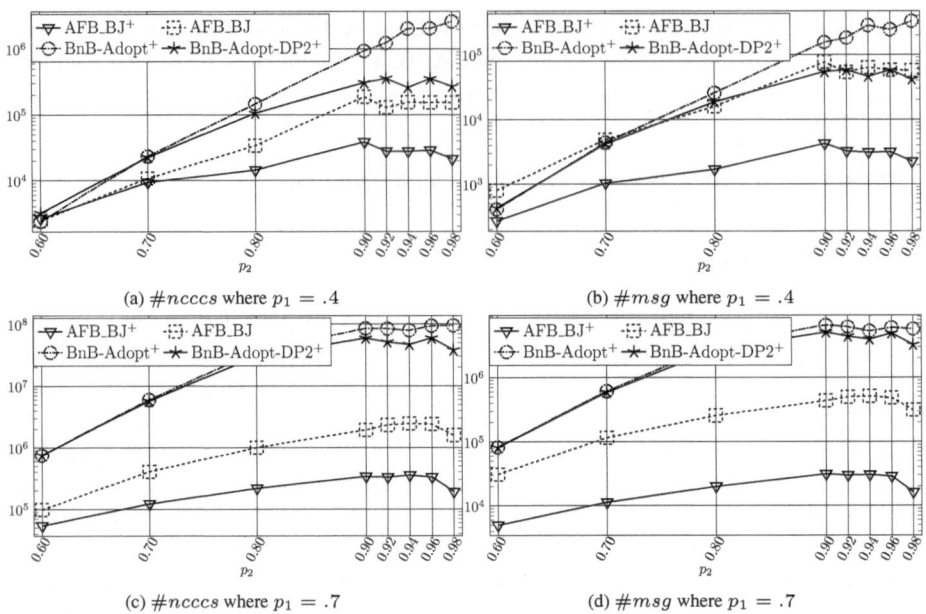

(a) #$nccs$ where $p_1 = .4$ (b) #msg where $p_1 = .4$

(c) #$nccs$ where $p_1 = .7$ (d) #msg where $p_1 = .7$

Fig. 3. Total number of messages sent and #$nccs$ performed on Max-DisCSP problems in logarithmic scale

Table 1. Total number of messages sent and #$nccs$ performed on binary random DCOPs where costs are randomly selected from 0 to 100

	#$nccs \times 10^3$					#$msg \times 10^3$				
p_1	0.4	0.5	0.6	0.7	0.8	0.4	0.5	0.6	0.7	0.8
AFB_BJ$^+$	31	77	148	299	554	3	7	14	27	48
AFB_BJ	122	308	654	1,601	3,442	54	111	186	379	658
BnB-Adopt$^+$	617	3,193	15,436	61,938	98,684	102	419	1,552	5,289	6,312
BnB-Adopt-DP2$^+$	180	958	5,266	25,869	75,414	34	151	636	2,549	5,864

meetings that share participants. We present here 4 cases each with different hierarchical scenarios [21].

The sensor network problem consists of a set of sensors that track a set of mobiles. Each mobile must be tracked by 3 sensors. Each sensor can track at most one mobile. The sensor network problems are encoded as follows. Variables/agents represent mobiles. The possible values of a variable/mobile are all combinations of three sensors that are able to track it. There are constraints between adjacent mobiles. Details are given in [1,11,2]. We present here 4 cases with different topology scenarios [21].

5.2   Results and Discussion

The results on instances of the first set of experiments (Max-DisCSPs) are illustrated in Fig. 3. In terms of computational effort (Figs. 3a and 3c), AFB_BJ$^+$ improves the AFB algorithms and performs faster than both BnB-Adopt$^+$ algorithms. The factor of

Table 2. Total number of messages sent and #$ncccs$ performed on Meeting Scheduling

cases	#$ncccs$				#msg			
	A	**B**	**C**	**D**	**A**	**B**	**C**	**D**
AFB_BJ$^+$	**4,987**	6,536	2,789	2,206	**373**	871	536	582
AFB_BJ	30,332	101,206	15,841	32,364	7,944	32,262	9,441	17,443
BnB-Adopt$^+$	272,490	63,352	51,134	30,030	15,507	10,472	8,717	8,278
BnB-Adopt-DP2$^+$	5,371	**4,224**	**2,165**	**1,647**	636	**749**	**511**	**485**

Table 3. Total number of messages sent and #$ncccs$ performed on Sensor Network

cases	#$ncccs$				#msg			
	A	**B**	**C**	**D**	**A**	**B**	**C**	**D**
AFB_BJ$^+$	5,599	6,182	2,395	4,869	2,043	1,999	325	1,430
AFB_BJ	167,862	190,423	12,084	33,988	127,544	145,421	7,853	33,280
BnB-Adopt$^+$	4,052	6,337	6,561	8,982	876	1,215	1,198	2,072
BnB-Adopt-DP2$^+$	**992**	**1,046**	**982**	**1,278**	**195**	**238**	**176**	**323**

improvements is 5 for sparse graphs and 7 for dense graphs. Concerning communication load (Figs. 3b and 3d), AFB_BJ$^+$ requires few messages compared to others algorithms. AFB_BJ$^+$ improves AFB_BJ by a factor of 20 (resp. 15) in sparse (resp. dense) instances. In dense instances, AFB_BJ$^+$ outperforms BnB-Adopt-DP2$^+$ by a large scale. BnB-Adopt$^+$ and BnB-Adopt-DP2$^+$ are the less efficient algorithms for solving Max-DisCSPs, and their performance dramatically deteriorates on dense Max-DisCSP problems. The DP2 heuristic improves the performance of BnB-Adopt$^+$. This improvement is clearer in the sparse problems than in dense ones.

For binary random DCOPs, the results are presented in Table 1. Both versions of BnB-Adopt$^+$ dramatically deteriorate compared to algorithms performing assignments sequentially. Again, the DP2 heuristic improves the performance of BnB-Adopt$^+$. AFB_BJ$^+$ improves the speed-up of AFB_BJ by a factor of 6 in dense instances. Regarding the #msg, the factor of improvement is 13.

Table 2 presents the results on meeting scheduling problems. Comparing AFB_BJ$^+$ to AFB_BJ, the obtained results show that AFB_BJ$^+$ reduces the number of #$ncccs$ by a factor of 10 and the number of required messages by a factor of 50 in all classes. AFB_BJ$^+$ outperforms BnB-Adopt$^+$ by a large factor on both considered measures. However, BnB-Adopt-DP2$^+$ benefits from its preprocessing step and performs faster than AFB_BJ$^+$.

For sensor networks, the results are presented in Table 3. Again, AFB_BJ$^+$ improves the performance of AFB_BJ by a large scale. Compared to AFB_BJ, AFB_BJ$^+$ reduces the #$ncccs$ by a factor of 15 and the number of messages by a factor of 50. BnB-Adopt$^+$ performs almost the same #$ncccs$ and the same number of messages as AFB_BJ$^+$. BnB-Adopt-DP2$^+$ outperforms all other algorithms since it needs very few messages and #$ncccs$ to resolves sensor networks instances.

Looking at all results together, we come to the straightforward conclusion that AFB_BJ$^+$ performs very well compared to its forward bounding counterparts. The reason for that amounts mainly to refined lower bounds and their use as value ordering heuristic. This guides the search first to promising assignments. The large gap in communication load can be explained by the fact that when an AFB_BJ$^+$ agent has the token to assign, it sends at-most one request to each lower agent, whereas other AFB algorithms need one message for each lower agent for each possible assignment. In addition, AFB_BJ$^+$ stores and maintains valid lower bounds to avoid redundant messages.

Both versions of BnB-Adopt$^+$ perform very poorly when solving Max-DisCSPs and random DCOPs. One possible reason is that in both algorithms, agents have a strongly asynchronous assignments policy. However, for structured problems, BnB-Adopt-DP2$^+$ has performance close to AFB_BJ$^+$. On some highly structured problems (sensor networks), it performs well. When we checked these instances, we found them very sparse with very few constraints. The constraint tree structure used in BnB-Adopt$^+$ combined with the very informed DP2 heuristic allows agents, in such very sparse instances, to initialize their lower bounds of values to a cost close to that of the solution.

Our experiments show that AFB_BJ$^+$ needs less messages than other algorithms. However, AFB_BJ$^+$ messages can be longer than those sent by other algorithms. The largest messages in AFB_BJ$^+$ (*lb* messages) are in $O(nd)$. To see the practical impact of these larger messages, we computed the total number of *bytes* exchanged by all algorithms.[5] AFB_BJ$^+$ is improved by BnB-Adopt-DP2$^+$ by a factor up to 2 on meeting scheduling and 47 on sensor networks. Except for these two cases, AFB_BJ$^+$ improves other algorithms in all benchmarks by a factor up to 94 (instead of factor 73 for $\#msg$) for AFB_BJ, 144 (instead of 236 for $\#msg$) for BnB-Adopt$^+$, and 80 (instead of 203 for $\#msg$) for BnB-Adopt-DP2$^+$.

In all our experiments, the longest message sent by AFB_BJ$^+$ was of size 366 bytes. The minimum datagram size that we are guaranteed to send without fragmentation of a message (in one physical message) is 568 bytes for IPv4 and 1,272 bytes for IPv6 when using either TCP or UDP [3]. Thus, counting the number of exchanged messages is equivalent to counting the number of physical messages.

6   Conclusion

We have proposed AFB_BJ$^+$, a revisited version of the AFB_BJ algorithm in which we refine the computations of lower bounds by future agents. In AFB_BJ$^+$, the lower bounds are computed for the whole domain of the last assigned agent providing him with a very informed value ordering heuristic. Our experiments show that AFB_BJ$^+$ improves the current state of the art in terms of runtime and number of exchanged messages on different distributed problems. The present work is a step forward in order to address real world applications in multi-agent coordination. Several directions need to be explored in the AFB family. A promising direction is that of variable ordering heuristics. Another direction will be to try to maintain consistencies stronger than forward bounding.

References

1. Ali, S., Koenig, S., Tambe, M.: Preprocessing techniques for accelerating the dcop algorithm adopt. In: Proceedings of the Fourth International Joint Conference on Autonomous Agents and Multiagent Systems, AAMAS 2005, pp. 1041–1048. ACM, New York (2005)

---

[5] In our implementation we do not perform any message compression.

2. Béjar, R., Domshlak, C., Fernández, C., Gomes, C., Krishnamachari, B., Selman, B., Valls, M.: Sensor networks and distributed csp: communication, computation and complexity. Artif. Intel. 161, 117–147 (2005)
3. Bessiere, C., Bouyakhf, E.H., Mechqrane, Y., Wahbi, M.: Agile Asynchronous Backtracking for Distributed Constraint Satisfaction Problems. In: Proceedings of the IEEE 23rd International Conference on Tools with Artificial Intelligence, ICTAI 2011, Boca Raton, Florida, USA, pp. 777–784 (November 2011)
4. Bonnet-Torrés, O., Tessier, C.: Multiply-constrained dcop for distributed planning and scheduling. In: AAAI Spring Symposium: Distributed Plan and Schedule Management, pp. 17–24 (2006)
5. Chechetka, A., Sycara, K.: No-Commitment Branch and Bound Search for Distributed Constraint Optimization. In: Proceedings of AAMAS 2006, pp. 1427–1429 (2006)
6. Gershman, A., Meisels, A., Zivan, R.: Asynchronous Forward-Bounding for Distributed Constraints Optimization. In: Proceedings of ECAI 2006, pp. 103–107 (2006)
7. Gershman, A., Meisels, A., Zivan, R.: Asynchronous Forward Bounding for Distributed COPs. JAIR 34, 61–88 (2009)
8. Gutierrez, P., Meseguer, P.: Saving redundant messages in bnb-adopt. In: AAAI 2010 (2010)
9. Gutierrez, P., Meseguer, P.: Removing redundant messages in n-ary bnb-adopt. J. Artif. Intell. Res (JAIR) 45, 287–304 (2012)
10. Hirayama, K., Yokoo, M.: Distributed partial constraint satisfaction problem. In: Smolka, G. (ed.) CP 1997. LNCS, vol. 1330, pp. 222–236. Springer, Heidelberg (1997)
11. Jung, H., Tambe, M., Kulkarni, S.: Argumentation as Distributed Constraint Satisfaction: Applications and Results. In: Proceedings of AGENTS 2001, pp. 324–331 (2001)
12. Léauté, T., Faltings, B.: Coordinating Logistics Operations with Privacy Guarantees. In: Proceedings of the IJCAI 2011, pp. 2482–2487 (2011)
13. Lynch, N.A.: Distributed Algorithms. Morgan Kaufmann Series (1997)
14. Maheswaran, R.T., Tambe, M., Bowring, E., Pearce, J.P., Varakantham, P.: Taking DCOP to the real world: Efficient complete solutions for distributed multi-event scheduling. In: Proceedings of AAMAS 2004 (2004)
15. Modi, P.J., Shen, W.M., Tambe, M., Yokoo, M.: ADOPT: Asynchronous Distributed Constraint Optimization with Quality Guarantees. Artif. Intel. 161, 149–180 (2005)
16. Nguyen, V., Sam-Haroud, D., Faltings, B.V.: Dynamic Distributed BackJumping. In: Faltings, B., Petcu, A., Fages, F., Rossi, F. (eds.) CSCLP 2004. LNCS (LNAI), vol. 3419, pp. 71–85. Springer, Heidelberg (2005)
17. Petcu, A., Faltings, B.V.: A Value Ordering Heuristic for Local Search in Distributed Resource Allocation. In: Faltings, B., Petcu, A., Fages, F., Rossi, F. (eds.) CSCLP 2004. LNCS (LNAI), vol. 3419, pp. 86–97. Springer, Heidelberg (2005)
18. Wahbi, M., Ezzahir, R., Bessiere, C., Bouyakhf, E.H.: DisChoco 2: A Platform for Distributed Constraint Reasoning. In: Proceedings of the IJCAI 2011 workshop on Distributed Constraint Reasoning, DCR 2011, Barcelona, Catalonia, Spain, pp. 112–121 (2011), http://dischoco.sourceforge.net/
19. Wahbi, M., Ezzahir, R., Bessiere, C., Bouyakhf, E.H.: Nogood-Based Asynchronous Forward-Checking Algorithms. Constraints 18(3), 404–433 (2013)
20. Yeoh, W., Felner, A., Koenig, S.: BnB-ADOPT: An Asynchronous Branch-and-Bound DCOP Algorithm. J. Artif. Intell. Res (JAIR) 38, 85–133 (2010)
21. Yin, Z.: USC dcop repository (2008), http://teamcore.usc.edu/dcop
22. Zivan, R., Meisels, A.: Message delay and DisCSP search algorithms. Annals of Mathematics and Artificial Intelligence 46(4), 415–439 (2006)

Optimizing STR Algorithms with Tuple Compression*

Wei Xia and Roland H.C. Yap

School of Computing
National University of Singapore
{xiawei,ryap}@comp.nus.edu.sg

Abstract. Table constraints define an arbitrary constraint explicitly as a set of solutions (tuples) or non-solutions. Thus, space is proportional to number of tuples. Simple Tabular Reduction (STR), which dynamically reduces the table size by maintaining a table of only the valid tuples, has been shown to be efficient for enforcing Generalized Arc Consistency. The Cartesian product representation is another way of having a smaller table by compression. We investigate whether STR and the Cartesian product representation can work hand in hand. Our experiments show the compression-based STR can be faster once the tables compress well. Thus, the benefits of the STR2 and STR3 algorithms respectively are retained while consuming less space.

1 Introduction

Table constraints are the most general form of finite domain constraints where the table defines the solutions (or non-solutions) of the constraint. Two state-of-the-art GAC approaches for non-binary table constraints, STR [1,2,3] and mddc [4], incorporate some form of compression. In particular, STR uses dynamic compression which compresses the table during search by removing invalid tuples. Tables can also be compressed with the Cartesian product representation [5,6] which was used before in the context of symmetry breaking and nogood learning. It was applied to compress table constraints in [7] and shown to improve the GAC-Schema+allowed algorithm.

Compression of a table constraint using the Cartesian product representation for the tuples, which we call c-tuple(s), gives a static compression of the table. Dynamic compression in STR, on the other hand, compresses by reducing the table size by *tabular reduction* [1] during search. However, the Cartesian product representation can inhibit tabular reduction. Thus, unlike GAC-Schema algorithms which use static tables where compression is likely to be beneficial, with STR there is interplay between both kinds of compression, changing their benefits respectively. A recent paper [8] proposed a more complex way to compress tables and applies STR on this compressed representation. However, their preliminary experiments showed that the revised algorithm to be competitive with STR1 [1], but not faster than STR2 [2]. Ideally we want a compression schemes for STR where the overall benefits can outweigh the costs.

In this paper, we return to the simple idea of Cartesian product compression and investigate whether static and dynamic compression approaches for GAC can be effectively combined on table constraints. We extend the STRx (STR2 [2] and STR3 [3])

---

* This work has been supported by grant MOE2012-T2-1-155.

algorithms to handle compression tuples [7]. We experiment with random and structured CSPs varying the degree of static and dynamic table compression. We find that compression is not always beneficial but when there is a reasonable amount of static compression, the compression algorithms, STR2-C and STR3-C are faster than STR2 and STR3 respectively. They also inherit the underlying properties of STR2 and STR3.

2   Background

A *constraint satisfaction problem* (CSP) $\mathcal{P} = (X, \mathcal{C})$ consists of a finite set X of variables and a finite set $\mathcal{C}$ of constraints. Variables $x_i \in X$ only take values from a finite *domain* $dom(x_i)$. An *assignment* (x_i, a) denotes $x_i = a$. An r-ary *constraint* $C \in \mathcal{C}$ on r distinct variables $x_1, \ldots, x_r$ is a subset of the Cartesian product $\prod_{i=1}^{r} dom(x_i)$, denoted by $rel(C)$, that restricts the values of the variables in C can take simultaneously. The *scope* is the set of variables denoted by $var(C)$ and r is the *arity* of C. A set of assignments $\theta = \{(x_1, a_1), \ldots, (x_r, a_r)\}$ *satisfies* C iff $(a_1, \ldots, a_r) \in rel(C)$, also called a *solution* of C or *tuple* of C and $\theta[x_i] = a_i$. Solving a CSP is finding an assignment for each variable from its domain so that all constraints are satisfied.

An assignment (x_i, a) is *generalized arc consistent* (GAC) to $\mathcal{P}$ iff for every constraint $C \in \mathcal{C}$ such that $x_i \in var(C)$, there is a solution θ of C where $(x_i, a) \in \theta$ and $a \in dom(x_i)$ for every $(x_i, a) \in \theta$. This solution is called a *support* for (x_i, a) in C. A variable $x_i \in X$ is GAC iff (x_i, a) is GAC for every $a \in dom(x_i)$. A constraint is GAC iff every variable in its scope is GAC. Finally, CSP $\mathcal{P}$ is GAC iff every constraint in $\mathcal{C}$ is GAC.

Definition 1. (C-tuple [5,7]). *Let C be an r-ary constraint. A compression tuple $\tau_c = (\{a_{1,1}, a_{1,2}, ..., a_{1,k_1}\}, ..., \{a_{r,1}, a_{r,2}, ..., a_{r,k_r}\})$ of C is the Cartesian product of a set of tuples. This compression tuple τ_c is also called a* c-tuple.

A c-tuple admits any set of assignments that assigns one of $a_{1,1}, ..., a_{1,k_1}$ to x_1, one of $a_{2,1}, ..., a_{2,k_2}$ to x_2, etc. Given a c-tuple τ_c, $\tau_c[x_i]$ denotes $\{a_{i,1}, ..., a_{i,k_i}\}$. Fig 1 (a) shows a table and Fig 1 (c) shows its compressed form. A c-tuple can potentially represent an exponential number of tuples. We extend the concept of *validity* to c-tuples.

Definition 2. (Validity of c-tuple). *Let C be an r-ary constraint. A c-tuple τ_c is valid on C iff $\forall x_i \in var(C)$, $\exists a_i \in \tau_c[x_i]$ such that $a_i \in dom(x_i)$.*

3   STR2-C: STR2 on Compression Tuples

STR2 [2], a refinement of STR, is shown to be one of the most efficient GAC algorithms for non-binary table constraints. In maintaining GAC during search, STR2 gets its efficiency by maintaining dynamically the table of valid tuples. When enforcing GAC on a table constraint, the validity of each tuple is checked. Once a tuple is found to be valid, the domains of the variables are updated with the consistent values belonging to the tuple. Otherwise, the tuple is removed from the table, thus, is not considered again as the search goes deeper. Upon backtracking, removed tuples will be added back to the table.

We now extend STR2 to work with table constraints on c-tuples which we call STR2-C. Like STR2, STR2-C maintains a table of valid c-tuples dynamically during search. The extension to STR2-C is straightforward, so we will describe it informally. When STR2-C identifies the validity of c-tuples, if a c-tuple is invalid, it will be removed from the table, otherwise, the values belonging to the c-tuples are used to update the domains of the variables. But unlike STR2, all the values belonging to a valid tuple are GAC-consistent, the valid c-tuples may contain inconsistent values. Thus additional value checks against the domains must be done when collecting the consistent values.

For the same reason, the domain updating phase of STR2 can cause rechecking of inconsistent values. In order to save some rechecks, a cursor can be used for each variable to separate the inconsistent values with the unchecked values, then the values which are already detected to be inconsistent can be skipped and only the unchecked values will be checked against the variables' domains. In STR2, one optimization is that once a variable's domain is found GAC-consistent, there is no need to seek supports for the variable. Such variables are skipped when the variables' domains are updated in the second phase of STR2. This optimization can be amplified in STR2-C, as the value checks against the domains in the second phase of STR2-C can also be skipped.

Compared to STR2, the potential runtime improvement of STR2-C is because the compressed table may be up to exponentially smaller than the original table. We could expect that gains in efficiency of STR2-C will depend on the size of the compression table to the original table. To illustrate this, consider the constraint:

$$C_{d,r}(x_1, x_2, ..., x_r) \equiv [\textstyle\bigwedge_{i=1}^{r} x_i = \{0, 1, ..., d-2\}] \vee [\textstyle\bigwedge_{i=1}^{r} x_i = d-1]$$

The domain of each variable is $0, 1, ..., d-1$. The table representation of $C_{d,r}$ has $(d-1)^r + 1$ tuples, while the c-tuple table has 2 c-tuples, which are $\{(\{0, 1, ..., d-2\}, ..., \{0, 1, ..., d-2\}), (\{d-1\}, ..., \{d-1\})\})\}$. Enforcing GAC on $C_{d,r}$ using STR2 takes $O(rd^r)$ time, while using STR2-C takes $O(rd)$ time.

However, there is also a drawback. Unlike STR2, which keeps valid tuples, STR2-C keeps the valid c-tuples that may still include invalid tuples. This will bring some rechecks of values which can slow down STR2-C compared with STR-2.

4   STR3-C: STR3 on Compression Tuples

STR3 [3] is a fine-grained table reduction based GAC algorithm, but uses a different table representation. Conceptually, in the STR3 representation, each variable-value pair is mapped to its set of tuples. STR3 is path-optimal, as it avoids unnecessary traversal of tables. STR3-C extends STR3 to work on c-tuples.

The STR3 and STR3-C representation is illustrated in Fig 1 (see [3] for details). Fig 1 (a) shows the original table representation and Fig 1 (b) is the equivalent STR3 representation. Compressing Fig 1 (a) with c-tuples gives Fig 1 (c) and the final STR3-C representation is Fig 1 (d).

We briefly summarize some ideas from the STR3 algorithm, and refer to the details in [3]. For each constraint C, $row(C, X, a)$ represents the set of tuples belonging to (X, a) in the equivalent table (e.g. Fig 1 (b)). $row(C, X, a)$ is associated with a cursor, represented by $row(C, X, a).curr$, to separate the untested and invalid tuples. In addition, each tuple τ is accompanied with a dependent list $dep(\tau)$ of variable-value

	X	Y	Z
1	a	a	a
2	a	a	b
3	b	b	c
4	b	c	c
5	c	b	c
6	c	c	c

(a) Standard table

X			Y			Z	
a	$\{1,2\}$		a	$\{1,2\}$		a	$\{1\}$
b	$\{3,4\}$		b	$\{3,5\}$		b	$\{2\}$
c	$\{5,6\}$		c	$\{4,6\}$		c	$\{3,4,5,6\}$

(b) Equivalent table

	X	Y	Z
1	$\{a\}$	$\{a\}$	$\{a,b\}$
2	$\{b,c\}$	$\{b,c\}$	$\{c\}$

(c) Standard compressed table

X			Y			Z	
a	$\{1\}$		a	$\{1\}$		a	$\{1\}$
b	$\{2\}$		b	$\{2\}$		b	$\{1\}$
c	$\{2\}$		c	$\{2\}$		c	$\{2\}$

(d) Equivalent compressed table

Fig. 1. The table representation of STR3 and STR3-C

pairs which treat τ as a valid support. STR3 works as follows. Once a value (X, a) is removed, all the unchecked tuples in $row(C, X, a)$ before the cursor will be checked . If a tuple is already invalid, there is no need to update its dependent list. Otherwise, the tuple τ is set to be invalid, then a new support should be found for each $(Y, b) \in dep(\tau)$. If a support τ_2 is found, (Y, b) will be shifted to $dep(\tau_2)$, otherwise (Y, b) is inconsistent and removed.

STR3-C works with c-tuples, i.e. the STR3 representation applied to the compressed c-tuple form of the original table. In STR3-C, $row(C, X, a)$ is replaced by a set of c-tuples containing value (X, a). The dependent list is based on c-tuples, but is still composed of the variable-value pairs. The key for STR3-C is the detection of the validity of c-tuples. This is the main difference from STR3, as when (X, a) is removed, the c-tuple in $row(C, X, a)$ may still be valid. The details of the algorithm is given in Fig 2. The $inv(C)$ in Line 1 is the set of invalid c-tuples of constraint C implemented as a sparse set. In the $inv(C)$ structure, $inv(C).members$ is the position of the last current element in $inv(C)$ and $inv(C).dense$ is the invalid c-tuples array. In Line 2, $comprTable$ is the standard compressed table (e.g. Fig 1c) and $comprTable[row(C, X, a)[k]][X]$ returns a set values of X appearing in the c-tuple $row(C, X, a)[k]$. Different from STR3, the standard compressed table is used to access the c-tuples. To check the validity of the c-tuple $row(C, X, a)[k]$, if a consistent value belonging to the c-tuple under variable X exists, the c-tuple is valid, so no need to update the dependent list of the tuple. Otherwise the c-tuple becomes invalid, and will be added into $inv(C)$. The save() function in Line 6 and Line 7 stores the states of $inv(C).members$ and $row(C, X, a).curr$ into the stack $stateI$ and $stateR$ respectively for backtracking. From Line 5 to Line 8, which is the same as STR3, STR3-C updates the dependent list of the invalid c-tuples, and the inconsistent values will be removed.

For example, consider a table constraint $C(x_1, x_2, x_3) = \{(0,0,0), (0,1,0), (0,2,0), (2,2,2)\}$ with domains $\{0,1,2\}$. In STR3, when $(x_2, 0)$ is removed, the first tuple $(0,0,0)$ becomes invalid. If $(x_1, 0)$ and $(x_3, 0)$ are in $dep((0,0,0))$, then $(x_1, 0)$ and $(x_3, 0)$ will be transferred to the second or third tuple by assuming $(x_2, 1)$ and

STR3-C($C : Constraint, X : Variables, a : Value$)
1   $prevMembers \leftarrow inv(C).members$
2   **for** $k \leftarrow 0$ *to* $row(C, X, a).curr$ **do**
        if $row(C, X, a)[k] \notin inv(C)$ **then**
3           $checkVal[] \leftarrow comprTable[row(C, X, a)[k]][X]$
            for $v \leftarrow 0$ *to* $checkVal[].size - 1$ **do**
                if $checkVal[v] \in dom(X)$ **then** break
            if $v = checkVal[].size$ **then**
4   add $row(C, X, a)[k]$ to $inv(C)$

5   **if** $prevMembers = inv(C).members$ **then** return *true*
6   save($C, prevMembers, stateI$)
    foreach $i \in \{prevMembers + 1, ..., inv(C).members\}$ **do**
        $k \leftarrow inv(C).dense[i]$
        foreach $(Y, b) \in dep(C)[k]$ *such that* $b \in D^C(Y)$ **do**
            $p \leftarrow row(C, Y, b).curr$
            while $p \leq 0$ **and** $row(C, Y, b)[p] \in inv(C)$ **do** $p \leftarrow p - 1$
            if $p < 0$ **then**
                removeValue(Y, b)
                if $D^C(Y) = \emptyset$ **then** return *false*
            else
                if $p \neq row(C, Y, b).curr$ **then**
7   save($(C, Y, b), row(C, Y, b).curr, stateR$)
                    $row(C, Y, b).curr \leftarrow p$
                move (Y, b) from $dep(C)[k]$ to $dep(C)[row(C, Y, b)[p]]$

8   return *true*

Fig. 2. STR3-C algorithm

$(x_2, 2)$ are consistent. However, for STR3-C, the first three tuples are compressed into one. When $(x_2, 0)$ is removed, the c-tuple $(\{0\}, \{0, 1, 2\}, \{0\})$ is still valid as $(x_2, 1)$ and $(x_2, 2)$ are consistent, thus the dependent list will not be checked or updated. But this c-tuple may be checked again when $(x_2, 1)$ or $(x_2, 2)$ is removed. Similar to STR2-C, the rechecks may potentially slow down STR3-C.

The difference between STR3 and STR3-C mainly lies in the cost of detecting the invalid tuples. STR3-C takes at most additional d value checks for each c-tuple when checking its validity. However, as the compressed table can be exponentially smaller than the original table, STR3-C can still have runtime improvement. As STR3 collects the invalid tuples incrementally, let's consider the cost for the constraint $C_{d,r}$, used in previous section, along a single search path of length m. For STR3-C, one c-tuple will become invalid after at most $r * d$ times validity checks along a single path, and each validity check is accompanied by at most d value checks. Assuming $O(1)$ cost at each search node, STR3-C will take $O(rd^2 + m)$ time, while STR3 will take $O(rd^r + m)$ [3] time for $C_{d,r}$.

5 Experiments

We prototype STR2-C and STR3-C in Abscon[1] which has implementations of STR2 and STR3. Experiments were run on an Intel i7/960 @3.20GHz on 64-bit Linux.

We investigate the performance of STRx algorithms under different table compression ratios with random and structured benchmarks given in Table 1.[2] In order to have sufficient variation in table compression, we also generate another two series of random CSPs. The first is MDD-p which generalises the *mdd-half* benchmark ($p = 0.5$) by building an MDD in a post-order manner with probability p that a previously created sub-MDD is reused [4]. Another series is modified from *rand-5-12* where the constraints have high tightness.[3] We decrease the tightness by randomly adding tuples such that the table is 2X (twice), 4X and 8X the size of *rand-5-12*. As some tables become quite large, half the constraints are removed. The average table size and table compression ratio based on series of benchmarks are given in Table 1 where $Cr = \frac{\#c-tuples}{\#tuples}$.

We use *dom/ddeg* and *lexico* as the variable and value ordering respectively when solving the CSPs to ensure the search space is the same [2]. Table 2 shows the average

Table 1. Statistics for the benchmarks

Series	#tuples	Cr	Series	#tuples	Cr	Series	#tuples	Cr
rand-3-20	2944	0.138	rand-5-12-2X	24884	0.451	mdd-0.5	39850	0.13
rand-8-20	78120	0.602	rand-5-12-4X	49768	0.243	mdd-0.7	39050	0.05
rand-5-12	12442	0.684	rand-5-12-8X	99536	0.124	mdd-0.9	39050	0.015
cril	1228	0.094	ramsey-a3	24	0.25			
ruler-25	238000	0.026	ramsey-a4	61	0.133			
ruler-34	850000	0.019	chessColor	78	0.115			

Table 2. The average runtime (in seconds) for random benchmarks

Series	#instances	STR2	STR2-C	STR3	STR3-C
rand-3-20	20	44.7	**27.6**	39.7	33.7
rand-8-20	20	**17.3**	25.5	280.0	245.6
rand-5-12	20	30.7	54.1	**9.3**	11.4
rand-5-12-2X	16	12.4	15.7	**8.2**	8.8
rand-5-12-4X	16	89.9	59.1	60.1	**43.7**
rand-5-12-8X	16	1061.3	**379.3**	1567.2	877.3
MDD-0.5	14	436.1	**212.7**	1628.5	596.7
MDD-0.7	9	1040.4	**172.3**	1580.7	304.6
MDD-0.9	10	262.4	**30.0**	383.0	39.0

---

[1] Available from http://www.cril.univ-artois.fr/~lecoutre/research/tools

[2] The benchmarks are available from http://www.cril.univ-artois.fr/~lecoutre

[3] The tightness of a table is defined as $1 - \frac{t}{d^r}$, where t is the number of tuples in the table, d is the domain size of the variables and r is the arity of the constraint.

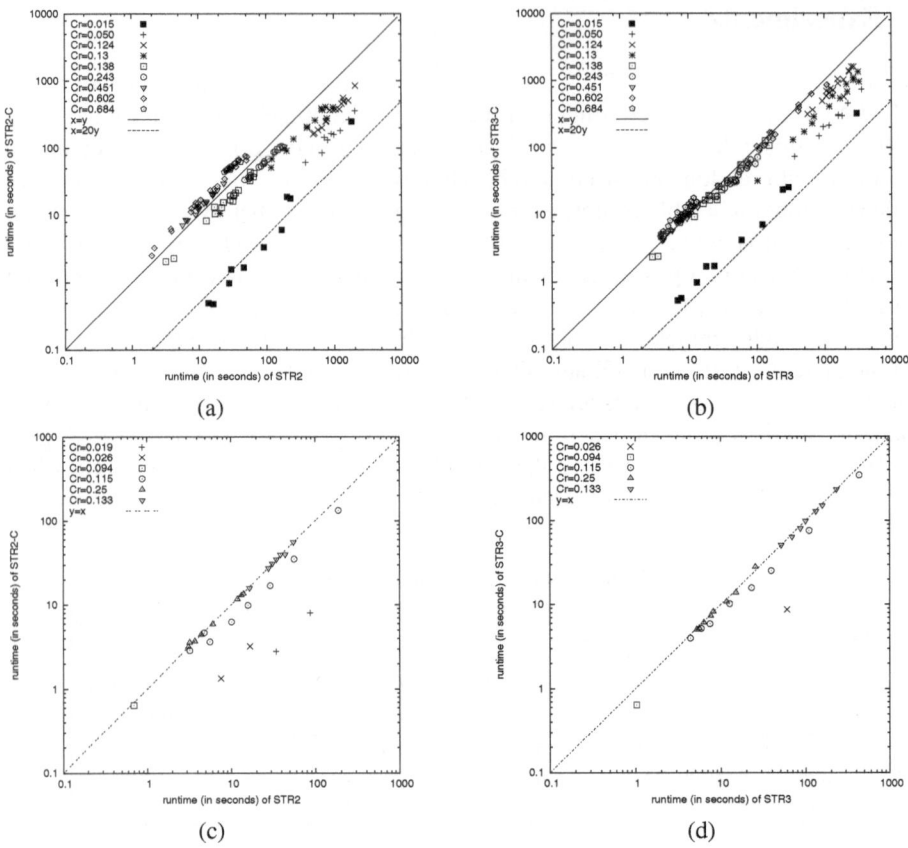

Fig. 3. (a) and (b) shows the runtime of STRx and STRx-C for random CSPs; (c) and (d) shows the runtime of STRx and STRx-C for structured CSPs with a max number of search nodes. (A smaller value of Cr indicates higher compression)

CPU time to solve the instances of different groups of random CSPs with the fastest across algorithms in bold and the fastest per column underlined. The cursor optimization is used in the STR2-C column which we found to give a small improvement on average. However, in some instances, the added overhead makes it slightly slower.

We found STRx-C, on average, is faster than the corresponding STRx algorithm. In particular, for the series *MDD-0.9*, STR2-C and STR3-C are up to 9 times faster than STRx respectively. For the structured CSPs, most of the problem instances take longer than the time out (1 hour), thus, we give the average runtime up to a maximum number of search nodes: 100000 nodes for ramsey, chessboardColor and cril, and 10000 nodes for golombRuler (in Fig 3 (c) and (d)).

Fig 3 details the runtime of STRx and STRx-C for each instance under different compression ratios. From Fig 3 (a) and Fig 3 (c), we see that STR2-C could be more than 20 times faster than STR2 for some instances. When the compression ratio is small

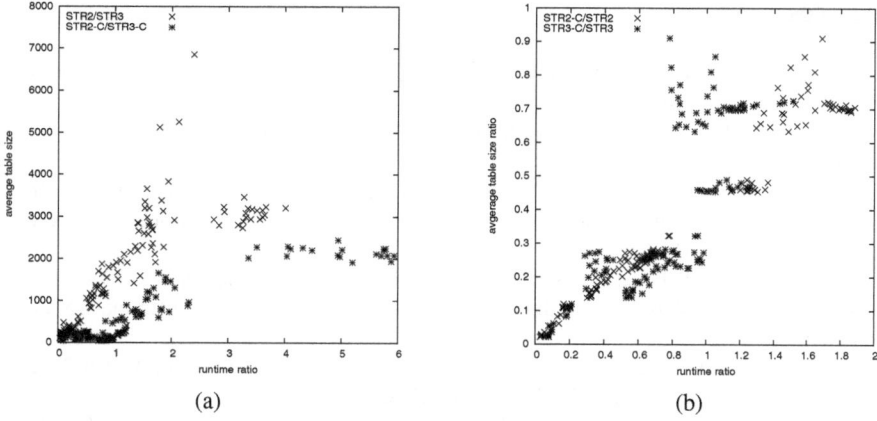

Fig. 4. (a) The runtime ratio of STR2/STR3 and STR2-C/STR3-C versus the average table size during search; and (b) The runtime ratio of STR2-C/STR2 and STR3-C/STR3 versus the average table size ratio during search

enough, STR2-C becomes faster than STR2. Thus, the more the table is compressed, the faster STR2-C gets. Fig 3 (b) and Fig 3 (d) show a similar comparison between STR3-C and STR3. We also find that STR3-C is faster when Cr is small. However, the slowdown or speedup of STR3-C on STR3 is less than STR2-C on STR2.

STR3 was shown to be faster than STR2 when table reduction does not drastically reduce the table during search [3]. We also investigate whether this property is inherited by STR2-C and STR3-C. As Fig 4 (a) shows, STR3-C is also faster than STR2-C when the average compressed table is large. We see that among the instances, STR2 is up to 4 times slower than STR3, while STR2-C is up to 6 times slower than STR3-C. This suggests that the difference between STR2 and STR3 is more pronounced after compression. Fig 4 (b) shows that STR2-C is fast as the average table size of the compression table is small enough, otherwise STR2-C is slower than STR2. We also observed that the average table size ratio during search of STR2-C over STR2 decreases as the compression ratio decreases. This is reasonable as when the table cannot be compressed much, the original table and the c-tuple table are close in size, so the table reduction during search becomes similar. Thus, the compression ratio becomes the basic determinant to the performance of the STRx-C algorithms and can be used to identify cases when STRx-C is beneficial. We also compared with c-tuples on GAC-schema+allowed. On average, the speedup or slowdown of c-tuples is greater with STR2 than with GAC-schema+allowed.

To summarize, our experiments show that our STR algorithms on compressed tables are competitive when the table can be compressed enough. For random CSPs, as the table compression ratio drops to 25%, the STRx-C algorithms become more efficient. This illustrates that static and dynamic table compression can cooperate well in practice even though the use of c-tuples has drawbacks in that it reduces the amount of table reduction and has some (small) overheads. We also show that the properties of STR2 and STR3 can be inherited by their compressed version algorithms.

References

1. Ullmann, J.R.: Partition Search for Non-binary Constraint Satisfaction. Information Science 177, 3639–3678 (2007)
2. Lecoutre, C.: STR2: Optimized Simple Tabular Reduction for Table Constraints. Constraints 16, 341–371 (2011)
3. Lecoutre, C., Likitvivatanavong, C., Yap, R.H.C.: A Path-Optimal GAC Algorithm for Table Constraints. In: Proceedings of the Twentieth European Conference on Artificial Intelligence (ECAI), pp. 510–515 (2012)
4. Cheng, K.C.K., Yap, R.H.C.: An MDD-based Generalized Arc Consistency Algorithm for Positive and Negative Table Constraints and Some Global Constraints. Constraints 15, 265–304 (2010)
5. Focacci, F., Milano, M.: Global Cut Framework for Removing Symmetries. In: Walsh, T. (ed.) CP 2001. LNCS, vol. 2239, pp. 77–92. Springer, Heidelberg (2001)
6. Katsirelos, G., Bacchus, F.: Generalized Nogoods in CSPs. In: Proceedings of the Twentieth National Conference on Artificial Intelligence (AAAI), pp. 390–396 (2005)
7. Katsirelos, G., Walsh, T.: A Compression Algorithm for Large Arity Extensional Constraints. In: Bessière, C. (ed.) CP 2007. LNCS, vol. 4741, pp. 379–393. Springer, Heidelberg (2007)
8. Gharbi, N., Hemery, F., Lecoutre, C., Roussel, O.: STR et Compression de Contraintes Tables. In: Journées Francophones de Programmation par Contraintes (JFPC), pp. 143–146 (2013)

Describing and Generating Solutions for the EDF Unit Commitment Problem with the ModelSeeker

Nicolas Beldiceanu[1,*], Georgiana Ifrim[2,**],
Arnaud Lenoir[3], and Helmut Simonis[2,***]

[1] TASC team (CNRS-INRIA), Mines de Nantes, France
nicolas.beldiceanu@mines-nantes.fr
[2] 4C, University College Cork, Ireland
{g.ifrim,hsimonis}@4c.ucc.ie
[3] EDF R&D, France
arnaud.lenoir@edf.fr

Abstract. We present an application to extract and solve constraint models from sample solutions of the Unit Commitment Problem of EDF, which computes the power output for each power plant in France as a 48 hour time series. Our aim is to describe and automatically generate the plant-specific model constraints common to the optimal solutions obtained over multiple days. The proposed system generates specific domains for each variable (i.e., time slot), binary constraints between consecutive time slots, and global constraints with functional dependencies over the entire time series. We employ time series clustering techniques for finding stronger constraints and we identify plant-specific time intervals, for which we add additional global constraints. A custom search routine and the generated models allow us to produce solutions corresponding to many overlapping global constraints. Our tool is based on the ModelSeeker [4], but specializes and extends that system for this specific application domain. Results indicate that useful models can be generated with this process.

1 Introduction

In this paper we present a first practical application inspired by ideas used in the ModelSeeker [4] tool. The ModelSeeker is a constraint acquisition system which generates global constraint models from example solutions of a problem. In [4], it was tested on a large variety of small puzzles and simple problems, exploiting many of the global constraints in the Global Constraint Catalog [3]. The ModelSeeker works for highly structured problems, where a given solution can be partitioned into regular subsets, for which the same constraints apply. In most of the problems, the domains of the variables are uniform, or can be computed as the result of a problem transformation.

Here, we extend the ModelSeeker to address an important, large scale optimization problem in electricity supply scheduling, the Unit Commitment Problem (UCP). This is

---

* This author benefited from the support of the *FMJH Program Gaspard Monge in Optimization and Operation Research*, and from the support to this program from EDF.
** The author is supported by the SFI 10/IN.1/I3032 project.
*** The author is supported by the FET/OPEN project ICON.

C. Schulte (Ed.): CP 2013, LNCS 8124, pp. 733–748, 2013.

a much less regular problem than those considered before, while having several unique features that can help in efficiently finding models and solutions. Each solution considered is a time series over a fixed time horizon, and we exploit this information to generate more accurate domains for our model. Consecutive variables represent solution values for consecutive time periods, therefore it also makes sense to discover binary constraints between them, e.g., using techniques described for the constraint acquisition tool CONACQ [5].

In Section 2 we describe the Unit Commitment Problem in general, and the specific one solved by the French electricity provider, Electricité de France (EDF) [10]. In Section 3 we give a high-level overview of our proposed approach, the **UCP-ModelSeeker** (Unit Commitment Problem ModelSeeker). In Sections 4, 5, and 6 we give details about each building block of our system. Section 7 presents results for the solutions generated from the model acquired with the UCP-ModelSeeker, showing that the overall approach is feasible, while Section 8 concludes and discusses future work.

2   The Unit Commitment Problem

The Unit Commitment Problem [13,6,12,16,17,11] is a core optimization problem in the electricity supply industry. Based on forecasted demand for a given time horizon, a complex model describes the possible power output values and operating cost of different power plants, and specific operating constraints for each plant. Depending on the type of plant (e.g., nuclear, hydro-electric, thermal), the possible power output levels, the min/max ramp-down/up constraints, and the number of shut-downs/start-ups in a given period may all be constrained. Due to its size and the complexity of constraints, the UCP can be very hard to solve, and a large variety of solution methods have been attempted [14]. As the problem is solved in an operational context, and constraints may change within a short time horizon, it is very useful to be able to rerun the model whenever the situation dictates. For many power systems, the computational effort restricts this more reactive use of the model. For example, solving the UCP using state-of-the-art techniques, takes 15 mins for 200 plants over a 24 hours time horizon, in a simulation of the Mexican Power System [14]. The UCP solving time is typically directly affected by the number of power plants, constraints, and length of the optimization horizon.

2.1   The EDF Unit Commitment Problem

EDF is the largest electric utility company in France, with a total of 98.8GW installed capacity mixing the following forms of energy: nuclear (85% of total, 58 power plants providing 63.1GW), thermal (5%, 47 plants, coal, oil, gas), hydraulic and renewable. The largest percentage of renewable energy is provided by hydro-electric plants (8%, 500 plants linked to 250 reservoirs) [8]. The day-ahead UCP model has 48 hours in half-hour periods (96 periods). The EDF high-level system overview is as follows [10]:

- Large size (10^6 variables, 10^6 constraints): all production plants are modeled with a large number of technical constraints and on a 48 hour horizon.
- Non-convex and non-continuous nature: some production costs have discontinuities and the production variables are discrete.

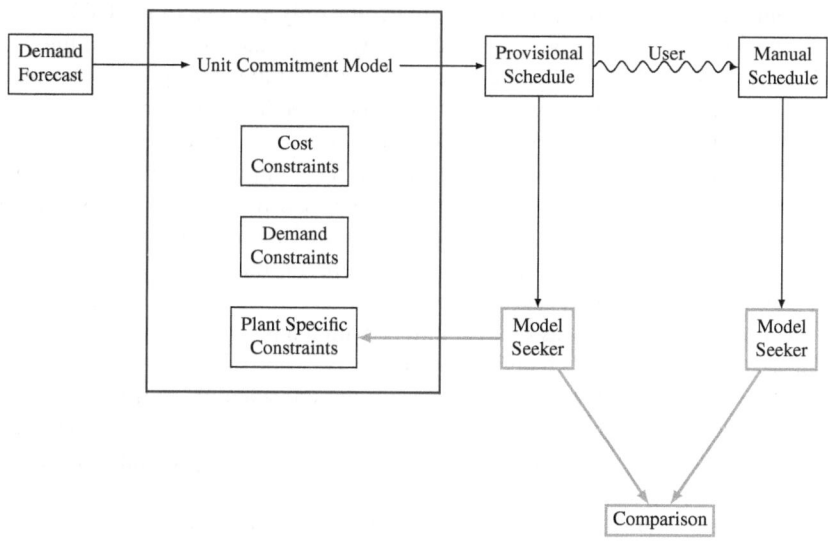

Fig. 1. The EDF Unit Commitment Problem

- Strict computational limits, due to a tight operational process: data collection ends at 12:30 and the feasible schedules have to reach the transport system operator by 16:30. Post-optimization involving human expertise is required, altogether leaving about 15 minutes for solving the optimization problem.

We consider **two use cases** associated with the EDF Unit Commitment Problem, as shown in Figure 1. The **first use case** derives plant-specific constraints from given (automated) solutions of the UCP. This serves multiple goals:

1. We want to see if the UCP-ModelSeeker can recover the plant-specific constraints which are part of the original EDF model. To make this study more realistic, we were not given the EDF real-world model. In this context, we are interested in discovering the constraints which govern the EDF schedule.
2. By analyzing optimal solutions of the problem, we may find new constraints which hold in (the majority of) optimal solutions, but are stronger than the constraints specified in the original model. Adding these constraints to the initial model may speed up the solving process, without affecting the solution quality.
3. We discover global constraints with functional dependencies, most of which can be expressed by automata with counters [2]. In the future we hope to extend the work described in [7] to generate linear constraints from these global constraints, that can then be directly added to the original MIP formulation of the optimization problem.
4. Some solving methods (e.g., column generation, genetic algorithms), require multiple feasible schedules for each plant as a starting point for the optimization procedure. Therefore, going beyond model discovery, we use the UCP-ModelSeeker-generated-model of each plant to *produce* new solutions for that plant.

Note that the plant-specific constraints are not the only element of the UCP. The over-all schedule must match the given demand exactly, which can be difficult if the demand is changing rapidly. Additionally, the overall objective of the model is to minimize gen-erating costs. This does not only cover the fuel cost needed to generate the electricity, but also fixed start-up/shut-down costs, and costs related to reserve capacity. In most de-regulated energy markets, the problem is further complicated by competing plant operators, each trying to satisfy their own objectives. Currently, the UCP-ModelSeeker does not consider explicitly demand and other cost as input data (implicitly, the pro-visional schedule is based on that data, see Figure 1), but uses only the discovered plant-specific constraints for generating new solutions.

The **second use case** takes as input solutions which have been *manually* modified by the network operators. These changes may be caused by problems in some plants, or by changing demand levels, or by operators modifying the schedule to make it more easy to implement (e.g., smoothing out power output curves). The resulting schedule must still satisfy most hard operational constraints, but may differ significantly from the au-tomatically generated schedule. We use the UCP-ModelSeeker to find the plant-specific constraints for the manually-adjusted schedules, and compare them to the constraints found for the original schedule. This can point to constraints and parameters of the problem which are currently not accurately modelled in the MIP model.

3   Overview of the UCP-ModelSeeker System

As input data, we are given the manually modified power output time series for all active power plants in France (261 plants in total) over a period of one month (April 2010). The solution for each day and plant consists of a time series of 96 integer values (given in MW), describing the output for the current and the next day in half-hour slots. The values can be negative, as the model also contains pumped-storage hydro systems, that can consume energy in low demand periods to pump water into storage reservoirs for later power generation. Some power stations also require significant power when shut down, this results in a constant negative value during the shutdown period.

Figure 2 shows our overall approach for describing and solving the model of each plant. The output of the plants can be quite different for different days in the study period. We therefore first cluster the plant-wise solutions into groups of similar profiles, and then generate constraints for each cluster. This step results in finding stronger (and more diverse) models than using all samples for a single model.

The output of a power plant is typically strongly correlated to the time of day. We use this information to generate specific domain constraints for each time period.

For each cluster, we try to learn constraints of different types. We begin with global constraints over the full time series, which express features of the series as functional dependencies. Other global constraints, without functional dependencies, may also hold. As we know that the solutions represent time series, it makes sense to generate possible binary constraints between consecutive time periods. In addition, the behavior of the time series may differ significantly over the optimization horizon. We have developed a technique which allows us to partition each time series into shorter intervals, for which stronger constraints can be applied (see Section 6 for details).

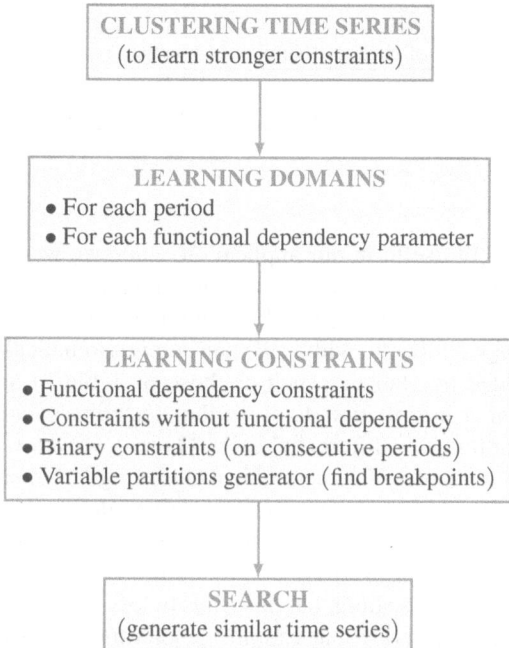

Fig. 2. Overview of UCP-ModelSeeker

Based on the UCP-ModelSeeker-generated-model, we then try to find solutions for the model. For this we specify some problem specific heuristics (custom search routine in Figure 2).

4   Clustering Time Series

Table 1 shows the combinations of *algorithms, metrics* and data representations (*features*) we have analysed for EDF plant-wise time series clustering. The Euclidean distance in the raw feature space does not consider the time dependency among time points. Similarly, the Manhattan distance treats time series as vectors, but is more robust to outliers [9]. Hoeffding's D statistic was recently suggested as another alternative to the Euclidean distance, to overcome the implicit Gaussian data distribution assumptions of that metric [15]. Dynamic Time Warping (DTW) is a time series metric that works even with time-shifted data (similar time series, possibly shifted, are given high similarity).

Table 1. Clustering Algorithms, Metrics and Features Analysed

Algorithm	Metric	Features
Hclust	Euclidean, Manhattan, Hoeffding's D, DTW	Raw, Stats
K-means	Euclidean	Raw, Stats
PAM	Manhattan	Raw, Stats

Table 2. Distribution of Power-Plants According to K-means Number of Clusters

Category (number of clusters)	Number of plants in category	Percentage
$K = 1$	10	3.80
$K = 2$	188	71.48
$K > 2$	65	24.71

DTW is nevertheless not useful in our application, since we want to differentiate, for example, between power peaks in the morning and in the afternoon.

We assessed the different clusterings both automatically using cluster-separation measures (e.g., average silhouette width [9]) and semi-automatically, by zooming into different plants (guided by statistics such as those in Table 2) and by checking the clusters against calendar information, to see if the clustering solution identifies workdays and weekends/holidays as different clusters. Overall, *k-means-euclidean* and *pam-manhattan* gave similar results. The hierarchical clustering *hclust* solution often agreed with *k-means*, but there were also cases where *hclust* settled for finer grained clusters. Regarding data representation, working in the original (*raw*) feature space seemed to provide reasonable clusterings. We have analysed different scaling of the data, for example row scaling or column scaling, but this leads to loss of information. For example, with row scaling, the time series that capture generating at constant level (e.g., 102 MW) or the turned-off (0 MW) state cannot be differentiated. The *stats* representation refers to using summary statistics of each time series as features (min, max, median,

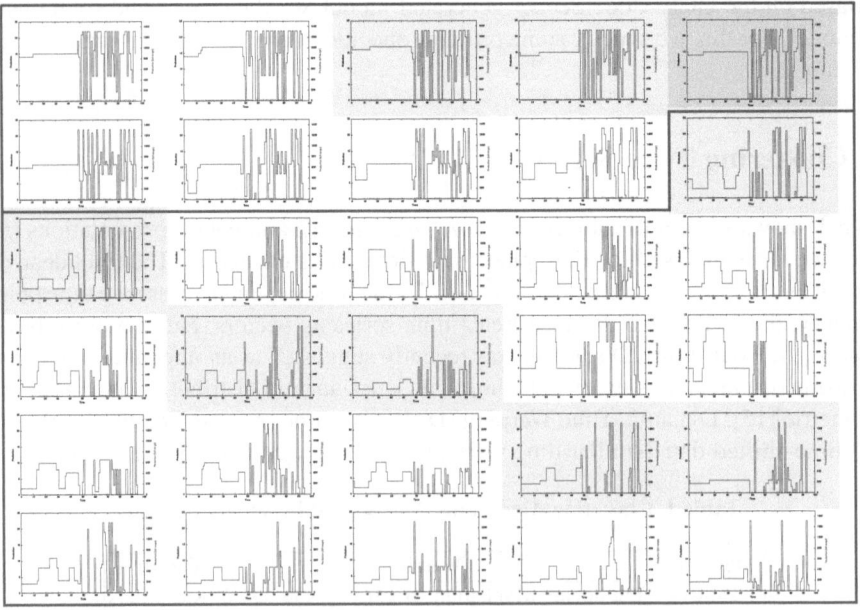

Fig. 3. EDF Power Output Profiles of Example Plant, 1-30 April 2010

mean, 1st and 3rd quartiles). Focusing on *kmeans-euclidean-raw* as our main clustering solution, we observed the following. The large majority of plants have 2 clusters, e.g., that separate workdays and weekends/holidays production profiles. A small percentage of plants have the same profile every day, and a third medium-sized category has plants with more than 2 clusters. Table 2 shows the number of plants in the three categories.

Figure 3 shows the 30 samples of April 2010 for one power plant, which serves as a **running example** in this paper. The *kmeans-euclidean-raw* clustering is shown in red. Weekends and public holidays are also highlighted, green and blue. For this particular plant, 2 clusters were found (01-09/04 and 10-30/04), that *did not* correspond to the workday/weekend split.

5 Variables and Domains

The ModelSeeker in [4] uses the range of values that occur in solutions as the initial domains of the variables. The default assumption is that in highly structured problems, like matrix models, all variables are quite similar and therefore share the same domain. For UCP, we can use more background knowledge. The decision variables denote the power output at different time periods, and output values at some time point may be quite different from the output at other time points, while we observe that the output levels at the same time on different days are quite similar. It then makes sense to treat each variable independently, and to only allow values in the domain which are already present in the given solutions. The output values are not ranges of integer values, instead each observed value corresponds to a specific operating mode of the plant.

Figure 4 shows the domains generated from the example solutions in Figure 3. The time periods are shown on the x-axis, the values on the y-axis. Each marker indicates

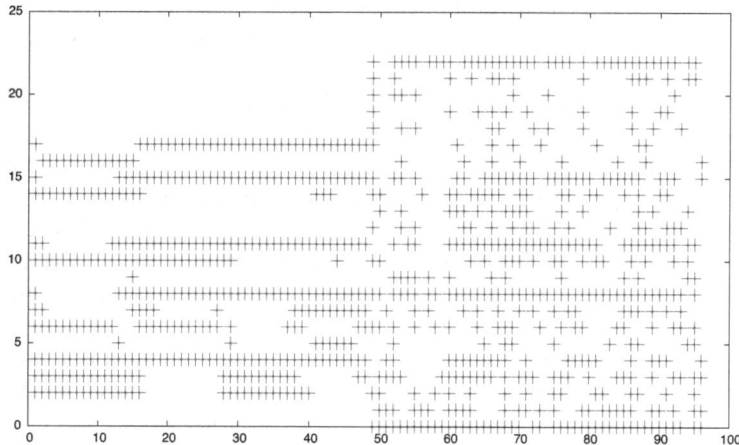

Fig. 4. Variable Domains for Example Plant

Table 3. Domain Sizes Over All Plants (Top: domain size, bottom: percentage of variables)

1	2	3	4	5	6	7	8	9	10	11	12	13	14	15	16	17	18	19-30
4.83	8.97	13.87	12.77	9.80	8.08	6.46	5.29	5.25	5.26	3.95	2.67	2.50	1.82	1.64	1.29	1.19	1.04	3.33

that a given value is used in some solution in this time period. The set of values occurring for each time period specifies the domain for the corresponding variable. This method of restricting the variables' domains generates tight domains. Table 3 shows aggregate results over all plants for April 2010, ignoring constant zero profiles, and without clustering of the solutions. The first row denotes the size of the domain, the second row the percentage of variables which have that domain size. Note that overall nearly 5% of all variables have constant value in all considered solutions.

6   Learning Constraints

In this section we discuss the steps of learning different types of constraints that can be used to describe features of time series.

6.1   Global Time Series Constraints

From a constraint perspective, computing features of a sequence of integer values can be related to *functional dependency constraints* for the following reasons:

1. By generalizing the sequence of integers to a sequence of variables, functional dependency constraints [1] extend the computation from a sequence of integers to a full constraint between variables. Since the Global Constraint Catalog [3] contains over one hundred such constraints, they are natural candidates for expressing plant specific features.
2. Values computed by many of the functional dependency constraints correspond to specific features in a time series, (e.g., *number of distinct values, largest value, number of peaks, maximum slope on the strictly increasing sequences*), by using a variety of such features we aim to accurately characterize the time series.
3. As mentioned in the introduction, the EDF production curves are the result of two distinct technical aspects. On one hand, the main driver is the overall electricity demand, which is similar to time series from econometrics or life sciences, i.e., price-change or growth of a being, that carry a degree of uncertainty. On the other hand, the EDF production curves also capture the underlying technological constraints of the production plants, which correspond to hard, physical constraints.
4. Since we not only aim to characterize the time series, but also to extract a set of constraints that can be added to a solver to enhance its performance, a natural choice is to select features that can be easily turned into constraints.

6.2   Examples of Constraints Used

The UCP-ModelSeeker uses the Global Constraint Catalog [3] as the set of candidate constraints. In the context of this new application, we have made the following changes to the ModelSeeker [4] and to the Global Constraint Catalog:

- Previously, the selection of global constraints in the catalog related to time series was quite limited. Motivated by capturing potential technological constraints of the production units, we have added 20 new constraints that represent many typical features of a production curve.
- We initially only search for constraints over the full time period of two days. This means that UCP-ModelSeeker doesn't use the partition generators available in the ModelSeeker of [4] (e.g., we do not try to interpret a sequence as a two or three dimensional matrix).
- Since the number of available EDF samples is quite large, we have written dedicated checkers for all automata constraints used in the UCP-ModelSeeker to improve performance. Most constraints without automata already had a checker available in the previous ModelSeeker system.

Below we show a set of simple constraints for a sequence $\mathcal{S} = s_1 \, s_2 \, \ldots \, s_n$.

- `among_diff_0`   : *number of values different from 0 in $\mathcal{S}$,*
- `maximum`   : *maximum value in $\mathcal{S}$,*
- `minimum`   : *minimum value in $\mathcal{S}$,*
- `minimum_except_0` : *minimum value in $\mathcal{S}$ discarding value 0,*
- `sum_ctr`   : *sum of the elements of $\mathcal{S}$,*
- `nvalue`   : *number of distinct values in $\mathcal{S}$,*
- `max_nvalue`   : *number of occurrences of the most used value in $\mathcal{S}$,*
- `min_nvalue`   : *number of occurrences of the least used value in $\mathcal{S}$,*
- `balance`   : *difference between the number of occurrences of the most and least used values,*
- `change`   : *number of consecutive values in $\mathcal{S}$ that are different.*

Most functional dependency constraints (features) are related to the peaks/valleys of the profile, where a peak corresponds to a value increase followed by a value decrease.

A variable s_p $(1 < p < n)$ of the sequence $s_1, s_2, \ldots, s_n$ is a *peak* if and only if there exists an i $(1 < i \le p)$ such that $s_{i-1} < s_i, s_i = s_{i+1} = \ldots = s_p$ and $s_p > s_{p+1}$. Similarly a variable s_v $(1 < v < n)$ is a *valley* if and only if there exists an i $(1 < i \le v)$ such that $s_{i-1} > s_i, s_i = s_{i+1} = \ldots = s_v$ and $s_v < s_{v+1}$. A peak variable s_p $(1 < p < n)$ is a *potential big peak* wrt. a non-negative integer Δ if and only if:

1. s_p is a peak,
2. $\exists i, j \in [1, n] \mid i < p < j$, s_i is a valley (or $i = 1$ if there is no valley before position p), s_j is a valley (or $i = n$ if there is no valley after position p), $s_p - s_i > \Delta$, and $s_p - s_j > \Delta$.

Let i_p and j_p be the largest i and the smallest j satisfying condition 2. A potential big peak s_p $(1 < p < n)$ is a *big peak* if and only if the interval $[i, j]$ does not contain any potential big peak that is strictly higher than s_p. Figure 5 illustrates the notion of big peak for a given sequence.

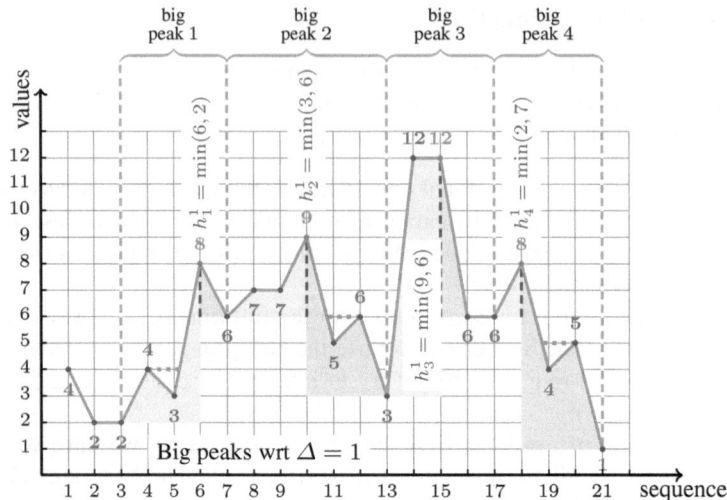

Fig. 5. Illustration of the `big_peak` constraint: A sequence $s_1\ s_2\ \ldots\ s_{21} =$ 4 2 2 4 3 8 6 7 7 9 5 6 3 12 12 6 6 8 4 5 1 and its 4 big peaks when $\Delta = 1$ with their respective heights $h_1^1 = 2,\ h_2^1 = 3,\ h_3^1 = 6,\ h_4^1 = 2$

We propose the following more complex functional dependency constraints related to profile inflexion:

- `peak` : *number of peaks of S,*
- `highest_peak` : *altitude of the highest peak of S,*
- `min_width_peak` : *smallest width of any peak of S,*
- `nvisible_from_start` : *number of peaks visible from the start of S,*
- `nvisible_from_end` : *number of peaks visible from the end of S,*
- `inflexion` : *number of peaks and valleys of S,*
- `min_dist_between_inflexion` : *minimum distance between consecutive inflexion of S,*
- `longest_increasing_sequence` : *range of the longest increasing subsequence of S,*
- `max_increasing_slope` : *maximum slope on the strictly increasing subsequences of S,*
- `min_increasing_slope` : *minimum slope on the strictly increasing subsequences of S,*
- `big_peak` : *number of big peaks of S.*

A further 8 constraints without functional dependencies were added to describe properties of all peaks in a time series, e.g. that all peaks have the same height.

6.3   Binary Constraints

The variables in our problem represent the power output in each time period, so that consecutive variables represent consecutive time periods. It is thus interesting to see how the power output changes from one time period to the next. For this, we introduce binary constraints between consecutive variables following the techniques introduced in CONACQ [5].

Figure 6 shows the lattice of binary constraints that we consider. For any two consecutive time periods of one sample, their relation is either equality, greater or smaller. To produce slightly stronger results, instead of just using $X > Y$ we consider $X \geq Y + C$

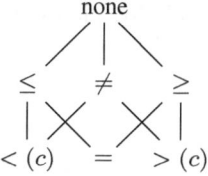

Fig. 6. Lattice of Binary Constraints

with the largest possible constant C. When combining multiple inequalities with constants C_1 and C_2, we use their minimum $\min(C_1, C_2)$ as the new value. If we combine different types of constraints in the lattice, we move up in the lattice. As soon as we have encountered all three possible outcomes between two variables in different samples, we drop the constraint between them. Figure 7 shows the constraints that are generated for our example plant. For our time series of 96 values, we can potentially produce 95 constraints. An entry marked with $-$ denotes that there is no constraint. In part (A) we see the constraints computed from the first cluster (days 1-9). Part (B) shows the time series from the second cluster (days 10-30), and part (C) the constraints corresponding to all 30 samples (no clustering). Note that the cluster size has a strong impact on the constraints generated: the fewer samples we use, the stronger the constraints will be. But even when using all 30 samples, we still find a significant number of constraints between consecutive variables. This indicates that these constraints capture an important regularity in the solutions, that our model should preserve. Table 4 shows that overall this technique is very useful, constraints are generated for more than three-quarters of

(A) Cluster 1 (Days 1-9)

(B) Cluster 2 (Days 10-30)

(C) All Samples (Days 1-30)

Fig. 7. Binary Constraints for Example Plant

Table 4. Binary Constraints Distribution for All Plants (No Clustering)

Constraint	none	=	$\geq$	$\leq$	$>$	$<$	$\neq$
Count	5852	9326	4388	5129	45	59	21
Percentage	23.58	37.21	17.68	20.66	0.18	0.24	0.08

all variable pairs, when considering all 261 plants. Note that strict inequalities are quite rare, but that equality constraints hold for over 37% of all consecutive variable pairs. By enforcing these constraints, we dramatically reduce the number of choices that need to be considered.

6.4   Learning Custom Variable Partitions

The ModelSeeker [4] does not only generate global constraints over the full set of variables, but considers many regular partitions of the variables to see if the same constraint with the *same* parameters holds for each of the smaller subsets. This allows finding identical constraints on rows or columns of matrix models, for example. We tried the same approach for the UPC, but the results were rather disappointing. While it is easy to partition the time series into smaller series of 24, 12 or 6 hours each, we only rarely find a constraint that holds with the same parameter value for each subsequence. The reason is that time series for the first and second day in the manually modified solutions are often very different, as most modifications are only applied to the first day of the schedule. This does not happen for all plants, and the change is not always exactly after 24 hours. But if we can recognize such a change, then we can state constraints with *different* parameters for the resulting subsequences. To compute time points when the largest change happens, we propose the following technique, shown in Figure 8 for a constraint with a functional dependency f_c. We consider how the values for the functional dependency values change if we were to split the sequence at time i or time $i+1$. a_i^{cd} denotes the functional dependency value for the front part of the sample solution time series with constraint c for day d at time i, while b_i^{cd} denotes the value for the back part. We can then see how the values change if we shift from i to $i+1$. If neither a nor b is affected, this is not a very interesting time point. But if the parameter values change for both a and b on many days at the same time i, we should consider splitting the time series here.

The overall interest in a split at time point i is determined by computing the change of the parameters for all constraints and all samples to produce a weight w_i. Time points with large weights are our candidates for splitting the time series into two sub sequences.

$$\forall_{i \in I}: \quad w_i = \sum_{c \in C} \sum_{d \in \text{Days}} |a_i^{cd} - a_{i+1}^{cd}| + |b_i^{cd} - b_{i+1}^{cd}| \qquad (1)$$

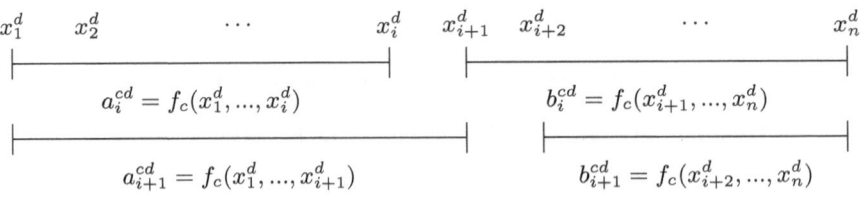

Fig. 8. Learning Custom Variable Partitions

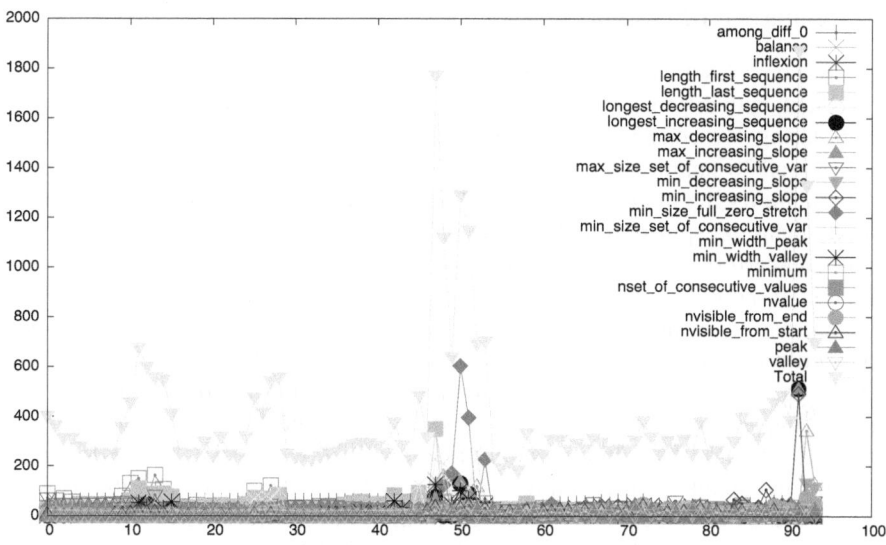

Fig. 9. Identified Custom Variable Partition for Example Plant

Figure 9 shows the resulting computation for our example plant. There is a clearly defined peak in the weight distribution around time period 48 (between the first and second day), and another peak at the end of the series. When either a or b are very short, we can expect rapid changes for most functional dependencies. We therefore ignore peaks that occur close to the start or the end of the time series, and only split the series into two parts, corresponding in this particular case to each day of the schedule.

Having identified this time series split-point for a plant, we can search for constraints which hold for each subsequence independently, thus we may find (potentially) different constraint parameters for each sub-period.

7   Generating Solutions with the UCP-ModelSeeker

Based on the previous steps, we have by now defined a finite constraint model, that has variables with individual domains, global constraints over the full set of time series (and smaller subsequences per each time series), and binary constraints between consecutive time periods. Now, we aim to use the acquired UCP-ModelSeeker model to generate new plant production profiles (i.e., find new solutions). The resulting model can be fairly difficult to solve without a custom search routine. In particular, we experienced that a first-fail variable selection is quite ineffective, compared to a sequential left-to-right selection. Furthermore, we found that trying values in increasing order is also ineffective. Instead, we either try values in *random order*, or try values by *frequency order* (by decreasing occurrence in the samples).

7.1   Results and Discussion

We show some results on the profiles generated using UCP-ModelSeeker, for our example plant, with different combinations of *clustering*, *constraints* and *search* routines. For clustering, we consider either cluster 1 (days 1-9), cluster 2 (days 10-30), or no clustering (all 30 samples). For search, we consider random and frequency-based value selection. We try the model with and without the custom split of time periods (for our example-plant the split is found at time 48). For all models, we include binary constraints and perform the variable selection left-to-right.

Figure 10 shows the generated profiles, while Table 5 describes the options used for each case and gives the mean absolute error (MAE) and mean squared error (MSE) of the generated solution wrt. the EDF samples considered for that scenario. A visual comparison with Figure 3 shows that the resulting profiles are quite "similar" to the original samples, and capture some of the properties of each cluster. Nevertheless, the visual inspection or the solution-quality measurement (e.g., MAE, MSE) do not fully determine the quality of the solutions produced. This will only be possible if we can compare the UCP-ModelSeeker solutions to the full real-world EDF-UCP model, or asses their utility in speeding up solving the original MIP model.

As a different experiment, Table 6 gives an indication of the impact of different search and constraint choices on the time required to find solutions for all 261 plants active in April 2010. We perform no clustering, and no custom splitting of the time intervals. We run each problem with a timeout of 10s, and report the percentage of problems solved, and the time required to find one solution for each plant. We see that the combination of left-to-right variable choice, frequent value selection, and using binary constraints is the only one which allows to find solutions for all plants in the given time limit.

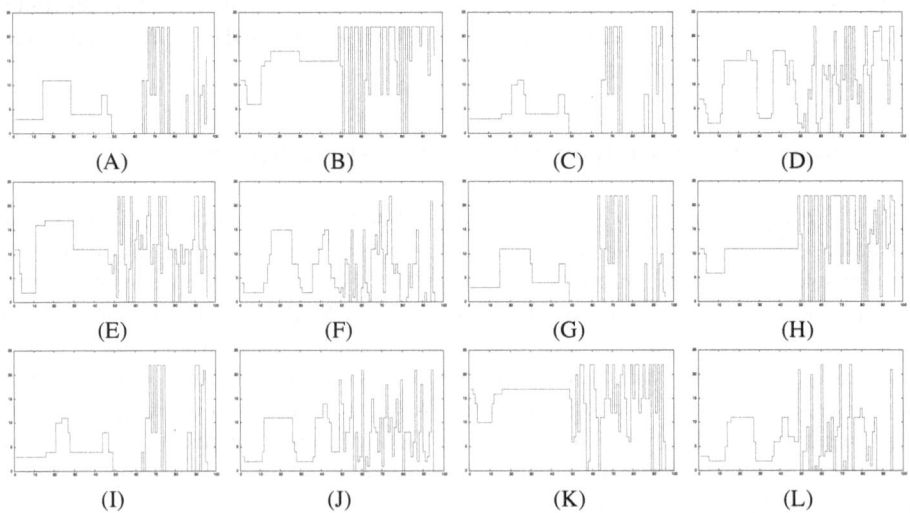

Fig. 10. Examples of UCP-ModelSeeker Generated Profiles for Example Plant

Table 5. Result Summary for Generated Profiles of Figure 10

Variant	Variable Selection	Value Selection	Binary Constraints	Cluster	Split	MAE	MSE	Time (sec)
A	left-right	Frequent	yes	All	no	452.30	**87.53**	1.10
B	left-right	Frequent	yes	1	no	449.67	104.62	0.53
C	left-right	Frequent	yes	2	no	298.43	70.90	0.87
D	left-right	Random	yes	All	no	649.97	114.20	1.43
E	left-right	Random	yes	1	no	492.90	106.68	0.54
F	left-right	Random	yes	2	no	422.33	82.48	0.89
G	left-right	Frequent	yes	All	yes	**445.10**	87.82	2.30
H	left-right	Frequent	yes	1	yes	**431.33**	**101.70**	1.03
I	left-right	Frequent	yes	2	yes	**294.00**	**70.70**	1.74
J	left-right	Random	yes	All	yes	547.37	97.23	2.32
K	left-right	Random	yes	1	yes	510.22	111.71	1.03
L	left-right	Random	yes	2	yes	397.86	78.38	1.73

Table 6. Solving Times for 261 Plants

Variable Selection	Value Selection	Binary Constraints	Cluster	Split	Percentage Solved	Time (sec)
first-fail	indomain	no	All	no	79.31	844.71
first-fail	indomain	yes	All	no	85.82	634.77
left-right	indomain	no	All	no	98.08	298.87
left-right	indomain	yes	All	no	98.85	264.94
left-right	Random	no	All	no	72.80	1098.31
left-right	Random	yes	All	no	89.27	572.28
left-right	Frequent	no	All	no	99.23	261.37
left-right	Frequent	yes	All	no	**100.00**	**232.63**

8   Conclusions and Future Work

In this paper we have presented the UCP-ModelSeeker system, which extends and specializes previous work on automated constraint acquisition for an important real-world optimization application, the Unit Commitment Problem, with a case-study on data made available by Electricité de France (EDF). Considering properties of the specific application, we have proposed partitioning of the production curves based on clustering, have added new functional dependency and binary constraints relevant for describing production curves, and defined a series of custom search strategies. Based on the acquired UCP-ModelSeeker model, we are able to generate new production profiles for each power plant. The aim is to compare these to the original EDF profiles, and use the insights gained in describing and generating the existing/new solutions, for speeding up solving. We are currently working with EDF on the assessment of the solution quality and a possible integration of some of these techniques into their existing MIP model.

References

1. Beldiceanu, N., Carlsson, M., Flener, P., Pearson, J.: On the reification of global constraints. Constraints 18(1), 1–6 (2013)
2. Beldiceanu, N., Carlsson, M., Petit, T.: Deriving filtering algorithms from constraint checkers. In: Wallace, M. (ed.) CP 2004. LNCS, vol. 3258, pp. 107–122. Springer, Heidelberg (2004)

3. Beldiceanu, N., Carlsson, M., Rampon, J.-X.: Global constraint catalog, 2nd edn. (revision a). Technical Report T2012-03, Swedish Institute of Computer Science (2012)
4. Beldiceanu, N., Simonis, H.: A model seeker: Extracting global constraint models from positive examples. In: Milano, M. (ed.) CP 2012. LNCS, vol. 7514, pp. 141–157. Springer, Heidelberg (2012)
5. Bessière, C., Coletta, R., Koriche, F., O'Sullivan, B.: Acquiring constraint networks using a sat-based version space algorithm. In: Proceedings of the Twenty-First National Conference on Artificial Intelligence and the Eighteenth Innovative Applications of Artificial Intelligence Conference, Boston, Massachusetts, USA, July 16-20, pp. 1565–1568. AAAI Press (2006)
6. Carrión, M., Arroyo, J.M.: A computationally efficient mixed-integer linear formulation for the thermal unit commitment problem. IEEE Transactions on Power Systems 21(3), 1371–1378 (2006)
7. Côté, M.-C., Gendron, B., Rousseau, L.-M.: Modeling the regular constraint with integer programming. In: Van Hentenryck, P., Wolsey, L.A. (eds.) CPAIOR 2007. LNCS, vol. 4510, pp. 29–43. Springer, Heidelberg (2007)
8. Dereu, G., Grellier, V.: Latest improvements of edf mid-term power generation management. In: Pardalos, P.M., Rebennack, S., Pereira, M.V.F., Iliadis, N.A. (eds.) Handbook of Power Systems I. Energy Systems, pp. 77–94. Springer, Heidelberg (2010)
9. Hastie, T., Tibshirani, R., Friedman, J.: The Elements of Statistical Learning. Springer Series in Statistics. Springer (2012)
10. Hechme-Doukopoulos, G., Brignol-Charousset, S., Malick, J., Lemaréchal, C.: The short-term electricity production management problem at EDF. Optima Newsletter - Mathematical Optimization Society 84, 2–6 (2010)
11. Hobbs, B.F. (ed.): The Next Generation of Electric Power Unit Commitment Models. Springer (April 2001)
12. Juste, K.A., Kita, H., Tanaka, E., Hasegawa, J.: An evolutionary programming solution to the unit commitment problem. IEEE Transactions on Power Systems 14(4), 1452–1459 (1999)
13. Kazarlis, S.A., Bakirtzis, A.G., Petridis, V.: A genetic algorithm solution to the unit commitment problem. IEEE Transactions on Power Systems 11(1), 83–92 (1996)
14. López, J.Á., Ceciliano-Meza, J.L., Moya, I.G., Gómez, R.N.: A MIQCP formulation to solve the unit commitment problem for large-scale power systems. International Journal of Electrical Power & Energy Systems 36(1), 68–75 (2012)
15. Musetti, A.T.Y.: Clustering methods for financial time series. PhD thesis, Swiss Federal Institute of Technology Zurich (ETH), Department of Mathematics (March 2012)
16. Takriti, S., Birge, J.R., Long, E.: A stochastic model for the unit commitment problem. IEEE Transactions on Power Systems 11(3), 1497–1508 (1996)
17. Virmani, S., Adrian, E.C., Imhof, K., Mukherjee, S.: Implementation of a lagrangian relaxation based unit commitment problem. IEEE Transactions on Power Systems 4(4), 1373–1380 (1989)

Solving the Agricultural Land Allocation Problem by Constraint-Based Local Search

Quoc Trung Bui[1], Quang Dung Pham[2], and Yves Deville[1]

[1] ICTEAM, Université catholique de Louvain, Belgium
{quoc.bui,Yves.Deville}@uclouvain.be
[2] SoICT, Hanoi University of Science and Technology, Vietnam
dungpq@soict.hut.edu.vn

Abstract. Agricultural land allocation is a problem that exists in most provinces in Vietnam. Each household owns many disconnected parcels, which reduces agricultural development. The solution to the problem is to repartition this agricultural land among the households, while satisfying some criteria. Historically, this problem has been approached neither using optimization technology nor computer science. The present paper describes the formulation of the problem and proposes a constraint-based local search algorithm for solving it. Experimental results on real data in Dong Trung village show that the solution computed by our algorithm is better than traditional solutions.

1 Introduction

In most provinces of Vietnam, agricultural land is still fragmented. One household has many parcels of different land categories (each category corresponds to a certain quality) from different fields. These parcels are very small and scattered. For example, in Vinh Phuc province, one household might have 47 parcels, each of which has the area of about ten square meters. The fact that each households has many separated small parcels leads to a lot of difficulties. First, households can not use machines for cultivating their small parcels. Second, fragmented parcels require a very high cost for visiting and controlling them. Third, the excessive number of tracks between the parcels results in a waste of agricultural land. Finally, projects of agricultural development are confronted with many difficulties caused by the huge number of small parcels.

The Vietnamese government promulgated a policy to overcomes this limitation. This consists of merging small parcels into large fields and then repartitioning these fields into larger parcels. In provinces where the policy was carried out, the results obtained have been very promising. After merging and repartitioning, the number of parcels held by a household markedly decreases (e.g., Bac Ninh [1], a reduction by a factor of 10) and the area of each parcel increases, with the rice output increasing considerably (e.g., Quang Nam [5], an increase of 20%–25%). Today, this land reallocation process has only been applied in few vietnamese provinces.

After merging the existing parcels (specified in the past), we have a set of fields of different categories (there might be several fields of the same category) and a set of households; each household has an *expected area* of agricultural land for each category. We need to specify, for each household, the parcel (the area and the position of the

C. Schulte (Ed.): CP 2013, LNCS 8124, pp. 749–757, 2013.
© Springer-Verlag Berlin Heidelberg 2013

parcel) in each field to avoid the above limitation. We consider fields of various shapes, i.e., rectangle, trapezium, quadrangle. But, for ease of presentation of the problem, we consider rectangular fields. Our solution however handles any quadrangular shape. We first present the traditional solution that people used in the past for allocating parcels to households, which does not use either computer science or optimization technology.

Traditional Solution. The Vietnamese government promulgated the following instructions to guide farmers in repartitioning the agricultural land. Fields are classified into categories (1–4) according to types of land quality, and determining a system of coefficients that presents an equivalence between categories. For example, a system of coefficients for three land categories $\langle 1.0, 1.2, 1.4 \rangle$ in which 1 m² of the first land category is equivalent to 1.2 m² of the second land category and 1.4 m² of the third land category. Fields of each land category are considered in turn for division into parcels, with the following rules:

1. The order of the fields is determined based on their land categories and geometrical positions and this is decided by the authorities.
2. The order of households is also determined by lot and households are assigned parcels with respect to this order. Suppose that this sorted list is $h_1, \ldots, h_n$.
3. Each field is divided into *zones* $z_1, \ldots, z_k$ of width 40–50 meters by lines which are parallel to one side of the field.
4. Each zone z_i is iteratively divided into parcels corresponding to a sequence of households $h_j, h_{j+1}, \ldots, h_p$ by parallel lines that are perpendicular to the parallel lines already used to separate the field into the zones. The next zone z_{i+1} will then be partitioned into parcels for households $h_{p+1}, \ldots$
5. At each step, suppose that household h_i under consideration, the current zone is z_j, and the current land category is c. The remaining area of z_j is R m². If R is greater than or equal to the expected area of h_i, then the next parcel of z_j will be allocated to h_i. Otherwise, h_i needs a supplementary area of S m². The following situations may occur:
   - R is smaller than 100, in which case the previous household h_{i-1} will receive this remaining area of R m². An equivalent (to R m² of category c) area of the next category $c + 1$ will be subtracted from the expected area of h_{i-1} for the category $c + 1$. The next zone z_{j+1} will then be considered for allocation to h_i.
   - Both R and S are greater than or equal to 100. The household h_i is allocated this remaining area of z_j (R m²), and will then receive a parcel of S m² in the next considered zone or field.
   - R is greater than or equal to 100, but S is less than 100. The household h_i is allocated this remaining area of R m² and an equivalent (to S m² of category c) area of the next category $c + 1$ will be added to the expected area of h_i for the category $c + 1$.

Limitations of the Traditional Solution. First, a lot of households have allocated areas which are different from their expected areas. The cause of this comes mainly from dividing a field into zones of fixed width, 40–50 m. Then, the threshold 100 m^2 may be not suitable for provinces where each household has thousands of square meters

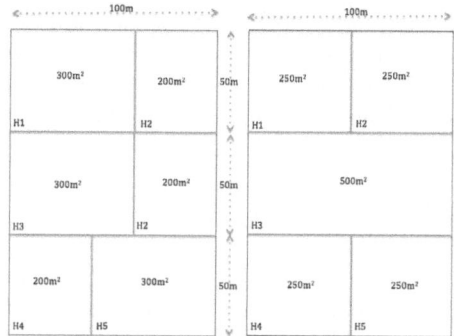

Fig. 1. A solution guided by the instructions of the government with five households $H1$, $H2$, $H3$, $H4$, and $H5$, and two fields of two different land categories with coefficients $\langle 1, 1 \rangle$. The expected areas of the first land category of the five households are 300, 400, 250, 300, and 250 m^2 (left part of the figure) and 250, 260, 560, 250, and $180 m^2$ of the second land category (right part of the figure). Applying the instructions of the government, each field is separated into three zones of width 50 m, and the allocated areas of the households are 300, 400, 300, 300, and 200 m^2 in the field on the left; and 250, 250, 500, 250, and 250 m^2 in the field on the right. In this solution, household $H2$ has three parcels of two land categories and households $H3$, $H4$, $H5$ are allocated parcels whose areas are different from their corresponding expected areas.

of agricultural land. In that case, there may be small parcels (with areas slightly more than 100 m^2) next to large parcels (with areas of some thousands of square meters). Finally, the instructions of the government do not take into account other expectations of the farmers, such as optimizing the number of parcels next to the canals, minimizing the distance from the house to the parcel, a nice form of the parcels, etc.

Objective of the Present Paper. The objective of the present paper is to improve on the traditional solution to the agricultural land allocation (ALA) problem by applying optimization technology. We hope that the developped method could be used in the forthcoming land reallocation in Vietnam. We propose a constraint-based local search algorithm for solving this problem. A local search algorithm typically starts from a solution and moves from one solution to a neighboring solution in the hope of improving an objective function that guides the search. A constraint-based local search algorithm is a local search algorithm that uses the violation of constraints and the evaluation of objective functions to guide the search [8]. ALA consists of two successive subproblems:

- **PArea** is the problem of computing the area of the parcel allocated to each household in each field. We do not consider any order between households.
- **PPos** is the problem of specifying the exact positions of the parcels allocated to the households. From the solution to **PArea**, we know, for each field, the set of households and their allocated areas from this field, and we have to determine the exact positions of the parcels for these households. This problem has been considered and solved in [3] but only for those cases in which the given field has the shape of a rectangle, triangle, or trapezium. We propose in this paper an approach for partitioning the fields with general quadrangular shape, which appear frequently in reality.

Contribution. We describe the problem formulation and a constraint-based local search algorithm for solving the ALA problem. We solve the **PArea** subproblem. For the **PPos** subproblem, we extend the results in [3] to partition a quadrangular field into parcels. We test the proposed algorithm on real data. The experimental results show the efficiency of the proposed algorithm compared with the traditional solution.

2   PArea: Computing the Parcel Areas

Input (1) A set of land categories $C = \{1, \ldots, m\}$ with a system of coefficients $(\alpha_1, \ldots, \alpha_m)$ that present the equivalences between the land categories. (2) A set of fields $\mathcal{F} = \{1, \ldots, p\}$, each field $f \in \mathcal{F}$ is associated with an area $A(f)$ and a land category $C(f) \in C$. (3) A set of households $\mathcal{H} = \{1, \ldots, n\}$, each household $h \in \mathcal{H}$ is associated with:

- a vector of *expected areas* $EA(h) = (A(h, 1), \ldots, A(h, m))$, in which $A(h, c)(h \in \mathcal{H}, c \in C)$ is the expected area of land category c.
- a vector of distances $(d(h, 1), \ldots, d(h, p))$, in which $d(h, f)(h \in \mathcal{H}, f \in \mathcal{F})$ is the distance from the house of household h to the center of field f.

Note that the equation $\forall c \in C, \sum_{f \in \mathcal{F}: C(f) = c} A(f) = \sum_{h \in \mathcal{H}} A(h, c)$ is always ensured.

Output. The output is the *allocated area* of each household in each field.

Ideally, we desire to have a solution in which each household receives a unique parcel with the corresponding expected area in some field for each land category if the corresponding expected area is greater than zero. Unfortunately, this often cannot be arranged. For example, if we have two fields of areas 300 m^2 and 700 m^2 of category 1, we have four households whose expected areas of land category 1 are respectively 200 m^2, 200 m^2, 400 m^2, and 200 m^2. In our approach, we prioritize the solutions in which each household receives a unique parcel in some field for each land category if the households expected area of this category is greater than zero. Otherwise, this household receives no parcel. In addition, we accept solutions in which some households have allocated areas different from their expected areas, and we try to minimize this difference.

In our approach, we iteratively allocate parcels to households for each land category, starting with land category 1 and finishing with land category m. The results computed for each land category c may change the input of the expected areas of the households for the next land category:

- For a household who receives an additional area a out of its expected area of land category c, its expected area for the land category $c+1$ will be lessened by $a \times \frac{\alpha_{c+1}}{\alpha_c}$
- For a household who receives area a less than its expected area of land category c, its expected area for land category $c + 1$ will be increased by $a \times \frac{\alpha_{c+1}}{\alpha_c}$

In the following, we give a mathematical formulation of the problem **PArea for a specific land category** c.

Input. (1) A set of fields $\mathcal{F}_c = \{f_1, f_2, \ldots, f_q\} \subseteq \mathcal{F}$ of land category c. (2) A set of households whose expected areas of land category c are greater than zero $\mathcal{H}_c = \{h_1, \ldots, h_k\} \subseteq \mathcal{H}$

Decision variables. Variable $a(h, f)$ $(h \in \mathcal{H}_c, f \in \mathcal{F}_c)$ presents the allocated area of household h in field f.

Invariant. $F(h) = \{f \in \mathcal{F}_c \mid a(h, f) > 0\}, \forall h \in \mathcal{H}_c$ represents the set of fields of land category c where the household h is allocated non-zero areas.

Constraints. **C1**: $A(f) = \sum_{h \in \mathcal{H}_c} a(h, f), \forall f \in \mathcal{F}_c$. **C2**: $|F(h)| = 1, \forall h \in \mathcal{H}_c$. Constraint **C1** states that the sum of areas allocated to households in each field is equal to the area of that field, and **C2** specifies that each household receives a unique parcel in some field.

Objectives. In order to respond to the expectations of the farmers, we propose the two following objectives, which are to be minimized: $F_0(c) = \sum_{h \in \mathcal{H}_c, f \in \mathcal{F}_c : a(h,f) > 0} d(h, f)$ and $F_1 = \sum_{h \in \mathcal{H}_c} |\sum_{f \in \mathcal{F}_c} a(h, f) - A(h, c)|$, in which $F_0(c)$ is the sum of distances from the houses of the households to the centers of the fields in which their allocated parcels are located, and F_1 is the sum of the differences between the allocated areas and the expected areas of the households.

A constraint-based local search algorithm has been developped for solving problem **PArea** for each land category c. The algorithm uses tabu lists in order to avoid revisiting solutions already explored. We always ensure the constraint **C2** holds by modeling a solution by a k-dimensional vector $(x(h_1), \ldots, x(h_k))$ in which $x(h_i) \in \mathcal{F}_c(h_i \in \mathcal{H}_c)$ indicates the field wherein the household h_i is allocated a parcel. During the search, there are invariants $\Delta(f) = |A(f) - \sum_{h_i \in \mathcal{H}_c} (x(h_i) = f) * A(h_i, c)|$ representing the difference between the area of the field f and the sum of the expected areas of the households allocated in this field. Our algorithm will find solutions minimizing $\sum_{f \in \mathcal{F}_c} \Delta(f)$ and F_0. Finally, the actually allocated areas of the households will be adapted based on the solution computed in order to satisfy the constraint **C1**.

The algorithm proceeds in two steps. Step 1 determines for each field of land category c a group of households who have parcels in this field. The solution is represented by a vector $S = (x(h_1), \ldots, x(h_k))$ in which $x(h_i) \in \mathcal{F}_c$ is a field of land category c in which a parcel allocated to household h_i is located, $\forall h_i \in \mathcal{H}_c$. Our algorithm considers two neighborhoods:

1. *Change-based neighborhood*
   $N_1(S) = \{(x(h_1), \ldots, x'(h_i), \ldots, x(h_k)) | h_i \in \mathcal{H}_c, x(h_i) \neq x'(h_i) \in \mathcal{F}_c\}$
2. *Swap-based neighborhood* $N_2(S) = \{(x(h_1), \ldots, x(h_{i+j}), \ldots, x(h_i), \ldots x(h_k)) | x(h_i) \neq x(h_{i+j}) \in \mathcal{F}_c, 1 \leq i < i + j \leq k\}$

In step 1, we suppose each household is allocated a quantity of area which is equal to its expected area. Due to this assumption, there exist, for each field, a difference δ between the area of the field and the sum of the allocated areas in this field. The goal of step 2 of our algorithm is to adapt the allocated area such that the sum of the allocated areas in the field fits the area of that field.

3 PPos: Finding the Positions of the Parcels

When problem **PArea** has been solved, we have, for each household, their allocated area in each field. The overall goal of the solution of **PPos** is to determine the exact position of the parcels in each field. In our previous paper [3], we specified the problem and proposed tabu search algorithms for **PPos** when the shape of fields was restricted to be a triangle, a trapezium, or a rectangle.

A field is a large quadrangle where hundreds of households must be assigned a parcel. The field will be decomposed using horizontal and vertical lines as illustrated in Figure 2. The field will be partitioned into *zones* by horizontal lines. Each zone is then partitioned into *parcels* by vertical lines which are allocated to households. Each household will be assigned only one parcel in the field. When one or more sides of the quadrangle are close to a canal, one has to maximize the number of parcels close to the canal (see Figure 2) as this is an important issue for Vietnamese farmers. The **PPos** problem is then divided into two independent subproblems.

- Problem **P1**: The given field has one or more canals that are located on its sides. This problem computes the zones and parcels next to the canals that are assigned to the househoulds.
- Problem **P2**: The remainder of the field, obtained by removing the zones computed by **P1**, has no canal on its sides. This remaining area, considered as a new field, is then partitioned into zones and parcels assigned to households.

The problem **P1** is viewed as a subset selection problem: we need to find a subset of households, satisfying some constraints while optimizing some objective function. The constraints are that each parcel must fit the exact area corresponding to a household, and that the ratio between the height and the width of a parcel must be bounded by some given constant in order to have parcels with an efficient shape. The objective function is to maximize the number of parcels next to the canal. A second, less important objective, is to have parcels with a shape close to a square. A tabu search is described in [3], where three neighborhoods are considered: removal-based, insertion-based, and swap-based neighborhoods.

The problem **P2** is viewed as a set partitioning problem. Given a set of households, we need to partition this set into subsets of households, one subset corresponding to one zone of the field. Each zone is then easily decomposed into parcels. Each parcel will thus fit the exact area corresponding to a household. The objective is to have parcels with a shape close to a square. A tabu search is described in [3], where two neighborhoods are considered: move-based and swap-based neighborhoods.

In [3], we restricted the shape of the field to be that of a triangle, trapezium, or a rectangle. However, most fields in reality are quadrangular. We here extend the problem to quadrangular fields. The idea is that a quadrangular field can be decomposed into smaller fields or can be approximated by a field of the shape of a trapezium, so that we can directly reuse the tabu search algorithms in [3].

When the quadrangular field has a shape close to a trapezium, the field is approximated by a trapezium with the same area. On the top of Figure 2, a quadrangle $ABCD$ (field of quadrangle $ABCD$ for short) is approximated by a trapezium $AB'C'D$, then the algorithms in [3] are applied to partition the trapezium $AB'C'D$ into the zones $B'E_1F_1C'$, $E_1F_1F_2E_2$, $E_2F_2F_3E_3$, and E_3F_3DA is divided into parcels.

When the quadrangle is not close to a trapezium, the field is decomposed into two parts: a triangular field and a pentagonal field that is approximated by a field with the shape of a trapezium with the same area. On the bottom of Figure 2, a quadrangle

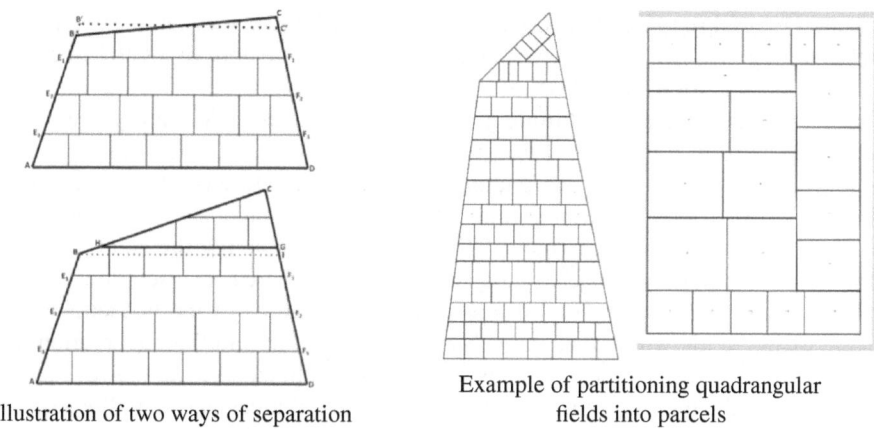

Illustration of two ways of separation

Example of partitioning quadrangular
fields into parcels

Fig. 2. Illustration and Example of partitioning quadrangular fileds

$ABCD$ is separated into a triangle CHG and a pentagon $ABHGD$. Then the pentagon $ABHGD$ is approximated by a trapezium. The algorithms in [3] can then be applied to partition these two fields into zones and parcels.

In the above decomposition of a quadrangular field into a pentagon and a triangle, we need to decide the position of the cut (line HG in Figure 2). A constraint is that the triangular field (CHG in Figure 2) should have an area corresponding to a subset of so that the sum of the allocated areas of a group of households. To identify this cut, we solve a subset selection problem [7] as follows: given a set $P = \{a_1, \ldots, a_p\}, P \subset \mathbb{R}^+$ and a threshold $\mathcal{T} \in \mathbb{R}^+$, the overall goal is to determine a subset $SP \subseteq P$ satisfying the constraint $\mathcal{T} \geq \sum_{a \in SP} a$ and minimizing the objective function $(\mathcal{T} - \sum_{a \in SP} a)$. In this problem, the threshold $\mathcal{T}$ is the maximum area of the triangular shape (in Figure 2), $\mathcal{T}$ is the area of the triangle BCI). To solve this subset problem, we developed a simple tabu search algorithm using the two neighborhoods of a solution $SP \subseteq P$: *Change-based neighborhood* $N_1(SP) = \{SP \cup a | a \notin SP, a \in P\} \cup \{SP \setminus \{a\} | a \in SP\}$ and *Swap-based neighborhood* $N_2(SP) = \{SP \cup a \setminus \{b\} | a \in P, a \notin SP, b \in SP\}$.

4   Experiments

We selected the 10th hamlet of Dong Trung village, Tien Hai district, Thai Binh province, Vietnam, to try our algorithms. We collected all the data before and after the land grouping of this hamlet. In this hamlet, there are 103 households, 3 land categories $\{c1, c2, c3\}$, and 24 fields adding to 204,744 m^2 (8 fields of $c1$, 7 fields of $c2$ and 9 fields of $c3$). In the 10th hamlet, the agricultural land allocation problem was already carried out by using the instructions of the government. The system of coefficients was set to $\langle 1.0, 1.0, 1.0 \rangle$ by the farmers. In the existing solution by the farmers guided by the instructions of the government, almost all households have three parcels, some households are allocated four parcels.

Our Tabu search algorithm was implemented in the COMET programming language [4] and always returns the best solution obtained in the search. The time limit for each execution was set to two minutes. The length of the tabu lists was set to 20. The experiments were performed on XEN virtual machines with one core of a CPU Intel Core2 Quad Q6600 @2.40GHz and 1GB of RAM.

Figure 2 illustrates the partitioning of two quadrangular fields into parcels by our extended algorithm for solving **PPos**, in which the field on the left of the figure does not touch a water source and the field on the right of the figure is close to three canals (blue lines). We compared the solutions computed by the tabu search algorithms with the existing solution by the farmers. The criteria for evaluating the solutions are the differences between the expected area and the allocated area for each household, for each land category, and the total distances from the houses of the households to the centers of the fields where their parcels are located. In the traditional solution, there are households whose expected area is equal to their allocated area. However, for the other households, the differences between the expected areas and the allocated areas are very high. This shows a greater inequality between households. In our approach, the solution computed ensures greater equality between the households in the sense that all households can tolerate the resulting differences. Moreover, the differences produced by our solution are very small.

Fig. 3. Comparing the traditional solution and our solution

where

$$diff(h, c) = A(h, c) - \sum_{f \in F : C(f) = c} a(h, f)$$

$$Max(c) = Max\{|diff(h, c)| \mid h \in \mathcal{H}\}$$

$$Avg(c) = \frac{1}{|\mathcal{H}_c|} \sum_{h \in \mathcal{H}} |diff(h, c)|$$

$$Var(c) = \sqrt{\frac{1}{|\mathcal{H}_c|} \sum_{h \in \mathcal{H}} (|diff(h, c)| - Avg(c))^2}$$

$$Max = Max\{|\sum_{c \in C} diff(h, c)| \mid h \in \mathcal{H}\}$$

$$Avg = \frac{1}{\mathcal{H}} \sum_{h \in \mathcal{H}} |\sum_{c \in C} diff(h, c)|$$

$$Var = \sqrt{\frac{1}{\mathcal{H}} \sum_{h \in \mathcal{H}} (|\sum_{c \in C} diff(h, c)| - Avg)^2}$$

$$F_0 = \sum_{c \in C} F_0(c)$$

Criteria	Trad. sol.	20 executions of tabu search		
		MIN	AVG	MAX
$Max(c1)$	88	5	7.75	15
$Avg(c1)$	39.51	2.06	2.39	3.21
$Var(c1)$	29.37	1.22	1.87	3.27
$Max(c2)$	98	6	8.45	17
$Avg(c2)$	32.93	1.95	2.39	3.16
$Var(c2)$	25.31	1.29	1.96	3.23
$Max(c3)$	89	1	3.35	6
$Avg(c3)$	13.96	0.28	0.85	1.68
$Var(c3)$	17.73	0.45	0.90	1.43
Max	89	1	1.15	2
Avg	13.65	0.04	0.17	0.33
Var	16.92	0.19	0.36	0.51
Added parcels	10	0	0	0
F_0 (km)	270.79	261.55	265.10	271.36

In figure 3, the last three columns present a summary of 20 executions of our tabu search algorithm. Column MIN (resp. AVG and MAX) presents the minimum (average and maximum) value of the 20 solutions. Other fairness criteria, such as in [2,6], could also be used.

Our solutions are clearly better than the existing solution obtained by the application of the instructions of the government. There does not exist any household that has more than two parcels. This means that our solutions always satisfy constraint **C2**. In the existing solution by the farmers, there are 10 households that have four parcels. The values of the first 12 criteria of our solutions are always much smaller than those for the existing solution by the farmers. The value of the objective function F_0 in most of our solutions is slightly smaller than that of the existing solution by the farmers.

This experiment has been presented to the Vice President of the Dong Trung village (Thai Binh province). This village is composed of 1460 households, and the land has already been reallocated using the traditional instructions of the government. The solutions produced by our algorithms were considered as a significant improvement over the traditional reallocation. The very low difference between the expected area and the allocated area for each household, the absence of added parcels and the form of parcels with a shape close to a square were particularly appreciated. His conclusion was that our tool should be used to the land allocation of the remaining Vietnamese districts and provinces.

5   Conclusion

In this paper, we solved the agricultural land allocation problem that has recently emerged in many provinces in Vietnam, using constraint-based local search algorithms. The problem comprises two subproblems. The first one is to compute the set of households that arc allocated parcels in each agricutural field, and the second one is to specify the exact position of each parcel allocated in each agricutural field. For the first subproblem, we described its formulation and proposed a constraint-based local search algorithm for solving it. For the second subproblem, we extended the results in [3] to compute the position of each parcel allocated in a field with the shape of a generic quadrangle. The experimental results show that our approach yields better solutions than the traditional solution. The experimental results were also validated by a Vietnamese authority in land allocation.

We are now reimplementing the algorithms as a C++ tool that could be used by the different vietnamese districts. This tool will also be extended to handle specific constraints and objective functions, to meet the expectations of the farmers in various provinces. Comparison with other optimization approches will also be investigated.

References

1. Bac Ninh News. Hieu qua don dien doi thua o yen phong (2010),
   http://www.bacninh.gov.vn
2. Bouveret, S., Lemaître, M.: Computing leximin-optimal solutions in constraint networks. Artif. Intell. 173(2), 343–364 (2009)
3. Bui, Q.T., Pham, Q.-D., Deville, Y.: Constraint-based local search for fields partitioning problem. In: Proceedings of the Second Symposium on Information and Communication Technology, SoICT 2011, pp. 19–28. ACM, New York (2011)
4. D. D. T. Inc. Comet Tutorial (2004)
5. Le Van. Hoan thanh don dien doi thua. Dan Viet (2011)
6. Ogryczak, W., Sliwinski, T.: On solving linear programs with the ordered weighted averaging objective. European Journal of Operational Research 148(1), 80–91 (2003)
7. Pruhs, K., Woeginger, G.J.: Approximation schemes for a class of subset selection problems. Theor. Comput. Sci. 382(2), 151–156 (2007)
8. Van Hentenryck, P., Michel, L.: Constraint-based local search. The MIT Press, London (2005)

Constraint-Based Approaches
for Balancing Bike Sharing Systems

Luca Di Gaspero[1], Andrea Rendl[2], and Tommaso Urli[1]

[1] DIEGM, University of Udine,
Via Delle Scienze, 206 - 33100 Udine, Italy
{luca.digaspero,tommaso.urli}@uniud.it
[2] Dynamic Transportation Systems, Mobility Department,
Austrian Institute of Technology
Giefinggasse 2, 1210 Vienna, Austria
andrea.rendl@ait.ac.at

Abstract. In order to meet the users' demand, bike sharing systems must be regularly rebalanced. The problem of balancing bike sharing systems (BBSS) is concerned with designing optimal tours and operating instructions for relocating bikes among stations to maximally comply with the expected future bike demands. In this paper, we tackle the BBSS by means of Constraint Programming: first, we introduce two novel constraint models for the BBSS including a *smart* branching strategy that focusses on the most promising routes. Second, in order to speed-up the search process, we incorporate both models in a Large Neighborhood Search (LNS) approach that is adapted to the respective CP model. Third, we perform a computational evaluation on instances based on real-world data, where we see that the LNS approach outperforms the Branch & Bound approach and is competitive with other existing approaches.

1 Introduction

Bike sharing systems are a very popular means to provide bikes to citizens in a simple and cheap way. The idea is to install bike stations at various points in the city, from which a registered user can easily loan a bike by removing it from a specialized rack. After the ride, the user may return the bike at any arbitrary station (if there is a free rack). This service is mainly public or semi-public, often initiated to increase the attractiveness of non-motorized means of transportation and is typically almost free of charge for the users. This, among other reasons, is why bike sharing systems have become particularly popular and an essential service in many European cities.

Depending on their location, bike stations have specific patterns regarding when they are empty or full. For instance, in cities where most jobs are located near the city centre, the commuters cause certain peaks in the morning: the central bike stations are filled, while the stations in the outskirts are emptied. Furthermore, stations located on top of a hilly region are more likely to be empty, since users are less keen on cycling up a hill and thus less keen on returning a

C. Schulte (Ed.): CP 2013, LNCS 8124, pp. 758–773, 2013.

bike to such a station. These differences in flows are one of several reasons why many stations have extremely high or low bike loads over time, which often causes difficulties: on the one hand, if a station is empty, users cannot loan bikes from it, thus the demand cannot be met by the station. On the other hand, if a station is full, users cannot return bikes and have to find alternative stations that are not yet full. These issues can result in substantial user dissatisfaction which may eventually lead users to abandon the service. This is why nowadays most bike sharing system providers take measures to *rebalance* them.

Balancing a bike sharing system is typically done by employing a fleet of trucks that move bikes between unbalanced stations overnight. More specifically, each truck starts from a depot and travels from station to station in a tour, performing loading instructions (adding or removing bikes) at each stop. After servicing the last station, the empty truck returns to the depot.

Finding optimal tours and loading instructions is a challenging task: the problem consists of a routing problem combined with the problem of distributing single-commodities (bikes) to meet the demand. Furthermore, since most bike sharing systems typically have a large number of stations (≥ 100), but a small fleet of trucks, the trucks can only service a subset of stations in a reasonable time, thus it is also necessary to decide *which* stations should be balanced.

In this work, we tackle the problem of balancing bike sharing systems in two steps. First, we formulate the problem as two different CP models: a *routing model* based on the classical Vehicle Routing Problem (VRP) formulation, and a *step model* that involves a planning perspective of the problem. We discuss both models in detail and compare their performance in a computational evaluation. In a second step, we employ each CP model in a Large Neighborhood Search (LNS) approach that is customized to the features of the respective CP model.

This paper is structured as follows: Section 2 provides the description of the BBSS problem, including our notation. Section 3 introduces our two CP formulations for the BBSS: the routing model in Sec. 3.1, and the step model in Sec. 3.2. Then we discuss our LNS approach in Section 4 and summarize our computational evaluation in Section 5. Section 6 concludes the paper.

1.1   Related Work

Balancing of bike sharing systems has become an increasingly studied problem in the last few years. Benchimol et al. [1] consider the rebalancing as hard constraint and the objective is to minimize the travel time. They study different approximation algorithms on various instance types and derive different approximation factors for certain instance properties. Furthermore, they present a branch-and-cut approach based on an ILP including subtour elimination constraints. Contardo et al. [5] consider the dynamic variant of the problem and present a MIP model and an alternative Dantzig-Wolfe decomposition and Benders decomposition method to tackle larger instances. Raviv et al. [10] present two different MILP formulations for the static BBSS and also consider the stochastic and dynamic factors of the demand. In the approach of Chemla et al. [4], a branch-and-cut approach based on a relaxed MIP model is used in combination with a tabu search that provides upper bounds. Rainer-Harbach et al. [9] propose a heuristic approach for the BBSS in which effective routes are calculated by a variable neighbourhood search (VNS) metaheuristic and the loading instructions are computed by a helper algorithm, where they study three different alternatives (exact and heuristic) as helper algorithms. Schuijbroek et al. [12] propose a new cluster-first route-second heuristic, in which the clustering problem simultaneously considers the service level feasibility constraints and approximate routing costs. Furthermore, they present a constraint programming model for the BBSS that is based on a scheduling formulation of the problem and therefore differs significantly from our formulations.

In [6] we presented a hybrid approach for the BBSS by combining CP with Ant Colony Optimization (ACO). In that work, we introduced a VRP-based CP formulation and integrate ACO into its search procedure to improve its performance. This work extends our previous work by introducing a novel CP model that is inspired by AI-planning. Moreover we revise the VRP-based model presented in [6] and we develop a problem-specific branching strategy. Finally, in this paper we attempt to enhance the two models by combining them into Large Neighbourhood Search (LNS).

2   Balancing Bike Sharing Systems

The problem of balancing a bike sharing system (BBSS) is concerned with finding *tours* for a fleet of vehicles and the respective *loading instructions* per stop such that the bike sharing system is maximally balanced after the vehicles finish their tour. Note, that we consider the static case of the BBSS where we assume that no bikes are moved independently between stations during the balancing operation (in other words, we assume that there are no customers using the service during balancing which can be a valid approximation for balancing systems at night).

Bike sharing systems consist of bike stations $S = \{1, \ldots, S\}$ that are distributed all over the city. Each station $s \in S$ has a maximal capacity of C_s bike racks and holds b_s bikes where $0 \leq b_s \leq C_s$. The target value t_s for station s states how many bikes the station should ideally hold to satisfy the customer demand. The values for t_s are derived in advance from a user demand model where $0 \leq t_s \leq C_s$.

An additional requirement of our project partner, Citybike Wien, was to ensure monotonic loading of bikes, i.e. to ensure that bikes are only added or removed from stations where necessary (without intermediate storage). Therefore, we introduce two different station types: bike 'sinks' and bike 'sources'. Bikes may only be added to sink stations, and bikes may only be removed from source stations. A station is either a sink or a source, depending on its respective demand (if the number of bikes in the station is larger than the demand, the stations is a source, otherwise it is a sink).

A fleet of vehicles $V = \{1, \ldots, V\}$ with capacity $c_v > 0$ and initial load $\hat{b}_v \geq 0$ for each vehicle $v \in V$, move bikes between stations to reach the stations' target values. The vehicles are associated with depot D where they start and end their tour. Thus, the set of possible stops in a tour is denoted $\mathcal{S}_d = \mathcal{S} \cup D$. We have a time budget of $\hat{t} > 0$ time units to complete the balancing operation (and after which every vehicle has to have reached the depot). The travel times between all possible stops are given by the matrix $travelTime_{u,v}$ where $u, v \in \mathcal{S}_d$, which includes an estimate of the processing times needed to serve the station, if $v \in \mathcal{S}$.

The goal is to find a tour for each vehicle including loading instructions for each visited station. The loading instructions state how many bikes have to be removed from, or added to the station, respectively. Naturally, the loading instructions must respect the maximal capacity and current load of both the vehicle and the station. Furthermore, each vehicle can only operate within the overall time budget and has to distribute all loaded bikes before returning to the depot (i.e., the truck has to be empty when returning to the depot).

After every vehicle has returned to the depot, each station $s \in \mathcal{S}$ has a new load of bikes, denoted b'_s. Obviously, the closer b'_s is to the the desired target value t_s, the better the solution. Thus, our objective is to find tours that manipulate the station states such that they are as close as possible to their target values. Furthermore, we are interesting in finding a low-cost route r_v for each vehicle $v \in \mathcal{V}$, so we also minimize the total travel time (which is equivalent to minimize the total traveling cost).

Therefore, we introduce an objective function f that contains two components: first, the sum of the deviation of b'_s from t_s over all stations $s \in \mathcal{S}$, and second the travel time for each vehicle:

$$f(\sigma) := w_1 \sum_{s \in \mathcal{S}} |b'_s - t_s| + w_2 \sum_{v \in \mathcal{V}} \sum_{(u,w) \in r_v} travelTime_{u,w} \qquad (1)$$

Note that Equation 1 defines a scalarization over a naturally multi-objective problem. Some points in the Pareto optimal set are hence neglected by construction. Our main reason for this choice is the need to compare with the current best approaches [9], which employ an equivalent scalarization. Furthermore, to the best of our knowledge, multi-objective propagation techniques are still a relatively unexplored research area.

3   Constraint Models for the BBSS

In this section we present two constraint formulations for the Balancing Bike Sharing Systems problem (BBSS).

3.1   Routing Model

The routing model is an adaption of the constraint model of the classical vehicle routing problem (VRP) that is described in [8]. The routing model uses successor variables ($succ_i$) to model the path of each vehicle and service variables ($service_i$) to represent the operations at each station.

The essence of the graph encoding underlying the routing model is depicted in Figure 1 along with the elements of a possible solution. According to the formulation proposed in [8], the graph structure of the original problem G (lower

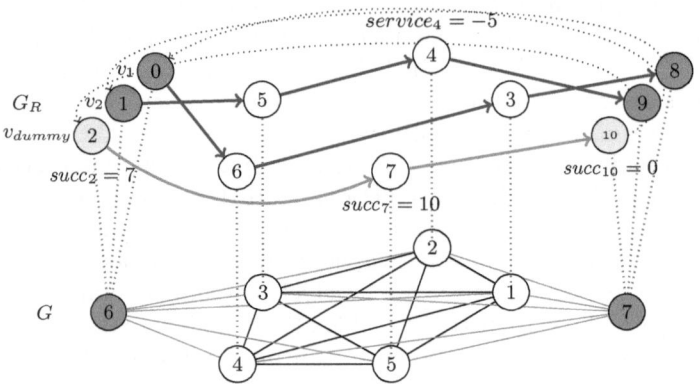

Fig. 1. Graph encoding of the BBSS problem employed in the routing CP model. The lower layer shows the original graph, whereas the upper layer shows the encoded graph, in the case of two vehicles, and the edges selected in a possible solution. The sub-path starting at node 2 and ending at node 10 corresponds to the set of unserved nodes.

layer) is encoded into an extended graph G_R (upper layer) by considering one replicate of the starting depot for each vehicle in order to identify each vehicle route as a sub-path in the graph starting at that node. The successor of the ending depot for a given vehicle is set to be the starting depot of the following vehicle (modulo the number of vehicles) so that we are searching for an Hamiltonian circuit in the extended graph. Moreover, for modeling service optionality, there is an additional vehicle v_{dummy}, whose sub-path comprises all the stations that are left unserved.

For brevity we do not give here a detailed specification of the variables and constraints that are involved in the model, which can be found in [6]. The main difference with respect to the cited paper, however, is the search strategy which will be outlined in the following section. Another difference is that the current version of the routing model employs the Hamiltonian `circuit` constraint instead of the `alldifferent` on the *succ* variables. Although the two formulations are equivalent, the `circuit` variant has a better sub-tour elimination behavior.

As a final remark, it is important to notice that the basic model does not allow visiting the same station more than once. Nevertheless, this limitation can be dropped by replicating each station in the extended graph G_R according to the number of revisits that will be allowed.

The dimensions of the routing model, given the size of the input, are reported in Table 3.

Search Strategy. The search strategy for the routing model attempts to incrementally construct the route for each vehicle by considering together the *succ* and the *service* variables, and employing a *smart* branching heuristic.

In detail, given a partial route for a vehicle, the next variable to be selected is the *succ* variable of the last node in the route. The possible values for this variable (i.e., the next station to be served) are ordered according to the contribution of the current vehicle load and the possible service at the next station in reducing its unbalance by preferring those that have a higher impact in this reduction. This value selection heuristic performs a one step look-ahead toward the next variable to be selected. Once the *succ* variable is set, the *service* variable of the next station is selected for branching. To be consistent with the look-ahead, the possible values for this variable are ordered according to their contribution in reducing the unbalance.

When the current route has to be ended because the time budget of the vehicle is finished or we have reached the vehicle's final depot, the next station to be considered for branching is the *succ* variable selection is the starting depot of the following vehicle.

Model Extensions. The model, and the solution methods built upon it, is quite flexible and make it possible to easily incorporate additional real-world aspects, which are not considered in the current problem statement.

For example it is quite trivial to allow for waiting times at stations (e.g., for avoiding the contemporary presence of two vehicles at the same station), by relaxing an equality constraint in the time accumulation formula. Also loading times at stations can be straightforwardly taken into account in the model.

Other possible extensions to the model include allowing the use of stations as *intermediate* depots by neglecting the 'sink' and 'source' concept or the relaxation of the requirement of having empty vehicles at the end of the route.

Finally, the model can be immediately adapted to consider the related problem of minimizing the working times in case of full rebalancing (similarly to [1]), just by imposing that the final unbalance of the stations should be zero.

3.2   Step Model

The step model considers the problem as a planning problem with a planning horizon of K steps, i.e., we try to find a route (with respective loading instructions) of maximal length K for each vehicle, where the first and the last stop is the depot. We introduce the set of steps $\mathcal{K} = \{0, \dots, K\}$ where 0 is the initial state and step K is the final state, thus each vehicle visits $K - 1$ stations. We set $K = \lceil \hat{t}/\tilde{t} \rceil + 1$, which is an estimated[1] upper bound of the number of steps.

In contrast to the routing model, this formulation allows us to directly represent the route of each vehicle by a sequence of stations of fixed length, as shown in Figure 2. This way we can formulate certain constraints more naturally, as we will see in the following description of the model.

Variables. All problem variables are summarized in Table 1: first, we introduce the routing variables *route*, where $route_{k,v}$ denotes the k-th stop in the tour of

---

[1] $\tilde{t}$ is the median of all travel times.

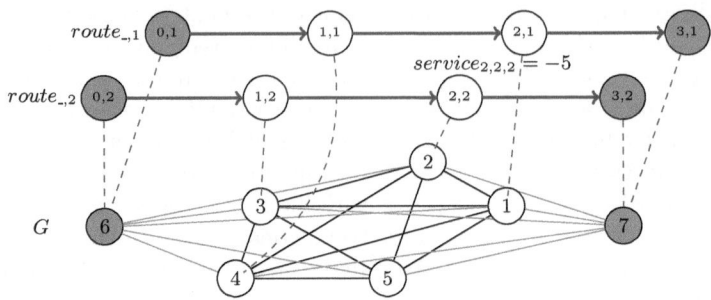

Fig. 2. Solution representation for the step CP model. The lower layer shows the original graph, whereas the upper layer shows the decision variables of the step model, i.e., the routes variables for two vehicles, and an example of the service variables.

vehicle $v \in \mathcal{V}$. Thus, the *route* variables range over the possible stops $\mathcal{S}_d$. Second, we introduce *service* variables where $service_{k,s,v}$ represents the number of bikes that are removed or added to station $s \in \mathcal{S}$ at step $k \in \mathcal{K}$ by vehicle $v \in \mathcal{V}$ and therefore ranges over $\{-C_{max}, C_{max}\}$, where $C_{max} = max_{s\in\mathcal{S}}\{C_s\}$ denotes the maximal capacity over all stations. The load of a vehicle is represented by the *load* variables where $load_{k,v}$ is the load of vehicle $v \in \mathcal{V}$ at step $k \in \mathcal{K}$. Furthermore, the variables $nbBikes_{k,s}$ state how many bikes are stored at station $s \in \mathcal{S}$ at step $k \in \mathcal{K}$. We introduce $\mathcal{K}_{-1} = \{0, \ldots, K-1\}$ for the set of steps excluding the last step and $\mathcal{K}_S = \{1, \ldots, K-1\}$ is the set of steps that concern stations, but not the depots (first and last step). We only search on the *route* and *service* variables.

Constraints. All constraints are summarized in Table 2 which we discuss in the following. First, we set up the initial state where step $k = 0$: the first stop of the route of each vehicle v is the depot (Eq. 2) and the initial load of v equals $\hat{b}_v$ (Eq. 3). The initial service is zero (Eq. 4), as well as the initial time (Eq. 5), and the initial number of bikes at station s equals b_s (Eq. 6).

Table 1. Variables of the step model

name [*dimension*]	domain	description
route [$\mathcal{K}$][$\mathcal{V}$]	$\mathcal{S}_D$	stop of vehicle $v \in \mathcal{V}$ at step $k \in \mathcal{K}$
service [$\mathcal{K}$][$\mathcal{S}$][$\mathcal{V}$]	$\{-C_{max}, C_{max}\}$	removed/added bikes at station $s \in \mathcal{S}$ by vehicle $v \in \mathcal{V}$ at step $k \in \mathcal{K}$
activity [$\mathcal{K}$][$\mathcal{S}$][$\mathcal{V}$]	$\{0, C_{max}\}$	movements at stop $s \in \mathcal{S}$ by vehicle $v \in \mathcal{V}$ at step $k \in \mathcal{K}$
load [$\mathcal{K}$][$\mathcal{V}$]	$\{0, c_{max}\}$	load of vehicle $v \in \mathcal{V}$ after step $k \in \mathcal{K}$
time [$\mathcal{K}$][$\mathcal{V}$]	$\mathcal{T}$	time when vehicle $v \in \mathcal{V}$ arrives at station at step $k \in \mathcal{K}$
nbBikes [$\mathcal{K}$][$\mathcal{S}$]	$\{0, C_{max}\}$	bikes at stop $s \in \mathcal{S}$ after step $k \in \mathcal{K}$

Table 2. Constraints of the step model

$$route_{0,v} = D \quad \forall\, v \in \mathcal{V} \tag{2}$$

$$load_{0,v} = \hat{b}_v \quad \forall\, v \in \mathcal{V} \tag{3}$$

$$service_{0,s,v} = 0 \quad \forall\, s \in \mathcal{S}, v \in \mathcal{V} \tag{4}$$

$$time_{0,v} = 0 \quad \forall\, v \in \mathcal{V} \tag{5}$$

$$nbBikes_{0,s} = b_s \quad \forall\, s \in \mathcal{S} \tag{6}$$

$$activity_{k,s,v} = |service_{k,s,v}| \quad \forall\, k \in \mathcal{K}, s \in \mathcal{S}, v \in \mathcal{V} \tag{7}$$

$$\texttt{atleast}(activity_{k,v}, 0, S-1) \quad \forall\, k \in \mathcal{K}, v \in \mathcal{V} \tag{8}$$

$$time_{k,v} \leq time_{k+1,v} \quad \forall\, k \in \{0,\ldots,K-1\}, v \in \mathcal{V} \tag{9}$$

$$service_{k,s,v} \geq 0 \quad \forall\, k \in \mathcal{K}, v \in \mathcal{V}, s \in \mathcal{S}:$$
$$b_s - t_s \leq 0 \tag{10}$$

$$service_{k,s,v} \leq 0 \quad \forall\, k \in \mathcal{K}, v \in \mathcal{V}, s \in \mathcal{S}:$$
$$b_s - t_s \geq 0 \tag{11}$$

$$load_{k+1,v} = load_{k,v} + \sum_{s \in \mathcal{S}} service_{k+1,v,s} \quad \forall\, k \in \mathcal{K}_{-1}, v \in \mathcal{V} \tag{12}$$

$$nbBikes_{k+1,s} = nbBikes_{k,s} - \sum_{v \in \mathcal{V}} service_{k+1,v,s} \quad \forall\, k \in \mathcal{K}_{-1}, s \in \mathcal{S} \tag{13}$$

$$time_{k+1,v} \geq time_{k,v} + travelTime_{route_{k,v},route_{k+1,v}} \quad \forall\, k \in \mathcal{K}_{-1}, v \in \mathcal{V} \tag{14}$$

$$(activity_{k,s,v} \geq 0) \Leftrightarrow (route_{k,v} = s) \quad \forall\, k \in \mathcal{K}, v \in \mathcal{V}, s \in \mathcal{S} \tag{15}$$

$$(route_{k,v} = D) \Rightarrow (route_{k+1,v} = D) \quad \forall\, v \in \mathcal{V}, k \in \{1,\ldots,K-1\} \tag{16}$$

$$(route_{k1,v1} = route_{k2,v2} \wedge route_{k1,v1} \neq D) \Rightarrow \quad \forall\, k_1, k_2 \in \{1,\ldots,K-1\},$$
$$time_{k1,v1} \neq time_{k2,v2} \quad v_1, v_2 \in \mathcal{V}, v1 \neq v2 \tag{17}$$

$$\texttt{count}(route_v, c), \texttt{dom}(c, 0, v_{max}) \quad \forall\, v \in \mathcal{V} \tag{18}$$

$$load_{K,s,v} = 0 \quad \forall\, s \in \mathcal{S}, v \in \mathcal{V} \tag{19}$$

$$route_{K,v} = D \quad \forall\, v \in \mathcal{V} \tag{20}$$

$$service_{K,s,v} = 0 \quad \forall\, s \in \mathcal{S}, v \in \mathcal{V} \tag{21}$$

Second, we continue with constraints that render the formulation consistent: first, the activity at station s for vehicle v at step k is always equal to the absolute value of the respective service (Eq. 7). Furthermore, every vehicle $v \in \mathcal{V}$ may only perform actions on at most *one* station at each step k, thus the activity is zero in at least $S - 1$ stations (Eq. 8). Moreover, we state that *time* is always incremental (Eq. 9) and ensure monotonicity ('sink' and 'source' stations) by stating that those stations that need to receive bikes to reach their target value must have *positive* services (Eq. 10), while stations from which bikes need to be removed to reach their target value, must have *negative* services (Eq. 11).

Third, we state the action constraints that describe how the state changes after each move: first, we update the load of vehicle v after servicing a station at step $k + 1$ (Eq. 12). Then we continue with updating the number of bikes at station s (Eq. 13) and updating the time at which vehicle v arrives at the station it services at step $k + 1$ (Eq. 14) where $travelTime_{route_{k,v},route_{k+1,v}}$ is expressed by an **element** constraint.

We link the *route* and *activity* variables (Eq. 15) and state that if vehicle v has returned to the depot before reaching the maximum number of steps, then it may not leave it anymore (Eq. 16). This way we add flexibility with respect to the tour length: vehicles may visit K stations or less. Moreover, we state that two different vehicles cannot visit the same station at the same time (Eq. 17). In Eq. 18 we use a `count` constraint and a temporary variable c to ensure that a station is visited at most v_{max} times in a solution. For the current formulation $v_{max} = 1$, however using a different value enables revisits on the step model.

For the final state, we constrain the load of each vehicle to equal zero (Eq. 19), the K-th stop is the depot (Eq. 20) which has zero service (Eq. 21).

Table 3 summarizes the dimensions of the step model as a function of the input size.

Search Strategy. In our search strategy, we try to construct feasible tours and corresponding loading instructions, for one vehicle after another. Therefore, we search upon the *route* and *activity* variables for each vehicle $v \in \mathcal{V}$: we begin with the *route* in a static order, i.e., $route_{0,v}, \ldots, route_{K,v}$, and continue with the loading instructions $activity_{0,v}, \ldots, activity_{K,v}$ using a dynamic variable selection, where we select the variable with the largest degree.

In order to obtain a good solution, the value selection for the route variables should return stations that are particularly in need of balancing. Therefore, we have implement a specialized value selection that returns those stations first that have a particularly high deviation from their target value. For the activity variables, we employ a dynamic max-value selection to achieve a maximal activity at each stop in the route.

Model Extensions. Similar to the routing model, the step model can easily be extended to incorporate additional real-world aspects, which are not considered in the current problem statement: waiting times at stations can be included, as well as variable loading times at stations. Furthermore, bike sharing system providers are often interested in a *minimal amount of service* at each station, i.e. a minimal amount of moved bikes per service. This can easily be expressed by applying a lower bound (α) on the *activity* variables.

Other possible extensions are to allow stations as *intermediate* depots by neglecting the 'sink' and 'source' concept by omitting the respecting constraint, or by allowing loaded vehicles to returned to (or leave) the depot. Finally, we can adapt the model to consider the minimizing the working times in case of full rebalancing by imposing that the final unbalance of the stations should be zero.

4　Large Neighborhood Search

Large Neighborhood Search (LNS) [13] is a local search metaheuristic based on the observation that exploring a *large neighborhood*, i.e., perturbing a significant portion of a solution, typically leads to *higher quality* local optima than the ones obtained with small perturbations. While this is an undoubted advantage in terms of search performance, it does not come without a price as exploring a large neighborhood structure can be *computationally impractical*. For this reason, LNS

Table 3. Dimensions of the CP models as a function of the number of vehicles V, the number of stations $S + D$, the number of steps K, and the time budget $\hat{t}_v$. Constraints are given as the total number plus a specific account of different constraint types.

Dimension	Routing Model	Step Model
Overall number of variables	$12V + 7S + 6$	$K \cdot (3V + 2VS + S)$
Number of fixed variables	$9V + 2$	$V(3S + 4) + 2S$
Size of the largest domain	$\hat{t}_v + 1$	$\hat{t}_v + 1$
Overall number of constraints	$14S + 3V + 5$	$K(3VS + 5V + S) + K^2V^2 + 2V$
`circuit` constraints	2	—
`element` constraints	$9S + 2V + 2$	KV
`iff` constraints	S	KVS
`imply` constraints	—	$KV + K^2V^2$
linear constraints	$4S + V + 1$	$K(2VS + 2V + S) + V$
`count` constraints	—	$KV + V$

typically involves *filtering* techniques that allow to keep the neighborhood size under control by removing unfeasible solutions. In particular, large neighborhood exploration has been successfully coupled with constraint-based propagation for tackling complex routing problems such as VRP with time windows [2,11].

The customary way of specifying a large neighborhood is to define two steps: (i) a *destroy* step, which takes a solution and relaxes a fraction $d \in [0, 1]$ (the *destruction rate*) of its variables, and (ii) a *repair* step, which takes the relaxed solution and reconstructs a feasible solution by assigning the *free variables*, usually through a greedy heuristic or an exhaustive search. Of course, different values of d originate different neighborhoods and imply different search efforts. For instance, at the most extreme cases, when $d = 1$ the original solution is completely replaced by a new one and local information is lost, while if $d \approx 0$ most of the solution is retained, and only a small neighborhood is explored. By adapting d during the solution process, e.g., based on the search performance, it is possible to codify more sophisticated behaviors, such as *stagnation avoidance*.

Similarly to most metaheuristics also LNS is a template method whose actual implementation depends on problem-specific details. In particular LNS requires to specify the following aspects:

- the way in which the *destroy* step is implemented, i.e., *which variables* are chosen for relaxation;
- the way in which the *repair* step is defined, i.e., random sampling, heuristic search (problem-specific greedy heuristics, ACO, ...) or complete search (depth-first, Branch & Bound, ...);
- whether the search for the *next solution* stops at the first feasible solution, at the first improving solution or continues until a local optimum is found;
- whether d *is evolved* during the search or not and its range of values;
- whether the *acceptance criterion* is strict improvement, equal quality or it is stochastic, e.g., as in Simulated Annealing;
- the *stopping criterion* employed.

In this work, most of these aspects are common to both CP models, however some components (in particular the *destroy* step) are model-specific because they depend on the modeling variables or on the branching strategy. We defer the description of these model-specific components to the last part of this section.

Solution initialization. The initial solution is obtained by a tree search with a custom branching strategy tailored for each model. The idea behind this strategy is to choose the station and the amount of service that will reduce most the unbalancing. Search is stopped after finding the first feasible solution.

Repair step. Similarly to the initialization, the repair step consists of a *Branch & Bound* tree search with a time limit, subject to the constraint that the next solution must be of better quality than the current one. The search starts from the relaxed solution and the time budget is proportional to the number of free variables ($t_{BAB} \cdot n_{free}$) in it. The tree search employs the same branching strategy used for solution initialization.

Acceptance criterion. A repaired solution x_t is accepted if it strictly improves the previous best x_{best}. If the repair step cannot find an improving solution in the allotted time limit, then a *idle iterations* counter ii is increased. When ii exceeds the maximum number of idle iterations ii_{max} a new initial solution is designated by using a random branching, and the search is restarted.

d update. The destruction rate d evolves during the search in order to implement an intensification/diversification strategy and to avoid stagnation of the search. In our implementation at each step its value is updated as follows:

$$d = \begin{cases} min(d \cdot 1.05, 0.8) & \text{if } x_t > x_{best} \\ d = d_{init} & \text{otherwise} \end{cases} \tag{22}$$

This update scheme will increase the *radius* of the neighborhood to allow solution diversification when the repair step cannot find an improving solution in a given neighborhood. When a new best solution is found, the original initial neighborhood radius is reset, so that the exploration of the newly discovered solution region is intensified.

Stopping criterion. We allow the algorithm a given timeout, when the time is up, the algorithm is interrupted and the best solution found is returned.

Destroy step. As mentioned before, the destroy step is the only model-specific component of our implementation. In fact, this is the most relevant aspect of LNS since it requires a careful selection of the variables that have to be relaxed in order to avoid unmeaningful combinations.

  Routing model: the relaxed solution is generated by selecting $d \cdot |R_i|$ stations from each route R_i and resetting the *succ*, *service*, and *vehicle* variables of these stations to their original domains. Moreover, also the *succ* variable of the stations preceding the relaxed ones are freed to allow for different routes. Note that since we are considering also these variables the final fraction of variables relaxed is in fact greater than d.

Step model: the relaxed solution is produced by selecting $d \cdot \sum_i |R_i|$ internal nodes (i.e., excluding the depots) from all the routes and resetting the *route* and *service* variables.

5   Computational Evaluation

In this section we report and discuss the experimental analysis of the algorithms. All the experiments were executed on an Ubuntu Linux 12.04 machine with 16 Intel Xeon CPU E5-2660 (2.20GHz) cores. For fair comparison, both the CP and the LNS algorithms were implemented in Gecode (v 3.7.3) [7], the LNS variant consisting of a specialized search engine and two specialized branchers.

The LNS parameters (ii_{max}, d_{init} and t_{BAB}) have been tuned by running an *F-Race* [3] with a confidence level of 0.95 over a pool of 150 benchmark instances from Citybike Wien. Each instance, featuring a given number of stations $S \in \{10, 20, 30, 60, 90\}$, was considered with different number of vehicles $V \in \{1, 2, 3, 5\}$ and time budgets $\hat{t} \in \{120, 240, 480\}$, totaling 900 problems. The tuning was performed by letting the algorithms run for 10 minutes. The best configurations were $ii_{max}s = 40$ and $t_{BAB} = 400$ for both models, $d_{init} = 0.05$ for the routing model and $d_{init} = 0.1$ for the step model. Note that the way the *destroy* step is designed determines the optimal value of d_{init}, as a consequence in both models a similar proportion of variables is relaxed (about $10 - 20\%$).

For benchmarking, we let the winning configurations for LNS and the pure CP models run for one hour, the results are summarized in Table 4.

5.1   Model and Solution Method Comparison

The main goal of this comparison is to understand and analyze the behavior of the CP Branch & Bound and LNS solution methods for the two problem models. Figure 3 shows exemplarily the evolution of the best cost within one search run on an instance from the Citybike Wien benchmark set featuring 30 stations. The pink and turquoise dashed lines represent the resolution using branch and bound respectively on the routing and the step model. The solid lines represent the median of 10 runs of LNS on the two models. The dark areas represent the interquantile range at each time step, while the light areas represent the maximum range covered by LNS over the 10 runs.

From the plot it is possible to see that, regarding the pure CP approaches (i.e., Branch & Bound), the routing model is clearly outperforming the step model. As for the LNS-based solvers, the situation is quite the opposite, with the step model outperforming the routing model on the median run. One must however consider that performance data collected on a single instance is of limited statistical significance. As for the comparison between pure CP approaches and LNS-based ones, the latter exhibit better *anytime* performance, reaching low areas of the objective function much faster then their Branch & Bound counterparts. Of course this comes at the price of *completeness*, and we expect CP approaches to rival with or outperform the LNS-based ones given enough time. It is worth noticing that this result is consistent across the whole benchmark set.

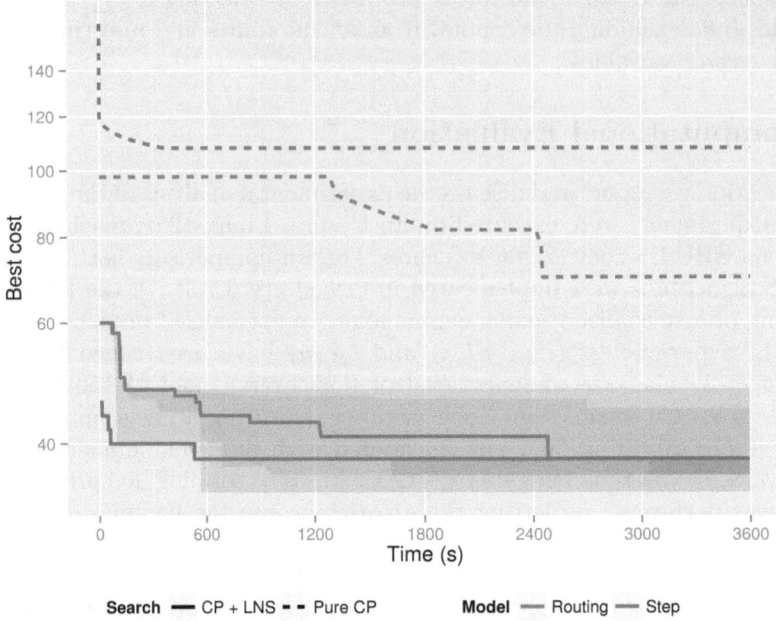

Fig. 3. Evolution of the best cost for the Pure CP and LNS solution methods for the routing and the step model (30 stations, 2 vehicles, time budget 480 minutes)

5.2 Comparison with Other Methods

In this second experiment, we compare our CP and LNS solution methods with the state-of-the-art results of [9], who solved the same set of instances using a Mixed Integer Linear Programming solver (MILP) and a Variable Neighborhood Search (VNS) strategy. The result of the comparison against the best of the three different VNS approaches in [9] are reported in Table 4. The reported results in each row are averages over 150 instances, grouped by size, number of vehicles and available time for the trucks to complete the tour.

Cells marked with a dash refer to instance classes for which the algorithm cannot reach a feasible solution within a hour. In these cases it makes no sense to compute a mean, thus for the CP models we report the number of instances in which the cost is inferior, equal or superior to the one obtained by the MILP solver in [9]. Of course, this may also happen in the case of LNS, since the initial solution is obtained through a Branch & Bound search.

In this table we did not report the results of our former ACO+CP solution approach [6], since it was outperformed by the methods proposed in this paper.

From the table it is possible to observe that the VNS heuristics proposed in [9] consistently outperform our pure CP and LNS-based solution methods on all the instance classes. On the other hand our LNS approach based on the routing model, has very close performances to [9] on almost all instances with 30 stations or less, although requiring more time to reach the same result.

Table 4. Comparison of our approaches with the MILP and the best VNS approach of [9]

Instance			CP Routing		CP Step		CP/MILP Routing	CP/MILP Step	MILP [9]			LNS Routing		LNS Step		VNS [9]	
S	V	t̂	f̄	t̄	f̄	t̄	< / = / >	< / = / >	ūb	l̄b	t̄	f̄	t̄	f̄	t̄	f̄	t̄
10	1	120	**28.3348**	149	35.2678	0	2/28/0	1/8/21	**28.3348**	**28.3348**	4	**28.3348**	53	35.4012	0	**28.3348**	2
10	1	240	4.2694	398	17.5353	334	2/28/0	0/1/29	4.2694	0.0042	3600	**4.1361**	148	17.5353	240	**4.2694**	10
10	1	480	**0.0032**	320	14.5356	462	9/21/0	0/0/30	0.0033	0.0028	3600	**0.0032**	128	13.1356	180	**0.0032**	17
10	2	120	10.4693	445	18.8022	52	0/22/8	0/4/26	9.8027	9.4377	911	10.2027	138	18.8689	23	**9.9360**	3
10	2	240	**0.0034**	56	13.6025	459	5/25/0	0/0/30	**0.0034**	0.0032	856	**0.0034**	65	13.1358	269	**0.0034**	19
10	2	480	**0.0032**	101	14.0691	644	2/28/0	0/0/30	0.0033	0.0028	1245	**0.0032**	101	13.1358	460	**0.0032**	15
20	2	120	63.4030	1042	72.1357	260	1/3/26	0/1/29	55.8029	26.4201	3600	57.9363	251	72.4024	131	**55.3363**	8
20	2	240	23.1388	1083	30.2712	389	11/0/19	4/0/26	19.7388	0.0038	3600	6.9391	292	27.5378	668	**4.2058**	58
20	2	480	14.6057	1089	20.0054	635	4/0/26	1/0/29	1.8091	0.0036	3600	0.0063	306	17.7386	721	**0.0061**	142
20	3	120	48.2043	1391	54.4369	310	3/0/27	1/1/28	37.3376	1.3478	3600	37.1377	250	52.2035	280	**31.7376**	13
20	3	240	20.3392	1347	20.6056	616.3439	4/0/26	5/0/25	6.1408	0.0040	3600	0.0067	285	17.1389	742.5992	**0.0065**	65
20	3	480	14.6724	1503	—	—	8/0/22	4/0/26	13.3419	0.0032	3600	0.0064	311	—	—	**0.0061**	114
30	2	120	114.8696	475	122.4692	250	2/3/25	1/2/27	106.9363	55.9491	3600	111.8030	264	123.4691	68	**104.7363**	12
30	2	240	59.3392	936	71.6047	439	23/1/6	14/0/16	74.9389	0.0049	3600	50.8060	256	69.7379	626	**34.6061**	109
30	2	480	22.9425	1320	22.4748	558.2628	24/0/6	25/0/5	69.7407	0.0046	3600	0.0097	338	20.0080	938.3446	**0.0093**	491
30	3	120	95.2044	612	104.6036	440	6/0/24	4/0/26	90.4042	16.3045	3600	90.6711	240	102.5369	265	**78.1377**	21
30	3	240	36.1417	1151	40.6069	645	26/0/4	24/0/6	61.6072	0.0046	3600	20.4752	272	39.4068	450.9957	**7.0752**	191
30	3	480	22.9425	1333	—	—	30/0/0	28/0/2	175.4000	0.0002	3600	0.0098	341	—	—	**0.0093**	399
60	3	120	284.0045	1311	287.7370	265.5874	5/0/25	7/0/23	274.2710	157.3735	3600	282.5379	228	287.2036	243.6891	**253.8046**	45
60	3	240	—	—	185.5412	146.4551	23/0/7	30/0/0	370.2000	0.0000	3600	184.3425	270	185.0743	587.3274	**126.7428**	521
60	3	480	67.2173	616	76.0806	541.0620	30/0/0	30/0/0	—	—	3600	55.5509	205	74.3471	625.0877	**6.6176**	3600
60	5	120	242.2741	1427	—	—	24/0/6	22/0/8	289.2711	34.6978	3600	242.4741	268	—	—	**196.6075**	99
60	5	240	115.8813	1323	107.3459	287.2886	30/0/0	30/0/0	370.2000	0.0000	3600	110.8813	273	106.1457	789.3186	**41.4816**	1556
60	5	480	55.5532	1259	—	—	30/0/0	30/0/0	—	—	3600	35.0875	259	—	—	**0.0190**	3600
90	3	120	480.3379	1761	364.4748	339.3373	13/1/16	14/0/16	492.2032	290.5999	3600	476.4046	335	364.2078	505.5610	**441.6047**	82
90	3	240	—	—	217.2814	561.5618	0/0/30	30/0/0	566.2667	0.0000	3600	370.1427	352	216.2811	548.0430	**294.4765**	985
90	3	480	—	—	—	—	0/30/0	30/0/0	—	—	3600	195.9516	367	—	—	**100.9522**	3600
90	5	120	436.8076	2071	262.7465	511.1937	30/0/0	29/0/1	566.2667	0.0000	3600	438.0076	429	262.4129	844.3994	**376.0743**	169
90	5	240	—	—	—	—	0/30/0	30/0/0	—	—	3600	271.0818	415	—	—	**174.2157**	3304
90	5	480	87.6287	752	88.7570	354.9604	30/0/0	30/0/0	—	—	3600	78.8955	350	85.3567	667.9110	**1.4285**	3600

As for the comparison with MILP, our CP models solved by Branch & Bound are able to match or outperform the upper bound solution found by MILP on the mid- and big-size instances ($S \geq 30$). Moreover, the routing model and the step model seems to have complementary strengths and weaknesses on the whole benchmark, with the step model being able to consistently find solutions on instances that are hard for the routing model and the other way round.

Overall, our LNS approach appears more robust on the largest instances, where pure CP often fails to find even a feasible solution. However, similarly to the Branch & Bound solution method, also in this case there is no clear winner.

6   Conclusions

In this paper, we have presented two novel CP models for the problem of balancing bike sharing systems (BBSS), and a Large Neighbourhuod Search (LNS) approach based on the propagation of the constraints defined in each model for obtaining good solutions in a reasonable time.

We have compared the results of our research against the state-of-the-art VNS and MILP solvers for BBSS proposed in [9]. Even though our approaches are not able to outperform the current bests, our results are reasonably close, and the two models we propose are based on a more general formulation of the BBSS problem that, for example, involves the possibility of visiting the same station repeatedly over the tour or to take into consideration the loading times.

Furthermore, we experimentally show that combining the power of constraint propagation with neighborhood search is a natural and effective way to trade *completeness* for *performance*. In fact, the LNS approaches based on our two CP models, consistently outperform their Branch & Bound counterparts by exploiting constraint propagation to limit the size of the neighborhood and reaching low-cost solutions very quickly.

As future work, we plan to consider different variants of LNS, e.g., employing different stopping conditions and acceptance criterions. Moreover, we are interested in solving the *dynamic* variant of the BBSS problem, where bikes are moved independently from station to station during the rebalancing, resulting in variable target values and variable station loads over time.

Acknowledgments. This work is part of the project BBSS, partially funded by the Austrian Federal Ministry for Transport, Innovation and Technology within the strategic program I2VSplus under grant 831740. Luca Di Gaspero and Tommaso Urli were supported by the Google Focused Grant Program on "Mathematical optimization and combinatorial optimization in Europe".

We thank Marian Rainer-Harbach, Petrina Papazek, Bin Hu and Günther Raidl from the TU Wien, and Matthias Prandtstetter and Markus Straub from the Austrian Institute of Technology, and Citybike Wien for the collaboration in this project, their comments and for providing the test instances and the results of the their methods.

References

1. Benchimol, M., Benchimol, P., Chappert, B., De la Taille, A., Laroche, F., Meunier, F., Robinet, L.: Balancing the stations of a self service bike hire system. RAIRO – Operations Research 45(1), 37–61 (2011)
2. Bent, R., Van Hentenryck, P.: A two-stage hybrid local search for the vehicle routing problem with time windows. Transportation Science 38(4), 515–530 (2004)
3. Birattari, M., Yuan, Z., Balaprakash, P., Stützle, T.: F-race and iterated F-race: an overview. In: Experimental Methods for the Analysis of Optimization Algorithms, pp. 311–336 (2010)
4. Chemla, D., Meunier, F., Calvo, R.W.: Bike sharing systems: Solving the static rebalancing problem. To Appear in Discrete Optimization (2012)
5. Contardo, C., Morency, C., Rousseau, L.M.: Balancing a Dynamic Public Bike-Sharing System. Tech. Rep. CIRRELT-2012-09, Montreal, Canada (2012)
6. Di Gaspero, L., Rendl, A., Urli, T.: A hybrid ACO+CP for balancing bicycle sharing systems. In: Blesa, M.J., Blum, C., Festa, P., Roli, A., Sampels, M. (eds.) HM 2013. LNCS, vol. 7919, pp. 198–212. Springer, Heidelberg (2013)
7. Gecode Team: Gecode: Generic constraint development environment (2006), http://www.gecode.org
8. Kilby, P., Shaw, P.: Vehicle routing. In: Rossi, F., Beek, P. (eds.) Handbook of Constraint Programming, ch. 23, pp. 799–834. Elsevier Science Inc., New York (2006)
9. Rainer-Harbach, M., Papazek, P., Hu, B., Raidl, G.R.: Balancing bicycle sharing systems: A variable neighborhood search approach. In: Middendorf, M., Blum, C. (eds.) EvoCOP 2013. LNCS, vol. 7832, pp. 121–132. Springer, Heidelberg (2013)
10. Raviv, T., Tzur, M., Forma, I.A.: Static repositioning in a bike-sharing system: models and solution approaches. To Appear in EURO Journal on Transportation and Logistics (2012)
11. Rousseau, L.M., Gendreau, M., Pesant, G.: Using constraint-based operators to solve the vehicle routing problem with time windows. Journal of Heuristics 8(1), 43–58 (2002)
12. Schuijbroek, J., Hampshire, R., van Hoeve, W.J.: Inventory rebalancing and vehicle routing in bike sharing systems. Tech. Rep. 2013-E1, Tepper School of Business, Carnegie Mellon University (2013)
13. Shaw, P.: Using constraint programming and local search methods to solve vehicle routing problems. In: Maher, M.J., Puget, J.-F. (eds.) CP 1998. LNCS, vol. 1520, pp. 417–431. Springer, Heidelberg (1998)

Constraint Based Computation of Periodic Orbits of Chaotic Dynamical Systems

Alexandre Goldsztejn[1], Laurent Granvilliers[2], and Christophe Jermann[2]

[1] CNRS, LINA (UMR-6241)
[2] Université de Nantes, LINA (UMR-6241),
2 rue de la Houssinière, 44300 Nantes, France
`firstname.surname@univ-nantes.fr`

Abstract. The chaos theory emerged at the end of the 19th century, and it has given birth to a deep mathematical theory in the 20th century, with a strong practical impact (e.g., weather forecast, turbulence analysis). Periodic orbits play a key role in understanding chaotic systems. Their rigorous computation provides some insights on the chaotic behavior of the system and it enables computer assisted proofs of chaos related properties (e.g., topological entropy).

In this paper, we show that the (numerical) constraint programming framework provides a very convenient and efficient method for computing periodic orbits of chaotic dynamical systems: Indeed, the flexibility of CP modeling allows considering various models as well as including additional constraints (e.g., symmetry breaking constraints). Furthermore, the richness of the different solving techniques (tunable local propagators, search strategies, etc.) leads to highly efficient computations. These strengths of the CP framework are illustrated by experimental results on classical chaotic systems from the literature.

Keywords: Chaotic dynamical systems, periodic orbits, topological entropy, numerical constraint satisfaction, symmetry breaking.

1 Introduction

A dynamical system is defined by a *state space* X (here $X \subseteq \mathbb{R}^d$, so a state is a vector of d reals) and an *evolution function*, which describes how the state $x_t \in X$ changes as time t passes. Continuous time dynamical systems (i.e., $t \in \mathbb{R}$) usually involve differential equations; in this case the evolution function is often called a flow. In this paper, we focus on discrete time dynamical systems (i.e., $t \in \mathbb{Z}$). They arise either from discrete models or discretizing continuous time dynamical systems (e.g., using the Poincaré map). In this case, the evolution function is a *map* $f : X \to X$, and the evolution of an initial condition $x_0 \in X$ is computed by $x_{k+1} = f(x_k)$, giving rise to *(forward) orbits* $(x_0, x_1, x_2, \ldots)$. Understanding the infinitely complex structure of orbits generated by very simple systems is the goal of the chaos theory.

The first evidence of chaotic behavior was found by Poincaré while partially solving seemingly simple three-body problem at the end of the 19th century: The two-body problem, consisting of computing the trajectory of two masses following Newton's gravitational laws, is easily solved and fully understood. Poincaré proved that infinitely

C. Schulte (Ed.): CP 2013, LNCS 8124, pp. 774–789, 2013.

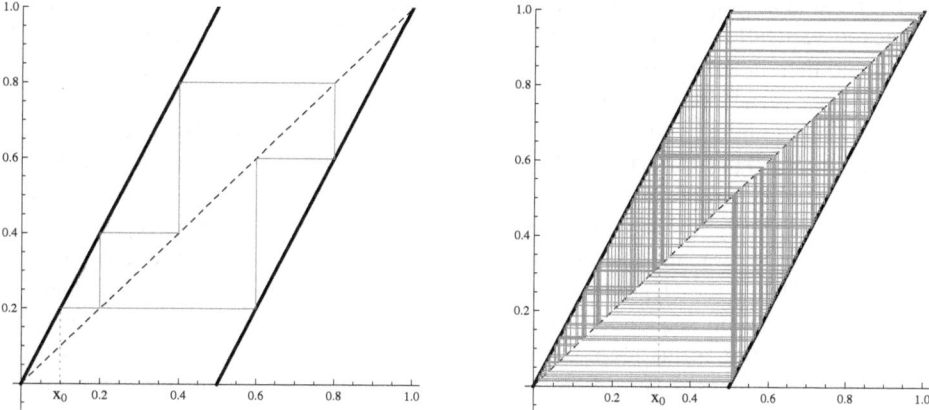

Fig. 1. The graph of the Dyadic map in thick black. Left: The orbit of $x_0 = 0.1$ in gray. Right: The orbit of $x_0 = \frac{1}{\pi}$ in gray.

complex trajectories arise when three bodies are considered, leading to the modern theory of chaos. One main discovery of Poincaré was the critical importance of periodic orbits in chaotic dynamical systems. Deep theoretical developments have followed during the 20th century (Markov partitions and corresponding symbolic dynamics, measure preserving maps and ergodicity, hyperbolicity, etc.) providing an increasingly accurate understanding of chaotic systems. Starting from the middle of that century, extensive simulations with computers (starting with Lorenz butterfly chaotic attractor) offered many illustrations of those chaotic behaviors, allowing these ideas to be disseminated toward the general public.

Formally defining a chaotic dynamical system is difficult: There exist several such definitions (Li-York chaos, Devanay chaos, positive topological entropy, see [20,6,21]), which are not equivalent and whose relationship is a current research topic. The common idea that chaos is the exponential sensitivity to initial conditions is wrong: Consider, e.g., the dynamical system defined by $X = \mathbb{R}^d$ and $f(x) = 2x$. Therefore $x_k = 2^k x_0$, so two different initial conditions diverge exponentially fast[1] while this simple linear system is definitely not chaotic. However, enforcing some kind of exponential divergence between the orbits of neighbor initial conditions within a *bounded* state space X leads to very complex systems. Such systems need to be simultaneously expanding (so as to show divergence) and contracting (since the state space is bounded). This leads to hyperbolic dynamical systems, which are consistently contracting in some directions and expanding in the other directions, the most well-understood chaotic behavior.

The Dyadic map is among the most simple systems showing hyperbolic chaos. It is defined by $X = [0, 1)$ and $f(x) = 2x$ mod 1. Multiplying by two is expanding, while taking modulo 1 enforces a contraction back to $[0, 1)$. Its graph is shown in Fig. 1, together with the orbit of $x_0 = 0.1$ (respectively $x_0 = \frac{1}{\pi}$) on the left (respectively

---

[1] Indeed $d(x_k, y_k) = d(2^k x_0, 2^k y_0) = 2^k d(x_0, y_0)$.

right) hand side graphic. Orbits of this map are easily visualized: An initial point x is moved vertically toward the graph of f (the thick line in Fig. 1) in order to reach $f(x)$, and then horizontally toward the line $x = y$ (the dashed line in Fig. 1) in order to reach $x = f(x)$; Repeating the process from the new point yields the orbit. On the left hand side graphic, $f(0.1) = 0.2$, and the orbit of 0.2 is periodic with period 4 (i.e., $f^4(0.2) = f(f(f(f(0.2)))) = 0.2$). This map has a striking interpretation when considering the binary representation of a real number $x = 0.b_1b_2b_3\cdots \in [0, 1)$, each b_i being a bit in $\{0, 1\}$: Indeed, multiplying x by two shifts left its binary representation (yielding $b_1.b_2b_3\cdots$), and the modulo 1 then removes the first bit on the left (yielding $0.b_2b_3\cdots$). Hence the Dyadic map is in direct correspondence with the shift map on one-sided infinite bit sequences. This is a simple example of the powerful tool symbolic dynamics represents for investigating chaotic dynamical systems. It has several important consequences here: First, it is well known that a real number is rational if and only if the binary representation of its fractional part is periodic after a given bit (the same actually holds in any base). For example, the binary representation of 0.1 is 0.0001100110011$\cdots$. Hence any rational number will converge to a periodic orbit after a finite number of applications of the Dyadic map; the period of this orbit is equal to the period of the binary representation. It follows there are exactly $2^n - 1$ initial states[2] yielding orbits of period n, called period-n orbits thereafter, and they are equally distributed within $[0, 1)$. On the contrary, irrational numbers are not periodic (the right hand side graphic of Fig. 1 shows the first 200 iterates of the orbit of $\frac{1}{\pi}$, which is seemingly random). Second, when computing an orbit using a computer, a finite binary representation has to be used. Therefore, any finite precision simulation has to converge toward zero. The 200 iterates of the orbit of $\frac{1}{\pi}$ shown in Fig. 1 have been computed using a 200-bit precision arithmetic.

The topological entropy is a real number associated to a dynamical system, which is meant to characterize the exponential divergence of orbits within a bounded state space. Suppose one can distinguish two points only if their distance is greater than $\epsilon > 0$, and consider a set $E \subseteq X \subseteq \mathbb{R}^d$ such that one can distinguish all points, so the mutual pairwise distance of the points in E is at least ϵ (such a set is called ϵ-separated). In this case, non intersecting balls of radius $\frac{\epsilon}{2}$ can be put around each point of E, entailing the cardinality of E to be at most $V_X/V_{\frac{\epsilon}{2}}$, where V_X and $V_{\frac{\epsilon}{2}}$ are respectively the volume of X and the volume[3] of the ball of radius $\frac{\epsilon}{2}$. A map f can improve the situation by separating initial points that were too close to be distinguished, leading to the definition of (n, ϵ)-separated sets: A set $E \subseteq X$ is (n, ϵ)-separated if two different points in E yield orbits that are separated by at least ϵ within n iterations of the map. Formally, for all $x, y \in E$ with $x \neq y$, $\max_{0 \leq k \leq n} d(f^k(x), f^k(y)) \geq \epsilon$. The maximal cardinality of (n, ϵ)-separated sets is denoted by $s(n, \epsilon)$. As mentioned above, $s(0, \epsilon) \leq V_X/V_{\frac{\epsilon}{2}}$, while iterating the map can only help distinguishing more points, so $s(n, \epsilon)$ is increasing with respect to n. The growth rate of $s(n, \epsilon)$ shows how quick the map separates points. In particular, whenever the growth rate is exponential for some

---

[2] There are 2^n binary representation of period n, but $0.1111\cdots$ is a periodic binary representation equal to 1 hence outside $[0, 1)$.

[3] Volumes are generalized by the Lebesgue measure in space of dimension greater than 3.

$\epsilon > 0$, i.e., $s(n, \epsilon) \approx ae^{bn}$ for some non-negative real constants a and b, the topological entropy of the map f is defined as $h_X(f) = b$. More precisely,

$$h_X(f) = \limsup_{\epsilon \to 0} \limsup_{n \to \infty} \frac{\log s(n, \epsilon)}{n}, \tag{1}$$

where the first limit is used because $s(n, \epsilon)$ is non decreasing in ϵ, and supremum limits are used in order to take into account irregular exponential growths. When the topological entropy is strictly positive, the cardinality of maximal (n, ϵ)-separated sets grows exponentially with n. Therefore, the minimal distance between points in a maximal (n, ϵ)-separated set decreases exponentially with n, while the map f still allows separating them by at least ϵ in at most n iterations. Hence the map induces an exponential expansion in spite of the bounded state space. Having a strictly positive topological entropy is the characterization of chaos that is most often used.

For the Dyadic map, one can easily see that the set of points that yield period-n orbits is $(n, 0.5)$-separated: Indeed, two such points differ in (at least) one bit among their n first bits, say the kth bit. Hence, iterating the map $k - 1$ times brings those two different bits at the first (fractional) place, so the distance between their $(k - 1)^{th}$ iterates is at least 0.5. Now, since there are $2^n - 1 \approx e^{n \log 2}$ such points yielding period-n orbits, the topological entropy of the Dyadic map is at least $\log 2$. As seen on this example, the topological entropy is closely related to the exponential growth of the number P_n of period-n orbits with respect to n. More generally, under the hypothesis that the system satisfies the axiom A hypothesis [20] (roughly speaking, it is hyperbolic), its topological entropy is equal to

$$h_X(f) = \limsup_{n \to +\infty} \frac{\log(P_n)}{n}. \tag{2}$$

Numerous techniques have been developed to provide computer assisted proofs of chaos related properties: E.g., the famous answer to Smale's 14th problem [29], and [1,19,3,30,27,13]. Proving that a dynamical system is chaotic is generally done by finding out a subsystem with known topological entropy (often by identifying some specific periodic orbits), leading to a certified lower bound on its topological entropy. Roughly speaking, the system is proved to be as complex as a known chaotic dynamical system. An upper bound on the topological entropy provides an estimate of the accuracy of the certified lower bound, but such upper bounds are difficult to obtain: [27] provides such an upper bound for one dimensional maps. On the other hand, [9,10,11] proposed to compute all periodic orbits up to a given period with certified interval techniques, hence inferring an approximation of the topological entropy using Eq. (2).

We show here that using CP for rigorously computing periodic orbits is a convenient and efficient approach: By benefiting of constraint propagation and symmetry breaking, a simple model can be used, while avoiding heavy preprocessing (Section 2). Furthermore, the CP framework allows tuning the propagation strength and the search strategy (Section 3) so as to achieve more efficient resolution (experiments on well-known chaotic systems are reported in Section 5). We show in particular that the solving process can be tuned for small periods, impacting the resolution for higher periods.

2   Modeling the Problem

After briefly recalling the basics of numerical constraint modeling, we introduce NCSPs whose solutions provide the periodic orbits of discrete time dynamical systems. Two standard models described in [9] are discussed with respect to numerical constraint solving. The flexibility of CP modeling allows considering alternative models.

2.1   NCSPs and Interval Arithmetic

Numerical constraint satisfaction problems (NCSPs) have variables representing real quantities, whose domains are thus subsets of $\mathbb{R}$. Their constraints are typically equations and inequalities on these quantities. For practical reasons, the domains are handled as intervals and the assignments are not enumerated. Instead, domains are split and filtered until a prescribed precision is reached. Interval arithmetic [24] allows enclosing the results of set-wise operations, and accounts for floating-point computational errors.

In this paper we denote $x = (x_1, \ldots, x_n)$ the variables, considered to be a n-dimensional vector for convenience. We also denote x a real assignment of the variables, i.e., a point $(x_1, \ldots, x_n) \in \mathbb{R}^n$. Intervals are denoted using bold-faced letters. Hence, the domains of x are denoted $\boldsymbol{x} = (\boldsymbol{x}_1, \ldots, \boldsymbol{x}_n)$, considered as a n-dimensional vector of intervals, also called a *box*. We denote $\boldsymbol{f}$ an interval extension of a function f, i.e., a function which computes an interval $\boldsymbol{f}(\boldsymbol{x})$ enclosing all the possible values of $f(x)$ for any real $x \in \boldsymbol{x}$. This definition naturally extends to function vectors f.

Interval arithmetic suffers from two problems: The *dependency problem* by which multiple occurrences of the same sub-expression are considered independent (e.g., $x - x$ evaluates to 0 for any real in $x \in [0, 1]$, but its interval evaluation at $\boldsymbol{x} = [0, 1]$ is $[-1, 1]$); And the *wrapping effect* by which the exact evaluation of an expression on an interval is in general poorly approximated using a single interval (e.g., $\frac{1}{x}$ evaluated at any $x \in [-1, 1]$ yields a real in $(-\infty, -1] \cup [1, +\infty)$ but its interval evaluation at $\boldsymbol{x} = [-1, 1]$ results in $(-\infty, +\infty)$). In addition, the practical use of floating-point computations induces the necessity of rigorous encapsulation of rounding-errors. These issues lead to potentially large over-approximations and must be carefully handled.

2.2   Folded Models of Periodic Orbits

Given a map f on a state space $X \subseteq \mathbb{R}^d$, we can characterize a period-n orbit with the fixed-point relation

$$x = \underbrace{f \circ f \circ \ldots \circ f}_{n \text{ times}}(x) = f^n(x). \tag{3}$$

Imposing it as a constraint on variables x with domains[4] X results in the NCSP *folded model* whose solutions are the initial states $x \in X$ of period-n orbits.

---

[4] In theory X may not be representable as a box, and the domains should be set to the smallest enclosing box. In practice however, the state spaces of classical chaotic maps are boxes.

Example 1. The famous *Logistic map* [23] is defined as $f(x) = rx(1-x)$ on $X = [0,1]$. It models the evolution of a population (x is the ratio to a maximum population) depending on a parameter $r \in \mathbb{R}^+$ representing a combined rate of reproduction and starvation. Despite its very simple formulation, this map has a chaotic behavior for some values of its parameter, e.g., $r := 4$. The folded model for period-2 orbits with this setting has a single variable x and a single constraint $x = f(f(x)) = -256x^4 + 512x^3 - 320x^2 + 64x$. Its four solutions are 0, $\frac{3}{4}$ and $(5\pm\sqrt{5})/8$, the two first ones being in fact fixed-points (period-1 orbits), the others constituting the only period-2 orbit.

Folded models present two major drawbacks when addressed with interval-based constraint solving methods. First, as soon as the map function contains more than one occurrence of a variable, the numbers of operations and occurrences of this variable in the constraint grow exponentially with the period. Though the factorized expression can still be compactly represented with a DAG, this cripples its interval evaluation by exacerbating both the dependency problem and the wrapping effect. This is even worse for the evaluation of the derivatives of the constraint, required to use interval Newton operators for proving the existence of real periodic orbits within boxes. Second, their solutions are the initial states of periodic orbits, but any point in a periodic orbit is an initial state for this orbit. Hence, as exemplified above, they have n solutions for each period-n orbit.

It is worth noting that the constraint can sometimes be simplified. For instance, that of the Dyadic map (see Section 1) can be rewritten $x = 2^n x \mod 1$, and the Logistic map function can be reformulated as $f(x) = \frac{r}{4} - r(x - \frac{1}{2})^2$. Such simplifications may reduce the over-approximations of interval arithmetic. Still, the intrinsic complexity of the model remains as an initial box forcibly grows exponentially in size with the iterations of the map due to its chaotic nature.

2.3 Unfolded Models of Periodic Orbits

The NCSP *unfolded model* aims at finding complete periodic orbits at once. Its variables $(x_0, \ldots, x_{n-1})$ represent the consecutive n states in an period-n orbit, each x_k being itself a vector of d variables with domains X. The constraints establish the links between consecutive points

$$x_{(k+1)\bmod n} = f(x_k) \quad k \in \{0, \ldots, n-1\}. \tag{4}$$

Example 2. The unfolded model for the period-2 orbits of the Logistic map with $r := 4$ is composed of the variables (x_0, x_1) and the constraints $x_1 = 4x_0(1 - x_0)$ and $x_0 = 4x_1(1 - x_1)$. Its four solutions are $(0,0)$, $(\frac{3}{4}, \frac{3}{4})$, $(\frac{5-\sqrt{5}}{8}, \frac{5+\sqrt{5}}{8})$ and $(\frac{5+\sqrt{5}}{8}, \frac{5-\sqrt{5}}{8})$. It is now obvious the first ones are fixed points, and the others represent the same orbit.

This model has the strong advantage that constraint expressions remain identically complex (as many operators and variable occurrences) when n grows, making it much more appropriate for constraint methods. However, it has $n \times d$ variables instead of d variables in the folded model, and its search space thus grows exponentially with the period n. This drawback must be balanced with the fact having the n states as variables allows connecting the states in the same orbit, defining more freely strong pruning operators involving several states, and splitting at any state during the search, definite advantages when taking into account the explosive nature of chaotic maps.

2.4   Other Models of Periodic Orbits

The flexibility of the CP framework makes it possible to consider alternative models to the two classical ones presented above. For instance, both folded and unfolded models naturally have a functional form, but it is sometimes interesting, e.g., in order to reduce variable occurrences, to manipulate symbolically each constraint as a relation. This can yield *relational* unfolded models of the form

$$F(x_{(k+1) \bmod n}, x_k) = 0 \quad k \in \{0, \ldots, n-1\} \tag{5}$$

whose interest will be illustrated in Section 5. It is also possible to reduce the search space by considering as variables only a fraction of the states in a periodic orbit, yielding *semi-unfolded models*. This could allow experimentally seeking an efficient trade-off between the folded and unfolded models, though in this paper we will focus only on those extremes in order to clearly illustrate their strengths and weaknesses.

2.5   Taking into Account Additional Properties

A nice feature of CP is its ability to include additional knowledge on the considered problem as constraints or within initial domains, yielding a variety of complemented models whose efficacy can then be tested.

Periodic orbits have an inherent cyclic state symmetry. It is difficult to handle it in folded models, but it naturally boils down to a cyclic variable symmetry in unfolded models, and can then be (partially) broken using the lex-leader constraints relaxation proposed in [14]:

$$x_{0,0} \le x_{k,0} \quad k \in \{1, \ldots, n-1\}, \tag{6}$$

where $x_{k,0}$ represents the first coordinate of state k. Note that the symmetry could be broken using any other coordinate. Though inducing only a partial symmetry breaking, these additional constraints reduce optimally the search space.

Example 3. The additional partial symmetry breaking constraint for the unfolded model whose solutions are the period-2 orbits of the Logistic map is $x_0 \le x_1$. It allows discarding the fourth solution, $(\frac{5+\sqrt{5}}{8}, \frac{5-\sqrt{5}}{8})$, as it is symmetric to the third one. It also halves the search space which is computationally very interesting.

Another property of the considered problem is that period-m orbits for any factor m of n are solutions of any NCSP model for period-n orbits. E.g., the two fixed-points of the Logistic map are solutions of models for any period n. In theory, additional constraints of the form $x_i \ne x_j$ for all $0 \le i < j < n$ would discard these *factor orbits*, but such constraints cannot be filtered with interval solving methods and are thus useless.

Many chaotic maps have been extensively studied and a lot of knowledge has been accumulated about them. For instance, the *trapping region* of a map f on X, i.e., the state subspace $X' \subseteq X$ whose image through the map $f(X')$ is strictly enclosed in X', may be known to be enclosed within an ellipsoid or a polytope. Since periodic orbits starting within a trapping region must belong entirely to this trapping region, we can restrict the search to the enclosing shape using some additional inequality constraints.

Another example is the *non-wandering part* (NWP) of a map f on X, i.e., the set of points $x \in X$ such that any neighborhood U of x verifies $f^n(U) \cap U \neq \emptyset$ for some $n > 0$. This set comprises all periodic orbits and can be approximated using a simple subdivision algorithm: Consider the directed graph whose vertices are the boxes in a regular ϵ-precise subdivision of X and whose arcs $x \rightarrow x'$ verify $f(x) \cap x' \neq \emptyset$; Removing iteratively sinks and sources in this graph yields an ϵ-precise approximation of the NWP of f. This paving can be used to setup the domain of the initial state of an orbit, as proposed in [9]. Its size however grows quickly with ϵ and it is difficult to predict the appropriate precision without a dedicated study of the considered map.

3 Solving the Problem

The standard complete constraint solving method is the branch&prune algorithm. It iteratively selects a box, prunes it using local consistency enforcing operators and interval methods (jointly designated as *contractors* in the following), checks if it contains a single solution and, otherwise, splits it into sub-boxes to be further processed. In this section we discuss the appropriate components of a branch&prune algorithm for solving NCSP models of periodic orbits of chaotic maps.

3.1 Pruning Periodic Orbits Domains

The basic pruning algorithm for NCSPs is an AC3-like fixed-point loop over simple, and inexpensive, contractors like, e.g., BC3-revise [5], HC4-revise [4] or MOHC-revise [2]. It is however sometimes needed to resort to stronger contractors in order to avoid too much splitting, on trade-off with more demanding computations at each node of the search-tree. This can be achieved using for instance a fixed-point of 3B (or more generally kB) [22] or CID [28] operators. Finally, it is essential in this work that the returned solutions are proven to enclose a unique periodic orbit of the considered map, otherwise no valid reasoning on the map (e.g., its topological entropy) could be derived. For this purpose, it is typical to use an interval Newton operator [25], providing in addition a more global consistency.

In this paper we consider essentially two pruning procedures: *BC5*, a fixed-point of HC4-revise and BC3-revise[5] contractors followed by an interval Newton application; and *BC5+CID(k)*, i.e., BC5 involving in addition CID(k) contractors[6] during the fixed-point phase.

---

[5] Typically generated for variables with multiple occurrences only.

[6] One CID(k) contractor for a variable x slices the domain of x into k parts, computes a fixed-point of HC4-revise contractors for all constraints and variables on each slice, and eventually takes the hull of all the pruned slices.

3.2 Splitting Periodic Orbits Domains

The standard splitting strategy for NCSPs is *round-robin* with bisection, which selects each time the next variable and splits its domain interval at its midpoint. Another typical strategy is *maxdom* which selects the variable with the largest domain.

The unfolded model for periodic orbits has a specific structure since variables are grouped into state coordinates and correspond to consecutive points in the orbit. We can thus consider dedicated splitting techniques, e.g., *initial-state* which splits only the coordinates of the initial state x_0 in the orbit, counting on pruning operators to reduce the domains of the other states. This idea was advanced in [9] as a mean of reducing the dimension of the search space.

Due to the explosive nature of chaotic maps, we think however that splitting all the states domains should pay-off. This will be confirmed in section 5 where we compare classical splitting strategies (*round-robin* and *maxdom*) on all variables to the dedicated *initial-state* splitting strategy.

3.3 Post-processing Solution Boxes

The branch&prune algorithm we have described outputs two types of boxes: Safe boxes which have been successfully certified to enclose a unique periodic orbit, and unsafe boxes which are not certified but have reached the prescribed maximum precision for the computation. When the partial symmetry breaking constraints (6) are used, boxes for which the corresponding strict inequalities are not certainly satisfied are also considered unsafe. Indeed such boxes may each contain a representative of the same periodic orbit. Note however that this never happened in the experiments reported in Section 5.

Unsafe boxes must be properly handled so as to allow rigorously counting the number of real periodic orbits. For this purpose, we apply a post-process that tries to certify them using a specific version of the interval-based Newton operator with *inflation* [18]. This operator acts like an interval local search algorithm, iteratively shifting and inflating slightly an initial box x so as to find a close box x' that can be certified. If it succeeds, x' replaces x in the solution set, after checking it does not enclose a periodic orbit already found in another safe solution box. Possibly symmetric boxes must be merged before applying this post-process. In case unsafe boxes remain after this process, only a lower bound on the number of real periodic orbits is obtained.

4 Related Work

In [9,10,11], an interval-based method dedicated to computing periodic orbits of chaotic maps is proposed. It amounts to a bisection algorithm which splits the domains of the orbit, using interval forward evaluation of the map along the orbit to discard boxes that provably do not contain any periodic solution, and applying an interval Newton operator to certify that a box contains a single solution. When the map is symbolically invertible, it uses both forward and backward interval evaluation along the orbit in a fixed-point manner. Several key ingredients are identified in [9] as essential to the efficiency of this method: The usage of the unfolded model and of the map symbolic inverse, the *initial-state* splitting strategy, and some preprocessing using the non-wandering part and some trapping region enclosure. Most of them are made unnecessary or even counterproductive by the CP framework, as illustrated by the experiments reported in the next section.

Non rigorous local methods for computing periodic orbits have also been investigated (see e.g., [8,26,7,12] and references therein). They usually work on the unfolded model, in a similar way as multiple shooting method for boundary value problems. Being incomplete, they are not used for estimating the topological entropy, although being useful for other purposes like computing longer periodic orbits.

5   Experiments

Constraint programming is implemented to handle several classical maps having different characteristics. Several issues are analyzed. What is the best way to model orbits? What are the good pruning and splitting techniques? Is it possible to take advantage of dedicated methods in the CP framework?

More precisely, we aim at comparing unfolded models with folded models, and the Cartesian form with the polar form of complex maps. Several splitting techniques (*max-dom*, *round-robin*, *initial-state*) are investigated. Different local consistency techniques are studied, in particular BC5 and BC5+CID(k). To this end, four standard maps are considered, namely Dyadic, Logistic, Hénon, and Ikeda. All techniques have been implemented in Realpaver [15] using default parameter settings. All experiments have been conducted on an Intel Core i7-620M 2.66GHz measured at 1666 MIPS by the Whetstone test.

We have implemented the previously introduced dedicated methods in our branch& prune algorithm, namely the non wandering part pre-paving, and trapping region constraints. In fact, we have observed that these methods do not change significantly the overall performances of the solving process. For instance, the solving time varies in proportion to $\pm 10\%$ (tested for Hénon and Ikeda) when the non wandering part is taken as input. It appears that propagation and split are together able to eliminate inconsistent regions of the search space without resorting to such methods.

For each problem, we found the theoretical number P_n of periodic orbits of the unidimensional maps, or the same number of periodic orbits as [9] for the Hénon and Ikeda maps. This number grows exponentially with n, i.e., $P_n \approx ae^{bn}$ where b approximates the topological entropy. As a consequence, the solving time t of the branch&prune algorithm must also grow exponentially with n. In fact, we aim at observing for a given map and a given strategy that $t \approx ce^{dn}$, where $d \geq b$ must hold since the solving process is complete. Therefore the difference $(d - b)$ quantifies the overall quality of this strategy. In the following, we will use a logarithmic scale on t and P_n to plot the results, the growth constants b and d corresponding to the slopes of the curves.

Remark: The experiments are carried out only for orbits of prime periods. Hence, possible issues of factor orbits and symmetry breaking are discarded, thus simplifying the post-processing phase and the interpretation of results. Following this approach still permits to compare the different techniques and to calculate accurate approximations of the topological entropy.

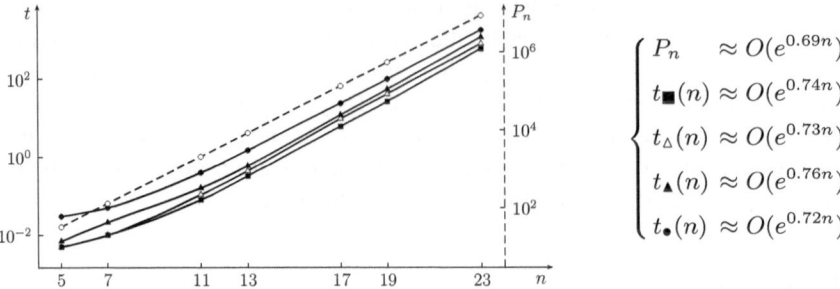

$$\begin{cases} P_n & \approx O(e^{0.69n}) \\ t_\blacksquare(n) \approx O(e^{0.74n}) \\ t_\triangle(n) \approx O(e^{0.73n}) \\ t_\blacktriangle(n) \approx O(e^{0.76n}) \\ t_\bullet(n) \approx O(e^{0.72n}) \end{cases}$$

Fig. 2. Finding orbits of Dyadic and Logistic using BC5 with *maxdom*. Left: ○ is the number of solutions P_n; the other curves represent the solving times of Dyadic's unfolded model and Logistic's unfolded model using the factorized expression (■), Dyadic's folded model (△), Logistic's folded model using the factorized expression (▲), and Logistic's unfolded model using the original expression (●). Right: Empirical asymptotic laws of these different techniques.

5.1 Unidimensional Maps

The two aforementionned unidimensional maps (Dyadic and Logistic) are interesting to illustrate the impact of modeling on the solving performance. Their folded models are simple enough, their number of operations growing linearly with n. The expression of Logistic can be factorized (the factorized form is used to generate the folded model). Dyadic is discontinuous due the modulo operation.

The topological entropy of these maps is equal to $\log 2$ since they have respectively 2^n (Logistic) and $2^n - 1$ (Dyadic) solutions. Their orbits are easily calculated by BC5 with *maxdom*, the number of splitting steps matching the number of solutions.

The results are depicted in Fig. 2. The topological entropy is the slope of the dashed line P_n. One can remark that the other curves corresponding to different models tend to become parallel to P_n, showing that the cost of calculating one solution is constant for all of them. Strikingly, the branch&prune algorithm behaves similarly when processing the folded models (curves △ and ▲) and the unfolded models (curve ■). In fact, the unfolded models exploit symmetry breaking constraints that reduce P_n by a factor n. However, pruning the folded models is easier since only one BC3-revise operator is applied at each node of the search tree, while pruning the unfolded models calculates a fixed-point of n HC4-revise operators (one per constraint) followed by an application of the interval Newton operator. Logistic's original unfolded model is worse (curve ●), since it requires applying BC3-revise operators due to the multiple variable occurrences.

Discontinuous or non differentiable functions, involving e.g., the modulo operation, are seemingly taken into account with no additional cost. However, they possibly interfere with the certification procedure. For instance, solving Dyadic's unfolded model produces two non certified boxes. The first box encloses the fixed-point $(0, \ldots, 0)$, which is located on the domain boundary. The second box $([1 - \epsilon, 1], \ldots, [1 - \epsilon, 1])$ contains no solution but it cannot be discarded by interval methods.

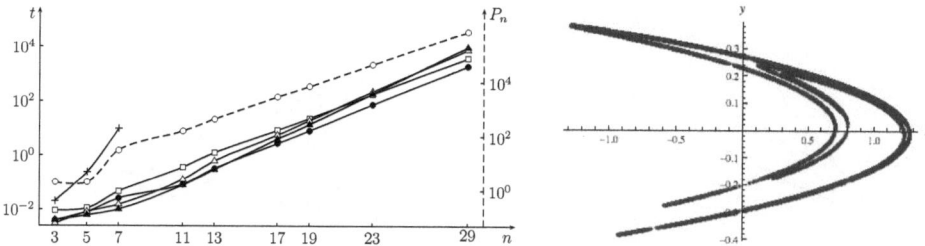

Fig. 3. Hénon map. Left: $\circ$ is the number of solutions P_n; $\bullet$ is the solving time t of BC5 with *maxdom*; $\triangle$ differs from $\bullet$ in the use of the *round-robin* strategy; $\blacktriangle$ differs from $\bullet$ in the splitting of the initial state alone; $\square$ differs from $\bullet$ in the use of CID(3) operators; $+$ differs from $\bullet$ in the use of the folded model. Right: Period-23 orbits, which clearly shows the well known strange attractor of the Hénon map.

5.2   Hénon Map

The Hénon map [16] is defined as $f(x, y) = (y + 1 - ax^2, bx)$, the standard parameter values $a := 1.4$ and $b := 0.3$ leading to a chaotic behavior. Given $x_k, y_k \in [-2, 3]$, $0 \leq k \leq n - 1$, the unfolded model is as follows:

$$\begin{cases} x_{(k+1) \bmod n} = y_k + 1 - ax_k^2 \\ y_{(k+1) \bmod n} = bx_k \end{cases} \tag{7}$$

The results are depicted in Fig. 3. The number of solutions P_n (dashed curve) gives an approximation of the topological entropy as $\log(P_n)/n \approx 0.46$. As expected, the folded model (curve $+$) is not tractable since its size grows exponentially with n. The other techniques are all able to isolate and certify all the solutions in reasonable time for the considered periods, corroborating the results in [9]. The best splitting technique is *maxdom* (curve $\bullet$), compared to *round-robin* and *initial-state* (curves $\triangle$ and $\blacktriangle$). Enforcing BC5+CID(3) (curve $\square$) seems to slow-down the solving phase but the growth constant is decreased from 0.55 to 0.51, demonstrating a better asymptotic behavior. In other words, we have $t_\bullet(n) \approx O(e^{0.55n})$ and $t_\square(n) \approx O(e^{0.51n})$.

We have also extracted from [9] the growth constant of the solving time t_G obtained from the best implemented method, which is approximatively equal to 0.58, i.e., $t_G(n) \approx O(e^{0.58n})$. Hence, on this problem the CP approach compares favorably in terms of complexity to the dedicated approach of [9].

5.3   Ikeda Map

The Ikeda map [17] is defined as

$$f(z) = a + b \exp\left(i\kappa - \frac{i\alpha}{1 + |z|^2} \right) z \tag{8}$$

where z is a complex number. The classical setting $a := 1$, $b := 0.9$, $\alpha := 6$ and $\kappa := 0.4$ yields a chaotic behavior. This map can be transformed into a two-dimensional

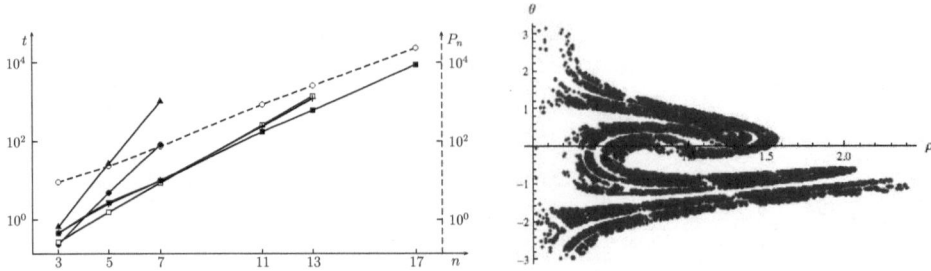

Fig. 4. Ikeda map. Left: ∘ is the number of solutions P_n; ▲ is the solving time t of BC5 with *max-dom* applied to the Cartesian model; • differs from ▲ in the use of the polar model; □ improves • with CID(3) operators; ■ improves • with CID(9) operators; + differs from ■ by *initial-state*. Right: Period-17 orbits, which clearly shows the well known strange attractor of the Ikeda map, although in polar coordinates here.

unfolded model[7] over the real numbers in two ways: The Cartesian form $z = x + iy$ yields

$$\begin{cases} x_{(k+1) \bmod n} = a + b(x_k \cos u_k - y_k \sin u_k) \\ y_{(k+1) \bmod n} = b(x_k \sin u_k + y_k \cos u_k) \\ u_k = \kappa - \alpha/(1 + x^2 + y^2) \end{cases} \tag{9}$$

for $k = 0, \ldots, n - 1$ and the polar form $z = \rho e^{i\theta}$ leads to the relational model

$$\begin{cases} \rho_{(k+1) \bmod n} \cos(\theta_{(k+1) \bmod n}) = a + b(\rho_k \cos(u_k)) \\ \rho_{(k+1) \bmod n} \sin(\theta_{(k+1) \bmod n}) = b(\rho_k \sin(u_k)) \\ u_k = \theta_k + \kappa - \alpha/(1 + \rho_k^2). \end{cases} \tag{10}$$

The domains can be defined as $x_k, y_k \in [-10, 10]$, $\rho_k \in [0, 10\sqrt{2}]$ and $\theta_k \in [-\pi, \pi]$ for every k. In both models, new variables $u_k \in (-\infty, +\infty)$ are added to share projections on common sub-expressions appearing in the constraints, hence augmenting the contraction power of interval constraint propagation. These variable domains are however never split, thus not increasing the size of the search space.

The results are depicted in Fig. 4. The number of solutions P_n (dashed curve) gives an approximation of the topological entropy as $\log(P_n)/n \approx 0.60$. We first compare the Cartesian model (curve ▲) with the polar model (curve •) both handled by BC5 with *maxdom*. The growth constants for these models are respectively equal to 1.83 and 1.46, i.e. $t_▲(n) \approx O(e^{1.83n})$ and $t_•(n) \approx O(e^{1.46n})$, promoting the use of the polar model. However, even using the polar model, the solving strategy BC5 with *maxdom* remains very inefficient with respect to the approximate topological entropy. This led us to enforce stronger consistency techniques in order to decrease the number of splitting steps by an exponential factor.

---

[7] The folded model of this map is far too complex to be tractable by interval solving methods.

The solving time is much improved by means of BC5+CID(3) (curve □ with growth constant 0.85) and especially BC5+CID(9) (curve ■ with growth constant 0.66, i.e. $t_\blacksquare(n) \approx O(e^{0.66n})$), considering the polar model. Finally, as observed for the Hénon map, the other tested splitting strategies are counterproductive. In particular, this phenomenon is illustrated by replacing *maxdom* with *initial-state* and solving the polar model with BC5+CID(9) (curve + with growth constant 0.81, i.e. $t_+(n) \approx O(e^{0.81n})$, to be compared to ■).

6 Discussion

Compared to the dedicated method proposed in [9,10,11], the CP framework offers a much more flexible, easy to deploy and to use environment. However, this high flexibility entails choosing the best combination of model and solving strategy. The results reported in Section 5 suggest that this choice can be performed as follows: The different combinations can be implemented to calculate period-n orbits for small values of n (e.g., with a timeout of a few minutes). On the basis of these results, the law $t \approx ce^{dn}$ can be approximated for each combination, by estimating the constants c and d, and the best combination can be used to solve the problem with greater periods.

A quantitative comparison of the respective efficiencies of the CP framework and the method of [9] is difficult to assess, since [9] does not focus on this aspect. Nevertheless, the asymptotic complexity, which does not depend on the computer, can be extracted from the results reported in [9] for the Hénon map: The time needed to compute all n-periodic orbits follows $t_G(n) \approx O(e^{0.58n})$. Our experiments on the Hénon map have shown an asymptotic time $t_\square(n) \approx O(e^{0.51n})$. This is a significant improvement with respect to the lower bound complexity $P_n \approx O(e^{0.46n})$.

On a qualitative perspective, the experiments reported in Section 5 allow arguing about several claims of [9]: First, the usage of local consistencies removes the necessity of symbolically inverting the map, which is critical for the efficiency of [9] but not always possible. Second, *initial-state* splitting strategy is not anymore a key ingredient for the efficiency, not even the best strategy in the CP framework. Finally, additional properties like the pre-computation of the non wandering part or some trapping region are not essential anymore: Local consistencies are able to efficiently remove boxes inconsistent with these additional properties using only the constraints $x_{(k+1)\bmod n} = f(x_k)$. In addition, the cost of their treatment may turn out to penalize the overall algorithm efficiency.

Future work shall tackle additional maps, including higher dimensional discrete time dynamical systems and ODE driven continuous time dynamical systems. One weakness of the approach, which is also pointed out in [9], is that the topological entropy approximation by counting the number of periodic orbits holds only for dynamical systems that satisfy the axiom A (although some exponential growth of the number of periodic orbits is a very strong hint of the presence of hyperbolic chaos in general). We shall investigate the possibility of providing some computer assisted proof of this property.

References

1. Arai, Z.: On hyperbolic plateaus of the Hénon map. Journal Experimental Mathematics 16(2), 181–188 (2007)
2. Araya, I., Trombettoni, G., Neveu, B.: Exploiting monotonicity in interval constraint propagation. In: AAAI (2010)
3. Banhelyi, B., Csendes, T., Garay, B., Hatvani, L.: A computer-assisted proof of σ_3-chaos in the forced damped pendulum equation. SIAM Journal on Applied Dynamical Systems 7, 843–867 (2008)
4. Benhamou, F., Goualard, F., Granvilliers, L., Puget, J.F.: Revising hull and box consistency. In: ICLP, pp. 230–244 (1999)
5. Benhamou, F., McAllester, D., Van Hentenryck, P.: CLP(Intervals) revisited. In: Procs. Intl. Symp. on Logic Prog., pp. 124–138. The MIT Press (1994)
6. Blanchard, F., Glasner, E., Kolyada, S., Maass, A.: On Li-Yorke pairs. Journal für die reine und angewandte Mathematik 2002(547), 51–68 (2002)
7. Crofts, J.J., Davidchack, R.L.: Efficient detection of periodic orbits in chaotic systems by stabilizing transformations. SIAM J. Sci. Comput. 28(4), 1275–1288 (2006)
8. Davidchack, R.L., Lai, Y.C., Klebanoff, A., Bollt, E.M.: Towards complete detection of unstable periodic orbits in chaotic systems. Physics Letters A 287(12), 99–104 (2001)
9. Galias, Z.: Interval methods for rigorous investigations of periodic orbits. International Journal of Bifurcation and Chaos 11(09), 2427–2450 (2001)
10. Galias, Z.: Rigorous investigation of the Ikeda map by means of interval arithmetic. Nonlinearity 15(6), 1759 (2002)
11. Galias, Z.: Computational methods for rigorous analysis of chaotic systems. In: Kocarev, L., Galias, Z., Lian, S. (eds.) Intelligent computing based on chaos. SCI, vol. 184, pp. 25–51. Springer, Heidelberg (2009)
12. Gao, F., Gao, H., Li, Z., Tong, H., Lee, J.J.: Detecting unstable periodic orbits of nonlinear mappings by a novel quantum-behaved particle swarm optimization non-Lyapunov way. Chaos, Solitons & Fractals 42(4), 2450–2463 (2009)
13. Goldsztejn, A., Hayes, W., Collins, P.: Tinkerbell is chaotic. SIAM Journal on Applied Dynamical Systems 10(4), 1480–1501 (2011)
14. Goldsztejn, A., Jermann, C., Ruiz de Angulo, V., Torras, C.: Symmetry breaking in numeric constraint problems. In: Lee, J. (ed.) CP 2011. LNCS, vol. 6876, pp. 317–324. Springer, Heidelberg (2011)
15. Granvilliers, L., Benhamou, F.: Algorithm 852: Realpaver: an interval solver using constraint satisfaction techniques. ACM Trans. Mathematical Software 32(1), 138–156 (2006)
16. Hénon, M.: A two-dimensional mapping with a strange attractor. Communications in Mathematical Physics 50, 69–77 (1976)
17. Ikeda, K.: Multiple-valued stationary state and its instability of the transmitted light by a ring cavity system. In: Opt. Comm., pp. 257–261 (1979)
18. Ishii, D., Goldsztejn, A., Jermann, C.: Interval-based projection method for under-constrained numerical systems. Constraints 17(4), 432–460 (2012)
19. Kapela, T., Simó, C.: Computer assisted proofs for nonsymmetric planar choreographies and for stability of the Eight. Nonlinearity 20(5), 1241 (2007)
20. Katok, A., Hasselblatt, B.: Introduction to the Modern Theory of Dynamical Systems. Cambridge University Press (1995)
21. Kolyada, S.F.: Li-Yorke sensitivity and other concepts of chaos. Ukrainian Mathematical Journal 56(8), 1242–1257 (2004)
22. Lhomme, O.: Consistency techniques for numeric CSPs. In: IJCAI, pp. 232–238 (1993)

23. May, R.M.: Simple mathematical models with very complicated dynamics. Nature 261, 459–467 (1976)
24. Moore, R.: Interval Analysis. Prentice-Hall (1966)
25. Neumaier, A.: Interval Methods for Systems of Equations. Cambridge University Press (1990)
26. Parsopoulos, K., Vrahatis, M.: Computing periodic orbits of nondifferentiable/discontinuous mappings through particle swarm optimization. In: Proceedings of the 2003 IEEE Swarm Intelligence Symposium, SIS 2003, pp. 34–41 (2003)
27. Sella, L., Collins, P.: Computation of symbolic dynamics for one-dimensional maps. J. Comput. Appl. Math. 234(2), 418–436 (2010)
28. Trombettoni, G., Chabert, G.: Constructive interval disjunction. In: Bessière, C. (ed.) CP 2007. LNCS, vol. 4741, pp. 635–650. Springer, Heidelberg (2007)
29. Tucker, W.: A rigorous ODE solver and Smale's 14th problem. Found. Comput. Math. 2, 53–117 (2002)
30. Wilczak, D., Zgliczynski, P.: Computer Assisted Proof of the Existence of Homoclinic Tangency for the Hénon Map and for the Forced Damped Pendulum. SIAM Journal on Applied Dynamical Systems 8, 1632–1663 (2009)

Laser Cutting Path Planning Using CP

Mikael Z. Lagerkvist, Martin Nordkvist, and Magnus Rattfeldt

Tomologic AB, Sweden
firstname.lastname@tomologic.com

Abstract. Sheet metal cutting using lasers is ubiquitous in the industry, and is used to produce everything from home decorations to excavator scoops. Metal waste is costly for the industry, both in terms of money, but also in terms of an increased environmental footprint. Tomologic develops a unique optimisation system that can reduce this waste drastically. This paper presents a CP approach to the *Laser Cutting Path Planning Problem (LCPPP)*, a very hard important sub problem within the Tomologic optimisation system. A solution to the LCPPP is, given a packing of some details on a metal sheet, an ordering of the cuts necessary to separate the details from the sheet. The problem is complicated by physical factors such as heat from the laser beam, or details moving or flexing. In the paper, we explain the problem in detail and present our CP approach that we developed for solving the problem. The possibility (in CP) of custom search heuristics turned out to be crucial to be able to solve the problem efficiently, as these could be made to guide the search to good first solutions.

1 Introduction

Most people have come across the problem of planning different shapes (hearts, Christmas trees, stars, etc) on gingerbread dough, and trying to minimise the dough waste that needs to be rolled out again. See Fig. 1 on the facing page for an example with hearts where, in 1(a), only three hearts fit but, when aligning the hearts together as in 1(b), one more heart can be made to fit. Now, replace the dough by metal sheets, and the technology to separate the shapes (or details) from those metal sheets by laser cutting machines. Then, aligning the details as in Fig. 1(b) is not trivial anymore, and the waste cannot simply be "rolled out" again, but the recycling process is very costly.

The sheet metal cutting market is huge: the number of active laser cutting machines is estimated to be around 50,000 globally, each such machine consumes around 1,500 tonnes of raw material each year, and the amount of metal waste is typically between 20 and 50 percent [1]. So *any* (general) decrease in waste means great savings!

Tomologic develops a unique optimisation system that can reduce this global metal waste considerably, by deploying a technology that makes alignments such as those in Fig. 1(b) possible. This is of great importance not only for the manufacturing industry, for which there are obvious cost savings, but also for the whole world, since the industry's environmental footprint can be made smaller.

C. Schulte (Ed.): CP 2013, LNCS 8124, pp. 790–804, 2013.

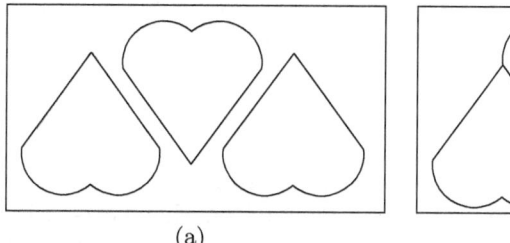 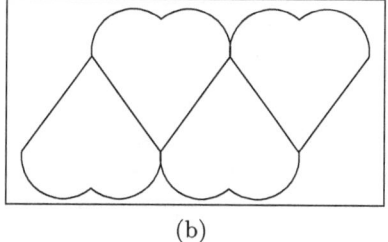

(a)   (b)

Fig. 1. How many hearts can be obtained from the gingerbread dough?

Tomologic's solution is based on technical knowledge of, given a packing of some details on a metal sheet, how to plan the cutting paths of the laser beam to separate aligned details, and still ensuring a high quality of the end products. In this paper, we formalise this very important and hard combinatorial sub problem that must be solved within the Tomologic optimisation system, and describe a constraint programming approach that we developed for solving it. The main contributions of this paper are:

- the introduction of a new problem domain in the context of a real life industrial problem of great importance;
- a constraint programming approach for the problem, including a formal model of variables and constraints, as well as customised search heuristics for solving the model.

In the following, we first discuss background and context in Sect. 2, after which we introduce the Laser Cutting Path Planning Problem in Sect. 3. We then present our constraint programming model in Sect. 4, where we start by describing the decision variables of the problem, followed by problem constraints as well as implied ones. Section 5 describes the search heuristics and optimisation goal, and Sect. 6 gives an overview of the implementation. Finally, in Sect. 7 we discuss current status and constraint programming impacts on the application development.

2   Optimisation for Sheet Metal Cutting

One of the large problems faced by the manufacturing industry today is metal waste. This is inevitable when, out of large metal sheets, using lasers or related techniques to produce anything from home decorations to excavator scoops. Such metal waste needs to be (i) transported from the manufacturing shops to metal recycling facilities (often overseas); (ii) melt down and restored to new raw material (for example new metal sheets); (iii) transported back to the manufacturing shops for further processing. This means increased costs, both in terms of money, but also in terms of increased environmental footprints for the end products. So the objective when optimising sheet metal cutting is very easy to understand:

> *Given a set of production details and a number of metal sheets, find*
> *a packing of the details on the sheets that minimises the overall metal*
> *waste.*

2.1   Current Technology

The traditional technology that is used for planning production details on metal
sheets is *nesting* [2], where the details are planned on the sheets using two-
dimensional irregular shape packing algorithms. Current state-of-the-art nesting software can produce sophisticated plans, but suffers from one important
limitation:

> *To ensure quality of the production details, any two adjacent details must*
> *be separated by a safety distance.*

This safety distance depends on the type and thickness of the metal sheets and,
of course, means that large amounts of waste in the form of metal skeletons are
unavoidable. For example, using the traditional nesting technology for solving
the hearts problem shown in Fig. 1 on the previous page, the solution in 1(b) is
not possible, as the laser cutting machine would not be capable of cutting those
aligned shapes safely.

However, by using a safety distance, the *only* condition (disregarding any optimisation criteria) that needs to be taken into account when developing nesting
algorithms, is the geometric non-overlapping constraint on all details. Given any
packing that fulfils this condition, the details are cut in isolation in some order,
without affecting each other.

2.2   The Tomologic Optimisation System

Tomologic introduces a completely new technology for planning production details on metal sheets. This technology is based on the observation that, under
some conditions, the safety distance between details can often be omitted. This
means that details can be aligned and separated by the width of the laser beam
only, and that cutting paths can be shared between several details. Tomologic's
knowledge of when this is safe to do is based on many years of hands on experience of manual production planning for, and operation of, laser cutting machines.

However, the alignment of production details complicates the problem considerably since (i) there are many more conditions to take into account in addition
to the geometric non-overlapping constraint, such as when and how two details
can be aligned; and (ii) the cutting path planning is much more complicated,
since the order of the cuts now depends on the packing.

Although complicating the problem, the alignment of production details
also means that the waste can be reduced considerably. For example, it is
often the case that waste in the form of metal skeletons (coming from the
use of a safety distance) is replaced by much less waste in the form of *metal*
frames (see Fig. 2 on the facing page, for example). Furthermore, the alignment

Fig. 2. Tomologic's technology (left) compared to the traditional nesting technology (right)

of production details also means that sophisticated cutting patterns can be deployed, which can decrease the time and energy necessary to drive the laser beam.

So the Tomologic optimisation system must solve two interacting problems, the first one being how to find a packing of the production details on the metal sheets, while the second one being how to plan the cutting paths given such a packing of details. In this paper we focus on the second problem, that we call the Laser Cutting Path Planning Problem, presented in the next section.

3   The Laser Cutting Path Planning Problem

Given a *packing* of a set of production details on a metal sheet, the *Laser Cutting Path Planning Problem* (*LCPPP*) is the problem of finding an order of the cuts necessary to separate the details from the sheet. In order to discuss this in greater detail, we need to introduce some terminology.

A packing consists of a number of *clusters*, each such cluster contains a number of details that are connected (directly or indirectly) to each other through *alignment cuts* (two sides of different details separated by the width of the laser beam only). Such clusters are separated by a safety distance. This is in contrast with the traditional nesting technology, where each cluster can contain at most one detail. A *pocket* is an area within a cluster that is not a detail, but completely surrounded by at least two connected details.

A *cutting path* describes the movement of the laser beam while it is turned on. This is analogous to paper pencil drawing, from the time that the pencil first touches the paper until it is lifted again. Given a cluster, we call a complete sequence (that separates each detail in the cluster from any other detail or the rest of the metal sheet) of such cutting paths a *cutting plan* for the cluster. A *piercing* is the process of creating a small hole in the metal sheet at the start of each cutting path. Due to additional heat produced by the laser beam in this process, there must be some space between piercings and details, or the details may suffer from defects. This means that after each piercing, and before starting the actual cut (that is, the cut separating the relevant detail from the rest of the sheet), there must be a short *lead-in* cut.

To reason about solutions to the LCPPP, we represent each cluster as a graph: the *cut graph* of the cluster. The edges of a cut graph represent cuts; either cuts separating details from the rest of the metal sheet, or cuts separating two details from each other (alignment cuts). The nodes of a cut graph represent the *connections* where two or more cuts meet (the *incoming cuts* of the connections). A cut graph is generated by identifying the cuts and connections of the cluster. In addition to natural connections that occur at the endpoints of alignment cuts, additional connections are introduced at positions that are well suited for piercings.

Example 1. Consider the instance of the LCPPP shown in Fig.3(a) on the facing page, consisting of one cluster containing four details (labeled d_1, ..., d_4), and one pocket (labeled p_1), to be separated from a metal sheet (its edges shown dashed). To separate the details from the metal sheet, thirteen cuts must be made, in some order. These cuts are labeled c_1, ..., c_{13} in the cut graph of Fig. 3(b), and should be interpreted as follows. Cut c_1 separates d_1 from the metal sheet; alignment cut c_2 separates details d_1 and d_2 from each other; alignment cut c_3 separates details d_2 and d_3 from each other; alignment cuts c_4 and c_5 separate details d_1 and d_3 from each other; cut c_6 and c_7 separate detail d_1 from pocket p_1; alignment cuts c_8 and c_{10} separate details d_1 and d_4 from each other; cut c_9 separates detail d_3 from pocket p_1; cut c_{11} separates detail d_4 from pocket p_1. Finally, cuts c_{12} and c_{13} separate d_4 from the metal sheet.

Each cut starts and ends in two out of nine connections labeled k_1, ..., k_9 (these connections are also shown on the details in (a) for clarity). Possible cutting path starting connections are identified in the cut graph by additional circles. The connection k_9 was introduced as an additional such possible starting connection.

A possible cutting plan for this instance is: c_4 starting in k_3; $c_5 \rightarrow c_2 \rightarrow c_3$ starting in k_4; $c_8 \rightarrow c_6 \rightarrow c_9 \rightarrow c_7$ starting in k_5; $c_{10} \rightarrow c_{11}$ starting in k_8; $c_{13} \rightarrow c_1 \rightarrow c_{12}$ starting in k_9.

A *solution* to the LCPPP is a cutting plan for each cluster that separates the production details from the rest of the metal sheet, and still ensuring production reliability of those details. This is achieved by imposing additional constraints on cutting plans. We may also impose optimisation criteria on cutting plans, for example with respect to improved detail quality or lower cutting time. These con-

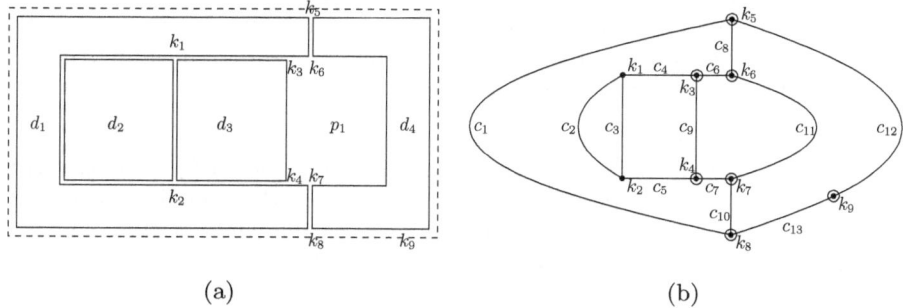

(a)   (b)

Fig. 3. An instance of the Laser Cutting Path Planning Problem

straints and optimisation criteria are discussed in the context of our constraint programming approach in the following three sections.

4 A Constraint Programming Model

4.1 Assumptions and Notation

We consider an instance of the LCPPP where cuts $\mathcal{C} = \{c_1, \ldots, c_n\}$ with connections $\mathcal{K}$ must be made to separate a number of production details from a metal sheet. For simplicity, we assume a single cluster; problems involving several such clusters are beyond the scope of this paper. Given this, we use

- *cut part* as a collective name for a detail or a pocket;
- incoming(k) to denote the incoming cuts to connection $k \in \mathcal{K}$; and
- arrays indexed by cuts as placeholders for our decision variables.

By abuse of notation we will sometimes use variable array names to denote sets or functions, and write formulas on elements, subsets, or function applications of such array names. For example, we use

- cutorder(P) to denote the cut order variables (defined below) of any of the cut parts in P; and
- cutparts(x) to denote the cut parts (at most two) that x is a cut order variable of.

This is exemplified further in Ex. 2 below.

4.2 Decision Variables and Their Domains

Cut Order Variables. We use an array cutorder$[c_1, \ldots, c_n]$ of *cut order variables* to represent the order in which the cuts are made, where the domain of each such variable is $1..n$. Furthermore, we let cutorder$[\bot] = -\infty$.

Cut Start variables. We use an array $\text{cutstart}[c_1, \ldots, c_n]$ of *cut start variables* to represent cutting path starting points, where the domain of each such variable is the starting connections of the corresponding cut, and $\perp$ (meaning that the corresponding cut does not start a cutting path).

Predecessor Variables. We use an array $\text{pred}[c_1, \ldots, c_n]$ of *predecessor variables* to represent the predecessors of the cuts, where the domain of each such variable is its adjacent cuts, and $\perp$ (meaning that the corresponding cut does not have a predecessor, since it starts a cutting path).

Example 2. Recalling the instance of Ex. 1 on page 794, the initial variable domains are as follows (only showing the domains for c_1, c_2, and c_{13}):

$$\text{cutorder}[c_1, \ldots, c_{13}] = \big[\, 1..13, 1..13, \ldots, 1..13 \,\big]$$
$$\text{cutstart}[c_1, \ldots, c_{13}] = \big[\, \{k_5, k_8, \perp\}, \{\perp\}, \ldots, \{k_8, k_9, \perp\} \,\big]$$
$$\text{pred}[c_1, \ldots, c_{13}] = \big[\, \{c_8, c_{10}, c_{12}, c_{13}, \perp\}, \{c_3..c_5, \perp\}, \ldots, \{c_1, c_{10}, c_{12}, \perp\} \,\big]$$

Now, the cutting plan given in Ex. 1 is equivalent to the assignments:

$$\text{cutorder}[c_1, \ldots, c_{13}] = [12, 3, 4, 1, 2, 6, 8, 5, 7, 9, 10, 13, 11]$$
$$\text{cutstart}[c_1, \ldots, c_{13}] = [\perp, \perp, \perp, k_3, k_4, \perp, \perp, k_5, \perp, k_8, \perp, \perp, k_9]$$
$$\text{pred}[c_1, \ldots, c_{13}] = [c_{13}, c_5, c_2, \perp, \perp, c_8, c_9, \perp, c_6, \perp, c_{10}, c_1, \perp]$$

Let $\text{cutorder}[c_i] = x_i$ for $1 \leq i \leq n$. The cut order variables of d_2 and $\{d_4, p_1\}$ respectively are :

$$\text{cutorder}(\{d_2\}) = \{x_2, x_3\}$$
$$\text{cutorder}(\{d_4, p_1\}) = \{x_6, \ldots, x_{13}\}$$

The cut parts of x_1 and x_3 respectively are:

$$\text{cutparts}(x_1) = \{d_1\}$$
$$\text{cutparts}(x_3) = \{d_2, d_3\}$$

4.3   Problem Constraints

We present the constraints first in English and then formally, possibly followed by an explanation.

Basic Graph Constraints. These constraints ensure that cutorder, cutstart and pred are correctly related.

(a) *Any given cut order can only be assigned once.*

$$\text{alldifferent}(\text{cutorder})$$

(b) *The cut order of a predecessor must be one less than the cut it precedes.*

$$\underset{c \in \mathcal{C}}{\forall} \text{cutorder}[c] = \text{cutorder}[\text{pred}[c]] + 1 \iff \text{pred}[c] \neq \perp$$

(c) *A starting cut must not have a predecessor.*

$$\underset{c \in \mathcal{C}}{\forall} \text{cutstart}[c] \neq \perp \iff \text{pred}[c] = \perp$$

(d) *A starting cut must have a correctly directed successor.*

$$\underset{c \in \mathcal{C}}{\forall} \left(\text{cutstart}[c] = k \wedge k \neq \perp \Rightarrow \underset{d \in \text{incoming}(k)}{\forall} \text{pred}[d] \neq c \right)$$

Each cut starting a cutting path in a connection k must not precede any of k's adjacent cuts. Otherwise, the cutting path would contain cuts with opposite directions (which is not possible, since a cutting path can have at most one start where it pierces the metal sheet).

Constraints Ensuring Production Reliability. These constraints ensure that important properties from the physical reality of laser cutting are maintained.

(e) *For some sets $K \subset \mathcal{K}$ of conflicting connections, at most one of those connections can start a cutting path.*

$$\left(\underset{k \in K}{\forall} \underset{c \in \text{incoming}(k)}{\forall} b_{ck} \iff \text{cutstart}[c] = k \right)$$
$$\wedge$$
$$\text{count}(b) \leq 1$$

The counting is done using additional boolean variables.

(f) *A cut separating two cut parts must not be the final cut for both parts.*

$$\underset{x \in \text{cutorder}}{\forall} \left(|\text{cutparts}(x)| = 2 \Rightarrow \max(\text{cutorder}(\text{cutparts}(x))) > x \right)$$

For each cut order variable x that corresponds to a cut c separating two cut parts p and q, the maximum cut order for *any* cut order variable of p or q must be greater than x. Otherwise, c is the final cut for both p and q.

(g) *For some pairs of sets of cuts $A, B \subset \mathcal{C}$, all cuts of A must be cut before the final cut of B.*

$$\max(\text{cutorder}(A)) < \max(\text{cutorder}(B))$$

(h) *For some sets of adjacent cuts $A \subset \mathcal{C}$ all sharing the same connection, no more than M pairs of those cuts may pass that connection consecutively.*

$$\left(\underset{c < d \in A}{\forall} b_{dc} \iff (\text{pred}[c] = d \vee \text{pred}[d] = c) \right)$$
$$\wedge$$
$$\text{count}(b) \leq M$$

The counting is done using additional boolean variables.

(i) *For some cut triplets (a, b, c) sharing a connection k, when a and b are cut consecutively, they must be cut after c.*

$$(\mathrm{pred}[a] = b \vee \mathrm{pred}[b] = a)$$
$$\Rightarrow$$
$$\mathrm{cutorder}[c] < \min(\mathrm{cutorder}[a], \mathrm{cutorder}[b])$$

4.4   Implied Constraints

(j) *The cut order of a predecessor must be strictly less than the cut it precedes.*

$$\forall_{c \in \mathcal{C}} \mathrm{cutorder}[\mathrm{pred}[c]] < \mathrm{cutorder}[c]$$

This constraint is implied by (b), and uses the property of $\mathrm{cutorder}[\bot] = -\infty$.

(k) *For all connections with exactly three incoming cuts, at most one pair of those cuts can pass through it consecutively.*

$$\forall_{k \in \mathcal{K}: |\mathrm{incoming}(k)| = 3} \left(\left(\forall_{c < d \in \mathrm{incoming}(k)} b_{cd} \iff (\mathrm{pred}[c] = d \vee \mathrm{pred}[c] = d) \right) \\ \wedge \\ \mathrm{count}(b) \le 1 \right)$$

For each connection k with exactly three incoming cuts, the number of distinct pairs of its cuts for which either is the predecessor of the other, can be at most one, since there are only three cuts in total. The counting is done using additional boolean variables for each distinct pair of cuts. This constraint is implied by the local properties around connections with three connected cuts.

(l) *The directed graph described by the predecessor variables consists of a set of simple paths.*

$$\mathrm{mirrored} = [m_o, \ldots, m_{n-1}] \tag{1}$$

$$\forall_{0 \le i < n} m_i = \begin{cases} x & \text{if } p_i = c_x \\ -(i+1) & \text{if } p_i = \bot \end{cases} \tag{2}$$

$$\mathrm{alldifferent}(\mathrm{mirrored}) \tag{3}$$

The implied constraint is a path constraint [3], and the above decomposition models the constraint. The variables used in (1) above are similar to the pred variables, the difference being that the cutting path starting point marker is indicated by unique negative values. The constraints in (2) can be implemented with element constraints since it is a total functional relation [4]. Replacing the cutting path starting point marker means that for any solution, all variables will be assigned different values, enforced by (3). This is in contrast with the pred variables, where all cuts starting paths are assigned $\bot$.

5   Optimisation and Search

Real world instances of the LCPPP can be quite large. It is not unreasonable to expect instances with thousands of variables and more than ten thousand constraints. This has several consequences for solving such instances to optimality, including high memory usage and long solving times. However, our goal is to find a good enough solution quickly, and not to find and prove the optimal solution. If no solution is found in a reasonable time frame, we consider that particular sub problem (or cluster) to be infeasible, and discard it as a potential solution. In our context, a reasonable time frame is a few seconds of running time.

In the following sub sections, we describe the general optimisation goal for the LCPPP, and the custom search heuristics that we developed.

5.1   Optimisation Goal

The overall goal of solving an instance of the LCPPP is to find a satisfying solution that has some combination of good properties in the context of sheet metal cutting using lasers. While the details of this goal is beyond the scope of this paper, we outline some general guide lines, in their order of importance.

- *Avoiding certain cut starts.* All cutting path starting connections are not equally good, but some starting connections may lead to undesirable marks.
- *Minimising the number of cutting paths.* Starting a new cutting path takes time, since it means that the metal sheet needs to be pierced.
- *Minimising the laser movement distance.* Moving the laser head between cutting paths takes time.

The first two goals are modelled as a minimisation problem over a sum using a valuation for each cutting path starting connection. While the third goal could be expressed in a similar way, our solution handles this more softly in combination with domain specific search patterns. These search patterns come from crucial domain knowledge of sheet metal cutting using lasers. Handling the third goal in this way works fine in our current approach, but it would likely have to be handled differently if a more general search heuristic was used, such as large neighbourhood search [5].

5.2   Custom Search Heuristics

The decision variables of the model in Sect. 4 are rather low level, while they are used to describe high level concepts such as graphs and their properties. So any single assignment to a variable has a low propagation impact, since it does not meaningfully constrain the solution space. In addition, most assignments have no or next to no impact on the optimisation goal. As a consequence, using standard constraint programming search techniques, either simple ones such as fail first, or more complicated ones such as weighted degree [6] or activity based search [7], is not effective enough. As a consequence of this, in order to find good enough solutions to the LCPPP quickly, we have implemented a set of custom search heuristics comprised by what we call *actions* and *strategies*. These custom search heuristics then drive the search towards good first solutions.

In the following, a *search node* is a partially instantiated solution, and a *choice* is a set of alternatives that restrict such a search node further.

Actions, Strategies, and Heuristics. An *action* is a function that accepts a search node, and returns a choice, or nothing if the action does not apply to the search node.

A *strategy* is a list of actions, and a specification of how to conduct the search among these actions. The specification indicates the maximum number of dispatches of each action, if the action should create choices or assignments (single alternative choices), or any limits that should be imposed on the search. For example, by limiting the number of times an action can be dispatched to one, we can create a sub list of strategies that must succeed on its first dispatch, or fail the whole strategy upon backtracking. A strategy will run the first applicable action that is available. If no action is applicable, the strategy is finished.

A *heuristic* is a list of strategies. Given a suitable set of strategies, an overarching heuristic that guides the search to good solutions can be defined.

Example Actions. We have defined over 40 different actions that perform meaningful choices for an LCPPP instance. Some examples are:

- *Extend alignment cut.* Given an open ended cutting path of alignment cuts, extend it with a successor alignment cut.
- *Start left bottom alignment cut.* Start a new cutting path of alignment cuts, choosing the left-most bottom-most possibility.
- *Start corner aligned contour.* Start a new cutting path in a graph contour corner.
- *Assign top left order.* Assign the top left unassigned cut order variable its minimum value.

The actions can roughly be classified into actions that start new cutting paths, actions that extend current cutting paths, and actions that assign cut orders. Most defined actions have some geometric meaning, and are derived from practical experience of how to plan cutting paths.

Example Strategies. New strategies that implement some desired behaviour are fairly easy to define by combining lists of actions. A typical such strategy is defined by an action starting a new cutting path, followed by a sequence of actions that extend the cutting path according to different properties. We have defined 15 different strategies so far for the LCPPP.

Example Heuristics. An example of a typical heuristic is as follows:

1. Build open cutting paths of alignment cuts that can be extended in both directions.

2. Assign good starting points.
3. Extend internal cutting paths from the chosen starting points.
4. Extend external cutting paths on the graph contour.
5. Assign remaining starting points, extend remaining cutting paths, and assign remaining cut orders.

When the final step is reached, the (partial) solution typically already has the interesting features defined already. This means that we are only interested in the existence of a single solution, which is similar to the radiotherapy planning [8] problem, as well as the use cases for the once-combinator [9]. At the moment, we have defined two main heuristics.

Example 3. Consider the instance of the LCPPP described in Ex. 1 on page 794, and recall that k_1 and k_2 are not possible cutting path starting connections. Following the general outline of a heuristic above, the search could perform the following steps to reach the described solution.

1. Set $\text{pred}[c_2] = c_5$ (speculative choice). Since k_1 is not a possible cutting path starting connection, set $\text{pred}[c_3] = c_2$ (all other alternatives at this point would force k_1 as a starting connection).
2. Connections k_3, k_4, k_5 and k_8 are identified as good starting points around pocket p_1.
3. Set $\text{cutstart}[c_5] = k_4$ and $\text{cutstart}[c_4] = k_3$ for paths from the pocket. Assign values for the paths $c_8 \rightarrow c_6 \rightarrow c_9 \rightarrow c_7$ starting in k_5 and $c_{10} \rightarrow c_{11}$ starting in k_8, finishing up the assigned starting points.
4. Assign the graph contour cuts, choosing k_9 (as best alternative among the available connections) as the starting point.
5. At this point, only the order remains to be set. Starting from top left among non graph contour cuts, cuts are ordered with k_4 starting the first cut, followed by cuts starting from (in order) k_3, k_5, k_8, and k_9

6   Implementation

The model and the search has been implemented using the Gecode [10,11] constraint programming system, version 3.7.3, as a stand-alone C++ application. The Tomologic optimisation system is implemented mostly in Java and Scala. Instead of integrating the C++ code using native calls, the model is run as a separate process. This ensures full separation between the Tomologic optimisation system and the CP application.

To facilitate the communication between Java and C++, a custom XML format is used for describing instances of the problem. In addition to the instance description, the XML definition also contains a list of the strategies that should be used. Each strategy is defined with the actions it contains and the search method to be used. This allows the application code to programmatically select the overall heuristic that is to be used for a particular instance, and to run the constraint model using different search strategies on the same instance.

The full implementation, including all supporting code such as visualisation and XML parsing consists of around 5000 lines of code and 1500 lines of documentation.

Variables. All variables are simple boolean or integer variables, and the largest domains for the integer variables are bounded by the number of cuts in the instance.

Constraints and Propagators. The constraints used in the model are mostly standard simple constraints. These include arithmetic, logical, counting, element, and min/max constraints. Such constraints are directly available as propagators in most constraint programming systems. The only global constraint in the model is alldifferent. After experimentation, we have concluded that the appropriate consistency level to use for the alldifferent propagation on the cutorder variables when solving LCPPP instances is bounds consistency [12].

Search. The search strategies are implemented as Gecode branchers [13]. Each brancher contains a list of implementations of actions that produce descriptions of the choices to be made. To run the strategies, a custom search function is used, where a normal Gecode depth first search engine is created for each strategy that is started. Since Gecode returns a partially instantiated solution when all currently installed branchers are exhausted, that returned solution can be used as the root node for the search engine for the next strategy.

Graphical Inspection Support. Invaluable for understanding the search process in large instances is to have good visualisation support. The Gecode Gist search tree visualiser [14,11] was used for understanding the search process. To understand partial solutions, a graphical visualisation was implemented that shows the currently assigned cutting paths and cut order domains. This gives a much more high level view of the current state, compared to just looking at the variables and their current domains. See Fig. 4 on the next page for an example Gist tree and visualised search state. The visualised search state shows the cutting paths under construction. For cuts that are known to be part of a specific cutting path, the cut is highlighted and their current cut order domain is displayed.

Parallel Search. In most cases the search trees are quite deep, and exploration only visits a very small part of the state space; the final number of leaf nodes visited is *much* lower than the number of nodes in the explored search tree. This means that using parallel search would not be very effective, since it does not significantly speed up getting to the first solution. However, in the surrounding context of the LCPPP, the machine is fully loaded by other tasks and, hence, not using parallelism for the LCPPP is not an issue.

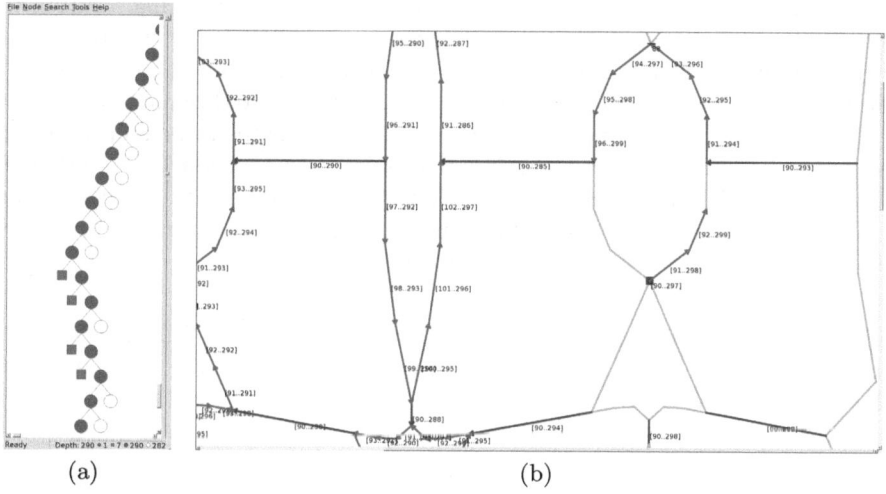

(a)   (b)

Fig. 4. Gist search tree (a) and visualisation of the current cutting paths (b). The cutting paths with assigned cut starts are shown as arrows indicating their directions. Green such cutting paths indicate pocket cuts, while red such cutting paths indicate alignment cuts. Grey cuts indicate cuts that do not yet belong to a cutting path. The ranges indicate the current domain for the respective cut order variables.

7   Constraint Programming Impact

Before the CP approach described in this paper was developed, we used a customised greedy algorithm for solving the LCPPP. At this time, the problem was not formalised, but based on the interaction between (non-CP) developers and our sheet metal cutting domain experts. The greedy algorithm quickly became difficult to maintain and, as more features in the form of additional constraints were introduced, the more often the algorithm had a hard time finding feasible solutions. It became clear that a more flexible and powerful approach was needed.

The formal modelling of the LCPPP that was necessary for the CP approach has been crucial for understanding the problem, and for gaining confidence in the generated solutions. Using CP as the vehicle for such a formalisation was very natural, since it allows the expression of the domain constraints and search heuristics in a reasonably high level.

Formalising, implementing, and testing a large and complicated constraint programming model such as the LCPPP requires a significant amount of time and experience with constraint technology. In total this took about four months, which included one constraint programming expert responsible full time, discussions with two sheet metal cutting domain experts, and one additional constraint programming expert, helping with constraint formulations and implementation.

In the process of formalising and understanding the LCPPP, it was possible to restructure and reimplement the previously used greedy algorithm by using an architecture inspired by the CP approach. (This was also a necessity since, during the development of the CP approach, having *some* reasonably working solution was crucial.) This has had the effect that the greedy algorithm can now handle many more cases and is much more robust, so its performance has increased drastically as a direct consequence of developing the CP approach. Due to this and to other practical reasons, the maintenance of the CP approach has stopped, and is currently not used in production. However, even though it is not used in production anymore for solving the LCPPP, we strongly believe CP to be a key factor in the *process* leading to our current solution. Keeping this in mind, constraint programming may very well be our first approach in future applications.

Acknowledgements. We thank Magnus Gedda and Jim Wilenius for many fruitful discussions about details of the LCPPP, as well as the anonymous referees for constructive reviews.

References

1. Tomologic AB. Market Survey (2010)
2. Bennell, J.A., Oliveira, J.F.: A tutorial in irregular shape packing problems. Journal of the Operational Research Society 60(S1), S93–S105 (2009)
3. Beldiceanu, N., Carlsson, M., Rampon, J.X.: Global constraint catalog, working version as of April 24 (2013), http://www.emn.fr/z-info/sdemasse/gccat/
4. Stuckey, P.J., Tack, G.: Minizinc with functions. In: Gomes, C., Sellmann, M. (eds.) CPAIOR 2013. LNCS, vol. 7874, pp. 268–283. Springer, Heidelberg (2013)
5. Shaw, P.: Using constraint programming and local search methods to solve vehicle routing problems. In: Maher, M.J., Puget, J.-F. (eds.) CP 1998. LNCS, vol. 1520, pp. 417–431. Springer, Heidelberg (1998)
6. Boussemart, F., Hemery, F., Lecoutre, C., Sais, L.: Boosting systematic search by weighting constraints. In: de Mántaras, R.L., Saitta, L. (eds.) ECAI, pp. 146–150. IOS Press (2004)
7. Michel, L., Van Hentenryck, P.: Activity-based search for black-box constraint programming solvers. In: Beldiceanu, N., Jussien, N., Pinson, É. (eds.) CPAIOR 2012. LNCS, vol. 7298, pp. 228–243. Springer, Heidelberg (2012)
8. Baatar, D., Boland, N., Brand, S., Stuckey, P.J.: CP and IP approaches to cancer radiotherapy delivery optimization. Constraints 16(2), 173–194 (2011)
9. Schrijvers, T., Tack, G., Wuille, P., Samulowitz, H., Stuckey, P.J.: Search combinators. Constraints 18(2), 269–305 (2013)
10. Gecode team: Gecode, the generic constraint development environment (2012), http://www.gecode.org/
11. Schulte, C., Tack, G., Lagerkvist, M.Z.: Modeling and Programming with Gecode (2012), Corresponds to Gecode 3.7.3
12. López-Ortiz, A., Quimper, C.G., Tromp, J., van Beek, P.: A fast and simple algorithm for bounds consistency of the alldifferent constraint. In: Gottlob, G., Walsh, T. (eds.) IJCAI, pp. 245–250. Morgan Kaufmann (2003)
13. Schulte, C.: Programming branchers. In: Schulte, C., Tack, G., Lagerkvist, M.Z. (eds.) Modeling and Programming with Gecode (2012), Corresponds to Gecode 3.7.3
14. Schulte, C.: Oz explorer: A visual constraint programming tool. In: Kuchen, H., Swierstra, S.D. (eds.) PLILP 1996. LNCS, vol. 1140, pp. 477–478. Springer, Heidelberg (1996)

Atom Mapping with Constraint Programming

Martin Mann[1], Feras Nahar[1], Heinz Ekker[5], Rolf Backofen[1,2,3,4],
Peter F. Stadler[5,6,7,8,9], and Christoph Flamm[5]

[1] Bioinformatics, Department for Computer Science, University of Freiburg,
George-Köhler-Allee 106, 79106 Freiburg, Germany
[2] Centre for Biological Signalling Studies (BIOSS), University of Freiburg, Germany
[3] Centre for Biological Systems Analysis (ZBSA), University of Freiburg, Germany
[4] Center for non-coding RNA in Technology and Health,
University of Copenhagen, Denmark
[5] Institute for Theoretical Chemistry, University of Vienna, Währingerstrasse 17,
1090 Vienna, Austria
[6] Bioinformatics Group,
Department of Computer Science, and Interdisciplinary Center for Bioinformatics,
University of Leipzig, Härtelstraße 16-18, 04107 Leipzig, Germany
[7] Max Planck Institute for Mathematics in the Sciences, Inselstraße 22, 04103
Leipzig, Germany
[8] Fraunhofer Institute for Cell Therapy and Immunology, Perlickstraße 1, 04103
Leipzig, Germany
[9] Santa Fe Institute, 1399 Hyde Park Rd., Santa Fe, NM 87501, USA
{mmann,backofen}@informatik.uni-freiburg.de,
{studla,xtof}@tbi.univie.ac.at

Abstract. Chemical reactions consist of a rearrangement of bonds so
that each atom in an educt molecule appears again in a specific position
of a reaction product. In general this bijection between educt and product
atoms is not reported by chemical reaction databases, leaving the Atom
Mapping Problem as an important computational task for many practical
applications in computational chemistry and systems biology. Elementary chemical reactions feature a cyclic imaginary transition state (ITS)
that imposes additional restrictions on the bijection between educt and
product atoms that are not taken into account by previous approaches.
We demonstrate that Constraint Programming is well-suited to solving
the Atom Mapping Problem in this setting. The performance of our approach is evaluated for a subset of chemical reactions from the KEGG
database featuring various ITS cycle layouts and reaction mechanisms.

1 Introduction

A chemical reaction describes the transformation of a set of educt molecules
into a set of products. In this process, chemical bonds are re-arranged, while
the atom types remain unchanged. Thus, there is a one-to-one correspondence,
the so-called *atom map* (or atom-atom mapping), between atoms in educts and
products. Atom maps convey the complete information necessary to disentan-
gle the mechanism, i.e. the bond re-arrangement, of a chemical reaction via the

C. Schulte (Ed.): CP 2013, LNCS 8124, pp. 805–822, 2013.

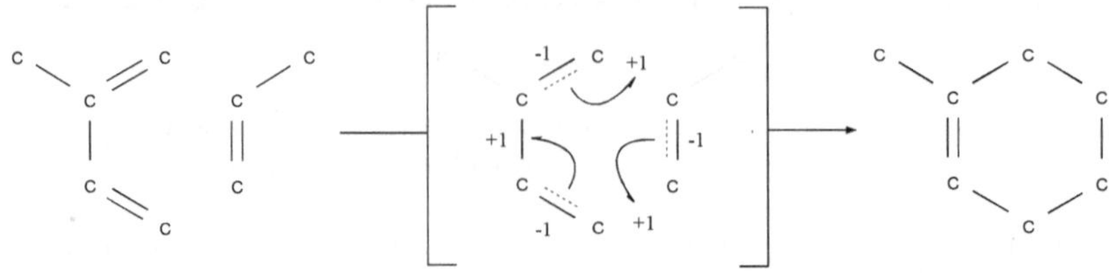

Fig. 1. Example of a Diels-Alder reaction. The imaginary transition state (ITS) is an alternating cycle defined by the bonds that are broken (dotted) and the bonds that are newly formed.

identification of bonds that differ in educt and product molecules. The changing parts of the molecules are described by a so called intermediate transition state (ITS) [17,24] that allows, for instance, a classification of chemical reactions [31,33,45]. Atom maps are a necessary requisite for computational studies of an organisms metabolism. For instance, the allow for consistency checks within metabolic pathway analyses [3] and play a role in the global analysis of metabolic networks [7,26]. Practical applications include, for example, the tracing or design of the metabolic break down of a candidate drug, which constitutes an important issue in in drug design studies [39].

For chemical reactions only the product and educt molecules are directly observable. The atom map therefore often remains unknown and has to be inferred from partial knowledge. Experimental evidence may be available from isotope labeling experiments. Here, special isotopes, i.e. atoms with special variations, are introduced into educt molecules that can then be identified in product molecules by means of spectroscopy techniques [44]. Such data, however, is not available for most reactions. The complete experimental determination of an atom map is in general a complex and tedious endeavor. Reaction databases, such as KEGG, therefore do not generally supply atom maps. The computational construction of atom maps is therefore an important practical problem in chemoinformatics.

Several computational approaches for this problem have been developed over the last three decades (for a recent review see [8]). The Educts and products are described as two not necessarily connected labeled graphs I and O, respectively. Vertex labels define atom types, while edge labels indicate bond types. The atom map is then determined as the solution of a combinatorial optimization problem resulting in a bijective mapping of all vertices of the educt molecule graph to corresponding vertices in the product molecule graphs. An illustration is given in Fig. 1.

The most common formulations are variants of the maximum common subgraph (isomorphism) problem [15]. Already the earliest approaches analyzed the adjacency information within educts and products [14,34]. The Principle of Minimal Chemical Distance, which is equivalent to minimizing an edge edit distance, was invoked in [28], using a branch and bound approach to solve the corresponding combinatorial optimization problem. Maximum Common Edge Subgraph (MCES) algorithms search for isomorphic subgraphs of the educt/

product graphs with maximum number of edges [13,22,23,33,40], an NP-hard problem. Furthermore, the use of specialized energetic [2,30] or weighting [32] criteria allows for the identification of the static parts of the reaction and, subsequently, of the atom mapping. A detailed investigation of the MCES from an Integer Linear Programming (ILP) perspective can be found in [6].

Akutsu [1] showed that the MCES approach fails for certain reactions. As an alternative, the Maximum Common Induced Subgraph (MCIS) problem was proposed as a remedy. This problem is also NP complete. Approximation results can be found in [27]. Algorithms for the MCIS iteratively decompose the molecules until only isomorphic sub-graphs remain [1,7,11,12]. Recently, an ILP approach incorporating stereochemistry was presented [16].

Neither the solutions of the MCES nor the MCIS necessarily describe the true atom map. Indeed, both optimality criteria are artificial and can not be derived from basic principles of chemical reactions. In fact, it is not hard to construct counter-examples, i.e., chemical reactions whose true atom maps are neither identified by MCES nor by MCIS. The re-organization of chemical bonds in a chemical reaction is far from arbitrary but follows strict rules that are codified e.g. in the theory of imaginary transition states (ITS) [17,24]. The ITS encodes the redistribution of bond electrons that occurs along a chemical reaction. Bond electrons define the atom-connecting chemical bonds and their according bond orders. Their redistribution is expressed in terms of the deletion or formation of bonds as well as changes of in the oxidation state of atoms, the latter resulting from non-bound electrons that are freed from or integrated into bonds. The ITS can be used to cluster, classify, and annotate chemical reactions [17,24,25]. These studies revealed, that only a limited number of ITS "layouts" are found among single step reactions and that these layouts represent a cyclic electron redistribution pattern usually involving less than 10 atoms [25]. In a most basic case, an elementary reaction, the broken and newly formed bonds form an alternating cycle (see Fig. 1) covering a limited even number of atoms [18], usually less than 8 [24]. In the case of homovalent reactions, i.e., those in which the number of non-bound electron pairs of all atoms (defining their oxidation state) remains unchanged, this cycle is elementary. That is, the transition state is a single, connected even cycle, along which bond orders change by ± 1 [25]. This property imposes an additional, strong condition of the atom maps that is not captured by the optimization approaches outlined in the previous paragraphs. Here, we explicitly include it into the specification of the combinatorial problem.

A *chemically correct* atom map is a bijective map between the vertices of the educt and product graphs such that:

1. The map preserves atom types
2. The total bond orders (including lone electron pairs) are preserved. Each broken bond thus must be compensated by a newly formed bond or a change in the oxidation number of an atom.
3. The broken and newly formed bonds constitute a chemically reasonable imtermediate transition state (ITS) following [25]. In the case of elementary chemical reactions, the transition state is an alternating cycle.

A formal definition of the combinatorial problem will be given in the following section. While cyclic transition states are very common, more "complex transition states" appear in non-elementary reactions, i.e., compositions of elementary reactions. Furthermore, even in elementary reaction, it is not true that the shortest ITS cycle is necessarily chemically correct. Empirically, transition states are most frequently six-membered cycles, while cycles of length 4 or 8 are less abundant [17,18,19,24]. As a consequence, we will consider several variants of the chemical reaction mapping problem:

1. **Decision Problem**: Is there an atom map with cyclic ITS? Of course one may restrict the question to ITS cycles of length k.
2. **Optimization Problem:** Find the minimal length k of an ITS cycle that enables an atom map.
3. **Enumeration Problem:** Find all atom maps with cyclic ITS (of length k).

Given a straightforward encoding of molecular graphs in terms of vertex indices, atom labels, and adjacency information, the atom mapping problem is naturally open to be treated as a constraint satisfaction problem with finite integer domains. This approach is particularly appealing when additional information on the ITS, e.g. its size or atoms involved in the ITS, are known.

2   Constraint Programming Formulation of the Atom Mapping Problem

We focus on the identification of the cyclic ITS. Once the ITS has been identified the overall atom mapping is easily derived. We formulate separate constraint satisfaction problems for different ITS layouts and cycle lengths. A fast graph matching approach is used subsequently to extend each ITS to a global atom mapping. In this section we follow closely [36]. We first formally define the problem, which is followed by a description of our constraint programming approach for identifying the cyclic ITS. Finally we discuss how to extend an ITS candidate to a complete atom mapping for the chemical reaction.

2.1   Problem Definition

Both educts and products of a chemical reaction are each represented by a single, not necessarily connected, undirected graph defined by a set of vertices V and a set of edges $E = \{ \{v, v'\} \mid v, v' \in V \}$. The educt (input) graph is denoted by $I = (V_I, E_I)$ and the product (output) graph by $O = (V_O, E_O)$. Here, each molecule corresponds to a connected component. Vertices represent atoms and are labeled with the respective atom type accessible via the function $l(v \in V_I \cup V_O)$. The principle of mass conservation implies $|V_I| = |V_O|$, i.e. no atom can dissolve or appear during a reaction. Edges encode covalent chemical bonds between atoms. For the CSP formulation we label each edge $\{x, y\} \in E_I \cup E_O$ with the number of shared electron pairs, i.e. its bond order: single, double or triple bonds are represented by a single edge with labels 1, 2, or 3, respectively. Non-bonding electron pairs of an atom, which define its oxidation state, are represented by loops labeled with the according number of unbound pairs.

We use an adjacency matrix $\mathcal{I}$ to encode the edge labels of the educt graph (and a corresponding matrix $\mathcal{O}$ for the products). The matrix elements $\mathcal{I}_{v,v'}$ denote the number of shared bond electron pairs for the edge between the atoms v and v' in the educt graph I. In practice $\mathcal{I}_{v,v'} \in \{0,1,2,3\}$. Non-bonding electron pairs (loops) are represented by the diagonal entries $\mathcal{I}_{v,v}$ and $\mathcal{O}_{v,v}$.

Consider a bijective function $m : V_I \to V_O$ mapping the vertices of I onto the vertices of O and a matrix $\mathcal{Q}$ with rows and columns indexed by V_I. Then $\mathcal{Q} \circ m$ is the matrix with entries $\mathcal{Q}_{m(x),m(y)}$, i.e. with rows and columns indexed by V_O. Thus the *reaction matrix* $\mathcal{R}^m = \mathcal{O} - (\mathcal{I} \circ m)$ is well defined and encodes the bond electron differences between educt and product.

Definition. An *atom mapping* is a bijective mapping $m : V_I \to V_O$ such that

1. $\forall_{x \in V_I} : l(x) = l(m(x))$ (preservation of atom types)
2. $\mathcal{R}^m \vec{1} = 0$ (preservation of bond electrons)

The reaction matrix $\mathcal{R}^m$ encodes the imaginary transition state (ITS) [17,24]. This definition of m is a slightly more formal version of the Dugundji-Ugi theory [14]. Our notation emphasizes the central role of the (not necessarily unique) bijection m. Since we consider I and O as given fixed input, the atom mapping m uniquely determines $\mathcal{R}^m$. The triple (m, I, O), furthermore, completely defines the chemical reaction. It therefore makes sense to associate properties of the chemical reaction directly with the atom map m.

Equivalently, the ITS can be represented as a graph $R = (V_R, E_R)$ so that E_R consists of the "changing" edges that lose or gain bond electrons during the reaction, i.e. $\mathcal{I}_{v,v'} \neq \mathcal{O}_{m(v),v(v')} \to \mathcal{R}^m_{v,v'} \neq 0$. The set of atom vertices $V_R \subseteq V_O$ covers all vertices with at least one adjacent edge in E_R. Each edge $\{v, v'\} \in E_R$ is labeled by the electron change $\mathcal{R}^m_{v,v'} \neq 0$, i.e. its change in bond order. See Fig. 2 for an example.

$\mathcal{I}$	v_1	v_2	v_3	v_4	v_5	v_6	v_7	v_8
v_1	0	1	0	0	0	0	0	0
v_2	1	0	1	2	0	0	0	0
v_3	0	1	0	0	2	0	0	0
v_4	0	2	0	0	0	0	0	0
v_5	0	0	2	0	0	0	0	0
v_6	0	0	0	0	0	0	2	1
v_7	0	0	0	0	0	2	0	0
v_8	0	0	0	0	0	1	0	0

$\mathcal{O}$	v'_1	v'_2	v'_3	v'_4	v'_5	v'_6	v'_7	v'_8
v'_1	0	1	0	0	0	0	0	0
v'_2	1	0	2	1	0	0	0	0
v'_3	0	2	0	0	1	0	0	0
v'_4	0	1	0	0	0	1	0	0
v'_5	0	0	1	0	0	0	1	0
v'_6	0	0	0	1	0	0	1	1
v'_7	0	0	0	0	1	1	0	0
v'_8	0	0	0	0	0	1	0	0

$\mathcal{R}^m$	v'_1	v'_2	v'_3	v'_4	v'_5	v'_6	v'_7	v'_8
v'_1	0	0	0	0	0	0	0	0
v'_2	0	0	+1	-1	0	0	0	0
v'_3	0	+1	0	0	-1	0	0	0
v'_4	0	-1	0	0	0	+1	0	0
v'_5	0	0	-1	0	0	0	+1	0
v'_6	0	0	0	+1	0	0	-1	0
v'_7	0	0	0	0	+1	-1	0	0
v'_8	0	0	0	0	0	0	0	0

Fig. 2. Adjacency matrices $\mathcal{I}$ for the reaction given in Fig. 1. The vertices $v_i \in V_I$ and $v'_j \in V_O$ are numbered in top-down-left-right order of their appearance in Fig. 1. The atom mapping $m(v_i) = v'_i$ defines $\mathcal{R}^m$ and thus the ITS graph R covers only vertices v'_2 to v'_7 since v'_1 and v'_8 do not show any bond electron changes.

It is important to note that the existence of an atom mapping m as defined above does not necessarily imply that $\mathcal{R}^m$ is a chemically plausible ITS.

We say that two edges $\{v, v'\}, \{v', v''\} \in E_R$ in R are *alternating* if $\mathcal{R}^m_{v,v'} \neq 0$ and $\mathcal{R}^m_{v,v'} + \mathcal{R}^m_{v',v''} = 0$. A *simple cycle* in R of size $k > 2$ is given by the vertex sequence $(v_1, v_2, \ldots, v_k, v_1)$ with $v_i \in V_R$, $\{v_i, v_{i+1}\} \in E_R$, $\{v_k, v_1\} \in E_R$, and $\forall i < j \leq k : v_i \neq v_j$. Such a simple cycle is called alternating if all successive edges as well as the cycle closure $\{v_2, v_1\}, \{v_1, v_k\}$ are alternating.

Definition. An atom map m is *homovalent* if $\mathcal{R}^m_{v,v} = 0$ for all $v \in V_R$. A homovalent reaction is *elementary* if its ITS R is a simple alternating cycle. Thus $\mathcal{R}^m_{v,v'} \in \{-1, 0, +1\}$ holds for all elementary homovalent reactions.

In the following we outline a novel algorithm for finding atom maps for a given ITS graph R that is guaranteed to retrieve all possible mappings given the educt and product graphs $\mathcal{I}$ and $\mathcal{O}$, respectively. To simplify the presentation, first only elementary homovalent reactions are considered. Generalizations are discussed in Sec. 3.

2.2 Constraint Programming Approach

The central problem to find an elementary homovalent atom mapping is to identify the alternating cycle defining the ITS R given the adjacency information of the educts $\mathcal{I}$ and products $\mathcal{O}$. This can be done via solving the Constraint Satisfaction Problem (CSP) as presented below. Note, due to the alternating edge condition within the ITS, we have to consider cycles with an even number of atoms only. In practice, the ITS of elementary homovalent reactions involves $|V_R| = 4$, 6, or 8 atoms [18].

Basic CSP Formulation: In the following, we will present a first basic CSP for an ITS of size $k = |V_R|$ that we already introduced in [36]. It is given by the triple (X, D, C) defining the set of variables X, according domains D_i, and the set of constraints C to be fulfilled by any solution.

We construct an explicit encoding of the ITS atom mapping using k variables representing the cycle in I and another set for the mapped vertices in O, i.e., $X = \{X^I_1, \ldots, X^I_k\} \cup \{X^O_1, \ldots, X^O_k\}$ with domains $D^I_i = V_I$ and $D^O_i = V_O$. Note, we do *not* directly encode the overall atom mapping problem but the identification of the two ITS subgraphs in the educts and products. Given this information, the overall atom mapping is easily identified as explained later.

To find a bijective mapping we have to ensure $\forall i \neq j : X^I_i \neq X^I_j$ and $\forall i \neq j : X^O_i \neq X^O_j$, i.e., a distinct assignment of all variables. To enforce atom label preservation we need arc consistency for $l(X^I_i) = l(X^O_i)$, i.e. we have to enforce $\forall e \in D^I_i : \exists p \in D^O_i : l(e) = l(p)$ as well as $\forall p \in D^O_i : \exists e \in D^I_i : l(p) = l(e)$. Analogously, homovalence is represented by $(\mathcal{I}_{X^I_i, X^I_i} - \mathcal{O}_{X^O_i, X^O_i}) = 0$. Due to the alternating bond condition, each atom can lose or gain at most one edge during a reaction. Thus, we can further constrain the variables with $|\text{degree}(X^I_i) - \text{degree}(X^O_i)| \leq 1$; where $\text{degree}(v)$ gives the out-degree of vertex v.

Finally, we have to encode the alternating cycle structure of the ITS in the mapping, i.e., for the sequence of bonds with indices 1-2-..-k-1. For all index pairs within the cycle (i, j) we therefore require pairs with even index i to correspond

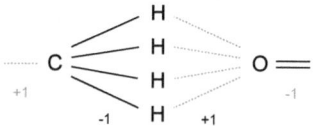

Fig. 3. Symmetries resulting from interchangeable hydrogens. The figure presents three successive atom assignments within an ITS mapping. Bonds present in I are given in black, bonds to be formed to derive O are dotted and gray. The ITS describes the loss of an hydrogen for the carbon (bond order decrease) and the bond formation between the decoupled hydrogen with the oxygen next in the ITS. It becomes clear that all 4 hydrogens are not distinguishable, which results in 4 possible symmetric ITS mappings.

the formation of a bond, i.e., we enforce $(\mathcal{O}_{X_i^O,X_j^O} \quad \mathcal{I}_{X_i^I,X_j^I}) = 1$, while all odd indices i are bond breaking $(\mathcal{O}_{X_i^O,X_j^O} - \mathcal{I}_{X_i^I,X_j^I}) = -1$ accordingly.

The homovalent ITS layout is rotation symmetric in itself (see Fig. 6). To partially counter this, we introduce order constraints on the input variables: $(\forall i > 1 : X_1^I < X_i^I)$; where $X_i < X_j$ denotes $\exists (x,y) \in D_i \times D_j : x < y$ using e.g. an index order on the vertices. This ties the smallest cycle vertex to the first variable X_1^I and prevents the rotation-symmetric assignments of the input variables. Note, since we constrain the bond $(1,2)$ to be a bond breaking $(\mathcal{O}_{X_1^O,X_2^O} - \mathcal{I}_{X_1^I,X_2^I} = -1)$, the direction of the cycle is fixed and all direction symmetries are excluded as well.

As we will show in the evaluation (Sec. 3), the basic CSP will produce many ITS candidates that do not enable an atom mapping over the whole educt and product graphs. Therefore, we introduce an extended version of this CSP that incorporates further constraints derived from the input.

Extended CSP Formulation: Investigating the given educt and product graph, we can exclude a large set of symmetric solutions that arise due to an exchange of hydrogens. The latter can form at most one single bond to other atoms. Thus, if a hydrogen participates in the ITS, its adjacent atom will do as well (since the bond is to be broken in the ITS). Most adjacent atoms are non-hydrogens, like carbon atoms, that can have multiple adjacent hydrogens. Since there is exactly one bond breaking and formation for each ITS atom, only one such adjacent hydrogen will be part of the ITS. This results in a combinatorial explosion due to the symmetries of adjacent hydrogen atoms. An example is given in Fig. 3.

To break this type of symmetry, we select for each non-hydrogen one adjacent "master" hydrogen and remove all other sibling hydrogens from the domains, both for educt and product variables X^I and X^O, respectively.

Furthermore, we can extend and tune the CSP formulation by comparing the graph structure of educts and products. To this end, we generate the sets N_I and N_O of local neighborhoods of all atoms (vertices) for the educt and product graph, resp., given by

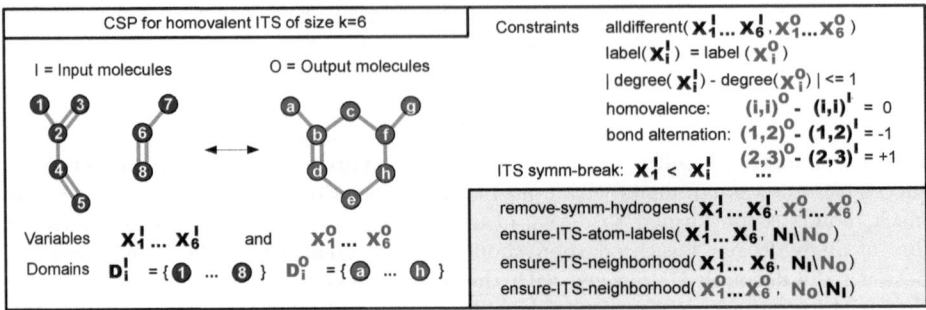

Fig. 4. Overview of the extended CSP for a homovalent ITS of size $k = 6$ where the extensions of the basic CSP are given in the gray box in the lower right

$$N_I = \{ N(v) \mid v \in V_I \} \text{ with} \tag{1}$$
$$N(v) = (l(v), \{ \mathcal{I}_{v,v'} \oplus l(v') \mid \text{where } v \neq v' \in V_I \wedge \mathcal{I}_{v,v'} > 0 \}) \tag{2}$$

where $N(v)$ is a tuple of the label of atom vertex v and an encoding of the set of all adjacent edges for this vertex. Note, $\oplus$ denotes string concatenation. N_O is derived accordingly. For example, the neighborhood sets for the reaction from Fig. 1 are

$$N_I = \{ 2 \times (\text{C}, \{1\text{C}\}), 3 \times (\text{C}, \{2\text{C}\}), 2 \times (\text{C}, \{1\text{C}, 2\text{C}\}), (\text{C}, \{1\text{C}, 1\text{C}, 2\text{C}\}) \}$$
$$N_O = \{ 2 \times (\text{C}, \{1\text{C}\}), 3 \times (\text{C}, \{1\text{C}, 1\text{C}\}), (\text{C}, \{1\text{C}, 2\text{C}\}), (\text{C}, \{1\text{C}, 1\text{C}, 1\text{C}\}),$$
$$(\text{C}, \{1\text{C}, 1\text{C}, 2\text{C}\}) \}$$

The subtraction $N_I \setminus N_O$ gives the local neighborhoods that are unique within the educts and thus are part of the ITS, i.e. have to be changed during the reaction. Therefore, we can derive a lower bound on the number of atoms of a certain type that are participating in the ITS. In the example this results in $N_I \setminus N_O = \{3 \times (\text{C}, \{2\text{C}\}), (\text{C}, \{1\text{C}, 2\text{C}\})\}$ revealing that at least 4 C-atoms of two types are ITS members.

Given this information, we formulate an extended version of the basic CSP. An arc-consistent global constraint on X^I is added, which enforces the occurrence of the identified ITS atom labels. This is automatically propagated on X^O via the atom label preservation constraints. In addition, we enforce that a valid assignment of the variables X^I and X^O preserves the ITS neighborhoods $N_I \setminus N_O$ and $N_O \setminus N_I$, respectively. To minimize propagation cost, this is ensured by a simple n-ary constraint propagating, which is propagated only after all variables have been confined to a single value. The full CSP is depicted in Fig. 4.

Although the CSPs from above are defined for domains of vertices $v \in V_I \cup V_O$, they can be easily reformulated using integer encodings of the atom indices allowing for the application of standard constraint solvers such as Gecode [42]. This enables the use of efficient propagators for most of the required constraints, such as the algorithm of Regin [41] for globally unique assignments. Only a few binary constraints, e.g. to ensure atom label preservation or the cyclic bond pattern, require a dedicated implementation as discussed in Sec 4.

All solutions for these CSPs are chemically valid ITS candidates. In order to check whether or not a true ITS is found we have to ensure that the remaining atoms, i.e., those that do not participate in the ITS, can be mapped without further bond formation or breaking. This is achieved using a standard graph matching approach as discussed in the following.

2.3   Overall Atom Mapping Computation

Given the CSP formulation from above, we can enumerate all valid ITS candidates. For a CSP solution we denote with a_i^I and a_i^O the assigned values of the variables X_i^I and X_i^O, respectively. Once the ITS candidate is fixed, we can reduce the problem to a general graph isomorphism problem with a simple relabeling of the ITS edges. Thus, we derive two new adjacency matrices $\mathcal{I}'$ and $\mathcal{O}'$ from the original matrices $\mathcal{I}$ and $\mathcal{O}$, resp., as follows: For all atom pairs (i, j) within the cyclic index sequence 1-2-..-k-1, we change the corresponding adjacency information to a unique label using $\mathcal{I}'_{a_i^I, a_j^I} = \mathcal{O}'_{a_i^O, a_j^O} \in \{f, b\}$ encoding if a bond between the mapped ITS vertices is formed (f) or broken (b). All other adjacency entries are kept the same as in $\mathcal{I}$ and $\mathcal{O}$, respectively.

Given these updated, "ITS encoding" adjacency matrices $\mathcal{I}'$ and $\mathcal{O}'$, the identification of the overall atom mapping m reduces to the graph isomorphism problem based on $\mathcal{I}'$ and $\mathcal{O}'$. Thus, all exact mappings of $\mathcal{I}'$ onto $\mathcal{O}'$ are valid atom mappings m of an elementary homovalent reaction, since the encoded ITS respects all constraints due to the CSP formulation.

2.4   Implementation Details

Our C++ implementation of the approach currently takes a chemical reaction in SMILES format [43], identifies chemically correct atom mappings, and returns these in annotated SMILES format. The latter provides a numbering of mapped atoms in the educts and products.

Molecule parsing, writing, and graph representation uses the chemistry module of the Graph Grammar Library (GGL) [35]. Note, we do an explicit hydrogen representation within the CSP formulation as in [16], since most homovalent elementary reactions involve the replacement of at least one hydrogen. Unfortunately, the compact string encoding of molecules in SMILES format does not explicitly represent hydrogens. Thus, we use the hydrogen correction procedures of the GGL to complete educt and product molecule input. The CSP formulation and solving is done within the Gecode framework on finite integer domains [42]. The final graph matching is done using the state-of-the-art VF2-algorithm [10], which is among the fastest available [9].

The CSP uses standard binary order constraints and the n-ary distinct and counting constraints provided by the Gecode library. Dedicated binary constraints propagating on unassigned domains have been implemented for preservation of atom label, degree, and homovalence. The alternating cycle is implemented by a sequence of k constraints propagating the edge valence change of ± 1. The ITS local neighborhood preservation to be enforced in the extended CSP is implemented by a dedicated n-ary constraint over all variables propagating on assignments only.

We are using a Depth-First-Search where the branching strategy chooses first variables with minimal domain size and first assigns non-hydrogen indices before hydrogen vertices are considered. The latter increases the performance to find the first solution since most reaction mechanism are constructed of at least 50% non-hydrogen atoms. Once a non-hydrogen is selected, propagation will ensure that adjacent hydrogens are considered for the neighbored variables within the ITS cycle encoding if appropriate.

For each ITS mapping identified, a full reaction atom mapping is derived via VF2-based graph matching. Therein, the discussed problem of hydrogen interchangeability (see extended CSP formulation) is faced again and would result in symmetric overall atom mappings. This is countered by first producing intermediate "collapsed" educt/product graphs, where all adjacent non-ITS hydrogens are merged into the atom labels of their adjacent non-hydrogens. This preserves the adjacency information and enables a unique mapping via VF2 excluding the hydrogen-symmetries. Furthermore, this compression speeds up the graph isomorphism identification since the graph size is approximately halved.

While not described here, the CSPs can be easily extended to find candidates for the entire atom mapping by introducing additional matching variables for all atoms participating in the reaction, all constrained to preserve atom label, vertex degree, and bond valence information. But first tests (not shown) revealed that the increase in CSP size and accordingly search and propagation effort needed does not repay due to the efficiency of the VF2 graph isomorphism approach used. Therefore, we omitted this approach from this work.

3   Application and Evaluation

In order to investigate the impact of our extended CSP formulation over the basic version, we selected a subset of homovalent elementary reactions from the KEGG LIGAND database [29]. The The reactions have been chosen to provide various ITS and reaction sizes for evaluation. The average size of the selected reactions, i.e. the average number of atoms, is about 30 (Tab. 2 column 2) while the whole KEGG database shows an average of 50 atoms per reaction. The example reactions cover homovalent ITS sizes of $k = 4$, 6, and 8 as introduced. Since there is no atom mapping information provided within the KEGG database, the example reactions had to be identified manually based on chemical knowledge. This again highlights the need for an automated identification of chemically feasible atom mappings as provided by our approach. The selected homovalent reactions are given in Tab. 1 with their respective KEGG ID, educts and products.

For each reaction, we applied our approach using both the basic and extended CSP formulation to evaluate the impact of the latter for various reaction and

Table 1. Elementary homovalent reactions from the KEGG LIGAND database [29] used for the evaluation of the approach. The educt and product molecules are given in SMILES notation [43].

Reaction	Educts	Products
R00013	C(=O)=O, C(C(=O)O)(C=O)O	2× C(=O)(C=O)O
R00018	N, N(CCCCN)CCCCN	2× C(CCN)CN
R00048	CC(O)CC(=O)OC(C)CC(O)=O, O	2× CC(O)CC(O)=O
R00059	N(C(=O)CCCCCN)CCCCCC(=O)O, O	2× C(CC(=O)O)CCCN
R00207	P(=O)(O)(O)O, O=O, CC(=O)C(=O)O	P(=O)(OC(=O)C)(O)O, OO, C(=O)=O

Table 2. Evaluation of the reactions from Tab. 1. Timings are given in seconds. For extended CSPs, the minimal set of ITS participating atoms is listed in column 3. Column "Sol. CSP" gives the number of CSP solutions (ITS candidates) tested via VF2 for final atom mappings.

Reaction	Atoms	CSP Type	k	Time 1st Sol.	Sol.	Sol. CSP	Time all Sol. CSP	VF2
R00013	14	Basic	6	0.03	1	346	0.8	0.03
		Ext. {2C}		**0.02**		76	**0.05**	0.02
R00018	36	Basic	4	10.4	1	73,924	2.62	19.9
		Ext. {2N}		**0.28**		36	**0.44**	0.01
R00048	30	Basic	4	0.1	2	26,178	1.44	6.1
		Ext. {2O}		**0.02**		24	**0.42**	0.03
R00059	44	Basic	4	0.34	1	194,210	9.45	63.15
		Ext. {H,C,N,O}		**0.03**		4	**2.08**	0.01
R00207	20	Basic	8	0.02	1	20,640	1.11	4.05
		Ext. {C,4O}		**0.01**		24	**0.56**	0.02

ITS cycle sizes. In Table 2 we report runtime, search, and solution details for the smallest ITS size k that yields a solution. For smaller values of k, the infeasibility tests were done within fractions of seconds and are therefore omitted.

Our atom mapping approach finds a first atom mapping for homovalent elementary reactions within milliseconds. It is clear that the additional constraints within the extended CSP formulation significantly increase the performance of the approach. This becomes even more striking when considering the timings for full solution enumeration. The extended CSP produces several orders of magnitude less ITS candidates (column "Sol. CSP"). Since the time consumption of the VF2 algorithm is about linear in the number of ITS candidates to test, this results in according speedups of the overall approach. Still there is room for optimization since the symmetry breaking within the CSP solution enumeration is not complete (see next section).

The strength of the extended CSP comes from the precomputed list of local neighborhoods to be part of the ITS candidate and the "hydrogen symmetry" breaking. For the reactions from Tab. 2, this list comprises on average about half the ITS resulting in the impressive impact of the constraint. For reaction

R00059, the list covers the whole ITS with an according immense reduction in ITS candidates.

As already expected based on the results from other approaches [16], only a single or very few reaction mechanisms, i.e. non-symmetric atom mappings, are identifiable, see Tab. 1 column "Sol".

4   Development and Future Work

The basic approach was implemented by a user not familiar with constraint programming within 1 month work time given the well documented and easily extendable Gecode library [42] and the chemoinformatics implementations provided by the GGL [35]. Extending the approach and adding the basic functionalities for symmetry exclusion required another week of implementation, such that we got a first prototype within 1.5 months. Given the current framework and available constraint implementations, we expect another month of implementation time to get the final atom mapping program that will cover most of the following features.

Branching strategies: The current CSP allows for further performance optimizations when solving the satisfaction problem. We are currently evaluating the impact of different branching strategies on the runtime of the approach. As a first result, a hierarchical value selection that first tries to assign vertices to the variables that are compatible to the neighborhoods participating in the ITS (see extended CSP formulation) and which selects hydrogen representing vertices last seems to allow for a good performance.

Symmetry breaking: As it can be seen from Tab. 2, the current CSP formulation still produces symmetric ITS solutions when enumerating all possible atom mappings. We are currently working on strategies to apply further symmetry breaking techniques during the solution enumeration of CSPs, i.e. symmetry breaking during search (SBDS) [4,21,5] (or the similar lightweighted dynamics symmetry breaking (LDSB) approach [37]), as well as symmetry exclusion in the final mapping phase. Both requires more sophisticated input analyses as e.g. done in [16].

CSPs for other ITS layouts: Of course, not all chemical transformations are based on a homovalent elementary ITS. This will in general be the case for multi-step reactions and for the so-called ambivalent reactions, in which the number of non-bonding electron pairs (and thus the oxidation number of some atoms) changes in the course of a reaction [25]. Figure 5, for example, shows a reaction for which it is not possible to find a simple homovalent circular ITS using the presented ITS encoding. Still the reaction shows a cyclic ITS with alternating bond electron changes for all but one bond [17].

We have extended the CSP-based framework outlined above to reactions with arbitrary cyclic ITS layouts, which allows for any defined bond and atom valence

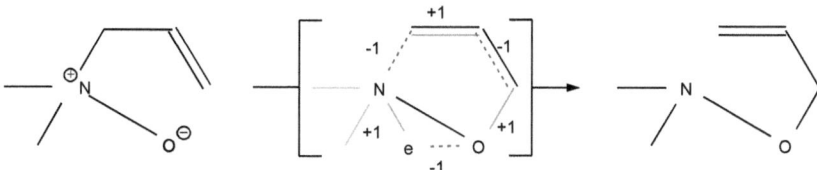

Fig. 5. The Meisenheimer rearrangement [38] transforms nitroxides to hydroxylamines. It does not admit a simple alternating cycle as ITS when molecules are represented as graphs whose vertices are atoms. An extended representation, in which the additional electron at the oxygen is treated a "pseudo-atom" can fix this issue. See Figure 6 for further details of such an ITS layout.

Table 3. Evaluation of CSPs for ambivalent reactions with an odd cycle ITS layout given in Fig. 6

Reaction		Atoms	k	Fig.	Sol.	Sol. CSP	Time all Sol.
Educts	Products						
O=[S--]=O.C=CC=C	O=S1(=O)CC=CC1	13	5	6 bottom	1	20	0.01
Cl[C--]Cl.C=C	ClC1(Cl)CC1	9	3	6 bottom	1	12	0
[O-][NH2+]CC=C	NOCC=C	12	5	5, 6 top	1	22	0

changes (i.e. charge changes) within the ITS. Figure 6 exemplifies odd ITS cycle layouts for ambivalent reactions [19]. The main difference to homovalent reaction CSP is the relaxation of the homovalence constraint, which is not enforced for all participating atoms [19]. Furthermore, the preservation of bond electrons for some ITS bonds instead of a change is enforced. The latter holds for instance for the bond connecting N^+ and O^- in Fig. 5.

Table 3 presents the timing results for our prototypical implementation of the ambivalent ITS layouts given in Fig. 6. The model is based on the extended CSP formulation for elementary reactions. Also for such ambivalent reactions, our CP-based atom mapping approach enumerates all possible atom mappings within milliseconds, as reported for homovalent reactions in Tab. 2. Note, the ambivalent CSPs require a different, ITS-specific symmetry breaking and thus have to enforce different static order constraints compared to the homovalent CSP. The ambivalent layouts given in Fig. 6 show no symmetry in itself such that actually no order constraint is needed here. The driving force of the CSP performance is the propagation of the oxidation state change for the atoms that get charged. This poses a very strong constraint for the ambivalent ITS identification.

We are currently identifying and verifying further ITS layouts, some of the already available layouts are given in Fig. 6. Considering the reaction classification work in [25,24,17,18], we expect a very limited number of possible ITS layouts within a few hundreds at most given the physics underlying chemical reactions. The overall approach will select, based on the provided input and the

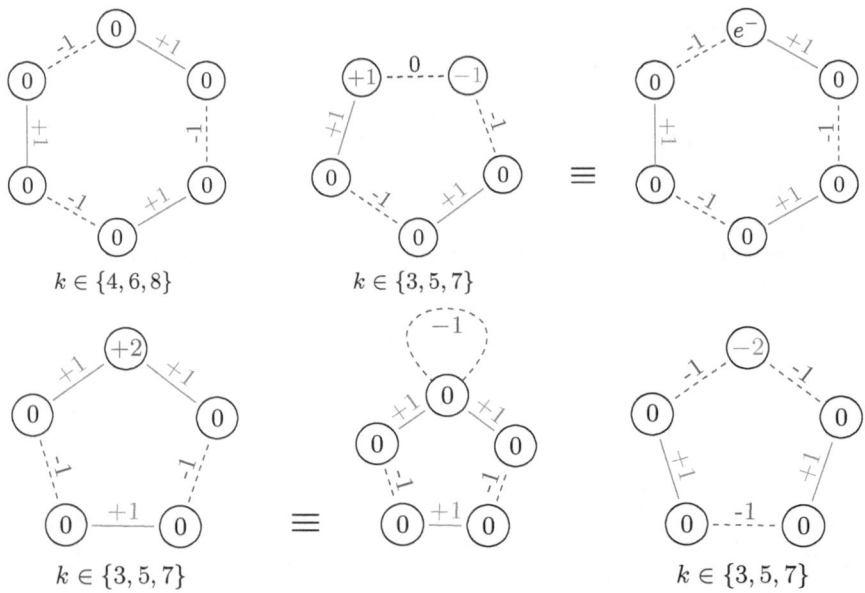

Fig. 6. Currently supported ITS layouts: The number within the vertices corresponds to atomic oxidation state changes, broken bonds are dotted given a negative bond label while formed bonds show positive numbers. (top) Homovalent elementary reactions result in even sized cycles with no oxidation state changes at the atoms (see Fig. 1). Note that odd cycles with two oppositely charged atoms separated by a non-changing pseudo bond (dashed edge labeled 0 see Fig. 5) are equivalent to the next larger even sized cycle with a virtual vertex for the moving charge (vertex label e^-). (bottom) Ambivalent elementary reactions involving non bonding electrons result in odd sized cycles and oxidation state changes of one atom. Note that this situation is equivalent to a non-elementary cycle with alternating bond labeling (bottom middle).

local neighborhood analyses presented for the extended CSP, the suitable ITS layouts and their respective CSPs and search for valid atom mappings.

Multi-step reactions: The current framework is designed to identify chemically feasible atom mappings for single-step reactions. Nevertheless, there cases where short-lived intermediate molecule structures are formed that are directly react further into the final products. Unfortunately, these intermediate structures are usually unknown, such that we cannot apply the presented approach.

As discussed by Hendriksen [24], often only two joint reactions with a single unknown intermediate are observed. We therefore plan to create "fused" ITS layouts based on our single-step ITS encodings that will allow for the correct identification of atom mappings for multi-step reactions and reveal the individual steps and intermediate structures. For the combination of ITS layouts, we are currently investigating the multi-step reaction analyses by Fujita [20] and Herges [25].

Webserver: The final atom mapping framework will be available both as stand alone tool as well as via a web front end including a visual depiction of the atom mappings. An according webserver framework ready for the integration is already available.

Graph Grammars and Atom Flow Network Generation: Atom mappings are the base to generate and analyze the atom flow in reaction networks [7,26]. Here, the chemical validity of the atom maps is of particular importance to ensure correct atom flow analyses. We will use our atom mapping approach to generate chemical graph rewrite rules that will be used within our GGL framework [35] to expand according reaction networks where molecular graph rewrite directly provides the atom flow information within the network.

5   Discussion

We have presented here the first constraint programming approach to identify chemically feasible atom mappings based on the identification of a cyclic inter-mediate transition state (ITS). The incorporation of the cyclic ITS structure within the search ensures the chemical correctness of the mapping that is not guaranteed by standard approaches that attempt to solve Maximum Common Edge Subgraph Problems [1]. To our knowledge, this is the first approach ex-plicitly incorporating the cyclic ITS structure into an atom mapping procedure.

The formulation of the CSP using only the atoms involved in the ITS results in a very small CSP that can be solved efficiently. Thus, it is well placed as a filter for ITS candidates for the subsequent, computationally more expensive graph matching approaches. The solutions of such an extended CSP are the desired chemically feasible atom mappings m. We apply advanced symmetry breaking strategies and thus can enumerate the different chemical mechanisms underlying a reaction for a given ITS cycle size.

The feasibility of the approach was demonstrated here for the special case of elementary, homovalent reactions, i.e., for reactions in which the transition state is an elementary cycle with an even number of atoms. The CSP formulation can be easily extended to arbitrary cyclic ITS layouts. Usually, such reactions are not homovalent, i.e., at least one atom participating in the ITS is gaining or losing non-bonding electrons, which requires some moderate changes in the formulation of the constraints. We are currently identifying all feasible ITS layouts and are developing a generic CSP formulations. This will result in a powerful approach to identify atom mappings with chemically valid ITSs.

Constraint programming was shown to be a very promising approach to solve atom mapping problems since it provides a very flexible framework to incorpo-rate combinatorial constraints determined by the underlying rules of chemical transformations.

References

1. Akutsu, T.: Efficient extraction of mapping rules of atoms from enzymatic reaction data. J. Comp. Biol. 11, 449–462 (2004)
2. Apostolakis, J., Sacher, O., Körner, R., Gasteiger, J.: Automatic determination of reaction mappings and reaction center information. 2. Validation on a biochemical reaction database. J. Chem. Inf. Mod. 48, 1190–1198 (2008)
3. Arita, M.: Scale-freeness and biological networks. J. Biochem 138, 1–4 (2005)
4. Backofen, R., Will, S.: Excluding symmetries in constraint-based search. In: Jaffar, J. (ed.) CP 1999. LNCS, vol. 1713, pp. 73–87. Springer, Heidelberg (1999)
5. Backofen, R., Will, S.: Excluding symmetries in constraint-based search. Constraints 7(3), 333–349 (2002)
6. Bahiense, L., Manić, G., Piva, B., de Souza, C.C.: The maximum common edge subgraph problem: A polyhedral investigation. Discr. Appl. Math. 160, 2523–2541 (2012)
7. Blum, T., Kohlbacher, O.: Using atom mapping rules for an improved detection of relevant routes in weighted metabolic networks. Journal of Computational Biology 15, 565–576 (2008)
8. Chen, W.L., Chen, D.Z., Taylor, K.T.: Automatic reaction mapping and reaction center detection. WIREs Comput. Mol. Sci. (2013)
9. Cordella, L.P., Foggia, P., Sansone, C., Vento, M.: Performance evaluation of the VF graph matching algorithm. In: Proceedings of the 10th International Conference on Image Analysis and Processing, ICIAP 1999, p. 1172. IEEE Computer Society Press (1999)
10. Cordella, L.P., Foggia, P., Sansone, C., Vento, M.: A (sub)graph isomorphism algorithm for matching large graphs. IEEE Transactions on Pattern Analysis and Machine Intelligence 26(10), 1367–1372 (2004)
11. Crabtree, J.D., Mehta, D.P.: Automated reaction mapping. J. Exp. Algor. 13, 1.15–1.29 (2009)
12. Crabtree, J.D., Mehta, D.P., Kouri, T.M.: An open-source Java platform for automated reaction mapping. J. Chem. Inf. Model. 50, 1751–1756 (2010)
13. de Groot, M.J.L., van Berlo, R.J.P., van Winden, W.A., Verheijen, P.J.T., Reinders, M.J.T., de Ridder, D.: Metabolite and reaction inference based on enzyme specificities. Bioinformatics 25(22), 83–2975 (2009)
14. Dugundji, J., Ugi, I.: An algebraic model of constitutional chemistry as a basis for chemical computer programs. Topics Cur. Chem. 39, 19–64 (1973)
15. Ehrlich, H.-C., Rarey, M.: Maximum common subgraph isomorphism algorithms and their applications in molecular science: a review. WIREs Comput. Mol. Sci. (2011)
16. First, E.L., Gounaris, C.E., Floudas, C.A.: Stereochemically consistent reaction mapping and identification of multiple reaction mechanisms through integer linear optimization. J. Chem. Inf. Model. 52(1), 84–92 (2012)
17. Fujita, S.: Description of organic reactions based on imaginary transition structures. 1. Introduction of new concepts. J. Chem. Inf. Comput. Sci. 26, 205–212 (1986)
18. Fujita, S.: Description of organic reactions based on imaginary transition structures. 2. Classification of one-string reactions having an even-membered cyclic reaction graph. J. Chem. Inf. Comput. Sci. 26, 212–223 (1986)
19. Fujita, S.: Description of organic reactions based on imaginary transition structures. 3. Classification of one-string reactions having an odd-membered cyclic reaction graph. J. Chem. Inf. Comput. Sci. 26, 224–230 (1986)

20. Fujita, S.: Description of organic reactions based on imaginary transition structures. 5. Recombination of reaction strings in a synthesis space and its application to the description of synthetic pathways. J. Chem. Inf. Comput. Sci. 26, 238–242 (1986)
21. Gent, I.P., Smith, B.M.: Symmetry breaking in constraint programming. In: Proceedings of ECAI-2000, pp. 599–603. IOS Press (2000)
22. Hattori, M., Okuno, Y., Goto, S., Kanehisa, M.: Heuristics for chemical compound matching. Genome Informatics 14, 144–153 (2003)
23. Heinonen, M., Lappalainen, S., Mielikäinen, T., Rousu, J.: Computing atom mappings for biochemical reactions without subgraph isomorphism. J. Comp. Biol. 18, 43–58 (2011)
24. Hendrickson, J.B.: Comprehensive system for classification and nomenclature of organic reactions. J. Chem. Inf. Comput. Sci. 37, 852–860 (1997)
25. Herges, R.: Organizing principle of complex reactions and theory of coarctate transition states. Angewandte Chemie Int. Ed. 33, 255–276 (1994)
26. Hogiri, T., Furusawaa, C., Shinfukua, Y., Onoa, N., Shimizua, H.: Analysis of metabolic network based on conservation of molecular structure. Biosystems 95(3), 175–178 (2009)
27. Huang, X., Lai, J., Jennings, S.F.: Maximum common subgraph: some upper bound and lower bound results. BMC Bioinformatics 7 (S4), S6 (2006)
28. Jochum, C., Gasteiger, J., Ugi, I.: The principle of minimum chemical distance (PMCD). Angew. Chem. Int. Ed. 19, 495–505 (1980)
29. Kanehisa, M., Goto, S., Sato, Y., Furumichi, M., Tanabe, M.: KEGG for integration and interpretation of large-scale molecular data sets. Nuc. Acids Res. 40(Database issue), D109–D114 (2012)
30. Körner, R., Apostolakis, J.: Automatic determination of reaction mappings and reaction center information. 1. The imaginary transition state energy approach. J. Chem. Inf. Mod. 48, 1181–1189 (2008)
31. Kotera, M., Okuno, Y., Hattori, M., Goto, S., Kanehisa, M.: Computational assignment of the EC numbers for genomic-scale analysis of enzymatic reactions. J. Am. Chem. Soc. 126, 16487–16498 (2004)
32. Latendresse, M., Malerich, J.P., Travers, M., Karp, P.D.: Accurate atom-mapping computation for biochemical reactions. J. Chem. Inf. Model (2013)
33. Leber, M., Egelhofer, V., Schomburg, I., Schomburg, D.: Automatic assignment of reaction operators to enzymatic reactions. Bioinformatics 25, 3135–3142 (2009)
34. Lynch, M., Willett, P.: The automatic detection of chemical reaction sites. Journal of Chemical Information and Computer Sciences 18, 154–159 (1978)
35. Mann, M., Ekker, H., Flamm, C.: The graph grammar library - A generic framework for chemical graph rewrite systems. In: Duddy, K., Kappel, G. (eds.) ICMB 2013. LNCS, vol. 7909, pp. 52–53. Springer, Heidelberg (2013), http://arxiv.org/abs/1304.1356
36. Mann, M., Ekker, H., Stadler, P.F., Flamm, C.: Atom mapping with constraint programming. In: Backofen, R., Will, S. (eds.) Proceedings of the Workshop on Constraint Based Methods for Bioinformatics WCB12, Freiburg, pp. 23–29. Uni Freiburg (2012), http://www.bioinf.uni-freiburg.de/Events/WCB12/proceedings.pdf
37. Mears, C.D., Garcia De La Banda, M.J., Demoen, B., Wallace, M.: Lightweight dynamic symmetry breaking. In: Proc. of the 8th International Workshop on Symmetry in CSPs, SymCon 2008 (2008), http://www.aloul.net/symcon/
38. Meisenheimer, J.: Über eine eigenartige Umlagerung des Methyl-allyl-anilin-N-oxyds. Chemische Berichte 52, 1667–1677 (1919)

39. Rautio, J., Kumpulainen, H., Heimbach, T., Oliyai, R., Oh, D., Järvinen, T., Savolainen, J.: Prodrugs: design and clinical applications. Nature Reviews Drug Discovery 7(3), 255–270 (2008)
40. Raymond, J.W., Willett, P.: Maximum common subgraph isomorphism algorithms for the matching of chemical structures. J. Computer-Aided Mol. Design 16, 33–521 (2002)
41. Regin, J.-C.: A filtering algorithm for constraints of difference. In: Proceedings of the 12th National Conference of the American Association for Artificial Intelligence, pp. 362–367 (1994)
42. Gecode Team. Gecode: Generic constraint development environment (2013), `http://www.gecode.org`, Available as an open-source library, from
43. Weininger, D.: SMILES, a chemical language and information system. 1. Introduction to methodology and encoding rules. J. Chem. Inf. Comp. Sci. 28(1), 31–36 (1988)
44. Wiechert, W.: C^{13} metabolic flux analysis. Meta. Eng. 3, 195–206 (2001)
45. Yamanishi, Y., Hattori, M., Kotera, M., Goto, S., Kanehisa, M.: E-zyme: predicting potential EC numbers from the chemical transformation pattern of substrate-product pairs. Bioinformatics 25(12), i179–i186 (2009)

Beyond Feasibility: CP Usage in Constrained-Random Functional Hardware Verification

Reuven Naveh and Amit Metodi

Cadence Design Systems, Israel
{rnaveh,ametodi}@cadence.com

Abstract. Constraint programming (CP) figures prominently in the process of functional hardware verification. The verification process is based on generating random tests according to given set of constraints. In this paper. we introduce INTELLIGEN, a propagation based solver, and the random generator of Cadence's Specman verification tool. INTELLIGEN is designed to handle several problems beyond the mere need to find a feasible solution, including: generating random tests with a 'good' distribution over the solution space; maintaining test reproducibility through different run modes and minor code changes; and debug of the solving process by verification engineers. We discuss the advantages of CP solvers over other solving technologies (such as BDD, SAT or SMT), and how INTELLIGEN overcomes the disadvantages of CP.

1 Introduction

Constraint programming (CP) is a major component in functional verification. Functional verification, tests a hardware device using a simulation of the design-under-test (DUT) behavior. Verification is performed by generating diverse random stimuli to produce interesting test scenarios, and collecting coverage to ensure that all aspects of the DUT have been tested. Constraint solvers are used to produce valid test cases that will satisfy the DUT's restrictions [6,5].

Typically, CP seeks a feasible solution as fast as possible. Functional verification changes this focus. Constraint problems in functional verification are not necessarily hard to solve, and often many solutions exist. Instead of seeking a single suitable solution, it is crucial to solve the problem many times with different solutions. Moreover, the random solutions should be well distributed over the (often enormous) solution space, and fulfill user-defined coverage requirements [7].

Several solving technologies are available to handle the random-constrained problems of verification environments. These include CP solvers, SAT solvers, and BDD solvers, as well as hybrid methods. BDD solvers are especially efficient in finding a solution with uniform distribution over the solution space [12]. CP solvers utilize the generation of a random value from a variable's domain and the randomization of variable selection ordering, to control the random selection [3]. SAT solvers are usually used to find one solution, but there are techniques that are used to generate random solutions such as XORSample [2].

C. Schulte (Ed.): CP 2013, LNCS 8124, pp. 823–831, 2013.
© Springer-Verlag Berlin Heidelberg 2013

An inherent part of the verification process is debugging. When generation leads to a conflict, or results do not satisfy the DUT expectations, the functional verification engineer must debug the constraint solving. This requires the solving process to be random stable: the generation must be reproducible. Typically, the functional verification engineer is not an expert in constraints and does not have actual knowledge in the underlying algorithms used by the solver. Hence, the debug tools should be designed and the solver chosen, with this in mind.

This paper focuses on two of the special needs in the domain of functional verification: distribution of the solutions, and random stability, and shows the advantages of using a CP solver instead of other solving techniques to address these (and other) needs. It also describes the disadvantages of a CP solver, and focuses on how INTELLIGEN, a Cadence-supplied constraint solver and random generator used by the Specman tool, overcomes those disadvantages.

2 INTELLIGEN

In the domain of functional verification, test scenarios are generated from user environments using constraint solvers. Cadence is an Electronic Design Automation (EDA) company which, among other things, supplies tools for functional verification. One of these tools, Specman, provides a framework and environment for the verification language e [10]. INTELLIGEN, a Cadence-supplied constraint solver and random generator used by the Specman tool, is a powerful solver that can handle a variety of constraints, including arithmetical, bit operations, and several global constraints (sum, count, all-different, etc.).

The user environments generated by INTELLIGEN may be quite large and include millions of lines of code, and tens of thousands of constraints and variables. The constraints may be on scalar fields, but can also be used to bind pointers or to decide which subtypes will be generated.

Because of the inter-dependencies inherent in the structure of the environment, it cannot be solved as a single-constraint problem. Therefore, INTELLIGEN analyzes the entire environment, and separates it into isolated solving problems, denoted as CFSs (Connected Field Sets). Dependencies between CFSs are computed to determine the correct order of solving, and for each CFS, a proper solving device is chosen, depending on the nature of the problem.

Test scenarios may contain thousands of CFSs, and may require that some of the CFSs be solved many times while others may be solved only once. The solving device therefore should ideally be reusable and have a relatively low build cost. In meeting this need, CP solvers have an important advantage over other solving techniques such as BDD or SAT.

INTELLIGEN also includes GenDebugger, a GUI debugging tool used to debug constraints [1,8]. GenDebugger is intended for use by the verification engineer, typically a software or hardware engineer with knowledge of procedural code debugging. GenDebugger allows debugging of constraint conflicts, unexpected generated results, or performance and distribution problems. It groups the debug information for each CFS, and represents the process of constraint solving as a set of relatively small steps. The steps present information about variables domains (before and after the step), and the relevant constraints that participated in the step. GenDebugger also includes breakpoints that let the user stop at specific points during the generation process.

Using a CP solver is particularly helpful for debug because it facilitates translation of the solver's main operations (such as "value-selection," "propagation," and "backtrack") into a sequence of elementary solving steps that can be described in terms known to the user ("randomization," "reduction," and "value cancellation" respectively). Because each solving step is directly related to specific constraints, the user can be informed which constraints led to which change.

Till now, we briefly discussed INTELLIGEN and some of the reasons (reuse, relative low-cost build, and easy operation translation for the users) for choosing a CP solver. In the next two sections, we describe two additional aspects (Distribution and Random Stability) of finding a good random solution, the advantage of a CP solver for handling these aspects, and how INTELLIGEN addresses them.

3   Distribution

The fundamental practice in functional verification is validating the DUT through many randomly generated tests, with a goal of achieving full coverage [7]. To achieve this goal, the tests should be varied, and the randomly generated values should be well distributed over the solution space.

A natural question is "What constitutes 'good distribution'?", and an intuitive answer might be uniform distribution over the solution space, an approach embedded in the SystemVerilog IEEE standard [9]. The concept of uniform distribution is clarified in [6]. There, the authors call for the tests to be distributed as uniformly as possible among all possible tests that conform to the constraint model, the intent being to reach a significantly different solution each time the same constraint model is solved. Thus, uniform distribution of all possible tests is not required, because two different solutions might still create very similar tests. Much more important than uniformly sampling the solution space is sampling all 'interesting' values of the different variables.

An important consideration is that generation problems in verification are usually asymmetric: a typical problem includes 'flags', which control the topology and conditions of the problem, and also as including generated scalar variables, whose domains may vary from very small to very large (e.g., all values of 32-bit unsigned integer). In these cases, the uniform distribution approach is hazardous because some values of the topology variables or small domain variables may not be generated because of their low probability. However, when the solution space is symmetric and comprised of variables with similar initial domains, a uniform or close to uniform distribution is the correct solution.

Consider Example 1, which demonstrates a case where an 'interesting' value for one of the variables will never be generated in a realistic test scenario.

Example 1. Assume two variables x and y with the domains $\{1..4\}$ and $\{1..2^{32}\}$ respectively, and a single constraint $(x = 1 \rightarrow y = 2)$.

The solution space of this problem is $\{ (1,2) \} \cup \{ (a,b) \,|\, a \in \{ 2..4 \}, b \in \{ 1..2^{32} \} \}$. When uniformly picking a random pair from this space, the probability for generating the value one for x is $1/(3 * 2^{32} + 1)$. In other words, an 'interesting' value for one of the variables will likely never be generated in a realistic test scenario. This will lead to unfilled areas in the user's coverage.

While unfeasible to generate all values of all variables, at the very least all four values for variable x should be generated. This will lead to non-uniform distribution of the values of y (which are 2 whenever x is 1). The user's expectations can thus be summarized as follows: For variables with a relatively small domain, all possible values are generated, while the distribution should be as uniform as possible for the rest of the model, as long as permitted by the constraint model.

Verification languages, such as e or SystemVerilog, include directives to control the distribution of specific fields, but even when no such directives are given, it is preferred that the generated values will vary and be distributed well. In this paper we do not discuss user directives, but only how generation results are naturally distributed depending on the solver technology.

In INTELLIGEN, a propagation-based solver, the distribution is the result of variable and value ordering. INTELLIGEN's variable ordering primarily picks a variable with the smallest domain (according to the first-fail principle of Haralick & Elliott [4]). If several variables have the same (or close enough) domain size, the solver chooses the variable randomly. Once a variable is chosen, the generated value is randomized uniformly from its domain. Though intended for finding a solution more quickly, variable ordering based on the first-fail approach also often leads to the desired distribution.

Finding a satisfying assignment for Example 1 when using INTELLIGEN's variable and value ordering is done in the following solving steps: (1) Choose the variable x (due to its smaller domain), and randomize a value from its domain (which is $\{ 1..4 \}$). (2) Propagate (reducing y to $\{ 2 \}$ if x was randomized to 1). (3) If y was not reduced to $\{ 2 \}$, choose the variable y, and randomize a value from its domain.

This flow (which we term 'randomization-propagation' flow) does not lead to a uniform distribution on the solution space, but it does ensure full coverage on the values of x (which are generated with equal proportions). As for the values of y, they will be varied enough once x was determined. This leads to the conclusion that CP heuristics, intended mainly for finding a solution efficiently, also contribute to achieving good, albeit not uniform, distribution.

As an enhancement to this behavior consider Example 2 which was extracted from a real life verification environment.

Example 2. Assume four variables *delay*, *kind*, *size*, and *lock* with the domains $\{ 0..2^{32} \}$, $\{ SINGLE, INCR, WRAP \}$, $\{ BYTE, HALFWORD, WORD \}$, and $\{ true, false \}$ respectively, and a set of constraints:

$(kind = WRAP \rightarrow lock = true)$ $(size = WORD \rightarrow kind = WRAP)$
$(kind \in \{ SINGLE, INCR \} \rightarrow lock = false)$ $(lock = true \rightarrow delay = 0)$

Experimental results of using INTELLIGEN to generate 10000 items with these constraints show 1437 generations of $(size = WORD)$, and 5141 generations of

($lock = true$). On the other hand, in 10000 generation using uniform-distribution BDD solver, no item at all has ($size = WORD$) or ($lock = true$). Therefore, INTELLIGEN's behavior meets the user's expectations, which would be that all possibilities of $kind$, $size$, and $lock$ will be generated.

A different principle regarding generation distribution can be defined as the unwillingness to generate values without a real cause for choosing them. If we look at the constraint ($x = 1 \rightarrow y = 2$) from Example 1, the value 1 for x should not be biased just because it is mentioned in a constraint. An unneeded repetition of values may lead to tests that are not varied enough, and to an uneven and biased distribution.

Example 3. Assume two variables x and y with the domains $\{1..2^{32}\}$ for both variables, and a single constraint ($x = 1 \rightarrow y \neq 1$) .

Applying 'randomization-propagation' flow to Example 3 will create the following steps: (1) Choose a variable (either x or y) and randomize its value. (2) If the chosen variable was randomized to one, omit one from the other variable's domain. (3) Choose the remaining variable and randomize its assigned value from its current domain. Observe that the impact of the constraint is minimal. The probabilities of x and y to be 1 are close to $1/2^{32}$, and therefore these values will be generated similarly to any other values in the variables' domains, and with very low probability.

A generation of biased values may occur when using an SMT solver over a CP theory. Most SMT solvers are designed to determine a single satisfying solution, but attempts have been made to adapt them generate random results [11]. The constraint ($x = 1 \rightarrow y \neq 1$) is equivalent to the SMT formula ($\neg(x = 1) \vee (y \neq 1)$)) which is translated to ($\neg b1 \vee b2$) where the atoms ($x = 1$) and ($y \neq 1$) are replaced by Boolean variables $b1$ and $b2$ respectively. The values for the Boolean variables are assigned by the SAT solver. Any of the three valid possibilities for ($b1, b2$), which are ($false, false$), ($false, true$) and ($true, true$), are acceptable. However, the value $true$ for $b1$ or the value $false$ for $b2$ enforces that x will be 1 or y will be 1, correspondingly. This differs from the CP behavior, which gave no additional weight to these values, and they were generated with very low probability, like the rest of the values in the variables' domains. This drawback of SMT solvers can also be seen in Example 4, which is a variation of Example 2.

Example 4. Suppose we add the constraint ($kind = SINGLE$) to Example 2.

Now there is no constraint enforcing $delay$ to become zero. Indeed, in 10000 generations using INTELLIGEN, no zeros were generated for this variable. However, the SMT solver, when the SAT layer has a literal for ($delay = 0$), might occasionally assign $true$ to this literal, leading to biased distribution. Thus, we see that the CP solver, and in particular the 'randomization-propagation' flow, maintains the principle of not generating biased values. The SMT behavior violates this principle and is much more sensitive to addition and removal of constraints, even when their influence on the environment is minor.

The examples thus far have demonstrated the positive aspects of CP solvers in achieving a good distribution when using first-fail variable ordering and uniform distribution value ordering. But as mentioned before, the 'randomization-propagation' flow does not always achieve uniform distribution. In cases of variables with symmetrical domains, such distribution might be the desired one, as can be seen in Example 5.

Example 5. Assume two variables, x and y, with the domains $\{0..9\}$ for both variables, and a single constraint $(x + y < 10)$.

Let us compare the probabilities of the solutions $(9,0)$ and $(0,0)$ for (x,y). The solution $(9,0)$ has 5.5% probability: 10% if x is chosen first, and 1% if y is chosen first. The solution $(0,0)$ has 1% probability. Since the solution space contains 45 solutions, a uniform randomization should choose each solution with 2.22% probability. In Example 5, there is no reason to prefer one solution over the other, and each combination of values can represent an interesting test-case. Assuming the solution space should be fully covered, this means that many data items will need to be generated until the coverage criterion is satisfied. The problem becomes much more serious when it involves many variables and not just two, as Example 6 illustrates.

Example 6. Assume a list m of variables with the domain $\{0..2^{32}\}$, and a single constraint $(\sum m_i = 10000)$.

Initial propagation over this constraint cannot narrow the initial domain of the variables beyond $\{0..10000\}$. Therefore, the first chosen variables will be generated relatively freely, enforcing the remaining variables to sum to a rapidly diminishing value. In a typical solution very few variables are randomized to values between zero and ten thousand, several more are small positive numbers, and the rest are zeros. This is definitely not the user's expectation of a solution.

Starting from Specman version 11.1, INTELLIGEN incorporated a more sophisticated randomization scheme regarding the sum global constraint, while still remaining loyal to the 'randomization-propagation' flow. While the main details cannot be shared, the main principle is that the variable's value is not randomized uniformly from the variable's domain, but rather it is chosen according to the expected solution for the whole list. The generated lists do satisfy the criterion of being uniformly distributed.

Using the latest Specman and INTELLIGEN, the experimental results for Example 6, with 10000 generations of a list of 100 variables, show: 6302 generations have a list which contains a zero, 2677 generations have a list which contains two zeros, and only six generations have a list which contains more than five zeros.

Contrast this with the naive 'randomization-propagation' flow prior to Specman 11.1: In 10000 generations, all results contain more than 70 zeros, and 4845 generations contain more than 90 zeros.

To summarize, the two main principles of distribution are to generate all the interesting test scenarios and to avoid biased generations of values. CP solvers using the 'randomization-propagation' flow satisfy these principles, unlike other solving techniques. However, constraint models which are not handled well by the CP solver should receive special attention by the solver.

4   Random Stability

Once a test involves picking a random solution, the requirement to maintain reproducibility of the generation results becomes crucial. This is especially true in the verification flow, which may look as follows: (1) A nightly regression catches a failure in a specific test case. (2) A verification engineer reproduces the test case and chooses its owner. (3) The owner runs the test case again, using debug or interactive tool. (4) A fix is implemented and validated by rerunning the test in the regression.

Each of these steps involves a different run mode: regression tests run in batch mode, while the user will often prefer to use GUI tools. In addition, when debugging the test, devices such as breakpoints or logging may be used. These different run modes might involve different settings (for example, the memory setting of the run might be influenced). Yet, it is highly important that the same generated results will be produced in each run. Any difference, even the smallest, may drastically change the flow of the test, leading to irreproducibility.

Maintaining reproducibility is a difficult task when using non-deterministic algorithms; naive parallel search or propagation algorithms do not ensure it. However, even in a totally deterministic generation, it is not a trivial matter to maintain reproducibility, because the machine environment and settings may influence random stability. The following scenario demonstrates a case in which a different behavior of garbage collection changes random stability:

A test generates a data item ten times. During the nightly run this test failed on DUT error in the tenth generation. The solving was done using a learning solver (e.g. SAT solver) and no garbage collections were performed. When debugging the test, the different settings increased its memory signature, so that garbage collection was issued before the tenth generation. The garbage collection may have removed some of the learned data, or even the entire solver.

In this scenario, where learning was used, part of the learned data was removed in the debug run, but in the nightly run all solvers and the corresponding learned data was kept. Therefore this scenario will produce different results for the tenth generation in each run mode of the test. The lesson is that even a deterministic solver is vulnerable to random stability problems, and that learning solvers are especially prone to such problems.

While absolute reproducibility is essential, verification tests have an even more pressing requirement: Code modifications that are irrelevant to the generation of a certain variable must not affect its generated values when using the same seed. However, using one seed will obviously make the generation very sensitive to code changes. To support the random stability requirement, many more seeds are required.

INTELLIGEN expands the concept of 'seed' and uses a multi-seed scheme. After breaking the constraint model into isolated solving problem CFSs, a unique seed is attached to each CFS and to each CFS variable. The CFS seed is used, for example, to randomize variable ordering when solving the CFS. This ensures that independent CFSs will keep random stability regarding each other. Each CFS variable has its own seed which is used to randomize values from its domain. This is done to improve random stability for the specific variable. Sometimes random stability may be kept even when code changes are done within the same variable's CFS, although it cannot be guaranteed, as Example 7 illustrates.

Example 7. Assume three variables x, y and z with the domains $\{1..2^{32}\}$ for all variables, and a constraint $(x \leq y)$.

Adding the constraint $(y \leq z)$ will affect the generation of x, although its domain is unchanged. Because the randomization results are dependent on variable ordering, if z is chosen first, the generated value of x is different. However, using propagation based solving ('randomization-propagation' flow used in INTELLIGEN as described in Section 2) is sometimes helpful in maintaining random stability even when code changes are made to closely related variables. The reason is that generation of values is done using the variable domains with their dedicated seeds. This is illustrated by Example 8.

Example 8. Assume three variables b, x and y, with the domains $\{0..1\}$, $\{1..2^{32}\}$, and $\{1..2^{32}\}$ respectively, and the constraints: $(b = 1 \rightarrow x < 100)$, and $(b = get\_flag\_value())$ where $get\_flag\_value()$ is a runtime method which may reflect the current status of the DUT.

Adding the constraint $(b = 1 \rightarrow y < 50)$ will not affect the generation of the variable x, although it is related to the same CFS with the variables y and b. This is because the domain of x remains the same after propagation, with or without the additional constraint. Note that a different randomization approach, for example one which generates a vector of solutions, will generate different values for x with or without the additional constraint. Translating the problem to a BDD is an example. The reason is that the additional constraint reduces the solution space, as it reduces the domain of y.

Maintaining random stability, in all its aspects, is one of the most important requirements INTELLIGEN is expected to support. Choosing a CP solver and using the multi-seed scheme enables INTELLIGEN to support this requirement.

5 Summary

This paper discusses the benefits of a CP framework to address the special requirements of functional verification. The requirements of the domain are beyond the need for finding a single feasible solution, and rely heavily on the randomness of the solutions. These benefits were one of the main motivations for the technology picked for Cadence's INTELLIGEN constraint solver. The considerations include: the ability of the solver to efficiently solve the constraint problems, the random quality and stability of the solutions, and the convenient way the results can be presented and reproduced for debug. CP solvers prove to be highly effective for these problems, and show advantages over other solving techniques.

References

1. Alexandron, G., Lagoon, V., Naveh, R., Rich, A.: Gendebugger: An explanation-based constraint debugger. 2010 TRICS, 35 (2010)
2. Carla, A.S., Gomes, P., Selman, B.: Near-uniform sampling of combinatorial spaces using xor constraints. In: Proceedings of NIPS 2006 (2006)
3. Dechter, R., Kask, K., Bin, E., Emek, R.: Generating random solutions for constraint satisfaction problems. In: AAAI/IAAI, pp. 15–21 (2002)
4. Haralick, R.M., Elliott, G.L.: Increasing tree search efficiency for constraint satisfaction problems. Artificial Intelligence 14(3), 263–313 (1980)
5. Lagoon, V.: The challenges of constraint-based test generation. In: Proceedings of the 13th International ACM SIGPLAN Symposium on Principles and Practices of Declarative Programming, pp. 1–2. ACM (2011)
6. Naveh, Y., Rimon, M., Jaeger, I., Katz, Y., Vinov, M., Marcu, E.S., Shurek, G.: Constraint-based random stimuli generation for hardware verification. AI Magazine 28(3), 13 (2007)
7. Piziali, A.: Functional verification coverage measurement and analysis. Springer (2004)
8. Rich, A., Alexandron, G., Naveh, R.: An explanation-based constraint debugger. In: Namjoshi, K., Zeller, A., Ziv, A. (eds.) HVC 2009. LNCS, vol. 6405, pp. 52–56. Springer, Heidelberg (2011)
9. Society, I.C.: IEEE Standard for System Verilog–Unified Hardware Design, Specification, and Verification Language. IEEE Std. IEEE (2005)
10. I. C. Society. IEEE Standard for the Functional Verification Language 'e'. IEEE Std. IEEE (2006)
11. Wille, R., Große, D., Haedicke, F., Drechsler, R.: Smt-based stimuli generation in the systemc verification library. In: Borrione, D. (ed.) Advances in Design Methods from Modeling Languages for Embedded Systems and SoC's. LNEE, vol. 63, pp. 227–244. Springer, Heidelberg (2010)
12. Yuan, J., Albin, K.: Simplifying boolean constraint solving for random simulation-vector generation. In: Proceedings of ICCAD, pp. 412–420 (2004)

Stochastic Local Search Based Channel Assignment in Wireless Mesh Networks

M.A. Hakim Newton[1], Duc Nghia Pham[1], Wee Lum Tan[2,3],
Marius Portmann[2,3], and Abdul Sattar[1]

[1] Institute for Integrated and Intelligent Systems, Griffith University
[2] Queensland Research Lab., National ICT Australia (NICTA)
[3] School of ITEE, The University of Queensland
{hakim.newton,d.pham,a.sattar}@griffith.edu.au,
{weelum.tan,marius.portmann}@nicta.com.au

Abstract. In this paper, we consider the problem of channel assignment in multi-radio, multi-channel wireless mesh networks. We assume a binary interference model and represent the set of interfering links in a network topology as a conflict graph. We then develop a new centralised stochastic local search algorithm to find a channel assignment that minimises the network interference. Our algorithm assigns channels to communication links rather than radio interfaces. By doing so, our algorithm not only does preserve the network topology, but is also independent of the network routing layer. We compare the performance of our algorithm with that of a well-known Tabu-based approach (by Subramanian et al.) on randomly generated sparse and dense network topologies. Using graph-theoretic evaluation and ns2 simulations (a widely used discrete event network simulator), we show that our algorithm consistently outperforms the Tabu-based approach in terms of both the network interference and the throughput obtained under various offered loads. In particular, for a practical setting of 3 radio interfaces per mesh node in a dense network topology with 12 channels available, our approach achieves 70% lower network interference and thus 15 times higher average throughput than those achieved by the Tabu-based approach.

1 Introduction

Wireless interference is one of the major factors that limits the performance of IEEE 802.11-based wireless mesh networks. There have been various approaches proposed to improve network performance by mitigating or taking into account the effects of interference in wireless networks. These approaches include optimising the transmission power used by the nodes in a wireless network [7], utilising network routing protocols that are interference-aware [9], and scheduling conflict-free transmissions that consider the physical interference in the wireless network [3]. In wireless mesh networks, a popular approach to minimise the effects of interference is to equip wireless mesh nodes with multiple radio interfaces, and assign non-overlapping channels to these interfaces. The key challenge in this

C. Schulte (Ed.): CP 2013, LNCS 8124, pp. 832–847, 2013.
© Springer-Verlag Berlin Heidelberg 2013

case is to design a channel assignment algorithm that assigns the available channels to the radio interfaces in such a way that minimises the network interference and thus maximises the network throughput, while at the same time preserving the network topology.

In this paper, we present a new centralised stochastic local search (SLS) algorithm for channel assignment in multi-radio, multi-channel wireless mesh networks. The objective is to minimise the network interference while satisfying the interface constraint. We assume that the interference model is binary. Consequently, two communication links are said to interfere with each other if they are assigned the same channel and are within interference range of each other. The *network interference* is the number of interfering pairs of links in the channel assignment. There is also an *interface constraint* for each mesh node to ensure that the number of channels being used at that node does not exceed the number of radios available at the node. Our algorithm assigns channels to communication links rather than radio interfaces. By doing so, our algorithm not only does preserve the network topology, but is also independent of the network routing layer.

Due to the constraints and optimisation issues in this application which moreover requires a reasonable quick deployment and allows incremental fine-tuning, we developed our algorithm on top of **Kangaroo**, a constraint-based local search system [13]. We designed a new constraint to model the interface constraint at each mesh node. We then represent the problem as a constrained optimisation problem and use stochastic local search to find a solution (i.e. channel assignment). Our search algorithm starts from a randomly generated initial solution; which may be infeasible. It then iteratively improves the feasibility and optimality metrics of the solution. The search attempts to improve the *feasibility metric* (i.e. the number of violated interface constraints) when the current solution is far from being satisfied; other times, it tries to improve the *optimality metric* (i.e. the network interference). When there is no improvement within a number of iterations, it restarts by randomly assigning values to a number of variables. For the selection of links that require changing its channel assignment, we use Novelty [12], a very well-known stochastic local search algorithm. For the selection of a channel to be assigned to a link, we pick the best possible channel.

We compare the performance of our algorithm with that of a well-known Tabu-based approach by Subramanian et al. [19] on randomly generated sparse and dense network topologies. It is worth noting that our industry partners in wireless mesh networks confirm our assumed network scenario (network topology, traffic pattern, etc.) and parameter choices to be realistic. Using graph-theoretic evaluation and ns2 simulations, we empirically show that our algorithm consistently outperforms the Tabu-based approach in terms of both the network interference and the throughput obtained under various offered loads. In particular, for a practical setting of 3 radio interfaces per mesh node in a dense network topology with 12 channels available, our approach achieves 70% lower network interference and thus 15 times higher average throughput than those achieved by the Tabu-based approach. Note that ns2 (a discrete event network simulator) uses a more complex and realistic interference model based on the Signal to Interference plus Noise Ratio (SINR) measured at the receiver nodes. It is the most widely used simulator in the networks community for the quantitative evaluation of network algorithms. Our ns2 evaluation strongly demonstrates that our simple interference model is an effective abstraction of the complex physical problem of wireless interference, and leads to very good network performance.

The rest of the paper is organised as follows: Section 2 reviews related work; Section 3 describes the model and problem formulation; Section 4 provides a brief review of the Tabu-based algorithm by Subramanian et al. [19]; Section 5 describes in detail our stochastic local search based approach; Section 6 presents our experimental evaluations. Finally, Section 7 summarises our conclusions and outlines the future work.

2   Related Work

The frequency assignment problems, also called channel assignment problems, have been a major research topic over the past years. Fast developments of wireless telephone networks and satellite communication projects have been the key factors behind this. Moreover, other applications like TV broadcasting and military communication have much inspired the interests in this research area. There have been various channel assignment algorithms proposed in the literature for various channel assignment problems and even for wireless mesh networks specifically. Interested readers can refer to a few good survey papers on various channel assignment problems [2] and on wireless mesh networks [5,18]. In this section, we briefly review a few relevant approaches.

Existing works on channel assignment algorithms can be divided into the *distributed* and *centralised* approaches. In distributed channel assignment approaches [8,17,16], individual network nodes compute its channel assignment based on locally gathered information about its network neighbourhood. Distributed channel assignment approaches are more suitable to be used once a network has been set up and is operationally running. This is because it is more adaptive to dynamic changes in local network topology (due to node failures or external interference) and any changes in the channel assignment can be confined to the local neighbourhood. In this paper, we are interested in the optimal channel assignment for mesh nodes in an initial network-wide deployment (before the entire network is operational running), which is better handled by a centralised channel assignment approach. Here, the centralised controller node is simply one of the mesh nodes configured to act in this role at deployment time. Even in a completely new and rapid deployment scenario, it is acceptable to wait for a few seconds for the channel allocation to complete.

In centralised channel assignment approaches [19,11,15], a central entity computes the optimal channel assignment based on global information about the network topology, such as the interference relationship between the nodes or communication links in the network. Typically, the interference relationship is represented as a conflict graph. The central entity then disseminates the channel assignment information throughout the network to every node.

Subramanian et al. proposed a Tabu-based approach in [19] in which the Tabu search based technique [6] is used to find a channel assignment that minimises the network interference. We will describe the Tabu-based approach in more detail in Section 4, and compare the performance of our proposed algorithm with this approach in Section 6. In [11], a greedy heuristic channel assignment algorithm called CLICA is proposed to find a connected and low interference network topology. Subramanian et al. compared the performance of their Tabu-based approach with CLICA in [19] and showed that their approach performs better. In the BFS channel assignment algorithm in [15], each mesh node has one radio interface configured to a default common channel in order to maintain network-wide connectivity. With typical mesh nodes having at most two or three radio interfaces each, this can lead to inefficient utilisation of the available channels in the network and poor network performance. In contrast, our algorithm in this paper preserves network connectivity by assigning channels to communication links, instead of using a dedicated radio interface configured to a common channel.

3   Model and Problem Formulation

A typical architecture of a wireless mesh network has two tiers: *backbone tier* and *access tier*. The backbone tier consists of stationary *mesh nodes* (or *nodes*) forming a wireless multihop backbone infrastructure, with one or more nodes also functioning as gateways to the Internet. The access tier sees the end-user client devices connecting to the mesh nodes in order to communicate with other client devices or to access the Internet. The 5GHz and 2.4GHz frequency bands are typically used for the communications in the backbone and access tiers respectively.

The mesh nodes usually have multiple radio interfaces, each of which might be configured with a channel $k \in \mathcal{K}$, where $\mathcal{K}$ is the set of available channels in the backbone tier. We assume all nodes to have the same number of r radio interfaces, each having omni-directional antennas with the same transmission range R_{tx}. Let D_{uv} denote the physical distance between two nodes u and v. There exists a *communication link* (or *link*) $l \equiv l_{uv}$ between nodes u and v, if $D_{uv} < R_{\text{tx}}$ and both nodes have a radio interface configured to a common channel. A given set of nodes V and the links E can be modelled as an undirected graph $G = (V, E)$. Let E_v denote the links incident on a node v.

In this paper, we are interested in the optimal assignment of channels to links in a multihop backbone infrastructure in order to minimise the interference between the links. Due to the inverse relationship between interference and network throughput, by minimising the interference in the network, we essentially maximise the network throughput. We model the interference between co-channel communication links with the range-based interference model. In this interference model, every node has an associated interference range, R_{int}, that is typically larger than the transmission range. A *unicast* transmission on a link l from node u to node v can be interfered with by another simultaneous transmission (on the same channel) from any node within the interference range of both nodes u and v.

We assume that the interference model is binary in the sense that the unicast transmission is successful if there is no interference present, and unsuccessful otherwise. We also assume that transmissions on different channels do not interfere. Note that our channel assignment algorithm does not depend on the choice of the interference model used in this paper, *i.e.* the range-based interference model. Our channel assignment algorithm will work with other interference models such as the hop-based and the protocol interference models. These interference models are called *pairwise* interference models after the fact that interference is defined on pairs of communication links. Interested readers can refer to [10] for a description of these and other interference models.

Given a pairwise interference model, we use a conflict graph G_c to model the set of interfering communication links in the wireless mesh network represented by a graph G. Each link l in G essentially becomes a vertex in the conflict graph G_c. An edge in G_c exists between a pair of vertices l and l', if the links l_{uv} and $l'_{u'v'}$ in the network are on the same channel and interfere with each other. With the range-based interference model that we adopt in this paper, an edge exists if any of the following is true: $D_{uu'} \leq R_{\text{int}}$, or $D_{uv'} \leq R_{\text{int}}$, or $D_{vu'} \leq R_{\text{int}}$, or $D_{vv'} \leq R_{\text{int}}$.

The objective of our channel assignment problem is to find a mapping ϕ that assigns unique channels to the links in such a way that minimises the network interference. Given a channel assignment ϕ, let $\phi(l)$ denote the channel assigned to a link l and $\sigma_\phi(v)$ the number of unique channels assigned to the links incident on a node v. The *network interference* $\eta(\phi)$ of a channel assignment ϕ is the number of edges in the conflict graph. A *feasible* channel assignment must satisfy all the *interface constraints* to ensure that for each node v, the number of unique channels assigned to all communication links incident on v is at most the number of radio interfaces available at v, *i.e.* $\sigma_\phi(v) \leq r$.

Note that by assigning channels to communication links rather than to radio interfaces, our channel assignment algorithm maintains the same network topology as in the case when a single channel is used for all communication links in the network. By doing so, our channel assignment algorithm is independent of the network routing layer.

4   Tabu-Based Algorithm

The Tabu-based algorithm by Subramanian et al. [19] comprises two phases, which can be viewed as optimisation and satisfaction phases. First, in the *optimisation phase*, the network interference is iteratively minimised without considering the interface constraints. Therefore, the best solution found in the first phase normally violates the interface constraints. Next, in the *satisfaction phase*, these violated interface constraints are heuristically fixed to obtain a feasible solution. Note that while fixing those violated constraints, the network interference may increase. However, the Tabu-based algorithm returns its one and only solution at the end of the second phase without attempting to improve the network interference further.

4.1   Optimisation Phase

In this phase, the Tabu-based algorithm starts with a random initial solution ϕ_0 wherein each link l is assigned to a random channel $k \in \mathcal{K}$ i.e. $\phi_0(l) = k$. Given a solution ϕ_i, it then obtains the next solution ϕ_{i+1} by selecting the best candidate (having the lowest network interference) from a number of randomly generated neighbouring solutions. A neighbouring solution ϕ_i' of a solution ϕ_i is generated by randomly selecting a link l and then assigning a random channel $k \in \mathcal{K}$ to the link. The neighbour generation process ensures that $k \neq \phi_i(l)$ and the pair (l, k) does not appear in a maintained tabu list τ, which is a first-in-first-out queue of a given length. When ϕ_{i+1} is obtained from ϕ_i, the modified link l and the new channel k assigned to the link are pushed into the queue; before that, one queue element is popped out, if the queue is full. The first phase of the algorithm terminates if there is no improvement in the network interference after $|E|$ iterations, where E is the set of links in the network [19].

4.2   Satisfaction Phase

In this phase, the Tabu-based algorithm mainly attempts to satisfy the violated interface constraints in the solution ϕ returned by the first phase. For this, it picks a node v that has the most violations in its interface constraint. The number of violations of the interface constraint at a given node v is $\max(\sigma_\phi(v) - r, 0)$. The algorithm then performs a merge operation that chooses two channels k and k', and makes a replacement $\phi(l) = k'$ for each incident link l to v such that $\phi(l) = k$. The replacement process has a recursive cascading effect on each node v' that l is also incident on. While satisfying the interface constraints, the merge operation may increase the network interference $\eta(\phi)$. Therefore, the choice of k and k' is greedily made so that the increase in $\eta(\phi)$ is minimised.

5   Stochastic Local Search Based Algorithm

We developed our stochastic local search based algorithm on top of Kangaroo, a constraint-based local search system [13]. We define the variables and functions such as constraints and objectives in the Kangaroo system. Kangaroo then efficiently propagates changes from variables to the dependant functions. In each iteration when the variables are assigned with new values, Kangaroo performs *execution* by updating the functions' values. It also helps to efficiently explore potential neighbouring solutions by performing *simulation*: computing the feasibility and optimality metrics temporarily due to the potential changes in the variables.

5.1   Constraint and Objective Functions

We designed a new constraint function AtMostCount in Kangaroo to model the interface constraints. The AtMostCount constraint maintains the degree of constraint violation and penalises the links for causing the violation. We also use NotEqual constraints to represent 'no conflict' between two given links, and Sum function to accumulate the constraints to form the top-level objective functions.

Each *function* $f(p_1, \cdots, p_n)$ has the parameters p_js that are either variables or other functions. A function f depends on a variable x, denoted by $f \to x$, if x is itself a parameter of f or f has a parameter $p \to x$. Each function f has a non-negative *metric* f^{m} denoting its evaluation. For each $x \leftarrow f$, it also has a non-negative *hint* $f^{\mathrm{h}}(x)$ denoting the preference level of changing x's value to improve f^{m}. A constraint f is satisfied when $f^{\mathrm{m}} = 0$ and in that case $f^{\mathrm{h}}(x) = 0$ for any x. This means a constraint's metric improves when it is minimised.

AtMostCount. The number of unique values occurred in the given n variables $x_1, \cdots, x_n$ must not exceed a specified limit m. This constraint, denoted by $C(x_1, \cdots, x_n)$, has its metric $C^{\mathrm{m}} = \mathsf{max}(c - m, 0)$ and for each variable x_j, a hint $C^{\mathrm{h}}(x_j) = n - c_j$; where c is the number of unique values used in the variables while c_j is the number of times x_j's value has occurred. Notice that C^{m} is the number of additional values used beyond the limit. Also, notice that $C^{\mathrm{h}}(x_j)$ captures the heuristic stated as 'the fewer the value of a variable occurs, the more the preference of the variable's value to be modified'.

NotEqual. The values of two given variables must not be equal. This constraint, denoted by $\bar{Q}(x, x')$, has its metric $\bar{Q}^{\mathrm{m}} = 1$ if x equals x', else 0. Its hints are defined as $\bar{Q}^{\mathrm{h}}(x) = \bar{Q}^{\mathrm{h}}(x') = \bar{Q}^{\mathrm{m}}$. When two variables have the same value changing either one's value could make them unequal and thus improve the metric.

Sum. This function, denoted by $S(f_1, \cdots, f_n)$, accumulates a given number of functions. We defined its metric $S^{\mathrm{m}} = \sum f_j^{\mathrm{m}}$ and for each variable $x \leftarrow S$, a hint $S^{\mathrm{h}}(x) = \sum_{x \leftarrow f_j} f_j^{\mathrm{h}}(x)$, where $1 \le j \le n$.

5.2   Constrained Optimisation Model

The constrained optimisation model used by our algorithm is defined in Procedure 1 ConstructModel. We first create a variable x_l for each link l in the network (Lines 1–2). The domain of each variable is $\mathcal{K}$, the set of available channels. For each node v, we then create an AtMostCount constraint C_v with the links incident on v (Lines 3–5). The limit given to each C_v is r which is the number of radio interfaces available at a node. The Sum function S_{sat} accumulates all the AtMostCount constraints (Line 6). The accumulated metric of S_{sat} must be 0 in order to get a feasible channel assignment for the network.

For each *potential* edge in the conflict graph G_c, we create a NotEqual constraint $\bar{Q}_{ll'}$ (Lines 7–9) with the variables x_l and $x_{l'}$ respectively associated

---

Procedure 1: ConstructModel

---

1 **foreach** $link\ l \in E$ **do**
2   Create a variable x_l with domain $\mathcal{K}$;
3 **foreach** $node\ v \in V$ **do**
4   Let X be the set of variables for the links E_v;
5   Create a constraint C_v with X and the limit r;
6 Create a function S_{sat} with all C_vs created above;
7 **foreach** potential $edge\ e\ in\ G_c$ **do**
8   Let l and l' be the vertices which e incidents on;
9   Create a constraint $\bar{Q}_{ll'}$ with variables x_l and $x_{l'}$;
10 Create a function S_{opt} with all $\bar{Q}$s created above;
11 Create a function S_{comb} with S_{sat} and S_{opt};

---

with the two links l and l'; note that these two links cause interference if they are assigned with the same channel. The Sum function S_{opt} accumulates all the NotEqual constraints (Line 10) and thus represents the network interference. The metric of this Sum function needs to be as small as possible. Finally, we have another Sum function S_{comb} that combines S_{sat} and S_{opt} (Line 11) to give an overall evaluation of the candidate channel assignment.

5.3 Local Search Method

Our search algorithm, shown in Procedure 2 PerformSearch, starts from a randomly generated initial solution; which may be infeasible (Line 6). It then iteratively improves the feasibility and optimality metrics of the solution (i.e. S_{sat} and S_{opt} respectively). In each iteration, our algorithm chooses one of the three options: *i)* satisfy: minimise the violation of interface constraints (Lines 20–23) when the current assignment is far from being feasible (i.e. $S_{\text{sat}}^{\text{m}} \geq$ Proximity); *ii)* optimise: minimise the network interference (Lines 16–19) if the current assignment is already feasible ($S_{\text{sat}}^{\text{m}}$ is 0) or very close to being feasible ($S_{\text{sat}}^{\text{m}} <$ Proximity); or *iii)* restart: randomly change a large portion of the current assignment to escape stagnation (Lines 12–15).

Function 3 selectVarValue selects a variable x greedily based on its hint obtained from VarObj. It then selects a new value k for x that minimises the metric of ValObj. The variable choice is limited to the subset of variables that the Constraint parameter depends on. Parameters VarObj and ValObj can be any Sum function (S_{sat}, S_{opt} or S_{comb}) defined above.

With a small probability, Function 3 selects a random variable in the subset (Line 4). Otherwise for most of the time, it selects a variable in the subset that has the maximum hint in VarObj, breaking ties on being assigned earlier (Lines 5–11). However, when that variable is the most recently assigned one in this subset, with a given small noise probability it will select the variable that has the second maximum hint (Novelty heuristic in Line 10).

Therefore, the first option (satisfy) in Procedure PerformSearch randomly selects a violated AtMostCount C with a view to improving the feasibility metric $S_{\text{sat}}^{\text{m}}$ by changing the value of a variable $x \leftarrow C$. Variable x and its value are selected by calling Function 3 with the following options Constraint=C, VarObj=S_{sat} and ValObj=S_{sat}.

Procedure 2: PerformSearch

1 BestSoln = $\emptyset$;
2 MinConflict = ∞;
3 Proximity = 1;
4 IterSinceRestart = 0;
5 RestartPeriod = $|E| \times 10$;
6 initialiseRandomly (CurrSoln);
7 **while** $S_{\text{comb}}^{\text{m}} > 0 \wedge \neg$TimeOut **do**
8   **if** $S_{\text{sat}}^{\text{m}} = 0 \wedge S_{\text{opt}}^{\text{m}} <$ MinConflict **then**
9   MinConflict = $S_{\text{opt}}^{\text{m}}$;
10   BestSoln = CurrSoln;
11   IterSinceRestart = 0;
12   **if** $++$IterSinceRestart $>$ RestartPeriod **then**
13   IterSinceRestart = 0;
14         $X =$ Select $\frac{|E|}{8}, \frac{2|E|}{8}$ or $\frac{3|E|}{8}$ variables randomly;
15   Assign random values to the variables X;
16   **else if** $S_{\text{sat}}^{\text{m}} <$ Proximity **then**
17   Select C randomly from all C_vs;
18         $(x, k) =$ selectVarValue($C, S_{\text{opt}}, S_{\text{comb}}$);
19   Assign the value k to the variable x;
20   **else** // $S_{\text{sat}}^{\text{m}} \geq$ Proximity
21   Select C randomly from C_vs such that $C_v^{\text{m}} > 0$;
22         $(x, k) =$ selectVarValue($C, S_{\text{sat}}, S_{\text{sat}}$);
23   Assign the value k to the variable x;
24 **return** BestSoln;

Function 3: selectVarValue(Constraint, VarObj, ValObj)

1 Noise = 0.01;
2 $X = \{x : x \leftarrow$ Constraint$\}$;
3 Select a variable $x \in X$ in the following way:
4   **if** bernoulii (Noise) **then** randomly;
5   **else** // use Novelty
6   Assume $x_1 \in X$ and $x_2 \in X$ have the first
7   and second maximum hint in VarObj
8   breaking tie on being assigned earlier;
9   **if** x_1 *is the last assigned variable* **then**
10   x= **if** bernoulii (Noise) **then** x_2 **else** x_1
11   **else** $x = x_1$
12 By simulation, select a value k for x
13   such that the metric of ValObj is minimised;
14 **return** (x, k);

For the second option (optimise), we could use a violated NotEqual constraint $\bar{Q}$ to limit the variable choices in Function 3. However, a $\bar{Q}$ depends only on two variables, and thus restricts the choice particularly in a stagnant situation. Notice that most of the time the variable selection is greedy and the value selection is also greedy with no tabu; which makes it harder to escape from a stagnant situation for any further improvement. The use of a random AtMostCount C gives more choices to select a variable from. Here, we set VarObj=S_{opt} to select a variable as we want to improve the network interference. However, we set ValObj=S_{comb} to select a value to find an assignment that might converge in the optimality metric but could even diverge in the feasibility metric.

The third option (restart) randomly assigns values to a number of variables, if there is no improvement after a given number of iterations.

6   Performance Evaluations

We compared the performance of the Tabu-based algorithm (using our implementation with the optimal settings as specified in [19]) and our SLS-based algorithm using both graph-theoretic evaluations and ns2 simulations [1]. In our evaluations, we used two different types of random network topologies: dense and sparse. We generated these networks by randomly placing 50 nodes respectively in 500×500 and 800×800 square meter areas. Using the default ns2 simulation settings for IEEE 802.11a, the transmission range R_{tx} is set to 163 meters, while the interference range R_{int} is set to 410 meters. With these settings, the average node degree is around 11 in the dense network while in the sparse network, the average node degree is around 5. The parameter choices are confirmed to be realistic by our industry partners.

6.1   Graph-Theoretic Evaluations

For both of the channel assignment algorithms, we used a computer equipped with Intel(R) Xeon(R) 64bit quad-core CPU X3470 @2.93GHz with 8MB L2 Cache and 8GB RAM running Ubuntu Linux version 12.04. We ran both the algorithms 25 times on each benchmark topology with a realistic 30 seconds cutoff time. Note that quick response time is important for an initial deployment of a wireless mesh network in a disaster situation. The choice of 30 seconds is based on user acceptance surveys undertaken by industry partners. For each run, we collected the best solution found within the cutoff time. We then used the median quality solution out of those 25 best solutions to run the ns2 simulations and to present our results and provide analysis based on them.

Note that the Tabu-based algorithm is a *one-off* algorithm meaning it produces only one solution as its output and then terminates. This is obvious from the two phases of the algorithm. The first phase iteratively minimises only the network interference while ignoring the interface constraints. Therefore, a feasible solution can only be found at the end of the second phase which fixes those constraint violations. Nevertheless, in most cases, the Tabu-based algorithm was able to return a solution within the stipulated cutoff time.

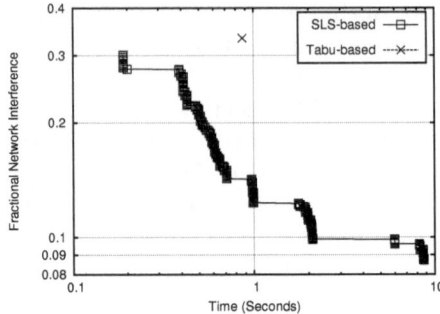

Fig. 1. Convergence of fractional network interference w.r.t. time for SLS-based and Tabu-based algorithms in the dense network with 12 channels and 3 radio interfaces per node

In contrast, our SLS-based algorithm is an *any-time* algorithm [21], meaning that one can terminate the algorithm at any time (after the initial time to find the first solution) and still has a valid solution. In other words, our algorithm is capable to quickly find a feasible solution. Once such a solution is found, our algorithm still keeps running in the quest of further improving solutions (in terms of the network interference) as long as the cutoff time permits (Procedure 2 Lines 7–10). Trading off with the time available, one can at any time take the best solution found so far.

We plot the typical convergence pattern of our algorithm in Figure 1 on a dense network with 3 radio interfaces at each node and 12 available channels. Notice that by the time the Tabu-based algorithm found its solution, our algorithm has already found a number of solutions that are much better. This was found true for all test cases. We observed the results found by the Tabu-based method over 25 runs on a test case to be very consistent, with only a small variance. This suggests that executing this method multiple times until the time cutoff would not make big differences. Conversely, our method found significantly better solutions than the Tabu method within the same time window, and further improved the solution quality substantially over time. If one were to terminate our algorithm early, still a sufficiently good solution would be returned. We used the optimal settings in [19] for our Tabu implementation used the optimal settings but we didn't tuned our method much.

For further comparison, we graphically show in Figure 2 the fractional network interferences of the solutions produced by the SLS-based and Tabu-based algorithms for the sparse network. These results were obtained by using 3 and 12 channels with the number of radio interfaces varied from 2 to 8. We show the same for the dense network in Figure 3. The *fractional network interference* is defined as the ratio of the number of edges in the conflict graph produced by a given channel assignment and the total number of conflicts in the single-channel network. From these results, it is clear that our algorithm obtained significantly better fractional network interference than the Tabu-based algorithm.

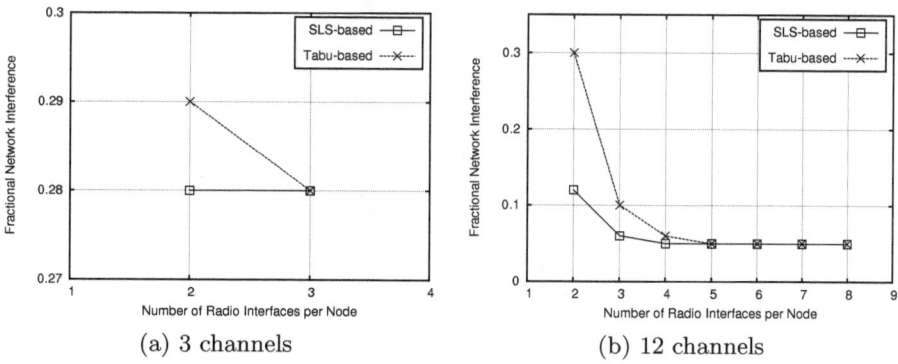

Fig. 2. Fractional network interference of the solutions produced by SLS-based and Tabu-based algorithms in the sparse network with (a) 3 or (b) 12 channels

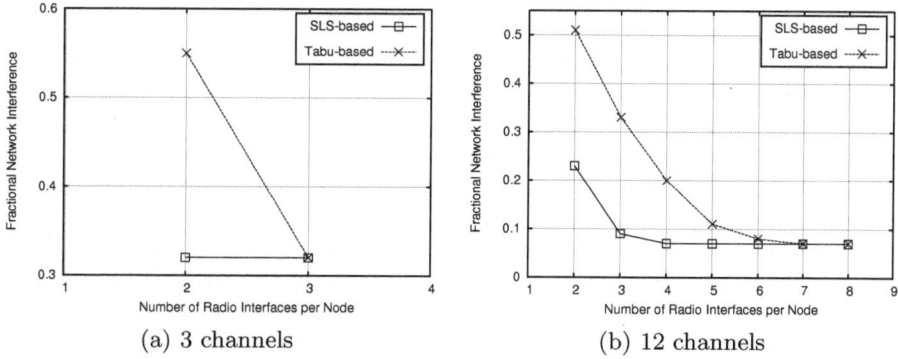

Fig. 3. Fractional network interference of the solutions produced by SLS-based and Tabu-based algorithms in the dense network with (a) 3 or (b) 12 channels

We sought a reasonable explanation for these performances. Note that our algorithm switches between the satisfaction and optimisation phases in an interleaving fashion. Moreover, both phases respect the interface constraints and the network interference in a separate or combined way. On the other hand, the Tabu-based algorithm has two separate phases where the satisfaction phase follows the optimisation phase. While the latter phase only minimises the network interference, the former phase just satisfies the interface constraints; no interaction between these two criteria is considered.

Our further observations suggest the solution quality of the Tabu-based algorithm greatly reduces by the fact that fewer than the total number of available channels are used by both of its phases. In particular, the satisfaction phase while addressing the violations of the interface constraints replaces one channel with another; which greatly reduces the number of channels being used in the end. As a result, the conflict graphs contain larger clusters and a higher number of edges

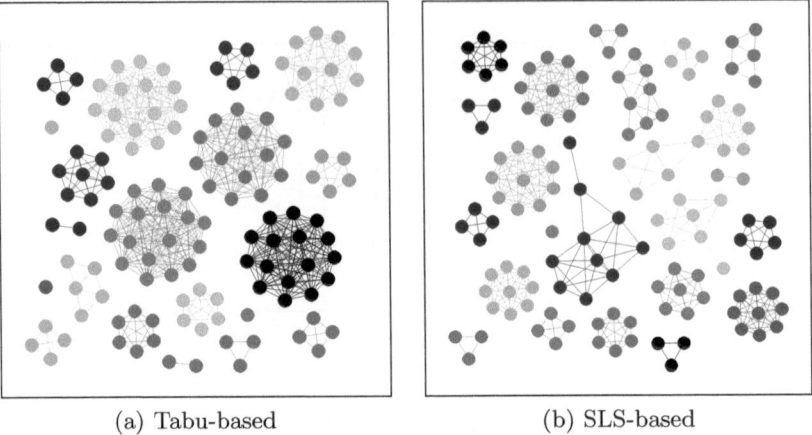

(a) Tabu-based   (b) SLS-based

Fig. 4. Conflict graphs for the solutions produced by SLS-based and Tabu-based algorithms in a sparse network with 12 channels and 3 radio interfaces per node

for the solutions produced by the Tabu-based algorithm, compared to those produced by the SLS-based algorithm. This can be observed in Figure 4 that shows the conflict graphs produced by the SLS-based and Tabu-based algorithms in the sparse network with 12 channels and 3 radio interfaces per node.

In Figures 2 and 3, notice that the fractional network interference produced by our SLS-based algorithm reaches the minimum after 2 radios for the 3-channels case and 4 radios for the 12-channels case. Further increase in the number of radio interfaces does not make any significant difference in the network interference. This means the number of radio interfaces at each node could effectively be reduced to these levels without increasing the network interference; which would result in lower hardware cost for the mesh nodes.

6.2   ns2 Simulations

In this evaluation, we investigate the network throughput performance of both our SLS-based algorithm and the Tabu-based algorithm. We use the IEEE 802.11a implementation in ns2 with its default settings, and added support for multiple interfaces and multiple channels [4]. Using the same two network topologies as in Section 6.1, we employ CBR traffic on every single-hop communication link in the network. We use this single-hop traffic model in order to evaluate the performance when all communication links in the network carry the same traffic load. We measure the throughput on every link as we slowly increase the offered traffic load on each link until the achievable throughput does not increase anymore. All transmissions are unicast transmissions with a packet size of 1000 bytes, and are using the IEEE 802.11a MAC protocol with the RTS/CTS feature turned off. The transmission bit rate is set to a fixed rate of 6Mbps, and each simulation run is 60 seconds long.

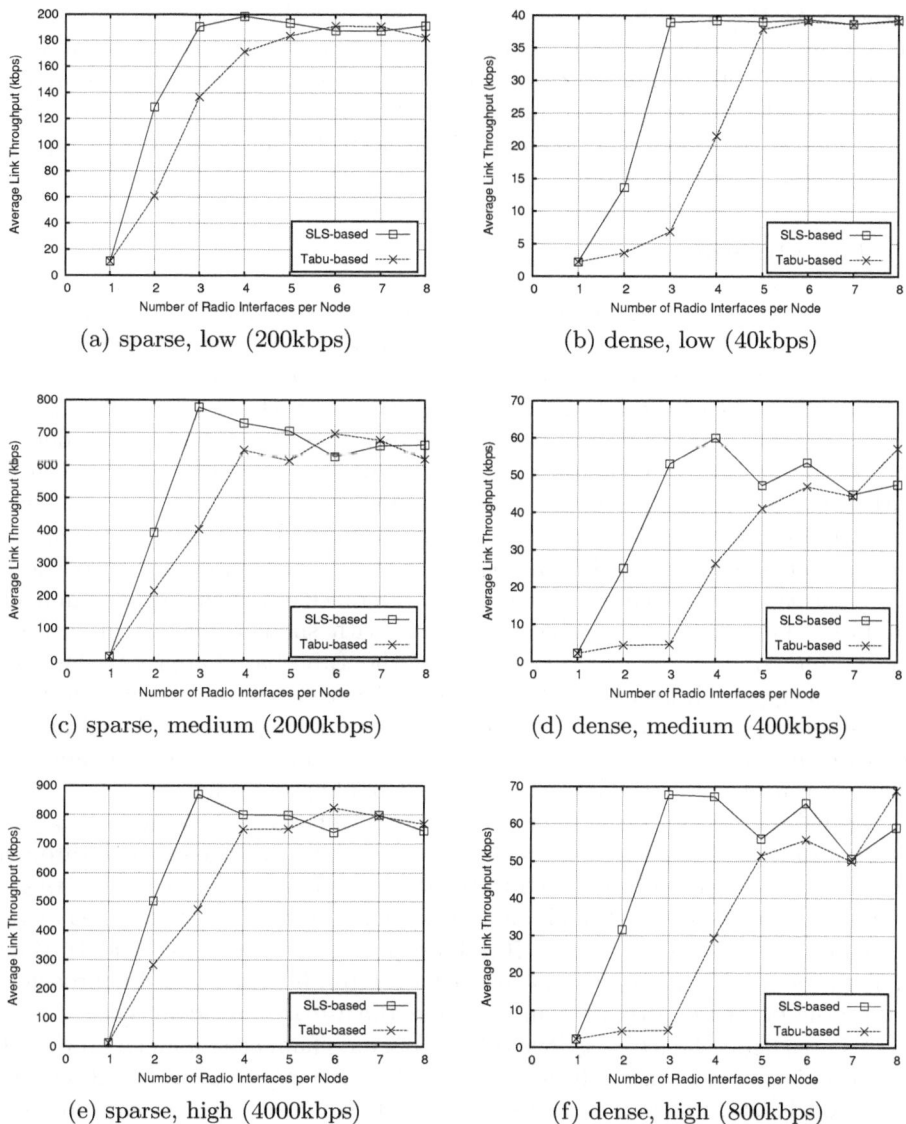

Fig. 5. Average link throughput in a sparse/dense network with 12 channels, and low/medium/high offered traffic load

Figure 5 shows the average link throughput achieved in the sparse and dense networks with 12 available channels, for different offered traffic loads as we vary the number of radio interfaces per node from 1 to 8. We observe that the average link throughput achieved by our SLS-based algorithm reaches a maximum after 3 or 4 radios. This is consistent with the graph-theoretic results shown earlier in Figures 2b and 3b. We also see that our SLS-based algorithm consistently outperforms the Tabu-based algorithm in terms of the achievable throughput obtained under various offered traffic loads. In particular, we see that for a practical network setting of 2 or 3 radio interfaces per node, our SLS-based algorithm achieves an average throughput that is markedly higher by as much as 2 times in the sparse network and 15 times in the dense network, compared to the Tabu-based algorithm.

7   Conclusion

In this paper, we have presented a new centralised stochastic local search algorithm to find a channel assignment that minimises the network interference in a multi-radio and multi-channel wireless mesh network. Using a binary interference model, we represent the interfering links in the network with a conflict graph. Our SLS-based algorithm preserves the network topology and is independent of the routing layer. We compared the performance of our SLS-based algorithm with the Tabu-based approach [19] on randomly generated sparse and dense network topologies by using graph-theoretic evaluation and ns2 simulations.

Our graph-theoretic results show that our approach significantly outperforms the Tabu-based approach in the network interference of the channel assignments produced. Our approach produces solutions with smaller clusters in the conflict graphs compared to those produced by the Tabu-based approach. Furthermore, our approach usually finds a number of better solutions even before the Tabu-based approach produces its only solution. Our approach still continues to improve the solution quality. Thus ours is an any-time algorithm meaning one could stop it at any time and still get a reasonably good solution.

The ns2 simulation results show that our SLS-based algorithm consistently outperforms the Tabu-based approach in terms of the average network throughput obtained under various offered traffic loads. Indeed, for a practical setting of 3 radio interfaces per mesh node in a dense network topology with 12 channels available, our approach achieves 70% lower network interference and 15 times higher average throughput than those achieved by the Tabu-based approach.

In terms of future work, we plan to explore the use of more advanced stochastic local search algorithms (e.g. gNovelty$^+$ [14]) in our channel assignment approach. We also plan to use other interference models, e.g. measurement-based interference models [20], in our channel assignment algorithm in order to capture more realistic network interference scenarios.

References

1. The network simulator ns-2, http://www.isi.edu/nsnam/ns/
2. Aardal, K.I., van Hoesel, S.P.M., Koster, A.M., Mannino, C., Sassano, A.: Models and solution techniques for frequency assignment problems. Annals of Operations Research 153(1), 79–129 (2007)
3. Brar, G., Blough, D.M., Santi, P.: Computationally efficient scheduling with the physical interference model for throughput improvement in wireless mesh networks. In: Proc. ACM MobiCom (2006)

4. Calvo, R., Campo, J.: Adding multiple interface support in ns-2. Tech. rep (2007), http://personales.unican.es/aguerocr/files/ucmultiifacessupport.pdf
5. Crichigno, J., Wu, M., Shu, W.: Protocols and architectures for channel assignment in wireless mesh networks. Ad Hoc Networks 6(7), 1051–1077 (2008)
6. Hertz, A., Werra, D.: Using tabu search techniques for graph coloring. Computing 39(4), 345–351 (1987)
7. Kawadia, V., Kumar, P.R.: Principles and protocols for power control in wireless ad hoc networks. IEEE Journal on Selected Areas in Communications 23(1), 76–88 (2005)
8. Ko, B., Misra, V., Padhye, J., Rubenstein, D.: Distributed channel assignment in multi-radio 802.11 mesh networks. In: Proc. IEEE WCNC (2007)
9. Liu, T., Liao, W.: Interference-aware QoS routing for multi-rate multi-radio multi-channel ieee 802. 11 wireless mesh networks. Trans. Wireless. Comm. 8(1), 166–175 (2009)
10. Maheshwari, R., Jain, S., Das, S.R.: A measurement study of interference modeling and scheduling in low-power wireless networks. In: Proc. ACM SenSys (2008)
11. Marina, M., Das, S., Subramanian, A.: A topology control approach for utilizing multiple channels in multi-radio wireless mesh networks. Computer networks 54(2), 241–256 (2010)
12. McAllester, D.A., Selman, B., Kautz, H.A.: Evidence for invariants in local search. In: AAAI/IAAI, pp. 321–326 (1997)
13. Newton, M.A.H., Pham, D.N., Sattar, A., Maher, M.: Kangaroo: An efficient constraint-based local search system using lazy propagation. In: Lee, J. (ed.) CP 2011. LNCS, vol. 6876, pp. 645–659. Springer, Heidelberg (2011)
14. Pham, D.N., Thornton, J., Gretton, C., Sattar, A.: Combining adaptive and dynamic local search for satisfiability. JSAT 4(2-4), 149–172 (2008)
15. Ramachandran, K., Belding, E., Almeroth, K., Buddhikot, M.: Interference-aware channel assignment in multi-radio wireless mesh networks. In: Proc. IEEE INFO-COM (2006)
16. Raniwala, A., Chiueh, T.: Architecture and algorithms for an IEEE 802.11-based multi-channel wireless mesh network. In: Proc. IEEE INFOCOM (2005)
17. Shin, M., Lee, S., Kim, Y.: Distributed channel assignment for multi-radio wireless networks. In: Proc. IEEE MASS (2006)
18. Si, W., Selvakennedy, S., Zomaya, A.: An overview of channel assignment methods for multi-radio multi-channel wireless mesh networks. Journal of Parallel and Distributed Computing 70(5), 505–524 (2010)
19. Subramanian, A., Gupta, H., Das, S.R., Cao, J.: Minimum interference channel assignment in multiradio wireless mesh networks. IEEE Transactions on Mobile Computing 7(11) (2008)
20. Tan, W., Hu, P., Portmann, M.: Experimental evaluation of measurement-based SINR interference models. In: Proc. IEEE WoWMoM (2012)
21. Zilberstein, S.: Using anytime algorithms in intelligent systems. AI Magazine 17(3), 73–83 (1996), http://rbr.cs.umass.edu/shlomo/papers/Zaimag96.html

Automatic Generation and Delivery
of Multiple-Choice Math Quizzes[*]

Ana Paula Tomás[1] and José Paulo Leal[2]

[1] DCC & CMUP, Faculdade de Ciências, Universidade do Porto, Portugal
[2] DCC & CRACS-INESC TEC, Faculdade de Ciências, Universidade do Porto
{apt,zp}@dcc.fc.up.pt

Abstract. We present an application of constraint logic programming to create multiple-choice questions for math quizzes. Constraints are used for the configuration of the generator, giving the user some flexibility to customize the forms of the expressions arising in the exercises. Constraints are also used to control the application of the buggy rules in the derivation of plausible wrong solutions to the quiz questions. We developed a prototype based on the core system of AGILMAT [18]. For delivering math quizzes to students, we used an automatic evaluation feature of Mooshak [8] that was improved to handle math expressions. The communication between the two systems - AgilmatQuiz and Mooshak - relies on a specially designed LATEX based quiz format. This tool is being used at our institution to create quizzes to support assessment in a PreCalculus course for first year undergraduate students.

1 Introduction

Mathematics assessment should help both student and teacher to understand what the student knows, and to identify areas in which the student needs improvement [14]. As a diagnostic means for assessing conceptual understanding and procedural fluency, multiple-choice tests are quite popular. They can be given and graded at low cost, in contrast to tests with open-answer questions. Nevertheless, their construction remains a time-consuming task. It can get easier when some authoring tool and a bank of questions can be used to produce the tests. There exist hundreds of exercise assistants. The use of a bank of questions, or of templates with parameters that can be randomly instantiated to create variants of the exercises is the most common approach in the design of systems that provide either interactive drills or multiple-choice tests for mathematics (e.g., [11,12,15,20,23,25]). The use of collections of semantically annotated mathematical learning objects is a trend [23,24]. Very often, the exercise systems provide worked out solutions for the drills or automatic feedback that is somehow hardwired to the problem model, even if encoded as a solution graph.

[*] Research funded in part by the ERDF/COMPETE Programme and by the Portuguese Government through the FCT - Fundação para a Ciência e Tecnologia under the projects PEst-C/MAT/UI0144/2011 and FCOMP-01-0124-FEDER-022701.

C. Schulte (Ed.): CP 2013, LNCS 8124, pp. 848–863, 2013.

When the intended answer is a mathematical expression, some systems give automatic feedback by making use of computer algebra systems or specific domain reasoners to diagnose the student answer [5,12,23] or to give hints [1,5,21]. This is useful for formative assessment (i.e., for self-assessment or assessment that is not directly contributing to the student grade). A similar feedback may be given for interactive multiple-choice questions, based on the analysis of individual responses and on the particular exercise model [5,12].

In [17], we proposed a novel approach for creating the drills, and adopted it for developing the AGILMAT prototype[1]. Instead of fixing the template of the parametric expression that is included in the question generator model [11], we focused on the algebraic procedures students know or learn in order to abstract and restrict the expressions in the questions. For that purpose, we tried to understand how the curricula contents condition the drills. This approach is feasible when we consider routine exercises about some topics, and their one-line solutions [17] or step-by-step solutions [1]. In AGILMAT, the expressions arising in drills are specified by constrained grammars and refined by some default profiles and possible user options (see Fig. 1). This is a distinguishing feature

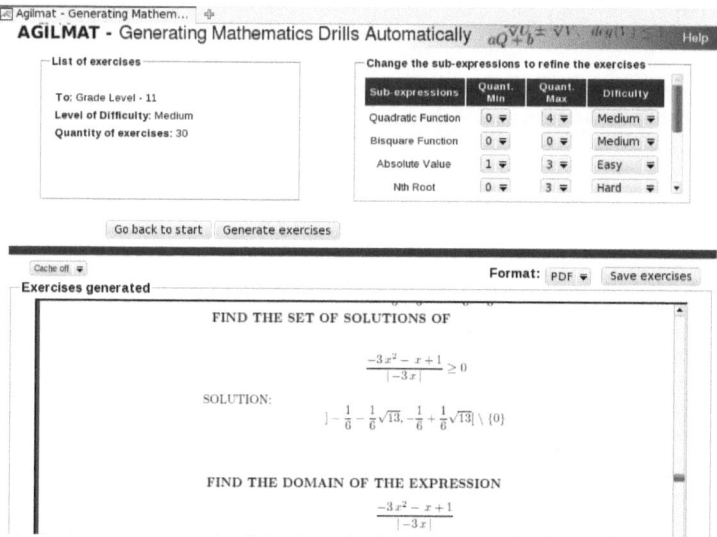

Fig. 1. The AGILMAT prototype available on the web. The notation $]a, b[$ and $[a, b[$ is employed instead of (a, b) and $[a, b)$ to represent intervals of real numbers

and an advantage of our work, making possible the generation of a large number of (non-trivial) examples. For a concise explanation on how this is done, please refer to section 2. A more detailed description can be found in [17,18]. Such

---

[1] http://www.dcc.fc.up.pt:8080/Agilmat

feature gives the user great flexibility to control the forms of the expressions. The system can produce automatically very different expressions or several expressions with the same pattern. The drills are always created and solved on the fly, if the cache option (see Fig. 1) is turned off.

1.1 AgilmatQuiz

In this paper, we describe AgilmatQuiz, the prototype we developed for producing multiple-choice questions for a Pre-Calculus course[2]. This course is being offered at our institution to the first year undergraduate students lacking the required mathematical background. AgilmatQuiz is based on an extension of the AGILMAT core system. Fig. 2 presents a screenshot of an exercise sheet, where questions 9 to 12, among some others, were produced using AgilmatQuiz.

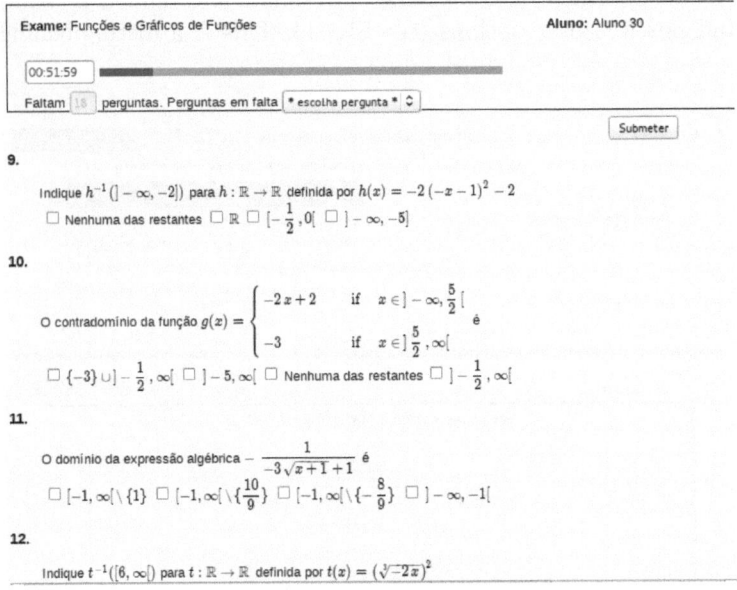

Fig. 2. Multiple-choice exercises about the notions of reciprocal image (9), range (10) and domain (11) of real valued algebraic functions, created by our system to a quiz

New types of expressions and of exercises were introduced. In particular, we were asked to create exercises about disjunctions and conjunctions of simple linear constraints, an extension that was quite easy. We were asked to create exercises involving piecewise-defined functions. This raises some difficulties that

---

[2] Please access `http://www.dcc.fc.up.pt/~apt/Research/AgilmatQuiz.html` to see more examples of questions created by our system.

we discuss in the paper. The questions created automatically cover essentially algebraic functions taught at high-school, the notions of domain, range, reciprocal image, piecewise-defined functions and the solution of equations and inequalities. Some questions about other topics were written by colleagues from the Mathematics Department.

For delivering math quizzes to students, we used Mooshak[3], a web based competitive learning system. Mooshak was originally developed for managing programming contests over the Internet under the ACM International Collegiate Programming Contest rules [8]. Quiz delivery is one of the educational features it supports currently. For AgilmatQuiz, it was improved to handle math expressions. The quiz questions are structured into groups and written in a specially designed LATEX based quiz format. LATEX is widely used in academia and therefore it made easier also the collaboration of our colleagues who were creating some questions by hand.

The rest of the paper is structured as follows. In section 2, we describe the main lines of the approach followed in AGILMAT and for the quiz generation. In section 3, we address the main changes introduced in the AGILMAT core system to be able to produce quiz questions. In section 4, we explain how the Mooshak system supports quiz delivery and grading. Section 5 concludes the paper.

2   Creating Drills Using AGILMAT

In the AGILMAT core system there are two main modular components – the *expression generator* and *exercise generator and solver* – which were implemented using Prolog based constraint programming languages. The *expression generator* processes the user constraints and produces a file with expressions and their *types* (i.e., templates). The *exercise generator and solver* processes this file and produces exercises (according to a specification) and their solutions. This module makes use of several submodules that handle arithmetic, set operations and symbolic constraints (to solve inequations, disequations and equations), along the lines we described in [17]. It uses also some modules for computing limits and derivatives of functions, performing simplications, and obtaining the image of a function when applied to a set (or an upper bound on this image). In addition, it uses a module for converting the internal representations of the exercises to LATEX, as well as their solutions.

Each exercise produced by the AGILMAT prototype has a *question* where a function expression is required. The expressions are built from polynomial functions, the absolute value function $x \to |x|$, and the power and radix functions $x \to x^n$ and $x \to \sqrt[n]{x}$, possibly using composition, addition, product and quotient operations. In the implementation of the expression generator, we follow the grammar proposed in [17] for the expressions. This grammar characterizes a wide range of algebraic expressions found in high school textbooks and whose zeros can be exactly computed by an algorithm students learn. To illustrate the main ideas, we present a fragment in Fig. 3. The category *sumexpr* denotes

---

[3] http://mooshak.dcc.fc.up.pt

$$
\begin{aligned}
prodexpr &\longrightarrow factor \mid factor * prodexpr \\
factor &\longrightarrow sumexpr \mid basic \\
basic &\longrightarrow ipol_2(\mathbf{x}) \mid bisqr \\
&\longrightarrow fbasic \mid fpol_1(fbasic) \mid fpol_1(\mathbf{x}) \\
fbasic &\longrightarrow \mathtt{abs}(basic) \mid \mathtt{pow}(basic, N) \mid \mathtt{rad}(basic, N) \\
ipol_2(T) &\longrightarrow \mathtt{pol}(T, [a, b, c]), \quad abc \neq 0 \\
fpol_1(T) &\longrightarrow \mathtt{pol}(T, [a, b]), \quad a \neq 0 \\
bisqr &\longrightarrow ipol_2(\mathtt{pow}(\mathbf{x}, N)), \quad N \geq 2
\end{aligned}
$$

Fig. 3. An fragment of the grammar proposed in [17]

some particular forms of sum expressions. We can see that $\sqrt[3]{(5x-1)^2}$ and $\sqrt{(2x+3)^5}$ are expressions of the category *basic* and instances of $\sqrt[N]{(ax+b)^M}$. This is rewritten as $\mathtt{rad}(\mathtt{pow}(\mathtt{pol}(x, [a, b]), M), N)$ and the expressions of this form are characterized by the pattern $\mathtt{rad(N)}$ o $\mathtt{pow(M)}$ o $\mathtt{p1}$ o $\mathtt{x}$. Composition, denoted by o, is the main operation. The generation of expressions for the exercises is driven by the generation of their patterns.

Each pattern, called *type*, is represented by a Prolog term with finite domain variables. These variables bind the exponents and, hence, can be constrained to tailor the expressions to specific needs. For example, if we need exercises about the quadratic function, we can constrain the degree of the expression to be two. The exponents are instantiated when the system creates an instance of the expression. Below, we show two other examples of couples of patterns and expressions produced by our generator, their internal representations and usual typesetting.

```
rad(3)o(abs o p1 o pow(2)o p1 o x + pow(2)o p1 o x)
rad(abs(pol(pow(pol(x,[3,-4]),2),[-4,-2])),3) + pow(pol(x,[1,-3]),2)
```

$$
\sqrt[3]{|-4(3x-4)^2 - 2| + (x-3)^2}
$$

```
pow(7)o ip(1)o(p1 o x/p1 o x)
pow(pol(pol(y,[-2,-1])/pol(y,[-3,4]),[-2,3]),7)
```

$$
\left(-2\frac{-2y-1}{-3y+4} + 3\right)^7
$$

Here, $\mathtt{ip(1)}$ is the internal pattern of expressions of the form $\mathtt{pol}(T, [a, b])$, with a and b non-null, whereas, $\mathtt{p1}$ corresponds to $\mathtt{pol}(T, [a, b])$, with b unrestricted. The variable in the expression is not relevant for the template and is passed as a parameter to the generator.

Each type can be used by the system to generate a single or several exercises of the same type. In this way, the difficulty level of the distinct versions would be similar as they are instances of the same template (only coefficients change). This can be important for grading. Nevertheless, for self-assessment or practice, drills with very different expressions favour the learning of concepts and properties instead of focusing on very specific methods for particular instances.

By setting parameters of the generator, the user can refine the types created. Some parameters are used to define constraints on the number of occurrences of each primitive function and of some categories, and also to enforce constraints on the difficulty level of the expressions. The latter is modeled as a weighted sum of the difficulty rates assigned to the primitive functions and some distinguished forms of constructs.

The prototype available on the web (see Fig. 1) allows some customization by default profiles but also by parameters that can be refined by the user. For further details, please refer to [17,18]. However, this is fairly less than what a user that knows CLP can do by interacting directly at the low level. The interface was kept simple because a preliminary version where users could tune several finer parameters was found too complex by a focus group of teachers.

The generator is implemented in a Prolog based constraint programming language and runs on top of the SICStus Prolog system [26]. The constraints act on finite domain variables associated to the types. In general, the grammar rules were implemented by predicates of the form

$$category(\texttt{Type,Degree,Rate,CountTypes,CountOps})$$

where `Degree`, `Rate`, `CountTypes`, `CountOps` are parameters used to constrain the generated `Type`. For example, for the *prodexpr* type, we can have:

```
prodstype(T,G,Rate,CTs,Ops) :-
    constrs(CTs,urestrs_factor),factorstype(T,G,Rate,CTs,Ops).
prodstype(Tb*T,G,Rate,CTs,Ops) :-
    rate_restr(prodstype,Rate,[RateB,RateT]),
    types_restr(prodstype,CTs,[CTsB,CTsT]),
    ops_restr(Ops,[1,OpsB,OpsT]), OpsT #=< OpsB,
    Gb #>= 1, Gt #>= 1, G #= Gt+Gb,
    constrs(CTsB,urestrs_factor),factorstype(Tb,Gb,RateB,CTsB,OpsB),
    prodstype(T,Gt,RateT,CTsT,OpsT).
```

where `constrs/2`, `rate_restr/3` and `ops_restr/2` impose new constraints on subtypes, rates and number of operators. Here, `OpsT #=< OpsB` is added to discard some symmetries. The user can define the rate value of the primitive functions (e.g., `p1`, `p2`, `abs`, `rad(_)`, `pow(_)`, ...) and of particular subexpressions (e.g., sums of radicals, quotients and products), through a predicate `user_rate/2`, used by `rate_restr/3`.

```
rate_restr(T,Rate,L) :- nonnegative(L), user_rate(T,Rt),
    sum([Rt|L],#=,Rate).
```

The parameter `CountTypes` is a list of finite domain variables, each one giving the number of occurrences of a given type and the user can define constraints on the values of these counters. Such constraints are imposed by `constrs/2` and can involve a single variable (e.g., to specify its domain), or any subset of them, and are specified by predicates.

```
constrs(CTs,Functor) :- Goal =.. [Functor,CTsConstr], call(Goal),
    single_vars_low_up_constrs(CTsConstr,CTs),
    user_other_restrs(Functor,CTs).
```

The `Goal` is a user-defined predicate that instantiates `CTsConstr` to a list of terms of the form `I-[Low,Up]`. This list is passed to `single_vars_low_up_constrs/2` to add new constraints on the lower and upper bound values of the variable associated to key `I`. The definition of `user_other_restrs/2` can be less trivial, and may be used to state more complex constraints on subsets of the variables. Still, the configuration constraints are usually simple value or arithmetic constraints, lower and upper bound cardinality constraints or conditional constraints, such as $l \leq x_i \leq u$, $l \leq \sum_{i \in I_1} x_i \leq u$, and $\sum_{i \in I_1} x_i \geq l \Rightarrow \sum_{i \in I_2} x_i \leq u$, where x_i denotes an integer variable, usually a counter.

From the definition of `prodstype/5`, we can see that the underlying CSP model is not a classic model, with a fixed static collection of variables and constraints. This happens very often in configuration problems [4,7]. In our application, new variables and constraints are added during the execution (e.g., the ones corresponding to `RateB`, `RateT`, `OpsB`, `OpsT`, `Gt`, `Gb`, and some variables in the lists `CTsB` and `CTsT`). We are not using global constraints in our application, although domain filtering algorithms for open global cardinality constraints have been investigated [9,19], for some dynamic CSPs. Considering the underlying execution model and the fact that the CSP model is rather dynamic, that will result in a burden without any payoff.

The expression generator produces a file of instances of expressions and their types. A call to `examples(File,DegreeI,RateMin-RateMax,X,NumbInst)` yields a `File` of expressions in the variable `X`, with `NumbInst` instances of each type. The degree of the expressions is `DegreeI` (and can be undefined) and the difficulty level is within `RateMin-RateMax`, which are positive integers.

In AGILMAT, such a file can be passed to the *exercise generator and solver* to create a sheet of exercises and their one-line solutions. We developed specific solvers to be able to handle some nonlinear constraints (in a real-valued variable) and compute exact solutions. A numerical approximation would not be a correct answer usually. Our solvers perform symbolic manipulations and the rules applied emulate steps students may take.

3   Extensions for AgilmatQuiz

The major extensions carried out in this work involved the two main modules, and consisted of:

- the definition of new forms of expressions;
- the modification of the symbolic solver to produce plausible wrong answers;
- the definition of new exercises and strategies for choosing the wrong answers.

In the design of AGILMAT and of this extension, we kept in mind the basic principle that we do not need to support full generality in order to obtain an useful tool. This allows us to partially circumvent some inherent theoretical difficulties of this work, including the ones due to the undecidability of some computer algebra problems [2,10]. We observe that, very often, the topics focused on in the literature are comparatively very *well-behaved*, with well-known canonical forms and solving algorithms (e.g., elementary school arithmetics [22], operations with fractions, linear constraints in a single variable, systems of linear equations, and so forth).

Since the type of the expression acts as a template, the AGILMAT prototype can be used to compute one or more expressions of each type. This feature makes easier the creation of several instances of the same question. This is useful for obtaining multiple-choice tests of the same difficulty level, although this is not too important or even desirable when the tests are used for self-assessment. The separation of the generation of types (templates) from the generation of the instances of the expressions provides the flexibility we need to deal with these two cases. By enforcing constraints on the difficulty level of the types of expressions, we can control the expressions that occur in the questions. This feature is important for multiple-choice questions since, usually, the student must find the answer to the question in a short time.

Besides some simple adjustments, such as the ones needed to create questions about conjunctive and disjunctive constraints, we focused on the generation of expressions for piecewise-defined functions. This is more challenging than the generation of simple expressions, as we need to split the function domain and control the way the different branches fit or do not fit.

3.1   Creating Expressions for Piecewise-Defined Functions

Although we could look at this problem as a constraint problem, devising its solution could be tricky, because we also have to create expressions that are interesting from the pedagogical point of view. This means that the numbers arising in the expression and the breaking points cannot be too scary.

We extended the generator to include a new type `piecewise(L)`, where L is the list of types of the branches. For the corresponding expression, we use a similar notation except that each branch is identified by a term $\mathrm{br}(Expr, DomExpr)$. Below, we show an example of a type and an expression of that type. Actually, the expression is still a *partial expression* as the domain of each branch is a free variable, represented by the underline character (adopting the Prolog notation).

```
piecewise([abs o p1 o x,p1 o x]).
piecewise([br(abs(pol(x,[-2,-1])),_),br(pol(x,[-5,-4]),_)]).
```

$$\begin{cases} |-2x - 1| & \text{if } x \in ? \\ -5x - 4 & \text{if } x \in ? \end{cases}$$

We decided to fix the coefficients of the functions first and then fix the domain of each branch, taking into account the points where the functions meet (although it is possible that the branches do not meet for some other functions). This made the extension of the generator easier. Besides and more importantly, in this way we avoid cumbersome coefficients in the resulting expressions, and therefore obtain expressions that resemble the ones defined by teachers. The quality and variability of questions created by AgilmatQuiz was one of the issues that our colleagues appreciated.

For affine and quadratic functions, the candidate breakpoints can be computed easily by solving an equation. For some other functions, to be able to guarantee that the resulting function is continuous at a breakpoint, we often need to compute lateral limits and, quite likely, to replace some coefficients (e.g., of the constant branches). For instance, $f_k(x)$, defined below, is continuous iff the constant k is zero, since $\lim_{x \to 1+} f_k(x) = 0$.

$$f_k(x) = \begin{cases} \frac{x-1}{\sqrt{x-1}} & \text{if } x > 1 \\ k & \text{if } x \le 1 \end{cases}$$

For the example given above, the complete expression yielded by the system was the following one, which means that the two branches actually meet.

```
piecewise([abs o p1 o x,p1 o x]).
piecewise([br(abs(pol(x,[-2,-1])),[a(-(infty)),a(rat(-1,1))]),
          br(pol(x,[-5,-4]),[f(rat(-1,1)),a(infty)])]).
```

$$\begin{cases} |-2x-1| & \text{if } x \in]-\infty, -1[\\ -5x-4 & \text{if } x \in [-1, \infty[\end{cases}$$

It is possible that the system selects another breakpoint. With some probability, fixed in the implementation, the points where the functions meet would not be selected as breakpoints. In our current implementation, the search for breakpoints that may guarantee continuity is supported for affine and quadratic functions, for example, but we are not reasoning about limits. For the remaining functions, the implementation ensures that the domain of every branch is nonempty by defining breakpoints in the intersection of the domains of the primitive functions, preferentially.

In the implementation, the difficulty rate of a piecewise-defined function is determined by the difficulty rate of the branch that has the highest rate and the total number of branches. A constraint is imposed on the weight of each new branch, so that it does not exceed the half of the previous one.

3.2 Solutions and Distractors

For generating a set of plausible wrong answers (i.e., distractors) for a quiz question, we modified the symbolic solver and some of its submodules to include buggy algebraic rules that translate known common errors or misconceptions.

For that purpose, we augmented the signature of some of the predicates with a new argument that acts as a constrained variable. By imposing constraints on this variable we can restrict and track the number of wrong rules applied in the derivation of an answer. In this way, the same predicate can be used to compute the correct solution if we restrict the value of the variable to be zero. To explain better what we mean, we give a fragment of the initial implementation of the predicate `domain_expr(Expr,X,Dom,DomExpr)`, which determines the domain `DomExpr` of an expression `Expr` in the variable `X` when `X` could only take values in the subset `Dom` of the real numbers.

```
domain_expr(X,X,Dom,Dom)  :- !.
domain_expr(pol(U,_L),X,Dom,Domf)  :- !, domain_expr(U,X,Dom,Domf).
domain_expr(rad(U,N),X,Dom,Domf)  :-  !,
  (even(N) ->
      (domain_expr(U,X,Dom,DomU),solve(DomU,U,geq,rat(0,1),X,Domf));
      domain_expr(U,X,Dom,Domf)).
```

The first rule is equivalent to saying that the identity function defined in the set `Dom` has domain `Dom`. The second rule basically says that the domain of the composition of a polynomial function and a function `U` of `X` is the domain of `U` in `Dom`. Finally, the third rule defines the domain of $\sqrt[N]{U}$ either as `DomU` if `N` is odd or as the solution set of $U \geq 0$ in `DomU` if `N` is even.

For the new version, `wrg_domain_expr(Expr,X,Dom,DomExpr,W)`, we added an extra argument `W`, that is a constraint variable, and add extra rules, which mimic frequent errors, known by experienced teachers.

```
wrg_domain_expr(X,X,Dom,Dom,W)  :- {W = 0}.
wrg_domain_expr(X,X,Dom,[a(-infty),a(infty)],Wf)  :- !,
    {Wf = 1},  Dom \= [a(-infty),a(infty)].
wrg_domain_expr(pol(U,_L),X,Dom,Domf,Wf)  :-
    {Wf >= 0},  wrg_domain_expr(U,X,Dom,Domf,Wf).
wrg_domain_expr(pol(U,_L),X,Dom,Domf,Wf)  :- !,
    {Wf = 1, Ok = 0},
    wrg_domain_expr(U,X,Dom,DomS,Ok),
    DomS \= [a(-infty),a(infty)], Domf = [a(-infty),a(infty)].
wrg_domain_expr(rad(U,N),X,Dom,Domf,Wf)  :- even(N), !,
    {Wi >= 0, Wii >= 0, Wf = Wi+Wii},
     wrg_domain_expr(U,X,Dom,DomU,Wi),
    ( wrg_solve(DomU,U,geq,rat(0,1),X,Domf,Wii);
       ({Wii = 1}, Domf = DomU) ).
wrg_domain_expr(rad(U,_N),X,Dom,Domf,Wf)  :- !, % N is odd
    {W >= 0, Wi >=0, Wf = W+Wi},
    wrg_domain_expr(U,X,Dom,Domi,W),
    ((({Wi=0}, Domf = Domi);
      ({Wi=1}, Domf = Dom, Domi \= Dom);
      ({Wi=1+Wii,Wii>=0}, wrg_solve(Domi,U,geq,rat(0,1),X,Domf,Wii))).
```

The constraint variable[4] is used to control the number of buggy derivation rules applied. If W is bounded to be zero, the predicate produces the correct answer, as before.

Now, the second rule is buggy since it ignores the given domain and yields $\mathbb{R}$ as the answer (the representation of $\mathbb{R}$ is [a(-infty),a(infty)]). For instance, if the stem asks for "the domain of $f : \mathbb{R}_0^+ \to \mathbb{R}$ given by $f(x) = 3x + 1$", the correct answer is $\mathbb{R}_0^+$ and a derivation applying the third clause and then the second one produces $\mathbb{R}$.

The fourth rule is buggy for a similar reason, as it ignores the restrictions imposed both by Dom and the domain of U, and returns $\mathbb{R}$ as the answer. For instance, for $g : \mathbb{R} \to \mathbb{R}$ given by $g(x) = 5\sqrt{x - 4} + 3$", the answer is $[4, \infty[$ but a derivation using the fourth clause yields $\mathbb{R}$.

In the fifth clause, the last branch is buggy and can produce a wrong answer (e.g., $\mathbb{R}$ for $\sqrt{x - 4}$ instead of $[4, \infty[$). Finally, the last rule is buggy and produces a distractor if the solution set of $U \geq 0$ in Domi differs from the correct answer (e.g., for $\sqrt[3]{x - 4}$, if Dom is $\mathbb{R}$, the correct answer is $\mathbb{R}$ and not $[4, \infty[$). If the problem asks for the solutions of, e.g., $\sqrt{1 + \sqrt[3]{x - 4}} = 0$, the application of this buggy rule can lead to a wrong solution.

In our experiments, for finding distractors, we often bound W to 1 when the predicate is called. In this way, we try to focus on more plausible distractors, resulting from a single error, and discard options that can give clues for looking somehow more absurd. Depending on the results of the computations, when W is not 0, the predicate can produce the correct answer as if it were a wrong answer. The set of all wrong solutions, computed by backtracking, is filtered out afterwards to discard repetitions and the "solutions" that are equal to the correct one. In a multiple-choice question about domains, the distractors for the question are selected from this final list, and possibly the item "None of the other ones". With some probability, this item can replace the correct solution also.

In general, the exact comparison of solutions is not straightforward, and can be undecidable [2,10]. In the implementation, we defined a canonical form for some expressions and constants and for the restricted sets manipulated by the system (which are unions of a finite number of intervals and isolated points), as in our previous work [17]. Our solver is not complete, as a domain reasoner. For comparing more complex constants (clearly, not rational numbers), the system sometimes performs a numeric comparison, after evaluating the constants as floating point numbers. In practice, by limiting the number of flaws to 1, we obtain more plausible distractors that are helpful for identifying a student error or misconception. In addition, we reduce the risk of finding equivalent solutions that are not syntactically equal (e.g., $2\sqrt{7 + 4\sqrt{3}}$ and $4 + 2\sqrt{3}$), by avoiding many alternative derivations and unnecessary computations that may yield complex

---

[4] The AGILMAT solver used the CLP($\mathbb{Q}$) module for supporting computations with rational numbers. This is the reason for W being not treated as a finite domain variable, as that avoid some (re-)implementation effort. We plan to fix that in a future version of the prototype.

constants. On the other hand, some more powerful procedures available in current computer algebra systems (e.g., Mathematica, Maple, Maxima, etc, to name a few) are too advanced for high-school or undergraduate students. Hence, given that our system can produce a huge number of exercise instances, the system can discard the ones it cannot solve exactly. Nevertheless, this is an issue that requires further research to understand the limits and possible improvements of our approach although, it is know that, in general, canonical forms cannot exist [10].

It is worth mentioning that there are other e-learning tools that make use of buggy rules for creating exercises or for diagnosis [21,22]. SLOPERT, for instance, is a reasoner and diagnoser for symbolic differentiation, developed in Prolog, used as a domain reasoner by LeActiveMath and MathBridge [5,21]. It is enriched with buggy derivation rules, implemented as clause predicates as in AgilmatQuiz, each one being annotated as buggy (wrong) or expert (correct) rule. A parameter keeps track of the history of the derivation, but there is not the same support for imposing constraints on the number of buggy rules that can be applied in a derivation.

3.3   New Exercises and Strategies for Selecting Distractors

The set of exercises was also enriched and the corresponding solving procedures were implemented. As we observe above, we try to create exercises that make some sense. This means that we sometimes can exploit the relationship between some notions, e.g. range and reciprocal image, to obtain exercises that are pedagogically acceptable. For instance, for creating the exercises about the notion of reciprocal image $f^{-1}(D)$, the system selects D as a subset of the range of f, and so it first tries to compute that range.

Poorly written distractors for a multiple-choice question can invalidate the question. Finding good distractors can be difficult even for teachers. For distractor development, we tried to attend to the following rules: "use plausible distractors; avoid illogical distractors; incorporate common errors of students in distractors; use true statements that do not correctly answer the item" [6]. Since the choices we consider are either wrong solutions or the correct one, we interpret the last goal as the inclusion of solutions that overlap, when the answer is a set.

Another guideline is to avoid or use sparingly "None of the above" [6]. However, the inclusion of this choice in some problems, either as distractor or correct answer, was a requirement from our colleagues (because the choices are shuffled by Mooshak, actually we use "None of the other ones" instead). This choice was intentionally used to make the guess of the correct answer by a simple analysis of the offered answers more difficult. We agree that this can be relevant for Mathematics. For instance, when the correct answer is a set, students may have to work out the solution if the other choices cannot be trivially discarded.

To prevent correct answers from being trivially guessed, the system tries to classify the distractors, for instance in terms of their intersection with the correct solution or the number of derivations that led to each one. This classification is sometimes used to introduce some bias on the selection procedure. In the implementation, the distractors are selected randomly for each problem, and the preference for distractors obtained by some rules can increase their weight. For instance, in problems about the reciprocal image $f^{-1}(D)$, a common error is the interpretation of f^{-1} as $1/f$. Distractors produced by the buggy rule that translates this misconception were given some higher weight. The number of times a distractor occurs is used in this case also.

The criteria for the selection of distractors from the computed list is an issue that deserves further investigation. In particular, a more accurate model for defining the preferences for some rules could be designed.

4   Quiz Delivery

In this section, we give further details on how the system is used to produce a quiz in the **mooshakquiz** style and how the quiz is delivered to students.

Mooshak Quizzes. Mooshak is a web based competitive learning system originally developed for managing programming contests [8]. It is used as an e-learning tool in several universities. Quiz delivery started as a complement for evaluation in programming courses, giving support for multiple-choice questions. It generates quizzes by randomly selecting questions and shuffling them and their items. Quizzes are graded automatically. Each correctly answered question adds its mark to the final grade. Incorrect answers are penalized so that a series of random answers to the quiz questions has an expected grade of zero. The system provides overall statistics per question. Quizzes are structured in groups. The number of questions in a group may be larger than the number of questions actually presented to students. A group may be regarded as a question bank. If this bank is created by AgilmatQuiz from a single template expression, the tests produced by selecting a single question from each group have similar difficulty level.

Quiz Format. Mooshak and AGILMAT use different formats. Mooshak uses its own XML based format to import and export data. AGILMAT uses LATEX as output format. To make the two systems interoperable with each other the main issue was the definition of a common quiz format. A natural candidate for this role is the Question & Test Interoperability (QTI) standard. This approach would require some implementation effort on the AGILMAT side. Moreover, QTI with MathML would be inappropriate for humans. Teachers must be able to produce quizzes in the selected format, as certain exercise types needed for the PreCalculus course are not yet covered by AgilmatQuiz.

The final decision was to create a new LATEX based format for quizzes – the **mooshakquiz** style. The quiz is defined as a document structured by LATEX environments defining groups, questions and choices. These environments are configured by parameters, such as the number of questions extracted from each group (typically 1) or the logical values of a choice (true or false). Text in these environments may contain math expressions. A quiz in this format may be processed as a LATEX document to produce an handout in PDF format, for instance.

On the AgilmatQuiz side, a predicate **groups**(p($Pred, Args, Nq$)$-File$) defines the groups of exercises to create. Here, $Pred$ is the name of the predicate that will generate the corresponding group using the expressions already saved in the file $File$. The definition looks like the following one.

```
groups(p(quizReciprocal,3)-p1)
groups(p(quizConjDisj,2,1)-p1).
groups(p(quizDomains,1)-rad1simple).
```

The sequence $Args$ is optional and defines parameters to this predicate $Pred$. Finally, Nq defines the number of questions that will be selected from each group to a test. The system produces a LaTeX file in the mooshakquiz style, that is used for quiz delivery.

```
\documentclass{article}
\usepackage[utf8]{inputenc}
\usepackage{mooshakquiz}
\begin{document}
\begin{quizgroup}{3}
\begin{quizquestion} Find $\displaystyle t^{-1}(]1,\infty[)$
            for $\displaystyle t :\mathbb{R}\rightarrow\mathbb{R}$
            given by $ \displaystyle t(x) = -6\,x-1$ \newline
\begin{quizchoice}{false}$ \displaystyle ]-\infty,-3[ $  \end{quizchoice}
\begin{quizchoice}{true}$ \displaystyle ]-\infty,-\frac{1}{3}[ $
                                \end{quizchoice}
\begin{quizchoice}{false}$ \displaystyle ]-\infty,-8[ $  \end{quizchoice}
\begin{quizchoice}{false}$ \displaystyle ]-\infty,0[ $  \end{quizchoice}
\end{quizquestion}
\begin{quizquestion} Find $\displaystyle t^{-1}(]4,\infty[)$
            for $\displaystyle t :\mathbb{R}\rightarrow\mathbb{R}$
            given by $ \displaystyle t(x) = 2\,(x-4)$ \newline
\begin{quizchoice}{false}$ \displaystyle ]-\infty,6[ $  \end{quizchoice}
\begin{quizchoice}{false}$ \displaystyle ]-1,\infty[ $  \end{quizchoice}
\begin{quizchoice}{false}$ \displaystyle ]-4,\infty[ $  \end{quizchoice}
\begin{quizchoice}{true}$ \displaystyle ]6,\infty[ $  \end{quizchoice}
\end{quizquestion}
...
\end{quizgroup}
\begin{quizgroup}{1} ... \end{quizgroup}
\begin{quizgroup}{1} ... \end{quizgroup}
\end{document}
```

Quiz Processing. Mooshak required minor changes to process quizzes. The function that imports quizzes in the mooshakquiz style converts the environment based structure of the document to XML, leaving text and math expressions unchanged. The document is then imported to the internal representation of Mooshak and processed as regular quiz, with text and math expressions inserted in HTML pages and presented on a web browser. Math expressions in LaTeX are converted on-the-fly in the browser using the MathJax [3] JavaScript display engine, which was crucial for a quick implementation.

5   Conclusions and Future Work

This work describes an approach for generating and delivering math quizzes using constraint logic programming. The main contribution is a novel approach for creating multiple-choice questions with a set of plausible wrong answers. We focused on a particular type of multiple-choice questions, but our approach could be exploited to support the generation of other types of questions or populate question repositories (if the output is written in some more standard exercise language). The work is an example of an application where the use of declarative languages was crucial for a rapid development of an useful tool. At the current stage, the generator and solver consist of about 7000 lines of code. But, it is not very easy to quantify the overall development effort of AgilmatQuiz, as it is an extension of AGILMAT. Our crude estimate is of about one month-person for the reported extension. We plan to improve the implementation and cover other topics. Constraint programming makes easier the re-usability and customization of the system. However, it would be interesting to study execution models where the propagation of constraints plays a more significant role in the program transformation. The prototype is currently used to support a remedial PreCalculus course for students entering the Faculty of Sciences at the University of Porto. A formal evaluation is planned. It would be already possible to draw some conclusions if we check whether experienced teachers can separate the exercises produced automatically from identical ones produced manually.

Acknowledgments. The authors would like to thank their colleagues responsible for the PreCalculus course and anonymous reviewers for helpful comments.

References

1. Beeson, M.: Design Principles of Mathpert: Software to Support Education in Algebra and Calculus. In: Kajler, N. (ed.) Computer-Human Interaction in Symbolic Computation, Texts and Monographs in Symbolic Computation, pp. 89–115. Springer, Heidelberg (1998)
2. Bradford, R., Davenport, J.H., Sangwin, C.J.: A Comparison of Equality in Computer Algebra and Correctness in Mathematical Pedagogy. In: Carette, J., Dixon, L., Coen, C.S., Watt, S.M. (eds.) Calculemus/MKM 2009. LNCS (LNAI), vol. 5625, pp. 75–89. Springer, Heidelberg (2009)
3. Cervone, D.: MathJax – A Platform for Mathematics on the Web. Notices of the AMS 59, 312–316 (2012)
4. Faltings, B., Macho-Gonzalez, S.: Open Constraint Programming. Artificial Intelligence 161, 181–208 (2005)
5. Goguadze, G.: ActiveMath – Generation and Reuse of Interactive Exercises using Domain Reasoners and Automated Tutorial Strategies. PhD thesis, Saarland University (2011)
6. Haladyna, T.M., Downing, S.M.: A Taxonomy of Multiple-Choice Item-Writing Rules. Applied Measurement in Education 2, 37–50 (1989)
7. Junker, U.: Configuration. In: Rossi, F., van Beek, P., Walsh, T. (eds.) Handbook of Constraint Programming, pp. 835–871. Elsevier (2006)

8. Leal, J.P., Silva, F.: Mooshak: a Web-based Multi-site Programming Contest System. Software – Practice and Experience 33, 567–581 (2003)
9. Maher, J.M.: Open Contractible Global Constraints. In: 21st International Joint Conf. on Artificial Intelligence, IJCAI 2009, pp. 578–583. Morgan Kaufmann Publishers, USA (2009)
10. Moses, J.: Algebraic Simplification: a Guide for the Perplexed. Communications of the ACM 14, 527–537 (1971)
11. Pinto, J.S., Oliveira, M.P., Anjo, A.B., Vieira Pais, S.I., Isidro, R.O., Silva, M.H.: TDmat-Mathematics Diagnosis Evaluation Test for Engineering Sciences Students. Int. J. Mathematical Education in Science and Technology 38, 283–299 (2007)
12. Sangwin, C.J., Grove, M.J.: STACK – Addressing the Needs of the "Neglected Learners". In: 1st WebAlt Conference and Exhibition, pp. 81–95 (2006)
13. Sangwin, C.: Computer Aided Assessment of Mathematics. Oxford University Press (2013)
14. Schoenfeld, A.H. (ed.): Assessing Mathematical Proficiency. Cambridge University Press (2007)
15. Snajder, J., Cupic, M., Basic, B.D., Petrovic, S.: Enthusiast: An Authoring Tool for Automatic Generation of Paper-and-Pencil Multiple-Choice Tests. In: ICL 2008, Villach, Austria (2008)
16. Sterling, L., Bundy, A., Byrd, L., O'Keefe, R., Silver, B.: Solving symbolic equations with Press. Journal of Symbolic Computation 7, 71–84 (1989)
17. Tomás, A.P., Leal, J.P.: A CLP-Based Tool for Computer Aided Generation and Solving of Maths Exercises. In: Dahl, V., Wadler, P. (eds.) PADL 2003. LNCS, vol. 2562, pp. 223–240. Springer, Heidelberg (2002)
18. Tomás, A.P., Leal, J.P., Domingues, M.: A Web Application for Mathematics Education. In: Leung, H., Li, F., Lau, R., Li, Q. (eds.) ICWL 2007. LNCS, vol. 4823, pp. 380–391. Springer, Heidelberg (2008)
19. van Hoeve, W.-J., Régin, J.-C.: Open Constraints in a Closed World. In: Beck, J.C., Smith, B.M. (eds.) CPAIOR 2006. LNCS, vol. 3990, pp. 244–257. Springer, Heidelberg (2006)
20. Xiao, G.: WIMS – An Interactive Mathematics Server. Journal of Online Mathematics and its Applications 1, MAA (2001)
21. Zinn, C.: Supporting Tutorial Feedback to Student Help Requests and Errors in Symbolic Differentiation. In: Ikeda, M., Ashley, K.D., Chan, T.-W. (eds.) ITS 2006. LNCS, vol. 4053, pp. 349–359. Springer, Heidelberg (2006)
22. Zinn, C.: Program Analysis and Manipulation to Reproduce Learners' Erroneous Reasoning. In: Albert, E. (ed.) LOPSTR 2012. LNCS, vol. 7844, pp. 228–243. Springer, Heidelberg (2013)
23. LeActiveMath: Language-Enhanced, User Adaptive, Interactive eLearning for Mathematics, EU project (2004–2006), http://www.leactivemath.org/
24. Math-Bridge: European Remedial Content for Mathematics, EU project (2009–2012), http://www.math-bridge.org/
25. PmatE – Mathematics Education Project. University of Aveiro, Portugal (1990), http://pmate4.ua.pt/pmate/
26. SICStus Prolog. SICS, Sweden, http://www.sics.se
27. STACK: System for Teaching and Assessment using a Computer algebra Kernel. University of Birmingham, UK, http://www.stack.bham.ac.uk/

Constrained Wine Blending

Philippe Vismara[1,2], Remi Coletta[1], and Gilles Trombettoni[1]

[1] LIRMM, UMR5506 Université Montpellier II - CNRS, Montpellier, France
{Philippe.Vismara,Remi.Coletta,Gilles.Trombettoni}@lirmm.fr
[2] MISTEA, UMR0729 Montpellier SupAgro - INRA, Montpellier, France

Abstract. Assemblage consists in blending base wines in order to create a target wine. Recent developments in aroma analysis allow us to measure chemical compounds impacting the taste of wines. This chemical analysis makes it possible to design a decision tool for the following problem: given a set of target wines, determine which volumes must be extracted from each base wine to produce wines that satisfy constraints on aroma concentration, volumes, alcohol contents and price. This paper describes the modeling of wine assemblage as a non linear constrained Min-Max problem (minimizing the gap to the desired concentrations for every aromatic criterion) efficiently handled by the Ibex interval branch and bound.

1 Introduction

Assemblage is the subtle blending of wines from different vineyard plots and/or different grape varieties, each contributing its own special flavor.

Wine blending is generally carried out by oenologists working for wineries. Oenologists can obtain wine blendings of the highest quality, but taste saturation entails a strong limit in the number of daily wine tasting sessions. Therefore the Nyseos company (www.nyseos.fr), which submitted the blending problem to us, provides chemical analysis tools to avoid a number of tasting sessions. These tools can analyze wine aromas by measuring a set of chemical compounds that impact wine taste [6]. These tools make it possible to develop a decision-support software for the following problem: given a set of target wines to be produced, which volumes must be taken from each base wine in order to make wines satisfying constraints on aroma concentrations, volumes, alcohol content, price, etc.

Moore and Griffin have shown that aroma concentrations of a wine blending satisfy linear constraints [11]. However, several other requirements lead to nonlinear constraints. For instance, the Nyseos company works on a model able to predict the color of a wine. The model will not be linear and the complexity of color modeling is confirmed by other researches [8]. Another critical point is that no less than a given amount of wine can be transferred from a tank to a target because of the loss of liquid in the pipes and the manipulation cost.

As we will see in this article, this requirement leads to a disjunctive constraint that can be modeled by boolean variables and nonlinear constraints.

C. Schulte (Ed.): CP 2013, LNCS 8124, pp. 864–879, 2013.

An interesting algorithmic research on wine blending has been presented in [8]. An artificial neural network approach has been used to select the wine quantities extracted from each base in order to elaborate a wine matching predefined aromatic criteria. In this work, aromatic criteria were not chemically analyzed. Instead, a panel of students carried out tasting to quantify predefined criteria. The neural network performed multicriteria optimization for adjusting each aroma. The comparison with our approach is difficult in terms of quality since we preferred to resort to monocriterion optimization. In addition, no performance (CPU time) results are shown in [8]. Another research dealt with the blending problem [7]. The main objective was to find the best matching between chromatograms of base and target wines. This problem was modeled by a *non constrained* nonlinear optimization solved by a local (Nelder-Mead) optimization method.

In this article, we present a mathematical modeling of the wine assemblage problem. The problem is modeled by a mixed (discrete and continuous) nonlinear program. We transform it into a non-linear (pure) continuous CSP handled by a rigorous interval Branch and Bound (B&B). We have built a constrained optimization model for minimizing in each target wine the gap between desired aromatic concentrations and obtained concentrations, while taking into account the minimal transfer disjunctive constraint. Absolute value and max operators have been removed from the obtained system.

2   The Wine Assemblage Problem

Figure 1 illustrates the definition of wine assemblage.

We consider a set of base wines numbered from 1 to $\mathcal{B}$. We denote by vol_b the volume of the base $b \in 1..\mathcal{B}$.

For different reasons, it is sometimes impossible to completely empty a tank. Let s_b^- be the minimum volume that must remain in tank b. (We have: $0 \leq s_b^- \leq \text{vol}_b$.) All base wines are analyzed in order to measure the concentration of selected key aroma compounds. These compounds are numbered from 1 to $\mathcal{A}$. We denote by $c_{b,a}$ the concentration of aroma a in base b.

A wine assemblage support tool should help to simultaneously build several target wines from a given set of bases. Hence, we consider a set of target wines, numbered from 1 to $\mathcal{W}$. For each wine w, we aim to produce an optimal volume $\widehat{\text{vol}}_w$. The final volume V_w of wine w should be as close as possible to $\widehat{\text{vol}}_w$ and must remain greater (resp. smaller) than a given lower bound vol_w^- (resp. an upper bound vol_w^+), i.e.:

$$\forall w \in 1..\mathcal{W}, \quad \text{vol}_w^- \leq V_w \leq \text{vol}_w^+ \tag{1}$$

These bounds are used to fulfill an order of a specific volume or to avoid producing an excessive volume.

Each target wine w is a blend of wines extracted from several tanks. We denote by $V_{w,b}$ the volume of wine w that has been pumped from base tank b.

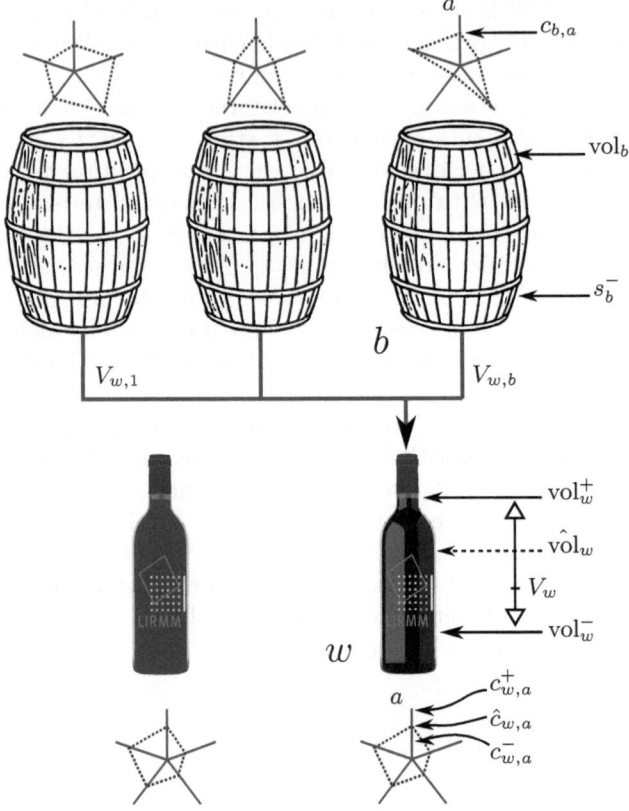

Fig. 1. Wine assemblage

We have a direct relation with V_w:

$$\forall w \in 1..\mathcal{W}, \quad V_w = \sum_{b=1}^{\mathcal{B}} V_{w,b} \tag{2}$$

Furthermore, all the volumes extracted from the same base tank b must leave a minimum volume s_b^- in the tank.

$$\forall b \in 1..\mathcal{B}, \quad s_b^- \leq \mathrm{vol}_b - \sum_{w=1}^{\mathcal{W}} V_{w,b} \tag{3}$$

When transferring wine between two tanks, a subpart is generally wasted in the pipes. Hence it is impossible to transfer very small volumes. If δ_V is the minimum volume that can be transferred between two tanks, we define the following *disjunctive constraint*:

$$\forall w \in 1..\mathcal{W}, \forall b \in 1..\mathcal{B}, \ (V_{w,b} = 0) \ \vee \ (\delta_V \leq V_{w,b}) \tag{4}$$

In addition to volume, each target wine is described in terms of aroma compound concentration. For a given wine w, we denote by $\hat{c}_{w,a}$ the desired concentration of aroma a.

The $C_{w,a}$ concentrations are to be as close as possible to $\hat{c}_{w,a}$ within an interval $[c_{w,a}^-, c_{w,a}^+]$, where $c_{w,a}^-$ (resp. $c_{w,a}^+$) denotes the minimum (resp. maximum) admissible concentration of aroma a in wine w. The relation between volumes and concentrations can be formulated as follows:

$$\forall w \in 1..\mathcal{W}, \ \forall a \in 1..\mathcal{A}, \ \ c_{w,a}^- \leq \frac{1}{V_w} \sum_{b=1}^{\mathcal{B}} (V_{w,b} \cdot c_{b,a}) \leq c_{w,a}^+ \tag{5}$$

In a similar way, we can model constraints on alcohol content or price per liter for the target wines. These can be treated like additional aromas.

3   A MINLP Formulation for Wine Blending

We can model the wine blending problem as a mixed nonlinear program (MINLP). We show in Section 4 how to straightforwardly transform the MINLP into a numerical CSP (NCSP) handled by interval methods. This explains why the bound constraints are directly modeled below by bounded domains, i.e., intervals.

3.1   Variables

For handling realistic volumes (due to (4)), for each wine $w \in 1..\mathcal{W}$ and each base $b \in 1..\mathcal{B}$, we create:

- a 0/1 variable $P_{w,b}$ and
- a variable $V'_{w,b}$ with a domain $D(V'_{w,b}) = [\delta_V, \min(\text{vol}_w^+, \text{vol}_b)]$ representing the volume coming from the base b in the wine w.
  (We have: $V_{w,b} \equiv P_{w,b} \cdot V'_{w,b}$. The introduction of these 0/1 variables of course avoids an explicit definition of the disjunctive constraint (4).)

For each volume of a wine $w \in 1..\mathcal{W}$, we also define a variable V_w of domain $[\text{vol}_w^-, \text{vol}_w^+]$ (see (1)).

3.2   Constraints

The system of constraints of our MINLP is described below.

- The channeling constraint (2) becomes:

$$\forall w, \ V_w - \sum_{b=1}^{\mathcal{B}} (P_{w,b} \cdot V'_{w,b}) = 0 \tag{2.i}$$

– The surplus constraint (3) also remains similar:

$$\forall b \in 1..\mathcal{B}, s_b^- \leq \text{vol}_b - \sum_{w=1}^{\mathcal{W}} (P_{w,b} \cdot V'_{w,b}) \tag{3.i}$$

– To enhance the performance results, we have added a constraint redundant to (3.i). This constraint simply ensures that the sum of the volumes of target wines is inferior to the sum of the base volumes:

$$\sum_{w=1}^{\mathcal{W}} V_w \leq \sum_{b=1}^{\mathcal{B}} \text{vol}_b \tag{6}$$

Aroma concentration requirements (see (5)) are decomposed into two constraints, and both parts of inequalities are multiplied by the positive volume V_w. $\forall w \in 1..\mathcal{W}$ and $\forall a \in 1...\mathcal{A}$, we have:
for the lower bound,

$$0 \leq \sum_{b=1}^{\mathcal{B}} V_{w,b} \cdot (c_{b,a} - c_{w,a}^-)$$

hence:

$$0 \leq \sum_{b=1}^{\mathcal{B}} P_{w,b} \cdot V'_{w,b} \cdot (c_{b,a} - c_{w,a}^-) \tag{5.i -}$$

and for the upper bound:

$$0 \leq \sum_{b=1}^{\mathcal{B}} V_{w,b} \cdot (c_{w,a}^+ - c_{b,a})$$

hence:

$$0 \leq \sum_{b=1}^{\mathcal{B}} P_{w,b} \cdot V'_{w,b} \cdot (c_{w,a}^+ - c_{b,a}) \tag{5.i +}$$

3.3   A Min-Max for Reaching the Highest Quality of Wines

In this application, the significant criterion is wine quality. Nevertheless, we could extend the definition of E below to errors on alcohol content or price. Bear in mind that a target wine is defined by a set of desired concentrations $\hat{c}_{w,a}$ for each of its aroma. Therefore, a way to optimize the quality of the target wines is to minimize a weighted sum of differences between the concentrations desired $\hat{c}_{w,a}$ and the concentrations obtained $C_{w,a}$ (see (5)). Furthermore, we want to minimize the maximal error on the set of target wines:

$$\max_{w \in 1..\mathcal{W}} \Omega_w (\lambda_{vol_w} \cdot e_{vol_w} + \sum_{a \in 1..\mathcal{A}} (\lambda_{w,a} \cdot e_{w,a})) \tag{7}$$

where:

- Ω_w is a parameter reflecting how important is a given wine w (Ω_w is assumed to be in $[0,1]$). This weight ensures a more accurate blending to the best wines among the targets.
- $e_{w,a}$ denotes the discrepancy between $\hat{c}_{w,a}$ and $C_{w,a}$.
- e_{vol_w} denotes the discrepancy between the volume $\hat{vol}_w$ of wine w desired and the volume V_w obtained.
- $\lambda_{w,a} \in [0,1]$ defines the weight of aroma a in the wine w. $\lambda_{vol_w} \in [0,1]$ weights the respect of the volume requirement of wine w compared to the satisfaction of aroma concentrations. For a given target wine w, we assume that $\lambda_{vol_w} + \sum_{a\in 1...\mathcal{A}} \lambda_{w,a} = 1$.

All the parameters, including the aroma concentrations, are measured with a given uncertainty. ε_a denotes the measure error related to the concentration of aroma a. We thus want to minimize the gap between $\hat{c}_{w,a}$ and $C_{w,a}$ within the limit given by this uncertainty ε_a. In other words, if the gap between the desired and obtained concentrations remains below the uncertainty, it will be considered as being null in the objective function. Thus, the variable $e_{w,a}$ describes the normalized concentration error in each aroma a for each wine w, as follows:

$$e_{w,a} = \max(\frac{|C_{w,a} - \hat{c}_{w,a}|}{\hat{c}_{w,a}} - \varepsilon_a, 0) \tag{8}$$

We can also describe the gap e_{vol_w} between the volume of a target wine V_w obtained and the volume $\hat{vol}_w$ desired with a similar expression:

$$e_{vol_w} = \max(\frac{\hat{vol}_w - V_w}{\hat{vol}_w} - \varepsilon_{vol}, 0) \tag{9}$$

Compared to the previous formula, the removal of the absolute value simply means that no error is taken into account if the volume V_w computed falls between the maximum volume vol_w^+ and the target $\hat{vol}_w$. This is illustrated by Figure 2.

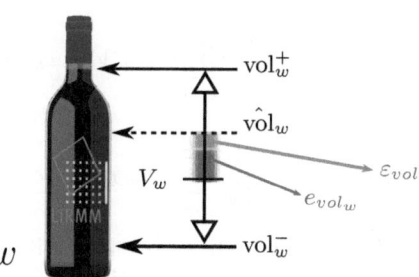

Fig. 2. Visualization of the gap e_{vol_w}

Following a usual way to define a Min-Max problem, we add a variable $E \in [0, +\infty]$ to be minimized and the following constraints:

$$\forall w \in 1..\mathcal{W}, \quad \Omega_w(e_{vol_w} \cdot \lambda_{vol_w} + \sum_{a \in 1..\mathcal{A}} (e_{w,a} \cdot \lambda_{w,a})) \leq E \tag{10}$$

Removing Max and Absolute Value Operators

In order to increase the performance, we attempt to remove max and absolute value operators. Observe that the maximum operator can be defined by

$$e = max(x, y) \equiv e \geq x \wedge e \geq y \wedge (e = x \vee e = y).$$

In addition, if the quantity e must be minimal for any reason, the last conjunct can be removed, thus simplifying the max operator. We can apply this simplification to (8). Indeed, $\lambda_{w,a}$ is positive so that minimizing E entails minimizing every variable $e_{w,a}$. Hence:

$$\forall w \in 1..\mathcal{W}, \forall a \in 1..\mathcal{A}, \quad ((e_{w,a} + \varepsilon_a) \hat{c}_{w,a} \geq |C_{w,a} - \hat{c}_{w,a}|) \wedge (e_{w,a} \geq 0) \tag{11}$$

The same simplification can be applied to (9), as follows:

$$\forall w \in 1..\mathcal{W}, \quad ((e_{vol_w} + \varepsilon_{vol}) \hat{vol}_w \geq (\hat{vol}_w - V_w)) \wedge (e_{vol_w} \geq 0) \tag{12}$$

We can also remove the absolute value operator above that can be transformed into a max operator as follows:

$$e = |x| \equiv e = max(x, -x)$$
$$\equiv e \geq x \wedge e \geq -x \wedge (e = x \vee e = -x)$$

Once more, if the quantity e must be minimal for any reason, the last conjunct can be removed, thus replacing the absolute value operator with two inequalities. We can apply this simplification to (11). Indeed, remember that every variable $e_{w,a}$ must be minimized and observe that $\hat{c}_{w,a}$ is positive. Thus, $\forall w \in 1..\mathcal{W}, \forall a \in 1..\mathcal{A}$,

$$(e_{w,a} + \varepsilon_a) \hat{c}_{w,a} \geq \frac{1}{V_w} \cdot \sum_{b=1}^{\mathcal{B}} (P_{w,b} \cdot V'_{w,b} \cdot c_{b,a}) - \hat{c}_{w,a}$$

$$(e_{w,a} + \varepsilon_a) \hat{c}_{w,a} \geq -\frac{1}{V_w} \cdot \sum_{b=1}^{\mathcal{B}} (P_{w,b} \cdot V'_{w,b} \cdot c_{b,a}) + \hat{c}_{w,a}$$

Multiplying both parts of these inequalities by the positive volume V_w, we finally obtain the following three categories of constraints: $\forall w \in 1..\mathcal{W}, \forall a \in 1..\mathcal{A}$,

$$e_{w,a} \geq 0 \tag{13}$$

$$V_w.(e_{w,a} + \varepsilon_a + 1).\hat{c}_{w,a} - \sum_{b=1}^{\mathcal{B}}(P_{w,b}.V'_{w,b}.c_{b,a}) \geq 0 \qquad (14)$$

$$V_w.(e_{w,a} + \varepsilon_a - 1).\hat{c}_{w,a} + \sum_{b=1}^{\mathcal{B}}(P_{w,b}.V'_{w,b}.c_{b,a}) \geq 0 \qquad (15)$$

As a result, we have succeeded in suppressing from our initial model all the absolute value and max operators. Although our interval nonlinear constraint solver can handle these operators, the performance is thus increased and the simplified model can also be implemented in other solvers.

3.4   Summary

In addition to the variables $P_{w,b}$, $V'_{w,b}$ and V_w defined in Section 3.1, we define new variables for the Min-Max: one variable $E \in [0, +\infty]$, $\mathcal{W}.\mathcal{A}$ variables $e_{w,a} \in [0, 1]$ (that absorb the unary constraints (13)) and $\mathcal{W}$ variables $e_{vol_w} \in [0, 1]$ that absorb the unary constraints of (12).

In addition to the constraints (2.i), (3.i), (6), (5.i-), (5.i+) defined in Section 3.2, we define new constraints for the Min-Max: (10), (12), (14), (15). The objective function simply consists in minimizing the value of the variable E.

4   Solving the MINLP with an Interval B&B

We wanted a free solver in order to embed it in the final dedicated tool for Nyseos. In addition, the MINLP detailed above could be handled by any MINLP solver such as Baron [13] or Couenne [2], but all of them are *not* rigorous (safe). This means that they sometimes miss the best solution due to round-off errors related to floating-point arithmetic. It is known that cases where the best solution is missed by unsafe solvers are rare but do occur in practice. Using a safe optimizer was reassuring for Nyseos. Furthermore, since:

- our modeling of the wine blending problem contains only one type of 0/1 variables,
- the interval solver Ibex features the very efficient IbexOpt interval B&B [14],
- the authors have a good command of Ibex [5,4],

we decided to simply encode the MINLP problem as an NCSP, i.e., a standard *continuous* system of nonlinear constraints (i.e., over the real numbers). To do so, the 0/1 variables are encoded by real-valued variables $P'_{w,b}$ of domain $[0, 1]$. To ensure these variables take 0/1 values, we simply add the following quadratic constraints:

$$\forall w \in 1..\mathcal{W} \text{ and } \forall b \in 1..\mathcal{B}, \ 4(P_{w,b} - \frac{1}{2})^2 = 1 \qquad (16)$$

This means that the initial disjunctive constraints that produce mixed constraints in the MINLP model are handled by continuous quadratic constraints.

Our good command of the interval solver `Ibex` enabled to produce an efficient strategy, but it would be interesting to compare in a future work our interval B&B with a MINLP solver like Couenne [2]. Quadratic solvers are not our first alternative choice because our model will probably be extended with other nonlinear (and non quadratic) constraints about color or wine varieties. To our knowledge, only one rigorous interval B&B, called `IBBA` [12], is endowed with a simple mechanism handling integral variables. `IBBA` could be compared with Ibex, although the two solvers are merging.

4.1   Constrained Global Optimization with an Interval B&B

A continuous constrained global optimization problem is defined as follows.

Definition 1 (Constrained global optimization)
Consider a vector of variables $x = (x_1, ..., x_n)$ varying in a domain $[x] = [x_1] \times \cdots \times [x_n]$, a real-valued function $f : \mathbb{R}^n \to \mathbb{R}$, vector-valued functions $g : \mathbb{R}^n \to \mathbb{R}^m$ and $h : \mathbb{R}^n \to \mathbb{R}^p$. We have $g = (g_1, ..., g_m)$ and $h = (h_1, ..., h_p)$.
Given the system $S = (f, g, h, [x])$, the constrained global optimization problem consists in finding:

$$\min_{x \in [x]} f(x) \quad subject\ to \quad g(x) \leq 0 \wedge h(x) = 0.$$

f denotes the objective function; g and h are inequality and equality constraints respectively.

Our `IbexOpt` constrained global optimizer [14] computes a floating-point vector x ϵ-minimizing[1]:

$$f(x)\ s.t.\ g(x) \leq 0 \wedge (-\epsilon_{eq} \leq h(x) \leq +\epsilon_{eq}).$$

Note that equalities $h_j(x) = 0$ are relaxed by "thick" equations $h_j(x) \in [-\epsilon_{eq}, +\epsilon_{eq}]$, i.e., two inequalities: $-\epsilon_{eq} \leq h_j(x) \leq +\epsilon_{eq}$. `IbexOpt` guarantees the global optimum of the relaxed system, although ϵ_{eq} can often be chosen almost arbitrarily small. (Most of the global optimizers like **Baron** [13] or **Couenne** [2] cannot offer any guarantee.)

In our wine blending problem, we set a constant ϵ_{eq}^1 equal to `1e-4` in the equality constraints (16). Another ϵ_{eq}^2 is set to `1e-1` in the equality constraints (2.i). This corresponds to 1 dl (deciliter), i.e., less than 0.1% of the target volumes (at least 500 liters). This means that the volumes are computed with an approximation significantly better than the ineluctable errors made during the actual blending, i.e., the errors induced by measures and loss of residual matter during the wine transfer from a base to a target tank.

---

[1] ϵ-minimize $f(x)$ means minimize $f(x)$ with a precision ϵ_{obj} on the objective, i.e., find x such that for all y, we have $f(y) \geq f(x) - \epsilon_{obj}$.

4.2   Algorithmic Features of Ibex

IbexOpt is implemented in Ibex (Interval Based EXplorer) and enriches this C++ library devoted to interval solving [5].

IbexOpt [14] follows an interval *Branch & Contract & Bound* schema. The process starts with an initial box $[x]$ that is recursively subdivided by a branching operator. The tree is traversed in best first search, in which a box with a smallest minimum cost is selected first. IbexOpt applies the following operators at each node (box) of the B&B:

Branch: A variable x_i is chosen and its interval $[x_i]$ is split into two sub-boxes.

Contract: A filtering process contracts the studied box, i.e., improves the bounds of its intervals, without loss of solutions.

Bound: The improvement of the lower bound is similar to a contraction (considering an additional variable corresponding to the objective cost). The lower bound guarantees that no feasible solution exists lower. Improving the upper bound amounts in finding a good (although generally not the best) feasible point, so as to cut branches in the search tree with a higher cost.

The process starts with an initial box $[x]$ and ends when the difference between the upper and lower bounds reaches a given precision ϵ_{obj} or when all the explored nodes reach a size inferior to a given precision.

At each node of the B&B, IbexOpt is called with efficient operators for reducing the search space and improving the lower bound of the objective function:

- The state-of-the-art HC4 [3,10] (continuous) constraint propagation algorithm is first used to contract the handled box.
- The operator X-Newton uses a specific interval Taylor to convexify the search space, contract the box and improve the lower bound [1].
- Two original algorithms are used to improve the upper bound by heuristically extracting an inner (entirely feasible) region that contains only solution points. This explains why equations are slightly relaxed. Roughly, the InHC4 algorithm is a dual algorithm of HC4 and InnerPolytope is a dual algorithm of X-Newton.

The default optimizer uses, as bisection heuristic, the SmearSumRel variant of Kearfott's branching heuristic using the *Smear* function [9]. The SmearSumRel and SmearMaxRel branching heuristics are described in [14].

5   Experiments on First Instances

We have modeled and solved several instances of wine assemblage. We also report in Section 5.5 a validation of our approach during a real tasting session.

For the ϵ-optimization, we have always required an accuracy ϵ_{obj} (goal precision) below 1e-4. The same precision is required for the solution (box) size: under this size, a box is not studied (and split) by the interval Branch & Bound. The goal accuracy is better than the errors ϵ_a made by the chemical tools when they measure the aroma concentrations (e.g, for the $c_{b,a}$'s).

5.1   Wine Blending Instances

We have modeled three instances of wine assemblage. The first one (WineBlending0) is a small and artificial instance. It was used to rapidly adapt the MINLP/NCSP model presented above until a rapid solving could be obtained. It contains 21 variables. WineBlending0 is solved in 0.18 seconds and only 6 branching nodes, independently from the ϵ_{obj} goal precision (1e-4 or 1e-8).

Real Instance 1

The second instance (WineBlending1) is a real instance provided by the Nyseos company. The instance consists in producing $W = 2$ target wines from $B = 7$ bases wines, taking into account $A = 11$ aroma.

The Min-Max problem, modeled as described in Section 3.4, contains 55 variables and 116 constraints:

- 2 volume (relaxed) channeling constraints,
- 7 base surplus constraints,
- 44 aroma concentration constraints,
- 49 constraints coming from the Min-Max encoding,
- 14 quadratic constraints modelling the disjunctive realistic volume constraints.

Real Instance 2

The second instance (WineBlending2) consists in assembling $W = 3$ target wines from $B = 6$ bases, taking into account $A = 7$ aroma.

The Min-Max problem contains 64 variables and 118 constraints:

- 3 volume (relaxed) channeling constraints,
- 6 base surplus constraints,
- 42 aroma concentration constraints,
- 49 constraints coming from the Min-Max encoding,
- 18 quadratic constraints modeling the disjunctive realistic volume constraints.

5.2   Results Obtained by the Default Optimizer

All the results reported in this paper have been obtained on a 2010 MacBook laptop with a 2.4 GHz Intel Core 2 Duo process.

We have first run the default optimizer of Ibex with a solution and goal precisions set to 1e-4 and with a timeout set to 5 minutes.

The optimizer reaches the timeout for WineBlending1 and WineBlending2 although rather good solutions are computed:

- In 5 min and 7894 branching nodes, an accuracy 0.002 is obtained on the goal (i.e., the maximum error on E) for WineBlending1.
- In 5 min and 8520 branching nodes, an accuracy 0.0054 is obtained on the goal for WineBlending2.

5.3   Algorithmic Analysis and Improvements

The results above have been obtained with the by-default constrained optimization strategy available in `Ibex` and briefly described in Section 4.2.

We have first analyzed which of the default features have an impact on performance and which ones do not. This analysis was fruitful.

Concerning the contraction/filtering part, the interval linearization operator `X-Newton` is surprisingly counterproductive. On the contrary, all the contraction is performed by the `HC4` constraint programming operator. Since the wine blending model built is mainly linear, this result is counterintuitive.

The opposite and more intuitive observation has been made on the inner regions extracted for improving the upperbound. The `InHC4` operator, issued from constraint programming principles (working constraint per constraint), appeared to be useless. On the contrary, the `InnerPolytope` algorithm (dual of *X-Newton*) that can extract an inner polytope from the feasible region is crucial. Without this feature, we could not obtain any answer from the optimizer.

Table 1. Comparison between the Smear-based branching heuristics. An entry is a multiline containing 3 information about the results obtained by a given branching strategy on a given instance. The first line of a multiline contains the CPU time, with a timeout set to 5 min; the second line gives the error obtained at the end (`1e-10` means that the last simplex call achieved by `InnerPolytope` finds a solution with no error (rounded to `1e-10`)); the third line reports the number of branching nodes.

		WineBlending1	WineBlending2
SmearSum	cputime (sec.)	> 300	57
	precision	0.00014	1e-10
	#nodes	21880	4146
SmearSumRel	cputime (sec.)	> 300	> 300
	precision	0.0018	0.00017
	#nodes	25864	24270
SmearMax	cputime (sec.)	> 300	111
	precision	0.0013	1e-10
	#nodes	25864	8851
SmearMaxRel	cputime (sec.)	0.35	27.4
	precision	1e-10	1e-10
	#nodes	23	2048

A second feature has a major impact on performance: the branching (bisection) strategy. We have observed that the basic bisection heuristics available in `Ibex` are inefficient or show a poor performance. The `largestFirst` heuristic, selecting a variable with the largest width, prevents the B&B from finding any feasible point during the tree search. The `roundRobin` also generally shows a poor performance, except if we rearrange statically the variables as preconised by a first analysis reported below. Only the Smear-based strategies behave well: the historical `SmearMax` and `SmearSum` heuristics [9]; the `SmearSumRel` variant called in the default optimization [14], and the `SmearMaxRel` variant that plays a fondamental role in our wine blending problem.

Based on this analysis, we have designed a dedicated strategy. The `X-Newton` operator has been first disconnected, thus bringing a speedup of a factor 4 on the two real instances. Second, we have compared the performance of the four Smear heuristic variants mentioned above. Table 1 gathers the results obtained.

We remark that `SmearSum` and, in particular, `SmearMaxRel`, show a good performance. Therefore `SmearMaxRel` has been chosen for our wine blending dedicated strategy.

We have also experimentally analyzed which variables are selected by the optimization process in slow and fast runs. We have empirically learnt that the CSP variables $P_{w,b}$ and $V'_{w,b}$ should often be selected, whereas the variables added for the Min-Max optimization should be rarely chosen. This study has led us to a dedicated heuristic that offers (only) the same performance as `SmearMaxRel` (23 seconds versus 27 seconds on `WineBlending2`). Of course, we need more instances to better learn from the data.

5.4   The Wines Produced

The layout of the solution computed for `WineBlending1` is shown in Figure 3. The details of the aroma concentrations $(C_{w,a})$ in the target wines are illustrated by $\mathcal{W} = 2$ radar graphs in Figure 4. The fact that the blue lines $(\hat{c}_{w,a})$ fall between the green lines $(\hat{c}_{w,a} - \varepsilon_a$ and $\hat{c}_{w,a} + \varepsilon_a)$ highlights visually that the best solution has been obtained within ε_a tolerances.

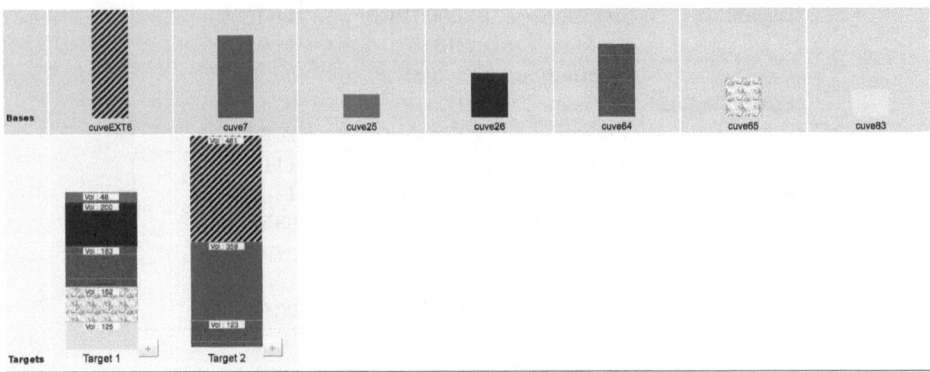

Fig. 3. Layout of the results for `WineBlending1`: volumes of the target wines (below) obtained by blending the bases (above)

The details of the aroma concentrations $(C_{w,a})$ in the target wines for `WineBlending2` are illustrated by $\mathcal{W} = 3$ radar graphs in Figure 5.

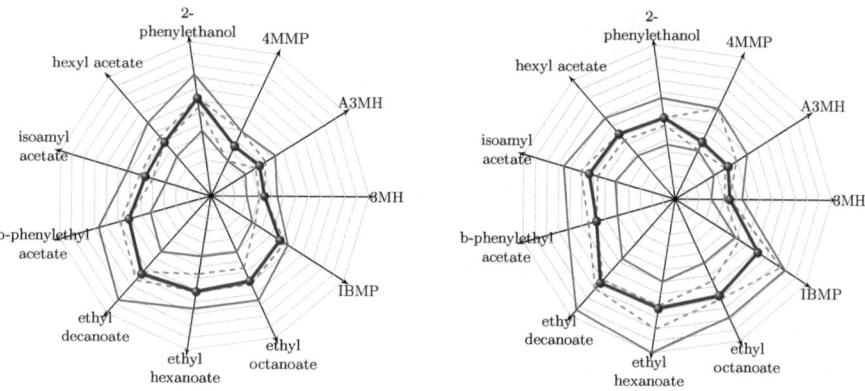

Fig. 4. Solution obtained by our Min-Max model on the two wines elaborated in instance `WineBlending1`. Every axis in a radar graph shows a computed aroma concentration $C_{w,a}$ (shown in thick line with balls), comprised within the imposed limits ($c_{w,a}^-$ and $c_{w,a}^+$) represented by solid lines, and as close as possible to the desired concentration $\hat{c}_{w,a}$. The 2 dashed curves represent the tolerances $\hat{c}_{w,a} - \varepsilon_a$ and $\hat{c}_{w,a} + \varepsilon_a$ on the desired concentration of aroma a in wine w.

5.5   Tasting Session

An interesting (qualitative) validation of our tool was carried out in collaboration with an oenologist. He was asked by the Nyseos company to elaborate a (target) wine by blending several given base wines. Nyseos wrote down the volumes the oenologist selected for the assemblage and carried out a chemical analysis of the final blend to measure its aromatic criteria. Then, using our tool, Nyseos created a similar blend with the same base wines, but using eventually different volumes extracted from each base wine. Nyseos finally compared the blendings used to obtain the human-made wine with the computer-made wine and asked the oenologist to carry out a blind-test on the two wines.

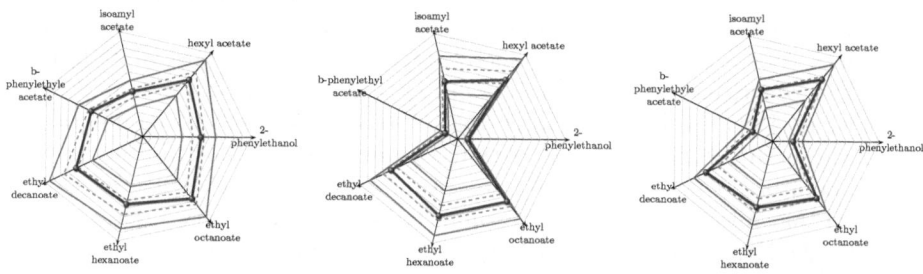

Fig. 5. Radar graphs obtained for the instance `WineBlending2`

The results were as follows: despite the blendings being significantly different, the oenologist could not distinguish between the two wines.

This one experiment is of course far from being representative, but is nonetheless a promising indication of the relevance of our tool.

5.6   A Future Configuration Tool for Wine Assemblage

The first experimental results are preliminary and were obtained on only two real instances. However, they suggest that the current strategy, maybe endowed with a dedicated branching heuristic, can handle efficiently most of the instances useful in practice. Therefore, to better fit the wishes of the client, we can imagine using our optimization algorithm interactively, inside a configuration tool. The user would be able to interact with the system via radar graphs corresponding to the different target wines, such as shown in Fig. 4 and Fig. 5.

A way to modify the blending is to increase (or decrease) the importance (weight) Ω_w of a wine. A slider under each radar graph could for instance be used for this purpose. An optimization process could then recompute a new solution with this specification.

Following the same idea, the user could modify the weight $\lambda_{w,a}$ of a given aroma in a wine (e.g., with a popup menu appearing when the mouse cursor position is on the corresponding axis of a radar graph), and the tool would run a new optimization.

A last and more intrusive possibility is to allow the client to strengthen the maximum admissible concentration $c_{w,a}^+$ (or minimum admissible concentration $c_{w,a}^-$) of aroma a in wine w. The client would simply select a bound and our tool would run two optimizations providing two information (assuming $c_{w,a}^+$ is selected). First, the smallest value of concentration $C_{w,a}$ for which there is a solution that respects all the constraints (such a solution is not necessarily optimal for the maximum error E). Second, we can also compute the smallest feasible value of $C_{w,a}$ that does not increase the maximum error E. These two bounds can be displayed as outstanding values for $c_{w,a}^+$.

6   Conclusion

We have reported in this paper a first attempt to handle the wine assemblage problem with constraint programming techniques. The problem can be modeled as a MINLP or a numerical CSP able to define disjunctive constraints that are critical in practice. These constraints ensure that a minimal amount of wines is transferred from a base to target wine. We have resorted to mono-criterion optimization and have worked to obtain a model with no max and no absolute value operators. We have designed an optimization strategy dedicated to wine blending that allows us to find in second the best solution to two real instances given by the Nyseos company (with no error in the volumes or the concentration requirements). A tasting session carried out by an oenologist has qualitatively validated our approach. These encouraging results offer the possibility to use our approach within an interactive configuration tool dedicated to wine assemblage.

References

1. Araya, I., Trombettoni, G., Neveu, B.: A Contractor Based on Convex Interval Taylor. In: Beldiceanu, N., Jussien, N., Pinson, É. (eds.) CPAIOR 2012. LNCS, vol. 7298, pp. 1–16. Springer, Heidelberg (2012)
2. Belotti, P.: Couenne, a user's manual (2013), http://www.coin-or.org/Couenne/
3. Benhamou, F., Goualard, F., Granvilliers, L., Puget, J.-F.: Revising Hull and Box Consistency. In: Proc. ICLP, pp. 230–244 (1999)
4. Chabert, G.: Interval-Based EXplorer (2013), http://www.ibex-lib.org
5. Chabert, G., Jaulin, L.: Contractor Programming. Artificial Intelligence 173, 1079–1100 (2009)
6. Dagan, L.: Potentiel aromatique des raisins de Vitis vinifera L. cv. Petit Manseng et Gros Manseng. Contribution à l'arôme des vins de pays Côtes de Gascogne. PhD thesis, École nationale supérieure agronomique, Montpellier (2006)
7. Datta, S., Nakai, S.: Computer-aided optimization of wine blending. Journal of Food Science 57(1), 178–182 (1992)
8. Ferrier, J.G., Block, D.E.: Neural-network-assisted optimization of wine blending based on sensory analysis. American Journal of Enology and Viticulture 52(4), 386–395 (2001)
9. Kearfott, R.B., Novoa III., M.: INTBIS, a portable interval Newton/Bisection package. ACM Trans. on Mathematical Software 16(2), 152–157 (1990)
10. Messine, F.: Méthodes d'Optimisation Globale basées sur l'Analyse d'Intervalle pour la Résolution des Problèmes avec Contraintes. PhD thesis, LIMA-IRIT-ENSEEIHT-INPT, Toulouse (1997)
11. Moore, D.B., Griffin, T.G.: Computer blending technology. American Journal of Enology and Viticulture 29(1), 50–53 (1978)
12. Ninin, J., Messine, F., Hansen, P.: A Reliable Affine Relaxation Method for Global Optimization. Technical Report RT-APO-10-05, IRIT (2010)
13. Tawarmalani, M., Sahinidis, N.V.: A Polyhedral Branch-and-Cut Approach to Global Optimization. Mathematical Programming 103(2), 225–249 (2005)
14. Trombettoni, G., Araya, I., Neveu, B., Chabert, G.: Inner Regions and Interval Linearizations for Global Optimization. In: AAAI, pp. 99–104 (2011)

The Berth Allocation and Quay Crane Assignment Problem Using a CP Approach

Stéphane Zampelli[1], Yannis Vergados[2], Rowan Van Schaeren[3], Wout Dullaert[4,5], and Birger Raa[3,6]

[1] Dynamic Decision Technologies, Providence
[2] Ph.D. Graduate from Brown University
[3] Antwerp Maritime Academy
[4] VU Amsterdam, Department of Informatics, Logistics and Innovation
[5] University of Antwerp, Institute of Transport and Maritime Management Antwerp
[6] Ghent University

Abstract. This paper considers the combination of berth and crane allocation problems in container terminals. We propose a novel approach based on constraint programming which is able to model many realistic operational constraints. The costs for berth allocation, crane allocation, time windows, breaks and transition times during gang movements are optimized simultaneously. The model is based on a resource view where gangs are consumed by vessel activities. Side constraints are added independently around this core model. The model is richer than the state of the art in the operations research community. Experiments show that the model produces solutions with a cost gap of 1/10 (7,8%) to 1/5 (18,8%) compared to an ideal operational setting where operational constraints are ignored.

Keywords: berth allocation, crane assignment, containers, terminal, constraint programming.

1 Introduction

A container terminal is a facility where cargo containers are transshipped between vessels and external trucks or trains. Cranes along the quay called 'quay cranes' are responsible for charging and discharging containers. Special trucks in the field move containers from quay cranes to container blocks in the yard. External trucks bring containers to the terminal and take away containers from the yard blocks. Many logistic problems arise in this context. We focus on two of them. On one hand, the berth allocation problem (BAP) positions vessels optimally, ensures security distances, and minimizes stay durations along the quay using a simplified model of the crane assignments. On the other hand, the quay crane assignment problem (CAP) considers the problem of detailed assignment of quay cranes to vessels in order to handle the incoming and outgoing containers where a feasible berth plan is already available. The challenge, proposed by our industrial partner, is to incorporate both problems together with those real-world constraints.

C. Schulte (Ed.): CP 2013, LNCS 8124, pp. 880–896, 2013.

Berth Allocation Problem. The BAP problem schedules the vessels by deciding the position of the vessel along the quay, estimating the duration needed to handle all the loading and discharging containers of the vessel, and avoiding vessel overlap along the quay. The difference with the CAP is that the stay duration along the quay is simplified by avoiding to compute the detailed crane assignment scheduling on each vessel. We review the detailed BAP problem constraints.

- The total length of all the vessels should be shorter than the quay length.
- Positions along the quay are represented by discrete bollards. The mooring ropes and wires used for securing the vessel along the quay length are attached to bollards. Every vessel is assigned a mooring place or berth that is a multiple of bollard distances. The distance between two bollards on the same quay length is equal. In Figure 1, the vessel uses bollard 2 to 5.
- Vessels along the quay should not overlap.
- The ideal berth of the vessel along the quay is computed by another yard optimization tool and is outside the scope of this paper. Ideal berth positions are an input in the context of this paper. Vessels discharge and load containers to and from containers blocks in the yard according to the yard planning. An ideal berth position can be precomputed for each vessel. The customer pays a fixed price for the container loading/unloading regardless of the yard position the container will occupy. The terminal wants to minimize the distance between the ideal berthing position and the precomputed one. Figure 2 represents a bad berth allocation.
- The computation of the handling time of the vessel depends on the cranes. In the BAP, this computation is simplified, not considering the detailed scheduling of the CAP. The handling time links the BAP and the CAP. Such a detailed scheduling is considered in the CAP below.
- Vessels have setup times. When a vessel arrives at a terminal and is safely moored alongside, the cranes can not immediately start to discharge the containers. The securing of the containers, called lashings, first need to be undone. The time needed for unlashing the containers differs per vessel and per stowage configuration. This time needs to be taken into account concerning the starting time for the cranes.

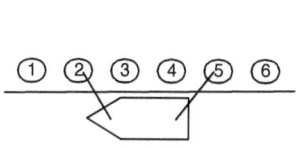

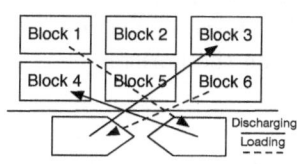

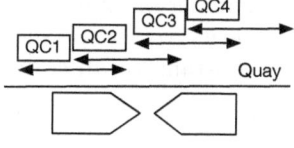

Fig. 1. Using bollards for defining the quay length occupied by a vessel

Fig. 2. A bad example of berth allocation regarding the yard distance cost

Fig. 3. An example of quay crane (QC) ranges on a container terminal with four quay cranes

Crane Assignment Problem. The CAP problem considers the detailed assignment of gangs to quay cranes and cranes to vessels. A gang is a team of workers consisting of a crane driver, a foreman, a person checking the container ids, two dockers to handle the containers and the driver moving the internal truck. Hiring a gang has a fixed cost per shift. Gangs have to be assigned to a crane but can be moved freely from one crane to another. A quay crane handles the containers to charge or discharge from or to a vessel. There are different types of cranes (Panamax, Post Panamax, Super-Post Panamax STS, ...). Vessels may accept only specific types of cranes. They also have a fixed arrival time at the terminal and leave as soon as all containers have been handled. A quay crane can be moved at any point in time from one vessel to another, creating a preemptive schedule. The overall goal is to minimize the operational cost of the terminal seen as a service from the point of view of the vessel operators. Let us review the detailed problem constraints.

- There is a maximum number of cranes available along the quay.
- Each vessel has a time window during which it needs to be handled alongside the quay length. The terminal operator will have to pay a fine, if the vessel arrives on schedule but cannot be handled within the agreed time frame.
- Repositioning cranes from one vessel to another takes 30 minutes.
- Gangs have breaks. Each gang works for eight hours. Each gang has a break of half an hour each four hours. During this break, crane repositioning is free, handled by a specialized team.
- The maximum number of cranes for each vessel is limited by the length and the number of bays of the vessel. Each quay crane has a fixed width, and hence a maximal number of cranes can work on one vessel simultaneously.
- Cranes operate on a common rail and have operating ranges. Cranes are electrically driven. The length of the source cables are chosen in such a way that an optimal coverage is given for the quay length (see Figure 3).
- Cranes are operated on a single rail, so they cannot cross each other.
- The gang cost per shift depends on the shift on which the gang operates. An example of relative gang costs is shown in Table 1. Note that gang costs must be paid in full, even if it only works during a part of a shift. There is always an integer number of gangs per shift.
- Crane productivity is measured in containers per hour and depends on a number of external factors (weather, crane driver, traffic, stowage plan, security vessel specific rule); however we consider the crane productivity as constant for all cranes along the horizon.

Table 1. Relative cost of a gang

	Weekday	Saturday	Sunday
Morning	1.05	1.50	2
Afternoon	1.15	1.50	2
Night	1.50	2	2

Table 2. Gang shifts and breaks

	Break 1	Break 2
Morning (06:00-14:00)	09:30-10:00	13:30-14:00
Afternoon (14:00-22:00)	17:30-18:00	21:30-22:00
Night (22:00-06:00)	01:30-02:00	05:30-06:00

Previous Work. We focus on literature about integrated BAP and CAP problems (BAPCAP). Readers interested in the abundant literature about BAP or CAP can refer to [4]. BAPCAP was first studied in [20]. They propose a two step approach. The first phase determines the berthing time and position of each vessel and the number of cranes to be allocated to the vessel. The second phase schedules the assignment of individual cranes, solved by a Lagrangian relaxation and dynamic programming respectively. Time is discretized in blocks of one hour. There is no detailed crane reallocation. Moreover, there is no consideration for gangs and gang cost. [19] gives a formulation of the BAPCAP, and solves the model in a two step procedures: a GA procedure (crossover and mutation) to assign ships in order to berth, and a heuristic to assign cranes to ships. [18] solves the BAPCAP by using a 4 steps GA-based approach, successively locating ship to berths, assigning quay cranes to berths, and designing berth and quay cranes scheduling. Note that there is no consideration for gang cost and cranes cannot be re-allocated. [17] decides on the berthing position, the berthing time, and the number of cranes that serve a vessel within the handling period, by taking into account drop of crane productivity due to interference. The BAPCAP model is then solved using heuristics and metaheuristics. Time is discretized in blocks of one hour. Detailed crane assignment and reallocation is not considered and authors suggest it should be post-processed. [16] proposes to optimize the cranes efficiency. The block periods consist of 12h and the horizon is limited to 6 blocks (3 days) because of the model complexity. It is extended by using a rolling horizon technique [7]. Among recent works, [15] considers many detailed constraints and studies the BAPCAP under uncertainty. Limited crane reallocation when a new vessel arrives can occur. The wharf is also segmented into fixed length segments. A non linear mixed integer model is solved using GA. Gang, shift costs and breaks are ignored. [6] proposes a pure MIP BAPCAP approach with time bucketized in periods of 2 hours and a rolling horizon. Recently, [14] splits the problem in two, with BAP on one side solved by GA and CAP is solved by a mixed-integer linear program. A Bi-Level Programming (BLP) approach is used to combine both subproblems. Running time is pretty high with 480 minutes reported for 3 vessels and 8 QCs, although the detailed scheduling of the crane is taken into account. [13] solves the two problems independently, although the BAP is continuous, meaning the quay length is not discretized in blocks. They use a nested loop-based evolutionary algorithm (NLEA), and two inner loops and one outer loop are suggested. Let us stress that in the OR literature, no papers can handle the set of constraints proposed in this paper in a single model. Our model schedules to the minute with dynamic crane reallocation and a computation of the actual gang and shift costs.

Paper Main Contributions. Paper contributions are threefold:

- The solved problem tackles a combination of many complicated technical constraints, such as setup times, transition times, time windows, shifts, breaks, smooth workforce allocation (work for large consecutive spans of time), spatial positioning, etc., in a large scale and realistic setting (5 days,

30 vessels). The crane allocation itself is done by a tractable subset inside the main model. It shows how useful CP is in tackling those OR problems.

- The proposed model shows how to solve a complex problems not only by identifying the main underlying structure in a form of a constraint but also by isolating several aspects into submodels and making them communicate through the variables. This is an interesting approach to tackle other challenging OR problems.
- This paper shows a case where CP can bring benefits to challenging problems inside the OR community that tends to be MIP or heuristics dominated. It has been agreed for a long time that combining detailed quay crane scheduling and berth planning is not an easy problem. MIP models become complex and heuristics are used to overcome this issue.

Benefits of CP. The scheduling features of CP are one of the keys against MIP or heuristics centric approaches as scheduling the cranes is the core of the problem. The declarativeness of CP was important, since more than 20 different models were tested, in a reasonable development cycle time (<2 men/month). It would have been difficult to test all those ideas using a heuristics approach, given the amount of coding and testing required. Our industrial partner did not believe all those constraints could be handled in a declarative way. CP declarativeness allowed not only to identify and state constraints but also to identify and integrate submodels of the problem at hand. The ability to easily add small side constraints also played a key role. For instance it is easy for a user to state in the model that a specific crane will go down in a given period of time or forbid certain crane/vessel assignments because of compatibility issues.

2   Model Description

Our CP model combines several submodels. The core model, described in Section 2.1, allocates gangs to vessel subtasks, minimizing the total gang cost and the lateness. The crane allocation and the berthing are added to the core model, in Section 2.2 and 2.3. Section 2.4 describes the objective function. A labeling procedure is proposed in Section 2.5. Section 3 discusses computational results.

2.1   Gang Allocation

Gangs need to be allocated to cranes, and cranes to vessels. Gangs can move freely from crane to crane and cranes can be reassigned at any time. The key idea is to view the gangs as a resource and use cumulative constraints. Viewing the cranes as a cumulative resource is a deadend since cranes are ordered and have exotic constraints like range and non crossing constraints. Each vessel is a set of activities that consumes a number of gangs in a preemptive way. The actual crane assignment is left to a separate submodel that makes sure the assignment is possible. Each gang delivers a certain amount of workforce that depends on the duration linearly. This workforce idea makes it easier to compute the shift gang cost and deal with side constraints such as breaks and setup times.

Notations. A range R is a consecutive finite sequence of integers; its minimum (maximum) is noted $\underline{R}$ (resp. $\overline{R}$). The range of input vessels is denoted $vessels$, and for each $b \in vessels$, the range of vessel activities is denoted Act_b. The time horizon $Horizon$ is a range of time units of 1 minute. The range $Shifts$ indexes the shifts. The shift duration (including breaks) is noted sd. The range $Gangs$ indexes the gangs. The ranges $Gangs_b = [0, mc_b]$ with $b \in vessels$ are the possible values for the number of cranes that can be allocated to a vessel. The ranges $Breaks$ is the ranges of breaks. We assume those ranges start at zero. The lower bound (resp. upper bound) of a finite domain variable x is denoted $\underline{x}$ (resp. $\overline{x}$).

Before declaring activities and constraints, we convert containers to the concept of workforce. A unit of workforce is the work of one gang during one unit of time (1 minute). This conversion is needed because the scheduling of gangs activities over vessels depend on gang units and time units, and know nothing about containers. Workforce is the link between the scheduling of gangs and the containers of vessel. Each vessel needs a minimum amount of workforce to leave. The following two definitions grasp those ideas:

Definition 1 (Crane Productivity). *The productivity of a crane is the number of containers it can handle per hour.*

Definition 2 (Workforce). *Given a crane productivity p, the workforce needed to handle c containers is defined by $(c * 60)/p$. The required workforce of a vessel, noted mw_b, is the workforce corresponding to its number of containers to handle.*

The only drawback is that a crane may be reassigned while a container is being moved, since only the required time is considered. However, this limitation has no impact on real operations: transition times can be shortened or extended to handle those cases in practice. We now define the set of activities $a_{b,i}$ with $b \in vessels$ and $i \in Act_b$.

Definition 3 (Activity). *An activity $a_{b,i}$ is defined by five variables:*

- $s_{b,i}$ *is the starting time,*
- $e_{b,i}$ *is the completion time,*
- $d_{b,i} = e_{b,i} - s_{b,i}$ *is the duration,*
- $cap_{b,i}$ *is the number of resources consumed by the activity between its starting time and its completion time.*
- *and $wkf_{b,i}$ is the workforce delivered by the task, with $0 \leq wkf_{b,i} \leq cap_{b,i} * d_{b,i}$.*

Our model creates one activity $a_{b,i}$ per vessel b and per index $i \in Act_b$. The capacity $cap_{b,i}$ is the constant number of gangs used by the activity. The equality of $wkf_{b,i}$ with $cap_{b,i} * d_{b,i}$ is not enforced because of breaks and transition times. For instance, if an activity overlaps with a break, the delivered workforce is inferior to its surface. An activity is an allocation of workforce to a vessel. Breaks and transition times are handled at the end of this section. Activities can be interrupted and are also optional (they can have a zero duration). Each vessel has its own time window. In the following, we abuse notations and use i instead of (b, i) when it is clear from context that we are speaking about a given boat.

Definition 4 (Time Window). *The time window of a vessel $b \in vessels$ is the couple (ta_b, td_b), where the integer ta_b denotes the arrival time of the vessel b and td_b the deadline of vessel b.*

Arrival time for each vessel $b \in vessels$ and each index $i \in Act_b$ is enforced:

Constraint 1 (Arrival). $\forall b \in vessels, i \in Act_b : s_{b,i} \geq ta_b$

Constraint 2 (Required Workforce). $\forall b \in vessels : \sum_{i \in Act_b} wkf_{b,i} \geq mw_b$

Let us ignore shifts for now. At any point in time, there is maximum $\overline{Gangs}$ gangs that can be hired. Given two variables s and d representing the starting time and the duration variables of an activity a_i, the mandatory part noted $mand(a_i)$ or $mand(s, d)$ is a range $[\overline{e} - \underline{d}, \underline{s} + \underline{d}]$ that can be empty if the mandatory range does not exist. This can be modeled by a cumulative constraint:

Definition 5 (Cumulative). *Consider a resource limited by a constant capacity c, and a set of activities $a_j \in A$. A constraint $cumulative(\{a_j \mid j \in A\}, c)$ ensures the following constraint: $\forall t \in Horizon \sum_{j \in I} cap_j \leq c$ where $I = \{j \in A \mid t \in mand(a_j)\}$.*

Activities may not exceed the maximum number of available gangs:

Constraint 3 (Global Cumulative). *The following constraint is added to the model: $cumulative(A, \overline{Gangs})$ where A is the set $\{a_{b,i} \mid b \in vessels, i \in Act_b\}$.*

Each vessel is also constrained on its maximum number of gangs at any point in time. An additional $|vessels|$ number of cumulative constraints are posted:

Constraint 4 (Local Cumulative). *For each $b \in vessels$, the following constraint is posted: $cumulative(A, \overline{Gangs_b})$ where A is the set $\{a_{b,i} \mid b \in vessels, i \in Act_b\}$ and $Gangs_b$ is the possible gang range for vessel b.*

Let us introduce shifts in the model. For each shift, a variable denoting the number of gangs used can be created:

Definition 6 (Gang Shift). *For all $sh \in Shifts$, $nbGangs_{sh}$ is the number of gangs used in shift sh.*

For each shift, a fake activity is created that spans over the whole shift and consumes the number of gangs that are not used during that shift.

Definition 7 (Fake Activities). *For all $sh \in Shifts$, a fake activity fa_{sh} is created with the following domains:*

- *starting time $s_{sh} = sh * sd$*
- *duration $d_{sh} = sd$*

- *capacity* $cap_{sh} = \overline{Gangs} - nbGangs_{sh}$
- *workforce* $w_{sh} = 0$.

Let us introduce breaks and transition time. Two break intervals are present in each shift sh, a first break

$$[\frac{se_{sh}}{2} - bd, \frac{se_{sh}}{2}]$$

and a second break:

$$[se_{sh} - bd, se_{sh}]$$

where se_{sh} is the ending time of the shift sh and and bd is the constant break duration. Each break $r \in Breaks$ can be associated with such an interval noted b_r. A variable bi_r is equal to time intersection between b_r and $[s_{b,i}, e_{b,i}]$. The total intersection between an activity and the breaks can be measured:

$$bi_{b,i} = \sum_{r \in Breaks} bi_r .$$

Regarding transition times, we considered a fixed and constant transition time denoted $transitionTime$ that is assigned to all activities. The transition time can be defined as

$$tt_{b,i} = max(0, transitionTime - fb_{b,i})$$

where $fb_{b,i}$ is defined as:

$$fb_{b,i} = bi_r \text{ where } r = min\{r \in Breaks \mid bi_r \neq 0 \wedge s_{b,i} \in b_r\}$$
$$= 0 \quad \text{if } r \text{ does not exist.}$$

The variable $fb_{b,i}$ denotes the intersection of a break with the beginning of a vessel operation. Indeed, cranes can be moved during breaks. Breaks occurring at the beginning of vessel operations hence shorten transition time. The actual workforce of the activity (b, i) can be defined.

Constraint 5 (Workforce). *For each activity (b, i), the workforce is*

$$wkf_{b,i} = (d_{b,i} - bi_{b,i} - tt_{b,i}) * cap_{b,i} .$$

Regarding the setup time, the transition time assigned to the first activity of the vessel stands for both the transition time of the cranes and the setup time. In this core model, gangs are assigned to vessels, using preemptive activities. Breaks and transition times are taken into account using the workforce variables. This first model is a relaxation of the problem as actual cranes along the quay are not assigned to vessels and vessel conflicting positions are ignored.

2.2 Space Allocation

Along the quay, the vessels should not overlap. The length of a vessel b is noted $length_b$. Let us define a vessel position along the quay:

Definition 8 (Position). *The position of vessel b along the quay is a finite domain variable and is denoted pos_b.*

Let us define the starting and ending time of vessel:

Definition 9 (Vessel Time Window). *The starting time of a vessel b is $s_b = \min_{i \in Act_b} s_{b,i}$, and its ending time is $e_b = \max_{i \in Act_b} e_{b,i}$.*

Non overlap between vessels is stated by enforcing that vessels overlapping in time should not overlap in space:

Constraint 6 (Non-overlap). $\forall\ (b, c) \in vessels \times vessels, b > c : (s_b \leq e_c \wedge e_b \geq s_c) \vee (s_c \leq e_b \wedge e_c \geq s_b) \Rightarrow (pos_c \geq pos_b + length_b) \vee (pos_b \geq pos_c + length_c)$

2.3 Crane Allocation

In this section a tractable submodel is presented for the crane allocation. This model can filter any inconsistent crane assignment value once the information is available from other submodels.

The first concept is the crane range. The assignment of cranes to a vessel can be represented as a range, because all cranes are consecutive along the quay and cannot cross each other, since they are each operated on a single rail.

Definition 10 (Crane Range). *The crane range of an activity (b, i) $(i \in Act_b)$ is a range $[sc_{b,i}, ec_{b,i}]$, where $sc_{b,i}$ is the starting crane and $ec_{b,i}$ the ending crane. The variable $nbCranes_{b,i}$ denotes the number of cranes assigned to vessel activity (b, i).*

The following constraint holds: $sc_{b,i} \leq ec_{b,i}$, and the number of cranes and the crane range are linked: $nbCranes_{b,i} = ec_{b,i} - sc_{b,i} + 1$.

Each crane has a certain span along the quay, because of physical constraints. This means that a crane can be assigned to a vessel if and only if the crane can reach the vessel along the quay. Given a vessel b, only a subset of crane ranges are available for vessel b. Let us define the *craneMin* array indexed by bollard positions. Since we focus on a given boat, we omit subscripts. The value $craneMin_p$ (resp. $craneMax_p$) is the leftmost crane (resp. rightmost crane) that can reach bollard range $[p, p + length_b]$. The consistency between crane positions and vessel positions can be added to the model:

Constraint 7 (Crane Position). $\forall\ b \in vessels, i \in Act_b : sc_{b,i} \geq craneMin[pos_b]$ and $ec_{b,i} \leq craneMax[pos_b]$.

The following set of constraints distribute the cranes between activities.

Constraint 8 (Crane Allocation). *For each pair of distinct tasks $((b, i)$, $((c, j))$ overlapping in time, their crane range follows their relative position:*

$$[(s_{b,i} \leq e_{c,j} \wedge e_{b,i} \geq s_{c,j}) \vee (s_{c,j} \leq e_{b,i} \wedge e_{c,j} \geq s_{b,i}) \wedge (pos_b < pos_c)] \Rightarrow ec_{b,i} < sc_{c,j}$$

and:

$$[(s_{b,i} \leq e_{c,j} \wedge e_{b,i} \geq s_{c,j}) \vee (s_{c,j} \leq e_{b,i} \wedge e_{c,j} \geq s_{b,i}) \wedge (pos_b > pos_c)] \Rightarrow sc_{b,i} > ec_{c,j}.$$

Once the position, the time span and the number of cranes of pairwise activities are bound, the right side constraints from Constraint 8 form a linear chain of inequality constraints. Given a time $t \in Horizon$, a total order is enforced upon crane range variables of activities intersecting in time t. Ignoring distinction between vessel and activity indexes, we have at a given time $t \in Horizon$:

$$sc_1 \leq_{k1} ec_1 < sc_2 \leq_{k_2} ec_2 < \ \dots \ \leq_{k_{n-1}} ec_{n-1} < sc_n \leq_{k_n} ec_n \quad (A)$$

where n is the number of vessel activities intersecting in time with t. $\leq_{k_i}$ is a notation for the binary constraint $s_i \leq e_i - k_i + 1$, k_i is the bound value of variable $nbCranes_i$, and $<$ is the binary inequality constraint.

It is well-known [Jeavons, 1995] that max-closed (or min-closed) constraints and arc-consistency detect at fixpoint if a constraint system is satisfiable. Both constraints $x < y$ and $x \leq_k y$ are max-closed and min-closed[1]. This implies the following property:

Property 1. Suppose the arc-consistent fixpoint has been computed for the chain of constraints (A) and the fixpoint does not fail. Then any value from any variable in the set of variables of (A) can be extended to a solution.

This last property implies that the labeling of the crane range variables can be skipped as propagation will ensure crane ranges can be instantiated.

2.4   Objective

The three components of the objective includes the lateness cost, cost induced by the distance with the ideal position, and the total gang cost. The lateness of a vessel $b \in vessels$ is easily defined:

Definition 11 (Lateness). *The lateness l_b of a vessel $b \in vessels$ is equal to* $\max(0, e_b - ta_b)$.

Lateness is the exceeded handling time with respect to the deadline of the vessel time window. Let pos_b be the position variable of vessel b. A position difference can be defined similarly:

Definition 12 (Distance Gap). *The distance gap dp_b of a vessel $b \in vessels$ with respect to its ideal position ip_b is equal to $|ip_b - pos_b|$.*

---

[1] We omit the proof due to lack of space. See [23] for the full proof.

The number of gangs used in each shift is defined by $nbGangs_{sh}$, see Section 2.1.

Constraint 9. *The objective variable obj is defined as*

$$obj = \sum_{b \in vessels} (l_b * lc_b) + \sum_{b \in vessels} (dp_b * dc_b) + \sum_{sh \in Shifts} (nbGangs_{sh} * gc_{sh})$$

where lc_b is the lateness cost per minute for vessel b, dc_b is the distance cost per meter for vessel b, and gc_{sh} is the cost of a single gang in the shift sh.

2.5 Labeling

The primary goal of the labeling is to minimize the total gang cost per shift while avoiding lateness. When the minimization of a resource is required in the cumulative constraint, a *fill hole* heuristic is used. The idea is to fill holes present inside the profile of the resource usage. A similar labeling has been used in the context of a soft cumulative [22]. The profile of a cumulative constraint can be defined as:

Definition 13 (Profile). *The profile of a cumulative constraint is a set of tuples (t_i, d_i, v_i), $i \in P$, such that:*

- *(non-overlap) $\forall\, i, j \in P$, $i \neq j : [t_i, t_i + d_i - 1] \cap [t_j, t_j + d_j - 1] = \emptyset$*
- *(usage reflection) $\forall\, t \in Horizon\ \exists\, i \in P : \sum_{k \in A} cap_k = v_i$ where $t \in [t_i, t_i + d_i - 1]$ and $A = \{j \in Act \mid t \in mand(a_j)\}$*
- *(cover) $\forall\, t \in Horizon\ \exists\, i \in P : t \in [t_i, t_i + d_i - 1]$*

The set Act denotes the set of all activities. Tuples of a profile are called *segments*.

Definition 14 (Minimal Profile). *A cumulative profile is minimal iff $\forall\, i, j \in P, i \neq j$, $v_i \neq v_j$, that is $|P|$ is minimal.*

In the following, we shall suppose that P is ordered with respect to t_i. We note invariably $i \in P$ and $(t_i, d_i, v_i) \in P$. Holes are defined with respect to left and right segments. The left (right) segment i of a profile P is the segment $i - 1$ (resp. $i + 1$). Its left (right) segment value is v_{i-1} (resp. v_{i+1}). The left and right segment of i may be undefined if $i = min(P)$ or $i = max(P)$. If they are undefined, their left or right segment value is equal to $\overline{Gangs}$.

A hole is an augmented segment. The profile segment is augmented with a depth information h:

$$h = \begin{cases} min(l - v_i, r - v_i) & \text{if } l - v_i > 0 \text{ and } r - v_i > 0 \\ l - v_i & \text{if } l - v_i > 0 \text{ and } r - v_i < 0 \\ r - v_i & \text{if } l - v_i < 0 \text{ and } r - v_i > 0 \\ 0 & \text{if } l - v_i \leq 0 \text{ and } r - v_i \leq 0 \end{cases}$$

where l and r are the left segment value and the right segment value resp. We say a segment is augmented by its hole value h.

The heuristic function uses a function called lmdh() for *leftmost deepest hole*. It returns an ordered sequence of holes based on the profile of the cumulative constraint that the next activity should try to fill. More specifically, considering the minimal profile P of the cumulative constraint, it returns a sequence O of augmented segments (t_j, d_j, v_j, h_j) such that:

1. O defines for C the same profile as P:
   $\forall t \in Horizon \; \exists \; j \in O : \sum_{k \in A} cap_k = v_j$ where $A = \{k \in Act \mid t \in mand(a_k)\}$.
2. Segments of O cannot cross shift boundaries
   $\forall j \in O, \exists \; sh \in Shifts : t_j \geq sh * sd \wedge t_j + d_j - 1 \leq ((sh + 1) * sd) - 1$.
3. h_j is the augmented hole value from the segment $i \in P$ for which $v_i = v_j$
4. the sequence O is sorted lexicographically on highest h_i and smallest t_i.

In other words, $lmdh()$ returns the same segments as P, except they are split at any shift beginning and they are ordered.

The labeling procedure is described in Algorithm 1. The vessels are scanned in increasing arrival time ta_b (line 1) and the activities of vessel b are scanned (line 3). The amount of workforce still to deliver is computed (line 4), and if no workforce is left, the remaining activities Act_b are assigned to a duration of zero so that they do not appear in the solution (line 4 to 7). If there is some work to do on the current vessel, the profile holes are then computed based on the information of the cumulative constraint, by calling lmdh() (line 8). The holes are ordered according to the gang cost corresponding to the shift they are in. The selected activity is forced to be included into the width of hole (line 9 to 11). The depth of the hole is adjusted if it is a border case. This can happen for instance if the left segment is undefined. Another possibility is that $h = 0$ because the segment is a hill. In both cases, h is set to the maximum possible number of gangs for the activity (line 13 to 15). The number of gangs, based on the augmented segment, tend to be the number of gangs that would fill the hole vertically, if any. Then the number of gangs is assigned, the activity is pushed leftmost, and the workforce delivered is maximized, maximizing the width of the activity (line 17 and 19). The current index of the activity is added to already used activities (line 23). When all activities of current vessel have been scheduled, line 25 and 26 assign a position to the vessel along the quay. It should be stressed that the crane allocation range variables are not labeled, as the crane allocation submodel is tractable, see Section 2.3.

The above labeling obtains good solutions. Using a naive labeling, where activities are pushed leftmost lead to worse results as demonstrated in the experiments. Moreover, we use large neighborhood search [9], where entire vessels are fixed with a 0.6 probability.

3   Computational Results

This section measures the performance of the proposed model on generated datasets and on a industrial dataset. To the best of our knowledge, most terminals schedule crane and berth by hand. Academic papers cover too few real-world

PROCEDURE label()
1: **for all** $b \in vessels$ by arrival order **do**
2:    $I \leftarrow \emptyset$ //I is the set of activities already used
3:    **for all** $i \in Act_b : i \notin I$ **do**
4:       int $lw \leftarrow mw_b - \sum_{i \in A_b} \underline{wkf}_{b,i}$ //workload left
5:       **if** $lw \leq 0$ **then** //if nothing to do for this vessel
6:          try constraint $d_{b,i} = 0$ //impose zero duration, as this activity is not used
7:       **else**
8:          **for all** $[t_i, d_i, v_i, h_i] \in$ lmdh() in increasing shift cost order **do**
9:             $h_1 \leftarrow t_i; h_2 \leftarrow t_i + d_i - 1;$
10:             try constraint $s_{b,i} \geq h_1$ //restrict activity to the segment $[h_1, h_2]$
11:             try constraint $e_{b,i} < h_2$
12:             $h \leftarrow h_i$
13:             **if** $h_i = 0$ **or** $h_i > \overline{nbCranes}_{b,i}$ **then** //if it is not a proper hole
14:                $h \leftarrow \overline{nbCranes}_{b,i}$ //set to max nbr of gangs for vessel b
15:             **end if**
16:             **for all** gangs g from h down to $\underline{nbCranes}_{b,i}$ **do**
17:                try constraint $nbCranes_{b,i} = g$ //impose nbr of cranes, starting from depth h
18:                try constraint $s_{b,i} = s_{b,i}$
19:                try constraint $wkf_{b,i} = \overline{wkf}_{b,i}$ //fix duration, as start and nbr of gangs are fixed
20:             **end for**
21:          **end for**
22:       **end if**
23:       $I \leftarrow I \cup \{i\}$
24:    **end for**
25:    try constraint $diffPos_b = \underline{diffPos}_b$ //label position close to the ideal position
26:    try constraint $pos_b = \underline{pos}_b$ //diffPos is an absolute value
27: **end for**

Algorithm 1. Dedicated labeling for the global model.

constraints. Each previous work has its own set of constraints and a comparison would not be fair. Commercial tools do not optimize globally and are a help to build the schedule by hand. Additional details can be found in [23].

Datasets Description. In order to validate the model, we generated datasets based on the authors' experiences and information found in various published academic papers. Industrial datasets were also used.[2] The onset for generating our instances meet client's operational requirements. Vessels are planned in advance with a time horizon of 5 days (7200 minutes). The total quay length is 2000 meters, matching the largest container terminal in the world, and there are up to 30 vessels. The average crane productivity is 35 containers per hour or 0.5833 per minute. The total amount of quay cranes available is set to 19. Crane

---

[2] Industrial datasets are available upon request.

width is 80 meters. This means that a vessel of 230 meters e.g. would have at most 3 cranes working on it simultaneously: $\lfloor 230/80 \rfloor$. Bollards are 20 meters apart. This distance is also used to add to the vessel's length around the vessel for safe mooring alongside the quay length. If a vessel stays longer than allowed by its commercial time window, the lateness cost is 5000€ per hour. Deviation with the ideal berth position costs one euro per meter of deviation. The gang costs use Table 1 and a base cost of 2600€. Shift details (working hours and breaks) are shown in Table 2. Setup leaving and arriving times and transition times for cranes are set to 20 minutes. The set Act_b is an input. The model uses 1 activity for barges with less than 35 containers. For other vessels, the number of containers (or workforce) is divided by a split threshold, typically a workforce of 100 containers for 4 cranes. More activities are useless (0 workforce) below this threshold.

MIP Relaxation. We need a measure of the gap with respect to optimality. The client uses MIP and the optimality gap is an expected output. We relaxed the gang allocation core submodel (see Section 2.1) into an integer program. This relaxation gives a lower bound to measure a gap with respect to an ideal operational setting. Crane allocation and space allocation submodels are ignored. Cranes can reach any vessel, can cross each other and can move instantly. Vessels can overlap along the quay. The MIP model considers cranes are helicopters and vessels can be positioned anywhere. Considering all vessels, the required mw_b has to be distributed into legal shifts (shifts intersecting with their vessel time windows) so that the total gang cost is minimized. The proposed MIP model is a lower bound relaxation of the gang allocation model from Section 2.1. A detailed description of this MIP model can be found in [23].

Results. The goal of our experiments is to measure the optimality gap between the CP model and the relaxed MIP model. All runs were performed on a 2,53Ghz Intel CPU with 1GB of RAM with a timeout of 10 minutes. The MIP solver is SCIP [8] and the constraint programming solver is Comet.

Three models were used. All models use an LNS procedure that randomly fixes vessels with a 0.6 probability. The first one is the *fill-hole* model that uses the *fill hole* labeling, denoted FH. The second model is the *naive* model where a naive labeling is used to assign activities in a leftmost manner ignoring the profile. The last one is the *fill-hole-relax* model (denoted FHR) where there is no crane range constraints, no non-overlap constraints, no transition time and time windows are relaxed to the boundary of the shift. The line FHR solves a simplified core model to compare the MIP relaxation and the CP approach.

Table 3 shows the results. Both MIP and CP approaches have a timeout of 600 seconds. If the MIP time column displays a time less than 600 seconds, optimality has been proven by the MIP. The CP time column displays the time of the last solution found. The distance in percentage with the MIP objective value is given in column GAP. The four columns under 'Objective Value' denotes the total objective value, the gang cost, the position cost, and the lateness cost.

Table 3. Results for all instances

H	Time (sec) CP	MIP	GAP	Objective Value Total	Gang	Pos.	L.	Extra Gangs	H	Time (sec) CP	MIP	GAP	Objective Value Total	Gang	Pos.	L.	Extra Gangs
Random1, 10 vessels									*Random4, 10 vessels*								
FH	504	600	7.8	20648	20589	59	0	5(67/62)	FH	582	600	13.6	29998	29473	525	0	6(86/80)
naive	600	600	-	-	-	-	-	-	naive	600	600	-	-	-	-	-	-
FHR	175	243	0.4	18522	18522	0	0	0(62/62)	FHR	211	600	0.4	26509	26509	0	0	0(80/80)
Random2, 10 vessels									*Industrial, 15 vessels*								
FH	483	8	11.0	20553	20446	107	0	6(65/59)	FH	458	2	11.9	15857	15666	191	0	4(48/44)
naive	385	7	27.8	25356	25321	35	0	7(66/59)	naive	428	3	23.3	18209	18078	131	0	8(52/44)
FHR	93	6	0.4	18314	18314	0	0	0(59/59)	FHR	501	2	0.9	14030	14030	0	0	0(44/44)
Random3, 10 vessels									*Industrial, 30 vessels*								
FH	542	343	18.8	36433	36265	168	0	12(104/92)	FH	60	12	16.5	29884	29050	834	0	11(90/79)
naive	600	356	-	-	-	-	-	-	naive	338	12	41.1	42335	41530	805	0	26(105/79)
FHR	364	600	0.7	28587	28587	0	0	0(92/92)	FHR	12	11	1.8	25878	25878	0	0	1(80/79)

Finally, the number of additional gangs hired with respect to the lower bound MIP approach is printed in column 'Extra Gangs'. A line marked '-' means the constraint programming model did not find any solution before the timeout.

Naive labeling performs poorly compared to the *fill hole* labeling used by the *fill-hole* model. The *naive* model did not find any solution before the timeout in 3 out of 4 random instances and uses two times the number of gangs in the industrial instances. The *naive* model tends to have a lower position cost. The *fill-hole-relax* CP approach is trapped in local optima, but finds good solutions up to 2%. This is expected as MIP is known to be stronger for flow-like problems. The overall performance of our proposed approach is 1/10 (7,8%) to 1/5 (18,8%) of additional cost compared to an ideal operational world (the MIP lower bound).

4   Conclusion

Container terminals are more and more automated and as a result optimization technologies are needed to efficiently solve the numerous logistics problems arising. This is also reflected in the operations research literature where recent works try to solve these integrated problems. The question is whether CP can help in this quest. We answer this question by considering the integration of two problems using a real world constraints with an industrial partner.

We have shown that operational and realistic constraints for BAPCAP can be successfully addressed in the context of a CP approach. This approach is modular in the sense that each set of operational constraints can be separated. The key idea is to use the gang allocation process as the main component, and view it as a resource. Other side constraints can be integrated around this basic model. Experiments show that the CP model can produce solutions close to 1/5th to 1/10th from an ideal operational world. Overall, this work shows that CP can be a technology of choice for tackling challenging problems in the maritime industry considered "out of scope" for the current approaches, even under complex operational and scale constraints.

Future research includes using alternative profile-centered labeling or additional LNS procedures. The resource view of the model opens the possibility to use many scheduling tools from the OR/CP community to improve performance or to integrate new types of side constraints. Integrating the yard management aspect by computing the ideal positions together with the scheduling would extend the integrated approach, for which CP may be the right optimization technology.

References

1. Drewry Shipping Consultants Annual Container Market Review and Forecast 2007/08.A London, 250 p. (2007)
2. Drewry Shipping Consultants Container Forecaster 1Q08 London, 218 p. (2008)
3. Grossmann, H., Otto, A., Stiller, S., Wedemeier, J.: Growth Potential for Maritime Trade and Ports in Europe. Intereconomics, 226–232 (2007)
4. Bierwirth, C., Meisel, F.: A survey of berth allocation and quay crane scheduling problems in container terminals. European Journal of Operational Research 202, 615–627 (2010)
5. Moorthy, R., Teo, C.-P.: Berth management in container terminal: the template design problem. OR Spectrum 28, 495–518 (2006)
6. Raa, B., Dullaert, W., Van Schaeren, R.: An enriched model for the integrated berth allocation and quay crane assignment problem. Expert Systems with Applications 38(11), 14136–14147 (2011)
7. Zhang, C., Zheng, L., Zhang, Z., Shi, L., Armstrong, A.J.: The allocation of berths and quay cranes by using a sub-gradient optimization technique. Computers and Industrial Engineering 58, 40–50 (2010)
8. Achterberg, T., Berthold, T., Koch, T., Wolter, K.: Constraint Integer Programming: A New Approach to Integrate CP and MIP. In: Trick, M.A. (ed.) CPAIOR 2008. LNCS, vol. 5015, pp. 6–20. Springer, Heidelberg (2008)
9. Shaw, P.: Using constraint programming and local search methods to solve vehicle routing problems. In: Maher, M.J., Puget, J.-F. (eds.) CP 1998. LNCS, vol. 1520, pp. 417–431. Springer, Heidelberg (1998)
10. Daganzo, C.F.: The crane scheduling problem. Transportation Research Part B 23, 159–175 (1998)
11. Peterkofsky, R.I., Daganzo, C.F.: A branch and bound solution method for the crane scheduling problem. Transportation Research Part B 24, 159–172 (1990)
12. Liu, J., Wan, Y.-W., Wang, L.: Quay Crane Scheduling at Container Terminals To Minimize the Maximum Relative Tardiness of Vessel Departures. Naval Research Logistics 53, 60–74 (2006)
13. Yang, C., Wang, X., Li, Z.: An optimization approach for coupling problem of berth allocation and quay crane assignment in container terminal. Computers & Industrial Engineering 63, 243–253 (2012)
14. Song, L., Cherrett, T., Guan, W.: Study on berth planning problem in a container seaport: Using an integrated programming approach. Computers & Industrial Engineering 62, 119–128 (2012)
15. Han, X., Lu, Z., Xi, L.: A proactive approach for simultaneous berth and quay crane scheduling problem with stochastic arrival and handling time. European Journal of Operational Research 207, 1327–1340 (2010)
16. Chang, D., Jiang, Z., Yan, W., He, J.: Integrating berth allocation and quay crane assignments. Transportation Research Part E 46, 975–990 (2010)

17. Meisel, F., Bierwirth, C.: Heuristics for the integration of crane productivity in the berth allocation problem. Transportation Research Part E 45, 196–209 (2009)
18. Liang, C., Huang, Y., Yang, Y.: A quay crane dynamic scheduling problem by hybrid evolutionary algorithm for berth allocation planning. Computers & Industrial Engineering 56, 1021–1028 (2009)
19. Imai, A., Chen, H., Nishimura, E., Papadimitriou, S.: The simultaneous berth and quay crane allocation problem. Transportation Research Part E 44, 900–920 (2008)
20. Park, Y., Kim, K.: A scheduling method for Berth and Quay cranes. OR Spectrum 25, 1–23 (2003)
21. Zhang, C.Q., Liu, J.Y., Wan, Y.W., Murty, K.G.: Storage space allocation in container terminals. Transportation Research Part B 37, 883–903 (2003)
22. De Clercq, A., Petit, T., Beldiceanu, N., Jussien, N.: Filtering Algorithms for Discrete Cumulative Problems with Overloads of Resource. In: Lee, J. (ed.) CP 2011. LNCS, vol. 6876, pp. 240–255. Springer, Heidelberg (2011)
23. Zampelli, S., Vergados, Y., Van Schaeren, R., Dulleart, W., Birger, R.: The berth allocation and quay crane assignment problem using a CP approach Technical Report at Universiteit Antwerpen (2013),
    http://www.ua.ac.be/main.aspx?c=wout.dullaert&n=9300&ct=005748

Author Index

Abío, Ignasi 80, 97
Akgun, Ozgur 107
Ansótegui, Carlos 117
Argelich, Josep 133

Bacchus, Fahiem 247, 273
Backofen, Rolf 805
Balafrej, Amine 143
Beldiceanu, Nicolas 529, 733
Bessiere, Christian 143, 159, 708
Bonet, Maria Luisa 117
Bouyakhf, El Houssine 143
Brockbank, Simon 175
Bui, Quoc Trung 749
Bulín, Jakub 184

Cai, Shaowei 481
Cambazard, Hadrien 47
Chakraborty, Supratik 200
Chu, Geoffrey 217
Clarke, Edmund 415
Cohen, David A. 230
Coletta, Remi 143, 864

Davies, Jessica 247
de Givry, Simon 263
Delić, Dejan 184
Delisle, Erin 273
Deville, Yves 749
Di Gaspero, Luca 758
Duck, Gregory J. 282
Dullaert, Wout 880

Ekker, Heinz 805
Ezzahir, Redouane 708

Fages, Jean-Guillaume 63
Fargier, Hélène 159
Feydy, Thibaut 628
Flamm, Christoph 805
Flener, Pierre 381, 529
Fontaine, Daniel 299
Francis, Kathryn 315
Frisch, Alan M. 107
Fukunaga, Alex 331

Gabàs, Joel 117
Gange, Graeme 340
Garcia de la Banda, Maria 432
Gaudreault, Jonathan 30
Gent, Ian P. 107
Goldsztejn, Alexandre 774
Granvilliers, Laurent 774
Gualandi, Stefano 356, 448
Gutierrez, Patricia 365

Habet, Djamal 678
Hamadi, Youssef 464
Hartert, Renaud 611
He, Jun 381
Hussain, Bilal Syed 107

Ifrim, Georgiana 733

Jackson, Marcel 184
Jaffar, Joxan 282
Janota, Mikoláš 415
Jeavons, Peter G. 230
Jefferson, Christopher 107
Jermann, Christophe 774
Jonsson, Peter 398

Kawas, Ban 587
Klieber, William 415
Koh, Nicolas C.H. 282
Kotthoff, Lars 107

Lagerkvist, Mikael Z. 790
Lagerkvist, Victor 398
Lapègue, Tanguy 63
Laumanns, Marco 587
Leal, José Paulo 848
Lecoutre, Christophe 159
Lee, Jimmy H.M. 365
Lei, Ka Man 365
Lenoir, Arnaud 733
Leo, Kevin 432
Levy, Jordi 117
Li, Chu-Min 133
Lombardi, Michele 356, 448
Loth, Manuel 464
Luo, Chuan 481

Mak, Terrence W.K.   365
Malapert, Arnaud   596
Mann, Martin   805
Manyà, Felip   133
Marinescu, Radu   497
Marques-Silva, Joao   415
Mears, Christopher   432
Meel, Kuldeep S.   200
Mehta, Deepak   47
Meseguer, Pedro   365
Metodi, Amit   823
Michel, Laurent   8, 299
Miguel, Ian   107
Milano, Michela   1
Moffitt, Michael D.   513
Moisan, Thierry   30
Monette, Jean-Noël   529

Nahar, Feras   805
Narodytska, Nina   545
Navas, Jorge   315
Naveh, Reuven   823
Newton, M.A. Hakim   832
Nieuwenhuis, Robert   80, 97
Nightingale, Peter   107
Niven, Todd   184
Nordh, Gustav   398
Nordkvist, Martin   790

Oliveras, Albert   80, 97
O'Sullivan, Barry   47, 263
Ouellet, Pierre   562

Pearson, Justin   381, 529
Pelsser, François   578
Pesant, Gilles   175
Pham, Duc Nghia   832
Pham, Quang Dung   749
Portmann, Marius   832
Prestwich, Steven D.   263, 587

Quimper, Claude-Guy   30, 562

Raa, Birger   880
Rattfeldt, Magnus   790
Razak, Abdul   497
Régin, Jean-Charles   578, 596
Rendl, Andrea   758
Rezgui, Mohamed   596

Rodríguez-Carbonell, Enric   80, 97
Rousseau, Louis-Martin   175

Sattar, Abdul   832
Schaub, Torsten   3
Schaus, Pierre   578, 611
Schoenauer, Marc   464
Schutt, Andreas   628
Scott, Paul   645
Sebag, Michèle   464
Simonis, Helmut   47, 733
Stadler, Peter F.   805
Stuckey, Peter J.   5, 97, 217, 315, 340,
    628
Su, Kaile   481

Tack, Guido   432
Tan, Wee Lum   832
Thiébaux, Sylvie   645
Thorstensen, Evgenij   230, 661
Tomás, Ana Paula   848
Toumi, Donia   678
Trombettoni, Gilles   864

Urli, Tommaso   758

van den Briel, Menkes   645
Van Gelder, Allen   694
Van Hentenryck, Pascal   7, 8, 299, 340,
    645
Van Schaeren, Rowan   880
Vardi, Moshe Y.   200
Vergados, Yannis   880
Vismara, Philippe   864

Wahbi, Mohamed   708
Walsh, Toby   545
Wilson, Nic   497
Wu, Wei   481

Xia, Wei   724

Yap, Roland H.C.   724

Zampelli, Stéphane   880
Zhang, Wei Ming   381
Zhu, Zhu   133
Živný, Stanislav   230